Essentials of
Human Anatomy

RUSSELL T. WOODBURNE, A.M., Ph.D.

Professor Emeritus of Anatomy in the University of Michigan Medical School

Seventh Edition
with 478 Illustrations

New York   Oxford
OXFORD UNIVERSITY PRESS
1983

Library of Congress Cataloging in Publication Data

Woodburne, Russell Thomas, 1904–
Essentials of human anatomy.

Bibliography: p.
Includes index.
1. Anatomy, Human. I. Title. [DNLM: 1. Anatomy.
QS 4 W884e]
QM232.W66   1983   611   82-7819
ISBN 0-19-503171-7   AACR2
ISBN 0-19-261435-5 (UK only)

Printing (last digit): 9 8 7 6 5 4 3 2 1

Printed in the United States of America

Preface to the Seventh Edition

As in previous revisions, the author's preparation of the seventh edition of this popular textbook has been guided by considerations of accuracy, relevance to clinical medicine, normal variation, and recent findings. To reduce the burden of memorization, detail has been reduced in the description of bones, branches of major blood vessels, and elsewhere. Electromyographic studies have again been consulted to reflect recent conclusions concerning muscle action and coordination. Embryological explanations are included where they appear useful. Useful new information has become available concerning the urogenital sphincters and is incorporated in this edition in both text and figures. Many illustrations have been improved and a number of new ones have been added.

Teachers of anatomy have recognized that it is helpful to the student to point out how direct are anatomical explanations for clinical entities. Accordingly, this edition presents under the heading of "Clinical Notes" many examples of direct anatomical relevance. In addition to these examples, which are usually juxtaposed immediately with the anatomy involved, two sections on the Applied Anatomy of the Upper Limb and the Applied Anatomy of the Lower Limb are placed at the end of the chapters devoted to the limbs. Many significant facts about dislocations, fractures, collateral circulation, and nerve injuries are discussed under these headings, and the anatomi-cal basis for the clinical problem is always brought out.

Radiologists in recent years have been able to examine more critically the anatomy of the living subject by computerized tomography. This technique produces radiograms of selected levels or cross-sections of the body (see figs. 184, 199, 332, 338). The radiologist's interpretation of a scan depends in large part on this knowledge of cross-sections of the normal body at various levels—note the correspondence of figs. 337 and 338. For this reason, as well as the value of cross-sectional views in understanding the location of structures and their anatomical relationships, it has become increasingly important for students to study cross-sections. As this textbook presents a much larger number of cross-sections than most do, it is especially helpful in this regard.

The book is concisely and clearly written; it conforms still to the original design of a shorter text, but it has sufficient detail to support the more extensive courses. It is the author's hope that it will continue to fill students' needs in their immediate study of Human Anatomy and that it will also serve them for reference when they are active practitioners.

Ann Arbor, Michigan   R. T. W.
December 31, 1981

Preface to First Edition

The pressure of steadily increasing knowledge in medicine and the other health sciences, and in anatomy itself, has been a stimulus for re-examination of teaching methods in anatomy, as it has in related subjects. A superficial response to such pressure is to reduce the time for and the content of anatomy courses. However, expanding frontiers in clinical and anatomical fields are not likely to be served by teaching less anatomy, though it is highly probable that advances can be made in the efficiency of presentation of the subject. It would appear that one area of anatomical teaching which would profit from improvement lies in the character of the textbooks available to the student. Enlarged by every accretion of knowledge through the decades, their systematic organization is an obstacle to learning when only isolated parts of the body are met by the student as systematic entities.

This text presents the basic concepts of the systems of the body and then examines the body, in detail, regionally. Within each region, the order of presentation is from superficial to deep—the only order in which the body can conveniently be dissected. This has the advantage of a concise and integrated description of the region under consideration, and much time and effort is saved in preparing for or reviewing each portion of the body. The regional method of description carries with it, however, the danger of dissociated learning—failing to relate the regional entity to the rest of the system of which it is a part. This danger is avoided by repeated stress on the continuity of parts, by numerous cross-references, and by illustrations that are designed to be synoptic as far as possible.

This is not an elementary text; in some respects its detail goes beyond that of the currently used textbooks. Its brevity comes from an adherence to the essentials of morphology presented functionally and concisely. The author recognizes that conciseness and brevity can lead to oversimplification and inaccuracy, and every effort has been made to be exact in description. Only the most frequent variations are discussed, for to be complete in the matter of variation places a textbook in the reference category. In the choice of such citations, vascular variations are given more attention than muscle and tendon variations, since information on the former is deemed to be of greater clinical utility.

A revision of anatomical nomenclature was adopted by the Sixth International Congress of Anatomists in 1955. This revision is followed in the present text. Its changes are in the direction of simplicity, consistency, and logic in terminology, and the last of the eponyms have been eliminated. No attempt is made, in the descriptions which follow, to carry the older terminology along in bracketed form. The beginning student has no prejudices or foreknowledge in the matter of names. Others accustomed to the traditional names should encounter no difficulty in making the simple transition in terms required by this revision.

It is a privilege and a pleasure for the author to acknowledge his indebtedness to his predecessors and colleagues in the field of anatomy. Most material contributed by previous workers has been incorporated into the general body of knowledge of the subject and cannot be specifically cited in a work of this kind. Indeed, only the more recent contributions are especially listed. In order, however, that the student may have available to him a list of specific citations and general references in which more detail may be found on such subjects as are especially interesting to him, selected references have been grouped at the end of each chapter. These lists are brief, in keeping with the character of the book, but should serve to lead the inquiring student into the more detailed sources.

I owe a real debt of gratitude to my colleagues in the Department of Anatomy of the University of Michigan for their careful reading of the manuscript and their constructive criticism of the illustrations; special thanks are due to Dr. Thomas M. Oelrich. I am indebted to my illustrators—Joanne C. Berger, William L. Brudon, David Sterrett, and Cecilia Graham—for the excellence of their work and for their tolerance of continual revision, and to my secretary, Esther L. Vowell, for her care in preparation of the manuscript. A real contribution to the work has been made by my wife, who has helped both in composition and in proofreading. The X-rays used were provided by the Department of Radiology of the University of Michigan Medical School. In grateful recognition of assistance from many, the author nevertheless accepts full responsibility for errors or omissions in the final result.

August 1957 *R. T. W.*

Foreword

To the student beginning his studies and observations in human anatomy: This text is designed as a teaching instrument dedicated to the development of an understanding of the human body. Details are present—and indeed the amount of detail in human anatomy is almost overwhelming—but the author has attempted at all times to simplify, to correlate, to explain, to integrate, and to describe in such a way that the student will understand the body as well as be informed about it. As a teaching instrument the book is not traditional but pragmatic. Its information is specific and accurate, related to the cadaver and checked by dissection. The illustrations are especially designed to illuminate the information in the text, and the student should refer to them constantly. It will be to the student's advantage if he colors the black and white illustrations as he goes along. This is advised in the belief that thoughtful and constructive coloration of the various tissues in any regional illustration adds to the learning process.

The text presents a logical analysis of regions, an analysis which follows the necessary sequence of dissection, but which gives emphasis to the central features of each region. The student will be aided if he recognizes this logical analysis and if he discerns the organization under which all description is made. Organization of material is an important aid to the learning of material, and the student should note and use organization at every step. Although the book is designed regionally, the student must also derive from his studies an adequate understanding of the systems of the body. To this end, the author has taken every opportunity to relate the specific regional entities to the systems of which they are parts. It will contribute to learning if the student will strive for both regional and systematic knowledge.

Anatomy is a laboratory subject. No one has ever been able to derive a satisfactory understanding of the human body by textbook reading alone. The subject is learned by the patient uncovering of structure after structure, cleaning each one to make a clear picture, and relating each to surrounding objects. A textbook should be used in the laboratory; information read should be correlated with information seen, and information seen should be used to check what is read.

Not all of the learning, however, can be done in the laboratory. There are important requirements for the review of material in the learning process; there are also important requirements for the introduction to that material.

There is a 'preview' phase of learning which is exceedingly useful for the learning of anatomy. This may consist of looking over in advance the topic to be dissected during the next laboratory period, examining the textbook organization, looking at its illustrations, seeing what these objects look like or in what terms they are described. This provides an overview of the material which will facilitate and enhance the laboratory examination. A lecture assists in the same preview sense, but pre-reading of the textbook serves as a useful introduction.

Review of the material, later, is another essential part of the learning process, and review within the textbook, within the atlas, and again on the body is important. Visualization should be stressed. One must see in the mind for both understanding and retention. Here the cadaver is paramount, but the atlas and textbook illustrations supplement the observations on the body. Learning is not done at once or in one step. It is produced by a number of contributory stages, and at every one of these this textbook will be useful.

Contents

Contents

Essentials of Human Anatomy

I

General Concepts in Anatomy

INTRODUCTION

Anatomy is the term usually applied to the study of the human body by the method of dissection. As observation of the tissues of the body has extended to more and more minute parts and as microscopes have been employed, the field of anatomy has been subdivided for convenience into **gross anatomy** and **microscopic anatomy.** Cytology, histology, and **organology** are segments of the general field of **microscopic anatomy,** having reference respectively to cells, tissues, and organs. The special study of the nervous system, **neuro-anatomy,** is pursued partly by gross but, for the most part, by microscopic techniques. The developmental aspects of the human organism fall under the heading of **embryology.**

All these branches of anatomy are concerned with various approaches to human morphology, but they are distinguished by special techniques and differing foci. There are no actual boundaries between the several parts of anatomy, and the discussions which follow will occasionally draw from all of them without regard to precise limitations. However, it is the anatomy as seen by the unassisted eye that is the focal point of this description, and it is to general aspects of the gross structure of the human organism that our attention is directed. Some of its special phases are designated **topographic,** or regional, **anatomy, radiographic anatomy, applied anatomy,** and **surgical anatomy.**

Description requires the body to have a standard orientation. Thus the **anatomical position** is an erect one with eyes forward and the palms of the hands to the front. This position of the hands is not an entirely natural one, it is a position of **supination;** the opposite, with the palms down or backward, is one of **pronation.** Description also uses standard reference terms (fig. 1) based on the **anatomical position.** The **median plane** is a vertical plane through the body reaching the surface at the midline in front and behind. This plane is also known as the **midsagittal plane** of the body and, with the exception of the unpaired viscera in the trunk cavities, divides the body into symmetrical halves. Other anteroposterior vertical planes parallel to the median, or midsagittal, plane are called **sagittal planes.** The **coronal plane** is a vertical one directed from side to side and thus is at right angles to the midsagittal plane. It gives its name to the coronal or frontoparietal suture of the skull and may also be designated as a frontal plane. The term **horizontal,** or transverse, **plane** refers to any plane at right angles to the vertical planes; it is a cross section. Fundamental terms for the front and back of the body are, respectively, **ventral** and **dorsal.** Since adult man stands erect, ventral is equivalent to **anterior** and dorsal is the same as **posterior. Cranial** and **caudal,** referring respectively to the head and tail regions of the trunk, are also useful directional terms. It is frequently necessary to stipulate that an object is **medial** and thus near or nearer the median plane of the body, or, conversely, **lateral** and thus farther away from the median plane. **Proximal** and **distal** contrast positions nearer the root of a limb and farther along its length. **Superficial** and **deep** are terms frequently used in describing a dissection and have their usual meaning of nearer or farther from the surface.

THE ORGANIZATION OF THE BODY

As one observes one's fellow man, his behavior is that of a sentient, reacting organism. He feels, sees, and hears; he moves and responds to stimuli and adapts to the conditions of his environment.

3

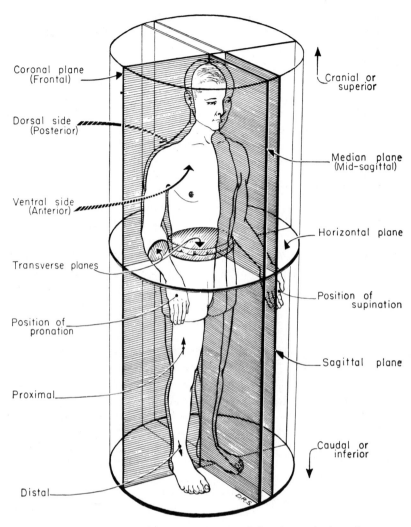

Fig. 1   The planes of the body and terms of direction and orientation.

He is thus a sensorimotor mechanism in outward behavior. He is also, however, an organism that is inconspicuously but incessantly preoccupied with maintaining his own life processes. The respiratory, digestive, and excretory needs of his body are carried on more or less automatically and continuously. He has an internal regulation that is due partly to endocrine gland secretions and partly to the activities of his nervous system. Finally, his genital apparatus provides for the continuity of the race.

Man is at once both complex and simple, for his manifold abilities and functions are made possible by rather extreme specialization in his tissues, and yet these tissues are fundamentally merely collections of single cells. Essentially, all the functions of the body are expressed in the qualities of the single cell. As **cells** of similar type are aggregated and become organized into **tissues** and as tissues of like and diverse character are collected and fabricated into **organs,** and organs into **systems,** the main subdivisions of the body take shape. The body can be described under headings designating its component systems with economy of space and homogeneity of subject matter. That this approach is not followed in this text is due to the fact that dissection is necessarily performed region by region and not system by system. The **system** represents, however, such a valuable organizing and simplifying concept that an initial acquaintance with the systems of the body is imperative. As usually listed these are: the skeletal system; the **muscles**; the **articulations**; the **circulatory system**; the **nervous system**; the **skin** and **organs of special sense**; and the **visceral systems**— respiratory, digestive, urogenital, endocrine. Most of these systems find representation in all or several regions of the body, although the visceral systems occupy the trunk almost exclusively. Among the **regions** of the body that may be usefully designated are the **upper limb**, the **head and neck**, the **back**, the **chest** or **thorax**, the **abdomen**, the **perineum**, the **pelvis**, and the **lower limb.**

As a conceptual basis for dissection in any of the regions of the body a general knowledge of the more basic systems is invaluable. In developing such an understanding of the whole organism it may be advantageous to return to the sensorimotor aspect of Man's organization and begin consideration of his systems with a description of the skin. This is also the initial site of dissection in any region of the body.

THE SKIN AND ITS APPENDAGES

The skin, or common integument (fig. 2), is a tough, pliable covering of the body which grades over into the more delicate mucous membranes of the body cavities at the mouth, the nostrils, the eyelids, and at the urogenital and anal openings. That it is infinitely more than a surface covering is shown by a consideration of its varied functions. The skin is an extensive **sensory organ,** supplied with a host of nerve endings which provide sensitivity to touch and pressure, temperature changes, and painful stimuli. Indeed, skin is the principal source of these 'general' sensations. The skin is a **protective** layer of considerable importance. Not only is it a strong, flexible covering but it also prevents loss of body fluids. Appreciation of this function comes especially as one considers the importance of skin grafts in covering raw, denuded, or burned areas of the body. The integument is especially significant in **temperature regulation.** As a warm-blooded (homeothermic) animal, Man's internal temperature is kept constant through external changes of 100° F. or more. Reduction of body temperature is a special function of skin, for heat loss through radiation, convection, and evaporation (from skin and lungs) accounts for approximately 95 per cent of the total heat dissipated from the body. Sweating is initiated by the effect of blood heat on brain centers and is mediated through cutaneous nerves reaching the sweat glands. Sweating also serves the **excretory functions** of the body, for if sweating is copious, up to one gram of nonprotein nitrogen may be eliminated per hour. The skin is concerned in the production of **vitamin D** through the action of the ultraviolet rays of the sun on its sterols. Studies of percutaneous transmission of drugs, vitamins, and hormones indicate that the skin has an important **absorptive** function when such substances are applied to it in a suitable vehicle.

Basic to effective functioning in all these metabolic aspects is the large **surface area** of the integument in the adult of 1.8 sq. meters (approximately a 6' by 3' sheet). The surface area increases about sevenfold from birth to maturity. The skin ranges in thickness from 0.5 mm. over the tympanic membrane and the eyelids to 6 mm. over the upper back, back of the neck, palm of the hand, and sole of the foot. It tends to be thicker on the posterior and extensor surfaces than on the anterior and flexor surfaces and generally approximates 1 to 2 mm. in thickness. The skin is loosely applied to underlying tissues and may be displaced and elevated in most regions of the body. Contrarily, it may be firmly attached to periosteum (as over the subcutaneous surface of the tibia) or to cartilage (as in the ear) or tightly bound to deep fascia or joint capsules (as seen in the flexion creases of the palm of the hand and digits).

The 'flesh color' of the skin is due to the blood color reflected through the epidermis. The color varies according to the thickness of the epidermal layers through which the reflected light rays pass, the state of constriction or dilation of the subpapillary vessels, and the degree of oxygenation of the blood. Variations in color from individual to individual and from race to race also depend on pigmentation. Pigmentation of skin is due to the presence in the deepest layer of the epidermis of so-called 'clear cells' which have branched processes extending into more superficial layers. Under appropriate enzyme action these cells form **melanin** and distribute this pigment as granules throughout their cell bodies and processes. Tanning from exposure to sunlight is due to a physiologic increase in pigment formation. Certain areas of the body exhibit constantly deeper pigmentation: the areola of the mammary gland, the external genital regions, and the axilla.

In areas transitional to mucous membrane, the skin lacks hair, has a ruddy color, and has a moistness or oiliness of surface which is evidence of a gradient toward mucous membrane. Such skin is typical of the lips, the nostrils, the external genitalia, and the anal region.

Observation of the skin, the dorsum of the hand as an example, clearly shows that **delicate creases** extend across the surface in various directions, intersecting one another and delineating irregular, diamond-shaped segments of the integument. Hairs typically emerge at points of intersection of these creases. Such creases represent flexion lines for the skin; they increase in frequency and depth as regions of free joint movement are approached. Certain of them are differentiated into definite **flexure lines.** These are lines of relative immobility and firm anchorage while the skin on either side of the line is folded passively toward it to accommodate the bending.

Clearly discernible on the pads of the fingers and toes, but extending over the palmar and plantar surfaces of the hand and foot, is a series of alternating **ridges** (cristae cutis) and **sulci** (sulci cutis). These friction ridges function to prevent slippage in the grasp and they result from the large size and specific arrangement of the dermal papillae under the epidermis. Ducts of sweat glands open along the summits of the ridges, and hairs and sebaceous glands are absent on these surfaces. The ridges and sulci are, in detail, highly individual and their whorls and patterns form the basis for identification through fingerprints (dermatoglyphics).

STRUCTURE

The skin is composed of a surface layer, the epidermis, and an underlying thicker lamina, the dermis (fig. 2). **Epidermis** is epithelium of the stratified squamous variety especially characterized by cornified surface layers. It is typically only a fraction of 1 mm. in thickness and is composed of many layers of cells. The cells in the deeper layers are living and proliferate actively; the cells produced pass gradually to the surface, becoming cornified as they approach it, and ultimately are shed by rubbing on clothing and other surfaces. The epidermis is nonvascular but is penetrated by sensory nerve terminals. On its deep surface the epidermis sends prolongations into the dermis, and irregularities of the dermis also interlock with the

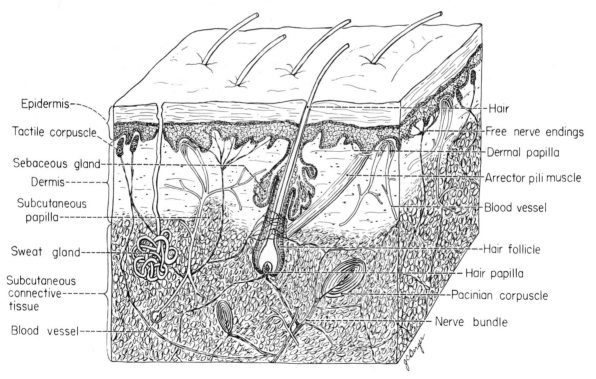

Fig. 2   The structure of the skin.

epidermis. These interlocking finger-like projections are called, respectively, **epidermal pegs** and **dermal papillae.**

Dermis is the deeper interlacing feltwork of connective tissue fibers which constitutes the greater part of the total skin thickness. It has a finely textured papillary layer, and a deeper, thicker, coarsely textured reticular layer which in turn grades over into the subcutaneous connective tissue. The **papillary layer** gives rise to the dermal papillae which may number 100 per sq. mm., their concentration varying from region to region. Most papillae enclose capillary tufts, thus bringing the blood into close relation with the epidermis. Some of them accommodate tactile corpuscles, numerous in areas of acute tactile sensitivity, scanty where such sensitivity is poor.

The deeper layer of the dermis, the **reticular layer,** is a dense mass of interlacing white (collagenous) and elastic connective tissue fibers. This layer accounts for the toughness and strength of skin and, when commercially processed, is the substance of leather. Its fibers run in all directions but are mainly tangential to the surface. The predominant orientation of fiber bundles in relation to the surface differs in different regions of the body, and study of these fiber arrangements has resulted in the description of patterns designated as **Langer's lines.** Surgical incisions in the direction of these lines run parallel to the principal fiber bundles and have less tendency to gape. The dermis contains a small quantity of fat, numerous blood vessels and lymphatic channels, nerves, and sensory nerve endings. Hair follicles, sweat and sebaceous glands, and smooth muscles are present in the layer. The underside of the dermis is invaginated by tufts of subcutaneous connective tissue similar to but larger and more dispersed than the dermal papillae of the papillary layer. The spacing and size of these invaginations is reminiscent of pig-skin, and the pig-skin appearance of the under surface of the dermis is a guide to the proper plane of separation between it and the subcutaneous tissues. These invaginations serve for the entrance into the skin of blood vessels and nerves.

The **subcutaneous connective tissue** (fig. 2) is composed of loose-textured, white fibrous connective tissue with which fat and slender elastic fibers are intermingled. It is of the type of 'areolar' connective tissue, so named by the ancients because of its gas-containing space in decomposing bodies.

In its relation to epidermis and dermis, it is correctly designated **hypodermis** (hypo = under). The subcutaneous fat varies in amount in different parts of the body but is absent in only a few regions, such as the eyelids, penis, scrotum, nipple, and areola. Where the fat layer is very prominent the hypodermis is designated **panniculus adiposus.** Fat is unequally distributed in the male and female and its local differences constitute a secondary sex characteristic. The fat is supported by strands and sheets of white fibrous connective tissue and in the scalp is firmly held in locules among the dense connective tissue fibers of the subcutaneous layer. The hypodermis varies in thickness but is generally much thicker than the overlying dermis. It contains blood and lymph vessels, the roots of hair follicles, the secretory portions of the sweat glands, cutaneous nerves, and sensory endings, especially Pacinian (pressure) corpuscles.

Subcutaneous bursae, single or multilocular spaces, exist in the subcutaneous tissue over joints that undergo marked bending, as at the elbow or knee. They contain a small amount of fluid and facilitate the movement of the skin.

The twitching of the skin of the horse and other four-footed animals gives evidence of **subcutaneous voluntary musculature.** Widely distributed in lower animal forms, this type of muscle is restricted in Man to the scalp, face, and neck where a subcutaneous sheet of muscle is differentiated into the **facial group of muscles.** It is concerned with the movements of facial expression. The palmaris brevis muscle in the hand is a vestigial, subcutaneous, voluntary muscle. Certain **involuntary,** or smooth, **muscles** also exist in the subcutaneous connective tissue. The dartos muscle of the scrotum and muscular tissue of the areola and nipple of the mammary gland are of this type. Subcutaneous muscle inserts into the overlying dermis.

HAIR

Hair (fig. 2) is distributed widely over the body, being absent only over the palm of the hand, the sole of the foot, the dorsum of the distal segment of the digits, the red portion of the lips, the glans and prepuce of the penis, the inner surfaces of the labia majora, the labia minora, and the nipple. These may be called glabrous surfaces. Hairs vary as to thickness and length. The very delicate **primary hair,** or **lanugo,** of the fetus and infant is succeeded by secondary hair or down hair. **Down hair** is, in turn, partially replaced by terminal hair. **Terminal hair** is present over much of the body, notably in the scalp, eyebrows, eyelids, vestibule of nose, and at the entrance of the external acoustic meatus. It is also prominent after puberty over the pubes and in the axilla, and as the beard in men.

The portion of the hair projecting from the surface of the skin is the shaft; the portion under the skin is the root. The root is enclosed in a tubular **hair follicle** which extends deeply into the dermis. A connective tissue papilla projects into the bottom of the follicle and conducts blood vessels into it. Hairs tend to be placed in groups, three being a common grouping, and they are implanted in the dermis obliquely to the surface rather than vertically. Such oblique directions are regional, forming hair-tracts in which the hairs have a common direction, as whorls in the scalp. The shaft of the hair is circular in cross section, although curly hair has a more oval cross-sectional form. Hair color is due to pigment in the hair cells (melanin and a soluble red pigment) and to air in the shaft of the hair. The life of a hair is from 2 to 4 years on the head but only from 3 to 5 months in the eyelashes; they are intermittently shed and replaced.

Each hair is associated with one or more **sebaceous glands.** These lie in the angle between the slanting hair follicle and the skin surface and their ducts open into the neck of the hair follicle. The angle in which the sebaceous gland lies is crossed by a slender smooth muscle, the **arrector pili** (fig. 2). This muscle attaches to the connective tissue sheath of the follicle at about its midlength and inserts at its other end in the papillary layer of the dermis. Contraction of the muscle erects the hair, squeezes out the secretion of the sebaceous gland, and, as a spasmodic phenomenon, creates the familiar 'goose flesh' or 'goose pimples.'

NAILS

The nail consists of an approximately rectangular horny plate on the dorsum of the terminal segment of the fingers and toes. The dense **nail plate** is composed of closely welded, horny scales or cornified epithelial cells. Its semitransparency allows the pink of the highly vascular **nail bed** to show through. The nail is partially surrounded by a fold of skin, the **nail wall,** and adheres to the subjacent nail bed, where the staunch fibers ending in the periosteum of the distal phalanx give the firm attachment necessary for the prying and scratching functions of the nails. The distal edge of the nail is free; the proximal edge constitutes its root. The nail is formed from the epithelium of the proximal part of the nail bed, an area in which the epithelium is particularly thick and which extends about as far distally as the whitened lunula. Developing from this **nail matrix** the nail moves out over the longitudinally parallel dermal ridges of the nail bed, growth being approximately 1 mm. per week. Sensory nerve endings and blood vessels are abundant in the nail bed.

SKIN GLANDS

The glands of the skin are the sebaceous, the sudoriferous, or sweat, glands, and the mammary glands. (The mammary glands are described with the pectoral region.)

Sebaceous Glands The sebaceous glands (fig. 2) are associated with the hairs and hair follicles. They exist throughout the body except in the skin of the palms and the soles and the dorsum of the distal segment of the digits. They are also found in certain areas where hair does not exist, such as the lips, the corner of the mouth and adjacent submucosa, the glans penis, the internal fold of the prepuce, the labia minora and clitoris, and the areola and nipple. In these locations the gland duct opens directly on the surface. The glands vary from 0.2 to 2 mm. in diameter. The largest exist on the ala of the nose where the size relationships are such that we might say that the hairs are accessory to the glands. Typically, sebaceous glands lie in the dermis with one to six glands grouped in relation to a hair and with the single duct opening into the neck of the hair follicle. The secretory segment is flask-shaped, the form being designated as alveolar. Several convergent alveoli may combine in a grape-like mass. The polyhedral cells of the sebaceous gland are continuously destroyed and renewed in the production of its oily secretion, which is known as **sebum.** Sebum provides a lubricant to the hair and skin and protects the skin from drying. Sebaceous glands do not appear to be under nervous control. The ciliary glands of the eyelid are modified sebaceous glands.

Clinical Note Acne is an affection of the sebaceous glands characterized by inflammation and accumulation of secretion. When the outlet is plugged, a blackhead (comedo) results. If plugging is permanent, a sebaceous cyst (wen) may be formed in the duct and follicles.

SWEAT GLANDS

The sweat glands (fig. 2) have a wide distribution in the skin of the body, exceptions being the nail beds, the margins of the lips, the concha of the ear, the nipple, the glans penis, the prepuce, and the labia minora. These eccrine sweat glands consist of simple, coiled tubes and generally have a frequency of from 100 to 400 per sq. cm. The secretory portion is a tube folded by several unequal coils into a ball of from 0.3 to 0.4 mm. in diameter. The duct begins at an abrupt narrowing of the tube and is spirally twisted in its course through the epidermis. On the palms and soles, where they are densest, and over the flexor surfaces of the digits the ducts open on the summits of the ridges (cristae cutis). Sweat is a clear fluid without cellular elements. Its importance in temperature regulation has been mentioned under the discussion of the function of the skin. Eccrine glands are supplied by the cholingeric fibers in sympathetic nerves, and anhydrosis may indicate a disturbance in the sympathetic nervous discharge.

In the armpit and about the anus there are large, modified sweat glands. These are from 3 to 5 mm. in diameter and lie deeply in the subcutaneous layer. Their ducts may open onto the skin surface directly or be associated with a hair follicle. The secretion includes disintegration products of the gland cells. The odor commonly associated with these glands is not inherent in the secretion product but is due to bacterial contamination from the skin. Pigment granules secreted with the axillary sweat provide its slight coloration. These apocrine glands vary with sexual development, enlarging at puberty, and in the female show cyclic changes correlated with the menstrual cycle.

BLOOD AND LYMPHATIC VESSELS

The **arteries** of the skin are derived from vessels in the subcutaneous connective tissue layer which form a tangential network or plexus at the boundary

between dermis and hypodermis. Branches from this network supply the fat, the sweat glands, and the deep parts of the hair follicles. Other branches in the dermis form a subpapillary network (fig. 2) from which is supplied twigs to the dermis and fine branches to the loops of the dermal papillae. The epidermis is avascular.

Arteriovenous anastomoses (p. 22) are found abundantly in the skin. The **veins** show an arrangement similar to the arteries, with a subpapillary plexus, a plexus at the interface between the dermis and hypodermis, and other subsidiary plexuses. The subcutaneous veins pass deeply with the arteries.

The **lymphatics** of the skin begin in the dermal papillae as blind outgrowths or networks and in the papillary layer form a dense, flat meshwork of lymphatic capillaries. From here lymphatic vessels pass to a deeper network at the boundary of dermis and hypodermis. The lymphatic vessels run centrally in company with the cutaneous blood vessels.

NERVES

Cutaneous nerves are of two types. Included are **afferent somatic fibers** mediating general sensation, such as pain, touch, pressure, heat, and cold. There are also **efferent autonomic** (sympathetic) **fibers** supplying the smooth muscle of the blood vessels, the arrectores pilorum muscles, and the sweat and sebaceous glands. **Afferent or sensory endings** have several forms. Free, or naked, nerve endings extend between cells of the basal layer of the epidermis and terminate around and adjacent to the hair follicles. They are receptive for painful stimuli and general tactile sensation. Encapsulated tactile corpuscles (fig. 2) lie in the dermal papillae and mediate the sensation of touch. Pacinian corpuscles (fig. 2) exist in the subcutaneous tissue and are plentiful along the sides of the digits. They appear to be pressure endings. Specific corpuscles for heat and cold have been described but there is not general agreement as to their identity. Further, in recent years considerable doubt has been expressed as to the above described specificity of nerve terminals.

THE MUSCLES

In the sensorimotor framework in which we observe man, movement is the second element. It is movement which characterizes animal life from the

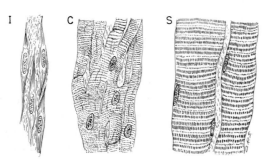

Fig. 3   The varieties of muscle; I—cells of involuntary muscle, C—cardiac muscle, S—skeletal muscle. (Redrawn after Patten.)

one-celled organism to the most complex. In the specialization of cells, contractility has become the particular property of muscle.

There are three varieties of muscle in the body (fig. 3): (1) **smooth muscle,** known also as **non-striated, involuntary,** and **visceral muscle;** (2) **cardiac,** or heart, **muscle;** and (3) **skeletal muscle,** also distinguished as **striated** and **voluntary.** Smooth, or involuntary, muscle typically forms the muscular layers of the walls of our hollow viscera, and of blood vessels. Its cells are elongated and fusiform, and each cell contains a single nucleus. This muscle contracts relatively slowly and less powerfully than striated muscle but can maintain its contraction longer. Cardiac muscle is also involuntary muscle; it is not under the direct control of the will. It exhibits cross striations, however, and is recognized as a separate variety. The fibers of cardiac muscle branch and anastomose and contraction waves spread through them. Physiologically we say that the heart action is myogenic; that is, its musculature has an automatic, rhythmic contraction which is modified but not initiated by nervous impulses. Skeletal muscle constitutes the largest category of muscle tissue in the body and is the principal subject of the following discussion.

Skeletal muscle constitutes some 40 per cent of the body mass in man. Muscles have various forms, some flat and sheet-like, some short and thick, others long and slender. In the abdominal wall certain sheet-like muscles extend from the bony framework of the back to the midline of the ventral abdominal surface and from the rib margin to the pelvis. The longest muscle of the body, the sartorius muscle of the thigh, is approximately 24 inches long. At the other extreme,

within the bone housing the ear there lie the tensor tympani muscle of 2 cm. length and the even shorter stapedius muscle. The length of a muscle, exclusive of its associated tendon, is closely correlated with the distance through which it is required to contract, for muscle fibers have been shown to have the ability to shorten to 57 per cent of their relaxed length. It must be clearly understoood that muscles produce movement by shortening; they pull, never push. The pull of a muscle is usually exerted across a joint and draws closer together the bones on either side of the joint, as in bending the elbow or flexing the knee (fig. 7).

STRUCTURE

The histological unit of skeletal muscle is a **muscle fiber** (fig. 3). A single muscle fiber is a long, cylindrical, multinucleated cell which varies from 10 to nearly 100 micra in thickness. Its length is given differently by various authors; teased

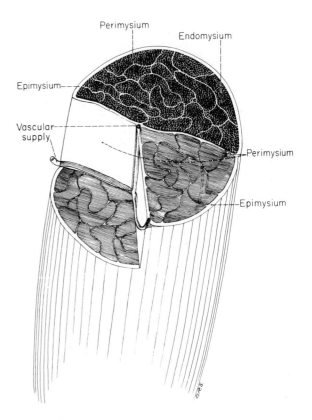

Fig. 4   The organization of muscle from bundles of muscle fibers and the investing and enclosing connective tissue layers.

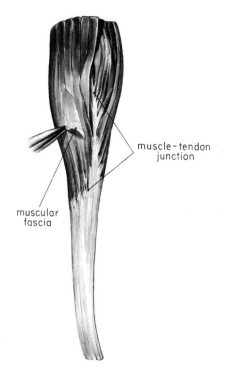

muscle-tendon
junction

muscular
fascia

Fig. 5   The muscle-tendon relationship; muscular fascia.

continuous with those of the periosteum, penetrate the bone, or blend with the fibers of dermis, joint capsules, or other connective tissue structures. Certain muscles are described as having 'fleshy' attachments, but this is no exception. Usually in such cases there is but a short span of connective tissue between a broadened fan of muscle and its attachment. In the 'tendinous' form of attachment the muscle tapers into apparent continuity with the discrete, shining, white **tendon** of closely packed collagenous fibers. The tendon is a flexible, immensely strong member of the muscle-tendon unit. It is estimated that a tendon with a cross-sectional area of one square inch (much larger than exists in the human body) would support a weight of from 10,000 to 18,000 pounds. In certain regions of the body, muscles are attached by means of **aponeuroses.** These are sheet-like tendons. Since the connective tissue elements (tendon, aponeurosis) are commonly included in the complete description of a muscle, the muscular, or contractile, part is often designated as the **'belly.'**

ATTACHMENTS

The two attachments of a muscle are usually described as an **origin** and an **insertion.** These are relative terms, for all parts of the body are capable of movement. The origin is the attachment of lesser movement, the insertion that of freer movement, and, usually, the origin is the more proximal attachment and the insertion the more distal. It is also apparent that even such definitions fail at times, for not only can one reach up and pull an object down but, in a different situation, one can reach up and pull oneself up to the object.

ARCHITECTURE OF MUSCLE (fig. 6)

In its simpler form, muscle is composed of parallel muscle fibers or parallel chains of muscle fibers. These may be collected into a discrete bundle, producing a 'fusiform' muscle, or dispersed as a broad, thin sheet. In either case the tendon lies at the end of the muscle. In such muscles contraction takes place through the maximal distance allowed by the length of the muscle but, in comparison with certain other muscles, is of limited power. A more powerful architectural type is one in which the muscle fibers approach the tendon obliquely and

specimens of up to 30 mm. in length have been isolated. Whole muscles are made up of bundles of these overlapping, interweaving, shorter, individual muscle fibers. The muscle fiber is covered by a thin membrane, the **sarcolemma,** and exhibits alternate light and dark cross striations. The end of the muscle fiber is bluntly rounded or tapered, and a delicate network of collagenous and elastic fibers, the **endomysium** (fig. 4), invests the entire muscle fiber. Parallel muscle fibers are organized into bundles and the whole is surrounded by a similar but heavier connective tissue covering, the **perimysium.** A still coarser connective tissue investment, the **epimysium,** surrounds and infiltrates the muscle itself. These intimate connective tissue relations are further exhibited at the muscle-tendon junction (fig. 5) where the sarcolemma at the end of the muscle fiber is fused with the collagenous fibers of the tendon and the collagenous bundles of the endomysium and perimysium pass directly over into those of the tendon. It is through such intimate connective tissue relationships that muscle produces traction on bone and other movable parts, for the dense connective tissue fibers of the tendon become

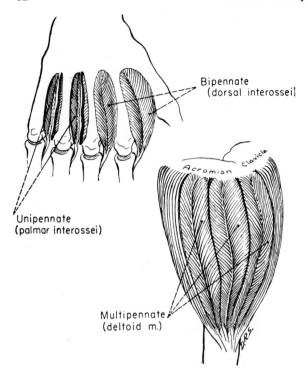

Fig. 6   The architecture of muscle.

insert in it in serial fashion along its length. If such muscle fibers approach the tendon from only one side, the form is called **pennate**. A **bipennate** arrangement of muscle and tendon is accurately visualized if one thinks of the barbs and central shaft of a feather. Certain muscles utilize the central tendon even more completely, with muscle fibers approaching it from all directions rather than simply the two directions of a flat plane. Such a muscle is designated as **circumpennate**. A muscle with a **multipennate** architecture exhibits multiples of one of the pennate forms. The advantage of the pennate type of muscle arrangement is that the power such a muscle can exert is considerably greater than that which a muscle of longitudinally parallel fibers is capable of producing. Muscle power is proportional to cross-sectional area— and cross-sectional area is measured at right angles to the long axis of the muscle fasciculi, not of the total muscle. Since the contraction of oblique fibers is not in the direction of the pull of the total muscle, the excursion of the tendon in this type of muscle is less than might be expected from the length of the muscle fasciculi contracting, but the power of the muscle is greater.

The length of a muscular belly is relative to the excursion permitted to the bone on which it attaches, and, with certain exceptions, any excess distance between origin and insertion is occupied by a noncontractile tendon. A muscle is a structure of high metabolic requirements and a tendon is one of low metabolism, and the economy of the body is such as to replace contractile elements by fiber elements wherever the former are unnecessary to the action required. This principle is expressed in the anatomy of a muscle with several heads of origin. Two heads of the triceps muscle act only in extension of the elbow; they exhibit an equal fascicular length and one nicely adjusted to the range of movement of the joint. The 'long head' of the muscle has a longer belly which permits additional movement at the shoulder.

It is of interest to examine the efficiency of muscle. As is the case with all machines, part of the energy going into muscular contraction is productive of movement and part of it is dissipated in heat. If muscular exertion is unproductive of motion, as in attempting too great a lift, all of the energy is expended in heat; but when a muscle is allowed to contract, 20 to 30 per cent of the energy going into it is productive of movement. This is actually a better efficiency factor than most machines.

CLASSIFICATION OF MUSCLE ACTIONS

It is useful to be able to describe the effect of muscle contraction on a joint in terms which may be applied throughout the body and which are related to the anatomical position. The descriptive terms express very poorly, however, the complexities of movement, since they resolve

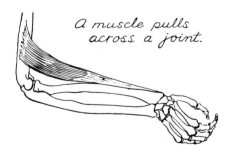

A muscle pulls across a joint.

Fig. 7

movement to arbitrary planes of action. **Flexion** (figs. 7, 8) is the term used to describe the bending of a part or the making of an angle, most easily visualized in the bending of the elbow or the knee. The opposite term **extension** fundamentally means a straightening. In the anatomical position most of the members of the body are in the extended position; the arms and legs are straight, as is also the back. In the case of the foot, these terms lead to some confusion since, in the erect position, the ankle is almost continuously semiflexed. From this position, further flexion is more readily understood under the term **dorsiflexion** and extension under the term **plantar flexion.** Plantar flexion is obviously bending in the direction of the sole; dorsiflexion, bending in the direction of the dorsum. For movement away from or toward the central axis of the body the terms **abduction** and **adduction** are applied. These terms lead to no confusion except in the hands and feet where there is movement of the digits away from and toward a plane wholly within these parts. Rotational movement is also recognized. Rotation of the anterior surface of a member toward the midplane of the body is **medial rotation;** rotation away from the midplane is **lateral rotation.** The rotary action of the forearm and hand, which can be readily observed as the hand is turned palm up or palm down, has a special designation. **Pronation** is rotation so as to turn the palm downward or backward; **supination** carries the palm upward or forward. There is a similar though less extensive movement of the foot in which rotation takes place in the tarsal joints. The rotation of the foot so that the sole turns outward is **eversion,** the opposite movement so that the sole turns inward is **inversion.** A special case of rotation is that which is seen in the very important opposing action of the thumb. This movement of rolling the thumb over onto the hand so that the pads of the digits converge into a firm grasp is termed **opposition;** it is exhibited to a lesser degree in the little finger and in the great and small toes. **Circumduction** is circular movement; to produce this type of motion, flexion and extension, abduction and adduction are combined in a proper sequence. Special terms, such as **protrusion** and **retraction, elevation** and **depression,** will be referred to in their proper context but are relatively self-evident.

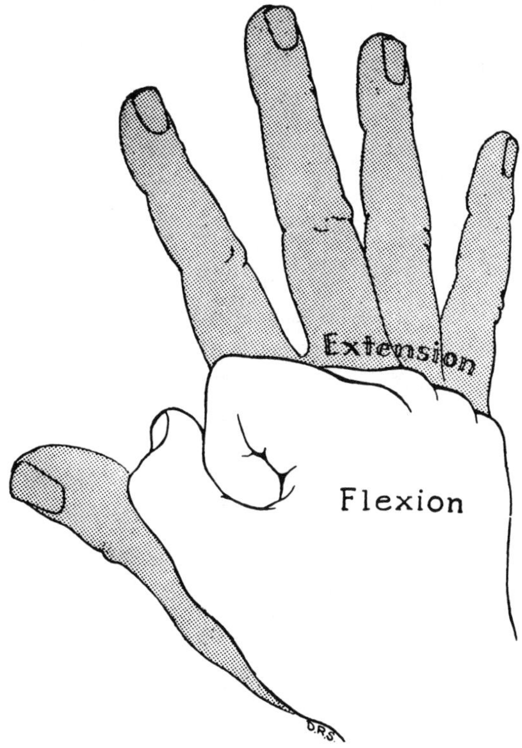

Fig. 8

NERVOUS CONTROL OF MUSCLES

Voluntary muscle is under the control of nerves and does not contract unless a nerve impulse reaches it. The paralysis of poliomyelitis is illustrative: this disease attacks the nervous system, but the muscles are just as effectively inactivated as if they had themselves been destroyed. So important is the nerve-muscle relationship that in dissecting the muscle the nerve is always displayed and positively identified. A muscle receives at least one nerve, but, if large, may well receive several. Such nerves may arise from different but adjacent segments of the spinal cord. The **motor end-plate** is the effector ending for striated muscle (fig. 20). It consists of a cluster of nerve branches terminating under the sarcolemma near an accumulation of nuclei of the muscle fiber located near the middle of the fiber. A muscular nerve is not entirely motor in function for about 40 per cent of its fibers conduct sensation from the muscle and tendon. Some pain fibers are supplied to muscle and to the surrounding connective tissue, but the principal type of afferent impulse from muscle and tendon is proprioceptive. Proprioceptive sensation provides information as to the degree of contraction of the muscle and the state of tension in the tendon. Combined with stimuli from joints it tells us the position of our members. This is an essential type of information for the co-ordination of movement; it is effective partly on the reflex plane and partly it is transmitted to consciousness. It is in large part responsible for the smoothness of muscle action and for our ability to produce complexly coordinated movement without careful analysis of each stage of the movement complex. Proprioceptive endings exist as spirally arranged nerve terminals in relation to special rather slender muscle fibers and as encircling terminals on tendon fibers.

The **motor unit** is composed of a nerve cell body plus its long axon and terminal branches, together with all the muscle fibers supplied by these branches. However, the number of muscle fibers in the unit varies greatly. Generally, muscles controlling fine or delicately modulated movements (of ossicles of ear, of eyeball or larynx, for example) have the smallest number of muscle fibers (under 10) per motor unit. Coarse-acting muscles (as of limbs) have larger motor units composed of over 100 and as high as 2000 muscle fibers to the single axon. It is clear then that each nerve fiber must branch repeatedly within the muscle, since a muscle fiber cannot contract without a nerve impulse reaching it. Study of contraction of individual muscle fibers, or small groups of them, shows that a muscle fiber contracts completely or not at all. So long as stimulation is sufficient to cause a contraction, there is only one degree of contraction and that is maximal. This is known as the 'all or none' law of muscle action. Since the individual muscle fiber and the motor unit act in this manner, it is apparent that we must look elsewhere for an explanation of the very delicate gradation that may be achieved in muscle action. Such graded movement is actually the result of variation in the number of nerve fibers carrying stimulation to the muscle. With only a few nerve fibers stimulated, a relatively small part of the muscle contracts; with all of the nerve fibers discharging, the entire muscle contracts maximally. There is also a gradient of nerve control in the opposite direction. The so-called 'lengthening reaction' refers to the controlled relaxation of a

muscle, and the progressive relaxation of certain components in any skilled act is an important part of the regulation of that act. While we speak of muscles and their actions we actually learn movements. This is one of the fortunate and impressive qualities of the nervous system—that in producing a skilled act, all of the regulation of tonus and of contraction of related muscles (antagonists, synergists, and fixators) is incorporated into the movement complex without our real attention or concern.

Electromyographic studies (Basmajian) show that muscles at complete rest show no electrical activity. That is, there is no random contractile activity as has been adduced to explain muscular 'tone.' The latter is thought to be determined by the passive elasticity or turgor of muscles and fibrous tissues and by the active but discontinuous muscular contractions following nerve stimuli. Further, under gravity conditions (or in supporting weights passively) muscles are not active as long as ligaments are adequate to maintain the body. If the ligaments are being overpowered or if voluntary movement is introduced, appropriate muscles immediately come into play. Man's erect posture (the weaknesses of which contribute to certain of his ills) is based on a line of gravity (p. 296) which intersects the curvatures of the vertebral column. With this balance, and aided by numerous ligaments, the erect posture is maintained with minimal muscle action.

INTEGRATION OF MUSCULAR ACTION

That movements are complexes has been indicated. We can better understand the results of muscle interaction if we analyze the part which individual muscles or muscle groups play in action patterns. It will be evident that in any situation certain muscles are acting as **prime movers.** For example, in the forearm there is a digital extensor muscle which is considered a prime mover for extension of the fingers, and, similarly, on the flexor side there are digital flexor muscles which are prime movers for flexion of the fingers. **Synergists** are muscles which assist a prime mover or complement its action. Thus, we find that while the digital extensor is a prime mover for extension of the fingers, complete straightening of the terminal segments of the digits requires the action of certain small muscles in the hand, the interosseous and lumbrical muscles. The interossei have other and different actions on the digits but also are synergists for complete digital extension. When we extend the digits, the same muscles tend to extend the wrist, but the wrist may be immobilized to produce a more discrete digital extension. Such immobilization of the wrist is produced by the wrist flexors (flexor carpi ulnaris, flexor carpi radialis, and palmaris longus) acting as **fixators.** These muscles are essential to discrete finger extension in preventing the wrist from being bent backward. The term **antagonist** is also used in analyzing muscle action; flexors are antagonistic to the extensors and vice-versa. So-called antagonists, however, are usually not active when the opposite prime movers are contracting. They are called into play at the end of a violent action to protect the joint and, at other times, serve as synergists to prevent undesired actions in a movement complex. The resistance of outside forces and the pull of gravity are other factors which serve as antagonistic elements to which muscles adjust in the production of smooth movement patterns.

SHUNT AND SPURT MUSCLES

Another aspect of muscular integration comes from the analysis of their theoretical capacities. A muscle whose pull operates largely along the line of the bones to which it is attached (as brachioradialis, fig. 7) exerts much of its force in maintaining firm contact of the articular surfaces in the joint across which it pulls. This type of muscle is called a 'shunt' muscle. If the muscle orientation is more transverse to the bone it moves (as brachialis or biceps brachii at the elbow) it is a 'spurt' muscle, capable of rapid and effective movement. Mac-Conaill and Basmajian ('69) have stated two laws in relation to muscle action, as follows.

(1) *The law of minimal spurt action.* No more muscle fibers are brought into action than are both necessary and sufficient to stabilize or move a bone against gravity or other resistant forces, and none are used in so far as gravity can supply the motive force for movement.

(2) *The law of minimal shunt action.* Only such muscle fibers are used as are necessary and sufficient to ensure that the transarticular force directed toward a joint is equal to the weight of the stabilized or moving part together with such additional

centripetal force as may be required because of the velocity of that part when it is in motion.

NOMENCLATURE OF MUSCLES

Throughout the study of anatomy the student should make use of the logical naming of structures to increase associations which will facilitate learning. The names of muscles almost always indicate some structural or functional feature of the muscle. **Geometric names** are common, as trapezius, quadratus lumborum. The **general form** of a muscle is indicated in the names gracilis, serratus anterior, longus capitis. **Location** is utilized in the terms temporalis, supraspinatus, tibialis posterior. The **number of heads** of origin is used for naming purposes, as in biceps, triceps, and quadriceps. A number of muscles are named according to their **principal action,** as flexor digitorum, extensor digitorum, flexor carpi ulnaris, levator scapulae. There are also names which combine several features, for example, action and shape in pronator teres and pronator quadratus.

STRUCTURES ACCESSORY TO MUSCLES

Subcutaneous bursae were mentioned under subcutaneous connective tissue. Similar fluid-containing spaces are found in relation to tendons. These **subtendinous bursae** are frequently located just proximal to the insertion of the tendon into a bone and reduce the friction that would otherwise occur as the tendon moves against it. Structures of related function are the synovial sheaths which closely surround certain tendons where they run across bone and under restraining ligaments or fascial bands. A **synovial sheath** (fig. 9) is a tubular sac wrapped around the tendon so that one layer is applied to the tendon and has its smooth inner surface in continuity with the enveloping or outer wall of the sheath. A small amount of synovial fluid lubricates these surfaces. Where the 'wrapped around' tube meets itself, the opposed layers form a complete or incomplete **mesotendon** through which tiny blood vessels are able to reach the tendon. Synovial sheaths, especially frequent in the wrist and ankle, are always a centimeter or two longer than the restraining ligament so that the tendon may move, unhindered, the full excursion provided by its muscle.

Highly organized, discrete connective tissue

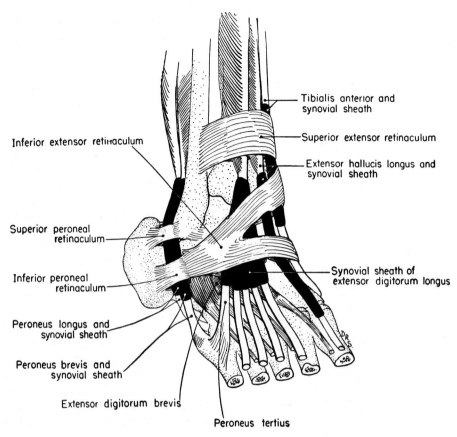

Tibialis anterior and
synovial sheath

Superior extensor retinaculum

Extensor hallucis longus and
synovial sheath

Inferior extensor retinaculum

Synovial sheath of
extensor digitorum longus

Superior peroneal
retinaculum

Inferior peroneal
retinaculum

Peroneus longus and
synovial sheath

Peroneus brevis and
synovial sheath

Extensor digitorum brevis

Peroneus tertius

Fig. 9   Synovial sheaths and retinacula at the ankle.

planes are the **deep fasciae** of the body. These are fibrous membranes devoid of fat and closely related to skeleton, ligaments, and muscles. Such fascia has no sharp distinction from epimysium as it invests a muscle but elsewhere it shows considerable specialization. **Intermuscular septa** (fig. 10) are laminae of deep fasciae which extend between and enclose muscle groups, frequently being continuous with the periosteum of the bones. Other representatives of the deep fasciae form cylindrical investments of whole regions of the body which are largely muscular. Such fasciae are made up of a lattice-like meshwork of connective tissue, the fascicles of which intersect one another at angles of 70° to 85°. They yield to increases in the circumference of muscle masses during contraction by increasing the angulation between the lattice-like fiber bundles. Such increases range from 5° to 13° (Im Obersteg, '48).

Other specialized fascial structures enclose and provide conduits for the passage of tendons. They are especially well seen in the fingers where the attachments of the **fibrous sheath of the digit**

to the margins of the phalanges complete a fibro-osseous tunnel or canal through which the tendon runs and by which it is prevented from bow-stringing as the finger is bent. **Retinacula** (fig. 9) are thickenings of the deep fasciae which serve to hold the tendons close to the bone as the joints are flexed.

The fascial structures associated with muscle are parts of the general differentiated and un-differentiated connective tissue matrix of the body. With or without fat, strongly sheet-like or very loose and diffuse, there is connective tissue around and between all the other body parts. In its looser and more fluid form it facilitates movements of muscle against muscle and everywhere it lies around blood vessels and nerves as a soft protective packing. Existing around and between all of the body structures, connective tissue elements are necessarily continuous with one another, and when we define discrete fascial planes that continuity is more or less disregarded. The degree of specialization and the extent to which the dissector divides connective tissue into discrete planes are among the

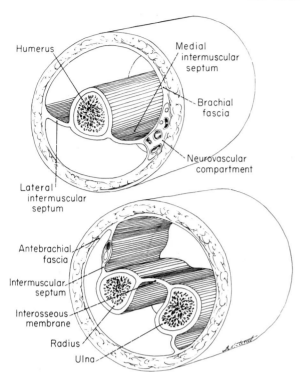

Fig. 10 Deep fasciae, intermuscular septa, and muscular and neurovascular compartments in the arm and forearm.

variables which make for confusion in the description of the fasciae of the body. The universal continuity of their planes and the intergrading of their composition must never be overlooked.

Blood vessels enter a muscle in close association with its nerve. Additional vessels, however, frequently enter extensive muscles without accompanying nerves. As a rule they are derived from the nearest major vessels of the region and it is unusual that individual names are given to blood vessels serving muscles alone. An anastomosing plexus of blood vessels is found within the muscle substance and, in correspondence with the high metabolic requirements of this tissue, the blood supply is rich. Capillaries and small vessels run parallel to the muscle fibers with frequent, short cross-communications between them. The requirement of muscle for blood is five or six times as great during exercise as during rest. The **lymphatics** of muscle have proved difficult to demonstrate but recently investigators have shown them to be present as a plexiform network in the perimysium and around the muscle fibers. Small lymphatic vessels in the connective tissue sheaths and intermuscular septa have been observed without question for many years.

THE CIRCULATORY SYSTEM

The one-celled organism lives in a watery medium. In this fluid environment the oxygen for its respiratory needs is dissolved, nutrient materials for its metabolism are provided, and the excretory products of the cell's interchange are carried away. This fundamental dependence on a fluid environment is no less a feature of multicellular organisms and holds true for Man. In the multicelled organisms, blood is the fluid that carries oxygen to the individual cells and carbon dioxide to the lungs for discharge. It dissolves the products of digestion and distributes them to the tissues, transports internal secretions from the cells that elaborate them to those that use them, and, with the lymph, carries the waste products of tissue metabolism to the kidneys, lungs, and skin for excretion. There is direct evidence that the life of a body part depends on a continuous and adequate circulation, for if there is serious interference with blood flow by blood vessel disease or occlusion, gangrene will set in with death of the part resulting.

The fluids concerned in supporting the metabolism of the cells and tissues are the **blood** and **lymph,** and the system of transport is composed of a pump, the **heart,** and tubes or vessels, the **arteries, capillaries,** and **veins.** The lymph is collected in **lymphatic capillaries** and vessels. The blood of the body has a volume of about 6 liters and is composed of a fluid portion, **plasma,** and a cellular portion, the **blood corpuscles.** The plasma is about 91 per cent water, the other 9 per cent of its volume being composed of proteins, salts, enzymes, and the products of digestion and excretion. Plasma also dissolves a fraction of the oxygen and most of the carbon dioxide of gaseous respiration. The red blood corpuscles combine with about sixty times as much oxygen as is carried in the plasma. The **red corpuscles,** or **erythrocytes** (fig. 11), are flattened biconcave discs averaging from 7 to 8 μ in diameter and are present in numbers of 5,000,000 to 6,000,000 per cu. mm. in the blood of adult Man. The normal life of a red blood cell is 120 days. An erythrocyte appears to consist of a highly elastic, semipermeable surface membrane surrounding a solution of hemoglobin in a structureless colloidal matrix. The hemoglobin readily combines with and releases oxygen under different conditions of gaseous pressure in lung and body tissue. Several types of **white blood corpuscles** (fig. 11) are present in circulating blood. All are nucleated cells, more or less ameboid in character,

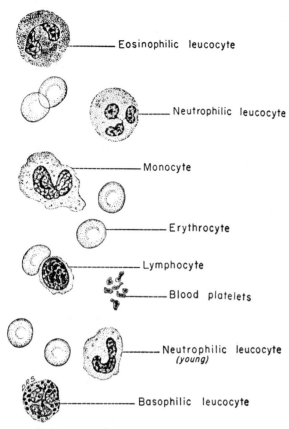

Fig. 11   The cells of circulating blood.

and probably all are concerned to a degree with the protection of the body against bacterial infection and toxic absorption. There are normally about 7000 white corpuscles per cu. mm. of blood. The white blood corpuscles exhibit some variety and their special description is not pertinent here. To be recognized among them are the granular cells—**neutrophilic leucocytes, eosinophilic leucocytes,** and **basophilic leucocytes;** and the non-granular cells—the **lymphocytes** and the **monocytes. Blood platelets,** or **thrombocytes,** tiny bits of granular cytoplasm of about $2\ \mu$ diameter, exist in rather large numbers (about 200,000 per cu. mm.) in circulating blood. They are considered to play an important role in the process of coagulation of the blood. **Lymph** is a fluid very similar to blood plasma with similar proteins and salts. In its principal collecting vessel, the thoracic duct, emulsified fat is also present after a meal. The cells of the lymph are almost entirely **lymphocytes** which vary from 2000 to 20,000 per cu. mm. in the thoracic duct of man.

An intimate fluid environment for the cells of the multicelled animal is provided by the distri-

bution of the blood through a diffuse and complex meshwork of tiny vessels. These are **capillaries** (fig. 15), composed almost entirely of endothelium and of just sufficient caliber to pass a red blood corpuscle (8 to 10 μ). The endothelium of the capillary is of flattened pavement cells, elongated in the direction of the vessel, each cell containing a centrally placed nucleus. There are no smooth muscle cells in the capillary wall and it is doubtful if there is any inherent contractility in the capillaries of mammals. The thin endothelial wall permits the transudation of blood plasma out of the capillary, thus allowing soluble food materials and oxygen to come into contact with the cells of the tissues, and the transfer of the waste products of their metabolism back to the plasma. The fluid plasma is again received into the capillary toward its venous end and returned to the circulating blood. A small proportion of the plasma is picked up by the lymph capillaries and returned to the blood stream through the lymph vessels. Normally, red blood cells do not pass through the capillary wall, but the more ameboid white corpuscles do pass out of the capillary to engulf degenerating cells and particulate matter or to form antibodies. Obviously, certain systems of capillaries are concerned primarily with the maintenance of normal function in muscle, skin, and other tissues, while other systems of capillaries are concerned with adding the products of digestion to the blood, eliminating the waste products of tissue metabolism, and effecting gaseous exchange of oxygen and carbon dioxide. The last-mentioned capillary system, that of the lungs, is served by a special vascular circuit.

Circulation of the blood is essential in support of the life and function of the tissues. In this circulation, the heart, the arteries, and the veins are necessary accessories. The heart provides the force for the movement of the blood through the system of larger vessels and capillaries. It is composed essentially of a pair of muscular pumps, one concerned with propelling blood into the capillary system of the lungs for oxygenation, the other concerned with forcing blood through the tissues of the remainder of the body. The propelling chamber of each pump, the **ventricle** (fig. 12), may be likened to a muscular sac. The muscular fasciculi of the sac have their connective tissue attachments in the region of its neck. The connective tissue at the neck of the sac also provides an inlet and an outlet for the muscular chamber with appropriate one-way valves. The receiving

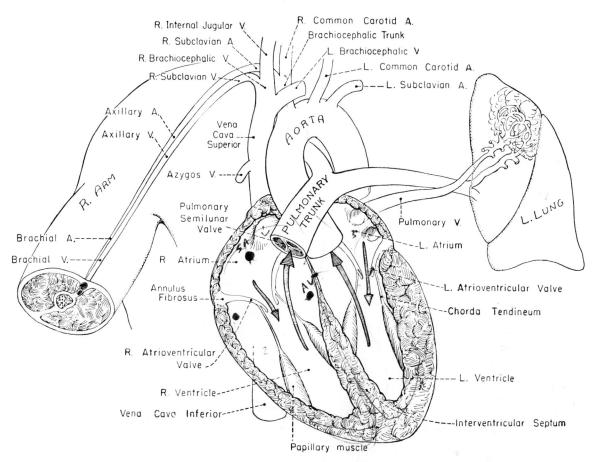

Fig. 12   The plan of the circulation; the chambers of the heart, the great vessels, and representative systemic and pulmonary circuits.

chambers, the **atria** (fig. 12), are less regularly shaped, have thinner muscular walls, and each receives a number of afferent blood vessels. The pumps, each composed of an atrium and a ventricle, lie fused side by side. In accordance with the capillary system to which each discharges, the right heart receives venous blood from the body mass and discharges it to the lungs for removal of carbonic acid and enrichment with oxygen. The left heart receives the enriched blood from the lungs and forces it out into the arteries and capillaries of the body. In view of the proportionately greater mass and resistance of the complex vascular system of the body, the muscular wall of the left ventricle is about three times as thick as that of the right ventricle. The muscle of the atria, being required only to propel the blood into the ventricles, is a relatively thin lamina. The chambers of the heart are lined by endothelium, a lining that characterizes the entire vascular system. Externally, the musculature of the heart is covered by a thin,

tough, connective tissue layer known as the **visceral pericardium.** The capacity of each chamber is approximately the same, 60 to 70 cc., and, at the average 72 per minute rate of heart beat, about 5 liters of blood passes through the heart each minute. This increases to 20 liters or more during exercise.

The **arteries** conduct the blood from the heart to the capillary bed. The arterial system begins in large vessels close to the heart. The largest, the aorta (fig. 12), carries the full volume of the blood expelled and is approximately an inch in diameter. The division of this and subsequent major branches leads to a continuous increase in their number and spread, with a concomitant progressive decrease in size. Thus one recognizes large, middle-sized, and small arteries and, finally, the tiny arterioles. Arteries (fig. 13) consist of a tube lined by endothelium similar to the capillary. The **tunica intima** has, like the capillary, its flattened surface cells longitudinally oriented. It includes subendothelial

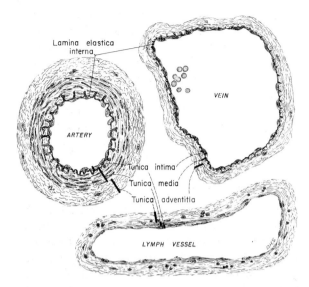

Fig. 13   Cross sections of histologic preparations of an artery, a vein, and a lymphatic vessel.

connective tissue and an elastic tissue layer designated the **lamina elastica interna.** Surrounding these inner layers is a relatively thick **tunica media** composed of smooth muscle and elastic tissue in varying proportions. The outer investment of arteries is the **tunica adventitia,** composed mainly of collagenous fibers. The principal structural differences between arteries of varying sizes are found in the tunica media. In the large vessels elastic tissue predominates, while in the smaller vessels smooth muscle forms most of its substance. These structural differences relate to the functional demands placed on such vessels. Larger arteries, receiving the full force of contraction of the heart, utilize a greater proportion of elastic tissue to smooth out the stroke of the heart and to act in a small way as an expansile reservoir for the blood. As a corollary, the recoil of the elastic tissue serves to prolong the thrust of the blood toward the periphery. The familiar pulsation which we feel in 'taking the pulse' is gradually smoothed out in the passage through ramifying arteries until a fairly smooth and continuous flow is present in the capillary bed. As the arteries branch and rebranch, producing smaller and smaller vessels, the proportion of smooth muscle increases until in the small arterioles the elastic tissue is essentially lacking. The arterioles and the small arteries are of greatest importance in regulating the flow of blood through the capillary bed. Their muscular walls are capable of contraction so as to shut off completely the flow to the capillary bed under

particular local conditions. This is part of a necessary control of blood flow, for the cross-sectional area of the entire capillary bed of the body is some one thousand times as large as the cross-sectional area of the aorta as it leaves the heart.

Veins (fig. 13) return the blood from the capillary bed to the heart. Generally they are thinner-walled and larger in diameter than their corresponding arteries. They also tend to be double or multiple, since an equal quantity of blood has to pass a given point as in the arterial outflow while at the same time the rate of flow is much reduced. The veins may also take separate courses toward the heart, some of them in positions quite remote from arteries, as in the superficial veins of the back of the hand, forearm, and arm. Veins form plexuses in certain regions, as around the vertebral column and in the pelvis. A vein has a structure similar to that of the artery with similar layers in its wall. The tunica media, however, is relatively poor in smooth muscle and elastic tissue and is, therefore, sometimes difficult to distinguish from the tunica adventitia. In large veins, the tunica adventitia is thick and includes some elastic tissue and smooth muscle among the predominantly white, fibrous tissue framework. Veins contain valves located at intervals along their course and so disposed as to permit flow only toward the heart. These are formed of simple duplications of the endothelial lining folded into the lumen of the vessel. A **valve** (fig. 14) is usually composed of two such infoldings opposite each other, a bicuspid arrangement. A valve cusp is rather crescentic in shape; its lower curvature is its attached margin; the upper curve is its free margin. Veins are frequently somewhat dilated above valves, and valves are especially frequent in the veins of the limbs and in muscle veins. Valves protect smaller veins from the higher pressure of the blood in larger veins and from pressure increases produced by muscle

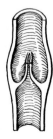

Fig. 14   A longitudinal section of a bicuspid venous valve.

action (Franklin, '37). They clearly serve to support the column of blood above them, dividing the column into smaller segments of less weight, assisting the flow of blood toward the heart, and preventing it from pooling in the areas of the body most affected by the force of gravity. Most importantly, muscular actions in the limbs and other parts of the body where valves are located work in conjunction with the valves to keep blood moving toward the heart. Valves are few in the veins of the abdomen, thorax, and neck; in the head and neck gravity assists the blood flow toward the heart.

HEART AND GREAT VESSELS (fig. 12)

The **heart** varies in size with the size of the individual and has been likened to the mass of the individual's two fists. The heart lies within the large **pericardial sac.** This is a serous sac formed of connective tissue and lined by a smooth layer of cells which secrete sufficient fluid to lubricate the adjacent surfaces. This mechanism allows for the movement of the heart within its connective tissue covering without friction. The heart is not actually located on the left side of the chest but it extends farther to the left of the midline than it does to the right. The readily identified position of maximal heart beat on the chest wall in the fifth interspace indicates the lower left limit of the heart. Large blood vessels enter and leave the heart in accordance with the general plan of circulation already discussed. The right atrium receives all the venous blood from the body, exclusive of the lungs, by means of three principal channels. The superior vena cava brings to the heart blood from the head, neck, upper limbs, and chest. The inferior vena cava collects the venous blood from the lower limbs, abdomen, and pelvis. The right atrium also receives the venous blood from the heart wall by means of a collecting vessel on its surface known as the coronary sinus. The right ventricle gives origin to the pulmonary artery (fig. 12) which discharges to the lungs all of the venous blood collected by the right atrium. The oxygenated blood from the lungs is returned to the left atrium of the heart by four pulmonary veins and this enriched blood is conveyed to the body and to the walls of the heart through the aorta which arises at the outlet of the left ventricle. The carotid arteries transmit blood to the head and neck, and the subclavian arteries to the upper limbs. As the aorta proceeds through the thorax and

abdomen, a number of arteries branch from it to distribute its blood to the tissues of the body. It will be observed that the term 'vein' is used for vessels carrying blood to the heart even though such blood has been enriched (arterial blood) by passage through the pulmonary capillary system, and that the term 'artery' is applied to the vessels leaving the heart for the lungs even though they do not transmit oxygenated blood. This is conventional usage.

NERVOUS CONTROL OF THE CIRCULATION

It has been pointed out how important the smooth muscle in the wall of small arteries and arterioles is for the regulation of blood flow in any peripheral region of the body. It is also evident that the rate and force of the heart beat are increased under certain stressful conditions and are decreased at rest. The nerve mechanism which controls these factors operates almost completely on an automatic and involuntary plane (p. 30). One cannot will an increase in heart rate, but such an increase occurs promptly during an emotional response or as a result of increased work on the part of the body. The heart beat may be described as myogenic, but whether it be rapid and forceful or more moderate in action is determined by nerve impulses. Arteries are also supplied by nerves, the nerves to the smooth muscle of blood vessels usually producing vasoconstriction, although active vasodilation has also been described. However, like the force of gravity acting on the somatic muscles of the body, pressure of the blood is one of the forces which is effective in vasodilatation. There are sensory mechanisms which respond to increases in pressure in the circulatory system and act reflexly to reduce the pressure toward more normal requirements. Such a responsive area is the carotid sinus at the beginning of the internal carotid artery. The carotid body, a collection of chromaffin tissue adjacent to the carotid sinus, acts as a chemoreceptor, responding to changes in the chemical constitution of the blood.

The presence of a pump in the circulatory system and of a complex system of minute vessels peripherally point to marked pressure changes in the system and the need to control and adjust to them. The usual pressure exerted by contraction of the ventricles is measured as a force sufficient to elevate a column of mercury to a height of 120 mm. This 'systolic' pressure is reduced by the resistance of the arterial tree to approximately 32 mm. of mercury at the arterial end of a capillary. At the venous end of a capillary loop the blood pressure has dropped to about 12 mm. of mercury, indicating that there is an appreciable resistance in the capillary bed. The blood begins its return to the heart with a force behind it of only about one-tenth that with which it left the heart, but, there being little further resistance in the circuit, there is only a small decrease of pressure in the subsequent venous return to the heart. Since, however, no further force is applied and, with progressive union of veins, the column of blood becomes heavier, the venous flow is much slower and requires much larger vessels to accommodate it. The obvious value of the venous valves is again demonstrated in these facts of blood pressure. Muscle action of the limbs, respiratory movements of the chest, and the excursions of the diaphragm are also thought to have a significant action in 'milking' the venous blood on toward the heart.

The pulmonary vascular circuit is much shorter and operates against much less peripheral resistance. Coupled with the reduced thickness of the muscular wall of the right ventricle ($\frac{1}{3}$ as thick as that of the left) is an initial blood pressure in the pulmonary circuit of only 25 mm. of mercury.

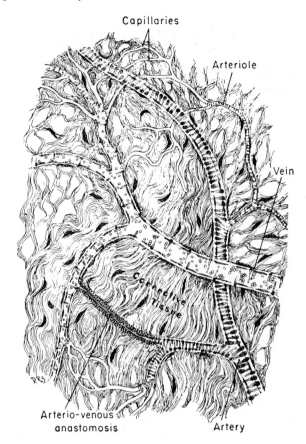

Fig. 15 Small arteries, veins, and capillaries; an arterio-venous anastomosis.

ARTERIOVENOUS ANASTOMOSES (fig. 15)

The cross-sectional area of the capillary bed is about a thousand times the cross-sectional area of the aorta. Even allowing for the increased capacity required as the blood moves more slowly, it is evident that we are not equipped to pass optimal amounts of blood through all of the capillaries at once. However, many parts of the body require a major blood supply only intermittently, such as the digestive system after a meal. The action of muscles is likewise local and intermittent, so that their capillary networks need to be wide open only at intervals. A means of control for local variations is the arteriovenous anastomosis. Such an anastomosis consists of a side branch of an arteriole leading directly into a venule. This short, shunting vessel is equipped with a well-developed muscular wall which is capable of completely closing this channel, thereby forcing blood to pass through an adjacent capillary system or, contrarily, of opening wide and short-circuiting the capillary bed. The regulation which such an anastomosis may effect on a capillary

circulation can obviously reduce the capillary blood flow in a part when blood is demanded elsewhere, or can contribute to an increase in capillary blood flow under conditions of high local requirements. These anastomoses also appear to be plentiful in the skin where they may be concerned with temperature regulation by permitting more rapid blood flow through the tissues, increasing the radiation of heat. They are especially noted in exposed parts, such as the lips, ears, nose, and digits.

THE LYMPHATIC SYSTEM (figs. 16, 17)

It has been noted that the blood plasma constitutes the fluid environment for the cells of the various tissues and is thereafter returned in large part to the veins of the body. A small proportion of this plasma, especially the fraction rich in macromolecules, large molecule proteins, and particulate

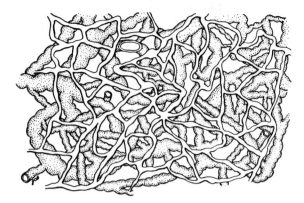

Fig. 16 A subcutaneous plexus of small and large lymphatic capillaries. (Modified from Teichmann.)

matter, is collected by the lymphatic capillaries. Between one and one-half and three liters of lymph are returned to the blood stream through the thoracic duct, the largest lymphatic channel of the body, each 24 hours.

The lymphatic system is the principal system of transport for assimilable fats in digestion, and approximately 95 per cent of its contents are received from the intestinal tract and the liver. Lymph is generally a clear, colorless fluid and usually contains no cells until lymphocytes are added in its passage through the lymphatic nodes. Lymph collected from the digestive tract may be, however, milky in appearance after a meal, due to contained fat particles. The lymphatic capillaries (fig. 16), like the blood capillaries, form the specific functional part of the system. They are simple endothelial tubes with closed ends which form richly anastomosing plexuses. They do not contain valves and they vary enormously in size from a few micra to almost a millimeter. In the dermal papillae and intestinal villi, lymphatic capillaries form blind pouches. The fluid pressure in the lymphatic capillaries of from $1\frac{1}{2}$ to 3 mm. of mercury is only a little less than that of tissue fluid pressure, but under increased activity tissue fluid pressure is elevated so that the differential pressure becomes greater. A pressure difference is required for the passage of plasma into the lymphatic capillary, and increased lymphatic flow accompanies muscular activity. Osmotic pressure also plays a role in the balance of forces determining differential pressures.

The distribution of lymphatic capillaries is uneven in the body. They are absent in tissues which have a scanty blood supply, as hyaline cartilage, epidermis, and the cornea of the eye. They are particularly rich in the inner and outer surface layers of the body, namely in the dermis and in the mucosa and submucosa of the alimentary tract. They are also numerous under serosal surfaces and in the capsules of joints. Muscles, bones, and fasciae exhibit a meager supply of lymphatic capillaries and they are unevenly distributed in glandular organs. There is no lymph in the brain or spinal cord, but lymphatic capillaries are present along peripheral nerves.

The collecting vessels of the lymphatic system contain valves, are rarely more than 0.5 mm. in diameter, and form a richly anastomosing plexus. Their structure resembles that of veins, but the specific tunics, or layers, are not sharply defined (fig. 13). The lymphatic vessels, communicating freely, run in loose connective tissue between organs, in the subcutaneous tissues, and along the course of arteries and veins. The drainage of lymphatic vessels tends to be similar to the drainage of veins. The thoracic duct, which drains the lymph from the lower half of the body and from the left side of the upper half of the body, is the largest of the lymphatic vessels. It is from 4 to 6 mm. in diameter and is evident in its course through the chest. It empties into the junction of the left subclavian and left internal jugular veins. Other lymphatic vessels which may be observed in the base of the neck are the jugular lymphatic trunk and the subclavian lymphatic trunk.

Lymphatic nodes (fig. 17) are organized collections of lymphatic tissue interposed in the course of the lymphatic stream. These bodies vary from the size of a pinhead to that of a small olive and are frequently even larger in disease. They have a flattened, oval form and are a pale pink in life but a rather indeterminate color in the embalmed cadaver. Lymph nodes located in the hilum of the lung may be quite black due to the accumulation within them of carbon particles taken from the lymphatic stream as it leaves the lung. The lymph nodes and vessels of the mesentery take on the milky whiteness of the absorbed fat after meals, for some 60 per cent of the fat absorbed through the intestinal tract is transported through the lymph. The lymph node has a fibro-elastic capsule from which connective tissue trabeculae extend into the center of the gland. Masses of

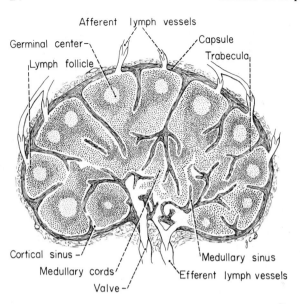

Afferent lymph vessels
Germinal center
Capsule
Lymph follicle
Trabecula
Cortical sinus
Medullary sinus
Medullary cords
Efferent lymph vessels
Valve

Fig. 17 A semi-schematic frontal section of a lymphatic node.

lymphocytes, as rounded lymph follicles, fill the periphery of the gland, and strings and cords of cells occupy the medulla. Along the supporting reticular tissue of the medulla lie reticulo-endothelial cells. These cells ingest foreign particles and debris floating past them in the sluggish lymphatic stream. The lymph node is characterized by the presence of numerous lymph vessels. Afferent collecting vessels penetrate the rounded surface of the node and within it form rich plexuses of wide capillary size surrounding the lymph follicles. The plexus of vessels extends into the medulla, or central part of the node, where vascular sinuses (medullary sinuses) occupy the spaces around and between the cords of cells. Lymph nodes generally exhibit a slightly indented hilum from which issue a much smaller number of larger efferent lymph vessels. One of the principal functions of the lymph node is the formation of lymphocytes. These blood cells are formed in great quantity in the germinal centers of the lymph follicles, are added to the lymph as it flows through the node, and are ultimately carried by this fluid to the blood stream. The reticulo-endothelial cells phagocytize particulate matter, bacteria, and cell fragments and impart to the node its commonly recognized filtration action. **Special lymphoid collections appear as solitary and aggregated nodules in the intestinal wall and** in the appendix. The palatine and pharyngeal tonsils and the spleen are lymphatic organs which exhibit an individual organization.

Clinical Note Lymph glands enlarge during infectious processes and may themselves become infected, swollen, and tender. Their enlargement and tenderness may be a diagnostic aid in disease. The lymphatic system is utilized by some kinds of cancer cells as a medium of spread through the body. This process of metastasis (change + position; thus migratory movement) accounts for the wide dissemination of cancer in the body in certain cases.

THE NERVOUS SYSTEM

In the sensorimotor framework in which we view Man, the nervous system is the adaptive and adjusting mechanism. It is through the nervous system that stimuli are received and interpreted and a selection of response made. Man's position at the head of the animal kingdom is fundamentally due to the capacity and effectiveness of his nervous system. His cerebral development has emancipated him from most of the more stereotyped reaction patterns dealing with the maintenance of life itself and provides, in its extra capacity, freedom for conceptual and abstract thought. It may be mentioned parenthetically that another feature of Man's pre-eminence is his highly mobile and skillful hand, a structure of advanced muscle-joint specificity coupled with a highly specific and abundant nervous innervation. The study of the detailed morphology of the brain is not a part of the present treatment, but the peripheral nervous system and the gross relations of the spinal cord are examined.

The peripheral nervous system (fig. 18) is composed of twelve pairs of cranial nerves and thirty-one pairs of spinal nerves. Each peripheral nerve is a collection of nerve fibers which are conductors for various types of sensation and for impulses passing to effectors, such as muscle or gland cells. The various functional entities found in a peripheral nerve are expressed as nerve components (fig. 19). Four components are present in any spinal nerve. **General somatic afferent** (incoming) fibers transmit sensation from the body to the spinal cord. Such sensations may be exteroceptive, as pain, temperature, touch, and pressure. Also included are proprioceptive sensations from muscles, tendons, and joint capsules, providing information

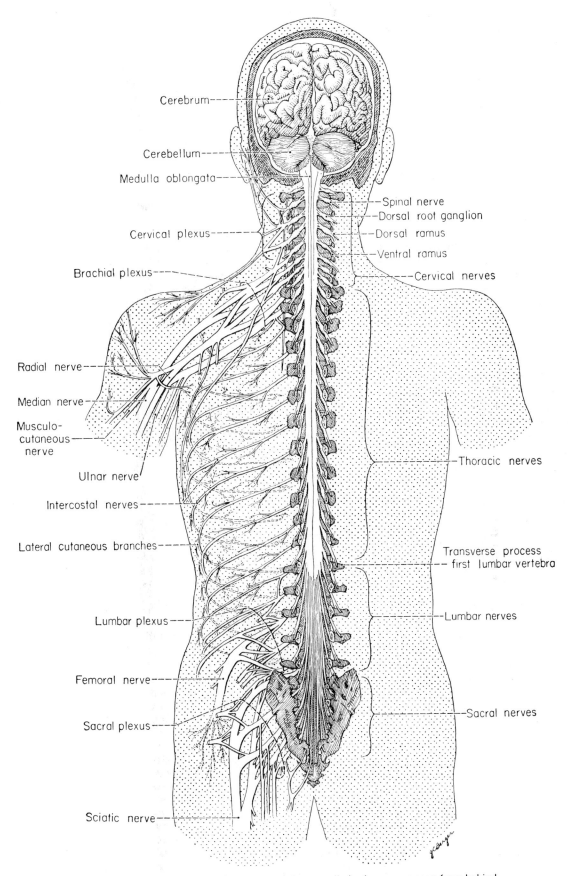

Cerebrum

Cerebellum

Medulla oblongata

Cervical plexus

Brachial plexus

Radial nerve

Median nerve

Musculo-
cutaneous
nerve

Ulnar nerve

Intercostal nerves

Lateral cutaneous branches

Lumbar plexus

Femoral nerve

Sacral plexus

Sciatic nerve

Spinal nerve
Dorsal root ganglion
Dorsal ramus
Ventral ramus

Cervical nerves

Thoracic nerves

Transverse process
first lumbar vertebra

Lumbar nerves

Sacral nerves

Fig. 18   The spinal cord, spinal nerves, and the great limb plexuses as seen from behind.

25

on joint position and the tension of tendons and muscle fibers. **General visceral afferent** fibers convey impulses from mucous membranes, glands, and blood vessels. **General somatic efferent** (outgoing) fibers carry motor impulses to skeletal muscles, and **general visceral efferent** fibers supply smooth muscle and glandular tissues.

Cranial nerves exhibit fundamentally the same components, but the type of sensation and certain of the muscle groups resident in the head region are designated as 'special.' Thus **special somatic afferent** fibers occur in those nerves serving the special senses of vision and hearing. **Special visceral afferent** fibers transmit impulses from special senses of visceral type, namely taste and smell. **Special visceral efferent** fibers distribute to muscles derived embryologically from branchiomeres, i.e., skeletal muscle which is embryologically related to the digestive tract, as the muscles of mastication or of facial expression. The **somatic efferent** component is composed of motor fibers to muscles of metameric origin in the head; this muscle is found in the orbit and tongue. Since the general skeletal muscle of the body is also metameric in origin, this component is essentially the same as general somatic efferent in spinal nerves. While the special components are found only in cranial nerves, certain cranial nerves also exhibit general visceral afferent and general visceral efferent components in their distribution to the smooth muscle and glandular structures of the head and

body, and general somatic afferent components are also represented.

There is nothing in the structure of the nerve fiber that determines the component to which it is related. Nerve fibers are simply conducting elements of the nerve cell, or neuron, and are capable of conducting an impulse in either direction and, artificially speaking, from any applied stimulus. The specificity which does exist in the intact functioning nerve fiber is a feature of its termination. A 'pain' fiber conducts painful impulses because of its freely branched ending in the epidermis of the skin; a 'touch' fiber conducts a touch stimulus because of its specialized encapsulated ending in a dermal papilla. Similarly, it is the motor end-plate termination of a motor nerve which gives it its specific function (fig. 19). The lack of unidirectionality in the conductivity of the nerve fiber is remedied by the presence of a minute gap between neurons known as the **synapse**. Neurons conduct in a chain type of sequence, and an impulse must pass from one neuron to the next at the synapse. The impulse will bridge this minute gap in only one direction, thus polarizing the neuron and determining the direction of conduction. Specificity of function and of conduction is provided by these terminal features of the nerve fiber, and the pattern organization of the synapses which a nerve impulse may follow in the central nervous system is the organization that gives the brain and spinal cord its special contribution to the body economy.

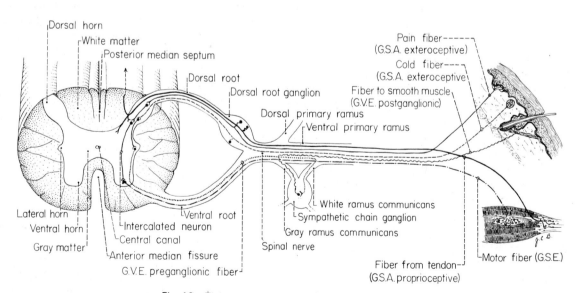

Fig. 19 The components of a typical spinal nerve.

THE NEURON (fig. 20)

A nerve fiber represents one of the processes of a neuron, the fundamental unit of nervous tissue. The neuron is composed of a cell body together with its processes, dendrites, and axons. The **cell body** of a neuron is most commonly angular in form. It contains a large, pale nucleus with a central nucleolus and cytoplasm characterized by stainable granules known as Nissl bodies. Appropriate staining methods also show the presence of mitochondria, Golgi apparatus, pigment granules, and neurofibrillae. The common multipolar form of the cell body is due to the radiation from the cell body of a number of delicate protoplasmic processes, the **dendrites.** These usually extend short distances from the cell and branch freely as they ramify in relation to other nerve processes. The polarity of the neuron is such that the dendrites and the cell body receive the neural impulse. The **axon** is a single process which usually emanates from the cell body at the side opposite the main dendrite and frequently from a somewhat protruding part of the cell body known as the **axon hillock.** The axon runs an unbranched course, although occasional collaterals may be seen, until it nears its destination, at which point it may divide into a number of branches. In certain neurons, the dendrites and axons emerge at opposite ends or sides of the cell, and the cell body has a bipolar, or spindle-shaped, form. In other neurons, the cell body is designated as unipolar, a description which applies when a single process of the cell body divides to form both the peripheral and central branch. In peripheral nerves, either dendrites or axons may be of considerable length and show much similarity of structure. The central part of the fiber is the **axis cylinder,** and it may be surrounded by a fatty sheath known as the **myelin sheath,** which is, in turn, covered by a transparent membrane, the **neurilemma sheath.** The whole constitutes a myelinated (or medullated) nerve fiber. The myelin is interrupted at about every half millimeter of length, and at these **nodes** the neurilemma dips in toward the axis cylinder. Under the neurilemma lies a flattened **neurilemma sheath cell** with an oval nucleus and reticular protoplasm. The **myelin sheath** may be thick or thin or may be essentially lacking. In the latter condition the nerve fiber is classified as an unmedullated one.

These fundamentals of neuron structure are

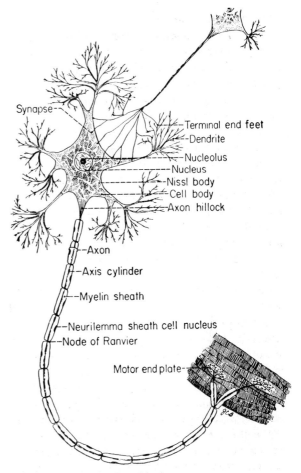

Fig. 20　A multipolar motor neuron.

exhibited by the two major types of nerve fibers in peripheral nerves. A **motor neuron** has a rather large, multipolar cell body with a richly branching ramification of dendrites and a long axon. The cell body is located in the **ventral horn** of the spinal cord gray matter, and its dendrites are relatively short, being largely confined to the ventral horn. Its axon leaves the ventral horn through the ventral white matter of the spinal cord and is organized with other axons of the same type into the **ventral root** of the spinal nerve. Continuing through the ventral root, the axon of the motor neuron may, in the case of those fibers destined for the muscles of the foot, run an over-all course of about three feet before its termination in a motor end-plate (fig. 19) under the sarcolemma of a striated muscle fiber.

The other primary type of nerve fiber in the peripheral nerve is the dendrite of a **dorsal root ganglion cell** (fig. 19). This arises as a freely branched terminal, or as an encapsulated ending, in the skin and may run an equally long course to reach its cell body in the dorsal root ganglion. This cell body is pseudo-unipolar with a single somewhat convoluted process resulting from a fusion within the ganglion of a dendrite and an axon. The axon of such a cell traverses the dorsal root of the spinal nerve and passes into the spinal cord through the **dorsal horn** of the gray matter. It may terminate within the dorsal horn or it may enter the white matter to ascend or descend to a different level of the spinal cord, or may even ascend to the brainstem. In this type of neuron, the dendrite may be a very long process and the axon a very short one. Although myelin is present on these nerve fibers within and without the central nervous system, neurilemma sheaths and neurilemma sheath cells are present only outside the central nervous system.

THE REFLEX ARC (fig. 19)

The axon of a dorsal root ganglion cell concerned in the transmission of a painful impulse may terminate in the ventral horn, in relation to the dendrites of a motor neuron. The impulse carried over the sensory neuron will then initiate a discharge through the motor neuron with resulting contraction in a muscle. This is the simple reflex arc which is basic to all neural function, although actually such a simple arc probably does not exist in Man. One's behavior, however, in withdrawing the hand when a finger is burned is so immediate and so effective a reaction as to give validity to the concept of functioning reflexes. At least one more cell is usually intercalated in the sensorimotor arc. This cell is located entirely in the gray matter of the spinal cord, is smaller in size, bipolar or multipolar in form, and is known as an **intercalated neuron.** Such a neuron can obviously serve to connect sensory and motor neurons, but it can also serve to bring stimulation from other sources to the same motor neuron and thus, while it introduces another step in the neural arc, serves to increase vastly the number of stimuli which can impinge on the motor neuron. The intercalated neuron is one of an exceedingly widespread type in the brain and spinal cord, and it is this sort of neuron which ultimately provides for the adaptability of neural responses.

The **spinal cord** exhibits in cross section a central,

butterfly-shaped area of gray matter surrounded by an irregular zone of white matter. The white matter of nervous tissue has whiteness imparted to it by the myelinated fibers it contains and is made up predominately of collections of ascending and descending nerve fibers. These may ascend or descend only short distances to accomplish a broader reflex integration within a few segments of the spinal cord. They may, however, ascend to carry sensory impulses to higher centers of the brain and thus to consciousness or descend from the brain to initiate willed action. This is a very incomplete statement of even the simplest variety of connections but serves to show that the spinal cord reflex is under the influence of other brain mechanisms at all times. Thus, there can be augmentation or facilitation of a simple reflex, or its completion may be inhibited by higher centers. Generally speaking, our more stereotyped reactions are managed on the basis of reflexes of varying degrees of complexity, and the more purposeful skilled actions depend on accurate control from the brain.

SPINAL CORD (figs. 18, 21, 243, and p. 303)

The spinal cord, enclosed in its coverings, lies in the vertebral canal formed by the superimposed vertebrae. When the vertebral canal is opened, a long tubular sheath, the tough, fibrous **dura mater,** comes into view. This extends from the foramen magnum of the skull to the second segment of the sacrum (S2) as a tube of relatively uniform caliber. Prolongations of the dural sheath extend along the spinal nerve at each vertebral segment. The dura mater narrows at the level of S2 into a close tubular investment of the filum terminale of the spinal cord and continues to the tip of the coccyx where, as the **coccygeal ligament,** it is attached. Directly under the dura mater is the second layer of the spinal meninges, the **arachnoid membrane.** This filmy (arachnoid = spidery) layer lies in close contact with the dura mater and is everywhere coextensive with the tubular cavity of the latter. Opening the arachnoid exposes the spinal cord with its intimately attached and coextensive **pia mater.** This third layer of the spinal meninges is inseparable from the surface of the spinal cord, invests its surface blood vessels, and is continued along its nerve roots. A space occupied by **cerebrospinal fluid** exists between the pial and

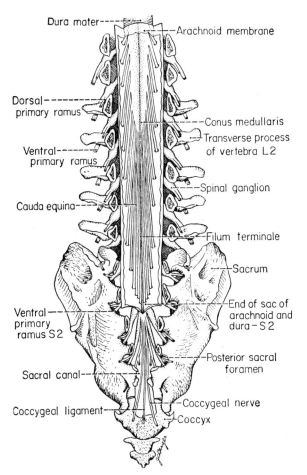

Fig. 21　The lumbosacral portion of the spinal cord and the cauda equina. The dura mater and arachnoid membrane have been incised and reflected.

21) and is continued in a mixed nervous and pial strand, the **filum terminale.** The filum terminale descends, becomes invested by the **coccygeal ligament** (filum) of dura mater, and, together with it, attaches to the tip of the coccyx. The relatively high termination of the spinal cord forces a descending course on the lumbar and sacral roots as they pass to their proper vertebral level of exit. The compact array of descending dorsal and ventral roots so formed within the lower end of the dural sac has an obvious likeness to a horse's tail, which accounts for its name, **cauda equina.**

Delicate, longitudinal grooves mark the dorsal surface of the cord, from several of which a continuous series of nerve rootlets emerges. These rootlets converge into a bundle for each segment, the **dorsal root** (fig. 19), and in their serial emergence give evidence of the segmental organization of the spinal cord. Ventrally, the spinal cord exhibits a deeper median sulcus, the **anterior median fissure,** lateral to which the ventral rootlets pass through the white matter to form a **ventral root** for each vertebral segment. The spinal cord enlarges somewhat at the levels of organization of the nerves to the limbs to provide the additional neural tissue required by the limb masses. At the respective levels **cervical** and **lumbosacral enlargements** exist (fig. 18).

The emerging dorsal and ventral rootlets combine regionally to form dorsal and ventral roots within the dural sac. Each root appears to perforate the dura mater separately, but actually the dura mater is prolonged as a tight sleeve along the more distal parts of each root. A swelling of the dorsal root at or within the intervertebral foramen constitutes the **dorsal root (spinal) ganglion.** It is the location of the cell bodies of afferent neurons of the spinal nerve. The ventral root joins the dorsal at the distal side of the dorsal root ganglion, thus forming a spinal nerve. The spinal nerve runs a very short course, for almost as soon as it emerges from the intervertebral foramen it divides into a dorsal and a ventral ramus (branch). The ventral ramus also has immediate connections with the sympathetic trunk, or a ganglion of it, located adjacent to the vertebral body.

THE PERIPHERAL NERVE

The **peripheral nerve** contains, as a rule, all the components of the typical spinal nerve, and thus it is almost never correct to speak of sensory or

arachnoid layers. This **subarachnoid space** is spanned by a scanty network of fibers of both pial and arachnoidal origin. A long, white fibrous band, the **denticulate ligament,** stretches along each side of the cord between the dorsal and ventral roots as an extension of the pia mater. Its twenty-one pairs of tooth-like attachments to the dura assist in stabilizing the spinal cord.

The spinal cord is about eighteen inches in length and extends from its continuity with the brainstem at the foramen magnum to the level of the upper border of the second lumbar vertebra. Its shortness in relation to the vertebral column is an expression of the early development of the nervous system and of the disparity in ultimate growth between the cord and the column. Just above the second lumbar level (L2) the spinal cord rapidly narrows as the **conus medullaris** (fig.

motor branches. One may designate a branch as a **muscular nerve,** recognizing that in addition to motor fibers to muscle it is also composed of sensory fibers for muscle-tendon (proprioceptive) sensation as well as both visceral efferent and visceral afferent fibers to the blood vessels of the muscle. Similarly, a nerve to the skin is a **cutaneous nerve** and contains many more fibers than its somatic sensory component requires. All peripheral nerves are mixed nerves resulting from a regrouping of fibers from the dorsal and ventral roots and of the somatic and visceral components of each. Thus, a nerve to a certain region carries all the nerve fibers required in that region, and several nerves are not needed to supply it. Similarly, there is a redistribution and regrouping of nerve fibers in the major nerve plexuses of the body to form nerves which carry all types of fibers to all sorts of structures in a certain region or part. Such plexuses are characteristic of the limb nerves in particular and make for economy in the branching and distribution of the peripheral nerves of these body parts. The brachial and lumbosacral plexuses (fig. 18) are the two largest formations of this type.

Much of the cross-sectional size of a peripheral nerve (fig. 22) is contributed by connective tissue. A relatively thick layer of **epineurium** is composed mainly of longitudinally arranged collagenous and elastic fibers. This layer surrounds all the nerve bundles composing a nerve and is supplemented by **perineurium,** made up of concentric layers of connective tissue enclosing and separating the separate nerve bundles. **Endoneurium** consists of strands of connective tissue extending in from the perineurium, separating and enclosing the nerve fibers.

DEGENERATION AND REGENERATION

If a peripheral nerve is cut, the portion disconnected from the cell body will die. Indeed, the cell body itself may exhibit degenerative changes known as chromatolysis but, after a period of several weeks, is likely to recover. The major change takes place in the distal parts of the nerve where the myelin is gradually transformed over a period of several weeks into a chain of lipoid droplets. The axis cylinder becomes disrupted and fragmented in a few days. The neurilemma sheath, however, persists and its sheath cells proliferate so that, in time, the degenerated nerve consists of a bundle of multinucleate tubes from which the degenerated fragments of myelin and axis cylinder have been removed. These neurilemma tubes represent a fortunate remnant in the nerve, for regenerating sprouts of the central nerve fibers entering such tubes are guided down them to the normal terminal distribution of the nerve, and re-establishment of effector or sensory connections can thus be made. Connective tissue infiltration producing scar tissue in the repair process of the surrounding tissues may seriously interfere with the distal growth of the regenerating sprouts and many turn backward or toward the side. Fortunately, the sprouts are produced in numbers greatly in excess of the axis cylinders of the central undamaged section of the nerve from which they come.

Clinical Note Great care is exercised surgically to provide as close an approximation of the sectioned surfaces of a cut nerve as possible, and under favorable circumstances many of the growing tips enter neurilemma tubes and effective regeneration is possible. If unimpeded, regenerating fibers proceed down the neurilemma tubes at the rate of 1 to 2 mm. a day.

THE AUTONOMIC NERVOUS SYSTEM (fig. 23)

There is a regulation over all our visceral structures, involuntary and automatic in nature, which is produced by the autonomic nervous system. The general visceral efferent neurons comprising this system have certain of their cell bodies inside the central nervous system (brain or spinal cord) and others of them outside it. The regulation of the

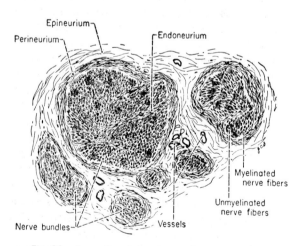

Epineurium
Perineurium
Endoneurium
Myelinated nerve fibers
Unmyelinated nerve fibers
Nerve bundles
Vessels

Fig. 22   A small peripheral nerve in cross section.

viscera is produced by a double innervation, the one set of neurons being organized as the **parasympathetic** (or craniosacral) portion of the autonomic nervous system, the other set constituting the **sympathetic** (or thoracolumbar) division of the system. The double innervation regulates or adjusts the action of the viscera by a balancing of their generally opposing actions. Parasympathetic stimulation slows the heart rate while sympathetic stimulation increases it, and the control of the heart at any one instant represents a balance of the two types of innervation.

As a general rule, parasympathetic stimulation acts to conserve or preserve the body as a vegetative organism. It slows the heart, constricts the pupil, increases peristalsis, and empties the bladder and the rectum. The sympathetic system, when in ascendancy, serves to prepare the body for crisis, as for the effort to defend itself. It serves the 'fight or flight' response. Its stimulation dilates the pupil, increases the heart rate, and raises the blood pressure to increase the blood flow to the somatic muscles, steps up respiration, and at the same time inhibits peristalsis and decreases the blood supply of the viscera. The general effect of sympathetic stimulation is the same as an injection of epinephrine into the blood stream, and there is evidence that an epinephrine-like substance, **noradrenaline,** is actually elaborated at most of the terminals of the sympathetic neurons and may be the medium whereby its effects are produced. In a similar way **acetylcholine** is liberated at the terminals of most of the parasympathetic effector neurons. Stimulation of the sympathetic portion of the system leads to a general and lasting discharge; by and large, all its effects happen together. In its organization one preganglionic neuron appears to activate from twelve to twenty postganglionic neurons. Contrarily, this ratio in the ciliary ganglion, a parasympathetic ganglion, is but 1:2, and there is much more rapid inactivation of acetylcholine than epinephrine. Parasympathetic activity is specific, discrete, and local, whereas sympathetic effects are widespread.

Structurally, each portion of the autonomic nervous system consists of two-neuron chains. The first neuron has its cell body in the visceral efferent column of the brain or spinal cord. The second neuron has its cell body in a ganglion outside the cerebrospinal axis. The first neuron synapses with the cell body in the ganglion and

is thus designated as the **preganglionic** neuron; the second neuron in the chain extends from ganglion to effector and is the **postganglionic** neuron. Origin in brain and spinal cord and location of the ganglion are specific in the two divisions, and the alternative name of each expresses these differences in origin. Thus, in the **parasympathetic** (craniosacral) division, preganglionic neurons arise in the nuclei of origin of cranial nerves III, VII, IX, and X and in the second, third, and fourth sacral segments of the spinal cord. In the **sympathetic** (thoracolumbar) division, the preganglionic neurons are located in the spinal cord beginning at the first thoracic level and ending at the second or third lumbar level. With reference to the ganglionic location of the postganglionic synapse, the **chain and collateral ganglia** belong to the thoracolumbar division. Ganglia in the craniosacral division are located more peripherally and are designated as either **peripheral** or **terminal ganglia.** The axons of preganglionic neurons are myelinated; axons of postganglionic neurons are unmyelinated, distribute through the visceral nerves and plexuses, and end in relation with involuntary and cardiac muscle or gland cells.

The sympathetic portion of the system distributes fibers to smooth muscle of blood vessels and to glands in every region or segment of the body but, as noted, has its origin restricted to the fourteen or fifteen segments of the spinal cord from the first thoracic to the second or third lumbar levels. The requirements for longitudinal spread of these preganglionic neurons is met by the **sympathetic trunk,** an accumulation of ascending and descending fibers located at the side of the vertebral column. **Chain ganglia** form enlargements of the sympathetic trunk opposite most spinal nerves. Into this system of chain ganglia and sympathetic trunk run the **white rami communicantes,** conducting preganglionic fibers from the spinal nerves at the levels of T1 to L2 or 3, and out of it pass **gray rami communicantes** (postganglionic fibers) to all thirty-one spinal nerves for peripheral distribution. Thus, the sympathetic trunk is entirely a dependency of the sympathetic division, and the 'white' and 'gray' designations of the rami express the predominant presence or absence of myelin on the nerve fibers. The **collateral ganglia** are cell stations located along the major abdominal blood vessels and serve abdominal and pelvic viscera.

In the parasympathetic division of the autonomic

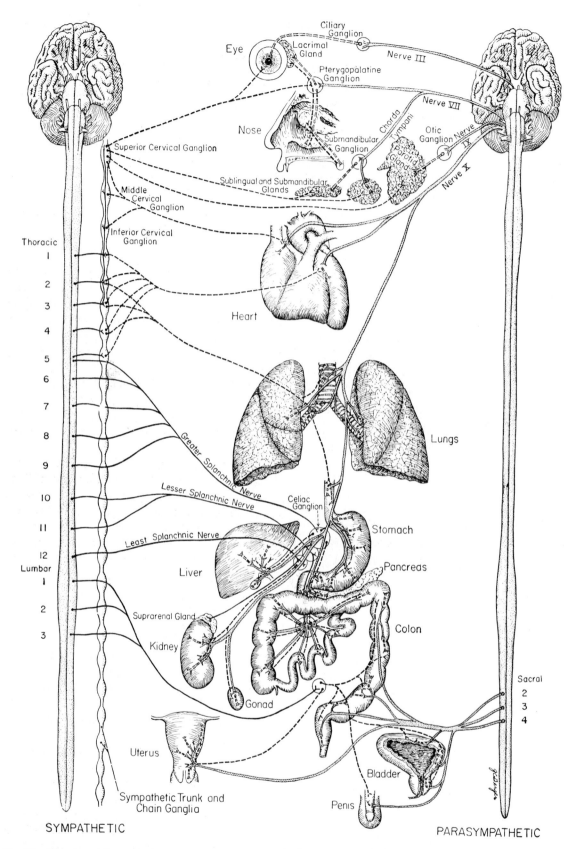

Ciliary
Ganglion
Eye
Lacrimal
Gland
Nerve III
Pterygopalatine
Ganglion
Nose
Nerve VII
Chorda Tympani
Submandibular
Ganglion
Otic
Ganglion
Nerve IX
Superior Cervical Ganglion
Parotid Gland
Sublingual and Submandibular
Glands
Nerve X
Middle
Cervical
Ganglion
Inferior Cervical
Ganglion
Thoracic
1
2
3
4
Heart
5
6
7
8
9
Lungs
10
Greater Splanchnic Nerve
Lesser Splanchnic Nerve
Celiac
Ganglion
11
Stomach
12
Least Splanchnic Nerve
Pancreas
Lumbar
1
Liver
2
Colon
Suprarenal Gland
3
Sacral
2
Kidney
3
4
Gonad
Uterus
Sympathetic Trunk and
Chain Ganglia
Bladder
Penis
SYMPATHETIC
PARASYMPATHETIC

Fig. 23 The plan of the autonomic nervous system; sympathetic nerves on the left, the parasympathetic innervation on the right. Postganglionic neurons are shown by dashed lines.

32

nervous system, the preganglionic neurons tend to follow the spinal nerves indistinguishably into their peripheral distribution, finally branching off to terminate in small **peripheral ganglia** in the neighborhood of the structures innervated. Alternatively, such a branch may run right into the hilus of a gland or have its terminal ganglion located between the muscular coats of the organ innervated. **Submucosal** and **myenteric ganglia** are terminal collections of such ganglion cells in the gastrointestinal tract. Postganglionic neurons are generally quite short in this portion of the autonomic nervous system.

In the foregoing discussion the autonomic nervous system has been described as a purely efferent system. This is in accordance with its original definition. However, it must be emphasized that visceral afferent neurons with cell bodies in dorsal root or cranial ganglia and with terminations in the spinal cord or brainstem make up one of the components of the autonomic nerves and represent the afferent portion of the neuron chains and arcs regulating visceral functions. The system is also under the dominance of higher brain centers. Investigations in recent years have called attention to visceral 'centers' in medulla, midbrain, hypothalamus, and cerebral cortex.

THE ARTICULATIONS OR JOINTS

A joint expresses the relation of two or more bones to one another at their region of contact. Some bones articulate at movable joints, others at only slightly movable joints, and still others at immovable joints. These differences are reflected in the structural details of the articulations. In joints in which the articulating bones are united directly by fibrous tissue or cartilage, little or no movement is allowed, whereas in the synovial type, the articulating bones move freely on one another.

THE FIBROUS JOINTS

The fibrous joints include sutures and syndesmoses. **Sutures** (fig. 24) are the typical form of junction of the flat bones of the skull. These bones, growing radially from ossification centers, make contact with one another along interlocking dentate or serrated edges or along overlapping or grooved

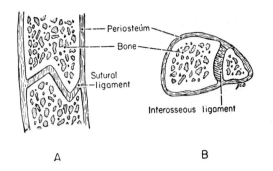

Fig. 24   The fibrous joints; A—a suture, B—a syndesmosis.

borders. The fibrous tissue trapped between the edges both separates the bones and joins them together and is continuous with the periosteum on their surfaces. Areas where these growing bones have not made contact at birth are represented by the **fontanelles** of the infant skull (fig. 30). These are usually all closed by the end of the eighteenth month of life. **Synostosis,** complete fusion of bone across the suture lines, begins on the inner side of the skull during the early twenties and progresses throughout life. All sutures are not obliterated until old age.

In the **syndesmosis** the apposed bones are simply joined together by intervening fibrous tissue. Such fibrous tissue is usually abundant and may be organized as an interosseous membrane or ligament (fig. 24). The inferior tibiofibular joint is an example of such a union. The attachment of the parallel borders of the radius and ulna and of the tibia and fibula by interosseous membranes are other syndesmoses.

THE CARTILAGINOUS JOINTS

The cartilaginous joints include transitional stages in complete bony union (synchondroses) and joints which retain at least a modified cartilage throughout life (symphyses).

Synchondroses (fig. 25) represent a temporary condition, later converted by ossification to bony continuity. The normal process of endochondral bone development involves a cartilaginous epiphyseal plate separating the shaft (diaphysis) and the end (epiphysis) of the bone. Through this type of joint is provided continuous growth in length of bone, for the hyaline cartilage of the epiphyseal plate is progressively replaced by bone

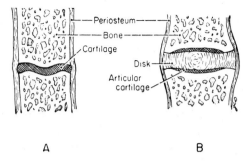

A   B

Fig. 25   The cartilaginous joints; A—a synchrondrosis,
B—a symphysis.

on its surfaces and elaborated in its middle. When maturity is reached, the cartilage is converted into bone by the extension of ossification from the diaphysial side, and the epiphysis becomes fused with the shaft. Another example of a synchrondrosis is seen in the cartilaginous union between the first rib and the sternum. This cartilage usually undergoes ossification during adult life.

The essential features of a **symphysis** (fig. 25) are the presence of hyaline cartilage on the bony surfaces concerned and the union of these cartilaginous surfaces by fibrous tissue or fibrocartilage. Such uniting tissues may form relatively thick pads capable of being compressed or displaced, thereby imparting some movement to the bones. The intervertebral disks are examples of symphyses and give, in their cumulative effect, a considerable flexibility to the vertebral column. In the center of each of the intervertebral disks is a semifluid, pulpy mass, the **nucleus pulposus,** which adds to the elasticity of the disk and also, being incompressible, contributes resiliency to the column. The symphysis pubis is another representative.

THE SYNOVIAL JOINTS (figs. 26-29)

The synovial joints provide for free movement. The apposed ends of the bones have thin coverings of **hyaline cartilage** over their full bearing surfaces, and a cavity exists in the joint. The cavity is lined by a **synovial membrane** which invests the inside of the capsule and covers all structures within the joint up to the edges of the hyaline cartilage. The synovial membrane is a sheet-like specialization of connective tissue, backed by unspecialized subsynovial connective tissue; it appears to elaborate the synovial fluid. Surrounding bones,

membrane, and fluid, the fibrous connective tissue of the **articular capsule** is attached to the bones just beyond the limits of the joint cavity. This type of joint is usually reinforced by strengthening **accessory ligaments** which are either separate or applied to the capsule; they limit motion in undesirable directions (fig. 27). Pads of fibrous tissue or fibrocartilage, **articular disks,** exist in certain joints, as do also subsynovial collections of fat.

Synovial joints occur in considerable number and variety in the body and are grouped according to the type of movement they permit. They are listed in an ascending order of movement as follows:

1. Joints providing simple gliding or sliding movement, **plane joints,** exist between the articular processes of the vertebrae. They are also present in the intercarpal, carpometacarpal, and intermetacarpal joints of the wrist and hand and in the comparable joints in the foot. The joint surfaces are flat, or of slight curvature, and the movement consists of a sliding of one surface on the other. Limitation of movement is provided by a tight, fibrous capsule.

2. The **hinge joint** (fig. 7) is one allowing movement around a single axis at right angles to the bones. This joint permits movements of flexion and extension only. Examples are seen in the elbow and in the joints between the segments of the fingers. In such joints the articular capsule is thin and lax on the surfaces where bending takes place, but strong collateral ligaments unite the bones at the margins of the joint.

3. The **pivot,** or trochoidal, **joint** also allows movement around a single axis, but around a longitudinal axis through the bone. This type of

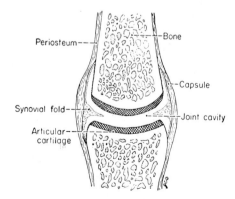

Fig. 26   A generalized synovial joint. Schematic.

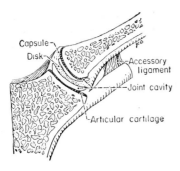

Fig. 27 The sternoclavicular articulation as a representative synovial joint. It has an articular disk which separates the joint cavity into two parts and a strong accessory (costoclavicular) ligament.

articulation occurs in the superior radioulnar joint and in that between the first and second cervical vertebrae. In each case a rounded process of bone rotates within a sleeve or ring composed of a bony fossa and a strong ligamentous band. The **annular ligament** fits the rounded process of bone so as to hold it in place as well as provide for its socket.

4. A **condyloid articulation** allows movement in two directions at right angles to one another. The shape of the articulation is ellipsoidal, and, like a segment of an egg, it has two curvatures which are unequal, one long and one short. Such a biaxial type of joint occurs in the wrist where movements of flexion and extension and of abduction and adduction occur, but rotation is prohibited by the unequal dimensions of the oval surface. Circumduction, a composite movement of flexion, abduction, extension, and adduction may also be produced. The metacarpophalangeal joints of the fingers and toes are other biaxial condyloid joints.

5. A second biaxial type of articulation is the **saddle joint.** Its two curvatures are, however, concavoconvex, or opposite to one another. The analogy of the saddle is apt and the carpometacarpal joint of the thumb is its clearest representative in the body.

6. The **ball and socket joint** (fig. 28) is the articulation of greatest freedom of motion. The ball and socket character allows motion in an almost infinite number of axes passing near the center of the 'ball.' In conventional descriptive terms, such joints as at the hip and shoulder allow flexion and extension, abduction and ad-

duction, medial and lateral rotation, and circumduction.

Articular Surfaces The articular surfaces of synovial joints present a number of points of interest. Hyaline cartilage covers the actual articulating surfaces. Articular cartilage is avascular and non-nervous and is apparently nourished by the synovial fluid. Varying in thickness from 0.5 to 6 mm. in different joints, the articular cartilage imparts the exceptional smoothness of surface characteristic of normal joints.

It is apparent in fitting together the bones of the skeleton that the articulating surfaces are not congruent but that the receiving fossa is generally of a somewhat flatter curvature. This incongruity serves to emphasize that very few joints of the body (only the pivot joints) are of the pin and sleeve type with the close contact seen in most bearing surfaces in mechanics. The surfaces of contact, or the bearing surfaces in weight-bearing joints, change with changing positions and do not constantly bear against any one part of the articular surface. All the weight or pressure exerted through the joint is exerted at one small area of contact, but the area is a steadily changing one as the bones move on one another. The movement of biological joints is predominantly gliding or sliding in character and, in accordance with the irregular curvatures of joint surfaces, movement takes

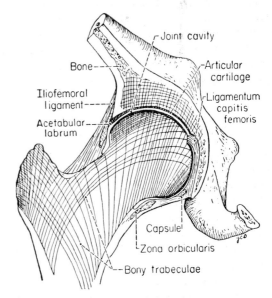

Fig. 28 The hip joint as a well-developed ball and socket joint.

place around continually shifting centers of rotation.

Synovial Fluid The synovial fluid is a clear, viscous fluid present in small quantity in synovial joints and thought to be produced by the synovial membrane. It resembles egg albumin in consistency, and its viscosity is related to its mucin content. The fluid serves as a lubricant for the joint surfaces and as a nutrient fluid for the articular cartilage. The high viscosity of synovial fluid is related to its lubricating function. The viscosity of a fluid lubricant should vary directly as the load and inversely as the velocity of movement. With the relatively heavy loading and slow rotary movement in biological joints, a high viscosity is indicated.

The incongruity of joint surfaces also appears to be a factor in efficient lubrication. As discussed by MacConaill ('32), lubrication studies on shafts and bearings indicate that a closely congruent fit allows only a very thin film of lubricant which tends to be forced from between the surfaces. Incongruous surfaces allow a wedge-shaped film of lubricant which exhibits a pressure gradient such that the lubricant is forced out of the acute, or leading, edge of the wedge, and thus between the bearing surfaces, and returns at the base of the wedge. This film has a counter-thrust sufficient to support the load and does effectively lubricate. Studies of mechanical bearing surfaces also show that flexible wedge-shaped pads assist in the formation of wedge-shaped films of lubricant. If these principles are applicable to biological articulations, menisci, disks, and fat pads should not be thought of as advantageously increasing congruity but rather as of value in conjunction with incongruity in furthering the development of efficient wedge-shaped films of lubricant in the joint.

Limitation of Movement Stability is a necessary attribute of biological joints. Freedom of movement and stability are usually inverse variables. The pull of muscles operating across a joint, the adjacent fasciae, and other soft tissues hold joint surfaces in apposition and contribute to their stability. Pressures exerted by superimposed weight, atmospheric pressure, ligaments, and articular capsules all contribute to stability. Strength is added in a joint by expanding the bearing surfaces so that a broader surface of support is provided. At the knee, the weight of the superimposed body is supported on bones making an end-to-end contact, an intrinsically unstable situation. Among the factors providing strength at this joint is the approximately threefold widening of the bony surfaces. Stability is frequently contributed by accessory ligaments which become tense at one or other extreme of motion. Hinge joints are characterized by a loose capsule on those sides of the joint toward which motion is directed and by strong collateral ligaments at the borders. Since the collateral ligaments pass across the axis of rotation and are never far from it, they can be relatively tense and thus impart stability in all positions of angular motion. At the extremes of motion, usually in extension, collateral ligaments are placed on maximum tension and stop the movement in a stable position. In the case of the knee joint, the cruciate ligaments are stabilizing structures acting in a manner similar to its collateral ligaments.

Accessory ligaments are complemented by muscles and tendons although electromyographic studies show that muscles never come into play if ligaments can carry the load. Close to almost all joints are tendons or collections of fibrous tissue serving for muscle attachments; they add strength to the joints. The importance of muscles and their tendons is well seen at the shoulder joint where a considerable number of short muscles hold the humerus closely against the articular fossa of the scapula and give a support to the joint not provided by its loose capsule. An extra advantage of this muscular support is that, as 'living ligaments,' muscles can relax as well as shorten and can provide a dynamic support, adjusted to the requirements of every changing situation.

The stability of the joints of the lower limb and the ready maintenance of the erect posture depend on a final 'screw-home' action of the bones and ligaments. At the knee (fig. 29) full extension involves a terminal rotation of the femur on the tibial articular surfaces. In this action the lateral condyle of the femur glides forward and the medial condyle backward, and the cruciate ligaments of the knee are made increasingly tense and hold the bones firmly against one another in the extended position. In the reverse of this action, one of the shorter flexors of the knee, the **popliteus muscle,** sends its tendon obliquely upward and forward to the lateral femoral condyle and serves to draw that condyle backward and thus to help initiate flexion.

At the hip joint the extremely strong capsular ligament is spirally arranged. Under the influence of this spiral ligament, as the thigh moves into

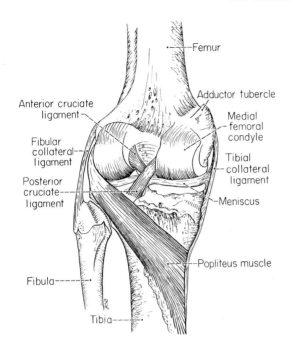

Femur

Adductor tubercle

Medial femoral condyle

Anterior cruciate ligament

Tibial collateral ligament

Fibular collateral ligament

Meniscus

Posterior cruciate ligament

Popliteus muscle

Fibula

Tibia

Fig. 29   The knee joint opened from behind to show its cruciate ligaments. The tendon of the popliteus muscle inserts into the lateral condyle of the femur.

The thumb may be flexed and extended, abducted and adducted, but its special feature, and one of the principal factors in Man's pre-eminence, is its ability to 'oppose' the other digits. The wrist adds the two actions of its condyloid type of articulation, and the forearm bones the rotary action of pronation and supination. At the elbow a firm hinge gives lift or push by flexion and extension. The ball and socket articulation of the shoulder provides a universal joint for action around any possible axis, and the combined joint and muscle suspension of the shoulder girdle allows the shoulder to be pointed up or down, forward or backward, or in any intermediate direction. These upper limb joints provide a whole series of variables in both position and action and allow the hand to operate at any angle of approach and through a maximum of movement patterns. Such a high degree of flexibility cannot be provided in a single joint. The wide range of the movements possible in the upper limb gives emphasis to the complexity of the neural and muscular organization that serves not only the immediate movement but also acts to fix and position all the joints in the series for their maximum contribution to the whole.

Nerve Supply of Joints

Joints are richly supplied with nerves. These are derived from the larger nerves that supply the overlying skin and the muscles that move the joints. Vasomotor and vasosensory fibers serve for the control of blood vessels. The principal type of sensation from joints is proprioception. From endings in the capsule, impulses pass to the spinal cord and brain to report on the position of the joint members and the degree and direction of movement. Such impulses may reach consciousness, but they also serve in spinal reflex mechanisms concerned in the control of muscles acting on the joint. This represents the special contribution of 'joint sense.' Pain fibers are also numerous in the capsule and accessory ligaments and respond especially to twisting and stretching. Compared with that arising in skin, joint pain is poorly localized, and may be referred to an overlying skin or muscle area or to another joint. Like deep pain elsewhere, there may be visceral disturbances associated with joint pain, and it may cause widespread reflex spasm in flexor muscles and inhibition of extensor muscles.

extension, the capsule is progressively twisted and shortened and the head of the femur is guided like a screw into its socket. When the thigh reaches full extension, the head is in its most completely congruent position in the acetabulum, firmly screwed into place by the shortening effect of the capsular ligament (close-packed position). At the ankle, the erect position brings the foot into dorsiflexion, a position in which the broadest part of the dorsum of the talus fits into the mortise formed by the other bones entering into the ankle joint. These mechanisms, producing firm, tight joints in the extended position of the lower limb, are of obvious value for an erect posture maintained with minimal muscle action. MacConaill ('69) believes that a slight rotary component contributing finally to the close-packed position is a characteristic of the stable position of many joints.

Serial Articulations Although the joints of the upper limb exhibit increased terminal stability also, the special contribution of these joints lies in the great range of mobility they impart to the hand. As the principal tool of the body, the hand is served by flexion and extension at the several joints of the fingers to which is added abduction and adduction at the metacarpophalangeal joints.

BLOOD SUPPLY OF JOINTS

Articular arteries are numerous and are derived from anastomosing vessels around joints. Articular capsules and ligaments are richly supplied, and branches to the synovial layer form a network of fine vessels. Veins accompany the arteries and lymphatic networks are present in the capsules.

THE SKELETAL SYSTEM

The skeletal system of Man consists of an internal set of bones which serves as the rigid supporting framework of the body. A small number of cartilages are also included in the system in adult Man. Certain parts of the supporting framework form chambers, such as the skull and the thoracic cage, of importance in protection of the contained soft parts. Bone everywhere serves for attachment of muscles, and in the joint system of the body the bones act as levers. The internal soft tissue of bone, the **bone marrow**, forms blood cells, and bone also acts as a storehouse for calcium and phosphorus from which these chemicals can be mobilized when needed. Varying somewhat with age and with the individual, there are usually recognized to be 206 bones in the complete skeleton. They are divided between axial and appendicular skeleton in accordance with following distribution:

Axial Skeleton		
Skull and hyoid	29	
Vertebrae	26	
Ribs and sternum	25	

		80
Appendicular Skeleton		
Upper limbs	64	
Lower limbs	62	

		126
Total		206

Many of these bones, particularly in the appendicular skeleton, are designated as **long bones,** meaning that their length is greater than their breadth, even though some of them are quite short, as in the fingers and toes. **Short bones** are bones of more nearly cuboidal character. They are found only in the wrist and ankle. Short bones are richly covered with articular surfaces, though there are always several nonarticular areas for the attachment of tendons and ligaments and for the entry of blood vessels.

Flat bones (fig. 30) exhibit a plate of compact bone on each flat surface separated by a thin marrow space. Flat bones are found for the most part in the vault of the skull and, therefore, are gently curved rather than flat. In the case of the skull bones, the marrow space between the tables is known as **diploe.** The sternum, ribs, and scapulae are other flat bones.

The classification of **irregular bones** includes bones of mixed shapes that do not fit into the other categories. Bones of the skull not of the flat type are irregular, and the vertebrae and the coxal bone of the hip are also included in the classification.

Sesamoid bones are small bones located where tendons play across the ends of long bones in the limbs. The sesamoid bone develops in the tendon and serves both to protect the tendon from excessive wear and to change the angle of approach of the tendon to its insertion. The improvement of this angle makes for a greater mechanical advantage at the joint.

Accessory, or supernumerary, **bones** occasionally cause difficulties in radiographic diagnosis and need to be recognized. Many bones develop from several centers of ossification, and it sometimes happens that, even in the adult, these parts of the bone have failed to make complete fusion. This gives the appearance of an extra bone, but

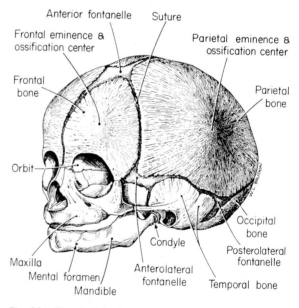

Fig. 30   The skull in the newborn. Ossification proceeds radially from central ossification centers. The fontanelles are areas of late junction of the bones.

it can usually be seen to be the missing part of one of the normally complete bones. **Wormian bones** are accessory bones which occur between the bones of the skull in the intervals where the converging borders of the ossifying flat bones come last into contact.

Clinical Note Another skull marking occasionally seen radiographically is a midline suture in the frontal bone. This bone normally develops in bilateral halves (fig. 30), but all trace of the suture between the halves usually disappears in adult life. An occasional persistence of this frontal **metopic suture** may be mistaken for a fracture line.

PHYSICAL PROPERTIES OF BONE

Mature bone combines an organic framework of fibrous tissue and inorganic salts—mainly calcium phosphate. The salts represent about 60 per cent of the weight of compact bone and are deposited in a fibrous structure of collagenous connective tissue. If the calcium salts are removed from a bone by immersion in acid, the decalcified bone retains an essentially normal form and appearance but loses much of its weight and all its rigidity. Such a bone can be tied in a knot without breaking. Bone requires strength to resist compression or crushing forces and tensile strength to resist disrupting forces. It is comparable to the stiffest woods in resisting tension and to concrete, common brick, or sandstone in its resistance to compression (Dempster & Liddicoat, '52). The tensile and compression strengths of bone are of an equivalent order, and it thus shows advantages over many structural materials, strong in one respect but weak in the other. The tensile strength of the human femur is shown in its maximum resistance before breaking of about 12,000 lbs. per sq. in. Under conditions of vertical loading, Evans and Lissner ('48) found that fresh adult human femurs suffered fractures of the neck and shaft at static pressures of between 1200 and 1800 pounds. This is twenty-five to thirty times that portion of the weight of the body applied to one femur in the standing position. For running and jumping, the factor of safety in the human femur is approximately five times.

The tubular character of the long bones of the body accords with good structural design. If a column, or a long bone, is supported at both ends and pressed on in its midlength, compression forces are resisted on its upper surface and tension

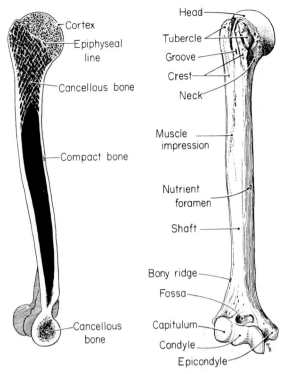

Fig. 31 Structural properties of a bone; A—compact and cancellous bone, B—processes and impressions.

forces on its lower surface. Since these forces are opposite and tend to neutralize each other in the core of the column, or bone, lightness may be achieved with no appreciable loss in strength by hollowing out this portion. Long bones also have thicker walls in their midlength with thinner cortical layers toward their ends.

The normally weight-bearing portions of bones have a thin surface layer, or shell, of compact bone and an internal zone of cancellous bone. The latter exhibits a pattern of bony trabeculae (fig. 32) so arranged as to resist, as struts, both compression and tension forces. These have been designated as 'pressure lamellae' and 'tension lamellae.' The lamellae lie at right angles to one another and are correctly placed in relation to lines of stress in the bone (fig. 32). That these lamellar patterns arise as a response to functional demands is clear. They do not achieve their final arrangement until locomotion is completely learned, and they become rearranged and reorganized whenever an abnormal set of stresses and strains remains after a badly healed fracture or the development of a deformity.

The typical long bone (fig. 31) has a shaft of

hard, smooth bone and ends which are expanded for articulation with neighboring bones. The articular surface is smooth and, in life, is covered by articular cartilage. It is designated as an **articular facet,** a **head,** or a **condyle,** depending on its

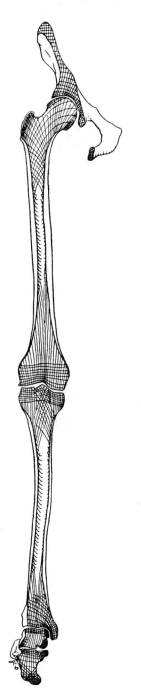

Fig. 32   The adaptation of trabecular structure to the weight-bearing function in the bones of the lower limb. Pressure and tension lamellae.

character. Near the ends are large foramina for veins and arteries, and piercing the shaft obliquely is a canal for the nutrient artery. Markings occur wherever fibrous tissue is attached to the bone. An elevation may occur as a **line,** a **ridge** or **crest,** a **tubercle,** a **tuberosity,** a **spine,** or a **trochanter.** Small, smooth, flat surfaces are called **facets.** A depression may be a **pit,** or **fovea,** or if larger, a **fossa. Grooves** and **sulci** are linear depressions. A **foramen** is a hole or opening; if it has length, it is a **canal.** Broad muscle attachments may make no mark on a bone, but the more discrete small areas of attachment of a tendon or ligament usually result in an elevation. The prominence of the marking is much greater in the highly muscular male than in the youth or female.

STRUCTURE OF BONE

Compact bone exists on the surfaces of bones and is thick in the shafts of long bones and in the surface plates of the flat bones of the skull. Spongy or cancellous, bone is characteristic of the articular ends of long bones and occurs in the short and irregular bones. The spaces in cancellous bone contain bone marrow and grade over into the large marrow cavity of the shafts of long bones. The trabecular area of the flat skull bones is called **diploe.** Elsewhere, cavitation may be exaggerated as in the pneumatic spaces in the bones of the head. These spaces, lined by mucous membrane, constitute the air sinuses of the head, contributing resonance to the voice and lightness to the bones. A bone is always covered externally by a connective tissue layer, **periosteum,** the deeper cell lamina of which is osteogenetic. The periosteum assists in the initial formation of the bone and is thereafter available to elaborate new bone in case of fracture or disease.

Bone is a blend of fibrous connective tissue and inorganic calcium salts. The inorganic substance of bone is deposited between the fibers of connective tissue, while the cells of the latter are transformed into bone cells lodged in cavities of the matrix, lacunae. The bone cells have anastomosing processes that extend out from the cells in connecting canals, the **bony canaliculi.** Fundamentally, bone consists of thin lamellae composed of these organic and inorganic materials, its thicker areas being produced by the formation of added lamellae apposed to the first. During development, the blood vessels of periosteum become surrounded by formed bone and thus constitute bony canals, the **Haversian canals.** These canals run predominantly longitudinally in a long bone, and the lamellae of bone form in the branching and anatomosing pattern of the canal.

BONE DEVELOPMENT (fig. 33)

Bone is formed either through a process of replacement of cartilage or by direct elaboration from periosteum. These are known, respectively, as **endochondral** and **intramembranous ossification,** and both processes are involved in the formation of a great many bones. In the case of a long bone, a cartilaginous model of the bone is first formed; the bone is 'preformed' in cartilage. The **perichondrium,** through its osteoblasts, forms a layer of bone on the surface of the cartilaginous shaft. Suceeding layers follow and the perichondrium is now called a **periosteum.** Meanwhile, calcification of the matrix of cartilage cells in the middle of the model has led, along with degeneration of interstitial cartilage, to a trabecular network of calcified cartilage. An invading **periosteal bud** of capillaries and osteoblasts penetrates the wall of the model at its midlength and lays down bone on the cartilaginous trabeculae present there. Formed on a network, this first 'center of ossification' is cancellous in nature. The combined processes of calcification and degeneration of the cartilage advance from the center toward the ends of the cartilage model, and with this advance goes the progressive formation of bone on the cartilaginous trabeculae. Ultimately, calcified cartilage is completely replaced by spongy bone. This is **endochondral bone development.** The process also involves destruction and absorption of bone, so that, as the whole bone enlarges both longitudinally and concentrically, a marrow cavity results which is many times the size of the original cartilaginous model (fig. 33). The simultaneous deposition of successive layers of bone on the surface by the periosteum (intramembranous ossification) forms a shaft of compact bone with characteristic lamellae.

Growth in length of a long bone is entirely due to longitudinal increase in the epiphyseal cartilaginous plate followed by progressive ossification through endochondral bone development. After birth, at various times in different long bones, an ossification center appears in the cartilage of each

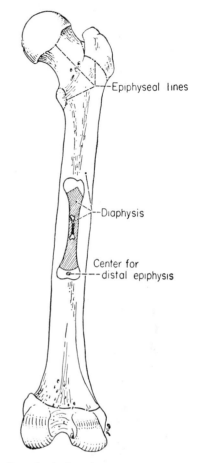

Epiphyseal lines

Diaphysis

Center for distal epiphysis

Fig. 33   Growth of a long bone, emphasizing the absorption of earlier stages in the matured bone. Inner figure, 13 weeks—26 mm.; middle figure, newborn—105 mm.; outer figure, adult—48 cm. femur. ($\times\frac{1}{4}$) (Drawn from data from W. J. Felts, *Am. J. Anat.,* 94: 1, 1954.)

end of the bone and begins to produce trabecular bone in this epiphysis. Bone is then undergoing development in the shaft, the diaphysis, and in the ends, the epiphyses, at the same time. Destruction of the intervening cartilage and its ossification does not catch up with its elaboration and growth until many years have elapsed, and thus the mature length of the long bone is attained. At an age typical for 'fusion' in the particular bone, ossification completely overcomes the elaboration of cartilage and **synostosis** takes place. Articular cartilage represents the only cartilage that remains on mature bone. Stature is the result of these nicely adjusted processes in bone development, **epiphyseal fusion** of the bones of the limbs occurring progressively from the age of puberty to maturity (about twenty-five years). Epiphyseal fusion takes place at nearly the same age for those ends of the several bones meeting in a particular joint, an age that usually does not correspond with fusion at the other end of the bones. The more actively growing end of a bone starts to ossify earlier and is the last to fuse with the diaphysis. Variation in growth rates at the two ends results in the nutrient artery becoming oriented so as to enter the bone in a direction away from the more actively growing end. The deposition of bone by the osteoblastic layer of the periosteum is typical of intramembranous bone development, and when no cartilage model occurs, as in the flat bones of the skull, the bone is deposited on the fibers and in the meshes of young fibrous connective tissue derived from mesenchyme. The tremendous increase in size between primordial and mature bone is effected by the concurrent processes of bone development and bone destruction. In the process of growth, the earlier stages of formed bone are represented by marrow cavity in the enlarged and mature bone, and modeling of bony contours is a continuous process (fig. 33).

The deposition of calcium salts in the cartilaginous model of each bone is detectable by X-ray examination, and the time of beginning ossification can be determined for all bones by this means. At a later period of life the closure of the metaphyseal intervals between the epiphyses and diaphyses of bone reaches an end-point in fusion of the epiphyses with the shafts. This is also detectable radiographically but is more difficult to time precisely. These data of ossification and fusion times are valid gauges of bone maturity but, especially in the case of fusion times, show wide individual differences.

BONE MARROW

Bone marrow is concerned in blood cell formation. The **red blood cell,** the **erythrocyte,** has a life of only 120 days so that continual replacement is necessary. Many of the varieties of **white blood corpuscles** are also the products of bone marrow. At birth, at which time **red bone marrow** fills the marrow cavity of all bones, red blood cells are produced exclusively by the bone marrow. As the ends of the long bones increase in size, the red bone marrow recedes from the medullary cavity, and, by puberty, red bone marrow is found only in cancellous bone. Further recession leads to its replacement by

yellow (fatty) **bone marrow** in later life in the distal parts of many long bones and in the lower segments of the vertebral column.

Clinical Note Red marrow persists in the sternum throughout life, and this is a convenient place for its aspiration for examination. Yellow bone marrow is almost pure fat.

VESSELS AND NERVES

Blood vessels are especially rich in those parts of bone which contain red bone marrow. Thus, the articular ends of long bones and the short bones exhibit many vascular foramina through which branches of the **articular arteries** enter. The **nutrient artery** divides into proximal and distal branches for the supply of the marrow, the metaphyses, and the inner portion of the compact bone of the shaft. **Periosteal twigs** enter the shaft at numerous points and supply the compact bone.

Veins accompany the several types of arteries. The importance of the blood supply for blood cell formation is indicated by the large and numerous veins that leave the foramina at the articular ends of the long bones and other marrow-containing bones. **Lymphatic vessels** are abundant in the periosteum, and perivascular lymph spaces occur in the Haversian canals. The periosteum is rich in sensory **nerves.** Other nerves, probably vasomotor in function, accompany the arteries into the bones.

CARTILAGE

Cartilage is a form of supporting tissue with a firm, resilient matrix but without the rigidity and strength of bone. It is characterized by a solid ground substance, has no blood or lymph vessels, and is not supplied with nerves. Three types are recognized—hyaline cartilage, white fibrocartilage, and elastic cartilage.

Hyaline Cartilage Hyaline cartilage has a bluish-white, translucent appearance. It is the type of widest distribution and appears in the cartilaginous model of developing bones. It persists in adult life as articular cartilage on the ends of bones, as costal cartilages, and as the cartilages of the trachea, the nose and nasal septum, the bronchi, and the larger laryngeal cartilages. Hyaline cartilage is enclosed by a perichondrium except over articular surfaces. The nonarticular representatives of hyaline cartilage have a tendency to ossify in later life.

White Fibrocartilage This type of cartilage consists of heavy collections of white, fibrous connective tissue in which a cartilage matrix is intermixed. It is less homogeneous than hyaline cartilage but is tougher and more flexible. It occurs in intervertebral and articular disks and in the glenoid lips of certain joints. It is present in the pubic symphysis and covers tendons where they bear on bones.

Elastic Cartilage In elastic cartilage the matrix is permeated by a rich network of elastic fibers. This imparts a yellow appearance and makes it highly resilient. Elastic cartilage occurs only in the external ear, the auditory tube, the epiglottis, and in a few small cartilages of the larynx.

VISCERAL SYSTEMS

The visceral systems of the body are the respiratory, digestive, urogenital, and endocrine. The parts of these systems are generally familiar, and conceptual material directed toward the understanding of their function is included in the descriptions of the organs concerned.

ANATOMICAL VARIATION

There is a wide-spread but faulty notion that human structure is standard and regular. This is far from the case; indeed obvious differences in facial conformation and general stature are outward indications of variability in all aspects of the organism. The same muscle in different individuals may differ as to extent of origin, detail of insertion, or even in nerve supply. A muscle may exhibit extra tendons, or it may be entirely missing. The blood supply of a structure or of a whole area may be furnished by one artery in one specimen and by a different vessel in another. Differences in size and branches of vessels are common. Even such stable elements as bone vary as to their processes, foramina, and articular surfaces. Variation is a phenomenon to be recognized and adjusted to in all aspects of human form and arrangement.

The fact of irregularity in anatomical structure raises several questions. What regularity is to be expected? Does variability occur in all tissues at the same or at differing rates? What is the significance of the variations encountered? The answers to such questions come largely from statistical analyses which themselves usually require interpretation. Regularity is commonly equated with normality, but the latter word tends to carry a connotation of correctness which the facts themselves do not support. Even worse, the term 'abnormal' carries overtones of 'deformity' and does not correctly describe the simple fact of differences in form and arrangement. These terms and their common implications are to be avoided in considering anatomical variation.

What is the magnitude of variation seen in anatomical structure? Actually, the closer one looks, the more variation one finds, for, generally, the farther one goes into minutiae, the more each specimen differs from its neighbor. If one asks only how many bones the adult body contains, a number can be found which will be relatively constant. If, on the other hand, one examines the patterns of the whorls in fingerprints, no two are alike. Between these extremes there is a great variety of examples. Variant structures of only 1–2 per cent occurrence, as Meckel's diverticulum, are rarities; the constant division of the gastroduodenal artery into gastroepiploic and pancreaticoduodenal arteries lies at the other extreme. A dorsal pancreatic artery arises from one source in 37 per cent of cases, in 33 per cent from another, and in 21 per cent and 8 per cent from other sources. What source is to be described as typical? Is the celiac source of the right hepatic artery in 83 per cent to be considered regular, or the 60 per cent termination of the inferior mesenteric vein in the splenic vein usual or a variant? Logically, if we are on the upper side of 50 per cent, we must consider the situation the more common or the usual. Regularity, then, occurs on the upper side of 50 per cent; variations are of lesser frequency. The more dependable regularities vary between 70 per cent and 100 per cent. This is the area in which descriptions are composed with confidence and there is no immediate need to qualify or stress variability.

With regard to the distribution of variation, bones have been cited as showing greater regularity; blood vessel examples are numerous on the side of variability. Muscle variants are readily identified in dissection but are not of marked frequency; variations of peripheral nerves, similarly, are infrequent. Ligaments share the stability of the osseous skeleton. Blood and lymphatic vessels arise in development by resolution of larger vessels out of a plexiform arrangement of channels and thus may be expected to show the greatest range of variation.

Bilateral symmetry is the rule in the vertebrate series. Are variants bilaterally symmetrical? It might be expected that an accessory tendon in one arm would be matched by one on the other side, or that a superficial brachial artery would appear in both limbs. But the variant is toward the opposite end of the scale from the usual, and significant variations are actually found twice as frequently on one side of the body as bilaterally.

All variations would appear to be related ultimately to the events of embryological development, and recent research points to critical stages in the process when certain anomalies of development are most likely to occur. The stimulus for anomalous development is unknown in most cases, though some infectious diseases and drug intake by the prospective mother are definitely implicated in certain abnormalities. Anomalies may appear as underdevelopment, such as cleft palate and harelip or spina bifida; excessive development is seen in extra fingers or toes. These errors of development are sufficiently striking to remove them from the category of normal variation. Certain vessels (median artery, ischiadic artery) are well developed during early embryonic stages and regress later relative to other vessels. They may persist as large functional and recognized vessels or they may totally disappear. These, however, are examples of occurrences which can readily be classified as normal variation. Much care in interpretation is needed in defining or assigning significance to the many variations that appear in the human specimen.

II

The Upper Limb

The upper limb, the 'arm' in the layman's usage, consists more exactly of shoulder, arm, forearm, wrist, and hand. It is an organ for grasping and manipulating. In the erect posture, the upper limb has become adapted into a body part having great freedom of movement and considerable utility in the performance of physical work. A number of highly mobile joints allow its muscles to so position and move the various segments of the limb that its terminal part, the hand, becomes the most effective tool in the animal kingdom. The hand can operate at any angle or in any position, and its opposable thumb gives it the ultimate in grasping and handling efficiency. Important contributions to the functional effectiveness of the limb are made by a blood vascular distribution rich enough to support its metabolic requirements and by a peripheral network of nerves and nerve endings that provides for it a high order of sensory acuity.

Serving in movement, position, and grasp, the upper limb is primarily composed of bones and joints and the muscles that move them. These structures, as well as its skin surfaces, are supplied by nerves and blood vessels. Lymphatic vessels are numerous, particularly as a superficial network in the subcutaneous tissues. Bursae and tendon sheaths facilitate movement by reduction of friction. The presence of only one point of bony union between the trunk and the limb makes for unhindered movement and manipulation, but results in an inevitable lack of security and stability. The sternoclavicular joint represents this single point of bony contact. Since the limb is an appendage of the trunk, all its nerves and blood vessels issue from the neck or the chest, and the more central parts of these systems come into view only by dissection of such areas.

THE BACK AND SCAPULAR REGION

SURFACE ANATOMY (fig. 34)

Certain muscles that move the shoulder are located in the back, others descend from the neck to the shoulder girdle, and still others arise on the chest wall. The position of these proximal muscles makes necessary the early consideration of the superficial parts of the back and of the back of the neck.

In the median line of the back is the **vertebral furrow,** deepest in the lower thoracic and upper lumbar regions. In the neck, the furrow ends above at the junction of the neck and scalp in the readily palpated **external occipital protuberance** of the occipital bone. The cervical spinous processes are, for the most part, short and deep; but at the base of the neck, the spine of the seventh cervical vertebra reaches the surface and gives to this vertebra its name of **vertebra prominens.** All the thoracic spines can be palpated; the third is opposite the root of the spine of the scapula. The vertebral furrow is deep in the lumbar region, and the spines of the lumbar vertebrae may be indicated by pits or depressions. The furrow ends at the top of the sacrum in a triangular, flattened area, the downwardly directed apex of which lies at the level of the third sacral spine. The **erector spinae muscle** makes a prominent vertical bulge on either side of the vertebral furrow; the bulge is especially heavy in the lumbar region where its lateral margin is indicated by a shallow groove about a hand's breadth from the midline. From the posterior superior iliac spine, the **iliac crest** can be followed lateralward and anteriorly. Its highest point is at the level of the fourth lumbar vertebra. Below the inferior angle of the scapula, the last five ribs can readily be felt lateral to the erector spinae muscle. The twelfth rib, however, may or may not be long enough to be palpated.

The **acromion** and **spine of the scapula** are subcutaneous. The acromion forms the point of the shoulder. Its tip is directed forward, and its medial border participates in the **acromioclavicular articulation.** The lateral border of the acromion can be followed backward into the **scapular spine,** the root of which lies opposite the third thoracic

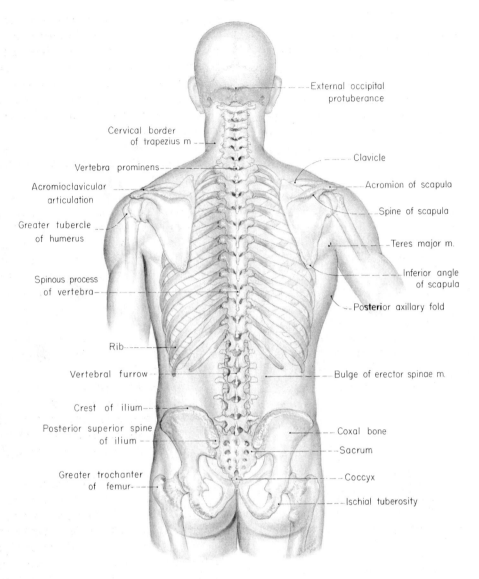

External occipital protuberance

Cervical border of trapezius m

Clavicle

Vertebra prominens

Acromioclavicular articulation

Acromion of scapula

Greater tubercle of humerus

Spine of scapula

Teres major m.

Inferior angle of scapula

Spinous process of vertebra

Posterior axillary fold

Rib

Vertebral furrow

Bulge of erector spinae m.

Crest of ilium

Posterior superior spine of ilium

Coxal bone

Sacrum

Greater trochanter of femur

Coccyx

Ischial tuberosity

Fig. 34   Surface anatomy and underlying bones of the back.

vertebral spine. The **inferior angle of the scapula** can be located under the skin and marks the level of the seventh thoracic vertebral spine when the arms are at the sides. The **trapezius** and **latissimus dorsi muscles** may be mapped out on the surface. The upper lateral border of the trapezius muscle is tensed and made prominent in a high shoulder shrug. Its lower lateral margin may be indicated by a line from the twelfth thoracic spine to the tubercle on the spine of the scapula. The upper border of the latissimus dorsi muscle is indicated by a line drawn from the seventh thoracic spine horizontally across the inferior angle of the scapula. Its lower lateral border forms the **posterior axillary fold.**

SUPERFICIAL TISSUES OF THE BACK

The **skin** of the back is thick, especially toward the nape of the neck and shoulder where its thickness may reach 5 to 6 mm. The **subcutaneous tissue** varies in thickness and fat content; it, also, is thicker and contains more fat in the shoulder region. The **subcutaneous nerves** of the back are branches of the dorsal rami of spinal nerves, one of the two fundamental divisions of such nerves.

A Typical Spinal Nerve (fig. 35)

Each **spinal nerve** represents the union of a dorsal and a ventral nerve root (fig. 35). These are formed by the gathering together of successive rootlets which emerge from the dorsolateral sulcus and the ventral root-zone of the spinal cord. The dorsal root is larger than the ventral, and a dorsal root ganglion, or **spinal ganglion,** forms a prominent swelling on it. The spinal ganglion is an ovoid mass of unipolar nerve cells, is from 4 to 6 mm. long, and is usually located in the intervertebral foramen where it is invested with a prolongation of dura mater. The dorsal root and ganglion of the first cervical nerve is usually much reduced or absent. The unipolar cell bodies in the spinal ganglion give rise to short central processes, which constitute the nerve fibers of the dorsal root, and to peripheral processes, which are continued distalward in the spinal nerve. At the outer margin of the ganglion, the dorsal and the ventral roots unite to form the spinal nerve.

There are thirty-one pairs of spinal nerves—eight cervical, twelve thoracic, five lumbar, five sacral, and one coccygeal. As a rule the nerves emerge below the vertebrae corresponding to them in number. The cervical region is an exception to the rule, however, for its eight nerves are associated with only seven cervical vertebrae, and the first cervical nerve emerges between the skull and the first vertebra. Thus the cervical nerves correspond in number to the vertebrae below them, whereas all more caudal nerves are numbered with the vertebrae above them.

The **spinal nerve** is very short. After providing a small **meningeal branch** to the membranes of the spinal cord, it emerges from the intervertebral foramen and almost immediately divides into **dorsal and ventral rami.** These rami are largely somatic in distribution, supplying the skin, the skeletal muscle, and the deeper tissues of the trunk and limbs. Each ramus contains fibers from both dorsal and ventral roots, and each of its branches is a 'mixed' nerve. The smaller dorsal ramus passes posteriorly between the transverse processes of adjacent vertebrae. It divides into medial and lateral branches, both of which supply muscles, and one of which migrates through the deep muscles of the back region to end as a cutaneous nerve. The details of this cutaneous supply follow. The muscular distribution of the dorsal ramus is a segmental one to the longitudinal muscles occupying the space between the spinous processes of the vertebrae and the angles of the ribs.

The larger ventral rami run ventrally and laterally, appearing to be the direct continuation of the spinal nerves. Close to their origin, the ventral rami are connected to **sympathetic chain ganglia** and to the associated sympathetic trunk by **rami communicantes.** The gangliated sympathetic trunk lies closely lateral, or anterolateral, to the bodies of the vertebrae at most vertebral levels, and the rami communicantes make short, direct connections between the spinal nerves and the ganglia. These rami are of two types. **White rami communicantes** conduct mainly axons of medullated (white) visceral efferent neurons originating in the spinal cord and terminating around the cell bodies in the sympathetic ganglia. The white rami arise from only the fourteen or fifteen spinal nerves of the segments between the first thoracic and the second or third lumbar spinal cord level. **Gray rami communicantes** pass from the sympathetic trunk to the spinal nerve and are composed principally of unmedullated (gray) neurons of the postganglionic sympathetic fibers. The cell bodies of these neurons

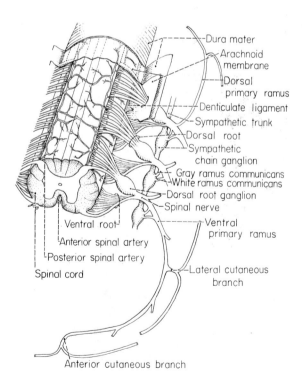

Dura mater
Arachnoid membrane
Dorsal primary ramus
Denticulate ligament
Sympathetic trunk
Dorsal root
Sympathetic chain ganglion
Gray ramus communicans
White ramus communicans
Dorsal root ganglion
Spinal nerve
Ventral primary ramus
Lateral cutaneous branch

Ventral root
Anterior spinal artery
Posterior spinal artery
Spinal cord

Anterior cutaneous branch

Fig. 35   The formation and branching of a typical spinal nerve.

lie in the sympathetic chain ganglia, and their axons reach smooth muscle and glands at the periphery of the body. The gray rami run to all spinal nerves and their visceral branches are included in all dorsal rami as well as ventral rami. The connections of the autonomic nervous system are included in a general discussion of the system on p. 30.

The distribution of the ventral ramus is generally segmental. Where segmentation of the body wall is most apparent (chest and abdominal wall), muscular branches innervate segmental musculature, such as the intercostal muscles, and cutaneous nerves are given off as lateral and anterior branches. The lateral cutaneous branch arises at the axillary line, and the anterior cutaneous branch at the border of the sternum. From these primary locations of entrance into the subcutaneous tissues, each cutaneous nerve distributes forward and backward to complete, with the help of the dorsal ramus, a circular band of cutaneous innervation around the body. The cutaneous nerves overlap onto adjacent bands of primary innervation, so that from any portion of surface skin the terminals of two adjacent nerves will mediate sensory impulses. Disregarding this 'nerve overlap,' the pattern of segmental innervation of the skin of the trunk and limbs describes an arrangement of segmental areas (dermatomes) of considerable regularity (figs. 65, 422). At the roots of the limbs, the ventral rami undergo some modification from this simple segmental pattern. Here redistribution and regrouping of nerve bundles lead to the formation of the great **brachial** and **lumbosacral plexuses.**

Cutaneous Branches of Dorsal Rami (figs. 35, 36) Each dorsal ramus divides into a **medial branch** and a **lateral branch.** These pass toward the surface, supplying the back muscles, and one or other of the branches terminates in the skin. Above the midthoracic level, the medial branches provide the cutaneous terminals; below this level, the lateral branches are sensory. Thus the cutaneous branches emerge from the muscle mass very close to the midline in the upper half of the body and several inches from the midline where the lateral branches are cutaneous. At lumbar levels, the cutaneous branches emerge lateral to the prominent bulge of the erector spinae muscle. The dorsal ramus of the first cervical nerve has no sensory distribution. The dorsal ramus of the second cervical

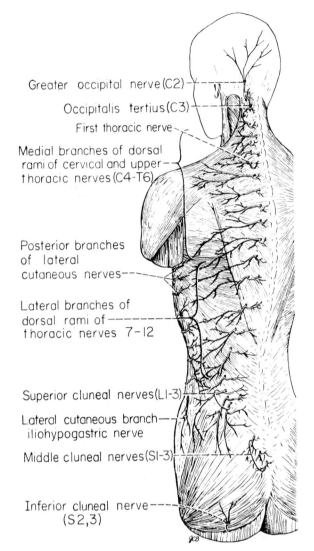

Greater occipital nerve (C2)
Occipitalis tertius (C3)
First thoracic nerve
Medial branches of dorsal rami of cervical and upper thoracic nerves (C4-T6)

Posterior branches of lateral cutaneous nerves

Lateral branches of dorsal rami of thoracic nerves 7-12

Superior cluneal nerves (LI-3)
Lateral cutaneous branch iliohypogastric nerve
Middle cluneal nerves (SI-3)

Inferior cluneal nerve (S2,3)

Fig. 36   The cutaneous nerves of the back.

nerve has a large medial branch which ascends to the vertex of the scalp as the **greater occipital nerve;** that of the third cervical nerve distributes in the upper part of the neck close to the median plane as the **third occipital nerve** (n. occipitalis tertius). It communicates with the greater occipital nerve. The dorsal rami of the other cervical nerves are generally small, and their sensory branches distribute to the back of the neck. Nerves C7 and 8 have little or no sensory distribution in the back. The cutaneous branches of the upper six or seven thoracic rami reach the surface near the spines of the vertebrae. That of the second thoracic nerve is large and distributes as far as the point of the shoulder; medial branches of the upper six or seven thoracic nerves supply the skin and sub-cutaneous tissues of the scapular region. The **lateral cutaneous branches** of the dorsal rami of **thoracic nerves** below the seventh run more obliquely downward and lateralward among the divisions of the erector spinae muscle. They pierce the latissimus dorsi muscle and distribute to the skin and subcutaneous tissue of the lower chest and loin regions. The oblique course of these and of the lateral cutaneous branches of the lumbar nerves are reflected in the downward slant of the dermatomes in the lower trunk. The lateral cutaneous branches of the first three lumbar nerves form the **superior cluneal nerves.** They pierce the thoracolumbar fascia at the lateral border of the erector spinae muscle, cross the iliac crest a short distance in front of the posterior superior iliac spine, and distribute into the gluteal region.

The dorsal rami of the fourth and fifth lumbar nerves have no cutaneous branches. The lateral branches of dorsal rami of the first, second, and third sacral nerves form the **middle cluneal nerves.** These become cutaneous on a line connecting the posterior superior iliac spine and the tip of the coccyx to supply the skin and subcutaneous tissues over the back of the sacrum and the adjacent area of the gluteal region. The dorsal rami of the fourth and fifth sacral and the coccygeal nerves do not divide into medial and lateral branches. They unite to form a cutaneous nerve which distributes in the neighborhood of the coccyx.

Muscles Connecting the Upper Limb to the Vertebral Column

Muscles of this group are superficially located in the back. They are the **trapezius, the latissimus dorsi, the levator scapulae, the rhomboideus minor,** and the **rhomboideus major muscles.** Ventral rami of spinal nerves supply these muscles, even though the distribution of cutaneous branches of dorsal rami overlie them and in spite of the position of the muscles in the back. The explanation lies in the developmental origin of the muscles from the ventrolateral sheet of trunk musculature and their subsequent migration to obtain an attachment to the vertebral column (fig. 64). Dorsal rami of spinal nerves innervate only deep back musculature, whereas muscles of the limbs and of the thoracic and abdominal walls proper are served entirely by ventral rami. Phylogenetic migration of muscles sometimes obscures their fundamental position

and relations, but the nerve supply of a muscle gives a reliable clue to its correct position and classification, for a nerve faithfully follows a muscle in any migration it may undergo.

Trapezius (figs. 37, 41, 42, 277, 280)

The trapezius is a triangular muscle, its name derived from the trapezoidal figure described by the muscles of the two sides. It is superficially situated in the upper back and back of neck. The muscle arises from the medial one-third of the superior nuchal line and the external occipital protuberance of the occipital bone, from ligamentum nuchae, from the spines of the seventh cervical and all thoracic vertebrae, and from the intervening supraspinal ligament.

The **ligamentum nuchae** is a sheet of mixed white fibrous and elastic tissue interposed between the cervical muscles of the two sides. It extends from the external occipital protuberance to the spine of the seventh cervical vertebra, and, representing the expanded supraspinal ligament, attaches deeply to the tips of the intervening spinous processes. Dorsal flexion of the neck necessitates shortness of the spinous processes of the cervical vertebrae, and the flexible ligamentum nuchae substitutes for bone in providing muscular attachments here.

The **trapezius muscle** is relatively flat and thin except in the low cervical and upper thoracic region where its increased thickness is matched by a distinct diamond-shaped accumulation of tendinous fibers of origin. The muscular fibers converge toward the bones of the shoulder. The occipital and upper cervical fibers of the muscle insert into the posterior border of the lateral one-third of the clavicle (fig. 41); the lower cervical and upper thoracic fibers reach the medial border of the acromion and the upper border of the crest of the spine of the scapula; the lower thoracic fibers converge to a triangular, flattened tendon which ends in the tubercle of the crest of the spine of the scapula.

The Trapezius assists in suspending the shoulder girdle. The upper thoracic part draws the shoulders strongly backward in squaring the shoulders and is important in pulling or extension movements of the arm. The lower segment of the muscle acts with the occipital and upper cervical portion to carry the point of the shoulder upward, for with the one part pulling downward and the other upward the scapula is rotated on the chest wall. In this 'force couple' the levator scapulae

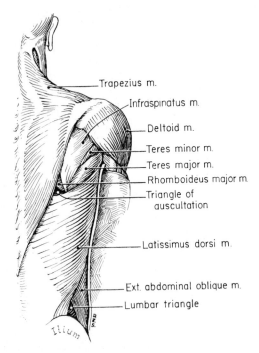

Trapezius m.
Infraspinatus m.
Deltoid m.
Teres minor m.
Teres major m.
Rhomboideus major m.
Triangle of auscultation
Latissimus dorsi m.
Ext. abdominal oblique m.
Lumbar triangle
Ilium

Fig. 37   Superficial upper limb muscles located in the back.

and the upper parts of the serratus anterior contract in concert with the upper part of the trapezius, and the lower, larger part of serratus anterior acts with the lower part of the trapezius. The lower and middle segments of the trapezius function together in pulling and in squaring the shoulders, and the middle fibers are active in abduction of the arm. A low level of activity in upper trapezius, levator scapulae, and upper serratus anterior is sufficient to suspend the shoulder girdle but the same muscles contract vigorously with loading of the shoulder or with weights in the hand.

The nerves to the trapezius are the **accessory nerve** (cranial nerve XI) and direct branches of ventral rami of the **third and fourth cervical nerves** (fig. 38). These nerves mingle to form the **sub-trapezial** plexus on the deep surface of the muscle, to which the cervical nerves run an almost horizontal course. The **accessory nerve** arises by lateral rootlets from the upper five cervical segments of the spinal cord. Successively joining, the rootlets form a nerve which ascends in the subarachnoid space of the vertebral canal, and passes upward through the foramen magnum into the cranium. The accessory nerve leaves the cranium through the jugular foramen and perforates the sternocleidomastoid muscle of the neck. Crossing the posterior cervical

triangle immediately under its fascial covering, the nerve reaches the trapezius about 5 cm. above the clavicle, thus providing the muscular innervation to both the sternocleidomastoid and trapezius muscles. It has been suggested that the cervical branches participating in the innervation of these muscles may be sensory (proprioceptive) in nature (Corbin & Harrison, '40) since the accessory nerve does not contain proprioceptive fibers. Stretch of the muscle fibers yields action potentials in the cervical nerve branches. A branch of the **transverse cervical artery** of the subclavian system joins the sub-trapezial plexus on the deep side of the muscle and ramifies with it (fig. 38) to supply the upper two-thirds of the muscle. The lower one-third receives its blood supply from the muscular per-forating branch of the dorsal scapular artery (p. 181 & fig. 38).

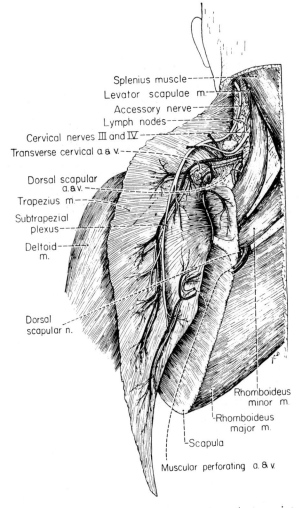

Splenius muscle
Levator scapulae m.
Accessory nerve
Lymph nodes
Cervical nerves III and IV
Transverse cervical a. & v.
Dorsal scapular a. & v.
Trapezius m.
Subtrapezial plexus
Deltoid m.
Dorsal scapular n.
Rhomboideus minor m.
Rhomboideus major m.
Scapula
Muscular perforating a. & v.

Fig. 38   Nerves and blood vessels supplying the trapezius muscle.

The trapezius is superficial in its entire extent. Its upper lateral border forms the posterior margin of the posterior triangle of the neck. In the neck, the trapezius is covered on both its superficial and deep surfaces by the **superficial layer of the cervical fascia** which, after blending with the periosteum of the clavicle and the scapula, continues beyond the neck as the fascia of the thoracic portion of the muscle.

Latissimus Dorsi (figs. 37, 57, 68, 277, 280)

Latissimus dorsi is a broad, triangular muscle situated in the lower part of the back. Its principal origin is aponeurotic, with attachments to the spinous processes of the lower six thoracic vertebrae and, through its fusion with the thoracolumbar fascia, to the lumbar and sacral spinous processes. The lower lateral portion of the muscle arises from the posterior one-third of the iliac crest. Additionally, it has muscular slips of origin from the last three or four ribs, which interdigitate (alternate on the bone) with slips of origin of the external abdominal oblique muscle. An attachment to the inferior angle of the scapula frequently occurs. Lateral to the inferior angle of the scapula, the much narrowed muscle curves spirally around the teres major muscle in the lower edge of the posterior axillary fold. It ends in a band-like tendon, 6 to 8 cm. long, inserted in the floor of the intertubercular groove of the humerus. Its spiral course takes its tendon ventral to the tendon of the teres major muscle, a bursa intervening, and turns it so that the inferiorly arising muscle fibers lie more cranially in the tendon and the dorsum of the muscle is represented ventrally.

Latissimus dorsi extends the humerus, drawing the arm downward and backward and rotating it medially. Its full action is seen in the crawl stroke in swimming, but it is used in all pulling movements. In climbing, it assists in drawing the trunk up toward the fixed humerus.

The blood and nerve supply of latissimus dorsi reaches the muscle from the axilla. Its nerve is the **thoracodorsal,** or middle subscapular, **nerve** from the posterior cord of the brachial plexus with fibers from C7 and 8. It reaches the muscle toward its tendinous end in a sheath of fascia which also encloses the **thoracodorsal artery** and a vein of the same name. The thoracodorsal artery is a branch of the subscapular artery from the axillary artery.

Latissimus dorsi is superficial in the lower part of the back where it overlies abdominal musculature and the posterior inferior serratus muscle. Its lower lateral border crosses the margin of the external abdominal oblique muscle but leaves a nonmuscular interval between them just above the iliac crest. This is the **lumbar triangle** (fig. 37), the floor of which is the internal abdominal oblique muscle. The upper part of latissimus dorsi is overlapped by the inferior portion of the trapezius and, in turn, covers a portion of the rhomboideus major muscle. The triangle formed by the borders of these three diverging muscles is the **triangle of auscultation** (fig. 37). Its floor is the chest wall. If the scapula is carried forward, parts of the sixth and seventh ribs and the interspace between them become uncovered in this triangle and are accessible for auscultation.

The Rhomboid Layer (figs. 38, 39, 42, 260, 277, 280)

Immediately under the trapezius lie three muscles which have a continuous insertion on the medial border of the scapula. These muscles also exhibit close similarities in action and innervation. From above down, they are the levator scapulae, the rhomboideus minor, and the rhomboideus major muscles.

The **levator scapulae** is a thick, strap-like muscle which arises by four separate tendons from the transverse processes of the first three or four cervical vertebrae. It inserts into the medial border of the scapula from the superior angle to the spine.

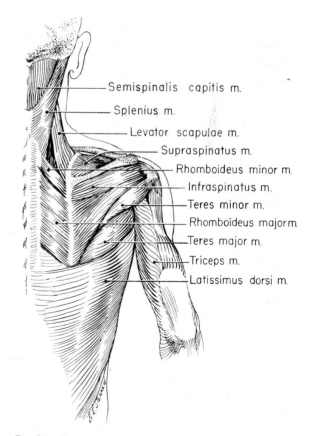

Fig. 39 The musculature exposed by removal of the trapezius and deltoid muscles.

It is overlapped and partially covered by the sternocleidomastoid and trapezius muscles.

The **rhomboideus minor** muscle is a slender slip parallel to and poorly separated from the rhomboideus major. It arises from the lower part of ligamentum nuchae, the spines of the last cervical and first thoracic vertebrae, and the associated portion of the supraspinal ligament. Directed downward and lateralward, it is inserted on the medial border of the scapula at the root of the scapular spine.

The **rhomboideus major** muscle is a flat, sheet-like muscle which arises from the spines of the second to the fifth thoracic vertebrae and the corresponding portion of the supraspinal ligament. Its fibers pass downward and lateralward to insert on the medial border of the scapula below the spine. The rhomboids draw the scapula upward and medialward and assist the serratus anterior muscle in holding it firmly to the chest wall. Their oblique traction also results in depression of the point of the shoulder. The levator scapulae assists in elevation, support, and rotation of the scapula but, with the scapula fixed, may also act in extension and lateral bending of the neck.

The **dorsal scapular nerve,** a branch of the brachial plexus from (chiefly) the fifth cervical nerve, is a part of the nerve supply to the levator scapulae and the sole innervation of the rhomboids. The dorsal scapular nerve approaches the levator at its anterior border and, passing under the muscle, descends inferiorly deep to (anterior to) this muscle and the rhomboids, adjacent to the medial border of the scapula (fig. 40). It supplies the rhomboids and the lower portion of the levator scapulae. The upper part of levator scapulae has an additional nerve supply from ventral rami of cervical nerves three and four.

The **dorsal scapular artery** (p. 181), a branch of the second or third part of the subclavian artery, crosses the neck to the lateral border of the levator scapulae muscle. From here it accompanies the dorsal scapular nerve inferiorly toward the inferior angle of the scapula (fig. 40). It sends branches into these muscles and, by small branches to both surfaces of the scapula, anastomoses with radicles of the subscapular and suprascapular arteries.

The levator scapulae and the rhomboid muscles underlie the trapezius muscle. The rhomboids cover and are closely applied to the posterior superior serratus muscle.

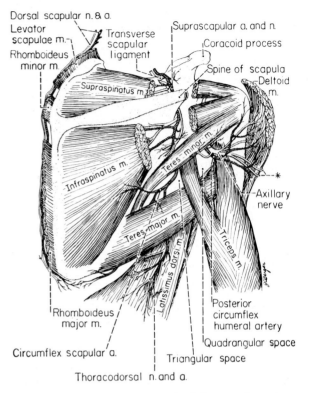

Fig. 40　Blood vessels and nerves of the scapular region; *—superior lateral brachial cutaneous nerve.

MUSCLES AND FASCIA OF THE SHOULDER

The muscles proper to the shoulder are the deltoid, the supraspinatus, the infraspinatus, the teres minor, the teres major, and the subscapularis.

Deltoideus (figs. 6, 37, 39, 41, 42, 43, 68, 69)

The deltoid muscle caps the point of the shoulder and is a coarsely fasciculated, multipennate muscle of triangular form. It has an extensive origin from the front of the lateral one-third of the clavicle, the lateral border of the acromion, and the lower lip of the crest of the spine of the scapula (fig. 41). The fibers converge to insert on the deltoid tuberosity situated on the lateral surface of the shaft of the humerus at its midlength. The more anterior (clavicular) part of the deltoid is formed of parallel fibers which may be separated from the remainder of the muscle. The spinous portion of the muscle is also composed of long fiber bundles and inserts into the posterior margin of the deltoid tuberosity.

The acromial, or central, part of the muscle exhibits a multipennate architecture (fig. 6). Four tendinous septa descend from the acromion into the muscle, and three septa ascend from the deltoid tuberosity. From the acromial bands, muscle fibers converge onto both sides of the intervening bands of insertion, resulting in the formation of a muscle of great power but short excursion. The increased power from the multipennate arrangement of fibers in this muscle is fortunate in view of the poor mechanical advantage at which it acts in abduction of the humerus.

The deltoid is the principal abductor of the humerus. In this action, the powerful central part of the muscle is primary. The deltoid requires assistance, however, from other muscles in its abduction action, and supraspinatus is its principal aid. Supraspinatus is especially important in the early phases of abduction; the activity of deltoid, minimal at first, increases progressively and is greatest between 90° and 180° of elevation. The supraspinatus muscle also, assisted by the infraspinatus, teres minor, and subscapularis muscles, holds the head of the humerus down and inward against the glenoid cavity of the scapula.

Both the position and the greater fiber length of the clavicular and scapular portions of the muscle signify actions differing in direction and degree from that of the central part. The clavicular fibers of the muscle assist in flexion and medial rotation of the arm, the scapular fibers assist in extension and lateral rotation.

Operating together, but without the central portion of the muscle, these marginal parts would adduct the humerus. In normal abduction movements their adduction potential is realized in stabilizing the shoulder joint.

The **axillary nerve** (C5, 6) from the posterior cord of the brachial plexus (p. 81) supplies the deltoid (fig. 40). An upper branch of the axillary nerve curves around the posterior aspect of the humerus and runs from behind forward on the deep surface of the muscle about 5 cm. from the acromion. The axillary nerve also supplies the teres minor muscle by a lower branch which ascends to its lateral and superficial surface. This branch provides a few twigs to the posterior part of the deltoid, and then becomes the **superior lateral brachial cutaneous nerve.** It curves over the posterior border of the deltoid muscle onto its surface and is cutaneous to the area over the lower half of the muscle. The axillary nerve is accompanied by the **posterior circumflex humeral artery** (fig. 40), a branch of the axillary artery (p. 77).

The deltoid muscle is superficial, overlies the greater tubercle of the humerus, and forms the rounded prominence of the shoulder. It is enclosed by a muscular fascia which is an extension of the superficial layer of cervical fascia from the trapezius, bony attachments intervening. The anterior border of the deltoid muscle consists of muscle bundles which are entirely parallel to the adjacent fasciculi of the pectoralis major muscle. These borders are separated by the cephalic vein and the deltoid branch of the thoracoacromial artery. Deep to the deltoid anteriorly is the coracoid process of the scapula. Posteriorly, the deltoid overlaps the long head of the triceps and the infraspinatus and teres minor muscles. Deep to the central portion of the muscle is the subacromial, or subdeltoid, bursa; and, in the floor of the bursa, the supraspinatus tendon.

The Subacromial Bursa Reflection of the deltoid muscle from its origin exposes the subdeltoid, or subacromial, bursa (fig. 43). This is a mucous bursa, approximately the size of a fifty-cent piece, situated directly under the deltoid muscle, and between it and the supraspinatus tendon and the joint capsule. It has an extension which runs deep to the acromion and the coracoacromial ligament. The bursa facilitates the movement of the deltoid muscle over the joint capsule and tendons.

Clinical Note Rupture of the supraspinatus tendon is common during middle and later life as a result of attrition or sudden strain. Due to the fusion of the supraspinatus tendon

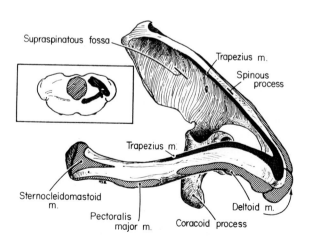

Fig. 41 The origin and insertion of muscles on the superior aspect of the clavicle and scapula. (Black dots—origin; white dots on black—insertion.)

Supraspinatous fossa

Trapezius m.

Spinous process

Trapezius m.

Sternocleidomastoid m.

Pectoralis major m.

Coracoid process

Deltoid m.

and the capsule of the shoulder joint, rupture of the tendon places the subacromial bursa into direct communication with the synovial cavity of the shoulder joint.

Supraspinatus (figs. 39, 42, 43, 68, 69)

Supraspinatus is a rounded muscle which occupies the supraspinatous fossa of the scapula. It arises from the medial two-thirds of the walls of this fossa and from the dense fascia which covers the muscle. The tendon is formed within the muscle, blends on its deep surface with the capsule of the shoulder joint, and inserts on the highest of the three facets of the greater tubercle of the humerus. The supraspinatus muscle acts early in abduction of the humerus and assists the deltoid in its full movement. Its medial traction resists downward displacement and possible dislocation of the head of the humerus. The **suprascapular nerve** (C5, 6) from the superior trunk of the brachial plexus (p. 80) enters the supraspinatous fossa through the scapular notch (fig. 40). Passing under the superior

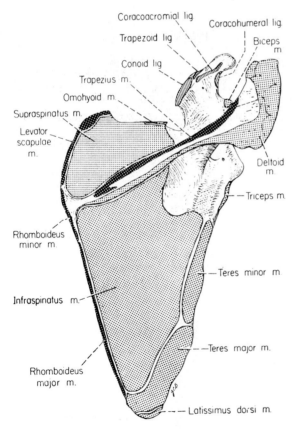

Fig. 42 The muscular and ligamentous attachments on the dorsum of the right scapula. (Black dots—origin; white dots on black—insertion; lines—ligaments.)

transverse scapular ligament, it is deep to the muscle and supplies it from its underside. The **suprascapular artery** from the thyrocervical trunk of the subclavian artery (p. 180) passes over the ligament and distributes with the suprascapular nerve. The supraspinatus muscle is covered by the trapezius. The acromion process of the scapula curves across its tendon and requires removal for clear observation of the muscle.

Infraspinatus (figs. 39, 42, 69, 272)

The **deep fascia** of the shoulder region is especially strong over the infraspinatus muscle. This fascia attaches to the scapula at the margins of the infraspinatous fossa and forms intermuscular septa between the infraspinatus and the teres minor and major muscles. The fascia gives origin to muscular fibers of the infraspinatus. One especially strong band of the infraspinatus fascia extends between the inferior angle of the scapula and the medial part of the spine of the scapula, and there anchors the lowest fibers of the trapezius. The infraspinatus fascia attaches to and is continuous with the fascia of the deltoid muscle.

The **infraspinatus muscle** arises from the whole of the infraspinatous fossa except its lateral one-fourth, and from the overlying infraspinatus fascia and the intermuscular septa. The tendon arises within the muscle and inserts on the middle facet of the greater tubercle of the humerus. It blends deeply with the capsule of the shoulder joint. The arm may be rotated through an angle of about 90°, and the infraspinatus is its chief lateral rotator. The muscle also assists in holding the head of the humerus in against the glenoid cavity of the scapula. The **suprascapular nerve and artery,** having traversed the supraspinatous fossa, pass through the notch of the scapular neck to enter the deep aspect of the upper part of the infraspinatus muscle (fig. 40). The muscle is also supplied by the **circumflex scapular artery** (axillary arterial system, p. 77). The infraspinatus muscle is partially covered dorsally by the trapezius and deltoid muscles. Teres minor lies parallel to it below and is separated from it by only an intermuscular septum.

Teres Minor (figs. 39, 40, 42, 44, 69, 272)

The teres minor is a narrow, elongated muscle which arises from the upper two-thirds of the lateral border of the scapula and from adjacent inter-

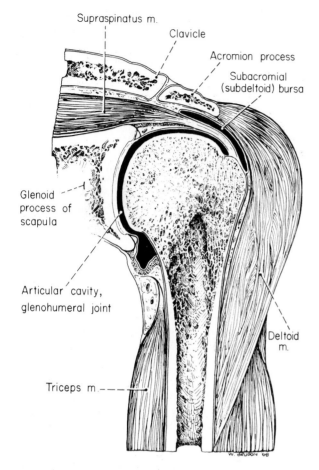

Fig. 43 A frontal section of the shoulder to show the subacromial (subdeltoid) bursa.

muscular septa. Its tendon is directed upward and lateralward to insert in the lower facet of the greater tubercle of the humerus; it blends deeply with the posterior part of the capsule of the shoulder joint. The muscle also reaches the back of the humerus for about 2 cm. below the tubercle. It is invested by the infraspinatus fascia and is sometimes inseparable from the infraspinatus muscle.

The teres minor muscle participates with infraspinatus in lateral rotation of the humerus and also fixes the head of the humerus to facilitate abduction and flexion of the arm. A branch of the **axillary nerve** reaches its lateral margin at about its midlength (fig. 40). The teres minor muscle is separated from the teres major by the long head of the triceps and by the axillary nerve and the posterior circumflex humeral vessels. It is pierced by the circumflex scapular vessels along the lateral border of the scapula.

Teres Major (figs. 39, 40, 42, 44, 68, 272, 277)

The teres major muscle is a thick, rounded muscle. It arises from the oval area on the dorsal surface of the inferior angle of the scapula and from the intermuscular septa separating it from adjacent muscles. The muscle is directed along the lateral border of the scapula toward the front of the humerus. Its fibers make a half twist, as do those of latissimus dorsi, and form a flat tendon, about 5 cm. wide, which inserts into the crest of the lesser tubercle of the humerus. Just before its insertion, the tendon lies against the latissimus dorsi which inserts in the floor of the intertubercular groove. A bursa intervenes between the two tendons.

Teres major is an adductor and medial rotator of the humerus. In these movements it acts with pectoralis major and latissimus dorsi. It also assists latissimus dorsi in extension of the arm at the shoulder. It is only recruited with resistance. A branch of the lower subscapular nerve (C5, 6) from

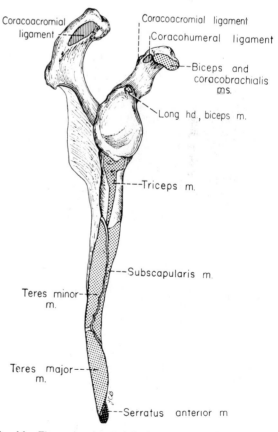

Fig. 44 The muscular and ligamentous attachments on the lateral border and processes of the right scapala. (Code as fig. 42.)

the posterior cord of the brachial plexus (p. 81) enters the anterior surface of the muscle near the middle of its upper border. The teres major muscle lies below the subscapularis muscle in the posterior wall of the axilla. The latissimus dorsi muscle crosses its axillary surface.

Quadrangular and Triangular Spaces (fig. 40)

The lateral divergence of the teres minor and teres major muscles produces a long, horizontally oriented triangle, which is bisected vertically by the long head of the triceps muscle. This division of the triangle results in a small, triangular space medially and a quadrangular space laterally. The humerus forms the fourth side of the latter. In the quadrangular space can be located the axillary nerve and the posterior circumflex humeral artery; in the triangular space, the circumflex scapular vessels.

Subscapularis

The subscapularis muscle lies on the costal surface of the scapula and forms much of the posterior wall of the axillary space. It is best seen from anteriorly after the clearing of the axillary space and is described in connection with the axilla (p. 84).

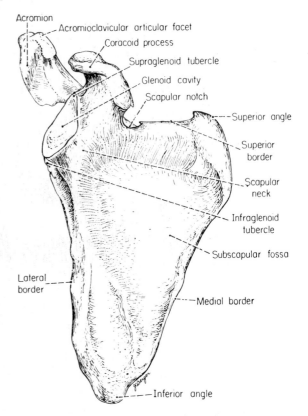

Fig. 45   The costal surface of the right scapula. ($\times\frac{1}{2}$)

THE SCAPULA (figs. 41-7)

The scapula lies against the posterior aspect of the thorax, overlying ribs two to seven, the subscapularis muscle and the muscles of the thoracic wall intervening. It is a flat bone of triangular form which, however, has certain prominent and rugged processes. The weight of the bone lies toward the shoulder in the spine and the acromion, the coracoid process, the glenoid fossa, and the lateral border (fig. 47).

The **body,** or blade, is thin and translucent. Its **costal surface** presents a large concavity, the **subscapular fossa** (fig. 45). The costal surface is thickened at its medial border and in triangular areas adjacent to the superior and inferior angles of the bone.

The **dorsum** of the body of the scapula (fig. 46) is convex and is separated by the spine into a larger infraspinatous and a smaller supraspinatous fossa. Near the **lateral border,** the dorsal surface has a sharp ridge. Its narrow upper portion is crossed about 3 cm. from the edge of the glenoid cavity by a groove for the circumflex scapular artery and vein. The ridge ends below in a broadened area at the inferior angle from which arises the teres major muscle. The **infraglenoid ridge** extends downward from the lower lip of the glenoid cavity along the lateral border for about 3 cm.,

terminating at the groove for the circumflex scapular vessels. The **medial border** is relatively straight and lies parallel with and adjacent to the vertebral column. It is the longest of the borders and is intermediate in thickness between the other two. Above the termination of the spine, there is a lateral inclination of the border toward the superior angle. The **superior border** is the thinnest and shortest, extending from the superior angle to the coracoid process. At its coracoid extremity the **scapular notch** forms a prominent indentation of the border (bridged by the **superior transverse scapular ligament**) which permits passage of the suprascapular nerve.

The **spine** of the scapula is a large, triangular process extending posteriorly from the dorsum of the scapula (fig. 46). Beginning at the medial border of the bone in a flattened projection, it rises and thickens laterally to end in a concave border which springs from the neck of the scapula just short of the middle of the glenoid cavity. The beginning of this concavity is the **notch of the neck** of the scapula. The spine is continuous with the acromion at the point of the shoulder. It separates the supraspinatous and infraspinatous fossae. A vascular foramen appears near the attachment of the spine to the body in the infraspinatous fossa. The free posterior border of the spine is thickened as the **crest of the spine.** The crest, blunt and roughened, presents two sharp lips.

The **acromion** is the continuation of the crest of the spine and forms the point of the shoulder. It overhangs the shoulder joint and projects forward to articulate with the outer end of the clavicle. The concave under surface of the acromion is smooth and is related to the subacromial bursa. The forward-directed extremity of the acromion presents, on its medial border, a smooth facet which participates in the acromioclavicular articulation.

The **coracoid process** (like a beak) is a thick, upward projection from the neck of the scapula (figs. 45, 47). Its root ascends from the neck and then bends into a horizontal part which projects forward and lateralward. The medial border of the root delimits the scapular notch and gives attachment to the superior transverse scapular ligament and, above this, to the conoid ligament. The lateral border of the root of the coracoid provides attachment for the coracohumeral ligament. The horizontal portion of the coracoid is broadened above and, near its base, gives attachment to the costocoracoid membrane and ligament. The lateral border is roughened for the coracoacromial and, at the root, the coracohumeral ligaments. The upper surface of the coracoid gives attachment to the trapezoid ligament (trapezoid portion of coracoclavicular ligament). The tip of the coracoid process provides the origin for the combined tendon of the coracobrachialis and the short head of the biceps brachii muscles.

The broadened **lateral angle** of the triangular scapula presents a shallow **glenoid cavity** (fig. 47) for reception of the head of the humerus. This is supported by a slightly constricted **neck.** The articular surface is directed forward and laterally, is somewhat oval with its long axis oriented vertically, and is broader below than above. The pinching-in of the upper part of this surface gives it somewhat the shape of a pear. The margin of the glenoid surface provides attachment for a fibrocartilaginous **glenoid labrum,** and, above, is elevated into a small prominence.

The scapula is ossified from seven or more centers, one for the body, two for the coracoid, two for the acromion, one for the medial border, and one for the inferior angle. Ossification in the body begins during the eighth week of intrauterine life. At birth a large part of the scapula is bony except the glenoid cavity and those parts listed above as having separate ossification centers. The ossification center for the coracoid process appears shortly after one year of age and fuses with the scapula at the age of fourteen to fifteen. Centers for the acromion, inferior angle, and medial border appear about puberty and fuse with the rest of the bone at sixteen to seventeen years for the acromion and twenty to twenty-one years for the others. Occasionally the union of the acromion and spine of the scapula persists in a fibrous condition throughout life and the line of separation may lead to misinterpretation as a fracture.

THE SUPERFICIAL FEATURES OF THE LIMB

SURFACE ANATOMY

Shoulder (fig. 48)

The **clavicle** is subcutaneous throughout its entire length. Medially, the sternal end of the clavicle forms a prominence bounding, on either side, the **jugular (suprasternal) notch.** The bone presents a double curve, projecting forward medially and receding backward laterally, thus accounting for the lateral concavity of the shoulder (infraclavicular fossa). The depth of this indentation marks the interval between the pectoralis major and deltoid muscles; it is the deltopectoral triangle. In the triangle the **coracoid process** of the scapula can be palpated 2 cm. below the clavicle. The acromial end of the clavicle is appreciably higher than the adjacent acromion, and the resulting palpable, or visible, elevation locates the **acromioclavicular articulation.** The **acromion** is also subcutaneous and forms the point of the shoulder. The rounded deltoid muscle covers the greater **tubercle of the humerus** which can be felt through the muscle. On abduction of the arm, the axillary folds are made prominent. The **anterior axillary fold** is formed by the border of the pectoralis major muscle; the lower margin of latissimus dorsi makes the **posterior axillary fold.**

Arm (fig. 48)

The skin of the arm is thin on its medial aspect and the hairs are fine. The biceps muscle forms a bulge in the

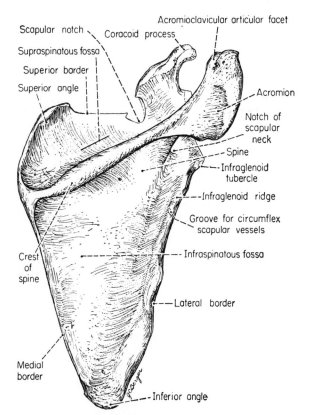

Scapular notch
Coracoid process
Acromioclavicular articular facet
Supraspinatous fossa
Superior border
Superior angle
Acromion
Notch of scapular neck
Spine
Infraglenoid tubercle
Infraglenoid ridge
Groove for circumflex scapular vessels
Infraspinatous fossa
Crest of spine
Lateral border
Medial border
Inferior angle

Fig. 46   The dorsal surface of the right scapula. ($\times\frac{1}{2}$)

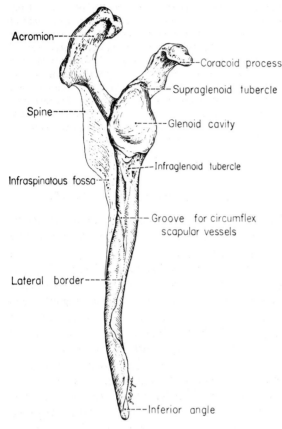

Acromion

Coracoid process

Supraglenoid tubercle

Spine

Glenoid cavity

Infraglenoid tubercle

Infraspinatous fossa

Groove for circumflex scapular vessels

Lateral border

Inferior angle

Fig. 47   A lateral view of the right scapula. ($\times \frac{1}{2}$)

front of the arm; the triceps muscle, one behind. Medial and lateral **bicipital grooves** separate these bulges and indicate the positions of medial and lateral **intermuscular septa** between the muscles of the anterior and posterior compartments of the arm. The **lateral sulcus** leads upward to the middle of the arm, ending at the deltoid tuberosity for the insertion of the deltoid muscle. The **medial sulcus** extends up to the coracobrachialis muscle. The **cephalic vein** runs upward in the lateral sulcus; the **basilic vein** lies in the medial sulcus.

Elbow (fig. 48)

The **epicondyles** of the humerus are palpable. The **medial epicondyle** is more prominent than the lateral and is located slightly higher and more anterior. The **head of the radius** can be palpated 2 cm. below and in front of the lateral epicondyle. The **olecranon** of the ulna forms the posterior prominence of the elbow; it gives attachment to the tendon of the triceps muscle. The region of the bend of the elbow anteriorly is designated as the **cubital fossa.** This triangular area is limited above by an arbitrary line between the epicondyles of the humerus. The downward-directed apex of the fossa is bounded on either side by the flexor and extensor muscle groups of the forearm. Pressure in the cubital fossa meets the hard, cord-like resistance of

the **biceps tendon.** This important landmark is lateral to the **brachial artery** which is, in turn, lateral to the **median nerve.** A large vein, the **median cubital vein** (fig. 51), lies under the skin of the cubital fossa, diagonally connecting the cephalic and basilic veins. This vein is a common site for withdrawal of blood. Behind the medial epicondyle, the **ulnar nerve** passes the elbow in the ulnar nerve groove. Its stimulation here produces tingling in the cutaneous area of distribution of the nerve—the ulnar digits and the ulnar side of the hand.

Forearm (fig. 48)

The skin of the forearm is thinner on its anterior aspect than on the dorsum. The **flexor muscle mass** forms a prominent bulge below the elbow on the medial side; an equally prominent bulge laterally is formed by the **extensor muscles.** The **ulna** is subcutaneous throughout its length and can be followed from the olecranon at the elbow to the **styloid process** at the wrist. The **radius** is subcutaneous in its lower half. Its styloid process can be located at the border of the wrist slightly beyond the familiar position of the radial pulse. It is against the expanded lower portion of the radius that the **radial artery** is compressed in feeling the pulse.

Wrist (fig. 48)

Location of the styloid processes of radius and ulna allows one to project the curved line of the **radiocarpal articulation,** for the styloids mark the ends of the line and the curve is somewhat convex toward the forearm. The styloid process of the radius extends 1 cm. lower than that of the ulna. One-third of the distance across the dorsum of the wrist from the radial side, one can palpate the sharp bony elevation of the **dorsal radial tubercle.** Two-thirds of the distance across marks the position of the **distal radio-ulnar articulation.** With strong extension of the thumb a tendon stands out diagonally on the back of the wrist. This is the tendon of the **extensor pollicis longus muscle** which passes the dorsal radial tubercle to enter the hand. The palmar aspect of the wrist exhibits either two or three transverse creases. The proximal one of two, or the middle of three, indicates the level of the **radiocarpal articulation.** The distal crease is at the **intercarpal joint.** Flexing the outstretched hand at the wrist brings into sharp relief a tendon entering the hand at the center of the wrist. This is the tendon of the **palmaris longus muscle.** A tendon 1 cm. to its radial side belongs to the **flexor carpi radialis muscle.** In the 13 per cent of limbs in which the palmaris longus muscle is lacking, the flexor carpi radialis tendon is the most prominent cord lateral to the center of the wrist. With the hand outstretched, the tendon of the **flexor carpi ulnaris muscle** forms a prominent subcutaneous ridge on the ulnar side of the wrist and leads to the **pisiform bone.** Just to the radial side of the palmaris longus tendon lies the **scaphoid bone** of the wrist, the tubercle of which is palpable.

Hand (figs. 48, 49)

The skin of the dorsum of the hand is delicate and exhibits a multitude of fine creases and criss-crossing lines that run generally transversely. Hair is absent over the distal phalanges, but elsewhere on the dorsum forms a distinctive pattern of growth. Under the loose subcutaneous tissue of the dorsum, the **metacarpal bones** can be palpated. In line with each metacarpal lie three **phalangeal bones,** except in the thumb where there are only two. The **knuckles** of the dorsum represent the distal ends of the more proximal bones in this series, for, in flexion of the digits, the proximal end of the distal bone turns down under the rounded end of the proximal bone. A prominent feature of the dorsum of the hand is the **dorsal venous arch** (p. 59). The skin of the palm of the hand is thick, and the subcutaneous tissue contains little fat. There are no hairs or sebaceous glands on the palmar aspect of the hand and fingers. The skin is firmly bound to the connective tissue planes and structures deep to it, thus forming many prominent skin creases. A pair of generally **transverse creases** at the middle of the palm (fig. 48) mark the palmar

aspect of the metacarpophalangeal joints. A radially located skin crease which partially encircles the root of the thumb (thenar eminence) serves as a flexion crease for movement of the thumb toward the palm. Three sets of transverse creases cross the digits. The proximal and middle sets tend to be double; the distal ones are single. For each digit the middle crease lies opposite the joint between the proximal and middle phalanges. The proximal crease is, however, well distal to the metacarpophalangeal joint line, and the distal crease is slightly proximal to the terminal joint of the digit.

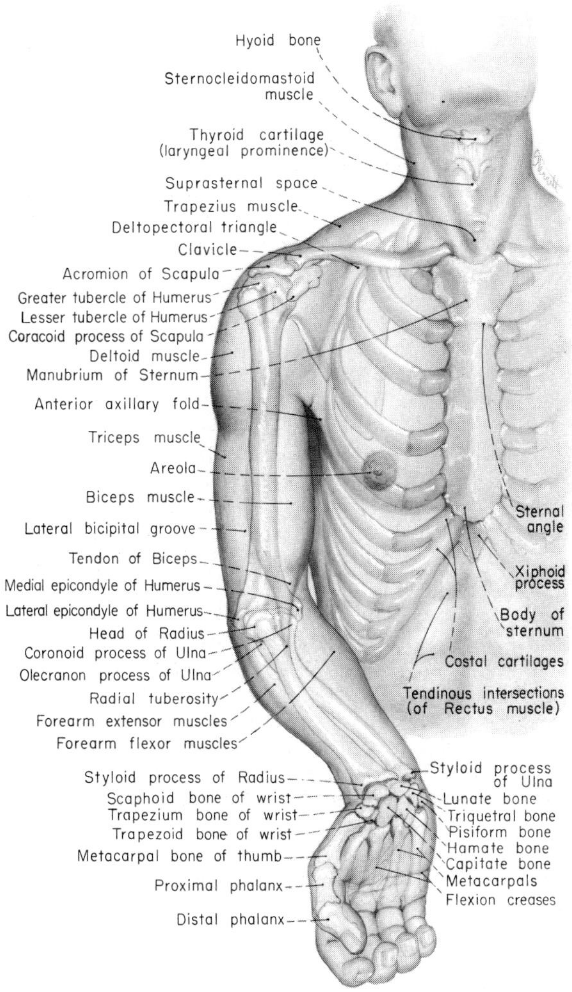

SUPERFICIAL VEINS (figs. 49, 51, 75)

Certain prominent subcutaneous veins, unaccompanied by arteries, are found in the limbs. The principal superficial veins in the upper limb are the **cephalic** and **basilic veins.** These veins originate by venous radicles in the digits and hand. Anastomosing longitudinal **palmar digital veins** empty about the webs of the fingers into similarly longitudinal **dorsal digital veins.** The adjacent dorsal veins of neighboring digits then unite to form relatively short **dorsal metacarpal veins** which end in the **dorsal venous arch,** a strong intercommunication proximal to the heads of the metacarpal bones (fig. 49). The radial continuation of the dorsal venous arch is the **cephalic vein.** It receives the dorsal veins of the thumb, passes across the tendon of the extensor pollicis longus muscle, and ascends at the radial border of the wrist. In the forearm, it tends to parallel the anterior border of the brachioradialis muscle, receiving tributaries from the dorsum of the forearm. In front of the elbow, the cephalic vein is connected to the basilic vein by an obliquely ascending **median cubital vein** (figs. 51, 75). This vein, present in 70 per cent of cases (Charles, '32), passes upward along the belly of the biceps muscle to terminate in the basilic vein above the elbow. It is frequently very large and may conduct so much blood from the cephalic vein to the basilic that the upper continuation of the cephalic vein is much reduced. The median cubital vein commonly receives one or more tributaries from the front of the forearm and is interconnected with deep veins of the forearm in the cubital fossa.

Leaving the cubital fossa (fig. 75), the cephalic vein (fig. 51) ascends in the lateral bicipital groove to the interval between the deltoid and pectoralis major muscles. It pierces the brachial fascia and passes upward and medialward between the parallel

Fig. 48 Surface anatomy and underlying bones of the chest and upper limb.

60   *The Upper Limb*

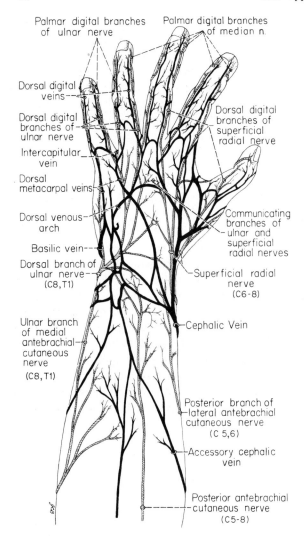

Fig. 49   Cutaneous nerves and superficial veins of the dorsum of the forearm and hand.

fibers of these muscles, accompanied by the deltoid branch of the thoracoacromial artery. Deep in the deltopectoral triangle, the cephalic vein perforates the costocoracoid membrane and empties into the axillary vein.

An **accessory cephalic vein** (fig. 49) arises in a venous plexus on the dorsum of the forearm or from the ulnar side of the dorsal venous arch. Ascending diagonally across the dorsum of the forearm it joins the cephalic vein at the level of the elbow.

The **basilic vein** (figs. 50, 51, 75) is the continuation of the ulnar end of the venous arch of the

dorsum of the hand. It ascends along the ulnar border of the forearm, receiving tributaries from both its anterior and posterior surfaces, and turns into the cubital fossa anterior to the medial epicondyle. After receiving the median cubital vein, the basilic vein continues upward in the medial bicipital groove and pierces the brachial fascia a little below the middle of the arm. Under the brachial fascia it enters the neurovascular compartment formed in the medial intermuscular septum, where it lies superficial to the brachial artery. It remains independent until the axilla is reached, and there it joins the brachial veins to form the axillary vein.

The **median antebrachial vein** is a frequent collecting vessel of the middle of the anterior aspect of the forearm. It begins in the palmar venous plexus and may terminate in the cubital fossa in the median cubital or basilic vein. It sometimes divides into a median basilic vein (which replaces in course and termination the median cubital vein) and a median cephalic vein, which borders the biceps laterally and joins the cephalic vein. In such cases (about 20 per cent according to Charles, '32) a clear M formation is produced by the veins of the cubital fossa. The median antebrachial vein may be large or absent.

Venous Drainage of the Limb

Deep veins accompany the major arteries of the limb and drain the areas supplied by the arteries. The deep veins of the arm are the **brachial veins.** These are paired accompanying veins (venae comitantes) of the brachial artery. At the lower border of the teres major tendon, the basilic vein joins with the brachial veins to form the large **axillary vein.** The cephalic vein empties into a more proximal portion of the axillary vein. The axillary vein becomes the **subclavian vein** as it passes over the first rib, and the subclavian joins the **internal jugular vein** behind the sternoclavicular joint to form the **brachiocephalic vein.** The right and left brachiocephalic veins unite to form the **superior vena cava** which terminates in the **right atrium** of the heart, to which all of the venous blood of the body returns except that from the lungs.

Clinical Note   Much use of these subcutaneous veins is made in medicine for withdrawal of blood samples, introduction of fluids, transfusions, and cardiac catheterization. This is especially true of the veins of the dorsum of the hand and around the wrist and of the cubital veins. Note that the basilic and median cubital veins at the elbow are frequently large and easily accessible for these purposes.

Cutaneous Nerves of the Upper Limb

The cutaneous nerves of the upper limb are, for the most part, branches of nerves of the brachial plexus (p. 78). The uppermost nerves, to the shoulder, are derived from the cervical plexus (p. 175).

Shoulder

Supraclavicular nerves (C3, 4) (figs. 50, 51)

The cervical plexus, composed of ventral rami of spinal nerves from the first through the fourth cervical, provides a number of cutaneous nerves to scalp, face, neck, and shoulder. Such cutaneous nerves all become superficial at the posterior border of the sternocleidomastoid muscle within the posterior triangle of the neck. The supraclavicular nerves descend, pierce the superficial layer of cervical fascia and the subcutaneous platysma muscle, and spread out under the skin from the midline to the point of the shoulder as medial, intermediate, and lateral supraclavicular nerves (p. 149).

Arm

Superior lateral brachial cutaneous nerve (from axillary, C5, 6)
Posterior brachial cutaneous nerve (from radial, C5 to T1)
Medial brachial cutaneous nerve (from brachial plexus, C8, T1)
Intercostobrachial nerve (lateral cutaneous branch of T2)

The **superior lateral brachial cutaneous nerve** (figs. 50, 51) is the termination of the lower branch of the axillary nerve of the brachial plexus (p. 81 & fig. 63). It turns around the posterior border of the deltoid muscle in its lower one-third, pierces the brachial fascia, and supplies the skin covering the lower half of the deltoid and the long head of the triceps muscles.

The **posterior brachial cutaneous nerve** (fig. 50) arises from the radial nerve separately, or in common with a muscular branch to the triceps

muscle. Traversing or passing on the medial side of the long head of the triceps muscle, the nerve penetrates the brachial fascia on the medial side of the arm near the axilla. It distributes below the deltoid and across the back of the arm in its middle one-third, above and behind the area supplied by the medial brachial cutaneous and the intercostobrachial nerves.

The **medial brachial cutaneous nerve** (figs. 50, 51, 63) arises from the medial cord of the brachial plexus (p. 80). It descends along the medial side of the brachial artery to the middle of the arm where it pierces the brachial fascia and distributes to the skin of the posterior aspect of the lower one-third of the arm as far as the olecranon.

The **intercostobrachial nerve** (figs. 50, 51) is the larger part of the lateral cutaneous branch of the second thoracic nerve. It emerges from the second intercostal space at the axillary line and penetrates the serratus anterior muscle to enter the axilla. Here it customarily anastomoses with the medial brachial cutaneous nerve. Crossing the axilla into the arm, the intercostobrachial nerve penetrates the brachial fascia just beyond the posterior fold of the axilla. Its cutaneous distribution is along the

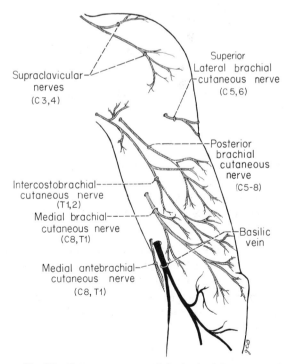

Fig. 50   Cutaneous nerves of the back of the arm.

medial and posterior surface of the arm from the axilla to the elbow. The size of the intercostobrachial nerve and the extent of its distribution tend to vary inversely with the size and distribution of the medial brachial cutaneous nerve with which it communicates.

Forearm

Lateral antebrachial cutaneous nerve (from musculo-
    cutaneous C5, 6)
Posterior antebrachial cutaneous nerve (from radial C5
    to 8)
Medial antebrachial cutaneous nerve (from brachial
    plexus C8, T1)

The **lateral antebrachial cutaneous nerve** (figs. 49, 51, 75) is the cutaneous terminal of the musculo-cutaneous nerve. The latter is appropriately named for it is muscular in the arm and cutaneous in the forearm. The lateral antebrachial cutaneous nerve arises from the musculocutaneous (p. 89 & fig. 63) in the interval between the biceps and brachialis muscles. It pierces the brachial fascia lateral to the biceps tendon just above the elbow, passes behind the cephalic vein, and, opposite the elbow joint, divides into an anterior and a posterior branch.

The larger **anterior branch** follows the cephalic vein into the forearm and distributes to the radial half of its anterior surface as far as the thenar eminence. Above the wrist it communicates with the superficial radial nerve. The **posterior branch** is small (fig. 49). Passing anterior to the lateral epicondyle, it distributes to the skin of the radial border of the forearm, and of its posterior aspect, as far as the wrist.

The **posterior antebrachial cutaneous nerve** (figs. 49, 51) arises from the radial nerve as it lies in its groove on the humerus (p. 91). The nerve has an upper and a lower branch. The smaller upper branch, the **inferior lateral brachial cutaneous nerve,** becomes superficial in line with the lateral intermuscular septum a little below the deltoid insertion. It accompanies the lower part of the cephalic vein and distributes to the skin and sub-cutaneous tissue of the lower half of the lateral and anterior aspect of the arm. The lower branch, the definitive **posterior antebrachial cutaneous nerve,** is large. It passes through the brachial fascia about 8 cm. above the elbow (slightly below the upper branch), descends behind the lateral epicondyle of the humerus, and distributes to the skin of the middle of the dorsum of the forearm as far as the wrist.

The **medial antebrachial cutaneous nerve** (figs. 49, 51, 63, 75) arises from the medial cord of the brachial plexus. A small branch arises in the axilla, pierces the axillary fascia, and descends to supply the skin of the anteromedial biceps area. The main nerve descends through the arm, anterior and medial to the brachial artery, in the neurovascular compartment formed in the medial intermuscular septum. At the junction of the middle and lower thirds of the arm and in company with the basilic vein, the nerve pierces the brachial fascia and divides into an anterior and an ulnar branch. The larger **anterior branch** passes superficial or deep to the median cubital (or median basilic) vein and distributes by several branches to the anterior and medial surfaces of the forearm as far as the wrist. The smaller **ulnar branch** passes downward and dorsalward in front of the medial condyle of the humerus and divides into branches that supply the skin on the postero-medial aspect of the forearm. Certain of its branches accompany the basilic vein.

Wrist and Hand

Superficial radial nerve (from radial C6 to 8)
Dorsal branch of ulnar nerve (C8, T1)
Palmar branch of ulnar nerve (C8, T1)
Palmar branch of median nerve (C6 to 8)
Digital branches of median nerve (C6 to 8)
Digital branches of ulnar nerve (C8, T1)

The **superficial branch of the radial nerve** (figs. 49, 51, 52) arises in the cubital fossa by the division of the radial nerve into deep and superficial branches. The superficial radial nerve (entirely cutaneous) courses through the forearm under cover of the brachioradialis muscle and in company with the radial artery. In the distal one-third of the forearm, the superficial radial nerve perforates the antebrachial fascia along the lateral border of the forearm and divides into two branches. The smaller **lateral branch** supplies the skin of the radial side and eminence of the thumb and communicates with the anterior branch of the lateral antebrachial cutaneous nerve. The larger **medial branch** divides into four dorsal digital nerves (fig. 49).

The first dorsal digital nerve supplies the ulnar side of the thumb; the second, the radial side of the index finger; the third, the adjoining sides of the index and middle fingers; the fourth communicates with a filament from the dorsal branch of the ulnar nerve and, with it, supplies the adjacent sides of the middle and ring fingers. The fourth branch may be lacking, in which case the superficial radial nerve supplies the digits only as far ulnarward as the middle of the third finger. The third finger is frequently a region of overlap between branches of the superficial radial nerve and of the dorsal branch of the ulnar nerve (fig. 52). Usually these two nerves anastomose on the back of the hand proximal to the third digit. Dorsal digital nerves extend to the base of the nail of the thumb, to the distal interphalangeal joint of the second digit, and not quite as far as the proximal interphalangeal joint of the third and fourth digits. The distal areas of the dorsum of the digits not supplied by the radial nerve receive an innervation by strong twigs of the palmar digital branches of the median nerve.

The **dorsal branch of the ulnar nerve** (figs. 49, 52) arises about 5 cm. above the wrist. It passes dorsalward beneath flexor carpi ulnaris and pierces the forearm fascia at the level of the styloid process of the ulna. At the ulnar border of the wrist, the nerve divides into three dorsal digital nerves.

The first branch courses along the ulnar side of the dorsum of the hand and supplies the ulnar side of the little finger as far as the root of the nail. The second branch divides at the cleft between the fourth and fifth digits and supplies their adjacent sides. The third branch may divide similarly and supply adjacent sides of the third and fourth digits, or, when the superficial radial nerve supplies this area, it anastomoses with the latter nerve. Sometimes, both nerves supply the intermediate zone. The dorsal branches to the fourth digit usually extend only as far as the base of the second phalanx, the more distal parts of the fourth and fifth digits being supplied by the palmar digital branches of the ulnar nerve (median nerve for the radial side of the fourth digit).

The **palmar branch of the ulnar nerve** (figs. 51, 52) arises about the middle of the forearm and descends under the antebrachial fascia in front of the ulnar artery. The nerve perforates the fascia just above the wrist and supplies the skin of the hypothenar eminence and of the medial part of the palm of the hand.

The **palmar branch of the median nerve** (fig. 51) arises just above the wrist. It perforates the palmar carpal ligament between the tendons of the palmaris longus and flexor carpi radialis muscles. Its branches supply the skin of the medial part of the thenar eminence and of the central depressed area of the palm.

The **digital branches of the median nerve** (figs. 49, 51, 52). Proper palmar digital branches may be identified at the margins of each of the digits. They are subcutaneous distal to the webs of the fingers

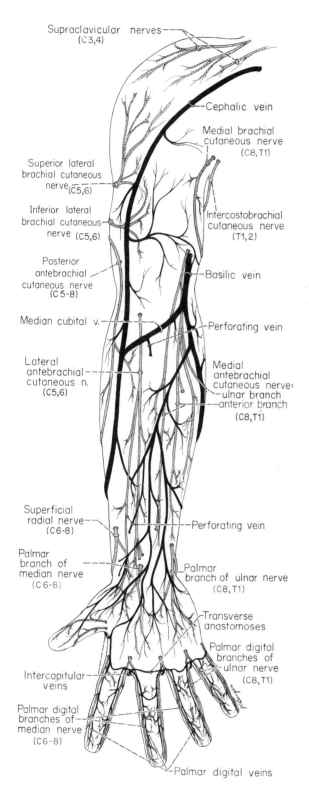

Fig. 51   Cutaneous nerves and superficial veins of the anterior aspect of the upper limb.

only, for they arise from common palmar digital nerves which lie under cover of the dense palmar aponeurosis in the central part of the palm. The first common palmar digital nerve gives rise to the principal muscular nerve to the short muscles of the thumb (motor or recurrent branch of median) and then divides into three proper palmar digital nerves. The lateral one of these runs along the radial border of the thumb to its extremity, giving numerous branches to the pad of the thumb and a strong twig to the dorsum to supply the matrix of the nail. The second proper digital supplies the medial side of the palmar aspect of the thumb and, likewise, sends a twig to the nail bed. The third branch supplies the radial side of the second digit. The second common palmar digital branch of the median provides two proper palmar digitals which supply adjacent sides of the second and third digits. The third common digital communicates in the hand with a digital branch of the ulnar nerve and divides into two proper palmar digitals for adjacent

sides of the third and fourth digits. The proper palmar digital nerves are large because of the density of nerve endings in the fingers, and many of their branches terminate in encapsulated endings. They lie superficial to the corresponding proper palmar digital arteries and veins, and each nerve terminates in the pad of the finger. As they pass along the margins of the fingers, each gives off branches for the innervation of the skin of the dorsum of the digits and of the matrices of the fingernails. These dorsal branches innervate the dorsal skin of the distal segment of the index finger, of the two terminal segments of the third finger, and an equal area of the radial side of the fourth finger.

The **digital branches of the ulnar nerve** (figs. 49, 51, 52). Proper digital branches of the ulnar nerve, like those arising from the median, lie under cover of the palmar aponeurosis until the webs of the fingers are reached. The proper palmar digital nerve to the ulnar side of the fifth digit and a common

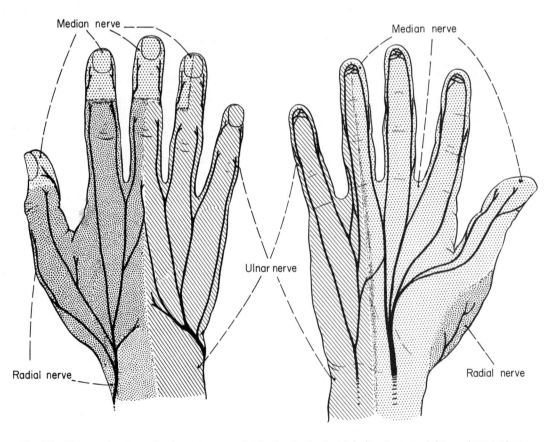

Fig. 52   The usual pattern of cutaneous nerve distribution in the hand. Left—dorsum; right—palmar surface.

palmar digital branch are terminals of the superficial branch of the ulnar nerve in the hand. The common palmar digital branch divides into two proper palmar digital nerves which supply the contiguous margins of the fourth and fifth digits and send branches to the dorsal surface of the second and third segments of these digits. The proper palmar digital branch to the ulnar side of the fifth digit also reaches the nail bed of the dorsum of this digit.

SUPERFICIAL LYMPHATICS (fig. 53)

The superficial lymphatic vessels of the upper limb begin in the hand. Here, a dense plexus of lymphatic capillaries pervades the skin and subcutaneous tissues; its meshwork is finest in the palm and on the palmar surfaces of the fingers.

The **digital lymphatic plexuses** are drained by channels which accompany the digital arteries along the margins of the digits. At the interdigital clefts (e.g., the veins) the collecting vessels of the palmar surfaces of the fingers pass onto the dorsum of the hand to join dorsal collecting vessels and empty into the **plexus of the dorsum of the hand.** Drainage of thumb, index finger, and the radial portion of the third finger is by collecting vessels which arise in this plexus and ascend along the radial side of the forearm; the channels draining the more ulnar fingers ascend along the ulnar side. Channels from the **lymphatic plexus of the palm** radiate to the sides of the hand and also upward through the wrist, making connection with two or three collecting trunks which ascend along the middle of the anterior surface of the forearm.

The **radial and ulnar lymphatic channels** from the fingers turn onto the flexor surface of the forearm and orient themselves on either side of channels ascending through its middle portion. Continuing subcutaneously through the arm, the ascending lymphatic channels reach the **axillary nodes.** These consist of several important groups of lymph nodes at the root of the limb which serve as collecting stations for lymph from the upper limb, shoulder, base of neck, and chest wall. They are more fully considered in connection with the axilla (p. 82).

Some of the ulnar lymphatic channels are interrupted by the **cubital lymph nodes.** This superficial group of one or two nodes is located 3 or 4 cm. above the medial epicondyle of the humerus, somewhat below the aperture in the brachial fascia for the basilic vein. The afferent vessels of these nodes include channels originating in the ulnar three fingers and the corresponding part of the forearm.

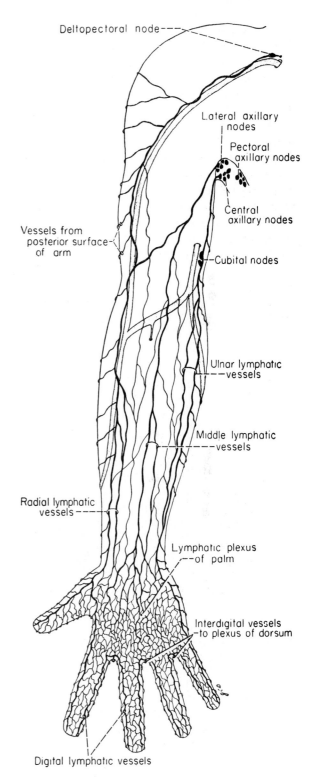

Fig. 53   The superficial lymphatics of the upper limb.

The efferent lymphatics of the cubital nodes accompany the basilic vein under the brachial fascia and terminate in the **lateral and central groups of axillary lymph nodes** (p. 82). Lymphatic channels of the posterior aspect of the forearm and arm drain toward the anterior aspect of the limb and empty into lymphatic vessels ascending on the anterior side of the limb. As a rule, one or more of the lateral collecting trunks of the arm follow the course of the cephalic vein to the deltopectoral triangle and perforate the costocoracoid membrane with the vein. Such lymphatic vessels usually terminate in an apical node of the axillary lymph group. In about 10 per cent of cases this channel is interrupted in the deltopectoral triangle by one or two small lymphatic nodules, the **deltopectoral nodes.**

A deep set of lymphatics also serves the limb. These drain joint capsules, periosteum, tendons, nerves, and, to a lesser extent, muscles. The collecting vessels accompany the major arteries of the limb, along which paths lie small, intercalated lymph nodes. The deep lymphatics of the limb end in the lateral and central axillary node groups.

Clinical Note Infection of the lymph vessels (lymphangitis) of the upper limb may result from a peripheral wound, usually of a penetrating nature. It will be evidenced by red streaks along the course of the lymphatic vessels. When the infection reaches the lymph nodes, they may become inflamed and enlarged (lymphadenitis) and will be tender to the touch.

SUBCUTANEOUS BURSAE

A subcutaneous **olecranon bursa** is nearly constant. It consists of an enlargement of the tissue spaces in the subcutaneous connective tissue over the olecranon and contains a small amount of fluid. Subcutaneous bursae are inconstantly found in the subcutaneous connective tissue over the medial and lateral epicondyles of the humerus.

THE ANTERIOR THORACIC WALL AND PECTORAL REGION

SURFACE ANATOMY (fig. 48)

In the midline of the chest there is a shallow **median furrow** over the subcutaneous sternum and between the sternal origins of the pectoralis major muscles of the two sides. Superiorly, the furrow broadens out over the **manubrium sterni.** At the upper border of the manubrium the **jugular (suprasternal) notch** is flanked by the prominences of the sternal end of each clavicle and by the cord-like sternal

attachments of the sternocleidomastoid muscles. The **nipple,** in the male and in the young female, usually lies opposite the fourth intercostal space. The **heart beat** is clearest in the fifth intercostal space on the left side, 8 or 9 cm. from the midline. In muscular subjects the **pectoralis major muscles** are evident. The lower lateral border of the muscle forms the **anterior axillary fold.** The digitations of origin of the **serratus anterior muscle** may bulge the skin; the origin from the fifth rib is the first that appears below the pectoralis major. The angle between the manubrium and body of the sternum forms a bony projection about 5 cm. below the jugular notch. This **sternal angle** is opposite the sternal end of the second rib and is the most reliable thoracic surface landmark. The **first rib** is, to a large extent, under cover of the clavicle, and ribs are best counted from the second at the sternal angle. The **seventh costal cartilage** is the last to reach the sternum. Attached to the lower end of the body of the sternum is the somewhat recessed and cartilaginous **xiphoid process.** The top of the manubrium is at the level of the disk between the second and third thoracic vertebrae; the sternal angle is on the horizontal plane of the disk between the fourth and fifth thoracic vertebrae. The **xiphosternal junction** is usually on the plane of the ninth thoracic vertebra; the tip of the xiphoid is one vertebral level lower. The xiphoid process lies at the junction of thorax and abdomen in the median plane and thus may serve as a guide to the diaphragm, to the lower border of the heart, and to the upper border of the liver.

CUTANEOUS NERVES OF THE CHEST (fig. 54)

With the exception of the first, second, and twelfth, the ventral rami of thoracic nerves retain the simplicity of typical spinal nerves (p. 47). They do not form plexuses, and each is distributed to a segment of the trunk but not to the limbs. They are divisible into two sets; an upper group composed of the third to the sixth nerves which distribute exclusively to the chest, and a lower group made up of the seventh to the eleventh, inclusive, which have a partially abdominal termination. The lateral and anterior cutaneous branches of the upper nerves supply the skin and subcutaneous tissue of the thoracic wall. The **lateral cutaneous nerves** pierce the musculature covering the ribs along the mid-axillary line and, in the subcutaneous tissue, divide into anterior and posterior branches. The posterior branches pass backward, supplying skin and subcutaneous tissue of the scapular and back regions, and meet the cutaneous distribution of the dorsal rami. The anterior branches pass over the pectoralis major muscle and distribute to the skin of the front of the thorax and, in the female, to the mammary gland. The latter nerves are known as **lateral mammary branches. The anterior cutaneous branches** of the upper group of nerves emerge from the thoracic musculature at the border of the sternum and provide short, medially running branches, which distribute to the midline, and a set of larger lateral cutaneous branches. The latter branches pass outward over the pectoral musculature, innervating skin and fascia, and in the female they supply the mammary gland by means of **medial mammary branches.** Their distribution meets that of the anterior branches of the lateral cutaneous nerves. The lateral and anterior cutaneous branches of the lower (thoracoabdominal) group of nerves have the same general relations but distribute partly over the thoracic wall and partly over the abdominal wall. The first thoracic nerve sends most of its nerve fibers into the brachial plexus and has no lateral cutaneous branch. Its anterior cutaneous branch terminates in the skin at the sternal end of the first intercostal space. The second thoracic nerve contributes most of its lateral cutaneous branch to the arm as the intercostobrachial nerve (p. 61). Its anterior cutaneous branch corresponds to those described.

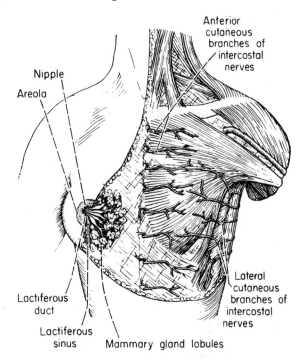

Fig. 54   Superficial features of the anterior thoracic wall; the mammary gland and cutaneous nerves.

THE MAMMARY GLAND (figs. 54, 55)

The mammary glands are accessory to the reproductive function in the female. They secrete milk for the nourishment of the infant, a process known as **lactation.** They are rudimentary and functionless in the male. Mammary glands are modified sweat glands, an origin which determines certain of the morphological features of the gland. It explains the absence of a special fibrous capsule or sheath, the lack of special blood vessels and nerves, and the generalized drainage of its lymph. The glands are situated on the anterior surface of the thorax. In the well-developed nulliparous female (one not having borne child), the breast extends from the second or third rib to the sixth or seventh costal cartilage and from the lateral border of the sternum to beyond the anterior axillary fold. The **sinus mammarum** is the median area between the breasts. The breasts are conical or hemispherical, but their shape varies with development and functional activity. The smooth, conical breasts of the nullipara become hemispherical with increase in fat, while in emaciation and in old age they may be reduced to flattened disks with irregular surfaces. After lactation, the breasts tend to become more pendulous, and, after

repeated pregnancies, they may be further elongated. The mammary gland in girls remains relatively undeveloped up to the age of puberty, when it shows a sudden, considerable increase followed by a long period of more gradual enlargement. The surface of the breast is smooth and convex. Its flattened upper surface shows no sharp demarcation from the anterior surface of the chest, but laterally and inferiorly its borders are usually well defined. At its greatest prominence, there is a pigmented, projecting **nipple** surrounded by a slightly raised and pigmented, circular zone, the **areola mammae.**

The mammary gland is a blend of its epithelial glandular tissue, the **parenchyma,** and the supporting and enclosing connective tissue of the tela subcutanea, the **stroma.** The essential part of each breast is a flattened circular mass of glandular tissue of a whitish or reddish-white color, thickest opposite the nipple and thinner toward the periphery. The gland is composed of from fifteen to twenty lobes, each lobe an irregular, flattened pyramid of glandular tissue, the apex of which is directed toward the nipple and the base toward the periphery of the gland. Each lobe has a single **lactiferous duct** which opens by a contracted orifice in a depression at the tip of the nipple. The ducts are parallel to one another in the nipple but diverge at its base. Beneath the areola, they dilate from a normal 1 to 2 mm. diameter to a size of 4 or 5 mm. The **lactiferous sinus** thus formed constitutes a reservoir for the contents of the duct system. Distally, the duct is reduced by division into branches, each terminal branch ending in a tubulo-saccular or spherical **alveolus.** A number of alveoli open into a common duct and constitute a lobule; all the lobules draining through the same excretory duct make up a lobe. The ducts do not anastomose.

The **stroma** consists of the connective tissue of the gland. Each of the parenchymal units—lobes, lobules, alveoli—is individually surrounded by loose connective tissue. The entire gland is similarly invested by connective tissue, which is not, however, sufficiently organized to form a true capsule. This investment is derived from the fibrous tissue of the tela subcutanea. The fatty component of the tela underlies the skin, interspersed between connective tissue strands; it imparts smoothness to the skin. As elsewhere in the body, the collagenous bundles of the subcutaneous tissue (or hypodermis) attach to and blend with the dermis. These bundles are some-times particularly well developed over the upper part of the breast and have been designated as the **suspensory ligaments.** There is no fat immediately beneath the areola and nipple, and the connective tissue here is loosely arranged. The connective tissue on the deep aspect of the mammary gland is loose, contains little fat, and is attached to the pectoral fascia. The spaces in the loose connective tissue attachments here are designated as **retromammary bursae.**

The **nipple** is conical, or cylindric, with a rounded, fissured tip which accommodates the fifteen or twenty openings of the lactiferous ducts. Its average height is from 10 to 12 mm. There is no fat in the nipple, nor does it have hairs or sudoriferous glands. Sebaceous glands are numerous, however, and their secretion protects the skin of the nipple. In the deeper parts of the dermis of the nipple, smooth muscle forms a loose layer continuous with the smooth muscle of the areola. The muscle fibers are principally circular though some also interlace with the lactiferous ducts. The circular muscle erects the nipple, making it narrower, harder, and more projecting. By continuous contraction, the muscle acts as a sphincter on the excretory system; by inter-mittent rhythmic contraction, it tends to empty the lactiferous ducts.

The **areola** is covered by delicate, pigmented skin, the color of which varies with the complexion but always darkens in pregnancy. The size of the areola shows considerable individual variation, its diameter ranging from 15 to 60 mm. It enlarges in pregnancy. Numerous sebaceous glands and the areolar glands produce elevations of the skin of the areola. The **areolar glands** are rudimentary milk glands and enlarge during pregnancy, as does the true mammary gland. The dermis of the areola lacks fat but contains a layer of smooth muscle. The muscle fibers are mainly circular and radial, and the layer is continuous with that in the nipple.

Variations occur in the mammary gland during both the life and sexual cycles. At puberty marked branching of ducts occurs, and alveoli develop. With each menstrual cycle the gland undergoes further proliferative changes with subsequent regression, but through the years progressive enlargement takes place, due in part to increased deposition of fat. During pregnancy, the breasts greatly enlarge with the formation of fresh glandular tissue. Proliferation in the mammary gland appears to be coincident with the persistence of the corpus luteum of pregnancy. The secretion of milk begins after delivery of the child. At this time the breasts are fully mature, and the suckling of the infant encourages and regulates their excretion.

The **arterial blood supply** of the mammary gland is abundant. The main vessels enter the breast from its superomedial or superolateral border; few vessels are found inferiorly. The arteries of the breast are derived mainly from subcutaneous vessels of the anterior thoracic region. Anterior perforating branches of the **internal thoracic artery** emerge from the thoracic wall musculature close to the border of the sternum. The second, third, and fourth perforating arteries supply the medial and deep surfaces of the mammary gland by **medial mammary rami. Lateral mammary arteries** are given off by the lateral thoracic artery (axillary system) and curve around the lower border of the pectoralis major muscle to reach the gland. **Intercostal arteries,** which course in the intercostal spaces, provide lateral cutaneous and anterior cutaneous branches which accompany the corresponding branches of intercostal nerves. The lateral cutaneous branches over the third, fourth, and fifth intercostal spaces provide **lateral mammary rami.** On and adjacent to the breast, the arteries course in the surface fat, sending penetrating anastomotic branches into the gland. A small part of its blood supply reaches the gland from its deep aspect by perforating branches of the pectoral rami of the thoracoacromial artery (p. 75).

The **veins** of the gland form superficial and deep plexuses and exhibit a pattern arrangement similar to that of the arteries. Venous drainage is principally to the internal thoracic, the lateral thoracic, and the upper intercostal veins, although certain small surface vessels may reach the veins of the base of the neck.

The **nerves** of the breast are the cutaneous nerves of the anterior thoracic region. The anterior branches of the lateral cutaneous branches of intercostal nerves two to six provide **lateral mammary rami.** The anterior cutaneous branches of nerves two to five or six supply **medial mammary rami** on the sternal side of the gland. The upper portions of the skin over the breasts are also supplied by terminals of the supraclavicular nerves (p. 61). The nerves of the breast include both sensory and sympathetic fibers. They reach the skin, the smooth muscle of the areola and nipple, the blood vessels, and the glandular tissue.

The **lymphatic drainage** of the mammary gland (fig. 55) is of much importance in view of the frequent development in this gland of cancer and the subsequent dissemination of cancer cells through the lymphatic stream. The anterolateral thoracic wall has a cutaneous lymphatic network continuous with that of the abdomen below and of the neck above. The **cutaneous lymphatics of the mammary region** are a part of this general system but exhibit a particular arrangement in relation to the nipple and the areola. A dense cutaneous network is continuous with the surrounding cutaneous lymphatic plexus and also discharges by small vessels into the subdermal **subareolar lymphatic plexus.** Toward the periphery of the areola, the subareolar plexus becomes progressively less dense, has larger meshes, and grades over into the **circumareolar plexus.** The latter is continuous with the general cutaneous lymphatic network of the skin over and around the breast. It is through this continuity of cutaneous lymphatic plexuses that cancer may metastasize to the skin of the opposite breast and to superficial abdominal or cervical regions.

The subareolar lymphatic plexus receives the converging lymphatic vessels from the gland proper. To this plexus pass perilobular and interlobular lymphatics as well as those which follow the lactiferous ducts. The **principal axillary lymph path** from the mammary gland has two parts. From the subareolar plexus a **lateral trunk** runs transversely outward toward the axillary lymph nodes. Tributary to it is a collecting vessel from the superior part of the gland. The **medial trunk** curves under the areola, receiving a tributary vessel from the inferior part of the gland, and runs lateralward to the base of the axilla. After winding around the anterior axillary fold, both of these trunks termin-

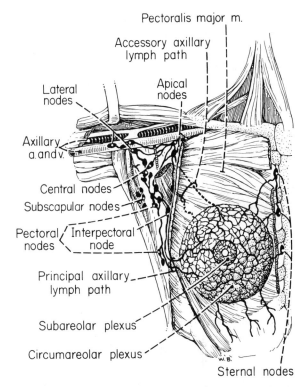

Fig. 55 The axillary lymph nodes and the lymphatic drainage of the mammary gland.

ate in the superior nodes of the **pectoral group of axillary lymph nodes.** Other collecting vessels of this principal path pass directly to the **central,** and still others to the **lateral groups of axillary nodes.**

Accessory axillary lymph paths are of two kinds. Certain lymphatic channels emerge from the periphery of the gland, and, perforating the pectoralis major muscle, follow the branches of the thoracoacromial blood vessels to the **apical axillary nodes,** or lateralward below the pectoralis minor to other **axillary nodes.** Other collecting vessels turn around the lower border of the pectoralis major muscle and ascend between the pectoral muscles to the apical axillary nodes. Small **interpectoral lymph nodules** may be found along these accessory lymphatic paths.

There is an additional lymphatic drainage of the mammary gland to the **sternal chain of nodes.** From the circumareolar plexus on the medial aspect of the gland, collecting vessels pass medialward in company with the anterior perforating branches of the internal thoracic vessels. They perforate the muscles at the edge of the sternum and penetrate the wall to end in the sternal lymphatic chain. These nodes are located deep to the sternum along the internal thoracic vessels, predominantly opposite the upper intercostal spaces. Efferent vessels from the chain reach the major lymph trunks or nodes of the base of the neck.

The lateral, the pectoral, the central, and the apical groups of axillary lymph nodes have been mentioned. They are more completely described, together with the subscapular group, in the section on the axilla (p. 82). Lymph passing through the axillary nodes is collected into channels which form the **subclavian lymphatic trunk** (p. 83).

PECTORAL MUSCLES AND FASCIAE

The pectoral region contains muscles and fasciae associated with the upper limb. It includes the pectoralis major muscle and the pectoral fascia, the pectoralis minor muscle and the clavipectoral fascia, the subclavius muscle, and the clavicle.

The Pectoralis Major Muscle (figs. 56–58, 60–61, 68, 272, 277, 280)

The pectoralis major muscle forms the fullness of the upper portion of the chest, and its lateral border is the anterior axillary fold. The muscle is covered by the adherent pectoral fascia, which encloses it, is attached at its origin to the clavicle and the sternum, and, below, is continuous with the fascia of the abdominal wall. The pectoral fascia leaves the lateral border of the muscle to form the **axillary fascia,** which constitutes the floor of the axillary space.

The pectoralis major muscle has a clavicular and a sternocostal portion. The clavicular part arises from the medial half of the anterior surface of the clavicle. The much larger sternocostal portion takes origin from the anterior surface of the manubrium and body of the sternum by tendinous fibers that decussate over the sternum with those of the other side. On the deep surface of the muscle, fascicles arise from the cartilages of the second to the sixth ribs. A small slip, sometimes separate as an abdominal part, arises from the anterior layer of the sheath of the rectus abdominis muscle. From this rather extensive line of origin, the muscle fibers converge toward the upper end of the humerus. As a means of concentrating the tendinous fibers onto a relatively restricted area of the humerus, the tendon

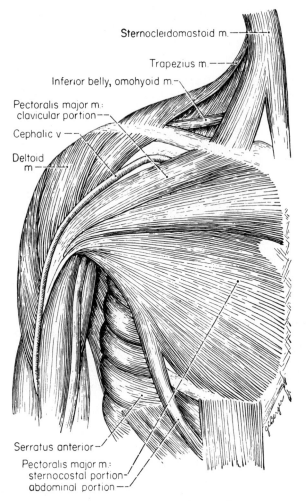

Fig. 56   The pectoralis major muscle.

is folded on itself, forming a 5 cm-wide, bilaminar (U-shaped) tendon with the fold of the laminae below (fig. 57). The fibers of the clavicular part of the muscle insert into the upper part of the anterior lamina; the upper sternal fibers, into the lower part of the same lamina and into its fold. The lower sternal and abdominal fibers reach the upper part of the posterior limb of the tendon. The tendinous fibers of both laminae insert into the crest of the greater tubercle of the humerus, those of the posterior lamina inserting higher and giving off a fascial expansion which spans the intertubercular groove and blends with the capsule of the shoulder joint.

The general action of the pectoralis major muscle is to flex and adduct the arm. The clavicular portion elevates the shoulder and draws the arm (or shoulder) forward and medialward. The sternocostal portion also produces forward and medial traction but draws the shoulder downward. The pectoralis major muscle is capable of medial rotation of the arm as well but usually becomes active only when this movement is resisted. When the arm is fixed, the muscle draws the chest upward toward it, a use made of the muscle in climbing and in forced inspiration. The nerves of the muscle are the **lateral** and **medial pectoral nerves** from the brachial plexus (fig. 57). They are derived from all the roots of the brachial plexus, C5 through T1. The pectoral branches of the thoracoacromial artery (axillary system, p. 76) supply the muscle in company with its nerves (fig. 57).

The pectoralis major muscle is superficial to the pectoralis minor and the clavipectoral fascia. Its upper lateral border is separated from the deltoid muscle by the loose connective tissue of the deltopectoral triangle. Distally, the separation of the parallel fibers of the pectoralis major and deltoid muscles is made by the cephalic vein and the deltoid branch of the thoracoacromial artery.

The Clavipectoral Fascia (fig. 60)

Deep to pectoralis major lies a plane of associated muscles and fascia—the subclavius and the pectoralis minor muscles and the clavipectoral fascia. The clavipectoral fascia invests the subclavius muscle and attaches to the under surface of the clavicle on both sides of the muscle. Extending down and lateralward from the clavicle to the upper border of the pectoralis minor muscle (clavipectoral), the fascia is also conveniently designated by its medial and lateral attachments as the **costocoracoid membrane** (fig. 60). Along the inferior border of the subclavius muscle, a distinct thickening of the membrane constitutes the strong **costocoracoid ligament** stretched between the coracoid process of the scapula laterally and the clavicle and first rib medially. The costocoracoid membrane is perforated by the cephalic vein, the thoracoacromial artery, and the lateral pectoral nerve. The clavipectoral fascia forms an investing fascia for the pectoralis minor muscle and, at its lower lateral border, reunites into a single sheet. This sheet extends downward and lateralward to blend with the **axillary fascia** of the floor of the axilla. Appearing to hold up the axillary fascia, it has been designated as the **suspensory ligament of the axilla.** Near the insertion of pectoralis minor, the fascia along the axillary border of the muscle forms a firm membrane which is continued into the fascia of the coracobrachialis and the short head of the biceps muscles.

The Pectoralis Minor Muscle (figs. 57, 60–61, 272, 277, 280)

The pectoralis minor muscle arises from the outer surfaces of the third, fourth, and fifth ribs near their costal cartilages, with a slip from the second rib being a frequent addition. Flat and triangular, the muscle converges to an insertion on the medial border and upper surface of the coracoid process. The pectoralis minor muscle draws the scapula forward, medialward, and strongly downward. It assists in pointing the shoulder downward and forward. When the scapula is fixed, the muscle aids in forced inspiration. The pectoralis minor is innervated by the **medial pectoral nerve** (C8, T1). After supplying the muscle, this nerve crosses the interpectoral space to reach the pectoralis major muscle. Pectoral branches of the thoracoacromial artery distribute to the muscle with the nerve.

The pectoralis minor muscle lies under cover of the pectoralis major. It is invested by the clavipectoral fascia, with the costocoracoid membrane filling the triangular area medial to it (between it and the subclavius muscle) and with the suspensory ligament of the axilla extending from its lateral border toward the axillary fascia. Deep to the tendon of pectoralis minor pass the axillary artery and the cords of the brachial plexus.

The Subclavius Muscle (figs. 58, 60)

This small, round muscle arises from the junction of the first rib and its cartilage. Its fibers are directed outward and slightly upward and insert in a groove on the under surface of the clavicle between the attachments of the conoid ligament laterally and of the costoclavicular ligament medially. The muscle assists in drawing the shoulder forward and downward. The **nerve to the subclavius** is a branch of the superior trunk of the brachial plexus, with fibers from C5, which reaches the upper posterior border of the muscle. The muscle receives its blood supply from the clavicular branch of the thoracoacromial artery (p. 76).

THE CLAVICLE (figs. 58, 59)

The clavicle is the strut of the upper limb. It props the shoulder out from the chest, so that the arm may have maximum freedom of motion. Situated above the first rib, its sternoclavicular joint constitutes the only point of bony union between the upper limb and the trunk. Laterally, the clavicle articulates with the scapula by means of the acro-

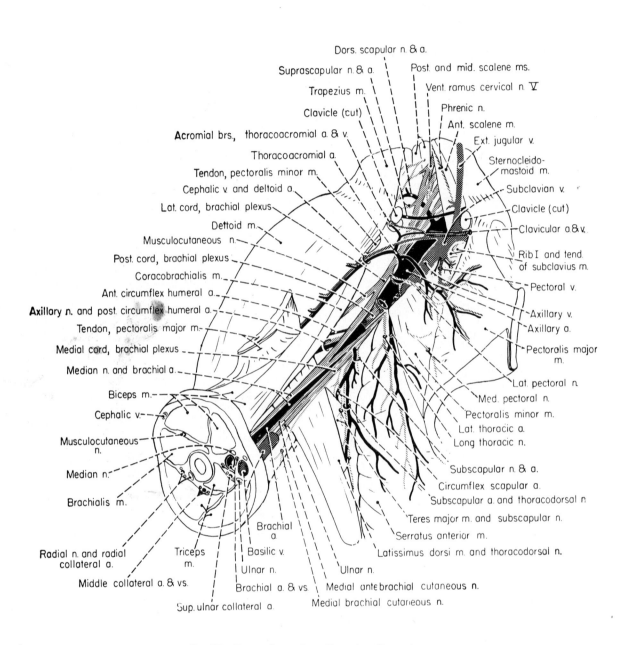

Dors. scapular n. & a.

Suprascapular n. & a.

Trapezius m.

Clavicle (cut)

Post. and mid. scalene ms.

Vent. ramus cervical n. V

Phrenic n.

Ant. scalene m.

Acromial brs., thoracoacromial a. & v.

Thoracoacromial a.

Tendon, pectoralis minor m.

Cephalic v. and deltoid a.

Lat. cord, brachial plexus

Deltoid m.

Musculocutaneous n.

Post. cord, brachial plexus

Coracobrachialis m.

Ant. circumflex humeral a.

Axillary n. and post. circumflex humeral a.

Tendon, pectoralis major m.

Medial cord, brachial plexus

Median n. and brachial a.

Biceps m.

Cephalic v.

Musculocutaneous n.

Median n.

Brachialis m.

Radial n. and radial collateral a.

Middle collateral a. & vs.

Sup. ulnar collateral a.

Triceps m.

Brachial a. & vs.

Ulnar n.

Basilic v.

Brachial a.

Medial antebrachial cutaneous **n.**

Medial brachial cutaneous n.

Ulnar n.

Ext. jugular v.

Sternocleido-mastoid m.

Subclavian v.

Clavicle (cut)

Clavicular a & v.

Rib I and tend. of subclavius m.

Pectoral v.

Axillary v.

Axillary a.

Pectoralis major m.

Lat. pectoral n.

Med. pectoral n.

Pectoralis minor m.

Lat. thoracic a.

Long thoracic n.

Subscapular n. & a.

Circumflex scapular a.

Subscapular a. and thoracodorsal n

Teres major m. and subscapular n.

Serratus anterior m.

Latissimus dorsi m. and thoracodorsal **n.**

Fig. 57   Pectoral muscles reflected, axilla, and arm.

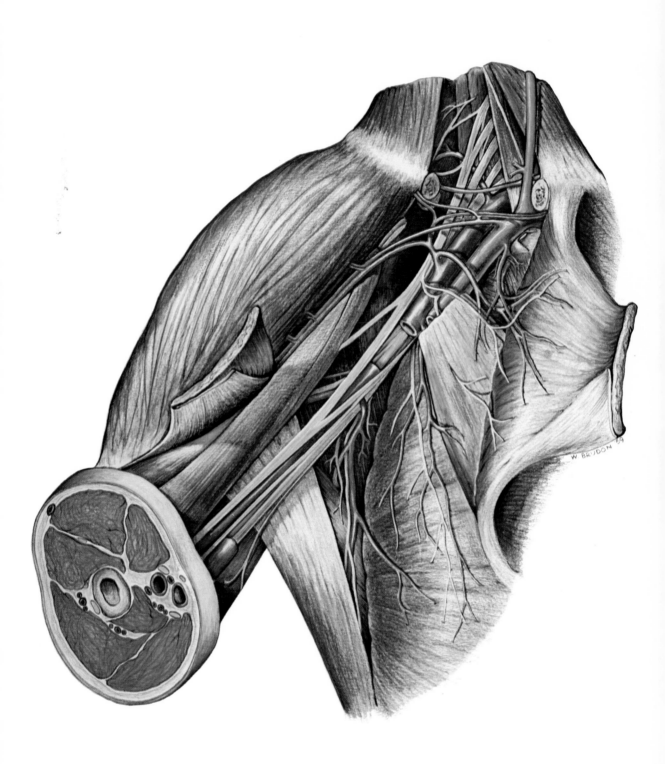

mioclavicular articulation. The bone has a double curve, so that its profile is like a very elongated capital S. Since it transmits shocks from the limb to the trunk, its curvatures have value in increasing its resilience.

The medial two-thirds of the clavicle is roughly triangular in section and is convex forward, whereas the lateral one-third is flattened and, viewed from the front, shows a marked concavity. The underside of the prismatic medial portion has a large, oval roughening medially; this is the **costal tuberosity** for the attachment of the costoclavicular ligament. Also on its underside is a foramen, directed laterally, for the nutrient artery.

The lateral one-third of the clavicle is broad, is flattened from above downward, and turns forward. On the underside posteriorly, at the junction of the medial two-thirds and the lateral one-third of the bone, is a prominent **conoid tubercle** for the conoid portion of the coracoclavicular ligament (fig. 59). Extending forward and laterally from the conoid tubercle is the **trapezoid line** for the attachment of the trapezoid portion of the coracoclavicular ligament.

The **sternal extremity** of the clavicle is triangular in shape and its articular surface is received into the clavicular fossa of the manubrium sterni, an articular disk intervening between the bony surfaces. The **acromial extremity** presents an oval facet directed lateralward and slightly downward for the acromion.

Though classed as a long bone, the clavicle has no medullary cavity. It consists of cancellous bone within a shell of compact bone. It is the bone that begins ossification earliest. Two primary centers, which later coalesce, appear near the center of the bone at the fifth week of fetal life. A third center for the sternal end appears about the seventeenth year and fuses with the shaft about the twenty-fifth year.

THE AXILLA

The axilla (figs. 60-61) is a pyramidal-shaped area at the junction of the upper limb and the chest and

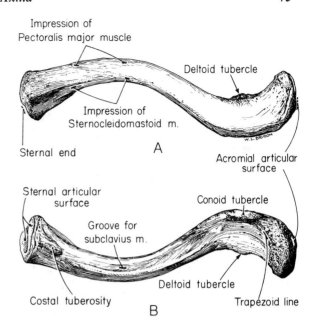

Fig. 59 The right clavicle; A—superior surface, B—inferior surface. ($\times\frac{1}{2}$.)

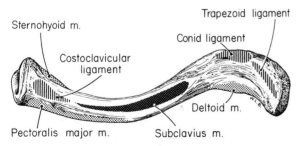

Fig. 58 The attachments on the inferior surface of the right clavicle. (Black dots—origin; white dots on black—insertion; lines—ligaments.)

has four sides, an apex, and a base. The **base** is the concave armpit, the actual floor of the space being the **axillary fascia**. The **anterior wall** of the pyramid is composed of the pectoralis major and minor muscles and their associated fasciae, and the lateral border of pectoralis major forms the **anterior axillary fold**. The **posterior wall** of the axilla is represented by the scapula and the scapular musculature. The lower member of this group, teres major, combines with the latissimus dorsi to form the **posterior axillary fold**. The chest wall covered by the serratus anterior muscle and its fascia forms the **medial wall**. The **lateral wall** is reduced by the convergence of the tendons of both the anterior and posterior axillary fold muscles and their insertions on the lips of the intertubercular groove of the humerus. The floor of the groove constitutes the lateral axillary wall. The **apex** of the axilla is blunted and triangular and is formed by the convergence of the bony members of the three major walls, the clavicle, the scapula, and the first rib. The interval between these bones is the entrance to the axillary space, through which pass all the neurovascular structures of the limb.

The **fasciae** of the axillary walls (fig. 60) are the actual boundaries of the space and exhibit continuities with one another. The **pectoral fascia** is attached above to the clavicle and medially to the sternum and encloses the pectoralis major muscle. The deeper **clavipectoral fascia** invests the subclavius and pectoralis minor muscles, is the **costocoracoid membrane** in the interval between them, and forms the **suspensory ligament of the axilla** lateral to the axillary border of the pectoralis minor muscle. The muscular fasciae of the scapular region are continuous with one another. The fascia of the infraspinatus and teres minor muscles is continuous with that of the teres major and, after attachment to the border of the scapula, with the fascia of the subscapularis. The fascia of teres major is continuous with that of the latissimus dorsi muscle as these muscles turn on one another approaching their insertions. There is also continuity of the fascia of the serratus anterior muscle with the pectoral and clavipectoral fasciae anteriorly, with the fasciae of the subscapularis and other scapular muscles posteriorly, and with the axillary fascia. The **axillary fascia,** thin and perforated, forms the fascial floor of the axilla. It is continuous anteriorly with the pectoral fascia and posteriorly with the fascia of latissimus dorsi. The **suspensory ligament of the axilla** attaches to the axillary fascia behind the lateral border of the pectoralis major muscle, and it is suggested that this attachment accounts for the sharp upward concavity of the armpit behind the anterior axillary fold. The axillary fascia is continuous medially with that covering the serratus anterior muscle and laterally with the brachial fascia investing the structures

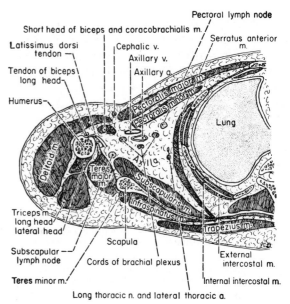

Fig. 61   A horizontal section through the trunk at the level of the shoulder to show the muscular boundaries of the axilla.

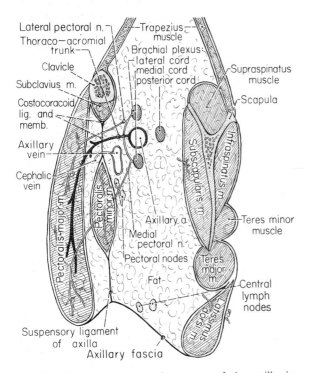

Fig. 60   The boundaries and contents of the axilla in vertical section.

of the arm. It is firmly attached to the skin and subcutaneous tissue of the armpit and is perforated by nerves and vessels, allowing continuity between the subcutaneous fat and connective tissue and the fat of the axillary space.

The skin of the axilla is pigmented. The down hair of the region is completely replaced at puberty by the coarse terminal hair characteristic of this region in the adult. Associated with the hairs of the axilla are special apocrine sweat glands. These glands are large (5 mm. diameter) and secrete large amounts of sweat. In the process of secretion, cell fragments of the glands are sloughed and, containing small amounts of pigment, give color to the sweat. The glands are closely associated with sexual development, enlarging at puberty and undergoing cyclic changes with the menstrual cycle in the female.

CONTENTS OF THE AXILLA

The principal contents of the axilla are blood vessels and nerves which pass from the neck and chest to the upper limb. These are the **axillary artery** and its branches, the **axillary vein** and its tributaries, and **branches of the brachial plexus.** The major vascular and nerve trunks are enclosed within the **axillary sheath,** a fascial extension of the prevertebral layer of cervical fascia covering the scalene muscles. The axillary sheath is adherent to the clavipectoral fascia behind the pectoralis minor muscle and is continued along the vessels and nerves as far as their entrance into the neurovascular compartment of the medial intermuscular septum of the arm. The **axillary lymph nodes** are lodged in the fat and loose areolar tissue of the axilla. The tendon of the long head of the biceps muscle occupies the intertubercular groove of the humerus and so is within the lateralmost part of the axilla. Also within the space are the short head of the biceps and the coracobrachialis muscles arising from the coracoid process of the scapula. Their origin and their fascial investment by the upper portion of the suspensory ligament of the axilla makes them **adjuncts** of the anterior wall.

The Axillary Artery (figs. 57, 62)

The term 'axillary artery' is applied to one portion of the main arterial stem of the upper limb. An appropriate regional name is given to the vessel in different parts of the limb. Ascending from the arch of the aorta, by which all systemic arterial blood leaves the heart, are a brachiocephalic artery on the right side and both a common carotid and a subclavian artery on the left. The subclavian artery (as a branch of the brachiocephalic artery on the right or by its independent origin on the left) supplies a number of important branches in the base of the neck and then passes over the first rib into the axillary space. It is then called the axillary artery. The axillary artery is defined by the limits of the axilla; it extends from the outer border of the first rib to the lower border of the teres major muscle, where its name is changed to brachial.

With the arm abducted, the course of the axillary artery can be indicated on the surface by a line extending from the middle of the clavicle to the groove just behind the coracobrachialis and biceps muscles. Proximally, it is deeply placed under the pectoral muscles and has the shoulder joint and neck of the humerus lateral to it. Its distal end is superficial, being covered only by skin, subcutaneous tissue, and the brachial fascia. The axillary artery is large, about 1 cm. in diameter and about 12 cm. in length. The axillary vein lies anterior and inferior to the artery, with the artery showing a gentle, downward curvature when the arm is at the side. With the arm fully abducted so as to bring the artery into a straight line, the vein rises and is more completely in front of the artery. The upper part of the axillary artery is crossed by the tendon of the pectoralis minor muscle. This allows the subdivision of the artery into a portion proximal to the pectoralis minor tendon, a portion behind it, and a portion distal to it. It is a memory help to observe that part one of the

artery has one branch; part two, two branches; and part three, three branches. For a discussion of collateral circulation in cases of ligation of the axillary artery, see p. 142.

For a discussion of collateral circulation in cases of ligation of the axillary artery, see p. 142.

The **first part of the axillary artery** is about 2.5 cm. long and lies behind the costocoracoid membrane, with the axillary vein below it. Above the artery lie the lateral and posterior cords of the brachial plexus, behind it is the medial cord. The artery is crossed by a communicating loop between the lateral and medial pectoral nerves. It has one branch.

1. The **superior thoracic artery** (figs. 57, 62) is a small vessel which arises at the lower border of the subclavius muscle. It runs downward and medialward behind the axillary vein to supply the intercostal muscles of the first and second intercostal spaces and the upper portion of the serratus anterior muscle. It anastomoses with the intercostal arteries.

The **second part of the axillary artery** lies behind the tendon of pectoralis minor and measures about 3 cm. in length. It is in reference to this part of the artery that the cords of the brachial plexus are named. The lateral cord is lateral to the artery, and the medial cord is medial to it. Posterior to the artery runs the posterior cord of the plexus, and behind the nerve is the loose areolar tissue which separates the artery from the subscapularis mucle. The branches of the second part are the thoracoacromial and the lateral thoracic arteries.

2. The **thoracoacromial artery** (figs. 57, 62) arises from the second part of the axillary artery beneath the upper border of the pectoralis minor muscle. It is a short trunk which issues from the anterior aspect of the axillary artery, turns the border of pectoralis minor to pierce the costocoracoid membrane, and then divides into four branches deep to the clavicular head of pectoralis major.

(a) The **acromial branch** (figs. 57, 62) passes lateralward across the coracoid process to the acromion. It gives branches to the deltoid muscle and participates, with branches of the anterior and posterior circumflex humeral and suprascapular vessels, in the formation of the acromial network of vessels on the surface of the acromion.

(b) The **deltoid branch** (figs. 57, 62) descends in the interval between the pectoralis major and deltoid muscles in company with the cephalic vein. It sends branches into these muscles, terminating opposite the insertion of the deltoid. It often arises with or as a branch of the acromial artery.

(c) The **pectoral branch** (figs. 57, 62) is a large branch

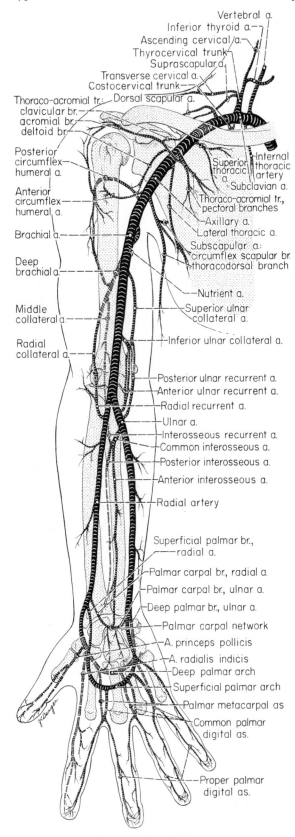

Vertebral a.
Inferior thyroid a.
Ascending cervical a.
Thyrocervical trunk
Suprascapular a.
Transverse cervical a.
Costocervical trunk
Thoraco-acromial tr.   Dorsal scapular a.
clavicular br.
acromial br.
deltoid br.
Posterior
circumflex
humeral a.
                        Superior   Internal
                        thoracic   thoracic
                        a.   artery
                            Subclavian a.
Anterior   Thoraco-acromial tr.,
circumflex   pectoral branches
humeral a.
                        Axillary a.
Brachial a.   Lateral thoracic a.
                        Subscapular a.:
                        circumflex scapular br.
Deep   thoracodorsal branch
brachial a.
                        Nutrient a.
                        Superior ulnar
Middle   collateral a.
collateral a.
                        Inferior ulnar collateral a.
Radial
collateral a.
                        Posterior ulnar recurrent a.
                        Anterior ulnar recurrent a.
                        Radial recurrent a.
                        Ulnar a.
                        Interosseous recurrent a.
                        Common interosseous a.
                        Posterior interosseous a.
                        Anterior interosseous a.

                        Radial artery

                        Superficial palmar br.,
                        radial a.
                        Palmar carpal br, radial a.
                        Palmar carpal br, ulnar a.
                        Deep palmar br, ulnar a.
                        Palmar carpal network
                        A. princeps pollicis
                        A. radialis indicis
                        Deep palmar arch
                        Superficial palmar arch
                        Palmar metacarpal as
                        Common palmar
                        digital as.

                        Proper palmar
                        digital as.

Fig. 62   The arterial system of the upper limb.

which descends between the pectoralis major and minor muscles. It gives branches to both of these muscles, anastomoses with intercostal and lateral thoracic arteries, and, in the female, supplies the mammary gland on its deep aspect.

(d) The **clavicular branch** (figs. 57, 62) is a slender vessel which runs upward and medialward, supplying the subclavius muscle and the sternoclavicular joint.

3. The **lateral thoracic artery** (figs. 57, 62) is rather variable. It may arise directly from the axillary artery, from the thoracoacromial, or from the subscapular artery and is frequently represented by several vessels. Typically (65%—Huelke), it arises from the axillary artery, descends along the axillary border of pectoralis minor, and sends branches to the serratus anterior and pectoral muscles and the axillary lymph nodes. In the female it provides lateral mammary branches which turn over the lower border of the pectoralis major muscle to reach the mammary gland. It anastomoses with branches of the aortic intercostal arteries, with anterior intercostal branches from the internal thoracic artery, and with radicles of the subscapular and thoracoacromial arteries.

An almost constant muscular artery to the subscapularis muscle enters its neurovascular hilum with the upper subscapular nerve. This **upper subscapular artery** (Huelke) is quite variable in origin, being most commonly a direct branch of the axillary artery but also frequently an offshoot of the subscapular or lateral thoracic artery.

The **third part of the axillary artery** extends from the axillary border of pectoralis minor to the lower border of teres major, a distance of about 6.5 cm. It lies against the coracobrachialis muscle laterally. The axillary vein is medial to the artery. The median nerve is formed proximally on the surface of this portion of the artery by the junction of contributing roots from the medial and lateral cords of the brachial plexus. The musculocutaneous nerve enters the coracobrachialis muscle lateral to the artery. The ulnar nerve and the medial brachial and medial antebrachial cutaneous nerves lie medial to it, and the radial and axillary nerves are posterior. The branches of the third part of the axillary artery are the subscapular, the anterior circumflex humeral, and the posterior circumflex humeral.

4. The **subscapular artery** (figs. 57, 62) is the largest branch of the axillary. It arises at the lower border of the subscapularis muscle and descends along its axillary border, supplying the subscapularis, the teres major, and the serratus anterior muscles. At 3 or 4 cm. from its origin the

artery divides into the circumflex scapular and the thoracodorsal branches.

(a) The **circumflex scapular artery** (figs. 57, 62), the larger branch of the subscapular, passes posteriorly through a triangular space defined by the borders of subscapularis and teres minor above, teres major below, and the long head of the triceps laterally (p. 56). It then turns beneath the teres minor muscle onto· the dorsum of the scapula which it grooves (p. 56). Ramifying in the infraspinatous fossa, it supplies the muscles of the dorsum of the scapula and anastomoses with the dorsal scapular artery and with the infraspinatous branch of the suprascapular artery. In the triangular space the artery gives off two branches. An anterior branch to the subscapularis courses deep to that muscle and supplies it; a second branch runs toward the inferior angle of the scapula between the teres major and minor muscles and supplies both of them.

(b) The **thoracodorsal artery** (figs. 57, 62) continues the general course of the subscapular artery to the angle of the scapula, supplying adjacent muscles and anastomosing with the circumflex scapular, the dorsal scapular, the lateral thoracic, and intercostal arteries. The thoracodorsal artery is the principal supply to the latissimus dorsi muscle, entering it ,on its deep axillary surface accompanied by the thoracodorsal (middle subscapular) nerve. It frequently has a thoracic branch which substitutes for the inferior portion of the complete distribution of the lateral thoracic artery.

5. The **anterior circumflex humeral artery** (figs. 57, 62), much smaller than the posterior circumflex, arises closely adjacent to the latter, or by a common trunk with it, and from the lateral side of the axillary artery. It runs anteriorly around the surgical neck of the humerus deep to the coracobrachialis and biceps muscles and anastomoses with the posterior circumflex humeral artery. It supplies adjacent muscles, one branch descending along the tendon of pectoralis major. Another branch ascends in the intertubercular groove to supply the long tendon of the biceps and the shoulder joint.

6. The **posterior circumflex humeral artery** (figs. 57, 62) arises at the level of the surgical neck of the humerus. It passes backward with the axillary nerve through the quadrangular space (p. 56) against the bone and deep to the deltoid muscle. Numerous branches supply this muscle; others reach the shoulder joint and anastomose in the acromial network on the acromion. A nutrient artery is supplied from the posterior circumflex humeral to the greater tubercle of the humerus. A descending branch follows the lateral head of the triceps, distributes to the long and lateral heads of this muscle, and anastomoses with an ascending branch of the deep brachial artery of the arm. Occasional enlargement of this anastomosis results in the origin of the deep brachial from the posterior circumflex humeral or of the posterior circumflex humeral artery from the deep brachial. The posterior circumflex humeral artery encircles the surgical neck of the humerus and anastomoses with the anterior circumflex humeral artery.

The Axillary Vein (figs. 57, 60, 61)

In connection with a description of the superficial veins of the upper limb (p. 59), mention was made of its general plan of venous drainage. As a rule, the deep veins of the limb are doubled, lying on either side of the corresponding artery, and may be designated as accompanying veins (venae comitantes) of the artery. Thus, in the forearm there are radial and ulnar veins and, in the arm, brachial veins which drain the regions supplied by the arteries of the same name. The brachial veins occupy, along with the brachial artery and certain nerves, the neurovascular compartment of the medial intermuscular septum into which the basilic vein also passes a little below the middle of the arm. Frequent cross anastomoses unite the brachial veins. At the lower border of either the teres major or subscapularis muscle, the lateral brachial vein crosses the axillary artery to join the medial brachial vein. At approximately the same level the basilic vein joins the medial brachial vein, and as a result of these somewhat variable junctions, the single axillary vein is formed. The **axillary vein** is a vein of large size and represents, as does the artery, a regional segment of the vascular system of the limb. It extends as far as the outer border of the first rib, where it becomes the subclavian vein. At first it lies on the medial side of the axillary artery which it partly overlaps; between them lie the median, the ulnar, and the medial antebrachial cutaneous nerves. In the upper part of the axilla the vein lies inferior and somewhat anterior to the artery and is separated from it by the medial cord of the brachial plexus. The medial pectoral nerve passes between the artery and the vein. One or two lateral axillary lymph nodes lie closely medial to the vein. It contains a pair of valves at the level of the lower border of the subscapularis muscle, and valves are also found at the ends of the subscapular **and the cephalic veins. Accidental wounds of the axillary vein are dangerous due to its exposed position, large size, and nearness to the thorax.**

The tributaries of the axillary vein correspond to the branches of the axillary artery with the exception of the thoracoacromial vein, the larger pectoral radicles of which usually empty into the distal portion of the subclavian vein. The **cephalic vein** perforates the costocoracoid membrane at the upper border of the tendon of the pectoralis minor muscle to terminate in the axillary vein. Before perforating the membrane, it usually receives all but the pectoral tributaries of the thoracoacromial vein. The **lateral thoracic vein** has certain tributaries which do not correspond with branches of the lateral thoracic artery. It receives numerous anastomosing **thoracoepigastric veins** from the lower thoracic and upper abdominal regions. It also receives the **costoaxillary veins** of the chest. These are, in part, anastomotic channels between intercostal veins and the lateral thoracic vein and provide an axillary linkage between the superior vena cava and the azygos system of veins to which the intercostal veins are tributary. The costoaxillary veins also constitute drainage channels from the lateral mammary and superficial axillary regions.

The Brachial Plexus (figs. 57, 63)

Peripheral nerves distribute to discrete regions of the body and provide a complete assortment of muscular, sensory, and sympathetic fibers to the region innervated. For the upper and lower limbs, the large brachial and lumbosacral plexuses allow the mingling of nerve fibers from several segments of the spinal cord and the union of the various components needed in each nerve territory. The brachial plexus does not originate in the axilla, and only part of its pattern of regrouping of nerve bundles is apparent in this region. The considerable axillary (infraclavicular) part of the brachial plexus necessitates, however, its complete description, so that the functional significance of its branches may be understood.

The **brachial plexus** is formed by the ventral rami of the fifth to the eighth cervical nerves and the greater part of the ramus of the first thoracic nerve. Small contributions may be made by the fourth cervical and the second thoracic nerves. The ventral rami, designated as the roots of the plexus, emerge from between the anterior and middle scalene muscles in line with the similar rami constituting the cervical plexus. The upper roots and trunks of the plexus pass downward and outward, whereas the ventral ramus of the first thoracic nerve has to ascend to pass across the first rib. The sympathetic fibers conducted by each root are added as they pass between the scalene muscles. Each of the ventral rami of the fifth and sixth cervical nerves receives a gray ramus communicans from the middle cervical ganglion. The cervicothoracic ganglion contributes gray rami to the seventh and eighth cervical and the first thoracic roots of the plexus.

The sources of the brachial plexus having been established, the steps in its complete formation may be anticipated. Five stages are recognized.

1. The undivided ventral rami.
2. The formation of three trunks.
3. The separation of each trunk into an anterior and a posterior division.
4. The union of the divisions to form three cords.
5. The derivation from the cords of the principal nerves of distribution.

After passing the margins of the scalene muscles, the ventral rami of the fifth and sixth cervical nerves unite to form the **superior trunk;** the seventh ramus continues alone as the **middle trunk;** and the first thoracic ventral ramus rises over the first rib to unite with that of the eighth cervical to form the **inferior trunk.** Stage three, the separation of each trunk into an anterior and a posterior part, is a division of much significance, for it indicates the separation of nerve bundles destined for the supply

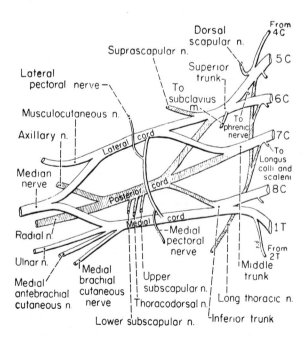

Fig. 63 The plan of the brachial plexus; postaxial divisions and nerves are cross-hatched, pre-axial cords and nerves are clear.

of the originally ventral parts of the limb from those which supply the dorsal parts. This fundamental nerve division is apparent at the limb-bud stage of development (fig. 64) and persists in essentially unaltered relations to the adult. Sometimes, in the plexus, the nerves divide before or during formation of the trunks, making identification of these stages difficult. The anterior and posterior divisions of the fifth, sixth, and seventh nerves are of approximately equal size. The posterior division of the eighth nerve is smaller, and that of the first thoracic is very small or may be entirely lacking. In the formation of the cords, stage four, the posterior divisions of all three trunks unite to form the **posterior cord**; the anterior divisions of the superior and middle trunk form the **lateral cord**; and the **medial cord** is the ununited anterior division of the inferior trunk. Thus, the posterior cord contains nerve bundles destined for the back of the limb from all nerves from C5 to T1; the lateral cord is formed of nerve bundles for the anterior portion of the limb from nerves C5 to 7; and the medial cord carries anterior nerve components from nerves C8 and T1. The cords of the brachial plexus are named by the relations they exhibit to the axillary artery—lateral, medial, and posterior to it—at the level of the crossing of that artery by the pectoralis minor muscle. A further stage of regrouping occurs below the level of the cords in the derivation from the lateral and medial cords of their three major terminal branches, the median, the ulnar, and the musculocutaneous nerves. In this regrouping a large segment of each of the two cords separates and joins the other on the anterior surface of the axillary artery to form the **median nerve** (fig. 57). The remainder of the lateral cord constitutes the **musculocutaneous nerve**; the rest of the medial cord is the **ulnar nerve**. Frequently, the contribution of the lateral cord to the median nerve is made in several separated bundles, a circumstance which makes it appear as though the median and musculocutaneous nerves were interconnected at a lower level in the arm. The posterior cord, lying posterior to the axillary artery, gives off the **axillary nerve** at the lower border of the subscapularis muscle. The remainder of the posterior cord is the large **radial nerve**. The five nerves thus formed from the three cords constitute the **terminal nerves** of the plexus, and their further description will be given in connection with the regions in which they distribute.

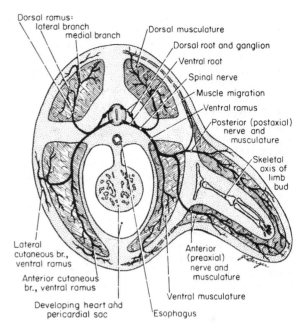

Fig. 64   The nerve-muscle relationships of the developing limb bud. (Adapted from Kollman.)

The nerves of distribution of the brachial plexus also include a number of nerves which arise from the rami, the trunks, and the cords at levels above the formation of the terminal branches. **Muscular twigs** to the anterior, middle, and posterior scalene and longus cervicis muscles arise from the lower four cervical nerves as they emerge from their intervertebral foramina. A contribution from the fifth cervical nerve is made to the phrenic nerve at the lateral border of the anterior scalene muscle. Other branches arising from proximal parts of the plexus are concerned with the innervation of the muscles of the shoulder. These nerves are divided topographically into supraclavicular and infraclavicular branches. Only the infraclavicular branches are approachable through the axilla.

1. The Supraclavicular Branches

(a) The **dorsal scapular nerve** (figs. 40, 63) is the nerve to the rhomboid and the levator scapulae muscles. Arising from the posterior aspect of the ventral ramus of C5 (with the frequent addition of a contribution from C4), it pierces the middle scalene muscle to reach the posterior triangle of the neck. It descends anterior to the levator scapulae and the rhomboid muscles adjacent to the medial border of the scapula.

(b) The **long thoracic nerve** (figs. 57, 63) supplies the serratus anterior muscle. Its three roots arise from the back of the fifth, sixth, and seventh cervical nerves. Those from the fifth and sixth nerves pierce the middle scalene muscle; they are joined by the root from the seventh at the level of the first rib. The nerve passes behind the brachial plexus and enters the axilla between the axillary artery and the serratus anterior muscle, on which muscle it ramifies in company with the lateral thoracic artery.

(c) The **nerve to the subclavius** (fig. 63) is a slender nerve which takes origin from the anterior aspect of the superior trunk. Its nerve fibers arise mainly from the fifth cervical nerve, with occasional additions from the fourth and sixth nerves. It descends across the anterior aspect of the plexus, in front of the subclavian artery and vein, to reach the subclavius muscle.

A frequent branch of the nerve to the subclavius muscle is the **accessory phrenic nerve**. This nerve, which usually represents the phrenic root from the fifth cervical nerve, passes medialward, either in front of or behind the subclavian or brachiocephalic vein, to join the phrenic nerve at a variable level. Its importance in the surgical procedure of 'phrenic interruption' stems from its frequency and from the fact that it may include a large proportion of phrenic nerve fibers.

(d) The **suprascapular nerve** (figs. 40, 57, 63) arises from the posterior aspect of the superior trunk, its nerve fibers coming from the fifth and sixth cervical nerves and, in about 50 per cent of cases, from the fourth. The nerve passes lateralward across the posterior cervical triangle, above the plexus, and along the anterior border of the trapezius muscle to reach the scapular notch. Passing through this notch, it is separated from the suprascapular artery by the superior transverse scapular ligament, the artery passing over the ligament. After entering the supraspinatous fossa, the nerve supplies branches to the supraspinatus muscle and to the shoulder joint; it then descends through the notch of the scapular neck to the infraspinatous fossa and terminates in the infraspinatus muscle.

2. The Infraclavicular Branches

(a) The **lateral pectoral nerve** (figs. 57, 60, 63) arises from the lateral cord and contains nerve fibers from the anterior divisions of the fifth, sixth, and seventh cervical nerves. It passes anterior to the axillary artery and forms a communicating loop with the medial pectoral nerve. Branches of the lateral pectoral nerve then pierce the costocoracoid membrane and are distributed to the pectoralis major muscle.

(b) The **medial pectoral nerve** (figs. 57, 60, 63) takes its origin from the medial cord and contains nerve fibers from the eighth cervical and first thoracic nerves. Passing forward between the axillary artery and vein, it forms a loop in front of the artery with the lateral pectoral nerve. The medial pectoral nerve penetrates the pectoralis minor muscle and supplies it. Branches also pass on, across the interpectoral space, into the pectoralis major. The medial and lateral designations of these two nerves refer to the cords from which they are derived. They are not topographic designations; indeed, the lateral nerve is the more medial on the chest wall.

(c) The **medial brachial cutaneous nerve** (figs. 50, 51, 57, 63) arises from the medial cord of the brachial plexus and contains fibers from the eighth cervical and first thoracic nerves in most cases. It descends through the axilla behind the axillary vein, pierces the brachial fascia about the middle of the arm, and distributes to the skin and subcutaneous tissue of the posterior aspect of the lower one-third of the arm as far as the olecranon. In the axilla the medial brachial cutaneous nerve forms a loop of communication with the intercostobrachial nerve. The medial brachial cutaneous nerve may be a branch of the medial antebrachial cutaneous nerve.

(d) The **medial antebrachial cutaneous nerve** (figs. 50, 51, 57, 63) arises from the medial cord closely adjacent to the ulnar nerve, and, like it, carries nerve fibers from the eighth cervical and first thoracic nerves. It first lies medial to the axillary artery and then anterior and medial to the brachial artery. At the junction of the middle and lower thirds of the arms, the nerve pierces the brachial fascia, emerging into the subcutaneous connective tissue with the basilic vein, and divides into anterior and ulnar branches. The distribution of these branches has been noted (p. 62).

(e) The **subscapular nerves** (figs. 57, 63) are branches of the posterior cord. They arise in close approximation to one another and are distinguished as an upper, a middle (thoracodorsal), and a lower subscapular nerve.

The **upper,** or short, **subscapular nerve** (fig. 63) is derived from the fifth and sixth cervical nerves. It may be represented by several short nerves and enters the upper portion of the subscapularis muscle which it supplies.

The **thoracodorsal (middle subscapular) nerve** (figs. 40, 57, 63) carries nerve fibers from the seventh and eighth cervical nerves, supplemented at times by fibers of the sixth nerve. It passes behind the axillary artery and accompanies the subscapular artery along the lower border of the subscapularis muscle. The nerve reaches the latissimus dorsi in the axilla and divides into branches on its deep surface to supply the muscle.

The **lower subscapular nerve** carries branches from the fifth and sixth cervical nerves. Arising from the posterior cord of the brachial plexus, it courses downward behind the subscapular vessels to the teres major muscle. It distributes to the teres major muscle, giving off one or two twigs to the axillary border of the subscapularis muscle in course.

(f) The **axillary nerve** (C5, 6) (figs. 40, 57, 63) arises from the posterior cord of the brachial plexus at the lower border of the subscapularis muscle. It passes to the back of the arm across the upper edge of the teres major tendon and enters the quadrangular space (p. 56), in company with the posterior circumflex humeral artery. The nerve provides an articular twig to the inferior portion of the capsule of the shoulder joint and then divides into a superior (anterior) branch and an inferior (posterior) branch. The superior branch winds around the surgical neck of the humerus as far as the anterior border of the deltoid, supplying the muscle, its branches ramifying with those of the artery. A few nerve filaments pass through the deltoid and end in the skin over the middle one-third of the muscle. The inferior (posterior) branch supplies the teres minor muscle and the posterior fibers of the deltoid; then, turning the posterior border of the deltoid in its lower one-third, ends as the **superior lateral brachial cutaneous nerve.** This nerve is cutaneous over the lower half of the deltoid and adjacent areas of the arm (p. 61).

Functional Analysis of the Brachial Plexus

The separation of the trunks of the brachial plexus into anterior and posterior divisions places the nerves derived therefrom into relation with muscles formed from primitive anterior and posterior meso-dermal masses during the development of the limbs. Figure 64 shows, in schematic form, the nerve-muscle relationships and the orientation of anterior and posterior muscle masses with respect to the central bony axis. These relations, once established, are never reversed, and the same fundamental cor-relation exists in the adult. Further, the anterior divisions of the trunks unite to form the lateral and medial cords of the brachial plexus, and the posterior divisions join to form the single posterior cord. Thus, all branches of the lateral and medial cords carry nerve bundles derived from the anterior divisions of the trunks, and the branches of the posterior cord conduct, exclusively, fibers from the posterior divisions.

It remains to determine how the primitive separa-tion in the musculature of the limb bud has been projected into the adult. In a cross section of the arm (fig. 10) the lateral and medial intermuscular septa, extending outward from the periosteum of the humerus, and the investing brachial fascia sur-rounding the muscle masses produce a clear separation of the arm into anterior and posterior compartments. Similarly, in the forearm the radius and ulna are interconnected by an interosseous membrane, and each bone is connected to the investing antebrachial fascia by an intermuscular septum. The discrete compartments thus defined lodge muscle groups of similar or related functions and the blood vessels and nerves which supply them. The concept is emphasized in the designation of **pre-axial** for the compartment and structures anterior to the bone and fascial plane, or axis of the limb, and of **postaxial** for the structures behind the bony and fascial axis. With the limb in the anatomical position, those parts anterior to the bony axis are all in a continuous plane down the front of the limb; the postaxial parts all lie con-tinuously down the back of the limb. Thus, a simple relationship exists between the nerves and muscle masses; the branches of the lateral and medial cords are all 'pre-axial' and innervate the pre-axial muscles of the limb, and posterior cord branches are 'postaxial' and innervate the post-axial musculature.

The radial nerve, being the only postaxial nerve below the shoulder, supplies all the postaxial musculature in the remainder of the limb. **The median, musculocutaneous, and ulnar nerves share the responsibility for pre-axial innervation in** accordance with a fairly simple pattern of alternation. *Thus, the musculocutaneous nerve is muscular in the arm and cutaneous in the forearm and it is the sole pre-axial muscular nerve in the arm. The pre-axial median and ulnar nerves are nerves of passage in the arm, but in the forearm and hand each contributes, the median nerve more heavily in the forearm, the ulnar nerve more heavily in the hand.*

In the shoulder region many of the supra-clavicular and infraclavicular branches of the plexus arise from recognizable pre-axial or postaxial cords or divisions. Other branches of undivided trunks or nerves arise from the front or back of these sources. Such origins have the same significance as the major anterior and posterior divisions of the trunks. The clavicle is an anterior and the scapula a posterior bone. One exception must be made with respect to the scapula, for its coracoid process has a phylogenetic history as a separate bone, the coracoid bone of lower vertebrates, and its fusion with the scapula is secondary. *Muscles arising from the scapula exclusive of the coracoid process belong, then, to a postaxial group at the shoulder and those from the clavicle and coracoid to a pre-axial group.* The nerve-muscle correlation is shown in the fact that *muscles of scapular origin are all supplied by postaxial branches of the plexus and that muscles taking origin from the clavicle and coracoid are supplied by pre-axial branches.* Throughout the limb only a very little muscle migration mars this simple plan of compartmentation and innervation, and its organizational value in learning nerve-muscle relationships is considerable. Figure 63, illustrating the brachial plexus, incorporates a shading difference to designate the pre-axial and postaxial divisions and nerves.

There is, further, an obvious serial order in the nerves of the brachial plexus as they arise from spinal nerves C5 to T1. This order is retained from the primitive serial morphology of the embryo, and, as in the limb bud, the fifth cervical nerve of the adult distributes to the cranial part of the limb and the first thoracic to its caudal part. The pattern of segmental cutaneous innervation as determined from the study of clinical cases is as seen in figure 65.

The Axillary Lymph Nodes (figs. 55, 60, 61)

The axillary lymph nodes number from twenty to thirty and are generally large. They consist of five subgroups, certain of which are related to the axillary walls.

1. A **lateral group** of three to five nodes lies medial and posterior to the distal portion of the axillary vein. These nodes receive the drainage of the upper limb except for that carried by the lymphatics along the cephalic vein. Efferent vessels of this group drain to the central and apical nodes and, in minor amount, to the inferior deep cervical nodes.

2. A **pectoral group** of nodes is located along the lateral thoracic artery adjacent to the axillary border of the pectoralis minor muscle. It consists usually of three to five nodes. They receive the drainage of the anterolateral part of the thoracic wall, including most of the drainage of the mammary gland, and of the skin and muscles of the supraumbilical part of the abdominal wall. Efferent lymphatic vessels from the pectoral nodes reach the central and apical groups.

3. A **subscapular group** of five or six nodes is distributed along the subscapular blood vessels from their junction with the axillary vessels to the point of contact of their branches with the chest wall. Afferent vessels of this node group drain the skin and muscles of the posterior thoracic wall and scapular region and of the lower part of the back of the neck. Their efferent lymphatics reach the central axillary nodes.

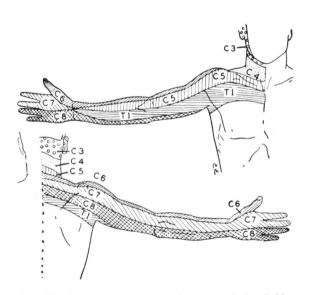

Fig. 65 Dermatome chart of the upper limb of Man outlined by the pattern of hyposensitivity from loss of function of a single nerve root (from original of fig. 9, J. J. Keegan and F. D. Garrett, *Anat. Rec.*, 102:417, 1948).

4. A **central group** of four or five nodes lies under the axillary fascia in the fat of the axilla. The intercostobrachial nerve crosses the axilla among these nodes. They are among the largest of the axillary nodes and, in any particular subject, their number is inversely proportional to their size. These nodes receive directly some lymphatic vessels from the arm and mammary regions, but primarily they receive lymph from all the preceding groups. The efferent channels of this group lead to the apical nodes.

5. An **apical group** of nodes represents the highest of the axillary lymph gland groups. Consisting of six to twelve nodes, it lies at the apex of the axilla along the axillary vein, adjacent to the superior border of the pectoralis minor muscle. This group of nodes receives the efferent vessels of all the other axillary lymph nodes. It also receives lymphatic channels that ascend along the cephalic vein and several of the accessory lymphatic vessels from the mammary gland. From the lymphatic channels interconnecting the apical nodes arises a lymphatic trunk which combines the lymphatic drainage of all the regions drained by the axillary nodes. This **subclavian lymphatic trunk** is single, as a rule, and usually empties into the junction of the internal jugular and subclavian veins. Occasionally, it may enter the jugular lymphatic trunk or, on the left side, the thoracic duct.

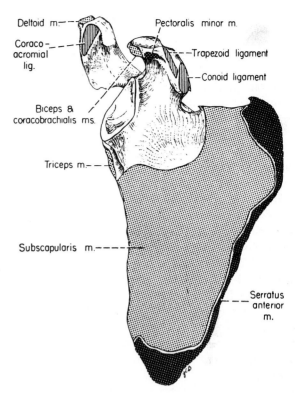

Fig. 66 The muscular and ligamentous attachments on the costal surface of the right scapula. (Black dots—origin; white dots on black—insertion; lines—ligaments.)

MUSCLES OF THE MEDIAL AND POSTERIOR WALLS OF THE AXILLA

Serratus Anterior (figs. 57, 66, 272, 277, 280)

Serratus anterior is a thin, muscular sheet overlying the lateral portion of the thoracic cage and the intercostal musculature. It arises by fleshy slips from the upper eight or nine ribs and inserts on the whole of the medial border of the scapula. The first digitation takes origin from the first and second ribs and the fascia covering the first intercostal muscle and inserts on the costal surface of the scapula at its superior angle. All other digitations spring from the corresponding ribs. The digitations from the second and third ribs form a thin, triangular sheet which inserts into nearly the whole length of the medial border of the scapula. The digitations from ribs four to eight or nine alternate in origin with slips of the external abdominal oblique muscle. Converging as the segments of a fan, they terminate in a large, oval area on the costal surface of the inferior angle of the scapula. The muscle keeps the scapula closely applied against the chest wall and draws the scapula forward around the curvature of the wall, as in a forward reach or pushing action. Its strong inferior portion rotates the scapula in elevation of the shoulder, especially in the shoulder component of flexion and abduction of the arm. Serratus anterior is not concerned in respiration. Its paralysis results in a 'winged scapula' and inability to elevate the arm above the horizontal. Serratus anterior is innervated by the long thoracic nerve from the brachial plexus (p. 80). The muscle is overlain by the long thoracic nerve and the lateral thoracic artery and is separated from subscapularis by loose areolar connective tissue.

Subscapularis (figs. 66, 68, 272, 277)

The subscapularis muscles fills the hollow of the costal surface of the scapula, arising from its medial two-thirds and from the intermuscular septa between it and the teres major and minor muscles. Tendinous bands arise from ridges on the bone and increase the area for muscular attachment. The robust tendon passes across the anterior face of the capsule of the shoulder joint and ends in the lesser tubercle of the humerus. It blends with the shoulder joint capsule as it nears its termination. The tendon is separated from the neck of the scapula by the large **subscapular bursa** which regularly communicates with the cavity of the shoulder joint through an aperture in the anterior portion of the capsule. The subscapularis muscle is the chief medial rotator of the arm. It acts as an extensor when the arm is at the side or in flexion. Subscapularis acts with infraspinatus and teres minor in holding the head of the humerus closely against the glenoid cavity; this is an important aid to supraspinatus and deltoideus in producing abduction of the arm. The muscle is innervated on its costal surface by the **upper** and **lower subscapular nerves** from the posterior cord of the brachial plexus (p. 81).

THE ARM

The arm, or brachium, extends from the shoulder joint to the elbow. The upper portion of the arm receives the tendons of the shoulder muscles. The arm muscles proper are few and are concerned primarily with providing flexion and extension at the elbow. The arm contains the terminal branches of the brachial plexus and portions of the great vascular channels of the limb. Its superficial aspects are considered elsewhere; its surface anatomy on p. 57, its superficial veins on p. 59, its superficial lymphatics on p. 65, and its cutaneous nerves on p. 61.

THE BRACHIAL FASCIA (figs. 10, 74)

The brachial fascia is a strong, tubular investment of the deeper parts of the arm. It is continuous above with the pectoral and axillary fasciae and with the fasciae of the deltoid and latissimus dorsi muscles. It is attached below to the epicondyles of the humerus and to the olecranon and is continuous with the fascia of the forearm. The brachial fascia contains both circular and longitudinal fibers, predominantly the former, and is stronger

on the dorsum of the arm. It is separated by loose areolar tissue from the muscles which it encloses. The fascia is perforated medially at the junction of the middle and lower thirds of the arm for the passage of the basilic vein and the medial antebrachial cutaneous nerve and elsewhere by other nerves and vessels in their passage to the subcutaneous tissues. Two **intermuscular septa** are prolonged upward from the epicondylar attachments of the brachial fascia. These blend with the periosteum of the humerus along its supracondylar ridges and, extending toward the margins of the arm, fuse with the under surface of the brachial fascia. The **lateral intermuscular septum** ends above at the insertion of the deltoid muscle, where it becomes continuous with the deltoid fascia. The **medial intermuscular septum** ends, similarly, in continuity with the fascia of the coracobrachialis muscles. The medial intermuscular septum is divided into two laminae, a stronger posterior and a weaker anterior, which, with the overlying brachial fascia, form the **neurovascular compartment** of the arm. This interfascial space provides for the passage of the brachial artery and its venae comitantes, the median nerve, and, except in the lower one-third of the arm, the basilic vein and the medial antebrachial cutaneous nerve. The medial and lateral intermuscular septa separate the anterior, or pre-axial, and the posterior, or postaxial, compartments of the arm.

THE MUSCLES OF THE ARM

The muscles of the arm are separated both in location and function into a pre-axial group and a postaxial group. The pre-axial muscles are the coracobrachialis, the biceps, and the brachialis. The postaxial muscles are the triceps and the anconeus. The anconeus muscle lies chiefly in the forearm but has morphological and physiological relations with the triceps. It is described with the muscles of the arm.

Coracobrachialis (figs. 57, 67, 68)

The coracobrachialis is a short, band-like muscle in the upper medial part of the arm. It arises from the tip of the coracoid process in common with the short head of the biceps. It inserts by a flat tendon into the medial surface of the humerus just proximal to the middle of its shaft. The musculocutaneous nerve passes diagonally through it at about its midlength.

Biceps Brachii (figs. 57, 67, 68, 72, 74, 75, 82)

The biceps is a long, fusiform muscle of two heads located on the anterior aspect of the arm. The **short head** arises by a thick, flattened tendon from the tip of the coracoid process, in common with coracobrachialis. The round tendon of the **long head** arises from the supraglenoid tubercle of the scapula. This tendon crosses the head of the humerus within the capsule of the shoulder joint and emerges from the joint anteriorly. It descends in the intertubercular groove of the humerus covered by a fascial prolongation of the tendon of the pectoralis major muscle. The bellies of the two heads unite immediately below the middle of the arm. The tendon of the insertion forms the strong vertical cord which is palpable in the center of the cubital fossa. Here it turns on itself so that its anterior surface faces lateralward and ends on the tuberosity of the radius. The **bicipitoradial bursa,** a synovial bursa of approximately 1 cm. diameter, lies between the tendon and the anterior part of the tuberosity. An **interosseous cubital bursa** frequently separates the biceps tendon from the ulna and the muscles covering the bone. From the medial and anterior part of the tendon, partly in continuity with the muscular fibers, a broad aponeurosis arises at the level of the bend of the elbow. This **bicipital aponeurosis** (figs. 67, 75, 82) passes obliquely over the brachial artery and median nerve and blends with the antebrachial fascia over the flexor muscles of the forearm. The pull of the bicipital aponeurosis is largely exerted on the ulna (Congdon & Fish, '53). The biceps brachii lies under the brachial fascia in front of the brachialis muscle. Its tendons of origin are concealed by the pectoralis major and deltoid muscles. At the medial margin of the muscle lie the coracobrachialis muscle and the brachial vessels and median nerve.

Brachialis (figs. 67–9, 72, 74, 75, 82)

The brachialis muscle arises from the lower half of the anterior surface of the humerus and from the medial and lateral intermuscular septa. Its upper extremity is represented by two pointed processes

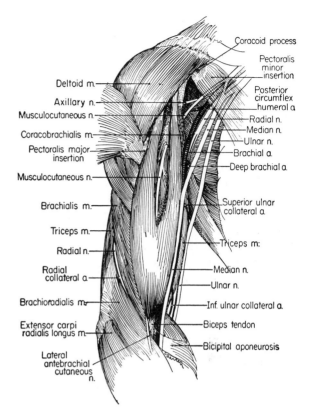

Fig. 67   The muscles, nerves, and vessels of the anterior aspect of the arm.

which lie on either side of the insertion of the deltoid muscle. The fibers of the muscle converge to a thick tendon which adheres to the capsule of the elbow joint and inserts on the tuberosity of the ulna and on the rough impression on the anterior surface of its coronoid process. The brachialis muscle is closely adjacent to brachioradialis distally, the two muscles being separated by the radial nerve. The muscle bulges beyond the biceps on each side, and in front of its medial border lie the brachial vessels and the median nerve.

Triceps Brachii (figs. 39, 40, 69, 73, 74)

The large triceps muscle occupies the entire dorsum of the humerus and arises by three heads of origin. The **long head** begins by a strong tendon attached to the infraglenoid tubercle of the scapula. Its belly descends between the teres major and minor muscles, lying medial to the lateral head, and joins the lateral and the medial heads in a common

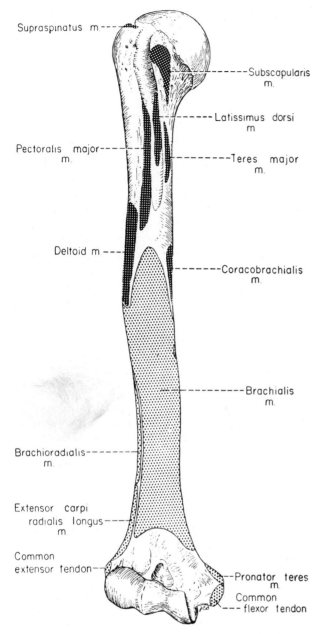

Supraspinatus m.

Subscapularis m.

Latissimus dorsi m.

Pectoralis major m.

Teres major m.

Deltoid m.

Coracobrachialis m.

Brachialis m.

Brachioradialis m.

Extensor carpi radialis longus m.

Common extensor tendon

Pronator teres m.

Common flexor tendon

Fig. 68 The muscular attachments on the anterior aspect of the right humerus. (Black dots—origin; white dots on black—insertion.)

the radial nerve and deep brachial vessels, its fibers join the common tendon of insertion. The **medial head** arises by fleshy fibers from the elongated triangular area on the back of the humerus between the teres major muscle and the olecranon fossa, from the entire length of the medial intermuscular septum, and from the lateral septum below the radial nerve groove. It will be observed that the lateral head arises wholly above and lateral to the groove and the medial head entirely below and medial to it. Further, the medial head is deep to the other two and is hidden by them. The tendon of the triceps forms a flat band covering the distal two-fifths of the muscle. The fibers of the medial head reach it on its deep surface. The tendon inserts on the posterior part of the olecranon and into the deep fascia of the forearm on either side of it. A **subtendinous olecranon bursa** frequently intervenes between the tendon and the olecranon just proximal to the insertion. The triceps is superficial in almost its whole length. The origin of the long head is concealed by the deltoid and teres minor muscles.

Anconeus (figs. 69, 73, 76, 79)

The anconeus is a small, triangular muscle that arises from the lateral epicondyle of the humerus. Its fibers diverge from this origin and insert into the side of the olecranon and the upper one-fourth of the posterior surface of the ulna. The muscle lies deep to the antebrachial fascia and extends across the elbow and the superior radioulnar joints.

Innervations

The **musculocutaneous nerve** innervates the preaxial muscles of the arm; coracobrachialis, brachialis, and biceps. The **radial nerve** supplies triceps and anconeus, the postaxial muscles, and sends a small twig into the inferior lateral part of brachialis. These nerves are more fully described on pages 89 and 90.

Muscle Actions

The principal movements produced by the muscles of the arm are flexion and extension of the forearm at the elbow. Brachialis and biceps brachii are the principal flexors, with brachialis always in use in this action. Biceps becomes active against resistance and is most effective when flexion is combined with supination. It is a powerful supinator

tendon of insertion. The **lateral head** arises from the posterior surface and lateral border of the humerus between the attachment of the teres minor muscle and the groove for the radial nerve, and from the lateral intermuscular septum. This head of the muscle is directed downward and medialward. Crossing the groove and concealing

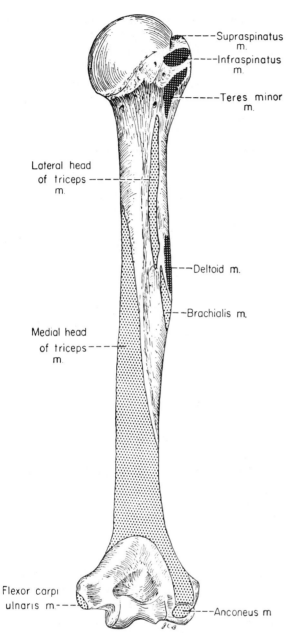

Fig. 69 The muscular attachments on the posterior surface of the right humerus. (Code as fig. 68.)

the arm at the shoulder and its tendon aids in stabilization of the joint. The long head of the triceps assists in extension and adduction of the arm. Coracobrachialis assists in flexion and adduction of the arm.

THE HUMERUS (figs. 68-71)

The humerus is the principal bone of the upper limb and, in length, equals approximately one-fifth of the body height. It is a long bone with a shaft and two extremities, both of which are articular. The upper end, or **head,** is approximately a hemisphere, its smooth articular surface facing upward, medialward, and backward. It articulates in the glenoid cavity of the scapula. The **anatomical neck** of the bone is only a slight indentation of the margin of the head for the attachment of the articular capsule. The junction of the head and shaft of the humerus is marked by its prominent greater and lesser tubercles.

The **greater tubercle** lies at the lateral margin of the bone and is the most laterally projecting part of the skeleton at the shoulder. It rises almost as high as the top of the head of the bone and has three flattened surfaces for muscle attachments. The **lesser tubercle** projects directly forward from the bone and is placed a little inferior to the greater tubercle. The **crest of the lesser tubercle** is a ridge of bone extending downward and medialward from the tubercle for about 5 cm. A prominent ridge also continues downward from the greater tubercle. This is the **crest of the greater tubercle.** It is almost continuous with the anterior roughening of the deltoid tuberosity and represents the upper part of the anterior border of the bone. Between the tubercles and their crests is a well-marked **intertubercular groove** that lodges the tendon of the long head of the biceps muscle. The **surgical neck** of the humerus is the narrowing of the shaft directly below the head and its tubercles. The name has significance in the frequency of fracture at this general level.

The **body,** or shaft, of the bone is somewhat rounded above and prismatic in its lower portion. An **anterior border** begins above in the greater tubercular crest, participates in the insertion of the deltoid below this, and then becomes ill-defined. Below, it is sharp again as the ridge between the radial and coronoid fossae and as the lateral lip of the trochlea. The **medial border** begins in the lesser tubercular crest and presents the **nutrient foramen.** The nutrient canal is directed distalward. Below this, the inferior one-third of the bone exhibits a sharp ridge, the medial supracondylar ridge, which ends in the prominent medial epicondyle. The **lateral border** begins above at the posterior extremity of the greater tubercle. It is indistinct above, where the lateral head of the triceps arises, and is interrupted in the middle of the shaft by the groove for the radial nerve. Below this it forms the sharp **lateral supracondylar ridge** and ends in the lateral epicondyle. Between the

of the forearm. Extension of the forearm is produced by the triceps, assisted by anconeus which generally initiates extension. The medial head of the triceps is usually active and the lateral and long heads are recruited for extra power.

Certain heads of these muscles are active at the shoulder joint. The long head of the biceps flexes

The Upper Limb

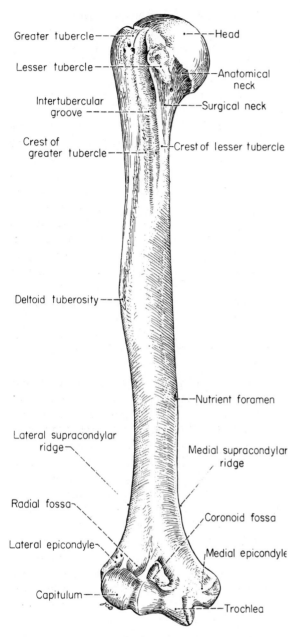

Greater tubercle — — Head

Lesser tubercle — — Anatomical neck

Intertubercular groove - - - Surgical neck

Crest of greater tubercle - - - Crest of lesser tubercle

Deltoid tuberosity - - -

— — Nutrient foramen

Lateral supracondylar ridge

Medial supracondylar ridge

Radial fossa

Coronoid fossa

Lateral epicondyle

Medial epicondyle

Capitulum —

— Trochlea

Fig. 70   The anterior aspect of the right humerus. ($\times\frac{1}{2}$)

The **inferior extremity** of the bone is flattened from before backward and is widened medio-laterally by the medial and lateral epicondyles. Its **articular surfaces** for the radius and ulna are directed forward, so that the bone appears to curve anteriorly. The epicondyles are continuous with the supracondylar ridges.

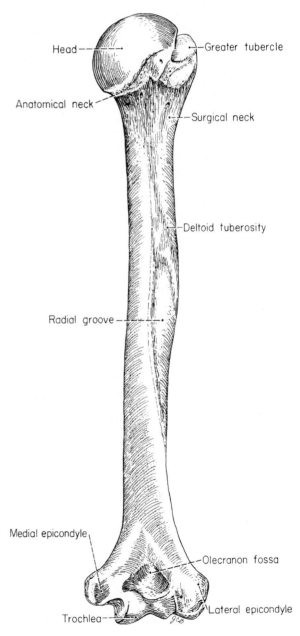

Head — — Greater tubercle

Anatomical neck — — Surgical neck

— Deltoid tuberosity

Radial groove — — —

Medial epicondyle

— Olecranon fossa

— Lateral epicondyle

Trochlea —

Fig. 71   The posterior surface of the right humerus. ($\times\frac{1}{2}$)

anterior and lateral borders of the bone and ending at its midlength is a prominent V-shaped roughening, the **deltoid tuberosity,** which marks the insertion of the deltoid muscle. The **groove for the radial nerve** begins on the posterior aspect of the bone and runs obliquely downward and forward, skirting the posterior margin of the deltoid tuberosity. In it descend the radial nerve and the deep brachial artery, and it separates the regions of origin of the lateral and medial heads of the triceps muscle.

The **lateral epicondyle** is a relatively inconspicuous process above the capitulum. Its posterior surface is subcutaneous. The **medial epicondyle** forms a marked medial projection above the elbow. It is higher than the lateral epicondyle, projects somewhat backward, and is grooved behind for the ulnar nerve. The articular surfaces are the trochlea, for reception of the ulna, and the capitulum, which articulates with the cup-shaped depression on the head of the radius. The **trochlea** (pulley) is shaped like a spool and has a deep depression between two well-marked margins. The depression is slightly spiral and receives the trochlear notch of the ulna. The medial rim of the trochlea is the more prominent, the lateral rim is a small projection which separates the trochlea and the capitulum. Above the trochlea are two fossae, the **coronoid fossa** in front and the **olecranon fossa** on the back of the bone. These receive the coronoid process and the olecranon of the ulna. The bone between these fossae is very thin and may be perforated. The **capitulum** (little head) is smaller than the trochlea and is globular in shape. It is directed anteriorly and inferiorly and articulates with the upper surface of the radius. Above it is a shallow **radial fossa** occupied by the edge of the head of the radius during full flexion of the elbow.

The humerus ossifies from eight centers of ossification, one for the shaft and seven for the processes: head, greater tubercle, lesser tubercle, trochlea, capitulum, lateral epicondyle, and medial epicondyle. Those of the proximal end coalesce to form a single epiphysis which fuses with the shaft at about the seventeenth or eighteenth year. Those of the distal end, except that for the medial epicondyle, also coalesce to form one epiphysis which joins the shaft between the thirteenth and the sixteenth years. The single epiphysis of the medial epicondyle joins the bone between the fourteenth and sixteenth years.

The Nerves of the Arm

The nerves of the arm are the terminal branches of the brachial plexus. These are the axillary, median, ulnar, musculocutaneous, and radial nerves.

1. The **axillary nerve** is almost entirely a nerve of the shoulder region, and its complete description is given in connection with the brachial plexus (p. 81).

2. The **median nerve** is formed, as described in the section on the brachial plexus (p. 79), by a root from both the lateral and the medial cords and contains fibers from the sixth, seventh, and eighth cervical and first thoracic nerves and sometimes from the fifth cervical. In this union the root from the medial cord passes across the medial aspect of the axillary artery, and the median nerve comes to lie anterolateral to the brachial artery in upper

levels of the arm. It is overlapped laterally by the coracobrachialis muscle (fig. 87). In wounds of the axilla, the median is the nerve most frequently injured. In the cubital fossa (fig. 75) the nerve has become medial to the brachial artery and, since both run relatively straight courses through the arm, the nerve crosses the artery anteriorly at about the midarm level. In the common case of a high division of the brachial artery into radial and ulnar arteries in the arm (p. 92), the median nerve tends to lie between them with one of the arteries superficial to the nerve. The median nerve occupies the neurovascular compartment of the arm adjacent to the brachial artery and has the basilic vein and the medial antebrachial cutaneous nerve superficial to it (fig. 74). The median nerve has no branches in the arm but merely passes through to reach its region of distribution in the forearm and hand. The occasional separation of the contribution of the lateral cord to the median nerve into several parts may give an appearance of intercommunication between the median and musculocutaneous nerves in the upper part of the arm.

3. The **ulnar nerve** is the largest nerve derived from the medial cord of the brachial plexus and carries nerve fibers from the eighth cervical and first thoracic nerves (figs. 63, 88). It arises at the lower border of the pectoralis minor muscle and descends through the axilla posterior to and in the angle between the axillary artery and vein. It is medial to the brachial artery in the neurovascular compartment in the upper half of the arm. At the level of insertion of the coracobrachialis muscle (middle of arm) the ulnar nerve pierces the posterior layer of the intermuscular septum and, in company with the superior ulnar collateral branch of the brachial artery, runs distalward behind the septum (fig. 74). The ulnar nerve has no branches in the arm unless one allocates to this region its articular branches at the elbow. As the nerve passes the back of the elbow between the olecranon and the medial epicondyle of the humerus, it gives off several fine filaments to the elbow joint.

4. The **musculocutaneous nerve** (figs. 57, 63, 72, 74) is the remainder of the lateral cord of the brachial plexus after the contribution to the median nerve is given off and consists of nerve fibers from the fifth and sixth cervical nerves. Below the pectoralis minor muscle the nerve is lateral to the axillary artery. The musculocutaneous nerve supplies the structures of the pre-axial compartment of the arm

and is muscular in the arm and cutaneous in the forearm. Close to its origin it gives off one or two nerves to the coracobrachialis muscle. The nerve pierces the coracobrachialis muscle at about its midlength (fig. 57), and continuing obliquely downward in the interval between the biceps and brachialis muscles, it inclines to the lateral side of the arm. In the interval between the biceps and brachialis, branches are given to these muscles. A filament of the nerve enters the humerus with the nutrient artery. An articular branch to the elbow joint is derived from the branch to the brachialis. The nerve pierces the brachial fascia lateral to the biceps tendon a short distance above the elbow and continues through the forearm as the lateral antebrachial cutaneous nerve (p. 62).

5. The **radial nerve** is the continuation of the posterior cord and is the largest branch of the brachial plexus (figs. 63, 89). It usually contains fibers from all five nerves contributing to the brachial plexus, but in some cases branches of the fifth cervical or first thoracic nerves may be absent. In the axilla it lies behind the axillary artery and in front of the subscapularis, teres major, and latissimus dorsi muscles. Entering the arm posterior to the brachial artery, the nerve lies medial to the humerus and in front of the long head of the triceps. Continuing posterolaterally, it passes diagonally across the back of the humerus through levels of the middle one-third of this bone. Here, accompanied by the deep brachial artery, it occupies a groove on the humerus and separates the lateral and medial heads of the triceps (figs. 73, 74). Reaching the distal one-third of the humerus at the lateral side of the arm, the nerve pierces the lateral intermuscular septum and runs between the brachialis and brachioradialis muscles across the front of the lateral epicondyle. In the cubital fossa (fig. 75) it is placed deeply between the aforementioned muscles, where it divides into its deep and superficial branches.

The branches of the radial nerve in the arm include cutaneous, muscular, and articular branches. Within the axilla are given off the muscular **branch to the long head of the triceps** and the **posterior brachial cutaneous nerve** (fig. 50). The latter crosses the tendon of latissimus dorsi ventrally, passes dorsal to the intercostobrachial nerve, pierces the deep fascia, and distributes to the skin of the middle of the back of the arm below the deltoid muscle. The upper branch to the medial head of the triceps arises at about the lower border of the teres major muscle. This nerve, frequently designated as the **ulnar collateral branch** of the radial nerve, runs with the

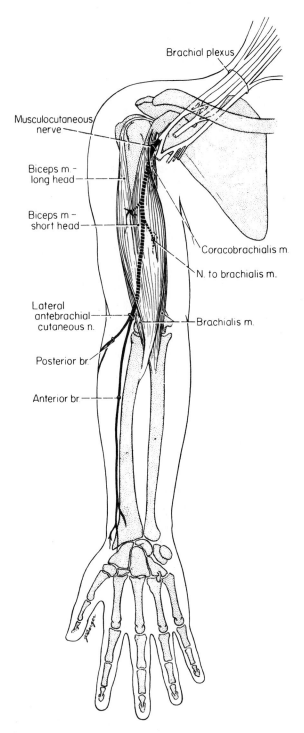

Fig. 72   The course and branches of the musculocutaneous nerve.

ulnar nerve through the middle of the arm to reach the distal part of the medial head of the triceps. The **nerve to the lateral head of the triceps** arises as the nerve

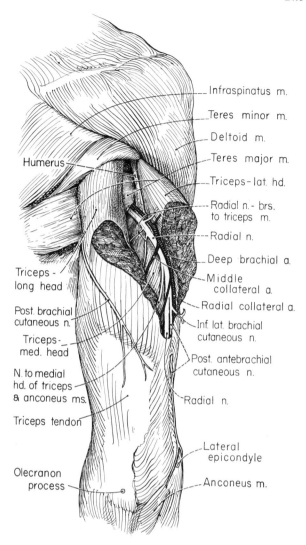

Fig. 73 The triceps brachii muscle and the course of the radial nerve and deep brachial artery.

nerve proper (figs. 51, 73) pierces the deep fascia in the line of the lateral intermuscular septum about 7 cm. above the elbow. It runs behind the lateral epicondyle and supplies the skin of the back of the forearm as far as the wrist. In the lower one-third of the arm the radial nerve lies anterior to the lateral intermuscular septum accompanied by the radial collateral artery. Here it supplies **muscular branches to brachioradialis** and to **extensor carpi radialis longus,** and to the lateral portion of brachialis. **Articular branches** to the elbow joint are given off from the radial nerve in this interval as well as from the ulnar collateral nerve and the nerve to the anconeus muscle. The radial nerve divides into its terminal branches, the superficial and deep radial nerves, in the interval between the brachialis and brachioradialis muscles.

THE BLOOD VESSELS OF THE ARM

The blood vessels of the arm are continuations of those in the axilla. The **brachial artery** is the axillary artery, renamed in the region of the brachium. The brachial veins unite to form the axillary vein.

The Brachial Artery (figs. 62, 74)

The brachial artery extends from the lower border of the teres major muscle to its bifurcation opposite the neck of the radius in the lower part of the cubital fossa. The course of the brachial artery is represented by a line connecting the middle of the clavicle and the midpoint between the epicondyles of the humerus when the limb is in right-angled abduction. The artery lies at first medial to the humerus and then in front of it before crossing the epicondyles at the elbow. It is superficial throughout its course. This fact is made use of in sphygmomanometry. The blood pressure cuff is applied at mid-arm to compress the artery and the pulse sounds are listened for in the cubital fossa where the artery is also superficial. Laterally, it lies against coracobrachialis and the biceps, the two muscles partially overlapping it. Posteriorly, it lies against the medial intermuscular septum and the medial head of the triceps. As it swings to the front of the arm, it has the medial border of brachialis behind it. Medially, the artery is in relation with the medial antebrachial cutaneous nerve and the ulnar nerve in the upper half of the arm; lower down, with the medial antebrachial and the median nerves. The median nerve, formed anterolateral to the axillary artery, is in front of the brachial artery in the middle

enters its groove, and on the posterior aspect of the humerus the **nerves to the medial head of the triceps and anconeus** are given off (fig. 73). The filament to anconeus represents a terminal part of one of the branches to the medial head which runs deeply through the length of this muscle. The **posterior antebrachial cutaneous nerve** (figs. 49, 51, 73) arises in the groove for the radial nerve and has an upper branch which is cutaneous in the arm and a lower branch which represents its principal forearm distribution. The upper branch, the **inferior lateral brachial cutaneous nerve,** perforates the lateral head of the triceps near its attachment to the humerus, pierces the deep fascia in line with the lateral intermuscular septum, and supplies the skin over the lower half of the arm on its anterolateral aspect. The posterior antebrachial cutaneous of the arm but lies medial to it at the elbow. In the cubital fossa the brachial artery is medial to the tendon of the biceps muscle and between this tendon and the median nerve. It is crossed by the bicipital aponeurosis at the bend of the elbow. The artery provides many unnamed, muscular branches in its course, principally from its lateral side. One, to the biceps muscle, may be large. The named branches of the brachial artery are the deep brachial, the nutrient, the superior ulnar collateral, and the inferior ulnar collateral arteries. Its terminal branches are the radial and ulnar arteries. For its collateral circulation, see p. 142.

The brachial artery is single in 80 per cent of specimens. In the other 20 per cent of cases a **superficial brachial artery** arises at upper brachial levels and descends through the arm superficial to the median nerve. Included in this 20 per cent group, the superficial brachial will constitute a high radial artery in 10 per cent, a high ulnar artery in 3 per cent and provide both the radial and ulnar arteries of the forearm in 7 per cent (Skopakoff). In the last case the brachial artery itself is likely to continue as the common interosseous artery, and the ulnar artery to form a superficial ulnar in the forearm.

1. The **deep brachial artery** arises from the medial and posterior aspect of the brachial, below the tendon of the teres major muscle. It is the largest branch of the brachial and accompanies the radial nerve in its course around the humerus (fig. 73). The deep brachial artery is a single branch of the brachial artery in 55 per cent of cases or of the axillary in 16 per cent. It is combined at its origin with the superior ulnar collateral artery in 20 per cent.

An **ascending branch** runs proximally between the lateral and long heads of the triceps muscle and anastomoses with the descending branch of the posterior circumflex humeral artery. Behind the humerus, the deep brachial artery divides into a middle collateral and a radial collateral artery. The **middle collateral artery** plunges into the medial head of the triceps, through which it descends to the elbow and for which it is an important muscular artery. At the elbow the middle collateral anastomoses with the interosseous recurrent artery and with other vessels contributing to the anastomosis about the elbow. The continuation of the deep brachial artery is the **radial collateral artery;** it accompanies the radial nerve in its course to the elbow. The radial collateral artery perforates the lateral intermuscular septum with the radial nerve and ends in front of the lateral epicondyle by anastomosing with the radial recurrent branch of the radial artery from the forearm.

2. The **nutrient humeral artery** arises about the middle of the arm and enters the nutrient canal on the anteromedial surface of the humerus.

3. The **superior ulnar collateral artery** arises from the brachial a little below the middle of the arm. It may take origin from the upper part of the deep brachial. The superior ulnar collateral artery immediately pierces the medial intermuscular septum and descends in company with the ulnar nerve behind the septum and on the surface of the medial head of the triceps. It passes, with the ulnar nerve, behind the medial epidoncyle of the humerus and between it and the olecranon, where it anastomoses with the posterior ulnar recurrent branch of the ulnar artery from the forearm and with the inferior ulnar collateral artery. The superior ulnar collateral also sends a branch in front of the epicondyle which anastomoses with the anterior ulnar recurrent branch of the ulnar artery and with the inferior collateral artery.

4. The **inferior ulnar collateral artery** arises about 3 cm. above the medial epicondyle. It divides on the brachialis muscle into a posterior and an anterior branch.

The **posterior branch** pierces the medial intermuscular septum and turns around the humerus deep to the triceps. It anastomoses with the posterior ulnar recurrent artery and with the interosseous recurrent and the middle collateral arteries, all of which participate in the network of vessels around the elbow. The **anterior branch,** muscular like the posterior, passes downward between the brachialis and the pronator teres muscles of the forearm and anastomoses in this interval with the anterior ulnar recurrent artery.

The Brachial Veins (fig. 74)

The brachial veins are a paired set of veins which accompany the brachial artery. They are formed at the elbow by the union of the venae comitantes of the radial and ulnar arteries. They have tributaries that run with the branches of the brachial artery and drain the same territories that the arteries supply. The brachial veins contain valves and make frequent cross anastomoses with one another. They occupy the neurovascular compartment on the medial side of the arm, lying on either side of the brachial artery and deep to the basilic vein. At the lower border of the teres major or subscapularis muscles, the more lateral of the brachial veins crosses the axillary artery to join the more medial brachial vein. At approximately the same level, the basilic vein joins

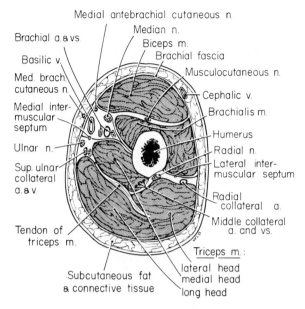

Fig. 74 A cross section of the right arm distal to the insertion of the deltoid muscle.

the medial brachial and, as a result of these somewhat variable junctions, the single axillary vein is formed.

The Brachial Lymphatics

The deep lymphatics of the upper limb drain its joint capsules, periosteum, tendons, nerves, and muscles. The vessels begin in the palm and are continued through the forearm with the blood vessels: radial, ulnar, anterior and posterior interosseous. In the cubital fossa five or six small lymph nodes lie at the confluence of these vessels. The **brachial lymphatics** arise in the cubital nodules and accompany the brachial vessels through the arm as two or three lymph trunks. Small, intercalated nodules may be encountered along the brachial lymph trunks. They receive or are joined by the superficial lymph vessels from the cubital nodes which enter the deeper part of the arm through the aperture of the deep fascia which accommodates the basilic vein and medial antebrachial cutaneous nerve. The brachial lymphatic trunks empty into the lateral and central groups of axillary lymph nodes (p. 82).

THE CUBITAL FOSSA (figs. 75, 82)

The **cubital fossa** is a triangular space at the bend of the elbow. It is bounded above by a line connecting the epicondyles of the humerus and its sides

are formed by the converging borders of the brachioradialis muscle laterally and the pronator teres muscle medially. These muscles are, respectively, the innermost members of the extensor group of forearm muscles, which arise from the lateral epicondyle and its supracondylar ridge, and of the flexor muscle group from the medial epicondyle. The floor of the fossa is composed of the brachialis muscle of the arm and the supinator muscle of the forearm. The central structure of this space is the **tendon of the biceps brachii muscle** which descends through it to its insertion on the tuberosity of the radius and spans the medial portion of the space with its fibrous expansion to the antebrachial fascia, the **bicipital aponeurosis** (p. 85). Medial to the tendon lies the **brachial artery** which bifurcates into the radial and ulnar arteries opposite the neck of the radius in the inferior portion of the fossa. The **median nerve** enters the forearm medial to the brachial artery. Although submerged between the brachioradialis and brachialis muscles, the **radial nerve** can be exposed in the interval between them by displacing brachioradialis lateralward. Superficial to the deep fascia, the **median cubital vein** crosses the area obliquely, overlying the bicipital aponeurosis. It is large and is commonly used for injections and transfusions. It is crossed by branches of the **medial antebrachial cutaneous nerve**. A **median cephalic vein** may lie toward the lateral side of the fossa in the subcutaneous tissues. The **lateral antebrachial cutaneous nerve** passes deep to this vein. The skin over the cubital fossa is thin and delicate.

THE FOREARM

The forearm extends from the elbow to the wrist and contains two bones, the radius and the ulna. It has many muscles, the tendons of which pass predominantly into the hand and its digits. The muscles are arranged as a flexor mass anteriorly and as an extensor mass posteriorly. Its nerves and blood vessels are continuations of those of the arm. The superficial structures—veins, nerves, and lymphatics—have been considered.

COMPARTMENTS AND FASCIA

The organization of the forearm, effected by an interosseous membrane connecting the forearm

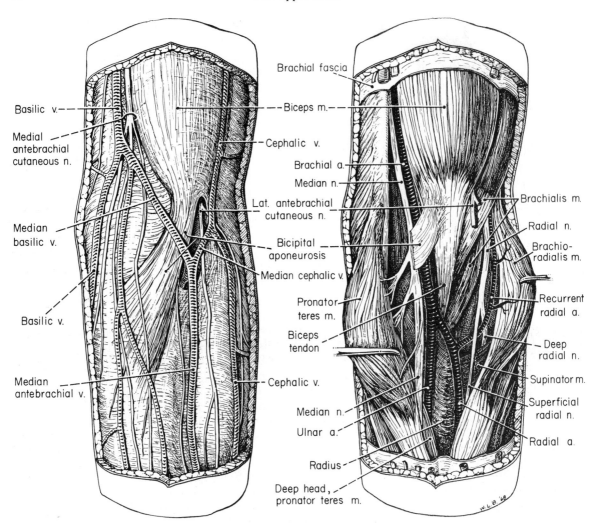

Fig. 75   The cubital fossa; superficial and deep views.

bones and by the antebrachial fascia, is into an anterior compartment and a posterior compartment (fig. 10). The anterior compartment contains the pre-axial flexor muscles together with nerves and vessels; the posterior compartment contains the postaxial extensor muscles with their accompaniments. The antebrachial fascia is continuous with the brachial fascia of the arm and with the deep fascia of the hand. Composed principally of circular fibers, but with additional oblique and longitudinal fibers arising in structures at the elbow (especially the bicipital aponeurosis), the fascia forms a cylindrical sheath enclosing all but the most superficial forearm structures. It is attached

to the posterior aspect of the olecranon and to the subcutaneous border of the ulna. This attachment provides one of the fibrous separations of the anterior and posterior compartments of the forearm; another intermuscular septum passes from the radius to the antebrachial fascia between the flexor and extensor groups of muscles. This septum (fig. 10) does not run directly lateralward to the antebrachial fascia, and thus the compartments are not as completely anterior and posterior as they are in the arm. In the proximal two-thirds of the forearm the antebrachial fascia provides fibrous origin for the underlying muscles. The strong interosseous membrane connects the interosseous borders of the radius and ulna throughout most of their length. It is more completely described with the articulations of the upper limb (p. 136). At the wrist, the antebrachial fascia is strengthened dorsally by oblique fibers that extend from the radius to the styloid process of the ulna and to the pisiform and triangular bones. This is the extensor retinaculum; it encloses the tendons of the extensor group of muscles. The palmar carpal ligament is a thickening of the antebrachial fascia anteriorly.

THE MUSCLES OF THE FOREARM

There are nineteen muscles in the forearm. Of this number, eleven are classified as extensor muscles; eight belong to the flexor group. These names designate simply a group characteristic, for certain of the muscles are primarily rotators of the forearm bones. It is an aid in organizing these muscles to divide them into functional groupings. Eighteen of the nineteen muscles can be arranged in functional groups of three. The muscle excluded is brachioradialis; this muscle has no action in the hand or digits but is actually an elbow flexor. With the exception of the first grouping, all others are composed of muscles moving the hand and digits.

1. **Muscles which rotate the radius on the ulna**
   a. Pronator teres
   b. Pronator quadratus
   c. Supinator
2. **Muscles which flex the hand at the wrist**
   a. Flexor carpi radialis
   b. Flexor carpi ulnaris
   c. Palmaris longus
3. **Muscles which flex the digits**
   a. Flexor digitorum superficialis
   b. Flexor digitorum profundus
   c. Flexor pollicis longus

All the muscles in these first three groups, with the exception of the supinator, are pre-axial muscles.

4. **Muscles which extend the hand at the wrist**
   a. Extensor carpi radialis longus
   b. Extensor carpi radialis brevis
   c. Extensor carpi ulnaris
5. **Muscles which extend the digits, except the thumb**
   a. Extensor digitorum
   b. Extensor indicis
   c. Extensor digiti minimi
6. **Muscles which operate in extension of the thumb**
   a. Abductor pollicis longus
   b. Extensor pollicis brevis
   c. Extensor pollicis longus

All the muscles of the last three groups are post-axial muscles.

There is both a morphological and a functional balance between these muscle groups. One group flexes the hand at the wrist; another group extends it. The structural requirements of their balance are met in their insertion on the opposite sides of the same bone; thus, the flexor carpi ulnaris inserts on the palmar side of the base of the fifth metacarpal bone, and the extensor carpi ulnaris inserts on the dorsal side of the base of the same bone. Extensor carpi radialis longus and brevis insert, respectively, on the dorsal side of the bases of the second and third metacarpal bones. Balance with these insertions is achieved in the termination of the tendon of flexor carpi radialis on the palmar aspect of the bases of the second and third metacarpal bones and in the continuity of the tendon of palmaris longus with the deep connective tissue (palmar aponeurosis) in the center of the palm of the hand. Balance between the digital flexors and extensors results from the insertion of the flexor tendons into the palmar aspect of the bases of the middle and distal phalanges and of the extensor tendons into the dorsum of the bases of the same bones. The extensors of the thumb insert in a logical sequence. There are three bones in the thumb, a metacarpal and two phalanges. The dorsal muscles of the thumb are also three, and their insertions, in sequence, are: the tendon of the abductor pollicis longus to the dorsum of the metacarpal; the tendon of the extensor pollicis brevis to the dorsum of the first phalanx; and the tendon of the extensor pollicis longus to the dorsum of the second phalanx.

The Extensor Muscles (figs. 68, 76, 78, 79)

The muscles listed in groups 4, 5, and 6 of the functional classification, the supinator of group 1, and the brachioradialis comprise the posterior antebrachial muscles. Six of these muscles compose a superficial set; five of them are more deeply located.

1. **Superficial Group** In the superficial group of posterior antebrachial muscles are the following,

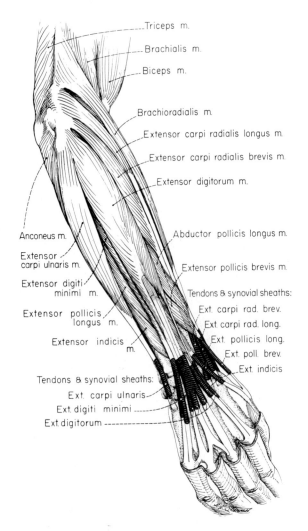

Fig. 76 The posterior antebrachial muscles and the synovial sheaths of their tendons at the wrist.

listed in the order in which they lie across the back of the forearm from the radial to the ulnar sides:

Brachioradialis
Extensor carpi radialis longus
Extensor carpi radialis brevis
Extensor digitorum
Extensor digiti minimi
Extensor carpi ulnaris

The supracondylar ridge of the humerus and the lateral epicondyle provide the origin for these muscles. The brachioradialis and extensor carpi radialis longus arise, respectively, from the upper two-thirds and the lower one-third of the supra-

condylar ridge. The other four members of the group utilize to a greater or lesser extent a tendon of origin common to all of them (**common extensor tendon**) which attaches to the lateral epicondyle. Large muscle masses in front and behind would seriously interfere with free flexion and extension at the elbow, whereas muscles placed at the margins are well out of the way. Thus, as here, the extensor origins are located laterally, and the origins of the flexor muscles are related to the medial epicondyle.

Brachioradialis (figs. 68, 76, 78) (origin noted above) inserts into the lateral side of the base of the styloid process of the radius. Its muscle fibers end about the middle of the forearm in a flat tendon. The muscle is superficially placed on the antero-lateral surface of the forearm, and its tendon of insertion is crossed by abductor pollicis longus and extensor pollicis brevis. The ulnar border of the muscle crosses the radial border of the pronator teres to define the apex of the cubital fossa. The superficial radial nerve and the radial artery descend through the forearm beneath the muscle and its tendon.

Extensor carpi radialis longus (figs. 68, 76, 78) (origin noted) exhibits a flat tendon which descends

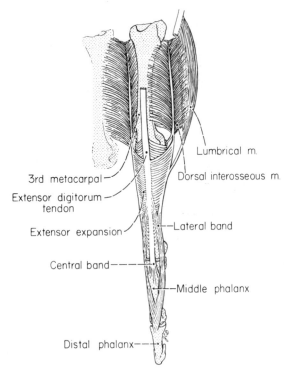

Fig. 77 The extensor expansion and the insertion of the tendon of the extensor digitorum muscle.

along the lateral border of the radius under abductor pollicis longus and extensor pollicis brevis. The tendon crosses the wrist, in company with the tendon of extensor carpi radialis brevis, and terminates on the dorsum of the second metacarpal bone of the hand.

Extensor carpi radialis brevis (figs. 76, 78) (common extensor tendon origin) has a tendon which becomes discrete in the lower one-third of the forearm, where it is closely applied against the tendon of extensor carpi radialis longus. The tendon of extensor carpi radialis brevis inserts on the dorsum of the base of the third metacarpal bone. In their passage under the extensor retinaculum, the extensor carpi radialis longus and brevis tendons are enclosed by a single synovial sheath.

Extensor digitorum (figs. 76, 78) divides above the wrist into four tendons which occupy a common compartment on the dorsum of the wrist under the extensor retinaculum (p. 110). Separating over the back of the hand, the four digital extensor tendons are joined together in a variable manner by obliquely running **intertendinous connections.** As the tendons pass over the back of the metacarpo-

phalangeal joints, they become flattened and are closely attached to the joint capsules, substituting as dorsal ligaments for these capsules.

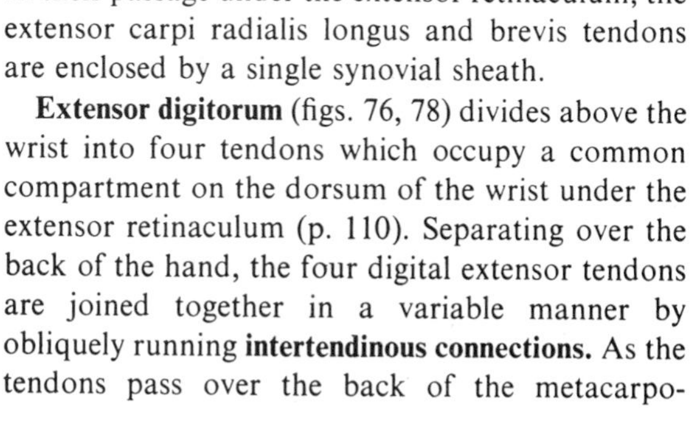

At the metacarpophalangeal joint and over the proximal two phalanges, an **extensor expansion** is formed by each of the tendons of the common extensor with the participation of tendons from the lumbrical and interosseous muscles of the palm of the hand (figs. 77, 100). Opposite the knuckles (metacarpophalangeal joints) a band of fibers passes from each side of the digital extensor tendon anteriorly on either side of the joint and attaches to the palmar ligament on the palmar aspect of the joint. This proximal spreading of the extensor expansion appears like a hood of fibers over the metacarpophalangeal joint. Over the dorsum of the proximal phalanx the digital extensor tendon divides into three slips. Of these, the central, broader one passes directly forward across the next joint and inserts on the base of the middle phalanx dorsally. The diverging bundles on either side receive and combine with the broadening fibers of a tendon of a lumbrical muscle on the radial side of the digit and of interosseous tendons on both sides of the digit. These tendons unite into a common band that proceeds distalward to the middle phalanx, over the dorsum of which the bands of both sides unite in a triangular aponeurosis. The apex of this aponeurosis attaches to the base of the distal phalanx.

Extensor digiti minimi is a slender muscle sometimes only incompletely separated from extensor digitorum. The tendon joins the ulnar side of the tendon of extensor digitorum to the fifth digit and helps to form the extensor expansion over the fifth digit. It provides independent extensor action for the fifth digit.

Extensor carpi ulnaris (figs. 76, 78) lies on the ulnar border of the forearm. It arises, like the previous muscles, from the lateral epicondyle of the humerus by the common extensor tendon and, in addition, from the middle two-fourths of the posterior border of the ulna. It inserts on the ulnar side of the base of the fifth metacarpal bone.

2. **Deep Group** The muscles of this group are generally covered by those of the superficial group, although certain of their tendons and parts of their fleshy bellies outcrop just above the wrist.

They are listed as follows:

Supinator
Abductor pollicis longus

Flexor carpi ulnaris m.
Flexor digitorum profundus m.
Ulnar n.
Basilic v. & medial antebrachial cutaneous n.
Palmaris longus m.
Flexor digitorum superficialis m.
Median antebrachial v.
Ulnar a.
Flexor carpi radialis m.
Median n.
Pronator teres m.
Ceph. v. & lat. antebr. cut. n.
Rad. a.;v., & superf. rad. n.
Brachioradialis m.
Extensor carpi radialis longus m.
Radius
Extensor carpi radialis brevis m.
Extensor digitorum m.
Dorsal antebrachial cut. n.
Supinator m.
Extensor digiti minimi m.
Deep radial n.
Anterior interosseous a. & n.
Posterior interosseous a. & v.
Extensor carpi ulnaris m.
Antebrachial fascia
Ulna

Fig. 78 A cross section of the upper portion of the forearm.

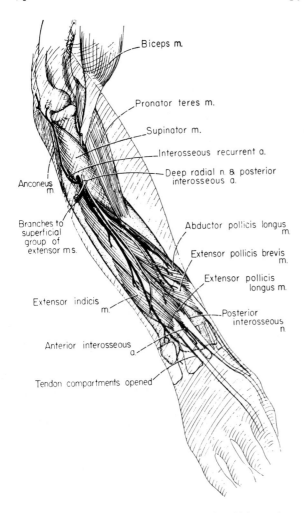

Fig. 79   The deep group of posterior antebrachial muscles; the associated blood vessels and nerves.

Extensor pollicis brevis
Extensor pollicis longus
Extensor indicis

The supinator is located in the upper one-third of the forearm, is completely hidden by superficial muscles, and is characterized by a wrap-around tendon attaching to the radius. The other four muscles arise from the radius and ulna below the supinator attachments and from the interosseous membrane between the bones. The abductor pollicis longus and extensor pollicis brevis are arranged from above downward on the radius and the interosseous membrane, the extensor pollicis longus and extensor indicis from above downward on the ulna and the interosseous membrane (fig. 80).

Supinator (figs. 78, 79, 80, 83, 84) The complex origin of the supinator includes the lateral epicondyle of the humerus, the radial collateral ligament of the elbow joint, the annular ligament of the radius, and the supinator crest and fossa of the ulna. Its fibers form a flat sheet which is directed downward and lateralward to wrap itself almost completely around the radius, inserting on the lateral surface of the bone in its upper one-third (from the radial tuberosity above to the insertion of the pronator teres muscle below). The supinator muscle is separable into superficial and deep layers;

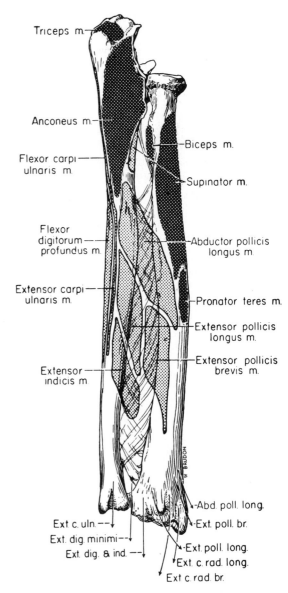

Fig. 80   The muscular attachments and the course of the tendons on the posterior aspect of the radius and ulna. (Black dots—origin; white dots on black—insertion.)

between these laminae runs the deep radial nerve from which it receives its innervation.

Abductor pollicis longus (figs. 79, 80) lies immediately below the supinator. It arises from the middle one-third of the posterior surface of the radius, and, extending across the forearm, from the lateral part of the posterior surface of the ulna below the anconeus. From this broad origin the fibers converge in a bipennate manner onto its tendon, the insertion of fleshy fibers continuing almost to the wrist. With the tendon of the extensor pollicis brevis muscle closely applied to its medial side, the tendon of the abductor crosses the tendons of extensor carpi radialis longus and brevis and inserts on the radial side of the base of the metacarpal bone of the thumb.

Extensor pollicis brevis (figs. 79, 80) (origin noted) inserts on the base of the proximal phalanx of the thumb. This muscle is a specialization of the distal part of the abductor pollicis longus muscle. Its fleshy part is medial to and closely connected with the latter, and their tendons may be intimately related as they cross the wrist.

Extensor pollicis longus (figs. 79, 80) arises as already noted. Distally, its tendon passes obliquely across the tendons of both radial extensors of the wrist and along the ulnar side of the first metacarpal bone to terminate on the base of the distal phalanx of the thumb.

On the dorsum of the wrist at its radial border, abduction and extension of the thumb bring into relief the tendons of extensor pollicis longus and extensor pollicis brevis to define a triangular interval known as the **anatomical snuff box** (fig. 81). In the depths of this depression the radial artery passes diagonally from the anterior surface of the radius distally onto the back of the hand.

Extensor indicis (figs. 79, 80) is a narrow muscle medial and adjacent to extensor pollicis longus. Its tendon passes under the extensor retinaculum with the tendons of extensor digitorum, and it is enclosed with them in a common synovial sheath. Opposite the second metacarpal bone, the indicis tendon joins the ulnar side of the tendon of the digital extensor for the index finger, participating with it in the formation of the extensor expansion over that digit.

The Flexor Muscles

The muscles of the second and third groups and the two pronator muscles of the first group in the functional classification of forearm muscles (p.

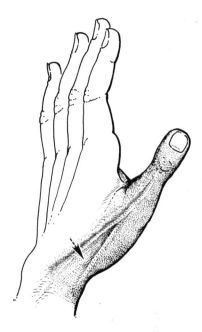

Fig. 81 The 'anatomical snuff-box' and its boundaries.

95) comprise the anterior antebrachial muscles. Five of these muscles belong to a superficial group, and three to a deep group.

1. **Superficial Group** The following list of the superficial muscles is in the order in which they lie from the radial to the ulnar side of the forearm, the flexor digitorum superficialis, however, actually lying deep to the first four muscles.

> Pronator teres
> Flexor carpi radialis
> Palmaris longus
> Flexor carpi ulnaris
> Flexor digitorum superficialis

These muscles have in common a tendinous origin from the medial epicondyle of the humerus. Intermuscular septa and the overlying antebrachial fascia are also used for muscle attachments, and certain muscles have additional bony origins.

Pronator teres (figs. 78, 79, 80, 83) has both a humeral and an ulnar head. The larger humeral head of origin arises from the medial epicondyle of the humerus by means of the common flexor tendon and from the adjacent septa and fascia. The small, deep ulnar head of origin arises from the medial side of the coronoid process of the ulna and joins the deep aspect of the humeral head. Between these two heads of origin passes the median nerve. The muscle is directed obliquely across the fore-

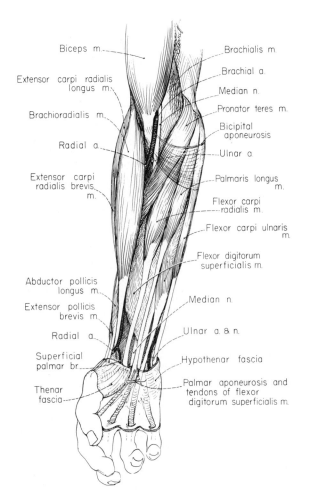

Biceps m.
Extensor carpi radialis longus m.
Brachioradialis m.
Radial a.
Extensor carpi radialis brevis m.
Abductor pollicis longus m.
Extensor pollicis brevis m.
Radial a.
Superficial palmar br.
Thenar fascia

Brachialis m.
Brachial a.
Median n.
Pronator teres m.
Bicipital aponeurosis
Ulnar a.
Palmaris longus m.
Flexor carpi radialis m.
Flexor carpi ulnaris m.
Flexor digitorum superficialis m.
Median n.
Ulnar a. & n.
Hypothenar fascia
Palmar aponeurosis and tendons of flexor digitorum superficialis m.

Fig. 82   The superficial group of anterior antebrachial muscles.

arm, and its tendon passes under the brachioradialis to insert on a rough impression on the shaft of the radius at the middle of its lateral surface. The radial margin of the pronator teres forms the medial border of the cubital fossa. Its tendon is crossed by the superficial radial nerve and the radial vessels and by the brachioradialis muscle.

Flexor carpi radialis (figs. 78, 82) (origin as noted) ends in a tendon which accounts for somewhat more than half its length. The tendon passes through the wrist in a fibro-osseous canal bounded by a split in the radial portion of the flexor retinaculum and, deeply, by the groove of the trapezium bone. It is inserted into the base of the second metacarpal and usually sends an additional slip to the third metacarpal bone. At the wrist the tendon

may be used as a guide to the radial artery which is immediately lateral to it.

Palmaris longus (figs. 78, 82) (origin as noted) is one of the more variable muscles of the body, being absent in about 13 per cent of cases. It terminates in a slender, flattened tendon which crosses superficial to the flexor retinaculum and inserts into the retinaculum and the palmar aponeurosis. Its spreading tendinous fibers constitute the chief portion of the palmar aponeurosis. At the wrist the median nerve lies to the radial side of its tendon.

Flexor carpi ulnaris (figs. 78, 82) has two heads of origin. A humeral head has the same origin as the

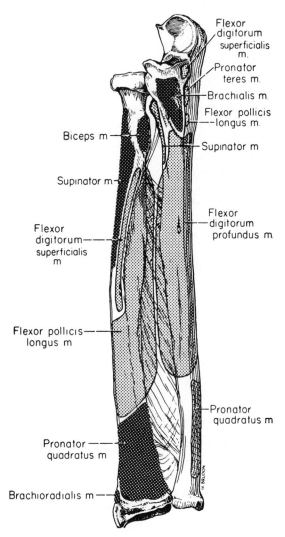

Flexor digitorum superficialis m.
Pronator teres m.
Brachialis m.
Flexor pollicis longus m.
Supinator m.
Flexor digitorum profundus m.
Biceps m.
Supinator m.
Flexor digitorum superficialis m.
Flexor pollicis longus m.
Pronator quadratus m.
Pronator quadratus m.
Brachioradialis m.

Fig. 83   The muscular attachments on the anterior aspect of the radius and ulna. (Code as fig. 80.)

other superficial flexors. The ulnar head arises from the medial border of the olecranon and the posterior border of the ulna in its upper two-thirds. The tendon of the muscle inserts on the pisiform bone and, through it by two ligaments, into the hook of the hamate and the base of the fifth metacarpal bone. The muscle overlies the ulnar nerve through its entire length and the ulnar artery in the lower half of the forearm.

Flexor digitorum superficialis (figs. 78, 82, 83) is deep to the other flexors of the superficial forearm group. The muscle arises by two heads of origin—humeroulnar and radial—connected by a fibrous band that crosses the median nerve and ulnar blood vessels. The larger **humeroulnar head** arises from the common tendon from the medial epicondyle of the humerus, from intermuscular septa, from the ulnar collateral ligament of the elbow, and, by a slender slip, from the medial border of the coronoid process. The **radial head** is a thin, fibromuscular layer arising from the upper two-thirds of the anterior border of the radius. The muscle forms two planes and the tendons of its superficial plane pass to the middle and ring fingers. The deep lamina, after sending a muscular slip to the superficial, divides into two parts which end in tendons for the index and the fifth digits. The four tendons pass through the wrist deep to the **flexor retinaculum** (p. 110), maintaining their paired and layered relationships. They begin to diverge toward the four digits in the palm, where they lie under the palmar aponeurosis and deep to the superficial palmar arterial arch and the digital branches of the median and ulnar nerves. Superficial to the tendons of flexor digitorum profundus and paired with them, the tendons of superficialis approach the digits, entering the proximal ends of the **fibrous sheaths of the digital flexors** opposite the heads of the metacarpal bones. Within each fibrous sheath (p. 123), the overlying superficialis and deeper profundus tendons pass distally against the palmar surfaces of the bones of the digits, a common synovial sheath investing them and lining the fibrous sheath.

The tendons of flexor digitorum superficialis insert into the palmar aspect of the shafts of the middle phalanges of digits two to five. This insertion would seem to be prevented by the tendon of the flexor digitorum profundus which intervenes between the superficialis tendon and the bone as it passes distally to insert on the base of the distal phalanx. The tendon of flexor digitorum super-

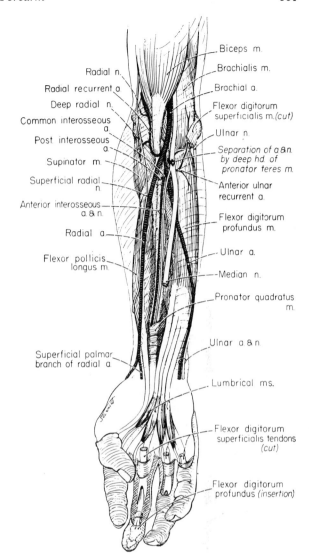

Fig. 84   The deep group of anterior antebrachial muscles.

Labels on figure:
Radial n.
Radial recurrent a.
Deep radial n.
Common interosseous a.
Post. interosseous a.
Supinator m.
Superficial radial n.
Anterior interosseous a. & n.
Radial a.
Flexor pollicis longus m.
Superficial palmar branch of radial a.

Biceps m.
Brachialis m.
Brachial a.
Flexor digitorum superficialis m.(cut)
Ulnar n.
Separation of a & n. by deep hd. of pronator teres m.
Anterior ulnar recurrent a.
Flexor digitorum profundus m.
Ulnar a.
Median n.
Pronator quadratus m.
Ulnar a & n.
Lumbrical ms.
Flexor digitorum superficialis tendons (cut)
Flexor digitorum profundus (insertion)

ficialis reaches the middle phalanx by splitting, allowing the profundus tendon to pass between its two halves, and then reuniting below the profundus tendon to attach to the periosteum of the phalanx. The division of the tendon takes place over the proximal phalanx. **Mesotendons** connect the tendons to the digital sheath. These **vincula tendinum** are folds of synovial membrane, strengthened by fibrous tissue, which conduct blood vessels to the tendon. They represent remnants of an original dorsal attachment of the long synovial sheath investing the paired superficialis and profundus tendons. The smaller **vinculum breve** exists as a triangular fold of synovial membrane between bone and the distal end of the tendons. **Vinculum longum,** frequently double, consists of a narrow strand between the tendon and the palmar surface of the proximal phalanx in the case of the superficialis tendon and between the profundus tendon and the inserting bundles of superficialis in the case of flexor digitorum profundus.

2. **Deep Group** The deep group of forearm flexor muscles has the following three members:

> Flexor digitorum profundus
> Flexor pollicis longus
> Pronator quadratus

These muscles have in common their deep position and their origins from ulna or radius and the interosseous membrane. The pronator quadratus covers the lower one-fourth of the forearm bones just as the supinator monopolizes the upper one-fourth (especially the radius). The two deep flexor muscles lie intermediate between them.

Flexor digitorum profundus (figs. 78, 83, 84) arises from the posterior border of the ulna by an aponeurosis common to it and flexor carpi ulnaris, from the proximal two-thirds of the medial surface of the ulna, and from the adjacent interosseous membrane. The muscle divides into four portions which produce, near the wrist, four discrete tendons. These pass under the flexor retinaculum where they lie side by side dorsal to the superficialis tendons. In the palm the tendons diverge and, opposite the heads of the metacarpals, both the superficial and deep tendons enter the ends of the fibrous digital sheaths. Opposite the proximal phalanx, the tendons of flexor digitorum profundus pass through the apertures produced by the splitting of the superficialis tendons and proceed to the base of the distal phalanx of digits two to five, on which they terminate (figs. 98, 100). The vincula tendinum of the profundus tendons have been described with those of the tendons of flexor digitorum superficialis.

Flexor digitorum profundus lies deep to flexor superficialis, flexor carpi ulnaris, the median nerve, and the ulnar blood vessels. It occupies the deepest muscular plane of the forearm. In the palm its tendons give origin to the lumbrical muscles, which are more fully described in the section on the hand (p. 123).

Flexor pollicis longus (figs. 82–84) arises principally from the anterior surface of the radius, from just below its tuberosity and oblique line almost as far as the pronator quadratus, and from the interosseous membrane. The tendon passes beneath the flexor retinaculum and continues downward medial to the thenar eminence. Passing between the two sesamoid bones of the metacarpophalangeal joint of the thumb, it enters a fibrous sheath and inserts on the base of the distal phalanx of the thumb.

Pronator quadratus (figs. 83, 84) is a small, fleshy, quadrilateral muscle located just above the wrist deep to the tendons of flexor digitorum profundus and flexor pollicis longus. It arises from the medial side of the anterior surface of the distal one-fourth of the ulna, and its fibers run transversely across the wrist to insert into the distal one-fourth of the anterior surface of the radius and the triangular area above the ulnar notch.

ACTIONS OF THE FOREARM MUSCLE

Rotary Movements

The rotary movements of the forearm are pronation and supination. With the elbow flexed, pronation carries the palm inward or downward; supination carries it upward and outward. The pronator muscles are pronator teres and pronator quadratus. Pronator quadratus is the principal mover, teres being added for speed and against resistance. The supinator muscle supinates the forearm (and with it the hand). It is supplemented by the more powerful biceps brachii muscle for speed and against resistance.

Brachioradialis has no action at the wrist. It is primarily an elbow flexor and thus is most allied functionally with the pre-axial arm muscles. It tends to return the forearm to its usual mid-pronation, mid-supination position especially against resistance and thus has some significance for rotary movements, but is principally concerned with establishing the close-packed position of the bones articulating in the elbow joint.

Wrist Movements

Flexion, extension, abduction, and adduction movements of the hand are accommodated at the wrist. Flexion is produced by the 'carpal flexors' (including palmaris longus) as well as by the 'digital flexors' after they have exerted their traction on the digits. The 'carpal flexors' are also synergists for the digital extensor muscles. Extension of the wrist is produced by carpal and digital extensors, the latter after exerting their traction primarily on the digits. The carpal extensors also stabilize the wrist for digital flexion, especially extensor carpi radialis longus. The ulnar and radially located carpal muscles produce, respectively, adduction and abduction, the flexor and extensor muscles of the same side co-operating in a single action. Abduction is assisted by the long abductor and the extensors of the thumb.

Digital Movements

Forearm muscles participate in digital movement, but the full understanding of digital movement must wait until the muscles of the hand have been considered. Flexor digitorum superficialis is a flexor of the proximal interphalangeal and meta-carpophalangeal joints of the medial four fingers. Flexor digitorum profundus primarily flexes the terminal phalanx but, continuing to act, flexes the middle and proximal phalanges also. Flexor digitorum profundus flexes the digits in slow action, superficialis being recruited for speed and against resistance. The extensor digitorum, assisted by the extensor of the index and fifth fingers, is the extensor of the fingers. Inter-connecting tendinous bands between the tendons of III, IV, and V prevent completely independent extension of these digits, but the index finger can be moved quite separately. The interosseous and lumbrical muscles of the hand are essential for full extension of the digits (p. 126). Free movement of the thumb is one of the important features of the human hand. The specific action of group 6 muscles (p. 95) will be detailed under the discussion of the actions of the muscles of the hand (p. 126).

THE ULNA (figs. 80, 83, 85, 86)

The ulna is the prismatic bone of the medial side of the forearm. Of the two forearm bones it is the more firmly connected to the humerus, but it is only indirectly articulated with the wrist and hand. It is, therefore, more robust proximally, and its principal processes at this end (olecranon and coronoid) project as open jaws to clasp the trochlea of the humerus. The radius is formed contrarily. Its lower end is broadened to take almost the full contact of the bones of the wrist; it is reduced proximally, where it has only a very shallow cup-shaped articulation for the capitulum of the humerus.

The ulna is a long bone. The proximal extremity, heavy and thickened, exhibits the opened jaws of the trochlear notch, the olecranon, the coronoid process, and the radial notch.

The **trochlear notch** is a concavity which describes about one-third of a circle and is divided by a longitudinal ridge into medial and lateral parts. The waist of

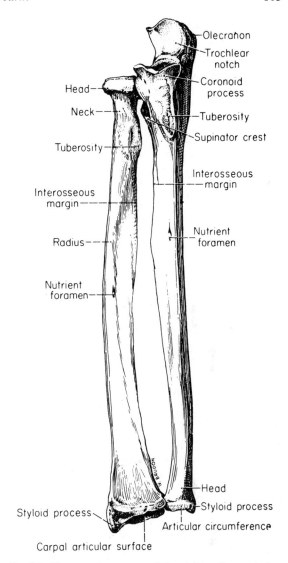

Fig. 84 The anterior aspect of the right radius and ulna. (×½)

the notch is constricted, and a roughness, or ridge, across the waist separates that part which comes from the olecranon from that part formed by the coronoid process. The notch receives the trochlea of the humerus. In extreme flexion the coronoid process enters the coronoid fossa of the humerus; the olecranon enters the olecranon fossa in full extension. The **olecranon,** the thick proximal projection of the shaft, contributes part of the trochlear notch and forms the point of the elbow. The **coronoid process** is a strong riangular projection from the front of the ulna, its upper end forming the inferior and anterior part of the trochlear notch. The anterior surface of the coronoid process is rough for the insertion of brachialis. On the lateral side of the coronoid process a shallow concavity constitutes the **radial notch** of the ulna; it receives the articular surface of the circumference of the head of the radius and its prominent edges give attachment to the ends of the annular ligament of the radius.

The **body** (shaft) of the ulna is thick in continuity with the coronoid and olecranon processes but becomes quite slender below.

Its **posterior border,** continuous with the dorsum of the olecranon is the subcutaneous border of the ulna. It leads to the styloid process below. Converging on the posterior border from the posterior lip of the radial notch is the **supinator crest** bounding the supinator fossa. The **interosseous margin** is a sharp ridge of the lateral side of the bone which gives attachment to the interosseous membrane. A **nutrient foramen** lies at the junction of the proximal and middle thirds of the anterior surface of the bone. It is directed toward the elbow and transmits a branch of the anterior interosseous artery.

The distal extremity of the ulna is small. It has a small, conical **styloid process** in line with the posterior border of the bone and the larger, rounded **head.**

The distal surface of the head is smooth and semi-circular in form and articulates with the articular disk of the distal radioulnar joint. The smooth articular surface of the circumference of the head of the ulna is received into the ulnar notch of the radius. In pronation of the hand the head of the ulna makes a rounded prominence on the dorsum of the wrist, and the styloid process can be palpated along the ulnar border.

Ossification of the ulna begins near the middle of the shaft in the eighth week of intrauterine life. A proximal epiphysis for the end of the olecranon appears about the eighth to the tenth year and fuses at the age of fourteen to sixteen years. An ossification center for the distal end of the ulna appears during the fifth to the sixth year and fuses with the shaft between the seventeenth and nineteenth years.

THE RADIUS (figs. 80, 83, 85, 86)

The radius is the shorter, laterally placed bone of the forearm. Its name comes from its resemblance to the spoke of a wheel. The radius has two ends and a shaft, and its smaller proximal end exhibits a head a neck, and a tuberosity.

The **head** is a thick disk, articular both on its free surface and on its circumference. The **upper surface** is a shallow cup for articulation with the capitulum of the humerus. The **articular circumference** of the head is wider medially for contact with the radial notch of the ulna and is narrower where it is held by the annular ligament. A smooth constriction below the disk-like head is the **neck** of the radius. The **tuberosity** is an oval prominence distal to the neck.

The **body** (shaft) of the radius is somewhat prismatic with a rounded surface laterally. It has a slight, lateral convexity which helps to keep the radius and

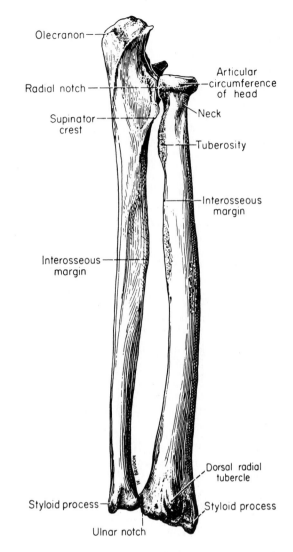

Fig. 86   The posterior aspect of the right radius and ulna. ($\times\frac{1}{2}$)

ulna separated during rotational movements of the forearm.

A sharp **interosseus border** marks the medial aspect of the shaft and gives attachment to the interosseous membrane. The anterior surface is broad and somewhat concave below as the bone expands at the wrist. Near the junction of the proximal and middle thirds of the bone, a **nutrient foramen** conducts a branch of the anterior interosseous artery into the bone. At its distal extremity, the lateral surface continues into the prominent pyramidal **styloid process.**

The distal end of the radius is quadrilateral. Its **carpal articular surface** makes the bony contact of the forearm with the hand and wrist. This surface is concave transversely as well as anteroposteriorly and is divided by a constriction and a marking on the articular surface into a more lateral, triangular-shaped portion for articulation with the scaphoid bone and a medial quadrangular part for articulation with the lunate bone of the wrist. The medial surface of the distal extremity is also concave and articular; as the **ulnar notch,** it receives the rounded head of the ulna. The **articular disk,** which separates the carpal and the distal radio-ulnar joints, attaches to the margin of the radius between the carpal and ulnar articular surfaces. To the anterior and posterior margins attach the fibers of the capsule of the wrist joint and the radiocarpal ligaments. The dorsum of the distal extremity is ridged and grooved for the passage of the extensor tendons across the wrist. Laterally, the pyramidal **styloid process** shows a flattened radial margin for the tendons of the abductor pollicis and extensor pollicis brevis muscles. On the dorsum of the bone radial to the **dorsal radial tubercle** is a broad, flat zone indistinctly divided into separate grooves for the tendons of the extensor carpi radialis longus and brevis muscles. On the ulnar side of the dorsal radial tubercle is a sharply defined oblique groove for the tendon of the extensor pollicis longus muscle. The broad area medial to this groove accommodates the tendons of the extensor digitorum and extensor indicis muscles.

The radius is ossified by means of a center of ossification appearing in the shaft at the eighth week of intrauterine life. A separate ossification center appears for the disk-like head in the fourth or fifth year, and this portion of the bone fuses with the shaft between the fourteenth and sixteenth years. The distal epiphysis appears about the end of the first year and fuses with the shaft between the eighteenth and nineteenth years.

THE NERVES OF THE FOREARM

The nerves of the forearm are the median, the ulnar, and the superficial and deep branches of the radial. The median and ulnar nerves supply the pre-axial muscular compartment in the forearm. *In summary, the ulnar nerve supplies the flexor carpi ulnaris and the ulnar portion of flexor digitorum profundus. All other pre-axial muscles of the forearm are supplied by the median nerve. The deep radial nerve is muscular to all postaxial muscles.* With the exception of the deep radial nerve all give rise to cutaneous branches a short distance above the wrist.

The **median nerve** (figs. 75, 78, 87) leaves the cubital fossa by passing between the two heads of the pronator teres muscle. Passing under the tendinous arch of the flexor digitorum superficialis muscle, the nerve descends through the middle of the forearm, adherent to the fascia on the underside of this muscle. Approaching the wrist, the nerve comes closer to the surface and is the most super-

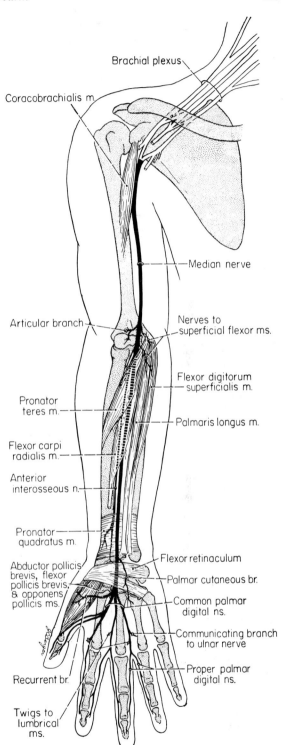

Fig. 87　The course and branches of the median nerve.

ficial structure under the flexor retinaculum. At the wrist it is located to the radial side of the tendon of the palmaris longus muscle; it enters the hand at the distal border of the flexor retinaculum.

In the forearm the median nerve gives rise to articular, muscular, anterior interosseous, and palmar cutaneous branches. One or two **articular twigs** are furnished to the elbow joint as the nerve passes it. Muscular branches are provided to the superficial flexor muscles of the forearm with the exception of the flexor carpi ulnaris. The **nerve to the pronator teres** usually arises at the bend of the elbow and enters the lateral border of the muscle. In common with or slightly below this nerve, a broad bundle of nerves pierces the superficial flexor mass to innervate the **flexor carpi radialis,** the **palmaris longus,** and the **flexor digitorum superficialis** muscles. The **anterior interosseous branch** arises from the back of the median nerve in the cubital fossa. It descends on the surface of the interosseous membrane in company with the anterior interosseous branch of the ulnar artery. Near its origin it furnishes branches to the radial portion of the flexor digitorum profundus muscle and to the flexor pollicis longus muscle. Descending to the upper border of the pronator quadratus muscle, the nerve supplies this muscle and then passes deep to it to end in sensory twigs to the wrist joint. The **palmar cutaneous branch** of the median nerve arises immediately above the flexor retinaculum, becoming superficial between the tendons of the palmaris longus and flexor carpi radialis muscles. It descends into the skin and subcutaneous tissue of the central depressed area of the palm.

The **ulnar nerve** (figs. 78, 88), having passed behind the medial epicondyle of the humerus (where it is susceptible to injury), enters the forearm between the two heads of the flexor carpi ulnaris muscle. At the elbow it lies against the ulnar collateral ligament of the joint in company with the posterior ulnar recurrent artery. At the wrist the ulnar nerve is immediately to the radial side of the pisiform bone, and a line connecting these proximal and distal locations marks its course through the forearm. In the upper part of the forearm the ulnar nerve lies on flexor digitorum profundus and is covered by flexor carpi ulnaris. Distally, the ulnar artery meets the nerve, and in the lower half of the forearm they lie side by side, the artery on the radial side of the nerve. As they approach the wrist, the nerve and its accompanying vessels emerge to the radial side of the flexor carpi ulnaris tendon and are then covered only by the antebrachial fascia. Along the radial aspect of the pisiform bone, the ulnar nerve divides into its terminal branches to the hand, the **superficial** and **deep branches.**

The ulnar nerve, like the median, has no branches in the arm. In the forearm it has articular, muscular, and

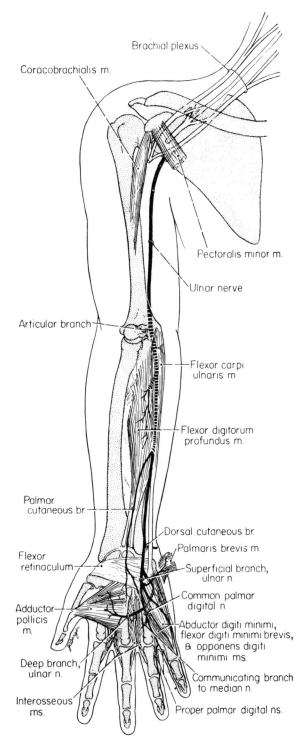

Fig. 88   The course and branches of the ulnar nerve.

cutaneous branches. The **articular branches** are several small filaments which are distributed to the elbow joint as the nerve lies in the groove between the olecranon and the medial epicondyle. In the upper part of the forearm **muscular branches** supply the flexor carpi ulnaris and the ulnar portion of the flexor digitorum profundus muscles. In the lower half of the forearm two cutaneous branches are given off. The **palmar cutaneous branch** arises from the ulnar nerve near the middle of the forearm and runs downward in front of the artery, accompanying it into the palm (see p. 63). The **dorsal cutaneous branch** is also more completely described under the discussion of the cutaneous nerves of the upper limb (p. 63).

The **superficial branch of the radial nerve** (figs. 78, 89) is the smaller of the two terminal branches of the radial nerve. A purely cutaneous nerve, it arises in the interval between the brachioradialis and brachialis muscles just above the elbow. In the forearm the superficial branch descends under cover of the brachioradialis muscle. In the upper one-third of the forearm it gradually approaches the radial artery, and in the middle one-third it lies laterally alongside the artery as both cross the insertion of the pronator teres muscle. In the lower one-third of the forearm the nerve deviates from the artery, passes dorsally under the tendon of brachioradialis, and appears on the dorsum of the forearm. Piercing the antebrachial fascia and descending across the extensor retinaculum, it divides into medial and lateral cutaneous branches to the dorsum of the hand and digits. The cutaneous distribution of this nerve is more completely described in the section on cutaneous nerves of the upper limb (p. 62).

The **deep branch of the radial nerve** (figs. 78, 79, 89) is the larger terminal division of the radial nerve. It is entirely muscular and articular. Descending across the elbow under cover of the brachioradialis and extensor carpi radialis longus muscles, it supplies the extensor carpi radialis brevis and supinator muscles before reaching the proximal border of the latter. Entering the supinator, the deep branch winds around the lateral side of the radius in the substance of the muscle. It passes through the muscle lengthwise between two thin laminae of muscle fibers. On the dorsum of the forearm the nerve emerges with the posterior interosseous artery in the interval between the superficial and deep muscles of the extensor group. Here it supplies branches to the extensor digitorum, the extensor digiti minimi, and the extensor carpi ulnaris muscles. Descending over the abductor pollicis longus muscle, the nerve gives branches in the lower half of the forearm to this muscle, to the extensor pollicis

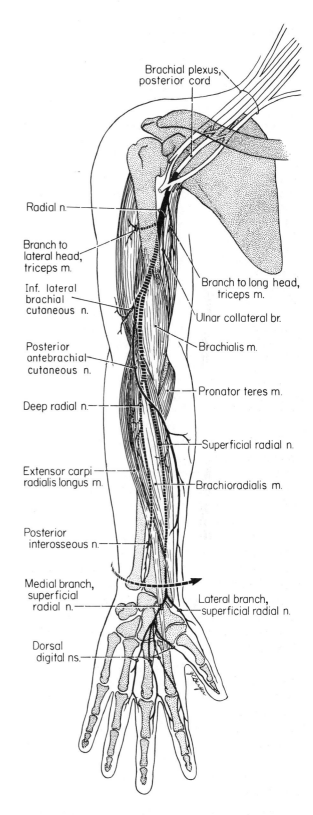

Fig. 89 The course and branches of the radial nerve.

brevis, the extensor pollicis longus, and the extensor indicis muscles. Now diminished in size and designated as the **posterior interosseous nerve,** it passes deep to the extensor pollicis longus and lies on the interosseous membrane where it is joined by the posterior interosseous artery. The nerve terminates at the wrist in a small gangliform enlargement from which twigs pass to the intercarpal joints.

BLOOD VESSELS OF THE FOREARM

The brachial artery (figs. 62, 75) usually divides opposite the neck of the radius into its terminal branches, the radial and ulnar arteries. In the forearm named branches of the arteries lie near the elbow and the wrist and only unnamed muscular branches arise through the middle regions of the forearm. Occasionally high origins of the radial and ulnar arteries occur at upper brachial levels (p. 92).

The Radial Artery

The radial artery (figs. 62, 78, 82, 84, 90), the smaller of the two branches, continues the direct line of the brachial trunk. The course of the artery in the forearm may be defined by connecting the position of the brachial artery next to the biceps tendon at the bend of the elbow with the position of the 'pulse' at the wrist. The artery lies in the intermuscular cleft of the lateral side of the forearm; in the upper one-third of the forearm it runs between the brachioradialis and pronator teres muscles. In the lower part of the forearm the artery lies under the antebrachial fascia with the superficial radial nerve lateral to it. Two venae comitantes accompany it. Its branches in the forearm are the radial recurrent, muscular, palmar carpal, and superficial palmar arteries. Its continuation into the hand and its branches there are described on p. 124.

1. The **radial recurrent artery** arises from the lateral side of the radial artery shortly below its origin. It ascends on the supinator, in the interval between the brachioradialis and brachialis muscles, and among the branches of the radial nerve. It supplies these muscles and the elbow joint and, in front of the lateral epicondyle, anastomoses with the radial collateral artery.

2. The **muscular branches** supply the muscles of the radial side of the forearm.

3. The **palmar carpal branch** is small and arises near the distal border of the pronator quadratus muscle. It runs across the wrist joint, anastomosing with the palmar carpal branch of the ulnar artery, the termination of the anterior interosseous artery, and recurrent branches of the deep palmar arterial arch to form the **palmar carpal network.**

4. The **superficial palmar branch** leaves the radial artery just before it turns from the radial border of the wrist onto the dorsum of the hand. Descending into the hand, the superficial palmar artery passes over or through the muscles forming the eminence of the thumb and sends branches into these muscles. It normally anastomoses with the superficial branch of the ulnar artery to form the superficial palmar arterial arch. The superficial palmar artery is usually slender, and its anastomosis with the ulnar artery may be quite tenuous.

The Ulnar Artery

The ulnar artery (figs. 62, 78, 79, 82, 84, 90) is the larger branch of the brachial artery. It begins in the cubital fossa, opposite the neck of the radius. The line of the ulnar artery may be defined in relation to the ulnar nerve, for the artery describes a gentle curve to the ulnar side of the forearm and takes its place on the radial side of the nerve at the junction of the proximal and middle thirds of the forearm. The artery, accompanied by its venae comitantes, passes deep to both heads of the pronator teres muscle and to all the other superficial flexors. It is crossed by the median nerve which passes between the two heads of the pronator teres. The artery has the flexor digitorum profundus deep to it. In the lower two-thirds of the forearm the artery lies to the radial side of the ulnar nerve and is covered, or overlapped, by the flexor carpi ulnaris muscle. The branches of the ulnar artery in the forearm are the anterior ulnar recurrent, the posterior ulnar recurrent, the common interosseous, muscular, the palmar carpal, and the dorsal carpal.

1. The **anterior ulnar recurrent artery,** arising just below the elbow joint, turns upward in the interval between the brachialis and pronator teres muscles. It sends branches to these muscles and, anterior to the medial epicondyle, anastomoses with anterior branches of the superior ulnar collateral and inferior ulnar collateral arteries.

2. The **posterior ulnar recurrent artery,** larger

than the anterior, arises a little below or in common with it. The posterior ulnar recurrent passes upward between the flexor digitorum superficialis and profundus muscles to reach the back of the medial epicondyle. Here it lies under the flexor carpi ulnaris muscle in company with the ulnar nerve. It supplies adjacent muscles and the elbow joint, anastomoses with posterior branches of the superior and inferior ulnar collateral arteries and with the interosseous recurrent artery, and participates in the formation of the vascular network of the olecranon.

3. The **common interosseous artery** arises from the radial side of the ulnar artery about 3 cm. from its origin. After a posterolateral and descending course of about 1 cm. between flexor pollicis longus and flexor digitorum profundus, the artery gains the upper border of the interosseous membrane. Here it divides into its branches, the anterior and posterior interosseous arteries.

(a) The **anterior interosseous artery** descends on the anterior surface of the interosseous membrane as far as the proximal border of pronator quadratus. It is accompanied by venae comitantes and the interosseous branch of the median nerve. In its course the artery supplies **muscular branches** to the adjacent muscles and **nutrient arteries** to the radius and the ulna. Proximally it gives off the **median artery**, a long, slender branch which accompanies the median nerve into the palm. The median artery is a reduced representative of one of the principal arteries of the forearm in development and, when large, makes a contribution to the superficial palmar arterial arch. At the upper border of pronator quadratus the anterior interosseous artery gives off a small palmar carpal branch which continues distalward on the interosseous membrane and contributes to the palmar carpal network. The anterior interosseous artery then perforates the interosseous membrane and passes to the dorsum of the forearm just above the wrist, where, as a dorsal terminal branch it anastomoses with the posterior interosseous artery. The dorsal terminal branch ends in the dorsal carpal network.

(b) The **posterior interosseous artery** (fig. 79), usually smaller than the anterior, passes to the back of the upper forearm between the interosseous membrane and the oblique ligament. It emerges between the supinator and abductor pollicis longus muscles in company with the deep radial nerve. Descending across the abductor pollicis longus, the artery lies between and sends branches to the superficial and deep extensor muscles of the forearm. Distal to the abductor pollicis longus, it leaves the deep radial nerve and continues across the other members of the deep extensor group of muscles to anastomose, above the wrist, with the dorsal terminal branch of the anterior interosseous artery. An **interosseous recurrent branch** is given off as the posterior interosseus artery lies beneath the supinator. It ascends on or through the supinator muscle and deep to the anconeus muscle to the interval between the lateral epicondyle of the humerus and the olecranon. Here the interosseous recurrent branch communicates with the middle collateral, the inferior ulnar collateral, and the posterior ulnar recurrent arteries and joins the vascular network of the olecranon.

4. **Muscular branches** of the ulnar artery distribute to the muscles of the ulnar side of the forearm.

5. The **palmar carpal branch**, a small vessel, arises at the upper border of the flexor retinaculum. It passes across the wrist deep to the tendons of flexor digitorum profundus to unite with the palmar carpal branch of the radial artery, the palmar carpal branch of the anterior interosseous artery, and the ascending branches of the deep palmar arch in the palmar carpal network.

6. The **dorsal carpal branch** takes its origin just above the pisiform bone. Winding around the border of the wrist under the tendons of the flexor and extensor carpi ulnaris muscles, it reaches the dorsum of the wrist deep to the extensor tendons. Here it anastomoses with the dorsal carpal branch of the radial artery. These, with the assistance of the dorsal carpal branches of the anterior and posterior interosseous arteries, form the dorsal carpal arterial arch. The branches of this arch are described with the anatomy of the hand (p. 116).

The Antebrachial Veins

Antebrachial veins are the venae comitantes of the arteries of the forearm. They arise in the hand, where both the superficial and deep palmar arterial arches are duplicated in **venous arcades.** At the radial side of these arcades the veins become the venae comitantes of the radial artery; from the ulnar side the venae comitantes of the ulnar artery arise. The **radial and ulnar venae comitantes** ascend along their respective sides of the forearm, having the same relations as the arteries. They receive numerous tributaries from the muscles between which they pass and they have frequent communications with the superficial veins. Venae comitantes of the interosseous arteries unite with those of the ulnar artery in the upper part of the forearm and the radial and ulnar veins unite at the bend of the elbow into the two venae comitantes of the brachial vein. In the cubital fossa the deep veins are connected with the median cubital vein.

THE LYMPHATICS OF THE FOREARM

The lymphatics of the forearm are represented by superficial and deep lymphatic channels and by minor intercalated nodes. The superficial lymphatics have been described under the superficial lymphatics of the upper limb (p. 65). The deep antebrachial lymphatic chains are continuous with the brachial lymphatic trunks (p. 93).

THE WRIST

The wrist is at the junction of the forearm and hand. Movement of the hand takes place at the wrist joint, an articulation between the expanded lower end of the radius and the articular disk of the distal radioulnar joint proximally and the first row of carpal bones distally. The mobility of the wrist is enhanced by the midcarpal joints between the three carpal bones of the first row and the four bones of the second, by articulations between the adjacent carpal bones, and by the joints between the distal row of carpals and the bases of the metacarpal bones. The description of these joints is given in the section on the joints of the upper limb (p. 138). The eight carpal bones of the wrist (figs. 91–93) are arranged in two rows. The proximal row consists of, from radial to ulnar sides, the scaphoid (boat-shaped), the lunate (moon-shaped), the triquetrum, and the pisiform (pea-shaped) bones. The pisiform bone is a sesamoid bone in the tendon of the flexor carpi ulnaris muscle and articulates only with the triquetral bone. The four carpal bones of the distal row are, radioulnarward, the trapezium, the trapezoid, the capitate (with a rounded head), and the hamate (with a hooked, or hamular, process). The trapezium and trapezoid bones are named for their shapes; they have a large number of surfaces (fig. 92). The complex joint structure of the wrist and the numerous tendons which cross it impart to the region an unusually sinewy and fibrous character. The ligaments of the wrist are numerous and strong; fibro-osseous canals are provided for the tendons; and synovial sheaths facilitate the movement of the tendons over the bones and joints.

COMPARTMENTS AND DEEP FASCIA

The antebrachial fascia is thickened at the wrist to form the extensor retinaculum and the palmar

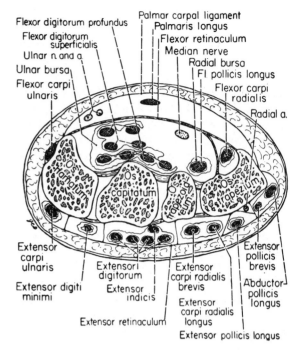

Fig. 90　A cross section of the wrist to show the tendon compartments and synovial sheaths.

carpal ligament. The **extensor retinaculum** (figs. 90, 94) is strengthened by the addition of oblique fibers that extend from the lateral margin of the radius to the styloid process of the ulna, the pisiform bone, and that carpal bone which lies distal to the end of the ulna, the triquetrum. The retinaculum has deep attachments to the ridges on the dorsum of the distal end of the radius. The **palmar carpal ligament** is a thickening of the antebrachial fascia over the palmar aspect of the wrist. It is attached to the styloid processes of both radius and ulna and crosses the tendons of the superficial flexor muscles and the ulnar nerve and blood vessels.

The **flexor retinaculum** (figs. 90, 96, 98) lies deep to the palmar carpal ligament, although the distal border of the latter has some continuity with the retinaculum. The retinaculum stretches between the ends of the concavity of the carpal bones and converts their arch into a fibro-osseous canal through which pass into the hand a considerable number of tendons and the median nerve. The flexor retinaculum is 2 to 3 cm. long and of almost the same width. It is attached radially to the tuberosity of the scaphoid and to both lips of the groove of the trapezium bone. On the ulnar side of the wrist the ligament attaches to the hook (hamulus) of the hamate and to the pisiform bones. The double attachment radially to both sides of the groove of the trapezium bone is by a Y-shaped split of the ligament. Between the arms of the Y and in the groove of the bone lies the tendon of the flexor carpi radialis muscle enclosed in its own synovial sheath. Through the larger fibro-osseous compartment deep to the retinaculum pass the tendons of the flexor pollicis longus, the flexor digitorum superficialis, and the flexor digitorum profundus muscles. The tendons of the superficialis destined for the third and fourth fingers lie superficially in the compartment; those for the second and fifth digits are more deeply placed. In the deepest part of the compartment the four tendons of flexor digitorum profundus lie side by side. The tendon of flexor pollicis longus passes through the canal to the radial side and deeply in the curve of the carpal bones. The median nerve also passes under the flexor retinaculum. It lies radial to the superficial row of flexor tendons.

Clinical Note Compression of these various structures under the flexor retinaculum, **carpal tunnel syndrome,** leads to paraesthesias and diminution of sensory acuity in the sensory distribution of the median nerve, loss of power and limitation of range of motion on abduction and opposition of the thumb and some wasting of substance of the thenar eminence.

Tendon Compartments of the Dorsum of the Wrist (figs. 80, 90)

The extensor retinaculum attaches deeply to the ridges on the dorsum of the radius. These and attachments to the ulna and to the capsular tissues of the joints separate the tendons which cross the dorsum of the wrist into the hand. Nine of the extensor muscles of the forearm send their tendons into the hand; they are accommodated in six compartments arranged longitudinally on the back of the wrist.

The tendons of the abductor pollicis longus and extensor pollicis brevis muscles occupy the most radial compartment, which is located on the lateral border of the styloid process of the radius. The second compartment occupies the rather smooth surface of the dorsum of the radius to the radial side of its tubercle and accommodates the tendons of the extensor carpi radialis longus and extensor carpi radialis brevis muscles. A slight ridge on the radius in the floor of this compartment guides the two tendons across the dorsum of the radius. The third compartment is occupied by the tendon of the extensor

pollicis longus muscle. This tendon occupies an oblique groove on the dorsal surface of the radius to the ulnar side of its tubercle. The dorsal radial tubercle and the ulnar lip of the groove for the tendon give attachment to deep processes of the extensor retinaculum which define the third compartment. Distal to the radiocarpal joint line, the tendon of the extensor pollicis longus muscle passes diagonally toward the thumb across the tendons of the extensor carpi radialis longus and brevis muscles. Thus, at carpal levels (fig. 90) the second and third compartments and their contents are reversed in their order from the radial side of the wrist. The fourth compartment is large, and the smooth ulnar one-third of the dorsum of the radius is its floor. In this compartment lie the four tendons of the extensor digitorum muscle and, deep to them, the tendon of extensor indicis. The small fifth compartment is located directly over the distal radioulnar articulation and transmits only the tendon of the extensor digiti minimi muscle. The sixth compartment overlies the head of the ulna and is bounded medially by the attachment of the extensor retinaculum to the styloid process of the ulna. The tendon of the extensor carpi ulnaris muscle occupies this compartment.

Synovial Sheaths (figs. 76, 90, 98)

Synovial sheaths, like bursae, reduce the frictional effects of the passage of the tendons through tight compartments or over bony processes. A sheath is formed like a double-walled tube, the delicate inner wall closely attached to the tendon and its outer wall lining the compartment in which the tendon lies. The two layers are continuous with one another at the ends of the tube. Their inner, or facing, surfaces are very smooth, and a small amount of synovial fluid allows easy sliding of one layer on another. Thus, as the tendon moves in its compartment or under a restraining ligament, the actual rubbing surfaces are the smooth, lubricated faces of the synovial sheath. Each of the compartments on the dorsum of the wrist contains a synovial sheath surrounding and investing the tendon or tendons included in the compartment. The upper ends of the sheaths on the dorsum lie at or about the upper border of the extensor retinaculum, and they extend variable distances below it. The synovial sheaths of tendons inserting on the bases of metacarpal bones (those of extensor carpi radialis longus and brevis and of extensor carpi ulnaris) end just short of the insertions of the tendons. The synovial investments of the tendons of the various digital extensor muscles terminate distally over the junction of the proximal and middle thirds of the metacarpal bones (fig. 76). On the palmar aspect of the wrist the tendons of

palmaris longus and flexor carpi ulnaris are not
provided with synovial sheaths. The digital flexors,
however, are protected at the wrist and in the
digits by rather complex synovial coverings. A long
synovial sheath for the tendon of flexor pollicis
longus extends along this tendon from several
centimeters above the flexor retinaculum to just
proximal to its insertion on the distal phalanx of
the thumb. This synovial sheath is called the 'radial
bursa.' The 'ulnar bursa,' the more complex covering of the digital flexor tendons, occupies the center
of the fibro-osseous tunnel of the wrist and extends
above the flexor retinaculum for several centimeters. This sheath exhibits the more primitive
'wrapped around,' or invaginated, character of
synovial sheaths, for the tendons of flexor digitorum superficialis are folded into it superficially
from the radial side and those of flexor digitorum
profundus are folded into the sac from the radial
side more deeply (fig. 90). The general sheath for
these eight tendons continues to about the middle
of the palm and terminates there, except for that
portion concerned with the fifth digit. This ulnar-
most part of the sac continues as far as the insertion of the profundus tendon on the base of the
distal phalanx of the fifth digit. Separate synovial
sheaths invest the digital parts of the tendons of
flexor digitorum superficialis and profundus in the
case of the second, third, and fourth digits (p. 123).

The tendon of flexor carpi radialis occupies a
synovial sheath as it lies in the separate compartment formed by the cleft of the Y-shaped ends of
the flexor retinaculum. This sheath begins several
centimeters above the proximal border of the flexor
retinaculum and extends distally as far as the
insertion of the tendon.

Nerves and Blood Vessels at the Wrist (fig. 90)

The **ulnar nerve and blood vessels** enter the wrist
together to the radial side of the pisiform bone.
They lie under the palmar carpal ligament and are
superficial to the flexor retinaculum as they pass
through the wrist into the palm of the hand. The
deep radial nerve ends in the intercarpal joints and
has no further representation in the hand, for there
is no postaxial musculature in the hand for it to
supply. The **median nerve** passes through the wrist
immediately under the flexor retinaculum, where it
lies to the radial side of the tendon of palmaris
longus and between it and the tendon of the flexor

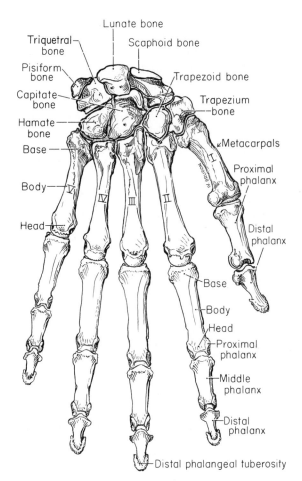

Fig. 91 The dorsal aspect of the bones of the right wrist
and hand. ($\times \frac{1}{2}$)

carpi radialis muscle. The **radial artery** at the wrist
moves from the palmar surface of the radius to
the back of the hand, where it passes deeply between the first two metacarpal bones. In its diagonal course between these points it occupies the
floor of the 'anatomical snuff-box,' the depression
on the radial border of the wrist defined by the
convergence of the tendon of the extensor pollicis
longus muscle and the superimposed tendons of
abductor pollicis longus and extensor pollicis brevis.

Bones of the Wrist and Hand
The Carpal Bones (figs. 91–93, 112)

The skeleton of the wrist consists of eight small
bones, closely held together by ligaments and
arranged in two rows, proximal and distal. The
bones of the proximal row, listed from the radial

to the ulnar side, are the **scaphoid**, the **lunate**, the **triquetrum**, and the **pisiform.** Those of the distal row, listed in the same order, are the **trapezium**, the **trapezoid**, the **capitate**, and the **hamate.** These are all short bones and consist of an outer shell of compact bone enclosing cancellous bony tissue. Fundamentally, they may be thought of as cubes, for each bone has six surfaces. Of these surfaces, the dorsal and palmar are nonarticular and give attachment to the dorsal and palmar ligaments which hold them together. The other surfaces are articular except for those surfaces of the bones of the ends of the rows that represent the borders of the wrist. These, too, serve largely for the attachment of ligaments. In conformity with the radiocarpal curvatures, the proximal articular surfaces are generally convex; the distal surfaces are concave. Nonarticular areas of the bones have foramina for the entrance of blood vessels.

1. The **scaphoid**, the largest bone of the proximal row, is named from its resemblance to a boat. It articulates with the radius proximally, the trapezium and trapezoid bones distally, and the capitate and lunate bones medially.

The smooth **radial articular surface** is convex and triangular in form. The smooth **distal surface** is triangular but concave and has a slight ridge separating a lateral articular surface for the trapezium and a medial one for the trapezoid bone. The **medial surface** presents two articular facets—a superior surface for articulation with the lunate and a larger inferior concavity for part of the head of the capitate bone. The prominent **tubercle** on the palmar aspect of the bone is palpable at the base of the thumb.

2. The **lunate bone** is crescentic in form. It articulates with the radius proximally, the capitate and hamate distally, the scaphoid laterally, and the triquetrum medially.

Its marked **proximal convexity** is for the more medial of the two carpal facets on the distal end of the radius. The **distal surface** presents a deep concavity for articulation with the capitate and for the hamate. The lateral, or radial, aspect of the lunate articulates with the scaphoid bone. A smooth, quadrilateral facet on the **medial surface** is for the base of the triquetral bone.

3. The **triquetrum bone** is pyramidal in form and is so oriented that its base is toward the lunate and its apex is down and ulnarward on the ulnar border of the wrist.

The **base** has a flat, quadrilateral facet matching that of the medial surface of the lunate. The **apex** is roughened for the attachment of the ulnar collateral ligament. The **ulnoproximal surface** presents a convex articular facet for contact with the articular disk of the distal radioulnar joint. The **inferior surface** articulates with the hamate. The **palmar surface** has a conspicuous, oval facet for the pisiform bone.

4. The **pisiform bone** is a small, pea-shaped bone with a single articular facet. It projects palmarward

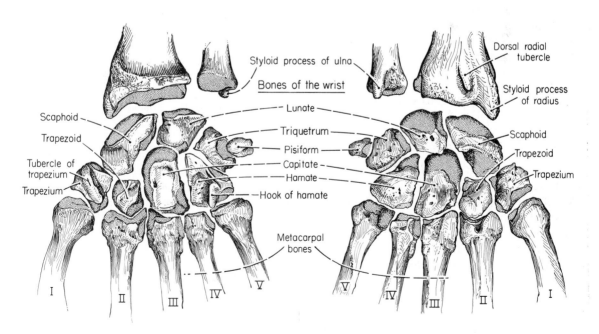

Fig. 92 · An exploded view of the wrist to show the articular surfaces of the bones. Left—palmar view; right—dorsum.

from the triquetral bone at the ulnar side of the wrist and is regarded as a sesamoid bone in the tendon of the flexor carpi ulnaris muscle.

It has a flat **dorsal surface** with a smooth, oval facet for the triquetrum. The palmar surface gives attachment to the flexor retinaculum. The lateral surface is grooved for the ulnar nerve and artery.

5. The **trapezium** is the radialmost bone of the distal row. It articulates with the scaphoid proximally, the first metacarpal distally, and the trapezoid and second metacarpal bones medially.

The **proximal surface** is concave for articulation with the scaphoid. The **distal surface** has a saddle-shaped facet for the base of the first metacarpal bone. The **palmar surface** of the trapezium presents a prominent **tubercle** with a deep groove on its medial side through which passes the tendon of the flexor carpi radialis muscle. The **medial surface** has a facet proximally for the trapezoid and articulates distally with the second metacarpal.

6. The **trapezoid bone** is somewhat wedge-shaped with the broader base of the wedge dorsally and the narrower part palmarward. Its four articular facets for the scaphoid, second metacarpal, trapezium, and capitate bones adjoin one another and are separated by sharp edges.

7. The **capitate** is the largest of the carpal bones and occupies the center of the wrist. It articulates with seven bones: the scaphoid and lunate proximally; the second, third, and fourth metacarpals distally; the trapezoid radially; and the hamate to the ulnar side.

The **proximal surface** of the capitate is its rounded head, which is received into a concavity formed by the scaphoid and the lunate bones. The **distal cuboidal extremity** of the bone is divided into three unequal parts for articulation with the bases of the second, third, and fourth metacarpal bones.

8. The **hamate** is characterized by its wedge-shaped form and by its flattened, hook-like process. The broad end of the wedge is its **distal surface** with two concave facets for the bases of the fourth and fifth metacarpal bones.

The apical **proximal surface** is narrow and convex and articulates with the lunate. The palmar surface gives rise to the curved **hamulus** which is directed forward and lateralward.

The Metacarpal Bones (figs. 91–93)

The metacarpal bones, five in number, form the skeleton of the hand proper. They are miniature 'long' bones, exhibiting a shaft, a head, and a base.

Two to three inches in length, the bones are palpable on the dorsum of the hand and terminate distally in the knuckles which are their heads. They are numbered from the radial to the ulnar side. The **shafts** are curved longitudinally so as to be convex dorsally and concave palmarward. They are approximately prismatic in cross section, presenting medial and lateral surfaces which are separated by a prominent anterior ridge, and are concave for the accommodation of the interosseous muscles. The dorsal surface of the shaft is flattened and triangular in its distal two-thirds, where it is covered by the tendons of the digital extensor muscles. The **head,** or distal extremity, of each metacarpal bone exhibits pits and tubercles at the sides for the attachment of ligaments; its rounded surface is smooth for articulation with the base of the proximal phalanx. The articular surface is continued to the palmar aspect of the head, for flexion of the proximal phalanx carries its distal end well under the head of the metacarpal (which then forms a knuckle of the clenched fist). The articular surface is also convex transversely, but less so than dorsopalmarward, so that the head fails to be a sphere but does allow both flexion and extension and abduction and adduction. The palmar aspect of the articular surface is grooved for the passage of the digital flexor tendons. The **base,** or carpal extremity, of the metacarpal is cuboidal in form and is broader dorsally than palmarward. Its dorsal and palmar surfaces, like the carpals, are rough for ligamentous attachments, and its ends and sides are articular.

The **first metacarpal** bone is shorter and stouter than the others, and its palmar surface faces inward toward the center of the palm. Its proximal surface is concavoconvex (saddle-shaped) for articulation with the trapezium. The head has two articular eminences on its palmar side for the sesamoid bones of the thumb. The **second metacarpal** bone is the longest, and its base is the largest of the metacarpals of the fingers. Its base has a deep groove for articulation with the wedge of the trapezoid bone. The distinguishing characteristic of the **third metacarpal** is its **styloid process,** a dorsally and radially placed proximal eminence. The **fourth and fifth metacarpal** bones are shorter and smaller, and both articulate proximally with the hamate bone.

The Phalanges (figs. 91, 93)

The phalanges, the bones of the fingers, are fourteen in number, two in the thumb and three in each of the other four fingers. Each bone is a

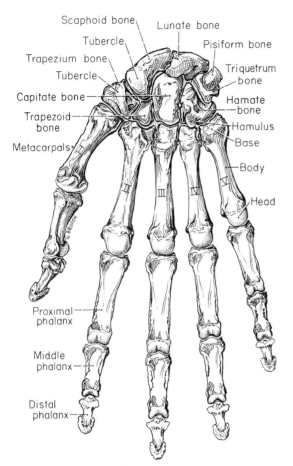

Scaphoid bone
Lunate bone
Tubercle
Pisiform bone
Trapezium bone
Triquetrum bone
Tubercle
Hamate bone
Capitate bone
Hamulus
Trapezoid bone
Base
Metacarpals
Body
Head
Proximal phalanx
Middle phalanx
Distal phalanx

Fig. 93 The palmar aspect of the bones of the right wrist and hand. (×½)

miniature long bone with a shaft and two extremities. The shaft tapers from its proximal to its distal end. Its dorsal surface is markedly convex from side to side; its palmar surface is nearly flat. The margins of the palmar surface are ridged for the attachment of the fibrous sheaths of the digits. The proximal end of the first phalanx of each digit is concave and oval and is broader from side to side for articulation with the head of a metacarpal. The proximal extremities of the bones of the second and third rows have two shallow concavities separated by an intervening ridge, the reverse of a pulley in shape. The distal extremities of the proximal and middle phalanges have the pulley shape to match the reverse-pulley of the proximal articulations just described; that is, two condyles separated by a median groove. The distal phalanges have roughened, elevated surfaces of horseshoe form on their palmar surfaces terminally. These support the pulp of the fingers. The sides of the bases of the

phalanges show tubercles for ligaments; the sides of the heads, except on the distal phalanx, exhibit shallow pits for ligamentous attachments.

Ossification in the Bones of the Wrist and Hand

Ossification of the **carpal bones** takes place from a single center in each bone. The beginning of ossification in these bones occurs in the following order: in the capitate and the hamate early in the first year of life, the capitate preceding; in the triquetrum, during the third year; in the lunate, during the fourth year; in the trapezium, trapezoid, and scaphoid in rather close sequence during the sixth year; and in the pisiform, in the eleventh year. Ossification starts earlier in the female and is completed between the fourteenth and sixteenth years.

The **metacarpal bones** are ossified from two centers: one for the body of the bone and one for the distal extremity in each of the four fingers but for the proximal extremity in the thumb. Ossification begins in the shafts in the eighth or ninth week of fetal life. During the second year the centers for the separate extremity epiphyses appear, and fusion takes place between the sixteenth and eighteenth years.

The **phalanges**, likewise, are ossified from two centers: one for the body and one for the proximal extremity. Ossification of the shafts of the bones begins about the eighth week of fetal life and in the epiphyses during the second and third years. Fusion of shafts and epiphyses occurs between the fourteenth and eighteenth years.

THE HAND

THE BACK OF THE HAND

Superficial Tissues

The **skin** of the back of the hand is thin and fine and has numerous, minute creases which are mainly transverse in orientation. The **subcutaneous connective tissue** is loosely arranged and contains little fat. Within the loose subcutaneous tissue the veins of the dorsum of the hand form the dorsal venous arch. A nonfatty membranous plane of subcutaneous connective tissue exists deep to the looser areolar tissue, and the cutaneous nerves lie in or on this deeper **membranous plane.**

The **fascia of the dorsum of the hand** (fig. 102) is continuous with the antebrachial fascia of the extensor surface of the forearm and of the extensor retinaculum at the wrist. Medially, it is attached to the dorsum of the fifth metacarpal bone and laterally, to the dorsum of the first metacarpal. The fascia also has an attachment to the dorsum of the second metacarpal bone, and over the first interosseous space it blends with the underlying dorsal interosseous fascia. The fascia of the dorsum encloses the tendons of the extensor muscles as they cross the hand toward the fingers and is continued into the extensor expansions on the dorsum of the digits.

Deep to the fascia of the dorsum is an interfascial cleft known as the **subaponeurotic space** (fig. 101). This interval separates the fascia of the dorsum from the deeper dorsal interosseous fascia, and through the space the radial artery runs its diagonal course onto the back of the hand. The **dorsal interosseous fascia** attaches to the dorsal surfaces of the second to the fifth metacarpal bones and covers the intervening dorsal interosseous muscles. It forms the dorsal limit of the interosseous-adductor compartment.

No intrinsic dorsal muscles exist in the hand. Thus the nerves of the dorsum of the hand are entirely cutaneous.

The **arterial system of the dorsum of the hand** (fig. 94) is formed under the extensor tendons, and its descending branches lie deep to the dorsal interosseous fascia. At the wrist the **radial artery** courses diagonally into the hand from a point a little beyond and medial to the styloid process of the radius. Passing under the tendons of the abductor and short extensor muscles of the thumb and over the radial collateral ligament and the scaphoid and trapezium bones it reaches the proximal end of the first dorsal interosseous space. Here it turns deeply into the palm of the hand, where it assists in the formation of the deep palmar arterial arch. As the radial artery passes under the tendon of the abductor pollicis longus muscle, it gives origin to its dorsal carpal branch and, just before it turns deeply through the interosseous muscle, to the first dorsal metacarpal branch.

The **dorsal carpal branch** passes medialward across the distal row of carpal bones under the extensor tendons and joins the dorsal carpal branch of the ulnar artery. This anastomosis, to which terminal twigs of the dorsal branch of the anterior interosseous artery contribute, constitutes the **dorsal carpal arterial arch.** From the arch three **dorsal metacarpal arteries** descend on the interosseous muscles of the second, third, and fourth intermetacarpal intervals. Opposite the heads of the metacarpal bones, the dorsal metacarpal arteries divide into **dorsal digital arteries** which proceed distally along the dorsal borders of contiguous digits. The dorsal metacarpal arteries are joined by branches of the palmar arterial arches in two locations: proximally, they anastomose between the bases of the metacarpals with perforating branches of the deep palmar arch; distally, they are connected by perforating branches with the junction of the palmar metacarpal and common palmar digital arteries of the corresponding interosseous spaces. The **proper dorsal digital arteries** are small and fail to reach the distal phalanges of the digits.

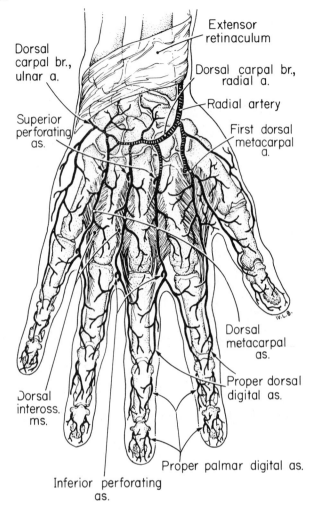

Fig. 94 | The dorsum of the hand; the dorsal interosseous muscles and the dorsal arteries.

They anastomose with proper palmar digital arteries which take the greater share in the supply of the digits. The **first dorsal metacarpal artery** divides similarly into two branches which supply the dorsal surface of the thumb and the radial side of the dorsal surface of the index finger. An arterial twig to the radial border of the thumb frequently arises from the radial artery in its diagonal course, and one to the ulnar border of the dorsum of the fifth digit is provided by the dorsal carpal branch of the ulnar artery.

THE PALM OF THE HAND

Superficial Structures

Attention has been called to the character of the skin and to the flexion creases of the palm of the hand (p. 59). The thick skin is richly supplied with sweat glands but contains no sebaceous glands. A moderate amount of fat underlies the skin of the hand and fingers, enhancing its pliability. All these features contribute to the effectiveness of the hand as a prehensile, or grasping, organ. Also important to this function are the ridges and furrows of the skin surface; they serve to reduce slippage. The ridges are oriented in arches, loops, and whorls, and their unique individual patterns are basic to the science of identification by finger prints (dermatoglyphics, fig. 96). The permanent skin (flexion) creases are due to the firm attachment between skin and deep fascia along the lines of the crease.

The principal cutaneous nerves of the palm of the hand are the **palmar cutaneous branches of the median and ulnar nerves** (p. 63). The **lateral antebrachial cutaneous nerve** (p. 62) terminates on the thenar eminence and communicates with the palmar cutaneous branch of the median nerve. The **superficial radial nerve** (p. 62) has a limited distribution over the thenar eminence and the root of the thumb. The **digital nerves** are more completely described in the section on the cutaneous nerves of the upper limb (p. 64). The **common palmar digital branches** of the median and ulnar nerves pass through the hand under the palmar aponeurosis and divide into proper digital nerves proximal to the webs of the fingers. The **proper digital nerves** are superficial to the corresponding digital arteries and are branches of the ulnar nerve for the fifth digit and the ulnar half of the fourth, and of the median nerve for the other three and one-half digits. Their distribution on the dorsum of the digits is detailed on p. 64.

In the subcutaneous tissue of the hypothenar eminence just beyond the wrist lies the **palmaris brevis muscle**. This is a flat, square muscle which arises at the

ulnar border of the palmar aponeurosis. Its parallel fibers insert into the skin of the ulnar border of the hand, and the muscle improves the grasp by drawing the skin of the ulnar side of the hand inward. The superficial branch of the ulnar nerve and the ulnar artery enter the hand deep to the muscle and supply it on its underside.

The **Compartments and Fascia of the Palm** (fig. 95) The fascia of the palm of the hand is continuous with the anterior antebrachial fascia and with the palmar carpal ligament of the wrist. At the borders of the hand it is continuous with the fascia of the dorsum at the attachments of the latter to metacarpals one and five. Sweeping from each side toward the center of the palm, the palmar fascia provides a tough membranous layer which overlies the muscles of the small finger and of the thumb. The fascia investing the muscles and associated vessels and nerves of the little finger constitutes the **hypothenar fascia** and, by means of a palmar attachment to the radial side of the fifth metacarpal bone, bounds the **hypothenar compartment** of the hand. In a similar way, the fascia over the thumb muscles and associated structures dips deeply to attach to the palmar aspect of the first metacarpal bone and bounds, with the metacarpal, a **thenar compartment** in the hand. This compartment contains the short muscles of the thumb except for the adductor pollicis. The **central compartment** of the palm is covered by a portion of the fascia of the palm which is continuous with the hypothenar and thenar fasciae but is reinforced superficially by the expansion of the tendon fibers of the palmaris longus muscle. This is the **palmar aponeurosis.**

Palmar Aponeurosis (fig. 95) Recognizable in the palmar aponeurosis are a superficial stratum of longitudinally running fibers continuous with the tendon of the palmaris longus muscle and a deeper layer of transverse fibers. These are continuous with the thenar and hypothenar fasciae and, proximally, with the flexor retinaculum and the palmar carpal ligament. The palmar aponeurosis broadens distally in the hand and divides into four digital slips. Some of its fibers attach to the overlying skin at the skin creases of the palm. The central parts of the digital slips pass into the digits, attaching superficially to the skin of the crease at the base of each digit and deeply to the fibrous sheath of the digit. The marginal fibers of the digital slips sink deeply between the heads of the metacarpal bones and attach to the metacarpophalangeal joint capsules, to the deep transverse metacarpal ligaments, and to the proximal phalanges of the digits. A digital slip to the

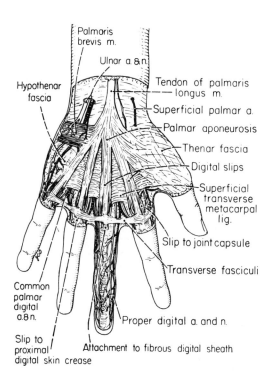

Fig. 95 The palm of the hand; the palmar fascia and the palmar digital arteries and nerves.

thumb is usually wanting, although longitudinal fibers of the palmar aponeurosis curve over onto the thenar fascia.

The deep attachments of the margins of the digital slips of the aponeurosis not only define the entrance to the fibrous sheath of each digit but are continued proximally into the hand for variable distances. These deep septa attach to the palmar interosseous fascia or to the shafts of the meta-carpal bones and provide communicating sub-compartments for each tendon and its lumbrical muscle. That septum passing to the third meta-carpal bone is stronger and more constant and separates a surgical 'thenar space' under the aponeurosis to its radial side and a 'midpalmar space' under the aponeurosis to its ulnar side (fig. 102). These are clinically important spaces because they may become infected and their boundaries will limit the spread of infective material in accordance with the attachments described.

At the approximate level of the transverse creases of the palm, transverse thickenings of the deeper aponeurotic layer appear between the diverging digital slips. Located opposite the heads of the metacarpal bones, these thick-enings are designated as the **superficial transverse meta-carpal ligament.** Distally, the webs of the fingers are reinforced by another accumulation of transverse fibers designated as **transverse fasciculi.**

Intrinsic Muscles of the Hand

All the intrinsic muscles of the hand are situated on its palmar side; all are pre-axial; and all are innervated by either the median nerve or the ulnar nerve. The thenar and hypothenar eminences are due, primarily, to the bulk of the musculature associated with the thumb and little finger, respec-tively, and these muscles occupy the thenar and hypothenar compartments. Both compartments contain an **abductor,** an **opponens,** and a **flexor** muscle for the digit concerned. These muscles also have similarities in origin and insertion in each compartment. *The flexor retinaculum and the bones to which it attaches, the scaphoid and trapezium radially and the hamate and pisiform on the ulnar side, provide the principal sites of origin for these muscles. The insertions of comparable muscles on the two sides of the hand are also similar; namely, the base of the proximal phalanx for the abductor and flexor muscles and the shaft of the metacarpal bone for the opponens.* The central compartment contains four slender **lumbrical muscles** associated with the

flexor digitorum profundus tendon. The **inter-osseous muscles** located in the intervals between the metacarpal bones occupy, with the **adductor pollicis muscle,** a deeply placed **interosseous-adductor com-partment** bounded by the dorsal and palmar inter-osseous fasciae. The general rule of innervation of the intrinsic muscles of the hand may now be stated. *The median nerve supplies the abductor pollicis brevis, opponens pollicis, flexor pollicis brevis, and and the radialmost two of the lumbrical muscles. The ulnar nerve supplies all other intrinsic muscles of the hand.*

The Hypothenar Compartment The hypothenar compartment contains three muscles concerned with the little finger, blood vessels, nerves, and the fifth metacarpal bone.

1. **Abductor digiti minimi** (figs. 96–98, 102) lies superficially along the ulnar border of the hand. It takes origin from the pisiform bone and from the tendon of the flexor carpi ulnaris muscle and ends in a flat tendon which inserts into the ulnar side of the base of the proximal phalanx of the little finger. A tendinous slip blends with the extensor expansion.

2. **Flexor digiti minimi brevis** (figs. 96–98, 102) lies adjacent to the abductor on its radial side. It arises from the hook of the hamate and the adjacent flexor retinaculum and inserts by a tendon which fuses with that of the abductor and ends on the ulnar side of the base of the proximal phalanx of the little finger.

3. **Opponens digiti minimi** (figs. 97, 98, 102) lies deep to the abductor and flexor muscles. It is triangular in form, with an origin from the hook of the hamate and the adjacent part of the flexor retinaculum. The muscle is inserted into the whole length of the ulnar margin of the fifth metacarpal bone.

4. **Vessels and Nerves of the Compartment** (figs. 96, 98) The ulnar nerve and artery cross the wrist against the radial border of the pisiform bone between the palmar carpal ligament and the flexor retinaculum. Here they divide into superficial and deep branches. The superficial branch of the ulnar nerve supplies the palmaris brevis muscle and sends a proper palmar digital branch down the ulnar side of the little finger. A common palmar digital branch for the sensory supply of adjacent sides of the fourth and fifth digits enters the central com-partment of the palm. A proper digital artery for the ulnar side of the fifth digit accompanies the

corresponding branch of the nerve, arising from the superficial part of the ulnar artery. The superficial portion of the ulnar artery then enters the central compartment, where it is the principal contributor to the superficial palmar arterial arch. The deep branches of the ulnar nerve and artery sink between the origins of the abductor digiti minimi and the flexor digiti minimi brevis muscles and perforate the origin of the opponens digiti minimi muscle. They supply these muscles and then pass into the central compartment, where they lie on the interosseous muscles.

The Thenar Compartment The thenar compartment contains three short muscles of the thumb which correspond to those in the hypothenar compartment, blood vessels and nerves, the tendon of the flexor pollicis longus muscle, and the first metacarpal bone.

1. **Abductor pollicis brevis** (figs. 96–98, 102) is a superficially placed, thin, relatively broad muscle which underlies the thenar fascia. It arises, frequently as two separated slips, from the flexor retinaculum, the tuberosity of the scaphoid bone, and the ridge of the trapezium. Its fiber bundles converge into a flat tendon which inserts on the radial side of the base of the proximal phalanx of the thumb and into the border of the tendon of the extensor pollicis longus muscle.

2. **Flexor pollicis brevis** (figs. 96–98, 102) lies to the inner side of and partly overlapped by abductor pollicis brevis. It takes origin from the flexor retinaculum and from the ridge of the trapezium bone and is inserted into the radial side of the base of the proximal phalanx of the thumb. There is a sesamoid bone in its tendon.

3. **Opponens pollicis** (figs. 97, 98, 102) underlies the abductor pollicis brevis and lies to the outer side of the flexor pollicis brevis muscle; it is frequently not well separated from the latter muscle. It arises from the flexor retinaculum and the ridge of the trapezium bone, passes more obliquely radialward than the other muscles, and inserts on the whole length of the radial border of the first metacarpal bone. This type of insertion for the whole length of the shaft of a metacarpal bone is unique to the opponens muscles and is well adapted to drawing the metacarpal toward the center of the palm; this rotates the digit and carries the pad of the thumb into opposing contact with the pads of the other digits.

4. **Nerves and Vessels of the Compartment** (figs.

96, 98) The median nerve passes through the wrist under the flexor retinaculum and enters the central compartment of the palm. At the distal border of the flexor retinaculum a stout branch is given off from its radial side which curves sharply, or 'recurves,' onto the muscles of the thenar eminence. This is the **motor branch of the median nerve** (recurrent nerve) in the hand, and it supplies the short abductor, the short flexor, and the opponens muscles of the thumb. Damage to this superficially placed nerve results in the loss of opposition in the thumb. Awkward and ineffective attempts to substitute for such loss usually involve flexion and adduction (p. 144). The recurrent nerve may arise from the median in common with the first common palmar digital nerve.

The muscles of the thenar eminence are supplied in part by the **superficial palmar branch of the radial artery.** The superficial palmar artery arises from the radial artery just before that vessel turns over the radial collateral ligament to the back of the hand. It passes across the muscles of the thenar eminence, or frequently deep to the abductor pollicis brevis, and supplies these muscles. The artery is usually small, but it contributes to the formation of the superficial palmar arterial arch.

The Central Compartment of the Palm The central compartment of the palm underlies the palmar aponeurosis and contains the superficial palmar arterial arch of the hand, branches of distribution of the median and ulnar nerves, the tendons of the digital flexors, their synovial coverings, and the associated lumbrical muscles. The structures immediately under the palmar aponeurosis are the superficial palmar arterial arch and the median and ulnar nerves.

1. The **superficial palmar arterial arch** (figs. 62, 96, 101) is formed by the distal curved portion of the ulnar artery (p. 118), usually completed by the superficial palmar branch of the radial artery (p. 108). Other vessels which occasionally complete the arch are the radialis indicis branch or the princeps pollicis branch of the radial artery. The arch is convex distalward and crosses the palm at the level of the line of the completely abducted and extended thumb. The artery is accompanied by venae comitantes and lies immediately under the palmar aponeurosis.

The branches of the superficial arch supply the medial three and one-half digits as a rule, the radial one and one-half receiving their blood supply from the

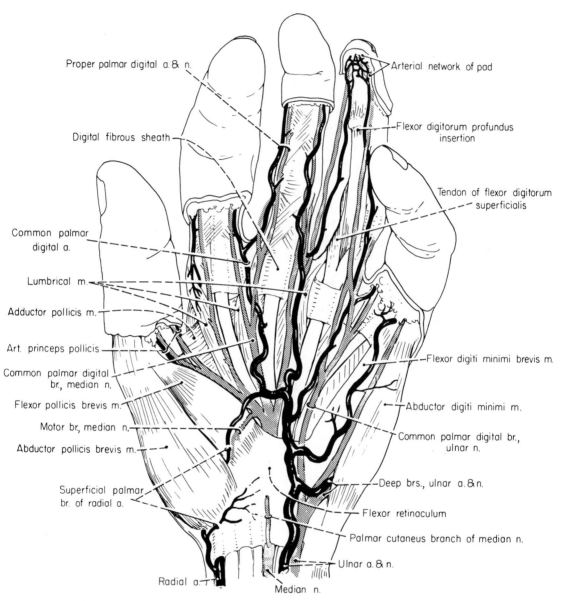

Proper palmar digital a. & n.

Digital fibrous sheath

Arterial network of pad

Flexor digitorum profundus insertion

Tendon of flexor digitorum superficialis

Common palmar digital a.

Lumbrical m.

Adductor pollicis m.

Art. princeps pollicis

Common palmar digital br., median n.

Flexor pollicis brevis m.

Motor br., median n.

Abductor pollicis brevis m.

Superficial palmar br. of radial a.

Flexor digiti minimi brevis m.

Abductor digiti minimi m.

Common palmar digital br., ulnar n.

Deep brs., ulnar a. & n.

Flexor retinaculum

Palmar cutaneus branch of median n.

Ulnar a. & n.

Radial a.

Median n.

Fig. 96   The palm of the hand; the muscles, superficial palmar arterial arch, and superficial nerves.

(Dissection by Prof. T. S. Chelvakumaran.)

120

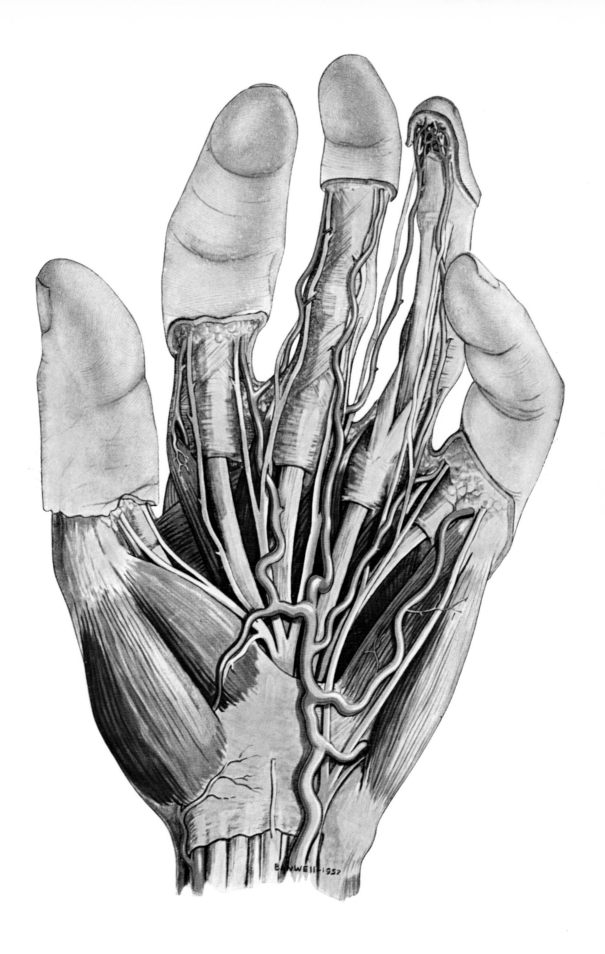

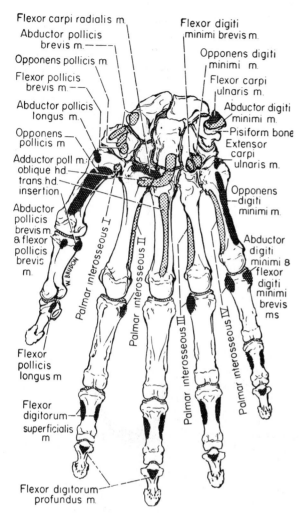

Flexor carpi radialis m.
Abductor pollicis brevis m.
Opponens pollicis m.
Flexor pollicis brevis m.
Abductor pollicis longus m.
Opponens pollicis m
Adductor poll m. oblique hd.
trans. hd.
insertion
Abductor pollicis brevis m. & flexor pollicis brevis m.
Flexor pollicis longus m
Flexor digitorum superficialis m
Flexor digitorum profundus m.

Flexor digiti minimi brevis m.
Opponens digiti minimi m.
Flexor carpi ulnaris m.
Abductor digiti minimi m.
Pisiform bone
Extensor carpi ulnaris m.
Opponens digiti minimi m.
Abductor digiti minimi & flexor digiti minimi brevis ms

Palmar interosseous I
Palmar interosseous II
Palmar interosseous III
Palmar interosseous IV

Fig. 97   The attachments of muscles on the palmar aspect of the bones of the wrist and hand. (Black dots—origin; white dots on black—insertion.)

deep palmar arterial arch. The superficial palmar arch gives origin to three **common palmar digital arteries.** Proceeding distalward on the flexor tendons and the lumbrical muscles and superficial to the digital nerves of the palm, the common palmar digital arteries unite at the webs of the fingers with the palmar metacarpal arteries, which descend from the deep palmar arch, and there also receive the distal perforating branches of the dorsal metacarpal arteries (p. 116). The short trunks so formed divide into **proper palmar digital arteries.** Two proper digital arteries run distalward along adjacent margins of the second and third, third and fourth, and fourth and fifth digits. The proper digital branch to the ulnar side of the fifth digit is a branch of the ulnar artery in the hypothenar compartment (p. 118). As the common palmar digital arteries pass toward the webs of the fingers, they are crossed by common palmar digital branches of the median and ulnar nerves, so that whereas these nerves are deep to the superficial palmar

arterial arch, the digital nerves lie superficial to the arteries at the webs of the digits. Thus, in the fingers the dorsal and palmar digital arteries lie between or in the span of the dorsal and palmar digital nerves.

The proper digital arteries supply the skin, subcutaneous tissue, tendons, joints, and bones of the digits and anastomose across the distal interphalangeal joints to form a terminal plexus in the pad of each digit (fig. 101). They also intercommunicate at other levels of the finger and send branches to the dorsum of the fingers which communicate with the smaller proper dorsal digital arteries and supply the tissues of the last two segments of the digits.

2. **Nerves of the Central Compartment** The nerves of the central compartment are terminal branches of the median and ulnar nerves. The **median nerve** (fig. 96) enters the palm of the hand under the flexor retinaculum radial to the tendon of palmaris longus, and its motor, or recurrent, branch supplies the muscles of the thenar compartment. This branch may arise at the radial side of the median nerve or as a branch of the first common palmar digital nerve. The median nerve divides, as it emerges below the flexor retinaculum, into three common palmar digital nerves.

The **first common palmar digital nerve** divides proximally in the hand into three **proper palmar digital branches.** The most lateral of these crosses obliquely over the tendon of the flexor pollicis longus muscle and runs along the radial border of the thumb. The second branch courses along the medial border of the thumb; the third supplies the radial side of the second digit and gives a twig to the first (radialmost) lumbrical muscle. The **second** and **third common palmar digital branches** of the median nerve pass distalward over the lumbrical muscles and, approaching the webs of the fingers, divide into **proper palmar digital branches** to the adjacent sides of the second and third and third and fourth digits. In the palm the second common digital nerve sends a twig to the second lumbrical muscle, and the third common digital nerve communicates with the neighboring common digital branch of the ulnar nerve. The proper palmar digital branches of the median nerve supply the radial three and one-half digits. Numerous offshoots reach the skin and soft tissues of the palmar portion of the fingers and form a plexus in the pulp of each digit. Dorsal branches curve onto the dorsum over the distal phalanx of the thumb and of the last two segments of the two and one-half fingers supplied (p. 64). The digital nerves are superficial to the digital arteries.

The **ulnar nerve** passes superficially through the wrist and divides into superficial and deep branches at the root of the hypothenar eminence (figs. 96, 98).

The superficial branch provides a proper digital branch to the ulnar side of the little finger (p. 64) and a common palmar digital branch which traverses the central compartment. It communicates with the third common palmar digital branch of the median nerve and divides, proximal to the web, into proper palmar digital branches to the contiguous surfaces of the fourth and fifth fingers. The terminal distribution of these nerves is similar to that described for the digital branches of the median nerve, and they, like the median branches, provide branches to the dorsum of the second and third segments of the fingers supplied. (For the deep branch, see the adductor-interosseous compartment, p. 125)

3. **Flexor Tendons and Sheaths** (figs. 90, 98–100, 102) The tendons of flexor digitorum superficialis and flexor digitorum profundus emerge from the wrist at the distal border of the flexor retinaculum and enter the central compartment of the hand. Here they fan out toward their respective digits, arranged in pairs, superficial and deep. They are invested by the **ulnar bursa** (p. 112, fig. 90) through the upper part of the palm. The general inferior limit of the ulnar bursa is the midpalm (fig. 98), but the portion investing the tendons of the fifth

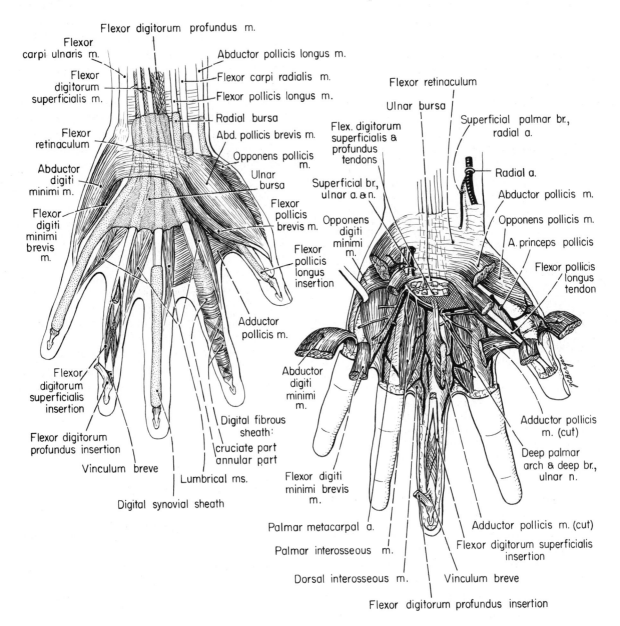

Fig. 98 The palm of the hand; synovial sheaths, tendon insertions, and deep structures.

digit continues to the base of its distal phalanx. The tendons for the other three digits have digital sheaths which, after a short gap in midpalm, continue the investment of these tendons to their insertions, being contained for almost their full length within the fibrous sheaths of the digits (figs. 98, 99). In accordance with the customary disposition of synovial sheaths, the digital sheaths begin a short distance (about 5 mm.) proximal to the entrance of the tendons into their fibrous sheaths.

The **fibrous sheaths of the digits of the hand** are strong coverings of the flexor tendons which extend from the heads of the metacarpals to the base of the distal phalanges and prevent the tendons from 'bowstringing' away from the bones of the digits (figs. 96, 98–100). They attach along the borders of the proximal and middle phalanges, to the capsules of the interphalangeal joints, and to the surface of the distal phalanx. Arching over the tendons, they form strong semicylindrical sheaths which, with the bones, produce fibro-osseous tunnels through which the tendons pass to their insertions. Over the shafts of the proximal and middle phalanges the sheaths exhibit thick accumulations of transversely running fibers (sometimes designated as 'pulleys'), whereas opposite the joints an obliquely crisscrossing arrangement is characteristic. These cruciate portions are thin and do not interfere with flexion of the joints. Proximally, the digital slips of the palmar aponeurosis attach to the digital sheaths.

Clinical Note Penetrating wounds of the fingers or hand may introduce infection into the synovial sheaths or the fibrous flexor sheaths, or the thenar or midpalmar spaces. Such infection will be limited, at least initially, by the boundaries of such specific sheath or space, or its spread may be facilitated by the communications such sheaths or spaces exhibit. Review these limitations or communications pp. 118, 122. **Dupytren's contracture** is a pathological thickening and contracture of fibers of the palmar aponeurosis. It produces marked tension on the digits with flexion, the digital tendons becoming like bowstrings; the hand may become practically useless.

4. **Lumbrical Muscles** (figs. 96, 98, 100, 102) The lumbrical muscles are four small, cylindrical muscles associated with the tendons of flexor digitorum profundus. The two lateral muscles arise just distal to the flexor retinaculum from the radial sides and palmar surfaces of the flexor digitorum tendons for the second and third digits. The two medial muscles arise from the contiguous sides of the tendons for the third and fourth and for the fourth and fifth digits. Each lumbrical tendon passes distalward on the palmar side of the deep transverse metacarpal ligament and shifts toward the dorsum past the metacarpophalangeal joint. Fanning out at the end, it inserts, at the level of the proximal phalanx, into the radial border of the expansion of the tendon of the extensor digitorum muscle (fig. 77).

The Interosseous-adductor Compartment (figs. 98, 100, 102) The interosseous-adductor compartment represents the deepest muscular plane of the palm of the hand. It is enclosed by two fascial membranes continuous with each other at the borders of the compartment. The **dorsal interosseous fascia** attaches to the periosteum of the dorsal surfaces of metacarpal bones one to five and invests the intervening surfaces of the dorsal interosseous muscles. The **palmar interosseous fascia** attaches to the same metacarpals and covers the interosseous musculature on the palmar side. The contents of the compartment are the deep palmar arterial arch and the deep branch of the ulnar nerve, the adductor pollicis muscle, the dorsal and palmar interosseous muscles, and the arterial arch of the dorsum of the hand.

1. The **adductor pollicis muscle** (figs. 96–98, 102) has two heads of origin separated by a gap through which the radial artery enters the palm of the hand. The **oblique head** arises from the capitate bone and from the bases of the second and third metacarpal bones. The **transverse head** arises from the palmar ridge of the third metacarpal bone. The two heads insert together by a tendon which ends in the ulnar side of the base of the proximal phalanx of the thumb. This tendon usually contains a sesamoid bone, which, with that developed in the tendon of

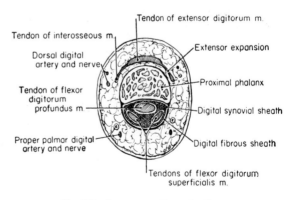

Fig. 99   A cross section of a finger.

the flexor pollicis brevis muscle, forms a pair of small sesamoid bones on either side of the tendon of flexor pollicis longus. The adductor pollicis lies deep to the long flexor tendons of the palm and overlies the interosseous muscles on the radial side of the third metacarpal bone.

2. The **deep palmar arterial arch** (figs. 98, 102) is formed by the terminal portion of the radial artery in conjunction with the deep branch of the ulnar artery. The radial artery enters the palm at the base of the first intermetacarpal space by passing between the two heads of origin of the first dorsal interosseous muscle. It then turns medially between the transverse and oblique heads of the adductor pollicis muscle, and, continuing with a slight downward convexity across the interosseous muscles, it joins the deep branch of the ulnar artery. The deep branch of the ulnar artery arises just distal to the pisiform bone and sinks deeply between the short abductor and flexor muscles of the fifth digit. Perforating the opponens digiti minimi muscle, it turns radially to anastomose with the radial artery. The deep palmar arterial arch lies deeper and more proximally in the palm than the superficial arch. It occupies the most proximal part of the cup of the palm and lies below the long flexor tendons under cover of the palmar interosseous fascia. The artery is accompanied by two venae comitantes and the deep branch of the ulnar nerve. The branches of the deep palmar arch are the

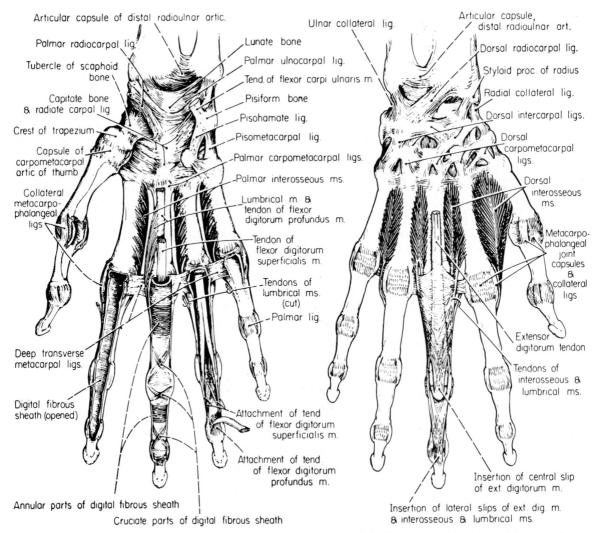

Fig. 100   The interosseous muscles, tendons and tendon sheaths, and the ligaments of the joints of the wrist, hand, and fingers.

princeps pollicis, the radialis indicis, three palmar metacarpals, the recurrent carpal, and the dorsal perforating.

(a) The **arteria princeps pollicis** (fig. 98) arises from the radial after that artery has traversed the interval between the two heads of origin of the first dorsal interosseous muscle. It descends along the ulnar border of the first metacarpal bone under the tendon of flexor pollicis longus, and at the head of the metacarpal it divides into two branches. These emerge at the border of the adductor pollicis and continue along the palmar borders of the thumb. They form a terminal network in the subcutaneous tissue of the pad of the distal segment of the thumb and distribute in other respects as do the proper palmar digital arteries of the fingers.

(b) The **arteria radialis indicis** (fig. 98) arises just beyond or in common with the princeps pollicis branch and descends between the first dorsal interosseous muscle and the transverse head of the adductor pollicis muscle. The artery runs along the radial side of the index finger, anastomosing in its terminal pulp and over its joints with the proper digital artery on the ulnar side of the digit. It frequently communicates at the lower border of the adductor pollicis with the superficial palmar arch, and sometimes it is concerned in the completion of this arch.

(c) The three **palmar metacarpal arteries** arise from the convexity of the arch and descend under the palmar interosseous fascia of the second, third, and fourth intermetacarpal intervals. They supply the interosseous muscles, the bones, and the lumbrical muscles and end near the webs of the fingers by joining the common palmar digital arteries from the superficial palmar arterial arch.

(d) The **recurrent carpal branches** are small. They ascend from the concavity of the arch and anastomose at the level of the wrist with the anterior interosseous artery and with the palmar carpal branches of the radial and ulnar arteries. This anastomosis forms the palmar carpal network.

(e) The **perforating branches,** three in number, pass dorsally between the heads of origin of the dorsal interosseous muscles of the second, third, and fourth interosseous spaces to anastomose with the dorsal metacarpal arteries from the arterial arch of the dorsum p. 116).

3. The **deep branch of the ulnar nerve** (figs. 96, 98), arising along the radial side of the pisiform bone, sinks into the palm between the origins of the abductor digiti minimi and flexor digiti minimi brevis muscles. It perforates the origin of the opponens digiti minimi and follows the course of the deep palmar arch across the interossei deep to the palmar interosseous fascia. In its proximal course it supplies the abductor digiti minimi, the flexor digiti minimi brevis, and the opponens digiti minimi muscles. As it crosses the hand, it supplies the third and fourth lumbrical muscles and all the interossei, both palmar and dorsal. Radially, it supplies the adductor pollicis muscle. The nerve also provides articular branches for the wrist joint.

4. The **interosseous muscles** (figs. 6, 94, 98, 100, 102, 103) are located in the intermetacarpal intervals and are of two types, palmar and dorsal. They are arranged side by side, one palmar and one dorsal muscle in each intermetacarpal space, but only the dorsal muscles are apparent on the dorsum of the hand. The four dorsal interossei are abductors of the digits and are bipennate in form. The four palmar interossei are adductors of the

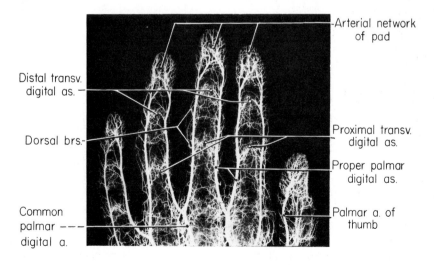

Fig. 101 Detail of small digital arteries. (Radiogram supplied by Dr. E.A. Edwards. Ref.—E.A. Edwards, 1960. 'Organization of the small arteries of the hand and digits.' *Am. J. Surg.,* 99:837.)

digits and are unipennate. The plane of reference for abduction and adduction of the digits is the midplane of the third digit. This is evident on simultaneously spreading and then approximating the extended digits.

Consideration of this plane of reference and the primary functions of the muscles allows the logical placing of each palmar and dorsal interosseous muscle. The requirements of abduction and adduction for five fingers are met in ten muscles. Of these, three special abductors and adductors exist in the abductor pollicis brevis, the adductor pollicis, and the abductor digiti minimi muscles. The requirement of seven other muscles is satisfied by the eight interossei, reduced by one, since the first palmar interosseous muscle parallels and inserts through the tendon of the adductor pollicis. The placing of the interosseous muscles may be begun with reference to the third digit. Since movement of this digit to either side takes it away from the plane of reference, abductor muscles are placed on either side. Thus, dorsal interosseous muscles in the second and third intermetacarpal intervals send their tendons to the base of the proximal phalanx of the third digit on both its radial and ulnar sides. The first dorsal interosseous is placed between the first and second metacarpal bones, inserting

on the radial side of the base of the proximal phalanx of the second digit; the fourth lies in the interval between the fourth and fifth metacarpal bones and has an insertion on the ulnar side of the base of the fourth digit. The palmar interossei are arranged to draw digits two, four, and five toward the plane of reference. The second inserts on the ulnar side of the base of the proximal phalanx of the second digit. The third and fourth palmar interossei insert on the radial side of the proximal phalanges of the fourth and fifth digits.

The bipennate **dorsal interossei** arise by two heads from the adjacent sides of the metacarpal bones between which they lie, but have a more extensive origin from the metacarpal bone of the digit into which they insert. The first dorsal interosseous muscle is large, and a fibrous arch between its heads allows passage of the radial artery into the palm. Between the heads of origin of the other dorsal muscles pass the dorsal perforating arteries, and in the crease formed by their central tendon run the dorsal metacarpal arteries from the arterial arch of the dorsum (fig. 94). These muscles are covered by the dorsal interosseous fascia.

The smaller, unipennate **palmar interossei** arise from the palmar surfaces of the first, second, fourth, and fifth metacarpal bones. Each arises from the entire length of the metacarpal and is inserted into the base of the first phalanx of the same digit. The palmar interossei are covered by the palmar interosseous fascia, and on them course the palmar metacarpal branches of the deep palmar arterial arch. The tendons of both the palmar and dorsal interosseous muscles pass dorsal to the deep transverse metacarpal ligaments and have two insertions. The insertion to the base of the proximal phalanx is concerned in the abduction-adduction function of the muscles. The other insertion, into the extensor expansion of the tendon of extensor digitorum (p. 97), produces flexion at the metacarpophalangeal joints and extension of the middle and distal phalanges at the interphalangeal joints.

5. The **arterial arch of the dorsum of the hand** and its branches of distribution underlie the dorsal interosseous fascia and thus belong in the interosseous-adductor compartment. They are described in the section on the dorsum of the hand (p. 116).

Actions of Muscles in Finger Movement

The participation of forearm muscles in finger movement was introduced in the discussion of the actions of these muscles (p. 102). In gross flexion and extension of all fingers together, the forearm muscles are in control, whereas the intrinsic hand muscles enter into activity in departures from simple opening and closing. The interosseous muscles act most effectively when there is combined metacarpophalangeal flexion and interphalangeal extension, principally producing interphalangeal extension. The lumbrical muscles are

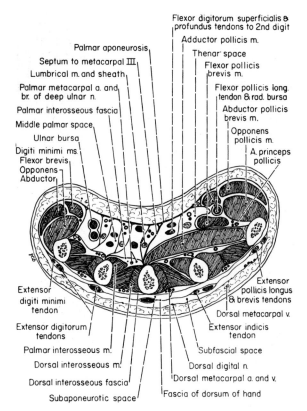

Flexor digitorum superficialis &
profundus tendons to 2nd digit
Adductor pollicis m.
Thenar space
Flexor pollicis brevis m.
Flexor pollicis long.
tendon & rad. bursa
Abductor pollicis brevis m.
Opponens pollicis m.
A. princeps pollicis

Palmar aponeurosis
Septum to metacarpal III
Lumbrical m. and sheath
Palmar metacarpal a. and br. of deep ulnar n.
Palmar interosseous fascia
Middle palmar space
Ulnar bursa
Digiti minimi ms.
Flexor brevis
Opponens
Abductor

Extensor digiti minimi tendon
Extensor digitorum tendons
Palmar interosseous m.
Dorsal interosseous m.
Dorsal interosseous fascia
Subaponeurotic space

Extensor pollicis longus & brevis tendons
Dorsal metacarpal v.
Extensor indicis tendon
Subfascial space
Dorsal digital n.
Dorsal metacarpal a. and v.
Fascia of dorsum of hand

Fig. 102 A cross section of the hand proximal to the separation of the thumb.

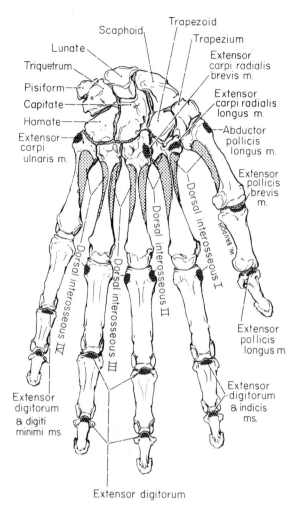

Fig. 103 The attachments of muscles on the dorsal aspect of the bones of the wrist and hand. (Code as fig. 96.)

rotated position of the metacarpal whereby its palmar surface is directed medially. The short abductor also assists in flexion. The opponens acts solely on the metacarpal of the thumb, drawing it across the palm and rotating it medially. The components of opposition are abduction, flexion, and medial rotation, the action bringing the tip of the thumb into contact with the pads of the other slightly flexed fingers. This is the single most distinctive digital movement and gives man his characteristic tool-making and tool-using capacity. In firm grasp, the flexor pollicis brevis is especially active. The motor or recurrent branch of the median nerve innervates the three muscles involved. Adductor pollicis adducts the thumb. The short muscles of the little finger are much less important than those of thumb. An abductor and a flexor produce their characteristic movements. The opponens digiti minimi rotates the fifth metacarpal bone medially and deepens the hollow of the hand.

silent during total flexion but are very active in extension of the proximal or distal interphalangeal joints and when these joints are being maintained in extension during metacarpophalangeal flexion (Basmajian).

Free movement of the thumb is most important in the more precise activity of the hand (fig. 104). The flexor pollicis longus flexes the thumb and extensor pollicis longus and brevis extend it. Abductor pollicis longus abducts and extends the metacarpal bone of the thumb. It is an accessory flexor of the wrist. The short muscles of the thumb provide flexion, abduction and adduction, and opposition. Abduction of the thumb carries it anteriorly out of the plane of the palm due to the

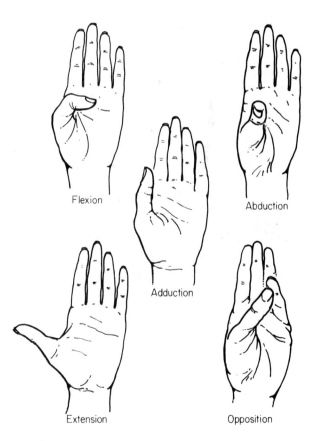

Fig. 104 The various movements of the thumb.

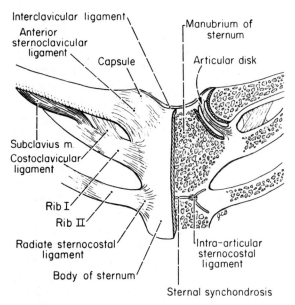

Fig. 105 The sternoclavicular joint and the articulations of the first two ribs with the sternum.

THE ARTICULATIONS OF THE UPPER LIMB

The articulations of the upper limb include those of the shoulder girdle as well as those of the limb proper. They may be listed as follows:

1. Sternoclavicular
2. Acromioclavicular
3. Shoulder
4. Elbow
5. Radioulnar
6. Wrist
7. Intercarpal
8. Carpometacarpal
9. Intermetacarpal
10. Metacarpophalangeal
11. Interphalangeal

THE STERNOCLAVICULAR JOINT (figs. 27, 105)

The sternoclavicular joint represents the only point of bony connection between the trunk and the upper limb. The scapula articulates with the clavicle at the acromioclavicular joint, but it is joined to the trunk by muscles only. The sternoclavicular joint has the movements, though not the form, of a ball and socket joint. Its bony surfaces are rather incongruent, but an articular disk interposed between them increases the capacity for movement in the joint. In this articulation the sternal end of the clavicle is applied to an articular fossa formed by the superolateral angle of the manubrium sterni and the medial part of the cartilage of the first rib. The clavicle is further united to the cartilage of the first rib by a strong ligament. The articular cartilage of this joint is partially fibrous in

character. An **articular capsule** surrounds the joint and is attached around the clavicular and the sternochondral articular surfaces. It is weak below but is reinforced above, in front, and behind by the capsular ligaments.

The **anterior sternoclavicular ligament** is a broad band of fibers applied to the capsule anteriorly. It is attached above to the upper and anterior aspect of the sternal end of the clavicle and passes obliquely across the joint downward and medialward to the front of the upper part of the manubrium sterni. It is a strong band reinforced anteriorly by the tendinous origin of the sternocleidomastoid muscle.

The **posterior sternoclavicular ligament** has similar attachments and a similar orientation. Behind it, lie the sternohyoid and sternothyroid muscles and certain of the major vessels arching into the base of the neck—the left brachiocephalic vein, the left common carotid artery, and the brachiocephalic artery.

The **interclavicular ligament** strengthens the capsule above. It passes from the sternal end of one clavicle to the sternal end of the other and is attached to the upper border of the sternum between the clavicles.

The **costoclavicular ligament** is a short, flat, dense band of fibers, attached below to the upper medial part of the cartilage of the first rib. It passes upward, backward, and lateralward to the costal tuberosity on the under surface of the clavicle. The tendon of the subclavius muscle inserts in front of the ligament; the subclavian vein is behind it.

The **articular disk** is a flat, circular disk of fibrocartilage which separates the joint into two synovial cavities. Occasionally, perforation of its thinner central part places the two cavities into communication. The disk is attached at its circumference to the capsule of the

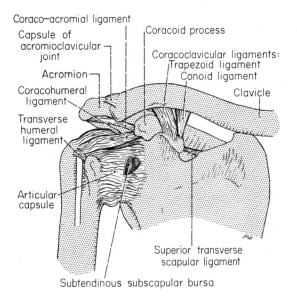

Fig. 106 The shoulder joint and the acromioclavicular articulation as seen from in front

joint, but, more importantly, it has a superior and posterior attachment to the upper border of the sternal end of the clavicle and ends below in the cartilage of the first rib. The disk exerts a cushioning effect on forces transmitted from the limb and compensates for the incongruity of the articulating surfaces. The disk also has an important ligamentous action. Fibrocartilaginous in the main, it is fibrous, or ligamentous, at its circumference and holds the sternal end of the clavicle down in its articular fossa. It resists the tendency for the sternal end of the clavicle to be displaced upward and medialward by strong, thrusting forces transmitted from the limb or induced when the limb is carried forcibly downward.

The **synovial membranes** of the joint are two in number. The looser lateral membrane reflects from the articular margin of the sternal end of the clavicle to the margins of the disk. The medial one lines the capsule between its sternal attachments and the disk. Synovial membranes are generally found as lining membranes of fibrous capsules but do not exist over the actual articulating surfaces of bones or disks.

The **arterial supply** of the sternoclavicular joint is provided by branches of the internal thoracic artery, of the superior thoracic branch of the axillary artery, of the clavicular branch of the thoracoacromial artery, and of the suprascapular artery. The **nerves** of the joint are derived from the anterior supraclavicular nerve.

THE ACROMIOCLAVICULAR JOINT (figs. 106, 107)

This is a small synovial joint in which the oval articular surfaces of the lateral end of the clavicle and of the medial border of the acromion articulate. The plane joint surfaces, covered by fibrous articular cartilage, slope downward and medially. The end of the clavicle rides higher than the adjacent surface of the acromion, and the slope of the surfaces tends to favor displacement of the acromion downward and under the clavicle. An **articular capsule** encloses the joint, attaching at the articular margins. It is relatively loose and is composed of strong, coarse fibers arranged in parallel fasciculi. The capsule is stoutest above, where it is reinforced by the fibers of the trapezius muscle. A wedge-shaped **articular disk** dips into the joint from the superior part of the capsule, partially dividing the joint cavity. A complete partition of the joint is rare. The **synovial membrane** lines the inner surface of the capsule.

The **coracoclavicular ligament** provides the real stability of the acromioclavicular joint. It is a very powerful, bipartite band which anchors the clavicle to the coracoid process medial to the joint and, since the coracoid and acromion have bony continuity, stabilizes the relationship of the clavicle and the acromion. Its two parts, named according to their shapes, are the posteromedial conoid ligament and the anterolaterally placed trapezoid ligament. The **conoid ligament** is triangular in form and has its apex downward, attached posteromedially to the base of the coracoid process. Its broadened base is fixed to the conoid tubercle on the under surface of the clavicle. The **trapezoid ligament** is strong, flat, and quadrilateral. It is attached below for about 2 cm. to a rough ridge on the upper surface of the coracoid process. Above, it is attached to the oblique **trapezoid line** on the under surface of the clavicle which runs anterolaterally from the conoid tubercle. The conoid ligament restrains backward movement of the scapula. The trapezoid ligament prevents excessive forward movement of the scapula and is especially important in resisting displacing forces which would cause the acromion to be carried down and under the clavicle.

Movements of the Shoulder Girdle

It has been pointed out that the clavicle forms a strut, holding the point of the shoulder well out from the trunk, thereby facilitating its free movement. It has also been noted that the sternoclavicular joint is the only point of bony union between the limb and the trunk. Thus, the clavicle represents a radius about which the shoulder may be moved, the sternoclavicular joint being its pivot point. A rigid union between clavicle and scapula would limit the shoulder, so that the glenoid cavity would

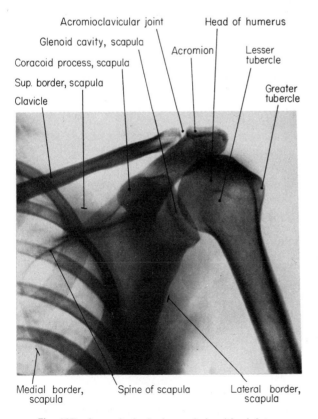

Acromioclavicular joint   Head of humerus

Glenoid cavity, scapula

Coracoid process, scapula   Acromion   Lesser tubercle

Sup. border, scapula

Clavicle   Greater tubercle

Medial border, scapula   Spine of scapula   Lateral border, scapula

Fig. 107   Acromioclavicular and shoulder joints.

always point in line with the clavicle. Actually, the 15° to 20° of movement allowed at the acromioclavicular joint is sufficient to permit the independent turning of the glenoid fossa in directions favorable for the humerus and the remainder of the limb. The acromion may glide forward or backward or may be turned upward or downward at the acromioclavicular articulation. Many of the axes around which movement takes place in this joint fall within the coracoclavicular ligament. At the sternoclavicular articulation the clavicle may be elevated or depressed, carried forward or backward, or rotated. The combination movement of circumduction is also possible. The freedom of movement of the clavicle is considerable. In full elevation of the upper limb, the clavicle may be raised to approximately a 60° angle from its usual horizontal position. The complex, muscular suspension and the numerous muscle attachments of the scapula make this bone primary in the movements of the shoulder girdle; the clavicle and its joints mainly accommodate passively to the positions and orientation of the scapula, but the last 30 degrees of scapular rotation is allowed by rotation of the clavicle on its long axis. The extension of certain muscle insertions from the scapula to the clavicle (trapezius), however, indicate the active participation of the clavicle in elevation of the shoulder. The principal movements of the shoulder and the muscles producing them may be summarized in tabular form as below.

Scapular Ligaments

Coracoacromial Ligament (fig. 106) This strong triangular ligament stretched between the acromion and the coracoid process completes, with the bones, a bony and ligamentous arch above the head of the humerus. Its broad base attaches to the lateral border of the coracoid process; a blunt apex is fixed to the tip of the acromion. It is related above to the clavicle and the deltoid muscles and, below, to the subacromial bursa and the tendon of the supraspinatus. The ligament is thicker at its margins, and thus anterior and posterior bands may be defined. Occasionally, an actual central gap in the ligament occurs.

Superior Transverse Scapular Ligament (fig. 106) This ligament converts the scapular notch into a foramen. It is a thin, flat band attached at one extremity to the base of the coracoid process and at the other to the medial border of the scapular notch. The suprascapular nerve runs beneath the ligament; the suprascapular vessels pass over it. The ligament is sometimes ossified.

Inferior Transverse Scapular Ligament This ligament, when present, extends from the lateral aspect of the root of the spine of the scapula to the margin of the glenoid process. With the bone, it forms a foramen for the passage of the suprascapular vessels and nerve into the infraspinatous fossa.

THE SHOULDER JOINT (figs. 43, 106–108)

The scapulohumeral joint is a ball and socket joint. It has the greatest freedom of movement of

Movement	Muscles Concerned
1. Elevation of scapula	Upper fibers of trapezius Levator scapulae Upper fascicles of serratus anterior
2. Depression of scapula	Passively, the weight of the limb Lower fibers of trapezius Actively, pectoralis minor Acting through the humerus, pectoralis major and latissimus dorsi
3. Upward rotation of the point of the shoulder and elevation of the scapula (movement usually combined with flexion or abduction of shoulder joint in raising hands above head)	Force couple of upper and lower fibers of trapezius carrying acromion upward and inferior angle outward Serratus anterior Levator scapulae
4. Downward rotation of the point of the shoulder and depression of the scapula	Pectoralis minor, rhomboids, and middle portion of trapezius Acting through humerus, pectoralis major and latissimus dorsi
5. Forward movement of the point of the shoulder	Serratus anterior, pectoralis minor; pectoralis major acting on the humerus
6. Retraction or backward movement of the point of the shoulder	Middle portion of trapezius, rhomboids Latissimus dorsi through action on humerus

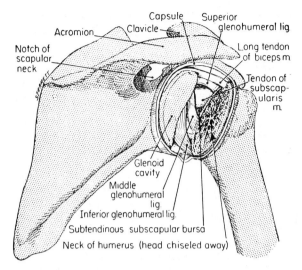

Fig. 108 The shoulder joint opened from behind; the glenohumeral ligaments.

all the joints of the body, a freedom which is inevitably accompanied by a considerable loss in stability. The large head of the humerus articulates against a glenoidal surface only a little more than one-third its size, and the articular capsule is loose. The head of the humerus is held into the glenoid cavity by an 'articular cuff' of short scapular muscles—supraspinatus, infraspinatus, teres minor, and subscapularis. The tendons of these muscles blend with the capsule and reinforce it, and when their attachments are cut, the humeral head is allowed to fall away from the glenoid surface by about 2 cm. The glenoid cavity of the scapula is slightly deepened and enlarged by a **glenoidal labrum,** or fibrocartilaginous rim, attached around its margin. The labrum is triangular in cross section with a free edge which is thin and sharp. The long tendon of the biceps muscle arises within the joint from the supraglenoid tuberosity and there blends with the fibrous tissue of the labrum.

The **articular capsule** of the shoulder joint forms a loose, cylindrical sleeve enclosing the articular parts of the bones. The capsule attaches to the scapula outside of the glenoid labrum and partly to the labrum itself, especially above and behind. On the humerus the capsule attaches to the anatomical neck immediately medial to the tubercles, except below where it extends onto the medial surface of the shaft of the bone a little below the articular head. With the arm at the side the lower part of the capsule is lax and redundant, but it becomes tense in full abduction. The tendons of the

short scapular muscles spread out over the capsule and blend with it proximal to their actual bony attachments, thus imparting strength to the capsule. The fibers of the capsule run generally in a longitudinal direction from bone to bone, but some oblique and transversely running fibers also occur. There are two openings in the capsule. One, at the upper end of the intertubercular groove, allows for the passage of the tendon of the long head of the biceps muscle. Here a fibrous band frequently stretches between the greater and lesser tubercles, arching over the tendon as it emerges from the capsule. The other opening of the articular capsule is a communication anteriorly between the synovial cavity of the joint and the subscapular bursa.

The **synovial membrane** extends from the margin of the glenoid cavity, lines the capsule, and reflects back onto the lower part and sides of the anatomical neck of the humerus to the limits of the articular cartilage. The tendon of the long head of the biceps muscle is enclosed in a tubular sheath of synovial membrane which is continuous with the general synovial lining of the joint at its attached end. This intertubercular synovial sheath extends along the biceps tendon through the intertubercular groove as far as the surgical neck of the humerus.

The **glenohumeral ligaments** (fig. 108) consist of three strengthening bands within the capsule. These thickenings do not appear on the outside of the capsule but may be seen on its inner aspect. All three lie in the anterior wall of the capsule and radiate from the anterior glenoid margin adjacent to and extending downward from the supraglenoid tuberosity. The **superior glenohumeral ligament** is slender, arises immediately anterior to the attachment of the biceps tendon, and parallels it to end near the upper surface of the lesser tubercle of the humerus. The **middle ligament** arises next to the superior and reaches the humerus at the front of the lesser tubercle just inferior to the insertion of subscapularis. It is usually well marked by virtue of its position obliquely under the opening of the subscapular bursa. The **inferior glenohumeral ligament** attaches to the scapula immediately below the notch in the anterior border of the glenoid process and descends obliquely to the humerus on the under surface of the neck. The middle and inferior ligaments are frequently only poorly separated.

The **coracohumeral ligament** is partly continuous with the capsule and is partly separated from it. It is a broad band which arises from the lateral border of the coracoid process. Flattening, it blends with the upper and posterior part of the capsule and ends in the anatomical neck of the humerus adjacent to the greater tubercle.

The **arterial supply** of the joint is composed of branches of the suprascapular branch of the subclavian artery and of axillary branches—anterior and posterior circumflex humeral and circumflex scapular. Its **nerves** are branches of the suprascapular, axillary, and subscapular nerves.

Movements of the Shoulder Joint

The architecture of the shoulder joint—its large humeral head, its small glenoid cavity, its loose capsule, and the numerous muscles acting on it —adapts the joint to great freedom of motion. As a ball and socket joint, movement can take place around an infinite number of axes intersecting in the globular head of the humerus. It is only due to the limitations of description that transverse, anteroposterior, and vertical axes are used to define the conventional movements of flexion and extension, abduction and adduction, medial and lateral rotation. Rotation of the humerus may be demonstrated by locking the elbow in flexion, holding the arm close to the side of the body, and moving the hand and forearm out away from the trunk and back again. In this position rotation of 170° is possible, but in other positions of the arm this motion may be much restricted. The combination movement of circumduction is also free. The principal muscles serving to move the humerus according to the conventional analysis follow in tabular form.

Flexion	Extension
Deltoid (anterior fibers)	Latissimus dorsi
Pectoralis major	Teres major
(clavicular fibers)	Deltoid (posterior fibers)
Biceps	Triceps (long head)
Coracobrachialis	Subscapularis

Abduction	Adduction
Supraspinatus	Pectoralis major
Deltoid (middle fibers)	Latissimus dorsi
	Teres major
	Subscapularis
	Coracobrachialis

Medial Rotation	Lateral Rotation
Subscapularis	Infraspinatus
Teres major	Teres minor
Latissimus dorsi	Deltoid (posterior fibers)
Pectoralis major	
Deltoid (anterior fibers)	

Such a table presents a rather simplified and arbitrary analysis of movement at this as at other joints, for the placement of the limb overhead or in markedly flexed, extended, or rotated positions makes additional muscles of importance in the action and, in addition, changes many of the above muscles from one listing to another. It is also worthy of note that muscles not listed in each classification act as fixators or synergists in the action under considera-

tion. Thus in abduction subscapularis, infraspinatus, and teres minor hold the head of the humerus closely in and down against the glenoid process and give important assistance to supraspinatus and deltoideus in effecting rotation of the head around the anteroposterior axis. The long tendon of the biceps muscle also assists in holding the humeral head in the glenoid fossa, prevents its being drawn too forcibly upward against the overlying acromion, and steadies the humerus in various movements of the arm and forearm. Running an intra-articular course nearly parallel to that of the coracohumeral ligament, it strengthens the upper part of the capsule.

The correlated activity of the shoulder girdle in movements of the shoulder must be clearly recognized. In abduction of the arm, a brief scapulohumeral phase up to 30° may be recognized, but between 30° and 170° the rotation of the shoulder girdle augments that of the scapulohumeral articulation in the ratio of 1:2. The total range of shoulder joint movement is 120°, and that of the scapula is about 60° (Inman, Saunders, & Abbot, '44). The description of scapular motion as it affects positioning of the arm is more completely given in reference to the joints of the shoulder girdle (p. 131).

THE ELBOW JOINT (figs. 109, 110)

The elbow joint is essentially a hinge joint (ginglymus). In it, the bones of the forearm articulate with the lower end of the humerus as the humeroulnar and humeroradial articulations. However, the elbow joint also includes within a common articular capsule the proximal radioulnar joint, an articulation which is properly described with the other radioulnar articulations (p. 135). The elbow joint exhibits clearly certain fundamental characteristics of hinge joints which may be looked for in other joints of this type. These are reciprocally convex and concave articular surfaces, a capsule which is loose on the aspects toward which movement takes place, strong collateral ligaments at the borders of the joint, and the grouping of muscles at the borders of the joint where they do not interfere with movement.

The **articular surfaces** are the spool-shaped **trochlea** of the medial portion and the spheroidal **capitulum** of the lateral part of the distal end of the humerus articulating with, respectively, the **trochlear notch of the ulna** and the slightly cupped upper surface of the **head of the radius.** The capitulum of the humerus is directed forward as well as downward, and the articular surfaces of the joint are most completely in contact when the elbow is flexed to a right angle with the forearm in a position midway between full pronation and supination. The shallow cup of the upper surface of the head of the radius clearly provides no firm grasp on the

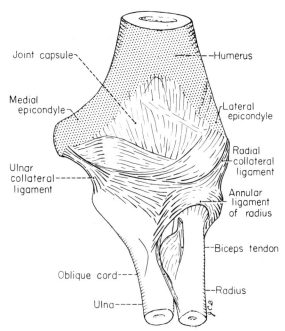

Fig. 109. The capsule and ligaments of the elbow joint, anterior view.

capitulum and, without the adjacent humeroulnar articulation, would form a weak ball and socket joint. Both the stability of the elbow joint and its limitation to the movements of flexion and extension are due to the ridged and grooved concavoconvex surfaces of the ulna and humerus. The reciprocal ridging and grooving of these surfaces, together with the strong collateral ligaments, completely prevent side-to-side movement in the joint.

The **articular capsule** encloses the articulation. It is weak in front and behind but is much strengthened at the sides by the ulnar and radial collateral ligaments. In front, the capsule has an attachment on the humerus along a line (like an inverted V) from the medial epicondyle upward to the limits of the coronoid and radial fossae and downward laterally to the lateral epicondyle (fig. 109). Distally, it is attached to the anterior border of the coronoid process of the ulna and to the annular ligament of the radius and is continuous on either side with the collateral ligaments. The posterior part of the capsule is thin and membranous. Its primarily transverse fibers extend loosely between the margins of the olecranon of the ulna below and the edges of the olecranon fossa of the humerus above. The posterior capsule also passes laterally behind the capitulum of the humerus to the lateral epicondyle and, below, to the posterior part of the annular

ligament and the posterior border of the radial notch of the ulna. The uppermost fibers of the posterior capsule tend to fall short of the upper limit of the olecranon fossa and to allow a small pouch of synovial membrane and a pad of fat to project from the upper part of the fossa when the joint is extended. This part of the capsule is well protected by the tendon of the triceps muscle.

The **collateral ligaments** are strong, triangular thickenings of the capsule which attach by their apices to the medial and lateral epicondyles of the humerus. They diverge below to broader distal attachments on the forearm bones and the annular ligament of the radius. The collateral ligaments lie at the sides of the joint and are so placed as to lie across the axis of movement of the joint in all positions. They are, therefore, relatively tense in all degrees of flexion and extension and place strict limitations on side-to-side displacements. The **ulnar collateral ligament** (fig. 110) fans out from its attachment to the medial epicondyle and has thick anterior and posterior bands united by a thinner intermediate portion. The **anterior band** reaches the medial edge of the coronoid process of the ulna, is closely associated with the common tendon of origin of the forearm flexor muscles, and gives rise to fibers of flexor digitorum superficialis. The **posterior band** attaches below to the medial edge of the olecranon. Between these bands is a thinner **intermediate portion** which is grooved by the ulnar nerve as it passes behind the joint. The intermediate portion ends below in a transverse band stretched between the attachments of the anterior and posterior bands to the coronoid process and the olecranon. The **radial collateral ligament** is a narrow, fibrous thickening, less distinct than the ulnar collateral ligament, which is attached superiorly to a depression below the lateral epicondyle of the humerus deep to the common tendon of origin of the forearm extensor muscles. Inferiorly, the ligament attaches to the annular ligament of the radius and to its bony

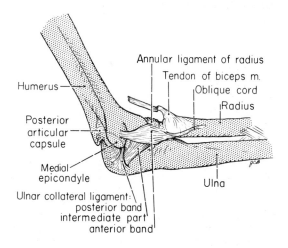

Fig. 110 The elbow joint from the ulnar side.

attachments, the margins of the radial notch of the ulna.

The **synovial membrane** of the elbow joint is extensive. It lines the capsule and is reflected onto the humerus over the radial and coronoid fossae in front and the olecranon fossa behind. Below, it reaches the attachments of the capsule and is continued into the proximal radioulnar articulation. Below the neck of the radius, a redundant fold emerges beyond the annular ligament to give freedom for rotation of the head of the radius. Synovial folds project into the recesses of the joint; especially constant is a semicircular fold which projects into the crevice between the periphery of the head of the radius and the capitulum of the humerus. Subsynovial fat pads lie outside the synovial membrane adjacent to the articular fossae. Such fatty accumulations project into the radial and coronoid fossae during extension and into the olecranon fossa during flexion.

The **blood supply** of the elbow joint is derived from the anastomotic connections of the collateral branches of the brachial system and the recurrent branches of the radial and ulnar arteries. The **nerves** of the joint are supplied anteriorly from the musculocutaneous, median, and radial nerves and, posteriorly, from the ulnar nerve and the radial branch to the anconeus muscle.

The **movements of the elbow joint** are flexion and extension. These are not produced in the exact line of the long axis of the humerus, for the articular surfaces of the lower end of the humerus form a plane which is not transverse to its long axis, the trochlear surface being distal to that of the capitulum. Thus, when the forearm bones are placed in coaptation with the humeral condyles, they extend downward and somewhat outward from the line of the humerus. The 'carrying angle' thus formed carries the hand away from the sides of the body in extension and supination, and the forearm bones form, with the humerus, an angle of about 170° (less in the female) rather than the 180° of a straight line. Full flexion of the elbow brings the hand up in front of the head of the humerus. A slight spiral orientation of the ridge of the trochlear notch and of the groove of the trochlea counters the obliquity of the articular plane of the elbow, so that the forearm bones flex against the humerus and not medial to it (as in a reversal of the 'carrying angle'). The freedom with which the hand is carried to the mouth in flexion at the elbow is due to the slight medial rotation of the humerus and the semipronated position of the forearm which is habitual. The 'carrying angle' is obliterated when the hand is pronated, and in that position the bones and joints of the limb extend to the straight line in which pulling and pushing movements are usually performed. As mentioned, the bony surfaces make their closest contact in right-angled flexion, with the forearm in a position midway between pronation and supination. This is the position of greatest stability of the elbow and is the position naturally assumed when fine manipulation is demanded of the hands and fingers. Flexion of the elbow is limited by opposition of the soft parts of the arm and forearm and by tension of the posterior muscles and ligaments. Extension stops with the straight position of the limb. It is limited by tension in the anterior muscles and by the anterior bands of the collateral ligaments. Flexion of the elbow joint is produced

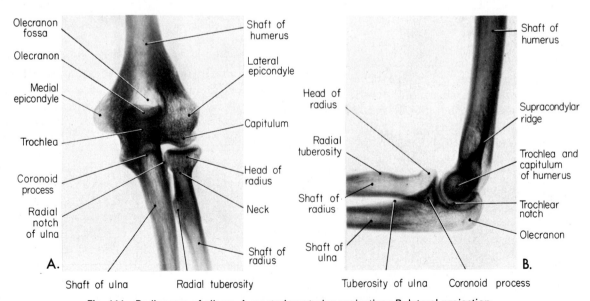

Fig. 111   Radiogram of elbow. A, posterioranterior projection; B, lateral projection.

by the action of the biceps and brachialis muscles with the assistance of brachioradialis and those forearm muscles arising from the medial epicondyle. Extension of the elbow is due to the pull of the triceps and anconeus muscles assisted by the superficial group of extensor muscles of the forearm which arise from the lateral epicondyle of the humerus.

THE RADIOULNAR ARTICULATIONS (fig. 112)

The bones of the forearm are united at their proximal and distal ends by synovial joints which produce the movements of pronation and supination. The radius rotates around a longitudinal axis which passes through the center of the head of the radius proximally and just inside the styloid process of the ulna distally (the apical attachment of the articular disk). These articulations are pivot joints. In addition, there is an intermediate union of the shafts of the radius and the ulna by means of an interosseous membrane.

The Proximal Radioulnar Articulation (fig. 112)

This articulation is enclosed within the articular capsule of the elbow joint and shares its synovial membrane. The **radial notch of the ulna** and the **annular ligament of the radius** form a ring within which the head of the radius rotates.

The head of the radius is covered by a layer of cartilage continuous from its upper surface onto its sides, the latter forming a smooth, circumferential articular surface for the radial notch and the annular ligament. The **annular ligament of the radius** is a strong, curved fibrous band which forms nearly four-fifths of a circle around the head of the radius and attaches by its ends to the anterior and posterior margins of the radial notch of the ulna. The ligament cups in under the head of the radius and thus serves as a restraining ligament, preventing withdrawal of the head of the radius from its socket as well as constituting a major share of the articular surface for the head. The annular ligament receives the radial collateral ligament and blends with the articular capsule of the elbow joint. Below the annular ligament a lax band of fibers, called the **quadrate ligament,** passes from the lower border of the radial notch of the ulna to the adjacent medial surface of the neck of the radius. The **synovial membrane** of this articulation is continuous with that of the elbow joint. Its lowermost reflection forms a loose sac around the neck of the radius, limited below by the quadrate ligament. The redundancy of the synovial membrane here accommodates to the twisting of the membrane that accompanies rotation of the head of the radius.

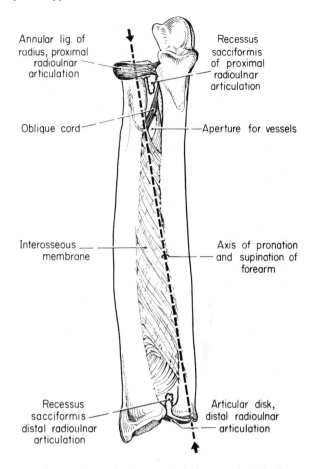

Fig. 112 The radioulnar articulations and their ligaments; the axis of rotation of the radius is indicated.

The blood vessels and nerves of the proximal radioulnar articulation are the same as the lateral aspect of the elbow joint.

The Distal Radioulnar Articulation (figs. 112, 113)

This articulation is a pivot joint formed between the head of the ulna and the ulnar notch on the lower end of the radius. Articular cartilage covers the bony surfaces and is present over both the lateral and distal surfaces of the ulna. The **articular capsule** is represented by transverse bands of no great strength which attach to the anterior and posterior edges of the ulnar notch of the radius and to the corresponding surfaces of the head of the ulna. Their distal edges blend with the margins of the articular disk. The fibrocartilaginous **articular disk** is the chief uniting structure of the joint. It is attached by its base to the sharp medial margin of the distal end of the radius and by its apex to the lateral side of the root of the styloid process of the ulna. The

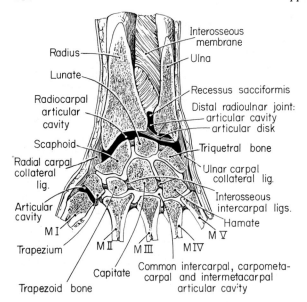

Radius
Lunate
Radiocarpal
articular
cavity
Scaphoid
Radial carpal
collateral
lig.
Articular
cavity
M I
Trapezium
M II
M III
Capitate
Trapezoid bone

Interosseous
membrane
Ulna
Recessus sacciformis
Distal radioulnar joint:
– articular cavity
– articular disk
Triquetral bone
Ulnar carpal
collateral lig.
Interosseous
intercarpal ligs.
Hamate
M V
M IV
Common intercarpal, carpometa-
carpal and intermetacarpal
articular cavity

Fig. 113 A vertical section through the distal radioulnar joint, the wrist joint, and the carpal joints.

disk is an essential part of the total bearing surface in the joint, for its upper surface articulates with the distal aspect of the head of the ulna. As a consequence, the joint cavity is L-shaped in vertical section, the vertical limb being formed between the ulna and the radius and the horizontal limb between the ulna and the articular disk.

The articular disk is thicker peripherally than centrally but is usually not perforated. The disk participates in the radiocarpal (wrist) joint through its distal surface, so that its perforation would lead to a communication between the distal radioulnar and the wrist joints. The **synovial membrane** is large in proportion to the size of the radioulnar joint and reaches proximally between radius and ulna, above the level of their articular surfaces, to form the **recessus sacciformis**. The **arterial supply** of the joint is from the anterior and posterior interosseous arteries and from the palmar and dorsal carpal networks. The **nerves** of the joint are the posterior interosseous branch of the radial nerve and the anterior interosseus branch of the median nerve.

The Interosseous Membrane (figs. 112, 113)

The shafts of the radius and ulna are connected by the oblique cord and the interosseous membrane. The **oblique cord** (figs. 109–112) is a slender, flattened fibrous band which extends from the lateral border of the tuberosity of the ulna to the radius distal to its tuberosity. The posterior interosseous vessels pass to the posterior side of the forearm in the interval between the oblique cord and the interosseous membrane.

The **interosseus membrane** (figs. 80, 83, 112) is a strong fibrous sheet which stretches between the interosseous

borders of the radius and the ulna. Proximally, it extends to within 2 or 3 cm. of the tuberosity of the radius; distally, it is continuous with the fascia of the posterior surface of pronator quadratus. An aperture in this distal portion of the membrane admits the anterior interosseous vessels to the posterior side of the forearm. The fibers of the interosseous membrane mainly run downward and medially from radius to ulna. The ulna is heavy and firmly articulated at the elbow, whereas the radius, with a rather weak relation to the humerus, is broadened at the wrist and is the essential forearm bone in the radiocarpal joint. Thus, the two bones require the firm fibrous connection provided by the interosseous membrane, and the predominant downward and medialward direction of its fibers serves to transmit forces carried from the hand up through the radius to the ulna, the bone of firm connection with the humerus. On the posterior side of the interosseous membrane there are a small number of fibrous bands which have a direction contrary to that of the majority of fibers. In addition to binding the forearm bones together in a complemental relationship, the interosseous membrane separates and gives fibrous attachment to the deep muscles of the anterior and posterior sides of the forearm and completes, with the bones and intermuscular septa, the separation of the forearm into pre-axial and postaxial compartments.

Movements at the Radioulnar Joints

The radioulnar joints contribute the movements of pronation and supination to the forearm. The longitudinal axis of this rotary movement passes proximally through the center of the head of the radius and distally through the apical attachment of the articular disk to the head of the ulna (fig. 112). This axis may be prolonged into the extended hand and is there represented by the fourth digit, around which the hand seems to turn as it follows the forearm. The ulna remains relatively stationary due to its fixed position with reference to the humerus, and the radius rotates, its head revolving within the circle of the annular ligament and the radial notch of the ulna. Distally, the radius travels around the relatively fixed lower end of the ulna. In full supination, the radius and ulna are parallel, and the interosseous space is at its widest. In full pronation, the radius and ulna are crossed in the form of an X, the crossing taking place in the upper part of the forearm. About 135° of rotation is possible in the full movement. Supination is the more powerful action, a fact utilized in the construction of screws and screwdrivers, the more effective force being used in inserting and seating the screw. The principal supinator muscles are the supinator and the biceps brachii; the principal pronators are pronator teres and pronator quadratus.

THE WRIST JOINT (figs. 100, 113, 114)

The wrist joint, or radiocarpal articulation, is formed by the distal extremity of the radius and the articular disk of the distal radioulnar articulation above with the proximal row of carpal bones below. These are the scaphoid, the lunate, and the triquetral bones, joined at their proximal edges by interosseous ligaments. The curvatures of these bones are combined into a convex, egg-shaped, or ellipsoidal, surface which fits into the concave surface of the radius and articular disk. These articular surfaces allow movement over either the long or the short curvature of the ellipsoid and thus provide for flexion and extension, abduction and adduction, and circumduction. Rotary motion is prohibited, since the length of the curved surface is greater than its width. In the straight position of the hand, the scaphoid bone lies below the lateral part of the radius, and the lunate is opposite the medial part of the radius and the articular disk. The triquetrum bone is then in relation to the disk and to the medial part of the joint capsule. Abduction and adduction change these relationships (fig. 114). The **articular capsule** encloses the joint and is strengthened by dorsal and palmar radiocarpal ligaments and by radial and ulnar collateral ligaments. The radiocarpal ligaments are composed of fibers running obliquely downward and ulnarward from the radius to the three carpal bones, their course and attachments determining that the hand shall follow the radius in its movements and displacements.

Due to the articular disk of the distal radioulnar joint and the completeness of the interosseous ligaments which unite the proximal surfaces of the carpal bones of the first row, the **synovial cavity** of the wrist joint is restricted to the radiocarpal space. When, as occasionally occurs, the articular disk is perforated, the radiocarpal joint communicates with the inferior radioulnar articulation. The lax **synovial membrane** lines the deep aspect of the capsule and presents numerous folds, expecially dorsally.

The **dorsal radiocarpal ligament** is a thin membrane extending between the posterior border of the lower end of the radius and the dorsal surfaces of the scaphoid, lunate, and triquetral bones. Its fibers are continuous with the dorsal intercarpal ligaments, and its principal band runs to the triquetral bone.

The **palmar radiocarpal ligament** is a broad band of fibers which extends from the palmar edge of the distal end of the radius and its styloid process to the palmar surfaces of the same three carpal bones that receive the

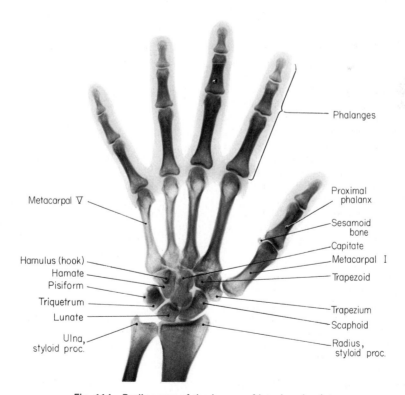

Fig. 114 Radiogram of the bones of hand and wrist.

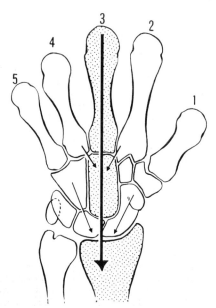

Fig. 115   Forces are gathered to the axis of the hand. (By permission, from *Muscles and Movements,* 2nd ed., by M. A. MacConaill and J. V. Basmajian, 1977. Robert E. Krieger Publishing Co., Inc.)

dorsal ligament, some fibers being prolonged to the capitate bone of the second row.

The **palmar ulnocarpal ligament** is formed of fibers which extend from the anterior edge of the articular disk of the distal radioulnar joint and from the base of the ulnar styloid process downward and lateralward to the carpal bones.

The **radial collateral ligament** is a thickening which passes from the tip of the styloid process of the radius to the radial side of the scaphoid and trapezium bones.

The **ulnar collateral ligament** is a rounded cord attached above to the styloid process of the ulna and below to the base of the pisiform bone and to the adjacent medial part of the flexor retinaculum. Other fibers reach the medial and dorsal surfaces of the triquetrum bone.

The **arteries** of the joint are derived from the dorsal and palmar carpal networks, and its **nerves** from the anterior interosseous branch of the median nerve, the posterior interosseous branch of the radial nerve, and the dorsal and deep branches of the ulnar nerve.

Movements of the wrist joint include flexion and extension, abduction and adduction, and circumduction. The limits of abduction (15°) and adduction (40°) of the hand are set by the styloid processes of the radius and ulna, and movement is freer in the ulnar direction because the radial styloid extends farther toward the hand than does the ulnar styloid. Flexion is freer than extension at the wrist, but the midcarpal joints combine with the radiocarpal joint in these actions and their contributions are unequal. Extension is actually freer in

the radiocarpal joint. The lack of rotary action at the wrist is made up by pronation and supination in the forearm; together these actions provide a second universal joint in the upper limb, adjacent to the hand. The muscles operating over the wrist joint have been described together with the compartmental relations which they exhibit at the wrist. No forearm muscles insert on the carpal bones. Flexor carpi radialis, flexor carpi ulnaris, and palmaris longus flex the hand at the wrist and stabilize it when the digital extensors are acting. Similarly, extensor carpi radialis longus, extensor carpi radialis brevis, and extensor carpi ulnaris extend the hand at the wrist. They stabilize the wrist when the digital flexors are active and cock the wrist dorsally when a fist is produced or when fine manipulative movements are required. In full dorsiflexion, the wrist joint is close-packed. Abduction of the hand follows simultaneous contraction of flexor carpi radialis and the two radial carpal extensors; adduction is produced by the combined actions of the flexor and the extensor carpi ulnaris.

JOINTS OF THE HAND AND FINGERS

Intercarpal Articulations (figs. 92, 100, 113–116)

The carpal bones are arranged in two transversely oriented rows. The arthrodial (gliding) intercarpal articulations may be classified as (a) those between the bones of the proximal row; (b) those between the bones of the distal row; and (c) those of the two rows articulating with each other. The ligaments of these joints are dorsal and palmar surface ligaments and interosseous ligaments. The bones of the proximal row have both **dorsal** and **palmar intercarpal ligaments** arranged transversely from scaphoid to lunate and from lunate to triquetrum. **Interosseous intercarpal ligaments** unite the proximal margins of the same bones and, extending completely through the depth of the intercarpal space, complete the distal limiting surface of the radiocarpal articulation. Similar **dorsal** and **palmar intercarpal ligaments,** generally transverse in orientation, unite the bones of the distal row—trapezium to trapezoid, trapezoid to capitate, and capitate to hamate. **Interosseous intercarpal ligaments** hold these bones in articular contact but do not extend through the full depth of the bones, so that the synovial cavities of their joints communicate around the borders of the ligaments. An interosseous ligament between the trapezium and the trapezoid bones may be lacking.

The articulation between the proximal and distal rows of carpal bones is the **midcarpal joint.** In its central portion the head of the capitate and the apex of the hamate are received into the cup-shaped cavity of scaphoid and lunate, a virtual ball and socket formation. To either side, the articulation of the trapezium and trapezoid with the scaphoid bone and of the ulnar surface of the hamate with the triquetral bone form gliding joints. The midcarpal articulation is strengthened by dorsal, palmar, and collateral ligaments.

The **dorsal intercarpal ligaments** connect the bones of the proximal with those of the distal row. The **palmar intercarpal ligaments** pass from the bones of the proximal row mainly to the capitate (**radiate carpal ligament**). The short **collateral ligaments** are placed on the radial and ulnar borders of the articulation. The stronger and more distinct radial ligament connects the scaphoid with the trapezium; the ulnar ligament connects the triquetrum and hamate bones. They are continuous with the collateral ligaments of the wrist joint. A slender interosseous ligament sometimes connects the capitate and scaphoid bones.

The articulation between triquetrum and pisiform forms a special intercarpal joint. A thin but strong articular capsule unites the bones, the pisiform being mounted on the palmar aspect of triquetrum. The pisiform is also anchored by the **pisohamate ligament** to the hook of the hamate and by the **pisometacarpal ligament** to the base of the fifth metacarpal bone. These ligaments resist the pull of the flexor carpi ulnaris muscle and form parts of its insertion.

The **intercarpal synovial cavity** is large and complex. It fills the midcarpal interval and extends between the adjacent carpal bones of each row, being continuous around the interosseous ligaments of the second row. The cavity extends over the distal aspect of the second row of carpal bones and thus includes the carpometacarpal articulations (except for the thumb). It is further prolonged into the intermetacarpal articulations between the bases of the second, third, fourth, and fifth metacarpal bones. It does not include the pisiform joint or the joint between the trapezium and the base of the first metacarpal bone, nor does it communicate with the wrist joint. The synovial membrane lines the articular capsule and covers the nonarticular surfaces within the capsule.

The **arterial supply** of the intercarpal joints is through the palmar and dorsal carpal networks. The **nerves** are twigs from the anterior interosseous nerve (median), the

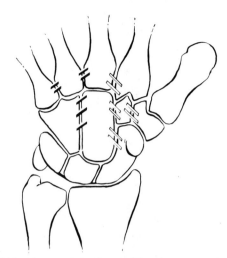

Fig. 116   Ligaments in the carpal region are so placed as to concentrate forces in the capitate. (By permission, from *Muscles and Movements,* 2nd ed., by M. A. MacConaill and J. V. Basmajian, 1977. Robert E. Krieger Publishing Co., Inc.)

posterior interosseous nerve (radial), and the dorsal and deep branches of the ulnar nerve.

Movements of the intercarpal articulations occur simultaneously with and in augmentation of the movements at the radiocarpal articulation. The principal region of intercarpal movement is that through the sinuous midcarpal joint. Here the movement of the head of the capitate in its socket and the gliding movements of the bones on either side contribute a considerable degree of flexion to the hand. Extension is freer in the radiocarpal joint, and flexion is freer in the midcarpal joint. The midcarpal joint also contributes to abduction of the hand, and it and the other intercarpal joints improve the grasp of the hand with slight rotary and gliding movements between the bones.

Carpometacarpal and Intermetacarpal Joints (figs. 92, 100, 113–116)

The **carpometacarpal joint of the thumb** is the separate, saddle-shaped articulation between the trapezium and the base of the first metacarpal bone. The articular surfaces are reciprocally concavo-convex, and a loose but strong articular capsule extends between the articular margins of the bones. Its **blood vessels** and **nerves** are twigs of those radiating toward the thumb. The bi-axial character of this joint is shown in its freedom of movement in flexion and extension, abduction and adduction, and circumduction. The looseness of its capsule also facilitates the movement of opposition of the thumb in which a small amount of rotary action is required (p. 119).

The **carpometacarpal joints of the four fingers** participate with the intercarpal and intermetacarpal joints in a common synovial cavity. The bones are united by dorsal, palmar, and interosseous ligaments.

Two **dorsal carpometacarpal ligaments** reach the second metacarpal from the trapezium and trapezoid; two pass to the third metacarpal from the trapezoid and the capitate; two reach the fourth metacarpal from the capitate and hamate; and the fifth metacarpal is connected by a single ligament with the hamate bone. The **palmar carpometacarpal ligaments** show rather similar connections, but the third metacarpal receives three ligaments: from the trapezium, the capitate, and the hamate. A short **interosseous ligament** is usually present between the contiguous inferior angles of the capitate and hamate bones and the adjacent bases of the third and fourth metacarpal bones.

The **intermetacarpal joints** are formed by the contiguous sides of the bases of the four metacarpal bones of the fingers. Such a joint does not exist between the first and second metacarpal bones. The four metacarpal bones are united by dorsal, palmar, and interosseous ligaments.

The **dorsal** and **palmar metacarpal ligaments** pass transversely from bone to bone. The **interosseous ligaments** connect their contiguous surfaces just distal to their articular facets. The **synovial cavity** of the intermetacarpal joints is an extension of the complex intercarpal and carpometacarpal articular cavity. The carpometacarpal and intermetacarpal joints receive arterial twigs from the dorsal and palmar metacarpal **arteries** and other branches of the dorsal and deep palmar arterial arches. Their **nerves** are derived from those supplying the intercarpal articulations.

Relatively little movement is possible between the carpal bones and the metacarpals of the four fingers; slight gliding movements characterize these joints. The metcarpal bone of the fifth digit has a flattened saddle articulation with the hamate bone so that it flexes appreciably during a tight grasp and rotates slightly under the influence of the opponens digiti minimi muscle.

The **deep transverse metacarpal ligaments** are short ligaments that connect the palmar aspects of the heads of the second, third, fourth, and fifth metacarpal bones. They are continuous with the palmar interosseous fascia and blend with the palmar ligaments of the metacarpophalangeal joints and with the fibrous sheaths of the flexor tendons. The tendons of the interosseous muscles pass dorsal to the deep transverse metacarpal

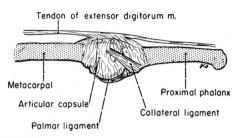

Fig. 117 The capsule and ligaments of the metacarpophalangeal joint.

ligaments, those of the lumbrical muscles to their palmar side. The ligaments limit the spread of the heads of the metacarpal bones.

Metacarpophalangeal Joints (figs. 100, 114, 117)

The metacarpophalangeal joints are condyloid in character, both the rounded head of the metacarpal and the oval concavity of the proximal end of the phalanx having unequal curvatures along their transverse and vertical axes. An articular capsule and collateral and palmar ligaments unite the bones in an end-to-end articulation. The **articular capsule** is loose and its attachment is closer to the articular cartilage on the dorsal than on the palmar surface. Dorsally, the expansion of the digital extensor tendon reinforces and attaches to the capsule.

The **palmar ligament** is a dense, fibrocartilaginous plate that increases the depth of the phalangeal articular surface anteriorly due to its firm attachment to the proximal palmar edge of the phalanx. It is loosely attached to the palmar surface of the metacarpal and, in flexion, passes under this surface of the head of the metacarpal, serving as part of the articular surface for the head. At the sides the palmar ligament is continuous with the collateral ligaments and with the deep transverse metacarpal ligaments. The **collateral ligaments** are strong cord-like ligaments attached proximally to the tubercle and the adjacent depression of the metacarpal bone and distally to the palmar aspect of the side of the base of the phalanx. Their fibers spread to attach to the palmar ligaments.

Movements of flexion and extension, abduction and adduction, and circumduction are permitted at these joints. With extension is associated abduction, as in spreading the fingers; with flexion is associated adduction, as in making a fist. The metacarpophalangeal joint of the thumb has only limited actions of abduction and adduction, for the head of its metacarpal is less convex and is wider from side to side. The thumb gets its freedom of motion at the carpometacarpal joint.

Interphalangeal Joints

Interphalangeal joints are structurally similar to metacarpophalangeal joints. They have a loose capsule and palmar and collateral ligaments, and are reinforced dorsally by the extensor tendon expansion. They are, however, hinge joints limited to the movements of flexion and extension due to the pulley-like form of their articular surfaces. Flexion is freer than extension and is especially facilitated at the proximal interphalangeal joint, where it may reach about 115°. Arteries and nerves of the metacarpophalangeal and interphalangeal articulations are derived from the adjacent digital branches.

APPLIED ANATOMY OF THE UPPER LIMB

In most clinical conditions and disabilities, anatomical features are of basic or ancillary importance. Cleft lip–cleft palate and congenital dislocation of the hip are obvious examples from the field of development. The following examples of dislocations, fractures, collateral circulation, and nerve injuries illustrate the determining influence of normal anatomical characteristics.

Dislocations

A joint is dislocated, or luxated, when its articular surfaces are wholly displaced one from the other so that apposition between them is lost. The joint is subluxated when its surfaces are partly displaced but retain some contact one with another. These conditions may be congenital, spontaneous, traumatic, or recurrent. Pain, swelling, deformity of the part, and marked limitation of movement are characteristic. There is always the danger of damage to major blood vessels and nerves in the vicinity—an example is the possible damage to the axillary artery or vein or to the cords or major nerves of the brachial plexus with anterior dislocation of the shoulder joint. Joint capsules are usually thin and weak in the direction of movement; they are stronger and may be reinforced by collateral and other ligaments at the sides. Dislocating forces obviously take advantage of the weaknesses of the capsules and are resisted by the strengthening bands applied at other locations.

Closed reduction of dislocations is usually possible; if not, open reduction may be resorted to. The joint capsule will usually be ruptured; in more severe cases the glenoid labrum may be torn from the bone, and joint ligaments may be stretched or be torn or ruptured. The fracture-dislocation is the most severe form of this injury—usually in the form of an intra-articular fracture along with the dislocation. These are often seen at the ankle or the hip joint. In the hip joint it may be associated with an avascular necrosis of the head of the femur due to loss of blood supply to the head. Numerous small arteries (p. 601) enter the neck of the femur, led to their bony foramina under the reflections of the synovial lining of the joint, and one vessel reaches the head of the bone through the ligamentum capitis femoris. In an intracapsular fracture of the femoral neck, most of these arteries suffer transection either inside or outside of the bone and the head fragment loses much of its blood supply. The artery of the ligament, having a different course, survives.

Injuries that produce joint dislocation in the adult produce epiphyseal separation in the child. These are most common in the distal radius, the proximal radius, the proximal humerus, and the distal tibia. Review bone development and the role of the epiphysis, pp. 41–2.

Dislocations are rare at the sternoclavicular joint for the clavicle fractures under lesser forces than produce dislocation. A fall on the point of the shoulder, however, may result in the clavicle being levered on the first rib as a fulcrum, the sternal end of the clavicle being forced upward and anteriorly out of the capsule. The intra-articular disk may be torn from the first rib. The strength of the costoclavicular ligament usually prevents a very wide displacement of the sternal end of the clavicle. Dislocation at the acromioclavicular joint is the "shoulder separation" of sports injuries. The distal end of the clavicle slides up the inclined plane of the acromion at the expense of the full integrity of the coracoclavicular ligament.

Dislocation of the shoulder joint is rather common, favored by those circumstances that make the joint so freely movable: shallow glenoid cavity, loose capsule, and only one-third of the head of the humerus in contact with the glenoid process. The humerus is long and has much leverage in dislocating forces. Anterior dislocation, the more common, may also be described as subcoracoid, the head of the humerus coming to lie under the coracoid process and producing an obvious bulge in the clavipectoral groove region. At the same time the deltoid bulge of the shoulder is lost. The head of the humerus goes out through a rent in the capsule over the long head of the triceps muscle and the inferior glenohumeral band. Posterior dislocation is more rarely encountered. The glenoid process of the scapula is directed somewhat forward and its posterior lip is better situated to resist posterior forces, and the strong rotator cuff muscles—infraspinatus and teres minor—located posteriorly splint the capsule there. Posterior dislocation is produced by medial rotation, abduction, and a force driving up the long axis of the humerus, the capsule is torn posteriorly in the region of the teres minor and the head of the humerus may become subspinous in location.

Most dislocations of the elbow are posterior ones through the relatively weak posterior capsule. Both forearm bones may be displaced together due to their firm union in the proximal radioulnar articulation. The olecranon process prevents anterior dislocation of the ulna unless it is fractured. The deformity is produced by combined hyperextension and a driving force upward from the hand. Dislocation of the radius forward is next common, usually produced by extreme pronation, levering the radius out of its normal position and tearing its restraining ligaments. The head of the radius will be palpated in the cubital fossa.

A fall on the outstretched hand may result in dislocation at the wrist—either at the radiocarpal joint or at the intercarpal interval. These can usually be reduced by manipulation but must not be confused with a distal radial (Colles) fracture (see below).

Finger dislocations occur at the metacarpophalangeal and at the interphalangeal joints and are usually amenable to reduction by manipulation.

Fractures

A fracture is a break in the continuity of a bone. It will be accompanied by tearing of periosteum and disruption and damage to the adjacent soft tissues. The jagged edges of the fragments can open blood vessels and lacerate nerves. The most common cause of fracture is violence—bone fractures when its capacity to absorb energy is exceeded. Also as we age, calcium is lost from our bones (osteoporosis), and fracture may occur with little violence or spontaneously. An open fracture is one associated with an open wound extending from the skin surface to the fracture; the fragments may be at or through the surface. A closed fracture is not associated with an open wound. An incomplete fracture retains bony continuity in part, as a "greenstick" fracture; the complete fracture exhibits a complete break. A comminuted fracture consists of three or more fragments and an impacted fracture has one fragment imbedded in another. Compression fractures occur, especially in vertebrae.

The fracture line may be longitudinal, transverse, oblique, or spiral. Such differences result partly from the type and direction of the trauma, as a direct blow or a twisting force, but are also affected by the structure of the bone involved. There will almost certainly be shortening of the member and overriding of the fragments; angulation or rotation of the fragments may occur.

Muscle pull is usually involved in these displacements. In fracture of the upper third of the shaft of the humerus, one can readily see that the traction of the pectoralis major will draw the proximal fragment sharply medialward, whereas the action of the deltoid muscle on the distal fragment will be to abduct that part of the bone, leading to wide displacement of the parts. Normally bones hold muscles out to their proper length and when that rigid architecture is lost, overriding of the fragments is almost inevitable.

The clavicle is the most frequently fractured bone of the body and fracture occurs most commonly in its middle third, the region of transition of its curvatures. In such a fracture, the weight of the limb and the pull of the pectoralis major muscle causes the distal portion of the clavicle to drop below the proximal fragment. The proximal fragment is also elevated by the sternocleidomastoid muscle as the shoulder collapses onto the chest. There can be serious damage to the major subclavian vessels and brachial plexus nerves from their laceration by the end of the distal fragment, although the subclavius muscle serves somewhat to protect them.

The scapula is pretty well encased in muscle and its fractures are not remarkable.

Fractures of the humerus through its mid or upper shaft carry the danger of serious damage to the radial nerve as it lies in its groove along the posterior and lateral surfaces of the bone, resulting in 'wrist drop' from paralysis of the forearm extensors. A supracondylar fracture may do damage to the vessels and nerves at the elbow.

A fall on the extended wrist frequently results in fracture of the distal end of the radius (Colles fracture). The radius will be fractured transversely in its lower one inch and the fragment will be displaced dorsally. The ulna will usually not be broken but its styloid process may be torn off. The hand will be displaced radially and is cocked dorsally in what has been called the 'silver fork' deformity. Fracture of the scaphoid bone of the wrist is not uncommon. This occurs at the isthmus of the bone and nonunion or vascular necrosis may result if viable blood vessels reach only one of the two fragments.

It has been said that 75% of all injuries resulting in disability occur in the hand and fingers. There are frequent fractures and dislocations of the bones here, but accidental amputation, burns, tendon lacerations and transections, penetrating wounds, and nerve injuries are the most common injuries.

Collateral Circulation

A collateral circulation is the result of the general plexiform character of peripheral blood vessels. One or two major arteries run through much of the limb, giving off large and small branches. These branches have primary responsibilities in supplying the tissues of certain regions or segments of the limb, but they and their finer branches come into very close spatial relationship with other vessels and into actual continuity in the form of anastomoses in many instances. Anastomosis of small branches is very common within the substance of muscles and on the capsules of joints. The vessels on joint capsules are usually named; those ramifying in muscle are not especially named but constitute a very rich anastomosing network. The plexiform nature of blood vessels provides numerous routes of blood flow through the limb. This is clearly of physiological value, for, while blood vessels generally run in protected locations, muscle action and the pressure of external objects can periodically restrict or stop the flow of blood through certain vessels. Aside from normal physiological variations, there are situations under which the limb may have to depend for its nourishment on its collateral circulation. Major vessels may be torn or severed when the bones are fractured; gunshot and other wounds, especially frequent under conditions of war, produce damage to major vessels which requires their ligation. The ligation of a large artery, abruptly stopping the flow of blood through it to more distal parts, places a considerable load on the anastomoses made by its more proximal branches. The nourishment of the distal parts of the limb under these conditions depends on filling the arterial tree beyond the ligature by blood flowing past the region of damage through vessels, usually small, having their origin above the level of the stoppage.

The collateral circulation in the upper limb is generally adequate to the emergency demands made upon it. Blood vessels are quite labile in early years and have some capacity for enlargement even in later years of life. When subjected to the pressure of an increased flow of blood accompanied by a constantly higher blood pressure, vessels take several weeks to increase in caliber, even under favorable conditions. In the child, vessel-shunting operations to increase blood flow to the lungs have shown that the subclavian artery of the neck and upper limb can be ligated without more than temporary impairment to the tissues of the limb. In more mature individuals such lability of blood vessels may not be expected, and under conditions of stiffening of the walls, as in arteriosclerosis, the collateral circulation of the limb will be unequal to emergency requirements. It should be recognized that, where there is a choice of location of a ligature, the collateral circulation will be aided by placing the ligature distal to a sizable branch rather than proximal to it. There are, because of

the branching pattern of a vessel, more favorable and less favorable locations for elective occlusions, and, therefore, a knowledge of the regions of anastomosis of the blood vessels of the limb is of value. The major regions of anastomosis of named vessels are considered below. The rich anastomoses of muscular arteries are not susceptible to ready analysis but nevertheless play an important role in establishing the collateral circulation of a part.

1. The **scapular anastomosis** represents the junction of numerous vessels from both subclavian and axillary sources. The dorsal scapular and the suprascapular arteries from the subclavian system and the circumflex scapular branch of the subscapular artery anastomose on the dorsum of the scapula. The first two of these vessels anastomose with anterior branches of the subscapular artery on the costal surface of the scapula. An **acromial network** is formed on the acromion of the scapula by the acromial branch of the thoracoacromial artery, by terminals of the anterior and posterior circumflex humeral arteries, and by the acromial branch of the suprascapular artery.

2. An **intercostobrachial anastomosis** is formed by the connections made between intercostal arteries, the lateral thoracic artery, the thoracodorsal artery, and pectoral branches of the thoracoacromial artery. Intercostal arteries may be fed from either or both the thoracic aorta and the internal thoracic arteries.

3. Anastomoses in the arm between shoulder and elbow are not numerous. If the upper part of the brachial artery is ligated, the collateral circulation is chiefly carried on by the anastomosis between the descending branch of the posterior circumflex humeral artery and the ascending branch of the deep brachial artery.

4. An anastomosis by the terminals of eight vessels is made around the elbow joint. Four descending vessels, all named 'collateral', participate. These are the superior ulnar collateral and inferior ulnar collateral branches of the brachial artery and the radial collateral and middle collateral branches of the deep brachial artery. They anastomose on the capsule of the joint with one another and with the anterior and posterior ulnar recurrent branches of the ulnar artery, the radial recurrent branch of the radial artery, and the interosseous recurrent branch of the common interosseous artery.

5. **Carpal networks** of anastomosing terminals characterize the joint capsules at the wrist. A palmar carpal network is formed by the junction of the palmar carpal branches of the radial and ulnar arteries, by a branch of the anterior interosseous artery from above, and by carpal recurrent branches of the deep palmar arterial arch below. The arterial arch of the dorsum of the hand serves both for the dorsal articulations at the wrist and for the supply of the dorsal aspect of the hand and fingers. It overlies the distal part of the carpus and is formed by dorsal carpal branches of the

ulnar and radial arteries and by the dorsal terminal branch of the anterior interosseous artery with which is combined the termination of the posterior interosseous artery.

6. The arterial arches of the hand are, themselves, important mechanisms of collateral circulation. Both the **superficial** and **deep palmar arches** unite branches of the radial and ulnar arteries. Other important communications in the hand are those of the palmar metacarpal arteries with the common palmar digital arteries, of the perforating branches of the dorsal metacarpals with the deep arch proximally and with the common palmar digitals at the webs of the fingers, and the anastomoses of the proper dorsal and palmar digital arteries in the fingers (fig. 101).

Nerve Injuries

There is no more instructive review of the nerve-muscle relations of the limb than a consideration of the effects of nerve injuries. Certain levels of injury are more common but all injuries have clear-cut sequelae. Certain of the more instructive cases follow.

Various injuries to the **brachial plexus** occur especially with excessive traction on its roots or branches as in difficult delivery of an infant or severe displacement in falls on the shoulder. The C5 and C6 roots can thus be damaged with subsequent paralysis of the supraspinatus and infraspinatus muscles, the subclavius, biceps brachii, deltoid, and teres minor, and partially brachialis and coracobrachialis. The limb will hang limply and will be medially rotated; the forearm will be pronated. There will be loss of sensation on the lateral aspect of the arm and forearm. Traction injuries may produce tearing of the lower roots, as T1. Its nerve fibers run in the median and ulnar nerves supplying the small muscles of the hand and paralysis and atrophy may result along with sensory changes. 'Cervical rib syndrome', 'scalenus anticus syndrome',

and 'thoracic outlet syndrome' are all terms reflecting compression of the lower trunk of the brachial plexus at its crossing of the first rib which produce deficits similar to tearing of the lower roots. These terms, however, imply a cervical rib as bone or as a fibrous strand, or excessive development of the anterior scalene muscle as causative agents.

The **radial nerve** may be injured in its groove on the posterior aspect of the humerus. By this level the nerve has given off all its branches to the triceps muscle so there is no loss of extension at the elbow. However, all postaxial muscles below this level will be paralyzed resulting in 'wrist-drop' and inability to extend the hand or digits (although the interphalangeal joints can still be extended by the unaffected interosseous and lumbrical muscles). There will be sensory loss in the areas of distribution of the posterior antebrachial cutaneous and superficial radial nerves.

Injury to the **median nerve** at the wrist emphasizes the importance of opposition in the activities of the hand. Loss of the motor or recurrent branch of the median nerve paralyzes the muscles of the thenar eminence with subsequent wasting of this area. The thumb can no longer be opposed to the other digits and the normal grasp of the hand is lost. A grasp will be attempted using adduction and flexion of the thumb toward the hand but this is an awkward and ineffective substitute and the hand's activities are markedly restricted. Loss of the short flexor and abductor muscles of the thumb results in this member being drawn into a position of close adduction and extension against the hand. There is also important sensory loss in the areas of distribution of the proper digital branches of the median nerve to the digits.

Injury to the **ulnar nerve** at the wrist produces a dramatically deformed 'claw-hand' attitude. The thumb is strongly abducted and all metacarpophalangeal joints are hyperextended. At the interphalangeal joints digits one and two are extended but three and four are partially flexed. There is marked wasting through the hand (hypothenar muscles, interossei, and lumbricals III & IV) and in the thenar web (adductor pollicis muscle). There are obvious sensory defects in the little finger and the medial half of the ring finger.

III

The Head and Neck

SURFACE ANATOMY

The Cranium The soft tissues of the scalp form a layer of relatively uniform thickness under which bony elevations may be visible. The **frontal** and **parietal eminences,** rounded projections of the corresponding skull bones, are usually evident. In the infant and child these eminences are so prominent as to give the skull a quadrilateral appearance. Posteriorly, at the junction of the neck and head, the **external occipital protuberance** is a readily palpable and useful landmark. Extending lateralward from it, the **superior nuchal line** leads to the **mastoid process of the temporal bone.** Located directly behind the external ear, the mastoid process projects downward and forward, its tip reaching the level of the lobe of the external ear.

The Face At birth the development of the face lags behind that of the cranium. The proportion of these parts is 1:8 at birth; it is 1:2 in the adult. The contours of the face are determined by its underlying bony conformation, the principal eminences being visible. The **superciliary arch** is the bulging part of the frontal bone above the eye. The bony **margin of the orbit** is also evident. The **zygomatic bone** forms part of the orbital margin laterally, and below the orbit it is the greatest prominence of the cheek. It joins a process of the temporal bone to form the **zygomatic arch** which ends posteriorly in front of the ear. Below the zygoma the **maxilla,** lodging the upper teeth, can be palpated through the cheeks. The lower part of the face is occupied by the **mandible.** The mandible is palpable through most of its extent, the prominence of the chin being the **mental protuberance.** The body of the mandible has a sharp inferior margin which ends posteriorly in the **angle of the mandible.** The **ramus** continues the bone upward toward the ear and ends in the **mandibular condyle** immediately anterior to the **external acoustic meatus.** Here the **temporomandibular joint** can be readily located, for in alternate depression and elevation of the lower jaw the condyle of the mandible can be felt to move within the **mandibular fossa** of the temporal bone. The act of mastication makes evident certain soft tissues of the face. As the jaw is clenched, the **masseter muscle** is indicated by a fullness between the angle of the mandible and the zygomatic arch. Under similar conditions, a fullness in the temporal region above the zygomatic arch marks the **temporalis muscle.**

The **facial artery** enters the face by crossing the mandible at the anterior border of the masseter muscle, and here its pulsations can be detected. The artery runs an irregular course toward the medial angle of the eye. In front of the ear and just above the zygomatic arch, the pulsations of the **superficial temporal artery** can be felt. This artery is frequently quite visible in older individuals, its frontal and parietal branches coursing in serpentine fashion over the temporal part of the scalp.

The Neck The localization of deeper structures of the neck is made partly in reference to palpable bones and cartilages and partly to triangles defined by superficial landmarks. The **hyoid bone** can be felt in the deep angle between the chin and the anterior part of the neck. The **body** of this bone is anterior; its **greater cornu** is palpable at the side of the neck. The **thyroid cartilage** forming the prominence of the larynx is visible in the median line of the neck 1 or 2 cm. below the hyoid bone. The upper margin of the cartilage is connected to the hyoid bone by the thyrohyoid membrane; its lower margin is attached to the cricoid cartilage by the cricothyroid ligament. Below the thyroid cartilage, palpation reveals the **cricoid cartilage,** followed by the rounded **trachea.** The cartilaginous rings of the latter can frequently be felt.

The principal muscular landmark of the neck is the sternocleidomastoid muscle. This bulging muscle bisects the neck diagonally, extending, as its name implies, from the sternum and clavicle below to the mastoid process and occipital bone above. The muscle separates an **anterior triangle** of the neck (defined by its anterior margin, the median line of the neck and the lower border of the mandible) from a **posterior triangle** behind the muscle (figs. 118, 121). The anterior cervical triangle is a region of approach to many important structures of the neck and is customarily subdivided for more precise localization. Above the hyoid, the suprahyoid portion of the triangle may be divided into two smaller triangles, the **submental** and the **submandibular** (fig. 118). The **digastric muscle** is a two-bellied muscle which attaches by an intermediate tendon to the hyoid bone. With the border of the mandible for a base, the digastric muscle completes the definition of the submandibular triangle, the apex of which is directed inferiorly. The largest structure in the triangle is the submandibular gland. Medial to the anterior belly of the digastric muscle and bordered

145

otherwise by the median plane of the neck and the body of the hyoid bone is the smaller **submental triangle** (fig. 118). Its floor is the mylohyoid muscle, and it contains the submental lymph nodes. In the infrahyoid part of the anterior triangle, the **omohyoid muscle** passes downward and lateralward from the hyoid bone to disappear behind the sternocleidomastoid muscle and thus subdivides this area of the anterior triangle into two triangles. The upper triangle, bordered by the omohyoid, the sternocleidomastoid, and the posterior belly of the digastric muscle, is known as the **carotid triangle** (fig. 118) due to its significance in approaching the carotid arterial system. It is also of importance for approaches to the internal jugular vein, the vagus and hypoglossal nerves, and the cervical sympathetic trunk. The inferior part of the anterior triangle is defined by the body of the hyoid bone above, the midline medially, and the borders of the omohyoid and sternocleidomastoid muscles laterally. It is designated as the **muscular triangle** (fig. 118) and contains the infrahyoid muscles and the thyroid gland. As a rule, the **isthmus of the thyroid gland** crosses the second, third, and fourth tracheal rings. The muscular triangle is also utilized for approaches to the lower levels of the carotid-jugular vascular system and to the trachea and esophagus.

The **posterior cervical triangle** (fig. 118) is bounded, anteriorly, by the border of the sternocleidomastoid muscle; posteriorly, by the adjacent margin of the trapezius; and below, by the middle one-third of the clavicle. The inferior belly of the omohyoid muscle crosses the triangle, running obliquely downward and backward several centimeters above the clavicle. The small subdivision of the posterior triangle thus defined is the **omoclavicular triangle.** The posterior triangle has as its floor the prevertebral layer of cervical fascia covering the splenius, levator scapulae, scalenus medius, and scalenus posterior muscles. The external jugular vein, descending vertically through the neck from a position behind the angle of the mandible, crosses the sternocleidomastoid muscle obliquely on a course toward the middle of the clavicle. Within the posterior triangle it pierces the deep fascia to reach the subclavian vein. The external jugular vein is frequently visible under the skin in older individuals.

THE SUPERFICIAL STRUCTURES OF THE NECK

SKIN AND SUBCUTANEOUS TISSUE

The **skin** of the neck, the scalp, and the face is thin and pliable. It is firmly attached to the underlying cartilages of the external ear and of the alae of the nose, but elsewhere it is generally underlain by loose, fatty areolar connective tissue. In the scalp the subcutaneous connective tissue is thick and is composed of a dense, closely felted layer of connective tissue in which fat lobules are imprisoned in well-formed locules. The skin of the lips is very delicate, and its rich capillary blood supply gives the lips their reddish character. Into the skin of all three regions (neck, scalp, and face) insert the terminal fibers of subcutaneous voluntary muscles.

The **subcutaneous connective tissue** of the head and neck contains the usual representation of cutaneous nerves and superficial veins. In the face and scalp, arteries distribute with the veins, and in the scalp all three elements course together through the subcutaneous connective tissue. The subcutaneous voluntary muscle, so well developed in these regions as the musculature of facial expression, takes origin from underlying bone and inserts into skin. The platysma muscle of this layer in the neck is superficial to the principal parts of the veins and nerves, but the representative in the scalp (epicranius muscle) generally underlies the subcutaneous nerves, arteries, and veins. In the face the vessels wind both superficial and deep to the muscles.

The Platysma Muscle of the Neck (figs. 121, 155)

The **platysma** is a broad, thin sheet of muscle

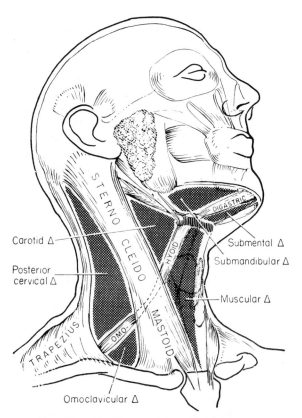

Carotid △

Posterior cervical △

Submental △

Submandibular △

Muscular △

Omoclavicular △

Fig. 118   The triangles of the neck.

located in the subcutaneous tissue of the neck. Its fibers originate from the fascia covering the upper parts of the pectoralis major and deltoideus muscles. Passing upward across the clavicle, the fascicles of the muscle incline medialward as they ascend and insert into the inferior border of the mandible and into the skin and subcutaneous tissues of the lower part of the face and the corner of the mouth. The more medial fibers interlace below the point of the chin with fibers from the opposite side. Fascicles of the muscle sometimes blend with depressor anguli oris or with the margin of orbicularis oris. The platysma muscle draws the corner of the mouth downward and widens the aperture as in expressions of sadness or fright. It can aid in depressing the lower jaw and in drawing upward the skin of the chest. The cervical branch of the facial nerve supplies the muscle on its deep surface. The platysma is superficial in the subcutaneous tissues of the neck, having the external jugular vein and the main cutaneous nerves of the neck deep to it.

SUPERFICIAL VEINS OF THE NECK

These consist of several longitudinally oriented veins, the external and anterior jugular veins together with their tributaries and communications. For their relations to other veins of the head region see p. 214.

External Jugular Vein

The external jugular vein (figs. 119, 121) crosses perpendicularly the superficial surface of the sternocleidomastoid muscle directly under the platysma muscle of the neck. It is formed a little below and behind the angle of the mandible by the union of the retromandibular vein and the posterior auricular vein. This junction occasionally takes place in the lower part of the parotid gland. Descending toward the middle of the clavicle, the vein pierces the superficial layer of cervical fascia about 2 cm. above the clavicle and ends in the subclavian vein. The vein is provided with two pairs of valves; one at its entrance into the subclavian vein, the other about 4 cm. above. As a rule, these valves are not entirely competent. As it passes the posterior border of the sternocleidomastoid muscle, the external jugular vein receives the **posterior external jugular vein** (fig. 119). This vein begins in the upper posterior part of the neck lateral to the cervical part of the trapezius muscle and drains the skin and superficial

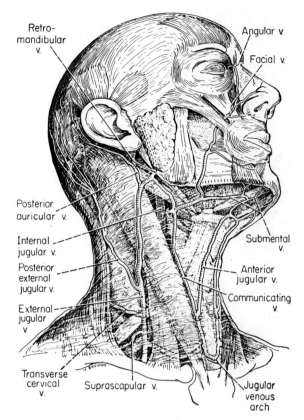

Fig. 119   The superficial veins of the neck.

muscles of the back of the neck. Descending toward the posterior border of the sternocleidomastoid muscle, it is there tributary to the external jugular vein. Near its termination in the subclavian vein, the external jugular vein receives the anterior jugular vein and, from the shoulder region, the suprascapular and transverse cervical veins. The **suprascapular** and **transverse cervical veins** (fig. 119) drain the muscles, bones, and other structures of the shoulder supplied by the corresponding arteries. They contain valves at their termination and sometimes empty directly into the subclavian vein.

Anterior Jugular Vein

This vein (figs. 119, 121) begins near the hyoid bone by the confluence of small veins of the submental and submandibular regions. It descends near the median line, sometimes on it as a single vessel, in the subcutaneous connective tissue. About 3 cm. above the manubrium of the sternum, the vein perforates the superficial layer of cervical fascia to lie in the suprasternal space between the two laminae of that fascia. Here it communicates by a venous arch with its fellow and then turns laterally, passes close behind the sternocleidomastoid muscle, and empties into the terminal portion of the external jugular vein. The anterior jugular vein varies considerably in size and has no valves. It may receive small laryngeal and thyroid tributaries. A **communicating vein,** unnamed but sometimes very large, frequently joins the anterior jugular vein above the sternum. It lies along the anterior diagonal border of the sternocleidomastoid muscle and usually connects above with the facial vein.

Cutaneous Nerves of the Neck

These have two sources: dorsal rami of cervical nerves, and ventral rami of cervical nerves two, three, and four from the cervical plexus. Terminal branches of dorsal rami of cervical nerves distribute in the back of the neck and on the back of the scalp. Cutaneous branches of the cervical plexus cover the entire lateral and anterior portion of the neck and extend into the head behind and over the lower part of the ear and over the parotid region in front of the ear.

Dorsal Rami of Cervical Nerves

The general distribution of the dorsal rami of cervical nerves is described on p. 47. The medial branch of this ramus of the second cervical nerve is the **greater occipital nerve** (figs. 120, 161). Turning upward over the obliquus capitis inferior muscle, it ascends between the muscles of the suboccipital triangle and the semispinalis capitis. The nerve pierces the latter muscle and the trapezius close to their attachments to the occipital bone and ascends on the back of the scalp in company with the occipital artery. The occipital artery, a posterior branch of the external carotid artery which arises high in the neck, converges on the greater occipital nerve by crossing the insertion of obliquus capitis superior and entering the scalp between the bony attachments of the trapezius and sternocleidomastoid muscles. The greater occipital nerve supplies the skin of the occipital part of the scalp for some distance lateral to the median line and as far superiorly as the vertex of the skull. Its lateral branches communicate with those of the lesser occipital nerve (see below). The greater occipital nerve gives off muscular branches to semispinalis capitis and, over the trapezius, communicates with

occipitalis tertius, the cutaneous branch of the third cervical nerve.

The dorsal rami of the third to the sixth cervical nerves divide into medial and lateral branches at the lateral border of the semispinalis cervicis muscle. The medial branches pass between semispinalis cervicis and capitis and supply these muscles. Piercing the trapezius muscle near its attachment to ligamentum nuchae, they supply the skin of the back of the neck in a serial manner from above downward. The medial branch of the third cervical nerve is **occipitalis tertius** (fig. 120). It communicates with the greater occipital nerve and supplies the skin of the upper part of the back of the neck near the median line and the skin of the scalp in the region of the external occipital protuberance.

Cutaneous Branches of the Cervical Plexus

The cutaneous branches of the cervical plexus are the lesser occipital, the great auricular, the transverse cervical, and the supraclavicular nerves. All these nerves approach the surface in the posterior triangle of the neck at about the middle of the posterior border of the sternocleidomastoid muscle. Their adjacent branches intercommunicate.

The **lesser occipital nerve** (C2, sometimes also 3) characteristically ascends in the neck along the posterior border of the sternocleidomastoid muscle (figs. 120, 161). It pierces the superficial layer of cervical fascia near the insertion of the muscle and divides into branches which supply the skin and subcutaneous tissue of the scalp behind and above the ear and of the upper portion of the cranial surface of the ear.

The **great auricular nerve** (C2, 3), appearing in the posterior triangle at the lateral border of the sternocleidomastoid just below the lesser occipital nerve, turns onto the muscle, and crosses it obliquely in a course toward the auricle and the angle of the mandible (figs. 120, 121). The great auricular nerve divides over the sternocleidomastoid muscle into three sets of branches. The **mastoid branch** reaches the skin over the mastoid process. The **auricular branches** supply the lower part of the auricle on both surfaces. The **facial branches** ascend in front of the ear among the superficial lobules of the parotid gland. They supply the skin and subcutaneous tissues over the parotid gland and over the angle of the mandible. Certain branches of the great auricular parallel the external jugular vein over the upper half of the sternocleidomastoid muscle.

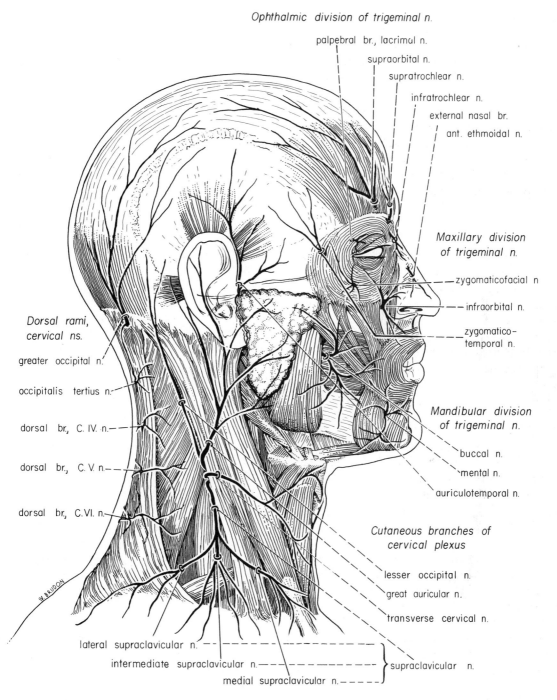

Ophthalmic division of trigeminal n.

palpebral br., lacrimal n.

supraorbital n.

supratrochlear n.

infratrochlear n.

external nasal br.

ant. ethmoidal n.

Maxillary division
of trigeminal n.

zygomaticofacial n

infraorbital n.

zygomatico-
temporal n.

Dorsal rami,
cervical ns.

greater occipital n.

occipitalis tertius n.

dorsal br., C. IV. n.

dorsal br., C. V. n.

dorsal br., C. VI. n.

Mandibular division
of trigeminal n.

buccal n.

mental n.

auriculotemporal n.

Cutaneous branches of
cervical plexus

lesser occipital n.

great auricular n.

transverse cervical n.

lateral supraclavicular n.

intermediate supraclavicular n.

medial supraclavicular n.

supraclavicular n.

Fig. 120   The cutaneous nerves of the head and neck.

The **transverse cervical nerve** (C2, 3) winds around the lateral border of the sternocleidomastoid muscle just below the great auricular nerve (figs. 120, 121). It crosses the muscle horizontally to reach the anterior triangle of the neck, passing deep to the platysma muscle and the external jugular vein. Near the anterior border of the sternocleidomastoid, the nerve divides into superior and inferior branches. These are distributed through the platysma to the skin and subcutaneous tissue of

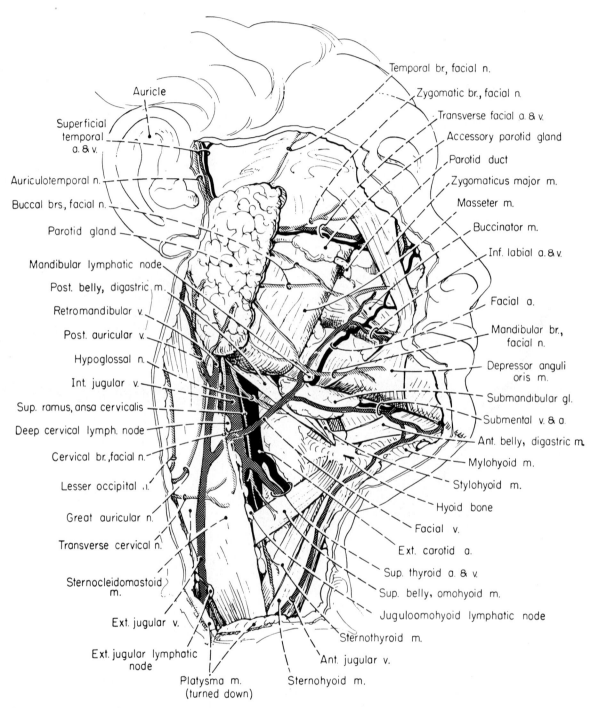

Auricle

Superficial
temporal
a. & v.

Auriculotemporal n.

Buccal brs, facial n.

Parotid gland

Mandibular lymphatic node

Post. belly, digastric m.

Retromandibular v.

Post. auricular v.

Hypoglossal n.

Int. jugular v.

Sup. ramus, ansa cervicalis

Deep cervical lymph. node

Cervical br.,facial n.

Lesser occipital n.

Great auricular n.

Transverse cervical n.

Sternocleidomastoid
m.

Ext. jugular v.

Ext. jugular lymphatic
node

Platysma m.
(turned down)

Sternohyoid m.

Ant. jugular v.

Sternothyroid m.

Juguloomohyoid lymphatic node

Sup. belly, omohyoid m.

Sup. thyroid a. & v.

Ext. carotid a.

Facial v.

Hyoid bone

Stylohyoid m.

Mylohyoid m.

Ant. belly, digastric m.

Submental v. & a.

Submandibular gl.

Depressor anguli
oris m.

Mandibular br.,
facial n.

Facial a.

Inf. labial a. & v.

Buccinator m.

Masseter m.

Zygomaticus major m.

Parotid duct

Accessory parotid gland

Transverse facial a. & v.

Zygomatic br., facial n.

Temporal br, facial n.

Fig. 121 Superficial anatomy of the face and neck. (The termination of the facial vein in the external jugular in this specimen is a variation).

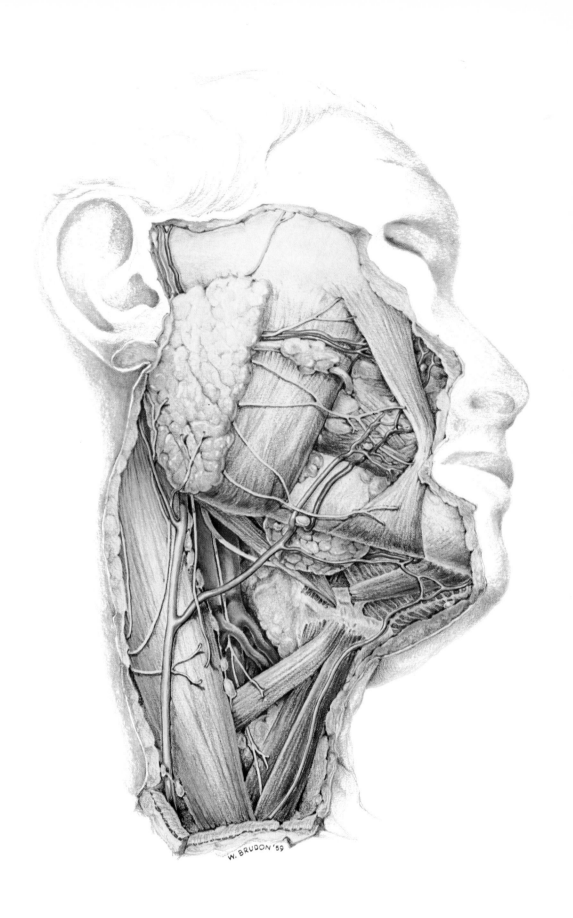

W. BRUDON '59

the anterior triangle of the neck from the mandible to the sternum.

The **supraclavicular nerves** (fig. 120) are branches of a large nerve trunk formed by contributions from the ventral rami of cervical nerves three and four. Like the other cutaneous branches of the cervical plexus, this nerve emerges at the posterior border of the sternocleidomastoid muscle. Descending through the inferior part of the posterior cervical triangle, it divides into three radiating branches which pierce the platysma near the clavicle.

The **medial supraclavicular nerves** are small, cross the clavicular head of the sternocleidomastoid, and terminate in the skin of the base of the neck and over the upper part of the sternum. They also contribute twigs to the sternoclavicular joint. The **intermediate supraclavicular nerves** cross the clavicle and supply the skin over pectoralis major as low as the third rib. Occasionally, one of these nerves passes through the clavicle. The **lateral supraclavicular nerves** pass across or through the insertion of the trapezius muscle and over the lateral part of the clavicle to the shoulder, supplying skin and subcutaneous tissues as far as the distal one-third of the deltoid muscle. Small twigs supply the acromioclavicular joint.

THE FASCIAL PLANES AND COMPARTMENTS OF THE NECK

As in the limbs and other parts of the body, the principal muscle masses, the viscera, and the main nerves and blood vessels of the neck are contained within fascial coverings composed of connective tissue organized in well-defined sheets and membranes. These fasciae have continuities and bony attachments which form fascial planes and compartments in which the deeper structures of the neck are confined.

The Superficial Layer of Cervical Fascia (figs. 122, 123, 124)

The superficial layer of cervical fascia encloses and covers the structures of the neck. It has a cylindrical form, corresponding to the column of the neck, so that it is circularly disposed in a cross section of the neck but also has longitudinal extent and attachments. As seen in cross section (figs. 122, 124), the superficial layer surrounds all the deeper parts of the neck. It splits into two sheets to enclose the sternocleidomastoid and trapezius muscles but exists as a single sheet over the anterior and posterior cervical triangles. In the posterior midline the layer is attached, from above downward, to the external

occipital protuberance, the ligamentum nuchae, and the spinous process of the seventh cervical vertebra. A relatively complete attachment to bone characterizes the uppermost extent of the layer—from behind forward to the external occipital protuberance, the superior nuchal line, the mastoid process, and the inferior border of the mandible. Beyond these lines of bony fixation, the layer continues into the face as the parotid fascia, investing the parotid gland and the masseter muscle, and attaching to the lower border of the zygomatic arch. The further ramifications of the parotid fascia are described on p. 207. The inferior attachments of the sternocleidomastoid and trapezius muscles largely determine the inferior extent of the superficial layer of cervical fascia. Following the sternocleidomastoid muscle, the fascia attaches to both anterior and posterior surfaces of the manubrium sterni and the medial one-third of the clavicle. The separation of the anterior and posterior laminae is prolonged upward above the manubrium for about 3 cm. between the convergent medial borders of the sternocleidomastoid muscles of the two sides. The interfascial space so formed, largely filled with fat but accommodating the transverse link between the two anterior jugular veins, is the **suprasternal space** (fig. 123). The superficial layer of cervical fascia attaches to the clavicle between the sternocleidomastoid and the trapezius and then, with anterior and posterior laminae on either surface of the trapezius, blends with the periosteum of the clavicle, the acromion, and the spine of the scapula. The fascia may be considered as ending at these bony attachments or it may be carried beyond into continuity with the pectoral and deltoid fascia.

In the upper part of the neck, the superficial layer of cervical fascia is attached to the whole extent of the hyoid bone (fig. 123). Above the hyoid bone the layer passes to its mandibular attachment, adhering to the fascia of the anterior belly of the digastric muscle and covering the submental and submandibular triangles. In the latter triangle the layer froms a sheath for the submandibular gland. No connective tissue invades this gland, so that the sheath loosely surrounds it. As noted in connection with the parotid gland, the deep layer of the parotid fascia (p. 207) arises from the deep lamina of the superficial layer on the sternocleidomastoid muscle. The deep layer of the parotid fascia blends with the fascia of the posterior belly of the digastric muscle and then is prolonged to the styloid process and the styloid muscles and over the deep part of the gland to the ramus of the mandible. This deep parotid fascia extends inferiorly and anteriorly on the digastric and

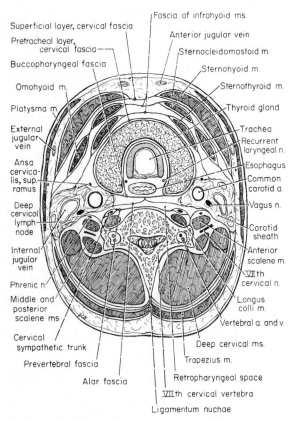

Superficial layer, cervical fascia
Pretracheal layer, cervical fascia
Buccopharyngeal fascia
Omohyoid m.
Platysma m.
External jugular vein
Ansa cervicalis, sup. ramus
Deep cervical lymph node
Internal jugular vein
Phrenic n.
Middle and posterior scalene ms.
Cervical sympathetic trunk
Prevertebral fascia
Alar fascia

Fascia of infrahyoid ms.
Anterior jugular vein
Sternocleidomastoid m.
Sternohyoid m.
Sternothyroid m.
Thyroid gland
Trachea
Recurrent laryngeal n.
Esophagus
Common carotid a.
Vagus n.
Carotid sheath
Anterior scalene m.
VIIth cervical n.
Longus colli m.
Vertebral a. and v.
Deep cervical ms.
Trapezius m.
Retropharyngeal space
VIIth cervical vertebra
Ligamentum nuchae

Fig. 122 A cross section of the neck at the level of the isthmus of the thyroid gland. Semischematic.

prevertebral fascia covers the scalene muscles, levator scapulae, splenius, and the other deep cervical muscles and forms, at the side of the neck, the floor of the posterior cervical triangle. Posteriorly, it attaches to the external occipital protuberance, the ligamentum nuchae, and the spine of the seventh cervical vertebra.

The superior limit of the prevertebral layer is determined by the muscles it encloses. Anteriorly, it ends in the region of the basilar process of the occipital bone, the jugular foramen, and the carotid canal. Posteriorly, the layer attaches to the inner edge of the superior nuchal line and to the mastoid process of the temporal bone. At the base of the neck the prevertebral fascia follows the prevertebral musculature into the chest—into the posterior mediastinum. Laterally, its investment of the scalene muscles carries the fascia into continuity with the fascia of the thoracic wall. The fascia on the deep aspect of the scalene muscles spreads over the cervical pleura, reinforcing it and giving it a superior support. This is known clinically as **Sibson's fascia.** As the lower cervical spinal nerves emerge between the anterior and middle scalene muscles and are extended lateralward over the first rib as the brachial plexus, they carry with them a prolongation of the prevertebral fascia. This prolongation, which covers the subclavian artery, also extends into the axilla as the **axillary sheath.**

stylohyoid muscles to the hyoid bone, where it blends with the attachment of the superficial layer of cervical fascia. Further, the stylomandibular ligament is a thickening of the deep parotid fascia extending from the tip of the styloid process to the angle of the mandible. This ligament establishes continuity with the deep layer of the sheath of the submandibular gland and, uniting the deep investments of the two glands, forms a separation between them.

The Prevertebral Layer of Cervical Fascia (figs. 122, 123, 124, 178)

The prevertebral fascia is also cylindrical in form and defines a smaller cylinder within the larger one formed by the superficial layer of cervical fascia. The cylinder of prevertebral fascia encloses the vertebral column and its associated musculature (fig. 122) and is an extension of the **vertebral fascia** of the back region. The prevertebral fascia crosses the midline anterior to the prevertebral muscles— longus colli and capitis, rectus capitis anterior and lateralis—and is then attached to the tips of the transverse processes. Continuing lateralward, the

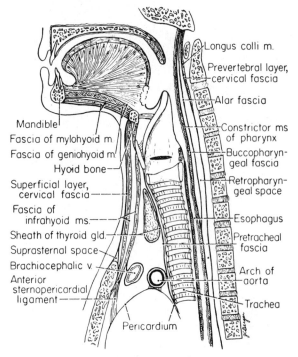

Mandible
Fascia of mylohyoid m.
Fascia of geniohyoid m.
Hyoid bone
Superficial layer, cervical fascia
Fascia of infrahyoid ms.
Sheath of thyroid gld.
Suprasternal space
Brachiocephalic v.
Anterior sternopericardial ligament

Longus colli m.
Prevertebral layer, cervical fascia
Alar fascia
Constrictor ms of pharynx
Buccopharyngeal fascia
Retropharyngeal space
Esophagus
Pretracheal fascia
Arch of aorta
Trachea
Pericardium

Fig. 123 The vertical disposition and attachments of the layers of cervical fascia. Semischematic.

The Fascia of the Infrahyoid Muscles (figs. 122, 123, 124)

The fascia of the infrahyoid muscles comprises two layers which, in a cross section of the neck, describe semicircles from side to side in the anterior half of the neck. Both layers blend with the superficial layer of cervical fascia along the median line. The more **superficial layer** of the infrahyoid fascia encloses the sternohyoid and omohyoid muscles of the two sides and, with these muscles, attaches above to the hyoid bone. This layer is carried along the omohyoid muscle behind the sternocleidomastoid muscle, where it fuses with the superficial layer of cervical fascia on the deep aspect of the latter muscle. Superiorly, the fascia ends at the lateral border of the omohyoid muscle. Below, the fascia follows the inferior attachment of the sternohyoid muscle to the back of the sternum medially and covers the omohyoid muscle lateralward. The fascia attaches to the clavicle behind the sternocleidomastoid muscle and thickens to form a pulley-like sleeve for the intermediate tendon of the omohyoid muscle (fig. 125). Still more lateralward, the fascia invests the inferior belly of the omohyoid muscle and, inferior to it, blends with the sheath of the subclavian vein. The **deeper lamina** of the infrahyoid fascia invests the sternothyroid and thyrohyoid muscles and ends superiorly at the hyoid bone. Lateralward, it extends to the carotid sheath with which it blends along the course of the internal jugular vein. Inferiorly, the layer follows the sternothyroid muscle to its attachment to the back of the sternum. Both the superficial and deep layers of the infrahyoid fascia are prolonged behind the sternum onto the left brachiocephalic vein and the pericardium. The fasciae of the infrahyoid muscles are definite and strong and are dealt with as specific entities in thyroid surgery.

The Cervical Visceral Fasciae (figs. 122, 123, 124)

In the central part of the neck lie the cervical viscera—pharynx, esophagus, larynx, trachea, thyroid and parathyroid glands—enclosed within a cylindrical fascial covering. This consists of two fasciae, the pretracheal layer of cervical fascia anteriorly and the buccopharyngeal fascia posteriorly. These fasciae blend with one another along the attachments of the pharyngeal constrictors to the hyoid bone and thyroid cartilage and at the posteromedial border of the thyroid gland. The **pretracheal fascia** (figs. 122, 123) closely covers the larynx and trachea and splits to enclose and form a sheath for the thyroid gland.

Superiorly, it attaches to the hyoid bone and, laterally, it blends with the buccopharyngeal fascia covering the insertion of the middle pharyngeal constrictor to the lesser and greater horns of the hyoid bone and to the stylohyoid ligament. Below, the pretracheal layer of cervical fascia attaches to the oblique line of the thyroid cartilage, where it also blends with the buccopharyngeal fascia. The sheath of the thyroid gland (figs. 122, 123) is a well-differentiated portion of the pretracheal fascia and encloses the gland on all sides. Below the thyroid gland, the pretracheal fascia covers the trachea and is continuous behind the sternum with the fascial sheaths of the aorta and with the fibrous layer of the pericardium in the thorax.

The **buccopharyngeal fascia** (fig. 178) is an external investment of the upper part of the alimentary tract. It covers the buccinator muscle and the pharynx and is prolonged downward over the posterior surface of the esophagus.

Superiorly, it attaches to the pharyngeal tubercle of the occipital bone and then extends lateralward over the pharyngobasilar fascia to the medial pterygoid plate. As an external covering of the superior constrictor muscle, the buccopharyngeal fascia attaches to the pterygomandibular raphe and then extends forward over the buccinator muscle, following it until its fibers become intermingled with those of other muscles of the lips. Inferiorly, the fascia covers the middle and inferior constrictor muscles of the pharynx and, at the bony and cartilaginous attachments of these muscles, blends with the pretracheal fascia. At the levels of the thyroid gland, the buccopharyngeal fascia of the lower pharynx and esophagus blends with the sheath of the gland at its posteromedial border. The inferior extent of the buccopharyngeal fascia behind the esophagus is uncertain, for the fascia gradually loses its membranous character.

The Carotid Sheath (figs. 122, 124)

The carotid sheath is a tubular investment of the internal and common carotid arteries, the internal jugular vein, and the vagus nerve. The cervical sympathetic trunk lies behind the sheath but is not included within it. Through middle neck levels the superior ramus of the ansa cervicalis lies in the sheath anterior to the carotid artery. Thin over the internal jugular vein, the areolar tissue of the carotid sheath both separates and invests the nerve and vessels. Its fibers adhere to and blend with the sheath of the thyroid gland anteromedially and with the fascia of the deep surface of the sternocleidomastoid muscle anterolaterally. Posteriorly, the carotid sheath is attached to the prevertebral fascia along the tips of the transverse processes of the

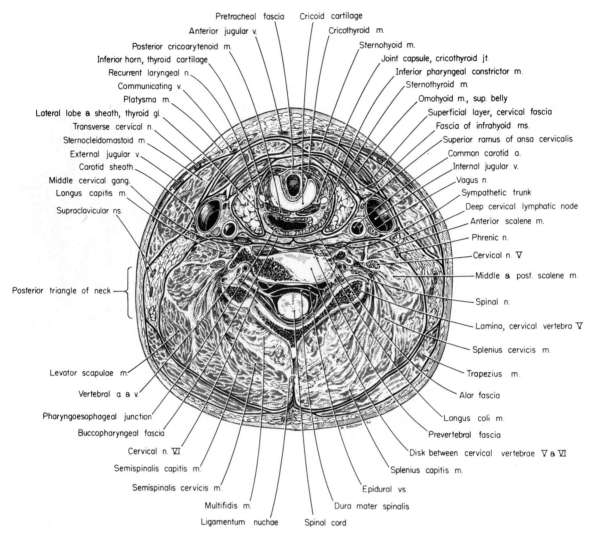

Pretracheal fascia   Cricoid cartilage
Anterior jugular v.   Cricothyroid m.
Posterior cricoarytenoid m.   Sternohyoid m.
Inferior horn, thyroid cartilage   Joint capsule, cricothyroid jt.
Recurrent laryngeal n.   Inferior pharyngeal constrictor m.
Communicating v.   Sternothyroid m.
Platysma m.   Omohyoid m., sup. belly
Lateral lobe & sheath, thyroid gl.   Superficial layer, cervical fascia
Transverse cervical n.   Fascia of infrahyoid ms.
Sternocleidomastoid m.   Superior ramus of ansa cervicalis
External jugular v.   Common carotid a.
Carotid sheath   Internal jugular v.
Middle cervical gang.   Vagus n.
Longus capitis m.   Sympathetic trunk
Supraclavicular ns.   Deep cervical lymphatic node
Anterior scalene m.
Phrenic n.
Cervical n. V
Middle & post. scalene m.
Posterior triangle of neck   Spinal n.
Lamina, cervical vertebra V
Splenius cervicis m.
Levator scapulae m.   Trapezius m.
Alar fascia
Vertebral a & v.   Longus coli m.
Pharyngoesophageal junction   Prevertebral fascia
Buccopharyngeal fascia   Disk between cervical vertebrae V & VI
Cervical n. VI   Splenius capitis m.
Semispinalis capitis m.   Epidural vs.
Semispinalis cervicis m.   Dura mater spinalis
Multifidis m.
Ligamentum nuchae   Spinal cord

Fig. 124   A cross section of the neck at the level of the cricoid cartilage.

vertebrae. The sternothyroid layer of the infrahyoid fascia ends in the carotid sheath over the internal jugular vein. In the upper part of the neck, the carotid sheath fuses with the fascia of the stylohyoid and posterior belly of the digastric muscles and, at the base of the skull, is attached around the jugular foramen and carotid canal. At the root of the neck, the sheath fuses with the scalene fascia and, in the chest, with the sheaths of the great vessels and the fibrous pericardium.

The Compartments and Interfascial Spaces of the Neck (figs. 122, 123, 124)

The cylindrical nature of the cervical fasciae gives emphasis to the fact that they are enclosures of longitudinal compartments. The superficial layer of cervical fascia encloses all except the most super-

ficial structures of the neck, but within it are other compartments. The prevertebral layer forms a complete enclosure for a compartment containing the cervical vertebrae and their associated longitudinal musculature, vessels, and nerves. The cervical visceral fasciae delimit a compartment which lodges all the visceral organs contained in the neck. The carotid sheath is a longitudinal neurovascular enclosure. The potential intervals, or clefts, between fascial planes constitute interfascial spaces of much importance in clinical medicine. The space of the posterior triangle of the neck is formed between the superficial and the prevertebral layers of cervical fascia. In the anterior cervical triangle there are potential interfascial spaces both superficial and deep to the infrahyoid muscle fasciae and also between them. The largest and most important interfascial interval in the neck is the **retropharyngeal space.** This is an areolar interval between the buccopharyngeal fascia anteriorly and the prevertebral fascia posteriorly. It is closed laterally by the carotid sheaths. The space extends superiorly to the base of the skull and inferiorly into the posterior mediastinum of the chest. Its loose areolar connective tissue accommodates the movements of the pharynx and associated parts during deglutition. The retropharyngeal lymph nodes lie in the lateral extremities of the space near the base of the skull. They constitute outlying nodes of the superior deep cervical chain of lymph nodes.

Forming a further subdivision of the retropharyngeal space is the **alar fascia.** The alar fascia is usually a delicate but sometimes a very definite plane of fascia which is attached along the midline of the buccopharyngeal fascia from the skull to the level of the seventh cervical vertebra. From this attachment, the fascia extends lateralward to terminate in the carotid sheath. Grodinsky and Holyoke ('38) showed it to be concerned in the formation of the carotid sheath.

The **lateral pharyngeal space** (figs. 157, 178) is a fat-filled area between the lateral aspect of the pharynx and the pterygoid muscles (medial wall of masticator and parotid spaces, see p. 215). It is limited posteriorly by the carotid sheath and below by the sheath of the submandibular gland. It extends above to the base of the skull. The space is traversed by the stylopharyngeus and styloglossus muscles. This area can be invaded by infections from the base of the tongue, the teeth, the tonsil, and the pharynx, and by its continuities with the retropharyngeal space, pass such material on as far as the posterior mediastinum.

THE ANTERIOR CERVICAL TRIANGLE

The anterior cervical triangle has been defined and its boundaries and subdivisions described (p. 145). The contents of the anterior triangle remain to be considered.

The Infrahyoid Muscles

The infrahyoid, or strap, muscles constitute a double-layered plane of muscle in the front of the neck superficial to the larynx, trachea, and thyroid gland. The group includes:

Sternohyoid   Sternothyroid
Omohyoid   Thyrohyoid

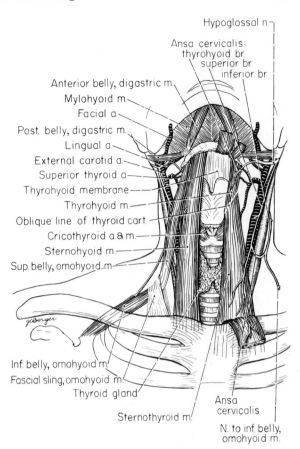

Fig. 125   The structures of the anterior cervical triangle.

The **sternohyoid** (figs. 121, 125) is a thin, narrow muscle which arises from the posterior aspect of the sternal end of the clavicle and from the back of the manubrium sterni. It ascends, with a slight medialward inclination, to insert on the lower border of the body of the hyoid bone. The sternohyoid muscle is innervated by one or more branches of the superior ramus of the ansa cervicalis which enter its lateral margin. The sternohyoid is the more medial muscle of the superficial sheet of infrahyoid muscles.

The **omohyoid muscle** (figs. 121, 125) is the lateral member of the superficial sheet of infrahyoid muscles. It consists of two bellies with an intermediate tendon. The **inferior belly** arises from the upper border of the scapula adjacent to and occasionally also from its superior transverse ligament. The inferior belly is narrow and passes obliquely upward and medialward through the lower part of the posterior cervical triangle. Under cover of the sternocleidomastoid muscle, the inferior belly ends in an intermediate tendon which is bound down to the clavicle by a strong process of the fascia of the infrahyoid muscles (p. 153). Turning abruptly upward from this fixation, the **superior belly** crosses the carotid sheath and lies lateral to the sternohyoid muscle to which it is joined by the superficial layer of the infrahyoid muscle fascia. The superior belly inserts into the lower border of the body of the hyoid bone lateral to the insertion of the sternohyoid muscle. The ansa cervicalis supplies both bellies of the muscle, the superior belly on its deep surface somewhat below its center and the inferior belly on the deep aspect of its proximal one-third.

The **sternothyroid muscle** (figs. 121, 125) is wider than the sternohyoid, under which it lies. It arises from the posterior surface of the manubrium sterni below the origin of the sternohyoid muscle and from the cartilage of the first rib. Lying closely behind the sternohyoid at the base of the neck, the sternothyroid muscle diverges lateralward from it as it ascends to insert on the oblique line of the thyroid cartilage (p. 184). One or two branches of the ansa cervicalis nerve enter the ventral surface of the muscle near its lateral border.

The **thyrohyoid muscle** (figs. 125, 126) arises from the oblique line of the thyroid cartilage and inserts on the lower border of the greater cornu of the hyoid bone. It appears to be a continuation of the sternothyroid muscle. The thyrohyoid is innervated from the hypoglossal nerve by a branch derived from the ansa cervicalis containing fibers of the first and second cervical nerves.

Actions. The infrahyoid muscles fix and steady the hyoid bone and are concerned, with the suprahyoid muscles, in movements of the tongue, hyoid bone, and larynx in both swallowing and phonation. The sternohyoid and omohyoid draw down on the hyoid; the sternothyroid, on the thyroid cartilage.

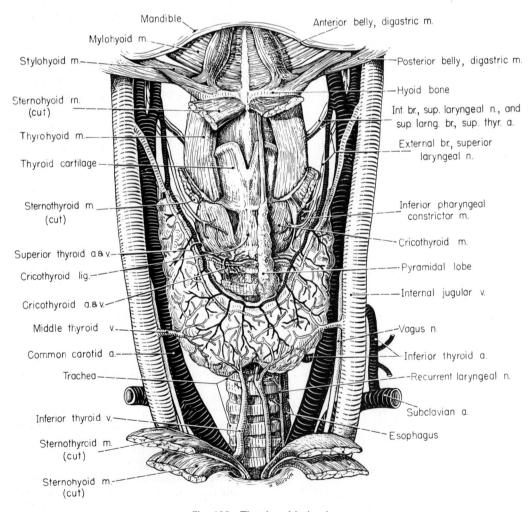

Fig. 126   The thyroid gland.

The latter muscle, acting with the thyrohyoid, also draws down on the hyoid. With the hyoid bone fixed from above, the thyrohyoid muscle draws the larynx upward in the act of swallowing and in the production of high notes in phonation. The sternothyroid draws down on the larynx during the production of low notes.

THE THYROID GLAND

The thyroid gland (figs. 122, 126, 130) is an endocrine gland which produces a hormone, **thyroxin,** of importance in controlling the rate of oxidation (metabolic rate) in the body. Its secretion is stored in small closed cavities, the follicles of the thyroid. The gland lies in the front of the neck. It has the general shape of an H, the middle bar of which, as the isthmus of the gland, arches from side to side over the upper rings of the trachea. The vertical limbs of the H are represented by two conical lateral lobes, blunted below and tapered above. The apices of the lateral lobes are limited by the oblique line of the thyroid cartilage to which attaches the overlying sternothyroid muscle. The lower rounded extremities of the lateral lobes usually descend to the level of the sixth tracheal ring. The **isthmus** of the gland unites the two lateral lobes across tracheal rings two, three, and four. A **pyramidal lobe,** of frequent occurrence but variable size, extends upward from the isthmus or from the junction of the isthmus and one of the lateral lobes, usually the left.

A band of connective tissue continues from the apex of the pyramidal lobe as far as the hyoid bone in most cases. The thyroid gland is developed largely from a median diverticulum of the ventral wall of the pharynx. The tubular primordium bifurcates and subdivides into a series of cellular cords from which the isthmus and the lateral lobes of the gland are formed. The connection with the pharynx is the **thyroglossal duct,** opening, in the embryo, at the foramen cecum of the tongue. The pyramidal lobe and its connective tissue continuation is a visible remnant of this development.

Clinical Note In rare cases there will be a failure of obliteration of the epithelium of the thyroglossal duct and a thyroglossal duct cyst may form. This may occur anywhere between the pyramidal lobe and the tongue.

The thyroid gland is overlain by the infrahyoid muscles and their associated fasciae. The gland arches across the trachea and esophagus (fig. 122), and the upper parts of the lateral lobes are molded against the cricoid and thyroid cartilages. The carotid sheaths containing the carotid arteries and internal jugular veins lie posterolateral to the gland,

in relation with its rounded posterior border. The anteromedial border of the gland is thin and inclines obliquely downward toward the midline. Along it lies a branch of the superior thyroid artery. The thyroid gland is invested by a proper, fibrous **capsule** which sends strands and septa deeply into its substance. A loose **sheath** is provided by the pretracheal layer of cervical fascia (p. 153 & fig. 122). This sheath is attached, along the posterior border of the gland, to the buccopharyngeal fascia, the two fasciae delimiting the cylindrical visceral compartment of the neck. The blood vessels ramifying on the gland do so between its sheath and capsule. Several ligaments are derived from the pretracheal fascia. Two lateral ligaments attach the medial surfaces of the lateral lobes to the cricotracheal membrane and to the cricoid cartilage. A less clear median ligament passes from the ventral surface of the cricoid cartilage to the dorsum of the isthmus. Closely applied to the posterior surface of the gland are the two pairs of parathyroid glands. They lie under the sheath of the thyroid gland and are sometimes embedded in the substance of the gland.

The **blood supply** of the thyroid gland (figs. 121, 126, 127, 139) is exceedingly rich, a characteristic of endocrine glands generally. Its **arteries** are the paired superior and inferior thyroid arteries. Its **veins** are variable but always include paired superior and inferior thyroid veins and usually middle thyroid veins as well. The **superior thyroid artery** is usually the first branch of the external carotid artery. The common carotid artery divides into its external and internal branches at the level of the upper border of the thyroid cartilage, and the external carotid gives rise to the superior thyroid artery opposite the thyrohyoid membrane. The superior thyroid artery (figs. 121, 126, 127, 139) runs downward and forward to the apex of the lateral lobe of the gland.

The superior thyroid artery sends a small **infrahyoid branch** along the lower border of the hyoid bone and a **sternocleidomastoid branch** downward and lateralward across the carotid sheath to the sternocleidomastoid muscle. A **superior laryngeal branch** arises opposite the thyrohyoid membrane which it pierces in company with the internal branch of the superior laryngeal nerve. Both ramify in the submucosa of the larynx. A small **cricothyroid branch** runs over the cricothyroid membrane, communicating with the artery of the opposite side, and **muscular branches** reach the inferior pharyngeal constrictor muscle and the esophagus. Reaching the superior pole of the gland, the superior thyroid artery divides into two branches. The larger **anterior branch** descends along the anterior border of the gland and ramifies on its anterior surface, anastomosing with the branch from the opposite side; the **posterior branch** descends on the posterior surface of the gland and anastomoses with the inferior thyroid artery.

Clinical Note The dissector should note the close spatial relationship between the superior thyroid artery and the external branch of the superior laryngeal nerve. The latter must not be included in a ligature placed on the superior thyroid artery lest the cricothyroid muscle be paralyzed, resulting in an alteration of the patient's voice.

The **inferior thyroid artery** is the largest branch of the thyrocervical trunk of the subclavian artery (p. 179 & figs. 126, 130, 139). The inferior thyroid artery ascends along the medial border of the anterior scalene muscle and, opposite the lower border of the cricoid cartilage, turns medialward behind the carotid sheath and the sympathetic trunk to lie behind the lateral lobe of the thyroid gland (fig. 130). The artery then descends to the inferior pole of the gland and finally turns upward and divides to supply the gland. As the inferior thyroid artery arches behind the carotid sheath, a slender **ascending cervical branch** runs superiorly in the interval between the anterior scalene and the longus capitis muscles. It supplies muscular, vertebral, and spinal twigs.

The inferior thyroid artery supplies adjacent muscles and gives off **pharyngeal, tracheal,** and **esophageal branches.** An **inferior laryngeal branch** ascends behind the larynx under cover of the inferior constrictor muscle of the pharynx and in company with the inferior laryngeal branch of the vagus nerve. It supplies the muscles and mucous membrane of the larynx and anastomoses with the superior laryngeal branch of the superior thyroid artery. The inferior thyroid artery usually divides into two terminal **glandular branches.** The inferior laryngeal branch of the vagus nerve passes frequently between but also variably in front of or behind these branches (Reed, '43) and is usually identified and avoided in thyroid surgery. The glandular branches of the artery distribute posteriorly and inferiorly, anastomosing with the superior thyroid artery and giving off small arteries to the parathyroid glands. An unpaired and variable **thyroidea ima artery** (10 per cent of cases) may arise from the brachiocephalic artery or from the arch of the aorta. When present, it ascends on the ventral surface of the trachea, which it supplies, and then distributes to the isthmus of the gland. It anastomoses with branches of the inferior thyroid arteries.

Superior thyroid veins accompany the superior thyroid arteries but are the only thyroid veins which do follow arteries. The **superior thyroid vein** (figs. 121, 126, 128) arises over the anterolateral surface of the gland. It receives tributaries corresponding to the branches of the superior thyroid artery—sternocleidomastoid, superior laryngeal, infrahyoid, and cricothyroid veins. The superior thyroid vein crosses the common carotid artery and empties into the internal jugular vein above the thyroid cartilage. Occasionally, it makes a common entry into the jugular with the lingual and facial veins. The **middle thyroid vein** (figs. 126, 128) arises from the venous plexus on the lateral surface of the gland. Leaving the gland near its lower pole, the vein crosses the common carotid artery and empties into the lower end of the internal jugular vein. The **inferior thyroid veins** (figs. 126, 128) arise in the venous plexus on the thyroid gland, communicating with the superior and middle thyroid veins. The two inferior thyroid veins descend on the surface of the trachea and, behind the manubrium sterni, empty into the left and right brachiocephalic veins, respectively; they have valves just proximal to their terminations. Occasionally, they unite to form a single vein which then empties into the left brachiocephalic vein. The inferior thyroid veins have esophageal, tracheal, and inferior laryngeal tributaries which correspond to the arteries of the same names. A high union of the right and left inferior thyroid veins constitutes the occasional **thyroidea ima vein.**

The **lymphatic drainage** of the thyroid gland is by a variety of routes. Its lymphatic channels begin in a subcapsular plexus from which vessels lead upward by both medial and lateral channels and downward by both medial and lateral channels. From the upper parts of the lateral lobes and the superior part of the isthmus, medial and lateral channels pass to the superior deep cervical nodes, sometimes interrupted in prelaryngeal nodes. Inferior to the gland are pretracheal and paratracheal nodes which receive the medial inferior drainage. The lateral inferior drainage reaches the inferior deep cervical nodes along the internal jugular vein. Dorsal to the gland, minor superior channels reach the retropharyngeal nodes; minor inferior channels, the paratracheal nodes along the recurrent laryngeal nerve.

The **nerves** of the thyroid gland are composed of nonmedullated postganglionic fibers which arise in the middle cervical and cervicothoracic ganglia. They reach the gland by way of cardiac, superior laryngeal, and inferior laryngeal nerves and along the arteries. Some of the nerve fibers are vasomotor, but others have been shown to end in close relation to the epithelial cells of the follicles of the gland.

Clinical Note Small, oval accessory thyroid glands are common near the hyoid bone, along the line of developmental migration as signified by the thryoglossal duct. Accessory glands may also occur in the superior mediastinum associated with the thymus gland.

THE PARATHYROID GLANDS (figs. 130, 139)

The parathyroid glands are from two to six small, flattened, oval bodies lying against the dorsum of the thyroid gland (Heinbach, '33). They are located under the sheath of the thyroid gland but have their own proper capsule. The parathyroids have only a spatial relationship with the thyroid, being very different in microscopic structure and being separate endocrine glands. They produce a hormone, **parathormone,** which maintains the normal relation between blood and skeletal calcium. Removal of all parathyroid tissue leads to the prompt development of tetany and death, but several of the glands may be lost without evident effect. The parathyroid glands usually measure from 5 to 7 mm. in length, 3 to 4 mm. in width, and 1 to 2 mm. in thickness. The superior glands lie, normally, at the level of the lower border of the cricoid cartilage; the inferior ones are near the inferior pole of the thyroid gland.

The parathyroid glands develop from the dorsal diverticula of the third and fourth pharyngeal pouches, and this circumstance has led to the expectation that there will be two pairs of glands, designated as parathyroids III and parathyroids IV. Parathyroids III are normally drawn by the developing thymus (also from the third pouch) lower in the neck and so constitute the inferior glands. They are less constant in position than the superior pair and are sometimes found below thyroid levels, even in the superior mediastinum of the thorax, in close relation to the thymus.

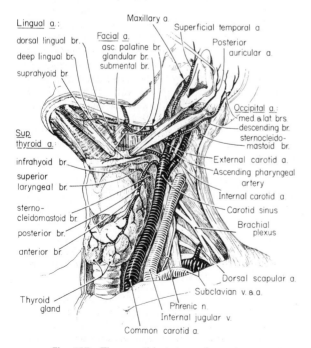

Lingual a.:
dorsal lingual br.
deep lingual br.
suprahyoid br.

Maxillary a.
Facial a.
asc. palatine br.
glandular br.
submental br.

Superficial temporal a
Posterior auricular a.

Sup.
thyroid a.:
infrahyoid br.
superior laryngeal br.
sterno-cleidomastoid br.
posterior br.
anterior br.

Occipital a.:
med. a lat. brs.
descending br.
sternocleido-mastoid br.
External carotid a.
Ascending pharyngeal artery
Internal carotid a
Carotid sinus
Brachial plexus

Thyroid gland

Dorsal scapular a.
Subclavian v. a a.
Phrenic n.
Internal jugular v.
Common carotid a.

Fig. 127   The carotid system of arteries.

The parathyroid glands are supplied by separate small branches of the **inferior or superior thyroid arteries** or by twigs arising from the longitudinal anastomosis between these vessels. The **venous drainage** is into the thyroid plexus of veins, and the **lymphatic channels** drain with those of the thyroid gland. The **nerve supply** is abundant; it comes from the thyroid branches of the cervical sympathetic ganglia and appears to be vasomotor in function.

THE CAROTID SHEATH AND ITS CONTENTS

The carotid sheath (p. 153 & figs. 122, 124) consists of aerolar tissue enclosing the carotid artery and the internal jugular vein. Its fibers fuse behind with the prevertebral layer of cervical fascia; anteromedially, with the pretracheal layer; and anterolaterally, with the superficial layer of cervical fascia on the deep aspect of the sternocleidomastoid muscle. The sheath is thin over the vein, thicker around the artery, and both separates and encloses the artery, the vein, and the vagus nerve. In the sheath the artery is medial and the vein is lateral with the nerve behind and between the vessels. The superior ramus of the ansa cervicalis nerve, conducting motor impulses to the infrahyoid muscles from the upper cervical nerves, lies on the sheath anterior to the artery but sometimes is embedded in the sheath. The sympathetic trunk lies behind the medial portion of the sheath.

Common and External Carotid Arteries (figs. 121, 127, 139)

The course of the carotid artery may be defined by a line beginning below at the sternoclavicular joint and terminating above midway between the angle of the mandible and the mastoid process of the temporal bone. This is the 'carotid line.' It represents the common carotid artery as high as the upper border of the thyroid cartilage and the external carotid above that level. The **common carotid arteries** are not of equal length. The right common carotid springs from the bifurcation of the brachiocephalic artery behind the right sternoclavicular joint, whereas the left common carotid is a direct branch of the arch of the aorta at a little distance along the arch. As a consequence, the left common carotid has a course of about 2 cm. in the superior mediastinum of the thorax before entering the neck. As variants, the left common carotid artery may be a branch of the brachiocephalic artery, or a short common stem may exist for it and the left subclavian artery (left brachiocephalic trunk).

The cervical portions of the common carotid arteries are about 8.5 cm. long. Separated below only by the width of the jugular notch, the two vessels diverge as they ascend, and above, the larynx and the thyroid gland project forward between them. At the root of the neck, the common carotid arteries are located deeply behind the sternocleidomastoid muscle and the infrahyoid muscles. Just above the clavicle, the anterior jugular vein crosses the artery as it passes behind the sternocleidomastoid muscle. Ascending behind the superior belly of the omohyoid muscle, each artery enters the carotid triangle bounded by the adjacent borders of the sternocleidomastoid and omohyoid muscles and the posterior belly of the digastric muscle. Here it is crossed by the sternocleidomastoid branch of the superior thyroid artery and by the superior and middle thyroid veins. At the level of the lower border of the cricoid cartilage, the inferior thyroid artery arches inward behind the common carotid artery and behind the prevertebral layer of cervical fascia.

The common carotid arteries have no collateral branches but divide terminally into the external and internal carotid arteries. The **internal carotid** (p. 195) has no branches in the neck but continues within the carotid sheath, in company with the internal jugular vein, to the base of the skull, where it enters the cranium (p. 281) to become the principal artery of the brain and of the structures within the orbit. Usually, it lies at first posterolateral to the external carotid artery but becomes internal to the latter as it ascends. The internal carotid artery occasionally (Faller, '46) arises anterior or anteromedial to the external carotid.

Carotid Sinus The terminal portion of the common carotid and the root of the internal carotid arteries are dilated for a distance of about 1 cm. This **carotid sinus** region is important in blood pressure regulation. Its walls are especially elastic and contain modified end organs which respond to changes in blood pressure. The peripheral nerve connected with these end organs is the carotid branch of the glossopharyngeal (cranial IX) nerve.

The External Carotid Artery (figs. 127, 139)

The external carotid artery, approximately equal in caliber to the internal carotid artery, extends from the upper border of the thyroid cartilage to the neck of the mandible, where it divides into the superficial temporal and maxillary arteries. It is relatively superficial in the carotid triangle, where it is covered by the superficial layer of cervical fascia and is overlapped by the anterior border of the sternocleidomastoid muscle. Here it is crossed by the hypoglossal nerve, by the facial and lingual veins, and sometimes by the superior thyroid vein. Leaving the carotid triangle, the artery is crossed by the posterior belly of the digastric and the stylohyoid muscles and separates these muscles from the stylopharyngeus muscle. Above, the artery grooves the deep surface of the parotid gland in company with the retromandibular vein. The facial nerve crosses both vessels. The external carotid artery has eight branches. Of these, four arise in the carotid triangle: the **superior thyroid,** the **lingual,** and the **facial arteries** from its anterior aspect; the **ascending pharyngeal artery** from its medial side. Two posterior branches, the **occipital** and **posterior auricular arteries,** arise at the level of the posterior belly of the digastric muscle; two terminal branches, the **superficial temporal** and **maxillary arteries,** take origin behind the neck of the mandible.

The **superior thyroid artery** is usually the first branch of the external carotid but frequently (36 per cent—Faller, '46) springs from the carotid bifurcation. It arises opposite the thyrohyoid membrane below the level of the superior horn of the hyoid bone (figs. 121, 126, 127, 130, 139). It may arch slightly upward but then turns downward to the thyroid gland. The artery is described in connection with the thyroid gland (p. 157).

The **lingual artery** (figs. 126, 139) arises from the external carotid opposite the tip of the greater horn of the hyoid bone. In the upper part of the carotid triangle, the lingual artery ascends past the greater horn of the hyoid bone and disappears under the posterior border of the hyoglossus muscle. It is crossed by the hypoglossal nerve on its way to the tongue (p. 231 & figs. 139, 180).

The **facial artery** (figs. 121, 127, 139) arises immediately above the lingual. It passes obliquely upward deep to the posterior belly of the digastric and the stylohyoid muscles over which it arches to lie in a groove on the deep surface of the submandibular gland. The artery curves over the border of the mandible at the anterior edge of the masseter muscle to enter the face, and the description of its facial branches is given on p. 203.

In the neck the facial artery gives off ascending palatine, tonsillar, glandular, muscular, and submental branches. The **ascending palatine branch** is small (fig. 180). It ascends between the styloglossus and stylopharyngeus muscles, turns downward over the upper border of the superior constrictor muscle of the pharynx, and enters the soft palate in company with the levator veli palatini muscle. It supplies the upper pharyngeal wall, the soft palate, the palatine tonsil, and the auditory tube. The small **tonsillar branch** (fig. 180) arises close to the ascending palatine. It ascends on the pharynx, pierces the superior constrictor muscle, and ends in the tonsil. **Glandular branches** are two or three twigs which enter the submandibular gland; **muscular branches** pass to the medial pterygoid and stylohyoid muscles. The **submental artery** (figs. 121, 127) arises from the facial artery near the lower border of the mandible. It runs toward the chin on the superficial surface of the mylohyoid muscle, giving branches to the muscle and to both the submandibular and sublingual glands. Turning over the border of the mandible, the artery anastomoses with the mental and the inferior labial arteries.

The **ascending pharyngeal artery** (figs. 127, 130) arises from the medial aspect of the external carotid artery in the carotid triangle. It is frequently the first or second branch of the artery. Long and slender, the vessel ascends at the side of the pharynx medial to the internal carotid artery and to the stylopharyngeus muscle and anterior to the prevertebral muscles.

Its branches are small and variable, but, aside from its named branches, it distributes twigs to the prevertebral muscles, the deep cervical lymph nodes, and the sympathetic trunk. **Pharyngeal branches** (fig. 130), three or four in number, supply the superior and middle constrictor muscles. A **palatine branch** (fig. 180) passes over the upper border of the superior constrictor muscle to the soft palate and its muscles and to the palatine tonsil. The course of this artery is like that of the ascending palatine branch of the facial artery, and it may replace the latter. The **inferior tympanic branch** accompanies the tympanic branch of the glossopharyngeal nerve to the tympanic cavity to supply the medial wall of the cavity and anastomose with other tympanic arteries. One or more **meningeal branches** enter the cranium through the hypoglossal canal, the foramen lacerum, or the jugular foramen to supply the dura mater. The one passing through the jugular foramen is the **posterior meningeal.**

The **occipital artery** (figs. 127, 139, 142, 156, 161) arises from the posterior aspect of the external carotid near the lower margin of the posterior belly of the digastric muscle. It courses upward and backward to the interval between the transverse process of the atlas and the mastoid process and then occupies the groove on the mastoid process. Proceeding between the muscles of the back of the neck, it pierces the superficial layer of cervical fascia between the sternocleidomastoid and trapezius along the superior nuchal line and enters the scalp.

Muscular branches of the occipital artery are given off for the supply of the digastric, the stylohyoid, and the posterior cervical muscles. A **sternocleidomastoid branch** generally arises from the occipital close to its origin but may be a separate branch of the external carotid. Descending over the hypoglossal nerve, it enters the muscle in company with the accessory nerve. The **meningeal branch** enters the skull through the jugular foramen to reach the dura mater of the posterior cranial fossa. An **auricular branch** ascends over the mastoid process to the back of the ear. It sometimes takes the place of the posterior auricular artery. The **mastoid branch** is a small twig that passes into the skull through the mastoid foramen for the supply of the dura mater, the diploe, and the mastoid air cells. The **descending branch** (fig. 142), the largest branch of the occipital, arises from it as it crosses the obliquus capitis superior muscle. At the lateral border of the semispinalis capitis muscle the descending branch divides into superficial and deep portions. The **superficial branch** descends between the splenius and semispinalis capitis muscles and anastomoses with the ascending branch of the transverse cervical artery. The **deep branch** descends between the semispinalis capitis and cervicis muscles and anastomoses with the deep cervical artery from the costocervical trunk of the subclavian and with branches of the vertebral artery of the same system. These anastomoses are important collateral paths in case of interference with blood flow in either the carotid or subclavian artery. The **terminal branches** of the occipital artery in the scalp are described on p. 210.

The **posterior auricular artery** (figs. 127, 139, 156, 161) arises from the posterior aspect of the external carotid at the level of the upper border of the posterior belly of the digastric muscle. It ascends through the parotid fossa to the notch between the external acoustic meatus and the mastoid process. Aside from muscular branches and a supply to the parotid gland, the posterior auricular artery provides stylomastoid, auricular, and occipital branches which distribute largely in the scalp (p. 211).

The **superficial temporal** and **maxillary arteries** (figs. 127, 139, 156, 161) are the terminal branches of the external carotid. They arise behind the neck of the mandible, and the superficial temporal artery continues upward into the scalp (p. 212). The maxillary artery has a wide area of distribution in the infratemporal fossa and the deeper parts of the face and is described in connection with those regions (p. 224).

The Internal Jugular Vein (figs. 128, 139)

This is the largest vein of the head and neck and collects the blood from the brain, from the superficial parts of the face, and from the neck. It begins, above, in the posterior compartment of the jugular foramen as a continuation of the sigmoid sinus of

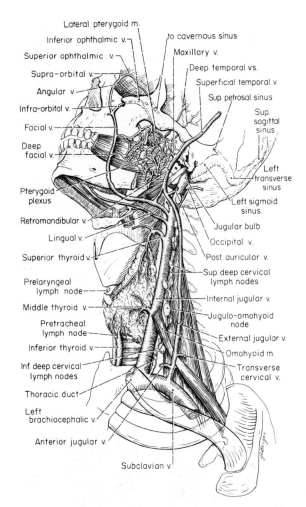

Lateral pterygoid m.
Inferior ophthalmic v.
Superior ophthalmic v.
Supra-orbital v.
Angular v.
Infra-orbital v.
Facial v.
Deep facial v.
Pterygoid plexus
Retromandibular v.
Lingual v.
Superior thyroid v.
Prelaryngeal lymph node
Middle thyroid v.
Pretracheal lymph node
Inferior thyroid v.
Inf. deep cervical lymph nodes
Thoracic duct
Left brachiocephalic v.
Anterior jugular v.
Subclavian v.

to cavernous sinus
Maxillary v.
Deep temporal vs.
Superficial temporal v.
Sup. petrosal sinus
Sup. sagittal sinus
Left transverse sinus
Left sigmoid sinus
Jugular bulb
Occipital v.
Post. auricular v.
Sup. deep cervical lymph nodes
Internal jugular v.
Jugulo-omohyoid node
External jugular v.
Omohyoid m.
Transverse cervical v.

Fig. 128   The deep veins of the face and neck. (Modified from Pernkopf.)

the dura mater and ends, below, behind the sterno-clavicular joint in the brachiocephalic vein. The brachiocephalic vein drains into the heart through the superior vena cava. Throughout its course, the internal jugular vein lies at the side of the carotid artery within the carotid sheath. Above, the vein is posterolateral to the artery and then becomes lateral to it; at the root of the neck, the vein lies somewhat anterior to the common carotid artery. As its beginning, the vein has a dilatation, the **superior bulb.** Another dilatation, the **inferior bulb,** occurs just above its termination. Near its termination, the internal jugular vein has a single bicuspid valve arranged to prevent upward flow of blood through the vein. The free margins of the valve flaps lie approximately 4 mm. from the end of the vein (Weathersby, '56). The **tributaries** of the internal jugular vein in addition to the sigmoid

sinus are the inferior petrosal sinus, the lingual, pharyngeal, facial, superior and middle thyroid veins, and, sometimes, the occipital vein.

The **inferior petrosal sinus** (p. 279) leaves the skull through the anterior part of the jugular foramen and empties into the superior bulb.

The **lingual veins** (p. 232) arise on the dorsum, the sides, and the under surface of the tongue and, accompanying the lingual artery, form a single vein emptying into the internal jugular at the level of origin of the lingual artery. The **vena comitans of the hypoglossal nerve** begins near the tip of the tongue. It may join the lingual vein or, passing backward on the hyoglossus muscle, it may terminate in the facial vein.

The **pharyngeal veins** arise from the venous plexus on the wall of the pharynx and empty into the internal jugular vein at the level of the angle of the mandible.

The course and communications of the **facial vein** are described on p. 204. It empties into the internal jugular vein opposite or just below the level of the hyoid bone. The facial vein may receive the superior thyroid, the lingual, or the sublingual veins.

The **superior** and **middle thyroid veins** are described in the consideration of the thyroid gland (p. 158).

The **occipital vein** is described with the veins of the scalp (p. 213).

The Vagus Nerve (figs. 129, 130, 139)

The vagus, or tenth cranial nerve, is the longest of the cranial nerves (vagus = vagrant). Arising from the medulla oblongata of the brainstem, it passes through neck, chest, and abdomen and gives off branches in each of these regions.

Its functions are indicated in the following listing of its components:

Special visceral efferent: motor innervation of the voluntary muscles of the larynx, pharynx, and palate (except the tensor veli palatini), and of the upper two-thirds of the esophagus. Fibers of this type include those of the cranial portion of the accessory nerve which are combined with the vagus.

General visceral efferent: parasympathetic preganglionic fibers to involuntary muscle and glands of the heart, esophagus, stomach, trachea, bronchi, intestines, and other abdominal viscera.

General visceral afferent: visceral sensation from the carotid body, base of tongue, pharynx, larynx, trachea, bronchi, lungs, heart, esophagus, stomach, and intestines.

Special visceral afferent: taste from the epiglottis and palate.

General somatic afferent: cutaneous sensation from the external ear and external acoustic meatus and sensation from the dura mater of the posterior cranial fossa.

The vagus nerve arises intracranially by the convergence of eight or ten rootlets which emerge from the medulla oblongata in the groove dorsal to the inferior olive. These rootlets are in line superiorly

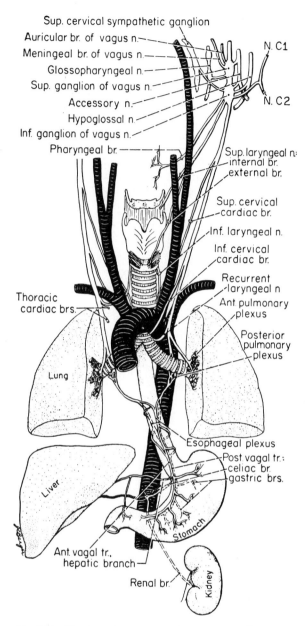

Sup. cervical sympathetic ganglion
Auricular br. of vagus n.
Meningeal br. of vagus n.
Glossopharyngeal n.
Sup. ganglion of vagus n.
Accessory n.
Hypoglossal n.
Inf. ganglion of vagus n.
Pharyngeal br.
N. C1
N. C2
Sup. laryngeal n.:
internal br.
external br.
Sup. cervical cardiac br.
Inf. laryngeal n.
Inf. cervical cardiac br.
Recurrent laryngeal n.
Ant. pulmonary plexus
Thoracic cardiac brs.
Posterior pulmonary plexus
Lung
Esophageal plexus
Post vagal tr.:
celiac br.
gastric brs.
Liver
Stomach
Ant. vagal tr., hepatic branch
Renal br.
Kidney

Fig. 129   The branches, communications, and distribution of the vagus nerve. (Redrawn and modified after Cunningham.)

nerve in the jugular foramen. The cell bodies of this ganglion are concerned primarily with the general somatic afferent (cutaneous) component of the nerve. After its exit from the foramen, the nerve exhibits a fusiform swelling of 2.5 cm. length which is the **inferior ganglion** (nodose ganglion). The cell bodies of this ganglion are concerned with the visceral afferent components of the nerve. In the region of the superior ganglion, a number of communications are made with other nerves: the cranial portion of the accessory nerve, the inferior ganglion of the glossopharyngeal, and the superior cervical sympathetic ganglion. The cranial part of the accessory nerve, the source of many of the motor fibers of the vagus, joins the vagus just proximal to the inferior ganglion. Further communications at the inferior ganglion are made by the vagus with the hypoglossal nerve, the superior cervical sympathetic ganglion, and the loop between

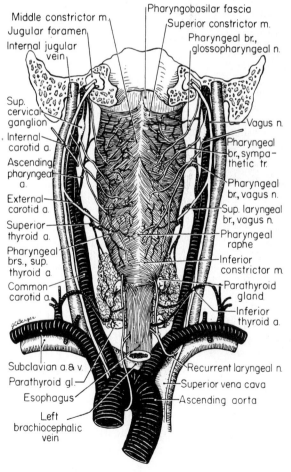

Middle constrictor m.
Jugular foramen
Internal jugular vein
Pharyngobasilar fascia
Superior constrictor m.
Pharyngeal br., glossopharyngeal n.
Sup. cervical ganglion
Internal carotid a.
Ascending pharyngeal a.
External carotid a.
Superior thyroid a.
Pharyngeal brs., sup. thyroid a.
Common carotid a.
Vagus n.
Pharyngeal br, sympathetic tr.
Pharyngeal br, vagus n.
Sup. laryngeal br, vagus n.
Pharyngeal raphe
Inferior constrictor m.
Parathyroid gland
Inferior thyroid a.
Subclavian a. & v.
Parathyroid gl.
Esophagus
Left brachiocephalic vein
Recurrent laryngeal n.
Superior vena cava
Ascending aorta

Fig. 130   The posterior aspect of the pharynx together with its blood vessels and nerves; the thyroid and the parathyroid glands.

with those which form the glossopharyngeal (cranial IX) nerve and inferiorly with those which form the accessory (cranial XI) nerve. The vagus nerve leaves the skull through the jugular foramen, contained in the same dural sheath with the accessory nerve. The glossopharyngeal nerve emerges through the jugular foramen anterior to the others and separated from them by a fibrous septum. A **superior ganglion** (jugular ganglion) forms a spherical swelling of about 4 mm. diameter on the vagus the first and second cervical nerves. The vagus nerve descends through the neck in the carotid sheath, lying in a separate compartment formed in the sheath behind and between the artery and vein. At the root of the neck, the right vagus passes anterior to the first part of the subclavian artery and behind the brachiocephalic vein and so enters the thorax. The left vagus descends between the left common carotid and the left subclavian arteries and, in the chest, passes to the left of the arch of the aorta.

The vagus nerve gives rise to meningeal and auricular branches while in the jugular fossa; in the neck there are pharyngeal, laryngeal, and cardiac branches.

The **meningeal branch** is a slender filament, arising from the jugular ganglion, which re-enters the cranium through the jugular foramen to distribute to the dura mater in the posterior cranial fossa.

The **auricular branch** enters the temporal bone through the mastoid canaliculus in the lateral wall of the jugular fossa. It communicates with the facial nerve within the bone and then emerges again through the tympano-mastoid fissure. The auricular nerve distributes to the back of the external ear, the floor of the external acoustic meatus, and the lower part of the tympanic membrane. It communicates on the ear with the posterior auricular branch of the facial nerve.

There are, usually, two **pharyngeal branches** of the vagus (figs. 129, 130). They arise from the upper part of the inferior ganglion, pass obliquely downward and medialward between the internal and external carotid arteries, and combine with the pharyngeal branches of the glossopharyngeal nerve and of the superior cervical sympathetic ganglion to form the pharyngeal plexus. From this plexus the vagal fibers supply the muscles of the pharynx and the soft palate, with the exception of the stylopharyngeus and tensor veli palatini muscles. The pharyngeal branch of the vagus gives origin to a **nerve to the carotid body,** a chemoreceptor sensitive to changes in oxygen tension of the blood, located at the carotid bifurcation. This flattened ovoid body lies behind the terminal portion of the common carotid artery and projects into its bifurcation.

The **superior laryngeal nerve** (figs. 126, 129, 139) arises from the lower part of the inferior ganglion. It passes obliquely downward and medialward toward the larynx, coursing medial to both the internal and external carotid arteries. It is joined by filaments from the superior sympathetic ganglion and from the pharyngeal plexus and gives a branch to the internal carotid artery. The nerve divides into a larger internal branch and a smaller external branch. The **internal branch** swings forward and pierces the thyrohyoid membrane under cover of the lateral border of the thyrohyoid muscle. Rami-

fying with the laryngeal branch of the superior thyroid artery, the nerve supplies sensory fibers to the mucous membrane and parasympathetic fibers to the glands of the epiglottis, base of tongue, ary-epiglottic fold, and greater part of the larynx. Filaments communicate with branches of the inferior laryngeal nerve. The **external branch** of the superior laryngeal nerve (fig. 126) is long and slender and descends on the inferior constrictor muscle of the pharynx to the lower border of the thyroid cartilage, where it terminates in the cricothyroid muscle. It also supplies the inferior constrictor muscle of the pharynx. This nerve descends medial and posterior to the superior thyroid artery and vein, but it is occasionally included in the connective tissue surrounding the vessels and is for this reason sometimes endangered in thyroid surgery.

Cervical cardiac branches of the vagus nerve (figs. 129, 139) are usually two in number. The **superior branch** arises sometimes as high as the superior laryngeal nerve and descends behind and medial to the carotid sheath. It frequently forms a common stem with the cardiac branch of the superior cervical ganglion. The **inferior cervical cardiac branch** of the vagus arises at the root of the neck just above the first rib. Both cardiac branches on the right side and the upper one on the left side pass into the thorax at the side of the trachea behind the major vessels to reach the deep cardiac plexus (p. 347). The **inferior cervical cardiac branch of the left vagus** descends across the left side of the arch of the aorta, inclining toward the midline, and ends in the superficial cardiac plexus to the right of the aortic attachment of the ligamentum arteriosum (p. 346). The cardiac branch of the left superior cervical sympathetic ganglion may run in a common trunk with it.

Recurrent laryngeal nerves of the two sides (figs. 129, 130, 139) have essentially the same distribution, but they 'recur' around different structures and at different levels on the two sides. These nerves spring from the vagus low in the neck or in the upper thorax and then pass upward to the larynx. On the right side, the nerve arises as the vagus crosses superficially the first segment of the subclavian artery (fig. 139). Looping under and behind this vessel, the recurrent nerve inclines medialward and ascends between the trachea and the esophagus. On the left side, it is given off at the level of the sternal angle as the vagal nerve passes the aortic arch. Here the recurrent nerve loops under the aortic arch and the ligamentum arteriosum and passes behind these structures to achieve the same relationship to the trachea and esophagus as on the right.

The recurrent laryngeal nerves turn under the sixth pair of aortic arches in development. The proximal parts of these arches form the pulmonary arteries on the two sides. On the left side, the distal portion of the sixth arch forms the ligamentum arteriosum; but on the right side, this segment and the fifth arch disappear, allowing the recurrent nerve to ascend to the fourth arch, the subclavian artery.

The recurrent laryngeal nerves ascend under the lower border of the inferior constrictor muscle of the pharynx to enter the larynx. Passing behind the articulation between the inferior horn of the thyroid cartilage and the cricoid cartilage, the nerves break up into their terminal branches. They are called the inferior laryngeal nerves above the cricothyroid articulation. **Cardiac branches** to the deep cardiac plexus are given off from the recurrent nerves as they loop behind the subclavian artery and the aorta respectively. As the nerves ascend, **tracheal and esophageal branches** are given off to the muscles and mucosa of the trachea and esophagus and **pharyngeal branches** to the inferior pharyngeal constrictor muscle. The terminal **inferior laryngeal branches** supply motor fibers to all the intrinsic muscles of the larynx except the cricothyroideus. The inferior laryngeal nerves have a minor role in the sensory innervation of the mucosa of the larynx, and there is a large area of overlap in the mucosal distribution of the inferior and superior laryngeal nerves. The recurrent nerves carry fibers of the bulbar portion of the accessory nerve (Rethi, '51).

Deep Cervical Lymph Nodes (figs. 128, 131, 139)

The deep cervical lymph nodes, as is generally true elsewhere in the body, are arranged along the blood vessels—arteries or veins. The deep cervical chain is the largest group of nodes in the neck. It consists of numerous large nodes, for the most part lateral and posterior to the internal jugular vein, although some are found anterior to it. The nodes are divided, arbitrarily but conveniently for description, into superior and inferior subgroups at the point of crossing of the omohyoid muscle over the vein. The **superior deep cervical nodes** thus occupy the carotid triangle of the neck. They are numerous and extend beyond the immediate neighborhood of the vein.

Medially lying nodes at the upper end of the chain occupy the lateral area of the retropharyngeal space and are separately designated as the **retropharyngeal nodes.** These nodes, one or two on each side, are especially concerned with the lymphatic drainage of the nasal fossae and the paranasal sinuses, the hard and soft palate, the middle ear, and the nasopharynx and oropharynx. The **deep parotid nodes** also represent marginal members of

the superior deep cervical chain. They lie on the lateral wall of the pharynx deep to or embedded in the deep surface of the parotid gland. The deep parotid nodes receive the afferent lymphatic drainage from the external acoustic meatus, the auditory tube and the tympanum, the soft palate, and the posterior part of the nasal cavity. Certain nodes of the superior deep cervical group are large. A prominent node, located anterior to the internal jugular vein just below the angle of the mandible and at the crossing of the posterior belly of the digastric muscle, has been designated as the **jugulodigastric node.** It is especially concerned with the lymphatic drainage of the tongue and the palatine tonsil, and its palpation is frequently an aid in diagnosis. The lowest node of the superior deep cervical series is located just above the omohyoid muscle as it crosses the internal jugular vein. This **juguloomohyoid node** receives some separate drainage channels from the submental region and from the tip of the tongue.

The **inferior deep cervical nodes** are in complete continuity with the superior deep cervical nodes, but, being overlapped by the sternocleidomastoid muscle and located inferior to the omohyoid muscle, they extend into and are approached through the omoclavicular portion of the posterior triangle of the neck. They are often designated as the **supraclavicular nodes** and are associated with other nodes and lymphatic channels of the posterior triangle. The lymphatic vessels linking the superior deep cervical nodes extend inferiorly, some becoming the afferent channels of the inferior deep cervical nodes, others bypassing the lower nodes to help form the **jugular trunk.** The jugular trunk is formed by the confluence of efferent channels from both superior and inferior deep cervical nodes. It is a short channel which empties into the thoracic duct on the left side and into the junction of the internal jugular and subclavian veins on the right.

THE GENERAL PATTERN OF THE LYMPHATICS OF THE HEAD AND NECK (fig. 131)

The deep cervical lymphatic chain represents the final common path for most of the lymphatic drainage of the head and neck. In the head, additional small groups of nodes are found superficially in association with blood vessels; in the neck, other node groups are located both superficially and deeply.

Superficial Lymph Nodes of the Head

These include the occipital, retroauricular, anterior auricular, superficial parotid, and facial nodes.

The **occipital nodes** (fig. 131) comprise two, three, or four nodes overlying the superior nuchal line in the interval between the attachments of the sternocleidomastoid and trapezius muscles. They are associated with the occipital artery and the greater occipital nerve. The area of lymphatic drainage of the occipital nodes is the occipital part of the scalp and the upper neck. A few of their efferent vessels pass to the superior deep cervical chain, but most of them end in the nodes along the accessory nerve.

The **retroauricular nodes** (fig. 131) are one or two behind the ear on the insertion of the sternocleidomastoid muscle. They drain the posterior parietal region of the scalp and the skin of the ear. Their efferent channels turn around the anterior margin of the sternocleidomastoid muscle to the superior deep cervical nodes and also descend in the posterior triangle to the accessory group of nodes.

The **anterior auricular nodes** are one or two in the subcutaneous connective tissue directly in front of the tragus. They drain the anterior parietal and frontal regions of the scalp, the anterior surface of the ear and external acoustic meatus, and the lateral portion of the eyelids. Their efferents pass to the superficial parotid and superior deep cervical nodes.

The **superficial parotid node group** (fig. 131) is composed of nodes lying over the parotid gland and under its fascia. There may be as many as ten nodes in this group. These nodes receive lymphatic drainage from the adjacent surface of the external ear and the external acoustic meatus, from the skin of the temporal and frontal regions, and from the eyelids, the lacrimal gland, the cheek, and the nose. The efferent lymphatics pass to the superior deep cervical nodes.

The **facial nodes** (fig. 131) include small nodes scattered along the course of the facial artery and vein. They tend to be grouped below the orbit (infraorbital nodes), at the angle of the mouth (buccal nodes), and over the mandible (mandibular nodes). The afferent lymphatics of the facial nodes arise in the skin and mucous membrane of the eyelids, nose, cheek, and lips. Their efferents pass to the submandibular nodes.

Superficial Lymph Nodes of the Neck

This group is composed of submental, submandibular, external jugular, and anterior jugular nodes.

The **submental nodes** (figs. 131, 171, 176) are two or three which lie on the mylohyoid muscle between the anterior bellies of the digastric muscles. They receive lymphatic drainage from the chin, the middle part of the lower lip, the cheeks, the incisor region of the gums and teeth, and the tip of the tongue. Their efferent channels pass principally to the submandibular nodes and also include one separate channel which descends diagonally to the juguloomohyoid node.

The **submandibular node group** (figs. 131, 171, 176) consists of three to six nodes along the inferior border of the mandible adjacent to the submandibular gland and even within its sheath. The submandibular nodes receive the lymphatic drainage of the chin, lips, nose and nasal fossae, cheeks, gums, lower surface of palate, and anterior portion of tongue. They also drain the submandibular and sublingual glands and receive the lymph collected by the submental and facial nodes. The efferent vessels of the submandibular nodes pass to the superior deep cervical nodes and also, by a frequent accessory path, to the juguloomohyoid node.

The **external jugular lymph nodes** (fig. 131) are one or two which lie along the external jugular vein over the upper portion of the sternocleidomastoid muscle. Their afferent lymphatics come from the lower part of the ear and the parotid region, and they drain into the superior deep cervical lymph nodes.

The **anterior jugular nodes** (fig. 131) accompany the anterior jugular vein in lower levels of the neck, lying between the superficial layer of cervical fascia and the infrahyoid muscle fascia. They drain the skin and muscles of the anterior infrahyoid region of the neck. Their efferent channels follow the anterior jugular vein lateralward to the inferior deep cervical nodes.

Deep Lymph Nodes of the Neck

In addition to the deep cervical lymph nodes already described (p. 165), there are other node groups which are classified among the deep lymph nodes of the neck. These are the accessory chain and the transverse cervical chain of the posterior triangle of the neck and the juxtavisceral nodes of the anterior triangle—the infrahyoid, prelaryngeal, pretracheal, and paratracheal groups.

The **accessory chain** (fig. 131) is composed of two to six nodes, confluent above with the superior deep cervical nodes, which follow the accessory nerve across the posterior triangle. Under the border of the trapezius muscle the efferent channels of this chain unite with the transverse cervical chain, and the lymph path turns medialward with the latter. The accessory nodes receive efferent vessels of the occipital and retroauricular node groups and drain the occipital and posterior parietal parts of the scalp. Lymph channels from the supraspinatous fossa and from the nape and lateral parts of the neck and shoulder also reach the accessory chain.

The **transverse cervical chain** of the posterior triangle (fig. 131) consists of from one to ten nodes extending medially above the clavicle along the transverse cervical blood vessels. These nodes are continuous with the lower nodes of the accessory chain laterally and with the inferior deep cervical nodes medially. The transverse cervical nodes receive the efferents of the accessory chain and sometimes some of the efferents from the apical group of axillary nodes. They also receive drainage from the lateral part of the neck, the anterior thoracic wall and the mammary gland, and, occasion-

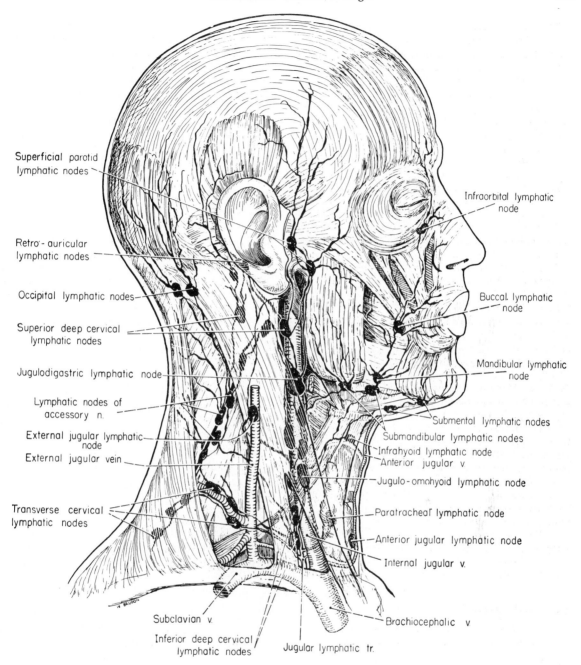

Superficial parotid lymphatic nodes

Retro-auricular lymphatic nodes

Occipital lymphatic nodes

Superior deep cervical lymphatic nodes

Jugulodigastric lymphatic node

Lymphatic nodes of accessory n.

External jugular lymphatic node

External jugular vein

Transverse cervical lymphatic nodes

Subclavian v.

Inferior deep cervical lymphatic nodes

Jugular lymphatic tr.

Infraorbital lymphatic node

Buccal lymphatic node

Mandibular lymphatic node

Submental lymphatic nodes

Submandibular lymphatic nodes

Infrahyoid lymphatic node

Anterior jugular v.

Jugulo-omohyoid lymphatic node

Paratracheal lymphatic node

Anterior jugular lymphatic node

Internal jugular v.

Brachiocephalic v.

Fig. 131 The general pattern of lymphatic drainage and the principal node groups of the head and neck.

ally, from the upper limb. The transverse cervical chain may empty its accumulated lymph into the jugular lymphatic trunk, into the thoracic duct or the right lymphatic duct, or independently into the internal jugular or subclavian veins.

The **juxtavisceral nodes** are those adjacent to the larynx, trachea, and thyroid gland. They include an **infrahyoid group** (fig. 131) located along the superior thyroid vessels on the thyrohyoid membrane, a **prelaryngeal group** on the cricothyroid ligament, **pretracheal glands** at the lower border of the thyroid gland along the inferior thyroid vessels, and **paratracheal nodes** (fig. 131) along the recurrent laryngeal nerves between the trachea and the esophagus. These lymphatic groups drain the larynx, the thyroid gland, the trachea, and the esophagus. The efferent channels of the infrahyoid nodes pass to the superior deep cervical nodes. Those of the prelaryngeal and pretracheal nodes reach the inferior deep cervical chain. The paratracheal nodes send their efferent channels along the recurrent laryngeal nerve to the lowest of the inferior deep cervical nodes or to the jugular trunk or the thoracic duct.

Lymphatic Channels

Lymphatic drainage through the various nodes of the head and neck follows a simple peripheral to central pattern. Lymphatic channels arising in the scalp pass to the nearest subjacent collection of nodes in the head region. Similarly, the lymphatic drainage of both the superficial and the deep parts of the face and neck is first to the nearest lymph node aggregation and then through more central nodes and channels. It is only by minute observation that one can identify macroscopically any but the larger lymphatic channels near the base of the neck. Here the jugular trunks, the subclavian trunks and the right lymphatic trunk have the size of small veins, and the thoracic duct may be as large as a medium-sized vein. The jugular trunk, the transverse cervical trunk, and the subclavian trunk on the right side frequently unite to form the **right lymphatic duct.** This duct is only 2 to 5 mm. long and courses along the medial border of the anterior scalene muscle at the root of the neck to end in the right subclavian vein at its angle of junction with the internal jugular vein. Its orifice is guarded by a bicuspid valve which prevents the passage of blood into the duct. The right lymphatic duct may also receive the bronchiomediastinal lymph trunk, or any of its usual tributary trunks may open separately into the subclavian vein.

The **thoracic duct** transports the greater part of the lymph and chyle. It is a common vessel for the collection of lymph from all of the body below the diaphragm and from the left side of the body above it. The thoracic duct begins in the abdomen in an occasional dilation, the **cisterna chyli** (p. 453), formed by the convergence of a number of large tributaries. Passing through the aortic hiatus of the diaphragm, the thoracic duct ascends through the thorax against the vertebral column, swinging somewhat to the left of the midline in the superior mediastinum. Emerging at the root of the neck between the left border of the esophagus and the left pleura, the thoracic duct arches behind the carotid sheath and the jugular lymphatic trunk and anterior to the prevertebral fascia and then turns down into the junction of the left internal jugular and the left subclavian veins. The highest point of the arch of the thoracic duct may be 3 or 4 cm. above the sternal end of the clavicle. The thoracic duct commonly receives the jugular and transverse cervical trunks of the left side and, less commonly, the subclavian trunk. The thoracic duct is developed bilaterally, but the large size of the arch of the aorta and of the thoracic aorta results in reduction or obliteration of the left primordium below the fifth thoracic vertebra and of the right primordium above that level. As remnants of its bilateral development, the thoracic duct is occasionally double in the neck or the chest or partially double with separated regions of junction. The cervical part of the duct exhibits double parallel trunks on the left side in as high as 64 per cent of cases (Rouviere, '32). The right cervical primordium may be retained, so that the duct empties on both sides, and in very rare cases the right termination may represent the entire duct.

THE CERVICAL PORTION OF THE SYMPATHETIC TRUNK (figs. 132, 133, 139)

The **cervical sympathetic trunk** is an upward continuation of the thoracic sympathetic trunk; it lies in the areolar tissue behind the carotid sheath and in front of the prevertebral muscles but is occasionally behind the prevertebral fascia. The trunk may appear as a solid cord or may be broken into several strands interconnecting the cervical sympathetic ganglia. The cervical sympathetic trunk consists largely of ascending preganglionic axons which traverse the white rami communicantes of the upper five thoracic nerves, mainly the second and third. The cervical sympathetic trunk receives no white rami communicantes. It also contains some postganglionic fibers and visceral afferent fibers with cell bodies in thoracic dorsal root ganglia. The ganglia of the cervical sympathetic trunk are usually three, formed by secondary fusion of the ganglionic primordia associated with the eight cervical nerves. As a rule, the superior cervical ganglion represents the upper four of these; the middle, the fifth and sixth; and the inferior, the seventh and eighth.

The **superior cervical ganglion** is the largest of the three and is from 25 to 35 mm. long. It is usually broad, flattened, and tapered at the ends and lies in front of the transverse process of the

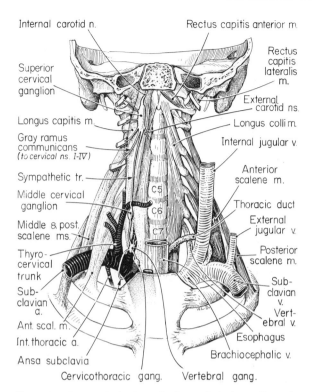

Internal carotid n.

Superior cervical ganglion

Longus capitis m.

Gray ramus communicans (to cervical ns. I-IV)

Sympathetic tr.

Middle cervical ganglion

Middle & post. scalene ms.

Thyro-cervical trunk

Sub-clavian a.

Ant. scal. m.

Int. thoracic a.

Ansa subclavia

Cervicothoracic gang.

Rectus capitis anterior m.

Rectus capitis lateralis m.

External carotid ns.

Longus colli m.

Internal jugular v.

Anterior scalene m.

Thoracic duct

External jugular v.

Posterior scalene m.

Sub-clavian v.

Vert-ebral v.

Esophagus

Brachiocephalic v.

Vertebral gang.

Fig. 132   The cervical sympathetic trunk and ganglia and the anterior and lateral cervical muscles.

second cervical vertebra. Its highest branch, the direct upward continuation into the head, is the **internal carotid nerve.** The branches of the internal carotid nerve distribute to smooth muscle and glandular structures within the cranium and are described more completely in connection with those structures (p. 284).

The superior cervical ganglion has delicate communications with the inferior ganglion of the glossopharyngeal nerve, the superior and inferior ganglia of the vagus, and the hypoglossal nerve. **Gray rami communicantes** (fig. 139) pass from the ganglion to the upper two to four cervical spinal nerves. These branches pass lateralward and dorsally and may reach the spinal nerves or the nerve loops between them. Those to the upper two nerves are constant; to the third nerve, frequent; to the fourth nerve, occasional. Four or five **pharyngeal branches** leave the medial aspect of the ganglion and, opposite the middle pharyngeal constrictor muscle, join the pharyngeal branches of the glossopharyngeal and vagus nerves to form the pharyngeal plexus. Two relatively large branches, **external carotid nerves,** leave the superior ganglion to join the

external carotid artery. These form the **external carotid plexus** which extends along the vessel and its branches. Perivascular fibers emanate from the plexus along the course of the arteries to structures served by the arteries and are vasoconstrictor, secretory, and pilomotor. The external carotid plexus is the source of the sympathetic innervation of the salivary and other oral and labial glands. Offshoots from the external carotid plexus reach the carotid body. The small **superior cervical cardiac nerve** arises from the ganglion or from the trunk just below it and descends posterior to the carotid sheath or in its posterior fascia. It usually lies ventral but may be dorsal to the inferior thyroid artery.

The superior cervical cardiac nerve has communications with the superior cervical cardiac branch of the vagus and with its inferior laryngeal branch, with the external branch of the superior laryngeal branch of the vagus, and with the middle cervical cardiac branch of the sympathetic trunk. At the root of the neck, there are differences in the course of the superior cervical cardiac nerves of the two sides. The right nerve passes either ventral or dorsal to the subclavian artery and then along the brachiocephalic artery to the deep cardiac plexus. The left nerve passes ventral to the common carotid artery and across the left side of the arch of the aorta to reach the superficial cardiac plexus. It may be combined with the inferior cervical cardiac branch of the left vagus which is directed to the same part of the cardiac plexus. The superior cervical cardiac nerve contains no afferent fibers from the heart.

The **middle cervical ganglion** is the smallest of the cervical ganglia and the most variable in form and position. It has an average length of 13.5 mm., but it may be double or absent. It commonly lies at the level of the cricoid cartilage in the bend of the inferior thyroid artery or superior to the arch of this artery. The ganglion and the trunk are also variably ventral or dorsal to the arching portion of the inferior thyroid artery. The ganglion gives rise to **gray rami communicantes** to cervical nerves five and six. The **middle cervical cardiac nerve** is the largest of the three cervical sympathetic cardiac nerves. It arises from the ganglion or from the trunk below it and descends behind the common carotid artery, frequently communicating with the superior cardiac nerve and the recurrent laryngeal nerve.

On the right side the middle cardiac nerve passes in front of or behind the subclavian artery and then along the brachiocephalic artery to the deep cardiac plexus. On the left side the nerve descends between the common carotid and subclavian arteries to the deep cardiac

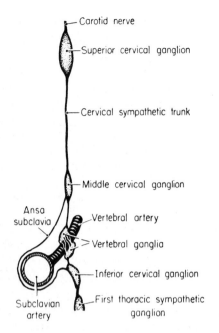

Fig. 133   The structural plan of the cervical sympathetic trunk.

plexus. The middle cervical cardiac nerve carries cardio-accelerator and visceral afferent fibers to the heart. One or more branches of the ganglion follow the inferior thyroid artery and provide an innervation to the thyroid gland.

The **cervicothoracic ganglion** (present as a fused ganglion in 82 per cent of specimens—Jamieson, Smith, & Anson, '52) represents a combination of the inferior cervical and the first or the first several thoracic ganglia. The inferior cervical portion of the cervicothoracic ganglion lies anterior to the base of the transverse process of the seventh cervical vertebra and is usually posteromedial to the origin of the vertebral artery. The sympathetic trunk between the middle and inferior cervical ganglia is separated into two parts, the larger strand passing directly between the ganglia. A smaller superficial strand, the **ansa subclavia,** descends from the middle ganglion, forms a loop over and around the subclavian artery, and rejoins the inferior ganglion behind the artery. Situated at the region of transition of the cervical and thoracic curvatures of the vertebral column, the cervicothoracic ganglion conforms to the change in direction of the sympathetic trunk and has its long axis oriented almost directly dorsoventrally. Frequently, the vertebral artery has a ganglionic mass anterior to it united to the cervicothoracic ganglion by encircling nerve

strands. This is the **vertebral ganglion,** thought to represent a low middle ganglion or a fusion of parts of the middle and inferior ganglia. The cervicothoracic ganglion supplies **gray rami communicantes** to the sixth, seventh, and eighth cervical and first thoracic spinal nerves.

The **inferior cervical cardiac nerve** arises from the cervicothoracic ganglion. It passes along the trachea to the deep cardiac plexus, communicating with the middle cervical cardiac nerve and the recurrent laryngeal nerve. Several large branches of the ganglion (vertebral nerves) pass to the vertebral artery and form the **vertebral plexus.** This continues along the artery into the cranial cavity and is prolonged onto the basilar, posterior cerebral, and cerebellar arteries. Branches of the cervicothoracic ganglion also distribute along the subclavian artery.

THE SUBMANDIBULAR TRIANGLE

The submandibular triangle (figs. 118, 132, 133) is the suprahyoid portion of the anterior cervical triangle. The inferior border of the mandible constitutes the base of the triangle, and the anterior and posterior bellies of the digastric muscle form the sides. The mylohyoid muscles provide a complete layer from one side of the mandible to the other, deep to the digastric muscles. They underlie the tongue and constitute a floor for the oral cavity. The submandibular and sublingual glands occupy the submandibular triangle; that part of the submandibular gland superficial to the mylohyoid frequently overlaps the muscular boundaries. Deep to the mylohyoid muscle, the lingual and the hypoglossal nerves pass toward the tongue. The lingual artery and vein are also approached through the triangle. The superficial layer of cervical fascia covers all but the subcutaneous structures. It attaches to the hyoid bone and to the mandible and forms a loose sheath for the submandibular gland. The anterior jugular vein takes origin in small venous radicles in the subcutaneous connective tissue, and the facial vein descends through the posterior portion of the triangle. Terminals of the transverse cervical and great auricular nerves and of the cervical branch of the facial nerve distribute over the triangle. The marginal mandibular branch of the facial nerve sometimes dips below the inferior border of the mandible and then lies in the triangle.

The Muscles of the Submandibular Triangle

The Digastric Muscle (figs. 121, 134, 135, 139)
The two-bellied digastric muscle has anterior and

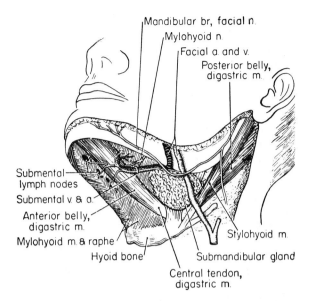

Mandibular br, facial n.
Mylohyoid n.
Facial a. and v.
Posterior belly,
digastric m.

Submental
lymph nodes
Submental v. & a.
Anterior belly,
digastric m.
Mylohyoid m. & raphe
Hyoid bone
Stylohyoid m.
Submandibular gland
Central tendon,
digastric m.

Fig. 134 The superficial structures of the submandibular triangle.

posterior portions. The **anterior belly** arises from the digastric fossa of the inner side of the lower border of the mandible close to the symphysis. The **posterior belly,** longer than the anterior, arises from the mastoid notch of the temporal bone. The two bellies descend toward the hyoid bone and are united by an intermediate tendon which is connected to the body of the hyoid bone by a loop of fibrous connective tissue. The digastric muscle elevates the hyoid bone and steadies it from above. The anterior belly is capable of drawing the hyoid forward; the posterior, of drawing it backward. With the hyoid bone held down by the infrahyoid muscles, the digastric muscles aid in opening the jaws. In association with the other muscles which raise the hyoid, they assist in elevation of the larynx and pharynx and, by raising the floor of the mouth, help to propel food into the pharynx in the act of swallowing. The two bellies of the digastric are fundamentally separate muscles, for they are supplied by different nerves. The anterior belly receives its innervation at its lateral border from the nerve to the mylohyoid, a branch of the mandibular division of the trigeminal nerve. The posterior belly is supplied by a branch of the facial nerve (p. 209) that enters the proximal one-third of its anterior margin.

The Stylohyoid Muscle (figs. 134, 139, 178) The stylohyoid muscle is medial and nearly parallel to the posterior belly of the digastric. It arises from the posterior border of the styloid process of the temporal bone near its root. Passing downward and forward and converging on the intermediate tendon of the digastric, the stylohyoid divides into two slips which pass on either side of the tendon to insert on the body of the hyoid bone adjacent to the greater horn. The stylohyoid muscle acts, with the posterior belly of the digastric, to draw the hyoid upward and backward. It is innervated by the branch of the facial nerve to the posterior digastric, the nerve entering the proximal one-third of the muscle on its deep surface.

The Mylohyoid Muscle (figs. 121, 134, 135, 139, 178) This muscle arises from the whole length of the mylohyoid line of the mandible from the symphysis in front to the last molar tooth behind. Its posterior fibers pass medialward and slightly downward to insert into the body of the hyoid bone. The middle and anterior fibers insert into a median fibrous raphe extending from the symphysis to the hyoid bone. The muscles of the two sides form a muscular floor for the oral cavity, supporting the tongue, the deep portions of the submandibular glands, the sublingual glands, ducts, nerves, and vessels. The mylohyoid muscle is innervated by the mylohyoid nerve from the mandibular division of the trigeminal nerve, the nerve entering the muscle on its superficial surface.

The Genoihyoid Muscle (figs. 144, 175, 177) This narrow muscle lies adjacent to the midline and superior to the mylohyoid. It arises from the mental spine on the posterior aspect of the symphysis menti of the mandible and inserts on the anterior surface of the body of the hyoid bone. It is innervated by a branch of the first cervical nerve conducted to it by the hypoglossal nerve. The muscles of the two sides are often blended together.

Group Actions The suprahyoid muscles serve in the swallowing reflex to elevate the tongue and the floor of the mouth, assisting in the rapid transfer of food into the pharynx. In the second phase of swallowing, as food starts down the pharynx, the stylohyoid and posterior digastric muscles both elevate and carry backward the tongue and hyoid, preventing any return of food to the mouth. These muscles also help open the jaw when the hyoid is held down by the infrahyoid muscles.

The Hyoglossus Muscle The hyoglossus muscle (figs. 121, 135, 139, 175, 177) is a muscle of the tongue which lies in the submandibular triangle. It arises from the whole length of the upper border of the greater horn of the hyoid bone and the adjacent surface of its body. Diverging somewhat, the fibers of the muscle ascend into the tongue. In the tongue the posterior fiber bundles run transversely, the middle obliquely, and the anterior bundles longitudinally, intermingling with the other muscle fasciculi that form the tongue. The hyoglossus muscle draws down the sides of the tongue and is a retractor. It is innervated by the hypoglossal nerve.

The Hyoid Bone (figs. 121, 134, 135, 139, 175, 176, 180)

The hyoid bone is a small U-shaped bone which can be palpated in the deep crease between the neck and the chin. It is composed of a body, two greater horns (or cornua), and two lesser horns. The **body** is the middle or bottom of the U, the **greater horns** are the limbs. They spring from the body about a finger's breadth from the median plane and extend backwards in the sides of the throat. The **lesser horns** are small pointed nodules that extend upward from the junction of the body with the greater horns and give attachment to the stylohyoid ligaments and to the middle pharyngeal constrictor muscles.

The body of the hyoid is convex forward and its anterior surface is divided by a transverse ridge into superior and inferior parts. The anterior surface provides attachment for muscles—geniohyoid in its superior part; mylohyoid, sternohyoid, and omohyoid in its inferior part. The posterior surface of the body is smooth and concave. It is separated by some fat and a bursa from the median thyrohyoid ligament which is, in turn, separated by a pad of fat from the epiglottis. The greater horns give attachment by their medial surfaces to the thyrohyoid membrane. Their lateral surfaces provide bony origin for the middle constrictor muscles of the pharynx and for the hyoglossus muscles, and, close to their junction with the body, they receive the fascial sling of the digastric and the insertion of the stylohyoid muscles. The thyrohyoid muscle attaches to the anterior two-thirds of the greater horn.

The hyoid bone is developed from the mesoderm of the second and third pharyngeal arches. From the third arch are derived the greater horns and most of the body. The lesser horn and the upper median part of the body come from the second arch, and it is between the two developmentally different parts of the body that the thyroglossal duct may continue to the foramen cecum of the tongue (p. 157). The stylohyoid ligament, the styloid process, and the stapes of the middle ear also arise from the second arch mesoderm. The hyoid is ossified from six centers, two for the body and one for each horn. Ossification begins during the first year and

is complete for the body and greater horn in middle life. The lesser horn may not achieve bony union with the rest of the hyoid until an advanced age.

The Submandibular Gland (figs. 121, 134, 135, 139, 177, 178)

This is a salivary gland lying below and in front of the angle of the mandible and occupying most of the space of the posterior part of the submandibular triangle. Its superficial surface is flattened and is covered by skin, subcutaneous connective tissue, the platysma muscle, and the superficial layer of the cervical fascia. In the subcutaneous tissue over it pass the facial vein and, occasionally, a branch of the facial nerve. The greater part of the gland is superficial; but it is folded around the free posterior border of the mylohyoid muscle, so that a small deep portion of the gland and its duct is found deep to the muscle (fig. 177). The superficial layer of cervical fascia forms a loose sheath for the gland from which it can readily be turned out. Submandibular lymph nodes lie on the gland and under the sheath but not within the proper capsule of the gland. The facial artery (fig. 134) reaches the face at the lower border of the mandible along the anterior edge of the masseter muscle and, in its course, grooves the posterior and superior surfaces of the gland. Posteriorly, the gland lies against the posterior digastric and stylohyoid muscles and, at the angle of the mandible, is adjacent to the apex of

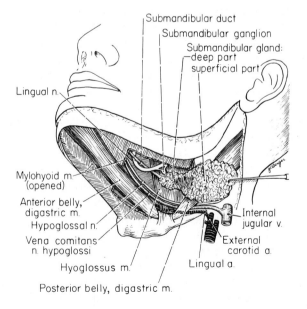

Fig. 135   The deep contents of the submandibular triangle.

the parotid gland. When large, the submandibular gland frequently overlaps the muscular boundaries of the triangle. The small, deep, tongue-like process of the gland and its duct lie on a different muscular plane from the superficial portion. They are located deep to the mylohyoid muscle and between it and the hyoglossus and genioglossus muscles (figs. 135, 177). The **submandibular duct** arises from the deep part of the gland, is about 5 cm. long, and has relatively thin walls and a constricted orifice (figs. 136, 139). It passes forward and medialward, lying at first on the hyoglossus and then on the genioglossus muscles, and opens at the summit of the **sublingual caruncle** at the side of the frenulum of the tongue. The duct has an intimate relation to the lingual nerve which crosses it twice in this intermuscular interval.

The **lingual nerve,** a branch of the mandibular division of the trigeminal, provides general sensation to the anterior two-thirds of the tongue—that portion of the tongue anterior to the circumvallate papillae. In its course to the tongue, it descends into the submandibular triangle, passes across the duct, and then, curving forward and upward into the tongue, spirals the underside of the duct and passes it again medially (figs. 135, 139). In the same intermuscular interval deep to the mylohyoid muscle, the **hypoglossal nerve** crosses the inferior portion of the submandibular triangle as it courses forward and upward to the tongue. This nerve supplies both the extrinsic and the intrinsic musculature of the tongue. Accompanying the hypoglossal nerve, usually inferior to it, runs the vena comitans of the nerve, draining the tongue and the sublingual region. On a still deeper plane and adjacent to the intermediate tendon of the digastric muscle, the lingual artery (figs. 135, 139) courses forward toward the tongue under the hyoglossus muscle. The artery is accompanied by small veins which contribute to the formation of the lingual vein.

The arterial supply of the gland is furnished by branches of the **facial artery,** and its veins enter the **facial vein.** The lymphatic drainage of the gland is to the submandibular lymph nodes (p. 166). Sympathetic and parasympathetic nerves reach the gland. Sympathetic filaments from the superior cervical sympathetic ganglion are conducted along the **facial plexus** on the facial artery. The parasympathetic fibers of the gland are carried in the chorda tympani branch of the facial nerve (fig. 137).

Peripherally, the **facial nerve** is motor to the

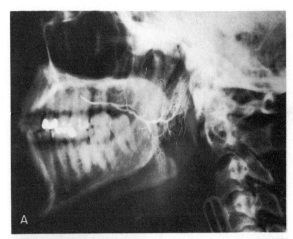
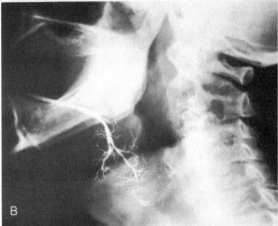

Fig. 136   Sialograms of (A) the parotid gland and (B) the submandibular gland.

muscles of facial expression. Intracranially, it has a portion which is sensory and parasympathetic and is designated as the **nervus intermedius.** Included in this nerve are fibers which convey the special sense of taste from the anterior two-thirds of the tongue and parasympathetic fibers associated with both the pterygopalatine ganglion deep in the head and the submandibular ganglion of the submandibular triangle. The fibers to the tongue and to the submandibular ganglion are collected into the **chorda tympani branch of the facial nerve.** This nerve leaves the petrous portion of the temporal bone through the petrotympanic fissure, crosses the medial surface of the spine of the sphenoid, and, high in the infratemporal fossa, joins the posterior aspect of the lingual nerve. This is a peripheral anastomosis only, the nerves maintaining their separate identities; but the chorda tympani is conducted by the lingual nerve to the tongue for its taste distribution and into the neighborhood of the submandibular gland for its parasympathetic distribution. The **submandibular ganglion** is a peripheral parasympathetic ganglion concerned with secretory and afferent nerves of the sublingual and submandibular glands. This triangular ganglion is suspended from the lingual nerve by two short roots and is connected below by nerve fibers to the deep portion of the submandibular gland. The ganglion lies on the hyoglossus muscle near the posterior border of the mylohyoid. The proximal root conducts preganglionic fibers to the ganglion and also, through it, to parasympathetic cell bodies in the stroma of the submandibular gland. The distal root conducts postganglionic fibers, after synapse in the submandibular ganglion, to the sublingual and small lingual glands. Small groups of ganglion cells in the stroma of the submandibular gland near the branches of the duct (Langley's ganglion) are considered to be the principal cell stations for synapses in the innervation of the submandibular gland.

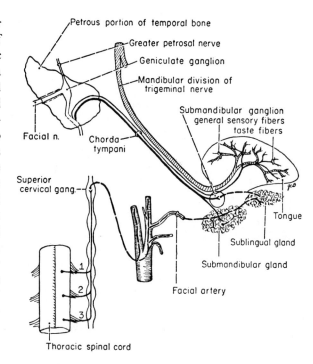

Fig. 137   The distribution of the chorda tympani and the innervation of the submandibular and sublingual glands.

The Sublingual Gland (figs. 177, 178)

This is the smallest of the three salivary glands. It is narrow and flattened, and lies beneath the mucous membrane of the mouth. Lateral to it is the mandible; medial to it are the genioglossus and geniohyoid muscles. The sublingual gland is located above the mylohyoid line and rests on the mylohyoid muscle. About twelve small sublingual ducts emanate from the superior border of the gland and open separately along the sublingual fold of the floor of the mouth. The sublingual gland has no fascial sheath but lies in the areolar tissue of the floor of the mouth. It receives its blood supply from the sublingual branch of the lingual artery and from the submental branch of the facial. The nerves of the gland are parasympathetic, as described in connection with the submandibular gland, and sympathetic, by way of a plexus along the facial artery.

THE POSTERIOR CERVICAL TRIANGLE

The **posterior triangle of the neck** (fig. 118) is defined by the diverging borders of the sternocleidomastoid and trapezius muscles and is limited below by the middle one-third of the clavicle. In the subcutaneous connective tissue the **platysma muscle**

(p. 146) spans most of the triangle. The **external jugular vein** (p. 147), formed over the upper portion of the sternocleidomastoid muscle, descends vertically into the lower portion of the triangle, where it pierces the superficial layer of cervical fascia. The cutaneous branches of the cervical plexus emerge at the posterior border of the sternocleidomastoid muscle just below the accessory nerve (p. 175). The **lesser occipital nerve** ascends along the posterior border of the sternocleidomastoid muscle. The **great auricular nerve** passes diagonally over this muscle toward the ear and the angle of the mandible in company with the external jugular vein, and the **transverse cervical nerve** crosses the muscle as it courses horizontally into the anterior cervical triangle. The **supraclavicular nerves** descend through the posterior cervical triangle to spread out as medial, intermediate, and lateral nerves (p. 151). All of these cutaneous nerves pierce the **superficial layer of cervical fascia** which stretches between the muscular and bony boundaries of the triangle and forms sheaths for both the sternocleidomastoid and trapezius muscles (p. 151). The floor of the triangle is formed by the middle and the posterior scalene, the levator scapulae, and the splenius muscles,

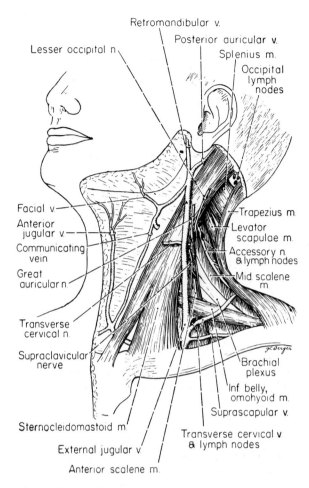

Lesser occipital n.
Retromandibular v.
Posterior auricular v.
Splenius m.
Occipital lymph nodes
Facial v.
Anterior jugular v.
Communicating vein
Great auricular n.
Transverse cervical n.
Supraclavicular nerve
Sternocleidomastoid m.
External jugular v.
Anterior scalene m.
Trapezius m.
Levator scapulae m.
Accessory n. & lymph nodes
Mid scalene m.
Brachial plexus
Inf belly, omohyoid m.
Suprascapular v.
Transverse cervical v. & lymph nodes

Fig. 138   The sternocleidomastoid muscle and the posterior cervical triangle.

covered by the prevertebral layer of cervical fascia (p. 152).

In the lower portion of the posterior cervical triangle, the diagonal inferior belly of the omohyoid muscle bounds a small **omoclavicular triangle** (fig. 118), through which pass the subclavian artery and vein. Here the superficial layer of the fascia of the infrahyoid muscles, covering the inferior belly of the omohyoid, descends behind the superficial layer of cervical fascia, separated from it by an areolar interval. The omohyoid fascia descends behind the clavicle and ends in the sheath of the subclavian vein.

The Sternocleidomastoid Muscle (figs. 121, 138)

This muscle bisects the neck diagonally. It is thick in its midportion, thinner and broader at either end, and arises by two heads. A narrow, tendinous **sternal head** springs from the anterior surface of the

manubrium sterni, and a broader, fleshy and aponeurotic **clavicular head** arises from the upper surface of the medial one-third of the clavicle. The two heads combine, shortly above their origin, into a robust muscle which inserts by a strong tendon into the lateral surface of the mastoid process from its apex to its superior border and along the lateral half of the superior nuchal line of the occipital bone. The sternocleidomastoid muscle draws the mastoid process down toward the shoulder of the same side and thus turns the chin upward and to the opposite side. Acting together, the muscles project the head forward and the chin upward. They are also respiratory muscles, drawing upward the clavicles and the sternum and, with them, the ribs. The nerve of the sternocleidomastoid is the **accessory**. A knowledge of the relations of this muscle contributes to the understanding of the anatomy of the entire anterior and lateral region of the neck.

The Accessory (cranial XI) Nerve (figs. 138, 140)

The accessory nerve has a cranial and a spinal portion. The spinal portion is the only part of the nerve that forms a discrete peripheral nerve.

The **cranial portion** consists of four or five small rootlets which arise from the side of the medulla oblongata below and in line with those of the vagus. The nerve formed by them passes to the jugular foramen, has a connection with the spinal portion but separates from it again, passes through the foramen, and then unites with the vagus just proximal to its inferior ganglion. The fibers of the cranial portion of the accessory nerve are incorporated in the vagus nerve and constitute, principally, its inferior laryngeal branch.

The **spinal portion of the accessory nerve** arises from motor cells in the lateral part of the ventral gray column of the first segments of the cervical spinal cord. The rootlets of the nerve emerge on the side of the cord between the ventral and dorsal roots and turn upward at the side of the cord, collecting as they ascend into a discrete nerve which passes between the denticulate ligament and the dorsal rootlets of the cervical nerves. This nerve enters the cranial cavity through the foramen magnum and penetrates the dura mater over the jugular bulb. In the jugular foramen it joins and then separates from or exchanges fibers with the cranial portion of the accessory nerve. Leaving the cranium by the jugular foramen, the accessory nerve at first lies between the internal carotid artery and the internal jugular vein. It then descends obliquely

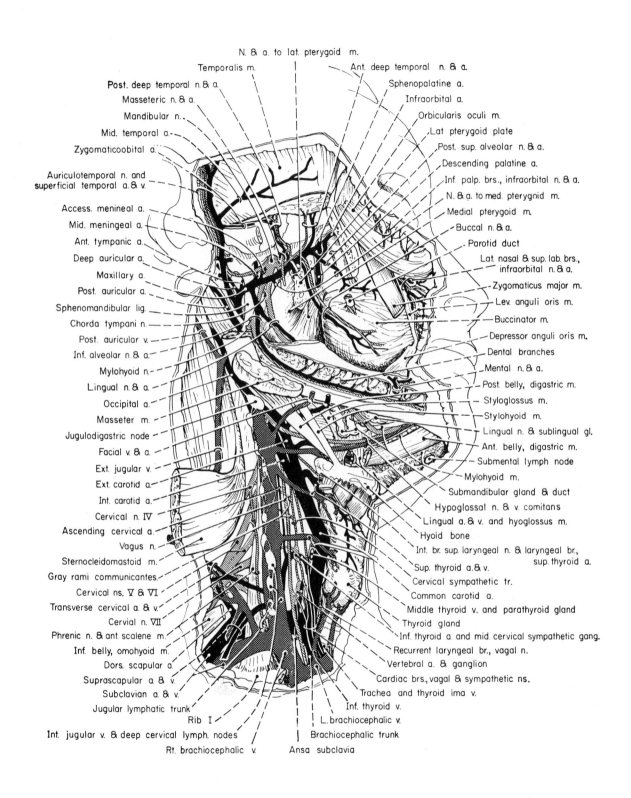

N. & a. to lat. pterygoid m.
Temporalis m.
Post. deep temporal n. & a.
Masseteric n. & a.
Mandibular n.
Mid. temporal a.
Zygomaticoobital a.
Auriculotemporal n. and
superficial temporal a. & v.
Access. menineal a.
Mid. meningeal a.
Ant. tympanic a.
Deep auricular a.
Maxillary a.
Post. auricular a.
Sphenomandibular lig.
Chorda tympani n.
Post. auricular v.
Inf. alveolar n. & a.
Mylohyoid n.
Lingual n. & a.
Occipital a.
Masseter m.
Jugulodigastric node
Facial v. & a.
Ext. jugular v.
Ext. carotid a.
Int. carotid a.
Cervical n. IV
Ascending cervical a.
Vagus n.
Sternocleidomastoid m.
Gray rami communicantes
Cervical ns. V & VI
Transverse cervical a. & v.
Cervial n. VII
Phrenic n. & ant. scalene m.
Inf. belly, omohyoid m.
Dors. scapular a.
Suprascapular a. & v.
Subclavian a. & v.
Jugular lymphatic trunk
Rib I
Int. jugular v. & deep cervical lymph. nodes
Rt. brachiocephalic v.

Ant. deep temporal n. & a.
Sphenopalatine a.
Infraorbital a.
Orbicularis oculi m.
Lat pterygoid plate
Post. sup. alveolar n. & a.
Descending palatine a.
Inf. palp. brs., infraorbital n. & a.
N. & a. to med. pterygoid m.
Medial pterygoid m.
Buccal n. & a.
Parotid duct
Lat. nasal & sup. lab. brs.,
infraorbital n. & a.
Zygomaticus major m.
Lev. anguli oris m.
Buccinator m.
Depressor anguli oris m.
Dental branches
Mental n. & a.
Post. belly, digastric m.
Styloglossus m.
Stylohyoid m.
Lingual n. & sublingual gl.
Ant. belly, digastric m.
Submental lymph node
Mylohyoid m.
Submandibular gland & duct
Hypoglossal n. & v. comitans
Lingual a. & v. and hyoglossus m.
Hyoid bone
Int. br. sup. laryngeal n. & laryngeal br.,
sup. thyroid a.
Sup. thyroid a. & v.
Cervical sympathetic tr.
Common carotid a.
Middle thyroid v. and parathyroid gland
Thyroid gland
Inf. thyroid a. and mid. cervical sympathetic gang.
Recurrent laryngeal br., vagal n.
Vertebral a. & ganglion
Cardiac brs., vagal & sympathetic ns.
Trachea and thyroid ima v.
Inf. thyroid v.
L. brachiocephalic v.
Brachiocephalic trunk
Ansa subclavia

Fig. 139 Deeper anatomy of the neck and structures of the infratemporal fossa.

176

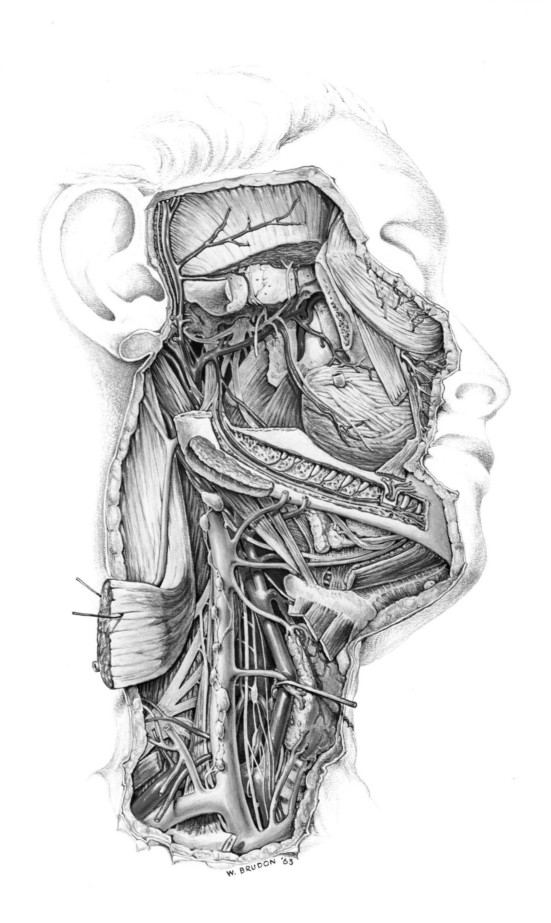

W. BRUDON '63

lateralward, usually anterior but sometimes posterior to the vein, and pierces the deep aspect of the sternocleidomastoid muscle. The nerve passes through the muscle, emerging into the posterior triangle at or below the junction of the upper and middle thirds of the length of the muscle. Continuing diagonally downward and backward under the superficial layer of cervical fascia, the nerve disappears under the anterior border of the trapezius muscle at about the junction of its middle and lower thirds. In or deep to the sternocleidomastoid muscle, the accessory nerve is joined by a branch of the second cervical nerve for the supply of this muscle. In the posterior cervical triangle the nerve is joined by branches of the third and fourth cervical nerves. The junction of these nerves and their distribution deep to the trapezius muscle is designated as the **subtrapezial plexus.** Experimentation on monkeys has indicated that the cervical nerves associated with the accessory nerve may supply the afferent proprioceptive innervation of the muscles supplied by it (Corbin & Harrison, '40). The accessory nerve is accompanied in the posterior cervical triangle by a chain of deep lymph nodes (p. 166).

The Cervical Plexus (figs. 121, 139, 140)

The ventral rami of the first four cervical nerves form the cervical plexus. The ventral ramus of the first nerve emerges from the vertebral canal above the posterior arch of the atlas, passes forward around the superior articular process, and descends in front of the transverse process of the atlas to form a loop with an ascending branch of the second cervical nerve. The ventral rami of the other cervical nerves emerge along the grooved upper surfaces of the transverse processes of the vertebrae posterior to the vertebral artery. Each of the first four cervical nerves is joined at its emergence by a **gray ramus communicans** from the superior cervical ganglion. Each nerve, except the first, divides into an ascending and a descending branch which unite to form loops. From the branches of these loops, the cervical plexus is formed opposite the first four cervical vertebrae ventrolateral to the levator scapulae and middle scalene muscles and deep to the sternocleidomastoid muscle. A **communication with the vagus nerve** is made between a branch from the loop of the first and second nerves and the inferior ganglion of the vagus. The components of the cervical plexus are (1) the cutaneous branches of the plexus, (2) the ansa cervicalis complex, (3) the

phrenic nerve, (4) contributions to the accessory nerve, and (5) direct muscular branches.

(1) The **cutaneous branches** of the plexus are the lesser occipital, great auricular, transverse cervical, and supraclavicular nerves. The first three arise from the second and third cervical nerves. They pass lateralward deep to the sternocleidomastoid muscle and enter the posterior triangle at the posterior border of the muscle just inferior to the accessory nerve. The supraclavicular nerves arise from the third and fourth cervical nerves and enter the posterior triangle slightly below the others. The further course and distribution of these nerves has had prior consideration (pp. 148, 151).

(2) The **ansa cervicalis** complex (figs. 121, 140) provides the innervation of the infrahyoid muscles and the geniohyoid muscle. From the loop between the first and second cervical nerves, a short nerve trunk joins the hypoglossal nerve just distal to its exit from the skull. Most of its fibers run with the hypoglossal nerve for only two or three centimeters and then leave it to descend in front of the internal and common carotid arteries as the **superior root of the ansa cervicalis.** The balance of these communicating nerve fibers pass off in two nerves farther anteriorly—the nerve to the thyrohyoid muscle and the nerve to the geniohyoid muscle. The term 'ansa cervicalis' refers to the loop formed between the superior root and a nerve which is formed by branches of the second and third cervical nerves, the **inferior root of the ansa.** The latter root descends behind the carotid sheath and, winding lateralward, joins the superior root either lateral to the internal jugular vein or between it and the common carotid artery. The loop so formed, **ansa cervicalis,** is quite variable as to level, lying anywhere between the angle of the mandible and the upper border of the clavicle. The superior root of the ansa gives off a slender branch to the superior belly of the omohyoid muscle. The nerves to the sternohyoid and sternothyroid muscles are larger and branch from the convexity of the loop. They frequently run for a distance as a common trunk. The nerve to the inferior belly of the omohyoid muscle passes from the distal part of the loop or from the inferior root to reach the muscle beyond its intermediate tendon. In spite of the association with the hypoglossal nerve, all the nerve fibers in the ansa cervicalis arise from the upper three cervical nerves.

(3) The **phrenic nerve** (figs. 139, 140, 261, 263, 271)

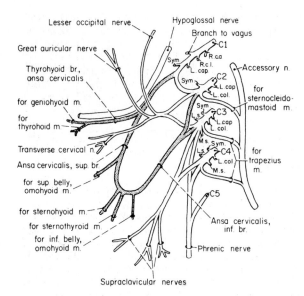

Lesser occipital nerve

Hypoglossal nerve

Branch to vagus

Great auricular nerve

Thyrohyoid br,
ansa cervicalis

for geniohyoid m.

for
thyrohoid m

Transverse cervical n.

Ansa cervicalis, sup. br.

for sup belly,
omohyoid m.

for sternohyoid m.

for sternothyroid m.

for inf. belly,
omohyoid m.

Supraclavicular nerves

Accessory n.

for
sternocleido-
mastoid m.

for
trapezius
m.

Ansa cervicalis,
inf. br.

Phrenic nerve

Fig. 140 The formation and branches of the cervical plexus; the ansa cervicalis is dotted.

is the sole motor nerve to the diaphragm and also provides sensation to its central part. The nerve arises by a large root from the fourth cervical nerve, reinforced by smaller contributions from the third and fifth nerves. It makes its appearance at the lateral border of the anterior scalene muscle, descends vertically over the ventral surface of this diverging muscle, and enters the chest along its medial border. The nerve lies behind the prevertebral layer of cervical fascia and is crossed by the transverse cervical and suprascapular vessels.

At the root of the neck the phrenic nerve passes between the first portion of the subclavian vein and artery and then in front of the internal thoracic artery and vein, where it is joined by their pericardiacophrenic branches. On the right side, the nerve continues into the chest along the right side of the superior vena cava; on the left side, it crosses the aortic arch to the lateral side of the vagus nerve. The thoracic course and termination of this nerve is described in the chapter on the Chest (p. 334).

Accessory phrenic nerves are common as aberrant contributions to the phrenic nerves. They frequently arise as branches of other cervical or brachial plexus nerves which normally receive contributions from cervical nerves three, four, or five. The most frequent accessory phrenic is a branch of the nerve to the subclavius muscle from the fifth cervical nerve (brachial plexus, p. 80). It courses medialward from the nerve to the subclavius and descends sometimes in front of or sometimes behind the brachiocephalic vein to join the phrenic nerve at the base of the neck or in the upper thorax. Other accessory phrenic nerves contain-

ing contributions from the third or fourth cervical nerves occur as branches of the medial supraclavicular nerve or of the ansa cervicalis.

(4) **Contributions to the accessory nerve** leave the cervical plexus at several points. The branch of the second cervical nerve for the supply of the sterno-cleidomastoid muscle frequently appears to arise from the lesser occipital nerve. Contributions to the accessory nerve for the supply of the trapezius arise from cervical nerves three and four. These usually run separate courses to join the accessory nerve in the posterior triangle or under the trapezius muscle.

(5) **Direct muscular branches** of the plexus supply the prevertebral muscles in the neck. From the loop between the first and second cervical nerves, small branches distribute to the rectus capitis lateralis, longus capitis, and rectus capitis anterior muscles. Additional branches from the second, third, and fourth cervical nerves reach the longus capitis and longus colli muscles. Muscular branches from the third and fourth nerves supply the middle scalene and the levator scapulae muscles.

The Brachial Plexus

The ventral rami of the last four cervical and the first thoracic nerves form the brachial plexus. Communications between the fourth and the fifth cervical and between the second and the first thoracic nerves may also contribute to the plexus. The plexus begins to form as the ventral rami pass between the anterior and middle scalene muscles. In this position, **gray rami communicantes** are added to the ventral rami of the fifth and sixth nerves from the middle cervical ganglion and to the ventral rami of the seventh and eighth cervical and first thoracic nerves from the cervicothoracic ganglion. The brachial plexus provides the innervation of the upper limb and shoulder region, and its formation, divisions, and branches are described in the section on the axilla (p. 80). The only branches of the plexus confined to the neck are muscular branches to the prevertebral musculature given off after the exit of the nerves from the intervertebral foramina. In addition to branches from the cervical plexus noted above, longus colli receives twigs from the fifth and sixth cervical nerves; the anterior scalene muscle, branches from the fifth, sixth, and seventh; the middle scalene, branches from nerves five to eight; and the posterior scalene, twigs from the seventh and eighth cervical nerves.

The Subclavian Artery (figs. 139, 141, 142)

The arterial system of the upper limb passes successively through the root of the neck and the axilla to reach the arm. It is the subclavian artery in the root of the neck and becomes the axillary artery after crossing the first rib. Although the subclavian arteries of the two sides have different origins, their course in the neck begins, for both, behind the sternoclavicular joints. From here, the vessels arch upward and outward and then downward to disappear behind the clavicles at their midlengths. The height to which the arteries rise in the neck varies from about 2 to 4 cm. In this course, the vessels are crossed anteriorly by the anterior scalene muscles and, for convenience in description, this muscle is used to define three parts to each artery—one medial, one behind, and one lateral to it. The arteries of the two sides differ in their relations in the first part but are alike in their second and third portions.

The **first part of the left subclavian artery** arises from the arch of the aorta behind and lateral to the left common carotid artery. It ascends through the upper part of the superior mediastinum on the left of the trachea to the root of the neck and then arches lateralward to the medial border of the anterior scalene muscle. The left vagus, phrenic, and cardiac nerves parallel the artery anteriorly, and the left vertebral and internal jugular veins cross it as they descend to the left brachiocephalic vein. The ansa subclavia arches in front of and below the artery (fig. 139). The thoracic duct arches across the artery just medial to the anterior scalene muscle to reach the confluence of the internal jugular and subclavian veins. The left recurrent laryngeal nerve ascends medial to the artery adjacent to the trachea and the esophagus. Laterally, the artery lies against the left lung and pleura, grooving the lung.

The **first part of the right subclavian artery** arises from the brachiocephalic artery behind the upper border of the right sternoclavicular articulation and passes upward and lateralward to the medial margin of the anterior scalene muscle. Its anterior relations with the internal jugular and vertebral veins and with the vagal, phrenic, cardiac, and ansa subclavian nerves are as on the left side. The right recurrent laryngeal nerve winds around the lower and posterior part of the vessel. Behind the artery are the apex of the right lung, the pleura, and the sympathetic trunk.

The **second part of the subclavian artery** represents the highest part of the arch of the vessel. It is short and lies between the anterior scalene muscle in front and the middle scalene muscle behind.

The **third portion of the subclavian artery** is its longest part. It extends downward and lateralward from the lateral margin of the anterior scalene muscle as far as the outer border of the first rib, where the vessel becomes the axillary artery. This portion of the artery is superficial and is contained within the omoclavicular triangle of the neck. It is crossed medially by the external jugular vein which receives the transverse cervical and suprascapular veins in front of the artery. Behind the veins, the nerve to the subclavius crosses the artery. The subclavian vein is parallel to the artery but lies below and anterior to it. Posteriorly, the artery lies against the inferior trunk of the brachial plexus; the middle and upper trunks of the plexus are above and lateral to the vessel.

Variation A variation of the right subclavian artery is its origin as the last branch of the aortic arch instead of as a branch of the brachiocephalic artery. With such an origin, the artery passes diagonally upward and to the right behind the trachea and esophagus and the right common carotid artery and, in such cases, the inferior laryngeal nerve of the right side does not 'recur' around the artery. This anomaly appears to follow the unusual disappearance of the right fourth aortic arch together with the persistence of the right dorsal aortic root between the origin of the seventh dorsal intersegmental artery and the usual site of fusion of the paired dorsal aortic roots. Such a vessel is apt to interfere with swallowing—'dysphagia lusoria'.

Branches of the subclavian artery are the vertebral, thyrocervical, and internal thoracic from its first portion; the costocervical trunk from its second part; and the dorsal scapular artery from its third portion.

(1) The **vertebral artery** (figs. 139, 141) arises from the dorsosuperior aspect of the ascending portion of the subclavian artery. It is surrounded by a plexus of nerve fibers from the upper part of the cervicothoracic sympathetic ganglion which lies posterior and medial to the artery. Inclining upward and medialward, the artery passes anterior to the ventral rami of the seventh and eighth cervical nerves and enters, from below, the costotransverse foramen of the sixth cervical vertebra. It ascends through the succeeding costotransverse foramina of the upper cervical vertebrae, lying anterior to the emerging spinal nerves. Winding behind the superior articular process of the atlas, the artery pierces the dura mater and passes through the foramen magnum into the cranial cavity. Within the cranial cavity the vertebral artery has a number of cranial branches and then unites with its fellow of the opposite side to form the basilar artery. The cranial relations and branches are considered in connection with that region (p. 280). The vertebral artery frequently enters a costotransverse foramen higher than the sixth cervical, and the left vertebral occa-

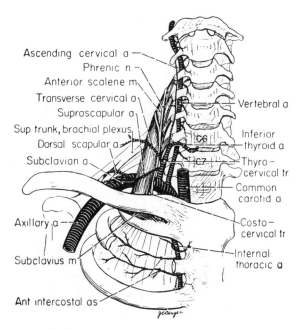

Ascending cervical a —
Phrenic n —
Anterior scalene m.
Transverse cervical a.
Suprascapular a.
Sup trunk, brachial plexus.
Dorsal scapular a.
Subclavian a.
Axillary a —
Subclavius m
Ant intercostal as

Vertebral a
Inferior thyroid a.
Thyro- cervical tr.
Common carotid a.
Costo- cervical tr
Internal thoracic a

C6
C7

Fig. 141   The right subclavian artery and its branches.

sionally (5 per cent) arises directly from the aortic arch.

The cervical portion of the vertebral artery provides spinal and muscular branches. The **spinal branches,** found at each segment, enter the vertebral canal through the intervertebral foramina. They give rise to radicular arteries which pass along the roots of the nerves toward the spinal cord and supply the cord and its meninges. Other branches ramify on the posterior surfaces of the bodies of the vertebrae, anastomosing with adjacent vessels above and below, and supplying the vertebral bodies and the intervertebral disks. **Muscular branches** are given off to the deep muscles of the neck. They anastomose with the descending branch of the occipital artery and with the deep cervical artery.

(2) The **thyrocervical trunk** is a short, thick stem which arises from the anterosuperior aspect of the subclavian just medial to the anterior scalene muscle and opposite the internal thoracic artery. After a short course, it divides into inferior thyroid, suprascapular, and transverse cervical arteries.

The **inferior thyroid artery** (figs. 139, 141) ascends along the medial border of the anterior scalene muscle behind the prevertebral fascia as high as the lower border of the cricoid cartilage. There, emerging from the prevertebral fascia, it turns medialward behind the carotid sheath and passes variably behind, in front of, or between strands of the cervical sympathetic trunk. Arching downward to the lower pole of the thyroid gland, the artery then

turns upward again and has inferior and ascending terminal branches. The distribution and branches of the inferior thyroid artery are considered in connection with the thyroid gland (p. 158). The slender **ascending cervical artery** springs from the arch of the inferior thyroid and ascends on the anterior scalene muscle medial to the phrenic nerve.

The **suprascapular artery** (figs. 139, 141) has a very distinctive course. It passes at first downward and lateralward across the anterior scalene muscle and the phrenic nerve, crosses the subclavian artery and the brachial plexus, and runs lateralward behind and parallel to the clavicle. Beneath the inferior belly of the omohyoid muscle, it arrives at the scapular notch and passes over the superior transverse ligament of the scapula to reach the supraspinatous fossa of the dorsum of the scapula. From a position deep to the supraspinatus muscle, the artery supplies it and then passes inferiorly through the notch of the scapular neck to the infraspinatous fossa. Here it ends by supplying the infraspinatus muscle and anastomosing with the circumflex scapular artery and the dorsal scapular artery.

The suprascapular artery provides a **nutrient branch** to the clavicle and an acromial branch to the skin over the acromion. As it passes over the superior transverse scapular ligament, a **subscapular branch** descends into the subscapular fossa to supply the subscapularis muscle and to anastomose with the subscapular and the dorsal scapular arteries. The suprascapular artery also supplies twigs to the acromioclavicular and shoulder joints.

The **transverse cervical artery** (figs. 139, 141) runs superficially lateralward across the posterior cervical triangle to the anterior border of the trapezius muscle. In its course, it is crossed by the inferior belly of the omohyoid muscle. The transverse cervical artery continues over the scapula lateral to the levator scapulae muscle and passes downward beneath the trapezius in company with the accessory nerve. It supplies the middle part of the trapezius muscle.

An occasional ascending branch (28 per cent) follows the anterior margin of the trapezius, supplying the trapezius, levator scapulae, and splenius muscles. When the dorsal scapular artery does not arise directly from the subclavian artery (30 per cent--Huelke, '62), it is a branch of the transverse cervical artery. This branch arises near the superior angle of the scapula, accompanies the dorsal scapular nerve, and supplies the levator scapulae, rhomboid, and serratus anterior muscles.

(3) The **internal thoracic artery** (fig. 141) takes its origin from the underside of the subclavian artery opposite the thyrocervical trunk. It descends, inclining forward and medialward, behind the sternal end of the clavicle, the subclavian and internal jugular veins, and the first costal cartilage. As it enters the thorax, the phrenic nerve crosses the artery from its lateral to its medial side. Below the first costal cartilage, the internal thoracic artery descends vertically about 1 cm. from the margin of the sternum and, at the level of the sixth intercostal space, divides into the musculophrenic and superior epigastric arteries. The cervical portion of the internal thoracic artery has no branches, and the further distribution of the artery is considered more completely in the section on the thorax (p. 318).

(4) The **costocervical trunk** (fig. 142) arises from the posterior aspect of the subclavian artery behind the anterior scalene muscle on the right side and, commonly, just medial to it on the left. It passes backward over the cervical pleura and the apex of the lung to the neck of the first rib, where it divides into the highest intercostal and the deep cervical arteries.

The **highest intercostal artery** descends in front of the neck of the first rib and, below it, gives off the **posterior intercostal artery** of the first intercostal space. The artery then continues to the second interspace to become the **second posterior intercostal artery**. These arteries end by anastomosing with anterior intercostal branches of the internal thoracic artery. The **deep cervical artery** passes dorsalward between the transverse process of the seventh cervical vertebra and the neck of the first rib. It turns cephalad in the intermuscular interval between the semispinalis cervicis and semispinalis capitis muscles, supplies these and adjacent muscles as high as the axis, and ends by anastomosing with the descending branch of the occipital artery.

(5) The **dorsal scapular artery** (figs. 139, 141) occurs in 70 per cent of specimens (Huelke, '62). It is usually a branch of the second or the third part of the subclavian artery, from which parts it arises with equal frequency. Rarely it arises from either of these parts of the subclavian artery in a common stem with the transverse cervical or the suprascapular artery. The dorsal scapular artery has an intimate relation to the brachial plexus, passing posteriorly through it, most frequently either above or below the middle trunk. Reaching the edge of the levator scapulae muscle, the artery frequently gives off a small ascending muscular branch and then descends along the medial border of the scapula

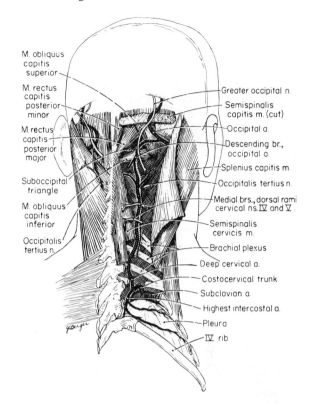

Fig. 142　The costocervical trunk of the subclavian artery and the descending branch of the occipital artery.

anterior to the rhomboid muscles. Here the artery accompanies the dorsal scapular nerve. It supplies the levator scapulae, rhomboid, and serratus anterior muscles, sends small branches to the dorsal and ventral surfaces of the scapula, and anastomoses with the suprascapular and subscapular arteries.

An almost constant **muscular perforating branch** of the dorsal scapular artery penetrates the rhomboid layer of muscles about 2 cm. medial to the base of the spine of the scapula. This vessel supplies the inferior portion of the trapezius muscle.

The Subclavian Vein (fig. 128)

The subclavian vein is the continuation of the axillary vein at the outer border of the first rib. It describes the same curving course through the inferior portion of the posterior cervical triangle as does the subclavian artery, but the vein lies anterior and inferior to the artery and, medially, passes in front of the anterior scalene muscle and the phrenic nerve. The subclavian vein joins the internal jugular vein to form the brachiocephalic vein behind the sternal end of the clavicle. In its course, the subclavian vein lies behind the clavicle and the subclavius muscle, and, below, it rests in a depression on the first rib and on the pleura. It has a bicuspid valve with anterior and posterior cusps, usually located just lateral to the opening of the external jugular vein.

The subclavian vein may receive only one venous tributary, the external jugular vein (p. 147). At its angle of junction with the internal jugular vein, the left subclavian vein receives the thoracic duct; in a similar location, the right subclavian vein receives the right lymphatic duct.

The **external jugular vein** enters the subclavian vein just above the clavicle after receiving the anterior jugular vein (p. 147) and the transverse cervical and suprascapular veins. The **transverse cervical** and **suprascapular veins** cross the lower portion of the posterior cervical triangle in company with the arteries of the same name and distribution. They usually terminate in the external jugular vein opposite the termination of the anterior jugular vein but occasionally open directly into the subclavian vein. They contain valves at their termination. A **dorsal scapular vein** occurs in the same proportion of cases as the artery of the same name and matches it in course and distribution. Like the artery, the vein passes between divisions of the brachial plexus and terminates in the subclavian vein lateral to the anterior scalene muscle. Its orifice is usually guarded by a bicuspid valve. In addition to these variable tributaries, the larger pectoral radicles of the thoracoacromial vein usually empty into the lateral portion of the subclavian vein.

The Anterior and Lateral Vertebral Muscles (fig. 132)

The anterior and lateral vertebral (prevertebral) muscles lie behind the prevertebral fascia in the floor of the anterior and posterior triangles of the neck. The anterior muscles are related to the anterior triangle; the lateral muscles, to the posterior triangle. The anterior group consists of longus colli and capitis and of rectus capitis anterior and lateralis; the lateral group consists of the anterior, middle, and posterior scalene muscles.

Longus colli (fig. 132) is located on the anterior surface of the vertebral column, extending from the atlas to the third thoracic vertebra. It has three parts, **vertical, inferior oblique,** and **superior oblique** portions being described. These tie together the bodies of vertebrae from T3 to C2 and also span to transverse processes (C5, 6) and to the anterior tubercle of the atlas. This muscle flexes the neck and assists in its rotation. It is supplied by branches from the cervical and brachial plexuses, with contributions from nerves C2 to 7.

Longus capitis (fig. 132) arises by four tendinous slips from the anterior tubercles of the transverse processes of vertebrae C3 to 6. Its fibers pass upward and medialward to insert on the under surface of the basilar part of the occipital bone in front of foramen magnum. The muscle assists in flexing the head and neck. It is innervated from the cervical plexus by branches of nerves C1 to 4.

Rectus capitis anterior (fig. 132) is a short, flat muscle immediately behind the upper part of longus capitis. It arises from the lateral mass of the atlas and inserts into the basilar part of the occipital bone immediately in front of the foramen magnum. The muscle flexes the head. It is innervated from the first (and sometimes the second) cervical nerves.

Rectus capitis lateralis, also short and flat, passes from the upper surface of the transverse process of the atlas to the under surface of the jugular process of the occipital bone. The muscle bends the head lateralward. It is supplied by the first cervical nerve.

Scalenus anterior (figs. 132, 139, 144) takes its origin from the anterior tubercles of the transverse processes of vertebrae C3 to 6. It descends with a lateralward inclination to end on the scalene tubercle and ridge of the first rib in front of the subclavian groove. This muscle is one of the essential muscular landmarks of the neck. The phrenic nerve is formed at its lateral margin, descends on the muscle under the prevertebral fascia, and enters the thorax by passing off its medial border a little above its insertion. The roots of the brachial plexus and the subclavian artery emerge between the anterior and middle scalene muscles, and the subclavian vein passes anterior to the insertion of scalenus anterior.

Scalenus medius (fig. 132) The middle scalene muscle is the largest and longest of the group. It takes origin from the posterior tubercles of vertebrae C2 to 7. Contributions from these tubercles form a compact muscle which inserts by a broad attachment into the upper surface of the first rib behind the groove for the subclavian artery. Scalenus medius is perforated by the dorsal scapular nerve and the upper two roots of the long thoracic nerve.

Scalenus posterior (fig. 132) is only incompletely separated from the middle scalene muscle. It arises from the posterior tubercles of the transverse processes of vertebrae C5 to 7 and inserts into the lateral surface of the second rib.

Group Actions When the scalene muscles act together from above, they elevate the first and second ribs; they are muscles of inspiration. Acting from below, they bend the vertebral column toward the side contracting; when both sides are in action, the vertebral column is flexed.

Nerves The scalene muscles are innervated through the brachial plexus: the anterior scalene by branches of nerves C5 to 7; the middle from nerves C3 to 8; and the posterior from nerves C7 and 8.

The prevertebral fascia on the deep aspect of the scalene muscles is reflected onto the cervical pleura as a supporting fibrous investment known as Sibson's fascia. Occasionally, a **scalenus minimus muscle,** an offshoot of scalenus anterior, passes under the subclavian artery to insert into the first rib behind the subclavian groove.

THE VISCERAL COMPARTMENT OF THE NECK

The visceral compartment of the neck is a cylindrical enclosure bounded by the pretracheal layer of cervical fascia in front and the buccopharyngeal fascia behind. The **pretracheal fascia** (p. 153), extending from the hyoid bone above to terminate below on the great vessels of the thorax, lies in front of the larynx and trachea. It splits around the thyroid gland, forming a sheath for that structure. The **buccopharyngeal fascia** (p. 153), essentially an external muscular fascia for the buccinator muscle of the cheek and the constrictor muscles of the pharynx, blends with the pretracheal fascia along the bony and cartilaginous attachments of the constrictor muscles. Thus, the pretracheal fascia forms the anterior and lateral, and the buccopharyngeal fascia forms the posterior and lateral limits of a space occupied by the cervical parts of the respiratory and digestive tubes. In addition to the thyroid and parathyroid glands, described under the anterior triangle of the neck, the visceral compartment contains the larynx and trachea and the pharynx and esophagus.

THE LARYNX (figs. 143–150)

The larynx and trachea are essential parts of the air passages, receiving inspired air from the nasopharynx above and passing it on to the bronchi and lungs below. The larynx is, however, more than a passageway for air; it acts as a valve to prevent the passage of solids and liquids into the airway below, and it is modified to control the expulsion of air from the lungs in the production of sound. In fulfillment of these functions, the larynx has cartilages to give it a relatively rigid form, muscles which act on the cartilages and ligaments to modify its aperture, and mucosal covered membranes and folds which provide a necessary lining and contribute to its air-controlling mechanism.

The larynx lies in the anterior portion of the neck, in front of the fourth, fifth, and sixth cervical vertebrae, and presents a median **laryngeal prominence,** colloquially termed the 'Adam's apple.'

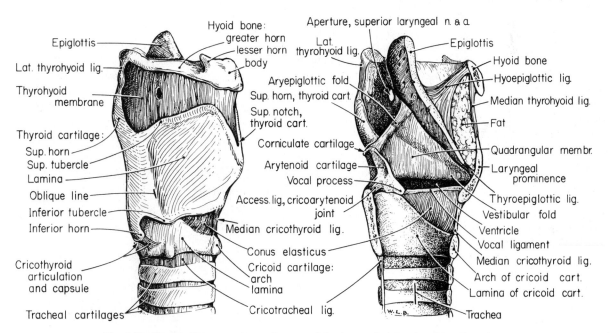

Fig. 143   The cartilages and membranes of the larynx in lateral and median views.

At the level of the prominence the larynx is triangular in cross section, reflecting the form of the thyroid cartilage. Inferiorly, at the level of the cricoid cartilage, the larynx is smaller and is circular in section. The larynx lies between the carotid sheaths laterally and is covered, anterolaterally, by the lateral lobes of the thyroid gland. The infrahyoid muscles are also anterior and lateral to it. The cartilages and muscles of the larynx are covered, posteriorly, by the mucous membrane of the pharynx which intervenes between the larynx and the prevertebral musculature and fascia and the vertebral column.

The Cartilages of the Larynx (fig. 143)

There are nine laryngeal cartilages; three are single, and six are in pairs. The three single cartilages are the thyroid, cricoid, and epiglottic cartilages. Of the paired cartilages, the arytenoid cartilages are the more important; the two corniculate and two cuneiform cartilages are small and of lesser significance.

The large **thyroid cartilage** is composed of two quadrilateral laminae fused together in the anterior midline to form the laryngeal prominence. This is more pronounced in the male, in which sex the laminae meet at a 90° angle (120° in the female). Immediately cephalic to the prominence, the thyroid laminae diverge to form a V-shaped **superior thyroid notch.** The posterior border of each lamina is thick and rounded. This border projects above the superior margin of the cartilage as the **superior horn,** the rounded end of which is joined to the tip of the greater horn of the hyoid bone by the lateral thyrohyoid ligament. The **inferior horn,** formed by a downward projection of the posterior border, is shorter and thicker. It has a slight forward and medialward inclination and, on the medial side of its tip, has a flat circular facet which articulates with an elevated articular surface on the side of the cricoid cartilage. The superior border of the thyroid cartilage has a sinuous curve which is continuous, anteriorly, with the superior thyroid notch and blends, posteriorly, into the superior horn. The superior border gives attachment to the thyrohyoid membrane by which the cartilage is suspended from the hyoid bone above. The generally concave inferior border of the cartilage is interrupted at about the junction of its anterior two-thirds with its posterior one-third by a projecting **inferior tubercle** which marks the lower extremity of the oblique line.

The external surface of each thyroid lamina is roughened by a **superior tubercle** immediately below and in front of the root of the superior horn and by an **oblique line** (fig. 149) which descends diagonally from the superior to the inferior tubercle. The oblique line provides attachment for the sternothyroid and thyrohyoid muscles and for the inferior constrictor muscle of the pharynx. The internal surface of each thyroid lamina is smooth, slightly concave, and covered by the mucous membrane of the interior of the larynx. Anteriorly, in the angle formed by the junction of the two laminae (inferior thyroid notch), are attached the stem of the epiglottis, the vocal and vestibular ligaments, and the thyroarytenoid, thyroepiglottic, and vocalis muscles.

The **cricoid cartilage** (figs. 142, 144, 146) is shaped like a signet ring, with the narrowed arch portion anteriorly and the expanded and flattened lamina posteriorly. The **lamina,** measuring about 2 cm. from above downward, has a median vertical ridge, to the lower part of which are attached the longitudinal fibers of the esophagus. On either side of the ridge is a broad, shallow depression for the posterior cricoarytenoid muscle. The narrow **arch,** 5 mm. in vertical dimension, completes the ring-like form of the cartilage. It gives attachment, anteriorly and laterally, to the cricothyroid muscle and, farther posteriorly, to the inferior constrictor muscle of the pharynx. On either side, at the junction of the arch and the lamina, there is a small, round articular surface by which the inferior horn of the thyroid cartilage articulates with the side of the cricoid (fig. 146). The lower border of the cricoid cartilage is relatively horizontal and is attached to the first ring of the trachea by the cricotracheal ligament. The upper border of the cartilage provides attachment for the median cricothyroid ligament, and more lateralward, for the conus elasticus and the lateral cricoarytenoid muscle. Rising posteriorly, the upper border is received between the spreading laminae of the thyroid cartilage. The superior aspect of the cricoid cartilage is notched in the posterior midline, and on either side of the notch it has a smooth, oval, convex surface directed upward and lateralward for articulation with a corresponding facet on the base of the arytenoid cartilage. The inner surface of the cricoid cartilage is smooth and is lined by mucous membrane.

The paired **arytenoid cartilages** (figs. 143, 147) are mounted, one on either side of the median plane, on the upper border of the lamina of the cricoid cartilage. They are situated between the posterior parts of the laminae of the thyroid cartilage. An arytenoid cartilage is shaped like a three-sided pyramid, its apex curving posteromedially. The base of the pyramid is triangular. Of the three borders of the pyramid, one forms the sharp, anteriorly directed **vocal process** to which the vocal ligament is attached. The laterally directed angle ends in the **muscular process** on which insert the posterior and lateral cricoarytenoid muscles. The broad base of each cartilage bears a concave articular facet, elongated in the anteroposterior direction, for articulation with the cricoid cartilage. Of the three other surfaces of the pyramid, the medial is triangular, flat, and vertical and faces the opposite cartilage. The posterior surface is smooth and concave and lodges the transverse arytenoid muscle. The anterolateral surface of the cartilage, convex and rough, provides attachment for the thyroarytenoid and vocalis muscles and bears, a short distance above the base, a small tubercle for the attachment of the vestibular ligament.

The **corniculate cartilages** (fig. 143) are two small, conical nodules of yellow elastic cartilage. They lie on the apices of the arytenoid cartilages and serve to prolong them backward and medialward. They are enclosed within the posterior parts of the aryepiglottic folds.

The **cuneiform cartilages,** also formed of yellow elastic cartilage, are small, rod-like bodies in the aryepiglottic fold anterior to the corniculate cartilages.

The **epiglottic cartilage** (figs. 143–146) is a single, spoon-shaped plate of yellow elastic cartilage which stands vertically behind the root of the tongue and the hyoid bone. Its broad, round, superior end is free; its tapered inferior extremity is attached by the **thyroepiglottic ligament** in the angle formed by the two laminae of the thyroid cartilage. The anterior surface of the cartilage is attached to the hyoid bone by the **hyoepiglottic ligament.** The anterior surface of the epiglottis is curved forward toward its tip. The mucous membrane covering this surface is reflected onto the root and sides of the tongue as one median and two lateral **glosso-epiglottic folds.** The depressions between these folds are the **valleculae epiglottica.** The mucosa-covered posterior surface of the epiglottis faces the vestibule of the larynx; its lower convex portion is the tubercle of the epiglottis. The cartilage is extensively pitted

for the reception of small mucous glands, and to its margins are attached the aryepiglottic folds.

Ossification The thyroid, the cricoid, and the basal parts of the arytenoid cartilages are hyaline in structure and undergo ossification as age advances. Ossification begins in the thyroid cartilage in the third decade, and these hyaline cartilages may, in old age, become completely converted to bone and thus become visible in radiographs. The other cartilages of the larynx, consisting of yellow elastic cartilage, show little tendency to ossification.

The Joints, Membranes, and Ligaments of the Larynx

The cartilages of the larynx are united by true joints. The **cricothyroid joint** of each side is formed between the elevated round facet on the side of the cricoid cartilage (figs. 143, 146) and the circular facet on the medial aspect of the inferior horn of the thyroid cartilage. Each joint has a ligamentous capsule lined by a synovial membrane; accessory ligamentous fibers strengthen the capsule posteriorly. At these joints a hinge action takes place between the two cartilages, the axis of rotation being a transverse one between the two joints. The **cricoarytenoid articulation** is characterized by two oval facets, concave on the base of the arytenoid and convex on the upper border of the cricoid cartilage. The long axis of the cricoid facet is transverse; that of the arytenoid is generally anteroposterior. As a result, the arytenoid cartilages are able to glide both forward and backward and in mediolateral directions. Due to the curvature and slope of the cricoid facet, backward movement of the arytenoid cartilage also tilts the vocal process of this cartilage upward, whereas forward movement carries it downward. Rotary movement around a vertical axis through the intersection of the joint surfaces allows for swinging of the vocal process medialward (adduction of the vocal folds) or lateralward (abduction of the vocal folds).

Several membranes unite the cartilages of the larynx. The **thyrohyoid membrane** suspends the larynx from the hyoid bone, attaching below to the upper border of the thyroid cartilage and to its superior horn. The upper extremity of the membrane passes behind the hyoid bone, from which it is separated by a bursa, and attaches to the upper margin of the bone and to its greater horn. This fibro-elastic membrane is thicker in its median part, where it forms the **median thyrohyoid ligament.** The posterior border of the thyrohyoid membrane extends from the tip of the superior horn of the thyroid cartilage to the end of the greater horn of the hyoid bone as a round, elastic cord. This **lateral thyrohyoid ligament** frequently contains a cartilaginous nodule. Hyoepiglottic and thyroepiglottic ligaments attach the epiglottis to the hyoid bone and the thyroid cartilage. The cricoid cartilage is connected to the first ring of the trachea by the **cricotracheal ligament,** a fibrous membrane similar to those connecting the cartilaginous tracheal rings.

The **conus elasticus** (figs. 143, 145) is an elastic membrane, the lower attachment of which is to the entire arch of the cricoid cartilage from one arytenoid articular facet to the other. The median part of this membrane is the **median cricothyroid ligament.** The paired lateral parts of the conus end above in parallel thickenings uniting the thyroid and the arytenoid cartilages. These elastic **vocal ligaments,** the supporting ligaments of the vocal folds, are stretched between the vocal processes of the arytenoid cartilages dorsally and the angle between the thyroid laminae ventrally, where they insert close to one another midway between the upper and lower borders of the laminae.

The **quadrangular membrane** (fig. 145) is a weak submucosal sheet of connective tissue ending in the aryepiglottic fold posterosuperiorly and, inferiorly, in a free margin which constitutes the vestibular ligament of the false vocal (vestibular) fold. Ventrally, the membrane has an attachment in the angle of the thyroid laminae and to the sides of the epiglottis. Dorsally, it ends in the corniculate and arytenoid cartilages. The quadrangular membrane underlies the thyroepiglottic and the aryepiglottic muscles in its posterior portion.

The Interior of the Larynx (figs. 144, 145)

The mucous membrane of the larynx is continuous with that of the pharynx above and of the trachea below. At the root of the tongue, reflections of the mucosa onto the anterior surface of the epiglottis form one median and two lateral **glosso-epiglottic folds.** From the side walls of the pharynx the mucosa passes forward between the thyroid cartilages laterally and the arytenoid and cricoid cartilages medially to line the **piriform recess.** This is the gutter which extends upward at the side of the epiglottis. From the piriform recess the mucous membrane passes over the aryepiglottic fold and descends into the larynx.

The **inlet of the larynx** (fig. 144), opening anteriorly from the pharynx, lies in an almost vertical plane. The projecting epiglottis, arytenoid cartilages, and aryepiglottic folds bound the inlet and carry it well into the cavity of the pharynx. The inlet is approximately triangular in shape, the upper free border of the epiglottis forming the base of the triangle. The apex of the triangular inlet is downward and backward between the two arytenoid cartilages. The aryepiglottic folds form the sides of the triangle and contain, posteriorly, the

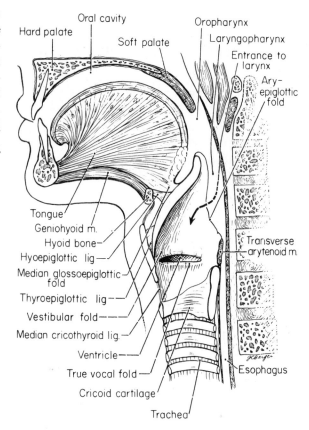

Fig. 144   A median section of the oral cavity, pharynx, and larynx.

corniculate and cuneiform cartilages which produce rounded tubercles in the folds.

Continuing from the inlet, the **cavity of the larynx** (fig. 144) presents, superiorly, a wide, triangular **vestibule** which ends below at the **rima vestibuli,** the aperture between the false vocal folds. The lateral walls of the vestibule are formed submucosally by the quadrangular membranes (p. 186) reinforced by the thyroepiglottic and aryepiglottic muscles. The free inferior margin of the quadrangular membrane constitutes the **vestibular ligament** which is attached, anteriorly, between the two thyroid laminae above the vocal ligament and, posteriorly, to the arytenoid cartilage slightly above its vocal process. The **vestibular folds** are thick pads of mucous membrane formed around the vestibular ligaments. The folds reach the angle between the thyroid laminae anteriorly but fall short of the dorsal laryngeal wall posteriorly. They contain many mucous glands and a few muscle fibers. The opening between the vestibular folds, the **rima vestibuli,** is wider than that between the true vocal folds, the rima glottidis, and consequently both sets of folds are normally visible by laryngoscopic examination. The **ventricle of the larynx** (fig. 145) represents a middle or intermediate portion of its cavity. It begins as a spindle-shaped interval between the vestibular and the vocal folds and continues upward, lateral to the overhanging vestibular fold, sometimes as high as the upper border of the thyroid cartilage. The wall of the ventricular appendix is packed with mucous glands and is the source of a lubricating mucous secretion for the vocal folds.

The **vocal folds** (true vocal folds) (fig. 145) are prominent, shelf-like folds stretched between the angle formed by the thyroid laminae anteriorly and the vocal processes of the arytenoid cartilages posteriorly. The folds contain, near their free margin, the vocal ligaments (p. 186), and over these liga-

ments the mucous membrane of the fold is thin and firmly attached. Lateral to the vocal ligaments lie the vocalis and thyroarytenoid muscles. The vocal folds and the space between them are designated as the **glottis** and are the part of the larynx most directly concerned in the production of sounds. The **rima glottidis,** the space between the apposed vocal folds and arytenoid cartilages, is the narrowest part of the laryngeal cavity. It averages 23 mm. in length in the male and 17 mm. in the female. The anterior three-fifths of this distance is intermembranous; the posterior two-fifths is intercartilaginous. The width and shape of the rima glottidis vary with the movements of the vocal folds and arytenoid cartilages during respiration and phonation (fig. 148). During quiet respiration, the opening is narrow and wedge-shaped with the medial surfaces of the arytenoid cartilages parallel to one another. During forced respiration, the glottic opening approximates the capacity of the trachea, becoming markedly widened. The broadest part of the fissure is then at the extremities of the vocal processes of the arytenoids, and, here, each side of the rima glottidis presents a marked angle. During the production of sounds, the vocal folds are closely approximated, so that the rima is reduced to a linear slit.

The **infraglottic portion of the larynx** extends to the inferior border of the cricoid cartilage. Its lumen changes from the slit-like form of the rima glottidis to become the circular form of the trachea.

The Muscles of the Larynx (figs. 145, 146, 147, 149)

The intrinsic muscles of the larynx maintain and control the airway through the larynx, control the expiration of air in phonation, and provide an effective sphincter against the entrance of foreign materials during the act of swallowing. The principal laryngeal muscles are the following:

Cricothyroid	Transverse arytenoid
Posterior cricoarytenoid	Oblique arytenoid
Lateral cricoarytenoid	Thyroarytenoid

The **cricothyroid muscles** (fig. 149) are located on the external surface of the larynx in the interval between the cricoid and thyroid cartilages. The fibers of each muscle arise from the anterior and lateral aspects of the arch of the cricoid cartilage. They radiate backward and upward and insert into the inferior border and medial surface of the thyroid

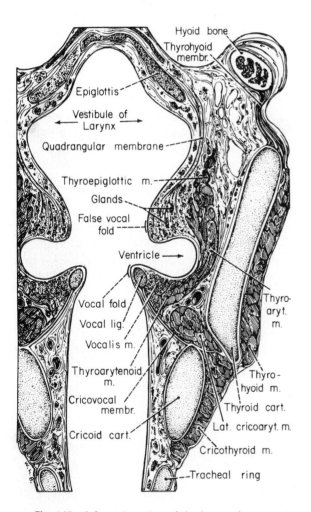

Fig. 145   A frontal section of the human larynx.

Hyoid bone
Thyrohyoid membr.
Epiglottis
Vestibule of Larynx
Quadrangular membrane
Thyroepiglottic m.
Glands
False vocal fold
Ventricle
Vocal fold
Vocal lig.
Vocalis m.
Thyroarytenoid m.
Cricovocal membr.
Cricoid cart.
Thyro-aryt. m.
Thyro-hyoid m.
Thyroid cart.
Lat. cricoaryt. m.
Cricothyroid m.
Tracheal ring

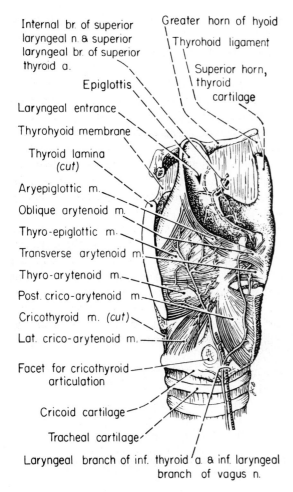

Internal br. of superior laryngeal n. & superior laryngeal br. of superior thyroid a.

Greater horn of hyoid

Thyrohoid ligament

Superior horn, thyroid cartilage

Epiglottis

Laryngeal entrance

Thyrohyoid membrane

Thyroid lamina (cut)

Aryepiglottic m.

Oblique arytenoid m.

Thyro-epiglottic m.

Transverse arytenoid m.

Thyro-arytenoid m.

Post. crico-arytenoid m.

Cricothyroid m. (cut)

Lat. crico-arytenoid m.

Facet for cricothyroid articulation

Cricoid cartilage

Tracheal cartilage

Laryngeal branch of inf. thyroid a. & inf. laryngeal branch of vagus n.

Fig. 146   The muscles and nerves of the larynx. (The left lamina of the thyroid cartilage has been removed.)

cartilage and into the anterior border of its inferior horn. By drawing the thyroid cartilage downward and forward toward the cricoid, the muscles increase the distance between the thyroid and the arytenoid cartilages (mounted on the cricoid). Thus, the anterior and posterior attachments of the vocal ligaments are carried farther apart and the tension of the vocal folds is made greater.

The **posterior cricoarytenoid muscles** (figs. 146, 147) lie on the dorsum of the lamina of the cricoid cartilage, covered by the mucous membrane of the laryngopharynx. The muscle fibers arise from the broad depression at either side of the median ridge of the cartilage and, passing lateralward and upward, converge to the back of the muscular processes of the arytenoid cartilage. The posterior cricoarytenoid muscles are the abductors of the

vocal folds; the drawing of the muscular processes backward turns the vocal processes lateralward.

The **lateral cricoarytenoid muscles** (figs. 146, 147) are small, rectangular muscles which arise from the upper border of the arch of the cricoid cartilage. Their muscle fibers pass backward and upward to insert into the front of the muscular processes of the arytenoid cartilages. These muscles abut against the lower borders of the thyroarytenoid muscles and are frequently inseparable from them. The lateral cricoarytenoid muscles draw the muscular processes of the arytenoid cartilages forward, causing their vocal processes to move medialward. These muscles are, therefore, adductors of the vocal folds and are the antagonists of the posterior cricoarytenoids.

The **transverse arytenoid muscle** (figs. 146, 147) is single and is lodged in the posterior concave surfaces of the arytenoid cartilages. Its fibers take origin from this posterior surface and the lateral border of one cartilage and insert into the same regions of the opposite cartilage. The muscle slides the arytenoid cartilage medialward on the oval cricoid facet and thus approximates (adducts) the vocal folds.

The **oblique arytenoid muscles** (fig. 146) consist of fascicles which overlie the transverse arytenoid muscle. Arising from the back of the muscular processes of the arytenoid cartilages, the muscles ascend medialward and cross each other, directed toward the apices of the opposite arytenoid cartilages. Many of the fibers are inserted there; others extend beyond and combine with fibers arising from the apices of the arytenoids to form the aryepiglottic muscles. The **aryepiglottic muscles** (fig. 146) pass forward and upward in the aryepiglottic folds to reach the thyroepiglottic ligament and the lateral margins of the epiglottic cartilage. The oblique arytenoid and the aryepiglottic muscles are sphincters of the vestibule of the larynx.

The **thyroarytenoid muscles** (figs. 146, 147) border the vocal ligaments. They arise from the inner surfaces of the thyroid laminae close to the laryngeal prominence and insert on the lateral borders of the arytenoid cartilages. These muscles draw the arytenoid cartilages forward, lessening tension in the vocal ligament. They are also important sphincters of the glottis, their forward traction on the arytenoids turning the vocal processes medialward. The thickening of the muscle that accompanies its shortening also contributes to sealing the glottis.

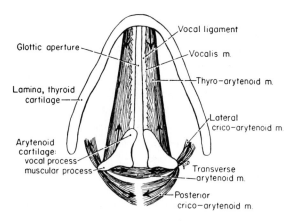

Fig. 147 A diagram illustrating the action of certain of the laryngeal muscles.

The thyroarytenoid muscles have as subsidiary parts the thyroepiglottic and the vocalis muscles. The **thyroepiglottic muscles** (fig. 146) continue the superior border of the thyroarytenoid muscles upward. From the same thyroid origin anteriorly, their fibers extend upward and backward to end in the quadrangular membranes and the lateral margins of the epiglottis. The thyroepiglottic muscles, with the superior margins of the thyroarytenoids, support the quadrangular membranes and act as sphincters of the vestibule of the larynx. The **vocalis muscles** are composed of those internal fibers of the thyroarytenoid muscles most closely related to the vocal ligament. Two muscular cones are described for each vocalis muscle (Galletti, '62). The broad bases of these muscular cones are represented in the attachments of their fibers into the vocal ligament, especially into its elastic tissue. The apices of the cones are: one in the angle between the thyroid cartilage and the vocal ligament; the other in the vocal process of the arytenoid cartilage. Such fibers are considered to be chiefly responsible for the control of pitch through their ability to regulate the vibrating part of the vocal ligaments.

The Action of the Laryngeal Muscles The vocal folds are abducted and the rima glottidis is widened by the posterior cricoarytenoid muscle; adduction of the folds and narrowing of the laryngeal aperture follow contraction of the lateral cricoarytenoid and the transverse arytenoid muscles. Adduction may be carried to complete glottic closure by the additional action of the thyroarytenoid muscles. Tension of the vocal ligaments is increased by the action of the cricothyroid muscles and lessened by the contracture of the vocalis portion of the thyro-

arytenoid muscles. The vocalis fibers produce the minute adjustments and segmental tensions or relaxations of the vocal ligament required in vocalization.

The posterior part of the glottic aperture is respiratory in function; the anterior part is concerned with phonation. Here, the escape of jets of air between the anterior parts of the two vocal cords produces vibrations, modified by the tongue, lips, and palate and affected by the resonating chambers of the head, which result in speech and other sounds. The cricothyroid and vocalis muscles are especially concerned.

As a reflex during the act of swallowing and to prevent entrance of foreign material, the laryngeal inlet is closed and the walls of the vestibule are approximated. In this action the arytenoid cartilages are tilted forward and medialward, the corniculate cartilages are carried forward to make contact with the tubercle of the epiglottis, the walls of the vestibule are flattened medialward, and the margins

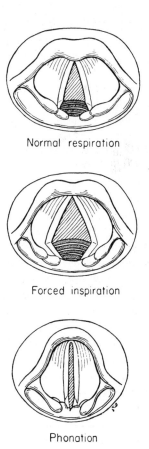

Normal respiration

Forced inspiration

Phonation

Fig. 148 The shape and size of the aperture of the larynx in different phases of its function.

of the epiglottis are drawn backward and medial-ward. In this way a T-shaped closure of the laryngeal inlet and approximation of the vestibular walls result. The muscles concerned in this protective action are the thyroarytenoid, the thyroepiglottic, the oblique arytenoid, and the aryepiglottic. Assistance is given to the T-shaped vestibular closure by the action of the tongue on the epiglottis. As the dorsum of the tongue is rolled backward, pitching a bolus of food into the pharynx, the upper portion of the epiglottis is bent backward and downward. Radiologic studies show that the tip of the epiglottis actually turns down over the arytenoids (Johnstone, '42). This provides additional protection against accidental entry of foreign material into the vestibule of the larynx.

Nerves and Vessels of the Larynx

The **nerves** of the larynx are the superior and the inferior laryngeal branches of the vagus (p. 164). The **internal branch of the superior laryngeal nerve** (figs. 146, 149), passing under the lateral margin of the thyrohyoid muscle, pierces the thyrohyoid membrane with the superior laryngeal artery. It supplies sensory fibers to the mucous membrane and parasympathetic fibers to the glands of the epiglottis, the base of the tongue, the aryepiglottic fold, and the greater part of the interior of the larynx. Taste buds are abundant in the mucosa of the posterior surface of the epiglottis and on the aryepiglottic folds. The superior laryngeal nerve is the principal sensory nerve of the larynx and sends branches to the surfaces of the epiglottis and to the aryepiglottic fold. A large medial descending branch to the back of the larynx overlaps and anastomoses with posterior branches of the inferior laryngeal nerve (fig. 146). The long, slender **external branch of the superior laryngeal nerve** (p. 164 & fig. 149) descends along the oblique line of the thyroid cartilage to the cricothyroid muscle which it supplies. All of the other intrinsic muscles of the larynx are innervated by the inferior laryngeal nerve.

The **inferior laryngeal nerve** (continuation of the recurrent laryngeal nerve, p. 165 & fig. 146) enters the larynx by ascending under the lower border of the inferior constrictor muscle of the pharynx. Passing behind the capsule of the cricothyroid articulation, the nerve divides into anterior and posterior branches. The anterior branch passes forward onto the lateral cricoarytenoid and thyroarytenoid muscles. It supplies these as well as the

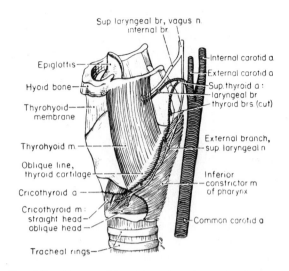

Fig. 149   The superior laryngeal nerve and laryngeal branches of the superior thyroid artery.

vocalis, the aryepiglottic, and thyroepiglottic muscles. The posterior branch supplies the posterior cricoarytenoid and transverse and oblique arytenoid muscles and anastomoses with the medial branch of the internal branch of the superior laryngeal nerve. The inferior laryngeal nerve has only a minor role in sensation, and considerable overlap exists in the distribution of the mucosal branches of the superior and inferior nerves.

The arteries of the larynx are the superior laryngeal, the inferior laryngeal, and the cricothyroid. The **superior laryngeal artery** (fig. 149) is a branch of the superior thyroid artery (p. 157) which pierces the thyrohyoid membrane in company with the internal branch of the superior laryngeal nerve. The **inferior laryngeal artery** (fig. 146) is a branch of the inferior thyroid artery (p. 158) which accompanies the inferior laryngeal nerve under the inferior constrictor muscle of the pharynx to the back of the larynx. The **cricothyroid artery** (fig. 149), a branch of the superior thyroid artery, frequently sends a twig through the cricothyroid membrane into the interior of the larynx. These arteries anastomose within the larynx, supplying the mucous membranes, muscles, and other tissues. **Superior** and **inferior laryngeal and cricothyroid veins** accompany the arteries; the superior and cricothyroid veins are tributary to the superior thyroid vein (p. 158), the inferior vein joins the inferior thyroid vein (p. 158).

The **lymphatics** of the larynx drain upward and downward from the true vocal fold and do not communicate across the median line. Drainage upward is by channels which emerge from the larynx in company with the superior laryngeal vessels and reach the **infrahyoid nodes** (p. 167) on the thyrohyoid membrane and the **superior deep cervical nodes.** The lymphatic channels draining inferiorly may pierce the cricothyroid membrane to reach the **prelaryngeal nodes.** Others follow the inferior thyroid vessels to the **pretracheal nodes** or, posteriorly, pass along the recurrent laryngeal nerves to the **paratracheal nodes.** Such paths lead, through these subsidiary nodes, to the inferior deep cervical chain.

THE TRACHEA (figs. 143, 144, 149, 154)

The trachea is the continuation of the airway below the larynx. It begins at the lower border of the cricoid cartilage at the level of the sixth cervical vertebra and terminates at the sternal angle (upper border of the fifth thoracic vertebra), where it divides into the right and left bronchi. The trachea is about 12 cm. long, of which 6 cm. is above the upper border of the manubrium sterni and about 6 cm. is in the chest. The corrugated cylindrical trachea has a maximum diameter of about 2 cm. In the neck (fig. 122) the trachea lies in the midline posterior to the infrahyoid muscles and the thyroid gland and is anterior to the cervical portion of the esophagus. The common carotid arteries ascend at the side of the trachea, separated from it above by the lateral lobes of the thyroid gland. The thyroid gland is wrapped around the tube, its isthmus portion overlying the second, third, and fourth tracheal rings anteriorly and the inferior poles of its lateral lobes lying lateral to the fifth or sixth ring. The recurrent laryngeal nerves ascend at the posterolateral borders of the trachea in the interval between the adjacent margins of the trachea and esophagus. The relations of the thoracic portion of the trachea are discussed in the chapter on the Chest (p. 348).

The lumen of the trachea is maintained against fluctuations of pressure by sixteen to twenty C-shaped cartilaginous rings. These hyaline rings occupy about two-thirds of the circumference of the trachea, their posterior ends being joined by fibrous tissue and smooth muscle. The cartilages are embedded in the fibro-elastic supporting membrane of the trachea which is continuous above with the perichondrium of the cricoid cartilage. The cartilages are from 3 to 4 mm. in breadth, and the spaces between them are about half as broad. The external surfaces of the cartilages are flat, but the internal surfaces are convex, so that each ring bulges slightly into the lumen of the tube. Two or more of the cartilages may partially unite, and their ends are sometimes bifurcated. The first cartilage is broader than the rest and is connected to the lower border of the cricoid cartilage by the cricotracheal mem-

brane. The last cartilage is thick and broad in the middle, and its lower border is prolonged downward and backward in a hooked process. This is the **carina** which forms a keel-like projection between the origins of the right and left bronchi. The smooth muscle joining the ends of the cartilaginous rings is the submucosal **trachealis muscle.** The submucosa is rich in mucous glands which maintain the moistness of its mucosal surface against the drying effect of the air passing through it. Like the mucous membrane of the bronchi and the lower portion of the larynx, that of the trachea has a ciliated epithelium. Airborne dust and foreign particles become trapped in the mucus of the airway, and the cilia sweep this material upward to the pharynx where it can be swallowed or expelled by coughing.

The **arteries** of the trachea are small vessels which reach the tube at various points along its length. They are branches of the inferior thyroid, the internal thoracic, and the bronchial arteries. Its **veins** terminate in the inferior thyroid veins. The **lymphatic channels** of the cervical trachea end in paratracheal, pretracheal, prelaryngeal, and inferior deep cervical nodes (p. 167). Those of the thoracic portion drain to the tracheobronchial nodes (p. 349). The smooth muscle and glands of the trachea receive a parasympathetic innervation through the **vagus nerve,** directly and by its recurrent laryngeal branches, and a sympathetic innervation by branches from the sympathetic trunk and its **ganglia.**

Clinical Note Tracheostomy is occasionally necessary when the airway is occluded. It is usually done by incising the tracheal cartilages in the midline just below the thyroid isthmus. Inferior thyroid veins represent the midline structures most likely to be encountered.

THE PHARYNX

The pharynx is the common chamber of the respiratory and digestive tracts and is located behind the nasal and oral cavities. Entered through these passageways, it is continued below into the separated larynx and esophagus. Here the digestive stream and the airway cross one another, the larynx lying anterior to the esophagus. The upper, or nasal, portion of the pharynx receives the terminations of the auditory tubes bilaterally. The pharynx is somewhat funnel-shaped in form. Its greatest breadth is about 5 cm. at its attachment to the base of the skull, and it tapers below to the 1.5 cm. width of the upper esophagus. The pharynx is approximately 12 cm. long; it begins at the base of the skull and terminates below at the level of the lower border of the cricoid cartilage. The pharyngeal wall is almost entirely posterior and lateral to its cavity, for the latter is broadly open in front to the nasal and oral cavities and to the larynx. Thus can be recognized a nasopharynx, which extends from the base of the skull to the level of the soft palate, and an oropharynx between the palate and the level of the hyoid bone. The lower laryngeal portion of the pharynx ends at the lower border of the cricoid cartilage, where the pharynx becomes continuous with the esophagus.

The **nasal part of the pharynx** (fig. 150) is entirely respiratory. Its walls are relatively rigid, and its chamber cannot be obliterated. It is entered through the posterior apertures of the nose which are separated by the posterior border of the nasal septum. The posterior wall of the nasopharynx abuts against the basilar part of the occipital bone, the anterior arch of the first cervical vertebra, and the body of the second vertebra. In the mucosa at the upper end of the posterior wall is the **pharyngeal tonsil,** a collection of lymphoid tissue which is called the **adenoids** when enlarged. In the median line above the pharyngeal tonsil a small diverticulum, the **pharyngeal bursa,** may extend deep to the mucosa as far as the basilar process of the occipital bone. The lateral wall of the nasopharynx lodges the opening of the auditory tube which communicates with the chamber of the middle ear. The opening of this tube (ostium pharyngeum tubae auditivae) is directed downward, and a hood-like overhang is provided for it by the projecting **torus tubarius** (fig. 150) representing the medial end of the tubal cartilage. Lymphoid tissue at the opening of the auditory tube is designated as the tubal tonsil. A vertical fold of mucous membrane, the **salpingopharyngeal fold,** covers the salpingopharyngeus muscle and extends downward from the posterior limb of the torus to lose itself in the pharyngeal wall. A smaller **salpingopalatine fold** descends from the anterior margin of the torus to the soft palate. An elevation of the pharyngeal orifice of the tube is caused by the presence of the levator veli palatini muscle as it approaches the lateral border of the soft palate under the mucosa. This is the torus levatorius. A laterally directed, slit-like recess behind the salpingopharyngeal fold is the **pharyngeal recess** (fig. 150).

The **oral portion of the pharynx** (figs. 150, 178) represents the digestive entrance of the chamber. Its anterior relationships are to the oral cavity and the dorsum of the tongue, and its inferior limit is at the level of the hyoid bone. The lateral walls of the oral cavity are occupied by the palatine tonsils bounded by the tonsillar pillars—the palatoglossal and the palatopharyngeal folds. Deep to the folds lie the corresponding muscles (p. 194). The palatine tonsil is more completely described in connection with the palate and tongue (p. 235).

The **laryngeal portion of the pharynx** (fig. 150) lies behind the aperture and posterior wall of the larynx. It is thus in relation to the epiglottis, the aryepiglottic folds, the interarytenoid notch, and the dorsum of the arytenoid and cricoid cartilages. Lateral to the laryngeal opening and internal to the thyroid cartilage and

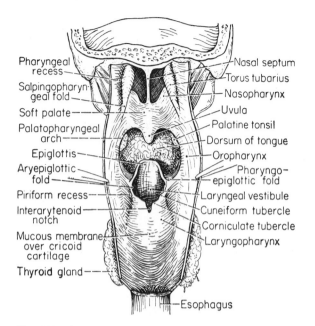

Fig. 150 The cavity of the pharynx, exposed by incision and retraction of its posterior wall.

thyrohyoid membrane is the deep depression of the **piriform recess.** It is limited above by the pharyngoepiglottic fold.

Structure of the Pharyngeal Wall

The pharyngeal wall is lined by a mucous membrane continuous with all the chambers with which it communicates. The submucosa is a strong fibrous sheet, known as the pharyngobasilar fascia, which is uncovered by the muscular layer of the pharynx in its uppermost part and is especially strong there. The **pharyngobasilar fascia** (fig. 151) blends with the periosteum of the base of the skull in front of the pharyngeal tubercle. Its attachment on the skull continues lateralward to the apex of the petrous portion of the temporal bone and then turns forward along the cartilage of the auditory tube. On either side the fascia attaches to the medial pterygoid plate, the pterygomandibular raphe, and the mylohyoid ridge of the mandible. The pharyngobasilar fascia thus defines the limits of the pharyngeal wall in its upper portion. It is less distinct below the mandible and tongue. The muscular coat of the pharynx lies external to the pharyngobasilar fascia, and the muscles are in turn covered externally by the buccopharyngeal fascia (p. 153). The pharynx is separated posteriorly from the vertebral column and the prevertebral muscles and fascia by the loose areolar tissue of the retropharyngeal space

(p. 155). This loose areolar tissue permits the elevation and depression of the pharynx which are entailed in swallowing movements.

The Muscles of the Pharynx (figs. 130, 151, 178)
The muscles of the pharynx are:

Inferior constrictor	Stylopharyngeus
Middle constrictor	Palatopharyngeus
Superior constrictor	Salpingopharyngeus

The inferior, middle, and superior constrictor muscles of the pharynx constitute an external, predominantly circular layer, each muscle of which overlaps, from below upward, the muscle above it. Thus they exhibit the overlapping relationship of stacked flowerpots. They are paired muscles which end in a median pharyngeal raphe, a fibrous band extending from the pharyngeal tubercle above to the lower limit of the inferior constrictor muscle. The stylopharyngeus, palatopharyngeus, and salpingopharyngeus muscles spread out as an internal

longitudinal muscular coat, being quite separate at their origins but blending with one another inferiorly in the pharynx.

The **inferior constrictor muscle** (fig. 124), the thickest of the constrictor muscles, arises from the oblique line of the thyroid cartilage and the area behind it, from the side of the cricoid cartilage, and from a tendinous arch between the inferior tubercle of the thyroid and the cricoid cartilages. Spreading backward and medialward, the muscle fibers end in the pharyngeal raphe. The inferior fibers of the muscle are continuous with the musculature of the esophagus. Most of the fibers of the inferior constrictor muscle ascend and, increasing in obliquity, overlap the middle constrictor.

The **middle constrictor muscle** arises from the upper border of the greater horn of the hyoid bone, from the lesser horn, and from the stylohyoid ligament. Fanning posteriorly and medially, its fibers both descend beneath the inferior constrictor muscle

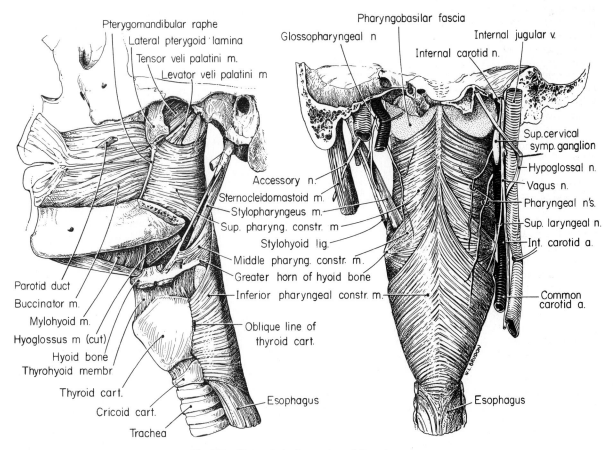

Fig 151   The external muscles of the pharynx.

and ascend to overlap the superior constrictor. The pharyngeal fibrous raphe is the insertion for the muscles of both sides.

The **superior constrictor muscle** arises from the lower one-third of the posterior border of the medial pterygoid plate and its hamulus, the pterygo-mandibular raphe, the mylohyoid line of the mandible, and the side of the root of the tongue. The muscle fibers radiate upward and downward, inserting in the pharyngeal raphe and, by the uppermost fibers, into the pharyngeal tubercle. The lower part of the muscle is overlapped by the middle constrictor; the upper border falls short of the base of the skull except at the midline. This leaves a crescentic nonmuscular area lateral to the pharyngeal tubercle in which the pharyngeal wall is composed largely of the pharyngobasilar fascia. In the gap between the origins of the middle and superior constrictor muscles, the stylopharyngeus muscle enters the pharyngeal wall (fig. 151).

The **stylopharyngeus muscle** arises from the medial aspect of the base of the styloid process. Directed downward and medialward, it insinuates itself between the upper border of the middle constrictor muscle and the superior constrictor at the lateral angle of the pharynx. The stylopharyngeus inserts into the posterior and superior borders of the thyroid cartilage and spreads out under the mucosa of the pharynx, its fibers becoming continuous with the neighboring fibers of the palato-pharyngeus muscle posteriorly. The glossopharyngeal nerve, having curved posteriorly around the lateral margin of the stylopharyngeus, enters the pharyngeal wall with this muscle.

The **palatopharyngeus muscle** is the substance of the palatopharyngeal fold, the posterior pillar of the palatine tonsil. In the palate the fibers of the muscle lie both posterosuperior and antero-inferior to the muscular bundles of the levator veli palatini and the musculus uvulae. The posterior superior lamina is thin; both laminae of fibers reach the midline, and the antero-inferior layer attaches to the palatine aponeurosis and the posterior border of the bony palate. Descending in the lateral wall of the pharynx (fig. 179), the palatopharyngeus muscle spreads out into a thin longitudinal layer which inserts, along with stylopharyngeus, into the posterior border of the thyroid cartilage and into the posterior wall of the pharynx. In the latter situation, its more medial fibers decussate with those of the opposite side.

The **salpingopharyngeus muscle** (fig. 179) consists of a thin muscular bundle which arises from the lower border of the auditory tube. It converges on the palatopharyngeus and blends with it in the pharyngeal wall.

Swallowing In the act of swallowing, the oral cavity and oropharynx are isolated from the respiratory passages by the closure of the laryngeal vestibule below (p. 189) and of the communication with the nasopharynx above. In the latter action the soft palate is elevated and drawn outward and backward by the levator veli palatini and tensor veli palatini muscles, and its middle one-third is brought into firm contact with the pharyngeal wall at about the upper border of the superior constrictor muscle. Contraction of the salpingopharyngeus muscles draws upward and inward on the lateral pharyngeal walls. As the tongue drives a bolus of food toward the back of the pharynx, the palatopharyngeus and stylopharyngeus muscles elevate the larynx and pharynx to receive it. Once food passes over the epiglottis, the serial contraction of the superior, middle, and inferior constrictor muscles of the pharynx drives it forcibly into the esophagus below.

In electromyographic studies of swallowing in the cat, the dog, and the monkey, Doty and Bosma ('56) observed a rather precise temporal sequence in the action of the muscles concerned. They noted that swallowing was initiated by the concurrent contraction of a group of muscles which include the superior pharyngeal constrictor, the palatopharyngeus, the palatoglossus, the posterior intrinsic muscles of the tongue, the styloglossus, the stylohyoideus, the geniohyoideus and the mylohyoideus. The middle pharyngeal constrictor, the thyrohyoid, and the thyroarytenoid muscles were inhibited at the onset but became active after varying delays, and action in the inferior constrictor did not start until the contraction of the intiating group of muscles was nearly over. The results are consonant with forcible passage of food from mouth to pharynx, closure of the laryngeal inlet and glottis, and progressive descending contractions in the pharyngeal musculature.

Nerves (figs. 130, 153) The pharyngeal plexus provides the innervation of the muscles of the pharynx, with the exception of stylopharyngeus. The inferior constrictor muscle receives an additional supply from the external branch of the superior laryngeal nerve (p. 164).

The **pharyngeal plexus** (fig. 130) is formed on the middle constrictor muscle, opposite the greater horn of the hyoid bone, by the pharyngeal branches of the glossopharyngeal and vagal nerves and of the superior cervical sympathetic ganglion. Several pharyngeal branches of the vagus (p. 164) pass downward and medialward between the internal and external carotid arteries, adjacent to several twigs from the glossopharyngeal nerve. Four to six filaments arising in the superior cervical sympathetic ganglion join these nerves on the wall of the pharynx. From the plexus thus formed, branches penetrate into the substance of the pharynx; the glossopharyngeal fibers are sensory; the vagal fibers, motor; and the sympathetic innervation, vasomotor. The nerve to the stylopharyngeus muscle (fig. 153) arises from the glossopharyngeal nerve as it passes behind and lateral to the muscle and penetrates it at about its midlength.

The vessels of the Pharynx The **arteries** of the pharynx (fig. 130) are the ascending pharyngeal (p. 161), the ascending palatine branch of the facial artery (p. 160), the descending palatine and pharyngeal branches of the maxillary artery (p. 225), and muscular branches of the superior thyroid artery (p. 157). The ascending pharyngeal (p. 161 & fig. 130) is frequently the first or second branch of the external carotid artery. Long and slender, it ascends at the lateral border of the pharynx and provides three or four pharyngeal branches. The **veins** of the pharynx form an external plexus and a plexus between the constrictors and the pharyngobasilar fascia. These plexuses drain to the pterygoid plexus above (p. 226) and to the internal jugular vein at the level of the angle of the mandible (p. 161 & fig. 128). The **lymphatic vessels** of the pharynx mainly emerge through its posterior wall near the midline, turn lateralward, and end in retropharyngeal and superior deep cervical nodes (p. 165).

Lateral Relations of the Pharynx

The pharynx is in relation laterally with the internal and common carotid arteries, the internal jugular vein, the cervical sympathetic trunk, and the last four cranial nerves. These structures are enclosed in or associated with the carotid sheath. Under the discussion of this sheath, the arterial and venous trunks and the cervical sympathetic and vagal nerves are considered (p. 162). The accessory nerve is described with the sternocleidomastoid muscle (p. 175).

The **internal carotid artery** ascends to the base of the skull, at first posterolateral to the external carotid but, increasingly, to the medial side of that artery. At the base of the skull the internal carotid artery enters the carotid canal. Through cephalic levels of the pharynx, it is separated from the more laterally coursing external carotid artery by the styloid process and the stylopharyngeus muscle, the

glossopharyngeal nerve, and the pharyngeal branches of the vagus nerve and the sympathetic trunk. At lower levels, the internal carotid artery is crossed superficially by the hypoglossal nerve, the digastric and stylohyoid muscles, and the occipital and posterior auricular arteries. Through pharyngeal levels, the **internal jugular vein** accompanies the internal carotid artery, lying posterolateral to it, and maintains the same positional relationship at the jugular foramen (fig. 152). The juxtaposition of the inferior aperture of the carotid canal, the jugular foramen, and the hypoglossal canal and the attachment of the carotid sheath at the base of the skull in the region of these foramina bring the glossopharyngeal, vagal, accessory, and hypoglossal nerves into the upper end of the carotid sheath. At a distance of one or two centimeters below the base of the skull, the **vagus** lies behind and between the vessels and the **sympathetic trunk** posterior and medial to the sheath as at lower levels of the neck (fig. 152). The **hypoglossal nerve** lies in the back of the sheath posterior to the vagus. At its emergence from the jugular foramen, the **accessory nerve** lies at first between the internal carotid artery and the internal jugular vein. It then descends obliquely lateralward, usually anterior but sometimes posterior to the vein, and enters the deep surface of the sternocleidomastoid muscle. The glossopharyngeal nerve (fig. 152) emerges from the jugular foramen lateral and anterior to the vagus and accessory nerves and descends posterior to the styloid process and its muscles.

The Glossopharyngeal Nerve (figs. 152, 153) As its name implies, the ninth cranial nerve is primarily related to the tongue and pharynx. A listing of its components, however, shows it to carry the

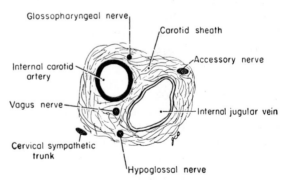

Fig. 152   A cross section of the carotid sheath one or two centimeters below the base of the skull.

same types of nerve fibers as does the vagus (p. 162).

> **Special visceral efferent:** motor innervation of the stylopharyngeus muscle of the pharynx.
>
> **General visceral efferent:** parasympathetic fibers for the parotid gland after synapse in the otic ganglion.
>
> **General visceral afferent:** visceral sensation from the parotid gland, carotid body and sinus, and from the mucous membrane of the pharynx, the middle ear, and the posterior one-third of the tongue.
>
> **Special visceral afferent:** taste from the posterior one-third of the tongue.
>
> **General somatic afferent:** cutaneous sensation from the external ear.

The glossopharyngeal nerve is formed by five or six rootlets which emerge from the side of the medulla oblongata in line with those which form the vagus nerve. The nerve passes through the jugular foramen in a separate sheath of dura mater and occupies a groove on the lower border of the petrous portion of the temporal bone. It may have a **superior ganglion** in this location, and, just before its emergence from the foramen, a larger and more constant **inferior ganglion** occurs. These contain the cell bodies for the afferent components of the nerve. The inferior ganglion of the

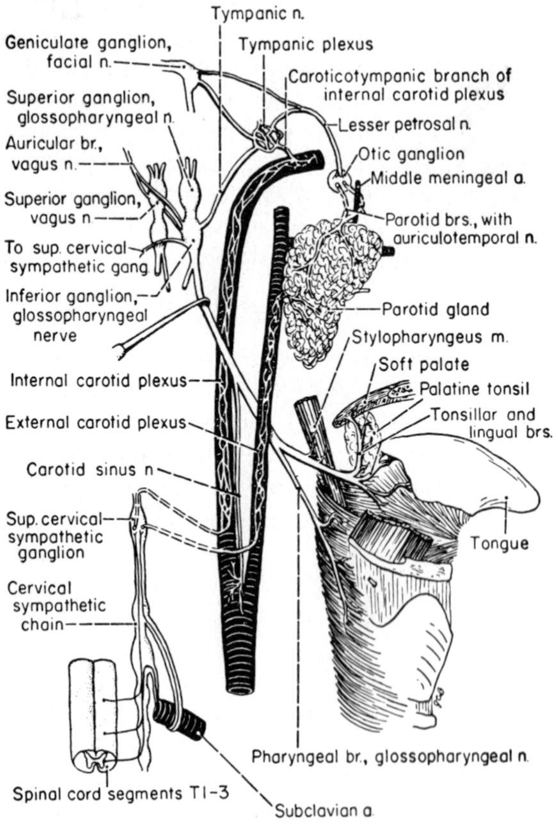

Fig. 153   The branches and distribution of the glosso-pharyngeal nerve and the innervation of the parotid gland.

glossopharyngeal communicates with the superior cervical sympathetic ganglion and with the auricular branch and superior ganglion of the vagus and gives rise to the tympanic nerve. A short distance below the ganglion, the glossopharyngeal nerve communicates with the facial nerve. It is through the communication with the auricular branch of the vagus that the somatic sensory (cutaneous) fibers of the glossopharyngeal nerve distribute.

The **branches of the glossopharyngeal nerve** are:

The **tympanic nerve** (fig. 153) supplies parasympathetic fibers, through the otic ganglion, to the parotid gland and sensory fibers to the mucous membrane of the middle ear. Arising from the inferior ganglion, the nerve ascends to the tympanic cavity through a canal in the bone separating the carotid canal and the jugular foramen. It enters the **tympanic plexus** on the promontory of the medial wall of the middle ear, forming this plexus with caroticotympanic branches from the internal carotid sympathetic plexus and a twig from the geniculate ganglion of the facial nerve. From the tympanic plexus sensory fibers distribute to the mucous membrane of the oval and round windows, the tympanic membrane, the auditory tube, and the mastoid air cells. The continuation of the tympanic nerve beyond the plexus is the **lesser petrosal nerve.** This nerve passes forward through the petrous portion of the temporal bone, is joined while in the bone by a branch from the geniculate ganglion of the facial nerve, and emerges into the middle fossa of the cranial cavity immediately lateral to the hiatus of the canal for the greater petrosal nerve. Leaving the cranial cavity through the fissure between the petrous bone and the great wing of the sphenoid or through a foramen in the latter bone, the lesser petrosal nerve ends in the **otic ganglion.** The latter lies medial to the mandibular portion of the trigeminal nerve at the origin of the nerve to the medial pterygoid muscle. The otic ganglion is a peripheral ganglion in the course of the parasympathetic innervation of the parotid gland; the postganglionic neurons arising in the ganglion distribute by way of the auriculotemporal branch of the trigeminal nerve to the gland (fig. 153).

The **carotid sinus nerve** (fig. 153) arises from the glossopharyngeal nerve just beyond its emergence from the jugular foramen. The nerve descends anterior to the internal carotid artery, has communications with the vagal and sympathetic branches to the carotid body, and ends in the wall of the carotid sinus, the dilated portion of the internal carotid artery. It is composed of afferent fibers from the chemoreceptors of the body and from the blood pressure receptors of the sinus.

Pharyngeal branches of the glossopharyngeal nerve (fig. 130) are three or four twigs which join the pharyngeal branches of the vagus and of the cervical sympathetic trunk on the middle constrictor muscle to form the pharyngeal plexus (p. 194).

The glossopharyngeal nerve descends along the posterior border of the **stylopharyngeus muscle** for several centimeters and then curves over the lateral aspect of the muscle. It supplies this muscle by branches which enter it at about its middle. This is the only muscular distribution of the nerve.

As the nerve continues toward the base of the tongue it gives off, deep to the hyoglossus muscle, **tonsillar branches** to the network around the palatine tonsil (fig. 153). From this network filaments pass to the soft palate and to the tonsillar pillars. In this portion of its course, the glosso-pharyngeal nerve lies deeply in the bed on the tonsil, coursing along the margin of the styloglossus muscle to reach the base of the tongue.

The **lingual branches** (fig. 153) provide afferent fibers for taste to the circumvallate papillae and general afferent neurons to the mucous membrane of the posterior one-third of the tongue.

The Hypoglossal Nerve (figs. 121, 139, 175)

The hypoglossal nerve (cranial XII) is the motor nerve of the tongue. It arises by a series of root-lets which emerge from the medulla oblongata in the anterolateral sulcus between the pyramidal and olivary eminences. These rootlets are collected into two bundles which perforate the dura mater separately but are united within the hypoglossal canal. **Meningeal twigs** arise in the hypoglossal canal and pass back to the dura mater of the posterior cranial fossa. These are thought to be sensory fibers derived from the loop between the first and second cervical nerves. Emerging from the hypoglossal canal, the hypoglossal nerve is medial to the internal carotid artery and internal jugular vein and then descends behind these vessels and to the lateral aspect of the artery. Passing posterior to the vagus nerve, the hypoglossal nerve communi-cates by numerous small branches with its inferior ganglion and, opposite the atlas, has a communica-tion with the superior cervical sympathetic ganglion. Also opposite the atlas, the hypoglossal nerve is joined by a nerve derived from a loop between the first and second cervical spinal nerves which brings

to it sensory fibers for its meningeal branches and provides muscular fibers for the suprahyoid and infrahyoid muscles which are innervated through the superior ramus of the ansa cervicalis (p. 177). The hypoglossal nerve turns forward near the angle of the mandible, loops around the occipital artery, and passes between the external carotid and jugular vessels (figs. 139, 175). Entering the sub-mandibular triangle deep to the posterior belly of the digastric muscle, the nerve passes above the hyoid bone on the surface of the hyoglossus muscle and under the mylohyoid muscle. It divides into terminal branches between the mylohyoid and genioglossus muscles. As the hypoglossal nerve crosses the occipital artery, a group of fibers derived from the loop between the first and second cervical nerves leaves it to form the superior ramus of the ansa cervicalis (see page 177). Farther along the hypoglossal nerve, branches to the thyrohyoid and geniohyoid muscles also represent neurons derived from the first cervical nerve. The terminal branches of the hypoglossal nerve distribute to the stylo-glossus, hyoglossus, genioglossus, and intrinsic muscles of the tongue. The nerve to the **styloglossus** arises near the posterior border of the hyoglossus, while the branches to the **hyoglossus** are supplied to the muscle as the nerve crosses it. The nerve to the **genioglossus** is given off as the hypoglossal nerve enters the substance of the tongue. The hypoglossal communicates freely with sensory branches of the lingual nerve.

THE ESOPHAGUS (fig. 154)

The esophagus is the portion of the digestive tract that extends from the pharynx to the stomach. It begins at the level of the lower border of the cricoid cartilage and ends below the diaphragm opposite the eleventh thoracic vertebra, a length of about 25 cm. It is a highly muscular tube and has an average diameter of about 2 cm. Beginning in the median line, the esophagus passes somewhat to the left so as to project about 5 mm. beyond the trachea at the base of the neck. It returns to the midline in the chest. The esophagus has three constrictions, being narrowest at its beginning (1.5 cm.). This is charac-teristic of tubes in the body, and its significance lies in the fact that material admitted to the tube will be allowed to pass through its entire length. The other constrictions lie in the thorax and are des-cribed on p. 358. The cervical esophagus lies directly behind the trachea and in front of the vertebral column and prevertebral musculature (figs. 122, 124). At its sides are the lateral lobes of the thyroid gland and the carotid sheaths. These relations are more intimate on its left side, for the inclination of the esophagus is in that direction. The recurrent laryngeal nerves ascend between the trachea and the esophagus. The esophagus has an inner circular and an outer longitudinal layer of muscle. The latter, thicker than on abdominal parts of the digestive tube, is covered by only a loose adventitial coat of connective tissue. Superiorly, the longitudinal coat is continuous with the muscular fibers of the pharynx and, in the anterior midline, is attached to the vertical ridge on the posterior surface of the lamina of the cricoid cartilage. Food passes through the esophagus with extreme rapidity under the influence of the peristaltic action of its musculature.

The blood vessels, lymphatics, and nerves of the esophagus are supplied regionally along its length. In the cervical region its arteries are branches of the inferior thyroid artery, and its veins are tributary to the inferior thyroid veins. The lymphatic drainage of the cervical esophagus is to the paratracheal and the inferior deep cervical lymph nodes. The nerves of the esophagus include an autonomic supply to its glands and smooth muscle and an innervation to the striated muscle of its cervical and upper thoracic parts. The recurrent laryngeal nerves provide both the parasympathetic and muscular innervations at cervical levels; sympathetic innervation is provided from the cervical sympathetic trunk.

THE SUPERFICIAL STRUCTURES OF THE FACE AND HEAD

The general character of the skin and subcutaneous connective tissue resembles and is described with the comparable tissues of the neck (p. 146). In the section on the neck, also, is described the platysma muscle which represents a cervical portion of the general 'facial' musculature. Subcutaneous veins of the neck, in part, arise in and are related to the subcutaneous veins of the head and face.

CUTANEOUS NERVES OF THE FACE AND HEAD

There is regional overlap between the cutaneous nerves of the neck and those of the face and head. Cutaneous branches of the cervical plexus extend upward behind and over the lower part of the ear (lesser occipital nerve) and over the parotid region in front of the ear (great auricular nerve). The greater occipital nerve (pp. 148, 213) arises in the neck but has its principal distribution in the back of the scalp. The principal nerves of the face and head are, however, derived from the trigeminal nerve.

Branches of the Trigeminal Nerve

The trigeminal nerve (cranial V) divides, before emerging from the cranium, into three primary divisions: the ophthalmic, maxillary, and mandibular nerves. These are named according to their principal areas of termination; respectively, the regions of the eye, the upper jaw, and the lower jaw. The first two divisions are wholly sensory; the mandibular division is largely sensory but also carries the fibers of the motor root of the trigeminal nerve. The largest cutaneous nerves of the three divisions are, respectively, the supraorbital, the infraorbital, and the mental. These emerge on the face at points along a single vertical line dropped from the site of the supraorbital notch or foramen— the junction of the medial and middle thirds of the superior orbital margin.

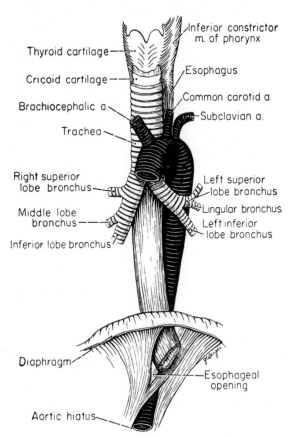

Fig. 154 The trachea, esophagus, and thoracic aorta.

Inferior constrictor m. of pharynx
Thyroid cartilage
Esophagus
Cricoid cartilage
Common carotid a.
Brachiocephalic a.
Subclavian a.
Trachea
Right superior lobe bronchus
Left superior lobe bronchus
Lingular bronchus
Middle lobe bronchus
Left inferior lobe bronchus
Inferior lobe bronchus
Diaphragm
Esophageal opening
Aortic hiatus

Cutaneous Branches of the Ophthalmic Division
The **supraorbital nerve** (figs. 120, 161) is the continuation of the frontal nerve which divides within the orbit into the supraorbital and supratrochlear nerves. The supraorbital nerve emerges through the supraorbital notch, or foramen, of the frontal bone, supplies small branches to the mucous membrane of the frontal sinus and to the upper eyelid, and then pierces the galea aponeurotica and the frontalis muscle of the scalp. Branches of the supraorbital nerve distribute to the skin and subcutaneous tissues of the forehead as far as the vertex of the skull. The **supratrochlear branch** of the frontal nerve (figs. 120, 161) emerges from the orbit at the upper medial angle of the eye, having passed above the tendinous pulley (trochlea) of the superior oblique muscle of the eye. It distributes to the skin and subcutaneous tissues of the medial portion of the upper eyelid and of the adjacent forehead. The **lacrimal nerve,** primarily concerned with the nerve supply of the lacrimal gland, has a small cutaneous distribution. Its palpebral branch (fig. 120) passes through the palpebral fascia of the upper eyelid near the lateral angle of the eye to distribute to the skin of the lateral portion of the lid. The **infratrochlear nerve** (fig. 120), a branch of the nasociliary portion of the ophthalmic nerve having its origin within the orbit, emerges as a slender nerve at the lower medial angle of the eye. It supplies the skin of the eyelids medially, the side of the nose, the lacrimal sac, the lacrimal caruncle, and adjacent conjunctiva. The **external nasal branches** of the anterior ethmoidal nerve (fig. 120) emerge on the surface of the nose between the nasal bone and the lateral nasal cartilage and supply the skin over the lower half of the nose.

Cutaneous Branches of the Maxillary Division
The **infraorbital nerve** (figs. 120, 139) is the largest terminal branch of the maxillary division. Having traversed the floor of the orbit and the infraorbital canal, it reaches the face at the infraorbital foramen deep to the levator labii superioris muscle. Under this muscle, the infraorbital nerve divides into three sets of branches. The **inferior palpebral branches** pass upward deep to the orbicularis oculi muscle and supply the skin and conjunctiva of the lower eyelid. The **external nasal branches** pass medialward to the skin and subcutaneous tissues of the side of the nose. **Superior labial branches,** three or four in number, descend to the skin and mucous membrane of the upper lip and gingiva and to the

labial glands. The infraorbital branches communicate freely with terminal branches of the facial nerve. The zygomaticofacial and zygomaticotemporal nerves are branches of distribution of the zygomatic nerve of the maxillary division. The **zygomaticofacial nerve** (figs. 120, 162) is a small nerve which appears on the face at the zygomaticofacial foramen in the zygomatic bone. It penetrates the fibers of the orbicularis oculi to supply the skin over the prominence of the cheek. The **zygomaticotemporal nerve** (figs. 120, 161) leaves the orbit through the zygomaticoorbital foramen (p. 241), skirts the anterior border of the temporalis muscle, and pierces the temporal fascia about 2 cm. above the zygomatic arch. The nerve is cutaneous to the anterior part of the temporal region.

Cutaneous Branches of the Mandibular Division
The mental foramen, located midway between the alveolar and basal borders of the mandible and below the second bicuspid tooth, conducts the large mental branch of the inferior alveolar nerve onto the face. The **mental nerve** (figs. 120, 139) divides under the depressor anguli oris muscle into three branches; one descends to the skin of the chin, the other two ascend to the skin and mucous membrane of the lower lip and gingiva. The **buccal branch** of the mandibular nerve (figs. 120, 139) is the principal cutaneous nerve of the cheek. It inclines from the infratemporal fossa to the surface along the anterior edge of the tendon of the temporalis muscle and enters the subcutaneous tissues at the anterior border of the masseter muscle. Certain of its branches distribute to the skin and subcutaneous tissues of the cheek; others perforate the underlying buccinator muscle and are sensory to the mucous membrane of the cheek and gums as far forward as the angle of the mouth. Over the buccinator muscle the buccal branches of the mandibular nerve communicate with buccal branches of the facial nerve, but only the branches of the latter nerve innervate the muscle. The **auriculotemporal nerve** (figs. 120, 161) reaches the subcutaneous tissues from the infratemporal fossa at a point between the condyle of the mandible and the external acoustic meatus. Turning upward across the root of the zygomatic arch, it is found posterior to the superficial temporal artery which it accompanies into the scalp. Several small **anterior auricular branches** supply the skin of the anterior upper part of the external ear, and the terminal superficial temporal branches pass to the skin and

subcutaneous tissues of the temporal region of the scalp as far as the vertex. As the auriculotemporal nerve passes horizontally outward through the infratemporal fossa, it supplies sensory branches to the temporomandibular joint, the external acoustic meatus, and the parotid gland.

Clinical Note The cutaneous and mucosal areas of distribution of the mandibular and maxillary divisions are the most common sites of the excruciating paroxysmal pain of trigeminal neuralgia (tic douloureux). This almost prostrating condition yields, though sometimes imperfectly, to alcohol injections of the nerve or to neurotomy of the division central to the semilunar ganglion.

THE FACIAL MUSCLES

The facial muscles, or muscles of facial expression, are subcutaneous voluntary muscles. In general, they arise from bones or fascia of the head and insert into the skin. Developed phylogenetically from a continuous sheet of musculature, they are sometimes indistinct at their borders. They are found in the neck, the face, and the scalp; the platysma muscle of the neck has been described (p. 146). Regardless of location, these muscles are all derived from the mesoderm of the hyoid branchial arch, and all are supplied by branches of the nerve of that arch, the facial (cranial VII) nerve.

The Superficial Muscles of the Face (fig. 155)

The superficial muscles of the face may be grouped in relation to the openings which they modify—the orbital opening, the nasal aperture, and the mouth.

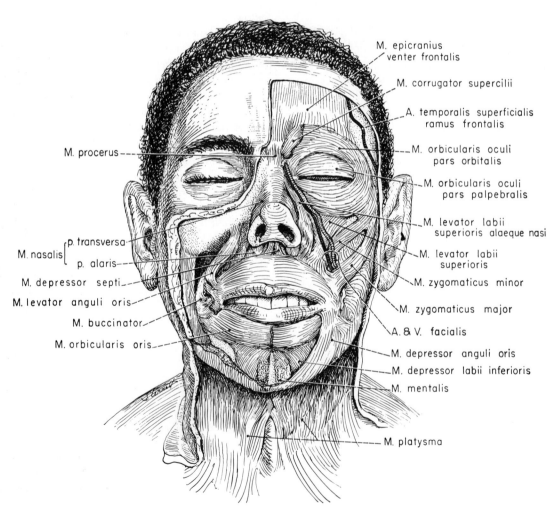

Fig. 155   The facial muscles.

The Muscles at the Orbital Opening. Orbicularis oculi (figs. 139, 155) is a broad, oval sheet of muscle which surrounds the orbit and occupies the eyelids. Its more peripheral **orbital portion** spreads widely onto the forehead and cheek; its **palpebral portion** constitutes the voluntary muscle of the eyelids. The orbital portion has a bony origin from the medial orbital margin between the supraorbital notch and the infraorbital foramen and a tendinous origin from the medial palpebral ligament. Its fibers form an ellipse, bowing above and below the eyelids to end in the cheek laterally. Peripheral fibers of the muscle overlie and occasionally blend with the muscles arising from the bones of the orbital rim.

The **medial palpebral ligament** (p. 244 & figs. 192, 193), about 5 mm. long and half as broad, arises from the frontal process of the maxilla anterior to the lacrimal groove. It extends lateralward into the eyelid in front of the lacrimal sac and divides, its parts being continuous with the tarsal plates of the upper and lower eyelids. An offshoot of the ligament leaves its posterior surface lateral to the lacrimal sac and attaches to the posterior lacrimal crest of the lacrimal bone.

The **palpebral portion of orbicularis oculi** (fig. 155) arises from the bifurcated portion of the medial palpebral ligament and spreads concentrically in the subcutaneous connective tissue of the upper and lower eyelids. Its fibers interdigitate laterally, superficial to the lateral palpebral ligament, to form the lateral palpebral raphe.

The **lacrimal portion of the orbicularis oculi** is associated with the posterior offshoot of the medial palpebral ligament. A small fascicle of muscle fibers covers the deep surface of this band, arising from the posterior crest of the lacrimal bone. Passing behind the lacrimal sac, the muscle divides into two slips for insertion into the medial parts of the tarsal plates of both lids; fibers also attach to the lateral wall of the sac.

The orbicularis oculi is the sphincter muscle of the eyelids. The lids are approximated by the palpebral portion in blinking and in sleep, but they are more forcibly brought together by the orbital portion, as in closing the lids tightly for protection against intense light or foreign bodies. The orbital part also contracts in winking. The lacrimal portion is essential in holding the lids close against the eyeball, thus facilitating the spread of the lacrimal secretion. It may also aid in dilating the lacrimal sac.

The **nerve supply** of orbicularis oculi is derived from temporal and zygomatic branches of the facial nerve. From just anterior to the notch of the ear, the zygomatic branch of this nerve is directed straight toward the lateral border of the orbicularis oculi. Temporal branches arching into the forehead above the muscle send twigs into its superior part.

The **corrugator** (fig. 155) is a narrow band of muscle lying deep to the superior portion of orbicularis oculi. Its origin is the medial part of the superciliary arch; its insertion is the skin of the medial half of the eyebrow. As its name implies, the corrugator draws the eyebrows medialward, producing vertical wrinkles in the forehead. It is supplied by the temporal branch of the facial nerve.

The Muscles of the Nose (fig. 155) The **procerus** is a small slip of muscle arising by a fibrous membrane from the lower part of the nasal bone and the lateral nasal cartilage. Its fibers ascend to intermingle with those of frontalis and to insert into the skin of the lower forehead between the two eyebrows. This muscle draws down the medial angle of the eyebrows and produces transverse wrinkles across the root of the nose. It is innervated by one of the more superior buccal branches of the facial nerve.

Pars transversa of the **nasalis muscle** (fig. 155) arises from the upper part of the canine eminence of the maxilla. Its fibers pass upward and medialward and end in an aponeurosis which is continuous over the cartilaginous portion of the nose with its fellow of the opposite side. **Pars alaris** of the **nasalis muscle** (fig. 155) arises from the maxilla above the lateral incisor tooth and inserts into the lateral part of the lower margin of the ala of the nose. Its more medial fibers insert into the lower end of the nasal septum and may be separately described as the **depressor septi muscle** (fig. 155). All three nasal muscles appear to be concerned in dilating the nasal aperture. In this action, the nose is flattened and widened; the greater alar cartilage is drawn downward and outward; and the lower end of the septum moves downward. The nasal muscles are supplied by the buccal branches of the facial nerve.

The Muscles of the Mouth (figs. 121, 139, 155) The musculature which modifies the oral opening passes into the lips from surrounding areas of the face. The fibers of insertion enter at or near the angles of the mouth and are lost in the substance of the lips, blending with fibers converging from all directions to form the oral sphincter, the orbicularis oris.

Three muscles enter the upper lip from above. The **m. levator labii superioris alaeque nasi** lies in the sulcus between the nose and cheek. Arising from the upper part of the frontal process of the maxilla, it descends to insert partly into the ala of the nose and partly into the skin of the lateral half of the upper lip. The **m. levator labii superioris** arises from the inferior margin of the orbit over the infraorbital foramen. It descends to insert into the lateral half of the upper lip. The slender **zygomaticus minor muscle** arises from the lower portion of the zygomatic bone and extends downward and medialward to the upper lip just medial to the angle of the mouth. These muscles are the principal elevators of the upper lip. M. levator labii superioris alaeque nasi also raises the ala of the nose; the zygomaticus minor is concerned in turning up the corner of the mouth. The muscles are supplied by buccal branches of the facial nerve.

M. levator anguli oris (figs. 139, 155) is thick and rounded and lies deep to the levator labii superioris muscle. Arising from the canine fossa of the maxilla just below the infraorbital foramen, its fibers descend, inclining lateralward, to the angle of the mouth. Here they blend with orbicularis oris, some fibers continuing into the lower lip. M. levator anguli oris elevates the angle of the mouth, at the same time drawing it medialward, and is innervated by a buccal branch of the facial nerve.

M. zygomaticus major (figs. 121, 155) arises from the zygomatic bone anterior to the zygomaticotemporal suture. Descending obliquely, the muscle reaches the angle of the mouth, where its fibers partly insert into skin and partly blend with orbicularis oris. **Zygomaticus major** turns the angle of the mouth upward and carries it outward, as in smiling and laughing. Its nerve is the zygomatic branch of the facial nerve. The muscle is quite superficial, and toward the angle of the mouth the facial artery and vein pass deep to it.

Risorius is a thin wisp of muscle which arises from the fascia over the parotid gland and may be partly fused with the upper fibers of platysma. It runs transversely to insert into the skin at the angle of the mouth and assists in widening the mouth, as in expressions of mirth. Risorius is innervated by a buccal branch of the facial nerve. The muscle may be absent.

M. depressor anguli oris (figs. 121, 155) lying below the angle of the mouth, is triangular in shape. Its fibers take origin from the oblique line of the mandible and, passing upward, converge at the angle of the mouth. They blend with orbicularis oris and insert into skin, many entering the upper lip. The muscle is prominent and superficial; it depresses the angle of the mouth. Buccal and mandibular branches of the facial nerve innervate it.

M. depressor labii inferioris (fig. 155) is small and quadrilateral. Its fibers arise from the mandible above its oblique line and medial to the mental foramen. The fibers of the muscle are directed upward and medialward and insert into the skin of the lower lip, the more medial fibers decussating with those of the opposite side. The muscle draws the lower lip downward and slightly lateralward. Its nerve is the mandibular branch of the facial nerve.

Mentalis (fig. 155) modifies the oral aperture only indirectly. It is a small, conical muscle arising from the mandible below the incisor teeth and inserting into the skin of the chin. It draws up the skin of the chin and thus assists in protrusion of the lower lip. The mandibular branch of the facial nerve innervates the muscle.

Orbicularis oris is an oral sphincter composed of interlacing fibers of the muscles described above, a deeper stratum of fibers derived from the buccinator muscle, and intrinsic bundles within the substance of the lips. The buccinator muscle provides a large part of the bulk of the orbicularis, its marginal fibers passing directly into the upper or lower lip from the upper or lower borders of the muscle, its central fibers decussating at the angle of the mouth. Superficial to the buccinator fibers, fascicles of levator anguli oris and depressor anguli oris cross each other at the angle of the mouth; fibers of the levator continuing into the lower lip, fibers of the depressor into the upper lip. These fibers end in the skin near the median plane of the lips. Other superficial fascicles of the orbicularis oris are derived from zygomaticus major, levator labii superioris, and depressor labii inferioris. As a rule, their fibers enter the lips obliquely. Intrinsic bundles are also oblique and pass through the substance of the lips from the skin to the mucous membrane. The deep stratum of the buccinator muscle is reinforced by short bundles of muscle passing from the alveolar borders of the maxilla and the mandible in the region of the incisor teeth toward the angles of the mouth. In the upper lip this muscle is the **incisivus labii superioris;** in the lower lip, the **incisivus labii inferioris.** The complex muscular structure of orbicularis oris provides for sphincter action, flattening of the lips against the alveolar arch, and protrusion. In addition, the muscles inserting into the lips from the surrounding face regions draw the lips radially according to the direction of their fascicles. Buccal branches of the facial nerve supply orbicularis oris. Superior and inferior labial branches of the facial artery run medialward in the substance of the lips between orbicularis oris and the mucous membrane. They supply the muscle, the labial glands, and the skin of the lips and anastomose with their fellows of the opposite side.

The **buccinator muscle** (figs. 121, 139, 155, 164, 165) lies on a deeper plane than the other muscles around the mouth and forms the principal substance of the cheek. Its form is quadrilateral, and it arises from the alveolar portions of the maxilla and the mandible lateral to the molar teeth and from the pterygomandibular raphe between these bones. (The **pterygomandibular raphe** is a ligamentous band which, stretched between the pterygoid hamulus superiorly and the posterior end of the mylohyoid line of the mandible inferiorly, separates the fibers of the buccinator from those of the superior pharyngeal constrictor muscle.) The fibers of the buccinator muscle pass toward the angle of the mouth, and those of the upper and lower borders of the muscle run directly into the upper and the lower lips. Fibers of the central part of the muscle decussate at the angle of the mouth; those from above enter the lower lip, those from below, the upper lip. The buccinator fibers form the deep stratum of orbicularis oris and end chiefly in the mucosa of the lips and, to a lesser degree, in the skin. The muscle compresses the cheek, draws the corner of the mouth laterally, and forces the lips against the teeth. It is an accessory muscle of mastication, keeping the food between the occlusal surfaces of the teeth and out of the cheeks. It assists in whistling and in blowing wind instruments. The buccal branches of the facial nerve are motor to the buccinator muscle.

The buccinator is covered deeply by the mucous membrane of the inside of the cheek. Externally, the muscle is invested by the buccopharyngeal fascia, and both muscle and fascia underlie the buccal pad of fat. The muscle is pierced by the duct of the parotid gland. The buccal branch of the mandibular division of the trigeminal nerve (sensory) and the buccal branch of the maxillary artery overlie the muscle and send twigs through it to the mucosa of the mouth (fig. 165).

The **buccal fat pad** (fig. 157) lies superficial to the buccinator muscle at the anterior border of the masseter muscle. It is encapsulated by the superficial layer of cervical fascia and is readily separated from the buccinator and other facial muscles. Prolongations of the fat pad extend deeply between the masseter and temporalis muscles and deep to the temporal fascia. The fat pad is prominent in infants,

and its apparent value in sucking has earned it the name of suctorial pad.

The Muscles of the Scalp The scalp musculature belongs to the facial group of muscles but is described in connection with the anatomy of the scalp (p. 215).

SUPERFICIAL BLOOD VESSELS OF THE FACE
(fig. 156)

Blood vessels in the subcutaneous tissues of the face, intimately related to the muscles of facial expression, include the facial artery and vein; the transverse facial artery and vein; the buccal, infraorbital, and mental arteries and veins; and other minor vessels.

The Facial Artery

This artery (figs. 121, 156, 178) arises from the external carotid artery in the carotid triangle of the neck. At or close to its origin, it is crossed by the posterior belly of the digastric and the stylohyoid muscles and by the hypoglossal (cranial XII) nerve. In the submandibular triangle the facial artery ascends deep to the submandibular gland, grooving its deep and superior aspect, and then passes superficially to gain the inferior border of the mandible. As it crosses the mandible at the anterior border of the masseter muscle, the artery is covered by skin and platysma muscle only, and its pulsations can here be felt. The facial course of the artery is an irregularly diagonal one, passing the angle of the mouth and inclining toward the medial angle of the eye. This portion of the artery is very tortuous, and the vessel lies irregularly superficial or deep to the muscles and nerves of the cheek and infraorbital regions. It is usually deep to platysma and zygomaticus major and superficial to buccinator and levator anguli oris. Its relation to the levator labii superioris muscle is variable. Ascending at the side of the nose, the artery terminates as the angular artery. The **facial vein,** which accompanies the artery, lies to its lateral side and is less tortuous and generally more superficial than the artery.

In the neck the facial artery gives origin to branches which do not reach the face: the ascending palatine artery, the tonsillar artery, branches to the submandibular gland and the muscles of the region, and the submental artery. The branches of the artery on the face are the inferior labial, the superior labial, the lateral nasal, and the angular.

The **inferior labial artery** arises opposite the angle of the mouth and runs in the lower lip within the orbicularis oris muscle close to the mucous membrane. It supplies the labial glands, the mucous membrane, and the muscles of the lower lip. Frequently, a second inferior labial, or 'infralabial', artery runs in the sulcus between the lower lip and the chin. These arteries anastomose with the corresponding vessels of the opposite side of the face, with the mental artery, and with one another.

The larger **superior labial artery** arises from the facial artery slightly above the inferior labial and, like it, courses tortuously in the upper lip between the orbicularis oris muscle and the mucous membrane. It supplies the skin, mucous membrane, muscle, and glands of the upper lip and gives off a small **septal branch** to the lower anterior part of the nasal septum. The superior labial artery anastomoses with its fellow of the opposite side.

The **lateral nasal artery** arises from the facial artery as it turns upward along the side of the nose. The lateral nasal artery supplies the ala and dorsum of the nose and anastomoses with its fellow of the opposite side, with the dorsal nasal branch of the ophthalmic artery, and with the infraorbital branch of the maxillary artery.

The **angular artery** is the terminal part of the facial artery. Ascending to the medial angle of the orbit, it supplies orbicularis oculi and the lacrimal sac. Its branches make connection with the infraorbital artery and with the dorsal nasal branch of the ophthalmic artery.

The Angular Vein and Facial Vein

The **angular vein** (fig. 156) is formed by the confluence of the supratrochlear and supraorbital

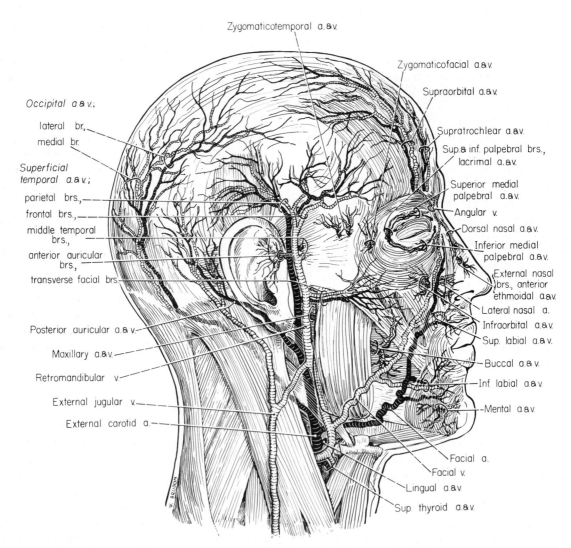

Fig. 156   The superficial blood vessels of the scalp and face.

veins of the forehead (p. 213) and is continued at the lower margin of the orbit into the facial vein. The angular vein receives small veins of the root of the nose and the superior palpebral vein and communicates by the nasofrontal vein with the superior ophthalmic vein. The angular, the facial, and the ophthalmic veins contain no valves. Blood can, therefore, pass either forward from the ophthalmic into the angular vein or backward from the facial or angular vein into the ophthalmic vein and to the cavernous and other venous sinuses of the cranium. At the side of the nose the **facial vein** receives external nasal veins and inferior palpebral veins from the lower eyelid. Superior labial and inferior labial veins drain the upper and lower lips, and small masseteric veins empty into the facial vein from the masseter muscle and the parotid region. The tributaries of the facial vein in the face communicate with the ophthalmic, the infraorbital, and the mental veins. There is also a large communication with the pterygoid plexus of veins (p. 226) around the lateral pterygoid muscle by way of the **deep facial vein** (fig. 128). Through the face, the facial vein follows the general course of the facial artery but lies to its lateral, or posterior, side. The vein is also less tortuous. Like the artery, the facial vein passes across the mandible at the anterior border of the masseter muscle. Unlike the artery, however, the vein has a superficial course through the submandibular triangle; where it receives the submental vein from the submental triangle, the external palatine and submandibular veins, and the accompanying vein of the hypoglossal nerve.

Just below the angle of the mandible, the facial vein usually receives a communication from the retromandibular vein. Passing deeply at the anterior margin of the sternocleidomastoid muscle, the facial vein crosses the external carotid artery and enters the internal jugular vein opposite to or below the level of the hyoid bone. It is also frequently connected to a **communicating vein** which runs along the border of the sternocleidomastoid muscle and joins the lower part of the anterior jugular vein. In the lower part of its course, the facial vein sometimes receives the superior thyroid, the pharyngeal, the lingual, and the sublingual veins.

The Transverse Facial Artery and Vein

These vessels (figs. 121, 156) lie transversely across the cheek about a finger's breadth below the zygoma and are, respectively, a branch of the super-

ficial temporal artery and a tributary of the corresponding vein (p. 213). They communicate with the infraorbital, the buccal, and the facial vessels.

The Infraorbital Artery and Vein

These vessels (figs. 139, 156) appear on the face at the infraorbital foramen. The artery is one of the terminal branches of the maxillary artery and reaches the face by traversing the infraorbital groove and canal in company with the infraoribtal vein and nerve. On the face, the artery divides under the levator labii superioris muscle into branches which supply the lower eyelid and lacrimal sac, the side of the nose, and the upper lip. Anastomoses are made with the labial and angular branches of the facial artery, with the dorsal nasal and lacrimal branches of the ophthalmic artery, and with the transverse facial artery. The **infraorbital vein,** small but in general corresponding to the infraorbital artery, is a tributary of the pterygoid plexus of veins and, through it, of the retromandibular vein.

The Mental Artery and Vein

The mental artery and vein (figs. 139, 156) appear on the face at the mental foramen. The mental artery is a terminal branch of the inferior alveolar artery which, as a branch of the first part of the maxillary artery, passes through a canal in the mandible to supply the lower teeth. The mental artery emerges from the bone under cover of the depressor anguli oris muscle and supplies the region of the chin. It anastomoses with the inferior labial artery and, below, with the submental artery. The **mental vein,** which communicates with similarly named veins, is a tributary of the inferior alveolar vein. The inferior alveolar vein empties into the pterygoid plexus and, thereby, into the retromandibular vein. The mental artery and vein distribute with the mental nerve (p. 199).

The Buccal Artery and Vein (figs. 139, 156)

The buccal artery is slender and, with the buccal nerve (p. 199) and vein, reaches the cheek over the buccinator muscle. As a branch of the maxillary artery, it emerges from behind the ramus of the mandible along the anterior margin of the tendon of the temporalis muscle. It supplies the buccinator muscle and the skin and mucous membrane of the cheek. The buccal artery anastomoses with the facial, infraorbital, and transverse facial arteries.

The **buccal vein** has a similar distribution and communication. It is a tributary of the pterygoid plexus and, through it, of the retromandibular vein.

Arteries and Veins of the Eyelids and Nose

Numerous small arteries and veins are found in the eyelids. These, together with certain vessels of the nose, are all terminals of the ophthalmic artery and vein. The ophthalmic artery is a branch of the internal carotid artery which enters the orbit in company with the optic nerve. It is the artery of the eye and its accessory structures, and its **lacrimal branch** supplies two terminal arteries to the upper and lower eyelids, the **lateral palpebral arteries** (fig. 156). These run medialward and anastomose with the medial palpebral arteries. The **medial palpebral arteries**, superior and inferior (fig. 156), arise from the ophthalmic artery opposite the pulley for the tendon of the superior oblique muscle. They emerge from the orbit on either side of the medial palpebral ligament and form, with the lateral palpebral arteries, **superior** and **inferior palpebral arches** superficial to the tarsal plates and near the free margins of the eyelids. The medial palpebral arteries anastomose with the supraorbital, infraorbital, angular, and transverse facial arteries. The lacrimal artery also provides **zygomatic branches** (fig. 156) which appear on the face over the zygoma and distribute with the zygomaticotemporal and zygomaticofacial nerves. The **dorsal nasal artery** is a small terminal branch of the ophthalmic which leaves the orbit at its medial angle above the medial palpebral ligament. It descends on the dorsum of the nose, provides a branch to the lacrimal sac, and anastomoses with the angular and lateral nasal branches of the facial artery. A terminal **external nasal branch** of the anterior ethmoidal branch of the ophthalmic artery (fig. 156) appears on the dorsum of the nose between the nasal bone and the lateral nasal cartilage and participates in the same communications as the dorsal nasal artery.

Clinical Note The **veins** which correspond to the arteries described above bear the same names and drain similar areas of the eylids and nose. They are all tributaries of the superior ophthalmic vein which drains the eye and its accessory structures in the orbit and empties intracranially into the cavernous sinus. A free communication exists between the superior ophthalmic vein and the supraorbital, supratrochlear, and angular veins by means of the **nasofrontal vein**. Free passage through these valveless veins and from the lips and face via the facial vein is commonly implicated in transmission of infection to the cavernous sinus which may result in a serious cavernous sinus thrombosis.

The Retromandibular Vein (figs. 156, 178)

This is a superficial vein of the face which is closely associated with the parotid gland. Its tributaries, however, include many veins draining structures of deeper location. The retromandibular vein is formed in the upper portion of the parotid gland, deep to the neck of the mandible, by the confluence of the superficial temporal vein (p. 213) and the maxillary vein. Descending just posterior to the ramus of the mandible through the parotid gland, or folded into its deep aspect, the vein is lateral to the external carotid artery. Both vessels are crossed by the facial nerve. Near the apex of the parotid gland, the retromandibular vein gives off an anteriorly descending communication which joins the facial vein just below the angle of the mandible. The retromandibular vein then inclines backward and unites with the posterior auricular vein (p. 214) to form the external jugular vein.

The **maxillary vein** is a short vessel which lies, with the first portion of the maxillary artery, between the sphenomandibular ligament and the neck of the mandible. The tributaries of the maxillary vein are combined in the pterygoid plexus of veins of the infratemporal fossa. The **pterygoid plexus** (fig. 128) lies on the lateral surface of the medial pterygoid muscle and surrounds the lateral pterygoid muscle. Its tributaries correspond with and accompany the branches of the maxillary artery; they are sphenopalatine, the pharyngeal, the vein of the pterygoid canal, the infraorbital, the posterior superior alveolar, the descending palatine, the buccal, two or three deep temporal, the pterygoid, the masseteric, the inferior alveolar, and the middle meningeal veins. Certain of these veins have a superficial distribution; however, most of them drain deep areas and are described more fully in other connections. The plexus has a number of deep communications. One superficial communication passes between the masseter and buccinator muscles and connects to the facial vein. This communication is the **deep facial vein.**

THE PAROTID GLAND (figs. 121, 157–159, 178)

The parotid gland is the largest of the salivary glands and occupies the side of the face in front of the ear. Its surface form is irregularly triangular. The superior border of the gland corresponds to the lower edge of the zygomatic arch; its posterior border lies in front of the external acoustic meatus and the sternocleidomastoid muscle and extends downward to the angle of the mandible. The diagonal anterior border of the gland connects the ends of the posterior and superior borders and overlaps the masseter muscle. The gland is overlain by fatty subcutaneous tissue in which distribute the

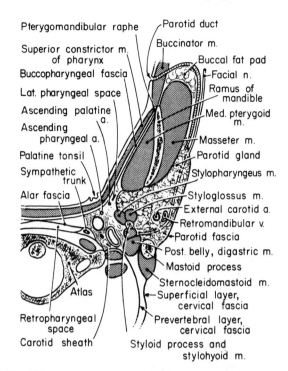

Pterygomandibular raphe
Superior constrictor m. of pharynx
Buccopharyngeal fascia
Lat. pharyngeal space
Ascending palatine a.
Ascending pharyngeal a.
Palatine tonsil
Sympathetic trunk
Alar fascia
Atlas
Retropharyngeal space
Carotid sheath

Parotid duct
Buccinator m.
Buccal fat pad
Facial n.
Ramus of mandible
Med. pterygoid m.
Masseter m.
Parotid gland
Stylopharyngeus m.
Styloglossus m.
External carotid a.
Retromandibular v.
Parotid fascia
Post. belly, digastric m.
Mastoid process
Sternocleidomastoid m.
Superficial layer, cervical fascia
Prevertebral layer, cervical fascia
Styloid process and stylohyoid m.

Fig. 157   A partial horizontal section of the parotid region of the face.

terminals of the great auricular nerve. The **anterior auricular lymph nodes** lie subcutaneously immediately in front of the ear. From the anterior border of the gland, the **parotid duct** extends forward about 1 cm. below the zygoma. Crossing the masseter muscle and the buccal fat pad, the duct turns deeply at the anterior border of the buccal pad of fat and penetrates the buccinator muscle, opening in the interior of the mouth opposite the second upper molar tooth. The location of the duct on the face may be given more exactly as the middle one-third of a line extending from the notch of the ear to a point midway between the red margin of the upper lip and the underside of the nose. The duct has a length of 4 to 6 cm. and is about 5 mm. in diameter; it ends in a small orifice. A separate accessory portion of the parotid gland is sometimes found lying above the duct.

While the surface form of the parotid gland is triangular, it exhibits an irregular wedge shape as it extends deeply into the face. Being composed of soft glandular material and having migrated to the surface after developing as a derivative of the oral cavity in the region of its duct opening, the parotid gland inosculates itself around and between the bones and muscles of the infratemporal fossa. A

horizontal section of the head (fig. 157) shows that the gland lies posteriorly against the sternocleidomastoid and the posterior belly of the digastric muscle and folds itself around the styloid process and the styloid muscles. The space in which it lies is constricted anteriorly by the ramus of the mandible and the associated masseter and medial pterygoid muscles. Deep relations of the gland are the carotid sheath (p. 159) and the lateral pharyngeal space. The waist-like constriction of the gland between the ramus of the mandible and the masseter muscle anteriorly and the posterior belly of the digastric muscle posteriorly has led to the designation of this portion of the gland as the **isthmus.** Deep to the isthmus, the gland is frequently expanded. The facial nerve splits into its temporofacial and cervicofacial divisions around the isthmus, and buccal branches of both divisions anastomose anterior to it.

The **parotid fascia** (figs. 157, 178) is an extension into the face of the superficial layer of the cervical fascia. Passing from the anterior border of the sternocleidomastoid muscle, the fascia splits to provide both a superficial and a deep layer on the parotid gland. The superficial layer covers the subcutaneous surface of the gland and extends upward to an attachment on the zygomatic arch. It sends deep extensions into the gland which become continuous with its stroma and form an intimate connection between the parotid gland and its covering fascia. The superficial layer of the parotid fascia is continuous, along the underside of the anterior border of the gland, with the fascia of the masseter muscle. The deep layer of the parotid fascia fuses with the fascial covering of the posterior belly of the digastric muscle and then attaches to the styloid process, becoming continuous with the fasciae of the styloid muscles. The portion of the gland adjacent to the lateral pharyngeal space is covered by a thin extension of the deep fascia which continues forward to the ramus of the mandible. The deep portion of the parotid fascia also forms a thickening, the **stylomandibular ligament,** which passes downward and lateralward from the tip of the styloid process to the angle of the mandible. This ligament separates the parotid from the submandibular gland.

The parotid gland has intimate relations with important blood vessels and nerves. The **external carotid artery** ascends, at first, in a groove on the deep surface of the gland. Then, frequently becoming surrounded by glandular tissue, it continues as far as the neck of the

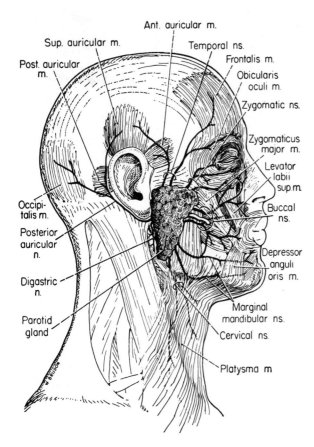

Fig. 158   The parotid gland and the facial distribution of the facial nerve.

arteries. Its drainage is to their accompanying veins and directly to the retromandibular vein. The lymphatic drainage of the gland is into the superficial and deep parotid nodes. Its nerves are the auriculotemporal and the external carotid plexus. The auriculotemporal nerve provides sensory fibers and also conducts to the gland parasympathetic impulses originating in centers associated with the glossopharyngeal (cranial IX) and facial (cranial VII) nerves. These signals are relayed through the otic ganglion (p. 196). Impulses mediated through the carotid plexus are sympathetic and are largely vasoconstrictor in function.

The Facial Nerve (figs. 121, 158, 159)

The facial nerve provides the motor innervation of the muscles of facial expression. It is also sensory, conveying taste from the anterior two-thirds of the tongue and general sensation from a small area of the external acoustic meatus. The facial nerve provides the parasympathetic innervation of the submandibular, sublingual, lacrimal, nasal, and palatine glands. The nerve enters the temporal bone by way of the internal acoustic meatus and leaves it through the stylomastoid foramen. It has a devious course in the bone and, within it, gives rise to several branches which emerge through separate openings. The detailed description of this portion of the nerve is reserved for discussion with the ear (p. 263). The branches of the facial nerve, after its emergence from the stylomastoid foramen, are primarily those concerned with the muscles of

mandible, where it divides into its terminal branches, the superficial temporal and the maxillary arteries. The **transverse facial artery** arises in the substance of the gland and emerges at its anterior border between the zygomatic arch and the parotid duct. The **retromandibular vein** runs a vertical course in the gland superficial to the external carotid artery. This vein is formed in the uppermost part of the gland by the union of the superficial temporal and the maxillary veins. The **facial nerve** has a close association with the parotid gland (p. 206). The **auriculotemporal nerve** arises high in the infratemporal fossa from the mandibular division of the trigeminal nerve. Passing between the lateral pterygoid muscle and the bone above, and superior to the upper part of the parotid gland, it curves over the root of the zygomatic arch and enters the scalp, accompanying and lying posterior to the superficial temporal artery. The **superficial parotid lymph nodes** are located under the superficial layer of the parotid fascia, embedded in the gland. A few, small **deep parotid lymph nodes** lie on the deep aspect of the gland or are embedded in it.

The blood supply of the parotid gland is provided by the external carotid artery and its branches; posterior auricular, superficial temporal, and transverse facial

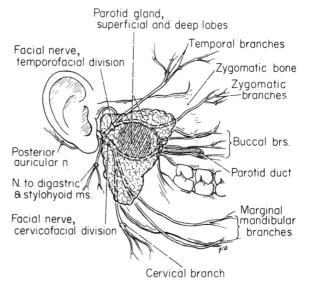

Fig. 159   The relation of the branches of the facial nerve to the superficial and deep lobes of the parotid gland.

facial expression. Given off very close to the foramen are the posterior auricular nerve and the nerve to the posterior belly of the digastric and the stylohyoid muscles. The facial nerve then approaches the parotid gland, passing forward superficial to the external carotid artery and the retromandibular vein. Just behind the posterior border of the ramus of the mandible and at one-third of the distance from the condyle to the angle of the mandible, the nerve divides into an upper **temporofacial division** (fig. 159) and a lower **cervicofacial division** from which are derived the nerves that emerge along the anterior border of the parotid gland. The temporofacial division passes superior to the isthmus of the parotid gland; the cervicofacial division, inferior to it. Buccal branches of each division anastomose with one another anterior to the isthmus so that this portion of the gland is surrounded by a ring of nerves.

The **posterior auricular nerve** (fig. 159) is the first extracranial branch of the facial nerve. Ascending between the external acoustic meatus and the mastoid process, the nerve divides into two branches, the **auricular branch** for the posterior auricular muscle and the **occipital branch** for the supply of the occipital belly of the epicranius muscle of the scalp. The posterior auricular nerve communicates with the lesser occipital and great auricular nerves and with the auricular branch of the vagus nerve.

The **nerve to the posterior belly of the digastric muscle** arises close to the stylomastoid foramen and enters the muscle near its mid-length. The **nerve to the stylohyoid muscle** is usually a branch of the nerve to the digastric but may arise from the facial trunk directly. It enters the upper part of the stylohyoid muscle.

The **temporal branches** from the temporofacial division are large (fig. 159). As a rule, two prominent communications unite the auriculotemporal nerve with a temporal branch of the facial. The temporal branches emerge along the superior border of the parotid gland, cross the zygomatic arch, and enter the temporal and frontal regions of the scalp. They supply the anterior and superior auricular muscles, the frontal belly of the epicranius muscle, the corrugator muscle, and the orbicularis oculi muscle. Besides the communications with the auriculotemporal nerve, more peripheral communications exist with the zygomaticotemporal, the lacrimal, and the supraorbital nerves.

The **zygomatic branches** of the temporofacial division (fig. 121) cross the zygomatic arch region of the face as they approach the lateral angle of the eye. They innervate the orbicularis oculi and the zygomaticus major muscles and communicate with the lacrimal branch of the ophthalmic division and with the zygomaticofacial branch of the maxillary division of the trigeminal nerve.

The **buccal branches** (figs. 121, 158, 159) of the facial nerve arise from both the temporofacial and cervicofacial divisions of the nerve and form an anastomosis in front of the isthmus of the parotid gland. Streaming medialward through the cheek, the buccal branches are directed toward the infraorbital region and the angle of the mouth. There is usually a prominent branch which runs just above the parotid duct. The buccal branches of the facial nerve communicate with the buccal branches of the mandibular nerve; other anastomoses with branches of the infraorbital nerve of the maxillary division of the trigeminal form an **infraorbital plexus**. The muscles supplied by the numerous buccal nerves are: zygomaticus major and minor, levator labii superioris, levator labii superioris alaeque nasi, procerus, risorius, buccinator, levator anguli oris, orbicularis oris, nasalis, and depressor anguli oris.

The **marginal mandibular branches** of the cervicofacial division of the facial nerve (figs. 121, 158, 159) emerge from the anterior border of the parotid gland near its apex and course forward over the masseter muscle and the mandible adjacent to the margin of the lower jaw. They run deep to the platysma and depressor anguli oris muscles and supply the latter muscle, the depressor labii inferioris, and the mentalis. The marginal mandibular branches (2 in 67 per cent) of the facial nerve communicate with the mental branch of the inferior alveolar nerve (trigeminal).

Clinical Note In about 19 per cent of cases, a marginal mandibular branch of the facial nerve curves below the border of the mandible and courses through the submandibular triangle, where it is endangered by incisions along the inferior margin of the mandible posterior to the facial artery.

The **cervical branch** of the facial nerve descends into the neck behind the angle of the mandible. It runs deep to the platysma muscle which it supplies. The cervical branch of the facial communicates deep to the platysma with the great auricular and transverse cervical nerves of the cervical plexus.

The significance of the facial-trigeminal communications in the face, neck, and scalp is uncertain. It is possible that terminals of the trigeminal nerve are conducted thereby into regions of distribution of the facial nerve; it has also been suggested that proprioceptive fibers from the muscles of the facial group utilize these communications to reach the major trigeminal proprioceptive centers of the brainstem.

Clinial Note Facial 'nerve' or facial muscle paralysis is distressing to the sufferer and is usually evident to a discerning observer in loss of normal facial contours. Weakness is particularly noticeable about the mouth, where there may be drooping of tissues or even drooling due to lack of control of the angle of the mouth. Asymmetrical muscular contraction shows up if the patient is asked to smile, show the teeth, frown, or forcibly close the eyes. Sagging of the lower eyelid may result in eversion of the lid and spilling of tears. These and other signs of paralysis are of value in the neurological examination.

THE SCALP (figs. 160–62)

The scalp is a five-layered structure composed of skin, subcutaneous connective tissue, the epicranius layer of muscle and aponeurosis, a scanty layer of loose subaponeurotic connective tissue, and the periosteum of the cranial bones. The skin of the scalp is thin, contains numerous sweat and sebaceous glands, and is firmly held to the next subjacent layer. The subcutaneous connective tissue is a thick, densely felted layer of connective tissue and fat in which the fat is firmly held in locules between the strong bundles of connective tissue fibers. The fibrous tissue of this layer is oriented in several directions—perpendicular to the surface and also obliquely between the skin and the underlying aponeurotic layer—so as to form a feltwork of fibers (fig. 160). Hair follicles extend into the second layer. The arteries, veins, and cutaneous nerves of the scalp are found in this layer and are firmly held in place due to the tight binding character of its collagenous bundles, which also accounts for the firm interconnection between the skin, the subcutaneous tissue, and the epicranius layer beneath. All three layers move together in wrinkling the

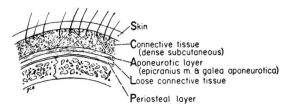

Fig. 160   The layers of the scalp as seen in cross section.

scalp. The third layer of the scalp is musculo-aponeurotic. It contains the frontal and occipital bellies of the **epicranius muscle** which are situated in the forehead and occipital regions, respectively, and are united by a strong membranous sheet, the **galea aponeurotica.** Associated with the extension of this sheet into the temporal region are the **auricular muscles.** The fourth layer of subaponeurotic connective tissue is represented by areolar tissue which is both loose and scanty and supports small blood vessels. Its loose character allows movement of the three superficial layers as one unit over the cranium and permits a wide spread of fluid accumulations in this layer. The fifth layer of the scalp is the pericranium. It is the periosteum of the cranial bones which adheres poorly over the surfaces of the bones of the skull and, due to this circumstance, is allocated to the scalp. The pericranium is firmly fixed to the bones only along the suture lines, for here the layer passes deeply to become continuous with the periosteum of their inner surfaces.

The Arteries of the Scalp (figs. 156, 161)

The external carotid artery gives rise to three branches which reach the scalp: the occipital, the posterior auricular, and the superficial temporal. The internal carotid artery provides, through its ophthalmic branch, the supraorbital and supratrochlear arteries. The terminal vessels anastomose freely with one another and provide a rich blood supply in the scalp. The blood vessels of the scalp are firmly held by the strong fibers of the subcutaneous connective tissue and are prevented thereby from retracting or contracting their lumina when severed. For this reason and because of the rich anastomosis of the vessels, bleeding is free and may be difficult to control in scalp wounds.

The large **occipital artery** (figs. 137, 161) arises from the back of the external carotid artery in upper levels of the neck. Passing upward and backward along the lower border of the posterior belly of the digastric muscle, the artery reaches and occupies the occipital groove on the mastoid portion of the temporal bone. From the groove it enters the scalp by piercing the cervical fascia in the interval between the attachments of the sternocleidomastoid and trapezius muscles to the superior nuchal line.

In the scalp the occipital artery divides into **medial** and **lateral** terminal **branches**. These ascend in the subcutaneous tissue of the posterior part of the scalp as far as the vertex, anastomosing with branches of the occipital artery of the opposite side and of the posterior auricular and superficial temporal arteries. The medial branch gives off a **meningeal twig** which enters the skull through its parietal foramen to supply the dura mater. Both the medial and the lateral branches of the occipital artery are accompanied by branches of the **greater occipital nerve** which joins the artery as it enters the scalp between the sternocleidomastoid and trapezius muscles. This nerve, the medial branch of the dorsal ramus of the second cervical nerve (p. 148), distributes with the occipital artery as far as the vertex of the skull. Its lateral branches communicate with radicles of the lesser occipital nerve.

The small **posterior auricular artery** (figs. 137, 161) arises from the external carotid immediately above the posterior belly of the digastric muscle. It runs upward and backward superficial to the styloid process to the groove between the mastoid

process and the cartilage of the external acoustic meatus. In its course, it supplies small twigs to the parotid gland and to the styloid and digastric muscles; its named branches are the stylomastoid, the auricular and the occipital.

The **stylomastoid branch** enters the stylomastoid foramen and ascends in the facial canal to supply the tympanic cavity, the tympanic antrum and mastoid cells, and the semicircular canals. The **auricular branch** ascends behind the ear beneath the posterior auricular muscle. It supplies the external ear on both surfaces, anastomosing with the auricular branches of the superficial temporal artery. It also supplies the scalp in the posterior temporal region, making connection with the superficial temporal and occipital arteries. The **occipital branch** passes backward to the scalp above and behind the ear. It supplies the occipitalis muscle and anastomoses with the occipital artery. The posterior auricular branch of the facial nerve (p. 209) accompanies the posterior auricular artery. The lesser occipital nerve

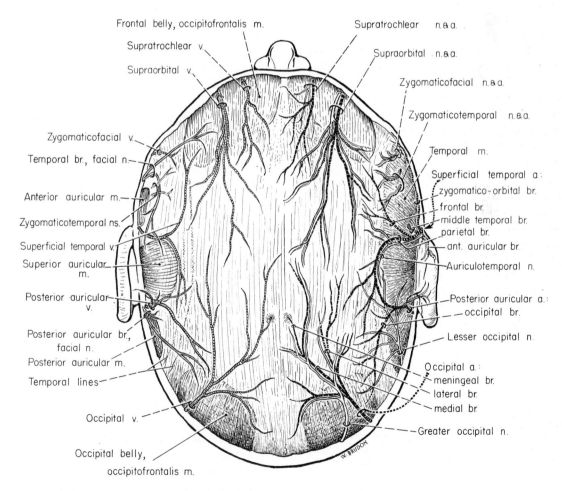

Fig. 161   Blood vessels, nerves, and muscles of the scalp; arteries and cutaneous nerves on the right, veins and muscular nerves on the left.

(p. 148) distributes in the region of supply of the artery, and the **posterior auricular vein** drains the area and participates in the formation of the external jugular vein of the neck.

The **superficial temporal artery** (figs. 139, 156, 161), though the smaller of the terminal branches of the external carotid, appears to be its direct continuation. It begins behind the neck of the mandible deep to the parotid gland, crosses over the posterior root of the zygomatic process of the temporal bone, and enters the subcutaneous tissues of the temporal region of the scalp. In front of the ear, it is very superficial and its pulsation is readily felt. The auriculotemporal nerve is directly behind the artery, and the **superficial temporal vein** accompanies it. About 5 cm. above the zygoma, the artery divides into its terminal **frontal** and **parietal branches.** Aside from twigs to the parotid gland, the branches of the superficial temporal artery are the transverse facial, the zygomaticoorbital, the middle temporal, the anterior auricular, the frontal, and the parietal.

The **transverse facial artery** emerges from the anterior border of the parotid gland near its upper end. The artery passes forward in the face across the masseter muscle between the parotid duct and the zygomatic arch. It may be accompanied by the zygomatic branch of the facial nerve. Branches of the artery are distributed to the parotid gland, the masseter muscle, and the skin of the face and anastomose with branches of the buccal, the infraorbital, and the facial arteries. The **zygomatico-orbital branch** may arise from the main artery or may be a branch of the middle temporal. It is a small vessel which passes forward along the upper border of the zygoma in the fat between the superficial and deep layers of the temporal fascia, the dense aponeurosis which overlies the temporalis muscle. The artery supplies the orbicularis oculi muscle and anastomoses with the lacrimal and palpebral branches of the ophthalmic artery. The **middle temporal artery** arises just above the zygoma, perforates the temporal fascia and temporal muscle, and ascends on the squamous portion of the temporal bone. It anastomoses with the posterior deep temporal artery (p. 225). The **anterior auricular branches** of the superficial temporal artery are three or four in number and are distributed to the anterior part of the auricle, the lobule, and part of the external acoustic meatus. The **frontal branch** of the superficial temporal artery runs tortuously forward and somewhat upward toward the frontal eminence. Its branches supply the muscles and skin of the anterior temporal and frontal regions and anastomose with lacrimal and supraorbital branches of the ophthalmic artery and with its fellow of the opposite side. The **parietal branch** courses upward and backward on the side of the head. It supplies muscles and the skin of the

parietal and posterior temporal regions and anastomoses with terminals of the posterior auricular and occipital arteries and with its fellow of the opposite side. The auriculotemporal nerve accompanies the superficial temporal artery but has a more restricted area of distribution in the scalp. It is described with the other cutaneous nerves (p. 199 & fig. 161).

The **supraorbital artery** (fig. 161) is a branch of the ophthalmic artery which arises as the ophthalmic artery crosses the optic nerve within the orbit (p. 253). Passing forward in the orbit with the frontal nerve, the supraorbital artery emerges onto the scalp through the supraorbital foramen, or notch. A tiny **diploic branch** frequently arises from the artery in the notch and there enters a minute foramen to supply the diploë and the frontal sinus. A deep branch of the supraorbital artery distributes between the frontalis portion of the epicranius muscle and the periosteum, but the superficial branch perforates the frontalis and supplies the skin and superficial structures of the forehead.

These branches anastomose with the angular branch of the facial artery and with the frontal branch of the superficial temporal artery. The supraorbital artery is accompanied by the supraorbital vein and distributes in company with the supraorbital nerve, a cutaneous nerve of the ophthalmic division of the trigeminal (p. 199 & fig. 161). The supraorbital vein communicates with the superior ophthalmic vein (p. 253) and empties into the angular vein (p. 204).

The **supratrochlear artery** is one of the terminal branches of the ophthalmic artery. It pierces the superior palpebral fascia at the upper medial angle of the orbit in company with the supratrochlear nerve. The artery supplies the skin and other superficial structures of the anterior and medial parts of the scalp and anastomoses with the supraorbital artery and with its fellow of the opposite side.

The supratrochlear vein unites with the supraorbital vein to form the angular vein. The supratrochlear nerve, a cutaneous branch of the ophthalmic division of the trigeminal nerve (p. 199), accompanies the supratrochlear artery into the scalp, distributing to the skin and subcutaneous tissue of the medial portion of the upper eyelid and of the adjacent forehead.

The Veins of the Scalp (fig. 161)

The veins that drain the superficial parts of the scalp are the venae comitantes of its arteries: the supratrochlear, the supraorbital, the superficial temporal, the posterior auricular, and the occipital. Venous drainage of the deeper parts of the scalp in the temporal region is by way of deep temporal veins which are tributary to the pterygoid plexus.

The **supratrochlear vein** begins in a plexus on the forehead through which it communicates with the frontal branch of the superficial temporal vein, the supraorbital vein, and its fellow of the opposite side. Descending near the midline, the two supratrochlear veins are joined at the root of the nose by a transverse communication, and each then unites, at the medial angle of the orbit, with the supraorbital vein of its side to form the **angular vein.**

The **supraorbital vein** begins in the forehead, where its radicles communicate with the supratrochlear vein and the frontal branch of the superficial temporal vein. The supraorbital vein communicates with the superior ophthalmic vein by means of an offshoot which passes through the supraorbital notch and which receives, while in the notch, the frontal diploic vein. At the medial angle of the orbit, the supraorbital and supratrochlear veins unite to form the angular vein. The angular vein descends obliquely at the side of the root of the nose, and at the lower margin of the orbit it becomes the facial vein (p. 204).

The **superficial temporal vein** is formed by tributaries which correspond to the frontal and parietal branches of the superficial temporal artery. These veins drain the side of the scalp, from the lateral part of the frontal region to the anterior part of the parietal region. They communicate with all other veins of the scalp and with the corresponding veins of the opposite side of the scalp. The frontal and parietal tributaries unite to form the superficial temporal vein in front of the ear and a little above the zygoma. Here the superficial temporal vein receives the **middle temporal vein** draining the temporalis muscle. The latter vein receives, as a rule, the **orbital vein,** which lies between the layers of the temporal fascia and corresponds to the zygomaticoorbital artery (p. 212). The superficial temporal vein descends subcutaneously across the root of the zygoma, receiving **anterior auricular veins, articular veins** from the temporomandibular joint, **parotid radicles,** and the **transverse facial vein** from the side of the face. Behind the neck of the mandible, the superficial temporal vein unites with the maxillary vein to form the retromandibular vein.

The **posterior auricular vein** begins in a plexus of veins in the subcutaneous tissues behind the ear which communicates with the parietal tributary of the superficial temporal vein and with the occipital vein. The posterior auricular vein receives tributaries from the back of the ear and the stylomastoid vein. It is larger than the corresponding artery and leaves it at the base of the scalp to descend across the upper part of the sternocleidomastoid muscle. The posterior auricular vein unites with the retromandibular vein to form the external jugular vein.

The **occipital vein** (fig. 161) begins in a plexus of veins in the occipital region of the scalp through which communications are made with the posterior auricular vein, the parietal radicles of the superficial temporal vein, and the occipital vein of the opposite side. The occipital vein descends with the occipital artery, pierces the cranial attachment of trapezius, and empties, in the suboccipital triangle, into a plexus drained by the deep cervical and vertebral veins. Occasionally, the occipital vein follows the course of the occipital artery and ends in the internal jugular vein. Sometimes the principal flow of the occipital vein is to the posterior auricular vein, through which it empties into the external jugular vein. The parietal emissary vein connects the occipital vein with the superior sagittal sinus, and one of the more lateral tributaries of the occipital vein receives the mastoid emissary vein.

The Cutaneous Nerves of the Scalp (fig. 161)

The cutaneous nerves of the scalp are described with the superficial nerves of the head and neck (p. 148). The close similarity in the course and distribution of the superficial nerves and blood vessels of the scalp is a reflection of the tendency for these structures to run together in all parts of the body. In the forehead the **supraorbital nerve** (p. 199) distributes as far as the vertex of the skull and lateralward as far as the temporal region. It is accompanied by the supraorbital artery and vein. The **supratrochlear nerve** supplies the medial region of the forehead. The **auriculotemporal nerve** (p. 199) enters the scalp in company with the superficial temporal artery and vein. The nerve, however, is small, and its area of distribution is less than that of the blood vessels (fig. 161). The **greater occipital nerve** (p. 148) distributes through the greater part of the posterior region of the scalp in company with the occipital artery and vein. The **lesser occipital nerve** (p. 148) is cutaneous to the scalp behind and above the ear and to the cranial aspect of the ear. Its main trunk has no companion vessel in the scalp, but certain of its branches are accompanied by the posterior auricular artery and vein. The

great auricular nerve (p. 148) has a very limited distribution in the scalp and is not in close association with blood vessels.

The Muscles of the Scalp (figs. 158, 161, 162)

The muscles of the scalp are the epicranius, or occipitofrontalis, muscle and the small auricular muscles.

Epicranius consists of two pairs of bellies— frontal and occipital—united by a broad, intermediate tendinous sheet, the **galea aponeurotica.** The **occipital bellies** arise from the lateral two-thirds of the highest nuchal line of the occipital bone. Ascending as thin sheets of parallel muscle fasciculi, the muscles end in the galea. The **galea aponeurotica** is a fibrous membrane composed mainly of sagittally running fibers. Posteriorly, it has a small bony attachment to the external occipital protuberance and the nuchal line between the two occipital portions of the muscle; anteriorly, it provides origin for the frontal bellies of epicranius. On the side of the scalp, the galea gives origin to and provides an investing fascia for the superior and anterior auricular muscles and is continued as a thinner layer over the temporal fascia to the zygomatic arch. The **frontal bellies** of epicranius have no bony attachments. They arise from the galea aponeurotica midway between the coronal suture and the superciliary arch and insert into the skin of the eybrows and the root of the nose. The medial fibers intermingle with procerus; the more lateral fibers, with orbicularis oculi and corrugator. The epicranius draws the scalp backward; the

special action of the frontal portion is elevation of the eyebrows, as in the expression of surprise. The occipital portion of the muscle is innervated by the posterior auricular branch of the facial nerve. The temporal branch of the facial nerve supplies the frontal portion of the muscle.

Clinical Note In case of transverse scalp incisions, the opposed pull of occipitalis and frontalis may lead to marked gaping of the wound edges, especially if the incision passes through the galea aponeurotica. In sagittally oriented incisions, gaping does not occur.

The **auricular muscles** are three: posterior, superior, and anterior. The **posterior auricular muscle** is narrow and takes origin from the mastoid portion of the temporal bone. It inserts into the cranial surface of the auricle. The posterior auricular vessels pass deep to this muscle. The **superior auricular muscle**, thin and fan-shaped, is the largest of the three auricular muscles. It arises from the galea aponeurotica and inserts by a thin, flattened tendon into the upper part of the cranial surface of the ear. The **anterior auricular muscle,** the smallest of the three, arises between two layers of the galea aponeurotica. It inserts on the front of the helix of the ear. The actions of the auricular muscles are indicated by their orientation. Though capable of training, they usually provide little specific movement in Man. The posterior auricular branch of the facial nerve supplies the posterior muscle; the anterior and superior auricular muscles are innervated by the temporal branch of the facial nerve.

THE PATTERN OF SUPERFICIAL VENOUS DRAINAGE OF THE HEAD AND NECK (figs. 119, 121)

The superficial veins of the head and neck have a pattern of organization that is inconstant but which is, nevertheless, of some reliability in plotting the usual course of blood flow through the system. The greatest variability occurs, as in all vascular patterns, in the peripheral ramifications of the veins, and most of the probable intercommunications have been noted in the descriptions of the veins already given. The **facial vein,** as the continuation of the **angular vein,** receives the venous drainage of the supraorbital and supratrochlear veins of the forehead and of the small palpebral and nasal veins. The facial vein usually receives, just below the angle of the mandible, a communication from the retromandibular vein and then penetrates deeply into the carotid triangle of the neck and empties into the **internal jugular vein.** Also beginning in the veins of the scalp, the **superficial temporal** and the **maxillary veins** unite behind the neck

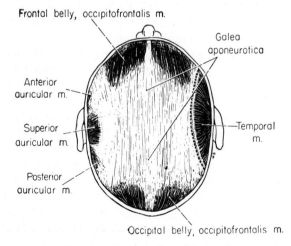

Frontal belly, occipitofrontalis m.

Galea aponeurotica

Anterior auricular m.

Superior auricular m.

Temporal m.

Posterior auricular m.

Occipital belly, occipitofrontalis m.

Fig. 162 The muscles of the scalp.

of the mandible to form the **retromandibular vein.** The **posterior auricular vein** joins the **retromandibular vein** over the upper end of the sternocleidomastoid muscle, and thus is formed the **external jugular vein.** The external jugular passes vertically downward across the sternocleidomastoid muscle, and in the posterior cervical triangle it receives additional tributaries. The **posterior external jugular vein** from the back of the neck and scalp empties into the external jugular as the latter crosses the posterior border of the sternocleidomastoid muscle. Just above the clavicle, the external jugular vein receives the **transverse cervical** and **suprascapular veins** from the region of the shoulder. The **anterior jugular vein,** arising in the submental region, descends subcutaneously along or close to the median line of the neck. Above the manubrium sterni, the vein turns lateralward close behind the sternocleidomastoid muscle to empty into the external jugular vein. The external jugular vein empties opposite the middle of the clavicle into the **subclavian vein.** Behind the sternoclavicular joint the subclavian vein joins the internal jugular vein to form the **brachiocephalic vein,** and the brachiocephalic veins of the two sides unite to form the **superior vena cava** leading to the heart. The **occipital vein** of the scalp usually drains into either the **deep cervical vein** or the **vertebral vein** but may empty into the internal jugular vein.

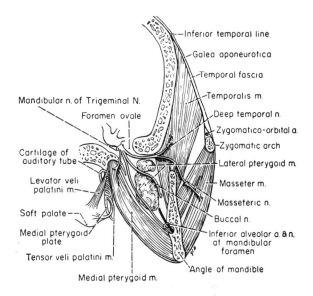

Fig. 163   The muscles of mastication and of the palate as seen in partial vertical section of the deep face; branches of the mandibular nerve.

THE DEEP FACE

The Muscles of Mastication and Their Fasciae (figs. 163–66)

The muscles of mastication—temporal, masseter, medial pterygoid, and lateral pterygoid muscles— are invested by fasciae which are continuities of the fasciae of the neck and which bound a **masticator space** enclosing these muscles and their associated vessels and nerves (fig. 163). The superficial layer of cervical fascia, after attachment to the inferior border of the mandible, continues into the face as both the parotid (p. 207) and masseteric fasciae. The **masseteric fascia** attaches to the inferior border of the mandible inferiorly and posteriorly. Anteriorly, after leaving the masseter muscle, it encircles the ramus to be continuous with the fascia of the medial pterygoid muscle deep to the bone. Superiorly, the masseteric fascia attaches to the zygomatic arch. The **pterygoid fascia** is continuous

at the borders of the mandible with the masseteric and cervical fasciae and with the stylomandibular ligament. Following the deep surface of the medial pterygoid muscle, the fascia reaches the medial pterygoid plate of the sphenoid bone. On the superficial surface of the muscle, the fascia ascends to the lateral pterygoid muscle which it splits to enclose.

Thickened between the two muscles, the fascia attaches to the base of the skull from the lateral pterygoid plate to the spine of the sphenoid and thus forms the pterygospinous ligament. Sometimes ossified, this ligament forms, with the skull, the **pterygospinous foramen** through which the branches of the mandibular nerve pass to the muscles of mastication. The approach of a needle to the mandibular nerve is occasionally blocked by a bony bar lateral to the foramen ovale. This is an ossified pterygoalar ligament.

The **sphenomandibular ligament** is a thickening of the pterygoid fascia which extends between the spine of the sphenoid bone and the lingula of the mandible. The **temporal fascia** is a strong aponeurotic sheet which covers the temporalis muscle and gives origin to many of its fibers. It is attached above to the entire extent of the superior temporal line and extends below to the zygomatic arch. Single above, it splits in its lower one-fourth into two laminae. The deeper lamina ends by attaching to the medial border of the zygomatic arch; the superficial layer, after attaching to the lateral border of the arch, continues below as the masseteric fascia. Between the layers there is a little fat and the zygomaticoorbital branch of the superficial temporal artery.

The Masseter Muscle (figs. 121, 163, 164, 178)

The masseter muscle is a thick, quadrangular muscle which overlies the angle of the mandible. The muscle arises from the zygomatic process of the maxilla and from the lower border of the zygomatic arch. A superficial portion utilizes the process and the anterior two-thirds of the arch; the smaller deep portion takes origin from the posterior one-third of the lower border and from the entire medial surface of the zygomatic arch. The muscle inserts into the lateral surfaces of the coronoid process, ramus, and angle of the mandible. The masseter is separated by the buccal pad of fat from the buccinator muscle anteriorly and is crossed by the parotid duct; it is partly overlapped by the parotid gland. The nerve to the masseter muscle is a branch of the mandibular division of the trigeminal nerve which reaches it by passing through the mandibular notch and penetrating its deep surface (p. 222).

The Temporalis Muscle (figs. 163, 164)

This is a fan-shaped muscle that occupies the

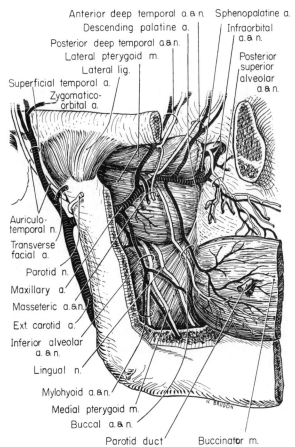

Fig. 165 The muscles, arteries, and nerves of the infratemporal fossa.

temporal fossa of the side of the head, an area outlined by the superior temporal line and the zygomatic arch. The muscle arises from the periosteum of the temporal fossa and from the temporal fascia which covers it. Thin at its periphery, the fibers of the muscle accumulate and form a thick tendon which passes medial to the zygomatic arch and inserts into the anterior border and medial surface of the coronoid process of the mandible. The most anterior fibers of the muscle extend down the anterior border of the ramus almost to the last molar tooth. Anterior and posterior deep temporal branches of the mandibular division of the trigeminal nerve enter the muscle on its deep aspect (p. 222).

The Medial Pterygoid Muscle (figs. 139, 163, 165, 167, 178)

The medial pterygoid muscle occupies the same position internal to the angle of the mandible as

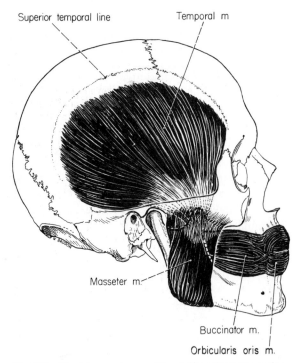

Fig. 164 The temporal, masseter, and buccinator muscles.

does the masseter on the outside. The medial pterygoid muscle arises from the medial surface of the lateral pterygoid plate and from the pyramidal process of the palatine bone. This larger portion of the medial pterygoid muscle lies deep to the lateral pterygoid muscle. A small muscular slip from the tuberosity of the maxilla arises superficial to the lateral pterygoid muscle and joins the main belly of the medial pterygoid muscle below. The fibers of the medial pterygoid are directed downward, backward, and lateralward to insert on the medial surface of the angle and ramus of the mandible as high as the mandibular foramen. Medially, the muscle is related to the tensor veli palatini muscle and to the superior pharyngeal constrictor. The nerve to the medial pterygoid is a branch of the mandibular division of the trigeminal nerve which arises just below the foramen ovale. It enters the muscle on its medial surface near its posterior border (p. 222).

The Lateral Pterygoid Muscle (figs. 163, 165–67)

The lateral pterygoid muscle is a short, thick muscle which is oriented horizontally high in the infratemporal fossa. Its main lower head of origin arises from the lateral surface of the lateral pterygoid plate. A smaller upper head arises from the infratemporal surface of the greater wing of the sphenoid. The two heads merge posteriorly but show a regional distinction at their insertion, the lower part ending in the pterygoid fovea of the neck of the mandible. The upper head of the muscle inserts into the articular disk of the temporomandibular joint and into the upper part of the neck of the condyle. The designation of **spheno-meniscus** is sometimes given to the uppermost portion of the muscle, reflecting its separate origin and special insertion. The lateral pterygoid muscle lies deeply in the infratemporal fossa almost completely under cover of temporalis. It is supplied by a branch of the mandibular division of the trigeminal nerve which enters at the upper medial border of the upper head of the muscle (p. 223).

THE TEMPOROMANDIBULAR JOINT (figs. 165–67)

The temporomandibular joint, by which the mandible articulates with the cranium, combines a hinge and a gliding joint. The sinuous articular surface of the temporal bone consists of the **mandibular fossa** and, anterior to it, the downwardly bulging **articular tubercle**. The **condyle** of the man-

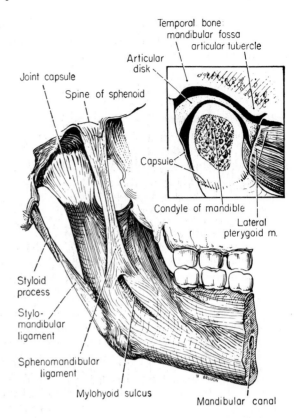

Fig. 166 The internal aspect of the temporomandibular joint and its medial ligaments. Inset—the lateral surface of the capsule removed to show the articular disk.

dible is rounded anteroposteriorly but resembles a narrow ellipsoid in being several centimeters in its longer transverse dimension. Its long axis is directed medialward and slightly backward, so that the planes of the two condyles, if extended, would meet at the anterior edge of the foramen magnum (fig. 167). The articular surfaces are covered by fibrocartilage instead of hyaline cartilage and are separated and made more congruent by an articular disk composed of dense fibrous connective tissue. The upper surface of the disk is concavoconvex to fit the mandibular fossa and articular tubercle, and its under surface is concave for reception of the condyle (fig. 166). The disk is thicker at its periphery, where it is attached to the articular capsule, than at its center, where it is sometimes perforated. The sphenomeniscus portion of the lateral pterygoid muscle inserts into the anterior portion of the disk. The **capsule** of the temporomandibular joint is thin and loose; it is attached above to the circumference of the mandibular fossa and the articular tubercle and below to the neck of the condyle of the mandible. There are two **synovial membranes** in the joint: one lines the capsule above the articular disk; the other, the capsule below the disk.

The capsule contains one strengthening ligament on its lateral surface; there are two separated accessory ligaments medially. The **lateral ligament** (fig. 165) is broad above where it attaches to the lower border of the zygoma and the tubercle. Its fibers are directed downward and backward to the lateral and posterior parts of the neck of the mandible. The **sphenomandibular ligament** (fig. 166) is thin and flat but very strong. It descends from the spine of the sphenoid bone (or the petrotympanic fissure) to the lingula of the mandible, separated from the medial aspect of the capsule by the maxillary vessels and the auriculotemporal nerve. The **stylomandibular ligament** is a thickening of the deep parotid fascia which extends from near the apex of the styloid process to the posterior border of the ramus of the mandible near its angle. It separates the parotid and submandibular glands.

The **nerves** of the temporomandibular joint are sensory twigs derived from the masseteric and auriculotemporal branches of the mandibular division of the trigeminal nerve. Its **blood vessels** are branches of the superficial temporal, middle meningeal, anterior tympanic, and ascending pharyngeal arteries. The **lymphatic drainage** of the joint is to the deep parotid nodes.

Movements and Muscle Actions The form of the articulating surfaces is such as to facilitate a hinge action between the articular disk and the mandibular condyle in the lower cavity of the joint and a gliding action between the disk and the temporal bone in the upper cavity. The concavoconvex form of the mandibular fossa and articular tubercle provides, furthermore, that when the mouth is opened, the jaw is depressed, so that the teeth are separated by direct downward movement of the jaw as well as by its hinge action. The jaws are opened by forward traction on the neck of the mandible by the lower portion of the lateral pterygoid muscles, assisted by the digastric, mylohyoid, and geniohyoid muscles (the hyoid bone being fixed by the infrahyoid muscles) (Moyers, '50). Electromyographic studies suggest that the sphenomeniscus portion of the lateral pterygoid muscle positions or stabilizes the condyle and disk against the articular eminence during closing movements of the mandible. The combined gliding and hinge action at this joint produces rotation of the mandible around a trans-

verse axis which might be projected between the two lingulae, a fact which gives significance both to the attachment of the sphenomandibular ligament at this point and to the entrance of the inferior alveolar nerve and vessels into the bone through the adjacent mandibular foramen. Closure of the jaws is a very powerful action produced by the combined action of the masseter, the medial pterygoid, and the temporalis muscles, the two former muscles holding the angle of the mandible in a muscular sling (fig. 163). The medial pterygoid muscle assists the lateral pterygoid in protrusion of the mandible; the posterior fibers of the temporalis retract it, assisted by the suprahyoid group of muscles. In chewing and grinding movements, the mandible is alternately protruded and retracted, the two sides usually moving in an opposite sense, so that the mandibular teeth are moved diagonally to the maxillary teeth. The strong medial traction of the medial pterygoid muscle is called into play to assist in producing the oblique direction of movement involved in this function.

THE MANDIBLE (figs. 164, 166, 167, 209, 213, 214)

The mandible forms the skeleton of the lower jaw and lodges the mandibular teeth. The fusion of its bilateral parts forms a U-shaped bone consisting of a horizontal body and, at the ends of the U, two vertically projecting rami. The body is composed of an inferior rim of dense basilar bone which gives support to a superior portion of alveolar bone hollowed out for the sockets of the teeth. Externally, the body exhibits a median ridge indicating the symphysis or line of fusion of the two mandibular halves.

The inferior border in front projects forward as the **mental protuberance,** and the elevations at either side of a central depressed area are the **mental tubercles.** Below the incisor teeth is the **incisive fossa.** The **mental foramen** is a prominent feature of the external surface and is located below the second bicuspid tooth midway between the alveolar and basal margins of the bone. It transmits the mental nerve and blood vessels onto the face. Beginning below the foramen is an **oblique line** which rises posteriorly to be continuous with the anterior border of the ramus. It provides attachment for muscles. Behind the symphysis the concave inner surface of the body has two sharp **mental spines.** Below these is a median ridge and at the inferior margin of the bone an oval depression at either side of the median line. The **mylohyoid line** arches upward from the outer limit of the digastric impression and, extending backward toward the ramus, becomes quite prominent as it nears the alveolar border. Above the line, just lateral to the mental spine, is a smooth oval fossa which lodges the sublingual gland; below the mylohyoid line, anterior to the angle, is a larger depression for the submandibular gland. The alveolar portion of the body of the mandible provides cavities for the sixteen lower teeth. The cavities vary in depth and size according to the tooth accommodated.

The **ramus** of the mandible is flat and quadrilateral. Several **oblique ridges** accommodate the tendinous attachments of the masseter muscle. The medial surface is heavily ridged at the angle for the insertion of the medial pterygoid muscle. The stylomandibular ligament attaches to the angle between the masseter and medial pterygoid muscles. The **mandibular foramen,** for the passage of the inferior alveolar nerve and vessels, perforates the medial surface of the ramus at about its center. This foramen is the beginning of the **mandibular canal** which ends below and anteriorly at the

mental foramen. From the canal, small channels pass to the tooth sockets and provide passage for the vessels and nerves to the teeth. The sharp **lingula** lies at the anteromedial edge of the foramen and gives attachment to the sphenomandibular ligament. Just behind and below the foramen, the **mylohyoid groove** runs downward and forward, lodging the mylohyoid nerve and vessels. The junction of the body and the ramus of the mandible is the roughened and everted **angle** of the mandible. These parts of the bone meet in the adult at an angle of from 110° to 120°. The thin anterior border of the ramus ends above in the **coronoid process.** Its posterior margin leads into the deep **mandibular notch** over which the nerves and blood vessels to the masseter muscle pass. The **condyle** of the mandible is an expansion of the posterior border of the mandible. It is received into the mandibular fossa of the temporal bone in the formation of the temporomandibular

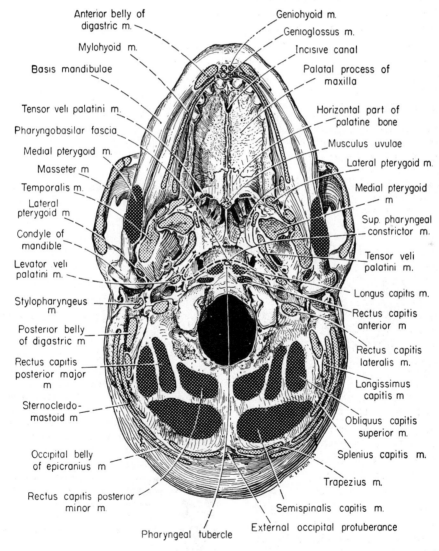

Fig. 167 The muscular attachments on the inferior aspect of the base of the skull and mandible. (Black dots—origin; white dots on black—insertion.)

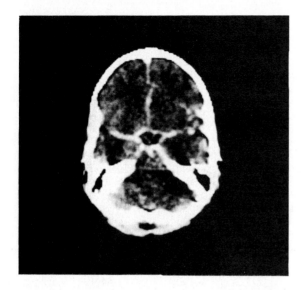

C.T. scan of head at level of arterial circle of base of brain.
(Courtesy of J. Seeger, M.D., University of Michigan)

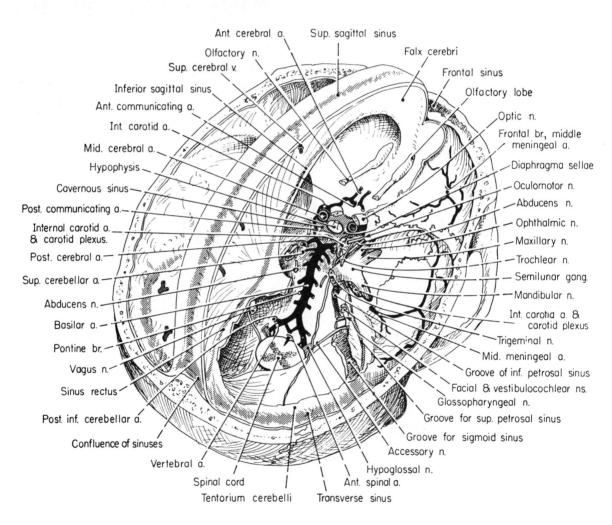

Fig. 168   The internal anatomy of the cranium; dura, dural venous sinuses, cranial nerves, arteries.

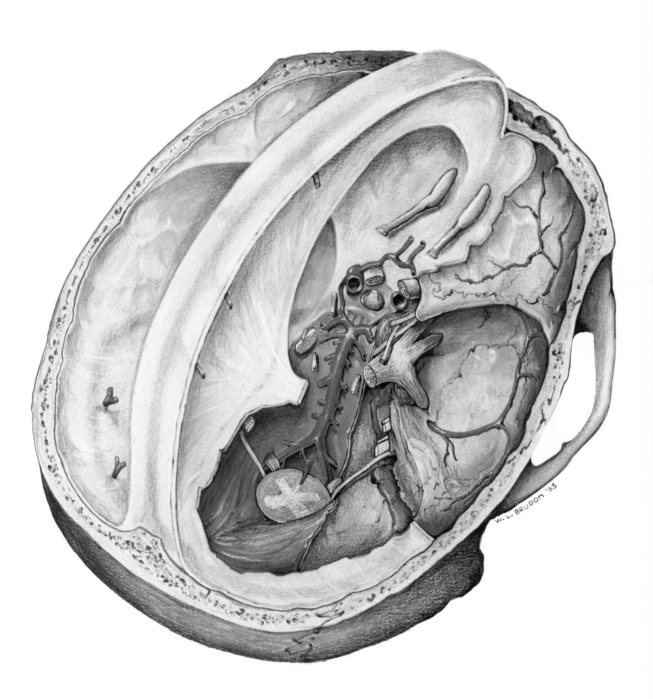

W. L. BRUDON '63

joint. The broadening of the condyle is in the mediolateral direction, its long dimension being about 2 cm. The long axis is directed medialward and somewhat backward and, if prolonged to the median line, would meet that from the opposite condyle at the anterior margin of the foramen magnum. The condyle is rounded, is narrow from front to back, and is curved forward, spoonlike. The **neck** of the condyle is flattened from before backward.

Age Changes in the Mandible. The life cycle of the mandible emphasizes the fundamental separation of the body of the bone into alveolar and basilar parts and the dependence of the form of the bone on the dental history of the individual. At birth, the body of the bone is a shell of alveolar bone containing the sockets of two incisor, one cuspid, and two deciduous molar teeth. The basilar part of the bone is a thin strip of denser bone along its inferior margin and, since the mental foramen lies at the junction of the alveolar and basilar parts, it is close to the inferior margin. The angle is obtuse (175°), the condyle being nearly in line with the body. The union of the two halves of the bone at the symphysis takes place during the first year, beginning inferiorly. After the first dentition, the body elongates and the basilar bone is increased, so that the mental foramen comes to lie midway between the superior and inferior borders of the bone opposite the second deciduous molar. The angle between body and ramus becomes 140° by about the fourth year. These changes continue to progress, so that, in the adult, the angle approaches a right angle (110° to 120°) and the mental foramen remains midway between the borders of the bone but, with permanent dentition, lies below the second bicuspid tooth. Old age is characterized by loss of teeth and absorption of alveolar bone, so that an edentulous mandible consists of a body composed entirely of basilar bone, and the mental foramen is then situated along its upper border. The angle is again increased to about 140°, and the neck of the condyle tends to be bent backward.

Ossification The mandible is formed in fetal development by ossification in the fibrous membrane covering the outer surface of Meckel's cartilage, the cartilaginous bar of the mandibular arch. From Meckel's cartilage are developed the malleus and incus (bones of the middle ear) and the sphenomandibular ligament. The cartilage largely disappears anterior to the ligament, although a small part below and behind the incisor teeth is ossified and incorporated into that part of the mandible. Ossification begins in the fibrous covering of the cartilage at about the sixth week of fetal life, each half of the mandible forming from a single center which appears near the mental foramen. The two halves of the mandible unite by ossification of the symphysis during the first year of life.

Clinical Note In spite of its general strength, the mandible is frequently fractured, especially in automobile accidents. Fractures in the region of its angle are less common than others due to the heavy muscular attachments on both its sides. Perhaps the most common fracture occurs at or just behind the mental foramen. Displacement of the fragments in such a fracture depends partly on the obliquity of the line of

fracture and largely on muscular traction. It will be remembered that all of the elevators of the jaw attach to the posterior fragment (temporalis, masseter, and medial pterygoid muscles) and that most of the depressors attach anterior to the fracture site (geniohyoid, mylohyoid, and digastric muscles). The medial pterygoid muscle also draws the posterior fragment medialward. A fracture of the mandible through its neck results in marked forward tilting of the condylar fragment due to the traction of the lateral pterygoid muscle. Bilateral anterior dislocation of the mandible is also relatively common with the condyles riding forward beyond the articular tubercles of the temporal bone.

THE INFRATEMPORAL FOSSA

The infratemporal fossa is the irregular space below and deep to the zygomatic arch and posterior to the maxilla. The space is limited above by the infratemporal surface of the greater wing of the sphenoid and its infratemporal crest (fig. 139) and below by the alveolar border of the maxilla. The medial limit of the space is the lateral pterygoid plate.

Between the upper part of the lateral pterygoid plate and the adjacent border of the maxilla is the **pterygomaxillary fissure** which leads from the infratemporal fossa to the triangular **pterygopalatine fossa. The foramen ovale** and the **foramen spinosum** open into the roof of the infratemporal fossa. The **inferior orbital fissure** is a horizontal cleft whereby the infratemporal fossa communicates with the orbit anteriorly. The infratemporal fossa accommodates the lower part of the temporalis muscle, the medial and lateral pterygoid muscles, the maxillary artery and pterygoid plexus of veins, and the mandibular division of the trigeminal nerve. The muscles are described on p. 215.

The Mandibular Division of the Trigeminal Nerve (figs. 120, 139, 165, 168, 169)

The **trigeminal** (cranial V) **nerve** is the largest of the cranial nerves. It is the principal cutaneous nerve of the face and scalp and is also sensory to the mucous membranes of the spaces of the head. In addition, it provides the innervation of the muscles of mastication and, thus, has the following three components:

General somatic afferent (exteroceptive): cutaneous sensibility in the face and head and sensation from the mucous membranes of the mouth, from the teeth, and from the meninges of the brain.
General somatic afferent (proprioceptive): proprioceptive sensation from the muscles innervated.
Special visceral efferent: motor supply of the four principal muscles of mastication, and of the mylohyoid, the anterior belly of the digastric, the tensor veli palatini, and the tensor tympani muscles.

The sensory fibers of the nerve enter the pons through the lateral part of its ventral surface; the motor root emerges adjacent to but rostral and medial to the sensory root. Both roots pass anteriorly in the posterior cranial fossa to reach the trigeminal ganglion which lies in a pocket of dura mater near the apex of the petrous portion of the temporal bone.

The **semilunar ganglion** (fig. 168) is flat, has a semilunar shape, and measures about 1 × 2 cm. The central processes of the nerve leave its concavity; the peripheral processes emerge from its convexity. The fibers of the small motor root pass deep to the ganglion and join the mandibular division of the nerve. From the convexity of the semilunar ganglion the peripheral processes pass outward, organizing into the three great divisions of the trigeminal nerve: **ophthalmic, maxillary,** and **mandibular.** *These are all predominantly sensory, and the ophthalmic and maxillary parts are exclusively so. Each of the divisions carries, on one or more of its peripheral branches, autonomic nerves destined for a gland or mucous surface to which the trigeminal nerve is sensory. No autonomic fibers arise in the central nuclei of the trigeminal nerve, and those that join and distribute with its branches are not truly parts of the nerve.* The ophthalmic nerve (p. 251) emerges from the anterior superior part of the semilunar ganglion and enters the orbit through the superior orbital fissure. The maxillary nerve (p. 240) arises from the middle portion of the semilunar ganglion and leaves the cranial cavity through the foramen rotundum.

The **mandibular nerve** is composed of a sensory and a motor root, the sensory root arising from the inferior angle of the trigeminal ganglion. The two roots leave the middle cranial fossa through the foramen ovale and unite just outside the skull into a common trunk of only 2 to 3 mm. length. **This trunk then divides into an anterior division, which is mainly muscular, and a posterior division, which is largely sensory.** The **branches of the trunk** are the meningeal and the nerve to the medial pterygoid muscle.

(1) The **meningeal branch** (fig. 169) re-enters the cranium through the foramen spinosum in company with the middle meningeal artery. It distributes, with this artery, by anterior and posterior branches to the dura mater; the posterior branch also supplies the mucous membrane of the mastoid air cells.

(2) The **medial pterygoid nerve** (figs. 139, 169) has a short course under cover of the pterygoid fascia immediately below the pterygospinous ligament and then penetrates the dorsomedial aspect of the medial pterygoid muscle. The medial pterygoid nerve is closely associated with the otic ganglion but is not functionally related to it. It provides the **nerve to the tensor veli palatini muscle,** which enters this muscle near its origin, and the **nerve to the tensor tympani muscle,** which penetrates the cartilage of the auditory tube to supply the tensor tympani muscle.

The **anterior division** *contains almost all the motor fibers in the mandibular nerve and one sensory branch, the buccal nerve.*

(3) The **masseteric nerve** (figs. 139, 165, 169) passes lateralward superior to the lateral pterygoid muscle. It goes through the mandibular notch with the masseteric artery and enters the deep surface of the masseter muscle near the zygomatic arch. It provides a branch to the temporomandibular joint.

(4) The **deep temporal nerves** (figs. 139, 165, 169) are usually two, sometimes three. Running close to the infratemporal surface of the cranium, they turn upward over the infratemporal crest and, with the corresponding arteries, enter the deep surface of the temporalis muscle. The **posterior deep temporal nerve** frequently arises with the nerve to the

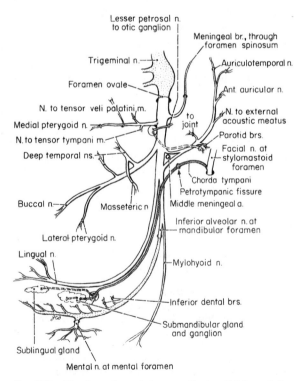

Fig. 169   The branches of the mandibular division of the trigeminal nerve.

masseter muscle; the **anterior deep temporal,** with the buccal nerve. A third deep temporal nerve is the **middle,** which distributes in the same manner but is intermediate in position between the other two.

(5) The **nerve to the lateral pterygoid muscle** (fig. 169) enters the muscle on its deep surface. Its origin may be combined with the buccal nerve.

(6) The **buccal nerve** (figs. 120, 139, 165, 169), frequently arising in common with the anterior deep temporal nerve and sometimes also with the nerve to the lateral pterygoid, emerges between the two heads of origin of the lateral pterygoid muscle. Descending obliquely toward the cheek, the nerve follows the tendon of the temporalis muscle, being bound by fascia to its anteromedial surface. In the cheek (p. 199) the branches of the nerve distribute sensory fibers to the skin and fascia and make communication with the buccal branches of the facial nerve. Other branches penetrate the buccinator muscle to supply the mucosa of the inner surface of the cheek and adjacent gingiva. *The buccal nerve is exclusively sensory and does not innervate the buccinator muscle, this innervation being a function of the buccal branches of the facial nerve.*

The large **posterior division** *of the mandibular nerve is almost entirely sensory, its only muscular branch being the mylohyoid.* The posterior division gives origin to the auriculotemporal nerve and then descends medial to the lateral pterygoid muscle. Here it divides into the lingual and inferior alveolar branches.

(7) The **auriculotemporal nerve** (figs. 120, 139, 165, 169) arises by two roots which encircle the middle meningeal artery just below the foramen spinosum and then unite. The nerve passes posteriorly medial to the lateral pterygoid muscle and between the sphenomandibular ligament and the capsule of the temporomandibular joint. Passing behind the neck of the mandible, the nerve runs lateralward through the uppermost part of the parotid gland or along its superior border. It turns upward over the zygomatic arch with and posterior to the superficial temporal artery and ascends into the scalp in front of the ear.

Several large **communicating branches** join the facial nerve in the substance of the parotid gland. **Communicating rami from the otic ganglion** join the auriculotemporal nerve close to its origin. These transmit the postganglionic fibers (with cells of origin in the otic ganglion) concerned with the parasympathetic innervation of the parotid gland. Such fibers (fig. 169) reach the gland along the parotid branches of the auriculotemporal nerve. The innervation of the parotid gland originates in the glossopharyngeal nerve which supplies preganglionic fibers through the lesser petrosal nerve (p. 196). Besides the **parotid branches** just mentioned, the auriculotemporal nerve supplies **articular twigs** to the back of the temporomandibular joint, several **anterior auricular branches** to the anterior superior part of the ear, and two **branches to the external acoustic meatus.** Entering the meatus between its bony and cartilaginous parts, these distribute to the skin lining the meatus and the tympanic membrane. The terminal **superficial temporal branches** accompany branches of the superficial temporal artery and supply the skin of the temporal region.

(8) The **lingual nerve** (figs. 139, 165, 169), arising medial to the lateral pterygoid muscle, descends anterior and medial to the inferior alveolar nerve between the medial pterygoid muscle and the mandible. Passing the medial pterygoid, the nerve crosses the mandibular attachment of the superior constrictor muscle and, traversing the lateral surface of the styloglossus and hyoglossus muscles, enters the submandibular triangle. The lingual nerve then descends across the duct of the submandibular gland, turns upward medial to it, and passes forward along the under surface of the tongue to its tip, lying immediately beneath its mucous membrane. The lingual nerve is joined high in the infratemporal fossa by the **chorda tympani branch of the facial nerve.** The association of the chorda tympani with the lingual nerve in the mediation of taste and for the parasympathetic supply of the submandibular and sublingual glands is discussed on p. 173. The lingual nerve proper is concerned with general sensation from the anterior two-thirds of the tongue and the mucous membrane of the floor of the mouth, including the lingual aspect of the lower gingivae.

(9) The **inferior alveolar nerve** (figs. 120, 139, 165, 169), larger than the lingual, is directed downward to the mandibular foramen behind the lingula. Traversing the mandibular canal, the nerve provides **inferior dental branches** which form a plexus in the bone and supply the molar and bicuspid teeth by minute filaments which enter the pulp cavities at the apices of the roots of the teeth. Opposite the mental foramen, the inferior alveolar nerve divides into its terminal incisive and mental branches. The **incisive branch** continues medialward to supply the cuspid and incisor teeth and the adjacent labial gingiva. The **mental nerve** emerges onto the face through the mental foramen. Deep to the depressor anguli oris muscle it divides into three branches: one descends to the skin of the chin;

the other two ascend to the skin and mucous membrane of the lower lip. The **mylohyoid nerve** arises from the inferior alveolar nerve just before the latter enters the mandibular foramen and descends in the mylohyoid groove of the mandible to the submandibular triangle. Here it lies on the inferior surface of the mylohyoid muscle deep to the submandibular gland. It provides the motor innervation of the mylohyoid muscle and the anterior belly of the digastric muscle.

Clinical Note   Nerve block of the inferior alveolar nerve at the entrance to the mandible (mandibular foramen) effectively anaesthetizes all the lower teeth. The needle should pass closely against the internal aspect of the ramus of the mandible, between it and the medial pterygoid muscle, and be directed toward the mandibular foramen.

The Maxillary Artery (figs. 139, 156, 165, 170)

Posterior to the neck of the mandible, the external carotid artery, having ascended through or deep to the parotid gland, divides into its terminal branches —the superficial temporal (p. 212) and maxillary arteries. The maxillary artery passes directly anteriorly, deep to the neck of the mandible, gives off branches in the infratemporal fossa, and divides into its terminal branches in the pterygopalatine fossa. Its branches are described in reference to these three regions, and mandibular, pterygoid, and pterygopalatine portions of the artery can be recognized.

The Mandibular Portion   This part of the artery extends horizontally forward from its origin. It lies between the neck of the mandible and the sphenomandibular ligament adjacent to the auriculotemporal nerve. Branches of this portion of the artery are the deep auricular, the anterior tympanic, the middle meningeal, the accessory meningeal, and the inferior alveolar.

The **deep auricular artery** ascends through the uppermost part of the parotid gland and penetrates the bony or cartilaginous wall of the external acoustic meatus. Giving a branch to the temporomandibular joint, it supplies the outer surface of the tympanic membrane and the skin of the meatus.

The **anterior tympanic artery,** ascending parallel to the deep auricular, enters the tympanic cavity through the petrotympanic fissure in company with the chorda tympani nerve. It supplies the lining membrane of the tympanic cavity and anastomoses with the other tympanic arteries, forming a vascular circle around the tympanic membrane with the

stylomastoid branch of the posterior auricular artery.

The **middle meningeal artery** is the principal artery of the cranial dura mater. It passes upward, superficial to the sphenomandibular ligament and between the two roots of the auriculotemporal nerve, and enters the middle cranial fossa by way of the foramen spinosum. It is accompanied through the foramen spinosum by a vein and the meningeal branch of the mandibular nerve. In the cranium (figs. 168, 214, 215) the artery runs forward for a short distance in a groove on the greater wing of the sphenoid bone, lying between bone and dura, and then divides into anterior and posterior branches.

The **anterior branch** continues forward in the groove in the sphenoid, reaching the parietal bone at its anterior inferior angle. Ascending near the anterior border of the parietal bone almost to the vertex of the skull, it gives off, in course, branches which are directed both forward and backward. The **posterior branch** curves backward over the squamous portion of the temporal bone to reach the parietal bone more posteriorly. It extends upward to the superior sagittal sinus and backward to the transverse sinus of the dura mater. The middle meningeal branches distribute partly to dura but principally to bone; they anastomose with the artery of the opposite side, with the accessory meningeal artery, and with the anterior and posterior meningeal arteries. Small branches of the middle meningeal artery within the cranium are the **ganglionic branches** to the semilunar ganglion and its dural covering, a **petrosal branch** that traverses the hiatus of the canal for the greater petrosal nerve and anastomoses with the stylomastoid branch of the posterior auricular artery, and a **superior tympanic branch** which reaches the tympanic cavity by way of the canal for the tensor tympani muscle, supplying the muscle on the way.

The **accessory meningeal artery** is frequently a branch of the middle meningeal artery. After supplying adjacent extracranial structures it enters the cranium through the foramen ovale and supplies the semilunar ganglion and the adjacent dura mater.

The **inferior alveolar artery** (figs. 139, 165) descends to the mandibular foramen posterior and lateral to the inferior alveolar nerve. Near its origin, it supplies a **lingual branch** which accompanies the lingual nerve to the mucous membrane of the mouth. Its **mylohyoid branch** arises immediately above the mandibular foramen. Piercing the sphenomandibular ligament, this branch descends with the mylohyoid nerve and supplies the mylohyoid muscle. In the mandibular canal the inferior alveolar artery supplies the teeth in the same manner as the corresponding nerve (p. 223). It gives rise to the **mental artery** (p. 205) which, on the face (fig. 156), anastomoses with its fellow of the opposite side and with the inferior labial (p. 204) and submental arteries (p. 161).

The Pterygoid Portion This part of the maxillary artery passes upward and forward through the infratemporal fossa. It may lie either lateral or medial to the lateral pterygoid muscle (lateral to it in two-thirds of the cases—Lurje, '46). Anteriorly, it sinks between the two heads of this muscle into the pterygopalatine fossa. The branches of this part of the artery are the masseteric, the anterior and posterior deep temporal, the pterygoid, and the buccal, *each of which distributes with a branch of the mandibular nerve.*

The **masseteric artery** (fig. 165) passes through the mandibular notch with the masseteric nerve and supplies the masseter muscle. It anastomoses with the transverse facial artery.

The **anterior** and **posterior deep temporal arteries** (figs. 139, 165) accompany the corresponding nerves over the infratemporal crest into the temporal muscle. They anastomose with the middle temporal artery.

The **pterygoid branches** are short vessels to the medial and lateral pterygoid muscles.

The **buccal artery** (figs. 139, 156, 165) is small and accompanies the buccal branch of the mandibu-

lar nerve downward and forward onto the buccinator muscle. It supplies branches to the muscle, the skin of the cheek, and the mucous membrane of the mouth. It anastomoses with the facial and the infraorbital arteries.

The Pterygopalatine Portion This part of the maxillary artery provides major arteries to the maxilla and maxillary teeth, to the nasal cavities, and to the palate. It terminates in the infraorbital region of the cheek. Its branches are the posterior superior alveolar, the infraorbital, the descending palatine, the artery of the pterygoid canal, the pharyngeal, and the sphenopalatine. *All of these distribute in company with branches of the maxillary division of the trigeminal nerve.*

The **posterior superior alveolar artery** (figs. 139, 165, 170) arises as the maxillary artery enters the pterygopalatine fossa. It descends on the tuberosity of the maxilla and provides branches to the molar and bicuspid teeth of the upper jaw and to the maxillary sinus by way of the alveolar canals. Other branches reach the maxillary gingivae and the buccinator muscle.

The **infraorbital artery** (figs. 139, 156) may arise in common with the posterior superior alveolar artery. It passes forward and upward and enters the orbit through the inferior orbital fissure. Proceeding forward in company with the infraorbital nerve, the artery occupies the infraorbital groove and canal and emerges on the face through the infraorbital foramen. The branches and communications in the cheek are described on p. 205. **Anterior superior alveolar branches** arise in the infraorbital canal and descend in the alveolar canals to the upper incisive and cuspid teeth and to the maxillary sinus. There is usually present a third alveolar branch, the middle superior alveolar, midway between the anterior and the posterior branches.

The **descending palatine artery** (figs. 139, 170), descends in the greater palatine canal in company with the greater palatine nerve. It reaches the oral surface of the hard palate by way of the greater palatine foramen. **Lesser palatine branches,** after emerging through the lesser palatine foramen, pass backward to the soft palate and the palatine tonsil. The **greater palatine artery** is the forward continuation of the descending palatine. From the greater palatine foramen, it passes anteriorly in a groove just internal to the alveolar process of the maxilla and distributes to the gingiva, the palatine glands, and the mucous membrane of the roof of the mouth.

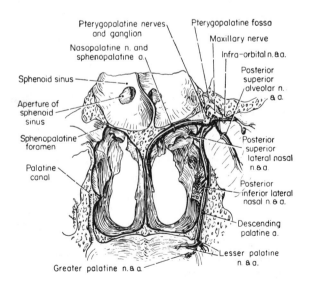

Fig. 170 The pterygopalatine fossa and branches of the maxillary nerve and artery. The portion of the skull illustrated has been sectioned vertically through the palatine canal and the direction of view is forward into the nasal cavities.

Reaching the incisive canal, the artery sends a branch through it to anastomose with the terminal branches of the sphenopalatine artery on the nasal septum.

The small **artery of the pterygoid canal** passes posteriorly through the pterygoid canal adjacent to the nerve of the same name. It distributes to the upper part of the pharynx and to the auditory tube and sends a small branch into the tympanic cavity to anastomose with other tympanic arteries.

The **pharyngeal artery** is a small branch which passes posteriorly in the pharyngeal canal with the pharyngeal branch of the maxillary nerve. It supplies the roof of the pharynx, the sphenoid sinus, and the lower part of the auditory tube.

The **sphenopalatine artery** (figs. 139, 170) enters the nasal cavity through the sphenopalatine foramen in company with the nasopalatine branch of the maxillary nerve. At its entrance it gives off **posterior lateral nasal branches** which supply the lateral nasal wall and the adjacent sinuses and anastomose with the posterior ethmoidal and anterior ethmoidal arteries. Crossing the roof of the nose, the sphenopalatine artery reaches the septum and runs forward and downward in a groove on the vomer toward the incisive canal, where it anastomoses with the termination of the greater palatine artery. The artery supplies the septum, and by **posterior septal branches,** reaches its uppermost part, where anastomoses are made with ethmoidal arteries.

The Maxillary Vein and Pterygoid Plexus (fig. 128)

The maxillary is a short vein that lies adjacent to the maxillary artery between the neck of the mandible and the sphenomandibular ligament. The vein joins the superficial temporal vein (p. 214) behind the neck of the mandible, thereby forming the retromandibular vein (p. 206). The tributaries of the maxillary vein are organized as the pterygoid plexus of veins. This plexus lies on the lateral surface of the medial pterygoid muscle and surrounds the lateral pterygoid muscle. Its tributaries, corresponding with and accompanying the branches of the maxillary artery, are the sphenopalatine, infraorbital, descending palatine, posterior superior alveolar, pharyngeal, vein of the pterygoid canal, deep temporal, pterygoid, masseteric, inferior alveolar, and middle meningeal veins. The plexus communicates with the cavernous sinus of the cranial dura mater by an emissary vein which passes through the fora-

men ovale and with the inferior ophthalmic vein by a vein traversing the inferior orbital fissure. The plexus also communicates with the facial vein by means of the deep facial vein (fig. 128) which passes between the masseter and buccinator muscles.

THE ORAL REGION

The oral region includes the mouth, the teeth and tongue, the hard and soft palate, and the region of the palatine tonsil.

THE MOUTH

The mouth consists of two parts: the vestibule and the mouth proper. The **vestibule** is the narrow interval between the lips and the teeth and gums. It communicates with the exterior through the orifice of the mouth. The **mouth** proper is limited laterally and in front by the alveolar arches and the teeth. The hard and soft palate forms its roof; the tongue and the sublingual mucosa, its floor. The mouth is continuous with the oropharynx at the faucial isthmus formed by the folds bounding the palatine tonsil.

Lips

The lips are two mobile, highly muscular folds covered externally by skin and internally by mucous membrane. The upper lip extends up to the nose medially and to the **nasolabial sulcus** laterally. The lower lip is separated from the chin by the **mentolabial sulcus.** The shallow, vertical groove of the upper lip is the **philtrum;** its bounding ridges end below in the labial tubercles of the free border of the lip. The **red margin** of the lip is an intermediate zone; it is covered by a dry and nonglandular mucous membrane continuous with the moist mucosa of the vestibule. Its mucous membrane is translucent and transmits the color of the blood in the underlying capillaries. The lips are richly vascularized and are well supplied with sensory nerves. Under the skin of the lip is a layer of fatty subcutaneous tissue continuous with that of the face. The **orbicularis oris muscle** has been described (p.\202). Many of its fibers and fibers of other facial muscles associated with it insert into the skin of the lips. In the submucosa between the muscular layer and the mucous membrane lies an almost continuous layer of **labial glands.** These small, nodular mucous glands, closely packed together, open through the mucous mem-

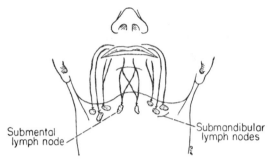

Submental lymph node

Submandibular lymph nodes

Fig. 171 The lymphatic drainage of the lips.

brane by minute individual ducts. The mucous membrane, covered by stratified squamous epithelium, is continuous with that of the gingiva. At the midline a fold of mucous membrane, the **frenulum,** connects the mucous membrane of each lip with the corresponding gingiva.

The lips receive a rich **blood supply** from the labial branches of the facial artery (p. 203). The nerves are branches of the infraorbital nerve for the upper lip (p. 199) and of the mental nerve (p. 199) for the lower lip. The **lymphatic vessels** of the upper lip and the lateral parts of the lower lip unite and pass to the submandibular lymph nodes (p. 166 & figs. 131, 171). Those from the medial part of the lower lip descend through the chin to the submental nodes (p. 166 & fig. 171), and some vessels close to the median plane may decussate.

The Cheeks

The cheeks resemble the lips in their structure but have the buccinator muscle as their principal muscular component. The fatty subcutaneous tissue of the cheeks is very loose and transmits the duct of the parotid gland. The **buccal pad of fat** (p. 203) is formed in the subcutaneous tissues. **Buccal glands,** similar in structure to but smaller than the labial glands, lie in the submucosa of the cheeks. The mucous membrane of the cheeks is reflected onto the gingiva.

The Gingivae

The gingivae consist of dense fibrous connective tissue which is firmly attached to the underlying alveolar bone and surrounds the necks of the teeth. They are covered by a smooth and vascular mucous membrane. This is continuous with the mucous layer of the lips and cheeks and is reflected into the alveolar sockets, where it is continuous with the periosteum lining these cavities.

The Teeth (figs. 172, 173, 178)

A prominent characteristic of human develop-

ment is the early and rapid increase in size and differentiation of the brain. With this goes a large cranial capacity, and the child is born with a large skull. In contrast, the face of the newborn child is small, and the subsequent growth and development of the face is approximately four times that of the cranial portion of the head. With this large postnatal change is correlated the development in Man of two sets of teeth. One set is composed of relatively few, small deciduous teeth; the later set consists of larger and more numerous permanent teeth. There are twenty deciduous teeth, ten in each jaw, and thirty-two permanent ones, sixteen in each jaw. The number of each of the types present are expressed in the following tables:

Deciduous Teeth

	Molar	Cuspid	Incisor	Incisor	Cuspid	Molar
Upper jaw	2	1	2	2	1	2
Lower jaw	2	1	2	2	1	2

Permanent Teeth

	Molar	Bicuspid	Cuspid	Incisor
Upper jaw	3	2	1	2
Lower jaw	3	2	1	2

	Incisor	Cuspid	Bicuspid	Molar
Upper jaw	2	1	2	3
Lower jaw	2	1	2	3

Form and Structure (figs. 172, 173) Each permanent tooth has a **crown** which projects above the gingival surface, a **root** embedded in the bone of the jaw, and a **neck,** the slightly constricted zone of junction of crown and root. The union of the root of a tooth and the bony wall of its socket is made by a vascular **alveolar periosteum** continuous at the margin of the tooth socket (alveolus) with the periosteum of the jaw and with the fibrous tissue of the gingiva. This is the **periodontal membrane.**

The surfaces of the tooth are designated according to the tissues to which they are related—**labial** or **buccal** are externally directed surfaces; **lingual** or **palatine** are inner surfaces. Because of the arch of the jaw, the teeth are placed next to one another in a curve, and directional terms that are strictly in accord with anatomical usage cannot be applied. With the median plane as a reference point, the adjacent surfaces of teeth are designated as medial and lateral for the incisors and cuspids and anterior and posterior for the bicuspids and molars. The biting surfaces are the occlusal surfaces.

The crowns of permanent **incisor teeth** (fig. 172) are chisel-shaped for cutting and tearing. The bevel of the

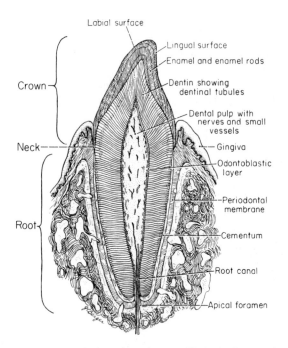

Fig. 172   A vertical section of a mandibular incisor tooth.

upper incisors is on the palatal aspect of the tooth; on the lower incisors it is on the labial surface. Thus, these teeth normally 'overbite,' with the incisor of the upper jaw in front of that of the lower. The roots of incisor teeth are single and conical. The **cuspids** (canines) are larger and stronger than the incisors, and their roots are long and form prominences on the surface of the jaw. The root of the upper cuspid is longer than that of any other tooth. The crown is large and conical and has a bluntly pointed cusp. The **bicuspid teeth** are characterized by two pyramidal cusps, a labial and a lingual, separated by a groove. The labial cusp is larger. The root is single, but its apex is frequently bifid. The **molar teeth** exhibit large crowns with three to five cusps on their occlusal surfaces. They are well adapted for grinding. The roots of the molars differ in the two jaws; there are three roots on the upper molars and only two on the lower. The roots of the late-appearing third molar (wisdom tooth) are compressed and more or less fused together. The deciduous teeth are smaller but generally resemble in form their counterparts in the permanent dentition.

The white crown of a tooth is formed of **enamel,** an extremely hard capping substance. The main mass of the tooth consists of a somewhat softer osseous tissue, the dentin (ivory) which, unlike enamel, has limited powers of regeneration. This material is hard and elastic and has a yellowish color. It has a finely striated appearance in section due to the **dental canaliculi,** fine tubules which traverse it. The **cement** is also a calcified material, disposed as a thin layer on the roots of the teeth. It increases toward the apex and is connected by

the periodontal membrane to the bone of the tooth socket. Each tooth has a central cavity which contains the **dental pulp** and which, continuous with the root canal, opens by an **apical foramen** at the apex of the root. The dental pulp is a loose areolar tissue containing a jelly-like matrix, and the numerous blood vessels, nerves, and lymphatics of the tooth.

Eruption of Teeth While individual variation is common, the teeth usually begin to appear at six months of age, and by the twenty-fourth month the deciduous teeth have all erupted. Although marked changes in the jaws and in the development of the permanent teeth are proceeding, no further visible change in the teeth takes place until the age of six years. During the sixth or seventh year, deciduous teeth begin to be replaced by the permanent incisors. This process is a continuous one, so that by the thirteenth year all deciduous teeth have usually been replaced by the permanent dentition. The third permanent molars (wisdom teeth) lag markedly behind, erupting irregularly and uncertainly between the seventeenth and twenty-first years. The chronology of dental eruption is expressed in greater detail in the following table adapted from Orban ('53):

Tooth Eruption	Deciduous Teeth		Permanent Teeth	
	Mandib-ular	Maxil-lary	Mandib-ular	Maxil-lary
	(mo.)	(mo.)	(yr.)	(yr.)
Central incisor	6	7½	6- 7	7- 8
Lateral incisor	7	9	7- 8	8- 9
Cuspid	16	18	9-10	11-12
First bicuspid			10-12	10-11
Second bicuspid			11-12	10-12
First molar	12	14	6- 7	6- 7
Second molar	20	24	11-13	12-13
Third molar			17-21	17-21

Nerves and Vessels The sensory nerves to the maxillary teeth are branches of the maxillary division of the trigeminal nerve (p. 241). The **posterior superior alveolar nerve** supplies the molar teeth; the **middle superior alveolar,** the bicuspids; and the **anterior superior alveolar** reaches the cuspid and incisor teeth. The mandibular teeth receive their innervation from the **inferior alveolar nerve** of the mandibular division of the trigeminal nerve (p. 223). The gingiva of the buccal surfaces related to the teeth of the upper jaw receive sensory branches from the **posterior superior, middle superior,** and **anterior superior alveolar nerves;** the part related to the bicuspids, cuspids, and incisors is also supplied by the infraorbital branch of the maxillary division of the trigeminal nerve. The palatine gingiva opposite the molar, bicuspid, and cuspid teeth receives an innervation from the **greater palatine branch** of the maxillary nerve (p. 241); that related to the incisive teeth is supplied by the **nasopalatine branch** (p. 241). The entire lingual surface of the mandibular gingiva is innervated by the **lingual branch** of the mandibular nerve (p. 223). The buccal portion opposite the molars is supplied by the **buccal branch** of the mandibular nerve (p. 223); that

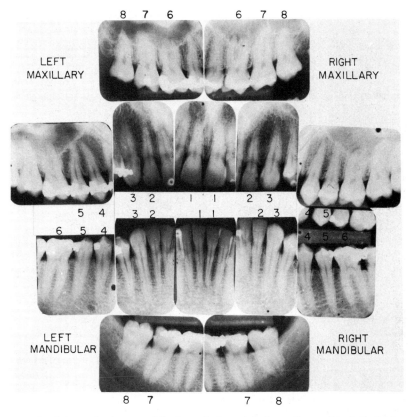

Fig. 173 A radiogram of adult teeth. 1, central incisor; 2, lateral incisor; 3, cuspid; 4, first bicuspid; 5, second bicuspid; 6, first molar; 7, second molar; 8, third molar. Courtesy. The University of Michigan Dental School.

opposite the bicuspids, cuspid, and incisors, by the **mental nerve** (p. 223).

The **arteries** of the teeth are derived from the alveolar arteries—the posterior, middle, and anterior superior alveolar branches of the maxillary artery (p. 225) for the maxillary teeth; the inferior alveolar artery (p. 225) for the mandibular teeth. These vessels provide gingival, alveolar, and dental rami. A dental ramus enters the root canal through an apical foramen and forms a rich capillary plexus within the pulp cavity. **Veins** of the same name and distribution accompany the arteries. The **lymphatic vessels** of the teeth and gingivae pass principally to the submandibular lymph nodes. A vessel from the anterior part of the mandibular gingiva occasionally reaches a submental node, and the more posterior palatal gingival drainage is to the superior deep cervical nodes.

THE TONGUE

The tongue is a highly mobile, muscular organ vital to the digestive functions of mastication, taste, and deglutition. It is also important in speech. The mobility of the tongue is enhanced by its suspension from three well-separated bilateral attachments: the mandible, the styloid process of the temporal bone, and the hyoid bone. From these bony attachments, the three principal extrinsic muscles of the tongue—genioglossus, styloglossus, and hyoglossus—enter the tongue to spread out into its substance. The muscular meshwork of the tongue is also contributed to by its intrinsic muscles, and the whole is covered by a mucous membrane containing papillae of various forms, taste buds, and the lingual tonsil. At its root the tongue is connected with the palate by the palatoglossal arches, with the pharynx through the attachments of the superior constrictor muscles, and with the epiglottis by the glossoepiglottic folds. The tip, sides, and dorsum of the tongue and part of its inferior surface are free. The tongue is partially separated into symmetrical halves by a septum of areolar tissue. On the dorsum the midline is marked by a depression, the **median sulcus.**

The tongue is composed of two parts, an anterior two-thirds and a posterior one-third, *which differ developmentally, structurally, and in nerve supply and appearance.* The anterior **oral part** is separated from the posterior **pharyngeal part** by a V-shaped **sulcus terminalis** on the dorsum of the tongue. The apex of the V is directed posteriorly and ends in a median pit, the **foramen cecum** of the tongue. This marks the site of the diverticulum from which the thyroid gland developed and may still, in the adult, retain its continuity with the upper end of the thyroglossal duct (p. 157).

The Mucous Membrane of the Tongue (fig. 174)

The mucous membrane of the dorsum is thin over the anterior part of the tongue. Here it is intimately attached to the underlying muscular tissue and contains numerous minute **lingual papillae.** Over the posterior one-third of the tongue, the mucous membrane is thick and freely movable, and the submucosa contains a large collection of lymphoid nodules forming the **lingual tonsil.** The thin mucous membrane of the under surface of the tongue reflects onto the lingual gingivae and onto the floor of the mouth where, in the median line, it is elevated into a distinct vertical fold, the **frenulum** of the tongue. The **papillae of the tongue** are projections of the

corium. They characterize the mucous membrane of the anterior two-thirds of the dorsum, imparting the characteristic roughness to the tongue. There are vallate, fungiform, and filiform papillae.

The **vallate**, or circumvallate, **papillae** (fig. 174) are large and only eight to twelve in number. They lie anterior to and correspond in arrangement to the V of the sulcus terminalis. Each flattened papilla is from 1 to 2 mm. wide and is surrounded by a moat-like circular sulcus. Taste buds are enclosed in both walls of the circular sulcus. **Fungiform papillae** are numerous on the sides and tip of the tongue but are irregularly spaced on the dorsum. They are globular and, their covering epithelium being thin and their blood supply rich, are easily recognized by their bright red color. **Filiform papillae** are minute, slender, conical papillae which cover the anterior two-thirds of the dorsum. They are arranged in rows which parallel the V of the sulcus terminalis but approach a transverse orientation at the tip of the tongue. Their epithelium is scaly and, in some animals, accounts for the raspy surface character of the tongue. **Foliate papillae,** found on the sides of the tongue in certain animals, are rudimentary in Man.

The Lingual Tonsil Behind the sulcus terminalis, the dorsum of the tongue is warty in appearance due to the projection of underlying nodular masses of lymphoid follicles (fig. 174). Each nodule surrounds a crypt which opens at the center of the nodule. Collectively, the nodules form the lingual tonsil, which is only indefinitely separated from the lower pole of the palatine tonsil. **Mucous glands** similar to the labial and buccal glands occur on the tongue. The **anterior lingual glands** are clusters of glands on either side of the frenulum. **Serous glands** are numerous in the region of the vallate papillae.

The Muscles of the Tongue (figs. 175, 177, 178)

The muscles of the tongue are the extrinsic muscles mentioned above and the intrinsic muscles—superior longitudinal, inferior longitudinal, transverse, and vertical. The muscles of the two sides are separated by the **lingual septum.**

The **hyoglossus,** as a muscle of the submandibular triangle, has been described on p. 171. Its fibers enter the side of the tongue between the styloglossus and inferior longitudinal muscles. The **chondroglossus muscle,** about 2 cm. long, is a separated portion of the hyoglossus. It arises from the medial side and base of the lesser horn and adjacent area of the body of the hyoid bone and enters the tongue between the hyoglossus and genioglossus muscles.

The **genioglossus** is a fan-shaped muscle lying vertically next to the median plane. Arising from the

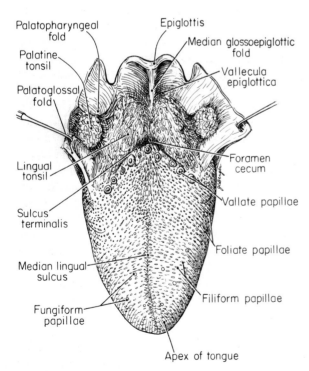

Palatopharyngeal fold
Palatine tonsil
Palatoglossal fold
Lingual tonsil
Sulcus terminalis
Median lingual sulcus
Fungiform papillae

Epiglottis
Median glossoepiglottic fold
Vallecula epiglottica
Foramen cecum
Vallate papillae
Foliate papillae
Filiform papillae
Apex of tongue

Fig. 174   The dorsum of the tongue and the palatine tonsil.

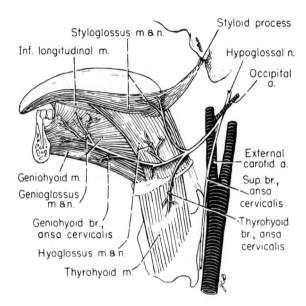

Fig. 175   **The muscles of the tongue and the muscular branches of the hypoglossal nerve.**

mental spine on the inner surface of the symphysis menti, the muscle diverges posteriorly to enter the whole length of the under surface of the tongue. Its superior fibers reach the tip of the tongue; the middle fibers, the dorsum; and the inferior fibers curve back to insert into the medial part of the hyoid bone.

The **styloglossus muscle** arises from the styloid process near its tip and from the stylomandibular ligament. It runs downward, forward, and medialward, its fibers diverging, and divides on the side of the tongue into two parts. The larger longitudinal fasciculus runs superficially along the lateral margin of the tongue to its tip. The smaller inferior oblique fasciculus penetrates the fibers of the hyoglossus to pass medialward into the depths of the tongue.

The intrinsic muscles (figs. 177, 178) have their attachments entirely within the tongue. The **superior longitudinal** is a superficial layer lying immediately under the mucous membrane of the dorsum from the base of the tongue to its apex. The **inferior longitudinal muscle** extends from the base to the apex on the inferior surface of the tongue between the genioglossus and hyoglossus. The **transverse lingual muscle** consists of bundles that pass transversely between the superior and inferior longitudinal layers. The fibers of this muscle arise from the septum and end in the submucous tissue of the side of the tongue. The fibers of the **vertical muscle**

are located at the borders of the tongue toward its tip. They pass from the superior to the inferior surface.

Movements of the Tongue Analysis of the simpler movements of the tongue may be made on the basis of the traction of the individual muscles described, whereas the understanding of its more complex actions involves various combinations of extrinsic and intrinsic muscles and differing groups on the two sides. The middle and inferior parts of the genioglossus are primarily concerned in protrusion of the tongue. The superior fibers of this muscle draw the tip back and down into the mouth. In addition, the middle fibers of the genioglossus depress the midregion of the tongue, producing a hollow in its dorsum for the reception of food. The styloglossus muscle retracts and elevates the whole tongue. The palatoglossus muscle, described with the palate, assists in elevation of the tongue. The hyoglossus muscles flatten the tongue and draw down its sides. The intrinsic muscles assist in the various actions mentioned and are especially concerned in deviations of the tongue to the side. In eating, the tongue forms itself into a trough-like receptacle, conducts the food between the teeth for their tearing and crushing actions, and prevents food from falling to the floor of the mouth. Finally it makes firm pressure against the palate above and forces the mixed food and saliva into the oropharynx. The tongue is an exceedingly mobile organ, and both its extrinsic and intrinsic muscles combine in all its actions. The nerve of both the extrinsic and intrinsic muscles of the tongue is the hypoglossal (p. 197 & fig. 175).

Vessels and Nerves The **lingual artery** (figs. 139, 180) provides the blood supply of the tongue. This artery (p. 160) arises from the external carotid opposite the tip of the greater horn of the hyoid bone. It ascends through the upper part of the carotid triangle and turns forward deep to the hyoglossus muscle. In this part of its course it is crossed by the hypoglossal nerve, the posterior belly of the digastric muscle, and the stylohyoid muscle. Turning upward under the hyoglossus muscle, it becomes the deep lingual artery and runs toward the tip of the tongue between the fibers of the genioglossus medially and the inferior longitudinal muscle laterally.

The lingual artery sends a small **hyoid branch** along the upper border of the hyoid bone. The **dorsal lingual branches** arise deep to the hyoglossus. They ascend to the back of the tongue and distribute to the tonsil, soft

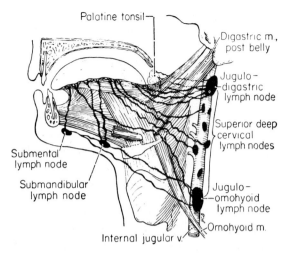

Fig. 176   The lymphatic drainage of the tongue.

palate, and epiglottis. A **sublingual artery** continues forward from the anterior border of the hyoglossus muscle as the 'lingual artery turns up toward the tongue. The sublingual artery supplies the sublingual gland, the mylohyoid muscle, and the mucous membrane of the floor of the mouth. The terminal **deep lingual artery** lies 5 mm. or less from the inferior surface of the tongue. It is connected near the apex by a branch with its fellow of the opposite side; elsewhere the lingual septum limits communication from side to side.

The **lingual veins (p 162)** accompany the lingual artery as two venae comitantes, receive the dorsal lingual vein, and terminate in the internal jugular vein. The **vena comitans nervi hypglossi** (figs. 135, 139) is usually larger than the lingual. Beginning below the tip of the tongue, the vena comitans passes backward on the hyoglossus muscle with the hypoglossal nerve and empties into the facial vein (p. 204).

The **lymphatics of the tongue** (fig. 176) arise in a rich submucosal lymphatic plexus and form a number of channels which drain to the entire series of superior deep cervical lymph nodes, the more anterior channels to the lower nodes and the more posterior to the higher.

The channels are divided into four groups. (1) Several **apical channels** on each side descend, one to the jugulo-omohyoid node, which is the lowest member of the superior deep cervical chain. Other channels end in the submental nodes (p. 166). (2) Four or six **marginal vessels** pass, either superficial or deep to the sublingual gland, downward and backward to the submandibular nodes and to the superior deep cervical nodes in the area between the digastric muscle and the omohyoid. (3) The **basal vessels** drain the posterior one-third of the tongue. They are large, are three or four in number, and end principally in the uppermost node of the superior deep cervical group, the jugulodigastric node. (4) The

central, or median, **vessels** descend between the genioglossus muscles, are joined by lymphatics from the deep part of the tongue, and turn lateralward to the superior deep cervical nodes below the digastric muscle. The more median of the lymphatic channels of the tongue decussate and terminate in nodes of the opposite side.

The **nerves of the tongue** are the fifth, seventh, ninth, tenth, and twelfth cranial nerves. They mediate general sensation and taste and supply its musculature; their distribution and function are given in the following table:

	General Sensation	Taste	Motor
Anterior two-thirds	Lingual nerve (V)	Chorda tympani (VII)	Hypoglossal (XII)—extrinsic and intrinsic muscles
Posterior one-third	Glossopharyngeal (IX)	Glossopharyngeal (IX) (includes vallate papillae)	
Epiglottic region of tongue	Superior laryngeal (X), internal branch	Superior laryngeal (X), internal branch	

The more complete description of these nerves is to be found as follows: the lingual nerve, p. 223; the chorda tympani, p. 173; the glossopharyngeal nerve, p. 195; the internal branch of the superior laryngeal nerve, p. 164 ; and the hypoglossal nerve, p. 197.

The Floor of the Mouth

Under the tongue the mucous membrane covering its root is reflected lateralward into continuity

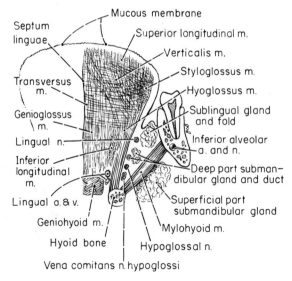

Fig. 177   A partial cross section of the tongue, mandible, floor of mouth, and submandibular gland.

with the gingiva of the mandibular arch. Support of the mucosa and the underlying structures is provided by the geniohyoid and especially by the mylohyoid muscles (oral diaphragm) of the submandibular triangle (p. 171). A median fold of mucous membrane forms the **frenulum of the tongue,** on either side of which lies the **sublingual caruncle.** This is the site of termination of the submandibular duct. The sublingual gland lies beneath the mucosa of the floor of the mouth and produces the **sublingual fold** (fig. 177), a longitudinal elevation at either side of the root of the tongue.

Clinical Note Incisions through the mucous membrane of the floor of the mouth lead into the paralingual space which is continuous with the space of the submandibular triangle. Such incisions give access to the sublingual gland, the submandibular duct, and the lingual nerve (fig. 177). Infiltration of the lingual nerve with an anesthetic can be done at the posterior

end of the mylohoid line of the mandible where the nerve passes between the superior pharyngeal constrictor and the mylohyoid muscles.

THE PALATE

The palate forms the arched roof of the mouth and consists of a hard and a soft palate. The **hard palate** is composed of the **palatine processes of the maxillae** and, united posteriorly to them, the horizontal portions of the **palatine bones.** The two sides unite in the midline to form a complete bony partition between the nasal and oral cavities. In the median plane of the hard palate, anteriorly, is the **incisive canal;** posteriorly, at the lateral extremity of the horizontal part of the palatine bone, lies the **greater palatine foramen.** The bony palate is covered by a dense **mucoperiosteum** composed of mucous

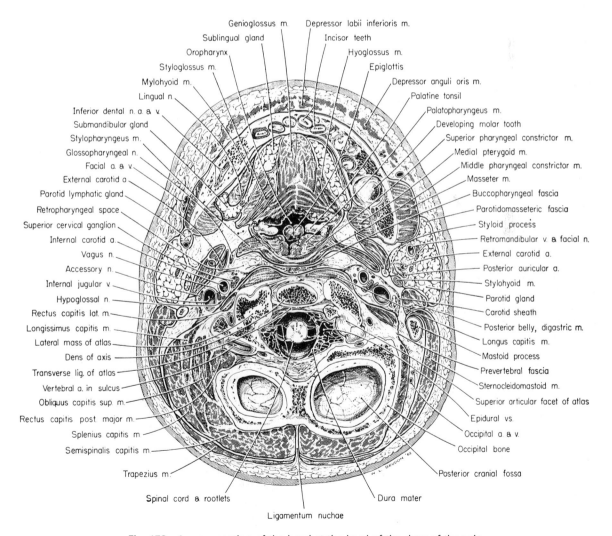

Fig. 178 A cross section of the head at the level of the dens of the axis.

membrane fused to the underlying periosteum. This mucoperiosteum contains, in the submucosa of the posterior palatal region, a layer of mucous palatine glands which empty through the mucosal surface. A midline ridge, the **palatine raphe,** ends anteriorly in the **incisive papilla** in the region of the incisive foramen. Transverse corrugations also mark the surface of the palate anteriorly.

The **soft palate** is a fold of mucous membrane enclosing an aponeurosis, muscle fibers, vessels and nerves, and mucous glands. The soft palate extends posteriorly from the hard palate, and its palatine aponeurosis is attached to the posterior border of the horizontal portions of the palatine bones. At its lateral margins the soft palate blends with the pharyngeal wall and gives rise to two mucous folds which overlie muscular bundles. These are the **palatoglossal fold** and muscle anteriorly and the **palatopharyngeal fold** and muscle posteriorly. The palatine tonsil is lodged in the triangular space between these folds (fig. 174). The soft palate arches posteriorly and inferiorly, and its under surface faces toward the mouth; its superior and posterior surface faces the pharynx. A soft conical projection at the middle of the free dependent border of the palate is the **uvula.** The **palatine aponeurosis** is a thin fibrous membrane continuous with the pharyngobasilar fascia of the pharyngeal wall. It is best developed in the anterior portion of the soft palate, where it is reinforced by the expanded tendon of the

tensor veli palatini muscle. The posterior part of the palate is freely movable and highly muscular, containing bundles of the levator veli palatini, palatoglossus, and palatopharyngeus muscles, and the musculus uvulae. Mucous glands are very abundant in the soft palate and are continuous with those of the posterior part of the hard palate. The **uvula** consists of mucous glands, areolar tissue, and the termination of the musculus uvulae, covered by mucous membrane.

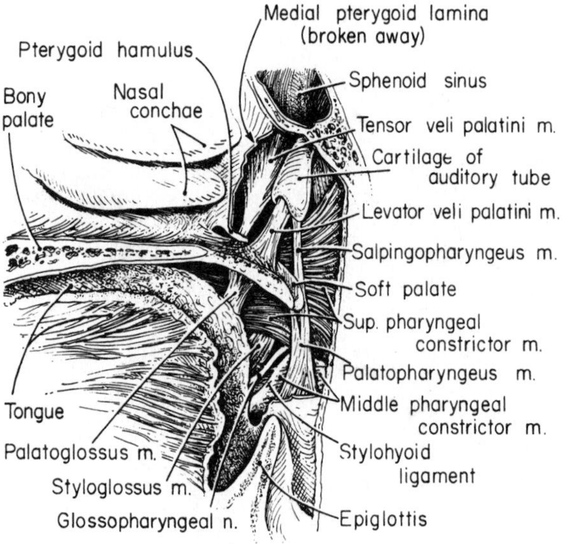

The **muscles of the palate** are the following:

Levator veli palatini	Palatoglossus
Tensor veli palatini	Palatopharyngeus
Musculus uvulae	

The **levator veli palatini muscle** (figs. 151, 163, 179) is a pencil-thick muscle which elevates the mucosa of the pharyngeal orifice of the auditory tube. Its origin is the under surface of the apex of the petrous portion of the temporal bone and the medial side of the posterior part of the cartilage of the auditory tube. Entering the pharynx above the upper concave border of the superior pharyngeal constrictor, the fibers of the muscle spread out into the soft palate. Most of them decussate with similar fibers from the other side, but the more anterior fibers end in the palatine aponeurosis.

The **tensor veli palatini muscle** (figs. 151, 163, 179) lies lateral and anterior to the levator. It arises from the scaphoid fossa at the base of the medial pterygoid plate, the spine of the sphenoid, and the lateral side of the membranous and cartilaginous wall of the auditory tube. The muscle fibers descend vertically between the medial pterygoid plate and the medial pterygoid muscle and converge on a tendon which winds around the pterygoid hamulus from which it is separated by a small bursa. The tendon, now directed horizontally medialward, inserts into the palatine aponeurosis and into the surface of the horizontal part of the palatine bone behind its transverse ridge.

The **musculus uvulae** arises from the posterior nasal spine of the palatine bones and the palatine aponeurosis. Consisting of two parallel bundles on either side of the median line, the muscle ends in the uvula.

The **palatoglossus muscle** (fig. 179) arises from the oral surface of the palatine aponeurosis and passes downward, forward, and lateralward, forming the substance of the palatoglossal arch, the anterior pillar of the tonsillar fossa. Its insertion is into the side of the tongue, some of its fibers spreading

Fig. 179   The muscles of the palate and the bed of the palatine tonsil.

over the dorsum and others blending with fibers of the transversus linguae muscle.

The **palatopharyngeus muscle** (figs. 178, 179) forms the posterior pillar of the tonsillar fossa. It assists in forming a thin, inner, longitudinal layer of muscle in the pharynx, with which it is described (p. 194).

Innervations The tensor veli palatini muscle is supplied by the mandibular division of the trigeminal nerve through the branch to the medial pterygoid muscle (p. 222). All others of the palatal muscles are supplied by the contribution of the vagus nerve to the pharyngeal plexus (p. 194).

Actions The essential function of the soft palate is to close off the naso-oral communication in swallowing (p. 194), in sucking, and during oral speech. Since the soft palate normally hangs downward into the pharynx, the action of both the levator and tensor muscles is to elevate it and to carry it somewhat posteriorly into contact with the pharyngeal wall, which is simultaneously drawn forward in this region by the upper fibers of the palatopharyngeus muscle (Passavant's ridge). The uvula appears to have no special function in swallowing but fills in the deepest part of the concavity of the pharynx. It has a minor function in speech in the production of the uvular 'r.' The palatoglossus muscle assists in the elevation and retraction of the tongue and in drawing the soft palate inferiorly. The palatopharyngeus and salpingopharyngeus muscles assist in elevation of the larynx and pharynx in the first stage of swallowing (p. 194). The contraction of the tensor veli palatini muscle during swallowing tenses the palate against the arching action of the tongue. It also tends to open the auditory tube. The tensor appears to be essentially inactive during oral speech at which time the levator veli palatini muscle is especially active.

Blood Vessels and Nerves The principal artery of the palate is the greater palatine branch of the maxillary artery (p. 225) which enters the palate by way of the greater palatine foramen. It runs foward, adjacent to the alveolar margin of the palate, to the incisive canal, where it anastomoses with a terminal branch of the sphenopalatine artery. The soft palate is supplied by small branches of the lesser palatine artery which emerge through the lesser palatine foramina. It is also reached by small branches of the ascending palatine branch of the facial artery (p. 160) and by the palatine branch of the ascending pharyngeal artery (p. 161). The **veins** of the palate are like the arteries and are tributary to the pterygoid plexus (p. 226). The sensory nerves of the palate are furnished by greater and lesser palatine branches of the maxillary division of the tri-

geminal nerve (p. 241). The greater palatine nerve anastomoses in the region of the incisive canal with the nasopalatine nerve (p. 241). An autonomic innervation to the glands and blood vessels of the palate is supplied over the same nerves by way of the greater petrosal and deep petrosal nerves (p. 242). The **lymphatic drainage** of the palate is by channels which pass lateral to the palatine tonsil to the superior deep cervical nodes.

Clinical Note Cleft palate occurs about once in 2500 births. It may or may not be associated with cleft lip. Fundamentally, it is due to failure of the lateral palatine processes to fuse with one another or failure of one or both of the lateral palatine processes to fuse with the primary palate. These fusions normally occur prior to the tenth week of development in man.

The Palatine Tonsil (figs. 150, 153, 174, 176, 178, 180)

The palatine tonsil is a collection of lymphatic tissue underlying the mucous membrane between the palatine arches. These are the palatoglossal and palatopharyngeal folds of mucous membrane and their corresponding muscles. The **palatoglossal arch** extends from the lower surface of the soft palate, near the root of the uvula, lateralward and downward to the lateral margin of the tongue, just behind its midregion. The **palatopharyngeal arch** arises at the posterior edge of the soft palate and ends inferiorly on the side wall of the pharynx. The oval palatine tonsil, about 2 cm. in its greatest dimension, bulges between the arches. It does not completely fill the space between them, for a small **supratonsillar fossa** exists at the uppermost extremity of the area. Part of the tonsillar mass is frequently embedded under the triangular fold of the

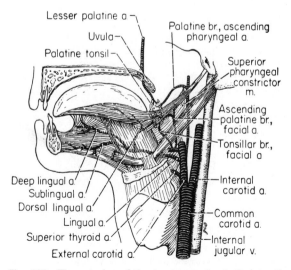

Fig. 180 The arteries of the tongue and palatine tonsil.

anterior arch. The free surface of the tonsil is marked by the openings of narrow **tonsillar crypts.** The lateral, or deep, surface of the tonsil is covered by a thin but firm fibrous capsule continuous with the pharyngobasilar fascia and is in contact externally with the superior pharyngeal constrictor muscle. External to the lower one-third of the tonsil, the styloglossus muscle descends toward the back of the tongue, having pierced the pharynx between the borders of the superior and middle constrictor muscles. Running inferior to the styloglossus muscle, the glossopharyngeal nerve reaches the base of the tongue (fig. 153).

The close spatial relationship of the palatine, lingual, and pharyngeal tonsils has been emphasized in the designation of a **tonsillar ring** at the oropharyngeal isthmus. The ring lies at the cephalic limit of endodermally derived epithelium and is presumed to have a protective significance.

Vessels and Nerves The tonsil is supplied by five small arterial branches (fig. 180). The facial artery (p. 160) sends an **ascending palatine branch** over the upper border of the superior pharyngeal constrictor and a **tonsillar branch** through the muscle into the tonsil. A **palatine branch** of the ascending pharyngeal artery (p. 161) also passes over the upper border of the superior constrictor muscle. The **dorsalis lingual branch** of the lingual artery (p. 231) provides a tonsillar branch at the antero-inferior aspect of the gland, and terminals of the **descending palatine branch** of the maxillary artery reach the gland from above. The **veins** form a plexus on the lateral aspect of the tonsil from which vessels empty into the pharyngeal plexus and into the facial vein. The **lymphatic vessels** of the tonsil (fig. 176) are numerous. They penetrate the pharyngeal wall and pass to the superior deep cervical lymph nodes. These channels terminate principally in the large node at the level of the posterior belly of the digastric muscle, the **jugulodigastric node,** which is for this reason frequently designated as the **tonsillar node.** The **nerves** of the tonsil are the **middle** and **posterior palatine branches** of the maxillary division of the trigeminal together with the **tonsillar branches** of the glossopharyngeal nerve. These nerves form a circular network, the **tonsillar plexus,** from which the tissue is supplied.

THE NASAL REGION

The External Nose

The skin of the nose is thin and movable over its root and upper portion. On the cartilaginous portion the skin is thick, adheres to the underlying cartilage, and accommodates many large sebaceous glands. The two elliptical **nostrils** are separated from each other by the movable portion of the septum. The margins of the nostrils exhibit a number of stiff hairs which trap particulate matter carried into the nose on the air stream. The nose has a bony substrate in its upper portion, which is provided by the two nasal bones meeting in the median line and making contact laterally with the frontal processes of the maxillae. The lower portion of the nose is supported by cartilages.

The small arteries of the nose are the lateral nasal branch of the facial artery, the nasal branch of the infraorbital artery, and the dorsal nasal branch of the ophthalmic artery. They are described under the superficial blood vessels of the face (p. 204). The venous drainage corresponds to the arterial distribution, the principal channel being the facial vein. Lymphatic channels from the nose drain to the submandibular nodes, by way of the facial nodes (p. 166), and to the superficial parotid group of lymphatic nodes (p. 166). The sensory nerves of the nose are branches of the infraorbital nerve (p. 199) and of the infratrochlear and external nasal branches of the nasociliary nerve (p. 199). Subcutaneous **muscles of the nose** are described on p. 201. They are supplied by buccal branches of the facial nerve.

The Cartilages of the Nose (fig. 181)

The cartilaginous framework of the lower part of the nose is provided by five major cartilages: two lateral nasal cartilages, two alar cartilages, and the septal cartilage. Several smaller pieces are found at the junction of the alar cartilages and the frontal processes of the maxillae at the sides of the nose. The cartilages are connected to each other and to the bones by tough fibrous connective tissue.

The **lateral cartilage** of the nose is flat and triangular and is located below the inferior border of the nasal bone. Its anterior border is continuous with the septal cartilage at the midline; its inferior border is connected by fibrous tissue with the alar cartilage. The **alar cartilage,** thin and flexible, lies in the ala of the nose. It is folded on itself so that it forms both the lateral and medial borders of the nostril. The oval lateral portion of the cartilage turns a sharp angle (about 30°) into the smaller, hook-shaped medial portion. The medial parts of the two alar cartilages are loosely connected to one another at the median line to form, with the thickened skin and subcutaneous tissue, the mobile portion of the septum. The **septal cartilage** forms the anterior portion of the septum of the nose. Irregularly triangular in form, it abuts below against the inclined edge of the vomer and behind against the edge of the perpendicular plate of the ethmoid bone. Its anterior margin, thickest above, is connected to the nasal bones and is continuous with the anterior margins of the lateral cartilages. Below, fibrous tissue connects the margin of the septal cartilage to the medial portion of the alar cartilages.

THE NASAL CAVITY (fig. 184)

The nasal cavity is entered through the nostrils and opens posteriorly into the nasopharynx through two choanae. The **choanae** (fig. 170), at either side of the posterior border of the nasal septum, measure approximately 2.5 cm. vertically and 1.5 cm. across. The general nasal cavity is divided into two individual nasal cavities by the septum. The medial wall of each nasal cavity is smooth and flat, whereas the lateral wall is irregular. Inside the aperture of the nostril is a dilatation, the **vestibule,** bounded by the medial and lateral portions of the alar cartilage. It is lined by skin which, in its lower part, gives rise to stiff hairs. The upper limit of the vestibule is formed by the inferior border of the lateral cartilage.

The **roof** of each nasal cavity is very narrow except posteriorly. The **floor** (fig. 181) is gently curved from side to side and almost horizontal anteroposteriorly. The bony floor is formed by the palatine process of the maxilla in its anterior three-fourths and by the horizontal portion of the palatine bone in its posterior one-fourth. The **medial wall** of the nasal cavity, the septum (fig. 181), is partly bony and partly cartilaginous. The upper part of the bony septum is formed chiefly by the perpendicular plate of the ethmoid and slightly by the crest of the nasal bone in front and by the crest of the sphenoid behind. The lower part is composed of the vomer and the nasal crests of the maxilla and palatine bones. The septal cartilage, previously

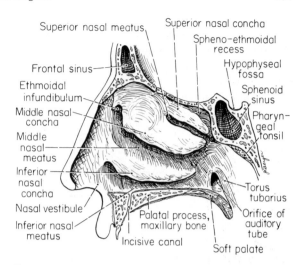

Fig. 182   The conchae and meatuses of the lateral nasal wall.

described, lies antero-inferiorly and fits into the gap between the diverging borders of the bones. In the adult, the nasal septum is frequently deviated to one or the other side, such deviations usually involving the septum and the ethmoid bone. Spurs of bone infrequently develop in the septum along the inclined edge of the vomer.

The **lateral wall** of the nasal cavity (figs. 182, 183, 184, 214) is made irregular by three projections, the superior, middle, and inferior **conchae.** Of these overhanging, scroll-like projections, the superior and middle are processes of the ethmoid bone, whereas the inferior concha is a separate bone of the skull. The conchae, almost reaching the septum, subdivide each nasal cavity into a series of groove-like passageways, the **meatuses** of the lateral nasal wall. The conchae provide additional areas of highly glandular and vascular nasal mucosa which modify the humidity and temperature of the inspired air. The meatuses are designated by the conchae above them; thus the superior meatus is inferior to the superior concha, and the middle and inferior meatuses are below the middle and inferior concha. The deep groove of the lateral wall above and behind the superior concha into which the sphenoid sinus empties is the **sphenoethmoidal recess.** The **superior meatus** is a short oblique passage above the posterior half of the middle concha. The opening for the posterior ethmoidal air cells is in the superior meatus. The large **middle meatus** is displaced into the cavity under the overhang of the middle concha due to the bulk of the middle ethmoidal air cells. The bulging mass is the **bulla ethmoidalis** (fig. 183), and the opening for the middle

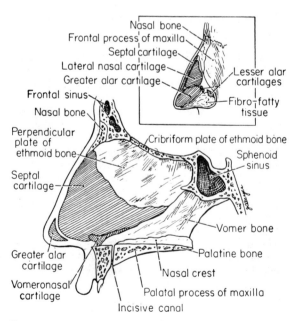

Fig. 181   The bones and cartilage of the septum of the nose. Inset—the lateral cartilage of the nose.

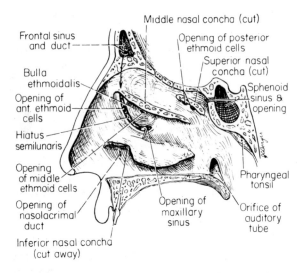

Middle nasal concha (cut)

Frontal sinus and duct

Opening of posterior ethmoid cells

Superior nasal concha (cut)

Bulla ethmoidalis

Opening of ant. ethmoid cells

Sphenoid sinus & opening

Hiatus semilunaris

Opening of middle ethmoid cells

Opening of nasolacrimal duct

Opening of maxillary sinus

Pharyngeal tonsil

Orifice of auditory tube

Inferior nasal concha (cut away)

Fig. 183 The openings of the paranasal sinuses and other apertures of the lateral nasal wall.

ethmoidal air cells is usually in its center. The uncinate process of the ethmoid bone forms a curved ridge inferior to the bulla, and the intervening deep groove anterior and inferior to the bulla is the **hiatus semilunaris.** Into a deep anterior portion of the hiatus drains the anterior ethmoidal air cells. Directly below the bulla and toward the posterior end of the hiatus semilunaris lies the **ostium of the maxillary sinus.** The **frontonasal duct** leads from the frontal sinus into the anterior end of the middle meatus or into the anterior portion of the hiatus. The **inferior meatus** of the lateral wall is placed between the inferior concha and the floor of the nasal cavity. It receives the slit-like termination of the nasolacrimal duct about 1 cm. behind the anterior end of the concha (fig. 183).

The Mucous Membrane

Mucous membrane lines the entire nasal cavity except the vestibule and is firmly bound to the periosteum and perichondrium of the supporting structures. It is continuous with the lining of all the chambers with which the nasal cavities communicate: the nasopharynx behind, the paranasal sinuses above and laterally, and the lacrimal sac and conjunctiva of the orbit. The mucous membrane is thick and very vascular over the conchae but is thin in the meatuses and in the sinuses. It contains a layer of glands with both mucous and serous alveoli represented.

Nerves (figs. 185, 186) The lining of the roof and adjacent surfaces of the septum and of the superior nasal concha and sphenoethmoidal recess is the ol-

factory epithelium. From the bipolar **olfactory cells** of this epithelium, central processes unite in fasciculi which pass through the cribriform plate to enter the under surface of the olfactory bulb of the brain. The **nerves of general sensation** in the nasal cavity are terminals of the nasociliary branch of the ophthalmic division of the trigeminal nerve and branches of its maxillary division. The nasociliary branch of the ophthalmic nerve (p. 252) distributes fibers to the anterior superior areas of the septum and lateral wall. The branches of the maxillary division of the trigeminal nerve (p. 240) are the nasopalatine to the septum, the posterior superior lateral nasal and posterior inferior lateral nasal to the conchae and meatuses of the lateral wall, and the anterior superior alveolar nerve to the floor and inferior meatus.

Vessels (figs. 185, 186) The arterial supply to the nasal cavity is principally from the sphenopalatine branch of the maxillary artery (p. 226). Its **posterior lateral nasal branches** ramify over the nasal conchae and lateral wall and provide twigs to the ethmoidal, frontal, and maxillary sinuses. **Posterior septal branches** of the sphenopalatine artery supply the upper part of the septum and anastomose with the ethmoidal arteries. The remainder of the septum is supplied by the continuation of the sphenopalatine artery. **Anterior and posterior ethmoidal branches** of the ophthalmic artery (p. 252) supply the anterior portions of the superior and middle conchae and meatuses and an adjacent area of the septum, as well as providing twigs to the frontal and ethmoid sinuses. The **veins** of the nasal cavity arise from a cavernous or mesh-like plexus beneath the mucous membrane, which is especially well marked over the middle and inferior conchae and the inferior portion of the septum. The veins drain posteriorly by the sphenopalatine vein to the pterygoid plexus; anteriorly, by

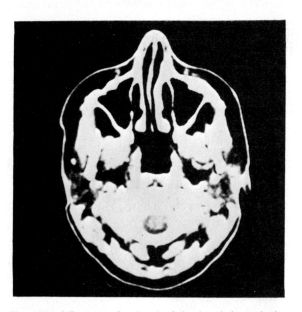

Fig. 184 C.T. scan of a level of the head through the nasal cavity, middle meatus, and maxillary sinus. (Courtesy of the Radiology Department of the University of Michigan Medical School.)

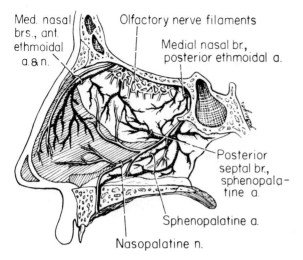

Fig. 185   The nerves and arteries of the nasal septum.

radicles which join the facial vein; and superiorly, by ethmoidal veins to the ophthalmic veins. One vein, draining superiorly, traverses the foramen cecum of the cribriform plate and empties into the beginning of the superior sagittal sinus of the cranium (p. 277). **Lymphatic channels** from the anterior nasal areas emerge between the nasal cartilages or below their free borders and end in facial and submandibular nodes. The lymphatic vessels of the remaining greater portion of the nasal walls drain posteriorly to a plexus in front of the torus tubarius and from there empty into the superior deep cervical node group (principally the jugulodigastric) and into the retropharyngeal nodes. Lymphatic drainge from the maxillary and ethmoid sinuses follows the same routes as from the posterior nasal areas.

The Paranasal Sinuses

The paranasal sinuses are pneumatic areas in the frontal, ethmoid, sphenoid, and maxillary bones, which are lined by mucous membrane continuous with that of the nasal cavity. Beginning in the fourth fetal month, they develop as evaginations of the nasal mucosa which invade the bones surrounding the nasal cavities. Absorption of bone around the invading mucosal sacs establishes the pneumatized areas known as the paranasal sinuses; their areas of evagination remain as the apertures by which communication is retained between the nasal cavity and the sinuses. Only the maxillary sinus exhibits a definite cavity at birth. Although the frontal and sphenoid sinuses are radiologically visible at six or seven years of age, the other sinuses are generally rudimentary until puberty after which time they develop adult proportions. The principal value of the paranasal sinuses appears to be as resonating chambers for the voice and as a means of lightening the bones of the head.

The Frontal Sinuses (figs. 181, 187, 188, 207, 214) These are located on either side of the median line within the frontal bone and behind the superciliary arches. They are rarely symmetrical, and the septum which separates them usually deviates to one side or the other. In a sagittal section of the cranium, the frontal sinus is roughly triangular, its greatest vertical dimension (about 2.5 cm.) being behind and above the superciliary arches. Its greatest anteroposterior depth of 2 cm. or less lies in the orbital process of the frontal bone. The frontal sinus communicates with the nasal cavity (fig. 183) by way of the frontonasal duct which empties into the anterior part of the middle meatus.

The Ethmoidal Air Cells (figs. 187, 188, 196) The ethmoidal cells are thin-walled spaces within the ethmoidal labyrinth of the lateral mass of the ethmoid bone, their peripheral walls being completed by the frontal, lacrimal, sphenoid, palatine, and maxillary bones. The number of cells varies from three to eighteen, the cells being larger when their number is smaller. The cells are subdivided into three groups, each group having a typical site of communication with the nasal cavity (fig. 183). The anterior group of ethmoidal cells opens into the depths of the anterior part of the hiatus semilunaris. The bulla ethmoidalis represents the bulge of bone over the middle group of ethmoidal cells, and the opening to them is usually on its summit. The posterior ethmoidal cells open into the superior meatus of the nose.

The Sphenoidal Sinuses (figs. 170, 187, 196, 211) These are spaces of approximately cuboidal form, having dimensions of about 2 cm., which are

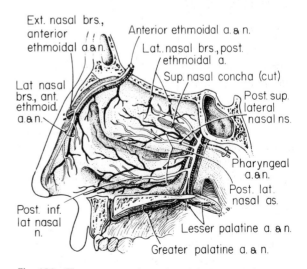

Fig. 186   The nerves and arteries of the lateral nasal wall.

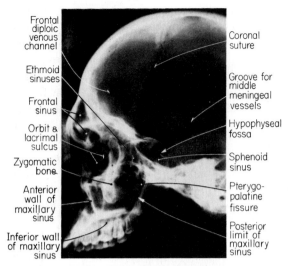

Frontal diploic venous channel
Ethmoid sinuses
Frontal sinus
Orbit & lacrimal sulcus
Zygomatic bone
Anterior wall of maxillary sinus
Inferior wall of maxillary sinus

Coronal suture
Groove for middle meningeal vessels
Hypophyseal fossa
Sphenoid sinus
Pterygo-palatine fissure
Posterior limit of maxillary sinus

Fig. 187 A lateral radiogram of the anterior portion of the skull to show the sinuses of the head.

located on either side of the median line in the body of the sphenoid bone. The intervening septum is often irregular and the cavities of the two sides asymmetrical. There may be extensions into the great wing or pterygoid process of the sphenoid bone or into the basilar portion of the occipital bone. Each sinus opens into the sphenoethmoidal recess by an aperture that begins in its upper anterior wall (fig. 183).

The pharyngeal branch of the maxillary artery and the corresponding branch of the maxillary division of the trigeminal nerve reach the sphenoid sinus. Each sinus is also supplied by the orbital branch of the maxillary division, by the posterior ethmoidal branch of the ophthalmic division of the trigeminal, and by the nerve of the pterygoid canal.

The Maxillary Sinuses (figs. 187, 188) These are the largest of the paranasal sinuses. Each maxillary sinus lies just external to the lateral nasal wall and extends up to the orbital surface of the maxilla, where its roof is ridged by the infraorbital canal. Its floor is formed by the alveolar process of the maxilla and is usually marked by conical elevations over the roots of the first and second molar teeth. The vertical dimension of the maxillary sinus is about 3.5 cm.; its other dimensions range between 2.5 and 3 cm. The communication of the sinus with the nasal cavity is by an opening which passes from its upper medial wall to the lower part of the hiatus semilunaris. In common with several other paranasal sinuses, the drainage of the maxillary sinus is very poor in the erect posture, and dependent drain-

age of the sinuses requires the laying of the head on one side.

The superior alveolar branches of the maxillary artery (p. 225), its infraorbital branch (p. 225), and the corresponding branches of the maxillary division of the trigeminal nerve supply the lining of the maxillary sinus.

The Maxillary Division of the Trigeminal Nerve (figs. 168, 170, 185, 186, 189, 190)

The maxillary division of the trigeminal nerve arises from the middle portion of the semilunar ganglion. *It is entirely sensory* and supplies the skin of the cheek, the side of the nose, the lower eyelid, and the upper lip; the mucous membrane of the nasopharynx, maxillary sinus, soft palate, tonsil, roof of mouth, and gingivae; and the roots of the upper teeth. Within the cranial cavity a **meningeal branch** accompanies the middle meningeal artery and supplies the dura mater. Emerging through the foramen rotundum, the maxillary nerve enters the deep, triangular pterygopalatine fossa from which branches of the nerve radiate to a major portion of its distribution.

The **Zygomatic nerve** (figs. 189, 190) enters the orbit through the inferior orbital fissure and divides into zygomaticotemporal and zygomaticofacial

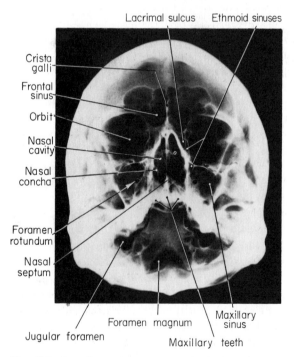

Lacrimal sulcus   Ethmoid sinuses
Crista galli
Frontal sinus
Orbit
Nasal cavity
Nasal concha
Foramen rotundum
Nasal septum

Foramen magnum
Jugular foramen
Maxillary sinus
Maxillary teeth

Fig. 188 A radiogram of the skull to demonstrate the sinuses of the head. Waters projection.

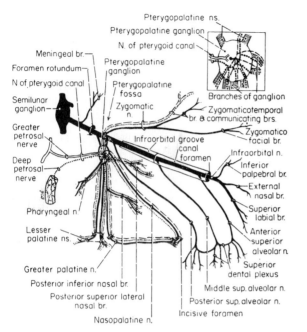

Fig. 189   The branches of the maxillary division of the trigeminal nerve. Inset—the nerve components of the pterygopalatine ganglion. (Dotted lines represent sympathetic nerves, dashed lines, parasympathetic nerves accompanying the zygomatic nerve and all branches of the maxillary nerve which supply mucous membrane.)

branches. The **zygomaticotemporal nerve** runs forward on the lateral wall of the orbit and emerges on the face through the sphenozygomatic suture or the zygomaticoorbital foramen. Its distribution on the face is to the anterior part of the temporal region (p. 199). A communicating branch to the lacrimal nerve, concerned with the transmission of autonomic nerve fibers to the lacrimal gland, is given off in the orbit (p. 251). The **zygomaticofacial nerve** runs forward in the inferior lateral angle of the orbit and emerges on the face at the zygomaticofacial foramen of the zygomatic bone. It supplies the skin over the prominence of the cheek (p. 199).

The **pterygopalatine nerves** (figs. 170, 189) are two short trunks which descend in the pterygopalatine fossa and receive connections from the pterygopalatine ganglion. Although the ganglion adds autonomic nerve fibers to the pterygopalatine nerves, they and their branches are predominantly composed of sensory fibers of the maxillary nerve.

(a) Several delicate **orbital branches** of the pterygopalatine nerves traverse the inferior orbital fissure and supply filaments to the periosteum of the orbit and the mucous membrane of the posterior ethmoid and sphenoid sinuses.

(b) The **greater palatine nerve** (figs. 170, 189, 190) descends through the greater palatine canal in company with the descending palatine artery. Emerging onto the oral surface of the palate at the greater palatine foramen, it passes forward medial to the alveolar process of the bone, giving branches to its mucoperiosteum as far as the incisor teeth. The nerve communicates in the region of the incisive canal with a branch of the nasopalatine nerve. **Posterior inferior lateral nasal branches** of the greater palatine nerve leave the nerve in the greater palatine canal and enter the nasal cavity, ramifying over the middle and inferior meatuses and the inferior concha.

(c) **Lesser palatine nerves** (figs. 189, 190) also descend in the greater palatine canal but emerge at the lesser palatine foramina. The lesser palatine nerves distribute to the palatine tonsil and to the soft palate. Some of the sensory fibers contained in these nerves belong to the facial nerve and have their cell bodies in the geniculate ganglion which they reach by traversing the greater petrosal nerve.

(d) **Posterior superior lateral nasal branches** (figs. 170, 186, 189, 190) pass through the sphenopalatine foramen into the posterior part of the nasal cavity and distribute to the superior and middle concha, the posterior ethmoidal air cells, and the posterior part of the septum.

(e) The **nasopalatine nerve** (figs. 170, 185, 189, 190) is one of the larger of the posterior superior nasal branches. It passes across the roof of the nasal cavity to the septum and runs obliquely downward and forward on the inclined edge of the vomer, supplying the septum. At the incisive canal, the nasopalatine nerves of the two sides descend to the palate where they communicate with the terminals of the greater palatine nerves.

(f) The **pharyngeal nerve** (figs. 185, 189, 190) leaves the posterior part of the pterygopalatine ganglion. It runs in the pharyngeal canal with the pharyngeal branch of the maxillary artery to reach the sphenoid sinus and the nasopharynx behind the auditory tube.

The **posterior superior alveolar nerve** (figs. 165, 189, 190) arises from the maxillary nerve just before it enters the infraorbital groove. The nerve descends over the tuberosity of the maxilla, supplies the adjacent gingiva and the mucous membrane of the cheek, and enters the posterior alveolar canals of the maxilla. Its branches supply the three molar teeth and the maxillary sinus and communicate with the middle superior alveolar nerve.

From the pterygopalatine fossa, the maxillary nerve enters the orbit by way of the inferior orbital fissure as the **infraorbital nerve** (figs. 189, 190). Lying successively in the infraorbital groove and infraorbital canal, the infraorbital nerve emerges onto the face through the infraorbital foramen. Its branches are the middle superior alveolar, anterior superior alveolar, and terminal.

The **middle superior alveolar nerve** (fig. 189) arises from the infraorbital in the posterior part of the infraorbital canal. Occupying a canal in the lateral wall of the maxillary sinus, the nerve descends to supply the two maxillary bicuspid teeth and adjacent gingiva and the mucous lining of the maxillary sinus. With the posterior superior and anterior superior alveolar nerves, it forms the superior dental plexus.

The **anterior superior alveolar nerve** (fig. 189) is given off from the infraorbital nerve just before the latter emerges from the infraorbital foramen. The anterior superior alveolar nerve descends in a canal in the anterior wall of the maxillary sinus and supplies the upper cuspid and incisor teeth and adjacent gingiva. Branches are also given to the

maxillary sinus, and a nasal branch enters the inferior meatus of the nose through a minute canal to distribute to the anterior part of the inferior meatus and floor of the nasal cavity.

Terminal branches of the infraorbital nerve in the face (fig. 120) are the inferior palpebral, external nasal, and superior labial. The **inferior palpebral branches** supply the skin and conjunctiva of the lower eyelid; the **external nasal branches,** the lateral part of the skin of the side of the nose. The **superior labial branches** are numerous. They descend deep to the levator labii superioris muscle and end in the skin and mucous membrane of the upper lip. These branches anastomose with buccal branches of the facial nerve to form the infraorbital plexus.

Clinical Note Since there are three superior alveolar nerves, anaesthesia of the upper teeth requires infiltration of an anaesthetic agent over the course of the nerve appropriate to the tooth concerned.

Autonomic Nerves Accompanying the Maxillary Nerve The **pterygopalatine ganglion** is a peripheral ganglion in the course of the greater petrosal nerve. About 5 mm. in length, it is attached to the pterygo-

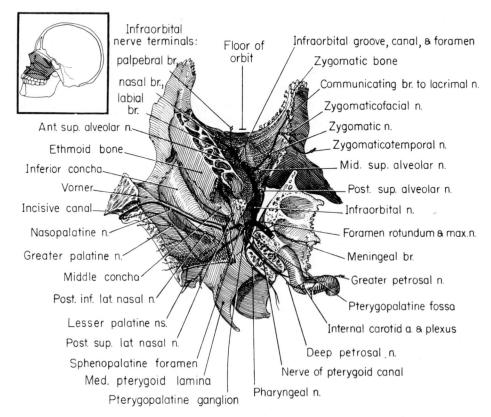

Fig. 190 Phantom of maxilla and adjacent bones to show course and branches of maxillary nerve.

palatine branches of the maxillary nerve. The **greater petrosal nerve** departs from the facial nerve adjacent to the geniculate ganglion. It is composed of general visceral efferent fibers for the parasympathetic supply of the lacrimal, nasal, and palatine glands, and of sensory fibers from the soft palate. Traversing the petrous portion of the temporal bone, it enters the middle cranial fossa at the hiatus of the canal for the greater petrosal nerve and runs forward between the dura mater and the semilunar ganglion, deep to the latter. Crossing the foramen lacerum lateral to the internal carotid artery, the greater petrosal nerve unites with the deep petrosal nerve to form the nerve of the pterygoid canal. The **nerve of the pterygoid canal** passes forward through the pterygoid canal in the sphenoid bone accompanied by the artery of the pterygoid canal (p. 226). The nerve supplies filaments to the sphenoid sinus while in the canal. The **deep petrosal nerve** is a branch of the internal carotid plexus, the continuation of the cervical sympathetic trunk in the cranium. The deep petrosal nerve consists of postganglionic sympathetic fibers which have their cells of origin in the superior cervical sympathetic ganglion and separate from the internal carotid plexus in foramen lacerum. At the pterygopalatine ganglion the fibers of the greater petrosal nerve synapse; contrarily, those of the deep petrosal nerve pass directly into the maxillary nerve branches. Carried indistinguishably in the branches of the maxillary nerve, these parasympathetic and sympathetic postganglionic fibers pass to the lacrimal gland and to the glands of the nasal cavity, palate, and upper pharynx. Fibers destined for the lacrimal gland reach it through the zygomaticotemporal branch of the zygomatic nerve and its communicating branch to the lacrimal nerve (p. 240). The other autonomic fibers accompany those branches of the pterygopalatine nerve (p. 241) which distribute to the nasal cavity, the palate, and the upper pharynx.

THE ORBITAL REGION

The Eyelids (figs. 191–93)

The **eyelids** are movable folds capable of closing in front of the eye and providing protection for it. The upper eyelid is the larger and, having an elevator muscle (levator palpebrae superioris), is the more movable. The aperture between the lids is the **palpebral fissure.** The eyelid is composed of five layers.

Lamination of the Eyelid

The layers of tissue composing the eyelid are the skin, the subcutaneous connective tissue, the muscular layer, the tarsofascial plane, and the conjunctiva.

(1) The **skin** of the eyelid is very thin. At the margin of the lid it assumes the character of a

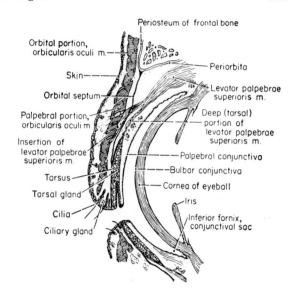

Fig. 191   A vertical section of the eyelids.

moist surface epithelium and is continued on the inner surface of the lid as the conjunctiva.

(2) The **subcutaneous tissue** is lax, scanty in amount, and rarely contains any fat. In the anterior edge of the lid occur short stiff hairs, the eyelashes, or **cilia.** These are longer in the upper lid, and in both lids they curve away from the palpebral fissure. Large sebaceous glands (glands of Zeis) are associated with the cilia, and nearby are modified sweat glands, the ciliary glands (of Moll). The cutaneous nerves of the eyelids are branches of the ophthalmic division of the trigeminal nerve (p. 199) and of the infraorbital branch of the maxillary divison (p.199). The arteries and veins of the eyelids are described with the superficial blood vessels of the face (p. 203). The blood supply of the eyelids is rich.

(3) The muscular layer of the eyelid consists mainly of the **palpebral portion of orbicularis oculi,** which arises from the medial palpebral ligament. Its fibers run elliptically toward the lateral palpebral raphe in which the muscle bundles of the two lids intermingle (p. 201). The **lacrimal portion of the orbicularis oculi** (p. 201) covers the posterior offshoot of the medial palpebral ligament. This small fascicle of muscle arises from the posterior crest of the lacrimal bone (also from the lateral wall of the lacrimal sac) and ends in the tarsus of each lid. Its posterior and medial traction improves the contact between the eyelid and the curved eyeball for better distribution of the lacrimal fluid, and it dilates the lacrimal sac.

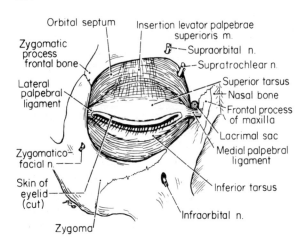

Fig. 192   The tarsofascial layer of the eyelids and the palpebral ligaments.

(4) The **tarsofascial layer** is an important plane of division in the eyelid between a superficial zone continuous with the subcutaneous tissues of the face and scalp and a deeper area continuous with the space of the orbit. The layer consists of a dense fibrous plate, the **tarsus,** and a membrane, the **orbital septum.** The tarsal plates, 2.5 cm. long and about 1 cm. in their greatest breadth (the inferior tarsus is considerably narrower), give support and form to the eyelid. They are semilunar in shape, with the straight edge at the lid margin. Medially, the tarsal plates are continuous with the bifurcated ends of the medial palepbral ligament (p. 201 & fig. 192); and laterally, they are attached to the zygomatic bone by the lateral palpebral ligament, deep to the corresponding muscular raphe. The **orbital septum** (the superior palpebral fascia in the upper lid and the inferior palpebral fascia in the lower lid) is continuous with the periosteum of the bones of the superior and inferior margins of the orbit and ends in the anterior surfaces of the tarsal plates, (figs. 191, 192). The superior palpebral fascia is perforated by tendinous terminals of the levator palpebrae superioris muscle which end among the bundles of orbicularis oculi and in the skin of the upper lid. Except for such minor perforations, the tarsofascial plane forms a complete barrier between the space of the orbit and that of the eyelid.

Embedded into the posterior surface of the tarsus in each lid are the **tarsal glands.** Vertically arranged and parallel to one another, they number about thirty in the upper lid and are somewhat fewer in the lower lid. Their length corresponds to the breadth of the tarsus, and their ducts open by

individual foramina on the free margins of the lids. The tarsal glands are modified sebaceous glands, each consisting of a single straight tube with many small lateral diverticulae. The oily secretion of these glands resists the overflow of tears at the lid margins.

(5) The **conjunctiva** lines the inner surface of each eyelid (palpebral conjunctiva) and is reflected over the anterior portion of the sclera and cornea of the eyeball as the bulbar conjunctiva. The lines of reflection from eyelid onto eyeball are the superior and inferior fornices. The **palpebral conjunctiva** is the thicker and is opaque and highly vascular. It becomes continuous at the margins of the lids with the lining of the tarsal glands, the lacrimal canaliculi, lacrimal sac, and nasolacrimal duct. The **bulbar conjunctiva** is thin and transparent and is loosely attached to the bulb of the eye. It constitutes the transparent epithelium of the cornea.

The **lacrimal lake** (fig. 194) is a shallow bay at the medial angle of the eye bounded laterally by the **semilunar fold** of conjunctiva, a rudimentary nictitating membrane. This depressed area constitutes a small reservoir for the lacrimal fluid and is largely filled by a rounded elevation, the **lacrimal caruncle.** Formed of moist skin, the caruncle exhibits minute hairs and sweat and sebaceous glands. The sebaceous glands are the source of the whitish secretion which collects at the medial angle of the eye.

Lymphatics have not been found in the eyeball or orbit. The lymphatic channels of the conjunctiva join those of the eyelids. From the lateral three-

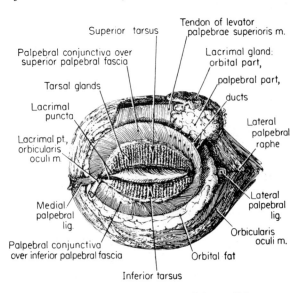

Fig. 193   The internal aspect of the eyelids.

fourths of the lids, lymphatic vessels pass to the anterior auricular and superficial parotid groups of lymph nodes, and those from the medial one-fourth accompany the facial vein to terminate in the facial and submandibular nodes.

THE LACRIMAL APPARATUS (fig. 194)

Tears are secreted by the lacrimal gland located behind the superolateral margin of the orbit. The fluid moves across the eyeball to the medial angle of the eye and is distributed in a uniform layer by the blinking of the eyelids. It is drained off by the lacrimal canaliculi which empty into the lacrimal sac, and from here the fluid passes into the nose through the nasolacrimal duct. When formed at a normal rate, the amount of lacrimal secretion reaching the nose is evaporated, but, when increased by emotion or other causes, it flows from the nose. The parts of the lacrimal apparatus are the lacrimal gland, the lacrimal canaliculi, the lacrimal sac, and the nasolacrimal duct.

The Lacrimal Gland

The gland occupies a hollow, the **lacrimal fossa,** in the orbital plate of the frontal bone behind the superolateral margin of the orbit. It is oval, about 2 cm. long, and divided by the expanded tendon of the levator palpebrae superioris muscle into a superior (orbital) and an inferior (palpebral) portion. The orbital portion of the gland is connected loosely to the periosteum of the orbit and rests on a fascial sheet extending between the superior and lateral rectus muscles. The smaller palpebral portion is connected to the orbital part

around the edge of the levator tendon and projects downward into the upper eyelid, with its deep surface in relation to the conjunctiva. The ducts of the gland number from six to ten. Those of the orbital part traverse the palpebral portion, and all empty into the palpebral conjunctiva just in front of its superior fornix. The vessels of the lacrimal gland are the lacrimal artery and vein of the ophthalmic system (p. 252). Its lymphatics drain to the anterior auricular nodes. The lacrimal branch of the ophthalmic division of the trigeminal nerve conducts sensory fibers to the gland and is also the final path for autonomic nerves to the gland which are provided by the greater petrosal and the deep petrosal nerves (p. 243).

The Lacrimal Canaliculi (fig. 194)

The lacrimal canaliculi occupy the margins of the eyelids from the lacrimal papillae to the lacrimal sac. They begin as **lacrimal puncta,** or pores, situated at the summit of the **lacrimal papillae** opposite the plica semilunaris and have their openings directed somewhat inward toward the lacrimal lake. Each canaliculus passes for a short distance perpendicular to the lid margin and then turns an approximate right angle to run nasalward to the lacrimal sac. The inferior canal is longer and the inferior punctum is larger. The canaliculi have firm walls and lumina of about 0.5 mm. diameter except at the turn, where they enlarge to about 1 mm.

The Lacrimal Sac

The sac is the upper dilated portion of the nasolacrimal duct. It is lodged in a vertical groove, the fossa of the lacrimal sac, formed by the lacrimal bone and the frontal process of the maxilla. The sac measures a little over 1 cm. in length and about 5 mm. in width and receives the lacrimal canaliculi individually just above the middle of its lateral side. The sac lies behind the medial palpebral ligament and in front of the lacrimal portion of the orbicularis oculi, where it may be dilated by the traction of fibers of this muscle which insert into its lateral wall. Its arched upper extremity extends only slightly above the medial palpebral ligament. The sac is weakest below the ligament.

The Nasolacrimal Duct (fig. 194)

The nasolacrimal duct, a continuation of the lacrimal sac, extends downward and slightly lateralward and backward to the inferior meatus of the

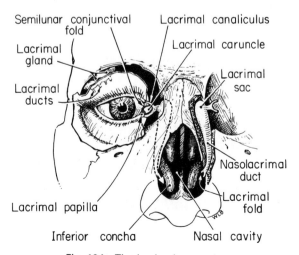

Fig. 194   The lacrimal apparatus.

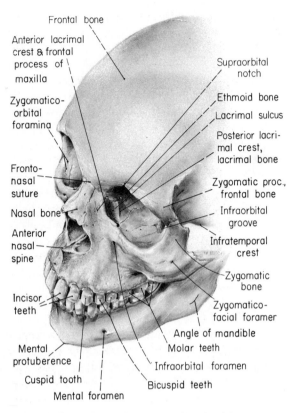

Frontal bone

Anterior lacrimal crest & frontal process of maxilla

Zygomatico-orbital foramina

Fronto-nasal suture

Nasal bone

Anterior nasal spine

Incisor teeth

Mental protuberance

Cuspid tooth

Mental foramen

Supraorbital notch

Ethmoid bone

Lacrimal sulcus

Posterior lacrimal crest, lacrimal bone

Zygomatic proc., frontal bone

Infraorbital groove

Infratemporal crest

Zygomatic bone

Zygomatico-facial foramen

Angle of mandible

Molar teeth

Infraorbital foramen

Bicuspid teeth

Fig. 195   The face, orbit, and lacrimal fossa.

nose. Its length is somewhat less than 2 cm., and it enters the nasal cavity about 1 cm. behind the anterior end of the inferior concha. The duct occupies the **nasolacrimal canal** formed by the maxilla, the lacrimal bone, and the inferior nasal concha but traverses the mucous membrane of the nose obliquely, so that its opening is partially guarded by the **lacrimal fold.**

THE BONY ORBIT (figs. 195, 196, 200, 211)

Each **orbit** is a pyramidal-shaped space, formed by seven bones of the skull, and has four walls and an apex. The medial walls of the orbits are parallel and 2.5 cm. distant from each other, and the intervening space is occupied by the ethmoidal air cells and the sphenoid sinus. The lateral walls diverge at 45° angles from the medial walls and are thus oriented at a 90° angle to each other. The orbital axes, projected through the middle of each orbit, diverge at a 45° angle from each other. The orbital aperture is quadrilateral; it has a vertical dimension of 3.5 cm. and a horizontal dimension of 4 cm. The walls of the orbit measure 5 cm. from apex to base, and

the divergence of the lateral wall is such that its anterior margin is 2 cm. posterior to that of the medial wall. The margins of the orbital aperture are strong; the bone of the margins is much heavier than that of the walls within the cavity.

The supraorbital arch of the frontal bone forms the superior margin and as much as one-third of both the lateral and medial margins. The zygomatic bone forms the remainder of the lateral margin and most of the inferior margin. The inferomedial angle of the orbital margin is provided by the frontal process of the maxilla. The **roof** of the orbit consists of the orbital plate of the frontal bone and, near the apex, the lesser wing of the sphenoid bone. The roof is concave, especially laterally, where the lacrimal fossa accommodates the lacrimal gland. The **lateral wall** is formed in front by the zygomatic bone and behind by the greater wing of the sphenoid bone. The **floor** slopes upward toward the medial wall. It is formed by the orbital surface of the maxilla, supplemented laterally and anteriorly by the zygomatic bone and medially and posteriorly by the palatine bone. Near the middle of the floor, the **infraorbital groove** extends forward from the inferior orbital fissure for several centimeters and then ends in the infraorbital canal. The **medial wall** is nearly vertical. It consists of the frontal process of the maxilla, the lacrimal bone, the orbital lamina of the ethmoid bone, and a small part of the body of the sphenoid bone. Anteriorly, the medial wall is indented for the lacrimal sac by the lacrimal groove, bounded posteriorly by the sharp posterior lacrimal crest. The medial wall forms only a thin partition between the orbit and the ethmoidal air cells and sphenoid sinus. The floor of the orbit is a bony separation between the orbit and the maxillary sinus; the frontal sinus frequently extends over the roof of the orbit nearly to its apex. The stronger lateral wall separates the orbit from the temporal fossa.

The principal openings of the orbit lie at the junction of its walls (fig. 200). The **optic canal** lies at the junction of the roof and the medial wall. It is an aperture between the two roots of the lesser wing of the sphenoid which transmits the ophthalmic artery and the optic nerve covered by the meninges. The **superior orbital fissure** occupies the upper lateral angle at the apex of the orbit and is bounded by the lesser and greater wings of the sphenoid bone. It transmits the oculomotor (cranial III), the trochlear (cranial IV), the ophthalmic division of the trigeminal (cranial V) and the abducens (cranial VI) nerves, sympathetic fibers from the cavernous plexus, and the ophthalmic vein. At the junction of the lateral wall and the floor, the **inferior orbital fissure** extends from the apex of the orbit two-thirds of the distance toward its base. The inferior orbital fissure accommodates structures which have only an indirect relation to the orbit

—the infraorbital nerve and artery, a communication between the inferior ophthalmic vein and the pterygoid plexus, and the orbital and zygomatic branches of the maxillary division of the trigeminal nerve.

Other apertures of the orbit are the **supraorbital notch,** or foramen, at the junction of the medial and middle thirds of the superior orbital margin, the zygomaticoorbital foramen for the zygomatic nerve in the lateral wall, and the canal for the nasolacrimal duct which leads inferiorly from the lacrimal groove. Along the frontoethmoidal suture lie the anterior ethmoidal and posterior ethmoidal foramina. The anterior opening transmits the anterior ethmoidal nerve and vessels; the posterior foramen, the posterior ethmoidal nerve and vessels.

The Fasciae of the Orbit (fig. 196)

The fasciae of the orbit are the periorbita, the bulbar sheath, and the muscular fasciae.

Periorbita

The periosteum of the bones forming the walls of the orbit (periorbita) is continuous through the optic canal and the superior orbital fissure with the periosteum of the bones lining the cranial cavity. The periorbita is also continuous at the orbital aperture with the superior palpebral and inferior palpebral fasciae of the eyelids (p. 244) and with the periosteum of the bones of the face. The periorbita is only loosely attached to the bones of the orbit and, in dissection, is readily displaced from the bones to form a conical investment of all the orbital contents. Laterally, from the roof of the orbit an investment of periorbita is given to the lacrimal gland. Medially, the periorbita attaches the trochlea, or pulley, of the superior oblique muscle of the eye to the trochlear fovea of the frontal bone.

Within the cranial cavity the periosteal layer is closely adherent to the meningeal layer of the dura mater, and these two layers constitute the cranial dura mater. As they pass through the optic canal, the periorbita follows the bony walls of the orbit, whereas the meningeal layer of the dura mater continues forward on the optic nerve as far as the junction of the nerve and eyeball and there fuses with the bulbar fascia. The space of the orbit is thus limited at the optic canal by the fusion there of the periosteal and meningeal layers of the cranial dura mater, and a complete separation between the orbit and the cranial cavity results.

The Bulbar Sheath (fig. 196)

The bulbar sheath is a thin membrane which covers all but the corneal portion of the eyeball. It separates the eyeball from the surrounding fat, muscles, and other tissues and, by its smooth inner surface, facilitates the movements of the eyeball. The bulbar sheath fuses, at the optic nerve entrance, with the sclera and with the sheath of the optic nerve. It is perforated by ciliary vessels and nerves. Anteriorly, the bulbar sheath is penetrated by the tendons of the ocular muscles and is reflected as tubular fasciae over the muscles. The sheath fuses with the ocular conjunctiva just in front of the insertions of the rectus muscles. Between the bulbar sheath and the sclera of the eyeball is a cleft known as the **intervaginal space.** This is largely a potential space crossed by numerous strands of areolar connective tissue.

The Muscular Fasciae (fig. 196)

The fasciae of the ocular muscles are continuous at their origins with the periorbita and, anteriorly, blend with the tubular extensions of the bulbar fascia reflected onto their tendons. From the muscular fasciae, fibrous expansions are given off which end in the periorbita and serve as check ligaments to restrict the action of the ocular muscles. The **medial check ligament** radiates from the sheath of the medial rectus muscle to attach to the posterior crest of the lacrimal bone; the stronger **lateral check ligament** extends from the sheath of the lateral rectus muscle to the zygomatic bone. The blending of these fascial extensions under the eyeball

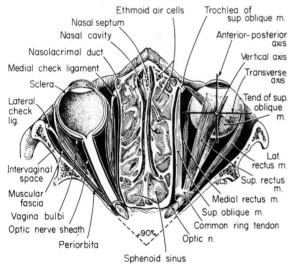

Fig. 196 The cavities, the fasciae, and the muscles of the orbit.

with the fasciae of the inferior rectus and inferior oblique muscles forms a hammock-like sling, the **suspensory ligament,** for the support of the eyeball.

THE OCULAR MUSCLES (figs. 196–200)

The ocular muscles are seven in number; six move the eyeball, and the seventh is an elevator of the upper eyelid. All except the inferior oblique muscle arise at the apex of the orbit and pass forward at the sides of the eyeball.

The Rectus Muscles The four rectus muscles are named according to their orientation with respect to the eyeball—**superior, inferior, medial,** and **lateral.** The rectus muscles arise from a **common ring tendon** which surrounds the upper, medial, and lower margins of the optic canal and encircles the optic nerve. The lateral part of the ring bridges across the superior orbital fissure (fig. 200). The lateral rectus **muscle has two heads of origin, one on either side** of the superior orbital fissure; they are separated by the nerves and the ophthalmic vein that enter the orbit through the fissure. Each muscle inserts by a tendinous expansion into the sclera about 6 mm. behind the margin of the cornea. The medial rectus is the broadest; the lateral rectus, the longest; and the superior rectus, the narrowest. The lateral rectus is innervated by the abducens nerve; the other three, by the oculomotor nerve.

The Superior Oblique Muscle This muscle lies in the upper medial portion of the orbit. It arises

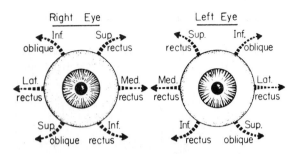

Fig. 198 Diagram of possible actions of the extrinsic muscles of the eye. Direction of arrow is direction of movement of cornea; curve of arrow reflects the direction of rotation around the anteroposterior axis.

immediately above the optic canal and runs forward to the trochlea which is attached in the trochlear fovea of the frontal bone. Through this fibrocartilaginous pulley, the tendon of the muscle passes and then turns lateralward and somewhat backward and downward under the tendon of the superior rectus muscle, ending in the sclera behind the equator of the eye. As the muscle which operates through the 'trochlea,' the superior oblique muscle is innervated by the trochlear (cranial IV) nerve.

The Inferior Oblique Muscle This is a small muscle located under the eyeball near the anterior margin of the orbit. It arises from the orbital surface of the maxilla just lateral to the lacrimal groove. Directed diagonally backward and lateralward below the inferior rectus muscle, the inferior oblique muscle inserts into the sclera beneath the lateral rectus muscle. It is innervated by the oculomotor (cranial III) nerve.

The Levator Palpebrae Superioris Muscle This muscle arises above and in front of the optic canal. As the uppermost of the ocular muscles, the levator expands beneath the roof of the orbit to end anteriorly in a wide aponeurosis. Superficial fibers of the aponeurosis descend through the orbital fat and penetrate the orbital septum to insert among the fascicles of orbicularis oculi and in the skin of the upper eyelid. The middle lamella of the muscle inserts into the upper border of the superior tarsus and consists largely of smooth muscle. The deepest layer of the muscle is represented by fibers that blend with the sheath of the superior rectus and end in the superior fornix of the conjunctiva. This muscle is innervated by the oculomotor nerve. Levator palpebae superioris is continuously active during waking hours except during closing of the lids. Simple lowering of the upper lid is accomplished by decrease of levator activity, but blinking is the result of contraction of orbicularis oculi.

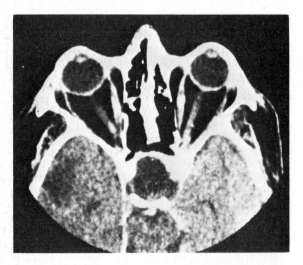

Fig. 197 C.T. scan of a level of the head through the orbits. Compare with fig. 196. (Courtesy of the Radiology Department of the University of Michigan Medical School.)

Actions of the Ocular Muscles (figs. 196, 198) The movements of the eyeball may be referred to three axes which pass through the center of the spheroid of the eyeball. Around a vertical axis through the center of the eyeball, lateral movements (abduction) and medial movements (adduction) take place. Around a horizontal axis through its equator, the eyeball is elevated (upward gaze) and depressed (downward gaze). The anteroposterior axis of the eye provides for rotation of the globe, its superior pole rotating nasalward or temporalward. The anteroposterior, or ocular, axis of the eye does not correspond to the orbital axis, which diverges from its fellow at a 45° angle (p. 246). For this reason, only the lateral and medial rectus muscles have a simple and single action on the eyeball. These muscles exert traction around only the vertical axis and, respectively, produce lateral movement (abduction) and medial movement (adduction). The other four ocular muscles approach the eyeball diagonally and move it on several axes. The superior and inferior rectus muscles predominantly elevate and depress the eye but also have a slight medialward traction, the superior rectus on the upper pole of the eye and the inferior rectus on the lower pole. Thus the superior rectus muscle elevates and adducts the eye and rotates the superior pole nasalward. The inferior rectus muscle depresses and adducts the cornea and rotates the superior pole temporalward. The superior and inferior oblique muscles also have actions on three ocular axes. The superior oblique muscle turns the cornea downward and outward and rotates the superior pole of the eye nasalward. The inferior oblique muscle turns the cornea upward and outward and rotates the superior pole of the eye temporalward. When the pupil is directed straight downward, the superior oblique and inferior rectus muscles act together. Direct upward gaze is produced by the combined action of the inferior oblique and the superior rectus muscles. Oblique movements require simultaneous action of two rectus muscles and one oblique muscle. Such combinations prevent orbital rotation and thus avoid disorientation of the retina with respect to the field of vision.

Innervation of the Muscles Three cranial nerves supply the ocular muscles, but two of these nerves innervate only one muscle each. *The abducens nerve innervates the abductor muscle, the lateral rectus; the trochlear nerve supplies the muscle which operates through a pulley (trochlea), the superior oblique. The other muscles are supplied by the oculomotor nerve*—the levator palpebrae superioris and superior rectus muscles by its superior branch, the medial rectus, inferior rectus, and inferior oblique muscles by its inferior branch.

Clinical Note In clinical testing of the capacity of each ocular muscle together with the integrity of its nerve supply, the patient is instructed to assume a gaze that will largely neutralize the similar action of other muscles and at the same time place the cornea on the axis of the muscle being tested (eye abducted neutralizes the action of the lateral rectus and leaves the inferior oblique with only a rotator function). Thus, as above, elevation of the abducted cornea, tests for the function of the superior rectus muscle and its oculomotor nerve supply. Similarly, depression of the abducted cornea, tests for the function of the inferior rectus muscle and its oculomotor nerve supply. With the eyes adducted, elevation of the cornea tests for the function of the inferior oblique muscle and the integrity of its oculomotor nerve supply, whereas depression tests for the superior oblique muscle and its trochlear nerve supply. The medial and lateral rectus muscles, having actions only on the vertical axis of the eyeball, are tested for, together with the integrity of their nerves (III & VI), in assuming the medial and lateral gaze positions in which the testing began.

THE NERVES OF THE ORBIT (figs. 168, 199–201, 219)

The nerves of the orbit are the three motor nerves to its muscles and the sensory ophthalmic division of the trigeminal nerve. The optic nerve arises in the

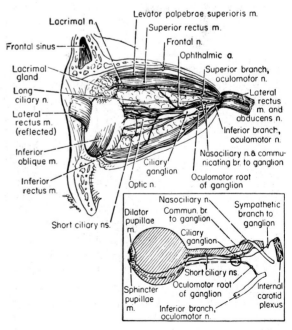

Fig. 199 The nerves and vessels of the orbit as seen after removal of its lateral wall. Inset—a schema of the innervation of the pupillary muscles.

retina of the eye and is described with the eyeball (p. 256). The other nerves all enter the orbit through the superior orbital fissure.

The Oculomotor Nerve

In addition to supplying most of the ocular muscles, the oculomotor nerve provides the parasympathetic innervation for the sphincter pupillae muscle of the iris and for the ciliary muscle of accommodation. It has the following three nerve components:

Somatic efferent: motor to the superior, medial, and inferior rectus muscles; the inferior oblique muscle; and the levator palpebrae superioris muscle.
General somatic afferent: proprioceptive fibers from the same muscles.
General visceral efferent to muscles of the iris and ciliary body with a synapse in the ciliary ganglion.

The oculomotor nerve is formed by rootlets which emerge from the midbrain in the oculomotor sulcus on the medial side of the cerebral peduncle of the brain. The nerve passes between the superior cerebellar and the posterior cerebral arteries near the termination of the basilar artery. Running anteriorly, it pierces the dura and lies in the lateral wall of the cavernous sinus (figs. 168, 199, 219) superior to the trochlear nerve and the ophthalmic and maxillary divisions of the trigeminal nerve. In the wall of the cavernous sinus, the oculomotor nerve receives a communication from the cavernous sympathetic plexus and one from the ophthalmic nerve.

The oculomotor nerve enters the superior orbital fissure between the two heads of the lateral rectus muscle as two branches, a superior and an inferior. The smaller **superior branch** of the oculomotor nerve passes medialward across the optic nerve. It supplies the superior rectus muscle on its underside and then turns around its medial border and terminates in the levator palpebrae superioris muscle. The **inferior branch** proceeds forward in the orbit below the optic nerve. It divides into branches which (1) pass under the optic nerve to the medial rectus, (2) end in the ocular surface of the inferior rectus, and (3) penetrate the posterior border of the inferior oblique muscle. The branch to the inferior oblique muscle is connected to the ciliary ganglion by a short thick root.

The **ciliary ganglion** is a pinhead-sized (2 mm.) ganglion located between the optic nerve and the lateral rectus muscle about 1 cm. from the posterior limit of the orbit. Into its lower posterior angle enters its oculomotor root from the inferior branch of the oculomotor nerve. The fibers contained in this root all synapse around the cells of the ciliary ganglion. The sensory root of the ganglion reaches it at its posterior superior angle and is the **communicating branch of the nasociliary nerve**

of the ophthalmic division of the trigeminal nerve. The sympathetic root from the cavernous plexus passes to the ganglion adjacent to the sensory root. The nerve fibers of both the sensory and sympathetic roots pass through the ganglion without synapse. The six to ten delicate **short ciliary nerves** leave the anterior part of the ganglion and course forward above and below the optic nerve. Accompanied by long ciliary branches of the nasociliary nerve, they pierce the sclera around the optic nerve to enter the eye. Indistinguishable in these nerves are the following components: sensory twigs of the ophthalmic division of the trigeminal nerve to intrabulbar structures, sympathetic postganglionic fibers from the superior cervical ganglion by way of the cavernous plexus to the dilator pupillae muscle of the iris and to blood vessels of the eye, and parasympathetic postganglionic fibers from the ciliary ganglion to the sphincter pupillae muscle of the iris and to the ciliary muscle of accommodation.

The Trochlear Nerve

The trochlear nerve is the smallest of the cranial nerves and *supplies only one muscle*, the superior oblique muscle of the eye. It contains somatic efferent fibers for this muscle and general somatic afferent fibers conducting proprioceptive impulses from the muscle. The trochlear is also distinguished as the only cranial nerve that emerges from the dorsal aspect of the brainstem (immediately caudal to the inferior colliculus). Descending toward the base of the brainstem, the nerve crosses the superior cerebellar peduncle and turns forward against the cerebral peduncle. It pierces the dura mater in the free border of the tentorium cerebelli just behind the posterior clinoid process and passes forward in the

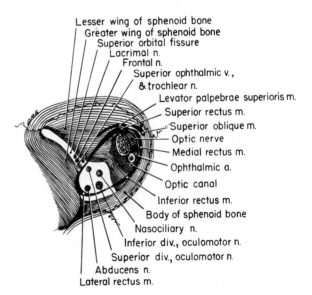

Lesser wing of sphenoid bone
Greater wing of sphenoid bone
Superior orbital fissure
Lacrimal n.
Frontal n.
Superior ophthalmic v.,
& trochlear n.
Levator palpebrae superioris m.
Superior rectus m.
Superior oblique m.
Optic nerve
Medial rectus m.
Ophthalmic a.
Optic canal
Inferior rectus m.
Body of sphenoid bone
Nasociliary n.
Inferior div., oculomotor n.
Superior div., oculomotor n.
Abducens n.
Lateral rectus m.

Fig. 200 The apex of the orbit to show the apertures, entering nerves, and origin of ocular muscles.

lateral wall of the cavernous sinus (fig. 168, 200, 219). At first below the oculomotor nerve, it crosses this nerve and enters the superior orbital fissure as the most superior of all the nerves entering the orbit. In the cavernous sinus the trochlear nerve has communications with branches from the cavernous plexus and from the ophthalmic division of the trigeminal nerve. In the orbit the trochlear nerve is medial to the frontal nerve and enters the superolateral border of the superior oblique muscle in its posterior one-third.

The Abducens Nerve (figs. 168, 200, 219)

Like the trochlear, the abducens nerve *supplies only one muscle* and has the same somatic efferent (motor) and general somatic afferent (proprioceptive) components. The abducens nerve emerges ventrally from the brainstem at the lower border of the pons. It pierces the dura mater on the dorsum sellae of the sphenoid bone and turns over a notch in the bone below the posterior clinoid process to enter the cavernous sinus. Passing forward within this sinus on the lateral side of the internal carotid artery, the nerve enters the orbit through the lower portion of the superior orbital fissure. Within the cavernous sinus the abducens receives communicating filaments from the cavernous plexus and from the ophthalmic division of the trigeminal nerve. At the apex of the orbit the abducens nerve passes between the two heads of origin of the lateral

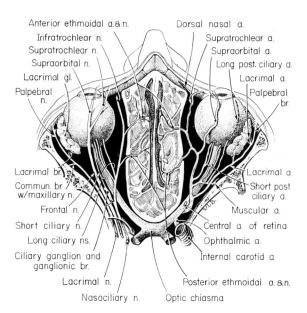

Fig. 201   The branches of the ophthalmic division of the trigeminal nerve (left) and of the ophthalmic artery (right).

rectus muscle, inferior to the other nerves in this location. It passes forward on the ocular surface of the lateral rectus muscle and terminates within it.

The Ophthalmic Division of the Trigeminal Nerve (figs. 168, 199–201, 219)

The ophthalmic nerve is the first and smallest division of the trigeminal *and is entirely sensory*. It distributes primarily to the eye and structures accessory to it in the orbit but also innervates the skin of the eyelids, nose, and forehead and the mucous membrane of the nasal cavity and some of the paranasal sinuses. Arising from the upper portion of the semilunar ganglion, the ophthalmic nerve passes forward in the lateral wall of the cavernous sinus inferior to the trochlear nerve. It is joined by filaments from the cavernous plexus and communicates with the oculomotor, trochlear, and abducens nerves. A **tentorial branch** turns backward with the trochlear nerve to end within the dura of the tentorium cerebelli. Just before reaching the superior orbital fissure, the ophthalmic nerve divides into its three branches: lacrimal, frontal, and nasociliary.

The Lacrimal Nerve The lacrimal nerve is the smallest of the three branches and enters the orbit through the outer limb of the superior orbital fissure. In the orbit it runs along the upper border of the lateral rectus muscle, receiving the communicating branch of the zygomaticotemporal nerve (p. 241) at about the midlength of the orbit. Through this communication it gains the autonomic nerves for the supply of the lacrimal gland. The lacrimal nerve enters the lacrimal gland, supplying it and the adjacent conjunctiva. It also sends a terminal branch through the superior palpebral fascia to the skin of the lateral portion of the upper eyelid (p. 199).

The Frontal Nerve This is the largest branch of the ophthalmic nerve. It enters the orbit through the superior orbital fissure and passes forward between the levator palpebrae superioris muscle and the periorbita. Midway through the orbit it divides into its supratrochlear and supraorbital branches.

The **supraorbital nerve** is the direct continuation of the frontal nerve. It emerges onto the forehead at the supraorbital notch and distributes by medial and lateral branches to the skin of the upper eyelid and the scalp (p. 199).

The **supratrochlear nerve**, smaller than the supraorbital, passes above the pulley of the superior oblique to distribute to the skin of the medial portion of the upper eyelid and the adjacent forehead (p. 199).

The Nasociliary Nerve The nasociliary nerve enters the orbit between the two heads of the lateral rectus muscle. Passing across the optic nerve, it runs below the superior rectus and superior oblique muscles toward the medial wall of the orbit. Along the border of the medial rectus muscle the nerve divides into its principal branches—the posterior ethmoidal, anterior ethmoidal, and infratrochlear nerves.

A branch of communication with the ciliary ganglion arises near the superior orbital fissure and runs forward on the lateral side of the optic nerve to the ciliary ganglion (p. 250). Although associated with the ganglion, this nerve conducts to the eyball sensory fibers whch are not different from those of the long ciliary nerves.

Several **long ciliary nerves** arise from the nasociliary nerve as it crosses the optic nerve. They accompany the short ciliary nerves from the ciliary ganglion (p. 248), pierce the sclera around the optic nerve, and are sensory to the eyeball.

The **posterior ethmoidal branch** passes through the posterior ethmoidal foramen and distributes to the mucous membrane of the posterior ethmoidal air cells and the sphenoid sinus.

The **infratrochlear nerve** springs from the nasociliary just before it enters the anterior ethmoidal foramen. The infratrochlear nerve continues forward, leaving the orbit just below the pulley of the superior oblique muscle. It supplies the skin of the eyelids medially, the side of the nose, the lacrimal sac, the lacrimal caruncle, and the adjacent conjunctiva (p. 199).

The **anterior ethmoidal nerve** is the terminal portion of the nasociliary. Passing through the anterior ethmoidal foramen (fig. 215), it supplies twigs to the anterior ethmoidal air cells and then enters the anterior fossa of the cranium. The nerve turns forward and runs on the cranial surface of the cribriform plate of the ethmoid as far as a slit at the side of the crista galli. Through this **ethmoidal fissure** the nerve descends into the nasal cavity. It supplies **internal nasal branches** to the mucosa of the anterior portion of the septum and lateral nasal wall (p. 238). The anterior ethmoidal nerve ends as the **external nasal branch** which emerges between the nasal bone and the lateral nasal cartilage to supply the skin of the lower half of the nose (p. 199).

THE OPHTHALMIC ARTERY (figs. 199, 201, 223)

The **ophthalmic artery** is a branch of the intracranial portion of the internal carotid artery which arises from it just as the internal carotid emerges from the cavernous sinus. The ophthalmic artery passes directly forward and enters the orbit through the optic canal, below and lateral to the optic nerve. It then curves across the optic nerve toward the medial side of the orbit, which it follows anteriorly.

Its terminal branches at the medial angle of the eye are the supratrochlear and dorsal nasal branches. The ophthalmic artery provides a series of branches to extra-ocular structures in the orbit, which distribute with the branches of the ophthalmic division of the trigeminal nerve, and a set of branches to the globe of the eye itself.

Ocular Vessels	Orbital Vessels
Central Artery of the Retina	Lacrimal
Short Posterior Ciliary	Supraorbital
Long Posterior Ciliary	Posterior Ethmoidal
Anterior Ciliary	Anterior Ethmoidal
	Medial Palpebral
	Supratrochlear
	Dorsal Nasal
	Muscular

The **central artery of the retina** is the first and one of the smallest branches of the ophthalmic artery. It arises close to the optic canal and pierces the optic nerve at about the middle of its intraorbital course. Passing forward in the center of the optic nerve accompanied by the central vein of the retina, the artery spreads over the surface of the retina. Its branches on the retina are the superior nasal, the superior temporal, the inferior nasal, and the inferior temporal.

Clinical Note The central artery is the sole supply of the retina. It has no significant collateral connections, and its obstruction leads to blindness and to atrophy of the retina.

The multiple **short posterior ciliary arteries** arise as the ophthalmic artery crosses the optic nerve. They pass forward with the short ciliary nerves, pierce the sclera around the optic nerve entrance, and supply the choroid and the ciliary processes. Two **long posterior ciliary arteries** are indistinguishable from the short vessels extra-ocularly but, within the eye, run forward between the sclera and choroid to the base of the iris. Here they divide into branches which form an arterial circle in the circumference of the iris (**greater arterial circle**) from which smaller branches pass to the pupillary margin to form another anastomotic ring (**lesser arterial circle**). **Anterior ciliary arteries** arise from muscular arteries. They accompany the tendons of the ocular muscles, form a vascular circle beneath the conjunctiva, and then pierce the sclera a short distance from the cornea and end in the greater arterial circle of the iris.

The **lacrimal artery** is frequently the largest branch of the ophthalmic. It accompanies the corresponding nerve to the lacrimal gland, providing on the way one or two zygomatic branches which accompany the zygomaticotemporal and zygomaticofacial nerves (p. 199). The terminal distribution of the lacrimal artery is by two **lateral palpebral arteries** which run medialward in the upper and lower eyelids, forming arterial circles with medial palpebral arteries (p. 206).

The **supraorbital artery** (p. 212) accompanies the supraorbital nerve, frequently giving rise to a diploic branch in the supraorbital foramen. It distributes to the skin and musculature of the forehead and anastomoses with the supratrochlear artery and the frontal branch of the superficial temporal artery (p. 212).

The **posterior ethmoidal artery** passes, with the corresponding nerve, through the posterior ethmoidal foramen to the posterior ethmoidal air cells and, entering the cranium, gives off a dural branch. The **anterior ethmoidal branch** of the ophthalmic artery duplicates the course of the anterior ethmoidal nerve (p. 252) and supplies the anterior and middle ethmoidal air cells and the frontal sinus. In the cranium it gives off the anterior meningeal artery to the dura mater and, in the nasal cavity, internal and external nasal branches. Superior and inferior **medial palpebral arteries** (p. 206) arise opposite the trochlea. They form, with the lateral palpebral arteries, the superior and inferior tarsal arches of the eyelids. The **supratrochlear artery** (p. 212), one of the terminal branches of the ophthalmic, runs with the supratrochlear nerve and distributes to the skin and muscles of the forehead. The **dorsal nasal artery** (p. 206), the other terminal of the ophthalmic artery, leaves the orbit just above the medial palpebral ligament. It supplies the lacrimal sac and the dorsum of the nose. **Muscular branches** of the ophthalmic artery vary in number and arrangement. A superior set supplies the superior oblique, levator palpebrae, and superior rectus muscles; an inferior set of vessels supplies the muscles below the eye. The anterior ciliary arteries are branches of the muscular branches.

THE OPHTHALMIC VEINS

There are superior and inferior ophthalmic veins. Neither vein contains valves. The **superior ophthalmic vein** begins in the **nasofrontal vein** which enters the orbit through the supraorbital foramen, or notch, after communicating with the supraorbital vein (p. 213). The superior ophthalmic vein has tributaries which correspond to the upper branches of the ophthalmic artery, and it is usually joined by the inferior ophthalmic vein at the medial end of the superior orbital fissure. It may leave the orbit between the two heads of the lateral rectus muscle but is usually above the muscular cone; it ends in the cavernous sinus (p. 278). The **inferior ophthalmic vein** begins on the

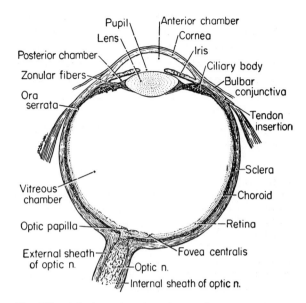

Fig. 202   A horizontal section of the eye through the optic nerve and the fovea centralis.

floor of the orbit in the confluence of muscular veins from the inferior oblique and inferior rectus muscles with the lower posterior ciliary veins. It proceeds to the back of the orbit, has a large communication through the inferior orbital fissure with the pterygoid plexus of veins, and empties into the superior ophthalmic vein. Occasionally it empties separately into the cavernous sinus.

THE EYEBALL (figs. 202, 203)

The eyeball is essentially a sphere, 2.5 cm. in diameter. Its form is modified by a small anteriorly bulging segment, the transparent cornea. The eyes look forward and an imaginary line connecting the anterior pole of the cornea with the posterior pole of the sclera is the ocular axis. In its essential structure the eyeball can be resolved into three layers: the outer fibrous supporting layer of sclera and cornea; the middle vascular layer of choroid, ciliary body, and iris; and the inner layer of nerve elements, the retina. A lens lies behind the cornea and is the principal refractive structure of the eye.

The Fibrous Tunic

(a) The **sclera** is a firm fibrous cup which constitutes the posterior five-sixths of the outer layer of the eye. Its greatest thickness of 1 mm. is in its posterior portion. It is continuous in front with the cornea at the corneoscleral junction; behind, it is perforated by the optic nerve, forming there a mesh-work, the **lamina cribrosa sclerae,** which permits the fascicles of nerve and the central artery and vein to pass through. Around the optic nerve entrance other small apertures transmit the ciliary nerves and vessels. The sclera is perforated in the region of its equator by the vorticose veins. The outer surface of the sclera is smooth and white and is separated from the bulbar fascia by loose connective tissue.

(b) The transparent **cornea** constitutes the anterior one-sixth of the fibrous tunic. It is a segment of a smaller sphere than the sclera and thus bulges forward from the latter. The cornea is dense and uniformly about 1 mm. thick. Its surface epithelium is the bulbar conjunctiva. The cornea is nonvascular, its nourishment coming from capillary loops of the anterior ciliary arteries at its margin. It is richly supplied with sensory nerves from the ciliary nerves.

The Vascular Tunic

(a) The **choroid** is a thin, highly vascular membrane which invests the posterior five-sixths of the eyeball, extending as far forward as the ora serrata of the retina. The choroid layer is brown in color due to the pigment cells of its outermost layer. It consists of a dense capillary plexus and of small arteries and veins held together in a meshwork of connective tissue. The choroid is loosely connected to the sclera except behind where it is pierced by the optic nerve; here it is firmly attached. Internally,

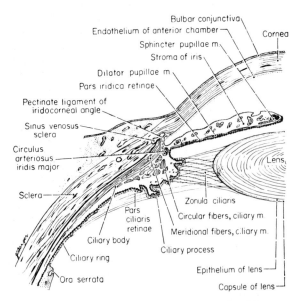

Fig. 203 A section of the corneoscleral angle of the eye.

the choroid is intimately attached to the pigment layer of the retina.

(b) The **ciliary body** (fig. 203) is an elevated zone of the anterior portion of the choroid layer. It has the form of a wedge-shaped ring, the outer flattened surface of the wedge being in apposition to the sclera and ending anteriorly at the scleral spur. The inwardly projecting angle of the wedge is directed toward the lens and is connected with it by the fibers of the suspensory ligament. The greatest bulk of the ciliary body is contributed by the ciliary muscle. Its inner surface is covered by the thin, pigmented layer of the retina.

Two regions are apparent in the ciliary body. One, the **ciliary ring,** is a smooth band about 4 mm. wide just anterior to the ora serrata of the retina. It is the narrow posterior taper of the wedge and contains only the thinnest part of the ciliary muscle. The other region is characterized by the **ciliary processes,** a circular series of from sixty to eighty projections of the choroid layer, each about 2 mm. in length. These have a meridional arrangement and give attachment to the suspensory ligament of the lens.

The **ciliary muscle** (fig. 203) occupies a zone of about 3 mm. breadth. The muscle is thickest anteriorly and consists of meridional and circular bundles. The **meridional fibers** arise from the **scleral spur** at the inner aspect of the corneoscleral junction and radiate backward to end in the ciliary processes and ciliary ring. **Circular fibers** of the muscle form bundles at the angle of the wedge, internal to the meridional fibers and adjacent to the root of the iris. The ciliary muscle is the muscle of accommodation. By its contraction the ciliary processes and ring are drawn toward the corneoscleral junction and the wedge-shaped ring is bulged toward its center. This reduces the tension on the fibers of the suspensory ligament, thereby allowing the natural elasticity of the lens to increase its curvatures, resulting in greater refractive power for vision of close objects. The muscle is innervated by parasympathetic fibers of the oculomotor nerve which arise in the nucleus of Edinger-Westphal in the floor of the aqueduct of the midbrain and which synapse in the ciliary ganglion (p. 250).

(c) The **iris** (fig. 203) is a thin, contractile membrane having a central aperture, the **pupil.** The color of the iris is imparted by pigment cells in its posterior layer, the pars iridica retinae, and in its stroma. If its pigment is confined to the pars iridica retinae, the iris is a shade of blue; if it is also prominent in the stroma, the color is brown or black. Within the loose stroma of the iris are two involuntary muscles. The **sphincter pupillae muscle** is formed of circular fibers arranged in a narrow band 1 mm. wide which surrounds the margin of the pupil. The **dilator pupillae muscle** consists of radiating fibers which, close to the posterior surface of the iris, converge from its circumference toward its center. The sphincter muscle has the same parasympathetic innervation as the ciliary muscle; the dilator is supplied by sympathetic fibers which arise in the superior cervical ganglion and reach the eye from the cavernous plexus and through the short ciliary nerves (p. 250). The greater and lesser arterial circles of the long ciliary arteries are formed in the circumference and in the pupillary margin of the iris.

The iris separates certain of the chambers of the eye (fig. 202). The **anterior chamber** lies behind the cornea and anterior to the iris. The **posterior chamber** is a narrow interval behind the iris and in front of the lens, the suspensory ligament of the lens, and the ciliary processes. These chambers are filled with **aqueous humor,** a clear, watery solution formed by the epithelium of the ciliary processes, which circulates forward into the anterior chamber, and is drained at the corneo-iridal angle of the eye through the mesh-like spaces of the pectinate ligament and the sinus venosus sclerae (fig. 203). The **vitreous chamber** of the eye lies behind the lens and ciliary processes. It is filled by the **vitreous body,** a body of transparent semigelatinous material, which is firmly adherent to the ora serrata region of the retina.

The **blood vessels and nerves** of the eyeball have been described. Long and short ciliary nerves (p. 252) conduct sensory fibers to structures of the fibrous and vascular tunics of the eye and autonomic nerves to the smooth muscle of the ciliary body and iris. Long and short posterior ciliary and anterior ciliary arteries, branches of the ophthalmic artery (p. 252), supply the parts of the fibrous and vascular tunics. The veins of the vascular tunic are almost all tributary to the **vorticose veins** which emerge from the eye at four to five points along its equator. They drain into the posterior ciliary and the ophthalmic veins. Anterior ciliary veins are also present and empty into the ophthalmic veins.

Clinical Note Glaucoma is a serious condition which results when the drainage of aqueous humor fails to keep up with its formation. There is then buildup of pressure in the fluid which causes severe pain and compression damage to the nerve fibers of the retina, especially at the optic disc.

Blindness can be the result but certain drugs can be used to reduce the rate of aqueous humor production and thus to hold intraocular pressure more nearly to normal.

The Nervous Tunic

In development, the vesicle that forms the retina undergoes an invagination, so that a two-layered cup results. This fundamental layering is retained in the nervous tunic of the eye, for the outer layer forms the outer pigmented epithelial layer, and the inner wall of the vesicle forms the light-receptive portion of the retina. Three regions are differentiated in the retina. The **pars optica retinae** is that portion characterized by nervous elements. Occupying the posterior part of the bulb, it ends in a jagged line, the **ora serrata** (fig. 203) a short distance behind the ciliary processes. At the ora serrata the nervous elements of the retina cease, and the **pars ciliaris retinae,** which extends forward over the ciliary body, is much thinner. It has outer pigmented and inner nonpigmented epithelial cell layers. The **pars iridica retinae** covers the posterior surface of the iris and ends at the pupillary margin. In the pars iridica retinae both cell layers are pigmented.

The **pars optica retinae** contains visual purple in the fresh state, but in the fixed cadaver it is a clouded, opaque membrane. It is firmly attached at the ora serrata and at the optic nerve entrance.

The **retina** is an outgrowth of the diencephalic portion of the brain, and its nervous portion is composed of neurons arranged in three layers: the light-receptive rods and cones, bipolar cells, and ganglion cells. The optic nerve is composed of the axons of the ganglion cells and is, strictly speaking, a brain tract rather than a cranial nerve. Separating these cellular layers in the retina are plexiform layers composed of the dendrites and axons of the cells, and supporting elements bind all together. Approaching the ora serrata, the typical layers are reduced and finally disappear. At the posterior pole of the eye there is an oval yellowish area, the **macula lutea,** and in its center, a depression, the **fovea centralis** (fig. 202). This depression is formed by a drawing aside of the ganglionic and bipolar neurons, so that light rays are unabsorbed on their way to the layer of cones which are exceedingly closely packed in this spot. The retina is here reduced to a thickness of 0.1 mm., and the dark color of the choroid is transmitted through it. Due to the narrow form and close packing of the cones and the removal of layers which normally intervene between the rods and cones and the light rays, the fovea centralis is the portion of the retina which provides the greatest visual acuity. It is that part of the eye on which the light rays are focused when the eyes are correctly accommodated and convergent.

About 3 mm. to the nasal side of the macula is the **optic** disk, the site of emergence of the optic nerve. At the margins of the disk the retinal cells are reduced, and the axons of ganglion cells which converge here from the entire retina pile up around its central depression. These axons pass through the lamina cribrosa sclerae and, receiving medullated sheaths, thicken to form the large optic nerve. The optic disk is the blind spot of the eye. From the center of the optic disk the arteria centralis retinae and its accompanying vein spread out on the inner surface of the retina (p. 252).

The Refractive Media

Light rays are bent at the interfaces of materials of differing densities. The refractive media in the course of light entering the eye are the cornea, the aqueous humor, the lens, and the vitreous humor. All of these have been described except the crystalline lens.

The **lens** (figs. 202, 203) is a transparent biconvex body, flatter on its anterior surface and more curved posteriorly. It is composed of layer on layer of transparent lens fibers. The transparent **capsule of the lens** surrounds it on all sides. The capsule is thicker in front than behind and is highly elastic. In middle age the equatorial diameter of the lens is 9 mm., and its thickness is 4 mm. The lens is nearly spherical in the fetus, but in old age it is considerably flattened. The shape of the lens is modified by the ciliary muscle as the eyes focus on objects at differing distances. Its surfaces, especially the anterior surface, become more rounded when the fixation is on near objects. The lens has a tendency to lose its transparency in later years, a condition known as cataract. The lens is held in place by a series of straight fibrils constituting the **suspensory ligament of the lens** (fig. 203). This is attached to the capsule from the equator of the lens forward for a short distance. The filamentous fibers arise from the epithelium of the ciliary body as far back as the ora serrata. When the ciliary muscle is relaxed, these fibers are under tension; when the ciliary muscle contracts, the suspensory ligament is relaxed and the lens surfaces become more rounded for close vision. This paradoxical effect results from the pull of the ciliary muscle fibers toward rather than away from the attachment of the suspensory ligament.

The Optic Nerve (figs. 168, 200, 202)

The optic nerve consists of the axons of the ganglion cells of the retina which converge from all the retinal quadrants to the optic disk located

3 mm. medial to the posterior pole of the eye. The bundles pierce the lamina cribrosa sclerae and then become myelinated and enclosed by the cranial meninges. The pia mater, the arachnoid, and the dura mater fuse with the sclera around the optic nerve entrance; from here they are prolonged on the orbital portion of the optic nerve into the cranial cavity to become continuous with their cranial counterparts. The nerve and its meningeal investments pass through the optic canal with the ophthalmic artery, the common ring tendon surrounding them all. The relations of the arteria centralis retinae and its accompanying vein to the optic nerve have been described (p. 252). The intracranial part of the optic nerve, about 1 cm. in length, passes backward and medialward above the ophthalmic artery, crosses the medial aspect of the internal carotid artery, and forms the optic chiasma over the diaphragma sellae. The chiasma lies immediately in front of the stalk of the hypophysis, blending with the base of the hypothalamus. The fibers from the nasal half of each retina decussate in the chiasma, resulting in a projection, through each optic tract, of impulses arising in the visual field of the opposite side.

THE EAR

The ear, or vestibulocochlear organ, consists of three parts: the external, the middle, and the internal ear.

THE EXTERNAL EAR (fig. 204A)

The external ear is composed of the oval **auricle,** or pinna, and the **external acoustic meatus,** which leads from the auricle to the middle ear.

The Auricle

This portion of the external ear consists of a cartilage connected to the skull by ligaments and muscles and covered by skin. The **skin** of the ear is thin and closely adherent to the cartilage and is prolonged inward along the external acoustic meatus as far as the tympanic membrane. Fine hairs cover it, and sebaceous glands are numerous in the concha. On the tragus and antitragus the hairs are longer and more numerous. The rim of the auricle, resembling the beginning of a spiral, is called the **helix.** Internal to the helix is a broader curved eminence, the **anthelix;** above and anteriorly, the anthelix forms two separated ridges, the **crura.**

The deep cavity of the auricle, the **concha**, is in front of the anthelix. It is bounded anteriorly by a small prominence, the **tragus**, which bears a tuft of hair. The tragus overhangs the beginning of the external acoustic meatus and has opposite it another small tubercle, the **antitragus**; the two are separated by the **intertragic notch.** Below the notch is the soft noncartilaginous **lobule** consisting of areolar tissue and fat. The cartilage of the ear is single and is irregularly ridged and hollowed to form the foregoing features. It is prolonged inward to form the external one-third of the acoustic meatus and is attached to the side of the skull by ligaments and muscles. Anterior and posterior ligaments attach, respectively, to the root of the zygomatic process of the temporal bone and to the outer surface of the mastoid process.

The **extrinsic muscles** are the anterior, superior, and posterior auricular muscles. They are described on p. 214. The **vessels** and **nerves** of the ear have also been described. Its cutaneous nerves are the lesser occipital nerve (p. 148) and the great auricular nerve (p. 148) of the cervical plexus, the auriculotemporal branch of the mandibular division of the trigeminal nerve (p. 199), and the auricular branch of the vagus nerve (p. 164). The arteries of the external ear are radicles of the posterior auricular and superficial temporal branches of the external carotid artery (p. 212). The veins correspond. The **lymphatics** of the external ear drain to the anterior auricular, retroauricular, and superficial parotid nodes (p. 166).

The External Acoustic Meatus (fig. 204A)

The external acoustic meatus is a canal, approximately 2.5 cm. long, which leads to the tympanic membrane. Its direction is first inward, forward, and upward; it then bends backward; and finally, forward and downward. The canal is straightened for examination by drawing the auricle upward, outward, and backward. Its lumen is not uniform, being somewhat narrowed toward the ear drum and having a constriction at the junction of the cartilaginous and osseous portions. The cartilaginous outer part of the canal constitutes about one-third of its length. It ends in the firm attachment of the cartilage to the circumference of the auditory process of the temporal bone. The osseous two-thirds of the canal is narrower than the cartilaginous part and is directed inward and a little forward. The inferior wall of the canal is about 5 mm. longer than the superior due to the obliquity of the tympanic membrane. The skin lining the acoustic

meatus is thin and closely adherent to the cartilage and bone. It covers the tympanic membrane, forming its outer layer. Numerous ceruminous glands lie in the subcutaneous tissues of the cartilaginous part of the canal. They do not exist in the osseous portion, and thus the ear wax which they secrete is deposited well away from the tympanic membrane. The external acoustic meatus lies behind the condyle of the mandible and the glenoid process of the parotid gland. The mastoid air cells are posterior to the bony meatus.

The **arteries** of the meatus are the superficial temporal and posterior auricular (p. 212) and the deep auricular branch of the maxillary artery (p. 224). The veins and lymphatics are those serving the external ear. The nerves of the meatus are the auricular branch of the vagus (p. 164) and the auriculotemporal nerve (p. 199). With the auricular branch of the vagus are combined a small number of peripheral process of the facial nerve (p. 263) and of the glossopharyngeal nerve (p. 196).

The Middle Ear (figs. 204–206)

The middle ear is a narrow cavity in the temporal bone; here the energy of sound waves is converted into mechanical energy through a chain of ossicles. The cavity lies just internal to the tympanic membrane, or ear drum, and communicates with the pharynx by means of the auditory tube. It also is continuous with the mastoid air cells. The vertical and anteroposterior dimensions of the tympanic cavity are both slightly over 1 cm. The transverse dimension is less and varies with the irregularities of the lateral and medial walls of the space. It is 6 mm. at the roof of the space, 4 mm. at its floor, and 2 mm. in its midregion, where the inwardly bulging tympanic membrane approaches the rounded promontory of the medial wall. The middle ear includes the **tympanic cavity** proper, the space directly internal to the membrane, and the **epitympanic recess,** the space above it. The mucous membrane which lines the middle ear is continuous with that of the mastoid air cells and the nasopharynx. The **roof** of the middle ear is formed by the thin, bony **tegmen tympani** which separates it from the cranial cavity. The **floor** of the tympanic cavity is a narrow chink between the medial and lateral walls. Here a thin bony separation exists between the middle ear and the jugular fossa, the layer of bone being perforated by the tympanic branch of the glossopharyngeal nerve.

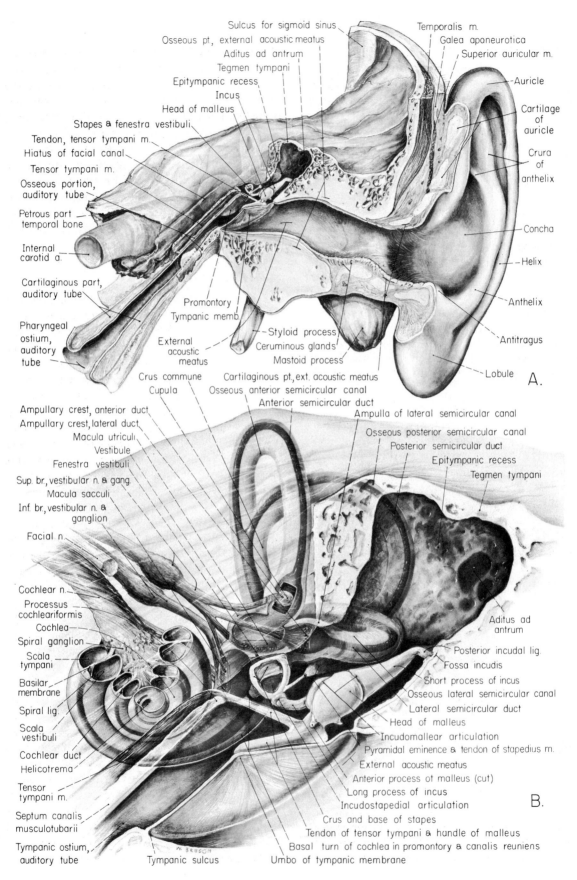

Sulcus for sigmoid sinus
Osseous pt, external acoustic meatus
Aditus ad antrum
Tegmen tympani
Epitympanic recess
Incus
Head of malleus
Stapes & fenestra vestibuli
Tendon, tensor tympani m.
Hiatus of facial canal
Tensor tympani m.
Osseous portion, auditory tube
Petrous part temporal bone
Internal carotid a.
Cartilaginous part, auditory tube
Pharyngeal ostium, auditory tube

Temporalis m.
Galea aponeurotica
Superior auricular m.
Auricle
Cartilage of auricle
Crura of anthelix
Concha
Helix
Anthelix
Antitragus
Lobule
A.

Promontory
Tympanic memb.
External acoustic meatus
Styloid process
Ceruminous glands
Mastoid process

Crus commune
Cupula
Ampullary crest, anterior duct
Ampullary crest, lateral duct
Macula utriculi
Vestibule
Fenestra vestibuli
Sup. br., vestibular n. & gang.
Macula sacculi
Inf. br., vestibular n. & ganglion
Facial n.
Cochlear n.
Processus cochleariformis
Cochlea
Spiral ganglion
Scala tympani
Basilar membrane
Spiral lig.
Scala vestibuli
Cochlear duct
Helicotrema
Tensor tympani m.
Septum canalis musculotubarii
Tympanic ostium, auditory tube

Cartilaginous pt., ext. acoustic meatus
Osseous anterior semicircular canal
Anterior semicircular duct
Ampulla of lateral semicircular canal
Osseous posterior semicircular canal
Posterior semicircular duct
Epitympanic recess
Tegmen tympani

Aditus ad antrum
Posterior incudal lig.
Fossa incudis
Short process of incus
Osseous lateral semicircular canal
Lateral semicircular duct
Head of malleus
Incudomallear articulation
Pyramidal eminence & tendon of stapedius m.
External acoustic meatus
Anterior process of malleus (cut)
Long process of incus
Incudostapedial articulation
Crus and base of stapes
Tendon of tensor tympani & handle of malleus
Basal turn of cochlea in promontory & canalis reuniens

Tympanic sulcus
Umbo of tympanic membrane
B.

Fig. 204 The petrous portion of the temporal bone and the ear. A—a cut-away specimen opening the external acoustic meatus, the middle ear, and the auditory tube. B—a cut-away and transparent representation of the middle and internal ear.

The Lateral Wall

The lateral wall of the cavity is formed by the **tympanic membrane** (figs. 204, 205). This is nearly circular and approximately 1 cm. in diameter. It is set obliquely into the external acoustic meatus, so that its outer surface faces outward, downward, and forward, and so that it forms an angle of about 55° with the floor of the meatus. The tympanic membrane is composed of three layers: the modified skin of the external meatus on its outer surface, the mucous membrane of the cavity internally, and an intermediate fibrous stratum composed of radial and circular fibers. This layer imparts strength to the membrane; the radial fibers diverge from the handle of the malleus, and the circular ones are more prominent at the periphery of the disk. The membrane has a fibrocartilaginous ring at the greater part of its circumference, which is fixed in the **tympanic sulcus** at the inner end of the external acoustic meatus. At the upper limit this sulcus is deficient and the membrane lacks the fibrous stratum which gives it its rigidity. Here the drum is designated as the **membrana flaccida,** a term applying to about one-sixth of its total area. The larger five-sixths, strengthened by the fibrous layer, is the **membrana tensa.** The manubrium (handle) of the malleus is firmly attached to the fibrous layer on its inner surface and extends to a little below the center of the membrane. The center point, the **umbo** (fig. 204B), is at the most indrawn part of the membrane. The epitympanic recess lodges the head of the malleus and the body of the incus. The tympanic membrane is set into vibration by sound waves which reach it through the external acoustic meatus.

The Medial Wall

The principal feature of the medial wall of the tympanic cavity is the **promontory,** a rounded bony eminence formed by the projection of the basal turn of the cochlea. The promontory is grooved by branches of the tympanic plexus which lie under its mucous membrane. Above the promontory is the oval **fenestra vestibuli,** an opening in the bone of the medial wall, closed in life by the footplate of the stapes. The long axis of this opening is horizontal and its greatest convexity is superior. Below and behind the promontory is the **fenestra cochleae,** or round window, closed in life by a membrane.

This membrane yields to the surge of fluid in the closed system of canals of the inner ear produced by the piston-like action of the footplate of the stapes at the vestibular window. The **prominence of the facial canal** is a curved bulge of bone forming the canal for the facial nerve (fig. 205). It lies superior to the vestibular (oval) window and then curves inferiorly and nearly vertically behind the promontory. The bone may be exceedingly thin over the nerve. Above the curve of the prominence of the facial canal, the lateral semicircular canal frequently bulges the bone so that it forms a prominence.

The Anterior Wall

The bone of the anterior wall separates the tympanic cavity from the carotid canal; it is perforated by the caroticotympanic nerves which leave the sympathetic plexus on the internal carotid artery to join the tympanic plexus. The anterior wall is incomplete, for at its superior extremity lie the openings of the auditory tube and (above it) the semicanal for the tensor tympani muscle. These canals are separated by a thin shelf of bone, the **septum canalis musculotubarii.** The **tensor tympani muscle,** 2 cm. in length, arises from the septum, the cartilaginous part of the auditory tube, and the adjacent part of the greater wing of the sphenoid. Its slender tendon enters the tympanic cavity and turns lateralward around a hook-like extremity of the septum on which the muscle lies, the **processus cochleariformis.** The tendon inserts into the manubrium of the malleus near its root (figs. 205, 206). The tensor tympani muscle is innervated by a twig of the medial pterygoid branch of the mandibular nerve (p. 222). It draws the handle of the malleus and the tympanic membrane toward the medial wall, increasing the tension and dampening the vibrations of the membrane.

The **auditory tube** is a communication between the nasal portion of the pharynx (p. 192) and the tympanic cavity. It measures 3.5 cm., of which one-third is osseous (at the tympanic end) and two-thirds is cartilaginous (related to the pharynx). The tube begins below the canal for the tensor tympani muscle and is directed downward, forward, and medialward, forming a 45° angle with the sagittal plane and one of 35° with the horizontal plane. The jagged extremity of the osseous portion of the canal serves for the attachment of its cartilaginous portion and is the narrowest portion of the tube. Through the mucous membrane of the tube the pharyngeal mucosa is continuous with that lining the tympanic cavity and mastoid air cells, affording a route for the passage of infectious material to the middle ear and mastoid area. In the cartilaginous portion the mucous membrane contains many mucous glands. A collection of lymphoid tissue lies near the pharyngeal orifice; this constitutes the **tubal tonsil.**

The Posterior Wall

The posterior wall of the tympanic cavity is open above, affording an entrance to the mastoid antrum. Below this, it presents the **fossa incudis** for the reception of the short process of the incus, and the projecting **pyramidal eminence.** The **aditus ad antrum,** the entrance to the mastoid antrum, is a large, irregular aperture which connects the epitympanic recess with the mastoid antrum. The **antrum** is the largest and most superior of the cavities in the mastoid bone. It communicates below and behind with the **mastoid air cells,** which vary in size and number in different individuals. The antrum and the mastoid cells are lined by mucous membrane continuous with that of the middle ear.

The **pyramidal eminence** (figs. 204B, 205) is a conical elevation of bone which juts forward from the posterior wall in front of the vertical portion of the facial canal. It is directed toward the vestibular window of the medial wall. The eminence is hollow, and its walls give rise to the fibers of the **stapedius muscle,** the central tendon of which emerges at an aperture on the summit of the eminence and inserts in the posterior surface of the neck of the stapes. A small branch of the facial nerve is conducted into the muscle at the base of the pyramidal eminence. The contraction of this muscle tilts the footplate of the stapes, tending to dampen its vibrations, and thereby serves a protective function.

The Auditory Ossicles (figs. 204–206)

Three small bones, the malleus, the incus, and the stapes, united by true joints, form a lever system which converts the vibrations in air impinging on the tympanic membrane into mechanical energy to oscillate the footplate of the stapes in the vestibular window. This lever system results in a decrease in amplitude but an increase

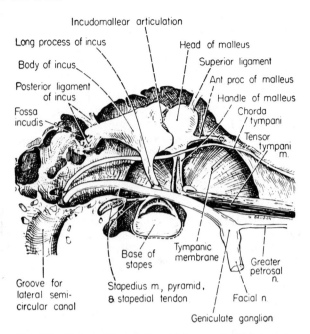

Fig. 205   The ossicles of the middle ear, the facial nerve, and the chorda tympani as seen from above and medially. (The bone of the facial canal has been removed.)

in power of the piston-like action of the stapes. The three bones are united by true joints and are suspended in the cavity of the middle ear. Their fixation in space depends on the attachments of the malleus to the tympanic membrane and of the stapes in the oval window and on ligaments which suspend the bones from the walls of the cavity.

The **malleus** receives its name from its resemblance to a hammer. Its tapered **manubrium** is firmly attached to the upper half of the tympanic membrane, and its rounded **head** extends into the epitympanic recess. The posterior aspect of the head is hollowed into an oval concave facet for the reception of the incus. A slender anterior process arises from the neck of the bone and is directed forward to the petrotympanic fissure, to the borders of which it is connected by ligamentous fibers.

The **incus** is named from its fancied resemblance to an anvil, but its **body** and two widely separated **crura** appear more like a lower molar tooth. The body of the incus has on its anterior surface a deep concavo-convex facet which articulates with the head of the malleus in the **incudomallear joint.** The two crura (processes) diverge at a right angle and the heavy, short process extends horizontally backward into the fossa incudis, where it is firmly attached by the posterior incudal ligament. The **long process** of the incus descends vertically, parallel to the manubrium of the malleus and, at its tip, bends medialward to form a rounded end, the **lenticular process.** This part of the incus articulates with the head of the stapes.

The **stapes** has an amazingly close resemblance to a stirrup. Its **head** is hollow for the lenticular process of the incus, the two forming the **incudostapedial articulation.** The head grades over into the **neck** of the bone into which the tendon of the stapedius muscle inserts. Two **crura** diverge from the neck and are connected by a flattened oval plate, the **base** of the stirrup, which is attached to the margin of the vestibular window by ligamentous fibers.

The incudomallear and incudostapedial articulations are synovial joints. The bones are covered by hyaline cartilage and the joint cavities are enclosed within articular capsules. The incudomallear is a saddle-shaped articulation; the incudostapedial, a ball and socket joint. The attachments of the malleus by its anterior ligament and of the incus by its posterior ligament have been described. In addition, each of these bones has a suspension from the roof of the epitympanic recess by delicate **superior ligaments.**

Blood Vessels and Nerves of the Middle Ear

The **arteries** of the middle ear are numerous. The two larger ones are the anterior tympanic branch of the maxillary artery (p. 224) to the tympanic membrane and the stylomastoid branch of the posterior auricular artery to the tympanic antrum and mastoid cells (p. 211). Smaller arteries include the inferior tympanic branch of the ascending pharyngeal artery (p. 161), the superficial petrosal branch of the middle meningeal artery (p. 224), a branch of the artery of the pterygoid

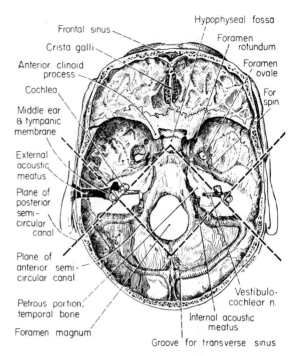

Fig. 207 The base of the skull with the cochlea and labyrinth represented as if in transparent bone. The planes of orientation of the vertical semicircular canals are shown by dashed lines.

canal (p. 226), and the caroticotympanic branch of the internal carotid artery (p. 282). The **veins** parallel the arteries and are tributary to the superior petrosal sinus (p. 279) and to the pterygoid plexus of veins. The **nerves** of the middle ear are branches of the **tympanic plexus,** formed by the tympanic branch of the glossopharyngeal nerve (p. 196) and by superior and inferior caroticotympanic nerves from the carotid sympathetic plexus which pass through the wall of the carotid canal. Branches of the tympanic plexus reach the mucous membrane of the oval and round windows, the tympanic membrane, the auditory tube, and the mastoid air cells. The **cranial portion of the facial nerve** traverses the petrous portion of the temporal bone. Its course and branches and the relations of the chorda tympani are described under the general discussion of the facial nerve (p. 263).

THE INTERNAL EAR (figs. 204, 206)

The internal ear provides the essential organs of hearing and of equilibrium. It consists of the cochlea, for the auditory sense, and a series of intercommunicating channels—the semicircular ducts, the utricle, and the saccule—for the sense of balance and position. These are tubular systems containing a fluid, **endolymph,** the displacements of which stimulate the sensory endings elaborated

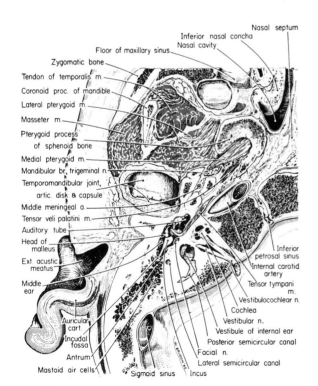

Fig. 206 A partial cross section of the head at the level of the temporomandibular joint and the ear.

in their epithelium. The tubules compose the **membranous labyrinth** which is contained within a fluid-filled series of bony canals, the **bony labyrinth.** The fluid of the bony labyrinth is the **perilymph.**

The Bony Labyrinth (figs. 204, 206, 207)

This consists of three parts: the vestibule, the semicircular canals, and the cochlea. The **vestibule,** the middle portion of the bony labyrinth, lies medial to the tympanic cavity with the cochlea anteriorly and the semicircular canals posteriorly. The **vestibular window** is an opening in its lateral wall. The movements of the footplate of the stapes produce surges in the fluid of the vestibule. The bony **semicircular canals** are three—anterior, posterior, and lateral. They lie above and behind the vestibule and behind the internal acoustic meatus. The canals describe the greater part of a circle, are about 1 mm. in diameter, and have an enlargement at one end, the **osseous ampulla,** which is twice the diameter of the rest of the tube. The semicircular canals are oriented at right angles to one another. The **anterior semicircular canal** is vertical and lies transverse to the long axis of the petrous portion of the temporal bone. Its lateral end is ampullated and empties into the vestibule; the opposite end joins the upper end of the posterior canal to form the **crus commune.** The **posterior semicircular canal** is also vertical and is disposed parallel to the long axis of the petrous bone. Its lower ampullated end has a separate opening into the vestibule; its upper end enters with the anterior canal by their crus commune. The **lateral canal** is horizontal and arches lateralward. Its ampullated end opens into the vestibule close to the corresponding end of the anterior canal; its nonampullated end joins the vestibule near the posterior ampulla. The lateral canal of one ear is nearly in the same plane as the other, whereas the anterior canal of one side is essentially parallel to the posterior canal of the other side.

The **cochlea** resembles a snail shell lying on its side. It is placed anterior to the vestibule and the internal acoustic meatus, and its apex is directed forward and lateralward and slightly downward. The bony cochlea measures about 5 mm. from base to apex and is almost 1 cm. broad across its base. A central axis of bone, the **modiolus,** gives origin to a thin, bony lamina, the **osseous spiral lamina,** which partially subdivides the windings of the spiral into two segments, the **scala vestibuli** above the spiral lamina and the **scala tympani** below it. The modiolus and the osseous spiral lamina both have many perforations for the passage of fibers of the cochlear part of the vestibulocochlear nerve. The bony cochlea makes two and three-quarters turns around the modiolus, and its turns measure about 3 cm. in length. The basal turn bulges the medial wall of the middle ear as the promontory, and the cochlear, or round window, represents a communication between the bony cochlea and the middle ear. The window is closed in life by a membrane. The basal turn of the cochlea has an opening into the vestibular part of the bony labyrinth and also presents a canal, the **aqueductus cochleae,** which leads to the inferior surface of the petrous portion

of the temporal bone. M. W. Young has described a 'saccus perilymphaticus' which closes the end of the aqueductus cochleae and prevents communication between the periotic and subarachnoid spaces.

The Membranous Labyrinth (fig. 204B)

The membranous labyrinth is composed of a series of modified tubules containing endolymph. The tubules are enclosed within the bony labyrinth and are surrounded by the perilymphatic fluid of the bony labyrinth. The membranous labyrinth has the same general parts and arrangement as the bony labyrinth, but its tubules are much smaller in diameter. The membranous cochlea and semicircular ducts follow faithfully the form of the bony parts corresponding to them, but within the bony vestibule lie two membranous sacs, the utriculus and the sacculus.

The **utriculus** is the larger and is irregularly oblong in form. It receives the five terminals of the semicircular ducts, the crus commune combining one end of the anterior and posterior ducts. From its anterior wall originates the utriculosaccular duct, which makes a communication with the sacculus by joining the endolymphatic duct. The utriculus has a specific area of sensory nerve terminals, the horizontal **macula utriculi** (fig. 204B).

The **sacculus** is smaller and more spherical. It gives rise to the endolymphatic duct from its posterior wall. By the junction of this duct with the utriculosaccular duct, the sacculus and utriculus are placed in communication. The endolymphatic duct continues through the vestibular aqueduct to end in a terminal dilatation, the **endolymphatic sac,** which lies in contact with the dura mater on the posterior surface of the petrous bone. The sacculus has a short, narrow connection to the vestibular end of the cochlear duct, the **canalis reuniens.** In its anterior wall lies a vertical area of sensory epithelium, the **macula sacculi.**

The **semicircular ducts** (fig. 204B) conform in shape to the osseous semicircular canals but are about one-third their diameter. Each duct has an ampulla, where a thickening of the wall projects transversely into the duct as the **ampullary crest.** Here its sensory nerves end. The semicircular ducts open into the utriculus.

The **cochlear duct** (fig. 204B) is the membranous part of the bony cochlea and is continuous with the remainder of the membranous tubules through the canalis reuniens connecting it to the sacculus. The cochlear duct is triangular in cross section, being bounded below by a fibrous extension of the osseous spiral lamina, the **basilar membrane,** and above by the more delicate **vestibular membrane.** The outer wall of the triangle is the thickened periosteum of that part of the bony cochlea, the **spiral ligament.** On the basilar membrane is mounted the thickened and highly specialized **spiral organ** (organ of Corti). Stimulation of the hair cells of the spiral organ results in auditory perception. The membranous cochlear duct lies between the segments of the perilymphatic space of the bony cochlea; the scala vestibuli is internal to it and toward the apex of the bony cochlea; the scala tympani is toward the base. The pulsations set up in the perilymph by the movements of the stapes at the vestibular window are propagated through the scala vestibuli, transmitted by a communication at the cochlear apex (the **helicotrema**) into the fluid of the scala tympani, and adjusted to by compensatory movements of the membrane closing the cochlear window of the bony vestibule. Regional stimulation of the hair cells of the organ of Corti results from this oscillatory movement of the perilymphatic fluid.

Blood Vessels of the Labyrinth The principal artery is the labyrinthine, a branch of the basilar artery (p. 281), which traverses the internal acoustic meatus and divides into branches that accompany the cochlear and vestibular nerves. The stylomastoid artery also provides some blood supply. The veins are similar and unite into the labyrinthine vein which drains into the inferior petrosal sinus (p. 279). Small veins are transmitted through both the vestibular aqueduct and the cochlear aqueduct.

CRANIAL NERVES VII AND VIII

The facial nerve (cranial VII) and the vestibulocochlear nerve (cranial VIII) traverse the internal acoustic meatus side by side. The facial nerve has a complex course through the petrous portion of the temporal bone and then emerges at the stylomastoid foramen, whereas the vestibulocochlear nerve distributes wholly to nerve endings in the labyrinth and cochlea within the bone.

The Cranial Portion of the Facial Nerve

The **facial nerve** (figs. 168, 205, 206) emerges from the brainstem at the inferior border of the pons in the recess between the olivary eminence and the inferior cerebellar peduncle. It is composed of two unequal roots: the larger motor root supplies all the muscles of facial expression (p. 200); the smaller root is known as **nervus intermedius.** It contains taste fibers from the anterior two-thirds of the tongue, fibers of general sensation from the external acoustic meatus, and parasympathetic and visceral afferent fibers for the submandibular, sublingual, lacrimal, nasal, and palatine glands. The two roots of the facial nerve enter the internal acoustic meatus with the vestibulocochlear (cranial VIII) nerve which, however, divides into branches within the meatus. The facial roots pass lateralward in the meatus between the cochlea and semicircular canals; at the lateral end of the internal acoustic meatus they fuse and the geniculate ganglion is formed. The **geniculate ganglion** is the sensory ganglion of the facial nerve. It is located at the abrupt bend (genu = knee) taken by the nerve as it turns from the internal acoustic meatus into the posteriorly directed facial canal. Adjacent to the geniculate ganglion arises the **greater petrosal nerve** which, after a short course in the bone, emerges at the **hiatus of the canal for the greater petrosal nerve** into the middle cranial fossa. Its further course and distribution are described on p. 242. A small **geniculotympanic branch** passes from the ganglion to the lesser petrosal nerve (p. 196). Distal to the geniculate ganglion, the facial nerve enters the bony facial canal, which passes posteriorly in the medial wall of the tympanic cavity above the vestibular window and then, behind the window, turns nearly vertically downward along the posterior wall of the cavity. The bony canal continues to the stylomastoid foramen, where the facial nerve emerges from the skull. In the descending portion of its course, the facial nerve gives rise to the branch to the stapedius muscle (which traverses a tiny canal into the base of the pyramid), a communication to the auricular branch of the vagus nerve (p. 164), and the chorda tympani.

The **chorda tympani** (fig. 205) arises from the facial nerve about 5 mm. proximal to the stylomastoid foramen and, turning sharply upward, enters a separate canal in the bone which conducts it into the tympanic cavity between the base of the pyramid and the posterior border of the tympanic membrane. The chorda passes forward over the medial surface of the tympanic membrane and under its mucous membrane, arching across the handle of the malleus lateral to the long process of the incus, and leaves the tympanic cavity near the anterior border of the membrane. This anterior aperture leads to the petrotympanic fissure, through which the chorda tympani emerges from the skull. The nerve contains taste fibers from the anterior two-thirds of the tongue and the parasympathetic innervation to the submandibular, sublingual, and lingual glands. Its peripheral course and relations are described with the mandibular nerve and the submandibular and sublingual glands (p. 173).

THE VESTIBULOCOCHLEAR NERVE (figs. 168, 204B, 206, 207)

The vestibulocochlear (cranial VIII) nerve consists of two functionally distinct and only incompletely united portions, the vestibular and cochlear nerves. These parts are united outside the brainstem and within the internal acoustic meatus but have distinctly different central connections and peripheral relations. Both portions are exclusively sensory; special somatic afferent (exteroceptive) for the cochlear nerve and special somatic afferent (proprioceptive) for the vestibular nerve.

The Cochlear Nerve (fig. 204B)

The peripheral processes of the cochlear nerve arise in the organ of Corti, their cell bodies forming the spiral ganglion of the cochlea which is situated in the osseous spiral lamina. The central processes converge and traverse the modiolus to form the cochlear nerve which, in the internal acoustic meatus, is combined with the vestibular nerve. At its entrance into the brainstem the cochlear nerve fibers are lateral to the vestibular root.

The Vestibular Nerve (figs. 204B, 206)

Peripherally, the vestibular nerve consists of three branches. The superior branch arises in the utricle and in the ampullae of the anterior and lateral semicircular ducts; the inferior branch, in the saccule; and the posterior branch, in the

ampulla of the posterior semicircular duct. These branches converge to the vestibular ganglion located in the outer part of the internal acoustic meatus. The central root ends in the vestibular centers of the medulla oblongata.

THE SKULL

The skull is the skeleton of the head. Here a series of flat bones are united by interlocking or overlapping sutures in the formation of the cranium, and a group of irregular bones participate in establishing the bony framework of the face and the base of the cranium. The cranium, enclosing a cavity accommodating the brain, consists of eight bones; the bones of the face total fourteen.

THE SUPERIOR ASPECT OF THE SKULL (fig. 208)

Viewed from above, the skull is somewhat oval in form. It is, however, broadened posteriorly by the parietal eminences, and it may exhibit similar projections anteriorly when the frontal eminences

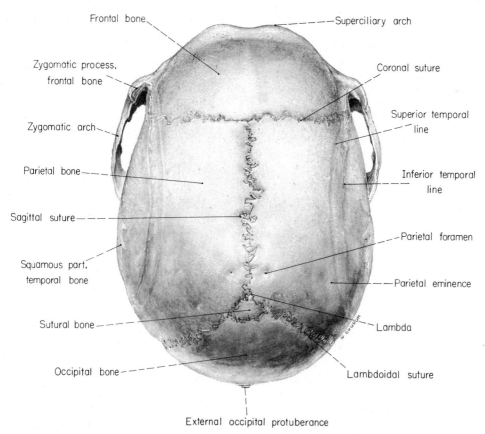

Fig. 208   The skull viewed from above. ($\times \frac{1}{2}$)

are large. In the child the prominence of both of these eminences gives the skull almost a square appearance. The **superciliary arches** of the frontal bone form the anterior limit of the skull in this view; the curvature of the occipital bone, the posterior limit. The **zygomatic arches** come into view laterally, except in broad skulls, in which they are hidden. The bones of the superior aspect of the skull are the frontal, the parietal, and the occipital. These form the **calvaria,** or skull-cap. The **sagittal suture** lies in the midline between the two parietal bones and ends anteriorly in a junction with the transverse **coronal suture** separating the frontal and the two parietal bones. It terminates posteriorly at the apex of the angular **lambdoidal suture** between the posterior borders of the parietal bones and the occipital bone. These junction points, representing the areas where the radial growth of the bones brings them last into apposition, are the sites of the anterior and posterior fontanelles of the infant (fig. 30).

The **posterior fontanelle,** located at the **lambda,** is triangular in form and closes during the first few months after birth. The **anterior fontanelle,** situated at the angle of junction of the two frontal and the two parietal bones, the **bregma,** is cruciate, or star-shaped. It is the largest of the fontanelles and is not filled in completely by bone until about the eighteenth month of life. The bilateral frontal bones, characteristic of developmental stages, are separated by a sagittal interfrontal suture during infancy. In most cases this **metopic suture** is completely obliterated. However, it persists in the adult in about 8 per cent of cases and, seen radiographically, may be mistaken for a fracture line. Small, irregular lateral fontanelles occur at the sphenoid and mastoid angles of the parietal bone. They are overlain by the temporal muscle, fuse early, and are of less significance than those of the median line. A **parietal foramen** lies close to the sagittal suture a short distance anterior to the lambda. It transmits the parietal emissary vein.

Clinical Note Premature closure of the fontanelles (synostosis) may substantially reduce the size of the cranial vault and thus prevent normal brain development.

THE POSTERIOR ASPECT OF THE SKULL (figs. 209, 213)

The back of the skull is round in outline, its roundness modified by the parietal eminences superolaterally and by the **mastoid processes** inferolaterally. In its center the **lambda** marks the junction of the interparietal **sagittal suture** and the highly serrated parietooccipital **lambdoidal suture.** The latter suture is continuous on either

side with the **parietomastoid** and **occipitomastoid suture.** One or more sutural bones may lie in the region of the lambda. Adjacent to the occipitomastoid suture is the mastoid foramen for the mastoid emissary vein. The **external occipital protuberance** is a marked process of the lower median line; the **superior nuchal line** extends outward from it on either side. Above the superior nuchal line is the faintly marked **highest nuchal line.** From the external occipital protuberance the **median nuchal line** descends toward the foramen magnum.

THE LATERAL ASPECT OF THE SKULL (fig. 209)

The lateral aspect of the skull is partly cranial and partly facial. The cranial portion consists of the temporal fossa and the external acoustic and mastoid regions. The facial portion is composed of the zygomatic arch, the infratemporal fossa, and the lateral aspect of the maxilla and mandible.

The Temporal Fossa (fig. 209)

The oval temporal fossa is bounded above and behind by the temporal lines and, in front, by the frontal and zygomatic bones. Inferiorly, the fossa is bounded superficially by the zygomatic arch and, deeply, by the **infratemporal crest of the sphenoid bone.** The floor of the temporal fossa is formed by parts of the frontal, parietal, and temporal bones and by the great wing of the sphenoid bone. These bones are separated from one another and from the zygomatic bone by the **coronal** (frontoparietal) **suture,** by the **squamosal suture** between the parietal and temporal bones, and by the **frontozygomatic** and **temporozygomatic sutures.**

The great wing of the sphenoid is separated from the bones surrounding it by the **sphenozygomatic suture** anteriorly, the **sphenofrontal** and **sphenoparietal sutures** above, and the **sphenosquamosal suture** posteriorly. The **pterion** is a point marked by the earlier position of the anterolateral fontanelle or by the junction of the great wing of the sphenoid bone with the adjacent frontal, parietal, and temporal bones. It lies about 3 cm. behind and a little above the zygomatic process of the frontal bone.

The two temporal lines run parallel to one another and are separated by a narrow band of smooth bone. The inferior one begins anteriorly at the temporal crest of the frontal bone, arches across the parietal bone, and ends in the supra-meatal crest of the temporal bone. It marks the superior limit of the temporal muscle.

The External Acoustic and Mastoid Regions (fig. 209)

The bony external acoustic meatus opens on the lateral aspect of the skull. The **aperture** of the meatus is bounded by the roughened free margin of the **tympanic portion of the temporal bone** to which attaches the cartilaginous part of the meatus. The **mastoid portion of the temporal bone** lies below the suprameatal crest and is separated from the external acoustic meatus by the **tympanomastoid suture.** It projects downward in the roughened conical **mastoid process.** The mastoid process is elaborated for the attachment of powerful neck muscles and is not present at birth.

The Zygomatic Bone and Arch (figs. 209–211, 213)

The zygomatic bone forms the prominence of the cheek. It has a broad connection with the maxilla and is also united with the frontal, the sphenoid, and the temporal bones. The large space internal to the zygomatic bone and arch accommodates the temporal muscle, and the temporal fascia attaches to the sharp superior border of the arch. Its roughened inferior border gives attachment to the masseter muscle and fascia. The **zygomaticofacial foramen** on the outer surface of the zygomatic bone transmits the zygomaticofacial nerve. The zygomatic bone and the slender zygomatic process of the temporal bone are united at an oblique suture to form the **zygomatic arch.** The upper border of the zygomatic process of the temporal bone is continued backward

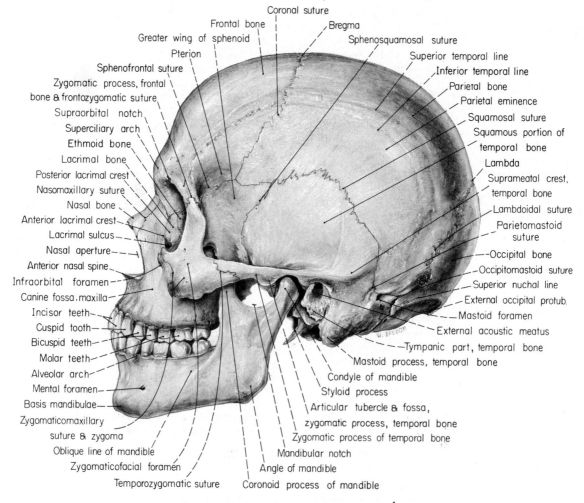

Fig. 209   The lateral aspect of the skull. ($\times \frac{1}{2}$)

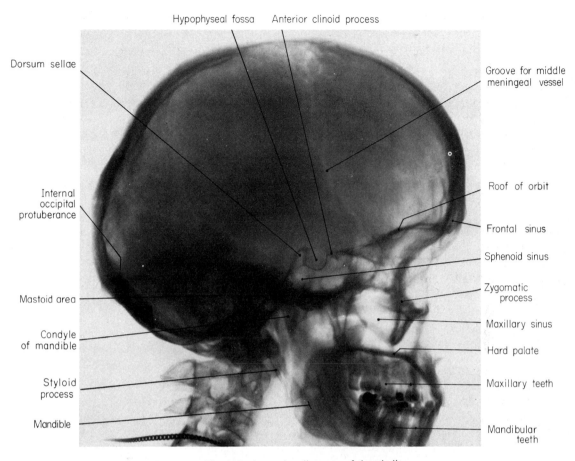

Fig. 210 Lateral radiogram of the skull.

above the external acoustic meatus as the supra-meatal crest; an enlargement of its lower border forms the **articular tubercle** of the temporomandibular joint. Behind the tubercle is the **mandibular fossa.**

The Infratemporal Fossa (figs. 139, 209, 213)

The infratemporal fossa is the irregular space below and deep to the zygomatic arch and posterior to the maxilla. It is bounded above by the infratemporal crest of the sphenoid bone and below by the alveolar border of the maxilla. The ramus of the mandible is its lateral boundary; the lateral pterygoid plate is its medial limit. The **foramen ovale** and **foramen spinosum** open into its roof. The **inferior orbital fissure** is a horizontal cleft between the infratemporal fossa and the floor of the orbit. Between the upper part of the lateral pterygoid plate and the adjacent border of the maxilla is the **pterygopalatine**

fissure which leads from the infratemporal fossa to the deep, triangular pterygopalatine fossa. The infratemporal fossa accommodates the principal muscles of mastication, the maxillary artery, and the mandibular division of the trigeminal nerve.

The Lateral Aspect of the Maxilla and Mandible

The infratemporal surface of the maxilla is convex and forms part of the anterior limit of the infratemporal fossa. This surface ends anteriorly in the zygomatic process of the maxilla and in a ridge extending upward from the socket of the first molar tooth. It is pierced about its center by the apertures of the alveolar canals for the posterior superior alveolar nerves and vessels. The **maxillary tuberosity** is a rounded eminence inferiorly which articulates medially with the pyramidal process of the palatine bone and, laterally, gives attachment to part of the medial pterygoid muscle. The description of the lateral aspect of the mandible is included with the general description of this bone (p. 218).

THE ANTERIOR ASPECT OF THE SKULL (figs. 195, 211)

Seen from in front, the orbits, nasal region, maxilla, mandible, and forehead are prominent features of the skull. The frontal bone of the forehead is smoothly convex; it turns into the orbits below, forming the strong supraorbital margins. These are thin and prominent laterally, heavy and more rounded medially. The smooth depressed area between the two superciliary arches is the **glabella.** It may show an inferior remnant of the metopic suture. **Supraorbital foramina,** or notches, are placed at the junction of the

medial and middle thirds of the supraorbital margins. At the root of the nose the **frontonasal suture** separates the corresponding bones, its midpoint being the **nasion.** Lateralward from the frontonasal suture runs the junction of the frontal bone with the maxilla and then with the lacrimal bone. Below the frontonasal suture the nasal bones, convex from side to side and united at the midline, form the bridge of the nose. They are supported deeply in the midline by the perpendicular plate of the ethmoid bone and join the maxillae along their lateral margins. Below the nasal bones appears the oval **nasal aperture,** its narrow end upward. In its interior the perpendicular plate of the ethmoid and the anterior end of the vomer lie vertically in the median plane. A jagged notch between these bones anteriorly

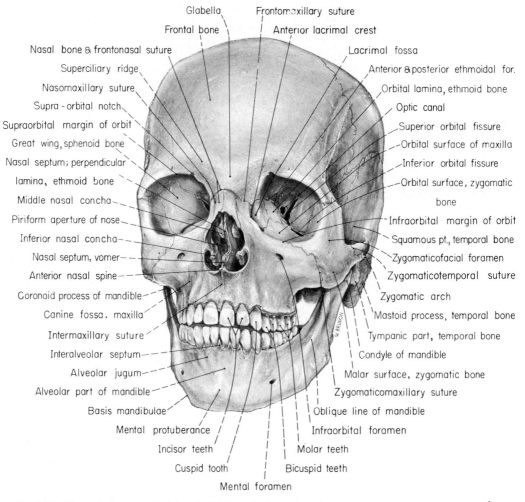

Fig. 211   The anterior aspect of the skull (viewed along the left anteroposterior orbital axis). ($\times \frac{1}{2}$)

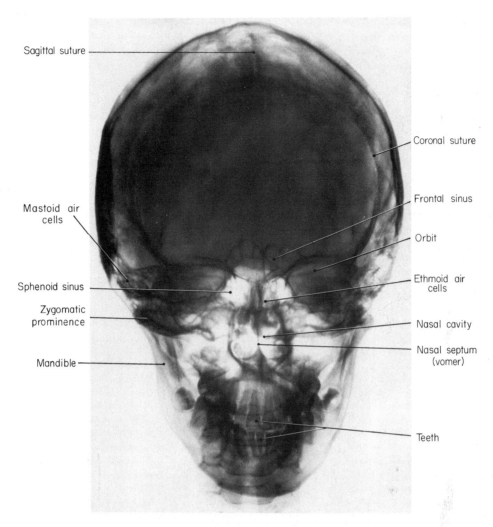

Sagittal suture

Coronal suture

Frontal sinus

Mastoid air
cells

Orbit

Ethmoid air
cells

Sphenoid sinus

Zygomatic
prominence

Nasal cavity

Nasal septum
(vomer)

Mandible

Teeth

Fig. 212   An anterior-posterior radiogram of the skull.

is the site of the septal cartilage. The sharp **anterior nasal spine** projects forward from the nasal aperture inferiorly in the median line.

The **maxillae** surround the nasal aperture, form the inferior orbital margin medially, and have a broad sutural connection with the zygomatic bones lateralward. Below, the alveolar processes of the maxillae expand to form sockets for the upper teeth. The **canine eminence** is an enlargement over the root of the cuspid tooth and separates the **incisive fossa** medially from the **canine fossa** laterally. The prominent **infraorbital foramen** lies just below the orbit and lateral to the nasal aperture. It transmits the infraorbital blood vessels and nerve.

The **zygoma** forms the prominence of the

cheek and the lateral border of the orbit and has sutural connections with the maxilla and the frontal bones. The **zygomaticofacial foramen** for the small nerve and vessels of the same name perforates the bone. The bony orbits and the mandible have been described. Prominent on the frontal aspect of the **mandible** is the median ridge indicating the position of the symphysis, the **mental foramen** below the second bicuspid tooth, and the **oblique line** running upward from the mental tubercle to become continuous behind with the anterior border of the ramus. The frontal view of the skull shows the supraorbital, infraorbital, and mental foramina all on a single vertical line dropped from the junction of the medial and middle thirds of the superior orbital margin.

THE BASE OF THE SKULL (figs. 167, 213)

The base of the skull exhibits anteriorly the alveolar arch of the maxilla, the hard palate, and the pterygoid plates and, posteriorly, the under surface of the occipital bone. In a middle zone occur the foramen magnum, the basilar process of the occipital bone, the under surface of the petrous portion of the temporal bone, and the styloid and mastoid processes, together with a number of foramina. Laterally, the infratemporal fossa is bounded by the under surface of the great wing of the sphenoid and by the zygomatic arch and mandibular fossa. The hard palate, together with its bones and foramina, is described on p. 233. Behind the hard palate the **posterior nasal apertures,** measuring 2.5 cm. in their vertical and 1.5 cm. in their transverse dimensions, are separated by the vomer. Also in the median line at the posterior margin of the palatine bone is the **posterior nasal spine** for the attachment of the

musculus uvulae. The vomer expands superiorly as alae, which abut against the rostrum of the sphenoid. Lateral to the alae of the vomer and at the roots of the pterygoid processes are the **pharyngeal canals.** The pterygoid processes appear as wings at either side of the nasal apertures. At their base and under projecting spines are the openings of the pterygoid canals. The **medial pterygoid plate** is long and narrow. Laterally, at its base it is hollowed out in the oval **scaphoid fossa** for the origin of the tensor veli palatini muscle; at the inferior extremity of the plate, the tendon of the tensor turns around its recurved **hamulus.** The **lateral pterygoid plate** is broad and wing-like and forms a large hollow cavity for the origin of the medial pterygoid muscle. The lateral surface of the lateral pterygoid plate forms the medial boundary of the infratemporal fossa and gives origin to the lateral pterygoid muscle. At the base of the lateral pterygoid plate lies the **foramen ovale** for the mandibular division of the trigeminal nerve and the accessory

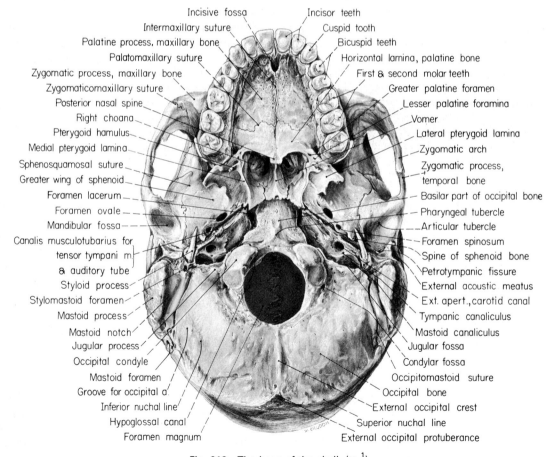

Incisive fossa
Intermaxillary suture
Palatine process, maxillary bone
Palatomaxillary suture
Zygomatic process, maxillary bone
Zygomaticomaxillary suture
Posterior nasal spine
Right choana
Pterygoid hamulus
Medial pterygoid lamina
Sphenosquamosal suture
Greater wing of sphenoid
Foramen lacerum
Foramen ovale
Mandibular fossa
Canalis musculotubarius for tensor tympani m. & auditory tube
Styloid process
Stylomastoid foramen
Mastoid process
Mastoid notch
Jugular process
Occipital condyle
Mastoid foramen
Groove for occipital a.
Inferior nuchal line
Hypoglossal canal
Foramen magnum

Incisor teeth
Cuspid tooth
Bicuspid teeth
Horizontal lamina, palatine bone
First & second molar teeth
Greater palatine foramen
Lesser palatine foramina
Vomer
Lateral pterygoid lamina
Zygomatic arch
Zygomatic process, temporal bone
Basilar part of occipital bone
Pharyngeal tubercle
Articular tubercle
Foramen spinosum
Spine of sphenoid bone
Petrotympanic fissure
External acoustic meatus
Ext. apert., carotid canal
Tympanic canaliculus
Mastoid canaliculus
Jugular fossa
Condylar fossa
Occipitomastoid suture
Occipital bone
External occipital crest
Superior nuchal line
External occipital protuberance

Fig. 213   The base of the skull. ($\times \frac{1}{2}$)

meningeal artery and, behind this, the small **foramen spinosum** which transmits the middle meningeal vessels and the meningeal branch of the mandibular nerve. Behind foramen spinosum is the **spine** of the sphenoid bone which gives attachment to the sphenomandibular ligament and the tensor veli palatini muscle. Medial to the spine is the deep **petrotympanic fissure** which continues into the mandibular fossa. The petrotympanic fissure separates the mandibular fossa into a smooth concavity anteriorly for the mandibular condyle and a nonarticular, rough posterior portion which is frequently occupied by a part of the parotid gland. The **styloid process** emerges between laminae of the tympanic part of the temporal bone. Posteriorly, at the base of the styloid process is the **stylomastoid foramen** which transmits the facial nerve and stylomastoid artery. Lateral to the stylomastoid foramen and between the tympanic portion of the temporal bone and the mastoid process is the **tympanomastoid suture** which accommodates the auricular branch of the vagus. The **mastoid process** is ridged and corrugated by muscle attachments; it has, medial to it, the deep **mastoid notch** for the posterior belly of the digastric muscle. Medial to the notch is a groove for the **occipital artery.** The **mastoid foramen** is close to the occipitomastoid suture near the posterior border of the mastoid portion of the temporal bone.

The under surface of the petrous portion of the temporal bone contains, in its midregion, the orifice of the **carotid canal** by which the internal carotid artery enters the skull. The **foramen lacerum** is the irregular area between the tip of the petrous bone, the great wing of the sphenoid, and the side of the basilar portion of the occipital bone. The inferior aspect of this gap in the dried bone is filled in, normally, by fibrocartilage. Across the cerebral surface of the cartilage the internal carotid artery passes toward the cavernous sinus. The pterygoid canal opens into the anterior edge of the foramen lacerum anteriorly. Between the petrous bone and the great wing of the sphenoid, anteromedial to the sphenoidal spine, is the deep groove of the **sulcus tubae auditivae** for the cartilaginous part of the auditory tube. The groove is continuous behind with the bony portion of this canal in the temporal bone and, at the forward end of the sulcus, ends in a notch in the free border of the medial pterygoid plate. Behind the orifice of the carotid canal lies the large **jugular foramen** located between the petrous portion of the temporal bone and the

occipital bone. This foramen may be compartmented. Posteriorly, it accommodates the jugular bulb of the internal jugular vein; anteriorly, it transmits the inferior petrosal sinus, and between these pass the glossopharyngeal, vagus, and accessory nerves. On the ridge of bone between the orifice of the carotid canal and the jugular foramen is the **tympanic canaliculus** for the passage of the tympanic branch of the glossopharyngeal nerve; in the lateral wall of the jugular foramen near the root of the styloid process is the **mastoid canaliculus,** through which the auricular branch of the vagus enters the temporal bone.

The **foramen magnum,** at the center of the base of the skull, is bordered anteriorly by the oval occipital condyles which articulate with the superior facets on the lateral masses of the first cervical vertebra. Through the foramen magnum pass the spinal cord and its meningeal coverings, the ascending rootlets of the accessory nerve, the vertebral arteries, the anterior and posterior spinal arteries, and the ligaments passing between the occipital bone and the axis. The condyles are roughened medially by the attaching fibers of the alar ligaments. Anterior to each occipital condyle is the **hypoglossal canal** for the hypoglossal nerve. Posterior to each condyle is a **condylar canal** for an emissary vein. The **median nuchal line** extends posteriorly from the foramen magnum to the **external occipital protuberance.** Extending lateralward from the protuberance are the **superior nuchal lines;** from the median nuchal line, at about its midlength, the **inferior nuchal lines.** These, as well as the surface of the occipital bone between them, are rough for the attachment of the muscles of the back of the neck. The thick basilar portion of the occipital bone lies anterior to the foramen magnum and the occipital condyles. About 1 cm. anterior to the foramen magnum, the basilar process exhibits a small midline elevation, the **pharyngeal tubercle,** to which is attached the median fibrous raphe of the pharynx.

THE INTERIOR OF THE CRANIUM (figs. 168, 214, 215)

The Inner Aspect of the Calvaria

Impressions of the meningeal arteries are apparent on the inner surface of the skull-cap. Of these, the grooves for the anterior and posterior branches of the middle meningeal arteries can be traced to their junction in a larger groove for the main stem of the artery. The groove for the anterior branch may be completely bridged over by bone in the region of the pterion. Along the median line, a longitudinal groove lodges the superior sagittal sinus, and its margins mark the attachments of the falx cerebri. The groove, beginning in the crest of the frontal bone, is narrow in front and broadens posteriorly. It ends in the region of the **internal occipital protuberance.** At the sides of the groove for the superior sagittal sinus, shallow depressions of the inner table of bone accommodate the arachnoid granulations. These are clumps of greatly folded arachnoid membrane which pierce the dura mater of the superior sagittal sinus. Through the foldings of the arachnoid membrane, the cerebrospinal fluid passes from the subarachnoid space around the brain into the venous stream of the superior sagittal sinus. Toward the posterior borders of the parietal bones, the **parietal foramina** open through the bone.

The Floor of the Cranial Cavity (figs. 168, 215)

The floor of the cranial cavity presents three distinct areas: the anterior, middle, and posterior cranial fossae.

The **anterior cranial fossa** is limited behind by the posterior borders of the lesser wings of the sphenoid bone and by the groove for the optic chiasma. Its floor is formed by the orbital plates

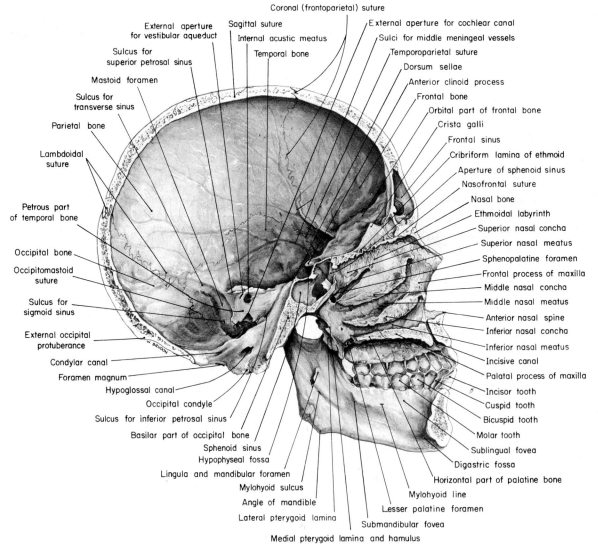

Fig. 214   A median section of the skull. ($\times \frac{1}{2}$)

of the frontal bone, the cribriform plate of the ethmoid, and the lesser wings and forepart of the body of the sphenoid bone. In the midline anteriorly, the **crest of the frontal bone** leads, at its base, to the **foramen cecum,** through which an emissary vein passes from the nasal cavity to the beginning of the superior sagittal sinus. Behind the foramen cecum, the **crista galli** (cock's comb) of the ethmoid bone gives attachment to the falx cerebri. On either side of the crista galli, the perforated **cribriform plate** of the ethmoid provides passage for the olfactory nerve fibers to the olfactory bulb which lies on the cribriform plate. At the posterior end of the frontoethmoidal suture is the **posterior ethmoidal foramen** for the posterior ethmoidal vessels and nerve. The **anterior ethmoidal foramen** lies at about the midlength of the suture and provides for a short passage of the anterior ethmoidal vessels and nerve along the floor of the cranial fossa. The **orbital plates** of the frontal bone are convex upward and

are ridged and furrowed in conformity with the convolutions of the orbital surface of the frontal lobes of the brain. Sphenoethmoid and sphenofrontal sutures pass across the posterior part of the fossa and mark the union of the corresponding bones.

The **middle cranial fossa** is composed of deep depressions at either side of the body of the sphenoid bone, extending from the posterior limit of the anterior cranial fossa backward to the superior angles of the petrous portions of the temporal bones. The fossa is continuous across the midline to include the chiasmatic groove and the sella turcica. The bones of this fossa are the body and the great wings of the sphenoid, the squamous portion of the temporal bone and the anterior surface of its petrous portion, and the sphenoidal angles of the parietal bones. The central part of the fossa contains anteriorly the chiasmatic groove which leads at either side into the

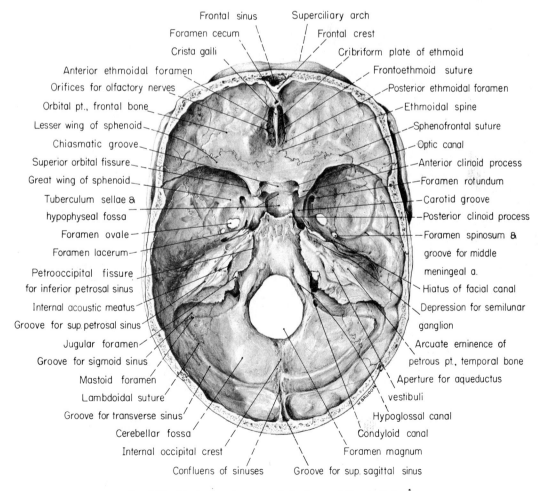

Fig. 215 The internal aspect of the base of the skull. ($\times \frac{1}{2}$)

optic canal for the optic nerve and ophthalmic artery. Behind the optic canal the **anterior clinoid processes** of each side project backward from the lesser wings of the sphenoid bone and give attachment to the tentorium cerebelli. Behind the chiasmatic groove the deep concavity of the **sella turcica** provides the **hypophyseal fossa** for the hypophysis. It is bounded in front by a pair of small projections, the **middle clinoid processes** and, behind, by a flat plate of bone, the **dorsum sellae,** on the upper angles of which are the **posterior clinoid processes.** Lateral to the sella turcica is the large **carotid groove** which begins behind and below at the foramen lacerum. It has the form of an S lying on its side, for anteriorly the groove turns upward and then backward to end at the medial edge of the anterior clinoid process. The occasional presence of bony union between the anterior and middle clinoid processes converts the end of the groove into a caroticoclinoid foramen. The groove marks the course of the internal carotid artery through the cavernous sinus.

The lateral portions of the middle fossa are deep and receive the temporal lobes of the brain, the bone being modeled in accordance with its gyri and sulci. Anteriorly, the **superior orbital fissure** provides communication with the orbit in the interval between the lesser and greater wings and the body of the sphenoid bone. This fissure transmits many of the nerves entering the orbit together with the ophthalmic vein (p. 246). Behind the superior orbital fissure is the **foramen rotundum** which leads to the pterygopalatine fossa and transmits the maxillary division of the trigeminal nerve. The larger, oval **foramen ovale** lies behind and lateral to the foramen rotundum. Through it pass the mandibular division of the trigeminal nerve and the accessory meningeal artery. The small **foramen spinosum,** behind and lateral to foramen ovale, transmits the middle meningeal artery and the meningeal branch of the mandibular nerve. The course and branches of the **middle meningeal artery** can be followed in grooves on the inner table of the skull. The main vessel passes anteriorly and somewhat lateralward for several centimeters over the medial portion of the squamous part of the temporal bone. It then divides into anterior and posterior branches, the **posterior branch** turning fairly sharply backward to ascend diagonally across the squamous parts of the temporal and parietal bones toward the lambda. The **anterior branch** continues forward and upward toward the anterior inferior angle of the parietal bone. Here, at the **pterion,** the groove is always very deep and may be roofed over by bone, a circumstance that frequently leads to tearing of the vessel in cases of skull fracture. Continuing in a normal groove above the pterion, the vessel is directed toward the vertex of the skull, providing numerous radiating branches. The **foramen lacerum,** medial to foramen ovale, transmits the internal carotid artery as it enters the carotid groove. The inferior aspect of the foramen lacerum is closed in life by fibrocartilage. A meningeal branch of the ascending pharyngeal artery pierces the layer of fibrocartilage. The pterygoid canal begins under the anterior edge of the foramen lacerum. The anterior surface of the petrous bone has, lateralward, a smooth, rounded **arcuate eminence** produced by the anterior semicircular canal and, medially and inferiorly, the slit-like **hiatus of the canal for the greater petrosal nerve.** Through this opening and along the groove leading from it pass the greater petrosal nerve and the petrosal branch of the middle meningeal artery. A smaller groove for the lesser petrosal nerve is frequently seen antero-inferior to that for the greater nerve and may lead to or toward the foramen ovale. On the anterior surface of the petrous pyramid near its apex a shallow depression lodges the semilunar ganglion of the trigeminal nerve.

The **posterior cranial fossa** is large and deep and accommodates the cerebellum, pons, and medulla oblongata. The fossa is formed by the dorsum sellae of the sphenoid bone, the occipital bone, and the petrous and mastoid portions of the temporal bones. The superior angle of the petrous temporal, forming the anterior limit of the fossa, gives attachment to the tentorium cerebelli and has a groove for the superior petrosal sinus. Posteriorly, the transverse groove on the occipital bone lodging the transverse sinus limits the posterior fossa. At the center of the fossa is the **foramen magnum** which presents anteriorly, at its sides, the hypoglossal canal for the hypoglossal nerve. Anterior to the foramen, the basilar portion of the occipital bone, fused with the base of the sphenoid, supports the pons and medulla oblongata. At the side of the basilar portion of the sphenoid, the petrooccipital fissure conducts the inferior petrosal sinus to the jugular foramen. The large, irregular, and sometimes compartmented jugular foramen is concerned with both cranial nerves and dural sinuses (p. 275). Above the jugular foramen

Foramina and other Openings of the Cranial Fossae and their Contents

Foramina	*Contents*
Cecum	Nasal vein to superior sagittal sinus
Cribriform plates	Filaments of olfactory nerves
Anterior and posterior ethmoidal	Vessels of same name and anterior ethmoidal nerve
Optic canals	Cranial N. II, ophthalmic artery
Superior orbital fissure	Ophthalmic vein, Ophthalmic div. of Cranial N. V, Cranial Ns. III, IV, and VI, sympathetic fibers
Rotundum	Maxillary div. of Cranial N. V.
Ovale	Mandibular div. of Cranial N. V and accessory meningeal artery
Lacerum	Internal carotid artery, greater and deep petrosal nerves, minor vessels
Hiatus of greater petrosal nerve	Greater petrosal nerve and petrosal br. of middle meningeal artery
Spinosum	Middle meningeal artery and vein and meningeal br. of mandibular N.
Internal acustic meatus	Cranial Ns. VII and VIII, labyrinthine artery. (Cranial N. VII also exits at stylomastoid foramen)
Jugular	Cranial Ns. IX, X, and XI, inferior petrosal and sigmoid sinuses, meningeal brs. of ascending pharyngeal and occipital arteries
Hypoglossal canal	Cranial N. XII
Magnum	Spinal cord, vertebral arteries, spinal roots of Cranial N. XI, dural veins, anterior and posterior spinal arteries
Condyloid	Condyloid emissary vein
Mastoid	Mastoid emissary vein

is the **internal acoustic meatus** for the facial and vestibulocochlear nerves and the labyrinthine artery. Behind and lateral to this is the **aqueductus vestibuli** transmitting the endolymphatic duct from the membranous labyrinth of the ear. Posterior to the foramen magnum an **internal occipital crest** separates the cerebellar fossae, gives attachment to the falx cerebelli, and lodges the occipital sinus. The deep grooves for the transverse sinuses begin at the internal occipital protuberance and are continued into the grooves for the sigmoid sinuses. These curve down and medialward across the mastoid portion of the temporal bone and end by turning forward into the jugular foramina. As the sigmoid groove reaches the mastoid portion of the temporal bone, the **mastoid foramen** opens into it, and the **condyloid canal** terminates in it just before the jugular foramen is reached.

The numerous details of foramina and contents may be summarized in a table.

THE CONTENTS OF THE CRANIUM

The cranial cavity is occupied by the brain, its meningeal coverings, and several systems of blood vessels which are largely accessory to the brain. The study of the brain and spinal cord is an extensive and special one, and the present description is limited to the meningeal coverings and the vascular systems.

Meninges of the Brain (figs. 168, 216–19)

Both brain and spinal cord are invested by three layers of protective and nourishing membranes: the pia mater, the arachnoid membrane, and the dura mater. These are continuous over the spinal cord and brain, but all features of the membranes are not the same in both regions.

Pia Mater This is the delicate, intimate, areolar investment of brain and spinal cord which enmeshes the blood vessels on their surfaces. It is thus described as a vascular membrane. The pia mater sends prolongations into the nervous tissue along perivascular spaces and as septa; it follows the contours of the brain and spinal cord faithfully and cannot be dissected from them.

Arachnoid This is a delicate, transparent membrane composed of a blend of collagenous and elastic fibers, covered on its inner surface by a layer of squamous mesenchymal epithelial cells. The arachnoid is separated from the pia mater by an interval which contains cerebrospinal fluid; this is the **subarachnoid space.** In the cranium the arachnoid membrane loosely invests the brain and does not dip into its sulci, except for the deep **longitudinal cerebral fissure.** The subarachnoid space is crossed by trabeculae derived from both the pia mater and the arachnoid. It is a very narrow interval over the convolutions, or gyri, of the brain but is more capacious in the sulci since the arachnoid layer does not dip into them. In regions where

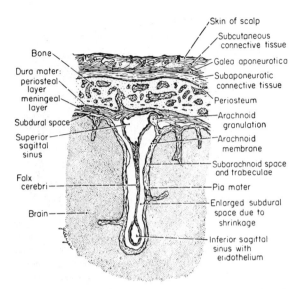

Fig. 216 A coronal section of the scalp, skull, and brain illustrating the meningeal relations and the formation of the dural venous sinuses.

the brain contours change markedly, the arachnoid bridges over the intervals between the brain parts, and subarachnoid cisterns occur. In these areas, greater quantities of cerebrospinal fluid are located. The **cisterna cerebellomedullaris** is the interval between the under surface of the cerebellum and the dorsum of the medulla oblongata. Cisternae at the base of the brain are the **cisterna pontis** at the border of that part of the brainstem called the

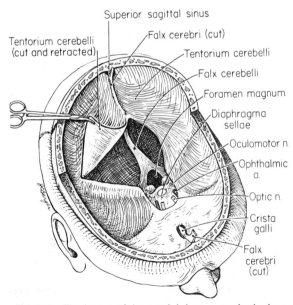

Fig. 217 The layers of the cranial dura mater in the base of the skull.

pons and the **cisterna interpeduncularis** on the ventral aspect of the midbrain where the arachnoid bridges across from the cerebral peduncle of one side to that of the other. Additional smaller cisterns exist. The **arachnoid granulations** are tuft-like collections of highly folded arachnoid which project through the dura mater of lateral bays of the superior sagittal sinus and of other dural sinuses. Through their thin membranes, the cerebrospinal fluid is passed into the blood stream.

Cerebrospinal fluid is a watery, alkaline fluid, similar in constitution to blood plasma. It is elaborated by (or through) the choroid plexuses of the lateral, the third, and the fourth ventricles of the brain. It occupies the intercommunicating ventricles and, being constantly formed, is drained from the ventricles by minute foramina in the roof of the fourth ventricle. These are the **median** and **lateral apertures of the fourth ventricle,** the latter pair being located at the extremities of the lateral recesses of the ventricle. Small additions to the cerebrospinal fluid are made through the perivascular channels of the brain surface and by the ependyma of the central canal of the spinal cord. The total volume of the fluid is from 130 to 150 cc. Emerging through the foramina into the subarachnoid space, the cerebrospinal fluid bathes the surface of the brain and spinal cord, providing a fluid suspension and a valuable shock absorber around these organs of the nervous system. The fluid has a pressure of about 100 mm. of water, which is intermediate between that of peripheral arterial and venous sinus pressures. Cerebrospinal fluid readily passes through the thinned-out membrane of the arachnoidal granulations and the endothelial lining of the dural sinuses and joins the venous blood of the sinus. A smaller part of the fluid is returned to the vascular system by way of the lymphatics of the cranial nerves.

Clinical Note Excessive elaboration or an obstruction to drainage of the cerebrospinal fluid may result in hydrocephalus (water on the brain). Obstruction can be the result of inflammation of the meninges closing off the minute median and lateral apertures of the fourth ventricle. The resultant back-pressure dilates the ventricles and compresses the cerebral cortex against the unyielding skull, causing a marked thinning and degeneration of the cortex with severe mental retardation. If the sutures of skull have not closed, as in the infant, there follows great enlargement of the ventricular system of the brain and of the total skull.

Dura Mater The dura mater is a thick, dense, fibrous layer which encloses and protects the brain and spinal cord. Its cranial and spinal portions are continuous at the foramen magnum, to the margins of which it is attached. The dura mater extends along the cranial and spinal nerves, as they leave the meningeal coverings, and becomes continuous with their sheaths. While the dura mater of the spinal cord is a long simple cylinder, the cranial dura has certain special features. The cranial dura mater is described as a two-layered structure, having a meningeal and a periosteal layer. The periosteum of the inner surface of the cranial bones adheres more closely to the meningeal dura than to the bone, except where it passes through the sutures. For this reason, the periosteum is classified as an outer layer of dura mater. The meningeal layer forms duplications which are prolonged between the parts of the brain as the falx cerebri, the falx cerebelli, the tentorium cerebelli, and the diaphragma sellae. Lastly, between areas of separation of the periosteal and meningeal layers or between duplications of the meningeal layers, dural venous sinuses are formed.

The **falx cerebri** is a sickle-shaped duplication of the meningeal layer of the dura which occupies the longitudinal cerebral fissure between the cerebral hemispheres. Narrow in front where it is attached to the crista galli of the ethmoid, it broadens as it passes backward and ends by becoming continuous with the tentorium cerebelli. Its upper convex margin corresponds to the sagittal suture and terminates posteriorly at the internal occipital protuberance. The superior sagittal sinus is developed within the convex attachment of this partition. The inferior margin of the falx cerebri arches over the corpus callosum of the brain and contains the inferior sagittal sinus.

The **tentorium cerebelli** separates the cerebellum from the cerebral hemispheres. It is elevated in the median line (like a tent) where the falx cerebri ends along its median ridge. The tentorium is attached at its circumference to the edges of the grooves for the transverse sinuses on the inner surface of the occipital bone and to the superior angle of the petrous portion of the temporal bones. This attachment ends anteriorly at the posterior clinoid processes. The anterior concave margin of the tentorium is free and arches closely around the brainstem, ending by an attachment to the anterior clinoid processes. In the line of junction of the falx cerebri and the tentorium lies the straight sinus.

The **falx cerebelli** is a small partitioning process of dura which separates the cerebellar hemispheres.

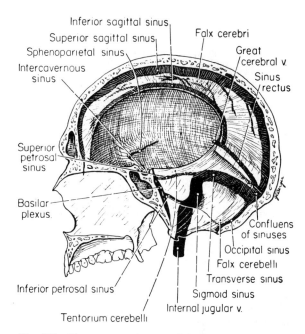

Fig. 218 The venous sinuses of the cranial dura mater.

It is attached to the internal occipital crest and, above, to the inferior aspect of the tentorium. The falx cerebelli contains the occipital sinus.

The **diaphragma sellae** is a horizontal duplication of meningeal dura which projects inward from the margins of the sella turcica. It roofs in the sella, covering the hypophysis, but leaves a central aperture for the stalk of the hypophysis.

The Venous Sinuses of the Dura Mater (figs. 168, 216, 218, 219)

The dural venous sinuses are endothelial-lined spaces between the layers of the dura mater. Some are formed between the periosteal and meningeal layers; others, within foldings of the meningeal layer. They represent collecting channels for the cerebral, meningeal, and diploic veins and communicate with external veins by means of emissary veins. They end directly or indirectly in the internal jugular vein. The dural venous sinuses are conveniently considered in two groups: (1) a group of midline sinuses and their continuations and (2) a group of paired sinuses related to the cavernous sinuses. The first group of sinuses includes the superior sagittal, the inferior sagittal, the straight sinus, the transverse and sigmoid sinuses, and the occipital sinus.

The **superior sagittal sinus** begins at the foramen cecum, where it receives a vein from the nasal cavity, and passes backward in the median plane to the region of the internal occipital protuberance. It lies in the root of the falx cerebri, occupying the triangular interval between the periosteal layer and the diverging meningeal sheets which form the falx. The sinus becomes larger as it is followed posteriorly. At parietal levels, laterally placed, irregularly-shaped bays of the sinus, the **venous lacunae,** open into it by somewhat restricted channels. The arachnoid granulations project into these venous lacunae, and many of the cerebral veins empty into them. The superior sagittal sinus receives the cerebral, diploic, and meningeal veins, and the parietal emissary vein. At the internal occipital protuberance the sinus ends in the confluens of sinuses or bifurcates to right and left to assist in forming the transverse sinuses (Browning, '53).

The **inferior sagittal sinus** occupies the inferior free margin of the falx cerebri. Beginning as a small channel approximately one-third of the distance behind the attachment of the falx, it ends posteriorly in the straight sinus. It has tributaries from the falx cerebri and from the medial surface of the hemispheres.

The **straight sinus** (sinus rectus) lies in the line of junction of the falx cerebri and the tentorium cerebelli and is the continuation of the inferior sagittal sinus in the tentorium. It receives, at its commencement, the great cerebral vein and, in its course, superior cerebellar veins. The straight sinus enters the confluens of sinuses or bifurcates to either side to end in the transverse sinuses.

The **transverse sinuses** are large and continue into the sigmoid sinuses. Beginning as continuations of the superior sagittal and straight sinuses, the transverse sinuses occupy the attached circumference of the tentorium cerebelli as far forward as the base of the petrous portion of the temporal

bone. From there a **sigmoid sinus** arches downward on each side, grooving the mastoid part of the temporal bone, and then turns forward over the occipital bone to the jugular foramen. In the jugular foramen it empties into the jugular bulb of the internal jugular vein. At the transition from the transverse to the sigmoid sinus, the superior petrosal sinus terminates. The transverse sinus receives inferior cerebral and superior cerebellar veins and a parietal diploic vein. The mastoid and condyloid emissary veins and inferior cerebellar veins empty into the sigmoid sinus.

The **occipital sinus** lies in the falx cerebelli. It begins by several small venous channels at the margins of the foramen magnum which communicate with the posterior internal vertebral venous plexus (p. 301). The sinus ends in the confluens of sinuses.

The **confluens of sinuses** is a dilatation at one side of the internal occipital protuberance which represents the junction of the superior sagittal, straight, and occipital sinuses with the right and left transverse sinuses. Sometimes it is a region of actual confluence, and sometimes the superior sagittal and straight sinuses (either or both) bifurcate here to form the right and left transverse sinuses (Browning, '53).

The paired sinuses related to the cavernous sinuses include the intercavernous sinuses, the sphenoparietal sinuses, the superior and inferior petrosal sinuses, and the basilar plexus.

The **cavernous sinuses** lie on either side of the body of the sphenoid bone and extend from the superior orbital fissure, in front, backward to the apex of the petrous portion of the temporal bone. They are formed between the meningeal and periosteal layers of the dura, and trabeculae from each layer cross the space, giving it a reticular or cavernous structure. This sinus is remarkable in having nerves in its outer wall and a nerve and a major artery coursing through it. In its outer wall, the oculomotor and trochlear nerves and the ophthalmic and maxillary divisions of the trigeminal nerve pass forward to the superior orbital fissure and the foramen rotundum. The internal carotid artery enters the sinus from foramen lacerum, arches forward and downward, and then turns upward and backward, leaving the cavernous sinus medial to the anterior clinoid process. The abducens nerve passes forward through the sinus, crossing the windings of the internal carotid artery externally. The sinus receives the superior

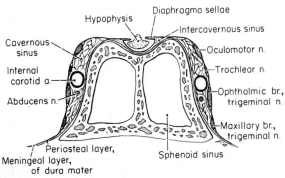

Fig. 219   A coronal section of the sphenoid and cavernous sinuses and of the hypophyseal fossa.

ophthalmic vein at the superior orbital fissure (p. 253), cerebral veins, and the sphenoparietal sinus. The cavernous sinus terminates posteriorly in the superior petrosal sinus, which ends in the transverse sinus, and in the inferior petrosal sinus, which ends in the bulb of the internal jugular vein. It communicates with the pterygoid plexus of veins by emissary veins through the foramen ovale and foramen lacerum and with the angular vein by means of the superior ophthalmic vein.

The **intercavernous sinuses** connect the two cavernous sinuses, passing anterior and posterior to the stalk of the hypophysis.

The **sphenoparietal sinus** is lodged in the dura mater on the under surface of the lesser wing of the sphenoid bone. It receives a radicle of the middle meningeal vein and veins of the dura mater and ends in the anterior part of the cavernous sinus.

The **superior petrosal sinus** lies in the margin of the tentorium cerebelli which is attached to the superior angle of the petrous portion of the temporal bone. It connects the posterior end of the cavernous sinus to the bend marking the transition between the transverse and sigmoid sinuses. It is small and narrow and is sometimes accommodated by a groove in the superior angle of the petrous bone. It receives cerebellar and inferior cerebral veins as well as veins from the tympanic cavity.

The **inferior petrosal sinus** lies in the petro-occipital fissure. It arises in the posterior inferior portion of the cavernous sinus and, passing through the anterior compartment of the jugular foramen, ends in the bulb of the internal jugular vein. It receives the labyrinthine vein and veins from the pons and medulla and from the underside of the cerebellum.

The **basilar plexus** is made up of interlacing venous channels in the dura of the basilar part of the occipital bone which connect the two inferior petrosal sinuses. The basilar plexus is also connected with the anterior vertebral venous plexus.

Clinical Note Cavernous sinus thrombosis may occur when inflammatory material is carried into the cavernous sinus through valveless veins which are tributary to it. Frequently implicated is the superior ophthalmic vein, the nasofrontal vein, and others. Note also the emissary veins described in the next paragraph and see the clinical note on p. 206.

Emissary Veins

Emissary veins connect the dural venous sinuses within the cranium with veins external to it. They are valveless and, as a consequence, may conduct blood either inward or outward in accordance with the pressures existing in the sinuses and in the external veins. Some are constant; others occur occasionally.

The **superior ophthalmic vein** is the largest vein of this type, though it is not always so classified. It connects the angular vein of the face with the cavernous sinus. The **mastoid emissary vein** unites the posterior auricular vein with the sigmoid sinus. The **parietal emissary vein** occupies the parietal foramen and connects the veins of the vertex of the scalp with the superior sagittal sinus. The **emissary vein of the foramen cecum** connects the veins of the nasal cavity with the superior sagittal sinus. The **condyloid canal,** when present, transmits an emissary vein which passes between the lower end of the sigmoid sinus and veins of the suboccipital triangle of the neck. Small, inconstant emissary veins accompany the hypoglossal nerve through the hypoglossal canal, the mandibular nerve through the foramen ovale, and the internal carotid artery through the foramen lacerum and the carotid canal.

Meningeal Arteries and Veins (figs. 168, 214, 215)

The arteries of the dura mater are the anterior, middle, and posterior meningeal arteries. The anterior and posterior meningeal arteries, from several sources, confine themselves to the region of the base of the skull. The middle meningeal artery supplies four-fifths of the dura mater.

The **middle meningeal artery** is a branch of the maxillary artery, under which description it has received consideration (p. 224). The tendency of its anterior branch to tunnel through the bone of the skull in the region of the pterion has been mentioned (p. 274). The middle meningeal artery supplies most of the supratentorial dura mater except for the floor of the anterior cranial fossa. The **accessory meningeal artery** (p. 224) is either a branch of the middle meningeal or a separate branch of the maxillary. It passes through the foramen ovale and supplies the semilunar ganglion and the adjacent dura mater.

The **anterior meningeal artery** is a branch of the anterior ethmoidal artery (p. 252). It supplies the floor of the anterior cranial fossa and is assisted by meningeal twigs of the posterior ethmoidal artery in the region of the cribriform plate.

The **posterior meningeal arteries** are branches of the ascending pharyngeal, the occipital, and the vertebral arteries. They vary in size and importance. The branches of the ascending pharyngeal artery (p. 161) enter the cranium by way of the jugular foramen, foramen lacerum, and hypoglossal canal. The principal branch enters through the jugular foramen. The meningeal branch of the occipital artery enters the cranium by the jugular foramen, and a mastoid branch enters through the mastoid foramen (p. 161). The meningeal branch of the vertebral artery (p. 179) arises from that artery as it pierces the dura mater to enter the cranium. The

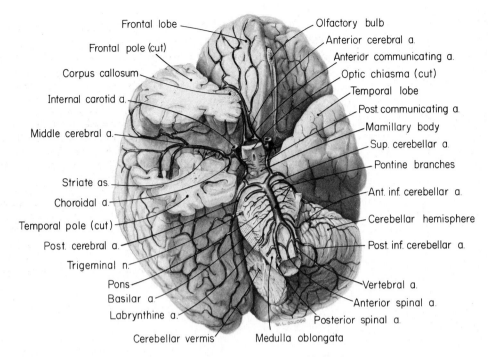

Frontal lobe — Olfactory bulb — Anterior cerebral a. — Anterior communicating a. — Frontal pole (cut) — Corpus callosum — Optic chiasma (cut) — Internal carotid a. — Temporal lobe — Post. communicating a. — Middle cerebral a. — Mamillary body — Sup. cerebellar a. — Pontine branches — Ant. inf. cerebellar a. — Striate as. — Choroidal a. — Cerebellar hemisphere — Temporal pole (cut) — Post. inf. cerebellar a. — Post. cerebral a. — Trigeminal n. — Vertebral a. — Pons — Anterior spinal a. — Basilar a. — Posterior spinal a. — Labrynthine a. — Cerebellar vermis — Medulla oblongata

Fig. 220   The major arteries of the brain.

posterior meningeal arteries distribute in the posterior cranial fossa, remaining below the tentorium cerebelli.

The **meningeal veins** correspond generally to the arteries with which they distribute. The anterior and posterior meningeal veins empty primarily into the dural venous sinuses of the anterior and posterior cranial fossae. The **middle meningeal vein** accompanies the middle meningeal artery, and it and its radicles lie external to the artery and its branches in the dura mater. The middle meningeal vein communicates with the sphenoparietal sinus and also terminates in the pterygoid plexus of veins.

Nerves of the Dura Mater

The sensory nerves of the dura mater are derived from all three divisions of the trigeminal nerve, from the vagus nerve, and from the hypoglossal nerve. The hypoglossal nerve has, centrally, only motor fibers, so that its sensory contribution to the dura is thought to arise from the two uppermost cervical segments. The convexity of the dura in the cranial vault is generally insensitive, except along the course of the meningeal blood vessels, whereas increased sensitivity (pain only) characterizes the floor of the cranial fossae. The various meningeal branches of the divisions of the trigeminal nerve (pp. 222, 240, 251) supply the dura mater of the anterior and middle cranial fossae. The meningeal branches of the vagus nerve (p. 164) and of the hypoglossal nerve (p. 197) distribute below the

tentorium cerebelli to the dura mater of the posterior cranial fossa. It is probable that all nerves to the posterior fossa arise from the upper cervical nerves (Kimmel, '61).

Diploic Veins

The cancellous bone (diploe) between the inner and outer tables of the cranium lodges the diploic veins. These are large and thin-walled and are formed of endothelium resting on a layer of elastic tissue. With the obliteration of the sutures of an adult skull, these veins anastomose and increase in size and are no longer related only to one bone. They communicate with the meningeal veins, the sinuses of the dura mater, and the scalp veins.

The **frontal diploic vein** ramifies primarily in the frontal bone and opens into the supraorbital vein and the superior sagittal sinus. The **anterior temporal vein** opens into the sphenoparietal sinus and a deep temporal vein; the **posterior temporal vein** largely drains the parietal bone and ends in the transverse sinus. The large **occipital diploic vein** is confined to the occipital bone. It empties into the occipital vein of the scalp or, internally, into the transverse sinus or the confluens of sinuses.

The Blood Vessels of the Brain (figs. 168, 218–23)

The Vertebral Artery The origin of the vertebral artery and its course in the neck are described in connection with the subclavian artery (p.179). The

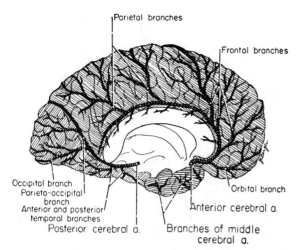

Fig. 221   The arteries of the median and inferior surfaces of the brain.

arteries of the two sides enter the cranium by way of the foramen magnum; they give off the meningeal, posterior and anterior spinal, and posterior inferior cerebellar branches and then unite with one another at the inferior border of the pons to form the basilar artery.

The **meningeal branch** arises in the foramen magnum and distributes in the cerebellar fossa, especially to the falx cerebelli. **Posterior spinal arteries** are small and multiple and arise from both the vertebral artery and the posterior inferior cerebellar artery. They descend on the spinal cord either medial or lateral to the emerging dorsal roots, branches anastomosing across the median line and participating in a chain-like series of connections with spinal branches of the vertebral artery and others at lower levels to form a long vessel on the dorsum of the cord. The **anterior spinal artery** springs from the vertebral shortly before the two vertebrals join in the basilar artery. Inclining medially as it descends, the anterior spinal unites with its fellow of the opposite side at the level of the foramen magnum. The single vessel descends in the anterior median fissure of the spinal cord and, reinforced by radicular branches of the vertebral and other arteries at lower levels, forms a longitudinal vessel running the whole length of the spinal cord. The **posterior inferior cerebellar artery** is the largest branch of the vertebral. It passes between the origins of the vagus and accessory nerves to the under surface of the cerebellum. A medial branch ramifies on the cerebellar vermis between the hemispheres, and a lateral branch anastomoses with radicles of the anterior inferior cerebellar and superior cerebellar arteries. The artery usually provides a posterior spinal branch.

The Basilar Artery The basilar artery, formed from the junction of the two vertebral arteries, proceeds forward in a shallow groove in the inferior midline of the pons. It ends at the superior border of the pons by bifurcating into the two posterior cerebral arteries. Its branches are the pontine, labyrinthine, anterior inferior cerebellar, superior cerebellar, and posterior cerebral.

Numerous small **pontine branches** spring from either side of the basilar artery to supply the pons. The **labyrinthine artery** enters the internal acoustic meatus with the vestibulocochlear nerve and is distributed to the internal ear. It is usually a branch of the anterior inferior cerebellar artery; less frequently, it arises from the basilar. The **anterior inferior cerebellar artery** distributes to the anterior part of the inferior surface of the cerebellum. The **superior cerebellar artery** arises near the end of the basilar. It passes lateralward, just behind the oculomotor nerve, and reaches the superior surface of the cerebellum. Like the posterior inferior cerebellar artery, it has medial and lateral branches which anastomose with the branches of that artery and also with the anterior inferior cerebellar artery.

The **posterior cerebral artery,** separated from the superior cerebellar artery by the oculomotor nerve, passes lateralward parallel to that artery. Receiving the posterior communicating branch of the internal carotid artery, it winds around the cerebral peduncle to reach the temporal and occipital lobes of the cerebrum. Like the other cerebral arteries, the posterior cerebral has small central branches to the brainstem and larger cortical branches to the temporal and occipital lobes of the cerebrum.

The Internal Carotid Artery The internal carotid artery has cervical, petrous, cavernous, and cerebral portions. The course and relations of the cervical portion are noted on p. 160. It has no branches. The internal carotid artery enters the carotid canal in the petrous portion of the temporal bone. Ascending, turning forward, and then ascending again, it traverses the windings of the carotid canal. The artery is accompanied by nerves of the carotid

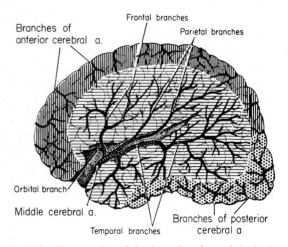

Fig. 222   The arteries of the lateral surface of the brain.

sympathetic plexus and is surrounded by a prolongation of the dura mater of the brain. It is
separated from the tympanic cavity by a thin bony
layer and, just short of the foramen lacerum,
another thin plate of bone separates the artery from
the semilunar ganglion of the trigeminal nerve.
While in the petrous bone, the internal carotid
artery gives off the small **caroticotympanic artery**
which enters the tympanic cavity through a minute
foramen in the carotid canal to anastomose with the
anterior tympanic branch of the maxillary artery
and with the stylomastoid artery.

The cavernous portion of the internal carotid
artery lies within the cavernous sinus, covered by
a layer of its lining endothelium. In the cavernous
sinus the artery describes another S-shaped series of
bends, first ascending toward the posterior clinoid
process and then turning forward at the side of the
body of the sphenoid bone. At the anterior end of
the sinus, the artery again turns upward and backward to the medial aspect of the anterior clinoid
process and perforates the dura mater of the roof of
the sinus. From the cavernous portion of the artery

arise minute twigs to the walls of the cavernous
sinus, to the hypophysis, and to the semilunar
ganglion. The ophthalmic artery (p. 252) arises just
as the internal carotid pierces the dura mater of the
roof of the sinus. The cerebral portion of the
internal carotid artery passes between the optic and
oculomotor nerves to the medial end of the lateral
cerebral fissure of the brain, where it divides into its
terminal branches, the anterior and middle cerebral
arteries.

The first branch of the cerebral part of the internal
carotid is the **posterior communicating artery.** This
vessel runs backward below the optic tract to anastomose with the posterior cerebral branch of the basilar
artery. The posterior communicating artery varies
much in size, being large when there is a marked inequality in the arterial supply of the brain through the
internal carotid and vertebral systems. It gives branches
to the optic tract, cerebral peduncle, internal capsule,
and thalamus.

The **choroidal artery** is a small vessel that arises lateral
to the posterior communicating artery. It follows the
optic tract and the cerebral peduncle as far as the lateral
geniculate body, where it enters the choroid plexus of the
inferior horn of the lateral ventricle.

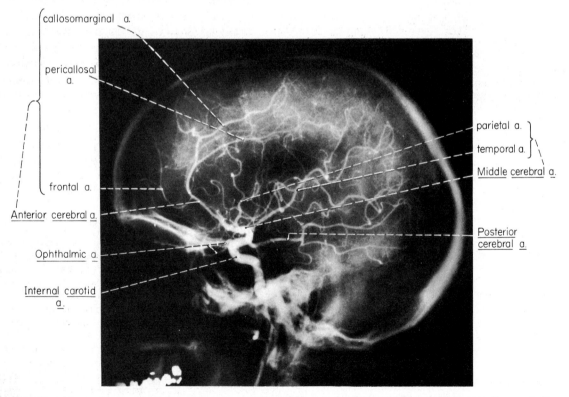

Fig. 223 Carotid arteriogram depicting the major arteries of the cerebral hemispheres. In this case the posterior
cerebral artery appears to arise from the internal carotid artery.

The **anterior cerebral artery** is the smaller of the two terminal branches of the internal carotid artery. It passes forward and medialward above the optic chiasma to enter the longitudinal fissure of the cerebrum. Swinging upward and backward over the rostrum and genu of the corpus callosum, the artery follows the dorsal surface of the corpus callosum as far as the parietooccipital sulcus. It has central and cortical branches. As the two anterior cerebral arteries reach the longitudinal fissure, they are united by a short (5 mm.), wide communication, the **anterior communicating artery.** The cortical branches of the anterior cerebral artery ramify on the medial surface of the hemisphere and over onto the superior gyri of the lateral surface.

The **middle cerebral artery** is the direct continuation of the internal carotid artery. It passes upward and lateralward in the lateral cerebral fissure and, on the surface of the insula, divides into numerous cortical branches. The **anterolateral central branches** penetrate the anterior perforated substance and comprise two sets, the medial striate and lateral striate arteries. The **medial striate arteries** supply the anterior parts of the lenticular and caudate nuclei and of the internal capsule. The **lateral striate arteries** supply the basal nuclei of the brain (most of putamen, part of the head and the body of the caudate, and the lateral part of the globus pallidus) and its internal capsule. Collectively, the striate arteries are known as the artery of cerebral hemorrhage. The cortical branches of the middle cerebral artery ramify over the insula and the lateral surface of the hemisphere.

The Circulus Arteriosus Cerebri The anterior and posterior communicating arteries indirectly connect the internal carotid arteries of the two sides and the basilar system, so that an arterial circle is formed at the base of the brain. These interconnections serve to equalize the blood supply to the various parts of the brain under conditions of fluctuating pressure through the major vessels. They are especially vital whenever a major source vessel has to be ligated. The terminal vessels in and on the brain anastomose, but the larger arteries do not, except in this arterial circle.

Clinical Note In view of the relatively minor terminal anastomoses of its vessels and the high vulnerability of nervous tissue when oxygen is lacking, the connections of the cerebral arterial circle are especially valuable. In spite of their presence, plugging of smaller brain vessels almost always leads to softening and necrosis of that brain tissue deprived of a normal blood supply. Most intracranial aneurisms occur in the vessels of or near the cerebral arterial circle.

The Veins of the Brain The veins of the brain have no valves, and their walls are very thin. They include the external cerebral, internal cerebral, and cerebellar veins. Among the external cerebral veins are the eight or ten **superior cerebral veins** which drain the superior, lateral, and medial surfaces of the hemisphere and terminate in the superior sagittal sinus. The more anterior veins run nearly at right angles to the sinus; the larger, more posterior veins pass obliquely forward to open into the sinus in a direction somewhat opposed to the current of blood within it (figs. 168, 218). The **middle cerebral vein** emerges from the lateral cerebral fissure and ends in the cavernous sinus or the sphenoparietal sinus. It may have anastomosing channels that reach the superior sagittal or the lateral sinuses. The small **inferior cerebral veins** drain the under surfaces of the hemispheres. They end in the sinuses at the base of the skull—the cavernous, superior petrosal, and transverse sinuses. A **basal vein,** beginning in the anterior perforated substance and communicating with the middle cerebral vein, winds around the cerebral peduncle and ends in the great cerebral vein.

The **internal cerebral veins,** two in number, are formed near the interventricular foramen by the union of the thalamostriate and choroid veins. After their formation the internal cerebral veins of each side pass backward parallel to each other between the layers of the tela choroidea of the third ventricle. They unite under the splenium of the corpus callosum to form the **great cerebral vein** which, as a short median trunk, curves around the splenium to end in the anterior end of the straight sinus.

Superior cerebellar veins end medially in the internal cerebral veins and the straight sinus or, laterally, in the transverse and superior petrosal sinuses. **Inferior cerebellar veins** end in the sigmoid, inferior petrosal, and occipital sinuses.

Clinical Note Fractures of the skull (sometimes lesser trauma) may lead to intracranial hemorrhage, the specifics of which are based on the layering of the meninges and the differing vessels of the layers. Noting these specifics is to review much intracranial anatomy. **Epidural hemorrhage** is due to rupture of the middle meningeal artery, especially its anterior branch as it grooves or tunnels the bone in the region of the pterion (p. 274). Such blood expands the potential space between the periosteal layer of the dura mater and the bone. Being under arterial pressure it results in increased intracranial pressure and symptoms requiring prompt attention. **Subdural hemorrhage** results from rupture of cerebral veins as they pass from the brain surface into the venous sinuses of the dura (especially the superior sagittal sinus). The bleeding is venous, may accumulate slowly, but the clots formed may require surgical removal or drainage. **Subarachnoid hemor-**

rhage is confined to the subarachnoid space and is thus due to rupture of cerebral vessels. Blood will be found in the cerebrospinal fluid.

The Cranial Portion of the Sympathetic Nervous System

The **internal carotid nerve,** the postganglionic cranial continuation of the superior cervical sympathetic ganglion, accompanies the internal carotid artery into the cranium. It provides two branches in the carotid canal, one on the medial and one on the lateral aspect of the artery. The lateral branch forms the internal carotid plexus, the medial follows the artery to the cavernous sinus to form the cavernous plexus.

The **internal carotid plexus** surrounds the lateral aspect of the artery, gives off the caroticotympanic nerves and the deep petrosal nerve, and has communications with the abducens nerve and the trigeminal ganglion. Two or three **caroticotympanic nerves** penetrate the bony wall of the carotid canal and join the tympanic plexus on the promontory of the middle ear (p. 196). The **deep petrosal nerve** passes through the cartilage which fills the foramen lacerum and joins the greater petrosal nerve to form the nerve of the pterygoid canal. The distribution of these nerves has been discussed (p. 243).

The **cavernous plexus** communicates with various nerves and continues as a plexus along the terminal branches of the internal carotid artery. It gives twigs to the oculomotor, trochlear, and ophthalmic nerves. The sympathetic fibers to the dilator pupillae muscle of the iris traverse the ophthalmic nerve, reaching the eyeball through the nasociliary and long ciliary branches. The sympathetic root of the ciliary ganglion is derived from the cavernous plexus. It enters the superior orbital fissure separately. The plexus also provides twigs which join the blood vessels of the hypophysis. The terminal fibers of the cavernous plexus follow the anterior and middle cerebral and ophthalmic arteries and their branches.

The nerve plexus on the **vertebral artery,** derived from the cervicothoracic ganglion, ascends along this artery into the cranium. This plexus distributes along the basilar, posterior cerebral, and cerebellar arteries.

The Hypophysis (fig. 224)

The hypophysis is a small endocrine gland

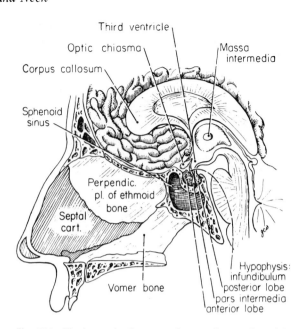

Fig. 224   The hypophysis as seen in a median section of the skull and brain.

located in the hypophyseal fossa of the sella turcica behind the optic chiasma. From its manifold effects on growth and development and on the metabolism of the body, it has been called the 'master endocrine gland.' Many of its effects are produced by its stimulation of other glands of the body, such as the thyroid gland, adrenal glands, and the gonads. Other of its hormones affect metabolism, and still others produce contraction of smooth muscle. Section of the supraopticohypophyseal tract leads to diuresis. The hypophysis is approximately spherical in shape and rather insignificant in size, being slightly over 1 cm. in anteroposterior and transverse diameters and less than 1 cm. in depth. It is roofed over by the diaphragma sellae except where this is perforated for the accommodation of the **stalk** which attaches the hypophysis to the anterior hypothalamic region of the brain. The periosteal layer of the dura mater lines the hypophyseal fossa. The hypophysis consists of an **anterior lobe,** developed from the ectoderm of the primitive oral cavity, and a **posterior lobe,** developed from the ectoderm of the floor of the diencephalic portion of the brain. **Pars tuberalis** and **pars intermedia** are small subdivisions of the anterior lobe which are not grossly distinguishable. The anterior lobe is concave posterosuperiorly, and the posterior lobe rests in the shallow cup formed by this concavity. The posterior lobe is connected to the brain by the stalk of the hypophysis through which runs a bundle of fine, unmedullated nerve fibers, the supraopticohypophyseal tract.

The blood supply of the hypophysis is provided (from below) by two inferior hypophyseal branches of the internal carotid arteries, and by one or two superior hypophyseal arteries which spring from the internal carotid artery just as it leaves the cavernous sinus. Branches of the superior arteries course along the hypophyseal stalk and enter both the stalk and the anterior lobe. The inferior arteries supply the posterior lobe and a large part of the anterior lobe. The venous drainage from both lobes is into capsular veins which drain into the cavernous and intercavernous sinuses. As far as is known, the hypophysis has no lymphatic circulation. Fine nerves reach it, by way of its arteries, from the internal carotid plexus and through its stalk (supraopticohypophyseal tract).

Table of Cranial Nerves

Number	Nerve	Component	Origin	Distribution and Function
I	Olfactory	Spec. visc. aff.	Olfactory epithelium	Smell from the nasal mucosa of the roof and upper sides of the septum and superior concha of nose
II	Optic	Spec. som. aff.	Ganglion cells of retina	Vision from the retina of the eye
III	Oculomotor	Somatic efferent	Midbrain	Motor to the superior, inferior, and medial rectus, inferior oblique, and levator palpebrae muscles
		Gen. visc. eff.	Pregangl.—midbrain Postgangl.—ciliary gang.	Parasympathetic innervation to the sphincter muscle of the iris and muscle of ciliary body
IV	Trochlear	Somatic efferent	Midbrain	Motor to superior oblique muscle of eye
V	Trigeminal	Gen. som. aff.	Semilunar ganglion	Sensation from the skin of the face and scalp, eyelids and cornea; mucous membranes of the mouth and tongue, palate, nose, teeth and gingiva, and meninges of the brain
		Spec. visc. eff.	Pons	Motor to the masticatory, mylohyoid, ant. belly of digastric, tensor veli palatini, and tensor tympani muscles
VI	Abducens	Somatic efferent	Pons	Motor to the lateral rectus muscle of the eye
VII	Facial	Spec. visc. eff.	Pons	Motor to branchiomeric muscles of facial expression, posterior belly of digastric, stylohyoid, and stapedius muscles
		Gen. som. aff.	Geniculate ganglion	Sensation from skin of external acoustic meatus
		Spec. visc. aff.	Geniculate ganglion	Taste from the anterior two-thirds of tongue via chorda tympani
		Gen. visc. eff.	Preganglionic—pons Postganglionic— pterygopalatine gang. Preganglionic—pons Postganglionic—sub-mandibular ganglion	Parasympathetic innervation to the lacrimal gland and the glands of the nasal cavity, palate, and upper pharynx Parasympathetic innervation to the submandib-ular and sublingual glands
VIII	Vestibulo-cochlear	Spec. som. aff. (exteroceptive)	Spiral ganglion	Hearing from the Organ of Corti
		Spec. som. aff. (proprioceptive)	Vestibular ganglion	Vestibular sensation from the semicircular ducts, utricle, and saccule
IX	Glossopharyngeal	Spec. visc. eff.	Medulla oblongata	Motor to the stylopharyngeus muscle
		Gen. visc. eff.	Preganglionic— medulla oblongata Postgang.—otic gang.	Parasympathetic innervation to the parotid gland
		Gen. visc. aff.	Superior ganglion	Visceral sensation from the parotid gland, carotid body and sinus, pharynx, and middle ear
		Spec. visc. aff.	Inferior ganglion	Taste from the posterior one-third of the tongue
		Gen. som. aff.	Inferior ganglion	Cutaneous sensation from the external ear

(Continued on following page)

X	Vagus	Spec. visc. eff.	Medulla oblongata	Motor to striated muscles of larynx, pharynx, palate (except tens. veli pal.), and upper two-thirds of esophagus
		Gen. visc. eff.	Preganglionic— medulla oblongata Postganglionic— cells on, in, or near viscus	Parasympathetic innervation to involuntary muscle and glands of heart, esophagus, stomach, trachea, bronchi, intestines, and other abdominal viscera
		Gen. visc. aff.	Inferior ganglion	Visceral sensation from carotid body, base of tongue, pharynx, larynx, trachea, bronchi, lungs, heart, esophagus, stomach, and intestines
		Spec. visc. aff.	Inferior ganglion	Taste from epiglottis and palate
		Gen. som. aff.	Superior glanglion	Cutaneous sensation from the external ear and external acoustic meatus and dura mater of posterior cranial fossa
XI	Accessory (cranial portion with vagus—principally the inferior laryngeal nerve)			
	spinal portion	Somatic efferent	Spinal cord ventral horn grey	Motor to the sternocleidomastoid and trapezius muscles
XII	Hypoglossal	Somatic efferent	Medulla oblongata	Motor to the extrinsic and intrinsic muscles of the tongue

IV

The Back

THE MUSCLES AND FASCIAE OF THE BACK

The more superficial muscular layers of the back are related directly to the upper limb and, as a consequence, the superficial features of the back are considered with the limb—its surface anatomy on p. 45, its cutaneous nerves on p. 48, and its muscles on p. 49. Deep to the limb muscles a thin layer of respiratory musculature arises from the thoraco-lumbar fascia of the back. It is represented by two muscles: the posterior superior serratus and the posterior inferior serratus. Under this pair of muscles the deep groove between the spinous processes of the vertebrae and the curving ribs is occupied by superimposed layers of longitudinal and oblique muscles. Each layer of muscles includes both long and short representatives. These muscles extend the vertebral column and the head and provide rotary and side to side bending movements for the back, neck, and head. They are complex in morphological detail but are presented here as simply as possible, for the understanding of their functions depends less on their anatomic minutiae than on their general plan.

Posterior Superior Serratus (fig. 226)

The posterior superior serratus consists of a thin quadrilateral plane of muscle and aponeurosis which lies directly under the rhomboid muscles at the junction of the neck and back. It arises by a broad aponeurosis attached to the ligamentum nuchae and to the spinous processes of the seventh cervical and the first two or three thoracic vertebrae. The aponeurosis ends in four flat muscular bellies which insert into the upper borders of ribs two to five, a

little lateral to their angles. The muscle is innervated by branches of the first four intercostal nerves.

Posterior Inferior Serratus (figs. 225, 226, 363)

The posterior inferior serratus muscle is, similarly, flat and quadrilateral but lies at the junction of the thoracic and lumbar regions. Its medial aponeurosis attaches, through the thoracolumbar fascia, to the last two thoracic and the upper two lumbar spinous processes. Passing upward and lateralward, the aponeurosis gives way to four flat muscular bellies which insert into the inferior borders of the last four ribs, lateral to their angles. The nerves of the muscle are branches of the ninth to the twelfth intercostal nerves.

The serrati are both inspiratory muscles. The posterior superior serratus elevates the upper four ribs, increasing the diameters of the thoracic cage and raising the sternum. The posterior inferior serratus draws down on the lower four ribs, holding them down against the upward pull of the dia-phragm as it contracts against the abdominal contents to enlarge the thoracic cavity in its vertical dimension.

The Nuchal and Thoracolumbar Fasciae (fig. 225)

The muscles of the back and of the back of the neck are enclosed by a fascial covering which attaches medially to the ligamentum nuchae, the tips of the spinous processes and the supraspinal ligament of the entire vertebral column, and the median crest of the sacrum. Underlying the super-ficial limb and respiratory musculature of the back, the layer attaches to the cervical transverse and lumbar transverse processes. Its lateral attachment

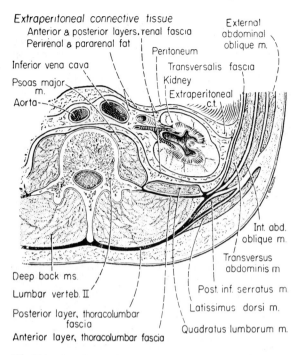

Extraperitoneal connective tissue
  Anterior & posterior layers, renal fascia
  Perirenal & pararenal fat
                                    Peritoneum
                                           External
                                           abdominal
                                           oblique m.
                                    Transversalis fascia
Inferior vena cava
                                    Kidney
Psoas major
  m.   Extraperitoneal
Aorta   c.t.

                                           Int. abd.
                                           oblique m.

                                    Transversus
                                    abdominis m

Deep back ms.
                                    Post. inf. serratus m.
Lumbar verteb. II
                                    Latissimus dorsi m.
Posterior layer, thoracolumbar
  fascia
Anterior layer, thoracolumbar fascia
                                    Quadratus lumborum m.

Fig. 225 A partial cross section of the lumbar region of the back, illustrating the thoracolumbar fascia, the deep back muscles, and the renal fasciae.

in the thoracic region is to the angles of the ribs lateral to the iliocostalis muscle and to the intercostal fascia. The layer in the neck, the **nuchal fascia,** is the posterior portion of the prevertebral layer of cervical fascia. It covers the splenius muscle and attaches to the skull below the superior nuchal line. The thoracic and lumbar portions of the layer constitute the **thoracolumbar fascia.** This is thin and transparent where it covers the thoracic portions of the deep muscles but is thick and strong in the lumbar region. Bulged by the thick mass of back muscles at lumbar levels, the thoracolumbar fascia extends beyond the line of the transverse processes and forms a strong enclosure for the erector spinae muscle mass. Thus, the **posterior layer of thoracolumbar fascia** extends from the midline lateralward over the erector spinae muscle and ends in the **aponeurosis of the transversus abdominis muscle.** An **anterior layer of the thoracolumbar fascia** lies deep to the erector spinae and ends medially in the transverse processes of the lumbar vertebrae. It combines laterally with the posterior layer of the thoracolumbar fascia in the aponeurosis of the transversus abdominis muscle. A thickening of the anterior layer between the twelfth rib and the transverse process of the first lumbar

vertebra forms the **posterior lumbocostal ligament.** Below, the thoracolumbar fascia attaches to the iliac crest and the lateral crest of the sacrum.

Splenius (fig. 227)

The splenius (a bandage) serves as a strap, covering and holding in the deeper muscles of the neck. It arises from ligamentum nuchae and the spinous processes from the seventh cervical to the sixth thoracic vertebrae. The muscle may be divided into two parts—**splenius capitis,** which inserts on the mastoid process and the lateral one-third of the superior nuchal line, and **splenius cervicis,** which ends in the posterior tubercles of the first two or three cervical vertebrae. The cervicis portion is the outer and lower portion of the unbroken width of the muscle. The splenius draws the head and neck backward and rotates the face toward the side of the muscle acting. Both sides contracting together extend the head and neck. The muscle is innervated by the lateral branches of the dorsal rami of the second to the fifth or sixth cervical nerves. The splenius lies directly under trapezius and is covered by the nuchal fascia; its mastoid insertion is deep to that of the sternocleidomastoid muscle. The splenius overlies the erector spinae and semispinalis muscles.

Erector Spinae (figs. 225, 226, 242, 272, 277, 280)

The erector spinae is a massive and complex muscle which occupies the vertebrocostal groove of the back and lies directly under the posterior layer of the thoracolumbar fascia. It begins below in a broad, thick tendon which is attached to the posterior aspect of the sacrum, the posterior portion of the iliac crest, and the lumbar spinous processes and supraspinal ligament. The muscular fibers, beginning on the anterior aspect of the tendon, split into three columns at lumbar levels: the lateral iliocostalis, the intermediate longissimus, and the more medial spinalis. The erector spinae muscle, as a whole, ascends throughout the length of the back, but its columns are composed of fascicles of shorter length. Each column is composed of a rope-like series of fascicles, various bundles arising as others are inserting; each fascicle spans from six to ten segments between bony attachments.

Iliocostalis, the lateral column, begins at the crest of the ilium and inserts on the angles of the ribs. Its higher fascicles arise in lower ribs and insert in upper ribs and, at cervical levels, end in the posterior tubercles of cervical transverse processes as high as the fourth

Erector spinae M. :

Longissimus capitis m.

Spinalis cervicis m.

Longissimus cervicis m.

Iliocostalis cervicis m.

Iliocostalis thoracis m.

Spinalis thoracis m

Ext. inter-costal m.

Longissimus thoracis m.

Iliocostalis lumborum m.

Transverso spinal complex:

Semispinalis capitis m.

Multifidus m.

Semispinalis cervicis m.

Posterior superior serratus m.

Semispinalis thoracis m.

Levatores costarum m

Posterior inferior serratus m

Quadratus lumborum m.

Multifidus m.

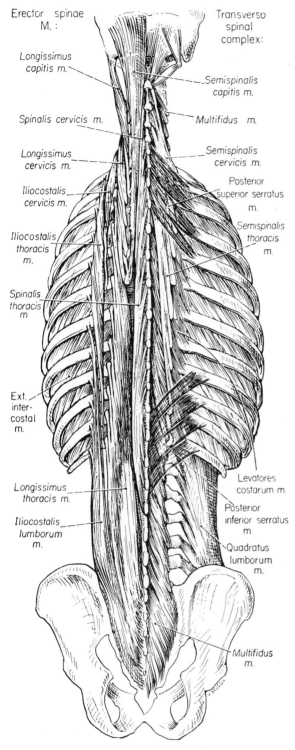

Fig. 226   The deep back muscles.

cervical vertebra. The regional differences in attachment are recognized in its three designations—iliocostalis lumborum, iliocostalis thoracis, and iliocostalis cervicis.

Longissimus, the large intermediate column, reaches the skull and thus has thoracis, cervicis, and capitis portions. Such regional names do not imply visible breaks in the muscle, however, for the continuous rope-like character of the erector spinae prevails in all its columns. The fibers of the longissimus arise on transverse processes at lower levels and insert into transverse processes at higher levels. The longissimus capitis inserts into the posterior margin of the mastoid process, beneath the splenius and sternocleidomastoid muscles.

Spinalis, the narrow medialmost column, arises from spinous processes and inserts into spinous processes. It has distinct thoracis and cervicis portions and a capitis portion which is usually inseparably blended with the semispinalis capitis.

The erector spinae extends the vertebral column and, acting on one side, bends the column toward that side. The capitis insertion of the longissimus serves to bend the head and rotate the face toward the same side. The erector spinae muscle also 'pays out' in slow flexion of the trunk but ceases activity when full flexion is reached, giving over then to ligamentous support. The muscle is innervated serially by branches of the dorsal rami of all spinal nerves. The erector spinae overlies the semispinalis and multifidus muscles of the transversospinal group.

The Transversospinal Muscle Group (figs. 226, 227)

Deep to the erector spinae lies a series of obliquely disposed muscles. Most of them have their origins in transverse processes and their insertions in spinous processes and are therefore characterized as transversospinal muscles.

The **semispinalis muscle,** as its name implies, occupies half the length of the vertebral column. It is divisible into capitis, cervicis, and thoracis portions, with the semispinalis capitis forming a superficial stratum completely covering the semispinalis cervicis. The **semispinalis capitis** is the largest muscle mass of the back of the neck. Its fibers arise from the transverse processes of the seventh cervical and the upper six or seven thoracic vertebrae and from the articular processes of the fourth to the sixth cervical vertebrae. The muscle inserts into the underside of the occipital bone, occupying the whole area between the superior nuchal and inferior nuchal lines. It is usually traversed by an imperfect tendinous intersection and incorporates medially the capitis portion of the spinalis part of the erector spinae. The semispinalis capitis is a very powerful extensor of the head. The **semispinalis cervicis** and **semispinalis thoracis** muscles form a continuous layer. They arise from the transverse processes of all the thoracic vertebrae, and their fascicles insert into spinous processes four to six segments higher than their origin. The uppermost insertion of the cervicis portion is into the spine of the axis (second cervical vertebra). The semispinalis cervicis and semispinalis thoracis muscles extend the upper vertebral column and rotate it toward the opposite side. The entire muscle complex is innervated by dorsal rami of the spinal nerves of the upper half of the back.

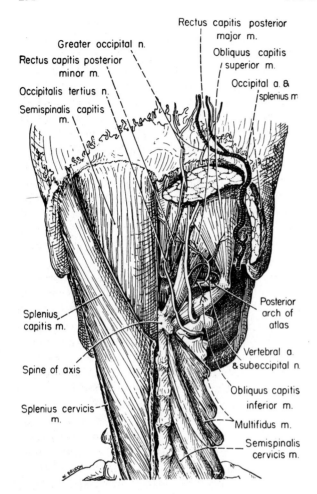

Rectus capitis posterior major m.

Greater occipital n.

Obliquus capitis superior m.

Rectus capitis posterior minor m.

Occipital a. & splenius m.

Occipitalis tertius n.

Semispinalis capitis m.

Splenius capitis m.

Posterior arch of atlas

Spine of axis

Vertebral a. & suboccipital n.

Obliquus capitis inferior m.

Multifidus m.

Splenius cervicis m.

Semispinalis cervicis m.

Fig. 227   The muscles of the back of the neck; the deeper suboccipital triangle is exposed on the right.

The **multifidus muscle** extends throughout the length of the vertebral column. It, too, is transversospinal in general character, although its transverse attachments are to transverse processes in the thoracic region, to articular processes at cervical levels, and to mammillary processes in the lumbar area. Its lowest fibers arise from the posterior surface of the sacrum and from the tendon of the erector spinae muscle; its highest origin is from the fourth cervical vertebra. The fibers of the multifidus insert two to four segments above their origin into the spinous processes of all vertebrae from the last lumbar to the axis. The muscle is heaviest in the lumbar region. It is innervated by dorsal rami of spinal nerves. The multifidus muscle extends the vertebral column and rotates it minimally toward the opposite side.

The **rotatores muscles** are the shortest representatives of the transversospinal group. They are similar in their attachments to the overlying multifidus but ascend to either the adjacent or the second vertebral spine above their origin (respectively rotatores breves and rotatores longi). Their innervation and action are the same as for the multifidus. Experimental evidence suggests that

multifidis and rotatores are more stabilizers of the vertebral column than prime movers, adjusting motion between individual vertebrae.

The Suboccipital Muscles (fig. 227)

The four suboccipital muscles, also components of the transversospinal muscle group, are the following:

> Rectus capitis posterior major
> Rectus capitis posterior minor
> Obliquus capitis inferior
> Obliquus capitis superior

These muscles lie deep to the semispinalis capitis, and three of them form the boundaries of the suboccipital triangle between the spine of the axis, the transverse process of the atlas, and the occipital bone below the inferior nuchal line.

The **rectus capitis posterior major** arises from the spinous process of the axis. Broadening as it passes upward, it inserts into the middle of the inferior nuchal line and into the occipital bone beneath this line.

The **obliquus capitis inferior** arises from the spinous process of the axis and, passing almost horizontally lateralward, ends in the transverse process of the atlas.

The **obliquus capitis superior** completes the formation of the suboccipital triangle. It arises from the transverse process of the atlas and, passing upward and medialward, inserts into the occipital bone above the outer part of the inferior nuchal line, where it overlaps the insertion of the rectus capitis posterior major.

The **rectus capitis posterior minor** lies medial to the rectus capitis posterior major and is more closely applied to the base of the skull. It arises from the posterior tubercle of the atlas and, widening as it ascends, inserts into the medial part of the inferior nuchal line and into the bone between it and the foramen magnum.

The suboccipital muscles extend the head and rotate it and the atlas toward the same side. They are all innervated by the suboccipital nerve, the dorsal ramus of the first cervical nerve.

The **suboccipital triangle** is the area between the two oblique muscles and the rectus capitis major muscle. Its floor is the posterior atlantooccipital membrane which is attached to the posterior margin of the posterior arch of the atlas. Dense fibrofatty tissue largely fills the triangle. Deep to the tough posterior atlantooccipital membrane, the **vertebral artery** occupies the groove on the upper surface of the posterior arch of the atlas as it passes medialward toward the foramen magnum. The **suboccipital nerve** emerges through the membrane, passing between the vertebral artery and the atlas, and divides in the fibro-fatty tissue of the triangle into branches to the suboccipital muscles. The **greater occipital nerve** (dorsal ramus of the second cervical nerve, p. 146), emerging below the obliquus capitis inferior, turns upward across that muscle and across the suboccipital triangle to reach the scalp by piercing the semispinalis capitis and trapezius muscles. The **occipital artery,** coursing medialward to join the greater occipital nerve on the back of the scalp, crosses the insertion of the obliquus capitis superior muscle.

Minor Deep Back Muscles

Minor deep back muscles are the interspinales, the intertransversarii, and the levatores costarum. The **interspinales** are short fasciculi placed in pairs between the spinous processes of adjacent vertebrae in the cervical and lumbar levels of the column. The **intertransversarii** are short pairs of muscles connecting adjacent transverse processes of the cervical and lumbar vertebrae. The two members of a pair lie one anterior and one posterior to an emerging ventral ramus of each spinal nerve.

The **levatores costarum** are deep back muscles which represent, in the thoracic region, the posterior intertransversarius muscles of the neck. Twelve in number on each side, they arise from the ends of the transverse processes of the seventh cervical and the upper eleven thoracic vertebrae and pass obliquely downward and lateralward to insert on the rib immediately below (levatores costarum breves) between its tubercle and angle. The four lowermost muscles divide into two fascicles, one like the above, the other descending to the second rib below its origin (levatores costarum longi). These muscles elevate the ribs, assisting in inspiration, and help to produce lateral bending of the vertebral column. They are innervated by branches of the dorsal rami of spinal nerves (Morrison, '54).

THE VERTEBRAL COLUMN

THE BONES

The vertebral column consists of 33 vertebrae superimposed on one another in a series which provides a flexible supporting column for the trunk and head. The vertebrae are separated by fibrocartilaginous intervertebral disks and are united by articular capsules and by ligaments. Of the fundamental 33 vertebrae, 5 are fused into the sacrum, and 4 combine in forming the coccyx in the adult. The remaining 24 vertebrae are classed as 7 cervical, 12 thoracic, and 5 lumbar. In accordance with their weight-bearing function, the vertebrae become larger and more massively constructed toward the lower end of the column, but all are built on the same fundamental plan.

Errors in development sometimes produce irregular columns. The first coccygeal segment may fuse with the sacrum to produce a sacrum of six segments. A similar numerical result occurs in the 5 per cent of cases in which the last lumbar vertebra is fused wholly or in part (sacralization) with the sacrum. Contrarily, in perhaps 5 per cent, the first sacral segment is lumbarized and separate. Other irregularities in cervical, thoracic, or lumbar parts of the column are occasionally recognized and, in old age, partial or complete fusion of the components of the vertebral column sometimes occurs.

General Characteristics of Vertebrae (fig. 228)

Vertebrae are composed of two parts—the anteriorly placed **body** and the posterior **vertebral arch** which encloses the **vertebral foramen.** The vertebral arch is formed of two pedicles and two laminae, from which arise four articular processes, two transverse processes, and one spinous process. The **body** is the largest part of the vertebra and has the form of a short cylinder. Its upper and lower surfaces are rough and flat except for a smooth rounded rim at their circumference. The bodies are separated and also bound together by the fibrocartilaginous **intervertebral disks** and are connected to one another by the long **anterior longitudinal** and **posterior longitudinal ligaments.** These are attached to the anterior and posterior surfaces of the bodies throughout the entire series of movable vertebrae. A nutrient foramen pierces the anterior surface of each body; on the posterior surface is a larger aperture for the exit of the basivertebral vein. The **pedicles** are short, stout processes which project backward from the junction of the posterior and lateral surfaces of the upper part of the body. **Vertebral notches** are formed by indentations of the rounded pedicles above and below; when the vertebrae are articulated, the notches of contiguous bones form the **intervertebral foramina.** The **laminae** are broad, flat plates of bone that continue backward and medialward from the pedicles. They close the vertebral foramen by uniting in the median plane, and the spinous processes arise from their junction. The laminae are rough along their upper and lower borders for the attachment of the ligamenta flava (p. 298).

The **spinous process** slopes downward and backward from the junction of the two laminae. The

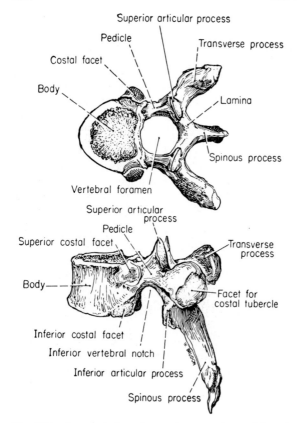

Fig. 228 A typical thoracic vertebra, superior and lateral surfaces. (×¾)

bilateral **transverse processes** project lateralward at the junction of the pedicles and laminae between the superior articular and inferior articular processes. These and the spinous process serve primarily for the attachment of muscles and ligaments. The **articular processes** are projections, two superior and two inferior, from the junction of the pedicles and laminae. Their articular surfaces tend to lie obliquely, the superior pair being directed posteriorly and somewhat upward and the inferior pair being directed anteriorly.

Special Regional Characteristics

Thoracic Vertebrae (figs. 228, 272, 277, 280) The thoracic vertebrae show the least modification of the basic pattern. The somewhat heart-shaped bodies are intermediate in size between those of the cervical and lumbar segments and exhibit two **costal facets** on each side at the junction of the body and the vertebral arch. One facet lies along the upper edge of the body; one indents its lower edge. The larger facet is at the upper border of the body. Each is a demifacet, adjacent demifacets of contiguous vertebrae forming a cupshaped articular socket for the head of a rib. These costal facets are found only on thoracic vertebrae, and the first such

vertebra has a complete costal facet at its upper border for the first rib and a demifacet at its lower border contributing to the articular surface for the second rib. The tenth, eleventh, and twelfth thoracic vertebrae have single, complete costal facets located at the upper border of each body for the head of the corresponding rib. The inferior notches of the pedicles of thoracic vertebrae are larger and deeper than in other regions of the column. The **laminae** are broad and sloping and overlap one another like shingles on a roof. The **spinous process** of a thoracic vertebra is long, slopes sharply downward, and terminates in a tubercle. The **vertebral foramen** is circular at thoracic levels and is smaller in this region than in either the cervical or the lumbar region. The **transverse processes** are thick, strong, and of considerable length. They extend posteriorly as well as lateralward and end in a clubbed extremity, on the anterior surface of which is a small concave facet for articulation with the tubercle of the corresponding rib. The **articular processes** are thin and triangular and have flat articular surfaces. The **superior articular processes** face backward, upward, and somewhat lateralward. The **inferior articular processes** project only slightly below the laminae; their facets are directed forward, downward, and somewhat medialward. The direction of the

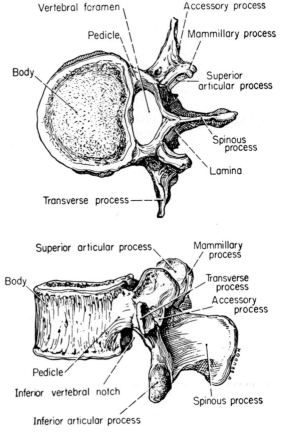

Fig. 229 A typical lumbar vertebra, superior and lateral surfaces. (×)

accessory, and transverse processes of the lumbar vertebrae.

Lumbar Vertebrae (fig. 229) The lumbar vertebrae have the largest and most massive bodies because the body weight supported by the vertebrae increases toward the lower end of the column. Their bodies are from 2 to 3 cm. deep and about 5 cm. wide. They are constricted at the sides and anteriorly. The **pedicles** are very strong, and the inferior vertebral notches are deep. The **laminae** are broad, short, and strong but do not overlap, a circumstance which allows lumbar puncture. The **vertebral foramen** is triangular in the lumbar region. It is larger here than in the thoracic vertebrae but smaller than at cervical levels. The **spinous process** of a lumbar vertebra is thick, broad, and hatchet-shaped. It may be notched inferiorly. The **articular processes** are strong and project markedly upward and downward. The facets on the superior processes are concave and face backward and medialward; those on the inferior processes are convex and are directed forward and lateralward. The articular surfaces of lumbar vertebrae facilitate flexion and extension and side to side bending. They do not permit rotation, the inferior articular processes being embraced by the superior processes of the vertebra next below. The **transverse processes** in lumbar vertebrae are long and slender. Mostly horizontal, they incline a little upward in the lower two segments. The lumbar transverse processes lie in front of the articular

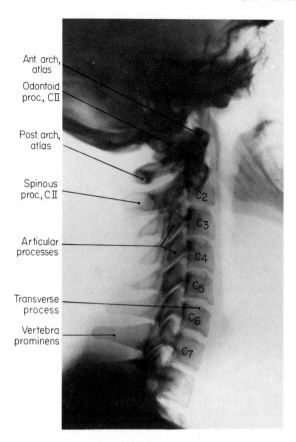

Fig. 230   A lateral radiogram of the cervical vertebrae.

articular processes of vertebrae largely determines the directions of movement of the trunk allowed in any particular region, the facets of the thoracic vertebrae favoring lateral bending and rotation of the column.

The vertebrae toward the ends of the thoracic region are necessarily transitional in form in the direction of the cervical and lumbar vertebrae. Variation in the costal facets has been noted. In addition to having one and a half costal pits, the **first thoracic vertebra** has a long and almost horizontal spinous process and long transverse processes. Its superior articular surfaces are wider and less upright; its body is broadened transversely and is slightly cupped on its upper surface like that of a cervical vertebra. The **eleventh thoracic vertebra** shows the beginning of transition to the lumbar type in its short and nearly horizontal spinous process, its very short transverse processes, and in the absence of articular facets at the ends of its transverse processes. The **twelfth thoracic vertebra** is distinctly transitional to the lumbar type. Its spine is short and broadened; its inferior articular process is turned outward and is convex; and its inferior vertebral notch is wide. The transverse process of the twelfth thoracic vertebra is rudimentary and exhibits three tubercles: superior, inferior, and lateral. These correspond to the mammillary,

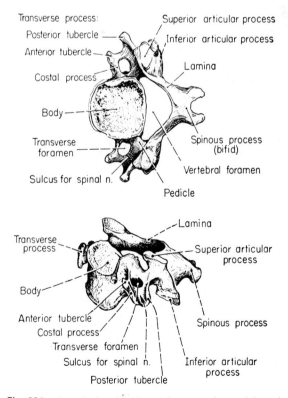

Fig. 231   A typical cervical vertebra; superior and lateral surfaces. ($\times \frac{3}{4}$)

processes (instead of behind them as in thoracic verte-brae) and are homologous with the ribs. On the back of each superior articular process there is a smoothly rounded projection, the **mammillary process**, and at the back of the root of the transverse process is an **accessory process**. These, together with the posterior part of the root of the transverse process, correspond to the trans-verse process of a thoracic vertebra. Progressive changes through the five lumbar vertebrae include widening of the bodies, thickening of the pedicles, and increased separation between articular processes. The **fifth lumbar vertebra**, in adjustment to its articulation with the sacrum, has the widest body and is much deeper in front than behind. Its spine and transverse processes are short, and the latter are thick and conical. Its inferior articular processes are as wide apart as its superior processes for articulation with the superior articular processes of the sacrum.

Cervical Vertebrae (fig. 231) The cervical vertebrae are the smallest of the series and are readily distin-guished by the presence of a large oval foramen in each transverse process. The **body** is small and somewhat broader from side to side than from front to back. The superior surface of the body is concave transversely with projecting lips on either side. The inferior surface is concave anteroposteriorly; its anterior lip projects downward to overlap the superior surface of the verte-bra below. The inferior surface is convex transversely

to fit the transverse concavity of the superior surface of the next adjacent vertebra. The **pedicles** project lateralward as well as backward and spring from the body midway between its superior and inferior surfaces, so that the superior vertebral notch is as deep as the inferior. The **laminae** are thinner above than below and enclose a large, triangular **vertebral foramen** for the accommodation of the upper and largest portion of the spinal cord. The **spinous process** is short and bifid. The **articular processes** form prominent lateral projections from the junction of pedicles and laminae and bear flat oval facets. The superior facets are directed upward, backward, and slightly medialward; the inferior facets face oppositely. Each **transverse process** is said to be perforated by the transverse foramen for the passage of the vertebral artery and vein and the vertebral nerve plexus. Actually, the 'costotransverse' foramen is located between the transverse process posteriorly and a costal process anteriorly. The robust transverse pro-cess springs from the junction of pedicle and articular process, is directed anteriorly and lateralward, and ends in a flattened posterior tubercle. The thin costal process shows its homology to a rib by arising from the body of the vertebra anteriorly; it ends in the anterior tubercle. The two processes are joined by a bar of bone, which together complete the formation of the costotransverse foramen, the connecting bar being deeply grooved superiorly for the emergence of the spinal nerve posterior to the vertebral vessels.

The second cervical vertebra (axis, fig. 232) **departs** considerably from the usual form of cervical vertebrae. The first and second cervical vertebrae are modified from the basic plan, the changes being essentially the incorporation of a part of the body of the first vertebra into the body of the second. This leaves only an anterior arch of bone in place of the body of the atlas (first cervical). In the **axis** (second cervical), the bony addi-tion forms the strong **dens** which rises perpendicularly from the upper surface of its body. This constitutes the pivot around which the atlas, carrying the head, rotates. To complete the pivot joint, the dens has concave, oval articular facets, anteriorly for the arch of the atlas and posteriorly for the transverse ligament of the atlas. The dens is pointed above and has a slight constriction where it joins the body of the axis. The **pedicles** of the axis are heavy and strong, as are also the **laminae**. The **spinous process** is broad and heavy and is bifurcated at its extremity. The **inferior articular surfaces** are similar to those of other cervical vertebrae, but the **superior articular surfaces** are large and are supported by the body, the pedicles, and the transverse processes. These articular surfaces are rounded and face obliquely up-ward and lateralward.

The first cervical vertebra (atlas, fig. 232), having lost part of its body to the second and having no spinous process, has the form of an oval ring of bone. It consists, therefore, of an anterior and a posterior arch and two lateral masses. The narrow **posterior arch** encloses a very large **vertebral foramen**. On its upper surface, behind the superior articular facet, it carries a sulcus for the verte-bral artery and, in the median line, a posterior tubercle

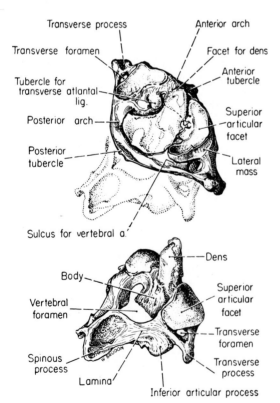

Fig. 232 The atlas and axis viewed obliquely. In the upper view, the atlas is fitted over a phantom of the axis.

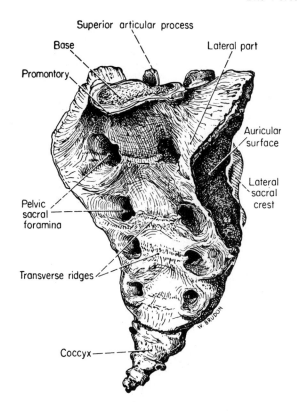

Fig. 233   The pelvic aspect of the sacrum.

and its transverse foramina are smaller than those of the other cervical vertebrae. They are occasionally absent.

The Sacrum (figs. 233, 234) **The sacrum, consisting** of the fusion of five vertebrae, is broadened by the amalgamation of large costal elements and transverse processes into heavy **lateral parts.** It is triangular in form, concave and relatively smooth on its pelvic surface, convex and highly irregular dorsally. The concave **pelvic surface** is marked by four **transverse ridges** separating the original bodies of the five sacral vertebrae. At the ends of the ridges are the large **pelvic sacral foramina** directed forward and lateralward and accommodating the ventral rami of the sacral nerves and branches of the lateral sacral arteries. The lateral parts, lateral to the foramina, are grooved for the sacral nerves and exhibit alternating ridges from which arise the fibers of the piriformis muscle.

On the **dorsal surface** of the sacrum appear the **dorsal sacral foramina,** medial to which the vertebral canal is closed over by the fused laminae. In the midline the reduced spinous processes form the **median sacral crest.** The shallow groove over the laminae lodges the sacral portion of the multifidus muscle. The laminae of the fifth segment, and sometimes of the fourth, fail to meet and thus is produced the **sacral hiatus,** an inferior entrance to the vertebral canal. Just medial to the dorsal sacral foramina, the fused articular processes form the indistinct **intermediate sacral crests.** The upper **articular**

which is the rudiment of the spinous process. Along the superior border of the arch posterior to the groove is attached the posterior atlantooccipital membrane. The **anterior arch** carries a prominent **anterior tubercle** for the attachment of the longus colli muscles. Posteriorly, this arch has a smooth oval concavity which receives the dens of the axis. The upper border of the anterior arch gives attachment to the anterior atlantooccipital membrane; the lower border, to the anterior atlantoaxial membrane. The **lateral masses** of the atlas support the skull and articulate below with the axis. The **superior articular facets** are large, oval, and concave. They face upward, medialward, and slightly posteriorly and each is the segment of a cup, in which rests the base of the skull, articulating by means of its occipital condyles. Below the medial margin of each superior articular facet is a tubercle for the attachment of the **transverse atlantal ligament.** The slightly concave surfaces of the **inferior articular processes** of the atlas conform in size and shape to the superior articular facets of the axis. They face downward and medialward. The transverse processes of the atlas are large and project lateralward over those of the axis. They provide attachment for the muscles which rotate the head, the greater part of this action occurring at the atlantoaxial joints.

The **seventh cervical vertebra,** transitional to the thoracic type, has the only long horizontal spine of the cervical group, which gives to this vertebra the name of **vertebra prominens.** Its transverse processes are large,

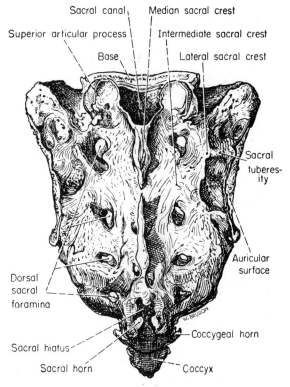

Fig. 234   The dorsal surface of the sacrum.

processes of the first sacral vertebra are large and oval. Their facets, for articulation with the inferior articular surfaces of the fifth lumbar vertebra, are concave from side to side and face backward and medialward. The tubercles of the inferior articular processes of the fifth sacral vertebra form the **sacral horns** and are connected to the horns of the coccyx. Lateral to the dorsal sacral foramina are the prominent **lateral sacral crests** of the sacrum, representing its transverse processes. These provide attachment for the dorsal sacroiliac ligaments and, inferiorly, for the sacrotuberous and sacrospinous ligaments.

The **lateral surface** of the sacrum has a narrow edge below, broadens above, and contains anteriorly the smooth ear-shaped **auricular surface.** This surface, covered with cartilage in life, is for articulation with the ilium. Behind it are three deep impressions for attachment of the massive dorsal sacroiliac ligament. The **superior surface** of the sacrum exhibits a projecting anterior border; this is the **sacral promontory.** The sacrum has a large oval body, separated from that of the fifth lumbar vertebra by a thick intervertebral disk. The **superior articular processes** of the sacrum are supported by short heavy pedicles, and its **laminae** enclose a flattened triangular vertebral canal. On either side of the body of the sacrum is the **ala,** formed of the costal and transverse processes of the first sacral vertebra. It is flattened anteriorly by the psoas muscle and is also crossed by the lumbosacral nerve trunk. In the articulated state, the ala of the sacrum is continuous with the ala of the ilium.

The Coccyx (figs. 233, 234) The coccyx may be a single bone fused from the four coccygeal elements, or the first segment may be separate from the rest. These vertebrae are much reduced, having no pedicles, laminae, or spinous processes. The pelvic surface is concave and relatively smooth like the sacrum and provides attachment for part of the coccygeus and levator ani muscles. On the dorsum the rudimentary articular processes present a row of tubercles; the larger superior pair form the **coccygeal horns** which articulate with the horns of the sacrum and enclose the fifth sacral intervertebral foramen. The base of the coccyx is oval for articulation with the sacrum.

The Vertebral Column as a Whole (fig. 235) The majority of vertebral columns range between 72 and 75 cm. in length, of which approximately one-fourth is accounted for by the intervertebral disks. The adult vertebral column has four curvatures— an anterior convexity in the cervical region; an anterior concavity in the thoracic region; an anterior convexity in the lumbar region; and the sacro-coccygeal curvature, concave anteriorly. Two primary curvatures exist in fetal life, both concave anteriorly and separated by the sacrovertebral angle where the curvature becomes abruptly deeper for the pelvic cavity. With the raising of the head after

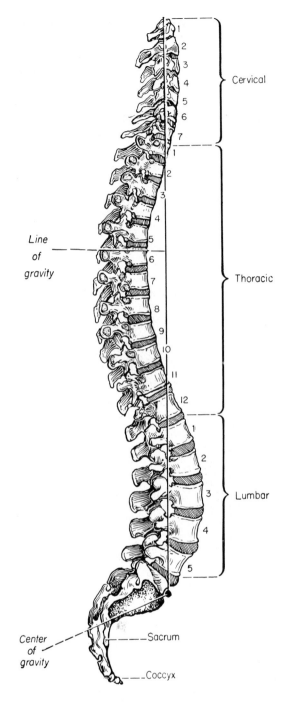

Fig. 235 A lateral view of the vertebral column illustrating its curvatures and the line and center of gravity of the body.

birth and with the assumption of the erect posture at a later period, two secondary curvatures develop at cervical and lumbar levels. The shallow cervical curve begins in the dens of the axis and ends at the second thoracic level. It can be obliterated by bending the head forward. The permanent thoracic curve is due to the shape of the vertebral bodies. It ends below at the twelfth thoracic vertebra. The prominent lumbar curve, deeper in women, is a secondary response to the upright position of the trunk and ends at the sacrovertebral angle. The curvature of the sacrum is permanent. It is a relatively smooth concavity in the male, deepest opposite the third sacral segment. In the female the sacrum is flatter above and more sharply bent forward below. The curvatures of the vertebral column make it a flexible support, and vertebrae curve back and forth about a line of gravity along which the weight of the upper body is projected into the lower limbs. This line of gravity passes successively through the dens of the axis, the bodies of the second and twelfth thoracic vertebrae, and the promontory of the sacrum. The center of gravity of the body is located just anterior to the sacral promontory. The vertebral foramina are large triangles in the cervical region, small circles in the thoracic vertebrae, large ovals at lumbar levels, and flattened triangles in the sacrum. The canal formed by these superimposed foramina is thus largest in the cervical and lumbar levels, which are the levels of origin of the large limb nerves and the regions of greatest movement in the column.

Ossification of the Vertebrae Each vertebra is ossified from three primary and five secondary centers. Primary centers, one in the body and two in the vertebral arch, appear in the cartilaginous model of the vertebra at about the twelfth week of intrauterine life. The centers for the arch begin near the bases of the superior articular processes and give rise to the pedicles, the laminae, and the articular, transverse, and spinous processes. At birth, a typical vertebra consists of three osseous parts— the body and two lateral arch pieces—joined together by hyaline cartilage. The line of union of body and arch, the **neurocentral synchondrosis,** is not obliterated until the third year in the cervical region; as late as the seventh year in the sacrum. The neurocentral synchondrosis falls so as to place the costal elements, and in the thoracic region the costal pit, on the side with the vertebral arch elements. The secondary centers appear in the cartilaginous extremities of the spinous and transverse processes and on the superior and inferior surfaces of the bodies. On the bodies they form two annular plates, thickest at the circumference and gradually thinning to a central deficiency. The secondary centers appear after the sixteenth year and fuse into the completed vertebra by the twenty-fifth year. Additional centers or modifications of this general sequence occur in the atlas, the axis, the sacrum, and the coccyx.

Clinical Note Failure of fusion of the two sides of a vertebral arch results in the gross developmental defect of *spina bifida.* The clinical importance of the defect is greatest when it occurs in the lower lumbar vertebrae and when spinal cord or meninges or both protrude through the defect. Abnormal curvatures of the column sometimes develop. **Kyphosis** (hunchback), an increase in the thoracic curvature, results from congenital or pathologic erosion of the anterior portion of one or more vertebral body. **Lordosis,** an abnormal increase in the lumbar curvature, accompanies any downward rotation of the pelvis and is fairly regularly seen in the later stages of pregnancy in compensation for the forward shift of the center of gravity. **Scoliosis,** or lateral curvature, also involves a rotation of the column in which the spinous processes of the verebrae turn toward the concavity of the curvature. Compensatory curves in the reverse direction also occur in order to keep the head directly over the feet. This complex deformity of the column may be congenital or may result from disease.

THE ARTICULATIONS OF THE VERTEBRAL COLUMN

The articulations of the vertebral column are the cartilaginous joints between the vertebral bodies, the synovial joints between the vertebral arches, the articulations between the axis and the atlas, and the articulation of the atlas with the skull.

The Articulations of the Vertebral Bodies (figs. 236, 239)

The bodies of the vertebrae are united by intervertebral disks and by the anterior and posterior longitudinal ligaments.

The **intervertebral disks** of fibrocartilage are interposed between the bodies of adjacent vertebrae from axis to sacrum. They vary in thickness and size in different regions of the column, but together they constitute about one-fourth of its length. In

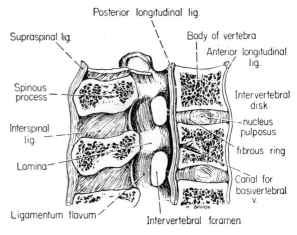

Fig. 236 A median section of several vertebrae to illustrate the intervertebral disk and the ligaments of the column.

the cervical and lumbar regions they are thicker in front than behind, thus accounting for the cervical and lumbar curvatures of the column. The disks are of uniform thickness at thoracic levels, where the curvature of the column is due to the slight wedge shape of the bodies of the vertebrae. They are thickest in the lumbar region of the column. The intervertebral disks are attached above and below to the thin layers of hyaline cartilage which cover the upper and lower surfaces of the vertebral bodies.

Each disk is composed of an outer fibrous ring, or **anulus fibrosus,** surrounding a **nucleus pulposus.** The fibrous ring consists of up to 20 concentrically arranged fiber bundles and of fibrocartilage. The fibers pass obliquely from the vertebral body above to that below and alternate in their diagonal direction in successive concentric rings. The more central concentric laminae tend to be incomplete and less distinct and they grade over into the more cartilaginous nucleus pulposus. The **nucleus pulposus** is a soft, pulpy, highly elastic, and compressible substance of a yellowish color. It has a fine fibrous matrix, a reticular structure, and a high water content. With increase in age, the nucleus exhibits an increase in fibrocartilage and a decrease in its fluid content.

The intervertebral disks are important shock ab-

sorbers. Under compression, they become flatter and broader and momentarily bulge from the intervertebral spaces. Ventrally and dorsally, their fibers are attached to the anterior longitudinal and posterior longitudinal ligaments.

The **anterior longitudinal ligament** is a broad collection of longer and shorter fibers and extends along the anterior surfaces of the bodies of the vertebrae, binding them together. It has a pointed attachment to the anterior tubercle of the atlas, widens as it descends, and ends below on the pelvic surface of the sacrum. Its superficial fibers are long, whereas its deeper fibers pass over only one or two vertebrae.

The **posterior longitudinal ligament** is a similar band of fibers common to all the vertebrae and is located in the vertebral canal on the posterior borders of the vertebral bodies. It is broadest above where, continuous with the tectorial membrane (p. 300), it is attached to the occipital bone. At thoracic and lumbar levels the ligament, broad over the intervertebral disks, narrows over the vertebral bodies so as to have serrated margins. It ends below in the sacral canal. Its deeper fibers attach to adjacent vertebrae; its superficial fibers pass over three or four vertebrae.

Clinical Note Protrusion or rupture of intervertebral disks, most common in lumbar or lumbosacral levels, result in pain and frequent disability. Usually the disk ruptures posterolaterally through the weaker lateral part of the posterior longitudinal ligament. Impinging and pressing on the spinal nerve descending from the segment above and coursing toward the intervertebral foramen, the protruded disk initiates pain which 'radiates' in the area of distribution of the spinal nerve affected.

The Joints between the Vertebral Arches (figs. 236, 239)

The vertebral arches are united by plane synovial joints between the articular processes, enveloped by articular capsules. The articular capsules are thin and loose, especially in the cervical region. They are attached to the margins of the articular processes of adjacent vertebrae. Accessory ligaments unite the laminae, the transverse processes, and the spinous processes. They include the ligamentum flavum, the supraspinal ligament, the ligamentum nuchae, the interspinal ligaments, and the intertransverse ligaments.

The **ligamenta flava** are yellow elastic ligaments which connect the laminae of adjacent vertebrae.

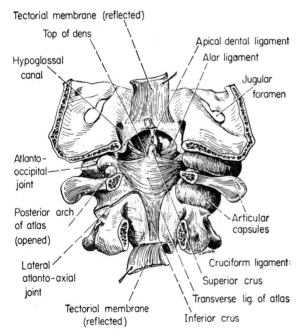

Fig. 237 The atlantoaxial and atlantooccipital joints, seen after removal of the vertebral arches.

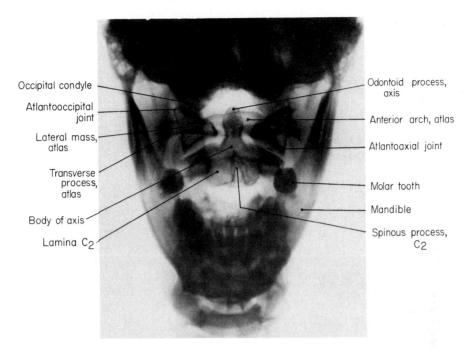

Occipital condyle

Atlantooccipital joint

Lateral mass, atlas

Transverse process, atlas

Body of axis

Lamina C₂

Odontoid process, axis

Anterior arch, atlas

Atlantoaxial joint

Molar tooth

Mandible

Spinous process, C₂

Fig. 238   A radiogram of the atlantooccipital and atlantoaxial joints.

Each is attached above to the front of the lower borders of a pair of laminae and below to the back of the upper borders of the next succeeding pair. The portions of the two sides meet at the root of the spine, separated by intervals for the passage of veins which communicate between the internal vertebral venous and external vertebral venous plexuses. The ligamenta flava extend lateralward as far as the articular capsules. These strong elastic ligaments assist in maintaining the upright posture and in resuming it after flexion of the column.

The **supraspinal ligament** is a band of longitudinal fibers interconnecting the tips of the spinous processes from the seventh cervical vertebra to the sacrum, where it ends in the median sacral crest. Its deepest fibers connect adjacent spines; more superficial ones extend over three or four vertebrae. It is continuous with the interspinal ligament and, above, with the ligamentum nuchae.

The **ligamentum nuchae** (p. 49) is a triangular, median, sheet-like, upward extension of the supraspinal ligament which reaches the external occipital protuberance and the median nuchal line of the occipital bone. Deeply, the ligamentum nuchae is attached to the posterior tubercle of the atlas and to the spinous processes of the cervical vertebrae. It is a rudiment of a highly elastic ligament in quad-

rupeds which assists in holding the head erect, but its principal advantage in Man is in providing for muscle attachments without limiting dorsal flexion of the neck as would long cervical spinous processes.

The **interspinal ligaments** are thin, membranous bands which unite adjacent spinous processes and extend from the supraspinal ligament to the ligamenta flava. They are longer and stronger in the lumbar region and insignificant at cervical levels.

The **intertransverse ligaments** connect adjacent transverse processes. Generally insignificant, they tend to be absent at cervical levels and are marked only in the lumbar region.

The Articulations between the Axis and the Atlas (figs. 237–40)

The joints between the axis and the atlas are of two types—bilateral gliding joints between the surfaces of their articular processes, and the median pivot joint between the dens of the axis and both the anterior arch of the atlas and the transverse atlantal ligament.

The **lateral atlantoaxial joints**, formed by the reciprocal surfaces of the articular processes of the atlas and the axis, are surrounded by a loose capsule lined by a synovial membrane. Applied to the back of the capsule is an **accessory atlantoaxial ligament** which passes from the back of the lateral mass of the atlas downward and medialward to the back of the body of the axis.

The **median atlantoaxial joint** is composed of two

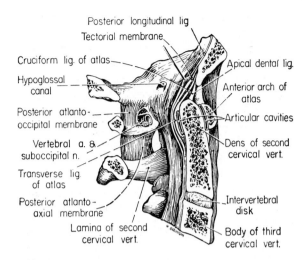

Fig. 239   A median section of the atlantooccipital region to illustrate the articular and ligamentous relations of the dens.

separate articulations: one between the anterior surface of the dens of the axis and the oval facet on the inner aspect of the anterior arch of the atlas; and one between the posterior surface of the dens and the fibrocartilaginous face of the transverse ligament of the atlas. Each cavity is enclosed by a thin capsule lined by a synovial membrane. The **transverse ligament of the atlas** is thick and strong. It arches from side to side behind the dens, attaching to small tubercles on the medial aspect of the lateral masses of the atlas (fig. 232). From the middle of the ligament a small band passes upward to the anterior edge of the foramen magnum, and another passes downward to the back of the body of the axis. These, together with the transverse ligament, constitute the **cruciform ligament of the atlas.** The large vertebral foramen of the atlas is posterior to the transverse ligament.

. Associated with the median atlantoaxial joint are certain ligaments which unite the axis with the occipital bone. The **tectorial membrane** covers the dens and its ligaments within the vertebral canal and is the upward extension of the posterior longitudinal ligament. From the body of the axis it expands as it passes upward and is attached to the occipital bone within the anterior edge of the foramen magnum. Here it blends with the dura mater. The **alar ligaments** are short but strong rounded cords which arise on either side of the apex of the dens. They pass upward and lateralward to the medial aspect of each occipital condyle. The **apical dental ligament** is a slender band seen between the diverging alar ligaments. It arises from the apex of the dens and attaches to the anterior margin of the foramen magnum.

The **anterior atlantoaxial membrane** is a strong membrane which passes between the anterior arch of the atlas and the front of the body of the axis between the lateral joints. In the median line it is overlain and strengthened by the uppermost narrow band of the anterior longitudinal ligament.

The **posterior atlantoaxial membrane** is a broad, thin membrane attached above to the inferior border of the posterior arch of the atlas and below, to the superior edge of the laminae of the axis. It is continuous with the ligamenta flava.

The Articulations of the Atlas with the Skull (fig. 237)

The atlantooccipital joints are formed between the convex occipital condyles and the concave superior articular facets of the atlas. Each of the pair is enclosed by an articular capsule which is thin and loose. The paired joint surfaces may be considered to be segments of a single ellipsoidal surface, the transverse diameter of which is the longer.

The **anterior atlantooccipital membrane** is composed of densely woven fibers which pass between the upper border of the anterior arch of the atlas and the anterior margin of the foramen magnum. Laterally, it is continuous with the articular capsules.

The **posterior atlantooccipital membrane** is connected above to the posterior margin of the foramen magnum. Below, it ends in the upper border of the posterior arch of the atlas. It reaches the articular capsules at each side. Below and laterally, the membrane arches over the vertebral artery and the suboccipital nerve as they cross the posterior arch of the atlas.

Movements of the Vertebral Column

Movement between vertebrae takes place in the resilient intervertebral disks and at the joints of the articular processes; the displacement between adjacent vertebrae is small, but the range of motion in the total column is considerable. Movements of the column are flexion, extension, lateral bending, and rotation. Extension is freer than flexion, the latter movement in the column being added to at the hips and at the atlantooccipital joints. The flexion-extension movement of the vertebral bodies on one another involves compression of the disks at one edge and stretching at the other. The anterior longi-

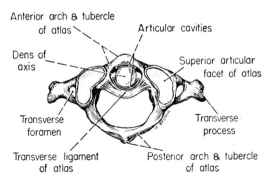

Fig. 240   The median atlantoaxial joint.

tudinal or the posterior longitudinal ligament becomes tense in limiting the motion. Generally speaking, the articular processes are set obliquely, so that lateral bending and rotation tend to be combined to a degree. In the neck all movements are permitted. With relatively thick disks and almost horizontally oriented articular processes, flexion and extension are free and lateral bending and rotation are allowed, each combined somewhat with the other. In the chest the obliquely set articular surfaces permit slight flexion and extension and free bending and rotation, but the thorax is a relatively stable region. This is due to the thin disks, the presence of the ribs and the sternum, and the overlapping of the long spinous processes, all of which tend to restrict the motion of the vertebrae. The lumbar disks are thick, and the vertical articular surfaces are sagittally oriented. Flexion and extension are free at lumbar levels, and lateral bending is permitted; but rotation is limited by the articular processes facing one another in a sagittal plane.

At the **atlantoaxial joints,** rotation of the skull and atlas together moves the face from side to side. The dens of the axis serves as a vertical pivot enclosed in the ring of the anterior arch of the atlas and its transverse ligament (fig. 240). As the ring turns around this pivot, the articular surfaces of the atlas glide forward and backward on the upper articular facets of the axis, their sloping and curved surfaces facilitating the turning movement. The alar liga-

ments are the chief check ligaments in the rotary movement.

The articular surfaces of the **atlantooccipital joints** resemble segments of an ellipse, the longer dimension of which lies transversely. Movements of flexion and extension (nodding of the head forward and backward) and lateral bending are permitted, the former movements being especially free. No rotary movement is possible at these joints.

Blood Vessels of the Vertebral Column

Arteries **Spinal arteries** are branches of vessels adjacent to the central axis of the trunk—vertebral and ascending cervical arteries in the neck; costocervical and posterior branches of intercostal arteries in the chest; dorsal branches of lumbar and iliolumbar arteries; and in the pelvis, branches of the lateral sacral arteries. Bilaterally, spinal arteries enter the intervertebral foramina with the spinal nerves of the corresponding segment and divide into three branches.

(a) A **postcentral branch** ramifies on the lateral part of the posterior longitudinal ligament. Its ascending and descending branches form anastomosing longitudinal channels, and transverse branches form interconnections from side to side.

(b) A **prelaminar branch** ramifies on the anterior aspect of the laminae and the ligamenta flava, forming similar but less regular longitudinal and transverse anastomoses.

(c) The **radicular branches** follow the nerve roots and contribute to the posterior spinal and anterior spinal arteries (p. 307).

Veins (figs. 241, 243) The veins of the vertebral column comprise rich venous plexuses which extend the whole length of the column both inside and outside the vertebral canal. All parts of the plexuses intercommunicate, and they are drained by a series of intervertebral veins. Internal vertebral plexuses are formed after the pattern of the arteries described above and receive the basivertebral veins. External vertebral plexuses are also interconnected with the internal plexuses and are drained in common with them. The **internal vertebral plexuses** are continuous networks which lie between the dura mater and the walls of the vertebral canal. Anteriorly, at the sides of the posterior longitudinal ligament are two wide **anterior longitudinal vertebral sinuses.** These extend the whole length of the column and are connected by transverse anastomoses which receive the basivertebral veins. They communicate above with the basilar and occipital sinuses, the end of the sigmoid sinuses, and the emissary vein of the hypoglossal canal. **Posterior longitudinal vertebral sinuses,** less marked like the corresponding prelaminar arteries, lie on the inner surface of the laminae and the ligamenta flava. They also exhibit cross connections. The **basivertebral veins** are wide endothelial-lined channels within the bodies of the vertebrae. They converge to the posterior aspect of the vertebral bodies and open into

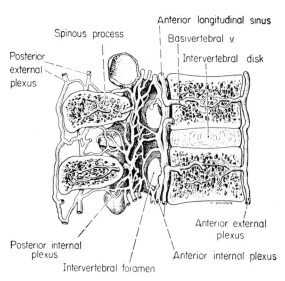

Anterior longitudinal sinus

Spinous process

Basivertebral v.

Posterior external plexus

Intervertebral disk

Posterior internal plexus

Anterior external plexus

Intervertebral foramen

Anterior internal plexus

Fig. 241 A median section of two vertebrae to illustrate the vertebral plexus of veins.

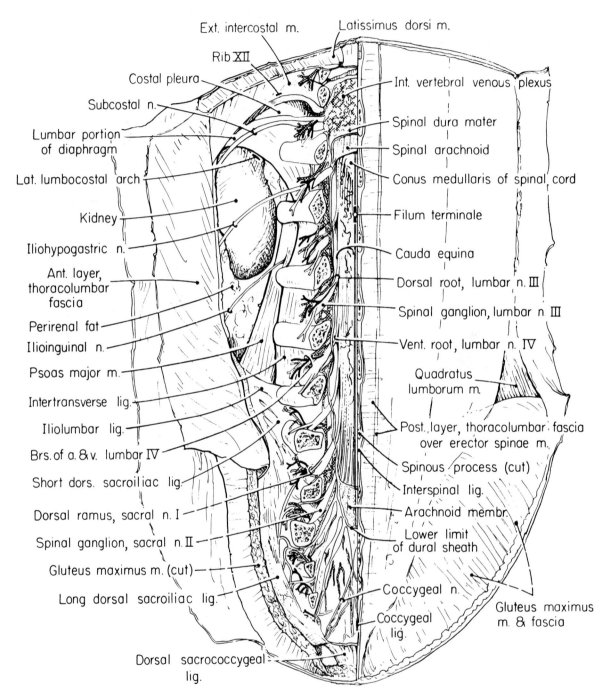

Ext. intercostal m.

Latissimus dorsi m.

Rib XII

Costal pleura

Subcostal n.

Lumbar portion
of diaphragm

Lat. lumbocostal arch

Kidney

Iliohypogastric n.

Ant. layer,
thoracolumbar
fascia

Perirenal fat

Ilioinguinal n.

Psoas major m.

Intertransverse lig.

Iliolumbar lig.

Brs. of a. & v. lumbar IV

Short dors. sacroiliac lig.

Dorsal ramus, sacral n. I

Spinal ganglion, sacral n. II

Gluteus maximus m. (cut)

Long dorsal sacroiliac lig.

Dorsal sacrococcygeal
lig.

Int. vertebral venous plexus

Spinal dura mater

Spinal arachnoid

Conus medullaris of spinal cord

Filum terminale

Cauda equina

Dorsal root, lumbar n. III

Spinal ganglion, lumbar n. III

Vent. root, lumbar n. IV

Quadratus
lumborum m.

Post. layer, thoracolumbar fascia
over erector spinae m.

Spinous process (cut)

Interspinal lig.

Arachnoid membr.

Lower limit
of dural sheath

Coccygeal n.

Coccygeal
lig.

Gluteus maximus
m. & fascia

Fig. 242   A dissection of the back exposing the spinal cord and nerve roots.

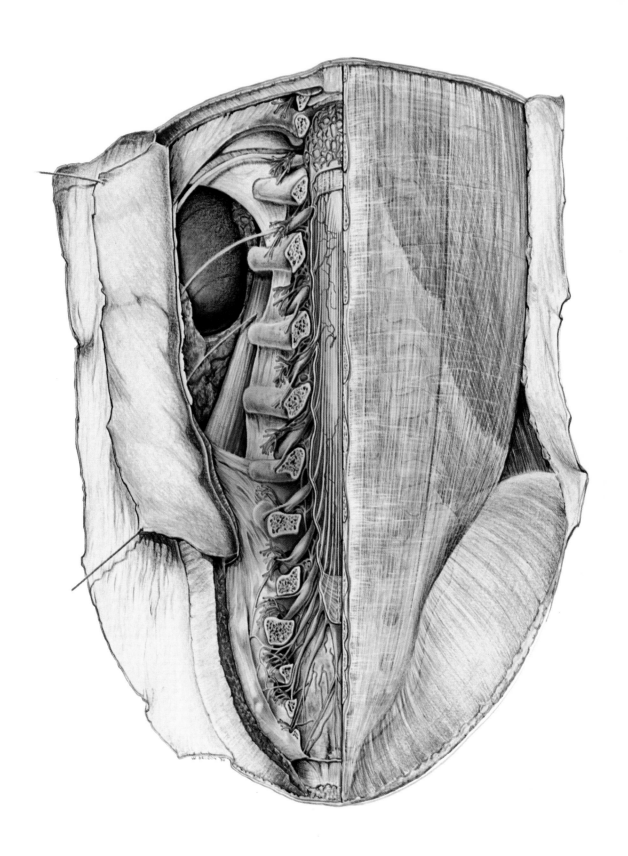

the transverse anastomoses between the anterior longitudinal sinuses.

The **external vertebral plexuses** are formed in several parts. The **anterior external vertebral plexuses** consist of anastomosing longitudinal venous and transverse venous channels along the front of the vertebral column. They communicate with the basivertebral and intervertebral veins. The **posterior external vertebral plexuses** lie around the spinous, articular, and transverse processes of the vertebrae. They communicate with the internal plexuses and the intervertebral veins.

The **intervertebral veins** arise in the longitudinal sinuses and pass outward through the intervertebral and pelvic sacral foramina. They collect blood from all the venous plexuses of the spinal canal. Varying with the region, these intervertebral veins are tributary to the vertebral, intercostal, lumbar, and lateral sacral veins.

The **vertebral veins** receive the intervertebral veins of the cervical region. The vertebral veins arise in the suboccipital triangle by the confluence of the intervertebral and muscular veins at that level. The vertebral veins descend through successive transverse foramina of the cervical vertebrae in company with and forming a plexus around the vertebral artery. They continue to receive intervertebral and muscular veins as they descend. Emerging from the transverse foramen of the sixth cervical vertebra, the vertebral vein empties into the posterior aspect of either the right or the left brachiocephalic vein. A bicuspid valve guards the mouth of each vertebral vein. The vein of the right side crosses the first part of the subclavian artery.

Clinical Note Batson ('40) has pointed out that the veins of the vertebral column form a system of great blood-carrying capacity extending, without valves, from the pelvis to the cranium and communicating at all levels with the other major venous channels of the abdomen, the chest, and the neck. He has noted its significance in the transfer of metastatic cancer cells to widely separated parts of the body under the influence of differences in venous pressure.

THE SPINAL CORD AND ITS COVERINGS

The Spinal Meninges (figs. 242–46)

The **spinal cord** is enclosed within three covering layers, just as is the brain. The **dura mater** of the spinal cord, the outermost covering, is a long tubular sheath of dense, white fibrous and elastic tissue which extends from the foramen magnum to the level of the second sacral vertebra. It is separated from the bony and ligamentous walls of the vertebral canal by epidural fat and areolar tissue and by the internal vertebral venous plexus. The spinal dura mater differs from the cranial dura mater, being composed of a single layer without duplications and without enclosed dural venous sinuses. It is continuous with the meningeal layer of the cranial dura mater at the foramen magnum, where it is attached to the periosteum of the occipital bone. It also is attached by areolar tissue to the periosteum of the second and third cervical vertebrae. The spinal dura mater sends prolongations outward over the dorsal root ganglia and spinal nerves and into the intervertebral foramina. These end by blending with the sheaths of all the spinal nerves and are adherent to the periosteum lining the foramina. When examined with the dural sheath opened, it is apparent that the dorsal and ventral roots are covered by separate prolongations of the dura mater which, however, blend into a single investment as the dorsal root ganglion is reached. The spinal dura mater is connected by fibrous slips to the posterior longitudinal ligament, especially toward the lower end of the vertebral canal. Opposite the second sacral vertebra the tubular dura mater tapers abruptly into a slender covering for the filum terminale of the spinal cord. This covering, the **coccygeal ligament** (filum of dura mater), descends to attach to the periosteum of the coccyx. The spinal dura mater forms a tube of varying caliber, being broadest in the cervical region and narrowest in the thoracic. Its inner surface is lined by squamous cells and is separated from the spinal arachnoid membrane by a capillary interval, the **subdural space.** This contains a film of fluid—just enough to moisten the apposed surfaces of the membranes.

The delicate spinal **arachnoid membrane** is continuous with the same layer investing the brain. It follows completely the limits of the spinal dura mater, being separated from it by only the minute subdural space. The arachnoid covers the nerve roots and the dorsal root ganglia and blends with the sheaths of the nerves. The arachnoid forms a voluminous covering for the cauda equina and follows the spinal dura mater as far as the termination of the dura mater in the coccygeal ligament at the second sacral level. The spinal **subarachnoid space** is a wide interval surrounding the cord and pia mater; it is largest in the lower part of the vertebral canal where it is occupied by the nerves of the cauda equina. It is continuous above with the cranial subarachnoid space and terminates below at the level of the second sacral vertebra.

The subarachnoid space is crossed by delicate trabeculae derived from both the arachnoid and the pia mater. These are condensed along the posterior median line as the subarachnoid septum. The subarachnoid space is partially subdivided at the side of the cord by the denticulate ligaments (see below), the points of attachments of which pierce the arachnoid to become fastened to the dura mater.

The spinal **pia mater** is the intimate connective tissue covering of the spinal cord which enmeshes its blood vessels as does the pia mater of the brain. The spinal pia mater is somewhat thicker, however, due to an outer layer of predominantly longitudinal connective tissue fibers. The pia mater is inseparable from the spinal cord and is prolonged over its roots and ganglia, continuing into the sheaths of the spinal nerves. It contributes to the trabeculae which bridge the subarachnoid space and to the subarachnoid septum. The pia mater provides a longitudinal, glistening band of fibers, the **linea splendens,** which ensheaths the anterior spinal artery occupying the anterior median fissure of the cord. At the conus medullaris, the pia mater continues into the **filum terminale,** gradually replacing all nervous elements within it. The filum terminale descends in the median line among the nerve roots of the cauda equina, enters the upper end of the coccygeal ligament, and ends with it by blending with the periosteum of the back of the coccyx.

The outer longitudinal fibers of the pia mater form two **denticulate ligaments,** one on either side of the spinal cord. These are narrow fibrous bands attached continuously along the spinal cord between the ventral and dorsal nerve roots. The lateral margin of each band is serrated, and about twenty-one triangular, tooth-like processes extend toward the dura. Impaling the arachnoid membrane against it, the tooth-like processes attach to the dura mater between the nerve roots. The attachments of the ligaments alternate with the passage of the nerve roots through the dura mater. The first attachment is at the level of the foramen magnum; the last one is between the last thoracic and the first lumbar nerve roots. The denticulate ligament provides an important fixation for the spinal cord, just as the cerebrospinal fluid is an essential buffer material. Both protect the cord from shocks and sudden displacements.

Clinical Note The absence of the spinal cord below the second lumbar level, the continuation of the subarachnoid space to the limit of the dural sac at sacral two, and the separation of the laminae and spinous processes of the lumbar vertebrae, combine to make the mid or low lumbar levels favorable sites for withdrawal of cerebrospinal fluid from the subarachnoid space. Such withdrawal provides an opportunity

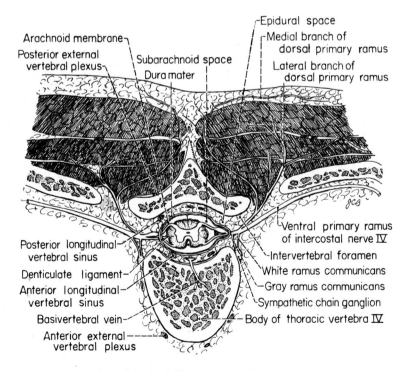

Fig. 243 A partial cross section of the back illustrating the formation and parts of an intercostal nerve.

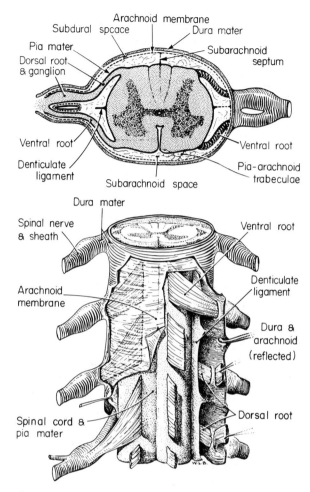

Fig. 244 The meningeal relations of the spinal cord and roots.

for diagnostic investigation of the fluid itself, for its replacement by air as in a pneumoencephalogram, or for the introduction of a radiopaque oil. In the latter case, filling defects of the subarachnoid space may locate tumors or other masses. Spinal anaesthetics can also be introduced by lumbar puncture. In lumbar puncture, the nerve roots of the cauda equina are seldom damaged by the needle, for, round and suspended in fluid, they roll away from its point.

THE SPINAL CORD (figs. 242–46)

The **spinal cord** (see also p. 28) is approximately cylindrical in form and has a length of from 43 to 45 cm. It is continuous with the medulla oblongata of the brain at the foramen magnum and terminates in the tapered **conus medullaris** at the lower border of the first lumbar vertebra or the upper border of the second. The conus is continued by a strand, the **filum terminale,** which extends to the base of the coccyx. Until the third month of fetal life the spinal cord is as long as the vertebral canal. Subsequently, the vertebral column lengthens at a more rapid rate than the cord. Thus, with early maturation in the nervous system and with regression of the coccygeal segments of the cord, its lower tip lies opposite the third lumbar vertebra at birth and is higher in the adult. The result of this positional disparity is that only the upper spinal cord segments are close to their corresponding vertebral levels. Toward the lower end of the spinal cord the differences are considerable. Thus, all the lumbar segments of the spinal cord are at the eleventh and twelfth thoracic vertebral levels, and all the sacral cord segments lie at the first lumbar vertebral level.

The dorsal surface of the cord has a number of longitudinal sulci, from several of which a continuous series of nerve rootlets emerge. These rootlets converge into a bundle, the **dorsal root,** and the portion of the cord which provides the rootlets of one dorsal root constitutes a segment of the spinal cord. The rootlets are crowded together in the cervical and lumbar regions, and the segments are a little over 1 cm. in length at those levels. The thoracic segments are over 2 cm. in length. Similarly, ventral rootlets emerge lateral to the anterior median fissure of the cord and combine into segmental **ventral roots.** The dorsal and ventral roots lie at first within the tubular sac of arachnoid and dura mater; but, as they proceed lateralward, they are invested by prolongations of dura mater and arachnoid, within which they also carry a covering of pia mater. The dorsal root has on it, as it reaches the intervertebral foramen, a large collection of cell bodies of its constituent neurons, the **spinal ganglion.** Just distal to the ganglion, the dorsal and ventral roots unite to form the **spinal nerve.** The spinal cord gives rise to thirty-one pairs of spinal nerves—8 cervical, 12 thoracic, 5 lumbar, 5 sacral, and 1 coccygeal. The spinal nerve has a very short course, almost immediately dividing into a dorsal ramus and a ventral ramus.

As a result of the inequality in ultimate length of the spinal cord and the vertebral column and of the emergence of all spinal nerves through the intervertebral foramina at their appropriate vertebral levels, there is a progressive descending obliquity of the dorsal and ventral roots within the subarachnoid space. The resulting compact array of roots within the lower end of the dura mater and the

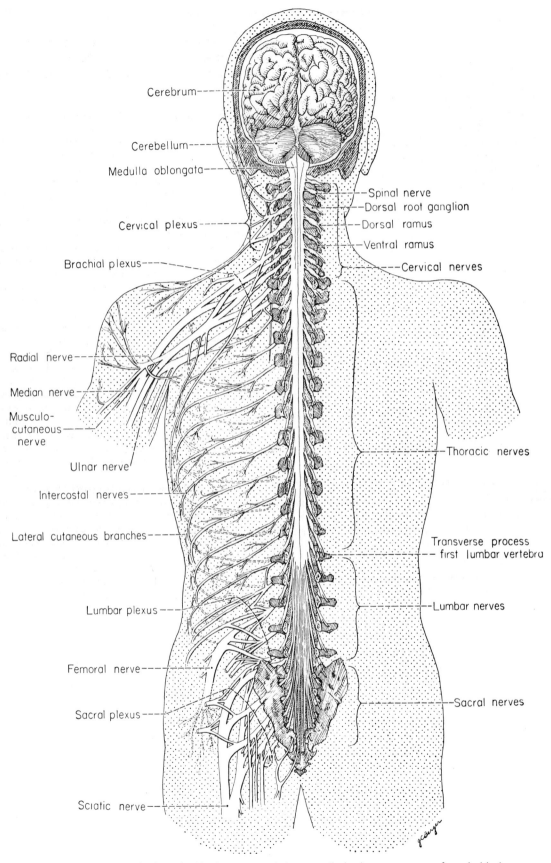

Cerebrum

Cerebellum

Medulla oblongata

Spinal nerve

Dorsal root ganglion

Dorsal ramus

Ventral ramus

Cervical plexus

Cervical nerves

Brachial plexus

Radial nerve

Median nerve

Musculo-
cutaneous
nerve

Ulnar nerve

Intercostal nerves

Thoracic nerves

Lateral cutaneous branches

Transverse process
first lumbar vertebra

Lumbar plexus

Lumbar nerves

Femoral nerve

Sacral nerves

Sacral plexus

Sciatic nerve

Fig. 245   The spinal cord, spinal nerves, and the great limb plexuses as seen from behind.

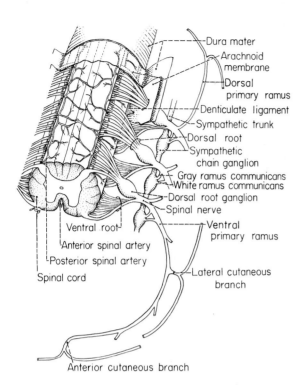

Fig. 246   The formation and branching of a typical spinal nerve.

arachnoid has an obvious likeness to a horse's tail which accounts for its name, the **cauda equina.**

The spinal cord does not have a uniform diameter. It is large at upper cervical levels and is smallest in its sacral and coccygeal segments. It is also markedly enlarged at the levels of organization of the nerves of the limbs. The **cervical enlargement,** giving rise to the nerves of the upper limb, includes the first thoracic and the lower four cervical segments. The **lumbar enlargement** gives rise to the nerves for the lower limb which arise from the second lumbar to the fourth sacral cord segments.

The Blood Supply of the Spinal Cord (figs. 242, 246) The spinal cord is nourished by an anterior spinal artery and posterior spinal arteries. The **anterior spinal artery** lies in the anterior median fissure of the cord and provides twigs which penetrate the cord, predominantly for the supply of the gray matter. The artery begins cephalically in the junction of the two anterior spinal branches of the vertebral artery (p. 281) and runs the length of the cord. The **posterior spinal arteries** also arise cephalically as the posterior spinal branches of the vertebral artery (p. 281). They descend on the cord along the line of emergence of the dorsal rootlets and are somewhat plexiform, their radicles winding between and on either side of the rootlets and anastomosing across the dorsal midline with those of the opposite side. Both the anterior spinal artery and the posterior spinal arteries are fed segmentally. The **spinal rami** of the vertebral, intercostal, lumbar, and lateral sacral arteries accompany their corresponding nerves through the intervertebral foramina, traverse the meningeal layers, and provide **posterior radicular** and **anterior radicular branches** (p. 301). The posterior radicular branches contribute to the posterior spinal arteries; the anterior radicular branches, to the anterior spinal artery. Not all nerves are accompanied by radicular branches, and perhaps only eight to ten anterior and ten to fifteen posterior radicular arteries are of significant size. A **great anterior radicular artery** has been recognized as making a major contribution to the anterior spinal artery variably at a lower thoracic or an upper lumbar level.

The **anterior external spinal** and **posterior external spinal veins** are also arranged longitudinally and are usually six in number. The three anterior veins lie in the anterior median fissure and on either side of the line of the emerging ventral roots. The three posterior external spinal veins lie, likewise, in the posterior median line and at the side of the emerging dorsal rootlets. The veins of the spinal cord are tributary to the internal vertebral venous plexus (p. 301).

V

The Chest

The anterior thoracic wall and pectoral region is a subsection of the chapter on The Upper Limb. Under this heading are considered the surface anatomy of the chest (p. 66), its cutaneous nerves (p. 66), the mammary gland (p. 67), the pectoral muscles and fasciae (p. 70), and the clavicle (p. 71).

THE BONY THORAX

The **bony thorax** (fig. 247) gives protection to the heart and lungs and provides attachment for many muscles besides those of the thorax itself—outwandered muscles of the upper limb and muscles of the back and the abdomen. The red marrow of the ribs and the sternum is one of the principal sites of red blood cell formation. The main function of the bony thorax, however, is to serve as the expansile but rigid walls of a bellows-like chamber, the interior capacity of which is alternately enlarged and reduced through the action of muscles upon it. By these changes the pressure of air within the lungs is alternately made less or greater than atmospheric pressure; as a consequence, air moves in and out of the lungs and aeration of the blood is carried on. The skeleton of the chest is formed by the thoracic vertebrae and the intervertebral disks, the ribs and their cartilages, and the sternum. The chest has a generally conical outline, being narrower above than below. The anteroposterior and transverse diameters of the chest of the infant or small child are approximately equal, the thorax being barrel-shaped; but with full development and use of the upper limb, the transverse diameter becomes increased. The transverse diameter of the adult chest is about one-fourth greater than its anteroposterior dimension, its greatest width being at the level of the eighth or the ninth rib. Associated with this shift in dimensions is the indentation of the cavity from behind by the thoracic vertebrae, so that in cross section the cavity of the chest is shaped like a kidney bean.

The **superior thoracic aperture** is small and kidney-shaped; it is oblique and has its uppermost limit posteriorly, where it rises to the level of the first thoracic vertebra. The superior thoracic aperture is bounded by the first pair of ribs and cartilages at the sides and by the manubrium of the sternum in front. The obliquity of the inlet is shown by the horizontal level of the upper border of the manubrium—the disk between the second and third thoracic vertebrae. The transverse dimension of the inlet is about 12 cm.; its anteroposterior dimension is about 6 cm. The **thoracic outlet** (inferior thoracic aperture) is large and irregular in outline. It is also oblique, the posterior wall of the chest being much longer than the anterior wall. The inferior aperture is bounded by the twelfth thoracic vertebra and the twelfth ribs behind, the cartilages of ribs 12 to 7 at the sides, and the xiphosternal junction in front. The aperture is closed by the diaphragm which arches upward into the chest and forms a movable partition between the thoracic and abdominal cavities. The sloping costal cartilages of ribs 7 to 10 in front form the **costal arch** into which the xiphoid process of the sternum descends in the median line.

The Sternum (figs. 247, 249)

The sternum is flat and elongated and is somewhat dagger-shaped. It is situated subcutaneously in the anterior midline of the chest and slopes downward and forward. The bone is slightly convex on its anterior aspect and a little concave posteriorly. Broader above and narrow below, it has an average length of 17cm. in the adult. The sternum consists of three parts: manubrium, body, and xiphoid process.

The **manubrium** (a handle) is generally quadrilateral but is much narrower inferiorly than superiorly. It is

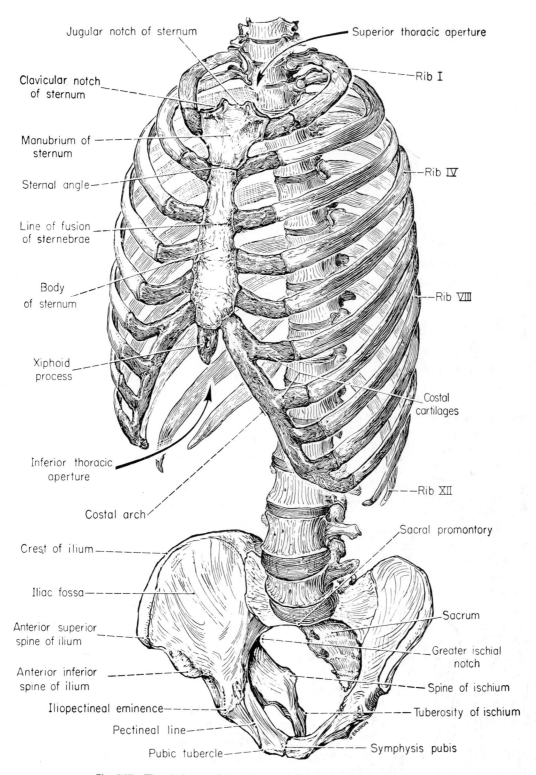

Jugular notch of sternum

Superior thoracic aperture

Rib I

Clavicular notch of sternum

Manubrium of sternum

Rib IV

Sternal angle

Line of fusion of sternebrae

Body of sternum

Rib VIII

Xiphoid process

Costal cartilages

Inferior thoracic aperture

Rib XII

Costal arch

Sacral promontory

Crest of ilium

Iliac fossa

Sacrum

Anterior superior spine of ilium

Anterior inferior spine of ilium

Greater ischial notch

Spine of ischium

Iliopectineal eminence

Tuberosity of ischium

Pectineal line

Pubic tubercle

Symphysis pubis

Fig. 247   The skeleton of the chest and abdomen; a three-quarter view.

also thicker superiorly. Its anterior surface is convex from side to side but concave from above downward. The thick **superior border** has a shallow concavity at its center which forms the floor of the readily palpable **jugular notch** at the root of the neck. At either side of this is a deep **clavicular notch** which provides an oval articular surface for the reception of the sternal extremity of the clavicle and the disk of the sternoclavicular joint. The **lateral borders** of the manubrium carry superiorly, a depression for the first costal cartilage and, at their inferior extremity, a notch for the reception of a part of the cartilage of the second rib. The **inferior border** of the manubrium, oval and rough, articulates with the superior border of the body by a synchondrosis, the fibrocartilage of which does not usually become ossified until old age. The manubrium and the body of the sternum do not articulate in a flat plane, but their line of junction projects forward as the **sternal angle**.

The **body** of the sternum is long, narrow, and thinner than the manubrium. It is widest near its lower end. Three transverse ridges cross its anterior surface opposite the articular depressions for the third, fourth, and fifth costal cartilages. These mark the lines of fusion of the segments of the sternum that form the body. An articular depression for the sixth costal cartilage is at the side of the fourth segment. In addition, the body has two demifacets—one at the angle of the superior and lateral borders for the second costal cartilage; the other at the angle of the inferior and lateral borders for the seventh cartilage.

The **xiphoid process** (sword-shaped) is thin and elongated. It is the smallest part of the sternum and is cartilaginous in youth; its upper part, especially, becomes ossified in the adult. The upper border of the xiphoid process articulates with the inferior border of the body of the sternum, receding from its plane in front but flush with it behind. The xiphoid process usually bears a demifacet for the articulation of the seventh costal cartilage at the junction of its superior and lateral borders. The xiphoid forms a synchrondrosis with the body of the sternum, and the xiphosternal junction marks the lower limit of the thoracic cavity in front. It is a landmark, in the median line, for the upper surface of the liver, the diaphragm, and the lower border of the heart. The xiphoid varies considerably in form, frequently being bifid and occasionally perforated.

The first sternal ossification centers appear in the manubrium during the fifth fetal month; three centers for the body appear bilaterally at that time, and the centers for the fourth and fifth sternebrae follow at about six-week intervals. The bilateral centers fuse to make single segmental centers, and the centers of the body fuse with one another from below upward from the fourth to the eighth year.

Clinical Note The sternal angle is readily palpable and represents the most reliable surface landmark of the chest. With respect to the ribs, the sternal angle is located at the sternal end of the second rib. All ribs are counted from here, palpating downward and lateralward to avoid confusion with their cartilaginous extremities.

The Ribs (figs. 247–50)

The ribs are thin, narrow, curved strips of bone. There are usually twelve on each side of the chest, but their number is occasionally increased by the development of a cervical or a lumbar rib, or it may be reduced. The first seven ribs are attached to the sternum by their costal cartilages and are thus characterized as **true ribs.** Of the remaining five **false ribs,** the eighth, ninth, and tenth have their cartilages connected to the cartilage above them. The cartilages of the eleventh and twelfth ribs end in the abdominal musculature; these are called **floating ribs.** The ribs increase in length from the first to the seventh and diminish thereafter. The first rib has the sharpest curvature; the curvature of each succeeding rib is more open. The ribs are twisted as well as curved, and the obliquity in the relationship of the shaft and its ends increases from the first to the ninth and thereafter diminishes. The point of greatest change in curvature of each rib is called the **angle** and occurs several inches from the head. The first, second, tenth, eleventh, and twelfth ribs have individual characteristics; the third to the ninth ribs can be considered to have a common type. The parts of a rib are the head, the neck, the tubercle, and the shaft.

The **head** has an articular surface composed of two facets separated by a **crest.** The two facets articulate with the demifacets of two adjacent vertebrae, and the crest is connected by an intra-articular ligament with the intervertebral disk. As a rule, the lower facet is the larger and articulates with the vertebra to which the rib corresponds in number. The **neck** of the rib extends lateralward

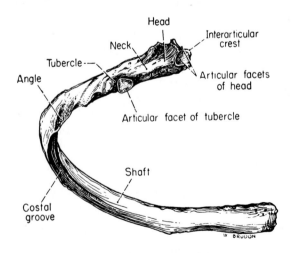

Fig. 248   A typical rib viewed from behind and below. ($\times \frac{1}{2}$)

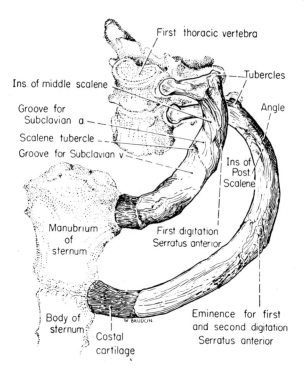

Fig. 249 Ribs one and two and their relations to the sternum and the first two thoracic vertebrae. ($\times\frac{1}{2}$)

from the head for about 2.5 cm. Smooth ventrally and rough dorsally, the neck has a sharp, cranially directed crest for attachment of the superior costotransverse ligament. The **tubercle** of the rib is a posteriorly directed eminence at the junction of the neck and the shaft. Its superior and lateral part is rough for the attachment of the lateral costotransverse ligament. On its inferior medial part is a small oval facet for articulation with the end of the transverse process of the lower of the two vertebrae to which the head is connected. The tubercle marks the beginning of the costal groove on the inferior surface of the shaft. The **shaft** (body) of the rib is thin and flat and strongly curved. A little beyond the tubercle the **angle** of the rib marks a rather abrupt change in curvature of the shaft. This is also the site of the greatest twist in the bone and bears an oblique line externally. The shaft is thick and rounded along its superior border but is thin and sharp on its inferior border where it forms the external bony margin of the **costal groove** for the intercostal vessels and nerve. The sternal end of the shaft is cupped for the reception of the costal cartilage.

The **first rib** (fig. 249) is the broadest, shortest, and most sharply curved of the series. It has no twist, and its

borders are directed inward and outward rather than up and down. The **head** has a single facet for articulation with the first thoracic vertebra. The **tubercle** is prominent. The superior surface of the rib has two shallow grooves—one anteriorly or medially for the subclavian vein; one just in front of its middle for the subclavian artery. The grooves are separated at the inner border of the rib by the **scalene tubercle** for the insertion of the anterior scalene muscle. Behind the groove for the artery are roughened areas for the attachment of the middle scalene and the serratus anterior muscles. The under surface has no costal groove, and the sternal end of the bone is thick.

The **second rib** is much longer than the first and is minimally twisted. Its head, neck, and tubercle are similar to the common type. Its outer surface faces outward and upward, and the special feature of this rib is a marked tuberosity at about the middle of this surface for the origin of a part of the first and the whole of the second digitation of the serratus anterior muscle.

The **tenth rib** (fig. 250) has a single articular facet on its head. Otherwise, this rib conforms to the common type.

The **eleventh rib** has a single facet on its head. Its tubercle is small and has no facet, and its angle is ill-defined.

The **twelfth rib** is small and slender and may be even shorter than the first. Its tubercle, angle, and costal groove are absent or poorly marked. The shaft tapers off at its sternal end and carries only a small costal cartilage.

Ossification of the Ribs Ossification of the ribs begins near the angle, a primary center appearing there at about the seventh week of intrauterine life. Ossification spreads rapidly through the rib. About the fourteenth year, a secondary center appears for the head of the rib, and shortly thereafter one develops for the tubercle in those ribs having tubercles. These secondary centers fuse with the shaft at about the twenty-fifth year. Frequently both parts of a tubercle have separate epiphyses.

Clinical Note A **cervical rib** (0.5–1% occurrence) is due to elongation of the transverse process of the seventh cervical

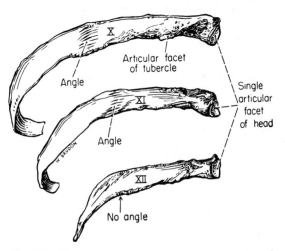

Fig. 250 Ribs ten to twelve viewed from behind. ($\times\frac{1}{2}$)

vertebra. The elongated process is most commonly continued by a fibrous strand which attaches to the first thoracic rib. These structures elevate the great vessels and the brachial plexus cords as they cross the upper rib cage and may produce symptoms referable to the tension thereby imposed on them. **Rib fracture** is common from direct blows or from the compression of the chest wall. If the ends of the fragments are forced inward, there may be laceration of the pleura and lung. However, in compression injuries, the fragments tend to point outward. The continuous movements of respiration prevent immobilization of the ribs but healing is relatively rapid with the deposit of a fairly large amount of callus. **Compression or crush injuries** to the driver's chest are now fairly common in automobile accidents, the chest being driven in by its impact on the steering wheel and its column. Ribs will be fractured outward and the sternum driven inward, in the extreme case, bottoming on the vertebral column. There may be rupture of the trachea and tearing of bronchi. Laceration of the great vessels may occur—the thoracic aorta possibly being torn at the attachment of the ligamentum arteriosum. The heart wall may be lacerated and, if the chambers are full at the instant of impact, their walls may be ruptured. Impact in the lower chest may result in damage to the liver or spleen.

Costal Cartilages (figs. 247, 249, 253)

The costal cartilages are bars of hyaline cartilage which prolong the ribs anteriorly and contribute materially to the elasticity of the walls of the chest. The upper seven cartilages join the sternum; the next three articulate with the cartilages just above them; the last two are mere tips which end in the wall of the abdomen. The cartilages increase in length through the first seven and then gradually become shorter. The cartilage of the first rib joins the manubrium sterni without a joint intervening, its perichondrium blending with the periosteum of both the rib and the sternum. Costal cartilages two to seven join the sternum by synovial joints. The first and second cartilages slope downward slightly; the third is horizontal; and the fourth inclines upward. The fifth, sixth, and seventh cartilages continue the downward slope of the ribs for about an inch and then turn upward with increasing pitch to reach the sternum. The eighth, ninth, and tenth do likewise but fail to reach the sternum, tapering to a point and articulating with the cartilages above. The inferior·borders of the sixth, seventh, eighth, and ninth cartilages present heel-like projections at their greatest convexity which carry smooth oblong facets for the articulation of these cartilages with those next below.

The Articulations of the Thorax

The ribs articulate by their heads with the bodies of the thoracic vertebrae and the intervertebral disks and by their tubercles with the transverse processes

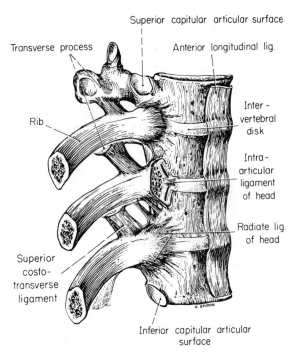

Fig. 251 The costovertebral junction, anterolateral aspect; the costovertebral articulations and ligaments.

of the thoracic vertebrae. Except for the first and the last two ribs, their costal cartilages form true joints with the sternum or with other costal cartilages.

The Costovertebral Articulations (figs. 246, 251) The articulations of the heads of the ribs are gliding joints between the heads of the typical ribs and the demifacets of two adjacent vertebrae.

The demifacets are separated by short, thick **intraarticular ligaments.** Each intraarticular ligament unites the crest between the facets on the head of the rib with the intervertebral disk. It is lacking in the joints of the first, tenth, eleventh, and twelfth ribs where the head has only one facet. The **articular capsule** of the capitular joint connects the head of the rib to the circumference of the articular cavity formed by the disk and the demifacets. The anterior part of the capsule is especially thickened as the **radiate ligament** of the head. This is composed of three flat fasciculi attached to the anterior part of the rib just beyond the articular surface. The upper fasciculus is connected to the vertebra above; the small middle one, to the intervertebral disk; and the lower fasciculus, to the vertebra below.

The Costotransverse Articulations (figs. 251, 252) Each typical rib articulates by its **tubercle** with the anterior surface of the transverse process of the inferior of the two vertebrae to which its head is joined. The small costotransverse joint is sur-

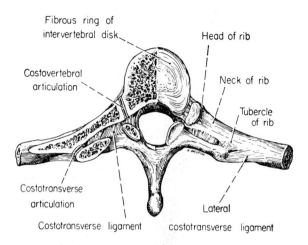

Fibrous ring of
intervertebral disk

Head of rib

Costovertebral
articulation

Neck of rib

Tubercle
of rib

Costotransverse
articulation

Lateral
costotransverse ligament

Costotransverse ligament

Fig. 252 Superior view of the costovertebral junction; the costovertebral and costotransverse articulations and ligaments.

rounded by a thin capsule which is strengthened by neighboring accessory ligaments.

The **costotransverse ligament** is a broad band of short fibers which connects the back of the neck of the rib with the anterior surface of the adjacent transverse process. The **lateral costotransverse ligament** passes obliquely from the apex of the transverse process to the non-articular portion of the tubercle of the rib. The **superior costotransverse ligament** is a broad band of fibers that ascends laterally from the crest on the neck of the rib to the caudal border of the transverse process above. This ligament may be divided into a stronger anterior costotransverse ligament and a feeble posterior costo-transverse ligament.

The Sternocostal Articulations (figs. 105, 247) The union of the first rib with the sternum is a synchondrosis in which the bones are firmly joined by cartilage, the perichondrium of which is continuous with the periosteum of the rib and the sternum. Synovial joints exist between cartilages two to seven and the facets on the side of the sternum. Their articular capsules are thin and are intimately blended with the radiate sternocostal ligaments.

The **radiate sternocostal ligaments** consist of broad thin membranous bands that diverge from the front and back of the sternal ends of the cartilages to end on the anterior and posterior surfaces of the sternum. On the front of the sternum such fibers mingle with those of the opposite side and with the tendinous fibers of origin of the pectoralis major muscle. An **intraarticular sternocostal ligament** (fig. 105), found constantly only between the second costal cartilage and the sternum, connects this cartilage to the fibrocartilage interposed between the manubrium and the body of the sternum. It separates this sternocostal articulation into two joint cavities. Occasionally the same separation occurs in the

joint of the third cartilage and even more rarely in that of the fourth. The **costoxiphoid ligaments** connect the anterior and posterior surfaces of the sixth and seventh costal cartilages to the front and back of the xiphoid process. They are in series with the radiate sternocostal ligaments.

The Interchondral Articulations Simple, gliding joints are formed between the costal cartilages of the sixth and seventh, the seventh and eighth, the eighth and ninth, and the ninth and tenth ribs. As each of the upper cartilages turns upward toward the sternum, it presents a broadened heel toward the cartilage below, and the oblong articular surface of each forms an interchondral joint. A thin articular capsule is strengthened laterally and medially by ligamentous fibers which pass from one cartilage to the next.

The Costochondral Articulations The lateral extremity of each costal cartilage is received into a cup-shaped depression at the sternal end of the rib. Here the two structures are bound firmly together by the continuity of the periosteum of the rib with the perichondrium of the cartilage.

Movements of the Ribs and the Sternum

The ribs articulate by their heads with the bodies and the intervertebral disks of the vertebrae and by their tubercles with the tips of the transverse processes. Their anterior extremities are related to the sternum through the costal cartilages. The joints of the head and the tubercle provide for movement that takes place around an oblique axis through the neck of the rib in the line of these two joints. This oblique axis, coupled with both the slope and curvature of the rib, requires that when the rib is elevated, its anterior extremity is carried forward as well as upward and its lateral margin is carried outward. Further, the relative fixation of the anterior end at the sternum forces the rib to elevate somewhat as a bucket handle fixed at both ends, and it swings both upward and lateralward. As a result of these factors, both the anteroposterior and the lateral diameters of the chest are increased as the ribs and sternum are raised, as in inspiration, and decreased with their fall.

THE MUSCLES, NERVES, AND VESSELS OF THE THORAX

MUSCLES AND FASCIA OF THE THORAX

The muscles of the thorax proper are the external intercostal, the internal intercostal, the subcostal,

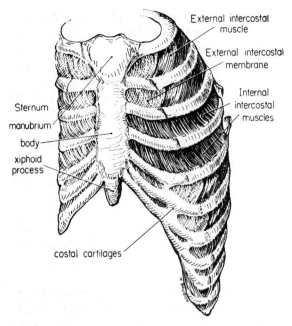

External intercostal muscle

External intercostal membrane

Internal intercostal muscles

Sternum

manubrium

body

xiphoid process

costal cartilages

Fig. 253   The thoracic cage to illustrate the external and internal intercostal muscles and the external intercostal membranes.

the transversus thoracis, the levatores costarum, the serratus posterior superior, the serratus posterior inferior, and the diaphragm. The serratus posterior superior and the serratus posterior inferior (p. 287) and the levatores costarum (p. 291) are described in the chapter on the Back. The diaphragm, which forms the muscular and highly movable partition between the thoracic and abdominal cavities, is the principal muscle of respiration; it is particularly effective in enlarging the vertical dimension of the thorax. The diaphragm is properly dissected, however, only from its abdominal side and is, therefore, more completely described in the chapter on the Abdomen.

The External Intercostal Muscles (figs. 253, 272, 277)

These are eleven in number, occupy the intercostal spaces, and extend from the tubercles of the ribs behind to the costochondral junctions anteriorly. Each muscle arises from the lower border of the rib which constitutes the upper limit of an intercostal space and inserts into the upper border of the rib below. The fibers of the muscle are directed downward and lateralward on the back of the chest and downward, forward, and medialward on the front. Ending at the costochondral junction for most interspaces, the muscle is continued to the sternal margin by a thin **external intercostal membrane.** The external intercostal muscles are a part

of the external sheet of thoracoabdominal musculature, and the seven lower muscles show a variable degree of continuity with fascicles of the external oblique muscle of the abdomen. The external intercostal muscles are innervated by branches of the intercostal nerves.

The Internal Intercostal Muscles (figs. 253, 272, 277)

Like the external intercostals, the internal intercostal muscles are eleven in number and occupy the intercostal spaces. However, they extend medialward to the border of the sternum and to the anterior extremities of the cartilages of the false ribs and terminate posteriorly at the angle of the rib. An **internal intercostal membrane** continues the plane of the muscle from the angle to the region of the tubercle of the rib. Each internal intercostal muscle arises from the upper border of the rib and costal cartilage which forms the lower limit of an intercostal space and is directed upward and forward to be inserted into the costal cartilage and into the upper edge of the costal groove of the rib above. Its fascicles course at right angles to those of the external intercostal layer and correspond in their orientation to those of the internal abdominal oblique muscle. The internal intercostal muscle is innervated by intercostal nerves. It lies directly under the external intercostal muscle, loose areolar tissue intervening.

The **subcostal muscles** are formed of those fasciculi of the internal intercostal muscle which, near the angle of the ribs, extend over two or more intercostal spaces. They correspond in direction and are continuous with the internal intercostal sheet of muscle.

The course of the intercostal nerves and vessels varies at different points in the thoracic wall (figs. 254, 255). Behind the anterior portion of the chest they run deep to the internal intercostal muscle and then between it and the transversus thoracis muscle, but posteriorly, they course between the internal intercostal (including the subcostal) and the external intercostal muscles. At the side of the chest they appear to lie between two laminae of the internal intercostal muscle. Since the typical course of the nerves and vessels in the thoracoabdominal layers is between the transversus and internal representatives (thoracis or abdominis), some authors have allocated the parts of the internal intercostal muscles deep to the nerves and vessels (mm. intercostales intimi), together with the subcostal muscles, to the deepest stratum.

The Transversus Thoracis Muscle (figs. 255, 256)

This muscle lies on the inner surface of the sterno-

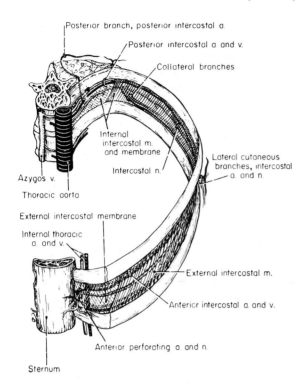

Posterior branch, posterior intercostal a.

Posterior intercostal a. and v.

Collateral branches

Internal intercostal m. and membrane

Intercostal n.

Lateral cutaneous branches, intercostal a. and n

Azygos v.

Thoracic aorta

External intercostal membrane

Internal thoracic a. and v.

External intercostal m.

Anterior intercostal a. and v.

Anterior perforating a. and n.

Sternum

Fig. 254 The anatomy of an intercostal space with special reference to the relations of intercostal vessels, nerves, and muscles.

chondral portion of the chest. Consisting of flat, thin, fasciculi, it arises from the dorsal surface of the lower half of the body of the sternum and the xiphoid process. Directed upward and lateralward, its fibers insert into the second to the sixth costal cartilages. The lowest fibers of the muscle are relatively horizontal and are continuous with fibers of the transversus abdominis muscle, the third and deepest plane of the thoracoabdominal musculature. The fasciculi of the transversus thoracis vary in size and number. The muscle is innervated by the second to the sixth intercostal nerves.

The Parietal Thoracic Fascia

The thoracic wall is covered internally by a parietal thoracic (endothoracic) fascia. This deep fascia invests the internal intercostal, subcostal, and transversus thoracis muscles. The fascia blends with the periosteum of the ribs and the sternum and with the perichondrium of the costal cartilages. It is pierced by the posterior intercostal arteries and veins along the ventrolateral curvature of the bodies of the vertebrae. The parietal thoracic fascia covers the thoracic surface of the diaphragm and constitutes the superior fascia of the diaphragm. It is con-

tinuous with the prevertebral layer of cervical fascia and with the scalene fascia (Sibson's fascia) along the inner border of the first rib and behind the sternum, with the fascia of the infrahyoid muscles. Inferiorly, the parietal thoracic fascia has continuity through the openings of the diaphragm with the parietal abdominopelvic (transversalis) fascia.

The Mechanism of the Thorax in Respiration

Under the action of a considerable number of muscles, the semirigid but expansile thorax enlarges its diameters, simultaneously reducing the pressure of air within, so that air under outside atmospheric pressure rushes in to fill the lungs. This is inspiration. Expiration is a response to the relaxation of these muscles, to the elasticity of the lungs, to the recoil of stretched fibrous and ligamentous structures of the joints and tendons of the thorax, and to increased intraabdominal pressure. Forcible expiration also requires muscle action, primarily of the muscles of the abdominal wall which, by compressing the abdominal contents, force the diaphragm upward into the chest. Transversus thoracis is also an expiratory muscle.

The thorax enlarges in three directions: anteroposteriorly, lateralward, and vertically. In the discussion of the joints of the ribs and costal cartilages (p. 313), the simultaneous enlargement of the anteroposterior and lateral diameters was related to the form and orientation of the ribs and cartilages together with their joints. The vertical diameter of the thorax is enlarged by the direct piston-like action of the diaphragm, assisted by those muscles which elevate the uppermost ribs and sternum. In quiet respiration there are moderate increases in all diameters. The vertical range of movement of the diaphragm is considerable—about 12 mm. in quiet respiration. and about 30 mm. in forced respiration. In the deepest possible inspiration it has been observed that the manubrium sterni may move 30 mm. in an upward and 14 mm. in a forward direction, that the width of the subcostal angle 30 mm. below the xiphosternal junction may be increased as much as 26 mm., and that the umbilicus may be retracted and drawn upward as much as 13 mm.

According to recent studies respiration is mainly carried out by the diaphragm and by those muscles that elevate the upper ribs and the sternum—anterior scalene, middle scalene, posterior scalene, and posterior superior serratus muscles. Assistance is given by the levatores costarum and intercostal muscles. In forced or labored respiration the shoulder girdle is fixed, and the muscles which pass from the trunk to the bones of the shoulder girdle may act on the thorax to enlarge its diameters. The sternocleidomastoid muscle powerfully elevates the sternum in full inspiration (Campbell, '55). Extension of the vertebral column by the deep back muscles also contributes to forced ventilation.

The function of the external intercostal and internal intercostal muscles has long been uncertain. Jones, Beargie, and Pauly ('53) have concluded, from electromyographic studies of costal respiration in Man, that the intercostal muscles are not separate in their actions. They act together to keep the intercostal spaces constant in width during the elevating action of the scalene muscles above and the depressing pull of the abdominal muscles below. They maintain rib spacing, transmitting an upward pull from rib to rib. They resist the blowing out or sucking in of the intercostal spaces with changes in intrathoracic pressures and are thus very necessary elastic and contractile elements of the flexible thorax. The intercostal muscles are more active in forced ventilation and also have a postural function in flexion and side to side bending of the trunk. Koepke, *et al.*, noted activity of the upper intercostal muscles during quiet inspiration with recruitment of lower intercostals only as deeper breathing is undertaken.

THE INTERCOSTAL NERVES (figs. 243, 246, 254, 255)

The **intercostal nerves** are the ventral rami of the first eleven thoracic spinal nerves. The ventral ramus of the twelfth thoracic is the **subcostal nerve.** The first intercostal nerve is the smaller part of the ventral ramus of the first thoracic spinal nerve, for its larger branch participates in the formation of the brachial plexus. In addition, this nerve usually has no lateral cutaneous branch. The intercostal nerves distribute to the thoracic and abdominal wall and do not form plexuses. They contribute preganglionic sympathetic fibers to the sympathetic chain through **white rami communicantes** and receive postganglionic neurons from the sympathetic chain ganglia through **gray rami communicantes.** These rami are connected to the spinal nerves near their exit from the intervertebral foramina.

The **thoracic intercostal nerves** are those which arise from the first six thoracic spinal nerves. They pass through the intercostal spaces of the chest in company with the intercostal vessels, supplying

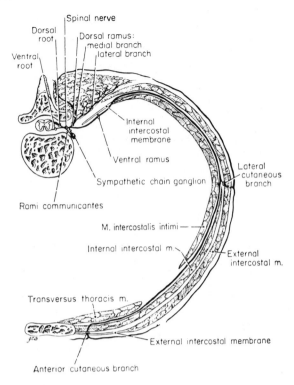

Fig. 255   A diagram of a typical spinal nerve, especially the relation of its intercostal branch to the intercostal musculature.

muscles and giving off cutaneous branches. Medial to the angles of the ribs, they lie between the pleura and the internal intercostal ligament. They then run in the interval between the internal and external intercostal muscles as far as the side of the chest, where they enter a separation between two laminae of the internal intercostal muscles. Toward the front of the chest, the nerves are wholly internal to the internal intercostal muscle and, deep to the costal cartilages, are found between that muscle and the transversus thoracis muscle. The intercostal nerves traverse the intercostal spaces inferior to the intercostal artery and vein of the same space. In their course, the nerves supply the external intercostal, internal intercostal, subcostal, serratus posterior superior, and transversus thoracis muscles.

Lateral cutaneous branches arise from the intercostal nerves in the axillary line. They pierce the external intercostal and serratus anterior muscles and divide into anterior and posterior branches. The anterior branches supply the skin of the lateral and ventral parts of the chest and of the mammary gland; the posterior branches pass backward to the skin over the scapular and dorsal chest areas, interdigitating in their terminal distribution with the medial branches of the dorsal rami of thoracic nerves. The lateral cutaneous branch of the second intercostal nerve is the intercostobrachial nerve (p. 61). The thoracic intercostal nerves terminate as **anterior cutaneous branches** which, at the side of the sternum, pierce the intercostal musculature and the pectoralis major muscle and fascia. They have short medial and longer lateral branches to the skin of the anterior aspect of the chest and of the mammary gland in the female.

The **thoracoabdominal intercostal nerves** (fig. 304) are the ventral rami of the thoracic spinal nerves seven to eleven inclusive. At the anterior ends of the intercostal spaces, these nerves pass behind the costal cartilages and proceed forward into the abdominal wall between the transversus abdominis and internal abdominal oblique muscles. They supply these and the external abdominal oblique and rectus abdominis muscles and end as anterior cutaneous branches. Besides the muscles just named, these nerves supply, in their course, the external intercostal, internal intercostal, and posterior inferior serratus muscles.

Lateral cutaneous branches, which reach the surface at about the axillary line, divide into anterior and posterior branches. The anterior branches supply the skin as far as the lateral border of the rectus abdominis muscle; the posterior branches pass dorsalward, and their terminals interdigitate with the lateral branches of the dorsal rami of the same spinal nerves. The **anterior cutaneous branches** of the thoracoabdominal intercostal nerves penetrate the anterior layer of the sheath of the rectus muscle near the midline. Their short medial and longer lateral branches distribute to the skin of the ventral surface of the abdomen.

The **subcostal nerve,** the large ventral ramus of the twelfth thoracic nerve, follows the lower border of the twelfth rib and has no thoracic distribution. It crosses the ventral surface of the quadratus lumborum muscle beneath the transversalis fascia, penetrates the transversus abdominis muscle, and then distributes like the lower intercostal nerves. It also supplies the pyramidalis muscle and communicates with the iliohypogastric nerve of the lumbar plexus.

Its large **lateral cutaneous branch** (figs. 304, 423) does not divide into anterior and posterior branches but passes downward over the crest of the ilium several centimeters behind the anterior superior spine of the ilium. This nerve distributes to the skin of the side of the hip, sometimes as low as the greater trochanter of the femur.

Areas of Distribution The description of a **typical spinal nerve** (p. 47) indicates the plan whereby the cutaneous branches of the ventral and dorsal rami of spinal nerves share the distribution of a segment, or dermatome, in the trunk region. It also includes mention of the phenomenon of 'nerve overlap,' whereby any particular point on the skin surface receives an innervation from two adjacent nerves. The ribs of the thoracic region lie at the boundaries of the dermatomes; thus the primary distribution of the cutaneous branches of each intercostal nerve is to the skin and subcutaneous tissue of the interspace in which that intercostal nerve runs. The thoracic intercostal nerves distribute to the skin and muscle of the thoracic wall. The thoracoabdominal intercostal nerves pass beyond the thoracic cage and continue their diagonal course downward into the abdominal wall. The dermatomes slope, therefore, in this region, so that the cutaneous and muscular terminals in the region of the umbilicus are those of the tenth intercostal nerve. Similarly, the subcostal nerve reaches, by its anterior terminals, as far as or beyond the half way point between the umbilicus and the pubis. The lateral cutaneous branch of this nerve passes into the hip region. Since the dermatomes are of approximately equal widths, it is possible to plot the area for each nerve from the above information.

THE INTERCOSTAL ARTERIES (figs. 254, 300)

Nine pairs of **posterior intercostal arteries** and one pair of subcostal arteries arise from the back of the thoracic aorta. The intercostal arteries of the first two intercostal spaces are provided by the costocervical trunk of the subclavian artery (p. 181), with additions from the superior thoracic branch of the axillary artery (p. 75). Since the thoracic aorta is slightly to the left of the median line (fig. 154), the right aortic intercostal arteries have a longer course than those on the left side. They cross the vertebral column and pass behind the esophagus, the thoracic duct, and the azygos vein. The posterior intercostal arteries pass lateralward, covered by parietal pleura, and are crossed by the sympathetic trunk opposite the heads of the ribs. At the proximal end of the intercostal space, each artery divides into an anterior and a posterior branch.

The **anterior branch** crosses the intercostal space obliquely toward the angle of the upper rib and then runs lateralward and forward in the costal groove. It lies first between the internal intercostal membrane and the pleura and then between the internal intercostal and external intercostal muscles. Like the intercostal nerve which it accompanies, the artery traverses the internal intercostal muscle as it passes forward around the chest and, anteriorly, lies between the internal intercostal and transversus thoracis muscles. The artery is accompanied by a posterior intercostal vein which lies above it and an intercostal nerve which lies below it. Each anterior branch gives rise to a **collateral intercostal branch** which, near the angle of the rib, descends to the upper border of the rib at the inferior boundary of the interspace and then continues forward in that location.

Both the collateral intercostal and anterior branches proper end at an indefinite point toward the anterior aspect of the chest by anastomosing with the anterior intercostal branches of the internal thoracic artery (p. 319). The first aortic intercostal artery anastomoses with the intercostal branch of the costocervical artery. The anterior and collateral branches of the posterior intercostal arteries of the seventh, eighth, and ninth spaces anastomose with the musculophrenic branch of the internal thoracic artery. Those of the tenth and eleventh spaces and of the subcostal artery (fig. 310) continue into the abdominal wall and anastomose with the superior epigastric branch of the internal thoracic artery, and with each other. **Muscular branches** of the posterior intercostal arteries supply the intercostal, pectoral, and abdominal muscles. A **lateral cutaneous branch** accompanies the corresponding lateral cutaneous nerve through the overlying muscles and, like the nerve, distributes by anterior and posterior branches to the skin and subcutaneous tissues. The anterior branches of these arteries in the third, fourth, and fifth spaces supply **lateral mammary branches** to the mammary gland.

The **posterior branch** of each posterior intercostal artery accompanies the dorsal ramus of the spinal nerve backward toward the intervertebral foramen and divides opposite this foramen into a spinal and a muscular branch. The **spinal branch** enters the foramen and distributes to the spinal cord, its meninges, and the vertebra (p. 307). The **muscular branch** continues with the dorsal ramus of the spinal nerve, divides like it into medial and lateral branches, and distributes to the muscles of the back and to the overlying skin and subcutaneous tissues.

The distribution of the **subcostal artery** is similar to that of the lower posterior intercostal arteries. It does not have a collateral branch. Anterior intercostal arteries are branches of the internal thoracic arterial system and are described below.

Clinical Note In coarctation of the aorta (p. 330), blood is fed into the distal thoracic aorta through the intercostal arteries and these may become so large as to erode the adjacent surfaces of the ribs. This circumstance is clearly seen in X-rays of the chest, giving evidence of the condition and a striking example of collateral circulation.

THE INTERCOSTAL VEINS (figs. 254, 297–300)

The intercostal veins accompany the intercostal arteries and nerves, lying highest in the costal groove (fig. 248). There are eleven **posterior intercostal veins** and one **subcostal vein** on each side. They are patterned after the arteries, having lateral cutaneous, muscular, intervertebral, and posterior tributaries. They contain valves which direct the flow of blood posteriorly, and they anastomose with **anterior intercostal veins** which are tributary to the internal thoracic vein.

The **first posterior intercostal vein** ascends across the neck of the first rib and arches forward over the apex of the pleura to end in either the brachiocephalic or the vertebral vein. The second, third, and fourth posterior intercostal veins form a **superior intercostal vein.** The superior intercostal vein of the left side is a tributary of the left brachiocephalic vein; that of the right side descends to empty into the arch of the azygos vein (p. 361). The posterior intercostal veins draining the fifth and the lower spaces on the right side drain segmentally into the azygos vein. Those of the left side frequently end in an accessory hemiazygos or a hemiazygos vein, or they may cross the vertebral column to the azygos vein (p. 361 & fig. 299). The **anterior intercostal veins** are described with the internal thoracic vein (p. 319).

Clinical Note The intercostal vein and artery are generally protected by their location in the subcostal groove; the intercostal nerve lies below the other two but still along the lower edge of the rib forming the upper limit of an intercostal space. A needle or trocar inserted through an intercostal space for the introduction of air to the pleural cavity (artificial pneumothorax) or removal of fluid from the pleural sac (paracentesis) should therefore be directed through the middle to lower portion of the intercostal space.

THE INTERNAL THORACIC ARTERY (figs. 141, 256, 272, 277)

The **internal thoracic artery** is a branch of the first part of the subclavian artery (p. 181). It descends behind the sternal end of the clavicle, the subclavian and internal jugular veins, and the first costal cartilage. Below this cartilage the artery runs vertically about 1 cm. from the margin of the sternum, dividing at the level of the sixth intercostal space into the musculophrenic and superior epigastric arteries. The artery descends directly behind the costal cartilages, the internal intercostal muscles, and the external intercostal membranes. Posteriorly, it rests on the pleura above and then on the transversus thoracis muscle below the third costal cartilage. The artery is accompanied by two veins which unite in the region between the first and third intercostal intervals. The branches of the internal thoracic artery are the pericardiacophrenic, the anterior mediastinal, the pericardial, the sternal, the anterior intercostal, the anterior

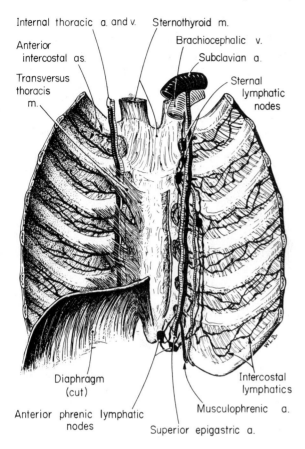

Internal thoracic a. and v.

Anterior intercostal as.

Transversus thoracis m.

Sternothyroid m.

Brachiocephalic v.

Subclavian a.

Sternal lymphatic nodes

Diaphragm (cut)

Anterior phrenic lymphatic nodes

Intercostal lymphatics

Musculophrenic a.

Superior epigastric a.

Fig. 256 The internal aspect of the sternum and ribs; the transversus thoracis muscle, the internal thoracic vessels, and the sternal lymphatics.

perforating, the musculophrenic, and the superior epigastric. An inconstant lateral costal branch may occur.

The **pericardiacophrenic artery** is a slender branch that accompanies the phrenic nerve to the diaphragm. It gives branches to the pleura and the pericardium and supplies the upper surface of the diaphragm.

The **anterior mediastinal arteries** are small twigs which supply the areolar tissue and lymphatics of the anterior mediastinum and the remains of the thymus gland.

The **pericardial branches** supply the upper portion of the anterior surface of the pericardium.

The **sternal branches** distribute to the transversus thoracis muscle and enter the nutrient foramina in the back of the sternum.

The **anterior intercostal arteries** supply the upper five or six intercostal spaces and are two to a space. They pass lateralward against the costal cartilages and ribs and anastomose with the anterior and collateral intercostal branches of the posterior intercostal arteries. They give off muscular branches to the intercostal, pectoral, and serratus anterior muscles.

The **anterior perforating branches,** one to an intercostal space, arise from the anterior aspect of the internal thoracic artery. They perforate the internal intercostal muscle in the first five or six spaces, supply the pectoralis major muscle, and end in the skin and subcutaneous tissues in company with the anterior cutaneous branches of the intercostal nerves. In the second, third, and fourth spaces these arteries supply **medial mammary branches** to the mammary gland.

The **musculophrenic artery** follows the costal arch behind the costal cartilages. It provides the anterior intercostal branches in the seventh, eighth, and ninth spaces, perforates the diaphragm, and ends, much reduced in size, in the tenth or the eleventh intercostal space. The musculophrenic artery gives branches to the lower part of the pericardium and also supplies the diaphragm and the peripheral parts of the abdominal muscles (fig. 310).

The **superior epigastric artery** is the direct continuation of the internal thoracic artery. It passes anterior to the transversus thoracis muscle to enter the sheath of the rectus abdominis muscle. It supplies the rectus abdominis and anastomoses within it with the inferior epigastric artery (figs. 306, 310). It gives small branches to the diaphragm, the peritoneum, and the skin and a twig which follows the falciform ligament to the liver.

The inconstant **lateral costal artery** is present in about 17 per cent of cases (Alexander, '46). It arises, when present, close to the first rib and descends behind the ribs lateral to the costal cartilages. Anastomosing with the upper intercostal arteries, it extends, variably, as far as the second to the seventh intercostal space.

THE INTERNAL THORACIC VEIN (figs. 256, 272, 277)

The **internal thoracic veins** are paired venae comitantes of the internal thoracic artery. Uniting in the region of the first to the third intercostal space, they form a single vein which ascends on the medial side of the artery. Each internal thoracic vein terminates in the brachiocephalic vein of the same side. The internal thoracic vein is formed by the confluence of the superior epigastric and musculophrenic veins. Other tributaries are those corresponding to the branches of the artery—six anterior perforating veins and twelve anterior intercostal veins from the upper six intercostal spaces; anterior mediastinal and thymic veins; and pericardial, muscular, and pericardiacophrenic veins. The valves of the internal thoracic veins prohibit blood from descending in this system.

THE SUPERFICIAL VEINS OF THE CHEST

The superficial veins of the chest form a plexus which has continuities into the abdominal wall, the shoulder region, and the base of the neck.

The **mammillary venous plexus** over the mammary gland drains medialward to the anterior perforating veins of the internal thoracic vein, lateralward to the lateral thoracic vein of the axillary system, and upward to tributaries of the anterior jugular and external jugular veins. A common route through the plexus of thoracic veins is the **thoracoepigastric vein** (fig. 302). This is a variably developed connection between the superficial epigastric vein of the subcutaneous abdominal wall belonging to the femoral system and the lateral thoracic vein of the axillary system (p. 78). The **lateral thoracic vein** also receives the costoaxillary veins, which conduct to it the venous drainage of the mammary gland and which, by their anastomoses, connect the lateral thoracic vein with intercostal veins. The various connections of the lateral thoracic vein provide a subcutaneous interconnection between radicles of the superior vena cava, the inferior vena cava, and the azygos and portal systems of veins.

THE PARIETAL LYMPHATICS OF THE CHEST

The lymphatic nodes of the chest wall are the sternal, the intercostal, and the phrenic nodes.

The **sternal nodes** (p. 69 & fig. 256) lie along the internal thoracic artery and consist of one or two nodes in each of the upper three or four intercostal spaces. They receive lymph from the medial and, less frequently, from the central and lateral parts of the mammary gland, the deeper parts of the anterior thoracic wall, the supraumbilical portion of the abdominal wall, and the upper surface of the liver by way of the anterior phrenic nodes. The efferent vessels of the sternal nodes usually unite to form a single trunk which may contribute to the bronchomediastinal lymph trunk or may empty independently into the confluence of the internal jugular and subclavian veins. It occasionally empties into the right lymphatic duct or into the thoracic duct on the left side.

The **intercostal nodes** (figs. 256, 297) lie near the heads of the ribs. One or two to an intercostal space, they accompany the intercostal vessels and drain the posterolateral portion of the chest wall. Their efferents in the five lower spaces unite to form a trunk which descends to empty into the cisterna chyli or into the beginning of the thoracic duct. The efferents of spaces three to six enter the thoracic duct either singly or in combination; those of the first two spaces ascend to the jugulosubclavian venous junction.

The **phrenic nodes** (p. 452 & figs. 256, 257) comprise

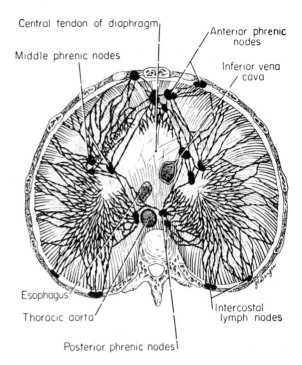

Fig. 257   The superior surface of the diaphragm, illustrating schematically the phrenic and intercostal lymphatics.

three sets—anterior, middle, and posterior—located on the thoracic aspect of the diaphragm. The **anterior phrenic nodes** include two or three nodes behind the base of the xiphoid process and one or two nodes on either side of it behind the seventh costal cartilages. These nodes receive lymph from the convex surface of the liver, the diaphragm, the ventral abdominal wall, and the middle phrenic nodes. Their efferents ascend to the sternal nodes. The **middle phrenic nodes** are two or three nodes on either side, close to the phrenic nerves as they penetrate the diaphragm. On the right side some of these nodes lie anterior to the termination of the inferior vena cava. The nodes of the middle group drain the middle portion of the diaphragm and, on the right side, receive drainage from the convex surface of the liver and from its deep substance by channels which follow the hepatic veins. The middle phrenic nodes send their efferents forward to the anterior phrenic nodes. The **posterior phrenic nodes** are situated on the back of the crura of the diaphragm adjacent to the aorta. They drain the posterior part of the diaphragm and receive a few efferent channels from the middle group of phrenic nodes. They are connected with the lumbar nodes and with the posterior mediastinal lymph nodes.

THE MEDIASTINUM, THE PLEURA, AND THE THYMUS

THE THORACIC MEDIASTINUM (fig. 258)

The lungs and their covering membranes, the pleurae, occupy the lateral portions of the thoracic cavity. All other structures contained in the chest are crowded into a thick partition of tissue at and on either side of the median plane. This is the **mediastinum,** more conveniently subdivided for descriptive purposes into a superior mediastinum, an anterior mediastinum, a middle mediastinum, and a posterior mediastinum.

The **superior mediastinum** (fig. 258) lies behind the manubrium sterni and the origins of the sterno-hyoid and sternothyroid muscles and anterior to the upper four thoracic vertebrae. It is bounded superiorly by the oblique plane of the first rib and inferiorly by a horizontal plane through the sternal angle. This latter plane reaches the vertebral column at the disk between the fourth and fifth thoracic vertebrae. Like the other subdivisions of the mediastinum, the lateral boundary is the parietal pleura. The superior mediastinum contains the great vessels related to the heart and peri-cardium—the arch of the aorta, the beginnings of the brachiocephalic trunk and the left common carotid and subclavian arteries, the brachio-cephalic veins, and the upper portion of the superior vena cava. The thymus occupies the superior mediastinum, and the thoracic parts of the trachea and the esophagus are included. Other structures of the area are the vagus and phrenic nerves, the left recurrent laryngeal nerve, and the thoracic duct. In the superior mediastinum the pericardium is loosely bound to the manubrium by

fibrous tissue which constitutes the superior sterno-pericardial ligament.

The **middle mediastinum** determines the defini-tion and contents of the anterior and posterior mediastina, all three mediastina extending from the plane of the sternal angle above to the diaphragm below. The middle mediastinum includes the peri-cardium and the heart and great vessels. It also contains the bronchi and other components of the roots of the lungs, the arch of the azygos vein, and the phrenic nerves.

The **anterior mediastinum** consists largely of areolar tissue anterior to the pericardium and posterior to the body of the sternum and the transversus thoracis muscles. It is bounded in-feriorly by the diaphragm. It includes some lymph nodes and lymph vessels and, below, the inferior sternopericardial ligament. The sternopericardial ligament consists of fibrous tissue which loosely attaches the pericardium to the upper end of the xiphoid process of the sternum.

The **posterior mediastinum** lies behind the peri-cardium, between the parietal pleura of the two lungs. It contains the thoracic portion of the descending aorta, the azygos and hemiazygos veins, the esophagus, the vagus and thoracic splanchnic nerves, the thoracic duct, and the posterior mediastinal lymph nodes and vessels.

THE PLEURAE (figs. 259, 260, 272, 289)

Each lung is invested by and enclosed within the **pleura,** which has the form of a closed, invaginated serous sac. If the fist is plunged into a balloon, it will invaginate a part of the skin of the balloon into the cavity formed by the remainder of the sphere, and around the wrist will be a circular region of con-tinuity of the outer and inner walls of the balloon. This is the essential analogy of lung and pleura, for, in development, the lung buds grow lateralward into the thoracic celom, invaginating and becoming invested by the mesothelial lining of the pleural portion of the celom. Thus are formed the **parietal pleura,** the portion which retains its original relation to the thoracic wall, and the **pulmonary,** or visceral, **pleura** which is invaginated by the growing lung bud. The visceral pleura intimately invests the lung, cannot be dissected from it, and follows all the fissures and indentations of the lobes of the lung. The apposed surfaces of the pleural layers are lined by a mesothelium which normally secretes a small amount of serous fluid. This lubricates the surfaces, so that respiration is carried on without friction or

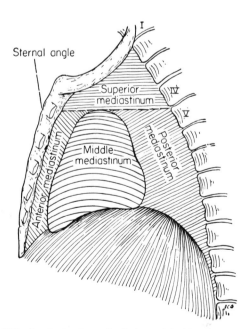

Fig. 258 A schematic sagittal view of the thoracic cavity; the thoracic mediastinum.

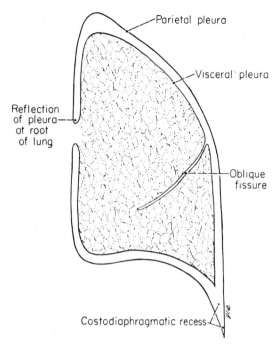

Fig. 259   A coronal section of a lung and its pleural coverings. Semischematic.

interference. Both the 'catch' in breathing and the pain accompanying inflammation of the pleura (pleurisy) are evidence of the value of this mechanism for normal respiration.

The parietal pleura lines the thoracic cavity and thus exhibits a **costal pleura** against the ribs and the intercostal muscles, a **diaphragmatic pleura** covering the thoracic surface of the diaphragm, and a **mediastinal pleura** against the mediastinum medially. The **cervical pleura** has its truncated apex at the plane of the first rib. These terms are designations of regional parts of a continuous sac which extends beyond the normal limits of the lung and provides space for the maximum expansion of the lung in forced ventilation. Beyond the lung margins, portions of the parietal pleura are in contact until expanded by excursion of the lung margins in inspiration. Such potential spaces are especially clear inferiorly where diaphragmatic and costal pleurae are in contact at the side of the diaphragm; this is the **costodiaphragmatic recess** (fig. 259). A similar potential cleft exists behind the sternum where costal and mediastinal pleurae are in contact, the **costomediastinal recess.**

The **mediastinal pleura** follows the curvature of the pericardium from in front to reach the root of the lung. Traced from behind, it overlies the bodies of the thoracic vertebrae, the thoracic aorta, and

the esophagus. The root of the lung is composed of the main bronchus, the pulmonary arteries and veins, the bronchial artery and veins, pulmonary lymph nodes, and the pulmonary nerve plexus. Over these structures the mediastinal pleura reflects onto the lung as the **visceral pleura.** If, in the analogy cited, the fist is driven straight into the balloon, the line of reflection around the wrist is a circular one. If, however, it is driven inward and upward, a teardrop-shaped line of entrance results, and the line of reflection of the layers corresponds to that form. Such is the case at the root of the lung, for with the considerable upward growth of the lung bud, the line of reflection surrounds the root of the lung but is tapered below. The tapered line of reflection extends toward the inferior border of the lung, and its two layers of pleura constitute the **pulmonary ligament**—a fold of serous membranes, not a fibrous ligament.

The relatively abrupt **lines of reflection** (figs. 259, 260) of costal pleura into mediastinal pleura both anteriorly and posteriorly and into diaphragmatic pleura inferiorly mark the limits of the pleural sac in all directions. This continuous line is conveniently referred to the surface of the chest, affording information on the greatest extent of the pleura under the thoracic cage. The **cervical dome** of the pleura projects slightly through the superior thoracic aperture but does not rise above the level of the neck of the first rib. Owing, however, to the obliquity of the first rib, the anterior projection of the neck of the rib is from 2.5 to 5 cm. above its sternal extremity or from 1.5 to 2.5 cm. above the sternal end of the clavicle. From this uppermost limit, the anterior margin of the pleura descends obliquely behind the sternoclavicular joint to reach the median line at the level of the sternal angle. Here the two pleural sacs come in contact and, in fact, may overlap one another slightly, each maintaining essentially a midline position down to the level of the fourth costal cartilage. At this level the two pleural sacs diverge gradually, the reflection on the right side curving only as far as is necessary to leave the sternum behind the seventh costal cartilage. First following the seventh costal cartilage, the line of reflection reaches the eighth costochondral junction (on the mid-clavicular line). It lies under the tenth rib at the axillary line, under the eleventh rib at the scapular line, and passes under the neck of the twelfth rib to reach the side of the body of the twelfth thoracic vertebra. From here the line of reflection turns upward along the sides of

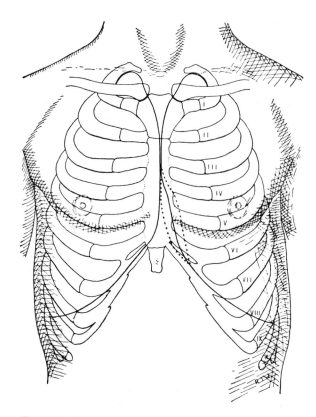

Fig. 260 The relations of the pleural reflections to the anterior chest wall. The solid line represents the mean. The interrupted lines (crosses) in the precordial area on the left represent variants between which lie 70 per cent of the cases studied (from original of fig. 2, R. T. Woodburne, *Anat. Rec.*, 97:210, 1947).

the bodies of the vertebrae to reach the cervical pleura above.

The line of reflection of the left pleura varies only a little from that of the right (fig. 260). In the precordial area the divergence of its reflection is slightly more marked, so that it leaves the sternum in the upper part of the fifth interspace. It crosses the lower border of the sixth cartilage, however, only 1 to 1.5 cm. lateral to the sternal margin and then follows behind the seventh costal cartilage, as on the right side. Although the divergence of the left reflection from the median line is rather more pronounced than on the right, the pleura does not exhibit a cardiac notch, as does the lung on the left side, but overlies the heart and pericardium to a greater extent. The inferior line of pleural reflection may be slightly lower on the left side but is not sufficiently different to require a separate description. The inferior margin of the pleura is relatively horizontal, its deepest point usually being behind the tenth rib at the midaxillary line. The cervical

pleura is maintained in position by a dome-like fascial expansion, **Sibson's fascia,** an offshoot of the prevertebral layer of cervical fascia on the deep side of the scalene muscles, which is attached in front to the inner border of the first rib (p. 152).

Clinical Note It is important in incisions over the kidney to realize that the pleura drops below the neck of the twelfth rib. Radiographic evidence in the living subject indicates that the pleura sometimes descends posteriorly as far as the lower margin of the second lumbar vertebra (Lachman, '42). Knowledge of the limits of the pleural sac in all directions is obviously important; among other reasons for the interpretation of fluid levels in the chest and in planning operative approaches through the chest wall. Further, a stabbing in the lower neck can open the pleural cavity or injure the apex of the lung. This is due to the obliquity of the first rib cited above.

Vessels and Nerves

The parietal pleura is supplied by twigs of arteries adjacent to it—posterior intercostal, internal thoracic, superior phrenic, and anterior mediastinal arteries. The veins of the parietal pleura correspond to the arteries. The visceral pleura is supplied by radicles of the bronchial and pulmonary arteries. Its veins are tributary to the pulmonary veins. The costal and peripheral portions of the diaphragmatic pleura are supplied by intercostal nerves; the central part of the diaphragmatic pleura and probably the mediastinal pleura are innervated by the phrenic nerve. Vagus and sympathetic twigs also reach the parietal pleura. The visceral pleura receives branches from the pulmonary plexuses. In contrast to parietal pleura, the visceral pleura is insensitive to contact, for it receives no nerves of general sensation.

The **lymphatics** of the visceral pleura combine with the superficial efferents of the lung. Those of the parietal pleura drain regionally—from the costal pleura into the intercostal and sternal nodes, from the diaphragmatic pleura into the phrenic nodes, and from the mediastinal pleura into the anterior mediastinal and posterior mediastinal nodes. A few superolateral channels reach the axillary nodes.

THE THYMUS (fig. 261, 262)

The **thymus** is composed of two flat flask-shaped lobes which occupy the superior mediastinum on either side of the median plane. The gland lies behind the manubrium sterni and is covered for the most part by the converging pleurae and their enclosed lung margins. The thymus lies anterior to the great vessels at the base of the heart—the arch of the aorta, the brachiocephalic veins, and the superior vena cava—and usually extends down onto the upper part of the pericardium. The gland occasionally reaches upward through the superior

thoracic opening into the base of the neck, sometimes as high as the lower pole of the thyroid gland.
The thymus develops as two entodermal diverticula
from the third branchial pouch. Descending ventrally, they meet and become joined to one another by
areolar connective tissue. No thymic tissue connects
them, however, and they form in the adult a bi-
lobed gland both connected and separated by connective tissue. The investing connecting tissue also
forms a distinct capsule for the gland. The thymus
is a prominent organ in the infant. It attains its
greatest relative size at about two years of age
but continues to grow until puberty. After puberty,
the thymus undergoes a gradual involution in which
the thymic tissue is usually replaced by fat. Thus, the
adult thymus retains approximately the form and
size of the gland of earlier years but is composed
largely of adipose tissue. The thymic tissue consists
primarily of lymphocytes. The gland apparently

produces immunologically potent cells that migrate
to the lymphoid organs of the body. There they
mature, reproduce, and differentiate into immunologically competent, antibody secreting cells.

The **arteries** of the thymus are derived from the
anterior mediastinal branches of the internal thoracic
arteries; the gland may also receive twigs superiorly
from the inferior thyroid arteries. The **veins** of the
thymus are mainly tributary to the left brachiocephalic
vein. This relationship results from the early development of the left brachiocephalic vein from a plexus of
veins which spanned the midline in the embryo and from
which the thymic veins also developed. Other thymic
veins may be tributary to the internal thoracic vein or
the inferior thyroid vein. The **lymphatics** of the gland
end in the sternal, anterior mediastinal, and tracheobronchial nodes. Small **nerves** are supplied to the gland
from the vagus nerve and the cardiac sympathetic
branches.

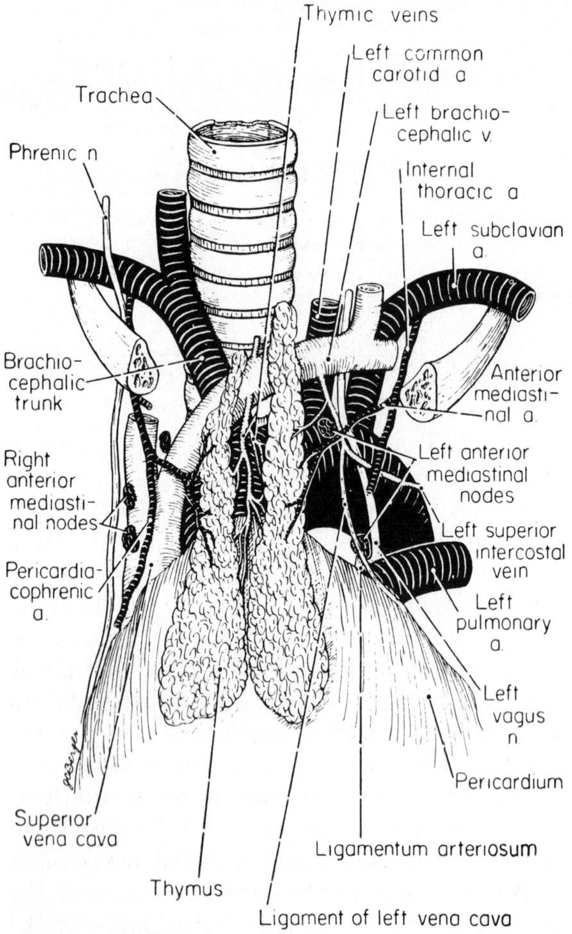

THE HEART AND THE PERICARDIUM

THE PERICARDIUM (figs. 261–65)

The **pericardium** is an invaginated serous sac
similar to the pleura, but the external surface of its
parietal layer is invested by a strengthening fibrous
coat. The pericardium encloses the heart and the
roots of the great vessels. Its base rests on the
diaphragm below, being molded by its curvatures,
and is located at the level of the xiphosternal
junction in the median line. The external fibrous
layer (fibrous pericardium) fuses with the central
tendon of the diaphragm and becomes continuous
with the superior fascia of the diaphragm. The pericardium is pierced by the inferior vena cava opposite
the caval aperture of the diaphragm. Superiorly, the
apex of the pericardium extends to the level of the
sternal angle. Here its fibrous layer becomes
continuous with the outer fibrous layer of the
aorta, the superior vena cava, and the pulmonary
arteries, and its serous layer reflects downward
over these vessels onto the heart. The fascia of the
infrahyoid muscles (p. 153) attaches to the fibrous
pericardium. The pericardium occupies the middle
mediastinum, and its anterior surface forms the
posterior boundary of the anterior mediastinum. It
is attached to the sternum by the **superior sternopericardial** and **inferior sternopericardial ligaments**
(p. 321). The upper portion of the pericardium is
separated from the sternum by the anterior portions
of the pleurae and the lungs. Below, it is in direct
relation with the left half of the lower portion of the

Fig. 261 The thymus, the pericardium, and the great
vessels at the base of the heart.

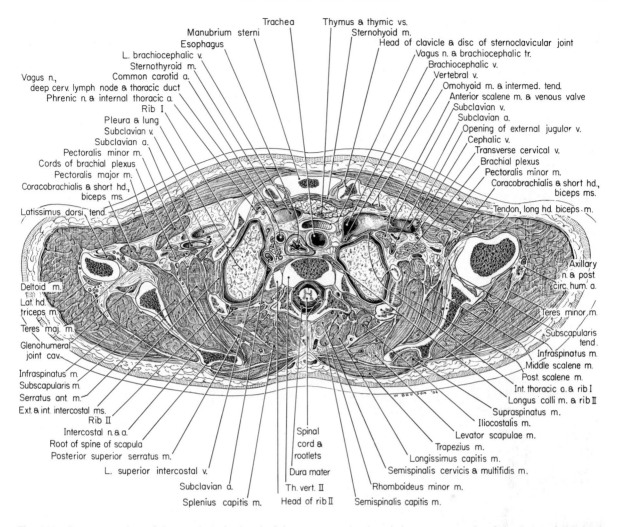

Fig. 262 A cross section of the trunk at the level of the thymus gland and the great vessels of the base of the neck.

body of the sternum and with the medial ends of the sixth and seventh costal cartilages and the left transversus thoracis muscle. The lateral surfaces of the pericardium lie against the mediastinal pleura, only the phrenic nerves and the pericardiaco-phrenic vessels intervening. The posterior surface of the pericardium forms the anterior boundary of the upper part of the posterior mediastinum; this surface is in direct contact with the esophagus and the thoracic aorta.

The serous face of the parietal pericardium is smooth and glistening and reflects onto the great vessels both at the apex of the pericardial sac and from its posterior wall. Following the vessels, the visceral pericardium becomes the outermost layer of the heart, the **epicardium.** A separate reflection of the visceral pericardium exists around the ascending aorta and the pulmonary arterial trunk;

this is the **arterial mesocardium.** The reflection about the veins, the **venous mesocardium,** includes the superior vena cava, two right and two left pulmonary veins, and the inferior vena cava. This departure from the simple invaginated condition to a state in which there are two lines of reflection comes about through the bending of the heart tube into its compound shape, the approximation of the arterial and venous ends of the tube, and a subsequent breaking through of the intervening visceral pericardium. The space between the arterial and venous mesocardia is the **transverse sinus of the pericardium;** it is situated deep to the great arterial channels and superior to the pulmonary veins. The **oblique sinus of the pericardium** is a box-like diverticulum of the pericardial cavity within the irregularities of the venous mesocardium between the right and left pulmonary veins. It is entered from below and

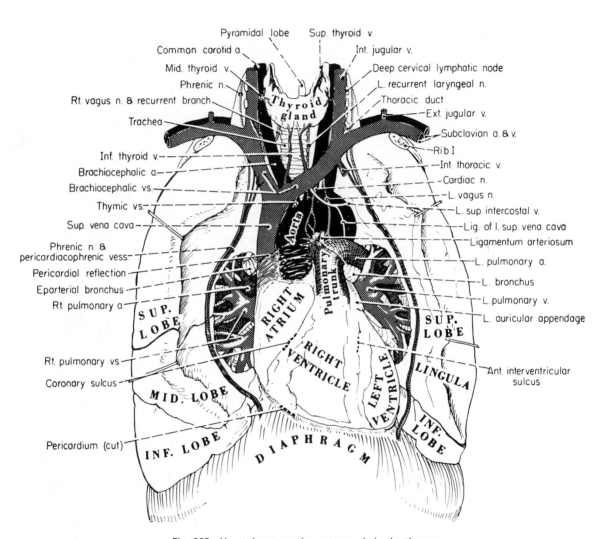

Pyramidal lobe — Sup. thyroid v.
Common carotid a.
Int. jugular v.
Mid. thyroid v.
Deep cervical lymphatic node
Phrenic n.
L. recurrent laryngeal n.
Rt. vagus n. & recurrent branch
Thoracic duct
Thyroid gland
Ext. jugular v.
Trachea
Subclavian a. & v.
Rib I
Inf. thyroid v.
Int. thoracic v.
Brachiocephalic a.
Cardiac n.
Brachiocephalic vs.
L. vagus n.
Thymic vs.
L. sup. intercostal v.
Sup. vena cava
Aorta
Lig. of l. sup. vena cava
Ligamentum arteriosum
Phrenic n. & pericardiacophrenic vess.
L. pulmonary a.
Pericardial reflection
Pulmonary trunk
L. bronchus
Eparterial bronchus
RIGHT ATRIUM
L. pulmonary v.
Rt. pulmonary a.
SUP. LOBE
SUP. LOBE
L. auricular appendage
RIGHT VENTRICLE
LEFT VENTRICLE
LINGULA
Rt. pulmonary vs.
Ant. interventricular sulcus
Coronary sulcus
MID. LOBE
INF. LOBE
Pericardium (cut)
INF. LOBE
DIAPHRAGM

Fig. 263 Heart, lungs, and great vessels in the thorax.

326

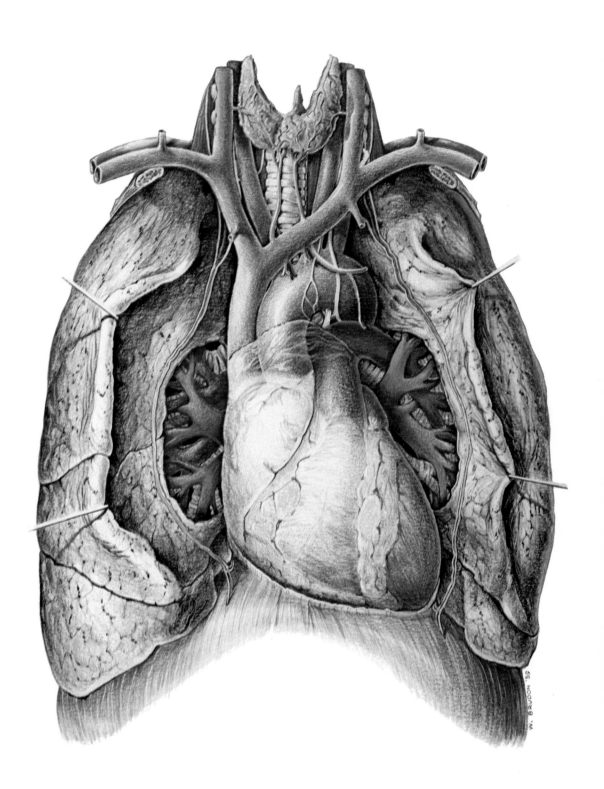

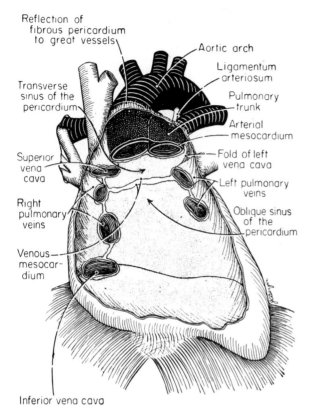

Fig. 264 The reflections and sinuses of the pericardium.

behind the heart and will admit several fingers. (Trace through these stages as illustrated in fig. 265).

Located at the left end of the transverse sinus between the left pulmonary artery and the superior left pulmonary vein is a small fold of serous pericardium, the **fold of the left vena cava.** It encloses a fibrous remnant of the left superior vena cava, which normally atrophies at an early stage of fetal life. From the lower end of the ligament, the **oblique vein of the left atrium** descends to end in the **coronary sinus** of the heart. The **ligament of the left superior vena cava** (fig. 261) is a strand of connective tissue which ends, as a rule, in the pericardium anterior and lateral to the fold of the left vena cava. It arises above in the left superior intercostal vein or the left brachiocephalic vein and may include a small vein. The continuity of these various structures has significance in the fact that the root of the left superior intercostal vein, the strand and vein of the ligament of the left superior vena cava, the oblique vein of the left atrium, and the coronary sinus of the heart represent parts of the embryonic left anterior and common cardinal veins (fig. 266). From these a persisting left superior vena cava occasionally is formed; when present, it ends at the heart in the coronary sinus.

The **arteries** of the pericardium are the pericardial, pericardiacophrenic, and musculophrenic branches of the internal thoracic artery, assisted by twigs from esophageal and bronchial arteries and from the thoracic aorta. The **nerves** of the pericardium are branches of the vagus and phrenic nerves and of the sympathetic trunks.

Clinical Note In pericardiocentesis the needle is inserted upward and backward at the upper end of the left xiphochondral junction. At this site the needle should miss the left pleura and lung and reach the cavity of the pericardium in its most dependent portion.

The Heart and Great Vessels (figs. 263, 267–72)

The heart and the roots of the great vessels occupy the pericardium. The heart is not centered in the

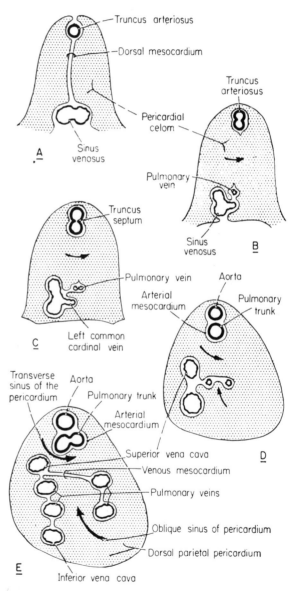

Fig. 265 A series of stages in the development of the reflections and sinuses of the pericardium. Redrawn and modified from Patten in Gould, S. E., 1960, *Pathology of the Heart*, 2nd ed., Thomas, Springfield.

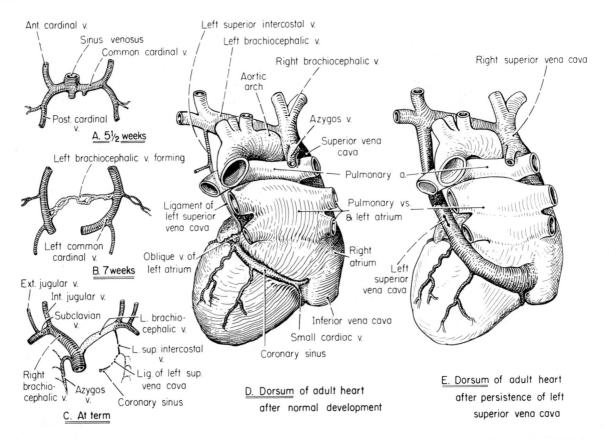

Fig. 266   The establishment of the great veins at the base of the heart. A-C, developmental sequence leading to a single (right) superior vena cava and to the coronary sinus. D, completed normal development. E, persistence of a left superior vena cava.

median plane but lies about two-thirds to the left of it and about one-third to its right. The heart, the pericardium, and the great vessels are related anteriorly to the sternum, the costal cartilages, and the ribs; the sternocostal projection of these parts represents the initial view of them during dissection.

Sternocostal Projection (figs. 263, 267)

The heart lies obliquely in the chest, with its **apex** downward and to the left, the site being readily recognized as the position of the 'heart beat.' The **base** of the heart is its most superior portion, and from it emerge the ascending aorta, the pulmonary arterial trunk, and the superior vena cava. The average outline of the heart, as projected anteriorly, may be defined in relation to its base, its apex, and the right end of its diaphragmatic surface. The base of the heart is represented by a horizontal or slightly oblique plane across the sternum at the level of the third costal cartilages. Such a line ends about 2 cm. from the left margin of the sternum

along the upper border of the third left costal cartilage and about 1 cm. from the right margin of the sternum along the lower border of the third costal cartilage. The point defining the apex of the heart is, on the average, about 8 cm. from the median plane in the left fifth interspace. The right end of the diaphragmatic surface lies under the chondrosternal junction of the sixth or the seventh cartilage on the right side. Appropriately curved lines incorporating these points outline the contour of the heart. From the right end of the base line, an outwardly curved line descends, has its greatest divergence (2 cm.) from the right sternal border in the fourth interspace, and ends behind the sixth or seventh chondrosternal junction. The diaphragmatic border of the heart is somewhat concave, conforming to the curvature of the central part of the diaphragm. It passes behind the xiphosternal junction and then curves downward and to the left to end at the apex of the heart. The left border of the heart is outwardly convex and connects the site of the apex with the left end of the base line. These are

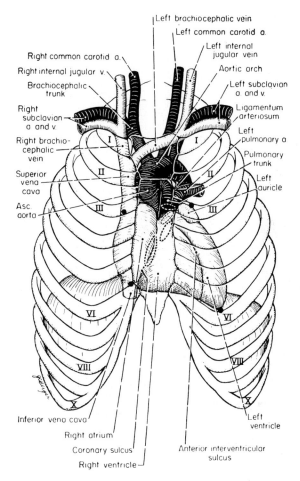

Fig. 267 The sternocostal projection of the heart and great vessels.

average locations of points and borders. The apex position varies in healthy young adults from 5.5 to 12 cm. to the left of the midline, and the greatest deviation from the sternal midline on the right side is between 3 cm. and 7 cm. The heart may be placed more to the right and thus lie more nearly in the midline or, conversely, may be even more markedly to the left than normal. Like all soft organs, the heart also conforms to the requirements of stature differences, being long and narrow in tall, thin individuals and squat and broad in short, stocky people (figs. 268, 269). It also responds to the effects of gravity, rising in recumbent positions and descending in the erect posture. Beyond the normal, displacement of the heart may be an indication of disease in it or surrounding organs.

The sternocostal projection of the pericardium differs from that of the heart only superiorly, where its apex ends on the great vessels at the level of the sternal angle (second costal cartilage), one

costal interval higher than the base of the heart. This upward prolongation places the apex of the pericardium superiorly; its base is inferior, where it is represented by its diaphragmatic surface.

The Sulci The sternocostal surface of the heart is marked by sulci which overlie the septa separating the chambers of the heart. The **coronary sulcus** (atrioventricular) begins on the base line of the heart in about the median plane and descends obliquely to the right to end at the right extremity of the diaphragmatic surface. This sulcus separates the right atrium, upward and to the right, from the right ventricle, downward and to the left. It lodges the right coronary artery and a tributary of the small cardiac vein. The coronary sulcus encircles the heart, separating both atria from both ventricles; on the back of the heart it lodges the coronary sinus, the terminal part of the great cardiac vein, and the circumflex branch of the left coronary artery. Beginning on the diaphragmatic border of the sternocostal surface about one-fourth to one-third of the distance in from the apex position is the **anterior interventricular sulcus.** Its position may be indicated by a slight notching of the diaphragmatic border of the heart. The sulcus is approximately vertical and ends above just to the left of the pulmonary trunk. It contains the anterior interventricular branch of the left coronary artery and the great cardiac vein. The anterior interventricular sulcus is the anterior indentation produced by the interventricular septum. It is continuous on the diaphragmatic aspect of the heart with the posterior interventricular sulcus which lodges the posterior interventricular branch of the right coronary artery and the middle cardiac vein.

The Chambers The sulci of the heart outline and delimit the surface projection of its chambers. Between the coronary sulcus and the anterior interventricular sulcus lies the triangular **right ventricle.** This occupies the greater part of the sternocostal surface of the heart and leads above into the pulmonary arterial trunk. Across the coronary sulcus from the right ventricle lies the **right atrium.** Its somewhat irregular surface is limited along the coronary sulcus by a wavy, or crenated, border. The right atrium occupies the right aspect of the heart and receives, toward its posterior surface, the superior vena cava above and the inferior vena cava below. The left atrium and the left ventricle lie predominantly at the back of the heart and have only limited representation on the sternocostal surface. The **left ventricle** bulges

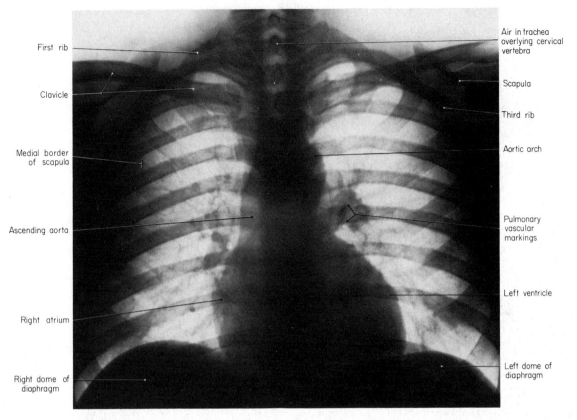

First rib

Clavicle

Medial border of scapula

Ascending aorta

Right atrium

Right dome of diaphragm

Air in trachea overlying cervical vertebra

Scapula

Third rib

Aortic arch

Pulmonary vascular markings

Left ventricle

Left dome of diaphragm

Fig. 268 A posterior-anterior radiogram of a normal chest of the sthenic type. It is characterized by a broad chest and heart. Compare fig. 269.

roundly to the left of the anterior interventricular sulcus and makes up the bulk of the ventricular part of the posterior surface of the heart, limited by the posterior interventricular sulcus. The **left atrium** occupies almost the entire posterior surface of the heart above the coronary sulcus, receiving there two left and two right pulmonary veins. The sternocostal projection includes only the auricular portion of the left atrium which curves over the left border of the heart to the left of the pulmonary trunk.

The Great Vessels (figs. 263, 267, 271, 272) At the base of the heart appear the beginnings of the ascending aorta and the pulmonary arterial trunk and the termination of the superior vena cava. Their roots are covered by visceral pericardium and are enclosed within the pericardial sac.

The **ascending aorta** takes its origin from the left ventricle at the lower border of the third left chondrosternal junction and is directed upward, forward, and to the right for about 5 cm. It ends at the level of the sternal angle by becoming the aortic arch. The diameter of the ascending aorta is 3

cm. on the average, and it conducts all the blood going into the systemic arterial system. The beginning of the ascending aorta is covered by the pulmonary trunk and by the right auricle. The right pulmonary artery turns behind it. At its beginning the aorta contains three localized dilatations, the **aortic sinuses.** These lie above and behind the aortic semilunar valves, and from the right and left sinuses arise the right coronary and left coronary arteries (p. 336).

Clinical Note The ascending aorta is more subject to **aneurism** than the succeeding parts. It receives the strongest thrust of blood from the ventricle and its wall is not reinforced by the apposition of the fibrous pericardium.

The **aortic arch** is the continuation of the ascending aorta, and emerges from the pericardium behind the right half of the sternum at the level of the sternal angle. The arch is directed first upward, backward, and to the left behind the lower half of the manubrium sterni. It then runs backward along the left side of the trachea and turns downward on the left side of the body of the fourth thoracic vertebra. The aorta arches, therefore, not only from

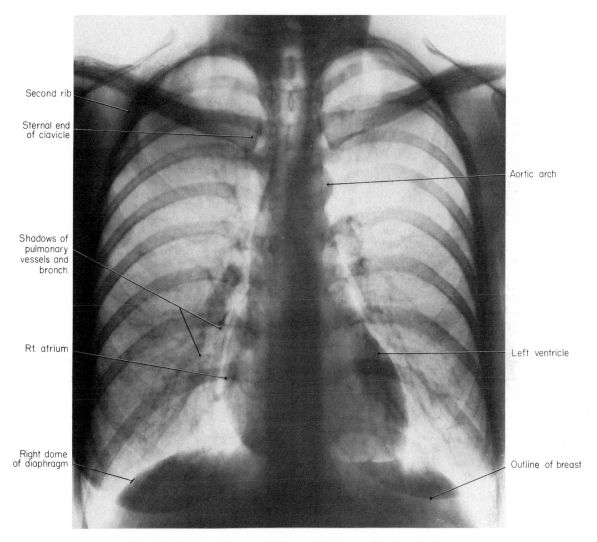

Second rib

Sternal end
of clavicle

Aortic arch

Shadows of
pulmonary
vessels and
bronch.

Rt. atrium

Left ventricle

Right dome
of diaphragm

Outline of breast

Fig. 269 A posterior-anterior radiogram of a normal chest of the asthenic type. The chest is narrower and the heart is oriented more vertically. Compare fig. 268.

right to left but markedly from anterior to posterior. The arch terminates at the level of the disk between the fourth and fifth thoracic vertebrae by becoming continuous with the thoracic aorta. The underside of the arch is at the level of the sternal angle. The thymus lies in front of the arch, succeeded anteriorly by the converging margins of the pleurae and the lungs. The trachea lies behind the arch and to its right side. Attached to the underside of the arch opposite the origin of the left subclavian artery is the **ligamentum arteriosum** (fig. 263). This is a band of connective tissue about 1.5 cm. long and from 3 to 5 mm. in diameter which represents the remains of the fetal ductus arteriosus, the dorsal part of the left sixth aortic arch. (The fetal **ductus arteriosus** constituted a necessary bypass of excess blood from

the pulmonary circuit into the systemic at a time when the lungs were merely developing as essential parts of the body, but were not concerned in the aeration of blood. The persistence of this channel after birth on the other hand results in a lack of separation of the two circuits and imperfect and incomplete oxygenation due to blood passing from the higher pressure of the systemic circuit into the lower pressured pulmonary system.) The ligament is directed anteriorly and inferiorly to the root of the left pulmonary artery. The left vagus nerve crosses the left side of the aortic arch under the left superior intercostal vein and, at the left side of the ligamentum arteriosum, gives off its recurrent laryngeal branch. The latter nerve turns under the ligament and passes up toward the neck behind and

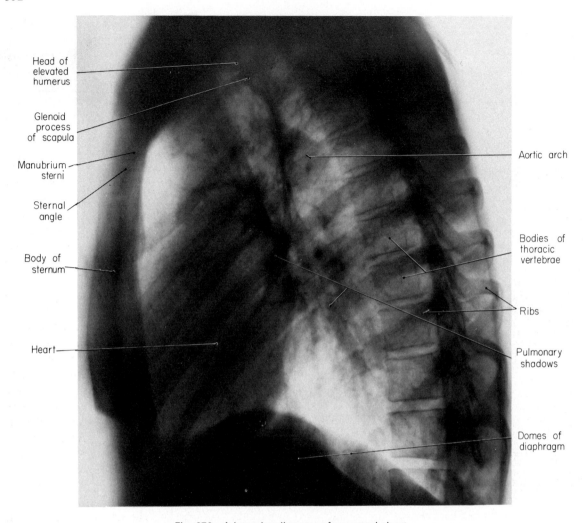

Head of
elevated
humerus

Glenoid
process
of scapula

Manubrium
sterni

Sternal
angle

Body of
sternum

Heart

Aortic arch

Bodies of
thoracic
vertebrae

Ribs

Pulmonary
shadows

Domes of
diaphragm

Fig. 270   A lateral radiogram of a normal chest.

to the right of the arch. One or more strands of nerve descend across the aortic arch to the right of the ligamentum arteriosum and turn under the arch. These are branches of the left vagus and the left sympathetic trunk to the superficial cardiac plexus. The pulmonary trunk bifurcates in the concavity of the arch, and behind the pulmonary trunk is the left bronchus. The aortic arch measures about 4.5 cm. in length. It is equal in diameter to the ascending aorta (3 cm.) until its branches are given off, after which its diameter is reduced to nearly 2 cm. The narrowing of the lumen at the level of origin of the left subclavian artery and the ligamentum arteriosum is the **aortic isthmus.**

Clinical Note   An abnormal narrowing that deprives much of the body of a normal circulation, **coarctation of the aorta,** takes place occasionally at the site of the aortic isthmus. It leads to the development of a large system of collateral circulation involving especially the scapular and intercostal arterial anastomoses to provide blood to the lower body.

The **branches of the aortic arch** (fig. 263) are the brachiocephalic trunk and the left common carotid and left subclavian arteries. The largest branch, the **brachiocephalic trunk,** is from 4 to 5 cm. in length. It arises from the beginning of the arch, behind the middle of the manubrium sterni. The trunk ascends obliquely upward, backward, and to the right to the upper border of the right sterno-clavicular articulation, where it divides into the subclavian artery (p.179) and the common carotid artery (p.159) of the right side. The brachiocephalic trunk usually gives off no branches, but occasionally the thyroidea ima artery (p. 158) arises from it.

The **left common carotid artery** springs from the highest part of the arᶜh, posterior to the brachio-

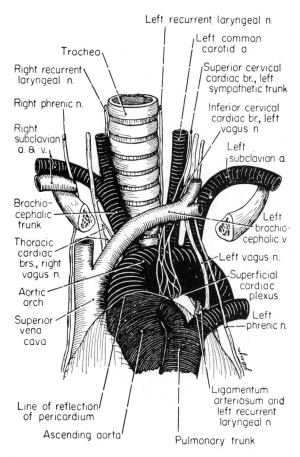

Left recurrent laryngeal n.
Left common carotid a
Trachea
Right recurrent laryngeal n.
Superior cervical cardiac br., left sympathetic trunk
Right phrenic n.
Inferior cervical cardiac br., left vagus n
Right subclavian a. & v.
Left subclavian a.
Brachio-cephalic trunk
Left brachio-cephalic v.
Thoracic cardiac brs., right vagus n.
Left vagus n.
Superficial cardiac plexus
Aortic arch
Left phrenic n.
Superior vena cava
Line of reflection of pericardium
Ligamentum arteriosum and left recurrent laryngeal n
Ascending aorta
Pulmonary trunk

Fig. 271 The great vessels and nerves at the base of the heart.

The **brachiocephalic veins** (fig. 263) arise behind the sternal ends of the clavicles by the union of the internal jugular vein (p. 161) and the subclavian vein (p. 181) of each side. They do not possess valves. The **right brachiocephalic vein,** only 2.5 cm. in length, descends almost vertically and joins the left brachiocephalic vein at the right border of the sternum just below the first costal cartilage. This junction forms the superior vena cava. The right brachiocephalic vein descends to the right of the brachiocephalic arterial trunk, the right vagus nerve lying between them. On the right side of the vein are the phrenic nerve and the right pleura and lung. The right brachiocephalic vein receives the vertebral vein (p. 303) which opens into its upper posterior portion, and, further down, the right inferior thyroid vein (p. 158) and the internal thoracic vein (p. 319). The posterior intercostal vein of the first intercostal space (p. 318) on the right side is also tributary to it. The **left brachio-cephalic vein,** arising behind the sternal end of the left clavicle, passes to the right behind the upper half of the manubrium sterni to unite with the right brachiocephalic vein in the formation of the superior vena cava. It has a length of about 6 cm. and crosses diagonally the left subclavian, the left common carotid, and the brachiocephalic arterial trunk together with the vagus and phrenic nerves. The left brachiocephalic vein is separated from the manubrium by the remains of the thymus, areolar tissue, and the origins of the sternohyoid and sternothyroid muscles. The tributaries of the left brachiocephalic vein are the left vertebral, internal thoracic, inferior thyroid, superior intercostal, the posterior intercostal vein of the first left intercostal space, pericardial, thymic, and mediastinal veins.

The **superior vena cava** is the final path to the right atrium of the heart for all the venous blood from the head and neck, the upper limbs, and the thoracic wall. It begins behind the first right costal cartilage in the union of the two brachiocephalic veins and descends to the level of the right third costal cartilage, where it ends in the upper part of the right atrium. It has a length of about 7.5 cm., curves slightly to the right in its intermediate portion, and is contained within the pericardium in its lower portion. The superior vena cava is over-lapped at its left margin by the ascending aorta and is covered anteriorly by the thymus, the right pleura, and the right lung. Posteriorly, it lies against the right bronchus, the right pulmonary artery, and the right superior pulmonary vein. The

cephalic trunk. It ascends behind the upper half of the manubrium sterni to the level of the left sterno-clavicular joint, where it enters the neck (p. 159). The trachea is behind and to the right of the artery in the superior mediastinum. Also behind it are the esophagus, the left recurrent largyngeal nerve, and the thoracic duct. To its left lie the left vagus and the left phrenic nerves, and the left subclavian artery is behind and lateral to it.

The **left subclavian artery** (p.179) arises from the arch about 1 cm. distal to the origin of the left common carotid and on the left side of the trachea. It ascends through the superior mediastinum, lying against the left lung and pleura laterally, and arches outward to the medial border of the anterior scalene muscle. The left vagus, phrenic, and cardiac nerves parallel the artery anteriorly, and the left vertebral and internal jugular veins cross it as they descend to the left brachiocephalic vein. The left recurrent laryngeal nerve ascends medial to the artery adjacent to the trachea and the esophagus.

azygos vein enters it posteriorly outside the pericardium just above the structures of the root of the right lung. The right phrenic nerve and the pericardiacophrenic vessels descend along the right border of the superior vena cava. The superior vena cava has no valves. It receives, in addition to the azygos vein, several small pericardial and mediastinal veins. The superior vena cava may be involved in a stab wound passing close to the sternum through the first or second interspace on the right side.

A left superior vena cava occasionally forms from the anterior and common cardinal veins of the left side. When present, it empties into the right atrium at the normal opening of the coronary sinus, replacing the coronary sinus and the oblique vein of the left atrium (fig. 266). A left superior vena cava drains the left brachiocephalic vein by way of the root of the left superior intercostal vein. When a left superior vena cava occurs, the right usually persists also, and the channels are present bilaterally. The ligamentous remnant of an undeveloped left superior vena cava is normally present (p. 326).

The **pulmonary trunk** (fig. 263) is the arterial continuation of the right ventricle and conducts nonaerated blood to the lungs for oxygenation. Short (5 cm.) and wide (3 cm.), it springs from the apex of the right ventricle and passes upward and backward across the origin of the ascending aorta. It lies to the left of the aorta as far as the under surface of the arch where, at the level of the sternal angle, it divides into right pulmonary and left pulmonary arteries. The pulmonary trunk is contained within the pericardium, being enclosed with the ascending aorta within the arterial mesocardium. The fibrous layer of the pericardium extends onto its branches, gradually becoming lost in their connective tissue. The pulmonary trunk lies behind the anterior end of the left second intercostal space. The right coronary and left coronary arteries pass toward the anterior surface of the heart along its sides. The left auricle lies along the left side of the pulmonary trunk, and the superficial part of the cardiac plexus lies above its bifurcation, between it and the aortic arch.

The **right pulmonary artery** runs horizontally to the right under the arch of the aorta, behind the ascending aorta and the superior vena cava, and in front of the right bronchus. At the root of the right lung it divides into two branches, a smaller upper branch to the superior lobe and a larger lower branch which supplies the middle and inferior lobes of the right lung. The further divisions of these arteries are segmental arteries which follow and correspond in name to the bronchopulmonary segments (p. 353).

The **left pulmonary artery,** shorter and smaller than the right, passes in front of the descending aorta and the left bronchus to the root of the left lung. It divides into two branches, one to the superior and one to the inferior lobe. The left pulmonary artery is connected to the underside of the aortic arch by the ligamentum arteriosum (p. 331). The fold of the left vena cava (p. 326) is a crescentic fold of serous pericardium reflected from the underside of the left pulmonary artery to the superior left pulmonary vein.

The Nerves of the Superior Mediastinum (figs. 271, 272, 286)

The **vagus nerves** descend in the neck (p. 162) posteriorly in the interval between the carotid artery and the internal jugular vein of each carotid sheath. At the base of the neck they cross the subclavian arteries, the recurrent laryngeal nerve arising at this level on the right side. Entering the chest on the right side between the brachiocephalic arterial trunk and the brachiocephalic vein, the right vagus lies at the side of the trachea. It passes dorsalward to the root of the right lung, where it forms the posterior pulmonary plexus. From the right recurrent laryngeal nerve and from the trunk of the right vagus as it lies at the side of the trachea, **thoracic cardiac branches** descend onto the tracheal bifurcation to contribute to the deep cardiac plexus. The **left vagus** enters the chest between the left common carotid and left subclavian arteries and behind the left brachiocephalic vein. It crosses the aortic arch with a dorsal inclination and is there crossed by the root of the left superior intercostal vein. It passes the ligamentum arteriosum, giving off its recurrent laryngeal nerve from its medial aspect as it does so. The nerve then runs behind the left pulmonary artery to the root of the left lung and forms the posterior pulmonary plexus on the back of the lung root. The **left recurrent laryngeal nerve** turns in under the ligamentum arteriosum and, ascending medial to the aortic arch, rises to the larynx (p. 190 & fig. 263). **Thoracic cardiac branches** arise from the left recurrent laryngeal nerve and run medialward to end in the deep cardiac plexus.

The **phrenic nerves** (figs. 261, 271, 272) are branches of the cervical plexus, and their cervical relations have been described (p. 177). At the base of the neck the phrenic nerves overlie the anterior

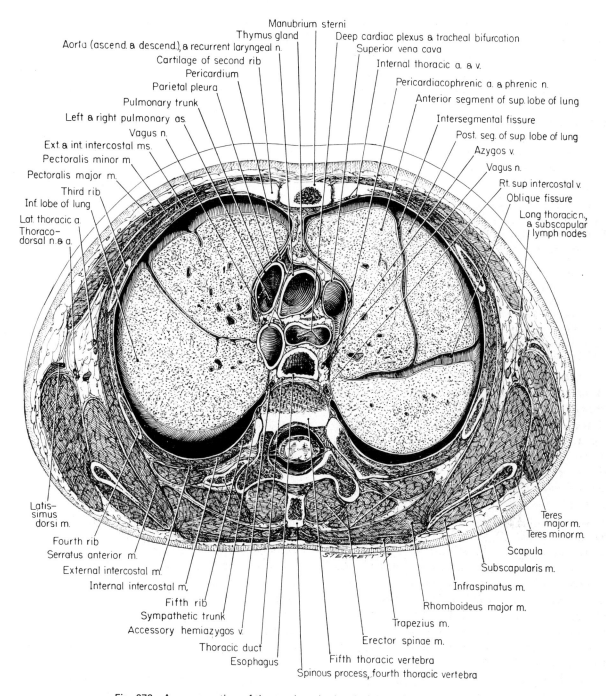

Manubrium sterni
Thymus gland
Deep cardiac plexus & tracheal bifurcation
Aorta (ascend. & descend.), & recurrent laryngeal n.
Superior vena cava
Cartilage of second rib
Internal thoracic a. & v.
Pericardium
Pericardiacophrenic a. & phrenic n.
Parietal pleura
Anterior segment of sup. lobe of lung
Pulmonary trunk
Intersegmental fissure
Left & right pulmonary as.
Post. seg. of sup. lobe of lung
Vagus n.
Azygos v.
Ext. & int. intercostal ms.
Vagus n.
Pectoralis minor m.
Rt. sup. intercostal v.
Pectoralis major m.
Oblique fissure
Third rib
Inf. lobe of lung
Long thoracic n., & subscapular lymph nodes
Lat. thoracic a.
Thoraco-dorsal n. & a.

Latis-simus dorsi m.
Teres major m.
Teres minor m.
Fourth rib
Serratus anterior m.
Scapula
External intercostal m.
Subscapularis m.
Internal intercostal m.
Infraspinatus m.
Fifth rib
Rhomboideus major m.
Sympathetic trunk
Accessory hemiazygos v.
Trapezius m.
Thoracic duct
Erector spinae m.
Esophagus
Fifth thoracic vertebra
Spinous process, fourth thoracic vertebra

Fig. 272 A cross section of the trunk at the level of the tracheal bifurcation.

scalene muscles under the prevertebral layer of cervical fascia. They enter the chest by passing superficial to the subclavian arteries and deep to the subclavian veins. Crossing the origin of the internal thoracic artery and joined by the pericardiaco-phrenic artery and vein, the nerves descend in front of the roots of the lungs between the parietal layers of the pericardium and the pleurae. The right nerve lies against the right brachiocephalic vein and the superior vena cava and has a direct course to the diaphragm. The left phrenic nerve lies between the left subclavian and common carotid arteries in the superior mediastinum and descends across the aortic arch posterolateral to the left vagus nerve. It is deviated by the mass of the heart from its straight line of descent and ends on the diaphragm below. Fine filaments of both phrenic nerves are distributed in the chest to the costal and mediastinal pleurae and to the pericardium. The terminal branches to the diaphragm are described with that structure (p. 452).

The Blood Vessels of the Heart (figs. 273, 274, 277, 280, 281)

The **arterial supply** of the heart is provided by the right coronary and left coronary arteries. The **right coronary artery** arises from the right aortic sinus and, passing forward, reaches the coronary sulcus between the pulmonary trunk and the right auricle. It descends in the coronary sulcus onto the diaphragmatic surface of the heart. A **marginal branch** passes toward the apex along the acute margin of the heart and ramifies over the right ventricle. Passing from right to left in the coronary sulcus on the back of the heart, the right coronary artery turns toward the apex in the posterior interventricular sulcus. This terminal part of the artery is known as the **posterior interventricular branch.** It supplies both ventricles and anastomoses, near the apex, with the anterior interventricular branch of the left coronary artery. The **left coronary artery**

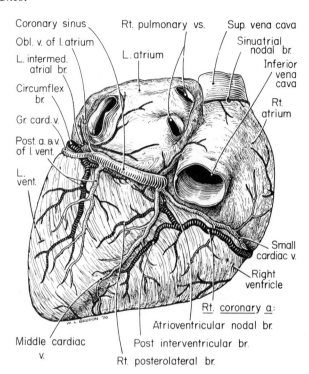

Fig. 274   The posterior aspect of the heart; arteries and veins.

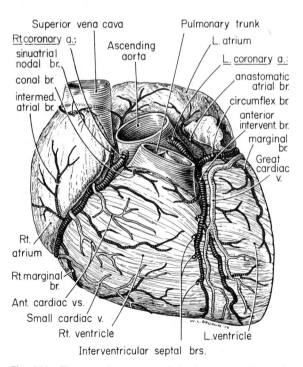

Fig. 273   The anterior aspect of the heart; arteries and veins.

arises from the left aortic sinus. Passing forward and to the left, the artery divides between the pulmonary trunk and the left auricle into an anterior interventricular and a circumflex branch. The **anterior interventricular branch** passes toward the apex of the heart in the anterior interventricular sulcus. It supplies both ventricles and the interventricular septum and anastomoses near the apex with the posterior interventricular branch of the right coronary artery. The **circumflex branch** circles toward the back of the heart in the coronary sulcus. It has a **marginal branch** which follows the left margin of the heart, supplying the left ventricle. The circumflex artery ends on the back of the left ventricle, anastomosing with the posterior interventricular branch of the right coronary artery. A **posterior artery of the left ventricle** frequently swings down the back of the left ventricle near the termination of the circumflex, matching the distribution of the posterior vein of the left ventricle.

Lesser branches of the coronary arteries (figs. 273, 274, 281). Both the right and left coronary arteries have small **conal branches** which arise close to the origins of the coronary arteries and mount onto the pulmonary conus. Branches to the surface of the right ventricle from each coronary also meet over the surface of that ventricle. **Atrial branches** are numerous though usually small. The important **sinuatrial nodal artery** usually (54%) arises from the right coronary artery distal to its conal branch, distributes on the deep aspect of the right atrium, circles the base of the superior vena cava and ends in the sinuatrial node at the cephalic end of the sulcus terminalis. An **intermediate atrial artery** arises from the right coronary artery nearly opposite its right marginal branch and ascends over the right atrial surface. An **anastomotic atrial branch** arises from the circumflex artery near its origin and passes toward the right on the internal surface of the left atrium. It may anastomose with terminals of the right sinuatrial artery and will usually be involved in those variant cases in which the sinuatrial artery is a left coronary branch. The circumflex artery also gives rise to an **intermediate atrial branch** which distributes along the left atrium above the coronary sulcus. The **atrioventricular nodal artery** arises from the right coronary artery opposite the origin of the posterior interventricular artery (85%) and penetrates deeply to the atrioventricular node at the base of the interatrial septum. **Interventricular septal arteries** arise from the anterior interventricular and posterior interventricular arteries. Of these, those from the anterior interventricular artery pass two-thirds of the way into the septum, those from the posterior interventricular one-third.

Clinical Note Terminal anastomoses between the cardiac arteries are usually small and require enlargement for survival in case of the occlusion of any of the larger branches of the coronary arteries. Interference with the coronary circulation results in ischemia and pain (angina pectoris) or a varying amount of permanent damage to the myocardium. A small area of damage usually allows recovery by connective tissue replacement in the infarcted area, but massive damage is likely to be fatal. The modern 'bypass' operation in which venous autografts are inserted to bridge the embolus area has saved many lives. The three most common locations of coronary occlusion, in descending order of frequency, are proximally in the anterior descending branch of the left coronary artery, proximally in the right coronary artery, and proximally in the circumflex branch of the left coronary artery.

The **veins of the heart,** for the most part, accompany the arteries but are not designated by the same names. The principal terminus of the veins is the **coronary sinus.** This is a short, wide venous channel situated in the posterior portion of the coronary sulcus between the left atrium and the left ventricle. It is frequently partially covered by muscular fibers from the left atrium. The coronary sinus opens into the right atrium between the orifice of the inferior vena cava and the atrioventricular opening. It has an imperfect, one-cusp valve, the **valve of the coronary sinus,** which is located at the right margin of its aperture. The tributaries of the coronary sinus are the great, middle, and small cardiac veins, the posterior vein of the left ventricle, and the oblique vein of the left atrium.

The **great cardiac vein** begins at the apex of the heart and ascends in the anterior interventricular sulcus, where it lies adjacent to the anterior interventricular branch of the left coronary artery. Reaching the coronary sulcus, the vein passes, in it, to the left to the posterior surface of the heart and ends in the left extremity of the coronary sinus. It has a valve at its entrance into the sinus. The great cardiac vein receives tributaries from the walls of both ventricles and the left atrium. When present, a left marginal vein empties into it.

The **middle cardiac vein** begins at the apex of the heart and ascends in the posterior interventricular sulcus. It opens near the right end of the coronary sinus and has a single valve at its termination. The middle cardiac vein accompanies the posterior interventricular branch of the right coronary artery.

The **small cardiac vein** usually begins as the marginal vein along the acute margin of the heart. Receiving tributaries from the right atrium, it turns to the back of the heart in the coronary sulcus in company with the right coronary artery. The vein ends in the right end of the coronary sinus.

The **posterior vein of the left ventricle** ascends on the diaphragmatic surface of the left ventricle to end in the coronary sinus. It is frequently paired on the ventricle with terminal branches of the circumflex branch of the left coronary artery.

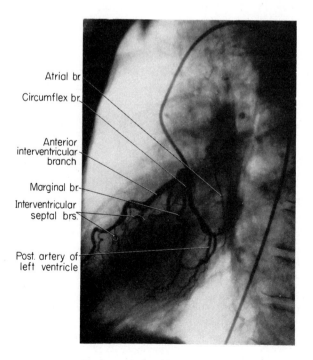

Atrial br.
Circumflex br.

Anterior
interventricular
branch

Marginal br.
Interventricular
septal brs.

Post. artery of
left ventricle

Fig. 275 A radiogram of the left coronary arterial circulation.

The **oblique vein of the left atrium** is a small vein which descends to the left end of the coronary sinus from the crevice between the left auricle and the inferior left pulmonary vein. It is continuous with the fold of the left vena cava (p. 326) and is part of the remains of the left common cardinal vein of the embryo.

Cardiac veins which do not empty into the coronary sinus are the anterior cardiac veins and the smallest cardiac veins. The **anterior cardiac veins** consist of three or four small vessels which arise on the anterior wall of the right ventricle, bridge across the coronary sulcus and the right coronary artery, and end by penetrating the wall of the right atrium. The **smallest cardiac veins (venae cordis minimae)** are numerous minute channels which begin in the myocardium of the heart and open directly into its chambers—principally the atria, but also the ventricles. They communicate with the capillary plexus of the myocardium but also directly with the arterioles and venules of the myocardium. Although designated as veins, they may function in carrying blood into the myocardium on occasion.

The Lymphatics of the Heart

A **subepicardial lymph plexus** receives lymphatic channels from the myocardium and the subendocardial connective tissue. Collecting vessels from the subepicardial plexus pass to the coronary sulcus and follow the coronary arteries. Vessels accompanying the right coronary artery pass over the arch of the aorta to empty into a left anterior mediastinal node (p. 365). Vessels accompanying the left coronary artery pass behind the arch of the aorta to a node of the tracheobronchial group—usually of the right side (p. 349).

The Chambers of the Heart (figs. 276–80)

The chambers of the heart are a right and a left atrium, and a right and a left ventricle. The separation of atria and ventricles is marked on the surface by the coronary (atrioventricular) sulcus. The ventricles are separated from each other by an interventricular septum, the position of which is indicated by the anterior interventricular and posterior interventricular sulci of the heart. The atria are separated by an interatrial septum. The right atrium receives all the systemic venous blood by way of the superior vena cava, the inferior vena cava, the coronary sinus, and the anterior cardiac veins. It passes this blood (fig. 12) through the right atrioventricular orifice into the right ventricle. The right ventricle, by its contraction, forces the blood out through the pulmonary arteries to the

lungs, the right atrioventricular valve closing to prevent reflux of blood into the right atrium. The oxygenated blood from the lungs is returned to the left atrium of the heart by the pulmonary veins. From here it enters the left ventricle through the left atrioventricular aperture. Contraction of the musculature of the left ventricle, together with concomitant closure of the left atrioventricular valve, drives the aerated blood into the arterial system of the body.

The cardiac cycle of alternating contraction and relaxation is repeated about seventy-five times per minute, the duration of one cycle being about 0.8 second. Three phases succeed one another during the cycle: atrial systole, or contraction of both atria—0.1 second; ventricular systole, simultaneous contraction of the ventricles—0.3 second; and diastole, complete relaxation or rest—0.4 second. The actual period of rest for each chamber is 0.7 second for the atria and 0.5 second for the ventricles; so that in spite of its continuing activity, the muscular parts of the heart are at rest longer than at work. During rest the chambers expand, and aided by respiratory movements, blood flows into the atria through their large venous tributaries and on through the atrioventricular apertures into the ventricles. Atrial systole, beginning with contraction

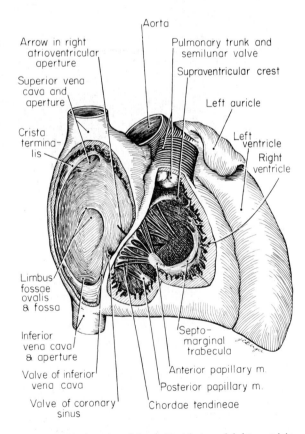

Fig. 276 The interior of the right atrium and right ventricle.

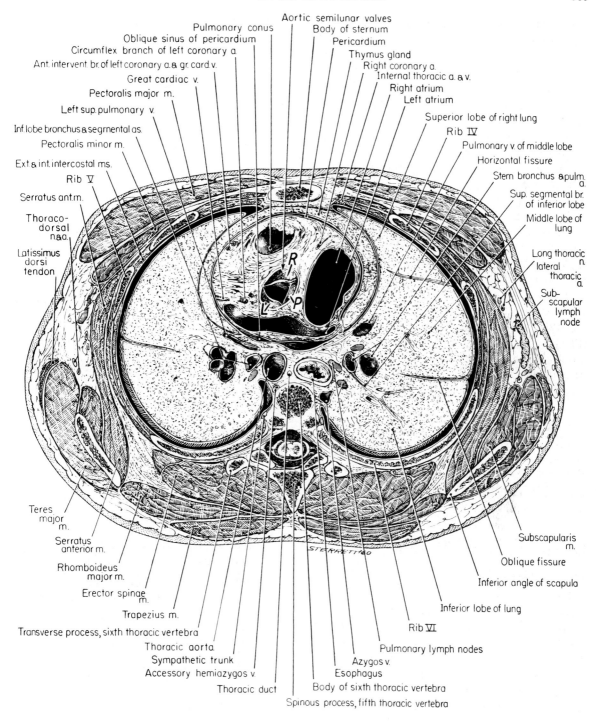

Aortic semilunar valves
Pulmonary conus
Body of sternum
Oblique sinus of pericardium
Pericardium
Circumflex branch of left coronary a.
Thymus gland
Ant. intervent. br. of left coronary a. & gr. card. v.
Right coronary a.
Internal thoracic a. & v.
Great cardiac v.
Right atrium
Pectoralis major m.
Left atrium
Left sup. pulmonary v.
Superior lobe of right lung
Inf. lobe bronchus & segmental as.
Rib IV
Pectoralis minor m.
Pulmonary v. of middle lobe
Horizontal fissure
Ext. & int. intercostal ms.
Stem bronchus & pulm. a.
Rib V
Sup. segmental br. of inferior lobe
Serratus ant. m.
Middle lobe of lung
Thoraco-dorsal n. & a.
Long thoracic n.
lateral thoracic a.
Latissimus dorsi tendon
Sub-scapular lymph node
Teres major m.
Subscapularis m.
Serratus anterior m.
Oblique fissure
Rhomboideus major m.
Inferior angle of scapula
Erector spinae m.
Inferior lobe of lung
Trapezius m.
Rib VI
Transverse process, sixth thoracic vertebra
Pulmonary lymph nodes
Thoracic aorta
Azygos v.
Sympathetic trunk
Esophagus
Accessory hemiazygos v.
Body of sixth thoracic vertebra
Thoracic duct
Spinous process, fifth thoracic vertebra

Fig. 277   A cross section of the trunk at the level of the aortic semilunar valve.

of the circular fibers around the openings of the great veins, empties the atria completely and distends the ventricles. Back pressure is minor, and the atria require only thin muscular walls. The force of ventricular systole is much greater but is unequal in the two chambers. The right ventricle has to overcome only the peripheral resistance of the pulmonary capillary bed, and its wall is only one-third as thick as that of the left ventricle. Correspondingly, the blood pressure in the pulmonary artery is only about one-third of that in the aorta. The atrioventricular and semilunar valves close in pairs—the atrioventricular valves, as the ventricles begin to contract; the semilunar valves, at the end of ventricular systole in response to back pressure from the elastic arterial bed. The closure of these pairs of valves gives rise to the characteristic heart sounds.

The **right atrium** has a posteriorly situated, thin-walled **sinus venarum** and an anterior more muscular portion. The **sinus venarum** receives the venae cavae. The separation of the sinus venarum and the remainder of the atrium is indicated on the right posterior surface of the atrium by a groove, the terminal sulcus. Curving to the right, this sulcus extends from the front of the superior vena cava to the front of the inferior vena cava and represents the line of union of the sinus venosus and the primitive atrium in the embryo. In the interior of the chamber, a smooth, vertical, muscular ridge, the **terminal crest,** separates the two parts and is continued to the postero-inferior wall of the atrium between the valve of the inferior vena cava and the valve of the coronary sinus. Behind the crest the atrium has a smooth surface, and in front of it the muscular fibers of the wall are raised in ridges (pectinate muscles). The opening of the **superior vena cava** is located in the superoposterior part of

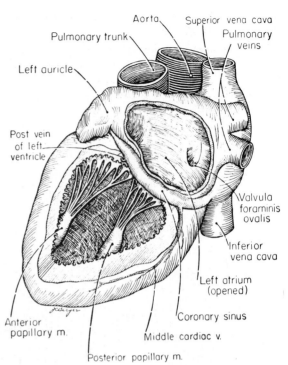

Fig. 279   The interior of the left atrium and left ventricle as seen through openings in the posterior wall of the heart.

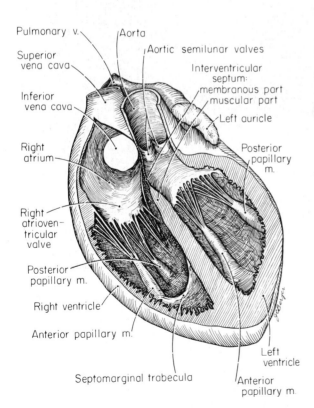

Fig. 278   A vertical section of the heart through the interventricular septum.

the atrium and is directed downward and forward toward the right atrioventricular opening. There is no valve in the superior vena cava. The **inferior vena cava** opens into the inferior extremity of the atrium. Its orifice is directed upward and backward toward the interatrial septum. The rudimentary **valve of the inferior vena cava** has a single, crescentic, vertical flap which fades away into the posterior atrial wall. The **opening of the coronary sinus** lies between the valve of the inferior vena cava and the atrioventricular aperture and is bordered by the semicircular **valve of the coronary sinus.** The posterior, or interatrial, septal wall contains an oval, depressed, thinned-out area, the **fossa ovalis.** This represents the position of the foramen ovale in the fetus. Its prominent oval margin, especially distinct above and at the sides of the fossa, is the **limbus fossae ovalis.** Under its margin at the upper limit of the fossa, a small valvular opening may lead obliquely upward into the left atrium. This is a remnant of the foramen ovale which persists into adult life in about 25 per cent of cases. This small oblique opening has only a little, if any, functional significance as a rule, for the greater pressure in the left atrium keeps the valvular margins in contact. The **right auricle** is the conical muscular pouch of the superior extremity of the atrium which lies against the root of the aorta. The auricle covers the first portion of the right coronary artery.

The **left atrium** has, like the right atrium, a larger smooth-walled portion and a smaller highly muscular auricle. Into the smooth-walled general cavity enter the four pulmonary veins—superior and in-

ferior left pulmonary and superior and inferior right pulmonary veins. On the interatrial septal wall a semilunar depression marks the fossa ovalis from this side. The depression is bounded inferiorly by a crescentic ridge, the **valvula foraminis ovalis.** Behind this lip, communication may be made with the

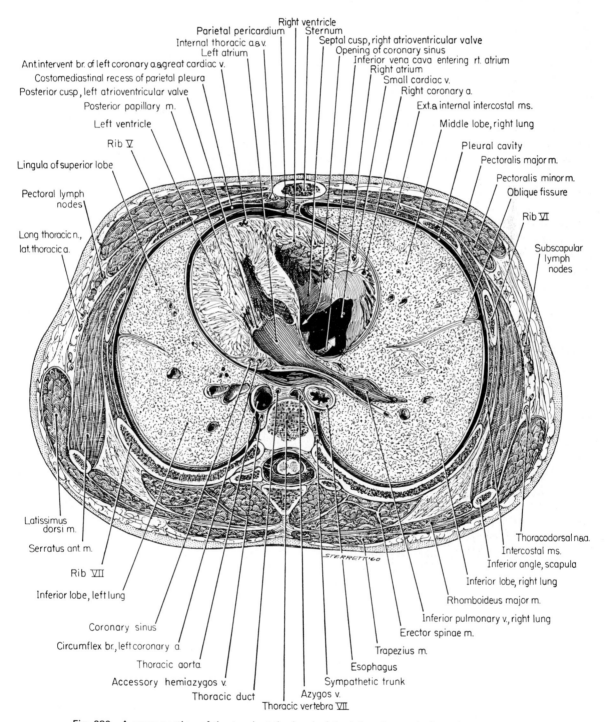

Fig. 280   A cross section of the trunk at the level of the left atrioventricular valve of the heart.

right atrium (see above). The **left auricle** is somewhat constricted at its junction with the atrium.
The auricle is narrow and is directed forward and
overlaps the root of the pulmonary artery.

The **right ventricle** is triangular in the frontal
plane, tapering superiorly into the root of the **pulmonary arterial trunk.** It forms a large part of the
sternocostal and a small part of the diaphragmatic
portions of the heart. The ventricle has a semilunar
outline in cross section, for its anterior wall is
convex forward, whereas its posterior wall, formed
by the interventricular septum, bulges into its
cavity. The muscular wall of the right ventricle is
only one-third as thick as that of the left ventricle.
Its endocardial surface is thrown into relief by criss-
crossing and anastomosing muscular ridges of the
myocardium. These fasciculi are designated as the
trabeculae carneae. At its right extremity the ventricle contains the large right atrioventricular orifice,
guarded by the **tricuspid valve.** To the underside of
the cusps of this valve are attached tendinous
strands, the **chordae tendineae.** These arise in conical
muscular projections of the wall of the ventricle
which are called papillary muscles. A large **anterior
papillary muscle** springs from the sternocostal wall
of the ventricle in partial continuity with the septomarginal trabecula; a much smaller **posterior
papillary muscle** arises from its diaphragmatic wall,
but chordae tendineae for the septal cusp arise
directly from the septal wall of the ventricle. A
muscular ridge, the **supraventricular crest,** arches
toward and over the anterior cusp of the tricuspid
valve and separates the muscular ventricle from
the pulmonary conus (fig. 276). The **septomarginal
trabecula** extends from the septum to the anterior
papillary muscle (fig. 278).

The **left ventricle** is longer and more conical in
shape than the right, and its wall is three times as
thick. It lies principally posteriorly and forms the
apex of the heart. In its interior it exhibits **trabeculae carneae** and two large **papillary muscles** together with their chordae tendineae. The papillary
muscles spring from the anterior and posterior walls
of the ventricle. The ventricle is entered through
the bicuspid (mitral) left atrioventricular valve. The
aortic opening, in front and to the right of the
bicuspid valve, leads into the aorta. The **interventricular septum** bulges into the cavity of the right
ventricle. It is thick and muscular except in its
upper and posterior part. Here the thin, fibrous
membranous part of the ventricular septum separates

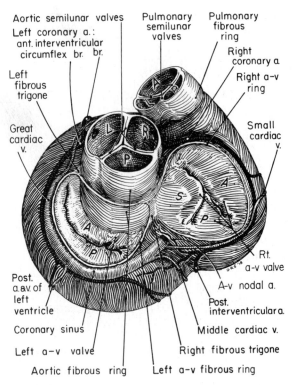

Fig. 281   A semischematic section of the heart through the
coronary sulcus.

the aortic ostium from the lower part of the right
atrium and the upper part of the right ventricle.

The Valves of the Heart (figs. 276–83)

At the coronary sulcus there is almost complete
separation of the myocardium of the atria and the
ventricles. Here, surrounding the valvular orifices
and providing attachment for the heart musculature, is a collection of dense connective tissue which
serves as the 'skeleton' of the heart. This fibrous
connective tissue exists as **fibrous rings,** giving circular form and rigidity to the atrioventricular apertures and to the roots of the aorta and pulmonary
trunk. The fibrous rings send thin, sheet-like extensions into the valves. The **aortic fibrous ring** is
massive and sleevelike (fig. 281). It furnishes semicircular attachments for the cusps of the aortic
semilunar valve and has a deep extension into the
left atrioventricular junction region. Here the atrial
musculature arises from the sleeve well above the
attachment of the anterior cusp of the left a-v valve.
The **septum membranaceum** is a continuation of the
aortic sleeve or cuff into the septum, and **right** and

left **fibrous trigones** are small offshoots of the cuff which provide for muscular origins. The **pulmonary fibrous ring** is smaller than the aortic, and is peaked like the latter for the attachment of the cusps of its valve.

The Atrioventricular Valves (figs. 276, 278–82) The right and left fibrous rings form the stiff margins of the atrioventricular valves and send into them a thin sheet of fibrous tissue. Covering both surfaces of this sheet of connective tissue is the lining membrane of the heart—the endocardium of the atria on the one side and of the ventricles on the other. The atrial surface of the valves is smooth and shining; the ventricular surface is attached by the chordae tendineae to the papillary muscles of the ventricle. The free margins of the valves are notched, the deepest notches separating the valves into their component cusps. The left atrioventricular (mitral) valve has two cusps, an anterior and a posterior. The right atrioventricular valve (the tricuspid valve) has three cusps—large anterior and septal cusps and a small posterior cusp. Each cusp receives chordae tendineae from more than one papillary muscle, and each papillary muscle sends chordae tendineae to more than one cusp. These muscles and tendinous cords combine to stay the atrioventricular valves against the pressure developed by the pumping action of the ventricles, for the valves have no intrinsic rigidity. A noncontractile tendinous cord would serve such a purpose if its distal attachment remained stationary, but the ventricular

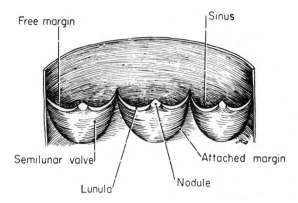

Fig. 283   The pulmonary trunk opened widely to show its semilunar valves.

walls and the heart apex move toward the fibrous axis of the heart during contraction. However, in the combination of tendinous cord and papillary muscle, the contraction of the papillary muscle compensates for the inward displacement of the ventricular walls, and the cords remain tense during systole. The endocardium of the ventricles covers the papillary muscle and the chordae tendineae and reflects to the underside of the valve cusps and then to the ventricular wall.

The Semilunar Valves (figs. 276–78, 281, 283) The semilunar valves are formed by a similar inward expansion of the fibrous rings at the roots of the aorta and the pulmonary trunk, covered on either side by epithelium—the endocardium of the ventricles proximally, the endothelium of the great vessels distally. These valves have no tendinous cords but depend, for their stability, on their semilunar form and their long attached margins. There are three valves in each orifice. With the heart oriented so that the ventricles are on the right and left of the interventricular septum, the aortic semilunar valve cusps are right, left, and posterior, and the pulmonary valve cusps are right, left, and anterior. This system agrees well with the coronary arteries, the right coronary artery arising in the aortic sinus behind the right aortic valve and the left coronary artery having a similar relation to the left aortic valve. Both the aorta and the pulmonary arterial trunk develop from one fetal channel (the truncus arteriosus) by an indentation and finally a pinching through of the common channel. In this process one of the developing valves is bisected, contributing to both sets of valves, and, thus, the attachments of the left and right valves are opposite one

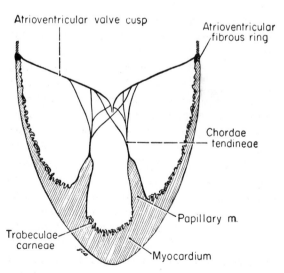

Fig. 282   A schematic vertical section through a ventricle and its atrioventricular valve.

another at the site of the constriction of the parent vessel (fig. 281).

The convex margins of the semilunar valves are attached to the wall of the artery, and their free borders are directed upward into the lumen of the vessel. The free margin of each valve exhibits a centrally placed **nodule** of dense connective tissue and, on either side of it, a crescentic thinned-out area, the **lunula.** The lunula is deficient in fibrous tissue and is thus formed principally of two layers of epithelium back to back. When the pocket-like valves are distended with blood due to back pressure in the arterial system after systole, the thin, flexible lunulae of the adjacent valves make intimate contact. At the same time, the hard, firm nodules come together to plug a tendency to a central deficiency between the curving edges of the valves. Behind the valves are the dilated sinuses of the aorta and the pulmonary trunk.

Clinical Note   Cardiac auscultation requires a knowledge of the surface projection of the valves of the heart, although the optimal sites for listening to their sounds are not all immediately over the valve in question. The projection of the valves is illustrated in figure 267. The valves (except for the tricuspid) are close to the left border of the sternum, the tricuspid being about midsternal. The pulmonary semilunar lies behind the third sternochondral junction and the aortic is at its lower border. The mitral and tricuspid valves are large but they center opposite the 4th rib for the mitral and the 4th interspace for the tricuspid. The mnemonic is P.A.M.T.-3, $3\frac{1}{2}$, 4, $4\frac{1}{2}$. **Anomalies of the heart and great vessels** occur. Attention has been called to the occasional retroesophageal right subclavian artery (p. 179), the embryologic events underlying persistence of a left superior vena cava (p. 326), location of aortic coarctation (p. 332), possible continuance of the ductus arteriosus (p. 331) and the relatively frequent persistence of a probe-patent foramen ovale in the interatrial septum (p. 340).

An **interatrial septal defect** larger than a probe-patent foramen ovale can result from inadequate development of the interatrial septum secundum or excessive resorption of the interatrial septum primum. Unless other conditions alter the relatively equal pressures on the two sides of the defect, little, if any, mixing of blood is expected. If, however, there are significant pressure differences, oxygenated and unoxygenated blood will mix with possible serious effects. **Interventricular septal defects** also occur and usually involve the membranous portion of the septum (p. 342 & fig. 278). This exposes the blood in the right ventricle to the higher pressure of the left ventricle, mixes the blood streams, and raises the pressure in the pulmonary circuit.

The valves of the heart may have deficiencies of two totally different sorts. **Stenosis,** or narrowing of the passageways, may develop either congenitally or as a result of disease. **Valvular incompetence** reflects failure of complete closure of the valve. The backrush of blood under high pressure in such cases can be recognized as a 'heart murmur.' These conditions interfere

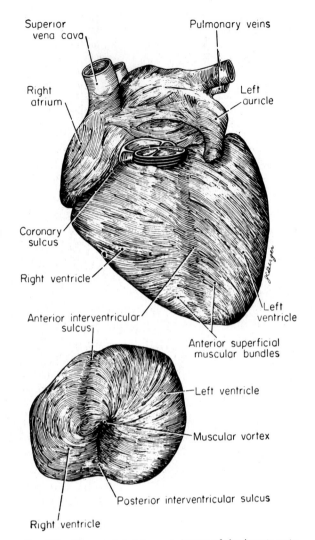

Fig. 284 The superficial musculature of the heart; anterior and apical views.

with the smooth onward flow of blood and usually necessitate compensatory changes to increase the force of propulsion.

The **'tetralogy of Fallot'** combines pulmonary stenosis and an interventricular septal defect with a shift of the ostium of the aorta over the septal defect and enlargement of the right ventricle. This congenital defect markedly reduces oxygenation of the blood. In recent years it has been corrected successfully by a bypass operation (Blalock procedure) which is roughly equivalent to producing an artificial ductus arteriosus. There are other complex deviations from the normal arrangement of heart valves and chambers and great vessels. The altered hemodynamics are instructive and provide the student with a test of his understanding of normal circulation.

The Heart Musculature (fig. 284)

The heart is essentially a muscular organ, and its fibrous connective tissue 'skeleton' serves as the origin of its musculature. The muscle fasciculi also 'insert' largely into the connective tissue substrate of the heart, including the chordae tendineae. The cardiac muscle is arranged in sheets and bundles which are susceptible of analysis.

In the **atria** the muscular fasciculi are arranged in superficial and deep layers. The superficial fibers, most numerous on the sternocostal surface and near the coronary sulcus, run transversely across the atria. A few of them turn into the interatrial septum. The deeper fibers of the atria are looping and circular in orientation and are generally confined to one atrium. Annular fibers surround the auricles and the extremities of the great veins ending in the atria.

The **ventricular musculature** arises from the fibrous trigones and the fibrous rings of the base of the heart. The spiral loops of which this musculature is composed embrace the cavities of either one or both ventricles, one limb of a loop lying on the outer surface of the heart and one in the interior. The superficial fibers of the sternocostal surface pass downward and toward the left; those of the dia-phragmatic surface pass toward the right. The principal anterior superficial bundle arises from the posterior aspect of the left atrioventricular and right atrioventricular fibrous rings and passes apicalward across the right ventricle. The bundle twists upon itself in a tight vortex at the apex and turns inward to spread out on the inner surface of the left ventricle, inserting by papillary muscles, into the septum, and into the opposite fibrous rings. Another prominent superficial bundle arises from the root of the aorta and from the left atrioventricular fibrous ring. It spirals toward the apex, making a double circle around the heart, somewhat like a figure 8 that is wider open at the top. The fibers of this bundle turn into the vortex of the left ventricle and insert into the opposite side of the tendinous structures from which the bundle arose. The deeper bundles also arise from the fibrous rings and form spirals and figure eights, but they do not reach the apex of the heart. One deep bundle makes a double circle of the cavity of the left ventricle, whereas others encircle only the right ventricle or loop around both ventricles.

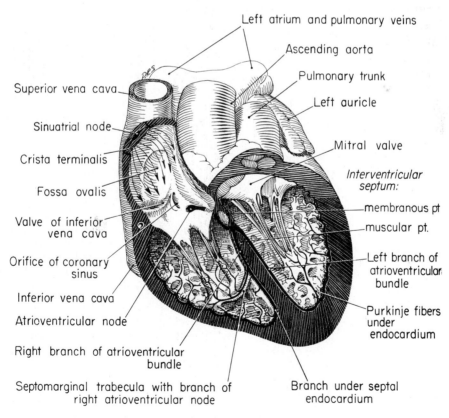

Fig. 285   The conduction system of the heart.

The configuration of these intermingling muscle bundles of the ventricular part of the heart is such as to compress the cavities by both a direct squeezing action (the deeper bundles) and a wringing motion (the more superficial spiral bundles). The combination of these actions powerfully empties the ventricles, and the pull of the spiral muscles causes the apex of the heart to impinge on the anterior chest wall, resulting in the characteristic apex beat.

The Conduction System of the Heart (fig. 285)

The heart has an automatic rhythmic beat. It also is under the influence of nerves which, however, serve only to change the force or frequency of the contractions of the heart muscle in accordance with the physiologic needs of the organism. Although the myocardium has an inherent rhythmicity, the orderly sequence of the cardiac cycle, in which the contractions of the ventricles follow those of the atria and proceed from the apex toward the base, is due to a conduction system composed of **nodal tissue** in two locations and a band of specialized cardiac muscle which conducts the stimulus for contraction to the ventricular muscle.

The **sinuatrial node,** a small collection of specialized myocardium, is located in the notch formed by the superior vena cava and the right auricle with the main atrial chamber. This is at the cephalic end of the sulcus terminalis. This node appears to initiate the heart beat, and from it the impulse to contraction is propagated over the atria. The **atrioventricular node** is a nodule of the same type of tissue that is situated in the septal wall of the right atrium immediately above the opening of the coronary sinus. This node initiates an impulse, in response to the stimulus reaching it through the atrial musculature, which controls the contraction of the ventricles. Since, for effective expulsion of blood, the ventricular contraction must progress from the apex toward the base of the heart, a conduction system is also required to carry the impulse for contraction from the atrioventricular node to the apical portions of the ventricles. The **atrioventricular bundle** (His) is a strand of specialized myocardium which passes from the atrioventricular node through the right fibrous trigone into the musculature of the ventricular septum. Its course carries it beneath the septal cusp of the tricuspid valve and along the lower margin of the septum membranaceum, immediately ventral to which it divides into **right and left branches.** (This course along the

lower margin of the septum membranaceum makes it very vulnerable when a septal defect has to be repaired.) The **right branch** follows under the endocardium of the septum until it reaches the **septomarginal trabecula,** a variable bundle of muscle which extends from the septum to the anterior papillary muscle of the right ventricle. The right branch usually traverses the septomarginal trabecula (moderator band) to reach the anterior papillary muscle and the sternocostal portion of the ventricular wall (Truex & Copenhaver, '47). It also gives off branches along its course through the septum. The **left branch** of the atrioventricular bundle is a flat, wide band which appears under the endocardium of the left ventricle a little below the septum membranaceum. It follows the septal wall toward the apex, dividing into branches to the anterior papillary and posterior papillary muscles and to the ventricular myocardium. The terminal strands of the atrioventricular bundle were discovered before the entire bundle was defined and may be designated as **Purkinje fibers** (their original name).

Clinical Note The role of the sinuatrial node in initiation and regulation of the heart beat is evident, and deficiencies in its stimulating action may lead to heart block. Fortunately, in recent years, the development of the cardiac pacemaker with its electrode introduced into the right ventricle via the superior vena cava and appropriate tributaries and its pulse generator implanted in a subcutaneous pocket has provided an alternate method of pacing the heart beat.

The Cardiac Plexus of Nerves (figs. 271, 286)

Modification of the intrinsic rhythmicity of heart action is produced by the cardiac nerves. These include both a sympathetic and a parasympathetic innervation. Stimulation through the sympathetic nerves increases the rate and force of the heart beat; slowing and reduction in force are the results of parasympathetic stimulation. The parasympathetic cardiac nerve is the vagus, which supplies three branches on each side. Three sympathetic branches on each side also reach the heart from the cervical sympathetic ganglia or the associated trunk and, in addition, direct branches from the upper four or five thoracic ganglia join the cardiac plexus.

The **cardiac plexus** is situated at the base of the heart behind and in the concavity of the arch of the aorta. It is divisible into superficial and deep portions, which, however, are interconnected, are separated only slightly, and are functionally unitary. The **superficial cardiac plexus** lies in the concavity

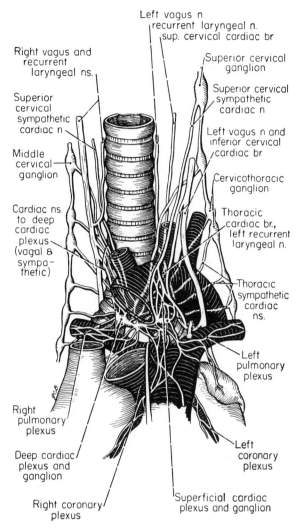

Right vagus and recurrent laryngeal ns.

Superior cervical sympathetic cardiac n

Middle cervical ganglion

Cardiac ns. to deep cardiac plexus (vagal & sympathetic)

Right pulmonary plexus

Deep cardiac plexus and ganglion

Right coronary plexus

Left vagus n.
recurrent laryngeal n.
sup. cervical cardiac br

Superior cervical ganglion

Superior cervical sympathetic cardiac n

Left vagus n and inferior cervical cardiac br

Cervicothoracic ganglion

Thoracic cardiac br., left recurrent laryngeal n.

Thoracic sympathetic cardiac ns.

Left pulmonary plexus

Left coronary plexus

Superficial cardiac plexus and ganglion

Fig. 286 The superficial and deep cardiac plexuses and the cardiac nerves.

of the aortic arch to the right of the ligamentum arteriosum. It usually contains a small parasympathetic ganglion. The superficial cardiac plexus receives the inferior cervical cardiac branch of the left vagus nerve (p. 164) and the superior cervical cardiac branch of the left sympathetic trunk (p.169). These nerves cross the aortic arch to the right of the ligamentum arteriosum, either separately or combined into a single strand, and turn under the aortic arch. The **deep cardiac plexus** is larger and lies behind the aortic arch in the areolar tissue over the tracheal bifurcation. It receives the superior cervical cardiac, middle cervical cardiac, and in-

ferior cervical cardiac nerves from the corresponding right cervical sympathetic ganglia and the middle and inferior cervical cardiac nerves of the left side (p.169). It is also joined by the superior and inferior cervical and thoracic cardiac branches of the right vagus and by superior cervical and thoracic branches of the left vagus nerve (p.164). From the upper four or five thoracic ganglia or from the thoracic sympathetic trunk, other visceral rami pass to the deep cardiac plexus. They are variable in number and size and approach the cardiac plexus in association with the intercostal vessels. Their size indicates that they make a considerable contribution to the cardiac plexuses.

The right half of the deep cardiac plexus gives branches to the right atrium, the right anterior pulmonary plexus, and the right coronary and left coronary plexuses. The left half of the deep plexus sends branches to the left atrium, the left anterior pulmonary plexus, and the left coronary plexus. The superficial cardiac plexus communicates with the left half of the deep plexus and contributes to the left anterior pulmonary plexus and the right coronary plexus. The **coronary plexuses** are right (or anterior) and left (or posterior). They accompany the coronary arteries and distribute into the territories served by the arteries.

The conduction system of the heart is richly innervated by these cardiac nerves—right vagal and right sympathetic branches ending chiefly in the region of the **sinuatrial node** and left vagal and left sympathetic branches terminating chiefly in the region of the atrioventricular node. The supply to the coronary arteries is also rich, vasoconstriction being produced by vagal stimulation; vasodilatation, by sympathetic. Visceral afferent fibers from the heart and coronary arteries are included in the sympathetic cardiac nerves, except for the superior cardiac branch, and in the vagal cardiac nerves as well. They are involved in the reflex arc concerned with slowing and acceleration of heart action. Heart rate and force of contraction appear to be controlled mainly through the inhibitory action of the vagus nerves; loss of vagal innervation results in an immediate acceleration of heart action. Visceral afferent fibers in the sympathetic cardiac nerves also mediate pain from the coronary arteries.

Clinical Note The pain of angina pectoris is frequently experienced in the tissues of the inner aspect of the left upper limb (as well as substernally). This ulnar border of the limb is supplied by the first and second thoracic nerves, and pain is 'referred' here because these spinal cord segments are also common to the visceral afferent terminations for the coronary arteries.

THE ORGANS OF RESPIRATION

RESPIRATION

External respiration is the interchange of gases (oxygen and carbon dioxide) between the air in the lungs and the blood in the capillaries of the pulmonary circuit. (**Internal respiration** is the gaseous interchange between the blood of the systemic capillaries and the cells of the tissues and organs of the body.) Through the action of the muscles of the chest, all diameters of the thorax are increased against the force of atmospheric pressure (p. 315). With this expansion, the pressures within the pleural cavity and within the spaces of the lung become less than that of the atmosphere, and, as a consequence, the lung is expanded and air rushes in. Expiration, in quiet respiration, is relatively passive and is a response to the elasticity of the lungs, the relaxation of the muscles of the chest, and atmospheric pressure acting on the thoracic wall. The quantity of air that can be exhaled by the deepest expiration after the deepest inspiration averages 3700 cc. for an adult male; this is the vital capacity. The amount of air breathed in and out during quiet respiration, the tidal air, is about 500 cc. The total epithelial surface area of the lungs available for respiration is about 70 square meters.

To facilitate the inflow and outflow of air, the air passages are kept open at all times by rigid or semirigid walls. This is the case with the nasal and oral cavities, the pharynx, the larynx, the trachea, and the bronchi. The lungs are soft, flexible, and elastic and respond passively to the changes in the diameters of the thorax. The lungs fill the pleural sacs most completely in areas of the chest where movement of bony parts is least; they have more room for expansion in areas where chest enlargement is not so limited. Thus in the apical and in the upper anterior mediastinal and posterior mediastinal portions, lung expansion is limited. It is much greater in lateral, anterior, and inferior directions. In full expansion, the lung roots move downward and forward, and this movement facilitates expansion in the apical and posteromedial regions. At the same time, the bronchi and pulmonary vessels elongate and increase in caliber (Macklin, '32). Respiration also takes place predominantly in that peripheral zone of the lung most immediately expanded by thoracic enlargement. In quiet respiration perhaps as little as 5 mm. in depth of

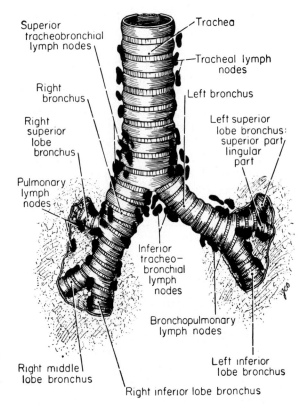

Fig. 287 The trachea and bronchi and associated lymphatic nodes.

the superficial zone is active. This will increase to 30 mm. in forced breathing.

TRACHEA (figs. 122–26, 152, 263, 272, 287)

A general description of the trachea and the special relations of its cervical portion has been given in the chapter on the Head and Neck (p. 191). The thoracic part of the **trachea** lies posteriorly in the superior mediastinum, where it is separated from the bodies of the upper four thoracic vertebrae by the esophagus. The trachea ends at the level of the sternal angle slightly to the right of the median line, being diverted to that side by the arch of the aorta. The respiratory shift of the bronchi and pulmonary vessels (Macklin, '32) carries the tracheal bifurcation as low as the sixth thoracic vertebra in full inspiration. Behind the lower half of the manubrium the aortic arch lies first in front of and then to the left of the trachea. The brachiocephalic trunk and the left common carotid artery arise from the arch anterior to the trachea and then diverge to either side of it as they ascend. In front of them are the left brachiocephalic vein and the remains of the thymus. On the right side the vagus nerve descends obliquely against the trachea, and on the left side the recurrent laryngeal nerve ascends between the trachea and the esophagus. Immediately above the bifurcation of the trachea, the deep cardiac plexus of nerves lies along its anterior and lateral surfaces. The trachea averages 1.7 cm. in diameter. It ends at the level of the sternal angle by dividing into the right and left bronchi. The last tracheal cartilage is broad, and its lower border is prolonged downward and backward in a hook-like process. This is the **carina,** which forms a keel-like projection on the inside between the origins of the right and left bronchi.

THE BRONCHI (figs. 263, 287, 290–94)

The two main bronchi pass to the roots of the lungs on either side. Like the trachea, they are maintained in a patent condition by C-shaped rings of cartilage in their walls. The **right bronchus** is shorter, straighter, and larger, and, being the more direct continuation of the trachea, foreign bodies are more apt to lodge in it. The right bronchus is about 2.5 cm. long and 1.3 cm. in diameter and enters the right lung at the level of the fifth thoracic vertebra. At the hilum of the lung the bronchus is superior and dorsal to the pulmonary artery, and the azygos vein arches over it. The **superior lobe bronchus,** its first branch, arises above the pulmonary artery and is frequently designated as the eparterial bronchus. A **middle lobe bronchus** and an **inferior lobe bronchus** complete the major divisions of the right bronchus. Each lobar bronchus further subdivides according to a pattern determined by the embryological budding of the bronchopulmonary primorida which results, in the adult, in the same segmental divisions of the bronchi and the lungs (see bronchopulmonary segments, p. 353).

The **left bronchus** is smaller in diameter (1.1 cm.) but is almost twice as long as the right bronchus (nearly 5 cm.). The subcarinal angle between the two main bronchi is 62°, the left bronchus diverging at a 37° angle, the right bronchus at 25°. The left bronchus passes under the aortic arch and proceeds lateralward anterior to the esophagus and the thoracic aorta. It is first superior to the pulmonary artery but divides into its lobar bronchi inferior to the artery; no eparterial bronchus occurs on this

side. The left **superior lobe bronchus** supplies the superior lobe, including the lingula; the left **inferior lobe bronchus** distributes to the inferior lobe. On the posterior surface of each main stem bronchus, the vagus nerve contributes to the posterior pulmonary plexus.

Blood Supply Bronchial arteries (fig. 296) are small and distribute along the posterior aspect of the bronchi. These arteries supply the lower trachea and the bronchi as far distal as the respiratory bronchioles. They also serve the tracheobronchial lymph nodes, a part of the esophagus, and some visceral pleura (Tobin & Zariquiey, '53). The **left bronchial arteries,** two in number, arise from the front of the thoracic aorta. The upper one arises opposite the fifth thoracic vertebra; the lower one, just below the left bronchus. The single **right bronchial artery** arises from the first right aortic intercostal or from the upper left bronchial artery. According to Latarjet ('54), the right bronchial artery and the first right aortic intercostal artery spring from a broncho-intercostal trunk from the aorta in about 80 per cent of cases. **Bronchial veins** drain the larger subdivisions of the bronchi and receive twigs from the tracheobronchial lymph nodes and other structures of the posterior mediastinum. The **right bronchial vein** enters the azygos vein as it arches over the root of the lung; the **left bronchial vein** empties into the accessory hemiazygos vein or into the left superior intercostal vein. That part of the blood from the bronchial arteries which reaches farthest along the bronchial tree is returned to the heart by the pulmonary veins.

Lymphatics (figs. 287, 294) The **tracheobronchial lymph nodes** form a continuous series along the trachea and along both the extrapulmonary and intrapulmonary bronchi; they total about 50 in number. Five groups can be recognized: **tracheal,** or **paratracheal,** on either side of the trachea; **superior tracheobronchial** in the angle between the trachea and the bronchi; **inferior tracheobronchial** in the angle between the two bronchi; **bronchopulmonary** at the hilum of each lung; and **pulmonary** along the larger bronchi in the substance of the lungs. This chain of nodes drains the lungs and the visceral pleura, the bronchi, the thoracic part of the trachea, the left side of the heart, part of the esophagus, and certain posterior mediastinal structures. The lymph flow through the chain is upward, and the efferents of the tracheobronchial nodes unite with those of the sternal and anterior mediastinal nodes to form the right and left **bronchomediastinal lymph trunks.** An exception to bilaterally symmetrical drainage occurs, however. The left inferior tracheobronchial lymphatic nodes send their efferent channels to join those of the right, rather than ascending on the left side (fig. 294). The right bronchomediastinal trunk may empty into the right lymphatic duct and the left trunk into the thoracic duct, but more commonly they open independently into the junction of the internal jugular and subclavian veins of their own side.

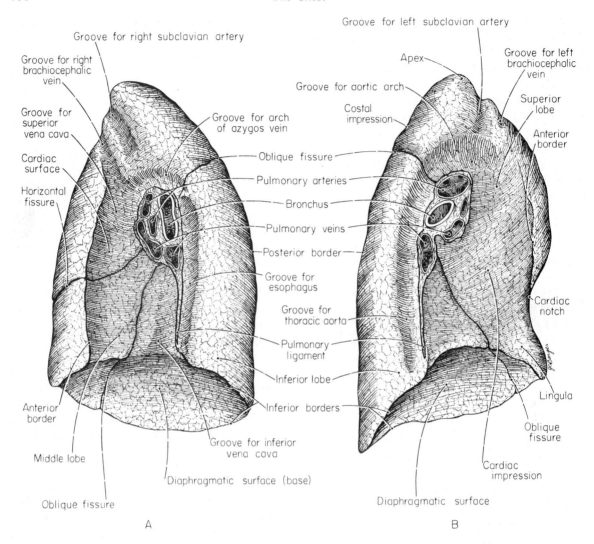

Fig. 288   The mediastinal surfaces of the right (A) and left (B) lungs.

THE LUNGS (figs. 263, 277, 280, 288–92)

The **lungs** are the essential organs of respiration. Here the air is brought into close relation with the blood of the pulmonary capillaries. In health, each lung lies free within its pleural cavity, attached only by its root and the pulmonary ligament. The adhesions so frequently seen in the cadaver between the visceral and parietal layers of the pleura are due to pleurisy. The right lung is the larger, but it is also shorter and wider. The liver, occupying the right side of the upper abdominal cavity, forces the dome of the diaphragm higher on that side, thus making for a shorter right lung. However, the heart occupies a greater proportion of the left side of the thorax, resulting in a smaller volume for the left lung. The

lung is light, spongy in texture, and highly elastic. The surface, covered by visceral pleura, is smooth and shining and is marked by numerous polygonal areas indicating the lobules of the organ. In color, the lung varies from pink to slate-gray or black, depending on the degree of impregnation of the lung and its interstitial areolar tissue by atmospheric dust particles.

Conforming to the outlines of the thoracic cage, each lung has an apex and a base; costal and mediastinal surfaces; and anterior, inferior, and posterior borders. The rounded **apex** reaches as high as the apex of the pleura, limited by the oblique plane of the first rib but rising as high as from 1.5 to 2.5 cm. above the sternal end of the clavicle. The **base** is broad and concave, and is molded by the arch of the diaphragm. The concavity of the right lung is deeper than that of the left. The sharp **inferior border** of the lung bounds the base and descends into the costodiaphragmatic recess of the parietal pleura in inspiration. The **costal surface** is large and convex in close adaptation to the curvatures of the ribs and the intercostal muscles. The **mediastinal surface** contains the root of the lung and, in front of this, the concave **cardiac impression.** This is larger and deeper on the left lung. Around the structures of the root of the lung, the reflection of parietal to visceral pleura takes place and, tapered inferiorly from the hilum, the prolongation of this reflection is the **pulmonary ligament.** In hardened specimens, a well-marked depression above and behind the hilum of the left lung is produced by the arch and the thoracic aorta. Running upward from the impression of the arch toward the apex of the lung is a groove formed by the subclavian artery, and just in front of this and indenting the anterior border is a groove for the left brachiocephalic vein. In front of the aortic furrow, near the base of the left lung, is a shallow depression for the lower part of the esophagus. Above the hilum of the right lung is the arched impression of the azygos vein connected to a groove running upward for the superior vena cava and the right brachiocephalic vein. Behind this vertical venous impression is a furrow for the subclavian artery. Behind the hilum of the right lung is an impression made by the esophagus, and, inferiorly, in front of the pulmonary ligament is the marking of the inferior vena cava. The posterior border of each lung is broad and rounded and rests in the deep concavity on either side of the vertebral column. It projects below into the costodiaphragmatic sinus. The anterior border of the lung is thin and sharp and overlaps the pericardium. On the left side the **cardiac notch** is a deep indentation of the anterior margin.

The Projection of the Lungs (fig. 289)

Since the lungs engage in a continuous cycle of expansion and contraction, their surface projection is not constant. However, it is useful to recognize a surface projection of their borders in quiet respiration from which their limits may be estimated in other conditions. Such a surface projection can be defined in the manner used for the pleurae. The apex and the upper anterior borders of the lungs conform closely to the pleural borders, descending obliquely behind the sternoclavicular joints and approximating the median plane at the sternal angle. Between the levels of the second and fourth costal cartilages, the anterior margins of the two lungs are on or nearly on the median line. The anterior margins of the two lungs differ below the fourth costal cartilage. That of the right side diverges gradually from the median line and leaves the sternum behind the sixth costal cartilage. The inferior margin is behind the sixth rib at the midclavicular line, the eighth rib at the axillary line, and the tenth rib at the scapular line. It ends posteriorly opposite the eleventh thoracic vertebra, continuing into the posterior border which ascends to the apex. The projection of the left lung is essentially the same except for its deep cardiac notch. This begins at the level of the fourth costal cartilage, the anterior margin of the left lung passing almost horizontally outward along that cartilage to about the parasternal line (midway between the sternal

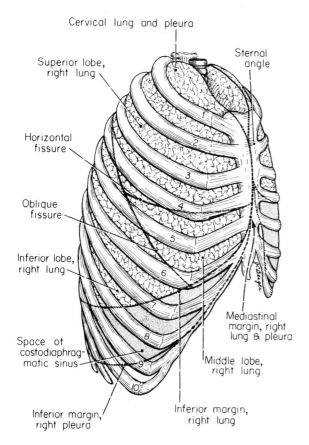

Fig. 289 The projection to the thoracic cage of the borders of the right pleura and of the lobes and fissures of the right lung.

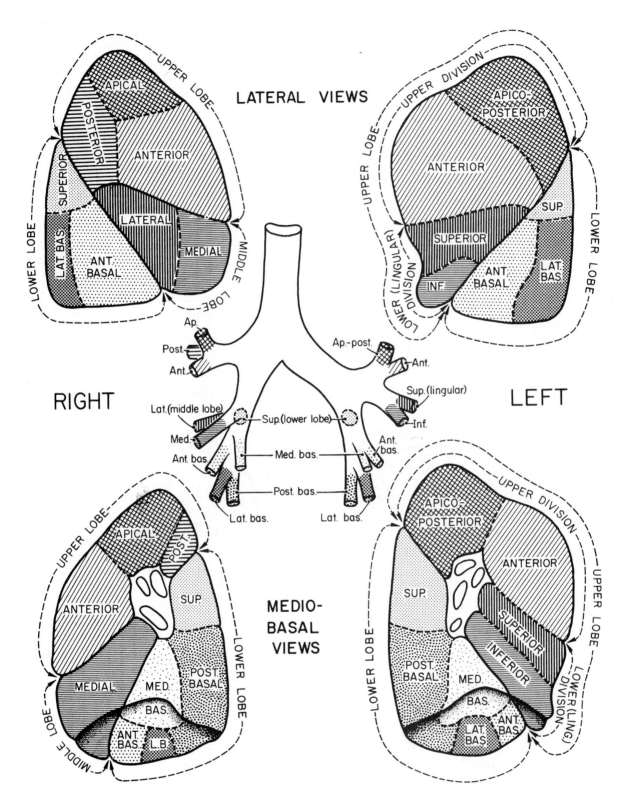

LATERAL VIEWS

RIGHT

LEFT

MEDIO-
BASAL
VIEWS

Fig. 290 Tracheobronchial branching correlated with the segmental subdivision of the lungs (from J. F. Huber, *J. Nat. Med. Assn.* 41:49, 1949).

border and the midclavicular line). Then descending and curving somewhat medialward again; the anterior margin of the left lung ends behind the sixth costal cartilage just medial to its chondrocostal junction.

Lobes and Fissures (figs. 288–90)

Each lung is cut diagonally by an **oblique** (interlobar) **fissure** which extends from the surface to the root of the lung. The **projection of the oblique fissure** is essentially the same for each lung. Beginning at the base of the spine of the scapula (opposite the spine of the third thoracic vertebra), its line runs obliquely downward and forward, crossing the fifth rib in the axillary line, and ends anteriorly behind the sixth chondrocostal junction (about 5 cm. from the border of the sternum). The oblique fissure separates a **superior lobe,** which is both superior and anterior, from an **inferior lobe,** which is inferior and posterior. A further subdivision of the superior lobe of the right lung is made by a

horizontal (accessory) **fissure.** It begins in the posterior axillary line under the fifth rib, immediately crosses the fourth intercostal space, and thereafter follows the fourth rib as far as the sternal margin. The horizontal fissure and the inferior part of the oblique fissure form the boundaries of the **middle lobe** of the right lung, a small, wedge-shaped segment located anteriorly. There is no horizontal fissure and no middle lobe in the left lung, although, by bronchial similarities, the **lingula** of the left lung is its homologue.

Bronchopulmonary Segments (figs. 290–92)

Small (lobular) units of the lung consist of a centrally placed lobular bronchus and a branch of the pulmonary artery surrounded by lung tissue, the whole enclosed by scanty areolar tissue, in the meshes of which arise the peripherally placed veins of the lobule. This fundamental arrangement follows the embryologic development of the bronchus and its associated pulmonary tissue. In

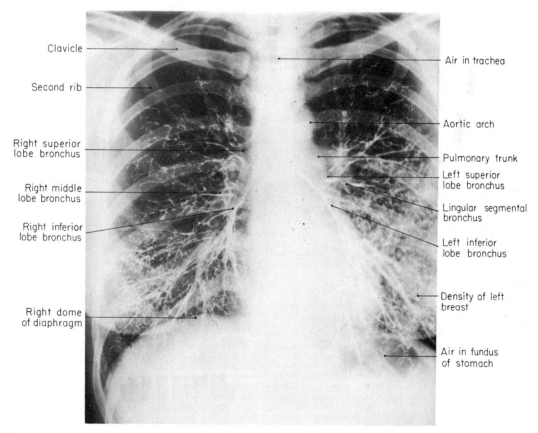

Fig. 291   A normal bronchogram in P-A projection. The bronchi are visualized with the aid of iodized oil.

practice, larger units, based on the direct branches of the lobar bronchi but having the same type of organization as the lobules, are recognized and named. These are the **bronchopulmonary segments.** Their delicate connective tissue boundaries can be demonstrated by dissection or inflation, and the occasional extra lung fissures usually follow their planes of separation.

In the right lung a **superior lobe bronchus** arises from the right bronchus about 2.5 cm. from the tracheal bifurcation. After a short course it divides into **segmental bronchi—apical, posterior,** and **anterior.** These give their names to the pulmonary tissue to which they distribute, and thus the superior lobe is divided into **apical, posterior,** and **anterior bronchopulmonary segments.** Five centimeters from the carina the right bronchus gives off the **middle lobe bronchus,** the segmental bronchi and corresponding pulmonary segments of which are **lateral** and **medial.** The remainder of the right bronchus is the **inferior lobe bronchus.** This bronchus gives rise to a **superior segmental bronchus** opposite or just below the origin of the middle lobe bronchus which serves the **superior bronchopulmonary segment of the inferior lobe.** The inferior lobe bronchus then continues its course toward the base of the lung, and its segmental bronchi and their pulmonary segments are all 'basal' in designation—**medial basal, anterior basal, lateral basal,** and **posterior basal.**

There is essential similarity on the left side. The **superior lobe bronchus** of the left lung arises from 4 to 5 cm. beyond the tracheal bifurcation. This lobar bronchus exhibits a further division into a superior part and an inferior, or lingular, division. From its superior part arise segmental bronchi essentially like those of the right superior lobe but numbering two rather than three. They are the **apicoposterior segmental bronchus** and the **anterior segmental bronchus.** From the lingular division of the superior lobe bronchus, **superior segmental** and **inferior segmental** bronchi arise. The lingular division of the left superior lobe bronchus is regarded as the homologue of the middle lobe bronchus of the right side and the lingula as representing a third part of the left lung comparable to the middle lobe of the right lung. As on the right, the left bronchus continues as the inferior lobe bronchus, and about 1.5 cm. farther along its course it gives rise to a **superior segmental bronchus** which serves a **superior bronchopulmonary segment** at the apical part of the inferior lobe. The 'basal' segmental bronchi and the bronchopulmonary segments of the left inferior lobe are like those of the right. The segmental bronchi and the segments of the lobe are, therefore, the **medial basal, anterior basal, lateral basal,** and **posterior basal.** The orientation of the lung segments in both lateral and mediobasal views is illustrated in fig. 290. For this scheme, for the terminology employed, and for emphasis on the basic concept we are indebted to Jackson and Huber ('43). More detailed analyses of branching and variation have been reported by Boyden ('45 and later).

Bronchial Divisions

Beyond the direct branches of the lobar bronchi which give their names to the bronchopulmonary segments are from twenty to twenty-five generations of branches, ultimately ending in terminal bronchioles. The first ten generations are **secondary bronchi.** They are widely spaced and fill two-thirds to three-fourths of the distance from the hilum of the lung to the periphery. They are lined by ciliated, pseudo-stratified epithelium containing goblet cells and have cartilage plates, mucous glands, and smooth muscle. The secondary bronchi are reached by radicles of bronchial arteries and veins. The remaining ten to fifteen generations of branches are **bronchioles.** These are more closely spaced, have ciliated columnar epithelium and goblet cells, lack cartilage in their walls, and have diameters ranging down from 3 to 1 mm. **Terminal bronchioles** are about 0.6 to 0.7 mm. in diameter, are lined by ciliated cuboidal epithelium, and are completely invested by smooth muscle. Each of these gives rise to several generations of **respiratory bronchioles** exhibiting scattered alveoli. Each respiratory bronchiole provides 2 to 11 **alveolar ducts** and each of these gives rise to 5 or 6 **alveolar sacs** lined by polyhedral **alveoli.** There are from 300 to 400 million alveoli, the functional lung unit, in each lung.

The Root of the Lung (fig. 288)

This is formed by the bronchus, the pulmonary artery, the pulmonary veins, the bronchial arteries and veins, the pulmonary plexuses of nerves, and the lymphatic vessels and nodes. These are held together by mediastinal connective tissue and surrounded by the reflection of the pleura. The superior vena cava and the right atrium lie in front of the root of the right lung, and the azygos vein arches above it. The root of the left lung is ventral to the thoracic aorta and inferior to its arch. Ventral to both roots are the phrenic nerves, the pericardiacophrenic vessels, and the anterior pulmonary plexuses of nerves. Dorsal to the roots lie the vagus nerves and the posterior pulmonary plexuses. There is a general correspondence on both sides in the position of entrance of the major structures into the lungs. The superior pulmonary veins are ventral, the bronchi are dorsal, and the pulmonary arteries are between them. The superior-inferior relations

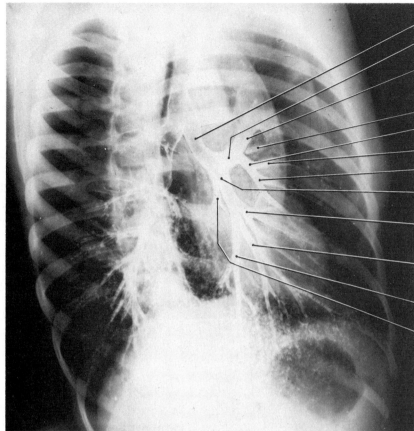

L. bronchus
Sup. lobe br.
apicopost. segm. br.
ant. segm. br.
Lingular br.
sup. segm. br.
inf. segm. br.
Inf. lobe br.
ant. med. basal segm. br.
lat. basal segm. br.
post. basal segm. br.
sup. segm. br.

A

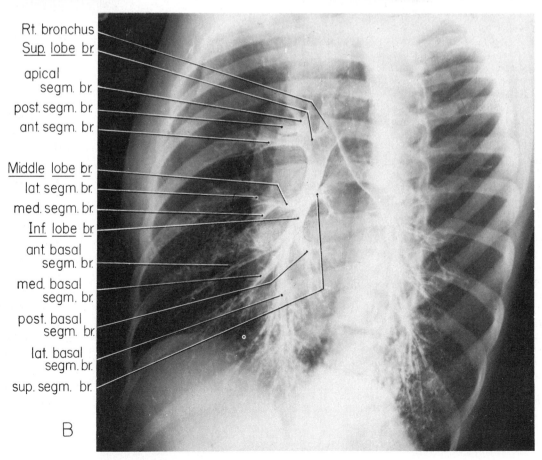

Rt. bronchus
Sup. lobe br.
apical segm. br.
post. segm. br.
ant. segm. br.

Middle lobe br.
lat. segm. br.
med. segm. br.
Inf. lobe br.
ant. basal segm. br.
med. basal segm. br.
post. basal segm. br.
lat. basal segm. br.
sup. segm. br.

B

Fig. 292 Normal bronchograms in oblique projections. A—right anterior oblique projection for favorable demonstration of the left bronchial tree. B—left anterior oblique projection for the right bronchial pattern.

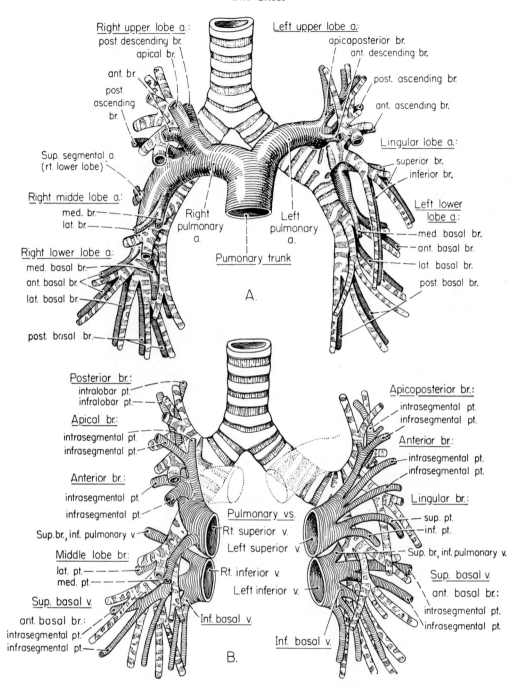

Fig. 293 A. The pulmonary arteries and their branches. Redrawn after Boyden in Luisada, A. A.; 1961, *Development and Structure of the Cardiovascular System.* Blakiston Div., McGraw-Hill Book Co. B. The pulmonary veins.

differ on the two sides, the eparterial superior lobe bronchus being most superior on the right side and the pulmonary artery most superior on the left side. The veins are generally inferior, and the inferior pulmonary vein is quite dorsal.

The Pulmonary Arterial Trunk This divides under the aortic arch into right and left branches (fig.

263). The **right pulmonary artery** is long and larger than the left and runs horizontally to the right behind the ascending aorta and superior vena cava and in front of the right bronchus to the root of the lung. Below and anterior to the superior lobe bronchus, the right pulmonary artery provides a trunk to the superior lobe of the right lung. The artery then arches across the main bronchus and follows its ramifications in a general posterolateral relation to the segmental bronchi of the middle and inferior lobes. Figure 293 locates and identifies the named branches. The **left pulmonary artery** passes horizontally anterior to the descending aorta and left bronchus to the root of the left lung. It arches over the left bronchus to attain a postero-lateral relation to the main and segmental bronchi. Its relation to the ligamentum arteriosum has been described (p. 331), as has also its relation to the fold of the left vena cava (p. 327). The branches of the left pulmonary artery follow and corres-pond with the segmental bronchi as represented in fig. 293. The **bronchial arteries** are described on p. 349.

The Pulmonary Veins Beginning in the pulmonary capillaries, the veins of the lungs unite into larger and larger channels. They are peripherally placed in relation to the lung lobule and maintain the same relationship to the bronchopulmonary segments. Thus, the veins run in the intersegmental connective tissue, and many drain adjacent segments. A single vein for each lobe is formed, and then the vein of

the middle lobe of the right lung usually joins that of the superior lobe, resulting in two pulmonary veins at the hilum of each lung. Here the **superior pulmonary vein** lies in front of and a little below the pulmonary artery, whereas the **inferior pulmonary vein** is at the lowest part of the hilum and dorsal to the superior vein. The right pulmonary veins pass behind the right atrium and the superior vena cava; the left veins pass in front of the thoracic aorta. The pulmonary veins end in the left atrium of the heart; they are without valves. Their tributary patterns are illustrated and specific names are supplied in figs. 263 and 293. The bronchial veins are described on p. 349.

The Lymphatics of the Lung (fig. 294) The lymphatic vessels of the lung originate in the super-ficial and deep plexuses. The **superficial plexus** lies under the visceral pleura, and its efferents follow the surfaces of the lobes and turn around their margins and borders to the bronchopulmonary nodes (fig. 287) situated at the hilum of the lung. The **deep plexus** arises in the submucosa of the bronchi and in the peribronchial connective tissue. There are no lymphatic vessels in the walls of the

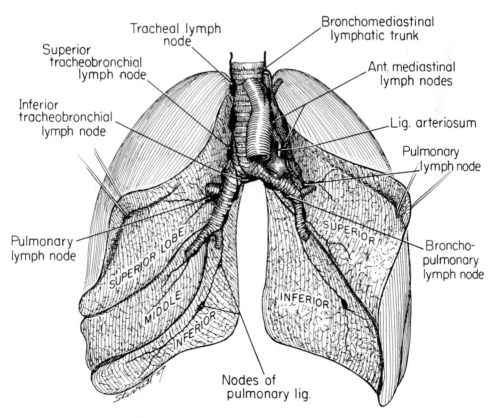

Fig. 294 The lymphatic drainage of the lungs.

pulmonary alveoli. The efferent vessels of the deep plexus follow the bronchi and the pulmonary arteries and veins to the hilum of the lung, ending in the **pulmonary** and **bronchopulmonary nodes** situated just inside and outside the lung substance. The superficial and deep networks communicate, have relatively few valves, and converge on the nodes at the hilum of the lung. The more central drainage of the lung is upward through the tracheobronchial lymph chain and to the blood stream by way of the mediastinal lymph trunk. There is also partial drainage of the superior lobe of the left lung to the left anterior mediastinal nodes. The left inferior tracheobronchial lymphatic nodes drain to the right side (fig. 294), and thus the inferior lobe of the left lung has a lymphatic drainage into the right bronchomediastinal lymph trunk.

The Pulmonary Nerve Plexuses (fig. 286) Afferent and efferent fibers of the vagus and sympathetic nerves supply the lung by way of the **anterior pulmonary** and **posterior pulmonary plexuses.** The sympathetic preganglionic fibers arise in the upper three or four thoracic spinal cord segments. Postganglionic sympathetic fibers reach the pulmonary plexuses by way of the cardiac plexus and from the upper thoracic ganglia directly. The **vagus nerves** are flattened out over the posterior aspect of the roots of the lungs and intermingle with the sympathetics to form the **posterior pulmonary plexuses.** Several small branches of the vagus on the anterior surface of the lung roots join with sympathetic filaments to form the **anterior pulmonary plexuses.** The anterior and posterior plexuses, located ventral and dorsal to the roots of the lungs, intercommunicate and are connected with the cardiac plexus. The posterior plexus is also continuous with the aortic and esophageal plexuses. The fibers of the anterior pulmonary and posterior pulmonary plexuses distribute to the bronchi and the pulmonary blood vessels.

On the bronchial tree the vagal innervation provides constrictor fibers to the bronchial musculature and secretory fibers to its mucous glands. Stimulation of the sympathetic innervation results in dilatation of the bronchi and reduction in glandular activity. On the pulmonary vessels the vagal innervation is vasodilator; the sympathetic innervation, vasoconstrictor, though the evidence here is not conclusive, and the control of the pulmonary vessels is thought to follow that of the systemic circulation. Afferent fibers in the vagus are

of two types. Fibers ending in neuromuscular terminations along the bronchial tree react to the stretching of the lung during inspiration; fibers from free nerve endings in the bronchial epithelium are concerned in the cough reflex and in transmission of pain.

THE POSTERIOR MEDIASTINAL STRUCTURES

THE ESOPHAGUS (figs. 277, 280, 295, 296)

The general description of the **esophagus** is included under the consideration of its cervical portion (p. 197). Lying slightly to the left of the median line at the base of the neck, it is returned to a midline position by the arch of the aorta. Below the arch, however, the esophagus inclines again to the left, passing through the diaphragm distinctly to the left of the median plane. The esophagus is narrowest at its beginning. It is also somewhat compressed as it passes behind the left bronchus. A dilatation just above the diaphragm is followed by a considerable constriction at its passage through the diaphragm. The position of the esophagus in the superior mediastinum is similar to that in the neck; it lies behind the trachea and in front of the vertebral column and prevertebral musculature. Passing behind and to the right of the arch of the aorta, it descends in the posterior mediastinum along the right side of the thoracic aorta. As the esophagus continues toward the left and as the thoracic aorta attains more nearly the median line over the vertebral bodies, the esophagus passes onto the front of the aorta at about the eighth thoracic vertebral level. Below the left bronchus the esophagus is in contact with the posterior aspect of the pericardium. It lies to the left of and partially overlaps the azygos vein, and the thoracic duct ascends between it and the azygos vein. The esophagus crosses the right posterior intercostal arteries in its descent.

At its passage through the diaphragm the esophagus is attached to the margins of the opening. This **phrenicoesophageal ligament** is an extension of inferior diaphragmatic fascia upward through the esophageal hiatus of the diaphragm for several centimeters and also downward toward the cardioesophageal junction. It allows independent movement of the diaphragm and of the esophagus during respiration and swallowing and it prevents herniation at the hiatus. The abdominal part of the

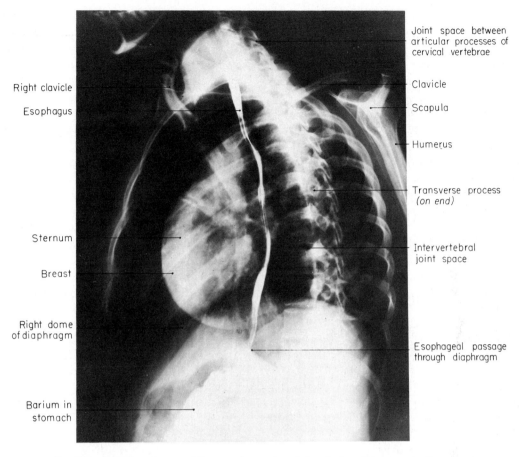

Right clavicle

Esophagus

Sternum

Breast

Right dome
of diaphragm

Barium in
stomach

Joint space between
articular processes of
cervical vertebrae

Clavicle

Scapula

Humerus

Transverse process
(on end)

Intervertebral
joint space

Esophageal passage
through diaphragm

Fig. 295   A barium image of the esophagus in a left anterior oblique projection.

esophagus is short and funnel-shaped. It grooves the left lobe of the liver and is covered by peritoneum only in front and on its left side. It ends in the stomach, after a course of about 1.25 cm., at the level of the eleventh thoracic vertebral body.

The Vessels and Nerves

In the cervical region an **arterial supply** to the esophagus is provided by the inferior thyroid artery. In the chest it receives esophageal branches from the bronchial arteries and from the thoracic aorta. Its lower portion is nourished from abdominal vessels—esophageal branches of the left gastric and left inferior phrenic arteries (Swigart, Siekert, Hambley, & Anson, '50). The **veins** of the esophagus are tributary to a number of systems. They drain into the inferior thyroid veins in the neck, into the azygos and hemiazygos veins in the chest, and into the left gastric (coronary) vein in the abdomen. The **lymphatics** of the esophagus arise mainly in

the mucosa. They drain regionally into the inferior deep cervical lymph nodes, the paratracheal nodes, the posterior mediastinal nodes, and the superior gastric nodes.

Clinical Note   Through the esophageal veins are established connections between the portal vein below and the azygos and superior caval systems above; this can be an important collateral drainage route for the portal blood. Being poorly supported in the submucosa and adventitia of the esophagus, these veins, when serving as a portal collateral, may become quite varicose, and, in the extreme case, may hemorrhage with serious consequences.

The **esophageal plexus** (fig. 129) of nerves is formed by the right vagus and left vagus nerves. Leaving the posterior pulmonary plexus, the right vagus passes principally onto the back of the esophagus but also has a branch which runs onto its anterior surface. Similarly, the left vagus below the root of the lung passes principally onto the front of the esophagus but has a branch which runs pos-

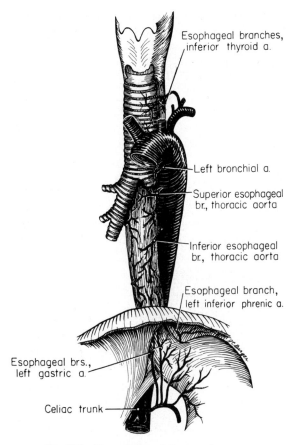

Esophageal branches,
inferior thyroid a.

Left bronchial a.

Superior esophageal
br., thoracic aorta

Inferior esophageal
br., thoracic aorta

Esophageal branch,
left inferior phrenic a.

Esophageal brs.,
left gastric a.

Celiac trunk

Fig. 296   The arteries of the esophagus.

teriorly to join the stem of the right vagus. All these branches redivide and form a network of nerves which enclose and infiltrate the muscular wall of the esophagus. The plexus also receives fine branches from the greater thoracic splanchnic nerve and from the thoracic sympathetic trunk or ganglia between the sixth and tenth thoracic levels. Just above the passage of the esophagus through the diaphragm, the esophageal plexus is again resolved into two vagal trunks. These enter the abdomen anterior and posterior to the esophagus and toward its right side. No longer representative of left vagus and right vagus nerves, they are now designated as the **anterior vagal** and **posterior vagal trunks.** The striated musculature of the upper one-third of the esophagus is supplied by the recurrent laryngeal nerves, and its glands are innervated by the same nerve and by branches from the cervical sympathetic trunk. The esophageal plexus, with its vagal and sympathetic contributions, constitutes the supply to the smooth muscle and glands of the lower two-thirds of the esophagus.

THE THORACIC AORTA (figs. 277, 280, 296, 300)

The **thoracic aorta** is the continuation of the aortic arch at the lower border of the fourth thoracic vertebra. Here it lies to the left of the vertebral column and has a diameter usually somewhat in excess of 2 cm. As the thoracic aorta descends, it approaches the median plane and terminates in front of the vertebral column by passing from the chest through the aortic hiatus of the diaphragm opposite the lower border of the body of the twelfth thoracic vertebra. Anterior to it, from above downward, lie the root of the left lung, the pericardium, the esophagus, and the diaphragm. Behind, it is crossed by the hemiazygos vein, the accessory hemiazygos vein, and the posterior intercostal veins. The thoracic duct ascends along the right side of the thoracic aorta, between it and the azygos vein. The esophagus is to the right of the aorta at middle thoracic levels but is in front of it below. The branches of the thoracic aorta, all small, are the following:

Pericardial	Mediastinal
Bronchial	Posterior Intercostal
Esophageal	and Subcostal
	Superior Phrenic

The **pericardial branches** are a few small vessels to the posterior aspect of the pericardium.

The **bronchial arteries** (fig. 296) consist of one right and two left bronchial arteries; only the left bronchial arteries arise from the aorta. These vessels are described with the bronchi (p. 349).

Several **esophageal arteries** arise from the front of the aorta; in 90 per cent of cases there is a superior smaller and an inferior larger vessel (Shapiro & Robillard, '50). These descend obliquely to the esophagus, where they communicate with each other and make up an anastomotic chain of vessels with esophageal branches of the inferior thyroid and bronchial arteries above and with those of the left gastric and left inferior phrenic arteries below.

The **mediastinal branches** are small and supply the lymph nodes and other tissues of the posterior mediastinum.

Nine pairs of **posterior intercostal arteries** and one pair of **subcostal arteries** arise from the back of the aorta. Their distribution to the intercostal and subcostal tissues is described with the chest wall (p. 317).

The **superior phrenic branches** are small and arise from the lowermost part of the thoracic aorta. They ramify on the posterior superior surface of the diaphragm, where they anastomose with the musculophrenic and pericardiacophrenic branches of the internal thoracic artery.

THE AZYGOS SYSTEM OF VEINS (Figs. 297–300, 331)

The Azygos Vein

The azygos vein is usually formed at the level of the diaphragm by the junction of the ascending

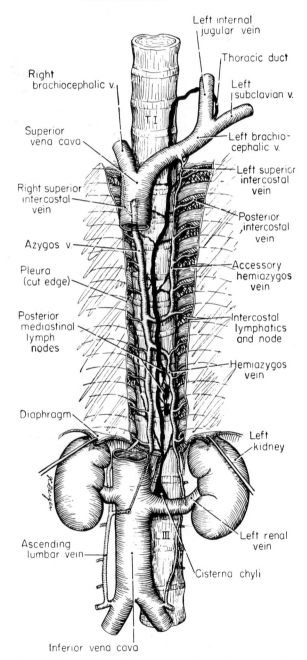

Fig. 297 Structures of the posterior mediastinum; the azygos system of veins and the thoracic duct.

lumbar vein (p. 464) and the right subcostal vein (lateral root of azygos vein, 94 per cent occurrence). However, the azygos vein is frequently in continuity with or added to by a posterosuperiorly directed connection of the inferior vena cava at the renal level (intermediate root of azygos vein). This origin or communication pierces the diaphragm lateral to the right crus, represents a persistence of the continuity of the supracardinal vein of the embryo, and occurs in about 45 per cent of specimens. A medial root of the azygos vein is prevertebral in location and passes through the aortic hiatus of the diaphragm. It may represent the coalescence of prevertebral veins or communicating connections with the inferior vena cava or upper lumbar veins and is found in about 30 per cent of cases. In the thorax, the azygos vein ascends along the right side of the vertebral column, overlying the posterior intercostal arteries and to the right of the aorta. The thoracic duct lies against the vertebral column between the azygos vein and the aorta. At the level of the fourth thoracic vertebra, the azygos vein arches anteriorly over the root of the right lung to end in the superior vena cava just before that vessel pierces the pericardium.

The azygos vein receives the right posterior intercostal veins as high as the fifth intercostal space directly and those of the second to the fourth intercostal spaces by way of the right superior intercostal vein which enters it as it arches over the root of the lung (p. 318). It usually receives the hemiazygos and accessory hemiazygos veins when these are present. The azygos vein receives the drainage of the esophageal veins, the mediastinal and pericardial veins, and the right bronchial vein. It frequently has some imperfect valves. When the azygos vein is a truly unpaired longitudinal thoracic vein, no hemiazygos or accessory hemiazygos veins occur, and the posterior intercostal veins of the left side then cross the vertebral column to enter it.

The Hemiazygos Vein

The hemiazygos vein arises, like the azygos vein, in the junction of the left subcostal and ascending lumbar veins. This lateral root is supplemented or substituted for by intermediate or medial roots in the same pattern of formation as for the azygos vein (see above) except that a posterosuperiorly directed connection of the left renal vein rather than of the inferior vena cava forms the intermediate root, and the medial root is rarer (10 per cent) than for the azygos system. The hemiazygos vein ascends on the left side of the vertebral bodies behind the thoracic aorta as far as the ninth thoracic vertebra. At this level the hemiazygos vein passes to the right across the vertebral column behind the aorta, the esophagus, and the thoracic duct and ends in the azygos vein. It receives the lower four or five posterior intercostal veins (p.318), the lower esophageal veins, the small left mediastinal veins, and sometimes the accessory hemiazygos vein. The hemiazygos vein may be lacking.

The Accessory Hemiazygos Vein

This vein occurs occasionally as a longitudinal venous channel in the upper portion of the chest on the left side. Interposed between the upper end of the hemiazygos vein and the left superior intercostal vein, it receives the posterior intercostal veins of the fourth to the seventh or eighth intercostal spaces (p.318). It descends superficial to the left posterior intercostal arteries, turns to the right across the eighth thoracic vertebra, and, passing behind the aorta, the thoracic duct, and the esophagus, empties into the azygos vein. It also receives the left bronchial veins and the small mediastinal veins. The upper part of the accessory hemiazygos vein is frequently connected with the left superior

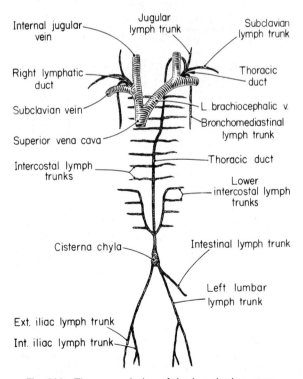

Fig. 298　The general plan of the lymphatic system.

intercostal vein, and its inferior terminus may be the hemiazygos vein.

THE THORACIC DUCT (figs. 272, 277, 280, 297, 298, 331)

The **thoracic duct** is the largest lymphatic channel of the body. It returns to the blood stream the lymph and chyle from all of the body below the diaphragm and from the left half of the body above the diaphragm. It arises in an occasional dilatation, the **cisterna chyli,** which lies on the surface of the second lumbar vertebra in the deep crevice formed by the aorta and the right crus of the diaphragm. The cisterna chyli (p.453), or the beginning of the thoracic duct, receives the **right lumbar** and **left lumbar trunks** which transmit the lymph from the lower limbs and from the wall and certain viscera of the abdomen and the pelvis. The **intestinal trunk,** which collects chyle and lymph from the digestive organs, is most commonly a tributary of the left lumbar trunk (70 per cent); only occasionally (25 per cent) does it enter the cisterna chyli directly.

The thoracic duct passes through the aortic hiatus of the diaphragm, lying between the aorta and the azygos vein, and ascends in this relationship through the posterior mediastinum. In its ascent, it lies against the vertebral column, the right posterior intercostal arteries, and the terminal portions of the hemiazygos and accessory hemiazygos veins as they pass across the vertebral column to the azygos vein. Approximately opposite the fifth thoracic vertebra (level of sternal angle), the thoracic duct inclines to the left side of the median plane and ascends behind the aortic arch and on the left side of the esophagus to the superior aperture of the thorax. In the base of the neck it passes lateralward behind the carotid sheath and in front of the vertebral artery and vein, where it is anterior to the prevertebral fascia and those structures which this fascia covers. Then, arching downward and anteriorly across the subclavian artery, the thoracic duct ends in the junction of the left subclavian and internal jugular veins, a constant valve being present in its last centimeter (Van Pernis, '49). This usual description of the thoracic duct will apply to two-thirds of specimens (Davis, '14). The arch of the thoracic duct may rise three or four centimeters above the termination of the duct. The thoracic duct is cream colored, irregular in its contours, and varies between 36 and 45 cm. in length. It may be plexiform in part of its course,

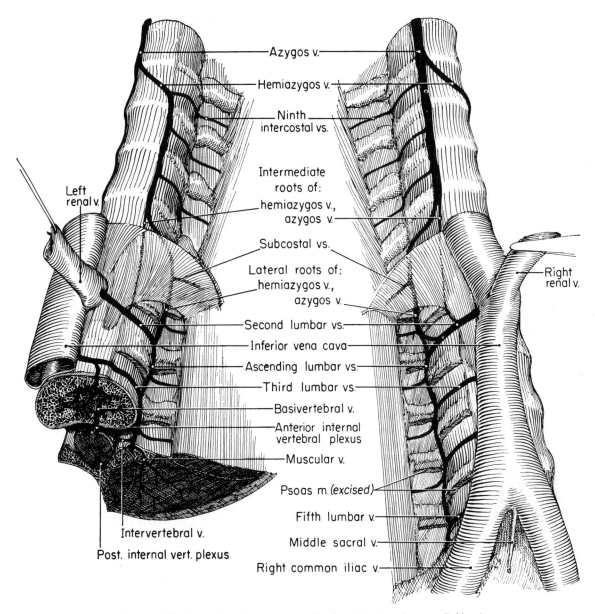

Fig. 299   The formation of the azygos (right) and the hemiazygos (left) veins.

and it occasionally divides, at the level where it usually crosses the vertebral column, to form a right and a left duct. Less commonly (5 per cent), it empties entirely on the right side, and a left terminus is then lacking.

The thoracic duct receives, near its beginning, a descending lymph trunk from the posterior intercostal nodes of the lower five intercostal spaces (p. 320). Higher in the thorax, the efferents of the posterior mediastinal nodes and of the posterior intercostal lymph nodes of the third to the sixth spaces enter it. At the base of the neck the thoracic duct usually receives the left jugular lymph trunk. It may also receive the left subclavian trunk, but that, like the left bronchomediastinal lymph trunk, usually has a separate opening into the subclavian vein adjacent to the venous confluence of this and the internal jugular vein.

Clinical Note   It is seldom realized how great a flow of lymph is accommodated by the thoracic duct. Accidental rupture or surgical separation of this channel allows spilling into the thoracic cage at rates ranging from 75 to 200 cc. per hour with resulting pressure on the lungs and heart (chylothorax).

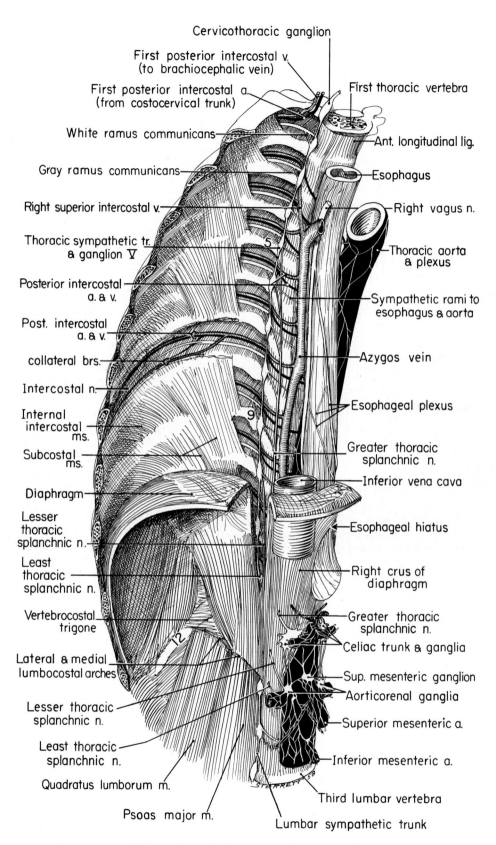

Cervicothoracic ganglion

First posterior intercostal v.
(to brachiocephalic vein)

First posterior intercostal a.
(from costocervical trunk)

White ramus communicans

Gray ramus communicans

Right superior intercostal v.

Thoracic sympathetic tr.
& ganglion V

Posterior intercostal
a. & v.

Post. intercostal
a. & v.

collateral brs.

Intercostal n.

Internal
intercostal
ms.

Subcostal
ms.

Diaphragm

Lesser
thoracic
splanchnic n.

Least
thoracic
splanchnic n.

Vertebrocostal
trigone

Lateral & medial
lumbocostal arches

Lesser thoracic
splanchnic n.

Least thoracic
splanchnic n.

Quadratus lumborum m.

Psoas major m.

First thoracic vertebra

Ant. longitudinal lig.

Esophagus

Right vagus n.

Thoracic aorta
& plexus

Sympathetic rami to
esophagus & aorta

Azygos vein

Esophageal plexus

Greater thoracic
splanchnic n.

Inferior vena cava

Esophageal hiatus

Right crus of
diaphragm

Greater thoracic
splanchnic n.

Celiac trunk & ganglia

Sup. mesenteric ganglion

Aorticorenal ganglia

Superior mesenteric a.

Inferior mesenteric a.

Third lumbar vertebra

Lumbar sympathetic trunk

Fig. 300 Structures of the posterior mediastinum; the thoracic sympathetic trunk and ganglia, the splanchnic nerves, the azygos vein, and the posterior intercostal vessels and intercostal nerves.

The Lymphatics of the Chest

The **parietal lymphatics** of the chest are described on p. 320. Of the visceral lymph node groups, the **tracheobronchial nodes** have been considered (p. 349) together with the lymphatic drainage of the pleura (p. 323), the heart (p. 338), the lungs (p. 357), and the esophagus (p. 359). There remain for consideration the anterior and posterior mediastinal lymph nodes.

The **anterior mediastinal nodes** (figs. 261, 294) are from two to five small nodes located, on the right, in the region of the superior vena cava and the right brachiocephalic vein and, on the left, along the left brachiocephalic vein and in front of the arch of the aorta. Of this latter group, the lowest node is constant and is situated anterior to the ligamentum arteriosum. It is frequently bound by connective tissue to the beginning of the left recurrent laryngeal nerve. The anterior mediastinal nodes receive the drainage of the thymus, the upper part of the pericardium and pleura, and the right side of the heart (p. 338) and partial drainage of the superior lobe of the left lung (p. 349). The efferents of the anterior mediastinal nodes join with those of the sternal and tracheobronchial node groups to form the bronchomediastinal lymph trunk (p. 349).

The **posterior mediastinal nodes** (figs. 297) lie behind the pericardium in relation to the esophagus and the thoracic aorta. They consist of 4 or 5 lymphatic nodes posterior to the lower portion of the esophagus and of 3 to 8 anterior and lateral to it. Other nodes lie along the thoracic aorta at levels below the lungs. The posterior mediastinal nodes receive lymph from the esophagus, the back of the pericardium, the back of the diaphragm, and the middle posterior intercostal spaces. Their efferents end mainly in the thoracic duct; some unite with the tracheobronchial nodes.

The Thoracic Sympathetic Trunk (figs. 277, 280, 300)

The **thoracic sympathetic trunk** and its associated ganglia represent a major part of the sympathetic (thoracolumbar) portion of the autonomic nervous system. The preganglionic neurons constituting the origin of the sympathetic innervation arise in the lateral horn gray matter of spinal cord segments thoracic one to lumbar two or three. These myelinated preganglionic neurons enter the sympathetic

trunk by way of the **white rami communicantes** which are present at each thoracic nerve level. Such neurons may terminate at the level of entrance into the trunk in the chain ganglion of that level, or they may ascend or descend in the trunk to a level more appropriate for distribution to the structure innervated. *The sympathetic trunk is, in fact, a structural response to the necessity for such fibers to reach levels either cephalic or caudal to their origin.* The **thoracic chain ganglia** are regions of synapse between preganglionic and postganglionic neurons concerned in the innervation of visceral tissues at predominantly thoracic levels. The **gray rami communicantes** of the thoracic nerves bring into the spinal nerves, for peripheral distribution, the postganglionic fibers concerned with smooth muscle and glands in the region of distribution of these nerves.

The **thoracic sympathetic trunk** is in continuity above with the cervical sympathetic trunk and, piercing the diaphragm, with the lumbar trunk. The thoracic sympathetic trunk lies against the neck of the ribs in the upper part of the chest; at midthoracic levels it rests against the costovertebral joints; and in the lower part of the thorax it is on the sides of the vertebral bodies. The trunk descends under cover of the parietal pleura and in front of the intercostal vessels and nerves. The thoracic sympathetic trunk turns sharply dorsalward from the position of the cervical trunk as it enters the curved thoracic cage. As it pierces the diaphragm and leaves the thorax to become continuous with the lumbar trunk, it makes a pronounced ventralward migration. The **thoracic ganglia** approximate the number of thoracic nerves, but they are frequently reduced by coalescence to less than twelve. The ganglia are connected to the spinal nerves by both white rami communicantes and gray rami communicantes and usually lie slightly inferior to their corresponding nerves. The first thoracic ganglion is larger than the rest, lies against the neck of the first rib, and is elongated dorsoventrally in accordance with the curvature of the uppermost part of the trunk. It is usually combined with the inferior cervical ganglion to form the **cervicothoracic (stellate) ganglion.** Sometimes the second or the second and third thoracic ganglia are also included in the common mass. The cervicothoracic ganglion occurs in about 80 per cent of specimens (Pick & Sheehan, '46; Jamieson et al., '52).

The distribution of fibers of the thoracic sympathetic trunk and ganglia is upward to the cervical trunk and ganglia, downward to the lumbar ganglia, into thoracic spinal nerves by way of gray rami communicantes, by visceral branches to thoracic viscera, and by the thoracic splanchnic nerves to collateral ganglia in the abdomen.

(1) Sympathetic fibers destined for structures in the head and neck arise in the upper two or three thoracic segments. They ascend as preganglionic fibers through the cervical sympathetic trunk and end in its ganglia. The peripheral distribution of the cervical sympathetic trunk is considered on p. 168. Part of the sympathetic innervation of the heart is also effected by this route, the superior, middle, and inferior cardiac sympathetic nerves being branches of the respective cervical sympathetic ganglia (p. 168). The cranial portion of the sympathetic trunk (p. 284) is an extension of the cervical part of the system.

(2) The connection between the thoracic and lumbar portions of the sympathetic trunk is a slender one, indicating a relative independence of the zones of distribution of these parts. The major contribution of the thoracic levels of the sympathetic system to the abdominal viscera is made by the thoracic splanchnic nerves (see below).

(3) Gray rami communicantes are bundles of postganglionic fibers which pass from each thoracic ganglion into the spinal nerve of the corresponding segment. They enter all branches of the nerve and of both its ventral and dorsal rami (Dass, '52) to distribute to smooth muscle and glands in the area of innervation of the nerve. By means of their gray rami communicantes, the first thoracic ganglion and frequently also the second thoracic ganglion contribute to the brachial plexus through the first thoracic nerve. Postganglionic fibers from the second ganglion may join the first thoracic nerve by means of a communicating branch between them (Kuntz, '35).

(4) Direct visceral branches to thoracic organs arise from the trunk and ganglia of the upper four or five segments. Branches to the **cardiac plexuses** (p. 346) occur in addition to the sympathetic innervation provided by the cardiac nerves from the cervical ganglia. **Pulmonary branches** follow the posterior intercostal arteries and reach the posterior pulmonary plexuses. **Esophageal branches** follow the arteries to the esophagus and join the plexus formed by the vagus nerves. Esophageal twigs may also be provided from the cardiac or aortic plexuses or by the greater thoracic splanchnic nerves. **Aortic branches** reach the aortic plexus from the lower five or six thoracic ganglia, from the cardiac plexus, and from the thoracic splanchnic nerves.

(5) The **thoracic splanchnic nerves** are preganglionic branches which pass through the diaphragm, synapse in collateral ganglia in the abdomen, and provide the sympathetic innervation for a large part of the abdominal viscera. The **greater thoracic splanchnic nerve** arises from the fifth to the ninth (or tenth) ganglia or intervening portions of the trunk. Contributions from these levels come off singly or in combination to form a large nerve trunk which descends along the ventrolateral aspect of the vertebral column. It perforates the crus of the diaphragm and ends in the abdomen in the lateral aspect of the celiac ganglion. Opposite the eleventh thoracic vertebra, a small **splanchnic ganglion** commonly occurs. Filaments of the greater thoracic splanchnic nerve are given off in the chest to the esophagus, the aorta, and the thoracic duct. The **lesser thoracic splanchnic nerve** arises from the tenth and eleventh ganglia (Edwards & Baker, '40; Reed, '51) or from the cord between them. It pierces the crus of the diaphragm near the greater thoracic splanchnic nerve and ends in the aorticorenal ganglion. The **least thoracic splanchnic nerve,** arises from the last thoracic ganglion or from the lesser thoracic splanchnic nerve. It, too, perforates the diaphragm and ends in the renal plexus.

VI

The Abdomen

SURFACE ANATOMY

The **abdomen** is that portion of the body bounded above by the diaphragm and below by the plane of the pelvic inlet. Its cavity is continuous with the pelvic cavity, and the abdominal wall grades over into the boundaries of the pelvis below and of the thorax above. The anterolateral wall of the abdomen is extensive and is the usual portal of surgical entry to the abdominopelvic cavity.

The abdominal wall (fig. 301) is bounded above by the cartilages of the seventh to the twelfth ribs and by the xiphoid process of the sternum. The

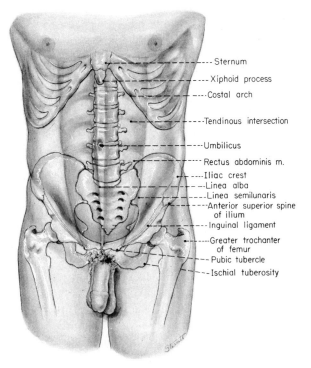

Fig. 301 The surface anatomy and underlying bony landmarks of the abdomen.

Labels on figure:
- Sternum
- Xiphoid process
- Costal arch
- Tendinous intersection
- Umbilicus
- Rectus abdominis m.
- Iliac crest
- Linea alba
- Linea semilunaris
- Anterior superior spine of ilium
- Inguinal ligament
- Greater trochanter of femur
- Pubic tubercle
- Ischial tuberosity

infrasternal angle is formed below the xiphosternal joint by the diverging cartilages of the seventh, eighth, and ninth costal cartilages. The lower border of the tenth costal cartilage is usually at the inferior limit of the **costal margin.** Inferiorly, the abdominal wall is bounded by the bones of the pelvis and by the inguinal ligament. The prominent **iliac crest** forms the upper limit of the region of the hip; the curve of the crest ends ventrally in the **anterior superior spine of the ilium,** dorsally in the **posterior superior iliac spine.** The highest point of the iliac crest is at the transverse level of the body of the fourth lumbar vertebra. In the pubic region the two coxal bones unite at the pubic symphysis. From here the **pubic crests** pass lateralward for about 2.5 cm., ending in the sharp **pubic tubercle.** Between the pubic tubercle and the anterior superior iliac spine is stretched the **inguinal ligament.** This ligament is the rolled-under inferior margin of the aponeurosis of the external abdominal oblique muscle and marks, in the groin, the separation between the abdominal wall and the lower limb. It is most readily visualized by dropping one leg to the floor while lying one one's back.

The **umbilicus** (figs. 301, 304) is a prominent midline marking located normally at the level of the intervertebral disk between the third and fourth lumbar vertebrae. It is an unreliable landmark for this level in a pendulous abdomen. Three vertical lines are visible on the anterior abdominal wall. The **linea alba** of the median plane marks the aponeurotic junction of the abdominal muscles of the two sides and separates the vertical bulges of the two rectus abdominis muscles. The linea alba, narrow below, widens to 1 cm. or more in its upper portion. Two **lineae semilunares** define the lateral margins of the rectus muscles. These lines begin at the pubic tubercles and end above at the costal margin in the region of the tips of the ninth costal cartilages. Three transverse lines cross the abdominal wall. These **lineae transversae** are produced by tendinous bands which interrupt the continuity of the fibers of the rectus abdominis muscle. One is located a little below the xiphoid process, one is at the level of the umbilicus, and the third is halfway between the other two (figs. 301, 305).

For localization of pain or swelling or for precise location of deep structures, the division of the abdomen into quadrants provides a convenient reference. Such quadrants (upper right and left, and lower right and left) are defined in relation to the vertical line of the linea alba and to an imaginary transverse plane through the umbilicus.

THE SUPERFICIAL STRUCTURES

SKIN AND SUBCUTANEOUS CONNECTIVE TISSUE

The **skin** of the abdomen is thicker in the lumbar region than it is ventrally. It is attached loosely to subjacent structures except at the umbilicus, where it adheres firmly. The **subcutaneous connective tissue** of the abdomen is characterized by the fat it contains, for this is one of the fat depots of the body. In the **tela subcutanea** of the inguinal portion of the abdominal wall can be defined two strata, a superficial fatty one and a deeper membranous plane (fig. 302). The membranous connective tissue plane is not differentiated at umbilical levels or above, but at the inferior limit of the abdominal wall it has definite attachments. It is attached along the iliac crest and, continuing medially, to the fascia lata of the thigh along a line about 1 cm. below the inguinal ligament. It then attaches to the pubic symphysis.

Clinical Note   Over the intervening superior border of the pubis, the membranous layer of subcutaneous connective tissue is not attached to bone, and a finger may be pushed behind it into the scrotum deep to the subcutaneous tissue (tunica dartos) of the scrotum. The unnatural communication thus enlarged, **abdominoscrotal opening,** is sometimes the pathway whereby extravasated fluid in the scrotum ascends into the abdominal wall. Retention of the fluid in the flanks and its failure to descend into the thighs is due to the other attachments of the membranous layer enumerated above.

Above the pubis a condensation of vertically arranged connective tissue strands of the membranous layer forms the **fundiform ligament of the**

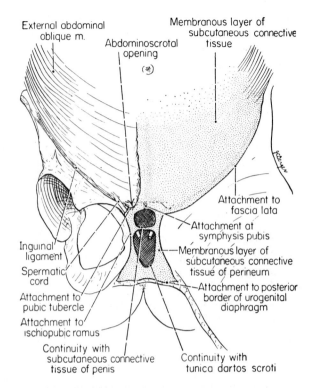

Fig. 302   The attachments and continuities of the membranous layer of subcutaneous connective tissue of the abdomen and perineum. (Modified and redrawn after Cunningham.)

penis. These fibers attach to the linea alba and pass downward on either side of the penis to enter the septum of the scrotum.

The genital swellings of the inguinal region of the fetal abdominal wall form the scrotum in the male and the labia majora in the female. The scrotum and the labia majora are, thus, cutaneous outpouchings of the abdomen. The **skin of the scrotum** in the adult exhibits scattered coarse hairs and has well developed sebaceous glands. The underlying subcutaneous tissue is known as **tunica dartos scroti** (fig. 304); it is directly continuous with the subcutaneous tissue of the abdominal wall. In the male the fatty and membranous layers of subcutaneous connective tissue blend into one layer in the scrotum and lose entirely their distinctive characteristics; tunica dartos scroti is without fat and contains smooth muscle intermingled with its areolar tissue. This muscle is the cause of the wrinkling of the skin of the scrotum. The cutaneous nerves of the scrotum are the anterior scrotal branch of the ilioinguinal nerve and posterior scrotal branches of both the pudendal and the posterior femoral cutaneous nerves. The **ilioinguinal nerve** is a branch of the lumbar plexus which supplies the lowermost portion of the abdominal wall. In the inguinal region this nerve traverses the inguinal canal to the **superficial inguinal ring,** where it emerges on the lateral aspect of the spermatic cord. It distributes to the skin of the upper and medial parts of the thigh, and its **anterior scrotal branch** (fig. 304) passes to the skin of the root of the penis and of the anterior portion of the scrotum. **Posterior scrotal nerves** reach the skin of the scrotum by way of the perineum and are described in connection with that region. The blood vessels and lymphatics of the scrotum are discussed in connection with the abdominal wall (p. 380).

The **labia majora** are two prominent longitudinal folds which bound the urogenital cleft in the female. They taper posteriorly, but, anteriorly, they blend into one another to form the rounded eminence of the mons pubis. The skin of the outer surface of each labium is pigmented and contains hairs. The inner surface of the labium is smooth and is provided with large sebaceous glands. Between the outer and inner surfaces of the labial folds is a considerable quantity of areolar tissue and fat. The fat is partially encapsulated as a tapering adipose cylinder which largely determines the form of the labium majus. The round ligament of the uterus emerges at the superficial inguinal ring and, entering the labium majus, ends in its areolar tissue and by attaching to its skin. **Anterior and posterior labial nerves** correspond in source and position to the anterior and posterior scrotal nerves.

CUTANEOUS NERVES (fig. 304)

The **cutaneous nerves** of the abdominal wall include the anterior and lateral cutaneous terminals of intercostal nerves seven to eleven. These are supplemented by the anterior terminals of the subcostal nerve and by a branch of the iliohypogastric nerve of the lumbar plexus. The abdominal distribution of the thoracoabdominal intercostal nerves and the subcostal nerve is described on p. 318. The lateral cutaneous branch of the subcostal nerve is not abdominal. It descends over the crest of the ilium several centimeters behind its anterior superior spine and ends in the skin of the hip region. The **iliohypogastric nerve** (L1), like the thoracoabdominal intercostal nerves, traverses the

abdominal wall from the lumbar region to the linea alba and serves both motor and sensory functions. Its **anterior cutaneous branch** pierces the aponeurosis of the external abdominal oblique muscle several centimeters above the superficial inguinal ring and distributes to the skin of the suprapubic region. The **lateral cutaneous branch** of this nerve crosses the iliac crest at the junction of its anterior and middle thirds and descends to supply the skin of the gluteal region, distributing posterior to the lateral cutaneous branch of the subcostal nerve.

It has been noted in relation to skin dermatomes (p. 318) that the anterior cutaneous branch of the seventh intercostal nerve ends immediately inferior to the xiphoid process of the sternum, and that the similar branch of the tenth intercostal distributes both above and below the umbilicus. The suprapubic region receives its cutaneous supply from the anterior cutaneous branch of the iliohypogastric nerve. The other intercostal nerves and the subcostal nerve distribute to intermediate regions in accordance with their order of sequence (fig. 304).

CUTANEOUS BLOOD VESSELS (fig. 303)

The tenth and eleventh posterior intercostal arteries and the subcostal artery reach the abdominal wall together with four lumbar segmental arteries. The intercostal and subcostal vessels provide lateral and anterior cutaneous branches which ramify in the subcutaneous tissue with the cutaneous nerves. The **musculophrenic branch** of the internal thoracic artery (p. 319) supplies twigs to the skin of the abdomen along the costal arch. The anterior cutaneous arteries are largely branches of the **superior epigastric artery** which, as a terminal of the internal thoracic artery, descends in the rectus sheath posterior to the rectus abdominis muscle. Other anterior cutaneous branches are terminals of the **inferior epigastric artery,** which enters the rectus sheath from below after arising from the external iliac artery. Veins accompany the arteries described.

In the region below the umbilicus, three small branches of the femoral artery enter the subcutaneous abdominal area from the groin. The **superficial epigastric artery** (fig. 303) arises from the anterior aspect of the femoral artery about 1 cm. below the inguinal ligament. Piercing the femoral sheath and cribriform fascia, the artery turns superiorly over the inguinal ligament and runs toward the umbilicus

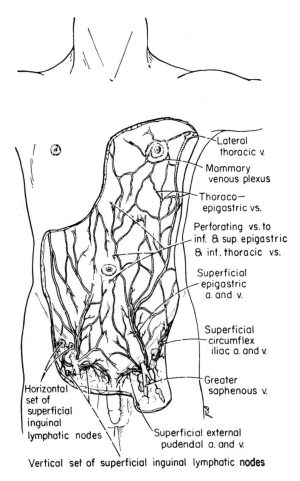

Vertical set of superficial inguinal lymphatic nodes

Fig. 303   The subcutaneous blood vessels of the antero-lateral wall of the trunk.

in the membranous layer of subcutaneous connective tissue. It supplies twigs to the inguinal lymph nodes and to the subcutaneous tissue and skin of the lower abdomen. The vessel anastomoses with branches of the inferior epigastric artery and with its fellow of the opposite side. The **superficial circumflex iliac artery** (fig. 303) also arises from the femoral artery about 1 cm. below the inguinal ligament. It pierces the fascia lata lateral to the saphenous opening and runs lateralward across the upper thigh below and parallel to the inguinal ligament. Branches are supplied to the skin of the abdomen and the thigh and to the inguinal lymph nodes. This artery anastomoses with the deep circumflex iliac, superior gluteal, and lateral femoral circumflex arteries. The **superficial external pudendal artery** (fig. 303) arises from the medial side of the femoral artery about 2 cm. below the inguinal ligament. It emerges through the saphenous opening and then

passes medialward and upward across the spermatic cord in the male (or the round ligament of the uterus in the female) to be distributed to the skin of the suprapubic region of the abdomen and to the penis and scrotum (or the labium majus in the female). It gives twigs to the inguinal lymph nodes and anastomoses with the deep external pudendal artery and with branches of the internal pudendal artery.

An additional arterial supply for the skin of the scrotum and labium majus is furnished by the **deep external pudendal branch** of the femoral artery. This artery arises inferior to the others and may be a branch of the medial circumflex femoral artery. It runs medialward across the pectineus muscle and then pierces the fascia lata to become subcutaneous. It supplies the pectineus and adductor longus muscles and anastomoses with branches of the superficial external pudendal and internal pudendal arteries and with the testicular artery in the male (or the artery of the round ligament of the uterus in the female).

The **veins** which correspond to the superficial epigastric, superficial circumflex iliac, and superficial external pudendal arteries course toward the groin where they terminate in the greater saphenous vein before it turns deeply through the saphenous opening. Occasionally these veins pierce the cribriform fascia and empty directly into the femoral vein. The external pudendal veins receive the superficial dorsal vein of the penis (or the clitoris) and the subcutaneous veins of the scrotum (or the labium majus). The superficial epigastric vein anastomoses in the region of the umbilicus with the paraumbilical vein and with radicles of the superior and inferior epigastric veins. It also communicates with the thoracoepigastric veins (fig. 303) which lead into the lateral thoracic vein of the axillary system (p. 78).

Cutaneous Lymphatics

A subcutaneous network, consisting principally of collecting vessels from the lymphatic capillary plexus of the dermis, forms an anastomosing system through the abdomen, the thorax, and the scrotum (or the labia majora). The vessels are particularly numerous in the scrotum and in the labia majora. The regional drainage of the subcutaneous plexus is such that the portion of the plexus above the umbilicus drains toward the chest and axillary regions—to the sternal nodes along the internal thoracic vessels and the pectoral nodes of the axillary group. Lymph from the subcutaneous plexus of the anterior and lateral parts of the abdominal wall below the umbilicus descends to the superficial inguinal nodes in channels which run with the subcutaneous blood vessels. The subcutaneous lymph of the scrotum and of the labia majora reaches the more medial superficial inguinal nodes.

The **superficial inguinal nodes** (fig. 303) are from twelve to twenty lymph nodes arranged in the form of a T in the subcutaneous tissue of the groin. The horizontal part of the T accounts for most of the nodes; they form a chain parallel to and about 1 cm. below the inguinal ligament. These nodes receive lymph from the lower abdominal wall, the buttocks, the penis and scrotum (or the labia majora), and the perineum. The larger but fewer nodes of the vertical limb of the T lie along the termination of the greater saphenous vein. Their afferent vessels drain chiefly the superficial lymphatic plexus of the lower limb, but they also receive some vessels from the skin of the penis, the scrotum, the perineum, and the buttocks.

THE ABDOMINAL WALL

Muscles and Fasciae of the Abdominal Wall (figs. 304–308, 363)

The muscles of the abdominal wall comprise two groups, anterolateral and posterior. The posterior muscle is the quadratus lumborum. The anterolateral muscles include two vertical muscles, rectus abdominis and pyramidalis, and three thin muscular layers which alternate in their fiber direction. These, named and oriented like the corresponding muscles of the thoracic wall, are the external abdominal oblique, internal abdominal oblique, and transversus abdominis muscles. These muscles have extensive aponeurotic insertions, their aponeuroses combining to form a sheath for the rectus abdominis and pyramidalis muscles. The alternation in the direction of their muscular and aponeurotic fascicles adds strength to the abdominal wall. The abdominal muscles are flexors of the trunk and the pelvis, side to side benders, and rotary muscles of the trunk. Acting together, they compress the abdominal viscera and elevate the diaphragm in respiration. They also 'bear down' in the acts of defecation and parturition (except for rectus abdominis).

Clinical Note The right-angled relationship between the muscular fascicles of the external and internal abdominal muscles is utilized in the small muscle-splitting surgical incisions in the abdominal wall. By successively separating and spreading the muscle fascicles an abdominal opening is produced with a minimum of incisional damage, and the fascicles are readily restored to their normal relations after the operation.

The External Abdominal Oblique Muscle (figs. 304, 363)

This is the most superficial of the muscles of the anterolateral group. Its fleshy part forms the lateral portion; its aponeurosis, the anterior portion of the layer. The muscle arises by eight fleshy digitations from the lower borders and outer surfaces of the lower eight ribs. The upper five slips of origin interdigitate with those of the serratus anterior muscle; the lower three digitations, with the costal attachments of the latissimus dorsi muscle. The muscular fibers arising from the last two ribs descend nearly vertically and insert into the anterior half of the outer lip of the iliac crest. This posterior margin of the external abdominal oblique is free and forms, with the converging border of the latissimus dorsi muscle and the iliac crest below, the **lumbar triangle** (p. 51). The muscular fasciculi from the remaining six ribs are directed obliquely downward and anteriorly to end in the external oblique aponeurosis. The muscle fibers terminate along a line which begins adjacent to the anterior superior spine of the ilium and ascends somewhat obliquely toward the costal margin. Aside from the fascicles from the last two ribs to the anterior half of the iliac crest, the insertion of the muscle is by means of its aponeurosis. Among other terminations, the fibers of the aponeurosis interlace with their fellows of the opposite side in the **linea alba** all the way from the xiphoid process to the symphysis pubis. The aponeurosis fuses with the underlying internal abdominal oblique aponeurosis in the formation of the **sheath of the rectus abdominis muscle.** This fusion takes place almost at the lateral border of the rectus muscle above the umbilicus but approaches progressively nearer the median line in passing from the umbilicus to the symphysis pubis. Additional specializations of the external oblique aponeurosis are the inguinal ligament, the lacunar ligament, the reflected inguinal ligament, the medial and lateral crura of the superficial inguinal ring, the intercrural fibers, and the external spermatic fascia.

The external oblique aponeurosis has, at its inferior extremity, a thickened, rolled-under border which is stretched between the anterior superior spine of the ilium and the pubic tubercle. This is

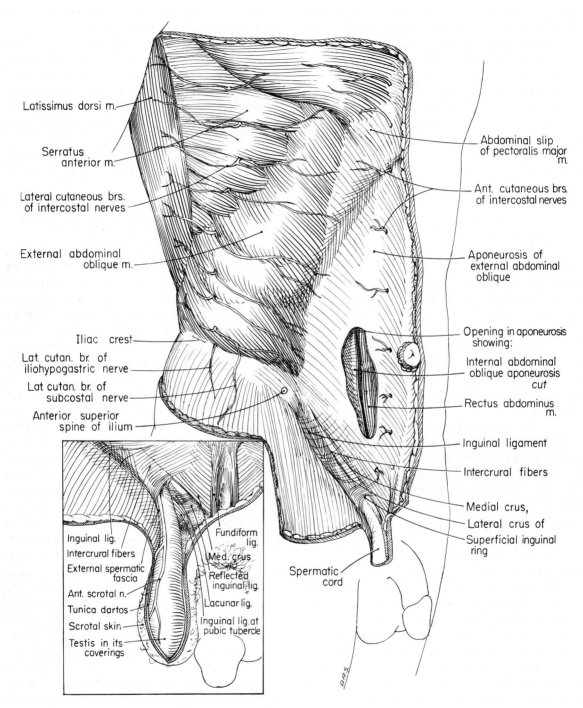

Fig. 304 The external abdominal oblique muscle layer and the overlying cutaneous nerves. Inset—the specializations of the external oblique layer in the inguinal region.

the **inguinal ligament** (figs. 304, 308), the turned-under edge of which gives attachment laterally to the lowermost muscular fascicles of the internal abdominal oblique and transversus abdominis muscles. Medially, the fibers of the inguinal ligament flatten down into the lacunar ligament. The fascia lata attaches to the inferior edge of the inguinal ligament. In its lateral one-third the ligament is attached internally to the iliacus fascia as this fascia passes into the thigh on the iliacus muscle.

The **lacunar ligament** (figs. 304, 308) represents the more medial rolled-under fibers of the inguinal ligament which, flattening down into a horizontal shelf, attach to the pecten of the pubis for about 2 cm. This ligament is continuous with the inguinal ligament from the pubic tubercle along the pecten and, as its fibers arch posteriorly to this receding line, presents a crescentic margin facing lateralward which constitutes the medial border of the femoral ring. The continuation of the fibers of the lacunar ligament lateralward along the pecten contributes to the **pectineal ligament.** The spermatic cord rests on the horizontal lacunar ligament and turns from it to emerge through the superficial inguinal ring.

The **reflected inguinal ligament** is minor. Certain of the fibers of the inguinal ligament are reflected from the pubic tubercle upward toward the linea alba, ending 2 or 3 cm. above the pubic symphysis. There is also some reflection from the lacunar ligament and the pubic crest. The small triangular sheet so formed lies behind the main layer of the aponeurosis. It is sometimes interpreted as a group of decussating fibers of the opposite side which descend to the pubic crest and tubercle.

The **medial and lateral crura of the superficial inguinal ring** are well marked in the lower portion of the aponeurosis. The **superficial inguinal ring** lies at the end of a triangular cleft in the external oblique aponeurosis and is located just above and lateral to the pubic tubercle. The long triangular cleft represents a weakness in the aponeurosis, but it is bordered on either side by a strong band of fibers. The lateral crus lies below the cleft. Its fibers run downward and medialward, forming the inferior margin of the superficial inguinal ring and blending with the inguinal ligament toward its insertion. The medial crus is a flat band of aponeurotic fibers which passes diagonally downward and medialward toward the symphysis pubis and forms the superior and medial boundary of the superficial inguinal ring. The ring transmits the spermatic cord in the male and the round ligament of the uterus in the female. Its superolateral margin is formed by intercrural fibers.

The **intercrural fibers** (fig. 304) are fiber bundles which cross the crura. These are scattered, reinforcing fibers which curve from the inguinal ligament upward and medialward over the triangular hiatus

of the aponeurosis and are lost beyond. They form the superolateral limit of the superficial ring.

The **external spermatic fascia** is a delicate tubulo-saccular sheath which represents the continuation of the intercrural fibers along the spermatic cord and over the testis. It is the outermost covering of the cord and testis and, as a fingerlike outpouching of the intercrural fibers, is clearly in continuity with the margins of the superficial inguinal ring.

The nerves and blood vessels of the muscles of the abdominal wall are considered together on p. 375.

The Internal Abdominal Oblique Muscle (figs. 305, 363)

This is the middle one of the three flat abdominal muscles. It is a thin muscular sheet which arises from the posterior layer of the thoracolumbar fascia, the anterior two-thirds of the iliac crest, the lateral two-thirds of the inguinal ligament, and, by deeper fibers, from the iliacus fascia for 5 or 6 cm. medial to the anterior superior spine of the ilium. From this posterior and inferior line of origin, the muscle fans out into a very extensive insertion. The posterior fasciculi ascend almost vertically to insert on the inferior borders of the lower three or four ribs and their cartilages. These fleshy bundles appear to be in continuity with the internal intercostal muscles. The rest of the muscle fibers spread fan-like forward and medialward and end in an aponeurosis as they approach the linea semilunaris. The ultimate insertion is the linea alba and the pubic bone; but in the upper three-fourths of the abdomen the aponeurosis splits at the linea semilunaris, sending one sheet anterior and one posterior to the rectus abdominis muscle. The posterior sheet attaches to the margins of the ninth, eighth, and seventh costal cartilages, and the two layers unite in the linea alba and at the margin of the xiphoid process. Above the umbilicus the posterior layer fuses with the aponeurosis of the transversus abdominis muscle; the anterior layer, with that of the external abdominal oblique muscle. Thus, an equal split of the abdominal aponeuroses forms the sheath of the rectus abdominis muscle (figs. 305, 307). In the lower one-fourth of the abdomen the internal abdominal oblique aponeurosis fails to split and passes to the median line entirely anterior to the rectus abdominis muscle. Muscle fibers arising from the inguinal ligament and the iliac fascia end in this part of the aponeurosis.

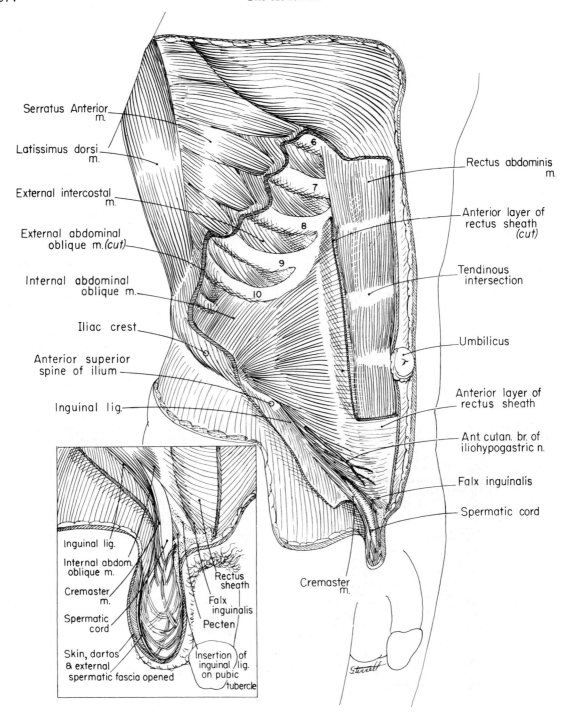

Fig. 305   The internal abdominal oblique muscle layer. The anterior layer of the rectus sheath has been partially removed. Insert—the cremaster muscle and fascia layer of the scrotum.

With the length of the abdominal wall considerably greater in the median line than toward its lateral side, the lower fibers of the internal abdominal oblique muscle necessarily arch downward to their insertion. Such fibers, arching over the spermatic cord, insert in the superior border of the pubis. They insert partially in common with the similarly arching fibers of the transversus ab-

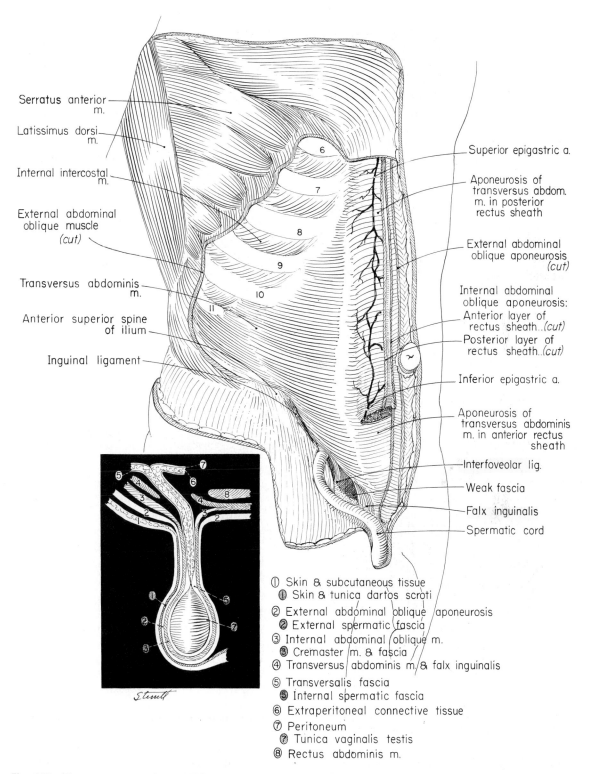

Serratus anterior m.
Latissimus dorsi m.
Internal intercostal m.
External abdominal oblique muscle *(cut)*
Transversus abdominis m.
Anterior superior spine of ilium
Inguinal ligament

Superior epigastric a.
Aponeurosis of transversus abdom. m. in posterior rectus sheath
External abdominal oblique aponeurosis *(cut)*
Internal abdominal oblique aponeurosis:
Anterior layer of rectus sheath..*(cut)*
Posterior layer of rectus sheath..*(cut)*
Inferior epigastric a.
Aponeurosis of transversus abdominis m. in anterior rectus sheath
Interfoveolar lig.
Weak fascia
Falx inguinalis
Spermatic cord

① Skin & subcutaneous tissue
❶ Skin & tunica dartos scroti
② External abdominal oblique aponeurosis
❷ External spermatic fascia
③ Internal abdominal oblique m.
❸ Cremaster m. & fascia
④ Transversus abdominis m. & falx inguinalis
⑤ Transversalis fascia
❺ Internal spermatic fascia
⑥ Extraperitoneal connective tissue
⑦ Peritoneum
❼ Tunica vaginalis testis
⑧ Rectus abdominis m.

Fig. 306 The transversus abdominis muscle. In its upper three-fourths, the rectus abdominis muscle has been removed to expose the blood vessels which lie against the posterior layer of its sheath. Inset—the layers of the abdominal wall (clear numbers) and their representatives in the coverings of the spermatic cord and testis (hatched numbers).

dominis muscle and, with them, form the sickle-shaped **falx inguinalis.** Due to the obliquity of the inguinal canal, by means of which the spermatic cord passes through the abdominal wall, the lowermost internal oblique muscle fibers arise from the inguinal ligament superficial to the cord but insert behind the cord at its medial extremity.

Among the coverings of the cord and testis, the **cremaster muscle and fascia** are the representatives of the internal abdominal oblique muscle layer. They form a layer which immediately underlies the external spermatic fascia. The cremaster muscle (fig. 305 inset) is formed by the lowermost fascicles of the internal abdominal oblique muscle arising from the inguinal ligament which, instead of arching over the spermatic cord, are applied to its lateral surface and descend with it into the scrotum. The muscle appears as a series of separated festoons, connected by fascia, which loop over the spermatic cord and testis. The highest fibers ascend on the medial side of the cord to insert into the pubic tubercle. The cremaster layer constitutes the middle and most readily recognized of the three covering layers of the cord and testis which are derived from the abdominal wall. The cremaster fascia is derived from the fascia of both the superficial and deep surfaces of the internal abdominal oblique muscle, and the cremaster layer represents a diverticulum of both the muscle and fascia of the lower border of this muscle. The cremaster muscle is innervated by the genital branch of the genitofemoral nerve. This nerve, a derivative of the lumbar plexus, joins the spermatic cord from within the abdomen at the deep inguinal ring, traverses the inguinal canal, and lies under the cremaster muscle. In the female only a very few muscle fibers are drawn down along the course of the round ligament of the uterus, and the cremaster layer is represented mostly by fascia.

The Transversus Abdominis Muscle (figs. 306, 363)

This is the third and innermost of the flat muscles of the abdomen. Its fibers have an extensive and varied origin. It arises from the deep surfaces of the costal cartilages of the lower six ribs (its fascicles interdigitating with fibers of origin of the diaphragm), from the fusion of the anterior and posterior layers of the thoracolumbar fascia, from the anterior three-fourths of the iliac crest, and the lateral one-third of the inguinal ligament, and from the iliacus fascia behind it. The muscular fibers end in an aponeurosis which is widest at the level of the iliac crest and narrows above and below this level. The aponeurosis is very narrow in the region of the xiphoid, the muscular fibers almost reaching the median line. The aponeurosis of the transversus abdominis muscle combines with the posterior layer of the internal abdominal oblique aponeurosis to form the posterior layer of the sheath of the rectus abdominis except in lower abdominal levels. About midway between the umbilicus and the symphysis pubis the aponeuroses of all three flat muscles pass anterior to the rectus abdominis muscle. Like the internal oblique, the lowest fibers of the transversus abdominis (those arising from the inguinal ligament) arch downward to the pubis. They insert into the superior border of the pubis and into the medial 2 cm. of the pecten. Together with the lowermost fibers of the internal oblique muscle, which end in the superior border of the pubis, they constitute the **falx inguinalis** (figs. 305, 306).

The inguinal arch of the transversus abdominis muscle carries its lowest fibers above the obliquely disposed spermatic cord to end behind, or deep to, the medial end of the cord. As the fibers arch over the cord, there arises from the deep surface of the muscle a few fascicles of muscle and tendon which drop more sharply downward just medial to the deep inguinal ring and across the course of the inferior epigastric vessels behind the muscle layers. This is the **interfoveolar muscle** or the **interfoveolar ligament** (figs. 306, 308); it is usually poorly represented but is occasionally a band of some strength. Its fibers descend behind the inguinal ligament to end in the anterior layer of the femoral sheath. The falx inguinalis and the interfoveolar ligament are united by a delicate sheet of fascia which represents the transversus abdominis muscle layer between these stronger bands (fig. 306).

The segmental arteries and nerves of the abdominal wall course between the internal oblique and the transversus abdominis muscles, and, where the muscles correspond in their fiber direction, the vessels and nerves give a useful indication of the plane of separation between them. The muscles are separated by sparse and loose connective tissue. The transversus abdominis muscle is covered on its deep aspect by the transversalis fascia.

The Rectus Abdominis Muscle (figs. 305–308, 331, 348, 378)

The rectus abdominis is the vertical muscle of the anterior abdominal wall. Lying on either side of the linea alba, it is narrow and thick inferiorly and broad and thin in upper abdominal levels. The muscle arises from the superior ramus of the pubis and from the ligaments of the symphysis pubis. It inserts on the anterior surface of the xiphoid process and into the cartilages of the fifth, sixth, and seventh ribs. In its anterior surface, but not extending through the thickness of the muscle, are three **tendinous intersections** (fig. 305). They are firmly adherent to the anterior layer of the rectus sheath. The lowest is at the level of the umbilicus; the highest is near the xiphoid process; and the third is halfway between these levels. The lateral border of the muscle is convex and coincides with the linea semilunaris, along which fusion of the aponeuroses of the flat muscle layers takes place. The **sheath of the rectus muscle** is formed by the fusion and separation of these aponeuroses. Above a line approximately midway between the umbilicus and the symphysis pubis, the aponeurosis of the internal oblique muscle splits into an anterior and a posterior lamella. Reinforced by the aponeurosis of the external abdominal oblique muscle externally and that of the transversus abdominis muscle internally, the anterior and posterior layers of the sheath are equal (fig. 307). In the lower one-fourth of the abdominal wall the aponeurosis of the internal oblique muscle does not split, and all three aponeurotic layers pass anterior to the rectus abdominis muscle. The line of shift in the arrangement of these aponeuroses is called the **arcuate line.** Since the shift of the posterior layer of the internal

oblique aponeurosis does not always take place at the same level as that of the transversus abdominis and since there are occasional partial and interrupted transitions, several lines frequently occur. As noted by McVay and Anson ('40), the transversus abdominis aponeurosis also splits and sends fascicles, interdigitating with those of the internal abdominal oblique muscle, into the anterior layer of the rectus sheath. This occurs only at levels between the arcuate line and the umbilicus. Below the arcuate line the rectus muscle lies against the transversalis fascia. The absence of the aponeurotic layers behind the rectus abdominis muscle below the arcuate line provides access to the muscle for the inferior epigastric vessels.

The Pyramidalis Muscle

This insignificant muscle, frequently absent, is also contained in the rectus sheath, where it lies anterior to the inferior portion of the rectus abdominis muscle. It is small and triangular and arises from the front of the pubis and the anterior pubic ligament. It ends in a pointed extremity in the linea alba a variable distance above the symphysis pubis. The muscle draws down on the linea alba; it is supplied by a branch of the twelfth thoracic nerve.

The Quadratus Lumborum Muscle (figs. 225, 226, 357, 368, 370, 377)

This four-sided muscle lies adjacent to the transverse processes of the lumbar vertebrae and is distinctly broader below than above. It arises from the iliolumbar ligament and from about 5 cm. of the posterior part of the iliac crest. The muscle inserts into the lower border of the last rib for about half the length of the rib and into the tips of the upper four lumbar transverse processes. An anterior lamina of the muscle is frequently seen with slips of origin from the transverse processes of the lower three or four lumbar vertebrae and an insertion into the lower border of the last rib.

The quadratus lumborum is innervated by branches of the twelfth thoracic nerve and of the first three or four lumbar nerves. The muscle fixes the last rib in relation to the pelvis, holding it down against the traction exerted by the diaphragm in inspiration. One side, acting alone, bends the trunk toward the same side.

Quadratus lumborum lies anterior to the lateral margin of the erector spinae muscle and is separated from it by the anterior layer of the thoracolumbar fascia. The medial portion of the muscle is anterior to the lumbar transverse processes and is overlapped by

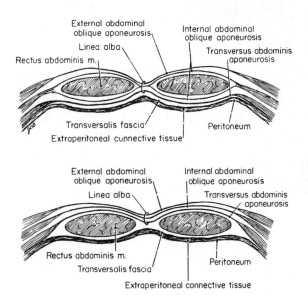

Fig. 307 The constitution of the rectus sheath. The upper figure represents the lamination in the upper three-fourths of the sheath; the lower figure, in its lower one-fourth.

the more anterior psoas major muscle. The transversalis fascia covers the quadratus lumborum muscle anteriorly. This fascia fuses with the anterior layer of the thoracolumbar fascia at the posterolateral border of the muscle and, at its medial border, attaches to the lumbar transverse processes. Between the fascia and the muscle the subcostal, iliohypogastric, and ilioinguinal nerves pass lateralward into the abdominal wall.

The Transversalis Fascia (fig. 308)

The fascia separating the flat abdominal muscles is loose, scanty, areolar connective tissue, for these muscles contract largely in unison and no bulk of intervening tissue is necessary. On the internal surface of the transversus abdominis muscle, however, there is a firm membranous sheet, the transversalis fascia. This fascia has continuities well beyond the transversus muscle and forms a lining fascia for the entire abdominopelvic cavity. It also displays thickenings and outpouchings which emphasize its independent character.

The transversalis fascia covers the internal surface of the transversus abdominis muscle and its

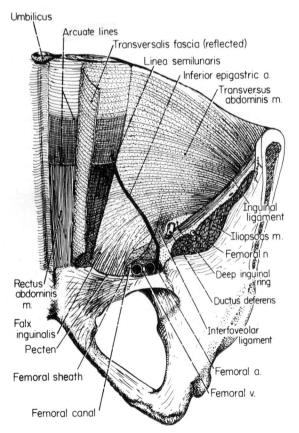

Fig. 308   The inguinal portion of the anterior abdominal wall as seen from within to illustrate the continuities of the transversalis fascia.

aponeurosis and is continuous from side to side across the linea alba. Below the arcuate line the transversalis fascia is the principal layer of the posterior sheath of the rectus abdominis muscle. At the lower border of the inguinal arch of the transversus abdominis muscle, the transversalis fascia blends with the weak fascia of this muscle which bridges between the falx inguinalis and the interfoveolar ligament. The strengthening here has been designated surgically as the **iliopubic tract,** said to run from the anterior iliac spine to the pubic tubercle, parallel to but behind the inguinal ligament. Along the costal arch the trasversalis fascia passes from the surface of the transversus abdominis onto the diaphragm and becomes the inferior fascia of the diaphragm. In the lumbar region it attaches to the anterior layer of the thoracolumbar fascia along the posterolateral border of the quadratus lumborum muscle and then sweeps across the anterior surface of that muscle to become its investing facia. Medial to the quadratus lumborum muscle, the fascia dips between it and the psoas major muscle, attaches to the lumbar transverse processes, and then continues over the psoas major and minor muscles. It covers the crura of the diaphragm and attaches to the anterior longitudinal ligament of the vertebral column behind the aorta and the inferior vena cava. Over the upper portions of the psoas and quadratus lumborum muscles the fascia is thickened to form the medial and lateral lumbocostal arches of the diaphragm. Below, the transversalis fascia is attached to the iliac crest and then descends upon the heavily aponeurotic iliac fascia. In the pelvis it covers the aponeurotic fascia of the obturator internus muscle and is then prolonged downward and medialward as the superior fascia of the pelvic diaphragm, thus completing the lining of the abdominopelvic cavity.

The transversalis fascia is characteristically continued along blood vessels and other structures leaving the abdominopelvic cavity. It forms a finger-like diverticulum along the first several centimeters of the femoral vessels. This constitutes the **femoral sheath** (fig. 308) which descends behind the inguinal ligament into the upper thigh. A similar but smaller prolongation is usually found over the obturator vessels and nerve as they pass into the obturator foramen. The principal outpouching of the transversalis fascia is the **internal spermatic fascia** (fig. 306), which invests the ductus deferens and the testicular vessels as they leave the abdominal cavity. The mouth of this outpouching constitutes the **deep inguinal ring**, located from 1 to 1.5 cm. above the middle of the inguinal ligament. The internal spermatic fascia traverses the inguinal canal and, in the scrotum, is the innermost of the three covering layers of the cord and testis.

The Extraperitoneal Connective Tissue

Between the layers of the abdominal wall and the peritoneum enclosing the peritoneal cavity is a blend of fibro-elastic connective tissue and fat. This extraperitoneal connective tissue is scanty in certain regions and thick in others. It separates the peritoneum and transversalis fascia by a clear layer of areolar tissue and fat in most individuals, but if, as in a very thin specimen, there is little or no fat in the layer, the peritoneum and transversalis fascia are bound closely together by the areolar tissue and appear as a single sheet. Fat is a conspicuous element, and the layer is thick around the kidneys and great vessels of the posterior body wall in most individuals, differences in amount depending on the degree of obesity of the specimen.

Nerves and Vessels of the Abdominal Wall

Nerves The innervation of the muscles of the abdominal wall is by ventral rami of spinal nerves thoracic seven through lumbar four. This is the same segmental sequence which provides the cutaneous nerves (fig. 304) in the region. Except for those supplying the quadratus lumborum and cremaster muscles, the nerves of the abdominal wall muscles are the intercostal nerves seven through eleven and the subcostal, iliohypogastric, and ilioinguinal nerves. The **thoracoabdominal intercostal nerves** enter the abdominal wall by passing dorsal to the costal cartilages. They then course medialward with a descending inclination in the interval between the transversus abdominis and internal abdominal oblique muscles. The subcostal, iliohypogastric (T12 & L1), and ilioinguinal (L1) nerves may be located in the lumbar region (p. 467) between the transversalis fascia and the quadratus lumborum muscle. Crossing that muscle, they perforate the transversus abdominis muscle—the iliohypogastric at the lateral border of the quadratus lumborum and the ilioinguinal near the anterior part of the crest of the ilium—and proceed forward between the transversus abdominis and the internal abdominal oblique muscles. The **iliohypogastric nerve** (fig. 305) pierces the internal abdominal oblique muscle several centimeters in front of the anterior superior spine of the ilium and continues medialward deep to the external abdominal oblique

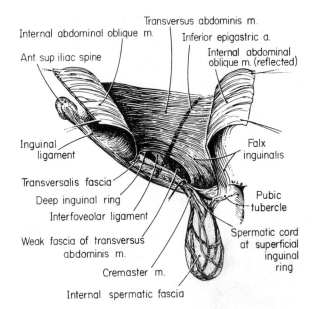

Fig. 309 The relations of the abdominal wall layers to the inguinal canal. The external abdominal oblique layer has been removed.

muscle. Its lateral cutaneous branch has been described (p. 369). The iliohypogastric nerve ends by an anterior cutaneous branch (p. 369) which penetrates the aponeurosis of the external oblique muscle several centimeters above the superficial inguinal ring. The **ilioinguinal nerve** pierces the internal abdominal oblique muscle near the deep inguinal ring and runs medialward through the inguinal canal. Becoming subcutaneous at the superficial inguinal ring, the ilioinguinal nerve distributes to the upper and medial parts of the thigh and, by the anterior scrotal nerve (fig. 304), supplies the skin of the root of the penis and the anterior part of the scrotum in the male. In the female the comparable anterior labial nerve supplies the skin of the mons pubis and the labium majus.

As the intercostal and subcostal nerves reach the lateral border of the rectus abdominis muscle, they penetrate the posterior lamina of the internal abdominal oblique aponeurosis and enter the rectus sheath. Here they provide lateral branches, which end in the rectus muscle, and medial branches, which pass behind the muscle, supply it, and then turn forward through it to pierce the anterior sheath as anterior cutaneous nerves (fig. 304). The tenth intercostal nerve enters the sheath just below the tendinous intersection at the level of the umbilicus. The higher nerves (T7 to 9) are above, and the lower nerves distribute, in sequence, below this level. The rectus abdominis muscle is innervated by the lower five intercostal nerves and the subcostal nerve. The external abdominal oblique muscle is innervated by intercostal nerves seven to eleven and the subcostal and iliohypogastric nerves; the internal abdominal oblique and transversus abdominis muscles are supplied by intercostal nerves eight to eleven and the subcostal, iliohypogastric, and ilioinguinal nerves.

Blood vessels (fig. 310) The blood vessels of the abdominal wall are derived from five systems, the terminal branches of each anastomosing with certain of the other systems represented. The arteries concerned are (1) the superior epigastric and musculophrenic branches of the internal thoracic artery; (2) the lower two posterior intercostal arteries, the subcostal artery, and the four lumbar arteries from the aorta; (3) the superficial epigastric, superficial circumflex iliac, and superficial external pudendal branches of the femoral artery; (4) the inferior epigastric and deep circumflex iliac branches of the external iliac artery; and (5) the iliolumbar branch of the internal iliac artery.

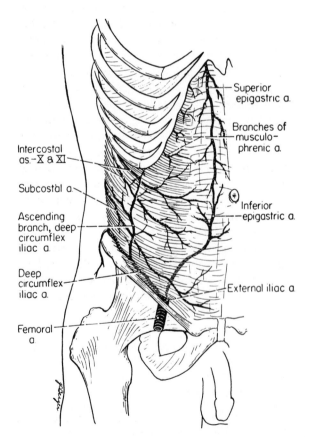

Intercostal
as.–X & XI

Subcostal a.

Ascending
branch, deep
circumflex
iliac a.

Deep
circumflex
iliac a.

Femoral
a

Superior
epigastric a.

Branches of
musculo-
phrenic a.

Inferior
epigastric a.

External iliac a

Fig. 310 The arteries of the anterolateral portion of the abdominal wall.

(1) The **superior epigastric** (fig. 306) and **musculophrenic arteries** enter the abdominal wall from above. They are described with the other branches of the internal thoracic artery (p. 319).

(2) The anterior and collateral branches of the **posterior intercostal arteries** of the tenth and eleventh intercostal spaces and the anterior branch of the **subcostal artery** continue into the abdominal wall. They supply the muscles of the wall and anastomose with the superior epigastric artery, with the upper lumbar arteries, and with each other. These arteries are more completely described in the chapter on the Chest (p. 318).

The **lumbar arteries** are in series with the posterior intercostal and subcostal vessels. Four in number, they arise from the back of the abdominal aorta at the levels of the bodies of the upper four lumbar vertebrae. They pass lateralward and posteriorly against the bodies of the vertebrae, posterior to the sympathetic trunk, and, on the right side, posterior to the inferior vena cava also. They lie behind the psoas muscle and the lumbar plexus and cross the quadratus lumborum, mostly posteriorly. The anterior branches of the lumbar arteries are small and extend but little beyond the lateral border of the quadratus lumborum muscle. Each artery also has a large posterior branch which, like the corresponding branch of the posterior intercostal arteries (p. 318), accompanies the dorsal ramus of the corresponding spinal nerve and divides into spinal and muscular branches. The spinal branch enters the intervertebral foramen and distributes to the spinal cord, the meninges, and the vertebrae. The large muscular branches supply the muscles of the back and send branches with the superior cluneal nerves to the overlying skin and subcutaneous tissues.

(3) The **superficial epigastric, superficial circumflex iliac,** and **superficial external pudendal branches** of the femoral artery are described among the superficial structures of the abdominal wall (p. 369).

(4) The **inferior epigastric** and **deep circumflex iliac arteries** are the principal branches of the external iliac artery. The abdominal aorta divides over the left side of the fourth lumbar vertebra into the two diverging common iliac arteries. After a course of about 5 cm., the common iliac arteries, in turn, divide into the external and internal iliac arteries, the division taking place at the level of the disk between the fifth lumbar vertebra and the sacrum. The external iliac artery furnishes the arterial supply of the lower limb; the internal iliac supplies predominantly the pelvis and the perineum. The com-

mon and external iliac arteries run a straight course of about 15 cm. to the midpoint between the anterior superior spine of the ilium and the symphysis pubis. Here the external iliac artery passes under the inguinal ligament and enters the thigh, its name changing to the femoral artery as it does so. In its descent the external iliac artery follows the medial border of the psoas major muscle, providing several small branches to it and to the neighboring lymph nodes. Its major branches arise just proximal to the inguinal ligament.

The **inferior epigastric artery** (figs. 306, 310, 378) arises from the external iliac immediately above the inguinal ligament and ascends obliquely toward the umbilicus in the extraperitoneal connective tissue of the anterior abdominal wall. Its origin is just medial to the deep inguinal ring and, consequently, the ductus deferens in the male (or round ligament of the uterus in the female) passes behind and then lateral to the artery to enter the inguinal canal. Nearing the arcuate line, the inferior epigastric artery pierces the transversalis fascia and enters the rectus sheath between the muscle and the posterior layer of the sheath. The artery provides numerous branches to the rectus abdominis muscle which anastomose with terminals of the superior epigastric and lower posterior intercostal arteries. Muscular branches originating proximal to the rectus portion of the vessel supply the flat abdominal muscles and the overlying skin and subcutaneous tissue.

Several small branches of the inferior epigastric artery arise near its origin. The **cremasteric artery** accompanies the spermatic cord and supplies the cremaster muscle and the other coverings of the cord. It anastomoses with branches of the testicular artery. In the female the small corresponding **artery of the round ligament** reaches the labium majus along the round ligament and anastomoses with the superficial external pudendal artery. A **pubic branch** of the inferior epigastric artery runs medialward along the inguinal ligament and then turns inward along the free border of the lacunar ligament to reach the back of the pubis. Here it anastomoses with the pubic branch of the obturator artery; it is the enlargement of this anastomosis which accounts for the frequent 'abnormal' obturator artery (p. 505).

The **deep circumflex iliac artery** (fig. 310) arises from the lateral aspect of the external iliac artery slightly distal to the inferior epigastric. It passes obliquely toward the anterior superior spine of the ilium in a fibrous canal formed between the transversalis fascia and the iliac fascia. Here it pierces the transversalis fascia and a little later the transversus abdominis muscle to take up the characteristic position of abdominal vessels between this muscle and the internal abdominal oblique. The deep circumflex iliac artery passes lateralward along the iliac crest, supplying the adjacent muscles and anastomosing with the superficial circumflex iliac, the iliolumbar, and the superior gluteal arteries. An **ascending branch,** frequently of considerable size, takes origin near the anterior superior iliac spine and distributes vertically within the muscular abdominal wall. It anastomoses with the subcostal and posterior intercostal arteries.

(5) The **iliolumbar artery** is a branch of the internal iliac artery, usually of its posterior trunk. Arising within the pelvis, the iliolumbar artery ascends anterior to the sacroiliac joint to the medial border of the psoas major muscle, behind which it divides into a lumbar and an iliac branch. The iliolumbar artery substitutes in part for a fifth lumbar segmental artery. Its **lumbar branch** supplies the psoas major and quadratus lumborum muscles, anastomoses with the last lumbar artery, and sends a small **spinal branch** into the vertebral canal to assist in the supply of the cauda equina. The **iliac branch** ramifies in the iliac fossa between the iliacus muscle and the ilium, supplying the muscle and providing a large nutrient artery to the bone.

Lymphatics The deep lymphatic vessels of the abdominal wall follow the blood vessels of the wall and may be divided into four principal groups: (1) A set of lymphatic vessels follows the inferior epigastric blood vessels and terminates in the external iliac nodes; (2) lymphatic vessels follow the deep circumflex iliac blood vessels to the same nodes; (3) a third set of channels follows the lumbar blood vessels to the nodes of the lumbar chain; and (4) the upper portion of the abdominal wall drains by channels which follow the superior epigastric blood vessels to the sternal nodes. Small lymphatic nodules are sometimes found along the inferior epigastric vessels and in the vicinity of the umbilicus.

SPECIAL ANATOMY OF THE INGUINAL REGION

Descent of the Testis (figs. 311, 313)

The testis develops in the extraperitoneal connective tissue of the upper lumbar region and, until nearly the end of intrauterine life, remains in the abdominal cavity. The testis forms in close association with the mesonephros, and their growing bulk pushes out the peritoneum covering them until, at 7 weeks of develop-

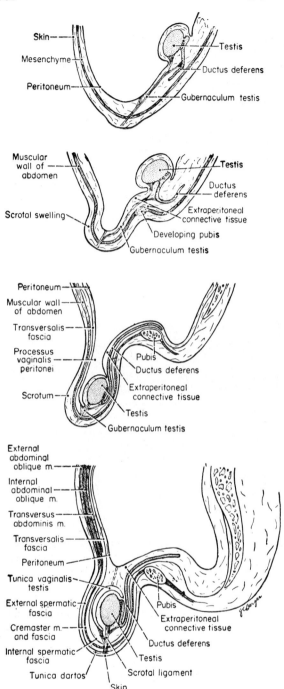

Fig. 311 The descent of the testis and the formation of its coverings and of the scrotum, diagrammed in four stages.

ment, the testis is intraperitoneal and is suspended by a mesorchium (Lemeh). At either end of the bulge the peritoneum is thrown into folds, the inferior one extending caudalward in the celomic cavity. At the same time a peritoneal fold progresses backward from the anterolateral abdominal wall. Fibrous connective tissue of the inferior fold from the testis and the mesonephros, the **genitoinguinal ligament,** coalesces with the connective tissue of the fold from the abdominal wall to form the **gubernaculum testis.** The gubernaculum thus becomes a connection between the developing testis in the lumbar region and the abdominal wall at the site of the future deep inguinal ring. The fold of peritoneum superior to the testis contains its blood vessels and may be designated as its vascular fold. Meanwhile, a celomic evagination begins in the inguinal region at the site of attachment of the gubernaculum. This peritoneal protrusion, the **processus vaginalis,** pushes into the scrotal swelling, the muscular and fascial layers of the abdominal wall necessarily being evaginated between the external skin and the internal peritoneum. As the processus vaginalis is deepened, its mouth becomes constricted. It would be incorrect to say that the gubernaculum draws the testis downward, for inequality in development and growth rates is probably fundamental to the apparent migrations of embryonic organs, but the net effect is much the same. By the third month of fetal life the testis is in the iliac fossa, and by the fifth month it lies close to the inguinal ring. The testis begins to pass through the ring during the seventh month of gestation and ordinarily lies in the scrotum by the eighth month. The passage of the processus vaginalis into the scrotal pouch thins out and carries down the inguinal portion of each of the abdominal wall layers and accounts for the presence in the adult scrotum of their derivatives, the internal and external spermatic fasciae and the cremaster muscle and its fascial layer. The gubernaculum testis becomes much reduced in the adult, its remnant constituting the **scrotal ligament** which extends from the inferior pole of the testis and the tail of the epididymis to the skin of the bottom of the scrotum. As development nears completion, the processus vaginalis becomes a thin tubular connection between the peritoneum of the abdominal cavity and that over the testis in the scrotal sac. Normally, both the abdominal and scrotal ends of this tube pinch off and the intervening part becomes atrophied, an indistinct connective tissue remnant being all that remains. The peritoneum covering the anterior and lateral aspects of the testis then forms a closed serous sac, the **tunica vaginalis testis,** and only a dimple in the abdominal peritoneum remains to indicate the site of the evagination. However, the processus vaginalis is said to be open in 50 per cent of infants until a month after birth, and irregularities in its obliteration have an important bearing on certain forms of inguinal hernia and hydrocele.

Clinical Note Hydrocele is a fluid accumulation within the tunica vaginalis testis due to secretion of abnormal amounts of serous fluid by the serous membrane. Occasionally the testis fails to descend or fails to make a complete descent. This condition is known as cryptorchidism. Abnormalities in descent may occur, as into the thigh or the perineum. Such testes are said to be ectopic in position.

Descent of the Ovary

Like the testes, the ovaries develop in the abdominal cavity and later migrate to the true pelvis. In the case of

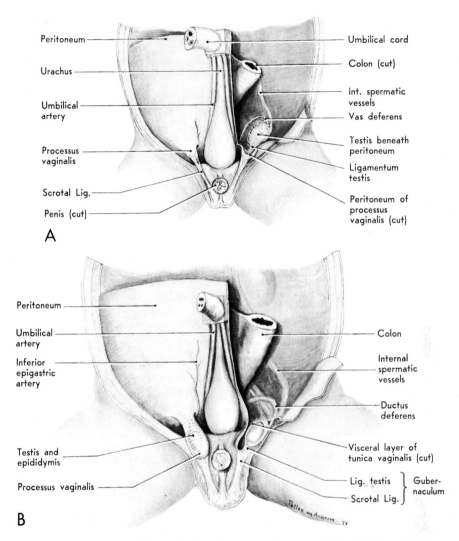

Fig. 312   The descent of the testis in relation to the development and growth of the bladder and abdomen. A—fetus of about 20 weeks age; B—fetus early in the seventh month. Internal spermatic vessels = testicular vessels. (By permission, from *Human Embryology,* 2nd ed., by B. M. Patten, 1953. Blakiston Div., McGraw-Hill Book Co.)

the ovary, too, a gubernaculum forms which connects it to the abdominal wall in the region where the labial swelling develops. There is also a vaginal process of peritoneum which enters the labial pouch. However, the gubernaculum in the female becomes attached to the paramesonephric ducts from which the uterus develops. As a result of this secondary attachment the ovary does not descend below the position of the uterus. The gubernaculum between the ovary and the uterus becomes the ovarian ligament, and between the uterus and the labium majus it forms the round ligament of the uterus. The round ligament of the uterus ends in the skin and connective tissue of the labium majus. The vaginal process of the peritoneum accompanies the round ligament of the uterus through the layers of the abdominal wall, but usually by the sixth month of development it is completely obliterated.

Clinical Note   Persistence of the vaginal process in the female (persistent 'canal of Nuck') leads to cysts and herniae of congenital origin.

The Inguinal Canal (fig. 309)

The evagination of the layers of the abdominal wall over the vaginal process of the peritoneum is originally straight dorsoventrally. However, with more mature development of the scrotal sac (or the labia majora), of the pelvis, and of the abdominal wall, the original straight line of passage through the wall becomes oblique, the point of entrance being more lateral than the point of exit. Thus, in the adult the abdominal wall contains an inguinal canal, about 4 cm. long, which lies above and parallel to the inguinal ligament. The entrance to the canal is the **deep inguinal ring,** located above the midpoint of the inguinal ligament; it is, instead of a ring-like opening, the mouth of a finger-like diverticulum of transversalis fascia. Into this sleeve-like process the spermatic cord (or the round ligament of the uterus in the female) passes to descend through the inguinal canal. The **superficial inguinal ring,** above and lateral to the pubic tubercle, constitutes the external opening of the canal.

The boundaries of the inguinal canal reflect its obliquity and depend on the behavior of the muscular and fascial layers in the inguinal region. The inguinal ligament forms the floor of the canal; the lacunar ligament underlying its medial end. The external oblique aponeurosis is superficial to the entire canal, its superficial inguinal ring marking the medial end of the canal. The transversalis fascia forms a posterior layer through the entire length of the canal medial to its formation of the deep ring; this layer is deep to the parts of the posterior wall contributed by both the internal oblique and transversus abdominis muscles. The internal oblique

muscle arises from the inguinal ligament in front of the deep ring and inserts behind and medial to the superficial ring. It thus forms part of the anterior wall of the canal lateralward and part of the posterior wall medialward and arches over the spermatic cord to assist in forming a roof for the canal in its intermediate portion. The lower fascicles of the transversus abdominis muscle arise from the iliacus fascia behind the lateral one-third of the inguinal ligament but fall short of covering the canal lateralward. Medialward, they have a broad posterior relationship to the canal through the interfoveolar ligament, the weak fascia, and the falx inguinalis.

The inguinal canal, in the male, is a passageway through the abdominal wall to the scrotum and testis for the elements of the **spermatic cord.** These are the ductus deferens, the deferential artery and vein, the testicular artery, the pampiniform plexus of veins, the lymphatics, and the autonomic nerves of the testis. Also traversing the inguinal canal are the **ilioinguinal nerve** and the genital branch of the **genitofemoral nerve.** The **cremasteric artery,** which arises from the inferior epigastric artery behind the

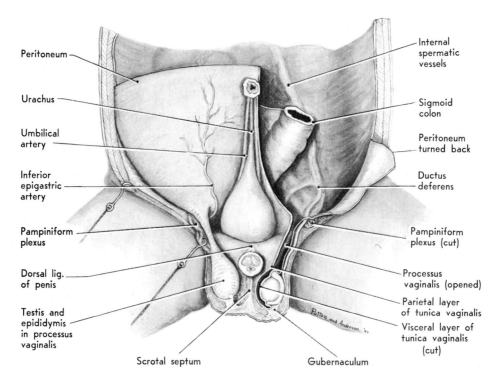

Fig. 313 The relations of the testis in a fetus of the ninth month. Internal spermatic vessels = testicular vessels. (By permission, from *Human Embryology,* 2nd ed., by B. M. Patten, 1953. Blakiston Div., McGraw-Hill Book Co.)

canal, penetrates its posterior wall and emerges with the spermatic cord at the superficial inguinal ring. The components of the spermatic cord are invested by the internal spermatic fascia all the way through the canal. They become covered also by the cremaster muscle and fascia in the middle of the canal but receive the external spermatic fascia, external to all, only as they emerge through the superficial inguinal ring. In the female the smaller inguinal canal accommodates the round ligament of the uterus, the artery and vein of the round ligament, and the ilioinguinal nerve. The homologues of the spermatic fasciae are difficult to delineate, but a few fibers of the cremaster muscle are usually observable along the round ligament. A tiny genital branch of the genitofemoral nerve can usually be traced to the cremaster fascicles.

The Scrotum and Coverings of the Cord and the Testis (fig. 306)

The scrotum is a cutaneous pouch formed of bilateral contributions which contain the testes and parts of the spermatic cord. Its bilateral development is seen in a median raphe which continues forward to the under surface of the penis and backward along the midline of the perineum to the anus. Internal to the raphe the **scrotal septum** divides the sac into its two chambers. The layers of the scrotum proper are the skin and the dartos. These, together with their nerves and blood vessels, are described on p. 368.

The coverings of the cord and testis are those derived from the abdominal wall in the descent of the testis. From without inward these are the delicate, membranous, **external spermatic fascia** representing the external abdominal oblique layer; the coarsely fasciculated **cremaster muscle** and its associated fascia, derived from the internal abdominal oblique layer; and the **internal spermatic fascia** derived from transversalis fascia. Internal to these layers, the structures of the spermatic cord lie embedded in areolar connective tissue which represents the extraperitoneal connective tissue of the abdomen.

The **tunica vaginalis testis** is the invaginated serous sac which partially covers the testis and represents the lower closed-off portion of the processus vaginalis of the peritoneum. The visceral layer is closely applied to the testis, the epididymis, and the lower part of the spermatic cord. It dips into the narrow interval between the body of the epidi-

dymis and the lateral surface of the testis, forming the slit-like **sinus of the epididymis.** The parietal layer is more extensive than the visceral, reaching onto the spermatic cord both above and below the testis and also covering its medial side. The apposed surfaces are smooth and are lubricated by a small amount of serous fluid.

Clinical Note **Inguinal Hernia** The passage of the spermatic cord through the abdominal wall produces a weakness in the inguinal region that predisposes it to inguinal hernia. Perhaps of even more direct significance is the frequent failure of obliteration of the vaginal process of the peritoneum. Herniation of abdominal contents into an unobliterated vaginal process and within the coverings of the spermatic cord results in an **indirect inguinal hernia.** This most common type of hernia, also producible after normal obliteration of the vaginal process, begins at the deep inguinal ring. The herniating structures traverse the inguinal canal, emerge at the superficial inguinal ring, and usually descend into the scrotum. The path of the indirect hernia is preformed; it is the path of evagination of the abdominal wall layers which occurred in the descent of the testis. The constituents of the spermatic cord are spread out over the peritoneal sac of the hernia, and these in turn are covered by the typical layers surrounding the cord and testis. Since such a hernia begins at the deep inguinal ring, the inferior epigastric artery is found medial to the neck of its sac. Electromyography indicates some protection for the

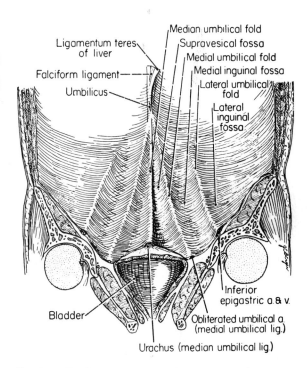

Fig. 314 The inguinal portion of the anterior abdominal wall as seen from behind.

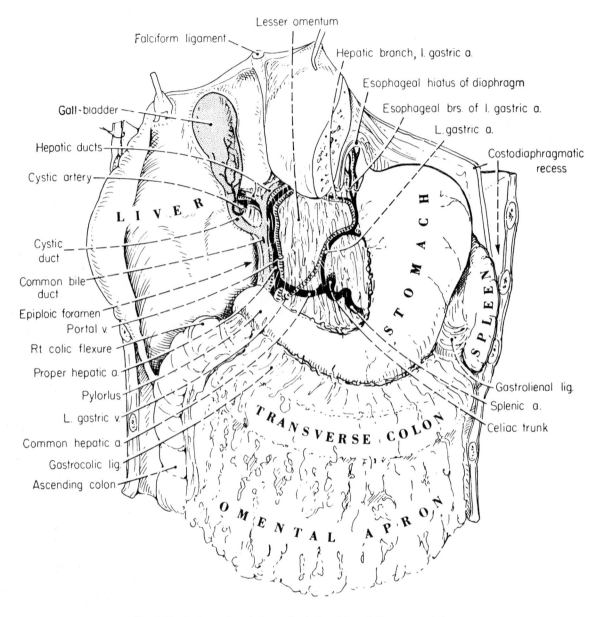

Falciform ligament

Lesser omentum

Hepatic branch, l. gastric a.

Esophageal hiatus of diaphragm

Gall-bladder

Esophageal brs. of l. gastric a.

L. gastric a.

Hepatic ducts

Costodiaphragmatic recess

Cystic artery

L I V E R

S T O M A C H

S P L E E N

Cystic duct

Common bile duct

Epiploic foramen

Portal v.

Rt. colic flexure

Proper hepatic a.

Pylorlus

L. gastric v.

Common hepatic a.

Gastrocolic lig.

Ascending colon

T R A N S V E R S E   C O L O N

O M E N T A L   A P R O N

Gastrolienal lig.

Splenic a.

Celiac trunk

Fig. 315 Peritoneal and visceral relationships of the upper abdomen.

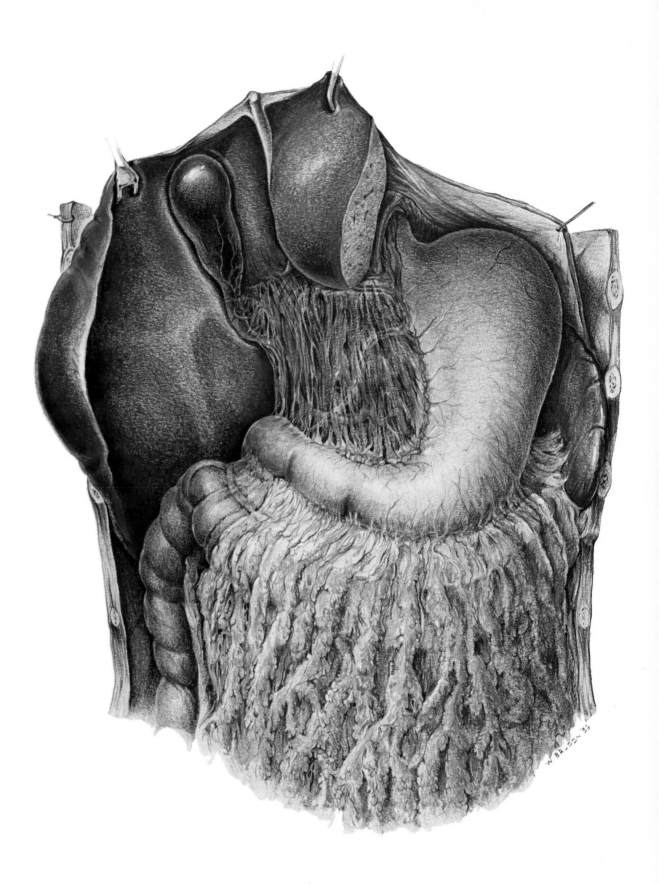

inguinal canal. The lower fibers of internal oblique and transversus abdominis are active in standing, and straining or coughing increases their activity.

Another type of inguinal hernia, the **direct inguinal hernia,** occurs less commonly (one-third as frequently as indirect herniae in males). The direct inguinal hernia bulges at the superficial inguinal ring and almost never descends into the scrotum. The weaknesses which predispose to hernia in this location are the superficial inguinal ring and the weakness of the abdominal wall lateral to the falx inguinalis. The inferior epigastric artery is lateral to the herniating mass in the direct type. The hernia begins internally between the lateral margin of the falx inguinalis and the interfoveolar ligament and, in contrast to the indirect type, pushes directly through the wall to the position of the superficial ring. The layers covering a hernia of this type are those of the abdominal wall in the region through which it passes: peritoneum, extraperitoneal connective tissue, transversalis fascia, the lateral portion of falx inguinalis or the weak fascia of the transversus abdominis muscle lateral to the falx, the cremaster muscle and fascia layer, intercrural fibers of the superficial inguinal ring, membranous and fatty layers of the subcutaneous tissue, and skin. On the rare occasion when a direct inguinal hernia turns down into the scrotum, it passes under the tunica dartos scroti and the skin of the scrotum.

The configuration of the peritoneal, or internal, aspect of the abdominal wall in the inguinal region (figs. 314, 378) also determines to some extent the typical locations and kind of inguinal herniae. In the median line the urachus, the remnant of the embryological allantoic duct, is represented in the adult by the fibrous **median umbilical ligament** and raises a **median umbilical fold** of peritoneum. Another degenerated embryological structure, the obliterated umbilical artery, remains in the adult as the **medial umbilical ligament.** This forms the **medial umbilical fold** of peritoneum. Farther lateralward, the inferior epigastric vessels elevate the **lateral umbilical fold** of peritoneum. The depressions which alternate lateral to each of these folds are, respectively, the supravesical fossa, the medial inguinal fossa, and the lateral inguinal fossa. The **lateral inguinal fossa,** lateral to the inferior epigastric vessels, is the site of evagination of the transversalis fascia which, as the deep inguinal ring, marks the starting point of indirect inguinal herniae. The **medial inguinal fossa,** between the inferior epigastric vessels and the obliterated umbilical artery, is the point of inception of direct inguinal herniae.

The inguinal region has a demonstrably firmer construction in the female, which is doubtless related to the passage of fewer and smaller structures through the abdominal wall. Accordingly, reports on the incidence of inguinal hernia in the female range from $\frac{1}{16}$th to $\frac{1}{20}$th the occurrence in males. Direct hernia is quite rare in the female, only $\frac{1}{12}$th as common as the indirect type. However, inguinal hernia, even though relatively infrequent, occurs two or three times as often as femoral hernia in the female.

The compressing force of the strong abdominal musculature may lead to rupture of the wall wherever a structural weakness occurs. Herniae through the umbilical scar, **umbilical herniae,** are fairly common in infants. **Femoral herniae** descend into the thigh behind the inguinal ligament, utilizing a tubular compartment of the femoral sheath, the femoral canal p. 534. **Obturator herniae** sometimes pass into the thigh at the obturator foramen p. 514. **Lumbar herniae** emerge in the muscular weakness of the lumbar triangle (p. 51). Regions of incomplete muscular closure of the diaphragm predispose to **diaphragmatic herniae** p. 450.

THE PERITONEUM

The **peritoneum** (figs. 315–323, 326), like the pleura and the tunica vaginalis testis, is an invaginated serous sac. The serous sac of the peritoneum is invaginated by and, therefore, forms a smooth covering layer for the gastrointestinal viscera of the abdominal cavity. The uninvaginated part of the peritoneum lines the inner surface of the abdominal wall and is separated from the transversalis fascia by loose, fat-containing extraperitoneal connective tissue. The smooth peritoneal covering of most of the organs of the abdominopelvic cavity provides for their gliding over one another freely in their normal movements and in displacements imposed by outside forces. The invaginated portions of the sac have a rather complex disposition in the adult, and various names are applied to their parts. The **lesser omentum** and the **greater omentum** are peritoneal sheets attaching the stomach to the body wall or to other abdominal organs. The **mesentery** is the peritoneal reflection from body wall to the small intestine, and the **mesocolon** is the similar attachment of the large intestine. Many peritoneal folds or sheets are designated as **ligaments** of the peritoneum. These totally lack the connective tissue character and strength of the usual ligaments of the body. **Folds of the peritoneum** are usually reflections of the peritoneum exhibiting more or less sharp borders.

The disposition of the peritoneum in the adult can appear meaningless and complex if it is examined without giving consideration to its developmental changes and modifications. It will, therefore, be described in relation to its embryology. In such a consideration the essential steps are (1) the primitive midline arrangement; (2) the counterclockwise rotation of the gastrointestinal tract and its consequences in the dorsal mesogastrium, together with the development of the omental bursa and the epiploic foramen; and (3) fusions and elaborations of peritoneal layers.

As an aid to identification, in both the adult pattern and the developmental history, the following listing of the principal parts of the peritoneum is presented. The peritoneal investment of intrinsically pelvic organs is covered in the chapter on the Pelvis.

I. **Derivatives of the dorsal common mesentery**

    The mesentery

    Transverse mesocolon

    Sigmoid mesocolon

    Ascending and descending mesocolon (not in the adult)

    Mesoappendix

II. **Derivatives of the dorsal mesogastrium**

    Greater omentum

        Gastrophrenic ligament

        Gastrolienal ligament

        Gastrocolic ligament

    Lienorenal (phrenicosplenic) ligament

    Phrenicocolic ligament

III. **Derivatives of the ventral mesogastrium**

    1. Lesser omentum

        Hepatogastric ligament

        Hepatoduodenal ligament

        Hepatocolic ligament (inconstant)

    2. Ligaments of the liver

        Falciform ligament

        Coronary ligament

        Right triangular ligament

        Left triangular ligament

The Primitive Midline Arrangement (figs. 316, 318)

The digestive tube forms in the midline by the sweeping together of the entodermal layers of each side of the body which are originally continuous with the extraembryonic membranes. The entoderm is backed by a layer of splanchnic mesoderm, so that the tube is double-layered. This outer covering of splanchnic mesoderm is separated by celomic space from a similar and continuous layer of somatic mesoderm which lines the body wall of the embryo. These mesodermal lining layers are primitive peritoneum and foreshadow in their relations to gut and wall the typical visceral and parietal layers of peritoneum. With the closure of the ventral body wall from the extraembryonic portions of the germ layers, the digestive tube lies as a midline structure attached, both dorsally and ventrally, to the body wall by the covering splanchnic mesoderm (figs. 316, 318). It is suspended and supported by both a dorsal and a ventral mesentery. The ventral mesentery is transient except in the region of the liver; it very early disappears inferior to the developing liver, the umbilical vein, and the common bile duct.

The stomach is indicated by a slight local dilatation of the gut at the end of the fourth week of development, and by six weeks its shape resembles that of the adult stomach. It is, however, a midline organ and its greater curvature faces dorsalward. Its peritoneal suspension, parts of the primitive common mesentery, may be designated as the **dorsal mesogastrium** from the dorsal body wall and the **ventral mesogastrium** from the ventral wall (fig. 316). At the end of the fourth week, the intestines are represented by the sagittally placed gut tube from the stomach to the cloaca. During the fifth week of development there is a rapid elongation of the intestinal segment, which results in a hairpin-shaped loop extending ventrally into the belly-stalk. This primary gut loop foreshadows the small intestine in its cephalic limb and through its U turn. The remainder of the caudal loop forms the large intestine. The intestine retains its dorsal body wall suspension as the **dorsal common mesentery.** The primordia of the pancreas, liver, and gall bladder appear as outgrowths of the gut tube just caudal to the stomach during the fourth week of development. The liver and the gall bladder are ventral outgrowths and develop in the **ventral mesogastrium;** the pancreas, although having two primordia, is definitively embedded in the **dorsal mesogastrium.** The original hepatic diverticulum grows rapidly, and the large liver mass which results is crowded ventrally and cephalically. The persistent proximal portion of the original diverticulum becomes the common bile duct, and an offshoot of it becomes enlarged to form the gall bladder (fig. 316). Thus, at approximately the end of the fifth week the gastrointestinal tract and its associated glands are recognizable. They all lie in the median line between the visceral layers of peritoneum which constitute the regionally designated dorsal common mesentery, dorsal mesogastrium, and ventral mesogastrium.

The Counterclockwise Rotation of the Tube (figs. 317–20)

A primary factor in the development of adult configurations in the gastrointestinal tube is a twist which occurs in the primary loop of the gut while it still extends out into the belly-stalk. Viewed from the ventral side, the twist is counterclockwise and

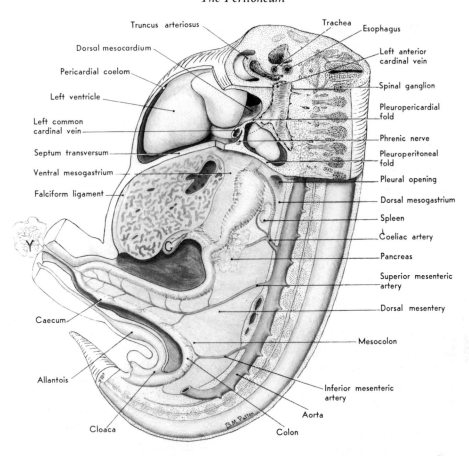

Fig. 316   The arrangement of the viscera and mesenteries of an embryo early in the seventh week of development. (By permission, from *Human Embryology,* 2nd ed., by B. M. Patten, 1953. Blakiston Div., McGraw-Hill Book Co.)

brings up cephalically and to the right the caudal part of the loop, representing the cecum and the large intestine, and turns the originally cephalic limb of the loop (small intestine) downward and to the left. The rotation of the loop takes place around an axis represented by the superior mesenteric artery, so that its branches to the gut simply rotate on the main line of the vessel. As the caudal part of the loop turns over the cephalic portion, the crossing segment becomes the transverse colon and the cecum terminates on the right side inferior to the liver. Exuberant growth of the small intestinal part of the tube begins immediately following the counterclockwise rotation of the tube and while the loop still projects into the belly-stalk. This results in elaborate coiling of this segment, and it is not until the tenth week of development that the abdomen has enlarged sufficiently to accommodate the growing intestinal tract. At that time the loop is pulled back into the abdominal cavity, the small intestinal coils returning first. They pre-empt the greater central part of the cavity and push the large intestine

to its left and superior margins. Since the exuberant growth of the small intestine is in its primary loop, its mesenteric attachment to the dorsal body wall remains restricted and its **mesentery** becomes a much folded, fan-like sheet.

The rotation of the gastrointestinal tract simultaneously throws a tight turn into the duodenum and turns the stomach from its median sagittal orientation to a frontal position. At the same time the stomach is carried caudalward. In consequence of these changes, the greater curvature of the stomach is directed toward the left and downward; the lesser curvature, to the right and upward. As the greater curvature swings to the left, the dorsal mesogastrium accommodates to this change by ballooning out to the left behind the stomach. This process carries the pancreas to the left and against the dorsal body wall and flattens it against the left kidney posteriorly. While the stomach is still in its early phases of rotation, the spleen begins to form between the layers of the dorsal mesogastrium in such a

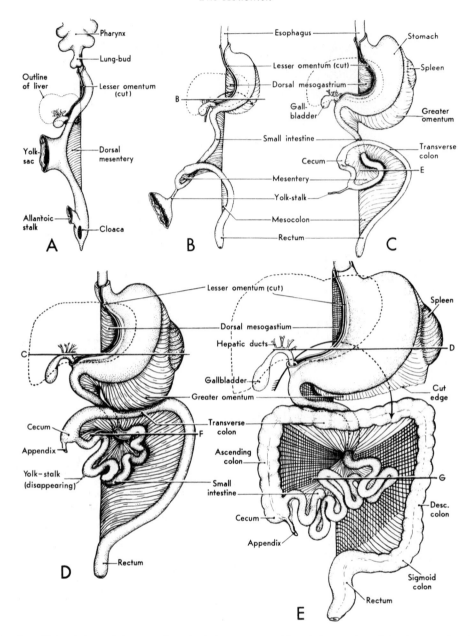

Fig. 317 A series of frontal views of the gastrointestinal viscera summarizing their growth and development in relation to the mesenteries of the abdominal cavity. Heavy horizontal lines, identified by letters, indicate the planes of the cross sections of fig. 314. Cross-hatched areas of the mesenteries in E indicate where fusion fasciae are formed. (By permission, from *Human Embryology*, 2nd ed., by B. M. Patten, 1953. Blakiston Div., McGraw-Hill Book Co.)

position that the ballooning out of the dorsal mesogastrium carries it far to the left. The rapid growth of the spleen itself causes it to expand the left face of the mesogastrium, and it nears its adult position under the lower ribs. The **greater omentum** is the term now applied to the dorsal mesogastrium as it leaves the stomach, and its local parts can be designated as the **gastrolienal ligament,** between the greater curvature of the

stomach and the spleen, and the **gastrophrenic ligament,** from the stomach to that part of the body wall represented by the diaphragm. The greater omentum grows especially freely from the inferior portion of the greater curvature of the stomach. Here the two layers of peritoneum sweeping from the two surfaces of the stomach grow caudalward in a duplication which extends ventrally and inferiorly beyond the transverse

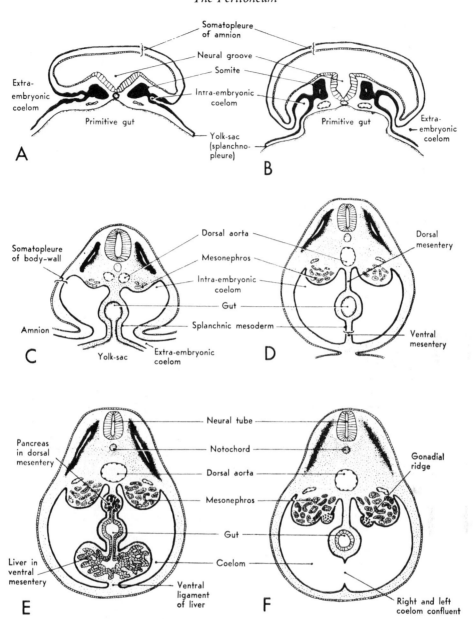

Fig. 318 Schematic transverse sections showing closing off of embryonic gut from primitive gut, separation of intra- from extra-embryonic coelom, and development of primary mesenteries. (By permission, from *Human Embryology*, 2nd ed., by B. M. Patten, 1953. Blakiston Div., McGraw-Hill Book Co.)

colon. Hanging down over the transverse colon, the double layers are still attached to the body wall dorsally; consequently, as they descend and then turn back superiorly, they come to lie against each other. This apposition constitutes the four-layered, fat-laden **omental apron** (figs. 315, 320), in which an actual or potential space still separates the two sets of layers in the adult. The layers returning from the fold to the body wall lie against the transverse colon and the superior layer of its mesocolon, and these apposed parts fuse. Thus, the inferior extension of the greater omentum becomes the **gastrocolic ligament** and helps form the **transverse mesocolon.**

As the gastrocolic ligament is traced to the left, it leaves the stomach and reaches the body wall over the diaphragm. This attachment of the dorsal meso-gastrium between the diaphragm and the left end of the transverse colon (splenic flexure) is the **phrenicocolic ligament.** It forms a horizontal shelf on which rests the spleen, a circumstance reflected in its alternative name of **sustentaculum lienis.**

With the assumption of a frontal orientation by the stomach, its lesser curvature faces to the right and the ventral mesogastrium is shifted to that side. The ventral mesogastrium (fig. 316) is the remains of the ventral mesentery after the very early disappearance of its caudal portion. The ventral mesogastrium is limited caudally by the umbilical vein and the hepatic diverticulum. The latter grows out from the entoderm of the duodenum during the fourth week of development and provides an offshoot which becomes the gall bladder. The hepatic diverticulum develops in a ventral and cephalic direction, and the liver mass formed from it spreads apart the two layers of the ventral mesogastrium. Cephalically, the liver cells grow into the septum transversum, the primordium of a part of the diaphragm, and this relationship persists in the adult in the adhesion of the two structures and in the presence of an area bare of peritoneum over the posterosuperior surface of the liver. The extreme growth of the liver spreads widely the two layers of the ventral mesogastrium, and the liver comes to occupy most of the right superior quadrant of the abdomen. The adult peritoneal relations are foreshadowed in this developmental sequence. The reflection of peritoneum from the liver to the diaphragm is determined by the partial fusion of the developing liver and the septum transversum. This originally circular reflection, the **coronary ligament,** is actually drawn out at its right and left extremities in the adult to constitute the **right triangular** and **left triangular ligaments** (fig. 321, 341). The **lesser omentum,** the double layer remaining in the interval between the stomach and the liver, encloses the common bile duct (hepatic diverticulum) and is a continuous sheet which includes both the **hepatogastric ligament** and the **hepatoduodenal ligament** (fig. 325). In some specimens this sheet continues farther to the right and invests the right flexure of the transverse colon as the **hepatocolic ligament.** This ligament is inconstant and results from an adhesion between the right flexure of the colon and the duodenum and an extension of peritoneum over both of them rather than following from a true developmental sequence. With the obliteration of its lumen, the umbilical vein is converted into a fibrous cord which is designated, in the adult, as the ligamentum

teres and the ligamentum venosum of the liver. The sagittally oriented, blade-like peritoneal reflection which encloses the ligamentum teres extends from the umbilicus to the liver as the **falciform ligament** (figs. 315, 316, 319–21). It is continuous with the coronary ligament on the upper surface of the liver and with the lesser omentum at the hilum of the liver.

The leftward expansion of the greater omentum creates a space behind the stomach which is continuous with the whole right side of the peritoneal cavity at this stage. This large bay is the **omental bursa** (figs. 319, 320); it has an inferior extension, or recess, behind the gastrocolic ligament between the duplicated layers of the omental apron. The omental bursa is retained in essentially these developmental relations to the adult stage, but its original wide communication with the right peritoneal space becomes constricted to an extreme degree. An important factor in this constriction is the growth of the liver in the ventral mesogastrium. The adult liver constitutes about one-fortieth of the total body weight and enlarges to this size between the layers of the ventral mesogastrium in the superior right quadrant of the abdomen. It thus fills and obliterates much of the right superolateral peritoneal cavity, and its caudate lobe forms the superior boundary of the entrance to the bursa omentalis, the **epiploic foramen** (figs. 315, 319D). The lesser omentum, like the stomach, becomes frontally oriented, and this sheet, carried posteriorly by its hepatic and duodenal attachments, bounds the epiploic foramen ventrally. Posterior to the foramen is the dorsal body wall, here represented by the inferior vena cava clothed by posterior body wall peritoneum. The inferior boundary of the epiploic foramen is the first part of the duodenum. The duodenum, carried posteriorly and into a frontal orientation by the intestinal rotation, has its first part tipped superiorly in the establishment of the oblique longitudinal axis of the stomach. Fusion of the duodenum to the body wall by resorption of its posterior (originally right) peritoneal surface fixes it in position and makes it an unyielding inferior boundary of the epiploic foramen. Such is the reduction of the originally wide entrance to the bursa omentalis that the adult epiploic foramen will admit only one or two fingers (fig. 319D at arrow & figs. 315, 325).

The adult **bursa omentalis** (lesser peritoneal sac) is a peritoneal space behind the stomach which is

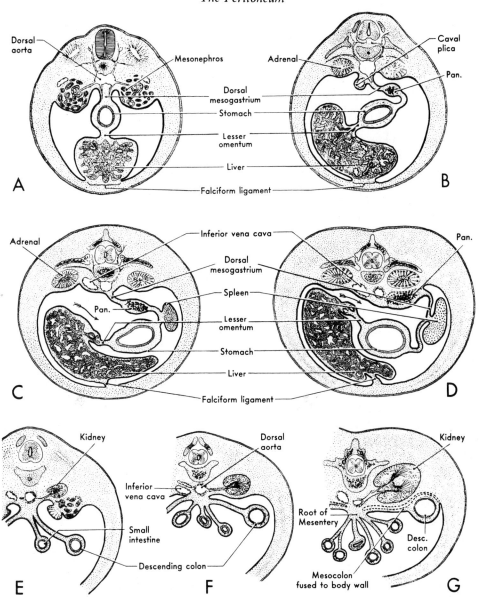

Fig. 319 The shifts of visceral position and the changing relations to the mesenteries as seen in cross sections. The planes are those indicated in fig. 317 by heavy horizontal lines and designated by the letters A to D for sections through the omental bursa and E to G for sections at the level of the kidneys. (By permission, from *Human Embryology,* 2nd ed., by B. M. Patten, 1953. Blakiston Div., McGraw-Hill Book Co.)

closed off from the major peritoneal cavity (greater peritoneal sac) except for the communication through the epiploic foramen (figs. 319, 320, 356). The bursa is bounded anteriorly, from above downward, by the lesser omentum, the posterior wall of the stomach and the gastrocolic ligament. Continuing through its inferior recess the limiting structures below are the transverse colon and its mesocolon and the anterior surface of the pancreas.

Ascending the posterior wall the bursa is bounded by posterior body wall peritoneum over the left kidney and suprarenal gland and is limited at its left extremity by the lienorenal, lienophrenic, and gastrophrenic ligaments and their enclosed nerves and blood and lymphatic vessels. The superior recess of the bursa is limited above by the left lobe of the liver and the posterior layer of the coronary ligament.

Fusions and Elaborations (figs. 317–21)

With the rotational changes in the abdomen and with the crowding of the abdominal space consequent to the growth of organs, some structures come to lie relatively immovably against one another or against the retroperitoneal organs of the posterior body wall. The serous layer of the peritoneum is, like the lining of a bursa, moist and slick and allows organs so covered to glide easily over one another. Where movement is not allowed by other factors, it is characteristic for the specific cells of the serous layer to degenerate, leaving behind the connective tissue substrate. This process has two results: (1) fusion of the adjacent structures by loss of the lubricating serous layer between them and (2) the formation of a **fusion fascia,** the double connective tissue plane which remains after loss of the serous cells.

Clinical Note   A fusion fascia has surgical importance, for being a peritoneal remnant along which or across which no blood vessels, nerves, or other accessory structures passed, it can be invaded in the adult body without fear of encountering any important vessels or nerves.

Peritoneal fusions are especially involved in the growth of the pancreas. The pancreas develops by outgrowths from the duodenum, a large primordium invading the layers of the dorsal mesogastrium and a small primordium associated first with the bile ducts. The smaller ventral primordium turns to the right and posteriorly around the bile ducts and becomes associated with the dorsal mesogastrium (fig. 339). As the sac of the omental bursa expands behind the stomach, the pancreas becomes frontally oriented in the posterior wall of the bursa and, with continued growth of the bursa, comes to lie relatively immovably against the vertebral column and the left kidney. The serosal covering of its posterior surface breaks down together with that of the opposing abdominal wall peritoneum; the pancreas now becomes retroperitoneal, and a fusion fascia is left behind it. As a consequence of this loss of peritoneal space behind the pancreas, the posterior part of the peritoneum of the omental bursa reaches the body wall not at the median line but lateralward over the left kidney and left half of the diaphragm (fig. 321). For this reason, the posterior portion of the dorsal mesogastrium becomes, in the adult, the **phrenicolienal ligament** or, as it reaches the kidney, the **lienorenal ligament.**

The counterclockwise rotation of the intestinal loop throws a tight turn into the duodenum and deviates that turn to the right and posteriorly, so that, with continued crowding from the overlying transverse colon, the duodenum comes to lie against the dorsal body wall. Here its posterior serosal surface is resorbed except for a small segment at its beginning. Since the head of the pancreas develops within the half-circle of the duodenum, the fusion fascia of both structures is in continuity as is also the peritoneum remaining on their ventral surfaces. This ventral peritoneum is also partially lost, for the transverse colon, on its return to the abdominal cavity, lies across both the duodenum and the pancreas. The transverse mesocolon (fig. 321) substitutes for the ventral peritoneum of these organs along the line of its superposition.

In the return of the intestinal loop to the fetal abdominal cavity, the large intestine is crowded to the left and superiorly (figs. 317-21). This establishes the descending colon and the sigmoid colon at the left margin of the abdominal cavity, the transverse colon across the pancreas and the duodenum, and the cecum on the right side inferior to the liver. The cecum later descends into the right iliac fossa, and a greater length of ascending colon results. The peritoneal sheet attaching the colon to the posterior midline becomes gathered in the region of the twist of the loop and is carried from that point across the duodenum and above the loops of the small intestine to the right side of the abdomen. It then grows downward with the cecum. As the ascending and descending parts of the colon become stabilized in the right and left lumbar fossae, their posterior serosal investment and that of the apposed peritoneal layers running to the midline are resorbed. These segments of the large intestine thus become retroperitoneal in position, and a fusion fascia exists behind them on both sides of the lumbar region of the abdomen (figs. 317, 319, 321). The cecum and the transverse and sigmoid segments of the large intestine retain their complete peritoneal investment. A rectal fusion fascia is described in the consideration of the rectum (p. 485).

The principal elaboration of the developing peritoneum is that of the omental apron (fig. 315). Its formation is described on p. 391. Although it normally hangs down over the transverse colon and the small intestine like an apron, it may be found tucked tightly into a recess or around one of the abdominal viscera, for it has the capacity of being drawn to an area of infection or inflammation and walling it off from the rest of the abdominal structures. The omental apron is also one of the fat

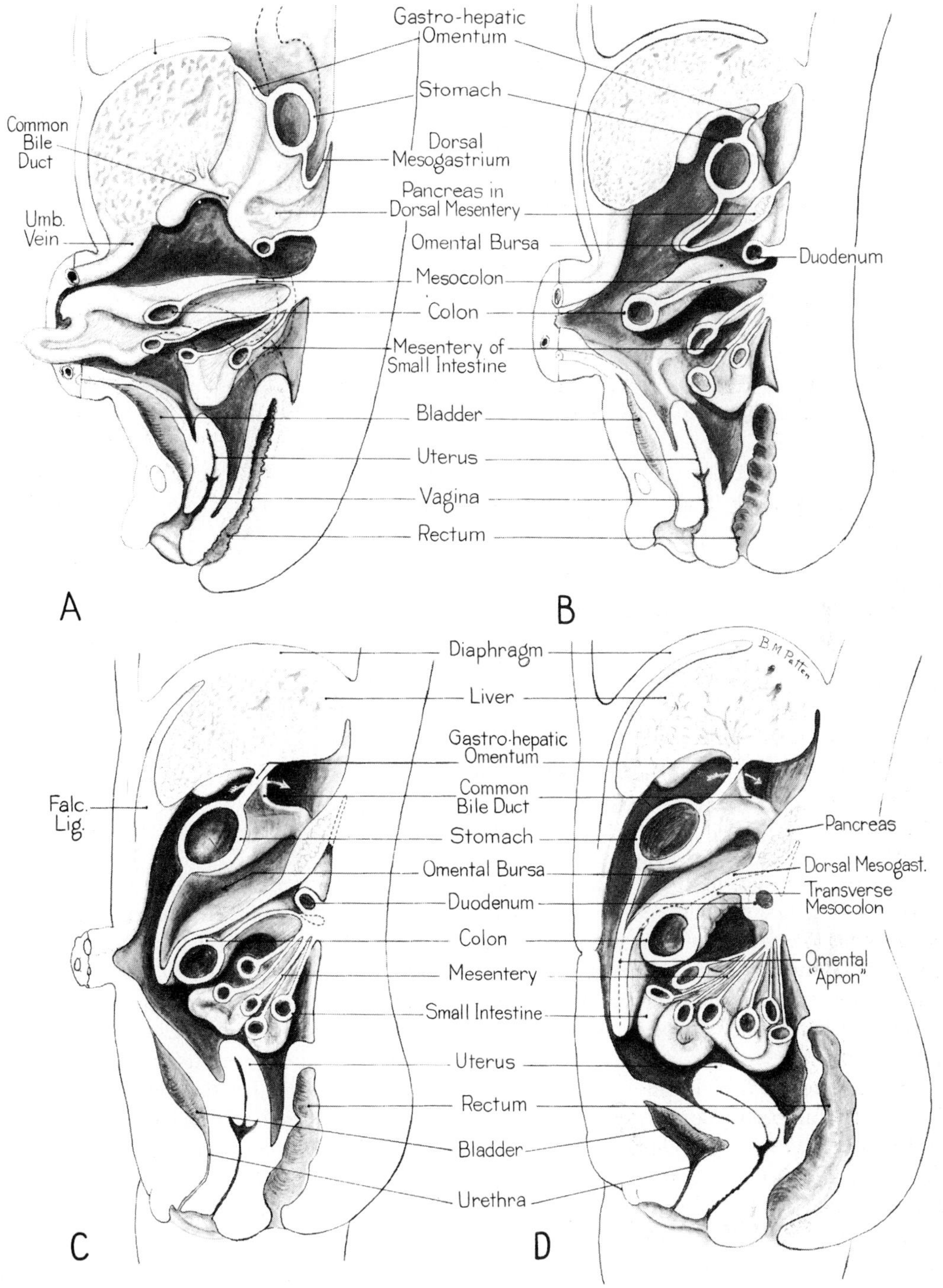

Fig. 320   The changes in the mesenteric relations of the gastrointestinal viscera as seen in schematized longitudinal sections of the body. White arrows in C and D lie in the epiploic foramen. (By permission, from *Human Embryology.* 2nd ed., by B. M. Patten, 1953. Blakiston Div., McGraw-Hill Book Co.)

depots of the body, and in all except very slender specimens it exhibits lobulated fat within its meshes.

The development of the peritoneum foreshadows the adult disposition, and the important layers and attachments have been listed. The adult configuration departs considerably from the original midline arrangement, and the observation of the attachments to the posterior body wall in the adult (fig. 321) gives a measure of the changes which have

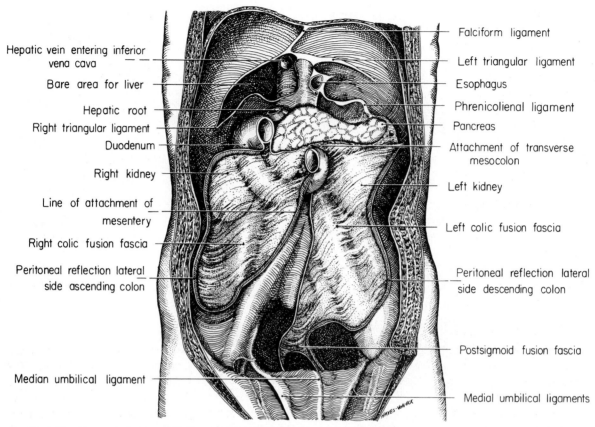

Hepatic vein entering inferior vena cava

Bare area for liver

Hepatic root

Right triangular ligament

Duodenum

Right kidney

Line of attachment of mesentery

Right colic fusion fascia

Peritoneal reflection lateral side ascending colon

Median umbilical ligament

Falciform ligament

Left triangular ligament

Esophagus

Phrenicolienal ligament

Pancreas

Attachment of transverse mesocolon

Left kidney

Left colic fusion fascia

Peritoneal reflection lateral side descending colon

Postsigmoid fusion fascia

Medial umbilical ligaments

Fig. 321   The lines of attachment of the gastrointestinal mesenteries which remain after removal of the viscera. The colic fusion fasciae are shown (from original of fig. 7. Mark A. Hayes *Am. J. Anat.,* 87:141, 1950.)

ensued. At the superior limit of the abdominal cavity, the parietal peritoneum of the falciform ligament spreads onto the superior surface of the liver and is continued to its posterior and inferior surfaces. The line of reflection from the diaphragm to the liver is the coronary ligament. The tapered poles of this originally circular reflection constitute the triangular ligaments. From the posterior leaves of the left and right triangular ligaments, the peritoneum passes to surround the esophagus and then becomes the attached parts of the dorsal mesogastrium—gastrophrenic, lienorenal, and phrenicocolic ligaments. The phrenicocolic ligament leads into the transverse mesocolon, the superior leaf of which may be traced to the left to the first part of the duodenum. The peritoneal investment of the rest of the duodenum is lost through fusion, but the mesentery reappears at the duodenojejunal flexure below the attachment of the transverse mesocolon to the left of the second lumbar vertebra. The mesenteric attachment is short and oblique. About six inches long, it extends from the left of the second lumbar

vertebra downward and to the right to end in the right iliac fossa. Here it becomes continuous with the broad posterior wall attachment of the ascending colon. The body wall attachment of the ascending colon goes into direct continuation with the root of the transverse mesocolon which is succeeded by the attachment of the descending colon, the root of the sigmoid mesocolon, and finally the attachment of the rectum. The disposition of the peritoneum in the pelvis is considered in the chapter on the Pelvis.

Folds and Fossae (figs. 322, 323)

Certain constant and a number of inconstant folds of peritoneum exist in the abdominal cavity. Some of these conduct blood vessels; others do not. In many cases their importance stems from the fact that recesses or fossae occur under them, into which it is possible for a knuckle of intestine to burrow and, with enlargment of the recess, to form an intraperitoneal hernia.

In the region of the ileocecal junction (fig. 322) several folds and fossae occur. From the cecum, 2 or 3 cm. below the ileocecal junction, the appendix arises. It is provided with a mesentery, the **meso-appendix,** which is a continuation of the mesentery extending to the appendix behind the terminal ileum. In the free margin of this small fold runs the **appendicular artery,** one of the terminals of the **superior mesenteric artery.** The appendicular artery sends small branches through the mesoappendix to the appendix. The **vascular cecal fold** is a constant fold which spans from the mesentery in the ileocecal angle across to the anterior surface of the cecum. In this fold runs the **anterior cecal artery,** another terminal branch of the superior mesenteric artery. Between the vascular cecal fold and the ileocecal junction is the **superior ileocecal fossa.** It is entered from below behind the vascular fold. An **ileocecal fold** is variable in this location. It extends from the surface of the terminal ileum to the cecum and crosses the root of the appendix; it carries no important blood vessels. The **inferior ileocecal fossa** lies between this fold and the mesoappendix and is entered from below. Behind the cecum the reflection of its peritoneum to the posterior body wall in the iliac fossa may form several sharp folds. These are the **cecal folds.** A shallow cul-de-sac between the cecal folds usually exists as a **retrocecal fossa.**

The region of the duodenojejunal junction is another area characterized by folds and fossae (fig. 323). Here at the transition of the duodenum into the jejunum, the small intestine emerges from a retroperitoneal into a peritoneal position, and at the top of its bend a fold of peritoneum passes upward and to the left. This is the **duodenojejunal fold.** A recess under it, entered from below, is the **superior duodenal fossa.** Inferiorly, extending to the

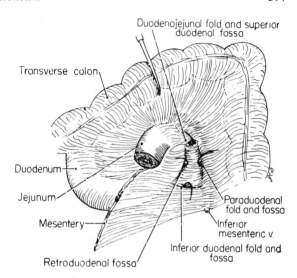

Fig. 323   Peritoneal folds and fossae at the duodenojejunal flexure.

left from the last segment of the duodenum, there is less commonly an **inferior duodenal fold.** This covers an **inferior duodenal fossa** which is entered from above. Although the last segment of the duodenum may be firmly held down by the peritoneal reflection across it, it is frequently covered on its underside by peritoneum, so that a peritoneal recess exists under it. This recess, the **retroduodenal fossa,** lies between the superior and inferior duodenal fossae. To the left of the duodenum, the inferior mesenteric vein passes cephalically behind the peritoneum and curves medialward above the duodenojejunal flexure to the back of the pancreas. This vein occasionally raises a fold of peritoneum over it, and a fossa of variable size may exist behind the fold. These are the inconstant **para-duodenal fold** and **paraduodenal fossa.** It will be noted that all four of these duodenal fossae could become confluent with each other to form a larger local sac of peritoneum. An **intersigmoid fossa** lies between the sigmoid mesocolon and the body wall peritoneum of the left iliac fossa. It represents the deepest part of the reflection of peritoneum between the body wall and the sigmoid colon.

Vessels and Nerves of the Peritoneum

The peritoneum is supplied with small blood vessels derived from various sources. The vessels of the abdominal wall serve the parietal peritoneum, whereas the peritoneum investing the organs is supplied through the visceral blood vessels. Lymphatic vessels also occur under the peritoneum. They drain regionally following

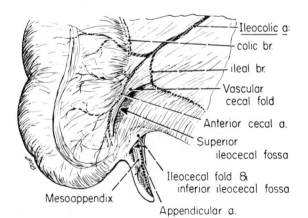

Fig. 322   Peritoneal folds and fossae in the region of the ileocecal junction.

the blood vessel pattern. The nerves of the peritoneum are relatively sparse. Sensory nerves are more frequent in the parietal peritoneum and are branches of the abdominal wall nerves—the seventh to the eleventh intercostal, subcostal, iliohypogastric, and ilioinguinal nerves. Vasomotor nerves are of sympathetic origin. The visceral peritoneum is insensitive to pain, touch, and temperature changes whereas the somatic sensory nerves supplying the parietal peritoneum provide acute, well localized sensation.

DIGESTION

Digestion is the process of breaking down ingested food into simpler chemical compounds. Such compounds are then absorbed into the body through the wall of the digestive tract, and the indigestible material is excreted. The system which produces these results extends from the oral cavity to the anus and is composed of the mouth, the pharynx, the esophagus, the stomach, the small intestine, and the large intestine (fig. 324). The tube is characterized by a moist surface epithelium containing millions of tiny glands producing mucus and a variety of enzymes for the breakdown of the food material. It has a relatively thick, muscular wall which is concerned with churning, mixing, and moving the food along the tract and a specialized surface for the absorption of the simpler chemical compounds resulting from enzyme action. Certain of the digestive enzymes are produced by the glands of the digestive tract; others are elaborated by organs accessory to it. Among such organs are the parotid, submandibular, and sublingual salivary glands, which empty their secretions into the mouth, and the liver and the pancreas of the abdomen. The elaboration of enzymes and the absorption of the assimilable chemical substances released is the essence of digestion. It requires, principally, a biochemical rather than an anatomical explanation.

The muscular layers of the digestive tract, an inner circular and an outer longitudinal layer, keep the mixed food and enzymes moving along the tract, an action known as peristalsis. **Peristalsis** has often been described as a wave of contraction spreading along the intestinal tube. Actually it is more than that. It has been found that a stimulus applied to the intestinal mucosa is followed by contraction of the longitudinal muscle at and below the stimulus and, three to five seconds later, by contraction of the circular muscle at and above it. Contraction of the longitudinal muscle distal to the food material shortens and thickens the longi-

tudinal fibers. Pressing against one another, they move the intestinal wall outward and enlarge the diameter of the lumen. The succeeding contraction of circular fibers proximal to the enlarged segment propels the intestinal contents downward, and this process, repeating itself at the next level, keeps the contained material moving onward.

The stomach receives mixed food and saliva from the esophagus. It is probable that the cardiac sphincter of the stomach is open when the stomach is empty but closed when food begins to be ingested. It then opens only in front of each oncoming peristaltic wave. Its closure at other times prevents the contractions of the stomach from forcing food back up the esophagus. When the stomach becomes partially filled, waves of contraction appear which originate about halfway along its length and progress toward the pylorus. These contraction waves become stronger as digestion proceeds, churning the food, mixing it with gastric juice, and reducing it to a semiliquid material known as **chyme.** Liquids pass the pyloric sphincter almost immediately, but solid material may remain in the stomach several hours before it is sufficiently liquefied to be passed on into the duodenum. The **pyloric sphincter,** formed by a greatly thickened circular layer at the termination of the stomach, relaxes ahead of contraction waves from the adjacent portion of the stomach but does not open with every cycle of contraction. Among other factors, it is under hormonal control from the duodenum.

Digestion is not completed in the stomach. As the chyme enters the duodenum, proteins and carbohydrates are only partially digested and fat shows little or no change. Fortunately for those patients who have to undergo surgical removal of the stomach, the digestion of all types of food can be completed in the small intestine. Progress through the small intestine is slow; it takes food from three to five hours to traverse the twenty feet of small intestine. Thus, ample time is allowed for complete digestion and for absorption as well. Peristalsis moves the food along the tract and brings it into contact with ever fresh epithelial surfaces. The waves move slowly (1 cm. per second) and at regular intervals. **Segmenting action** of the muscles of the small intestine does not move the contents through the tube but mixes and churns it. This action is produced by local contractions of the circular muscles which appear and disappear at the rate of from ten to thirty times a minute. **Pendular movements,** ring-like contractions of the circular

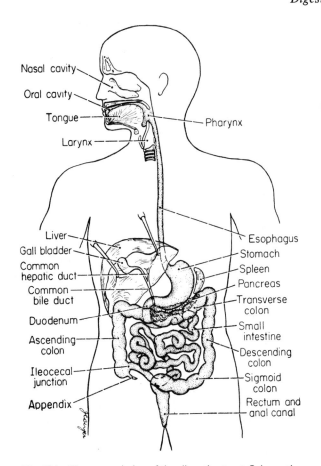

Nasal cavity

Oral cavity

Tongue

Larynx

Pharynx

Liver

Gall bladder

Common
hepatic duct

Common
bile duct

Duodenum

Ascending
colon

Ileocecal
junction

Appendix

Esophagus

Stomach

Spleen

Pancreas

Transverse
colon

Small
intestine

Descending
colon

Sigmoid
colon

Rectum and
anal canal

Fig. 324   The general plan of the digestive tract. Schematic.

muscle which travel down the intestine for short distances and occasionally reverse themselves, also serve to mix the intestinal chyme.

The movements of the large intestine are similar to those of the small intestine, but they are slower. The colon shows long periods of inactivity and regular peristalsis is rare. **Mass,** or rush, **peristalsis** is the type of movement characteristic of the large intestine. It occurs usually after meals and is reflexly produced by the entrance of food into the stomach. It is characterized by a strong peristaltic wave which sweeps over most of the length of the colon, moving almost all the contained material into the terminal colon and the rectum. This sudden filling of the rectum stimulates the desire to defecate. Rectal pressure, usually low, rises with the arrival of fecal material and is further increased as additional peristaltic rushes reach the rectum. Powerful contractions of the longitudinal muscles of the colon increase the pressure. In the anal canal a considerable thickening of the circular muscle layer forms the **internal anal sphincter,** and the

anal aperture is further guarded by a powerful skeletal muscle, the **external anal sphincter.** Both of these must be relaxed before defecation can take place. The internal sphincter reacts to the rectal pressure, but the external sphincter is under the control of the will. Voluntary contractions of the diaphragm and the abdominal wall musculature serve to increase intra-abdominal pressure and to facilitate the act of defecation.

The **autonomic nerves** of the alimentary tract have specific secretory effects in its various parts but are especially concerned with motor activity. The parasympathetic nerves increase motility in the tract and are important in keeping the chyme moving along. The sympathetic nerves slow peristalsis and thus oppose the emptying mechanism. The parasympathetic supply is effected by preganglionic fibers which reach the tube and synapse with the cell bodies of neurons located in the wall of the tract. In the intestines there is a nerve and ganglionic plexus between the circular and longitudinal muscle layers, the **myenteric plexus,** and another in the submucosal connective tissue, the **submucosal plexus.** The cell bodies within these plexuses are those of the postganglionic neurons in the two-neuron parasympathetic innervation of the intestinal tract; their axons are very short and are contained within the wall. The sympathetic neurons reaching the intestinal tract are already postganglionic and contribute to the myenteric and submucosal plexuses only as terminal nerve fibers; they have no cell bodies there. The myenteric plexus is, then, composed of mixed preganglionic and postganglionic parasympathetic and postganglionic sympathetic elements concerned with the regulation of the muscular layers of the intestines. The submucosal plexus, of similar constitution, is concerned with the regulation of the various glands of the mucosa and of the smooth muscle of the muscularis mucosae. Peristalsis depends on the integrity of the myenteric plexus, but the muscular tone of the layers and the segmenting movements of the tube seem to depend only on the intrinsic contractility of the smooth muscle fibers.

Clinical Note   Variation exists in the number of ganglion cells of the myenteric and submucosal plexuses. Increase in the cells of the myenteric plexus appears to result in increased muscular activity in the tube and to underlie, in the extreme case, ulcerative disease entities. Conversely, too few ganglion cells are thought to result in stasis and failure of normal evacuation. Pain afferents from the abdominal viscera are carried in sympathetic nerves.

THE STOMACH

GENERAL RELATIONS (figs. 315, 324–27, 331)

The **stomach** is the first abdominal representative of the digestive tract (for terminal esophagus, see p. 358). It lies principally in the left superior quadrant of the abdomen but terminates across the median line and frequently descends below the plane of the umbilicus. The position of the stomach is variable, for it is firmly fixed only at its beginning and at its end and is invested in its intermediate parts by the broad and movable sheets of the greater and lesser omenta. The stomach receives the termination of the esophagus 1 or 2 cm. below the esophageal hiatus of the diaphragm. Here the esophageal attachments of the diaphragm and the peritoneal reflections (gastrophrenic ligament) give a relatively stable location to the cardiac end of the stomach. The esophagogastric junction is on the horizontal plane of the tip of the xiphoid process and to the left of the eleventh thoracic vertebral body. Below, the terminal, or pyloric, portion of the stomach is in continuity with the retroperitoneal duodenum, and the fixation of the latter to the posterior body wall determines the level of the pyloric end of the stomach. The first part of the duodenum ascends as high as the body of the first lumbar vertebra, so that the termination of the stomach is usually (and most reliably in the supine position of the body) at the level of the body of the first or the second lumbar vertebra. Between these fixed points the stomach varies in contour and level of descent and, the organ being J-shaped, its lowest point is usually below the vertebral level of its pyloric termination.

The parts of the stomach are the cardia, the fundus, the body, and the pylorus. The **cardia** of the stomach is the portion immediately surrounding the esophageal opening. The upper end of the stomach expands upward, filling the dome of the diaphragm on the left side. This is the **fundus** of the stomach (fig. 331); it is limited below by the horizontal plane of the cardiac orifice. The expanded fundus is in complete continuity with the descending **body** portion. The upturn of the J of the stomach is separated from its vertical portion by the **angular notch,** a sharp indentation about two-thirds of the distance along the lesser curvature. The angular notch marks a vertical separation between the body of the stomach to its left and the **pyloric portion** of the stomach to its right. A less definite notching of the greater curvature sometimes assists in this

division, but the indentations of the curvatures and surfaces produced by peristaltic movements of the stomach are more marked than any other notchings except the angular notch. The pyloric portion of the stomach consists of the **pyloric antrum** to the right of the angular notch and the markedly constricted terminal **pylorus.** The latter is composed of a greatly thickened muscular wall and a narrowed **pyloric canal.**

Clinical Note **Variations in Position** (figs. 326–28) Differences in the size and position of the stomach seen in the cadaver give only a partial indication of the positional changes in this organ. Variations are common in response to differences in stature, body position, contents, respiratory movements, and degree of filling of adjacent organs. **Stature differences** affect the stomach to a considerable degree but in a very simple manner. It will be almost self-evident that the long, slender abdomen of the tall, thin individual can only contain viscera of the same long, thin form. Similarly, the broad, stocky body form will necessarily be characterized by broader, more transversely oriented organs. It is typical for individuals of the first stature type to possess a stomach, the lower limit of which is not far from the pubis in the erect posture. Contrarily, the stomach of a short, stocky person usually descends only a little, if any, below the level of the pyloric termination in the duodenum. The lower limit and form of the stomch also vary considerably between the erect and supine **positions of the body,** for the movable viscera move downward under the influence of gravity in the erect position. A normal variation of one or two vertebral levels between these two body positions is to be expected. Such is the elevating and normalizing effect of the supine position, however (Colton, Anson et al., '47), that the transpyloric line, determined originally in the fixed cadaver, has considerable validity for all stature types. The transpyloric line may be marked on the surface of the body at the midpoint between the suprasternal notch and the symphysis pubis and is at the level of the first lumbar vertebra.

The effect on the stomach of its **contents** is considerable. All hollow viscera of the body collapse with their walls in contact when they are empty and become enlarged and more globular when filled. The weight of the contents also has a dragging-down effect, so that both the form and the inferior limit of the stomach are affected. Like all abdominal viscera, the position of the stomach is also varied by the **movements of respiration.** In all these variations a fundamental relationship between the stomach and the transverse colon remains constant. The inferior curvature of the J of the stomach lies superiorly and is matched just below it by a similar curvature of the transverse colon. This relationship sometimes requires the transverse colon to undergo a considerable downturn to the brim of the pelvis and then a steep ascent to the position of its left (splenic) flexure.

Curvatures and Surfaces (figs. 325)

The axis of the stomach is oblique and extends from the fundus downward, to the right, and ventralward. The **lesser curvature** marks the right

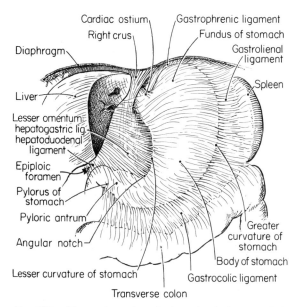

Fig. 325. The peritoneal and visceral relations, and the curvatures and parts of the stomach.

border of the organ and prolongs the line of the esophagogastric junction in a curve concave to the right as far as the pylorus. The **greater curvature** is the left and inferior border of the stomach. It begins at the cardiac notch at the superior aspect of the esophagogastric junction and follows the superior curvature of the fundus and then the convex curvature of the body down to the pylorus. The greater curvature is four or five times as long as the lesser. As a consequence of the frontal orientation of the stomach, the organ has essentially anterior and posterior surfaces, although the anterior surface also faces somewhat superiorly and the posterior surface somewhat inferiorly. The left half of the anterior surface of the stomach is in contact with the diaphragm above and with the abdominal wall below. The right side of this surface lies against the left and quadrate lobes of the liver. The posterior surface of the stomach forms the anterior wall of the omental bursa. The structures lying in the floor of the omental bursa form a shallow **stomach bed** on which the organ rests in the supine position. These are the diaphragm, the spleen, the left suprarenal gland, the upper portion of the left kidney, the pancreas, the left flexure of the colon, and the transverse mesocolon.

Peritoneal Relations (figs. 314, 325)

The stomach is entirely covered by peritoneum except where the blood vessels course along the

curvatures and except for a small triangular space behind the cardiac orifice. Here the stomach is left bare by the gastrophrenic peritoneal reflection. The layers covering the stomach sweep off it at its curvatures. At the lesser curvature the two layers extend to the liver as the gastrohepatic portion of the lesser omentum. From the greater curvature the greater omentum spreads widely—to the diaphragm as the gastrophrenic ligament, to the spleen as the gastrolienal ligament, and to the transverse colon as the gastrocolic ligament.

STRUCTURE (fig. 329)

The stomach is composed of four coats: mucous, submucosal, muscular, and serous. The **mucous membrane** is thrown into a series of coarse **gastric folds.** These are oriented chiefly longitudinally and rather regularly along the lesser curvature, where they tend to define a channel graphically designated in German as the *magenstrasse* (literally, the street of the stomach). These folds partially disappear when the stomach is distended. On and between the folds the mucous surface is thrown into hillocks and separating hollows and resembles the surface of a sponge. This surface is in turn the site of numerous **gastric foveolae** (gastric pits), into which open the gastric glands concerned with digestion. The **submucosal layer** is a loose, areolar, vascular layer that permits the wrinkling of the mucosa. The **muscular coat** of the stomach is composed of the longitudinal and circular layers typical of the gastrointestinal tract but also contains an internal oblique layer. The

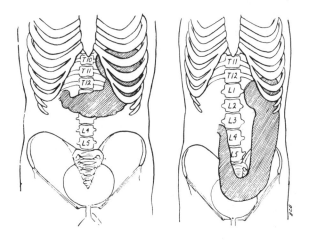

Fig. 326 Variations in shape and position of the stomach that can be correlated with stature. Traced from radiographic projections.

outer longitudinal layer is not as uniform as the circular layer, being concentrated particularly along the curvatures. The circular muscle is the principal part of the muscular coat. The inner oblique layer is composed of muscle fasciculi which are continuous with the deepest circular fibers of the esophagus; they radiate from the cardiac orifice of the stomach over the anterior and posterior surfaces. The **serous coat** is the peritoneum.

The **cardiac orifice** marks the transition from the esophagus to the stomach. It is oval in form, being compressed from side to side, and the layers of the esophagus are here continuous with those of the stomach. On the left of the junction a deep notch, **the cardiac notch,** separates the esophagus and the fundus and produces a fold in the interior. This fold, together with decussating fibers of the diaphragm and strengthened circular fibers of the lower esophagus, forms a sphincter for the orifice which prevents regurgitation under normal conditions. The **pyloric orifice** of the stomach is narrow and is

surrounded by a thick ring of circular muscle constituting the pyloric sphincter. The thickened sphincter is very evident to the touch and marks the termination of the stomach and its transition to the duodenum.

VESSELS AND NERVES OF THE STOMACH

The stomach receives its arterial blood supply from all branches of the celiac trunk. Its venous drainage is into the portal vein, in common with the venous drainage of all other parts of the gastrointestinal tract. The lymph drainage of the stomach is to nodes of the celiac group. The principal nerves of the stomach are derived from the anterior and posterior vagal trunks which arise by resolution of the esophageal plexus.

The Celiac Trunk (figs. 315, 330, 340)

The aorta begins its abdominal distribution by passing through the aortic hiatus of the diaphragm

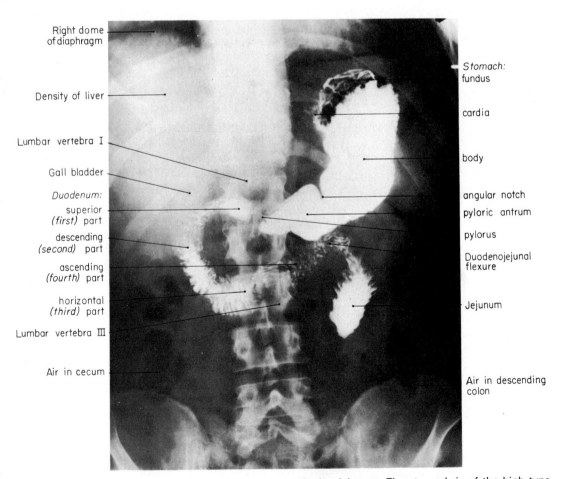

Fig. 327 A barium image of the stomach, duodenum, and beginning jejunum. The stomach is of the high type.

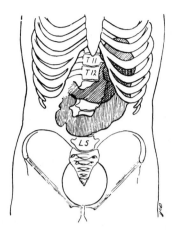

Fig. 328 The higher position of the stomach in the recumbent position and the downward shift which accompanies the erect position. Traced from radiographic projections.

in front of the lower border of the twelfth thoracic vertebra. The abdominal aorta is short, ending by a division into the left and right common iliac arteries anterior to the body of the fourth lumbar vertebra. The aorta has a number of paired branches to the bilaterally represented abdominal glands and to the wall of the abdominal cavity but, of special importance for the gastrointestinal tract, it provides three ventral unpaired branches of large size for the supply of this system. These are the celiac trunk and the superior mesenteric and inferior mesenteric arteries.

The **celiac trunk** arises from the front of the abdominal aorta just below the aortic hiatus and at the level of the upper portion of the first lumbar

vertebra. Only from 1 to 2 cm. long, it passes horizontally forward above the upper margin of the pancreas and then divides behind the posterior body wall peritoneum into the left gastric, common hepatic, and splenic arteries. In addition, it often gives rise to the inferior phrenic artery, and it is frequently the source of the dorsal pancreatic artery. Close around the celiac trunk lie the celiac lymph nodes and the celiac plexus of nerves and their ganglia.

The Left Gastric Artery This is the smallest of the celiac branches. It courses upward and to the left toward the cardiac end of the stomach. The vessel lies behind the body wall peritoneum in the floor of the bursa omentalis, where it frequently raises a fold of peritoneum, the **left gastropancreatic fold.** It reaches the stomach in the bare area behind the cardia, here giving off **esophageal branches** which ascend along the esophagus to provide an important part of its arterial supply (p. 359). At this point the left gastric artery also has an **hepatic branch** in an appreciable number of cases (18 per cent—Moosman & Coller, '51) which passes to the left lobe of the liver between the layers of the lesser omentum. The left gastric artery turns onto the lesser curvature of the stomach between the layers of the hepatogastric ligament and follows this curvature as far as the pylorus. It provides branches to both surfaces of the stomach and anastomoses terminally with the right gastric artery.

The Common Hepatic Artery The common hepatic artery, larger than the left gastric, runs forward and to the right along the upper border of the pancreas, and, at the upper margin of the duodenum, passes from behind the posterior body wall peritoneum into the lesser omentum. It frequently raises a **right gastropancreatic fold** in reaching the lesser omentum. The branches of the common hepatic artery are the gastroduodenal and the proper hepatic arteries, and, rarely, the right gastric artery.

(a) The **gastroduodenal artery** is a short, thick trunk which takes its origin from the common hepatic artery at the upper border of the first part of the duodenum. It descends behind this portion of the duodenum and divides at its inferior border into the right gastroepiploic and the anterior superior pancreaticoduodenal arteries. In its course it passes between the duodenum and the head of the pancreas. Proximal branches of the gastroduodenal artery are the supraduodenal artery, the large posterior superior pancreaticoduodenal artery, and the small retroduodenal twigs.

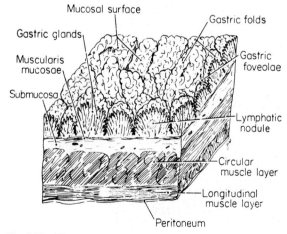

Fig. 329 The structure of a segment of stomach. (Modified from Braus.)

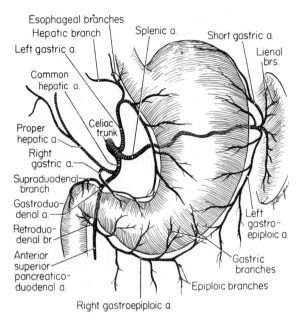

Fig. 330   The arteries of the stomach.

The **supraduodenal artery** is a slender branch which usually descends from the beginning of the gastroduodenal artery onto the upper border of the first part of the duodenum. Its branches ramify over both its anterior and posterior surfaces, more extensively over the anterior surface. The supraduodenal artery is not constant but occurs in the majority of cases (Wilkie, '11). The **posterior superior pancreaticoduodenal artery** arises behind the duodenum from the first 2 cm. of the gastroduodenal artery. Its origin is usually to the left of the descending common bile duct, but it may arise in front of the duct. The posterior superior pancreaticoduodenal artery passes to the right anterior to the common bile duct, sinks between the duodenum and the head of the pancreas, and describes an arch directed downward and to the left across the posterior aspect of the head of the pancreas. It gives off both duodenal and pancreatic branches and anastomoses with the posterior inferior pancreaticoduodenal branch of the superior mesenteric artery. It is more fully described in connection with the arteries of the pancreas (p. 414). The gastroduodenal artery provides small, multiple **retroduodenal branches** to the inferior aspect of the first part of the duodenum just proximal to its terminal division.

The **right gastroepiploic artery** passes from right to left along the greater curvature of the stomach but several centimeters from it between the layers of the gastrocolic ligament. It anastomoses with the left gastroepiploic branch of the splenic artery to form a vascular arch along the greater curvature from which pass gastric branches to both surfaces of the stomach and omental branches into the layers of the greater omentum. The **anterior superior pancreaticoduodenal artery** descends across the head of the pancreas near the sulcus between the pancreas and the duodenum. It provides duodenal and pancreatic branches and turns in under the inferior border of the head of the pancreas to anastomose with the anterior inferior pancreaticoduodenal branch of the superior mesenteric system.

(b) The **proper hepatic artery** is the continuation of the common hepatic artery distal to the gastroduodenal artery. It ascends between the layers of the hepatoduodenal portion of the lesser omentum, lying to the left of the common bile duct and anterior to the portal vein. The proper hepatic artery usually gives off, near its beginning, the right gastric artery.

The small **right gastric artery** descends through the lesser omentum to the pyloric end of the lesser curvature of the stomach. It supplies branches to both surfaces of the pyloric portion of the stomach and anastomoses with the left gastric artery. It is most commonly a branch of the proper hepatic artery but may arise, in order of frequency, from the left hepatic, the gastroduodenal, or the common hepatic arteries.

The right and left hepatic arteries arise by division of the proper hepatic artery sometimes near the porta hepatis, sometimes more proximally in the lesser omentum. The **right hepatic artery** passes to the right, usually behind the common hepatic duct, to gain the right end of the liver hilum. Here it breaks up into several branches which enter the right lobe of the liver with radicles of the portal vein and hepatic duct. As it passes between the hepatic duct and the cystic duct, it gives rise to the small **cystic artery** which follows the cystic duct to the gall bladder. The right hepatic artery is frequently (14 per cent) a branch of the superior mesenteric artery (Moosman & Coller, '51, fig. 349). Such a branch, the most common variation in the hepatic arterial supply, ascends to the liver in the hepatoduodenal ligament posterior to the portal vein. The **left hepatic artery,** smaller but longer than the right branch, runs to the left end of the porta hepatis. It provides a branch to the caudate lobe and frequently one to the quadrate lobe and then enters the substance of the left lobe of the liver. The more detailed distribution and the variations of the hepatic and cystic arteries are considered in the description of the liver (p. 424).

The Splenic Artery This is the largest branch of the celiac trunk. It arises from the left side of the trunk distal to the left gastric artery. The splenic artery runs a highly tortuous course, partially embedded in the superior border of the pancreas, behind the peritoneum of the floor of the bursa omentalis. It passes horizontally to the left, enters the lienorenal ligament over the superior portion of the left kidney, and is conducted through this ligament to the hilum of the spleen. The splenic artery may extend as a single entity to the hilum of the spleen and there break up into its terminal branches, or it may divide into superior and inferior terminal branches as it overlies the pancreas some distance from the spleen (Michels, '42). As a consequence, the short gastric and the left gastroepiploic arteries may arise from the main vessel or from its terminal branches.

The splenic artery gives off a number of pancreatic branches in its undulating course along the upper border of the gland. The relatively large **dorsal pancreatic artery** (p. 415) is most commonly a branch of its first several centimeters. Small **superior pancreatic branches** pass from the artery into the substance of the pancreas. The **arteria pancreatica magna** is a large superior pancreatic artery which usually enters the pancreas at about the junction of the medial two-thirds and the lateral one-third of the gland. Within the gland this artery distributes along the pancreatic duct. A **caudal pancreatic artery** arising from the distal portion of the splenic artery distributes to the tail of the pancreas.

The **short gastric branches** are four or five small vessels which pass through the gastrolienal ligament to reach the fundus of the stomach. Here they anastomose with branches of the left gastric and left gastroepiploic arteries. The **left gastroepiploic artery** arises from the splenic artery or an inferior terminal branch and passes toward the greater curvature of the stomach through the gastrolienal ligament. Its branches (gastric and epiploic) are distributed to both surfaces of the stomach and to the greater omentum, like those of the right gastroepiploic with which they anastomose. The superior and inferior terminal branches of the splenic artery usually divide into four splenic branches which enter the spleen along its hilum and which establish a vascular segmentation for the organ (p. 409).

Veins of the Stomach

The veins of the stomach correspond in position and in course to the arteries of the stomach. They are tributary to the portal vein (figs. 361, 362) or its radicles. Venous blood from the stomach, as from all of the gastrointestinal tract, flows to the liver, where it passes through a second capillary network before passing on to the heart.

The **left gastric vein** arises from tributaries on both surfaces of the stomach. It passes from right to left along the lesser curvature to the cardia, where it receives esophageal veins. The left gastric vein then turns to the right and posteriorly and descends in company with the left gastric artery behind the posterior body wall peritoneum. Passing beyond the celiac arterial trunk, the left gastric vein ends in the portal vein. This circular course, first along the lesser curvature and then inferiorly on the body wall, is expressed in the old name 'coronary vein.'

The small **right gastric vein** is formed from tributaries of both surfaces of the pyloric region of the stomach. It accompanies the right gastric artery between the layers of the lesser omentum and, passing from left to right, ends directly in the portal vein. A **prepyloric vein** ascends over the pylorus to the right gastric vein. This vein, more clearly evident in life, is an assistance to the surgeon in identifying the pylorus.

The **right gastroepiploic vein** accompanies the right gastroepiploic artery within the layers of the gastrocolic ligament. It receives tributaries from the inferior portions of both the anterior and posterior surfaces of the stomach and from the greater omentum. Passing from left to right, the right gastroepiploic vein is joined by the anterior superior pancreaticoduodenal vein in front of the head of the pancreas. The vein crosses the uncinate process of the pancreas and ends in the superior mesenteric vein. The middle colic vein frequently joins the confluence of the right gastroepiploic and anterior superior pancreaticoduodenal veins so as to form a **gastrocolic vein** with a termination in the superior mesenteric vein.

The **left gastroepiploic vein** completes the arch of veins along the greater curvature of the stomach and has the same pattern of drainage as the right gastroepiploic vein. It is directed to the left in the folds of the gastrocolic ligament. Entering the gastrolienal ligament, the left gastroepiploic vein ends in the beginning of the splenic vein.

The **short gastric veins,** four or five in number, drain the fundus and the superior part of the greater curvature of the stomach. They pass between the layers of the gastrolienal ligament toward the hilum of the spleen, where they terminate in the splenic vein.

The Lymphatics of the Stomach (fig. 333)

The lymphatic vessels of the stomach have a pattern which is similar to that of its arteries and veins. The vessels drain the anterior and posterior surfaces toward the curvatures where the lymph nodes of the stomach are concentrated.

Four lines of lymphatic drainage characterize the stomach.

(1) A large group of vessels drain the surfaces toward the lesser curvature of the stomach to end in the **left gastric nodes.** These form a chain of from ten to twenty nodes extending from the angular notch of the

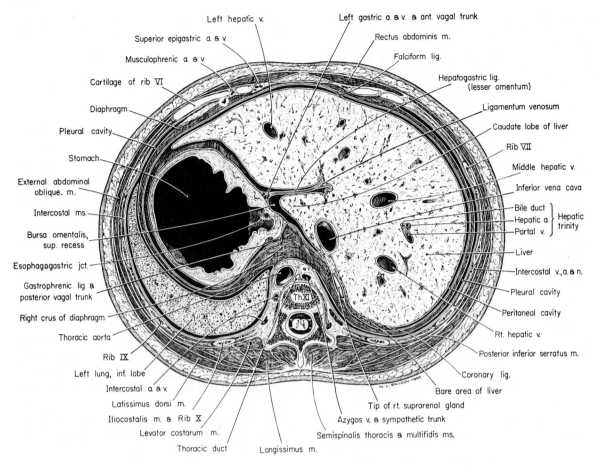

Left hepatic v.
Superior epigastric a. & v.
Musculophrenic a. & v.
Cartilage of rib VI
Diaphragm
Pleural cavity
Stomach
External abdominal
oblique. m.
Intercostal ms.
Bursa omentalis,
sup. recess
Esophagogastric jct.
Gastrophrenic lig. &
posterior vagal trunk
Right crus of diaphragm
Thoracic aorta
Rib IX
Left lung, inf. lobe
Intercostal a. & v.
Latissimus dorsi m.
Iliocostalis m. & Rib X
Levator costarum m.
Thoracic duct
Longissimus m.

Left gastric a. & v. & ant. vagal trunk
Rectus abdominis m.
Falciform lig.
Hepatogastric lig.
(lesser omentum)
Ligamentum venosum
Caudate lobe of liver
Rib VII
Middle hepatic v.
Inferior vena cava
Bile duct
Hepatic a. } Hepatic
Portal v.   } trinity
Liver
Intercostal v.,a. & n.
Pleural cavity
Peritoneal cavity
Rt. hepatic v.
Posterior inferior serratus m.
Coronary lig.
Bare area of liver
Tip of rt. suprarenal gland
Azygos v. & sympathetic trunk
Semispinalis thoracis & multifidis ms.

ThXI

W. L. BRUDON—1962

Fig. 331   A cross section of the abdomen at the level of the esophagogastric junction.

stomach upward to the cardia. At the cardia a small subsidiary group of five or six **paracardial nodes** surround the esophagogastric junction. The efferent vessels of the left gastric nodes follow the left gastric blood vessels to end in the celiac nodes.

(2) A few lymph vessels drain the pyloric portion of the lesser curvature to the two or three **right gastric nodes** along the corresponding artery. These in turn, discharge to the **hepatic nodes** along the common hepatic artery.

(3) The left halves of both surfaces of the stomach drain by lymphatic channels directed toward the greater curvature. Lymphatic vessels from the fundus and the upper portion of the body of the stomach meet no lymph nodes at the greater curvature but follow the short gastric and left gastroepiploic blood vessels to the **pancreaticolienal nodes.** These are three or four nodes situated along the posterior superior border of the pancreas on the splenic blood vessels. One or two members of the group may be found in the gastrolienal ligament. The pancreaticolienal nodes drain the stomach, the spleen, and the pancreas and discharge to the celiac nodes.

(4) From the lower portion of the left half of the stomach, lymphatic vessels drain to the six to twelve **right gastroepiploic nodes,** the efferents of which pass to the right to the pyloric nodes. The **pyloric lymph nodes** are from six to eight nodes located in the angle between the first and second parts of the duodenum in close relation to the terminal division of the gastroduodenal artery. Their efferents pass along this artery to the **hepatic nodes** along the common hepatic artery. The latter nodes are tributary to the celiac nodes.

Lymph draining from the stomach by means of these various routes passes through the lymph node groups described and ultimately reaches the **celiac nodes,** from three to six large nodes located adjacent to the celiac trunk. The celiac nodes also receive lymph from the liver and the gall bladder by means of the hepatic nodes and from the pancreas and the spleen by way of the pancreaticolienal nodes. The efferent channels of the celiac nodes help form, with efferents from the superior mesenteric nodes, the **intestinal trunk.** This trunk represents the final common path of lymph from the gastrointestinal

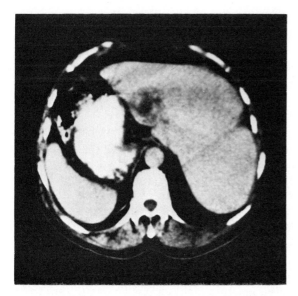

Fig. 332 C.T. scan of the abdomen at a level approximately that of fig. 331. (Courtesy of the Radiology Department of the University of Michigan Medical School.)

tract and its associated glands. The trunk empties most commonly into the left lumbar lymph trunk, just before that channel reaches the cisterna chyli (p. 439).

Nerves of the Stomach (fig. 334)

The nerves of the stomach are both parasympathetic by way of the vagus nerve and sympathetic by means of perivascular plexuses from the celiac plexus.

The right and left vagus nerves above the diaphragm form the esophageal plexus (p.359). In this plexus branches of the right and left vagus nerves intermingle. Just above the diaphragm the plexus is resolved into two nerve trunks which, however, lie anterior and posterior to the esophagus and both of which receive fasciculi from the originally right and left vagus nerves. The **anterior and posterior vagal trunks** pass through the esophageal hiatus of the diaphragm anterior and posterior to the terminal esophagus, and both lie toward its right side. They contain preganglionic visceral efferent and general visceral afferent fibers. Both supply **gastric branches** to the stomach which arise at the cardiac end of the stomach. From four to six in number, the gastric branches diverge from the lesser curvature over the anterior and posterior surfaces of the stomach,

penetrating the wall to reach the myenteric and submucosal plexuses. Each anterior and posterior vagal trunk gives off one longer branch, 'the principal nerves of the lesser curvature,' which descend on the anterior and posterior surfaces of the stomach as far as the pyloric antrum.

In addition to the gastric branches, the anterior vagal trunk has an **hepatic branch** which passes from the stomach to the liver in the upper part of the hepatogastric ligament in company with the hepatic branch of the left gastric artery. The hepatic branch contributes to the plexus on the hepatic artery and to the gall bladder and bile ducts. It also has a small offshoot which descends through the hepatogastric ligament to the pyloric portion of the stomach, the duodenum, and the pancreas (Jackson, '49). The **posterior vagal trunk** makes a major contribution to the celiac plexus by means of a celiac branch which follows the left gastric artery and vein. This branch loses its identity in the celiac plexus, and its radicles are not distinguishable in the branches of the plexus. They provide, however, the preganglionic parasympathetic innervation to all of the organs of the abdomen beyond the stomach that are supplied by the arterial branches of the celiac trunk and the superior mesenteric artery. In addition, vagal fibers reach the kidneys by way of the celiac plexus and its branches.

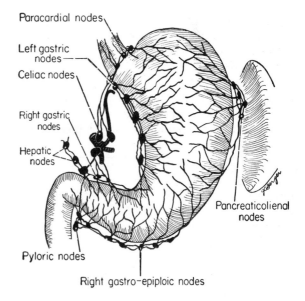

Fig. 333 The lymphatic drainage of the stomach. Semischematic.

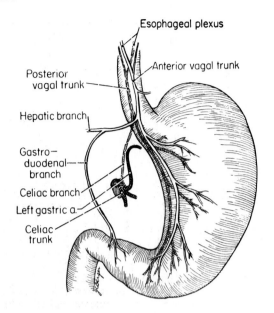

Fig. 334 The abdominal branches of the anterior and posterior vagal trunks.

The sympathetic supply of the stomach is by perivascular plexuses (left gastric, hepatic, and splenic) which emanate from the celiac plexus. These are composed primarily of postganglionic sympathetic fibers, the cell bodies of which form the celiac ganglia. The preganglionic fibers concerned reach the celiac ganglia by way of the greater thoracic splanchnic nerve (p. 366) from thoracic cord segments five to nine or ten. The perivascular plexuses may also conduct to the stomach some parasympathetic fibers reaching the celiac plexus through the posterior vagal trunk.

The nerves to the stomach appear to have somewhat mixed functions. However, the parasympathetic innervation initiates or enhances muscular movements, and the sympathetic innervation is important in vasomotor control. The parasympathetic fibers exert the greater influence on the secretion of water and hydrochloric acid in the fundus; the sympathetic innervation has the major influence in the secretion of enzymes in the stomach. Afferent impulses principally accompany the sympathetic system.

THE SPLEEN

The **spleen** is a soft lymphatic organ of purplish color located posterior to the upper part of the stomach and between it and the diaphragm (fig. 352). The diaphragmatic surface of the spleen is convexly curved to fit into the concavity of the diaphragm. The ninth, tenth, and eleventh ribs are in relation to the spleen, the diaphragm and the pleura intervening. The tenth rib lies along the long axis of the spleen. About 12 cm. long and 7 cm. broad in the average case, the spleen is entirely surrounded by peritoneum. It develops in the dorsal mesogastrium, the left sheet of which it expands, and is attached to the stomach by the gastrolienal ligament. Its body wall attachments are the phrenicosplenic ligament above and the lienorenal ligament below. The spleen normally does not descend below the costal margin but rests on the left flexure of the colon and the phrenicocolic ligament (sustentaculum lienis). Thus, the spleen is seldom palpable through the abdominal wall unless it is enlarged.

The anteromedially directed visceral surface of the spleen (figs. 315, 335) is triangular in outline. The apex is directed superiorly and medialward; the base is a flattened surface which makes contact with the left flexure of the colon. A slight ridge descends from the apex down the middle of the spleen, superolateral to which is a marked concavity that receives the adjacent surface of the stomach. Just posterior to the gastric impression, a series of indentations characterize the hilum of the spleen; these are the sites of entrance of the splenic arteries and veins. Inferomedial to the ridge, a less well-marked concavity is the renal impression. Toward its lower end, the hilum is occasionally in contact with the pancreas. The laterally directed superior

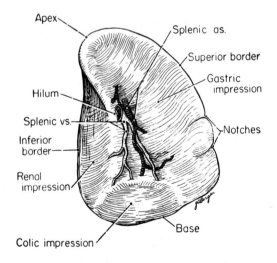

Fig. 335 The borders, the visceral relations, and the hilum of the spleen.

border of the gland is thin and is often notched, especially toward its lower end. The inferor border faces also medialward and is smooth and rounded.

The size and weight of the spleen show considerable variation. The gland normally contains a large quantity of blood, and the size of the splenic artery (or the vein) is an indication of the volume of blood which passes through its capillaries and sinuses. The capsule and trabeculae of the spleen contain smooth muscle, and the organ, besides showing rhythmic contractility, is capable of expelling large quantities of blood into the circulatory system. Splenic pulp contains abundant red blood cells, monocytes, and macrophages, and its arterioles are surrounded by lymphatic nodules. Knisely ('36) observed the living spleen under the microscope and reported that the arterioles ended in venous sinuses marked off at both the arteriolar and venous ends by contractile sphincters. The functions of the spleen are only incompletely known. Its cell and tissue constitution appear to indicate that it collects waste products from the blood, removes dead or dying red blood cells, and elaborates lymphocytes. Accessory spleens are occasionally found between the layers of the gastrolienal ligament. They vary from the size of a pea to that of a plum.

Vessels and Nerves

The blood vessels of the spleen are the splenic artery and vein. The **splenic artery** is described on p. 404. Its branches are usually four, which enter the spleen separately along the hilum. The lack of anastomosis of these vessels within the spleen establishes a vascular segmentation of the organ. Numerous observers have determined the pattern as **superior polar, superior middle, inferior middle,** and **inferior polar arteries** and **segments.** The splenic vein arises by the union of proper splenic veins which emanate from the hilum in correspondence with the arterial arrangement. Receiving the short gastric and left gastroepiploic veins, the splenic vein passes toward the median plane behind the pancreas (fig. 337). It lacks the tortuosity of the splenic artery to which it is inferior. The splenic vein receives the caudal pancreatic veins, the vena pancreatica magna, and, in 60 per cent of cases, the inferior mesenteric vein (p. 439). The **lymph vessels** of the spleen arise only from the capsule and the trabeculae; they reach the **pancreaticolienal lymph nodes.** The nerves of the spleen are derived from the celiac plexus, especially from the greater thoracic splanchnic nerve through which it receives fibers from thoracic spinal cord segments six to eight. Some of its nerves are vasomotor; some supply the nonstriated muscle of the capsule and trabeculae; and some are afferent.

THE DUODENUM

At the pylorus the stomach empties into the intestinal tract. This is divided into a longer portion, the small intestine, and a shorter terminal large intestine. The small intestine is, generally speaking, highly convoluted and is suspended freely by a mesentery. It averages about 21 feet in length when measured in the unembalmed state and may be as short as 11 feet or as long as 25 feet (Underhill, '55). In contrast to its general arrangement, the first and shortest part of the small intestine is not freely suspended by a mesentery. This is the **duodenum.** The remainder of the tube is divided into two parts —**jejunum** and **ileum.** These names were applied by the ancients on the basis of quite superficial characteristics (jejunum = empty; ileum = twisted), and now they have only limited utility in differentiating the upper from the lower part of the small intestine. The proximal 8 feet of the tube may be designated as the jejunum; the distal 12 feet, as the ileum.

General Relations

The **duodenum** (figs. 315, 336, 337) is the fixed and retroperitoneal portion of the small intestine. It receives its name from the fact that its length is equal to the breadth of twelve fingers (about 25 cm.). The developmental events which lead to its

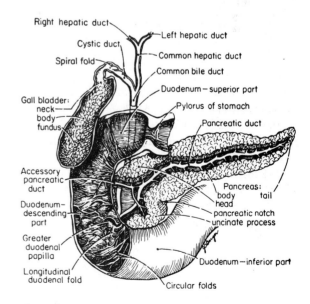

Fig. 336 The duodenum and pancreas; the biliary and pancreatic ducts.

retroperitoneal position and the formation of a fusion fascia behind it have been enumerated under the discussion of the peritoneum (p. 392). The settling of the transverse colon across the duodenum and the pancreas as the umbilical loop of intestine returns to the abdominal cavity is followed by the replacement of a portion of the original surface peritoneum of these parts by primitive mesocolon. The definitive line of attachment of the transverse mesocolon thus lies across the duodenum and the pancreas and leaves between its layers a linear strip of these organs uncovered by peritoneum. At its extremities, however, the duodenum is continuous with the completely peritonealized stomach and jejunum, and thus its first and fourth parts tend to be within the peritoneum.

The duodenum has the form of an incomplete circle, a form established by the twisting it experienced in the counterclockwise rotation of the primitive intestinal loop. In this incomplete circle, its termination is not far removed from its beginning. It is useful to designate the parts of the circle of the duodenum as first, second, third, and fourth; and it is an aid to memory to be able to relate the first part of the duodenum to its normal level opposite the first lumbar vertebra, the second part to the second vertebra, and the third part to the third vertebra. Occasionally the circular form of the duodenum is replaced by a U- or V-shaped type due to differences in the length of the parts.

The **first (superior) part of the duodenum** is short (5 cm.), is in direct continuity with the pylorus, and is completely peritonealized. The hepatoduodenal portion of the lesser omentum surrounds this segment of the duodenum. It makes contact with the neck of the gall bladder and is usually stained with bile in the cadaver. The first part of the duodenum is related above to the quadrate lobe of the liver as well as to the gall bladder. The gastroduodenal artery, the common bile duct, and the portal vein pass behind the first part of the duodenum; below and behind it is the head of the pancreas. This part of the duodenum passes posteriorly along the right side of the body of the first lumbar vertebra and there turns down into the descending (second) part of the tube. It is thin walled in consequence of the absence of circular folds in its interior.

The **second (descending) portion of the duodenum** (from 7 to 10 cm. long) descends retroperitoneally along the right side of the first, second, and third lumbar vertebrae. Its middle one-third is crossed ventrally by the transverse colon (fig. 321). Here

the colon is not covered posteriorly by peritoneum, and a triangular area of the anterior duodenal surface is attached to it by extraperitoneal connective tissue. Both the supracolic and infracolic surfaces of the duodenum are peritonealized. The supracolic portion is in relation to the right lobe of the liver; the infracolic part, to the coils of the small intestine. The second part of the duodenum overlies the hilum of the right kidney posteriorly; it also rests on the renal vessels, the inferior vena cava, and the psoas muscles. The medial surface of the duodenum is in contact with the pancreas. At the junction of the middle and lower thirds of this part of the duodenum, the common bile duct and the pancreatic duct penetrate the posteromedial wall to empty on the summit of the greater duodenal papilla. The consequent interruption in the integrity of the duodenal wall explains the frequency of duodenal diverticuli occurring in this location. An accessory pancreatic duct sometimes pierces the anteromedial wall at the junction of its upper and middle thirds.

The **third (inferior) portion of the duodenum** (from 6 to 8 cm. long) crosses the lower part of the body of the third lumbar vertebra from its right to its left side. It is separated from the vertebral column by the right psoas muscles, the inferior vena cava, the aorta, and the right testicular (or ovarian) vessels. This portion of the duodenum is retroperitoneal. Its ventral surface is covered by peritoneum except toward its left extremity where the superior mesenteric vessels and the root of the mesentery cross the duodenum. It is of topographic value to note that the superior mesenteric artery arises above and crosses the anterior aspect of the third part of the duodenum, whereas the inferior mesenteric artery arises from the aorta directly below the duodenum.

The **fourth (ascending) portion of the duodenum** is short. It ascends on the left side of the vertebrae as far as the upper border of the body of the second lumbar segment, a distance of about 5 cm. In its course it passes along the left side of the aorta and anterior to the left psoas muscles and the left testicular (or ovarian) vessels. Terminally, the fourth part turns abruptly forward to become continuous with the jejunum at the duodenojejunal flexure. The end of the duodenum is covered by peritoneum and is movable, but most of the fourth segment is retroperitoneal. The root of the mesentery begins at the duodenojejunal flexure, and its attachment passes diagonally downward and to the right over this segment, leaving the duodenum at the junction of its third and fourth parts. The duodenojejunal flexure is stabilized by a fibromuscular band, the **suspensory muscle of the duodenum** (Jit, '52), the connective tissue of which arises around the celiac arterial trunk and from the right crus of the diaphragm (Low, '07). Its smooth muscle, well developed in the adult, is an upward extension of the nonstriated muscle of the duodenum. The suspensory muscle passes superomedialward from the duodenojejunal flexure. Passing superolateralward from the flexure is the superior duodenal peritoneal fold, under which is the superior duodenal fossa. These and other folds and fossae in the region of the duodenojejunal flexure are described under the peritoneum (p. 396).

Interior of the Duodenum

The internal surface of the small intestine is greatly folded, thus increasing its absorptive and secreting surface. **Circular folds** project a centimeter or less into the lumen, most of them extending for one-half or two-thirds of the circumference of the tube. Others complete the circle or even produce a spiral of several complete turns. The folds are composed of mucosa and submucosa, the latter being dense and firm enough so that these are permanent foldings of the interior. The surface of the circular folds and the mucous membrane between them is also covered by tiny finger-like projections which, rather than being seen distinctly, impart a plush-like appearance to the mucous membrane. These are the **villi** of the mucosa. They are irregular in size and shape but increase the epithelial surface of the intestinal tract tremendously. They are formed only by the mucosa.

In the duodenum, circular folds are not found in the first 3 to 5 cm. beyond the pylorus. They begin in the descending segment, and distal to the papilla for the common bile and pancreatic ducts, they become especially large and numerous. The villi of the duodenal mucosa are also thickly set and have an elongated form. They have been likened to a page of a book set on one of its long edges. Duodenal glands are also characteristic of the duodenum. These large glands extend into the submucosa. They are most numerous near the pylorus of the stomach.

The **greater duodenal papilla,** on which are the terminal openings of the common bile duct and the pancreatic duct, is found at the junction of the middle and lower thirds of the second part of the duodenum. Here the posteromedial wall exhibits a somewhat tubular projection overlain by a hood-like fold. The papilla is continued below by the tapered **longitudinal fold of the duodenum.** The occasional **lesser duodenal papilla** marks the termination of the accessory pancreatic duct. It lies in the anteromedial wall at the junction of the superior and middle thirds of the second part of the duodenum. The blood vessels, lymphatics, and nerves of the duodenum are described with the pancreas (p. 414).

THE PANCREAS

GENERAL RELATIONS (figs. 335–40)

The **pancreas** is an accessory digestive gland. It has been called the abdominal salivary gland because histologically it resembles the salivary glands of the oral region. Its secretion, the **pancreatic juice,** is an important digestive fluid having an action on proteins, fats, and carbohydrates. The pancreas also has an internal secretion concerned with sugar metabolism. This secretion, synthesized as **insulin,** is elaborated by pancreatic islet tissue and is taken up by the blood stream. There are perhaps one million clusters of islet cells scattered through the organ; in man, they appear to be more numerous in the tail of the gland.

The pancreas lies transversely across the posterior abdominal wall from the duodenum to the spleen and is behind the stomach. The gland is divided into a head, a neck, a body, and a tail. The **head** is the more expanded part of the pancreas, occupying the area defined by the C-shaped curve of the duodenum and separated from it by only a shallow sulcus. Continuing to the left and somewhat upward from the head are successively the neck, the body, and the tail of the gland. The transverse mesocolon crosses the anterior surface of the head, its layers reflecting above and below over the pancreas and duodenum. Superior to the attachment of the mesocolon the pancreas is related to the pylorus; inferior to it are the coils of the small intestine. Posteriorly, the head overlies the second and third lumbar vertebrae, the inferior vena cava, the renal veins, the right renal artery, and (uncinate process only) the aorta. The common bile duct descends diagonally across the back of the head partially embedded in its substance. The lower left portion of the head of the pancreas is inserted behind the superior mesenteric vessels, forming the **uncinate process** (fig. 339).

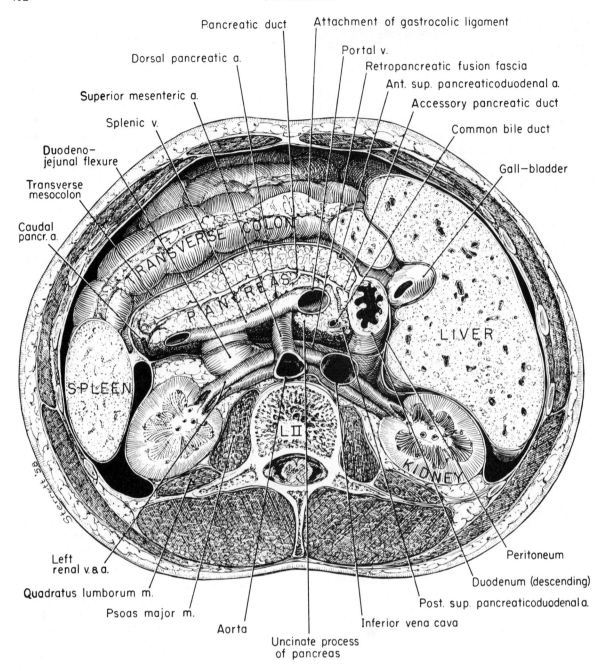

Fig. 337 A cross section of the abdomen at the level of the pancreas.

The **neck** of the pancreas is continuous with the upper left portion of the head. This somewhat constricted portion is grooved posteriorly by the superior mesenteric artery and vein. The superior mesenteric vessels emerge through the pancreatic notch below the neck and descend across the uncinate process and the third part of the duodenum to enter the mesentery. Behind the neck of the pancreas the superior mesenteric and splenic veins unite to form the portal vein. The anterior surface of the neck, continuous with the corresponding surface of the body of the pancreas, is covered by peritoneum and lies in the floor of the omental bursa. It rises superior to the pyloric portion of the stomach and is related, through the lesser omentum, with the left lobe of the liver.

The **body** of the pancreas continues from the neck toward the left and somewhat upward. In

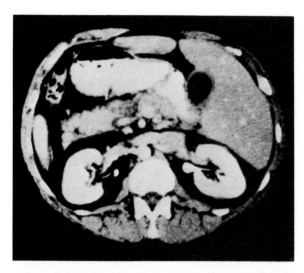

Fig. 338   C.T. scan of abdomen at a level approximately that of fig. 337. (Courtesy of the Radiology Department of the University of Michigan Medical School.)

cross section this portion of the pancreas is somewhat triangular, the apex of the triangle being an anteriorly projecting ridge along which is attached the transverse mesocolon. The ridge separates an anterior surface above from an inferior surface below. The posterior surface of the gland is the base of the triangle.

The borders of the triangular body are superior, anterior, and inferior. The **anterior surface**, in continuity with the neck of the pancreas, lies in the floor of the omental bursa and forms part of the stomach bed. The **posterior surface** is flat and is covered by the fusion fascia existing behind the pancreas. This surface crosses, from right to left, the aorta, the left suprarenal gland, and the left kidney. The splenic vein courses from left to right along the posterior surface. The narrow **inferior surface** is covered by the inferior reflection of the transverse mesocolon and is in relation with the duodenojejunal flexure and the coils of the jejunum. The **superior border** is invaded by the convoluted splenic artery. The **anterior border** is the ridge along which the transverse mesocolon attaches. The **inferior border** of the gland separates the inferior and posterior surfaces; the inferior pancreatic artery lies along or close to this border.

The **tail** of the pancreas is usually blunted and turned upward. It enters the lienorenal ligament and frequently makes contact with the spleen; inferiorly, it is in relation with the left flexure of the colon.

The Pancreatic Ducts

The pancreatic ducts are two—a main and an accessory duct. The **pancreatic duct** begins in the tail by the union of small ducts; other subsidiary ducts join it in a herringbone pattern as it passes through the gland. The duct courses to the right midway between the superior and inferior borders and lies toward the posterior surface of the gland. In the head of the pancreas it turns both inferiorly and toward the posterior surface. The duct then curves to the right to join the common bile duct, with which it passes obliquely through the wall of the duodenum. The greater duodenal papilla marks the termination of these two ducts.

The **accessory pancreatic duct** is variable. Typically, it empties at the lesser duodenal papilla, about 2 cm. proximal to the greater papilla, in the anteromedial duodenal wall. Traced to the left into the head of the pancreas, it has several branches. One branch descends anterior to the pancreatic duct and drains the lower portion of the head. Another branch usually connects with the main duct at the point where, on entering the head of the gland, the main duct turns inferiorly and posteriorly. The accessory duct may lack a patent connection to the bend of the pancreatic duct (40%), or, contrarily, it may be large and constitute the direct continuation of the duct of the body and tail (8%). Rarely (2%), the termination of the accessory duct in the duodenum is lacking, and it then empties completely into the pancreatic duct. The variations of the ducts are explainable by their fusion or lack of fusion in development (see below).

Development of the Pancreas

The pancreas develops in two parts from a dorsal and a ventral primordium. The dorsal primordium arises as a bud from the dorsal aspect of the duodenum a short distance above the hepatic diverticulum and grows upward and backward between the layers of the dorsal mesogastrium. Its central stalk is its duct, and its distal expanded portion forms the upper anterior part of the head, the neck, the body, and the tail of the pancreas. The duct of the dorsal primordium runs straight through the head to the opening recognized in the adult as that of the accessory pancreatic duct. The ventral pancreatic primordium is a diverticulum from the primitive bile duct and arises ventrally from this part of the hepatic diverticulum. It forms the posterior inferior part of the head and the uncinate process of the gland, and its duct has, from its first development, an intimate association with the common bile duct. With the rotation of the stomach and the duodenum (p. 392) and with the twisting of the duodenum on its long axis, the ventral primordium turns to the right and then behind and comes into contact with the dorsal primordium. It is this circling behind of the ventral primordium that brings the uncinate process posterior to the superior mesenteric vessels. The ducts of the two primordia fuse in the upper part of the head of the gland, and the principal path of the pancreatic secretion becomes established through the region of fusion. Thus, the pancreatic duct of the adult is formed from the distal part of the duct of the dorsal primordium and the proximal part of the duct of the ventral primordium; at the point of their fusion the duct takes a rather abrupt turn inferiorly and posteriorly. The remaining proximal part of the duct of the dorsal primordium becomes the accessory pancreatic duct in the adult.

Clinical Note Heterotopic pancreatic tissue (accessory pancreas) has been reported variably from 0.6% to 5.6% in routine autopsies. In 70% of the cases in which such nodules were found, they were located on the stomach, duodenum, or jejunum. Annular pancreas is a rare anomaly. It consists of a complete ring of pancreatic tissue around the descending portion of the duodenum. It is considered to be due to a doubling of the ventral primordium of the pancreas, or as a splitting of it, one part going to the left as the other part follows the twist of the duodenum to the right and dorsalward. The duct of the leftward migrating portion is usually still in continuity with that of the more normal remainder but occasionally they empty separately. An annular pancreas can constrict the lumen of the duodenum.

VESSELS AND NERVES OF THE DUODENUM AND THE PANCREAS

The close positional relationship of the duodenum and the pancreas results in their blood and lymph vessels being the same in whole or in part and in their nerves reaching them by similar courses. They are therefore described together.

Arteries of the Duodenum and the Pancreas (figs. 337, 340)

The duodenum and the pancreas are situated between the celiac trunk and the superior mesenteric artery and receive major blood vessels from both. Furthermore, their branches anastomose in distinct arcades both anterior and posterior to the pancreas and thus form arterial connections between these two great ventral branches of the abdominal aorta. The principal arteries from the celiac trunk to the duodenum and the pancreas are branches of the gastroduodenal artery—anterior superior pan-creaticoduodenal and posterior superior pan-creaticoduodenal arteries. The principal arteries from the superior mesenteric artery are corres-ponding branches of its inferior pancreaticoduo-denal artery—anterior inferior pancreaticoduo-denal and posterior inferior pancreaticoduodenal arteries. Of these paired sets, the anterior vessels anastomose openly with one another and the posterior vessels do likewise, and clear and definite anterior and posterior arcades are formed on the surfaces of the pancreas. The **anterior arcade** is formed along the right border of the head of the pancreas within about 1 cm. of the groove between the descending part of the duodenum and the pancreas. It enters this groove where the pancreas abuts against the third part of the duodenum, and the arcade is completed on the posterior surface of the uncinate process. The **posterior arcade** is shorter and more superiorly located. It unites, on the back of the pancreas, the posterior superior and posterior inferior pancreaticoduodenal arteries. The posterior arcade being shorter and higher, its duodenal branches are longer than those of the anterior arcade. A third, small, prepancreatic arcade is usually seen over the head of the pancreas (fig. 340).

The division of the gastroduodenal artery into the right gastroepiploic branch and the **anterior superior pancreaticoduodenal artery** is a constant one. The **posterior superior pancreaticoduodenal artery** is also virtually constant, and in almost all cases it arises from the first 2 cm. of the gastroduodenal artery. Its origin is usually to the left of or anterior to the common bile duct. This vessel passes to the right, crosses the bile duct, and sinks dorsally between the duodenum and the pancreas. It then crosses the posterior surface of the pancreas posterior to the common bile duct. Thus, the artery spirals the common bile duct, a point of surgical interest in any approach to the duct. The artery gives off small arterial twigs to the inferior part of the duct.

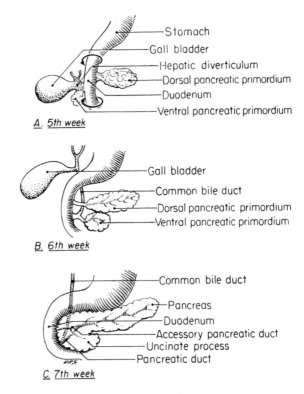

A. *5th week*

Stomach
Gall bladder
Hepatic diverticulum
Dorsal pancreatic primordium
Duodenum
Ventral pancreatic primordium

B. *6th week*

Gall bladder
Common bile duct
Dorsal pancreatic primordium
Ventral pancreatic primordium

C. *7th week*

Common bile duct
Pancreas
Duodenum
Accessory pancreatic duct
Uncinate process
Pancreatic duct

Fig. 339 Three stages in the development of the pan-creas to show the fusion of the primordia and the establishment of the definitive duct system.

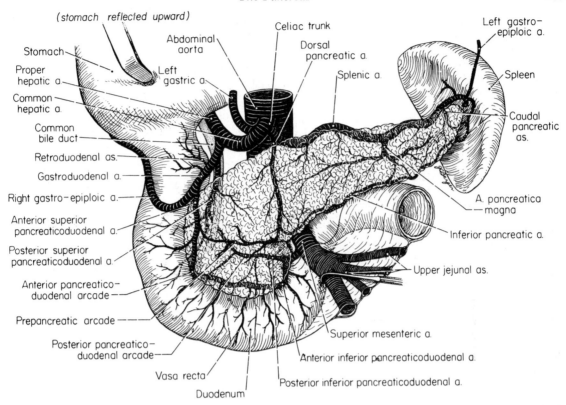

(stomach reflected upward)

Stomach
Proper hepatic a.
Common hepatic a.
Common bile duct
Retroduodenal as.
Gastroduodenal a.
Right gastro-epiploic a.
Anterior superior pancreaticoduodenal a.
Posterior superior pancreaticoduodenal a.
Anterior pancreatico-duodenal arcade
Prepancreatic arcade
Posterior pancreatico-duodenal arcade
Vasa recta
Duodenum

Abdominal aorta
Left gastric a.
Celiac trunk
Dorsal pancreatic a.
Splenic a.

Left gastro-epiploic a.
Spleen
Caudal pancreatic as.
A. pancreatica magna
Inferior pancreatic a.
Upper jejunal as.

Superior mesenteric a.
Anterior inferior pancreaticoduodenal a.
Posterior inferior pancreaticoduodenal a.

Fig. 340 The arteries of the duodenum and pancreas

The origin of the **anterior and posterior inferior pancreaticoduodenal arteries** is variable, although their occurrence is nearly constant. They most frequently arise through the division of the inferior pancreatico-duodenal branch of the superior mesenteric artery; less frequently they come individually from the superior mesenteric artery. The first jejunal artery is sometimes the origin of the single trunk or of the separate inferior pancreaticoduodenal vessels. When these vessels take origin from a jejunal artery, they pass posterior to the superior mesenteric artery and vein in reaching their area of distribution on the back of the pancreas.

The anterior and posterior arcades provide both pancreatic branches to the head of the pancreas and vasa recta to the duodenum. The **vasa recta** of the anterior arcade supply the anterior surface of the second, third, and fourth parts of the duodenum even from the part of the arcade which lies dorsally behind the uncinate process. The vasa recta of the posterior arcade, longer than those from the anterior arcade, reach the posterior surfaces of the same parts of the duodenum. They branch out over the surface of the duodenum, their radicles penetrating the wall to the submucosa. Shapiro and Robillard ('46) found the vasa recta of the duodenum to be virtual end arteries, with less anastomosis between adjacent vessels than similar arteries of the jejunum and ileum. Vasa recta to the fourth part of the duodenum pass posterior to the superior mesenteric vessels to reach this segment. The fourth part of the duodenum receives an additional arterial supply by

branches of the uppermost jejunal arteries. The first part of the duodenum lies superior to the superior pancreaticoduodenal vessels. It receives only recurrent branches of the highest vasa recta of this system. However, its main supply is by the supraduodenal and retro-duodenal branches of the gastroduodenal artery (p. 403) and by twigs from the right gastric and right gastro-epiploic arteries. Its blood supply does not appear to be as rich as that provided to the remainder of the duodenum through the vasa recta.

The **dorsal pancreatic artery** is named for its dorsal origin and its dorsal course in relation to the pancreas. It is almost always present but shows variability in its origin. The vessel arises most commonly (37%) from the proximal portion of the splenic artery (its first 2 cm.) or as a fourth branch of the celiac trunk (33%). It has a fairly frequent occurrence as a high branch of the superior mesenteric artery (21%); it is less frequently (8%) a proximal branch of the hepatic artery (Woodburne & Olsen, '51). The dorsal pancreatic artery descends behind the neck of the pancreas, dividing toward its inferior border into right and left branches. The right branch turns to the right at the inferior border of the gland and runs along the upper margin of the pancreatic notch. It sends a branch down onto the anterior surface of the uncinate process and then passes across the head of the pancreas to anastomose with a small left branch of the anterior superior pancreatico-duodenal artery. This anastomosis forms a **prepancreatic arterial arcade**; it supplies only pancreatic tissue and is usually of small caliber. The left branch of the dorsal pancreatic artery forms the inferior pancreatic artery.

The constant **inferior pancreatic artery** runs along the inferior border of the pancreas, frequently embedded in the dorsal aspect of that border of the gland. It is typically a continuation of the left branch of the dorsal pancreatic artery. When the dorsal pancreatic artery is lacking, the inferior pancreatic artery is apt to be the direct continuation of the left branch of the anterior superior pancreaticoduodenal artery. The inferior pancreatic artery anastomoses with radicles of the caudal pancreatic and pancreatica magna arteries.

The **pancreatica magna artery** is the largest of the series of superior pancreatic branches of the splenic artery (p. 405). It plunges into the substance of the pancreas in the region of the junction of the middle and left thirds of the gland. Its branches tend to be oriented along the main pancreatic duct, and they anastomose with branches of the caudal pancreatic, inferior pancreatic, and dorsal pancreatic arteries.

Caudal pancreatic branches are distal branches of the splenic artery or of one of its terminal branches, frequently the left gastroepiploic artery. They are present in most specimens and anastomose with radicles of the inferior pancreatic and pancreatica magna arteries.

Veins of the Duodenum and the Pancreas (figs. 361, 362)

The veins of the duodenum and the pancreas correspond with the arteries and usually lie superficial to them. Anterior and posterior venous arcades end in two superior pancreaticoduodenal and two inferior pancreaticoduodenal veins named like the arteries. The **anterior superior pancreaticoduodenal vein** unites with the right gastroepiploic vein, passes through the attachment of the transverse mesocolon across the head of the pancreas, and ends in the superior mesenteric vein. One or more colic veins from the right flexure of the colon so frequently join the right gastroepiploic vein that they make this tributary a gastrocolic vein. The **posterior superior pancreaticoduodenal vein** empties directly into the adjacent portal vein. The **anterior inferior pancreaticoduodenal** and **posterior inferior pancreaticoduodenal veins** are tributaries, either singly or as a common trunk, of the uppermost jejunal vein or the superior mesenteric vein. A **dorsal pancreatic vein** corresponding completely to the artery of the same name is seen only occasionally. A vein which arises on the ventral surface of the neck of the pancreas, the **pancreatic cervical vein** (Kirk, '32), descends to terminate in the superior mesenteric vein as it disappears behind the neck of the gland. The **inferior pancreatic vein** corresponds to the artery of the same name. It is a tributary of either the superior or the inferior mesenteric vein; less commonly, of the splenic vein. **Caudal pancreatic** and **pancreatica magna veins** empty into the splenic vein on the back of the pancreas.

Lymphatic Drainage

The lymph node groups associated with the duodenum and the pancreas are the **pancreaticolienal nodes** along the splenic artery and the **pyloric nodes**

along the gastroduodenal artery at the junction of the first part of the duodenum and the pancreas. These nodes are described on p. 406 (fig. 333). The **superior mesenteric nodes** around the superior mesenteric vessels and scattered, small **pancreaticoduodenal nodes** along the pancreaticoduodenal arcades are also concerned. The lymphatic drainage of the duodenum is by channels which follow the blood vessels; they are interrupted in the pancreaticoduodenal nodes and end in the pyloric nodes of the gastroduodenal artery above or in the superior mesenteric nodes below. The efferents of the pyloric nodes pass to the hepatic nodes along the common hepatic artery. In addition, the channels of the posterior aspect of the duodenum and the head of the pancreas communicate with the lymphatic vessels of the common bile duct. The lymphatic drainage of the head of the pancreas is similar to that of the duodenum. The lymphatic drainage of the neck, the body, and the tail of the pancreas follows the blood vessels, ending in the pancreaticolienal nodes along the splenic vessels and in the superior mesenteric nodes at the roots of the corresponding vessels. The line of attachment of the transverse mesocolon appears to constitute a watershed, drainage above it being upward to the celiac nodes; below it, downward to the superior mesenteric nodes.

Nerves of the Duodenum and the Pancreas

The nerves of the duodenum and the pancreas are autonomic nerves from the celiac and superior mesenteric plexuses. Included are preganglionic vagal, postganglionic sympathetic, and afferent fibers. They distribute to the duodenum and the pancreas by way of the hepatic, splenic, and superior mesenteric plexuses (p. 457). The sympathetic supply is vasomotor and is accompanied by afferents, especially for pain. The vagal innervation to the pancreas is said to reach both acinar and islet cells. This innervation is presumed to influence pancreatic enzyme formation, but the pancreatic secretion is largely controlled by hormones.

THE LIVER

GENERAL RELATIONS

Size, Location, and Projection

The **liver** is the largest gland in the body, accounting for one-fortieth of the body weight in the adult. It represents one-twentieth of the weight of the body at birth and produces most of the bulge of the abdomen in the newborn. It has a number of important functions; the contribution of the liver to digestion is its continuous production of bile. The liver occupies much of the right upper quadrant of the abdomen and conforms to the concavity of the diaphragm. Following the curvature of the diaphragm, the liver rises to its highest point behind the fifth rib in the right mammary line. Its superior outline (in surface projection) is continued to the median line in the plane of the xiphosternal junction and ends on the left just below the apex of the heart (the left fifth interspace approximately 8 cm. from the median line). The inferior margin of the liver is oblique (fig. 352). Leaving the diaphragm on the left side just below the apex of the heart, it descends diagonally, passes the left costal margin at the junction of the costal cartilage of the eighth rib with that of the seventh, and, in the median line, is about midway between the xiphoid process and the umbilicus. The inferior border of the liver reaches the right costal margin at the tip of the ninth costal cartilage and then follows the costal margin downward and posteriorly.

Surfaces and Borders (figs. 315, 341, 352)

The **diaphragmatic surface** of the liver (fig. 341A) is dome-shaped and conforms to the concavity of the underside of the diaphragm. This surface is extensive and is divisible into superior, anterior, right, and posterior parts. The **superior surface** is related, through the diaphragm, with the base of the right lung, the pericardium and the heart, and (on its extreme left) with the base of the left lung. The heart accounts for a shallow fossa, the **cardiac impression,** where it rests upon the liver. The **posterior** and **right parts** of the diaphragmatic surface are large and are in contact with the diaphragm and the lower ribs. The **posterior surface** includes most of the bare area between the reflections of the coronary ligament and contains a sulcus for the inferior vena cava. The anterior and superior surfaces are subdivided by the falciform ligament. The **anterior surface** lies against the diaphragm, the costal margin, the xiphoid process, and the abdominal wall.

The **visceral surface** of the liver (figs. 315, 341B) is directed downward, backward, and to the left. It is separated from the diaphragmatic surface by the **inferior border,** which is sharp anteriorly but blunt and rounded posteriorly. The inferior border is notched by the ligamentum teres just to the right of the median plane (**notch of the ligamentum teres**). Farther to the right, there may be a shallow notch in the border for the fundus of the gall bladder. The visceral surface of the liver is in relation to the stomach and the duodenum, the right flexure of the colon, the right kidney, and the right suprarenal gland. It lodges the gall bladder and is deeply indented posteriorly by the inferior vena cava. At about the middle of the visceral surface, the **porta** of the liver is the region of branching and entrance

of the hepatic artery and the portal vein and of exit of the hepatic bile ducts. These structures reach the liver surrounded by the layers of the lesser omentum, and at the margins of the porta the peritoneal layers reflect onto the liver. Extending from the inferior border to the porta, there is a deep **fissure for the ligamentum teres**; the ligamen-

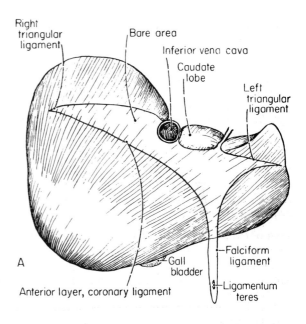

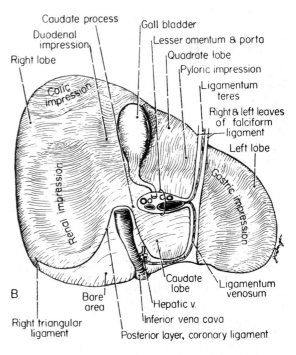

Fig. 341 The peritoneal and visceral relations of the liver A—diaphragmatic surface, B—visceral surface.

tum teres, the obliterated remains of the umbilical vein, passes through this to end in the left branch of the portal vein. From the porta to the posterior surface of the liver is a deep **fissure for the ligamentum venosum.** This lodges the ligamentous remains of the ductus venosus of the fetus. The ligamentum venosum extends from the left branch of the portal vein to the left hepatic vein adjacent to its termination in the inferior vena cava.

Lobes of the Liver

The porta of the liver, the fossae for the gall bladder and for the inferior vena cava, and the deeper fissures for the ligamentum teres and the ligamentum venosum combine to form an H-shaped figure on the visceral surface of the liver (figs. 315, 341B). The limbs and crossbar of the H separate the various liver lobes. The external marking which corresponds most nearly to the fundamental internal separation of the liver into lobes is made by the combined gall bladder and vena cava fossae. To the right of this limb of the H is the **right lobe** of the liver. Between the gall bladder fossa and the fissure of the ligamentum teres is the **quadrate lobe,** and between the caval fossa and the fissure of the ligamentum venosum is the **caudate lobe** of the liver. The caudate lobe extends onto the posterior part of the diaphragmatic surface. A ridge of liver substance which continues the caudate lobe toward the right between the porta and the fossa for the vena cava is known as the **caudate process.** To the left of the deep fissures for the ligaments is the **left lobe** of the liver. According to the internal morphology of the liver (p. 419), the quadrate lobe and part of the caudate lobe belong to a larger left lobe, more nearly equal in substance to the right lobe.

Peritoneal Relations (figs. 315, 341)

The liver develops in the ventral mesogastrium, and the reflections of its peritoneum in the adult are those of the ventral mesogastrium to the anterior abdominal wall and diaphragm on the one hand and to the stomach and the duodenum on the other. The **falciform ligament** represents the inferior limit of the ventral mesogastrium and encloses the ligamentum teres of the liver in its free border. This double-layered fold extends from the umbilicus upward and slightly to the right, reaching the liver at the notch for the ligamentum teres. Here its layers pass over both the diaphragmatic and visceral surfaces of the liver. The layers of the diaphragmatic surface

spread to the right and left as anterior layers of the **coronary ligament** (fig. 341A) until they reach the sharp reversals of peritoneal reflection to the diaphragm known as the right and left triangular ligaments. The **left triangular ligament** is located near the extremity of the superior surface of the left lobe. The **right triangular ligament** lies to the right and posteriorly and locates the right end of the bare area of the liver. The posterior layer of the coronary ligament passes medialward from the right triangular ligament adjacent to the posterior part of the inferior border of the liver and then swings around the inferior vena cava to reflect onto the caudate lobe. The posterior layer of the coronary ligament bounds the bare area of the liver posteriorly and is reflected as a broad sheet from the liver onto the right kidney (hepatorenal ligament). The line of attachment of the coronary ligament passes along the left margin of the fossa for the vena cava and, circling the caudate lobe, meets the posterior layer of the left triangular ligament in the fissure for the ligamentum venosum. On the visceral surface of the liver (fig. 341B) the layers of the falciform ligament reflect onto the liver along the line of the fissure for the ligamentum teres as far as the porta hepatis. Similarly, the lines of reflection of the posterior layers of the coronary ligament and of the left triangular ligament come forward on either side of the fissure for the ligamentum venosum as far as the porta hepatis. At the porta the peritoneal reflections become continuous. The right sheet of reflection is carried around the structures of the porta (portal vein, hepatic artery, and common hepatic duct) and is prolonged over them downward and to the left to the stomach and the duodenum. This double layer of peritoneum is the **lesser omentum.** It fans out to attach to the whole of the lesser curvature of the stomach and to the first part of the duodenum.

Visceral Relations (figs. 315, 331, 337, 341B)

The liver suspends and overhangs the stomach, the pylorus, and the duodenum and makes contact with other abdominal viscera. The left lobe presents a concave **gastric impression** on its visceral surface. Following the continuity of the gastrointestinal tract, the pylorus rests against the quadrate lobe; the first part of the duodenum, against the gall bladder. (This portion of the duodenum is frequently stained with bile after death.) The descending portion of the duodenum lies to the right of the gall bladder and the porta at the margin of the large **renal impression.** This shallow concavity of the visceral surface of the right lobe marks the contact area of the upper part of the right kidney and the right suprarenal gland. A shallow **colic impression** in the anterior portion of the same surface receives the right flexure of the colon.

INTERNAL MORPHOLOGY OF THE LIVER
(figs. 341–45)

The hepatoduodenal ligament transmits the hepatic artery, the portal vein, and the common bile duct. The **common bile duct** is the excretory duct of the liver and empties into the second portion of the duodenum. It appears to be formed near the hepatic porta by the junction of the common hepatic duct from the liver and the cystic duct from the gall bladder. Actually the cystic duct and the

gall bladder constitute a reservoir adjacent to a main bile stream composed of the common hepatic and common bile ducts. The **hepatic artery** conducts arterial blood for the nourishment of the liver tissues, whereas the **portal vein** carries to the liver blood from the gastrointestinal tract containing certain of the products of digestion. These are the raw material on which the liver acts. The venous drainage of the liver is by the **hepatic veins**—short, wide channels which open into the inferior vena cava as it lies in the caval fossa of the posterior surface.

The radicles of the hepatic artery, portal vein, and hepatic duct distribute together through the liver, and the tributaries of the hepatic veins lie at the opposite extremity of the functional liver lobule. The presence of both a nourishing arterial system

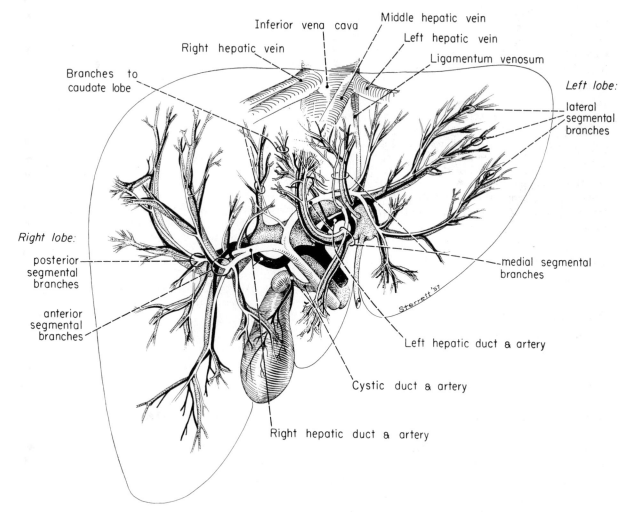

Fig. 342   Composite drawing illustrating the usual pattern of branching of the bile (white), the hepatic artery (black) and the portal vein (stippled). (By permission of *Surg., Gyn., & Obs.)*

420 *The Abdomen*

and vessels carrying in blood to support the function of the liver is fundamentally similar to the pulmonary circuits. Studies of intrahepatic anatomy have revealed segmentation of the liver which is suggestive of that in the lung, being based in both organs on their larger functional units. The segments of the liver include radicles of the hepatic duct, hepatic artery, and portal vein, which behave alike, and radicles of the hepatic veins which have a more individual pattern. Although much of the basic information involved is in the older literature, the more recent studies of Healey and Schroy ('53) and of Healey, Schroy, and Sorenson ('53) have placed our knowledge of the segmentation of the liver on a firmer basis.

The Bile Ducts (figs. 342, 343)

Injection studies of the intrahepatic duct radicles reveal that the fundamental division of the liver into right and left lobes falls on a line which may be roughly projected to the visceral surface through the gall bladder bed and the fossa for the inferior vena cava. No communication exists within the liver between the right and left hepatic duct systems. The **right lobe** is divided further into an **anterior segment** and a **posterior segment**. The plane of separation runs obliquely, so that, projected to the surface, the anterior and superior parts of the diaphragmatic surface of the right lobe belong to the anterior segment; the posterior part of the diaphragmatic surface and the whole of its visceral surface belong to the posterior segment. The **left lobe** is divided into **medial** and **lateral segments** along the line of the fissures for the ligamentum teres and ligamentum venosum. Thus, the traditional left lobe of the liver is the lateral segment, and the quadrate lobe and part of

the caudate lobe fall in the medial segment of the true left lobe. Each segment is defined by the area of distribution of a major, or segmental, division of the lobar duct, and the intrahepatic fissures which separate the segments are not crossed by branches of either the bile ducts, the hepatic artery, or the portal vein. Each of the four segments so defined (two right and two left) may be divided by its biliary drainage into **superior** and **inferior areas.** The caudate lobe is split internally into a part belonging to the right lobe and a part belonging to the left lobe with the caudate process draining to the right bile duct. The constancy of pattern in the bile ducts is high, although variations do exist. So-called extrahepatic accessory ducts are actually not accessory but are aberrant segmental or area ducts. The segmental and area divisions of the liver and the pattern of branching of the bile duct may be seen in figs. 342 and 343.

The Hepatic Artery and the Portal Vein (figs. 342, 344)

The proper hepatic artery divides in the hepatic porta into right and left hepatic arteries. The intrahepatic course of these arteries corresponds closely to that of the biliary ducts, and they distribute to the right and left lobes and their subdivisions as defined by the biliary duct distribution. The **right hepatic artery** joins the duct of the right lobe and divides, like it, into **anterior segmental** and **posterior segmental branches.** These accompany the corresponding segmental ducts, running along their lower borders. Further, like the ducts, each segmental artery divides into **superior area** and **inferior area branches.**

At the hilum of the liver, the **left hepatic artery** is located well below the left hepatic duct. Passing a short distance to the left, the artery divides into **medial segmental** and **lateral segmental branches,** each branch providing **superior area** and **inferior area branches.** The order of branching of the left hepatic artery does not appear to be quite as regular as that of the right hepatic artery. Just as the drainage of the caudate lobe is divided between the right and left biliary ducts, so also is its arterial supply provided by both the right and left hepatic arteries. *There is no evidence, from injection studies, of arterial anastomoses within the liver, each segment or area being supplied by a specific branch and by it alone.*

The pattern of distribution of the **portal vein** and its branches is similar to that of the ducts and arteries. Its right branch has the same subdivision and follows the same pattern of division. The left branch distributes to the same segments and areas as do the ducts and arteries but has a largely unbranched transverse portion proximal to its highly branched umbilical part (fig. 344). There are usually three portal branches to the caudate lobe.

The Hepatic Veins (figs. 344, 345)

Like the pulmonary veins of the lungs, the hepatic veins of the liver constitute its functional venous drainage. Like the pulmonary veins also, the hepatic veins lie in the intersegmental planes and drain adjacent segments. Their radicles interdigitate with those of the

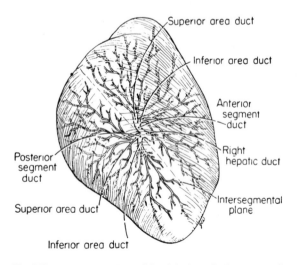

Fig. 343 The subdivisions of the right hepatic duct as seen in transparency through the right diaphragmatic surface of the right lobe of the liver. (Based on fig. 2, J. E. Healey, Jr. and P. C. Schroy, *A.M.A. Arch. Surg.,* 66:599, 1953.)

hepatic arteries, portal vein, and biliary ducts. Three main hepatic veins are distinguishable: a left, a middle, and a right. The **left hepatic vein** is located in the upper end of the left segmental fissure, where it receives the drainage of the entire lateral segment and of the superior area of the medial segment of the left lobe. The **middle hepatic vein** courses in the lobar fissure which bisects the liver deep to the gall bladder and the inferior vena cava. This vein drains the inferior area of the medial segment of the left lobe and the inferior area of the anterior segment of the right lobe. The **right hepatic vein** lies in the intersegmental plane of the right lobe and receives the venous blood of the entire posterior segment and of the superior area of the anterior segment of the right lobe. The hepatic veins converge on the inferior vena cava as it lies in its deep fossa on the surface of the liver. They have no valves and empty individually as a rule, although the middle and left veins frequently unite for about 1 cm. before entering the vena cava. Small, but sometimes numerous, hepatic

veins drain separately into the inferior vena cava caudal to the entrance of the main veins. These lower veins on the left side drain the caudate lobe, but, on the right, there is a larger vein which drains the visceral surface of the posterior segment of the right lobe. It may be visible through the capsule of the liver.

EXCRETORY APPARATUS OF THE LIVER (figs. 315, 336, 337, 338, 341–44)

In the porta of the liver, the right and left hepatic ducts unite to form the **common hepatic duct,** which then passes downward and to the right between the layers of the lesser omentum for about 4 cm. It is joined at an acute angle by the cystic duct, the two forming the common bile duct. The common hepatic duct is accompanied by the proper hepatic artery and by the portal vein.

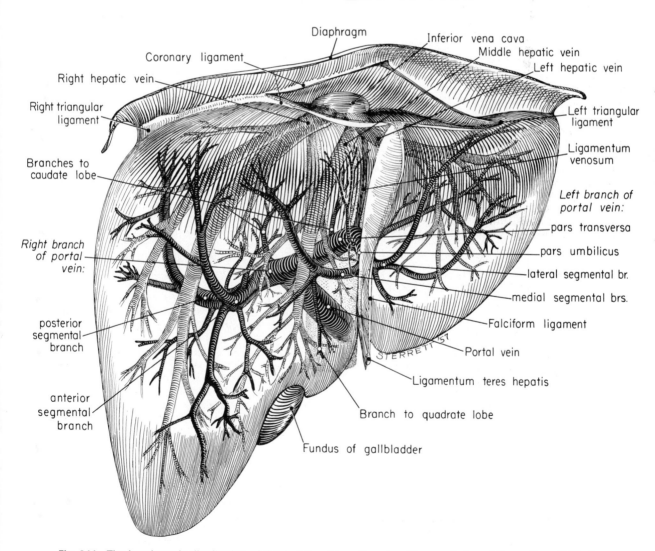

Fig. 344 The intrahepatic distribution of the portal and hepatic veins. (By permission of *Surg., Gyn., & Obs.*)

The Gall Bladder and the Cystic Duct (figs. 315, 336, 341, 346–48)

The gall bladder and the cystic duct constitute a diverticulum of the biliary tract, the gall bladder serving as a reservoir for bile secreted in the intervals between active phases of digestion. In addition to storing bile, the gall bladder concentrates it by extraction of water. The pear-shaped gall bladder is from 7 to 10 cm. long, is 2.5 cm. broad at the fundus, and has a capacity of about 35 cc. It lies in the gall bladder fossa of the visceral surface of the liver, its bulbous end (fundus) making contact with the anterior abdominal wall at the junction of the right semilunar line and the costal arch (tip of the ninth costal cartilage). Positional variation is frequent in the living, however, as seen in roentgenograms. The **fundus** of the gall bladder lies downward, forward, and to the right; the **body** and the tapered **neck** are directed upward and backward and to the left. The peritoneum of the visceral surface of the liver sweeps across the gall bladder but does not ordinarily suspend it as by a 'mesocyst.' The relatively broad area of attachment of the gall bladder to the gall bladder fossa of the liver is sometimes crossed by small veins and by small hepatocystic bile ducts. The fundus of the gall bladder rests on the transverse colon, the neck is related to the duodenum. The **neck** of the gall bladder is directed toward the porta hepatis, makes an S-shaped curve, and is continuous with the cystic duct. Crescentic folds of mucous membrane in its interior are spirally arranged and constitute the **spiral fold.** The gall bladder has serous, fibromuscular, and mucosal layers. The mucous membrane has a honeycomb appearance due to the elevation of folds in low criss-crossing ridges. Its

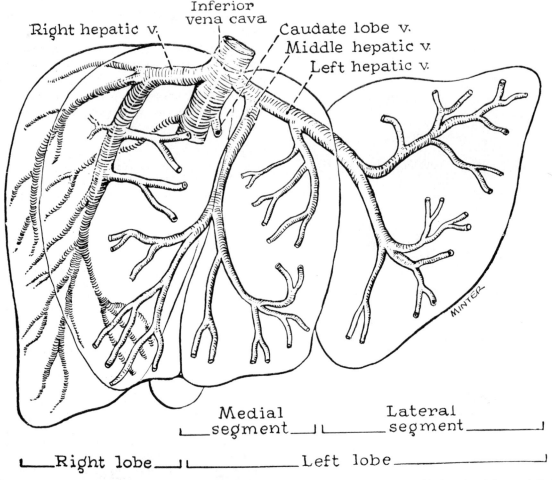

Fig. 345 The prevailing pattern of the hepatic veins in the human liver. (Courtesy of J. E. Healey, Jr.: *J. Internat. Coll. Surg.* 22:543, 1954.)

epithelium is capable of concentrating the contents of the gall bladder six to ten times.

The neck of the gall bladder passes without interruption into the **cystic duct.** The cystic duct is about 4 cm. long and is directed backward, downward, and to the left to join the common hepatic duct in the formation of the common bile duct. The **spiral fold** of the neck of the gall bladder continues into the upper portion of the cystic duct. It is said to be an aid in keeping the lumen of the duct open. It is also a common site of impaction of gallstones.

The Common Bile Duct (figs. 336, 337, 346, 347)

The common hepatic and cystic ducts unite at an acute angle to form the common bile duct which descends in the free border of the lesser omentum to the duodenum. About 8 cm. long, it lies to the right of the hepatic artery and anterior to the portal vein in its relatively short supraduodenal segment. The duct descends behind the first portion of the duodenum and then crosses the posterior surface of the head of the pancreas. Frequently embedded in the back of the pancreas (actually overlapped

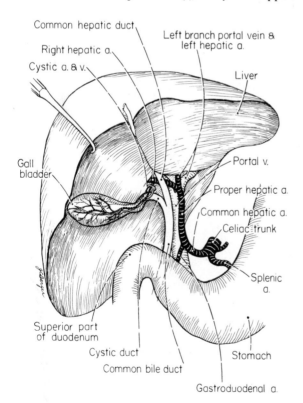

Fig. 346 The excretory apparatus of the liver, the gall bladder and bile ducts; the proper hepatic artery.

by a lingula of pancreatic tissue), it lies in front of the inferior vena cava. The posterior superior pancreaticoduodenal artery and vein spiral around the pancreatic portion of the duct in forming the posterior vascular arcade of the pancreas. The common bile duct and the pancreatic duct meet one another and, together, pass obliquely through the posteromedial wall of the second part of the duodenum at the junction of its middle and lower thirds. In this oblique passage (about 1.5 cm. in length) the ducts pass through the longitudinal and transverse layers of muscle and then lie, for most of this distance, in the submucosal layer. The two ducts open side by side but separately in most (80 per cent—Sterling, '54) cases. However, the common wall between their passageways is thin and it is short or nonexistent in about 17 per cent of individuals. Where the channel is a completely common one (3 per cent), constriction of the terminal opening on the papilla can lead to backflow of bile into the pancreatic duct. Deflection of the thin common wall of the ducts to one side or the other (or its absence) forms an enlarged chamber just proximal to the terminal opening; this is the **hepatopancreatic ampulla.** Since the opening of the **greater duodenal papilla** is the narrowest part of the biliary passage, the ampulla is a common site for the impaction of gallstones.

Clinical Note Accessory bile ducts occur in about 16 per cent of specimens (Moosman & Coller, '51). They are to be looked for in the angle between the cystic and common hepatic ducts, for they usually empty into one of these ducts adjacent to their junction.

The Finer Anatomy of the Choledochoduodenal Junction (fig. 348)

The passage of the common bile and pancreatic ducts through the wall of the duodenum is an oblique one involving a hiatus in the longitudinal duodenal muscle and a transversely oval opening in the circular layer. Muscular fascicles pass from the circular layer of the duodenum onto the ducts as reinforcing and connecting fibers which are variable in position and arrangement but serve to anchor the ducts in place. The specific duct sphincters are two. The **bile duct sphincter** (sphincter choledochus) consists of a funnel-shaped muscular sheath which surrounds the bile duct for about $\frac{1}{2}$ cm. just proximal to its passage through the duodenal wall (pars superior) and an additional $\frac{1}{2}$ cm. band of fibers which line and encircle the bile duct from its penetration of the wall to the confluence of the bile and pancreatic ducts (pars inferior). The **sphincter of the hepatopancreatic ampulla** encloses and is capable of constricting the ampulla. In those cases in which the bile and pancreatic ducts remain separate through the length of the papilla this thin layer of muscle encloses both and is then more properly termed the **sphincter of the papilla** (Boyden, '57). The duodenal musculature does not compress the bile duct, nor does it control the flow of bile. The sphincter of the bile duct is, however, capable of stopping the passage of bile.

VESSELS AND NERVES OF THE LIVER AND ITS DUCT SYSTEM

Arteries of the Liver and the Gall Bladder (figs. 346, 349, 350)

The **proper hepatic artery** (p. 404) ascends to the porta of the liver in the free border of the lesser omentum on the left of the common bile duct. Here it divides into a right and a left branch. This course and division after an origin from the celiac trunk is the normal arrangement and occurs in 85 per cent of specimens. In 14 per cent the right hepatic artery arises behind the pancreas as the highest branch of the superior mesenteric artery or from a common hepatic artery from the superior mesenteric. Such a vessel ascends through the lesser omentum posterior to the portal vein. This is the commonest variation in the origin of the right hepatic artery. Nearing the porta of the liver, the right hepatic artery passes posterior to the common hepatic duct in three-fourths of those cases having

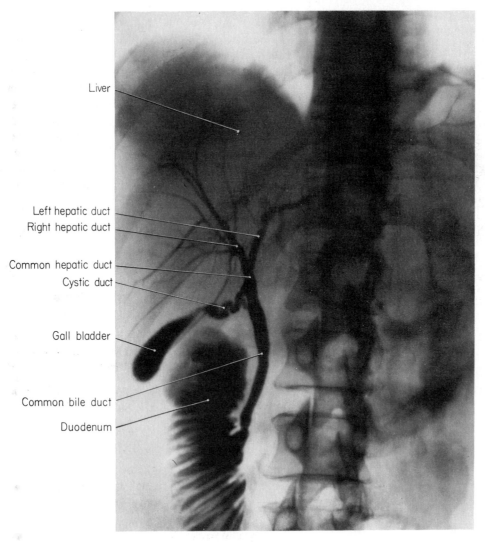

Liver

Left hepatic duct
Right hepatic duct

Common hepatic duct
Cystic duct

Gall bladder

Common bile duct
Duodenum

Fig. 347   A percutaneous cholangiogram.

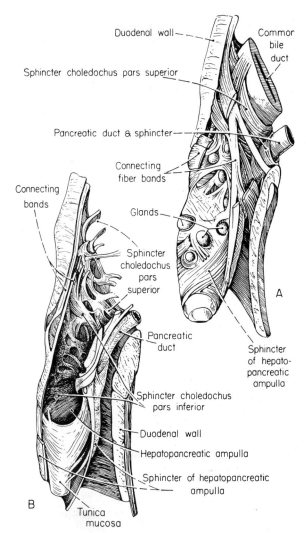

Duodenal wall

Common bile duct

Sphincter choledochus pars superior

Pancreatic duct & sphincter

Connecting fiber bands

Connecting bands

Glands

Sphincter choledochus pars superior

A

Pancreatic duct

Sphincter of hepatopancreatic ampulla

Sphincter choledochus pars inferior

Duodenal wall

Hepatopancreatic ampulla

Sphincter of hepatopancreatic ampulla

B

Tunica mucosa

Fig. 348 The sphincters of the choledochoduodenal junction. Redrawn after Boyden. A—greater duodenal papilla as seen from anterior aspect with mucosa removed. B—sagittal section of bile duct exposing cavity of bisected papilla.

the normal origin. In most of the remaining cases it passes anterior to the hepatic duct. As it passes through the cystohepatic angle, the right hepatic artery usually gives rise to the cystic artery. The **left hepatic artery** passes to the left end of the porta and enters the left lobe of the liver. It frequently (40 per cent—Healey, Schroy, & Sorenson, '53) provides, in the porta, a special artery for the medial segment of this lobe. This is the **middle hepatic artery** of Michels ('55) and others. The intrahepatic course and distribution of the hepatic arteries have been described (p. 420).

The **cystic artery** usually (72 per cent of cases)

arises from a normal right hepatic artery as that vessel crosses the cystohepatic angle. The cystic artery passes downward and forward along the neck of the gall bladder and provides two branches —one to the peritoneal surface, the other to the attached surface of the gall bladder. As its most frequent variations, the cystic artery may arise as a branch of either a right hepatic or a common hepatic artery from the superior mesenteric artery (13 per cent), or it may spring from a normal (celiac) proper hepatic artery (6 per cent—Moosman & Coller, '51). Cystic arteries are usually single, but they are occasionally double.

A **left hepatic branch of the left gastric artery** (p. 403) occurs in 18 per cent of cases. The **common hepatic artery** with its entire complement of branches arises from the superior mesenteric artery in 4 per cent of specimens (Daseler, Anson, Hambley, & Reimann, '47). The studies of the intrahepatic distribution of the arteries of the liver emphasize that there are no accessory hepatic arteries in the sense of extra and expendable vessels. All hepatic arteries, regardless of source, distribute to discrete portions of the liver, and there are no appreciable intrahepatic anastomoses. The small anastomoses which do occur among the hepatic arteries are formed in the porta and in the fossa for the ligamentum teres. The collateral circulation of the hepatic arteries afforded outside the liver by the interconnections of gastric, gastroepiploic, pancreatic, and intestinal vessels is, however, very considerable (Michels, '55).

Arteries of the Common Bile Duct

The blood vessels of the common bile duct are, in the main, short twigs which approach the duct at right angles and supply, individually, from 2 to 3 cm. of its length. The **cystic artery** supplies twigs to the upper part of the duct; the **right hepatic artery,** branches to its upper and central parts. The retroduodenal portion of the common bile duct is supplied by twigs from the gastroduodenal artery. The **posterior superior pancreaticoduodenal artery** spirals the pancreatic part of the duct and provides it with from three to five branches, some of which ascend along the duct.

Veins of the Liver

The **hepatic veins** drain into the inferior vena cava. They are entirely intrahepatic, and their distribution has been considered (p. 420).

The **portal vein** (figs. 359, 362), formed behind the neck of the pancreas by the junction of the superior

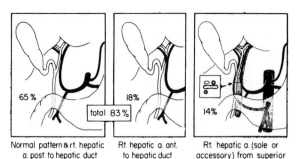

Normal pattern & rt. hepatic   Rt. hepatic a. ant.   Rt. hepatic a. (sole or
a. post. to hepatic duct   to hepatic duct   accessory) from superior
                                                       mesenteric a.

Common hepatic a. from   Left hepatic a. (or part) from
superior mesenteric a.   left gastric a.

Fig. 349 Variational anatomy of the hepatic arteries.

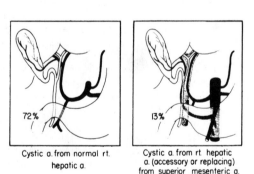

Cystic a. from normal rt.   Cystic a. from rt. hepatic
hepatic a.   a. (accessory or replacing)
                                from superior mesenteric a.

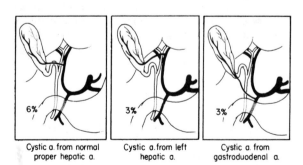

Cystic a. from normal   Cystic a. from left   Cystic a. from
proper hepatic a.   hepatic a.   gastroduodenal a.

Fig. 350 Variational anatomy of the cystic arteries.

mesenteric and splenic veins, ascends to the liver in the free margin of the lesser omentum posterior to the common bile duct and the proper hepatic artery (fig. 346). Its full description follows the consideration of its tributaries from the intestines (p. 440). Arriving at the hepatic porta, the portal vein divides into a right and a left branch which convey portal blood to the fundamental right and left lobes of the liver. The intrahepatic distribution of the vessel has been described (p. 420). The left branch, longer but smaller than that of the right side, receives anteriorly the ligamentum teres (obliterated umbilical vein) and is united to the left hepatic vein by the ligamentum venosum (obliterated ductus venosus). The left branch also receives the paraumbilical veins which arise in the umbilical region of the anterior abdominal wall and accompany the ligamentum teres to the liver.

Veins of the Biliary Ducts and the Gall Bladder

The **vein of the cystic duct** accompanies the duct and anastomoses with the veins of both the gall bladder above and the common bile duct below. The **vein of the common bile duct** and the **vein of the hepatic duct** lie along these ducts. They are connected below with the posterior superior pancreaticoduodenal vein and terminate above by penetrating directly into the liver. There are several **veins of the gall bladder neck.** They communicate with the veins along the cystic and common bile ducts but, in the main, they penetrate directly into the liver opposite the neck of the gall bladder (Petrén & Karlmark, '32). Small veins traverse the gall bladder bed to pass directly into the liver.

Lymphatics of the Liver (fig. 351)

The lymphatics of the liver are of interest both for the large amount of lymph they conduct and for their ramification to a large number of node groups. The lymphatic vessels of the liver are both superficial and deep, the two sets anastomosing freely near the surface of the gland.

The **deep lymphatics** are in two sets. (1) Starting in capillaries around the terminals of the portal vein, lymphatic vessels accompany the portal vein and converge onto the hilum of the liver. Here a number of lymph vessels emerge and end in the hepatic nodes along the hepatic artery. (2) A smaller group of channels arises along the radicles of the hepatic veins and follows these veins to the caval opening of the diaphragm. They end in the middle group of phrenic nodes of the right side of the chest (p. 320).

The **hepatic nodes** are from three to six nodes

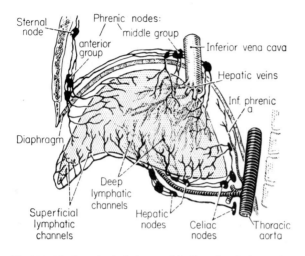

Fig. 351 The lymphatic drainage of the liver. Semischematic.

which lie along the proper hepatic and common hepatic arteries and along the bile ducts. They vary in number and position, but two are constant. One, the **cystic node,** lies in the curve of the neck of the gall bladder; the other is located along the upper part of the common bile duct and may be called the **node of the epiploic foramen.** Efferents from the hepatic nodes pass to the celiac group and thus contribute to the lymph of the intestinal lymph trunk (p. 439).

The **superficial lymphatics of the liver** arise in the subperitoneal areolar tissue over the entire surface of the organ. They can be analyzed according to the surfaces from which they arise. The lymphatics of the **diaphragmatic surface** of the liver drain in several directions: (1) From the anterior, right, and superior diaphragmatic surfaces, channels pass through the sternocostal hiatus of the diaphragm and end in the **anterior group of phrenic nodes** (p. 320). (2) Five or six vessels from the superior and posterior surfaces pass through the caval foramen and end in the **middle phrenic nodes** (p. 320). (3) From the posterior surface of the left lobe, a few vessels pass toward the esophageal hiatus of the diaphragm and end in the **paracardial group of the left gastric nodes.** (4) From the posterior surface of the right lobe, lymph vessels pass to the **nodes of the inferior phrenic artery** and thence to the celiac group. (5) The drainage of the lower 2 or 3 cm. of the anterior part of the diaphragmatic surface turns around the inferior border of the liver and ends in the **hepatic nodes** at the porta. The lymph collected

from the **visceral surface** of the liver mostly ends in the **hepatic nodes** of the hepatic porta.

Lymphatics of the Gall Bladder and the Bile Ducts

The gall bladder has a rich mucosal and subserous lymphatic network from which lymph vessels pass to the **cystic node** at the neck of the gall bladder and to the **node of the epiploic foramen.** The lymph flows through these nodes to more proximal hepatic nodes and then to the celiac group.

The biliary ducts possess both mucosal and external lymphatic networks; their drainage is like that of the gall bladder. The nodes concerned are the cystic, the node of the epiploic foramen, and the more proximal hepatic and celiac nodes.

Nerves of the Liver and the Biliary System

Nerves are prominent along the hepatic artery, portal vein, and common bile duct. The **anterior hepatic plexus** surrounds the hepatic artery and consists of branches of the left celiac plexus and of the hepatic branch of the anterior vagal trunk (p. 407). The **posterior hepatic plexus** is related to the portal vein and common bile duct. Its component nerves arise in the right celiac plexus and from the celiac division of the posterior vagal trunk. The posterior plexus is larger than the anterior, but ganglia are numerous in both. The region of the junction of the common bile duct and the duodenum receives an abundant innervation through the gastroduodenal nerve, an offshoot of the hepatic plexus with branches from both halves of the celiac plexus and from the celiac division of the posterior vagal trunk.

The finer branches of these nerves form terminal networks about the liver cells. No ganglion cells occur in the liver. The nerves of the gall bladder and the ducts distribute to their blood vessels and form plexuses in their outer connective tissue layers, distributing to the mucous membrane and the muscle layers. The outer plexus contains many ganglion cells.

In spite of the rich innervation of these organs, the specific action of the nerves is uncertain, for both hormonal and nervous controls exist in the liver and the biliary system. The secretion of bile by the liver is largely independent of nervous control. The gall bladder is capable of rapid periodic emptying by the contraction of its muscular layers, but there is practically no evidence that nervous mechanisms are in any way involved (Boyden, '53). The sympathetic nerve supply is vasomotor in both the liver and the biliary system.

THE JEJUNOILEUM

GENERAL RELATIONS (figs. 352–54, 357)

The **jejunoileum** is the portion of the small intestine which is suspended from the posterior body wall by the mesentery (the duodenum is the

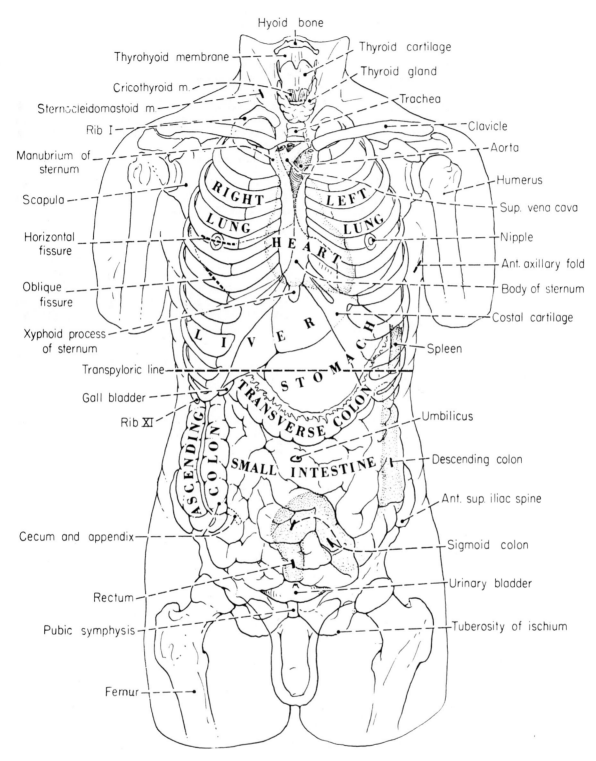

Fig. 352 The thoracoabdominal viscera; especially the gastrointestinal tract.

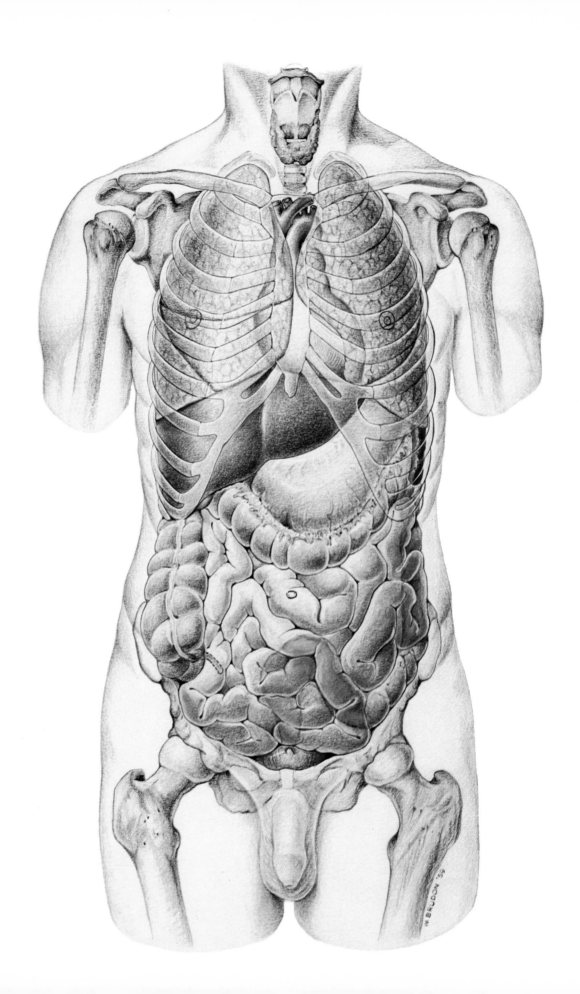

retroperitoneal portion of the small intestine). The combination term 'jejunoileum' is an expression of the fact that there is no clear line of demarcation between the two parts, such differences as there are being the results of gradual changes along the tube. The upper two-fifths (8 feet) is considered jejunum; the lower three-fifths (12 feet) is ileum. The diameter of the jejunum is slightly greater, and its wall is thicker, more vascular, and of a deeper color than that of the ileum.

Clinical Note Intestinal localization is sometimes a matter of practical importance. The upper one-third of the free intestine occupies the left upper quadrant of the abdomen, the middle one-third is located in the middle of the abdominal cavity, and the lower one-third lies in the pelvis and in the right iliac fossa. The distal portion of the ileum is almost always in the pelvis, from which it ascends over the right psoas muscles and the right iliac vessels to end in the medial aspect of the cecum. The amount of fat between the layers of the mesentery is also an indicator of distance along the tract, for the mesenteric fat increases progressively from above downward. Thus, little fat exists in the mesentery of the upper jejunum, and 'windows' of translucency between the blood vessels of the mesentery are numerous. Such 'windows' become progressively less clear in the ileum, for here the fat usually occupies the full width of the mesentery to the intestinal wall, making it thicker and more opaque. There is also a progression in the arcades formed by the intestinal branches of the superior mesenteric artery. The arcades, beginning singly in the jejunum, increase to four or five in succession in the ileum.

Meckel's diverticulum is an occasional feature of the ileum. It is a finger-like pouch which springs from the free border of the intestine; its diameter is similar to that of the ileum, and it averages about 5 cm. in length. It occurs in about 1 per cent of individuals, somewhere in the last meter of the ileum. Representing the remains of the vitelline duct of early fetal life, the diverticulum may be free or may be connected with the umbilical region of the abdominal wall by a fibrous band. Rarely it has a fistulous opening at the umbilicus.

Structure (fig. 353)

The wall of the intestine, like that of the stomach, is composed of four coats: mucosal, submucosal, muscular, and serous. The mucous membrane exhibits circular folds and intestinal villi. The **circular folds** are described under the duodenum (p. 411). They are well developed in the jejunum but become very small or absent in the ileum. The difference in thickness of the jejunum and the ileum is largely dependent on the gradual loss of the circular folds as the tube is followed downward. The **intestinal villi,** closely set and shaped like pages of a book in the duodenum, become less numerous and finger-like in the ileum. Between these extremes there is a gradual transition in their frequency and form, the middle region of the tract exhibiting villi of a flat, triangular shape. The strong **submucosa** is composed of fibro-elastic and areolar connective tissue. It forms the core of the circular folds. **Solitary lymph nodules** occur in the submucosa and bulge the mucous membrane in all parts of the small intestine. **Aggregated lymph nodules** occur in patches along the antimesenteric border of the ileum. The **muscular coat** of the jejunoileum is composed of complete circular and longitudinal layers. The circular layer is thicker than the longitudinal, and both of them become thinner toward the lower ileum. Between the two layers lies the myenteric nerve plexus of unmedullated nerve fibers and ganglion cells; a similar plexus occurs in the submucosa. The **serous coat** is composed of the peritoneum and the connective tissue connecting it to the muscular layer. The jejunoileum is covered by peritoneum except along the narrow strip between the attachment of the layers of the mesentery; through this narrow interval it receives its blood vessels and nerves.

The Mesentery

The mesentery is defined and partially described under the peritoneum (p.387). It is a double layer of peritoneum which attaches the jejunoileum to the posterior wall and which conducts to the tube the mesenteric blood and lymphatic vessels and nerves (fig. 357). These, plus fat and areolar connective tissue, comprise the contents of the mesentery. The mesentery is a highly folded fan, for its intestinal border is the length of the jejunoileum—twenty feet —whereas its root is attached over a distance of only six or seven inches of the posterior body wall. This transition in the length of the borders is achieved in a width of eight or nine inches. The body wall attachment of the mesentery begins over the fourth part of the duodenum to the left of the second lumbar vertebra and ends in the right iliac fossa anterior to the right sacroiliac joint. Between these points the attached border of the mesentery crosses the third part of the duodenum, the aorta, the inferior vena cava, the right ureter, and the right psoas muscles.

VESSELS AND NERVES

The Superior Mesenteric Artery (figs. 322, 337, 340, 354)

The superior mesenteric artery is the second of the ventral unpaired branches of the abdominal aorta (celiac, superior mesenteric, and inferior mesenteric arteries). The superior mesenteric artery supplies all of the small intestine except the proximal part of the duodenum; it also supplies the cecum, the ascending colon, and most of the transverse colon, the embryonic midgut. The artery arises from the aorta behind the neck of the pancreas in front of the lower one-third of the first lumbar vertebra. (For variations in level of

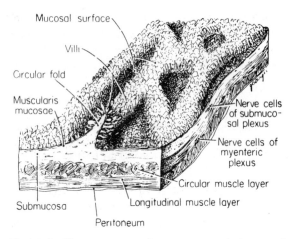

Fig. 353 The structure of a segment of the small intestine. (Modified from Braus.)

origin of this and other visceral branches of the abdominal aorta, refer to Cauldwell and Anson, '43.) The superior mesenteric artery descends across the uncinate process of the pancreas and the third part of the duodenum to enter the mesentery. The course of the artery is oblique toward the right iliac fossa where its last intestinal branch anastomoses with a higher branch, the ileocolic. The artery occupies the mesentery and is bowed to the left and downward in a curve from which its intestinal branches arise. The superior mesenteric vein is to its right and anterior, and the superior mesenteric plexus of nerves surrounds the artery. The branches of the superior mesenteric artery are the inferior pancreaticoduodenal, the intestinal, the ileocolic, the right colic, and the middle colic. It also may exhibit certain inconstant branches. In 21 per cent of specimens the dorsal pancreatic artery constitutes the highest branch of the superior mesenteric. In 14 per cent of cases a right hepatic branch springs from the superior mesenteric artery behind the neck of the pancreas; in 4 per cent the common hepatic artery arises in this manner.

(1) The **inferior pancreaticoduodenal artery** arises from the superior mesenteric or from its first jejunal branch at the level of the pancreatic notch. It passes to the right and divides into two branches: the anterior inferior pancreaticoduodenal and posterior inferior pancreaticoduodenal arteries. These participate in forming the anterior and posterior pancreaticoduodenal arcades as more fully described on p. 414. Sometimes, the anterior and posterior branches arise separately from either the superior mesenteric or the first jejunal artery.

As branches of the first jejunal artery, they pass posterior to the superior mesenteric vessels to reach the pancreas and the duodenum.

(2) **Intestinal arteries,** from twelve to fifteen in number, arise from the convex left side of the superior mesenteric artery. They run nearly parallel to one another, and each vessel divides into two arteries which unite with adjacent branches to form a series of arches convex toward the intestine. From this first series of arches other branches arise which unite with similar branches above and below to form a second series of arches. Only one or, at most, two successive arches are formed at the upper jejunal end of the intestine, but in the ileal region, three, four, or even five arches may succeed one another before the intestinal wall is reached. The arches become smaller as the intestine is approached, and from the last arch in the series small, straight vessels (vasa recta) arise which pass alternately to one or the other side of the small intestine. Occasionally a terminal branch divides and passes to both sides. The vasa recta are not end-arteries, but the longitudinal anastomosis between them depends on small vessels (Doran, '50).

Clinical Note Occlusion of a series of vasa recta (commonly the veins) reduces nutrition of the tissues of the tube. This may result in gangrenous obstruction due to faulty circulation which not only stops peristalsis but releases toxic substances.

(3) The **ileocolic artery** (figs. 322, 354) is the lowermost of the branches which arise from the right side of the superior mesenteric artery. It is a separate branch of the superior mesenteric in 60 per cent of specimens and has a source in common with the right colic artery in 40 per cent. Directed toward the cecum, the ileocolic artery diverges but little from the course of the terminal intestinal branch of the superior mesenteric artery. It descends behind the peritoneum toward the right iliac fossa, where it gives off an **ascending branch** which anastomoses with the right colic artery. The ileocolic artery terminates at the upper border of the ileocecal junction, here dividing into four or five branches.

(a) a **colic branch** ascends along tne medial aspect of the ascending colon anastomosing with downwardly directed terminals of the ascending branch; (b) an **anterior cecal branch** passes through the vascular cecal fold onto the anterior surface of the cecum; (c) a **posterior cecal branch** crosses the back of the ileocecal junction to the posterior aspect of the cecum; (d) an

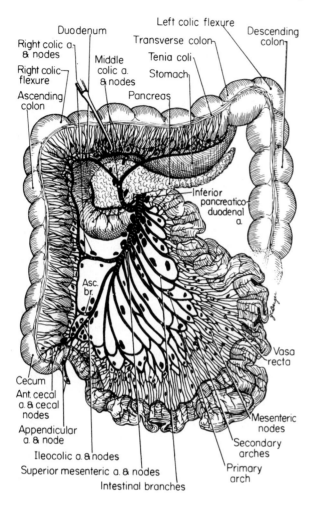

Duodenum
Left colic flexure
Transverse colon
Descending colon
Right colic a. & nodes
Right colic flexure
Middle colic a. & nodes
Tenia coli
Stomach
Ascending colon
Pancreas
Inferior pancreatico-duodenal a.
Asc. br.
Cecum
Ant. cecal a. & cecal nodes
Appendicular a. & node
Ileocolic a. & nodes
Superior mesenteric a. & nodes
Intestinal branches
Vasa recta
Mesenteric nodes
Secondary arches
Primary arch

Fig. 354 The branches and distribution of the superior mesenteric artery; the mesenteric lymph nodes.

appendicular artery descends behind the terminal part of the ileum and enters the mesoappendix; and (e) an **ileal branch** turns back onto the terminal ileum and anastomoses with the last intestinal branch of the superior mesenteric artery. The appendicular artery runs near the free margin of the mesoappendix and gives off a series of small branches which supply the appendix from its root to its tip. It frequently is a branch of the ileal, the anterior cecal, or the posterior cecal artery, and it may be double (Shah & Shah, '46).

(4) The **right colic artery** usually (27%) arises from about the middle of the right side of the superior mesenteric artery. It passes to the right behind the posterior body wall peritoneum, crossing in its course the testicular (or ovarian) vessels, the right ureter, and the right psoas muscles. Reaching the middle of the ascending colon, the artery divides into a descending branch, which anastomoses with the ascending branch of the

ileocolic artery, and an ascending branch, which anastomoses with the middle colic artery. From the loops formed by these anastomoses, branches are distributed to the ascending colon and the beginning of the transverse colon.

The right colic artery exhibits considerable variability. It may spring from the root of the ileocolic artery (26%) or it may be a branch of the middle colic (24%). It may be absent altogether (13%), in which case its area of supply is covered by the ascending branch of the ileocolic artery and the right branch of the middle colic.

(5) The **middle colic artery,** present and single in 88 per cent, takes origin from the front of the superior mesenteric artery immediately below the neck of the pancreas in 60 per cent of cases. Entering the transverse mesocolon, the artery passes downward and forward between its layers and divides into right and left branches. The right branch anastomoses with the right colic artery; the left branch, with the left colic branch of the inferior mesenteric artery. From the arcades so formed, secondary and tertiary loops sometimes develop, especially in the regions of the right and left flexures of the colon, and terminal branches distribute to the transverse colon. The anastomotic channels along the large intestine are frequently so large as to constitute a **marginal artery** which follows the arch of the colon. The middle colic arises in common with the right colic in 27 per cent of cases.

Clinical Note The middle colic artery must be carefully avoided when a surgical passage is being made through the transverse mesocolon for short-circuiting operations such as gastroenterostomy.

The Superior Mesenteric Vein (figs. 361, 362)

The superior mesenteric vein accompanies the superior mesenteric artery and lies anterior and to its right in the root of the mesentery. Like the artery, the vein crosses the third part of the duodenum and the uncinate process of the pancreas. It terminates behind the neck of the pancreas by joining the splenic vein to form the portal vein. The portal vein is more completely described following the consideration of the inferior mesenteric vein (p. 440).

The tributaries of the superior mesenteric vein include veins which correspond to the branches of the artery and some which do not. Beginning at the ileocecal junction, the **anterior cecal vein,** the **posterior cecal vein,** and the **appendicular vein** unite to form the **ileocolic vein. Intestinal veins,** generally anterior and to the right of their corresponding arteries, pass to the root of the mesentery to join the superior mesenteric vein. A **right colic vein** enters its right side, and a **middle colic vein** reaches it at the pancreatic notch.

The **right gastroepiploic vein** runs from left to right along the greater curvature of the stomach between the layers of the gastrocolic ligament and receives both gastric and epiploic tributaries. Turning onto the head of the pancreas at the right extremity of the gastrocolic ligament, the right gastroepiploic vein receives the anterior superior pancreaticoduodenal vein and ends in the superior mesenteric vein at the pancreatic notch. It also receives colic veins from the right flexure of the colon so frequently as to form a gastrocolic vein. A single **inferior pancreaticoduodenal vein** may empty into the superior mesenteric vein behind the neck of the pancreas. There may be both anterior and posterior inferior pancreaticoduodenal veins or, as is most common, either the single trunk or the pair of veins may empty into the highest jejunal vein. Other pancreatic veins which are usually tributary to the superior mesenteric vein are the **pancreatic cervical vein** and the **inferior pancreatic vein.** The superior mesenteric vein occasionally receives the inferior mesenteric vein.

The Superior Mesenteric Lymph Nodes (fig. 354)

The superior mesenteric lymph nodes form a system in which the large central nodes at the root of the superior mesenteric artery receive the drainage from four subsidiary groups: mesenteric, ileocolic, right colic, and middle colic. Further, there is a common pattern of distribution within all of these subsidiary groups—smaller, more numerous nodes are located adjacent to the intestinal tract along the anastomosing branches of the terminal arches, and larger, less numerous nodes are arranged along the named arteries. Drainage through the lymphatic channels connecting these nodes is from peripheral to central parts of the system.

(1) The **mesenteric nodes** number in excess of two hundred and lie between the layers of the mesentery. The smaller peripheral nodes lie along the vasa recta and in the terminal loops of the arterial system; the larger intermediate nodes are found in relation to the primary arcades and along the intestinal arteries.

(2) The **ileocolic nodes** lie along the ileocolic artery.

They show a tendency to clump into one group near the origin of the artery and a second group more distally located. At the ileocecal junction, small nodes wander out on the arterial branches—**cecal nodes** accompany the anterior cecal and posterior cecal arteries, and **appendicular nodes** are located along its artery and in the mesoappendix.

(3) The **right colic nodes** include peripheral nodes along the colic branches of the right colic artery or along the marginal artery and intermediate nodes along the right colic artery itself. The nodes associated with the cecum and the ascending colon average seventy-five in number.

(4) The **middle colic nodes,** about forty in number, include the nodes of the transverse colon from the right flexure as far as the left one-third of this segment of the colon. The smaller, more numerous peripheral nodes lie along the wall of the transverse colon and between the wall and the anastomosing loops of the artery; the larger, less numerous intermediate nodes lie along the middle colic artery.

The **superior mesenteric nodes** proper are large and numerous. They lie about the main stem of the superior mesenteric vessels in front of the duodenum and the head of the pancreas. These nodes receive the ultimate efferents of all the subsidiary groups as well as part of the drainage of the pyloric nodes and channels from the head of the pancreas and the duodenum. The efferent vessels of the superior mesenteric nodes partly drain to the celiac nodes and partly assist in forming the intestinal lymph trunk (p. 440).

The Superior Mesenteric Nerve Plexus

The superior mesenteric artery and all its branches are invested by a perivascular nerve plexus through which parasympathetic preganglionic, sympathetic postganglionic, and afferent nerve fibers are conducted to the parts of the small and large intestines served by the artery. Its parasympathetic fibers are derived from the celiac division of the posterior vagal trunk, and its sympathetic fibers arise in the superior mesenteric ganglion after synapse with preganglionic neurons of the thoracic splanchnic nerves. Both the vagal parasympathetic innervation and the thoracic splanchnic sympathetic innervation of the gastrointestinal tract terminate with the end of the distribution of the superior mesenteric artery—at about the junction of the middle and left thirds of the transverse colon. More distal parts of the large intestine are innervated from other parts of the autonomic nervous system (p. 439). The peri-

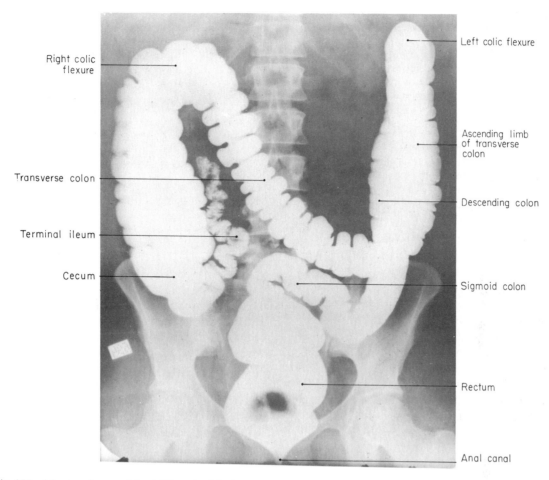

Right colic flexure

Transverse colon

Terminal ileum

Cecum

Left colic flexure

Ascending limb of transverse colon

Descending colon

Sigmoid colon

Rectum

Anal canal

Fig. 355 A barium image of the full length of the large intestine. Note segmentation, especially in the transverse colon.

vascular nerve plexus follows all the branches of the superior mesenteric artery. Its vagal fibers end in the submucosal and myenteric plexuses of the intestinal tract where synapse is made with short, intramural postganglionic neurons.

The perivascular plexus is a continuation of the **superior mesenteric plexus.** This is essentially the lower portion of the celiac plexus although it may be more or less detached from that plexus. It frequently contains a **superior mesenteric ganglion.** Extending downward from the celiac plexus along the aorta, the superior mesenteric plexus surrounds the root of the superior mesenteric artery and accompanies it and its branches in their distribution to the duodenum, the pancreas, the jejunoileum, the cecum, the ascending colon, and the transverse colon.

THE LARGE INTESTINE

GENERAL RELATIONS (figs. 352, 355, 357, 378)

The **large intestine** is continuous with the ileum. It is about 5 feet long and is composed of the cecum and the vermiform appendix, the ascending colon, the transverse colon, the descending colon, the sigmoid colon, the rectum, and the anal canal. These parts succeed one another in an arch which lies first to the right of the small intestine, then successively above or in front of it, to the left of it, and finally below it. The large intestine is only larger than the small intestine in a general sense, for the diameters of the parts of the tube are so nearly alike that local dilatations or constrictions make 'large' and 'small' meaningless.

Clinical Note Three surface features serve to distinguish isolated loops of the small or the large intestine presenting through an abdominal opening (fig. 356). (1) The longitudinal musculature of the large intestine is not so complete as that of the small intestine. It is separated into three bands, the **teniae coli.** (2) The longitudinal bands of muscle of the large intestine are about one-sixth shorter than the colon, so that the wall is forced to bulge between the teniae. The colon, therefore, presents three rows of **sacculations, or haustra coli,** alternating between the teniae. (3) The third characteristic of the large intestine is the occurrence along its length of **epiploic appendages.** These are fat-filled tabs, or pendants, of peritoneum that project from the serous coat of the large intestine except on the rectum. They are numerous and are attached adjacent to the teniae, but they are not found along the tenia omentalis.

The **teniae coli** (fig. 356), about 1 cm. in width, are approximately equally spaced around the tube. They arise at the root of the appendix, which has a complete longitudinal muscular coat, and separate there to spread over the surface of the cecum and onto the ascending colon. The names applied to the teniae coli are derived from their relations to the peritoneal attachments of the transverse colon—one lies in the transverse mesocolon, the **tenia mesocolica;** one lies in the attachment of the greater omentum, the **tenia omentalis;** and one lies on the inferior surface of the transverse colon free of any peritoneal fold, the **tenia libera.** The attachment of the mesocolon has continuity with the line of peritoneal reflection on the ascending colon and the descending colon, and thus the tenia mesocolica is posteromedial in position on these segments. Following out the correspondence of surfaces of the ascending colon, the transverse colon, and the descending colon, the tenia omentalis is posterolateral and the tenia libera is anterior on the ascending colon and the descending colon. Similar positional relations follow on the sigmoid colon, but over the rectum the longitudinal muscular coat is more uniform. Here it forms two wider muscular bands, one anterior and one posterior to the rectum.

Between the sacculations of the wall of the large intestine produced by the teniae, the wall is puckered and thrown into folds in the interior. These folds have a crescentic form and are designated as the **semilunar folds** in contrast with the circular folds of the small intestine. The mucous coat of the large intestine is smooth, lacks villi, and covers the haustra and semilunar folds uniformly (fig. 358). The mucosa lodges numerous, closely set, straight, tubular mucous glands, and the mucosal surface

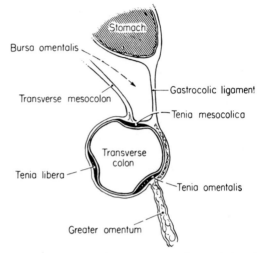

Fig. 356 A section of the transverse colon and stomach to show teniae and attachments.

of the colon is characterized by their regularly arranged circular openings.

THE CECUM (figs. 352, 355, 359, 378)

The **cecum** is the blind pouch of the large intestine which lies below the junction of the ileum and the colon. It is broader (7.5 cm.) than long (6 cm.) and occupies the right iliac fossa. Its closed end is directed downward, and it opens above into the ascending colon. The cecum is usually covered completely by peritoneum, although cecal folds and a retrocecal fossa are defined behind it by the continuity of its peritoneal coverage to the posterior body wall (p. 397). Its relative freedom in the abdominal cavity results in occasional migration or displacement. The cecum develops its adult form from an originally tapered or funnel-shaped diverticulum and exhibits some variations in final form. The normal cecum (90 per cent of cases) is markedly sacculated to the right, deviating the anterior and posterolateral teniae to the left and displacing the apex of the cecum to the left and posteriorly.

The Ileocecal Valve (figs. 355, 359, 378)

The ileocecal valve is formed by the infolding of the end of the ileum into the cavity of the cecum. As seen from within, the valve appears as a projecting fold of mucous membrane showing two thick rounded lips. The substance of the lips is a duplication of the circular muscle layer of the ileum which projects into the cecum. The peritoneum and the longitudinal coat of the ileum are not duplicated but are stretched across the crease between the ileum and the cecum. The upper lip of the valve somewhat overhangs the lower, and the two are continued to either side of the opening in a more or less distinct ridge, the **frenulum** of the valve. The configuration of the valve is frequently more circular (fig. 378). The ileocecal valve is a questionable sphincter. It does not prevent barium which has filled the colon per rectum from entering the terminal ileum. The position of the ileocecal valve may be projected to the anterior abdominal wall at the junction of the right semilunar line and the spino-umbilical line (anterior superior iliac spine to umbilicus).

The Vermiform Appendix (figs. 322, 352, 359)

The appendix is the originally tapered tip of the cecum which, due to the unequal growth of the right wall of the cecum, has become displaced onto its medial surface. A much narrowed tube in the adult, the appendix varies between 2 and 20 cm. in length; its average length is aout 8 cm. The appendix has a complete peritoneal investment and a small mesentery, the **mesoappendix.** This fold is derived from the left leaf of the peritoneum and is a continuity of the mesentery. It is triangular in form and is attached along the whole length of the appendix. The appendicular vessels run near the free margin of the mesoappendix. The lumen of the appendix is narrow but is usually complete. Occasionally, especially in later life, the lumen becomes obliterated, and then the appendix is converted into a fibrous cord. The blood vessesl, lymphatics, and nerves of the cecum and the appendix have been considered (p. 430).

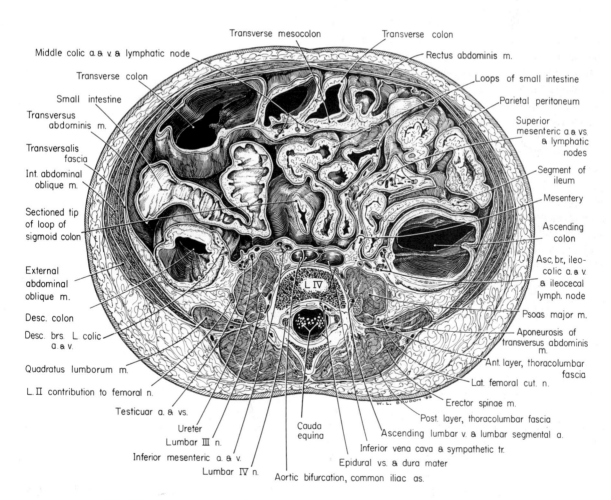

Fig. 357 A cross section of the abdomen at the level of the bifurcation of the aorta.

Clinical Note The root of the appendix is its only fixed point. This is located about 3 cm. below the ileocecal valve or, projected to the surface of the body, at the junction of the right semilunar line and the interspinous line (anterior superior iliac spine to anterior superior iliac spine). The body and the tip of the appendix may be directed anywhere in the large semicircle between the root of the mesentery and the peritoneal fixation of the ascending colon. Most commonly, the tip lies retrocecally or hangs over the brim of the pelvis. When hidden, the appendix can be located by following down the anterior tenia of the cecum to the root of the appendix, for the appendix has a complete coat of longitudinal muscle.

THE ASCENDING COLON (figs. 352, 354, 355, 357)

The **ascending colon** is that portion of the large intestine which extends between the ileocecal valve and the right flexure of the colon. It is a narrower segment than the cecum and lies retroperitoneally along the right side of the posterior abdominal wall. Posteriorly, it is joined by loose connective tissue with the fasciae over the iliacus, quadratus lumborum, and transversus abdominis muscles and the lower lateral aspect of the right kidney. The ascending colon makes contact at its right flexure with the right lobe of the liver lateral to the gall bladder (colic impression of right lobe). Extending medialward behind the body wall peritoneum as far as the median line is the fusion fascia of the ascending colon.

THE TRANSVERSE COLON (figs. 337, 352, 354, 357, 360)

The **transverse colon** is the largest and most movable part of the large intestine. It begins in the **right (hepatic) flexure** of the colon and terminates in the left flexure opposite the spleen. The transverse colon

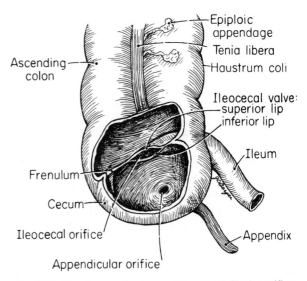

Fig. 359 The ileocecal valve and the appendicular orifice as seen through a window in the wall of the cecum.

is attached posteriorly by the transverse mesocolon on a line which crosses the second part of the duodenum and passes across the pancreas except for its distal extremity. The transverse colon is further attached along its whole length to the greater curvature of the stomach by the **gastrocolic ligament.** The **omental apron** is fixed to the anterior surface of the transverse colon and hangs below it. The transverse colon usually descends in its midportion but always conforms to the curvature and position of the inferior border of the stomach. In tall, thin individuals the transverse colon may reach the pelvis. The **left colic** (splenic) **flexure** represents the highest point attained by the transverse colon. Here the phrenicocolic ligament holds it into contact with the rounded inferior border of the spleen. Below and behind the transverse colon lie the coils of the small intestine. Above, it is in relation to the greater curvature of the stomach and to the liver and the gall bladder. The blood vessels, lymphatics, and nerves of the ascending and transverse segments of the colon are described on p. 430.

THE DESCENDING COLON (figs. 352, 355, 357, 360, 378)

The **descending colon** passes inferiorly from the left colic flexure to end in the sigmoid colon at the brim of the pelvis. The descending colon is usually much narrower than the other colic segments. It is retroperitoneal, like the ascending colon, but is more apt to have a partial mesocolon, especially in the iliac

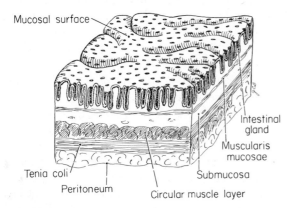

Fig. 358 The structure of a segment of large intestine. (Modified from Braus.)

fossa. Passing at first along the lateral border of the left kidney, the descending colon descends over the quadratus lumborum muscle to the iliac fossa. In the iliac fossa the colon curves medialward and downward in front of the iliacus and psoas muscles and ends at the pelvic brim in the sigmoid colon.

THE SIGMOID COLON (figs. 352, 355, 357, 360, 378)

The **sigmoid colon** is characterized by its S-shaped loop and by the presence of a mesocolon. The sigmoid colon begins at the brim of the pelvis on the left side and passes across to the right and upward and then back and down to end at the median line opposite the third segment of the sacrum. Losing its free mesocolic attachment at this level, the tube is continued as the rectum. The **sigmoid mesocolon** diminishes in breadth from the middle toward the ends of the loop where it disappears. Thus, the sigmoid colon has a greater range of motion in its middle part. The attachment of the sigmoid colon is along a V-shaped line, the apex of the V bounding the **intersigmoid fossa** of the peritoneal cavity (p. 397). The loop of the sigmoid colon usually occupies

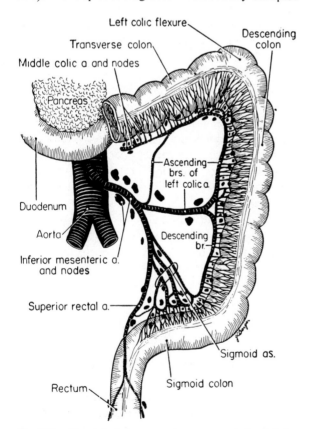

Fig. 360   The inferior mesenteric artery and the inferior mesenteric system of lymph nodes.

the pelvis. It may rest on the bladder in the male or on the uterus in the female. Frequently the lower coils of the small intestine are in front of the sigmoid colon.

Clinical Note   A long, mobile sigmoid segment with a long mesentery creates a situation susceptible to a twisting of the sigmoid colon on its mesentery (volvulus).

Diverticulosis is fairly common in the sigmoid colon. This is a herniation of the lining mucous membrane through the circular layer of muscle between teniae coli. Weakness of the wall where the blood vessels penetrate is thought to be a predisposing cause.

THE RECTUM AND THE ANAL CANAL

These parts of the large intestine are described in the chapter on the Pelvis.

VESSELS AND NERVES

The arteries, veins, lymphatics, and nerves of the cecum and the appendix, the ascending colon, and the greater part of the transverse colon have been described. These parts are supplied through the superior mesenteric systems of blood and lymph vessels (p.429), and their nerves reach them through the superior mesenteric plexus (p. 432). The inferior mesenteric system supplies the distal portion of the colon and the rectum.

The Inferior Mesenteric Artery (figs. 357, 360)

The inferior mesenteric artery is the third of the ventral unpaired branches of the abdominal aorta related to the gastrointestinal tract; developmentally, it is the artery of the hindgut. It arises from the aorta about 3 to 4 cm. above its bifurcation. The origin of the inferior mesenteric artery is just below the third portion of the duodenum and at the level of the third lumbar vertebra. The artery supplies the left one-third of the transverse colon, the descending colon, the sigmoid colon, and the greater part of the rectum. It passes downward to the left of the aorta behind the peritoneum of the posterior abdominal wall, giving off as branches the left colic, sigmoid, and superior rectal arteries.

The **left colic artery** passes to the left, behind the peritoneum, directed toward the middle of the descending colon. Approaching that segment, the artery divides into an ascending and a descending branch. The **ascending branch** passes upward across the left kidney toward the left flexure of the colon. It supplies branches to the upper part of the descending colon, enters the left end of the transverse colon, and turns to the right within it. It supplies the left one-third or more of the transverse colon and anastomoses with the left branch of the middle colic artery. One or more secondary arches may be formed in the left colic flexure. An **ascending branch** of more medial position frequently accompanies the inferior mesenteric vein upward. It then enters the transverse mesocolon and turns to the left to reach the area of anastomosis of the left colic and middle colic arteries. This vessel, or a branch of it, may anastomose with the superior mesenteric artery (or middle colic) to establish a **central intermesenteric anastomosis** in 11 per cent of specimens (Williams and Klopp). The **descending branch,** likewise supplying branches to the descending colon, ends by anastomosing with an ascending branch of the sigmoid artery.

The **sigmoid arteries,** two or three in number, descend across the left psoas muscle, the ureter, and the testicular (or ovarian) vessels and enter the sigmoid mesocolon. They give off ascending and descending branches and form several interconnecting loops. From the loops, vasa recta supply the sigmoid colon. The ascending branch of the first sigmoid artery anastomoses with the descending branch of the left colic; the descending branch of the lowest sigmoid has only a weak or no anastomosis with the superior rectal artery (Basmajian, '54).

The **superior rectal artery** is the continuation of the inferior mesenteric artery. Crossing the left common iliac vessels, it descends into the pelvis between the layers of the sigmoid mesocolon. At the upper end of the rectum, opposite the third segment of the sacrum, the artery divides into two branches which follow along the sides of the rectum. The two branches redivide, and halfway down the length of the rectum their radicles pierce the rectal wall to ramify within the submucosa of the rectum and the upper part of the anal canal. Here they form a series of small longitudinally oriented vessels which anastomose with each other and with terminals of the middle rectal and inferior rectal arteries from the internal iliac system.

The anastomoses along the large intestine between adjacent branches of the inferior mesenteric artery, between adjacent branches of the superior mesenteric artery, and between adjacent branches of the two systems sometimes result in a long patent arterial arch. This is the **marginal artery;** it helps to equalize blood supply in the large intestine when for any reason the flow is impeded through regular channels.

The Inferior Mesenteric Vein (figs. 361, 362)

The inferior mesenteric vein corresponds in pattern to the artery which it accompanies. The vein begins in the **rectal plexuses** which lie in the submucosal layer of the anal canal and the rectum and on the outer surface of their muscular coats. The rectal plexuses interconnect radicles of the inferior rectal, middle rectal, and superior rectal veins. The **superior rectal vein** arises by the union of tributaries from the rectal plexuses and accompanies the corresponding artery across the common iliac vessels. Here it becomes the **inferior mesenteric vein;** it receives as tributaries the **sigmoid veins** from the sigmoid colon and the **left colic vein** from the descending colon and the distal portion of the transverse colon. The inferior mesenteric vein lies to the left of its artery and crosses the branches of the artery in its ascent. Departing from the artery at the level of the left colic branch, the inferior mesenteric vein

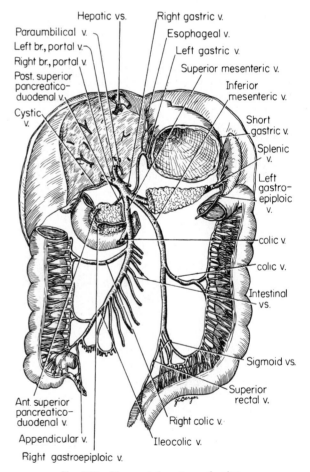

Fig. 361   The portal system of veins.

continues to ascend behind the peritoneum on the left side of the fourth part of the duodenum. Here it may raise the peritoneum and form a paraduodenal fold (p. 397). The vein turns above the duodeno-jejunal flexure and, in most cases (60 per cent), ends in the splenic vein behind the pancreas. It frequently (40 per cent) empties into the superior mesenteric vein.

The Inferior Mesenteric Lymph Nodes

The lymph nodes associated with the distal one-third of the transverse colon, the descending colon, and the sigmoid colon are distributed like those along the more proximal part of the large intestine (p.432). Peripheral nodes lie close to the colon (in the epiploic appendages and along the vasa recta), and larger intermediate nodes lie along the superior rectal, sigmoid, and left colic arteries. The central **inferior mesenteric nodes** lie around the root of the inferior mesenteric artery. They receive the drainage from the outlying node groups, and their efferents pass upward to the superior mesenteric nodes and also end in the neighboring lumbar chain.

The counting of lymph nodes in cleared preparations reveals many small nodes not recognized on dissection or palpation (Kay, '42). When reliably counted, there appear to be an average of forty nodes associated with both the descending and sigmoid portions of the colon and from fifty to sixty along the rectum and the anal canal.

The lymphatic drainage of the distal one-third of the transverse colon and the left colic flexure has an alternate path along the ascending radicle of the left colic vein and thence by way of the upper segment of the inferior mesenteric vein to the central superior mesenteric nodes.

The Intestinal Lymph Trunk It has been seen that the lymphatic drainage of the gastrointestinal tract passes through smaller, more numerous peripheral nodes, less numerous but larger intermediate nodes, and then to the relatively few large central nodes at the roots of the great visceral arteries. The lymph collected by the inferior mesenteric nodes turns partly into adjacent lumbar nodes but is also partly transmitted to the higher superior mesenteric nodes. The superior mesenteric nodes drain partly to the celiac nodes above them. The final common path of this lymphatic collection is the **intestinal trunk.** This trunk is formed by the union of efferents from both the celiac and superior mesenteric nodes. The

intestinal trunk turns under the inferior border of the left renal artery and ends in the left lumbar trunk in 70 per cent of specimens and in the cisterna chyli in 25 per cent; it crosses to the right lumbar trunk in only 5 per cent (quoted from Morris).

The Inferior Mesenteric Nerve Plexus

The inferior mesenteric plexus is a perivascular network of nerves which follows the inferior mesenteric artery and its branches. It usually contains a visible ganglion, and, although less evident grossly, it also contains ganglionic cells in small collections along its nerves. The inferior mesenteric plexus is derived from the aortic plexus through which pass preganglionic and postganglionic sympathetic fibers from the visceral branches of the upper two or three lumbar nerves. The inferior mesenteric plexus is inferior to the region of distribution of the thoracic splanchnic nerves. It contains afferent nerves, but it is unlikely that any parasympathetic fibers reach the colon by way of this plexus.

The Parasympathetic Nerves (fig. 409 & p. 513)

The parasympathetic supply of the distal colon and of the pelvic viscera arises in sacral segments of the spinal cord. From sacral nerves two, three, and four, the **pelvic splanchnic nerves** pass forward on either side of the rectum to the inferior hypogastric plexus. The preganglionic parasympathetic fibers of the pelvic splanchnic nerves distribute to the viscera of the pelvis and the perineum as components of this plexus. However, their supply of the left one-third of the transverse colon, the left colic flexure, the descending colon, and the sigmoid colon is by independent routes. Emerging from the inferior hypogastric plexus and from the hypogastric nerves just below the promontory of the sacrum, discrete nerves pass on either side upward in the sigmoid mesocolon and then behind the posterior body wall peritoneum of the lumbar region. These nerves run obliquely to the course of the vessels, communicate with but do not join the perivascular plexuses on the main arterial branches, and ascend out of the pelvis to reach the parts of the distal colon. The nerves follow the arteries only along the vasa recta. Several investigators have followed their separate course as far as the left flexure of the colon. While the junction of the middle and left thirds of the transverse colon appears to be the principal area of transition between the vagal innervation and the sacral parasympathetic innervation, physiologic evidence indicates that there is a considerable overlap in the intrinsic plexuses. The parasympathetic supply to the colon and the rectum serves the emptying reflex. The sympathetic innervation inhibits the emptying mechanism and is vasomotor.

THE PORTAL VEIN (figs. 361, 362)

The **portal vein** is the large collecting vein into which empties all the venous blood from the gastrointestinal tract (except the lower part of the anal canal), from the spleen, and from the pancreas. The vein conducts this blood to the liver; here it branches like the hepatic artery and ends in capillary-like vessels—liver **sinusoids**. The portal vein is formed behind the neck of the pancreas by the union of the superior mesenteric vein and the splenic vein. Since the inferior mesenteric vein empties into one or the other of these veins just before their junction (splenic vein—60 per cent; superior mesenteric vein—40 per cent), the portal vein, at its beginning, represents the venous drainage of these three major veins. The portal vein ascends obliquely to the right behind the first part of the duodenum and, in the free margin of the hepatoduodenal ligament, lies behind the common bile duct and the proper

hepatic artery. At the porta of the liver, it divides into right and left branches which accompany the corresponding branches of the hepatic duct and the hepatic artery. The portal vein is surrounded by the posterior hepatic plexus and is accompanied by lymphatic vessels. The left branch of the portal vein is joined anteriorly by the ligamentum teres (obliterated umbilical vein); posteriorly, by the ligamentum venosum (obliterated ductus venosus). Before entering the hepatoduodenal ligament, the portal vein receives the left and right gastric veins and the posterior superior pancreaticoduodenal vein. The paraumbilical vein unites with the left branch of the portal vein at the hilum of the liver. The descriptions of many of the tributaries of the portal vein have been given—inferior mesenteric vein (p. 438); superior mesenteric vein (p. 431) splenic vein (p. 409); left gastric vein (p. 405); right gastric vein (p. 405); posterior superior pancreaticoduodenal vein (p. 416); and paraumbilical vein (p. 426).

Collateral Circulation of the Portal Vein

Clinical Note **Collateral Circulation of the Portal Vein** The portal vein and its tributaries have no valves. As a consequence, blood may flow centrifugally through the system if there is an impediment to its flow in a central direction. This occurs in certain diseases of the liver, and it then becomes essential that there be alternate routes for the return of this large volume of blood to the heart. Such alternate routes are the regions of anastomosis of its terminal radicles with veins which drain toward the heart by way of the superior vena cava or the inferior vena cava.

The esophageal tributaries of the left gastric vein communicate with esophageal veins which empty into the azygos vein of the chest (p. 361).

The rectal plexuses (fig. 392) of the anal canal and the lower rectum allow communication between tributaries of the superior rectal, middle rectal, and inferior rectal veins. The middle rectal and inferior rectal veins transmit their blood to the inferior vena cava.

The paraumbilical veins anastomose with small veins of the anterior abdominal wall which, as radicles of the superior epigastric, inferior epigastric, thoracoepigastric, and segmental vessels, connect with the superior or the inferior vena cava.

The retroperitoneal veins draining the colon (ileocolic, right colic, middle colic, and left colic veins) have anastomotic connections with testicular (or ovarian) veins and especially with small veins of the pararenal fat. These are usually tributary to the suprarenal and renal veins or to a circumadrenorenal vascular circle which involves these together with the testicular (or ovarian) veins. The inferior vena cava is the termination of venous blood entering this retroperitoneal network. Surgically, portocaval and splenorenal shunts have been developed to deal with extreme cases of portal obstruction.

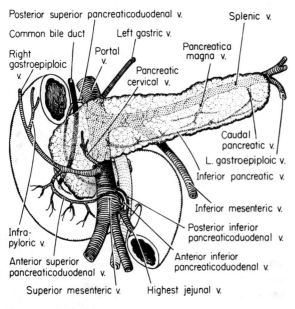

Fig. 362 The formation of the portal vein; pancreatic veins.

Posterior superior pancreaticoduodenal v.
Common bile duct
Left gastric v.
Right gastroepiploic v.
Portal v.
Pancreatic cervical v.
Splenic v.
Pancreatica magna v.
Caudal pancreatic v.
L. gastroepiploic v.
Inferior pancreatic v.
Inferior mesenteric v.
Posterior inferior pancreaticoduodenal v.
Infra-pyloric v.
Anterior superior pancreaticoduodenal v.
Anterior inferior pancreaticoduodenal v.
Superior mesenteric v.
Highest jejunal v.

THE KIDNEYS

GENERAL RELATIONS (figs. 225, 363, 364, 368, 369, 372)

The retroperitoneal **kidneys** lie embedded in fat and fibrous connective tissue on either side of the vertebral column through lower thoracic and upper lumbar levels. The anteriorly projecting lumbar portion of the vertebral column, with the psoas muscles, forms deep gutters on either side of the column in which the kidneys lie. The superior extremities of the kidneys reach the upper border of the body of the twelfth thoracic vertebra; their inferior extremities lie at the level of the third lumbar vertebra and farther from the median plane than are the superior poles. The left kidney is usually somewhat longer than the right kidney, and the right kidney is usually slightly lower than the left kidney due to the mass of the liver above it. The kidneys are from 10 to 12 cm. long, from 5 to 6 cm. wide, and about 3 cm. thick. Lying against the sloping sides of the psoas muscles, their **anterior surfaces** face anteriorly and slightly lateralward; their **posterior surfaces,** posteriorly and slightly medialward. Their convex **lateral borders** are directed posteriorly as well, and their **medial borders** also face anteriorly. The medial border of the kidney is indented, exhibiting a **hilum** for the entrance of the renal blood vessels and for the accommodation of the **renal pelvis** (which continues as the **ureter**). Thick lips of the anterior and posterior surfaces of the kidney overhang the hilum and enclose the renal sinus, a fat-filled space around the blood vessels and the pelvis of the kidney.

Renal Fascia and Fat (figs. 225, 364)

The kidneys are embedded in a thick mass of adipose and fibrous connective tissue derived from the extraperitoneal fat and areolar tissue. Around the kidney the two elements form discrete layers—fatty layers separated by a thin membranous sheet, the **renal fascia.** These structures surround both the kidney and the suprarenal gland, but a thin extension of the renal fascia separates the compartments occupied by the two organs. Arising from the extraperitoneal connective tissue along the lateral border of the kidney, the renal fascia provides anterior and posterior laminae which pass medialward across the corresponding surfaces of the kidney. The laminae unite medially except where they are in relation to the renal artery and vein. Medial to the hilum of the kidneys, the anterior and posterior laminae continue across the vertebral column anterior and posterior to the aorta and the inferior vena cava but adhere so closely to the adventitia of the vessels that no side to side communication between the spaces of the kidneys is permitted. The posterior layer of renal fascia (somewhat thicker than the anterior layer) blends with the transversalis fascia over the vertebral column. Elsewhere at the margins of the kidney (or suprarenal gland where it intervenes), the renal fascial layers blend and then lose their identity in the extraperitoneal connective tissue. An inferior 'conal' extension of the perirenal space has been demonstrated radiographically. Inferomedially, a delicate extension of the renal fascia is prolonged along the ureter as the **periureteric fascia.** The **renal fat** lies both outside and inside the renal fascia. The layer within is frequently designated as the **perirenal fat;** that outside the fascial envelope is known as the **pararenal fat.**

Relations of the Kidneys

Each kidney lies in a paravertebral gutter which is bounded medially by the psoas muscles. Behind the kidney lies the quadratus lumborum muscle and, at its lateral border, the transversus abdominis muscle (fig. 363). Behind the approximate junction of the upper and middle thirds of the kidney, the

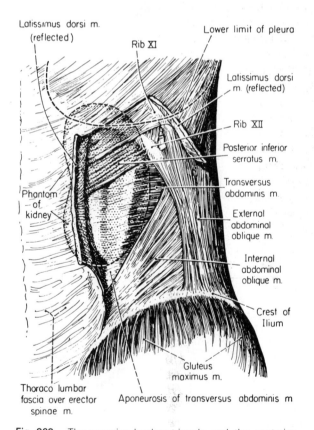

Latissimus dorsi m.
(reflected)

Rib XI

Lower limit of pleura

Latissimus dorsi m. (reflected)

Rib XII

Posterior inferior serratus m.

Transversus abdominis m.

External abdominal oblique m.

Internal abdominal oblique m.

Crest of Ilium

Phantom of kidney

Gluteus maximus m.

Thoraco lumbar fascia over erector spinae m.

Aponeurosis of transversus abdominis m

Fig. 363 The superior lumbar triangle and the posterior relations of the kidney.

fasciae of the quadratus lumborum and psoas muscles are thickened into the lateral and medial lumbocostal arches (fig. 370). These give rise to muscular fibers of the diaphragm and, as a consequence, the superolateral one-third of the kidney lies in front of the diaphragm. The subcostal nerve and its accompanying blood vessels descend diagonally across the back of the kidney, conforming to the slope of the twelfth rib and lying between the quadratus lumborum muscle and its covering fascia. At lower levels but in the same plane, the iliohypogastric and ilioinguinal nerves diverge from the psoas muscles toward the transversus abdominis muscle.

The **anterior relationships** of the kidneys are with the other abdominal organs (fig. 373) and differ on the two sides with the exception that the suprarenal glands cap the superomedial surfaces of each kidney. The second portion of the duodenum lies anterior to the medial border of the **right kidney,** and, lateral to the duodenum, the superior two-thirds of the kidney is in contact with the right lobe of the liver. Over the inferior one-third of the renal surface lie the right flexure of the colon laterally and coils of the jejunum medially. Across the hilar region of the **left kidney** lies the pancreas, its contact area ending lateralward in the superiorly extending splenic area. In the triangular zone between the suprarenal

gland, the spleen, and the pancreas, the stomach makes contact with the upper part of the left kidney. Inferior to the areas of contact of the pancreas and the spleen, the left kidney is in relation to the left flexure of the colon laterally and to coils of the jejunum medially.

The Renal Tubule (fig. 365)

The unit of renal structure is the tubular **nephron.** There are over one million of these in each human kidney. Each nephron begins in a double-walled cup, the **glomerular capsule,** which encloses a capillary tuft, the **glomerulus.** The glomerular capsule leads into the **proximal convoluted tubule.** After several turns, this tubule straightens out and runs toward the inner portion (medulla) of the kidney; it has a total length of about 14 mm. Tapering abruptly, it goes into continuity with the thin-walled **straight tubule** (Henle's loop) which descends somewhat into the medulla and then returns to the cortex of the kidney. The straight tubule may be 20 mm. long. This tubule passes into the **distal convoluted tubule** which is about 5 mm. long. The latter, which exhibits a specialized **macula densa** region as it passes near the glomerulus, empties into the **collecting duct.** The collecting duct, after receiving many ductal tributaries, is directed toward the hilum of the kidney, where it ends on the surface of a pyramidal eminence, the **renal papilla.**

The **glomerulus** is a network of capillaries which arises from an **afferent arteriole,** a branch of an interlobular renal artery. The capillary network ends not in a vein

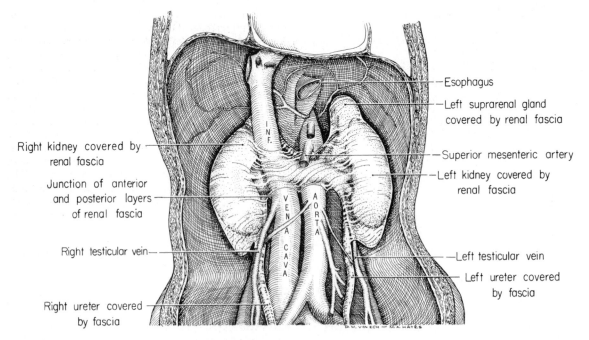

Fig. 364 The posterior abdominal region, illustrating the renal fascia and its extensions (from original of fig. 3. Mark A. Hayes. *Am. J. Anat.,* 87:137, 1950).

but in an **efferent arteriole** which descends to form another capillary plexus in close relation to the straight and convoluted tubules. This double capillary network in relation to the tubular nephron forms the essential mechanism of the kidney, for the glomerulus and the glomerular capsule provide for filtration of the blood plasma, and the second capillary plexus around the tubules provides for selective reabsorption of water and dissolved materials back into the circulation.

Clinical Note Glomerular filtration pressure is about 50 mm. of mercury (75 mm. of blood pressure minus an osmotic pressure of 25 mm.), which is sufficient to force a considerable fraction of blood plasma through the glomerular membrane. About one-sixth of the blood plasma which enters the kidney is filtered into the capsular space (about 100 liters per day for each kidney). Proteins and other blood colloids do not normally pass the double barrier of the capillary wall and the glomerular capsule. The osmotic pressure of the blood in the capillaries distal to the capsule is the driving force for reabsorption of water and and dissolved materials. Reabsorption is such that excreted urine in Man is normally from one to one and a half liters per day.

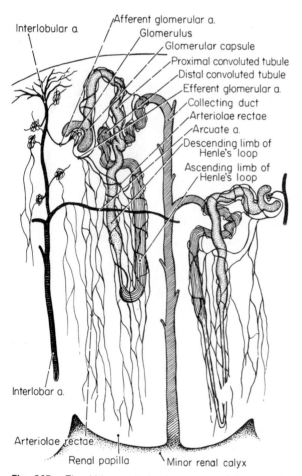

Fig. 365 The kidney tubule; its parts and its vascular relations. Schematic.

Labels on figure:
Interlobular a.
Afferent glomerular a.
Glomerulus
Glomerular capsule
Proximal convoluted tubule
Distal convoluted tubule
Efferent glomerular a.
Collecting duct
Arteriolae rectae
Arcuate a.
Descending limb of Henle's loop
Ascending limb of Henle's loop
Interlobar a.
Arteriolae rectae
Renal papilla
Minor renal calyx

GROSS STRUCTURE (fig. 366A)

On examining a section of the kidney, it is apparent that a marginal granular zone is distinguishable from a more central, striated region. The outer zone is the **cortex;** its granularity is due to its high content of glomeruli. The inner zone is the **medulla,** striated by its straight tubules and collecting ducts. Both zones contain the various parts of the nephron in a mixture. In spite of the radial orientation of the nephron in the kidney, glomeruli occur in the outer medulla as well as in the cortex, and tubules are also found in the cortex. Indeed, the inner portion of the cortex has a partially medullary appearance, its radiating tubules being designated as the **pars radiata** of the cortex. The **medullary pyramids** of straight tubules and collecting ducts are separated by columns of cortical substance which extend to the renal sinus. These are the **renal columns.** Traversing each renal column is a principal branch of a renal artery, the interlobar artery; its name is a reflection of the lobar character of a single medullary pyramid and the cortical tissue on all sides of it. In the subprimate mammalian kidney and in the human fetus, the surface of the kidney is indented along the lines of separation of these lobes (lobular kidney).

The **interlobar arteries** pass from the renal sinus peripheralward through the renal columns and then, at the line between the cortex and the medulla, undergo a branching resembling the stays of an umbrella arching away from its shaft. These are the **arcuate arteries.** From them, a number of small **interlobular arteries** pass into the cortex to redivide into the **afferent arterioles** of the glomerular capsules. The **efferent arterioles,** emerging from the glomeruli, break up into second capillary networks around the tubules of the kidney (arteriolae rectae). Near the cortex-medulla boundary, arterial systems occur in which the glomerulus is reduced or absent. Such vessels provide only the second of the usual two sets of capillaries. They are not numerous, but they provide a shunting mechanism for renal blood which is of value when the vasomotor mechanism of the afferent and efferent arterioles has reduced the blood flow through the glomeruli. The veins of the kidney substance have a pattern like that of the arteries and drain mainly into the renal veins in the sinus of the kidney. Some small veins in the superficial part of the cortex communicate through the fibrous capsule with minute veins in the renal fat.

THE EXCRETORY DUCTS (figs. 357, 366-68, 373, 378)

The **renal pyramids** taper to form, by their apices, the eight to twelve **renal papillae** on which open the major collecting ducts of the kidney. The renal papillae are received into usually eight cup-shaped **minor calyces** which represent the beginning of the duct system of the kidney. Some of the minor calyces receive several renal papillae. The minor calyces empty into the two or three larger **major calyces** which drain the urine from the superior, middle, and inferior portions of the kidney. Still within the sinus of the kidney, the major calyces empty into a common chamber, the **renal pelvis,** which leads inferomedialward beyond the borders of the hilum of the kidney. Toward the lower pole of the kidney, the extrarenal renal pelvis narrows to form the **ureter** which conducts the urine to the bladder.

The epithelium of the calyces becomes continuous with that covering the renal papillae, and this in turn is continuous with the epithelium of the large collecting ducts emptying on the papillae. The connective tissue capsule of the kidney turns in to line the renal sinus and is there continuous with the connective tissue wall of the calyces. The calyces, the pelvis, and the ureter each possess a muscular wall, and the flow of urine is not left to dependent drainage but is conducted by peristaltic action in the duct system. Jets of urine enter the bladder two or three times a minute.

The **ureter** is a thick-walled muscular tube with a small lumen. Each ureter has a length of 25 cm.; half of this length is abdominal, and half is pelvic. The pelvic portion is described in the chapter on the Pelvis. The abdominal portion of the ureter lies in the extraperitoneal connective tissue of the lumbar region. Periureteric fascia is prolonged downward on it from the renal fascia and tends to bind it to the overlying peritoneum. The ureters descend on the psoas muscles until, crossing the bifurcation of the common iliac arteries or the beginning of the external iliac arteries, they enter the pelvis. Each ureter is crossed obliquely by the testicular (or ovarian) vessels. The right ureter is also crossed anteriorly by the right colic and ileocolic vessels; the left ureter, by the left colic vessels.

The **arteries** of the abdominal portion of the ureter are branches of the renal artery and of the testicular (or ovarian) artery. Its veins drain to the comparable veins. The **nerves** of the ureter are derived from the renal and testicular (or ovarian) plexuses in its abdominal portion and from the inferior hypogastric plexus in its pelvic portion. Afferent fibers from the

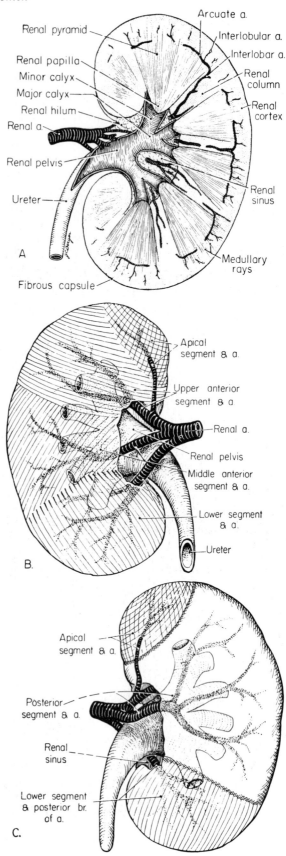

Fig. 366   A—a semischematic coronal section of the kidney; B—vascular segmentation of the kidney, anterior view; C—vascular segmentation of the kidney, posterior view.

ureter probably reach the spinal cord by way of the eleventh and twelfth thoracic nerves and the first lumbar nerve.

Clinical Note **Doubling of the ureter** is an occasional developmental anomaly (3 per cent). Embryologically, it appears to be due to a very early division of the primary renal pelvis with a subsequent extension of this division inferiorly. The division may stop somewhere between the kidney and the bladder, producing a Y-shaped ureter, or, as more often

happens, it may be complete. When two complete ureters occur on one side, the one arising from the superior renal pelvis terminates in the bladder by an orifice which is lower than the one arising from the inferior pelvis.

VESSELS AND NERVES

Blood Vessels of the Kidney (figs. 337, 366, 368, 373)

The **renal arteries** arise, one on each side of the aorta, at the level of the upper border of the second

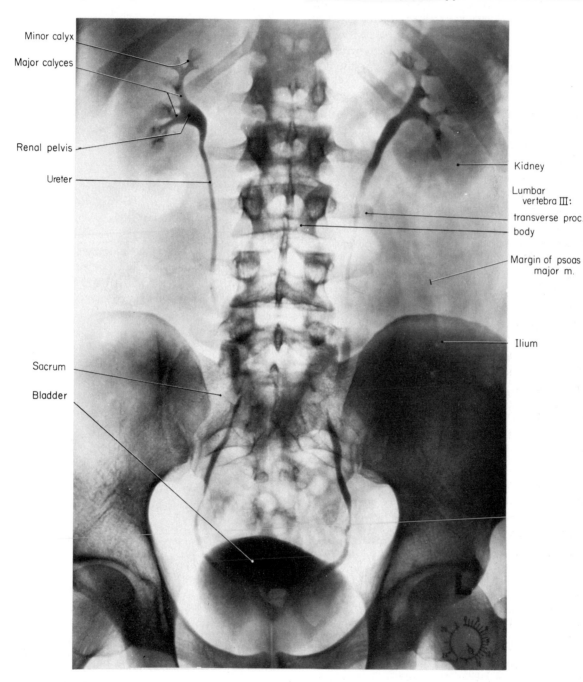

Fig. 367   An intravenous urogram demonstrating the parts of the duct system of the kidney.

lumbar vertebra. Their origin is about 1 cm. below that of the superior mesenteric artery. The renal arteries are of large size in accordance with the vascular requirements of the kidneys, and they pass directly transversely to the hilum of the kidneys, coursing anterior to the crura of the diaphragm and the psoas muscles. The **right renal artery,** longer and a little lower than the left, passes behind the inferior vena cava, the head of the pancreas, and the second part of the duodenum. The **left renal artery** lies behind the left renal vein, the pancreas, and the splenic vein. The renal veins pass lateralward anterior to the arteries, but at the hilum of the kidney the arteries and the veins break up into branches which enter the sinus of the kidney both anterior and posterior to the renal pelvis. Within the sinus these major branches divide into anterior interlobar and posterior interlobar branches which pass into the kidney substance within the renal columns. Their intrarenal distribution is described on p. 443. The anterior interlobar branches are larger and distribute somewhat past the midcoronal plane of the kidney. Extrarenally, each renal artery gives off an **inferior suprarenal artery,** which ascends to the lower part of the suprarenal gland, and **ureteric branches** to the ureter.

Clinical Note   Accessory renal arteries are frequent (23 per cent of cases—Anson, Cauldwell, Pick, & Beaton, '47). They are more common on the left side and below the main artery than above it. Accessory vessels tend to pass directly to the surface of the kidney, frequently to the inferior pole. They commonly arise from the aorta but are also seen taking origin from the common iliac, the internal iliac, and other abdominal vessels. The kinking of the ureter over an accessory renal artery may lead to dilatation of the pelvis and the calyces of the kidney (hydronephrosis). The back pressure so produced in the pelvis and calyces may be sufficient to cause ischemia of the medulla of the kidney, the vessels of which are under greatly reduced blood pressure in the second capillary plexus of the kidney.

Graves ('54, '56) has analyzed the intrarenal distribution of the anterior and posterior divisions of the renal arteries and has found that their branches demarcate constant 'vascular' renal segments (figs. 366B & C). The segments are the apical, the upper anterior, the middle anterior, the lower, and the posterior. The **apical segment** involves both surfaces of the kidney in the approximate suprarenal contact area. It is served by an apical artery which is usually a branch of the anterior division of the renal artery but which may have a more proximal origin from the renal artery or the aorta. With this artery is usually associated the inferior suprarenal artery. The **upper anterior segment** and the **middle anterior segment** are defined and vascularized by separate branches of the anterior division of the renal artery. These segments involve only the anterior portion of the kidney and are radially disposed below

the suprarenal area. Each one occupies a little more than one-fourth of the anterior surface of the kidney. The **lower segment** is the inferior one-third of the kidney substance and includes both the anterior and posterior surfaces. It is supplied by a lower branch of the anterior division of the renal artery or by a vessel having a more proximal origin from the renal artery or the aorta; the artery of this segment may arise in common with the testicular (or ovarian) artery. The **posterior segment** represents the posterior parenchyma of the kidney opposite the upper anterior and middle anterior segments. It is supplied by the posterior division of the renal artery, which has no other distribution except, occasionally, to the apical segment. Each of the five segments has its own artery, and each artery provides several interlobar branches. Injection studies show that there is no collateral circulation between the segments. Graves has also shown that so-called accessory arteries are, in fact, normal segmental arteries, the origin of which is more proximal than usual. This principle also explains the variations of normal renal arteries. The segmentation of the kidney and the specificity of its vessels accord with similar findings with respect to the liver (p. 418).

The **renal veins** arise from tributaries which issue from both the anterior and posterior sides of the sinus of the kidney. They are large and generally lie anterior to the renal arteries. The **left renal vein** is longer (7.5 cm.) than the right and passes in front of the aorta just below the superior mesenteric artery to end in the inferior vena cava. It receives the left suprarenal vein and the left testicular (or ovarian) vein. It frequently receives the left inferior phrenic vein also. The **right renal vein,** 2.5 cm. long, lies behind the second part of the duodenum. It ends in the right side of the inferior vena cava. The left renal vein frequently receives the left second lumbar vein on its dorsal aspect. This communication provides the intermediate root of the hemiazygos vein (about 25% of cases), which ascends lateral to the left crus of the diaphragm to join the lower intercostal or subcostal vein in the thorax.

Lymphatics of the Kidney and the Ureter

The lymphatic vessels of the kidney form three plexuses: one in the substance of the kidney, one beneath its fibrous capsule, and one in the perirenal fat. The last two intercommunicate. From the substance of the kidney, four or five lymph trunks issue at the hilum, where they are joined by vessels from the capsule, and follow the renal vein to the **lumbar lymph nodes.** The perirenal plexus is drained directly to the same nodes. The lumbar lymph nodes drain by way of the lumbar trunks to the cisterna chyli (p. 453). The lymphatic drainage of

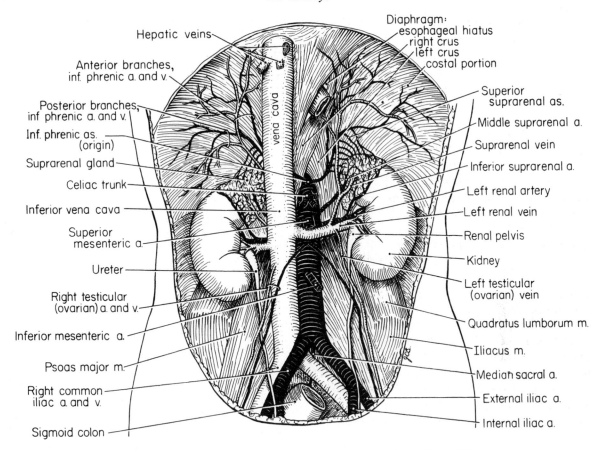

Fig. 368   The blood vessels of the diaphragm, suprarenal glands and kidneys.

the upper portion of the abdominal ureter is with the renal vessels to the lumbar nodes. From the lower portion of the abdominal ureter, lymphatic vessels pass to the lower nodes of the lumbar chain and to the common iliac nodes.

Nerves of the Kidney (fig. 373)

An autonomic and afferent innervation is provided for the kidney by the nerves of the renal plexus. Its nerves are derived from the celiac plexus, the lesser thoracic splanchnic and least thoracic splanchnic nerves, and the aorticorenal ganglion. The renal plexus sends offshoots to the testicular plexus and the inferior vena cava on the right side. The principal efferent innervation of the kidney is vasomotor, autonomic nerves being supplied to the muscular coats of the afferent and efferent arterioles. Nerves appear to influence urine formation only by changing the blood supply to the kidney; it is doubtful if there are any true secretory nerves to the

organ. Afferent fibers probably reach the spinal cord through the tenth, eleventh, and twelfth thoracic nerves.

Clinical Note The **kidney** develops in the pelvis, and through a normal migration, due primarily to inequalities in growth rates in the trunk, it finally achieves its typical upper lumbar position. Occasionally the migrational change is interfered with in whole or in part, and the kidney may then be found in the pelvis, at the sacral promontory, or at any intermediate position below the normal level. Such kidneys are apt to be misshapen or smaller than normal and to receive their blood supply from the lower end of the aorta or from the iliac vessels. The **horseshoe kidney** is also usually ectopic in position. Such a kidney, not too infrequent, is characterized by a fusion across the midline of the lower poles of the two primordia. Horseshoe kidneys are unrotated—their hila are directed forward, and their ureters cross their lower portions. Their blood supply is usually derived from vessels from the lower part of the aorta or from the iliac arteries, and their interconnecting band is often crossed by the inferior mesenteric artery.

THE SUPRARENAL GLANDS

General Relations (figs. 364, 368, 369, 373)

Two **suprarenal glands** are situated on the superomedial aspect of the kidneys. These are endocrine glands; in fact, each gland is divisible into a cortex and a medulla, and these parts are separate endocrine glands, differing from one another in their developmental origin and in their function. The **cortex** is essential to life. It secretes hormones of the steroid type, and interference with its function causes great disruption of fluid and electrolyte balance in the body. It is also concerned with carbohydrate metabolism and is important in normal bodily reactions to stress. The **medulla** of the suprarenal gland is developed from the same cells which produce the sympathetic nervous system. Indeed, it may be thought of as a postganglionic sympathetic organ, for it receives only a preganglionic nervous innervation. It secretes epinephrine and norepinephrine which, introduced into the blood stream, have effects identical to those which follow activation of the sympathetic portion of the autonomic nervous system (p. 30).

The suprarenal glands are approximately triangular in form, from 3 to 5 cm. in height, from 2 to 3 cm. in breadth, and less than 1 cm. in thickness. They are concave on their under surfaces where they fit against the kidneys. Each gland has a **hilum** on its anterior surface; from this furrow emerges the central suprarenal vein. The **right suprarenal gland** lies in front of the diaphragm and makes contact anteriorly with the inferior vena cava medially and the liver laterally. The **left suprarenal gland,** somewhat semilunar in form, rests against the left crus of the diaphragm posteriorly. Anteriorly, it is in relation with the cardia of the stomach superiorly and with the pancreas inferiorly. The two glands are separated by the celiac arterial trunk and the celiac plexus of nerves. The suprarenal glands are enclosed within the renal fascia but are separated from the kidneys by a delicate sheet of fascia. They are usually embedded in fat. Small accessory suprarenal glands are sometimes found in the connective tissue around the main glands.

Vessels and Nerves

The Suprarenal Arteries (figs. 368, 369, 373)

The suprarenal glands, as endocrine glands, are highly vascular. More blood passes through these glands relative to their size than through any other organs of the body, with the possible exception of the thyroid glands. Many variations in the detail of distribution of their blood vessels exist (Gagnon, '56) but the general pattern is relatively standard. Three types of suprarenal arteries supply the glands. An average of from six to eight **superior suprarenal arteries** are derived from the inferior phrenic artery (principally from its posterior branch). They descend to the superior border of the suprarenal, some of them redividing before penetrating the gland. Laterally, one or more end in the pararenal fat. A **middle suprarenal artery** (or several) arises from the abdominal aorta, in most cases at or above the level of the renal artery. It runs to the medial aspect of the gland, breaks up into numerous twigs to this surface of the gland, and anastomoses with the superior suprarenal and inferior suprarenal arteries. One or more **inferior suprarenal arteries** arise from

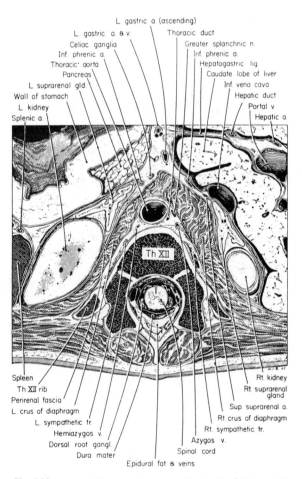

L. gastric a. (ascending)
L. gastric a. & v.
Celiac ganglia
Inf. phrenic a.
Thoracic aorta
Pancreas
L. suprarenal gld.
Wall of stomach
L. kidney
Splenic a.

Thoracic duct
Greater splanchnic n.
Inf. phrenic a.
Hepatogastric lig.
Caudate lobe of liver
Inf. vena cava
Hepatic duct
Portal v.
Hepatic a.

Th XII

Spleen
Th XII rib
Perirenal fascia
L. crus of diaphragm
L. sympathetic tr.
Hemiazygos v.
Dorsal root gangl.
Dura mater
Epidural fat & veins

Rt. kidney
Rt. suprarenal gland
Sup. suprarenal a.
Rt. crus of diaphragm
Rt. sympathetic tr.
Azygos v.
Spinal cord

Fig. 369 A partial cross section at the level of the twelfth thoracic vertebra to show the relations of the kidney and suprarenal glands.

the renal artery. They pass lateralward along the renal border of the suprarenal gland and give rise to numerous twigs which penetrate the inferior surface of the gland. Their terminals may make connection with those of the superior suprarenal at the lateral edge of the gland.

The three types of suprarenal arteries give rise to a variable number of twigs which approach the gland perpendicular to its surface. As many as sixty such twigs have been counted penetrating the gland, and the average is twenty-five. The contribution of the superior suprarenal vessels to the supply of the gland is about equal to the combined contributions of both the middle suprarenal and inferior suprarenal arteries. Many small offshoots are given to the perirenal fat and to the adjacent nerve plexuses and ganglia. Frequently, a **circumadrenorenal vascular arcade** is formed which connects lateral branches of the suprarenal arteries above with the renal, the intercostal, the lumbar, and with other arteries below the kidneys.

The Suprarenal Veins

Small suprarenal veins, corresponding to the arteries, follow the arteries to the gland. The principal venous drainage of the gland is, however, by way of the prominent, single **suprarenal,** or central, **vein.** This vein emerges at the hilum on the anterior surface of the gland. It is usually about 5

mm. in diameter. On the left side (after union with the inferior phrenic vein) it is a tributary of the renal vein, but on the right side it empties directly into the posterior surface of the inferior vena cava.

Lymphatics of the Suprarenal

The lymphatic vessels of the suprarenal gland arise from a plexus under the capsule and from another in the medulla of the gland. They follow the vessels of the gland, predominantly the suprarenal vein, and end in the upper nodes of the lumbar chain.

Nerves of the Suprarenal (fig. 373)

The suprarenal gland is provided with many fine nerves which stream toward it from the celiac plexus and the greater thoracic splanchnic nerve. These are principally medullated preganglionic neurons from the greater thoracic splanchnic nerve which end on the secretory cells of the medulla. There appears to be only a vasomotor nerve supply to the cortex.

THE DIAPHRAGM

GENERAL RELATIONS (figs. 331, 368–73)

The **diaphragm** is the great musculofascial sheet that separates the thoracic and abdominal cavities. It has the shape of a dome with its convexity bulging into the thoracic cavity. In moderate expiration it reaches as high as the fifth rib on the right side and the fifth interspace on the left. The diaphragm has an irregularly shaped, central tendinous portion which underlies the heart and the medial portions of the lungs. Its musculature is peripheral and radiates from the sternum, the ribs, the costal cartilages, and the lumbar vertebrae toward the central tendon. The muscle may be divided into sternal, costal, and lumbar portions.

Structure (fig. 370)

The diaphragms of the body consist of a central sheet of muscle covered on both surfaces by a membranous layer of fascia. For the thoraco-abdominal diaphragm, the inferior fascia is provided by the upper portion of the parietal abdomino-pelvic fascia, the transversalis fascia (p. 378). Its superior fascia is the parietal thoracic fascia (endothoracic fascia) which lines the interior of the thoracic cavity covering the internal intercostal, subcostal, and transversus thoracis muscles. The fibrous pericardium blends with the superior fascia

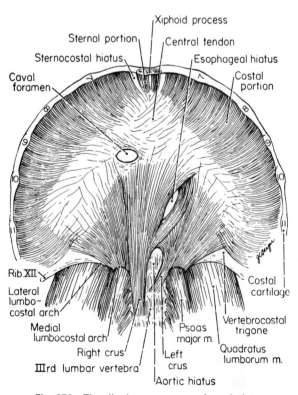

Fig. 370   The diaphragm as seen from below.

of the diaphragm beneath the heart. There is continuity of the superior and inferior diaphragmatic fasciae through the openings of the diaphragm.

The **sternal part** of the diaphragm consists of two fleshy slips from the back of the xiphoid process. The **costal portion** consists of muscular slips which arise from the inner surfaces of the costal cartilages and adjacent bone of the lower six ribs. Such slips interdigitate with fascicles of origin of the transversus abdominis muscle. The **lumbar portion** is composed of two crura, which arise from the lumbar vertebrae, and of muscle fibers which spring from aponeurotic arches overlying the psoas and quadratus lumborum muscles.

The Crura The musculotendinous crura arise from the anterior surfaces of the bodies of the lumbar vertebrae, from the anterior longitudinal ligament, and from the intervertebral disks. The right crus is larger and longer than the left crus. It takes origin from the upper three lumbar vertebrae, whereas the left crus arises from only the upper two. The right crus usually splits to enclose the esophagus as it pierces the diaphragm. The fibers of the right crus decussate beyond the elliptical opening, making the **esophageal hiatus** a complete muscular sphincter. The left crus is smaller and lies farther from the median plane than the right crus. It may send a slip under the right crus directed toward the vena caval foramen (15%), and it participates in some fashion in the formation of the esophageal hiatus in 60 per cent of cases. The medial tendinous margins of the crura meet in the median line, forming a more or less distinct arch across the front of the aorta and defining the **aortic hiatus.** The muscle fibers of the lateral margins of the crura are continuous with those arising from the medial lumbocostal arches. The crura are usually pierced by the thoracic splanchnic nerves on both sides.

The Lumbocostal Arches Between the vertebral column and the last rib, the musculature of the diaphragm arises from the lumbocostal arches. The **medial lumbocostal arch** is a thickening in the fascia over the psoas major muscle and extends from the side of the body of the second (or first) lumbar vertebra across the psoas to the tip of the transverse process of the first lumbar vertebra. The **lateral lumbocostal arch** is a thickening of the fascia of the quadratus lumborum muscle and extends from the transverse process of the first lumbar vertebra to the tip of the lower margin of the twelfth rib.

From the series of origins represented by the sternal, costal, and lumbar portions of the diaphragm, the muscular fibers converge to insert into the central tendon. The fibers from the xiphoid process are short and are separated from those of the costal portion by a small sternocostal hiatus. The fibers from the ribs, the costal cartilages, and the lumbocostal arches are long and curve as they ascend. The fibers of the crura diverge to either side as they enter the diaphragm.

Clinical Note A gap in the musculature of the diaphragm frequently occurs medial to the lowermost fibers arising from the twelfth rib. Here, muscular fibers may be lacking over a part or the whole of the lateral lumbocostal arch. The muscular deficiency in this **vertebrocostal trigone** is of variable size and occurs in 80 per cent of cases (quoted from Cunningham). With the muscle layer absent, the superior and inferior fasciae represent the full thickness of the diaphragm, and the triangular defect extends from the lumbocostal arch to the central tendon. The vertebrocostal trigone is in contact with the upper one-third of the kidney and represents a weakness through which congenital diaphragmatic hernia may take place and through which perinephric abscess may extend into the thorax. The most common congenital diaphragmatic defect is a persisting **pleuroperitoneal canal** located at the extremity of the left leaf of the central tendon. Varying amounts of abdominal viscera may herniate through this route. Esophageal **hiatal hernia** is more likely to be acquired. In such defects the cardia and fundus of the stomach (variable amounts) slide upward through an enlarged esophageal hiatus.

The Central Tendon This strong aponeurosis is the tendon of insertion of all the muscular fibers of the diaphragm. It is situated closer to the front of the partition, so that the posterior muscular fibers are longer. The central tendon lies immediately below the pericardium; indeed, the external fibrous coat of the pericardium blends with the superior fascia of the diaphragm and the central tendon. The central tendon is crescentic in outline but bulges forward and is incompletely divided into three lobes, or leaves. The right lobe is the largest; the middle lobe is intermediate in size; and the left lobe is the narrowest. The central tendon consists of interlacing tendinous bundles continuous at the margins with the muscular fibers. The strongest interlacement is found in the region of the caval opening at the posterior junction of the right and middle lobes of the tendon.

The diaphragm is the principal muscle of inspiration. The muscular fibers draw the central tendon downward, increasing the vertical dimension of the chest and decreasing the vertical dimension of the abdominal cavity. This action increases the volume and at the same time decreases the pressure in the thoracic cavity and both decreases the volume and increases the pressure in the abdominal cavity. In inspiration, the abdominal wall muscles are forced outward to allow for these changes in the thoracic volume. In defecation, as also in parturition, the abdominal wall muscles are contracted along with the diaphragm to increase markedly the intra-abdominal pressure.

The Openings of the Diaphragm (fig. 370)

The **aortic hiatus** is a passageway behind the diaphragm and is not properly an opening in it. The aorta emerges below the diaphragm at the lower border of the twelfth thoracic vertebra or at the level of the disk below it. The hiatus is bounded at the sides by the crura of the diaphragm, the fibrous material of which unites across the aorta. On the right side of the aorta the thoracic duct and, sometimes, the azygos vein pass from below upward through the aortic hiatus.

The elliptical **esophageal hiatus** lies in the muscular part of the diaphragm at the level of the tenth thoracic vertebra. It is located above, in front of, and a little to the left of the aortic hiatus and is formed by the divergence and subsequent decussation of the bundles of the right crus, together with a less constant participation of the left crus (p. 450). In 70 per cent of cases both margins of the hiatus will be provided by bundles of the right crus, alone in 40 per cent, or with the addition of a superficial bundle from the left crus to the right margin in 30 per cent. Besides the esophagus, esophageal branches of the left gastric artery and vein and the anterior and posterior vagal trunks occupy this opening.

The **vena caval foramen** is an opening in the central tendon at the posterior junction of its right and middle leaves. It is the highest of the three openings and lies at the level of the eighth thoracic vertebra or the disk below this. It transmits the inferior vena cava together with branches of the right phrenic nerve and lymphatic vessels ascending to the middle phrenic nodes.

The **sternocostal hiatus,** the small interval between the adjacent sternal and costal muscular fasciculi, allows passage of lymphatic channels from the convex surface of the liver to the anterior phrenic nodes. The crura of the diaphragm are pierced by the greater splanchnic and lesser splanchnic nerves; the sympathetic trunks pass behind the diaphragm under the medial lumbocostal arches. The hemiazygos vein communicates with the ascend-ing lumbar vein behind the diaphragm on the left side in the depths of the psoas muscle or deep to the medial lumbocostal arch.

VESSELS AND NERVES

Blood Vessels of the Diaphragm

The **arteries** of the diaphragm ramify on both its superior and its inferior surfaces. Vessels to the superior surface are the **pericardiacophrenic** and **musculophrenic branches** of the internal thoracic artery (p. 319) and the **superior phrenic branches** of the thoracic aorta (p. 360). The blood supply of the inferior surface is provided by the inferior phrenic branches of the abdominal aorta.

The **inferior phrenic arteries** typically arise from the front of the abdominal aorta just below the aortic hiatus and immediately above the origin of the celiac trunk. However, they arise almost as frequently from the celiac trunk either separately or by a single stem common to both sides. The inferior phrenic arteries may also arise from the renal arteries, and other aberrancies occur. The inferior phrenic arteries diverge across the crura of the diaphragm, the right one passing behind the inferior vena cava and the left one passing behind the esophagus. Each artery divides into an anterior and a posterior branch.

The **anterior branch** ramifies anteriorly and medial-ward on the under side of the diaphragm, anastomosing with the same branch of the other side and with the terminals of the musculophrenic and pericardiaco-phrenic arteries. The **posterior branch** runs posteriorly and lateralward toward the ribs and anastomoses with the terminals of the intercostal and superior phrenic arteries. The inferior phrenic arteries provide **superior suprarenal branches** as described on page 448.

The **veins of the diaphragm** correspond to the arteries. From the superior surface of the diaphragm, **pericardiacophrenic** and **musculophrenic tributaries** empty into the internal thoracic veins. From the posterior curvature, venous twigs end in the azygos and hemiazygos veins. **Inferior phrenic veins** of the abdominal surface correspond to the inferior phrenic arteries. The right inferior phrenic vein ends in the inferior vena cava. The left inferior phrenic vein is often double. Its posterior portion joins with the left suprarenal vein to become tributary to the left renal vein. The more anterior radicle passes in front of the esophageal hiatus to empty into the inferior vena cava.

Lymphatics of the Diaphragm

The lymphatics of the diaphragm are formed in two plexuses, one on its thoracic side and one on its abdominal side. These plexuses are richer where the diaphragm is covered by pleura or peritoneum, and they communicate freely. Lymphatic vessels of the superior surface of the diaphragm converge on the three subgroups of **phrenic nodes**—anterior, middle, and posterior (p. 320 & fig.¹351). Many offshoots of the plexus of the abdominal side perforate the diaphragm to join the superior plexus and distribute with it. Other lymphatic channels of the abdominal surface follow the course of the inferior phrenic vessels and end in the upper nodes of the lumbar chains of each side (fig. 371).

Nerves of the Diaphragm

The **phrenic nerve** of the cervical plexus (C3 to 5) is the sole motor nerve of the diaphragm. It also carries sensory fibers from the central portion of the diaphragm. The phrenic nerve is described in its cervical relations in the chapter on the Head and Neck (p. 177). Its thoracic course has also been considered (p. 334).

At the thoracic surface of the diaphragm, or a centimeter or two above it, the phrenic nerve divides into three branches. These are a sternal and an anterolateral branch, radiating into the anterior half of the diaphragm, and a posterior branch to the posterior half. The latter nerve shortly divides into posterolateral and crural branches. The sternal branches of the two sides sometimes anastomose in front of the pericardium. The phrenic branches distribute largely on the abdominal surface of the diaphragm, the right posterior branch passing through the vena caval foramen. The subperitoneal branches of the phrenic nerve distribute with the inferior phrenic vessels and, along these vessels, communicate with the phrenic sympathetic plexus, an offshoot of the celiac plexus. By way of the phrenic plexus sympathetic nerves are provided to the diaphragm and its blood vessels. Branches to the margins of the diaphragm are supplied by intercostal nerves nine through eleven. These are sensory nerves to the peripheral part of the diaphragm.

THE PARIETAL LYMPHATICS OF THE ABDOMEN

The visceral lymphatics of the abdomen—celiac, superior mesenteric, and inferior mesenteric nodes together with the lymphatic drainage paths of the nonintestinal viscera of the abdomen—have been described. The **parietal lymph nodes** (figs. 371, 372) lie in a continuous chain along the aorta and its subdivisions; they are the external iliac, common iliac, and lumbar nodes.

The **external iliac nodes** lie about the external iliac vessels, especially toward their lower ends.

From eight to ten large nodes of this group are located predominantly on the lateral and posteromedial aspect of the vessels. The lower posteromedial nodes are continuous with the deep inguinal nodes of the femoral canal, and an upper constant member of the group lies on the obturator nerve. The lateral nodes tend to be lodged between the artery and the psoas muscle. The external iliac nodes receive the drainage of the superficial inguinal and deep inguinal nodes and deep lymphatic vessels of the part of the anterior abdominal wall below the umbilicus. Small, outlying inferior epigastric and circumflex iliac nodes are interposed in the lymphatic drainage along the correspondingly named blood vessels. Afferent lymphatic vessels direct from

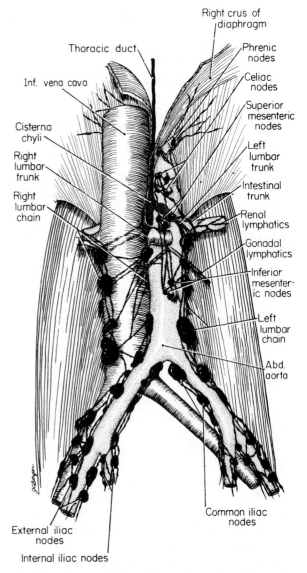

Fig. 371   The lymphatics of the lumbar region. Schematic.

the pelvis reach the more medial members of the external iliac group; these nodes receive part of the lymphatic drainage of the bladder, the prostate, the ductus deferens, the seminal vesicles, the prostatic and membranous portions of the urethra, the uterine tube, the uterus, and the vagina. The external iliac nodes discharge to the common iliac nodes and are also interconnected by lymphatic channels with the internal iliac nodes.

The **common iliac nodes,** about six in number, lie laterally, medially, and posteriorly along the course of the common iliac vessels. The lateral and posterior nodes of the group are in continuity with the external iliac nodes. The medial nodes lie over the promontory below the bifurcation of the aorta. The common iliac nodes receive the efferent lymphatic vessels from the external iliac and internal iliac groups and thus combine lymph from the pelvic wall and viscera with that from the lower limb, the perineum, and the lower abdominal wall. The efferents of the common iliac nodes pass to the lumbar chain.

The **lumbar chain of nodes** extends from the aortic bifurcation to the aortic hiatus of the diaphragm. These nodes are large and numerous. On the right side, they lie partly in front of the inferior vena cava and partly behind it on the psoas major muscle. The nodes of the left side lie at the side of the abdominal aorta on the psoas. The lumbar nodes receive (1) the efferent lymphatic vessels of the common iliac nodes; (2) the lymphatic vessels from the testis in the male and from the ovary, the uterine tube, and the fundus of the uterus in the female; (3) vessels from the kidneys, the suprarenal glands, and the abdominal surface of the diaphragm; and (4) the efferent vessels from the lateral abdominal wall which accompany the lumbar arteries. The left lumbar chain also receives part of the efferents from the inferior mesenteric nodes. The interconnecting vessels pass the lymph upward through the lumbar chain, the vessels from the uppermost nodes uniting to form a single **lumbar trunk** on each side.

Cisterna Chyli At about the level of the renal blood vessels, the lumbar trunks converge posterior to the great vessels and end in the lower portion of the thoracic duct. An occasional (25 per cent— —Nelson) dilatation of this portion of the thoracic duct is designated as the cisterna chyli. The cisterna lies on the bodies of the upper two lumbar vertebrae between the right crus of the diaphragm and the abdominal aorta and represents a dilated receptacle for the lymph gathered from the lower part of the body. It may be very irregular in form, or it may be replaced by a plexus of vessels formed by the intestinal and lumbar lymph trunks. (The intestinal trunk is, in 70 per cent of cases, a tributary of the left lumbar trunk.) When, as frequently happens, the lumbar trunks unite at levels above the second lumbar vertebra, no cisterna chyli exists. Into the cisterna chyli, or the lower end of the thoracic duct, also empty a pair of descending intercostal lymph trunks formed by the efferent vessels from the lower intercostal nodes; they descend through the aortic hiatus of the diaphragm. The complete course of the thoracic duct is described in the chapter on the Chest (p. 362 & fig. 298). Rarely there are communications from the cisterna chyli or thoracic duct to the azygos vein, the inferior cava, or the lumbar veins at low thoracic and upper lumbar levels.

THE AUTONOMIC NERVES OF THE ABDOMEN

As is true elsewhere in the body, the autonomic innervation of the abdominal viscera is by both sympathetic and parasympathetic nerves. In the sympathetic supply, the thoracic splanchnic nerves penetrate the crura of the diaphragm to terminate in the celiac ganglia and its subdivisions, and the lumbar sympathetic trunk contributes visceral branches to the plexuses along the aorta and its branches. The parasympathetic innervation in the abdomen is represented by the anterior and posterior vagal trunks, which descend through the esophageal opening of the diaphragm, and by ascending branches of the pelvic splanchnic nerves. Nerves combining both parasympathetic and sympathetic neurons distribute to the viscera, predominantly by perivascular plexuses (exclusively by these plexuses for sympathetic fibers). In these nerves of distribution it is not possible to distinguish grossly between the parasympathetic and the sympathetic nerve bundles.

The Sympathetic Nerves

(1) The **thoracic splanchnic nerves** (fig. 300) constitute the principal source of sympathetic nerves in the abdomen. They are described in the chapter on the· Chest (p. 366). The greater thoracic splanchnic nerve (T5 to 9 or 10) terminates in the abdomen in the lateral border of the celiac ganglion. The lesser thoracic splanchnic nerve (T10, 11) terminates in the aorticorenal

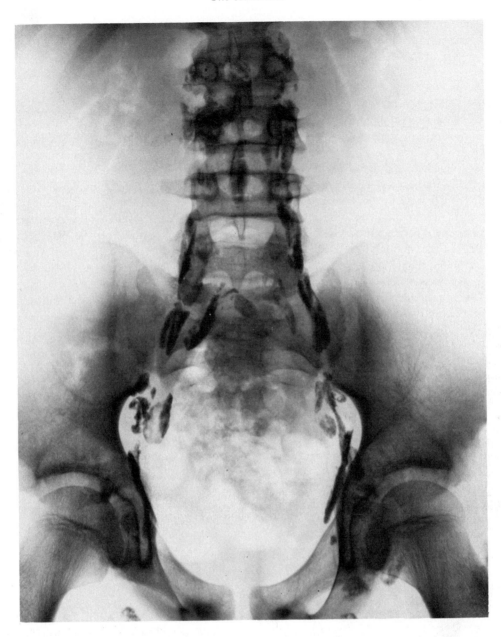

Fig. 372 Lymphogram of lumbar levels. (Compare fig. 371.)

ganglion; whereas the least thoracic splanchnic nerve (last thoracic ganglion), when present, ends in the renal plexus. These are preganglionic and visceral afferent nerves; their preganglionic fibers end around the ganglion cells in the ganglia and the associated plexuses. Postganglionic neurons continue through the subsidiary plexuses to the viscera.

(2) The **abdominal portion of the sympathetic trunk** (fig. 373) contributes visceral branches (lumbar splanchnic nerves) to the perivascular

plexuses and also gray rami communicantes to all of the lumbar spinal nerves. The **lumbar sympathetic trunk** lies more ventralward than does the thoracic portion of the trunk, being located against the ventrolateral curve of the bodies of the lumbar vertebrae internal to the bulging psoas muscles (fig. 357). The cord connecting the thoracic and lumbar portions of the trunk is slender, indicating that not many nerve fibers ascend or descend between the thoracic and lumbar levels. The sympathetic trunk passes under the medial lumbo-

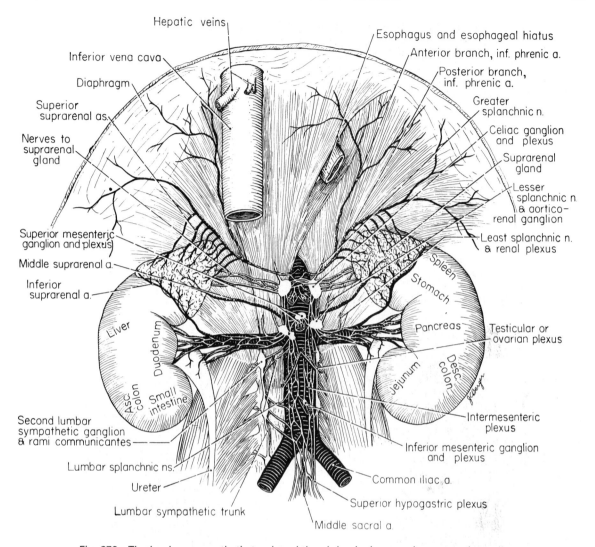

Fig. 373   The lumbar sympathetic trunk and the abdominal nerve plexuses and ganglia.

costal arch and then curves rather abruptly ventralward into its lumbar location. On the right side it lies under the inferior vena cava; on the left side it is overlapped by the lumbar lymph nodes along the aorta. The **ganglia of the lumbar sympathetic trunk** are irregularly fused, and many studies have emphasized their variability. There may be as few as two ganglia on the trunk or as many as six. Tsouras ('49) found that three or four ganglia were present in 80 per cent of the cases. When individually represented, the ganglia lie on the bodies of the corresponding vertebrae or the intervertebral disks below them. Thus, the first ganglion, frequently partly concealed by the medial lumbocostal arch, lies at the level of the first lumbar vertebra or the

disk below it. The ganglion of the second lumbar vertebra is said to be the largest and most constant. The last lumbar ganglion commonly lies deep to the common iliac vessels at the side of the fifth lumbar vertebra. Inferior to it, the sympathetic trunk continues into its sacral portion. There are but few interconnections between the lumbar sympathetic trunks of the two sides.

The lumbar sympathetic trunks receive **white rami communicantes** from the upper two or three lumbar spinal nerves, give rise to **gray rami communicantes** to all lumbar spinal nerves, and provide the **lumbar splanchnic nerves** which participate in the supply of certain abdominal and most of the pelvic viscera. The prevertebral position of the lumbar sympathetic trunk and its association with the massive psoas muscle, arising from the sides of the lumbar vertebrae, removes the trunk some distance from the emerging lumbar spinal nerves, so that both the white and gray rami are long. The white rami communicantes arise from only the first two or three lumbar nerves and descend obliquely to the lumbar ganglia, running deep to the fibers of the psoas muscle. Preganglionic neurons arising in the upper two or three segments of the lumbar spinal cord and destined for distribution by way of the lower lumbar, sacral, and coccygeal nerves descend to the level of those nerves in the sympathetic trunk. A **gray ramus communicans** for each lumbar spinal nerve arises from the sympathetic trunk or from one of its ganglia. The gray rami take a more transverse path to the spinal nerves than do the white rami; the gray rami commonly accompany the lumbar arteries under the fibrous arches of the psoas, frequently splitting, doubling, and rejoining (Kuntz & Alexander, '50) as they pass toward the spinal nerves.

The **lumbar splanchnic nerves** (fig. 373) are three or four short branches of the lumbar sympathetic trunk which arise predominantly at the levels of the first, second, and third lumbar vertebrae. The upper two nerves pass forward, medialward, and inferiorly to end mainly in the intermesenteric plexus; a few strands of the first nerve reach the celiac or the renal plexus. The lumbar splanchnic nerves of the right side pass behind the inferior vena cava to reach the intermesenteric plexus. The lower two lumbar splanchnic nerves pass either in front of or behind the common iliac vessels and contribute, over the sacral promontory, to the superior hypogastric plexus. The lumbar splanchnic nerves, for the most part, provide the sympathetic innervation for the pelvic viscera; their principal abdominal representatives join the inferior mesenteric plexus for the supply of the distal part of the large intestine.

Vascular, osseous, and **articular branches** of the lumbar sympathetic trunk are generally small twigs which distribute to the adjacent vertebrae and their ligaments, to the lumbar lymph nodes, and to the arteries of the abdominal cavity. Vascular branches to the abdominal aorta and the inferior vena cava are contributed by the lumbar sympathetic trunk and by the lumbar splanchnic nerves. Similar branches to the common and external iliac arteries arise in the lumbar trunk and lumbar splanchnic nerves or are branches of the ureteric and testicular (or ovarian) nerves. **Thoracolumbar sympathectomy** may be performed to produce vasodilation of the arteries of the lower limbs. The postganglionic

fibers concerned reach the vessels of the limb as branches of the lumbar and sacral nerves.

THE PARASYMPATHETIC NERVES

(1) The **anterior vagal** and **posterior vagal trunks** are described in connection with the stomach (p. 407). Aside from gastric branches, the anterior vagal trunk has an hepatic branch which reaches the liver through the upper part of the hepatogastric ligament. This branch contributes to the plexus on the hepatic artery and reaches the gall bladder and bile ducts. It also has an offshoot which descends through the hepatogastric ligament to the pyloric portion of the stomach, to the duodenum, and to the pancreas. The posterior vagal trunk supplies the stomach but also sends a celiac branch along the left gastric artery to the celiac plexus. This branch loses its identity in the plexus, but its radicles accompany the subdivisions of the celiac and superior mesenteric plexuses and distribute to all the organs of the abdomen beyond the stomach that are supplied by branches of the celiac trunk and the superior mesenteric artery. They are, therefore, components of the perivascular plexuses described hereafter and reach as far along the gastrointestinal tract as the last one-third of the transverse colon. Vagal fibers also reach the kidneys by way of the celiac and renal plexuses and are probably also continued along the testicular and ovarian plexuses.

(2) That portion of the large intestine which is supplied by the inferior mesenteric artery and most of the pelvic viscera are innervated by the **pelvic splanchnic nerves.** These are preganglionic parasympathetic nerves that have their cells of origin in the second, third, and fourth segments of the sacral spinal cord and arise from the corresponding sacral spinal nerves. Passing forward on either side of the rectum, the pelvic splanchnic nerves join the inferior hypogastric plexus for the supply of the pelvic viscera. For the supply of the distal colon, however, the pelvic splanchnic nerves give off discrete branches which, like others that spring from the hypogastric nerves below the sacral promontory, run independent courses under the peritoneum along the rectum and in the sigmoid mesocolon toward the left flexure of the colon (p. 513, fig. 409). These nerves do not join the perivascular plexuses of the inferior mesenteric artery until they reach the vasa recta, although they anastomose with perivascular plexus nerves as they cross them.

PLEXUSES AND GANGLIA OF THE AUTONOMIC
SYSTEM OF THE ABDOMEN

The sympathetic and parasympathetic nerves to the
abdominal viscera distribute by means of a plexus of
nerves and ganglia prolonged along the abdominal
aorta. The principal mass of this system is formed
by the celiac plexus and its ganglia located around
and on either side of the celiac arterial trunk at the
level of the upper part of the first lumbar vertebra.

The Celiac Plexus and Ganglia (figs. 369, 373)

The dense celiac plexus surrounds the root of the
celiac arterial trunk and contains two large ganglia,
one on either side. The plexus lies between the
suprarenal glands on the ventral surface of the
crura of the diaphragm and on the abdominal
aorta. The two **celiac ganglia** are about 2 cm. in their
long dimension and lie lateral and ventral to the
aorta. They are irregular in shape, may be broken
into smaller masses, and are interconnected with
each other by a dense network of nerves which pass
principally inferior to the celiac trunk. The right
ganglion, covered by the inferior vena cava, receives
the right greater thoracic splanchnic nerve. The
left ganglion receives the left greater thoracic
splanchnic nerve. The celiac branch of the posterior vagal trunk ends in the celiac plexus and
frequently in both ganglia. The celiac plexus is
continued down the abdominal aorta and is
prolonged along all its branches. It gives rise to
subsidiary perivascular plexuses named according
to the blood vessels along which they pass.
Several ganglia close to the celiac ganglia and
only incompletely detached therefrom are the
aorticorenal ganglia and the **superior mesenteric
ganglion.** The aorticorenal ganglia lie below the
celiac ganglia and above the level of origin of the
renal arteries, typically in the angle between these
arteries and the aorta. The lesser thoracic
splanchnic nerves end in the aorticorenal ganglia.
The superior mesenteric ganglion is usually
single. It lies on the aorta just above or occasionally below the root of the superior mesenteric
artery. The celiac plexus is continuous below with
the intermesenteric plexus of nerves which is, in
turn, continuous with the superior hypogastric
plexus.

The Intermesenteric Plexus

This plexus consists of from four to twelve
nerves on the anterior and anterolateral aspects of
the aorta between the mesenteric arteries. On the
left side the outermost nerves actually lie lateral
to the aorta; on the right side the outermost nerve
is in contact with the inferior vena cava. Above,
the outer nerves are united with the lower parts of
the celiac and aorticorenal ganglia; the intermediate
ones cross under the superior mesenteric artery.
Below, the outer nerves descend to enter the superior
hypogastric plexus, whereas the central ones form
the inferior mesenteric plexus. The descending
nerves are connected by oblique fibers to form an
open network. The intermesenteric plexus receives
contributions from the first two lumbar splanchnic
nerves and gives off renal, testicular (or ovarian),
and ureteric branches, as well as occasional
branches to the pancreas, the duodenum, the aorta,
and the inferior vena cava.

The Superior Hypogastric Plexus

This plexus is continuous with the intermesenteric
plexus below the origin of the inferior mesenteric
artery and lies anterior to the lower part of the
abdominal aorta, its bifurcation, and the median
sacral vessels. It is anterior to the last two lumbar
vertebrae and the first sacral vertebra. The plexus
consists of a flattened band of intercommunicating
nerve bundles which descend over the aortic
bifurcation. Broadening below, it divides over the
sacral promontory into a right and a left hypogastric nerve (p.511). In addition to its continuity
with the intermesenteric plexus, the superior hypogastric plexus receives the lower two lumbar
splanchnic nerves.

The Subsidiary Plexuses

The subsidiary plexuses are perivascular or consist of
nerves running along the branches of the abdominal
aorta. It is by these plexuses that the autonomic innervation is conducted to the viscera.

The **inferior phrenic plexuses** arise from the superior
part of the celiac plexus and accompany the inferior
phrenic arteries to the diaphragm. The right plexus
communicates at the vena caval foramen with the right
phrenic nerve. A small **inferior phrenic ganglion** is
found in this communication. Branches of both plexuses
supply the peritoneum and the diaphragm and provide
filaments to the gastroesophageal junction and the
cardiac end of the stomach.

The **celiac plexus** sends offshoots along the three
branches of the celiac arterial trunk, thus forming the
hepatic, splenic, and gastric plexuses. The **hepatic
plexus** gives off secondary plexuses along the right
gastric, gastroduodenal, and cystic arteries. The main
plexus consists of fibers arranged in two principal but
interconnected networks, one anterior to the hepatic
artery and one posterior to the portal vein. These sub-

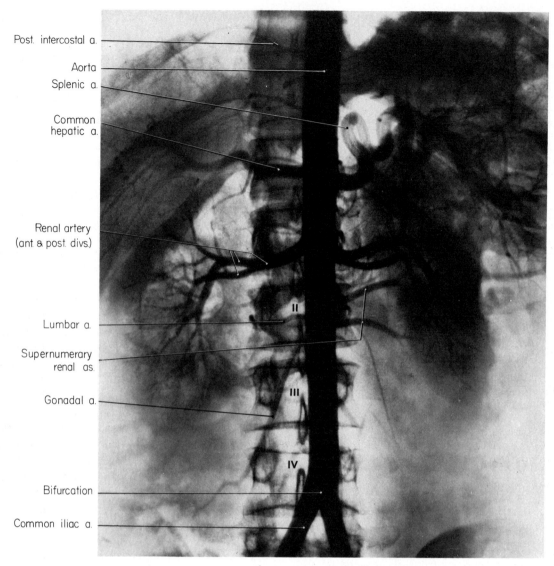

Post. intercostal a.

Aorta

Splenic a.

Common
hepatic a.

Renal artery
(ant & post. divs)

Lumbar a.

Supernumerary
renal as.

Gonadal a.

Bifurcation

Common iliac a.

II

III

IV

Fig. 374   A normal abdominal aortogram (gastrointestinal vessels not filled).

divide and follow the vessels into the substance of the liver, but this organ is less profusely innervated relative to its size than are the stomach, the kidneys, and the suprarenal glands. The anterior hepatic plexus supplies the common bile duct. The **cystic plexus** is small; it innervates the gall bladder. The **right gastric** and **gastro-duodenal plexuses** follow these arteries and their branches and reach the stomach, the pancreas, and the duodenum. A large plexus follows the **right gastroepiploic artery** to supply the stomach along its greater curvature, communicating with the splenic plexus from the left.

The **splenic plexus** consists of from four to six filaments which follow the artery to the spleen. Extensions are sent along the pancreatic, short gastric, and left gastroepiploic arteries to supply the corresponding structures.

The **gastric plexus** consists of from one to four twigs which accompany the left gastric artery and supply the stomach from its lesser curvature. Radicles of the gastric plexus do not quite reach the pylorus but communicate with terminals of the hepatic plexus along the right gastric artery. The gastric plexus also communicates with the gastric branches of the vagal trunks. Nerves to the greater curvature of the stomach are terminals of the plexuses along the right gastroepiploic and left gastroepiploic arteries.

The **superior mesenteric plexus** is the largest offshoot of the celiac plexus and is a continuation of its lower part. A **ganglion** is frequently found in the plexus above the root of the superior mesenteric artery. The nerve fibers of the plexus divide to follow the branches of the artery—inferior pancreaticoduodenal, jejunal, ileal, ileocolic, right colic, and middle colic. They conduct sympathetic and parasympathetic fibers to the organs supplied by the corresponding arteries. The parasympathetic fibers in the superior mesenteric plexus are derived from the celiac branch of the posterior vagal trunk.

The **inferior mesenteric plexus** is derived from the intermesenteric plexus. It also receives contributions from the lumbar splanchnic nerves but probably does not contain parasympathetic fibers. The plexus may contain an **inferior mesenteric ganglion,** but, more frequently, small collections of ganglion cells occur in the thickened bundles of the proximal part of the plexus. The plexus divides in accordance with the branching of the inferior mesenteric artery into the secondary left colic, sigmoid, and superior rectal plexuses. The sympathetic fibers which distribute through the inferior mesenteric plexus arise in the upper lumbar splanchnic nerves; the parasympathetic innervation to this part of the large intestine reaches it by independent pathways which have been described (p. 456).

The **suprarenal plexuses** consist of a number of nerves which radiate to each gland from the homolateral celiac ganglion and the greater splanchnic nerve. The phrenic nerves also send twigs to the glands. Relative to their size, these glands have a richer innervation than any other organ. The fibers of the plexuses are mostly preganglionic sympathetic fibers which end directly around the medullary cells; these cells are homologous with postganglionic sympathetic neurons. Nerve fibers in the cortex are sparse and are confined to the vicinity of the blood vessels.

The **renal plexuses** consist of an open meshwork of nerves along the renal arteries, supplemented by nerves which run both superior and inferior to the arteries. The plexuses receive from four to eight branches from the celiac and aorticorenal ganglia which convey both sympathetic and parasympathetic (vagal) fibers, branches from the first (sometimes second) lumbar splanchnic nerve, renal branches from the intermesenteric nerves, and contributions from the superior hypogastric plexus. The contributions from the superior hypogastric plexus are thought to be the pathways for the parasympathetic innervation of the renal pelvis and the upper end of the ureter (Mitchell, '53). In addition, the lesser thoracic splanchnic nerve ends in the aorticorenal ganglion and the least thoracic splanchnic nerve in the renal plexus; contributions from these nerves join the renal plexus on either side. **Renal ganglia** are usually found along the posterosuperior aspect of the renal artery but, occasionally, are anterior to it. There are communications between the renal and suprarenal plexuses and also with the ureteric plexus below.

The **ureteric plexuses** are composed of nerves which interlace along the ureters; the plexuses are both abdominal and pelvic in location. Nerves to the upper portions of the ureters are derived from the renal plexuses, the intermesenteric plexus, and the superior hypogastric plexuses. The latter contributions probably provide parasympathetic nerves from the pelvic splanchnic nerves. Tiny nerve filaments to the midportions of the ureters are derived from the superior hypogastric plexuses and from the hypogastric nerves. The nerve supply of the lower portions of the ureters comes from the hypogastric nerves and the inferior hypogastric plexuses and is associated with the innervation of the

ductus deferens, the seminal vesicle, and the urinary bladder.

The **testicular** and **ovarian plexuses** are closely related to the ureteric plexuses and regionally receive branches from the same sources. These plexuses lie on and follow the testicular and ovarian arteries. The testicular plexus supplies the spermatic cord, the epididymis, and the testis; the ovarian plexus supplies the ovary, the broad ligament, and the uterine tube and communicates with the uterine plexus. It may be predicated, on embryologic and reflex grounds, that the parasympathetic fibers of the testicular and ovarian plexuses are vagal in origin.

THE ABDOMINAL AORTA

The **aorta** enters the abdomen in front of the lower part of the body of the twelfth thoracic vertebra by passing through the aortic hiatus of the diaphragm. It is only about 13 cm. in length, for it ends by dividing into the two common iliac arteries in front of the lower part of the fourth lumbar vertebra slightly to the left of the median line (figs. 356, 374). The level of its bifurcation can be marked on the anterior surface of the body as that of the intercrestal line (a line across the highest level of the iliac crests) or as a point from 2 to 3 cm. below and slightly to the left of the umbilicus (fig. 375). Posteriorly, the abdominal aorta lies against the lumbar vertebrae, the intervertebral disks, and the anterior longitudinal ligament. The left lumbar veins cross behind it to reach the inferior vena cava. On the right the aorta is in relation above with the azygos vein, the cisterna chyli and the thoracic duct, the right crus of the diaphragm, and the right celiac ganglion. Below, it lies against the inferior vena cava. On the left of the aorta are the left crus of the diaphragm, the left celiac ganglion, and the left lumbar chain of lymph nodes. The aorta is posterior to all the viscera which cross the midline; it is crossed by the left renal vein. The intermesenteric plexus lies in close contact with the vessel. The diameter of the aorta is about 2 cm. as it passes through the aortic hiatus. It diminishes considerably in size after giving rise to the celiac arterial trunk and the superior mesenteric artery, and then it retains a fairly uniform caliber to its termination.

Clinical Note Aneurism of the abdominal aorta frequently involves the segment below the renal arteries and may extend into the common iliac vessels also.

BRANCHES OF THE ABDOMINAL AORTA

The branches of the abdominal aorta may be classified as visceral, parietal, and terminal. Some of these are paired and some are unpaired. With the

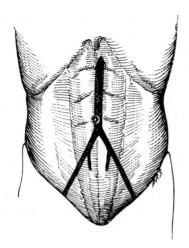

Fig. 375   The surface projection of the abdominal aorta and the iliac arteries.

exception of the testicular (or ovarian), median sacral, and common iliac arteries, all the branches have been described. They are listed below in the order in which they arise.

1. Inferior phrenic arteries (p. 451)
2. Celiac trunk (p. 403)
3. Middle suprarenal arteries (p. 448)
4. First lumbar arteries (p. 380)
5. Superior mesenteric artery (p. 429)
6. Renal arteries (p. 445)
7. Testicular (or ovarian) arteries (p. 460)
8. Second lumbar arteries (p. 380)
9. Inferior mesenteric artery (p. 437)
10. 11. Third and fourth lumbar arteries (p. 380)
12. Median sacral artery (p. 460)
13. Common iliac arteries (p. 461)

The Testicular Arteries (figs. 357, 368, 378)

The testicular arteries are two long, slender vessels which arise from the anterior aspect of the abdominal aorta just below the renal arteries. They descend retroperitoneally and in an oblique direction across the psoas muscles to the deep inguinal ring. In its descent, the right testicular artery crosses the inferior vena cava and goes behind the middle colic and the ileocolic arteries. The left vessel passes behind the left colic and sigmoid arteries and the sigmoid colon. The testicular arteries cross obliquely over the ureters and the lower parts of the external iliac arteries to reach the deep inguinal ring. Emerging from the ring are two testicular veins which accompany

each artery; these then join to form a single vein which usually lies lateral to the artery. In its abdominal course, each testicular artery gives off **ureteric branches** to the abdominal part of the ureter as well as twigs to the perirenal fat, the peritoneum, and the lumbar lymph nodes. Entering the inguinal canal, each testicular artery is accompanied by the ductus deferens and is enmeshed by the pampiniform plexus of veins. It descends to the testis as one of the constituents of the spermatic cord and then breaks up into several branches. Two or three of these branches accompany the ductus deferens, supply the epididymis, and anastomose with the artery of the ductus deferens and the cremasteric artery; others pierce the posterior part of the tunica albuginea of the testis and supply the substance of the gonad.

The Ovarian Arteries

The course and relations of each ovarian artery are just like those of the testicular vessels as far as the brim of the pelvis. Here the ovarian artery turns medialward to enter the pelvis by crossing the external iliac artery from 2 to 3 cm. from its origin. The ovarian artery passes through the suspensory ligament of the ovary to reach the broad ligament of the uterus. In the broad ligament it runs medialward below the uterine tube, then turns posteriorly in the mesovarium. Over the ovary it breaks up into terminal branches which enter the ovary through the hilum in its anterior border. Branches are also given to the uterine tube and to the ovarian ligament of the uterus; these anastomose with branches of the uterine artery. A small branch is given to the ureter during the abdominal course of the artery.

The Median Sacral Artery (fig. 404)

This is a small vessel which may be regarded as the caudal continuation of the abdominal aorta. It arises from the back of the aorta just above its bifurcation. The artery descends in the median line over the lower two lumbar vertebrae and the sacrum and the coccyx, ending over the pelvic surface of the coccyx in branches which form loops with the lateral sacral arteries. Opposite the fifth lumbar vertebra, the left common iliac vein crosses the median sacral artery; the artery is accompanied by a vein which empties into the left common iliac vein. The median sacral artery gives origin to the fifth pair of **lumbar arteries** in front of the last lumbar vertebra. These distribute like the other lumbar arteries (p. 380). As it descends over the sacrum, the median sacral artery gives rise to small parietal branches which anastomose with the lateral sacral arteries laterally. Small visceral branches pass to the rectum and anastomose with the superior rectal and middle rectal arteries.

The Common Iliac Arteries

The common iliac arteries (figs. 373, 375, 404) are the terminal branches of the abdominal aorta; they begin opposite the middle of the body of the fourth lumbar vertebra, a little to the left of the median plane. The common iliac arteries divide into the external iliac and internal iliac arteries opposite the intervertebral disk between the fifth lumbar vertebra and the sacrum. The common iliac arteries lie in straight-line continuity with the external iliac arteries, which pass under the inguinal ligaments midway between the symphysis pubis and the anterior superior iliac spines of either side. Of the combined length of these arteries of 15 cm., the common iliac accounts for 5 cm.; the external iliac, 10 cm. The internal iliac artery is the artery of the pelvis; the external iliac artery supplies the lower limb.

The superior hypogastric plexus is formed in front of the upper ends of the common iliac arteries, and the ureters cross them at their terminations. The left common iliac artery is crossed in addition, by the superior rectal vessels. Behind each artery are the bodies of the fourth and fifth lumbar vertebrae and the intervertebral disk between them, the sympathetic trunk, and the psoas muscles. The right common iliac artery is, however, separated from these structures by the terminations of the two common iliac veins and the beginning of the inferior vena cava. The common iliac arteries give rise to the external iliac and internal iliac arteries. Other small and inconstant branches pass to the peritoneum, the psoas muscles, the ureter and adjacent areolar tissue.

The **external iliac arteries** are the direct continuations of the common iliac arteries and lie along the medial borders of the psoas major muscles. Passing under the inguinal ligament midway between the symphysis pubis and the anterior superior spines of the ilia, the external iliac arteries become the femoral arteries, the major vessels of the lower limbs. The ureters cross the arteries at their origins, and in the female the ovarian vessels cross them 2 or 3 cm. beyond. Near the lower end of

each artery, the testicular vessels and the genital branch of the genitofemoral nerve lie on the artery, and these are all crossed by the deep circumflex iliac vein. The ductus deferens crosses this portion of the artery in the male; the round ligament of the uterus does so in the female. External iliac lymph nodes lie along and at the sides of the artery, and one node is usually directly in front of the artery at its termination. The external iliac vein is medial to the artery. The external iliac artery gives rise to small branches to the psoas major muscle and to the neighboring lymph nodes. Its principal branches are the inferior epigastric and deep circumflex iliac arteries (p. 380).

THE INFERIOR VENA CAVA

The **inferior vena cava** (figs. 368, 376), formed on the right side of the fifth lumbar vertebra by the junction of the two common iliac veins, collects the venous blood of the lower limbs and the nonportal blood of the abdomen and the pelvis. The vena cava is longer (20 cm.) than the abdominal aorta (13 cm.) and leaves the abdomen through the central tendon of the diaphragm at the level of the eighth thoracic vertebra. At that level it pierces the fibrous pericardium and ends in the postero-inferior portion of the right atrium of the heart. The thoracic course of the inferior vena cava is very short, and within the pericardium the serous layer of the pericardium reflects onto it except posteriorly. Its atrial orifice is marked anteriorly by the **valve of the inferior vena cava,** a crescentic fold which does not form a competent valve in the adult but which, in the fetus, serves to direct the blood entering by the vena cava across the atrial septum into the left atrium.

In the abdomen the inferior vena cava lies to the right of the abdominal aorta, the right crus of the diaphragm, and the caudate lobe of the liver. It begins below, posterior to the right common iliac artery, but is more anteriorly placed at the level of the kidneys where the left renal vein crosses the aorta anteriorly. The vena cava passes through the diaphragm in its most anterior opening. These differing relations are the result of the formation of the inferior vena cava from several fetal channels: the supracardinal vein below the renal level, the subcardinal plexus at the renal level, the liver sinusoids within the liver, and an outgrowth of this hepatic portion between the liver and the subcardinal plexus (fig. 376). Anomalies of the inferior vena cava

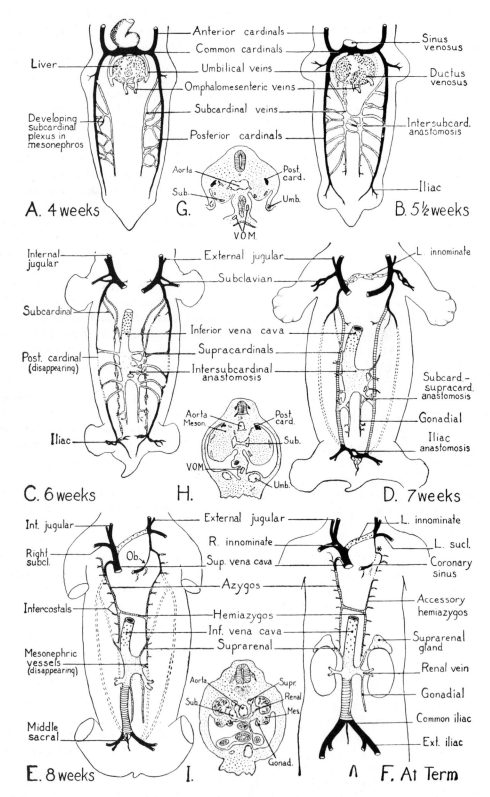

Fig. 376 Schematic ventral diagrams showing the major steps in the development of the inferior vena cava. (By permission, from *Human Embryology*, 2nd ed., by B. M. Patten, 1953. Blakiston Div., McGraw-Hill Book Co.)

are usually due to irregularities in persistence or atrophy of the fetal channels concerned in its formation. The supracardinal veins above the renal level are represented in the azygos and hemiazygos veins, an arrangement which is reflected, in the adult, in the rather large communication between these veins and the dorsal aspect of the inferior vena cava and left renal veins, respectively (p. 361). The vena cava is in relation anteriorly, from below upward, with the right common iliac artery, the mesentery, the right testicular artery, the third part of the duodenum, the pancreas, the common bile duct, the portal vein, and the posterior surface of the liver.

TRIBUTARIES

Formed by the junction of the common iliac veins, the inferior vena cava receives veins which correspond with the arterial branches of the abdominal aorta with the exception of such veins as empty into the portal vein. From below upward, the inferior vena cava receives the lumbar veins, the right testicular (or ovarian) vein, the renal veins, the right suprarenal vein, the right inferior phrenic vein, and the hepatic veins. It also receives a few smaller tributaries including one or two direct from the ureter.

The Common Iliac Veins

The common iliac veins are formed by the union of the external iliac and internal iliac veins in front of the sacroiliac joint. They ascend obliquely toward the right side and unite to form the inferior vena cava over the right half of the fifth lumbar vertebra behind the right common iliac artery. The **right common iliac vein** is much shorter than the left and is more vertical in its course. The **left common iliac vein,** longer than the right and more obliquely disposed, courses first on the medial side of its artery and then passes behind the right common iliac artery. Each common iliac vein receives the external and internal iliac veins and the iliolumbar vein. The left common iliac also receives the median sacral vein. The internal iliac vein is described in the chapter on the Pelvis (p. 507).

The **external iliac vein** is the companion of the external iliac artery and begins at the level of the inguinal ligament as a continuation of the femoral vein. The right vein is situated at first on the posteromedial aspect of the external iliac artery and then

becomes directly posterior to it. The left external iliac vein lies continuously along the medial aspect of the corresponding artery. Each vein crosses the lateral surface of the internal iliac artery just before it terminates, separating that artery from the psoas muscle. Each vein is anterosuperior to the obturator nerve. About one of four external iliac veins contains a bicuspid valve, which is competent in most cases (Basmajian, '52). The tributaries of the external iliac vein are the inferior epigastric, deep circumflex iliac, and pubic veins. The **inferior epigastric** and **deep circumflex iliac veins** are formed by the union of the accompanying veins of the corresponding arteries. The inferior epigastric vein joins the external iliac vein about 1 cm. above the inguinal ligament; the deep circumflex iliac vein is tributary to the external iliac vein about 2 cm. above this ligament. The **pubic vein** is a communication between the obturator and the external iliac veins. It begins at the obturator canal and accompanies the pubic branch of the inferior epigastric artery. When of large size, it is likely to form the main termination of the obturator vein.

The **iliolumbar vein** is named for its drainage. It receives the fifth lumbar vein and tributaries from the iliac fossa as well as from the vertebral canal. It accompanies the distal parts of the iliolumbar artery but, proximally, passes behind the psoas major muscle and ends in the back of the common iliac vein.

The **median sacral vein** represents the union of the accompanying veins of the corresponding artery. These arise on the pelvic aspect of the sacrum by veins (sacral venous plexus) that communicate with the lateral sacral veins and receive tributaries from the sacral canal. The median sacral vein is usually a tributary of the left common iliac vein.

The Lumbar Veins

The lumbar veins are usually five on each side, one vein accompanying each lumbar artery from the abdominal aorta and one running with the lumbar branch of the iliolumbar or of the median sacral artery. Their pattern of distribution corresponds to the lumbar arteries (p. 380), and they drain the posterior wall of the abdomen, the vertebral canal, the spinal cord, and the meninges. The lumbar veins communicate in the abdominal wall with the subcostal vein and other abdominal wall veins and also with the external and internal vertebral plexuses by way of the intervertebral vein. The lumbar veins lie on the bodies of the vertebrae deep to the psoas major muscles and end in the posterior aspect of the inferior vena cava. The left lumbar veins are longer than the right and pass posterior to the abdominal aorta, except when ending in the left renal vein as does the second left lumbar vein in the majority of cases (Anson & Kurth, '55). The fifth vein is a tributary of the iliolumbar vein or of the common iliac vein; the others usually end in the inferior vena cava, although the upper one commonly ends in the ascending lumbar vein. The lumbar veins are all united by the longitudinal ascending lumbar vein.

The **ascending lumbar vein** lies between the psoas major muscle and the roots of the transverse processes of the lumbar vertebrae. Beginning in the common iliac or the iliolumbar vein, it serially interconnects the lumbar veins dorsal (and sometimes ventral) to the emerging lumbar nerve roots. Cranially, the ascending lumbar vein unites with the subcostal vein to form the lateral root of the azygos vein on the right or of the hemiazygos vein on the left (p. 361). The ascending lumbar veins are thinned out between the second and third lumbar levels and blood flow appears to increase upward and downward on the cranial and caudal sides of this level respectively.

The Testicular Veins

The testicular veins arise from the testis and the epididymis. They form the **pampiniform plexus** which, consisting of from eight to ten anastomosing veins, ascends along the ductus deferens, mainly anterior to it. This plexus, as a constituent of the spermatic cord, traverses the inguinal canal and ends near the deep inguinal ring by forming two accompanying veins of the testicular artery. In the abdomen tributaries from the ureter join the veins. The two veins unite into one, and the resulting **testicular vein** ends on the right side in the inferior vena cava at a lower level than the level of origin of the corresponding artery. The left testicular vein is longer; it ends in the left renal vein, joining it at a right angle. The testicular veins are usually provided with valves at their terminations, and their relations are like those of the corresponding arteries (p. 460).

The Ovarian Veins

The ovarian veins arise from the ovary and form a pampiniform plexus in the broad ligament which reaches the brim of the pelvis. From the plexus are formed two veins which subsequently form the single ovarian vein. The relations and manner of termination of the ovarian veins are like those of the testicular veins. Within the broad ligament the ovarian veins communicate with the uterine plexus, and, like the uterine veins, they become greatly enlarged during pregnancy.

The Renal, Suprarenal, Inferior Phrenic, and Hepatic Veins

These veins are described under the structures which they serve. References to these descriptions are as follows:

Renal veins (p. 446)
Suprarenal veins (p. 449)
Inferior phrenic veins (p. 451)
Hepatic veins (p. 420)

THE ABDOMINAL STRUCTURES RELATED TO THE LOWER LIMB

The psoas major, psoas minor, and iliacus muscles of the lower limb have their origins within the abdomen. The lumbar plexus, which provides a number of important nerves of the lower limb, is also formed in the abdominal region.

THE FASCIAE

The **transversalis fascia** (p. 378) forms an internal parietal lining of the entire abdominopelvic cavity. In the lumbar region it covers the quadratus lumborum and psoas muscles constituting the **quadratus lumborum fascia** and the **psoas fascia**. Attachments to the anterior layer of the thoracolumbar fascia, to the lumbar transverse processes, and to the anterior longitudinal ligament of the vertebral column define the specific portions of the transversalis fascia forming these muscular fasciae. The transversalis fascia attaches to the inner lip of the iliac crest and then descends over but does not form the iliacus fascia. The **iliacus fascia** (Hayes, '50) is aponeurotic like the internal obturator fascia and the heavy fasciae of the muscles of the shoulder (similarity in pectoral and pelvic girdle musculature) and continues deep to the inguinal ligament with the iliopsoas tendon. In many specimens, fat and areolar tissue are found between the iliacus fascia and the overlying transversalis fascia. The iliacus fascia is also attached to the inguinal ligament later-

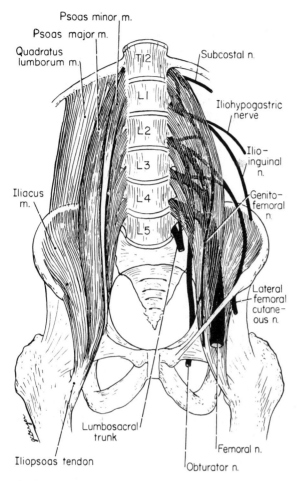

Fig. 377 The lumbar and iliac musculature and the nerves of the lumbar plexus.

and the intervertebral disks of the last thoracic and all five lumbar vertebrae; (2) from the transverse processes of all the lumbar vertebrae; and (3) from membranous arches which pass over the sides of the bodies of the four upper lumbar vertebrae permitting the lumbar arteries and veins and the rami communicantes to pass beneath them. Narrowing as it passes into the thigh, the muscle goes under the inguinal ligament just lateral to the iliopubic eminence. It crosses the capsule of the hip joint, separated from it by the iliopectineal bursa, and ends in the **iliopsoas tendon** which inserts in the lesser trochanter of the femur. This tendon also receives most of the fibers of the iliacus muscle.

The psoas major muscle flexes the thigh or, acting from below, flexes the lumbar vertebral column and bends it toward the same side. Its rotary actions on the thigh are equivocal and unimportant (Basmajian). With iliacus, it shows some flexion activity in relaxed standing, for only in hyperextension of the hip is the joint in the close-packed position with all ligaments taut. Its nerves are derived from the ventral rami that compose the lumbar plexus, principally lumbar nerves two, three, and four but sometimes from the first lumbar as well.

THE PSOAS MINOR MUSCLE (figs. 377, 379)

The **psoas minor muscle** is absent in about 40 per cent of specimens. When present, it arises from the sides of the bodies of the twelfth thoracic and first lumbar vertebrae and from the intervertebral disk between them. Its muscle fibers give rise to a long flat tendon which ends in the iliopubic eminence, the arcuate line of the ilium, and the iliac fascia. The muscle flexes the pelvis on the trunk. Its nerves are offshoots of the first and second lumbar nerves which often run in company with the genitofemoral nerve.

THE ILIACUS MUSCLE (figs. 377–80)

The **iliacus** is a fan-shaped muscle that occupies the iliac fossa. It arises from the iliac crest and the greater part of the iliac fossa, from the iliolumbar and ventral sacroiliac ligaments, and from the ala of the sacrum. The fibers of the muscle converge to insert into the lateral side of the tendon of the psoas major (iliopsoas tendon) and end in the lesser trochanter of the femur. Some of the fibers are prolonged onto the shaft of the femur for several

ally and participates with it in giving origin to fibers of the internal abdominal oblique and transversus abdominis muscles. Following the medial border of the iliopsoas muscle, the iliacus fascia attaches to the iliopubic eminence and then extends along the superior pubic ramus and over the pectineus muscle, blending with the deep layer of fascia lata behind the saphenous opening. It thus separates the muscular and vascular lacunae under the inguinal ligament. The femoral sheath is a diverticulum of transversalis fascia and is not contributed to by the pectineus or iliacus fasciae.

THE PSOAS MAJOR MUSCLE (figs. 377–80)

The **psoas major muscle,** long and tapered at its ends, lies at the side of the bodies of the lumbar vertebrae. It arises (1) from the sides of the bodies centimeters below the lesser trochanter. The iliacus muscle flexes the thigh or, acting from below, flexes the pelvis on the thigh. Like psoas major, the rotary actions of iliacus are insignificant. The nerves of the iliacus muscle arise from the femoral nerve (L2 to 4) and pass into the iliacus muscle in its midregion.

THE LUMBAR PLEXUS (figs. 377–80)

The **lumbar plexus** is formed from the ventral rami of the first three lumbar nerves and part of the fourth lumbar nerve. In about 50 per cent of cases, a small branch of the twelfth thoracic nerve joins the first lumbar. The lumbar plexus is located anterior to the transverse processes of the lumbar vertebrae deeply within the psoas muscle. The ven-

tral rami are connected with the sympathetic trunk and ganglia, as has been described (p. 454), and then form branches which unite with others to constitute the definitive nerves of the plexus.

The first lumbar nerve, usually after receiving a contribution from the twelfth thoracic nerve, splits into superior and inferior branches. The superior branch redivides into the iliohypogastric and ilio-inguinal nerves; the smaller inferior branch unites with a small superior branch of the second lumbar nerve to form the **genitofemoral nerve.** The remaining greater portion of the second ventral ramus and the third and fourth ventral rami divide into smaller anterior and larger posterior portions. These fundamental divisions unite to form the pre-axial (anterior, p. 81) **obturator nerve** and the postaxial

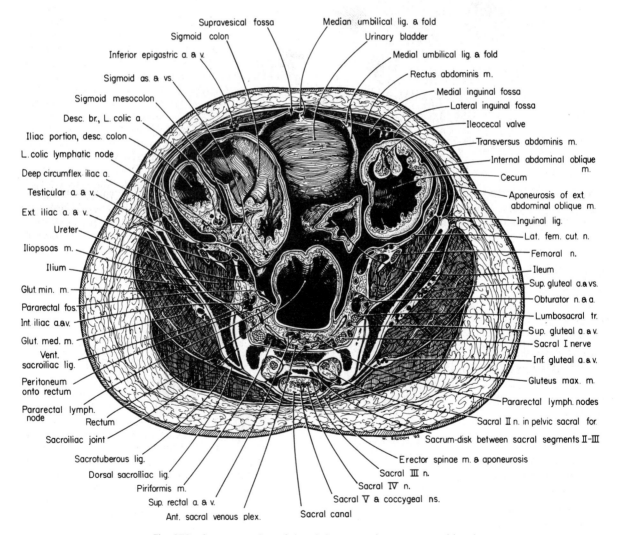

Fig. 378   A cross section of the abdomen at the upper rectal level.

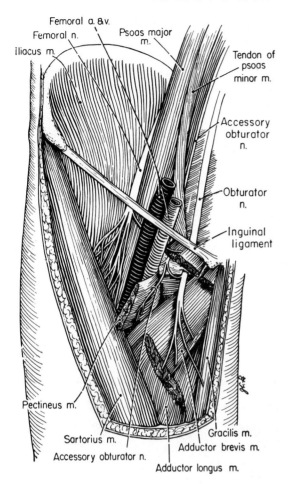

Fig. 379   The femoral, obturator, and accessory obturator nerves (from original of fig. 1, R. T. Woodburne, *Anat. Rec.*, *136:367, 1960*).

(posterior, p. 81) **femoral** and **lateral femoral cutaneous nerves.** A portion of the fourth lumbar nerve joins the sacral plexus. The **accessory obturator nerve (L3, 4)**, present in 9 per cent of cases, is an offshoot of the anterior branches which give rise to the obturator nerve. The nerves of the lumbar plexus also provide muscular branches to the psoas and quadratus lumborum muscles. The branches of the lumbar plexus may be summarized as follows:

Iliohypogastric-L1 (T12)
Ilioinguinal-L1
Genitofemoral-L1, 2 (pre-axial)
Lateral femoral cutaneous-L2, 3 (postaxial)
Obturator-L2, 3, 4 (pre-axial)
Accessory obturator-L3, 4 (pre-axial)
Femoral-L2, 3, 4 (postaxial)

The Iliohypogastric Nerve

This nerve arises from the first lumbar nerve, together with a frequent contribution from the twelfth thoracic nerve. The iliohypogastric nerve emerges from the psoas major muscle along its lateral border, crosses the quadratus lumborum muscle between the muscle and its internal fascia, and penetrates the transversus abdominis muscle near the crest of the ilium. This nerve is motor to the abdominal musculature and ends in an anterior cutaneous branch to the skin of the suprapubic region and a lateral cutaneous branch in the hip region. Its course and terminal branches have been described (p. 379).

The Ilioinguinal Nerve

The ilioinguinal nerve emerges from the psoas muscle and has a course similar but slightly inferior to that of the iliohypogastric. Indeed, these two nerves are frequently joined into a common trunk for part of their course through the abdominal wall. The ilioinguinal nerve traverses the inguinal canal and ends in the skin over the upper medial part of the thigh and the anterior part of the scrotum in the male (anterior scrotal branch, p. 369) and of the mons pubis and the labium majus in the female (anterior labial branch, p. 369).

In 35 per cent of cases (Oelrich and Moosman, '77) the ilioinguinal nerve is combined with the genitofemoral nerve (see below) and has a course within the abdomen. Passing in the typical course of the genitofemoral nerve along the psoas major muscle toward the anterior abdominal wall, the common nerve divides as it approaches the wall, one branch continuing into the thigh on the surface of the external iliac artery. This constitutes the femoral branch of the genitofemoral nerve. The other branch traverses the deep inguinal ring, where it lies dorsal to the structures of the spermatic cord. This branch covers the distribution of both the normal ilioinguinal nerve and the genital branch of the genitofemoral nerve, supplying the cremaster muscle as well as the cutaneous distribution of the ilioinguinal nerve.

The Genitofemoral Nerve

The genitofemoral nerve arises by the union of branches from the anterior portions of the first and second lumbar nerves. It passes through the psoas muscle, emerging on its ventral surface medial to the psoas minor at the level of the third or fourth lumbar vertebra. Descending under the psoas fascia, the genitofemoral nerve divides at a variable level, sometimes before its emergence from the psoas, into genital and femoral branches. The **genital branch** is small, passes in front of the lower part of the external iliac artery, and enters the inguinal canal at the deep inguinal ring. It supplies the cremaster muscle (p. 376) and gives small branches to the skin and fascia of the scrotum and the adjacent part of the thigh. The **femoral branch** is the more medial of the two and continues under the inguinal ligament on the anterior surface of the external iliac artery. Below the inguinal ligament it

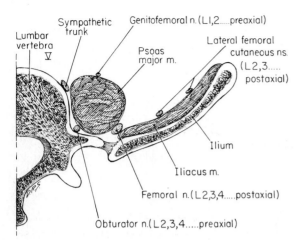

Fig. 380   The major nerves of the lumbar plexus.

pierces the femoral sheath and passes through the saphenous opening (or directly through the fascia lata) to supply the skin over the femoral triangle lateral to that supplied by the ilioinguinal nerve. In the triangle the femoral branch gives off a small twig to the femoral artery and communicates with an anterior cutaneous branch of the femoral nerve.

The Lateral Femoral Cutaneous Nerve

This nerve arises from the posterior branches of the second and third lumbar nerves. Appearing at the lateral border of the psoas muscle, the nerve—single or in several branches—crosses the iliacus muscle deep to its fascia directed toward the anterior superior spine of the ilium. It passes under the lateral end of the inguinal ligament superficial or deep to the sartorius muscle and, about a palm's breadth below the anterior superior spine, divides into anterior and posterior branches. Their distribution is discussed with the cutaneous nerves of the thigh (p. 529).

The Obturator Nerve

The obturator nerve is the principal pre-axial nerve of the lumbar plexus. It arises from the anterior branches of lumbar nerves two, three, and four and descends along the medial border of the psoas muscle. The obturator nerve passes behind the common iliac vessels and then runs forward lateral to the internal iliac vessels and the ureter. Lying against the obturator internus muscle, the nerve accompanies the obturator artery and vein to the obturator foramen and passes with them through the obturator canal into the thigh. In the obturator canal the nerve divides into an anterior and a posterior branch which supply the adductor group of muscles, the hip and knee joints, and the skin of the medial aspect of the thigh above the knee (p. 529).

The Accessory Obturator Nerve (fig. 379)

This nerve is small and is present in only 9 per cent of specimens. It arises from the anterior branches of the third and fourth lumbar nerves. The nerve descends along the medial border of the psoas muscle and passes across the superior pubic ramus adjacent to or under the femoral vein. In the thigh it lies deep to the pectineus muscle and supplies this muscle on its posterolateral surface. The nerve frequently also communicates with the anterior branch of the obturator nerve and sends a branch to the capsule of the hip joint.

The Femoral Nerve

The femoral nerve is the largest branch of the lumbar plexus. It is formed by the posterior branches of the second, third, and fourth lumbar nerves. The nerve emerges through the fibers of the psoas muscle at its lower lateral border and descends at the line of junction of the psoas and iliacus muscles. It passes under the inguinal ligament within the groove formed by these muscles. Entering the femoral triangle of the upper thigh, the femoral nerve breaks up immediately into numerous branches. These are described in the chapter on The Lower Limb (p. 555). Within the iliac fossa, muscular branches of the femoral nerve supply the iliacus.

Muscular Branches of the Lumbar Plexus

The nerves to the quadratus lumborum muscle arise independently from the first three or four lumbar nerves and the subcostal nerve; those to the psoas major, from the second, third, and fourth and sometimes from the first lumbar nerve. The nerves to the psoas minor arise from the first and second lumbar nerves.

VII

The Perineum

The **perineum,** defined anatomically, is the entire outlet of the pelvis (fig. 381). This diamond-shaped space is bounded anteriorly by the pubic symphysis and anterolaterally, on either side, by the inferior ramus of the pubis and the ramus and tuberosity of the ischium. Posteriorly, the perineum ends at the coccyx, and its posterolateral limit is the sacrotuberous ligament which stretches from the sides of the sacrum and the coccyx to the ischial tuberosity. The rhombus enclosed within these bony and ligamentous limits is arbitrarily divided, for convenience in description, by a line passing from side to side between the anterior parts of the tuberosities of the ischia. The **urogenital triangle** is anterior to this line; the **anal triangle** is behind it.

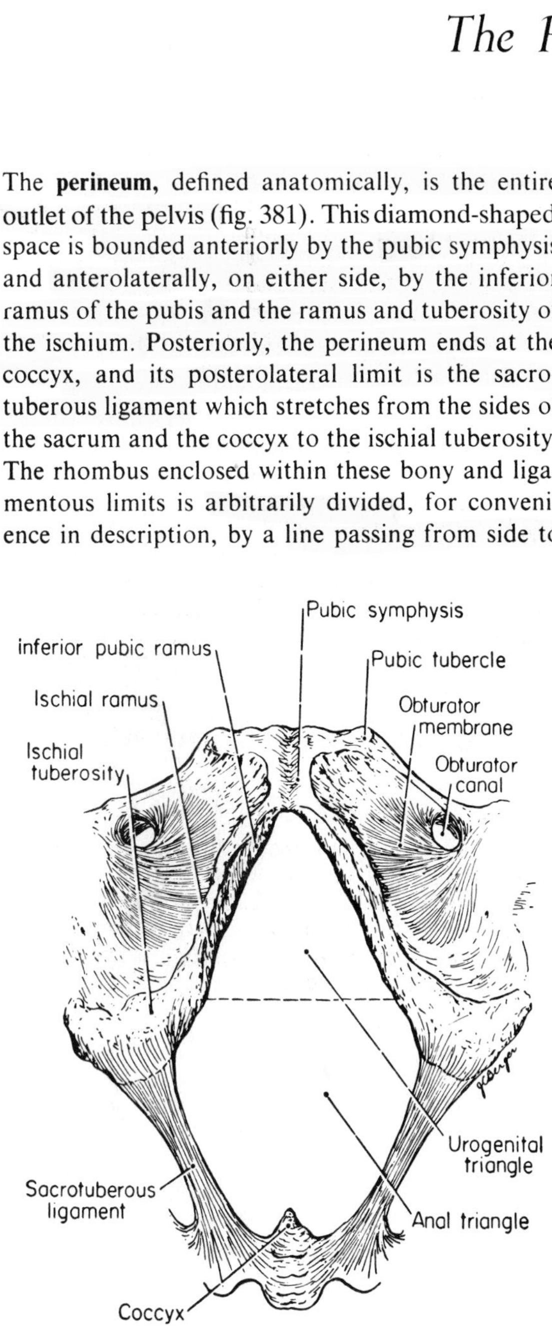

Aside from the cutaneous and subcutaneous structures, the urogenital triangle may be described in superficial and deep compartments. A membranous layer of subcutaneous connective tissue similar to and continuous with the membranous layer of the subcutaneous connective tissue of the abdomen attaches, on either side of the triangle, to the inferior ramus of the pubis and to the ramus of the ischium and, behind, to the posterior free margin of the perineal membrane. Between this layer and the membrane is the **superficial perineal space,** a compartment which lodges the roots of the penis (or clitoris in the female) and their associated muscles. Existing superior (deep) to the perineal membrane is the **deep perineal space** which, in the male, is occupied by the membranous urethra and the bulbourethral glands; in the female, by the urethra and the lower portion of the vagina. Associated with these structures are the sphincter muscles related to them and their blood vessels and nerves. It is convenient to describe the urogenital and the anal triangles separately.

Fig. 381 The boundaries and subdivisions of the perineum or outlet of the pelvis.

THE SUPERFICIAL STRUCTURES OF THE UROGENITAL TRIANGLE

It is an aid in understanding the perineum in the two sexes, to note that the fundamental structures have remained bilateral with a midline cleft in the female, but the bilateral primordia have fused into midline structures in the male (penis, scrotum). There is nevertheless representation of the same structures in both sexes.

The superficial structures of the urogenital tri-

angle are the penis and the scrotum in the male and the homologous external genital parts in the female.

THE PENIS (fig. 382)

The **penis** is composed of three cylindrical masses of cavernous tissue bound together by fibrous tissue and covered by skin. Two of the cavernous bodies, the **corpora cavernosa penis,** lie side by side on the dorsum of the penis, whereas the third lies ventrally in the median plane. This is the **corpus spongiosum penis.** Distally, the penis exhibits a conical extremity, the **glans penis.** It is formed by an expansion of the corpus spongiosum penis which fits over the blunt terminations of the corpora cavernosa penis. The prominent margin of the glans, the **corona,** projects backward beyond the ends of the corpora cavernosa penis. The corpus spongiosum penis is traversed by the urethra which opens near the summit of the glans in a slit-like opening, the **external urethral orifice.**

The **skin** of the body of the penis is thin, hairless, and dark; it is only loosely connected with the deeper parts of the organ. On the ventral, or urethral, surface the skin exhibits a median raphe which is continuous with the raphe of the scrotum. Along the base of the glans the skin forms a free fold, the **prepuce** (foreskin), which overlaps the glans to a variable extent. The skin of the inner surface of the prepuce is continuous with that covering the glans and resembles mucous membrane. The skin over the glans is firmly attached to the underlying erectile tissue. From the deep surface of the prepuce a small median fold, the **frenulum of the prepuce,** passes to a point immediately below the external urethral orifice. Numerous sebaceous glands are present in the skin, especially on the urethral surface of the penis. Circumcision, amputation of the prepuce, a traditional rite of certain religions, is also common as a simple hygienic measure.

The **subcutaneous connective tissue** of the penis is loose areolar tissue devoid of fat but reinforced by a prolongation of smooth muscle fibers from the tunica dartos scroti. In the dorsal midline of the penis the **superficial dorsal vein** passes toward the root of the penis, where it communicates with the superficial external pudendal veins of the upper thigh. The superficial external pudendal veins empty into the greater saphenous vein of each side. A **deep fascia** encloses the cavernous bodies and the main vessels and nerves of the penis. It is attached

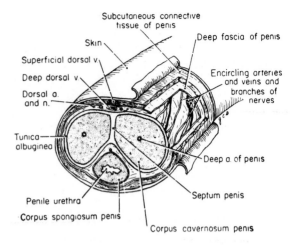

Fig. 382 The structures of the penis as seen in a cross section of its body.

distally in the groove between the shaft of the penis and the corona of the glans and extends backward over the roots of the cavernous bodies and their muscles in the superficial perineal space. The cavernous bodies, their muscles, and the vessels and nerves are more completely considered in connection with the superficial space of the perineum (p. 473). The **cutaneous vessels and nerves** of the penis are branches of those which lie along the dorsal midline of the penis under its deep fascia: the dorsal artery of the penis (p. 475); the deep dorsal vein (p. 476); and the dorsal nerve of the penis (p. 475). The **lymphatic drainage** of the penis is to the superficial inguinal lymph nodes.

THE SCROTUM

The **scrotum** is a cutaneous pouch, formed of bilateral contributions, which contains the testes and parts of the spermatic cords. It is developed from the skin of the abdominal wall in the region of the genital swellings. The bilateral formation of the scrotum is evident in a midline **scrotal raphe** which continues forward to the under surface of the penis and backward along the median line of the perineum to the anus. Internal to the raphe, the scrotal sac is divided into two chambers by the scrotal septum. The layers of the scrotum are the skin and the tunica dartos. These layers, together with the cutaneous nerves entering the scrotum from in front (anterior scrotal nerves), are described in the chapter on the Abdomen (p. 368).

The **posterior scrotal nerves** arise from the pudendal nerve; they are supplemented by perineal branches of the posterior femoral cutaneous nerve. The pudendal is the principal nerve of the perineum, supplying both muscular and sensory nerves in this area. The **pudendal nerve** is a branch of the sacral plexus and enters the perineum from the gluteal region. After supplying the skin and muscles of the anal triangle, it divides into the perineal nerve and the dorsal nerve of the penis (figs. 384, 386). The perineal nerve divides at the posterior border of the perineal membrane into posterior scrotal branches and a deep branch, which is mainly muscular to the muscles of the urogenital triangle but also supplies the bulb of the penis. The posterior scrotal nerves pass forward in the plane of the membranous layer of subcutaneous connective tissue providing branches to the skin and superficial tissues of the perineum and the scrotum. The lateral branch of the posterior scrotal nerve communicates with the perineal branch of the posterior femoral cutaneous nerve. The posterior scrotal nerves are accompanied by branches of the internal pudendal artery and vein. The large **posterior femoral cutaneous nerve,** also a branch of the sacral plexus, descends as a sensory nerve in the back of the thigh. At the lower border of the gluteus maximus muscle a **perineal branch** arises and runs medially across the origin of the hamstring muscles of the thigh toward the perineum. It enters the perineum by piercing the fascia lata along the ischiopubic ramus (fig. 386) and distributes laterally in the perineum to the skin and subcutaneous tissue of the scrotum and the base of the penis. Its branches communicate with the posterior scrotal nerves and the inferior rectal nerve of the pudendal system.

EXTERNAL GENITAL STRUCTURES IN THE FEMALE
(fig. 383)

The genital swellings of the inguinal region of the fetal abdominal wall do not fuse across the midline in the female as they do in the male in the formation of the scrotum. Consequently, there remains a midline cleft in the female bounded at the sides by the bilaterally represented labia minora and labia majora. Both the urethral and vaginal openings come to the surface in the vestibule defined by these folds. Also included in the female pudendum are the mons pubis, the clitoris, an erectile mass known as the bulb of the vestibule, and the greater vestibular glands.

The **labia majora** are two prominent longitudinal folds which bound the urogenital vestibule. They taper posteriorly where they approach the anus but blend into one another anteriorly to form a rounded fatty eminence, the **mons pubis.** The skin of the outer surface of each labium is pigmented and contains hairs. Their inner surfaces are smooth and are provided with large sebaceous glands. Within the labial fold is a considerable quantity of areolar tissue and fat. The fat is partially encapsulated in the form of a tapering cylinder which largely determines the form of the labium. The labia meet across the midline in front in the **anterior labial commissure.** Posteriorly, they are not actually joined, but a slight transverse fold in front of the anus can frequently be identified. This is the **posterior labial commissure.** The round ligament of the uterus emerges at the superficial inguinal ring of the abdomen and, entering the labium majus, ends in its areolar tissue, and is attached to its skin. Anterior and posterior labial nerves correspond in source and position to the anterior scrotal (p.369) and the posterior scrotal nerves (p. 369). The blood vessels of the labia are the external pudendal branches of the femoral artery and the labial branches of the internal pudendal

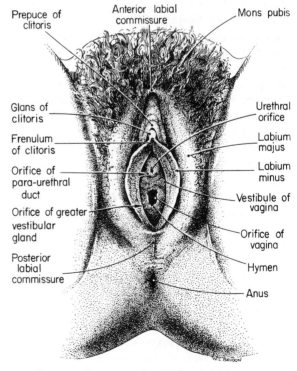

Fig. 383 The external genitalia of the female.

artery. Its lymphatic channels pass to the superficial inguinal lymph nodes.

The **labia minora** are two small cutaneous folds between the labia majora. Their hairless skin is smooth and moist like the medial surfaces of the labia majora, but, unlike the greater folds, there is no fat in the labia minora. Sebaceous glands open on their surfaces. The labia minora extend from the clitoris backward for about 4 cm. In the young female the posterior ends of the labia minora are connected by a small fold, the **frenulum pudendi.** In the region of the clitoris each labium divides—the upper division meets its fellow of the opposite side above the clitoris, forming the **prepuce of the clitoris;** the lower division passes beneath the clitoris and unites with its under surface. This and the same structure of the other side form the **frenulum of the clitoris.**

The **clitoris** is the morphological equivalent of the penis but does not contain the female urethra. Like the penis, the clitoris arises by two robust crura in the superficial space of the urogenital triangle. Its two **corpora cavernosa clitoridis,** divided by an incomplete septum, are enclosed in a dense fibrous membrane to constitute the body of the clitoris. The body is about 2.5 cm. in length and, like the root of the penis, is angulated toward the perineum. It is supported at its angle by a **suspensory ligament** which passes from the pubic symphysis to the clitoris. The free extremity of the clitoris (glans clitoridis) is a small rounded tubercle of erectile tissue. Like the glans penis, it is covered by a very sensitive epithelium. The **internal pudendal artery** provides branches to the clitoris—the deep branch to the crus and the dorsal artery of the clitoris to the body and glans. The **pudendal nerve** provides the dorsal nerve of the clitoris and also conducts sympathetic fibers to the cavernous tissue.

The **vestibule of the vagina** is the cleft between the labia minora. In its floor are the openings of the urethra, the vagina, and the ducts of the greater vestibular glands. The **external urethral orifice** lies from 2 to 3 cm. behind the glans clitoridis and immediately in front of the vagina. The opening usually appears as a vertical slit, the margins of which are prominent and in contact with one another. On either side of the urethral opening may be identified the minute openings of the **paraurethral glands.** The vaginal orifice is also a median slit; it is located below and behind the urethral opening. The **hymen** is a thin, membranous fold which represents the edges of the vaginal orifice. When stretched, the hymen makes a crescentic fold across the posterior border of the vaginal orifice, and after rupture of the hymen, the vaginal orifice is enlarged. Fragments of the hymen around the margins of the vaginal orifice may form small rounded elevations, the **hymeneal caruncles.** The **greater vestibular glands** lie in the superficial space of the perineum, but their openings appear posterolateral to the vaginal opening in the grooves between the hymeneal caruncles and the labia minora. Numerous minute openings of mucous glands (the lesser vestibular glands) are located between the urethral and vaginal orifices.

THE SUBCUTANEOUS CONNECTIVE TISSUE

The **subcutaneous connective tissue** of the penis is loose areolar tissue which is without fat and is continuous with the tunica dartos scroti. On the other hand, the subcutaneous connective tissue of the labia majora is heavily laden with fat which is partly encapsulated. In the urogenital and anal triangles of the perineum much fat is present in the subcutaneous connective tissue; in fact, in the anal triangle the fat and connective tissue of this layer is thickened into a mass which fills the ischiorectal fossae and constitutes a resilient pad of compressible material on either side of the anal canal. The **fatty layer** of the subcutaneous tissue of the perineum is also continuous with the subcutaneous fat of the thighs and, at the sides of the scrotum and the penis, with the superficial fatty layer of the abdomen.

As in the abdomen, there is in the urogenital triangle of the perineum a distinct **membranous layer** of subcutaneous connective tissue. The membranous layer is attached at the sides to the inferior pubic rami, to the ischial rami, and, posteriorly, to the posterior border of the perineal membrane. In the median plane the layer attaches to the median raphe of the bulbospongiosus muscle and to the septum of the scrotum. Anteriorly, the membranous and fatty layers blend in the formation of the tunica dartos scroti and the subcutaneous tissue of the penis. Like the fatty layer, the membranous layer of the perineum is continuous at the sides of the scrotum and the penis with the similar membranous layer in the abdomen. The membranous layer of the subcutaneous connective tissue of the urogenital triangle is the surface boundary of the superficial space of the perineum. The deep limit of this space is the perineal membrane.

Clinical Note Due to its posterior and lateral attachments, the membranous layer of the urogenital triangle forms a compartment which is open only anteriorly into the scrotum, the subcutaneous space of the penis, and the lower abdominal wall. Accidental rupture of the male urethra may lead to accumulations of urine in this compartment. Because of the connections of the limiting membranous plane, there is no passage of urine backward or into the thigh; but it may fill the scrotum and the subcutaneous space of the penis and rise into the abdominal wall behind its membranous layer (see also abdominoscrotal opening, p. 368). The attachment of the layer in the abdomen to the fascia lata just below the inguinal ligament prevents fluid from descending into the thighs, and it tends to accumulate in the flanks.

THE SUPERFICIAL SPACE OF THE PERINEUM

The superficial perineal space lies between the perineal membrane (inferior fascia of the urogenital diaphragm) and the membranous layer of the subcutaneous connective tissue. The space contains, in the male, the crura of the corpora cavernosa penis, the corpus spongiosum penis, and the bulb of the penis; the ischiocavernosus, bulbospongiosus, and superficial transverse perineal muscles; the deep fascia of the penis; and the vessels and nerves associated with these structures. Homologous structures are found in the female, their arrangement being somewhat modified by the presence of a midline cleft.

THE ERECTILE BODIES (fig. 384)

The three erectile bodies of the penis are described and their arrangement noted on p. 470. Erectile tissue is composed of a meshwork of interlacing and intercommunicating spaces lined by an endothelium directly continuous with that of the veins which drain the meshwork. Each corpus cavernosum penis is enclosed in a dense, white fibrous coat (**tunica albuginea**) which, where the bodies are in contact, fuses with that of the other side to form the median **septum of the penis.** This septum is incomplete, and the spaces of the two cavernous bodies communicate. The structure of the corpus spongiosum penis resembles that of the corpora cavernosa, but the fibrous tunica albuginea is thinner and more elastic and the cavernous meshwork is finer. The glans penis is also composed of erectile tissue but does not have a strong tunica albuginea. The skin of the glans is thin and firmly adherent to the erectile tissue beneath. The urethra is expanded vertically within the glans as the **fossa navicularis.** The **deep fascia of the penis** surrounds loosely the three cavernous bodies of the penis and is continued backward into the perineum as the fascia of the muscles related to them. It ends distally at the junction of the body and the glans of the penis, and, dorsally, it encloses the dorsal artery and nerve of the penis and the deep dorsal vein.

At the root of the penis, the two corpora cavernosa penis diverge as the **crura of the penis.** Slightly swollen at first, they taper posteriorly and end just in front of the ischial tuberosities. Each crus is firmly attached to the periosteum of the medial side of the ischiopubic ramus, and each is covered by an ischiocavernous muscle. The corpus spongiosum penis continues backward in the median line but becomes enlarged posteriorly and thus forms the **bulb of the penis.** The bulb is firmly attached to the perineal membrane through which the urethra passes to enter the bulb and traverse the corpus spongiosum. The bulb of the penis is covered by the bulbospongiosus muscle.

The root of the penis is firmly held to the underside of the pubic arch by the **suspensory ligament of the penis.** This median, triangular band of strong fibrous tissue extends from the symphysis pubis and the arcuate pubic ligament to the deep fascia of the penis. A more superficial **fundiform ligament** (p. 368) is formed of longitudinally arranged strands of the membranous layer of the subcutaneous connective tissue of the abdomen, reinforced by fibers of the external oblique aponeurosis and the linea alba. Its fibers pass down on either side of the penis to enter the septum of the scrotum.

THE MUSCLES (fig. 386)

The Bulbospongiosus Muscle

This muscle, in the male, overlies the bulb of the penis and is formed of two symmetrical halves united along the midline by a fibrous raphe. Its muscle fibers arise from the central tendinous point of the perineum and from the median raphe anterior to it. The fibers diverge from the median raphe forward and lateralward. The most posterior fibers cover the end of the bulb of the penis and terminate in the perineal membrane; the middle fibers encircle the forepart of the bulb and the corpus spongiosum and end in a strong aponeurosis on the dorsum of the corpus spongiosum. The most anterior fibers of the muscle are longer and more divergent. They surround all three erectile bodies at the root of the penis and insert into the deep fascia on the dorsum of the penis. The **central tendinous point of the perineum,** more commonly referred to as the **perineal body** in the female, is a region of inter-digitation of the fibers of the bulbospongiosus muscle in front, the sphincter ani externus muscle behind, and the two superficial transverse perineal muscles at the sides. It is located at the center of the perineum just in front of the anus and is firmly attached to the posterior border of the perineal membrane by the fascia and tendons of the super-ficial transverse perineal muscles. The bulbospong-iosus muscle, exerting forward compression, expels the last of the urine in the urethra. It also serves as an accessory muscle of erection for the penis by compressing the bulb and impeding the venous re-turn; it expels the semen in ejaculation. In the female the muscle compresses the bulb of the vestibule and constricts the vaginal orifice. The bulbospongiosus muscle is innervated by the pudendal nerve.

The Ischiocavernosus Muscles

The ischiocavernosus muscles cover the crura of the penis (or clitoris). Each muscle arises from the medial side of the ischial tuberosity and from the ischiopubic ramus on either side of the crus. Its fibers pass forward over the crus and insert into the sides and under surface of the crus and into the corpus cavernosum. The ischiocavernosus muscles compress the crura, impede the venous return, and maintain erection of the penis (or clitoris). The muscles are supplied by the pudendal nerve.

The Superficial Transverse Perineus Muscles

These muscles arise on either side from the an-terior and medial portions of the ischial tuberosity. They pass transversely inward to end in the central tendinous point of the perineum. To the posterior aspect of the muscles are added fibers from the deep portion of the external sphincter ani muscle. The slender superficial transverse perineus muscles lie in the posterior border of the superficial space of the perineum, the membranous layer of subcutaneous connective tissue curving over them to attach to the edge of the perineal membrane. Their contraction serves to fix the central tendinous point, assisting, by this stabilization, the action of the other muscles.

They are supplied by the pudendal nerve. These muscles may be missing or very weak.

The **deep fascia of the penis** continues from the penis both superficial and deep to the muscles of the superficial space of the perineum and constitutes an investing fascia for these muscles. Superiorly, it attaches to the suspensory ligament of the penis and to the perineal membrane.

Clinical Note An extravasation of urine can be limited by an intact deep fascia, Buck's fascia clinically. Such a collection of fluid will form a spindle-shaped enlargement of the penis as far distally as the junction of its glans and body and proximally will bulge the fascia over the muscles of the superficial perineal space to their limits.

Nerves and Vessels of the Perineum (figs. 386, 403, 431)

The Pudendal Nerve

The pudendal nerve is almost the sole somatic nerve of the perineum. It supplies all the perineal musculature and is sensory to most of its skin sur-faces.

The pudendal nerve arises in the pelvis by roots from sacral nerves two, three, and four. It passes through the greater sciatic foramen into the gluteal region, where it lies medial and inferior to the sciatic nerve. It then crosses the spine of the ischium medial to the internal pudendal artery and vein. Entering the lesser sciatic foramen between the sacrotuberous and sacrospinous ligaments and accompanied by the internal pudendal vessels and the nerve to the internal obturator muscle, the pudendal nerve passes into the **pudendal canal.** This is a fascial canal formed by a split in the obturator internus fascia on the lateral wall of the ischiorectal fossa. The canal crosses the obturator internus muscle about 4 cm. above the lower margin of the ischial tuberosity and conducts the pudendal nerve and the internal pudendal vessels along the margin of the ischio-rectal fossa to the urogenital triangle of the perineum.

The pudendal nerve has three branches. The **in-ferior rectal nerve** arises from the pudendal nerve early in its course in the pudendal canal and, leaving the canal with the inferior rectal branches of the internal pudendal vessels, crosses the ischiorectal fossa toward the anus. It supplies the external sphincter ani muscle and the skin around the anus and communicates with the posterior scrotal and perineal nerves.

The **perineal nerve** arises from the pudendal nerve within the pudendal canal, and leaves it toward the posterior border of the perineal membrane. Its posterior scrotal (or posterior labial) branches become superficial and distribute to the skin and fascia of the perineum and the scrotum (or labium majus). They communicate with the inferior rectal nerve and with the perineal branch of the posterior femoral cutaneous nerve. The deep branch of the perineal nerve is mainly muscular, supplying branches to the superficial transverse perineus, the bulbospongiosus, the ischiocavernosus, the deep transverse perineus, and the sphincter urethrae muscles. The nerve to the bulb is given off from the muscular nerve to the bulbospongiosus muscle. It pierces this muscle to supply the bulb of the penis, the corpus spongiosum, and the mucous membrane of the urethra as far as the glans penis.

The remainder of the pudendal nerve is the dorsal nerve of the penis (or clitoris). Arriving at the posterior border of the perineal membrane with the internal pudendal vessels, this nerve proceeds forward deep to the perineal membrane (fig. 384). It sends a slender branch through the fascia to supply the erectile tissue of the crus and corpus cavernosum of the penis (or clitoris) and then pierces the perineal membrane near its anterior border. Continuing forward in company with the dorsal artery of the penis, the nerve lies on the dorsum of the penis (or clitoris). It distributes to its distal two-thirds, sending branches around the sides to reach the under surface of the organ (fig. 382). This nerve is much smaller in the female than it is in the male.

The Internal Pudendal Artery

This is one of the terminal branches of the internal iliac artery which has its origin within the pelvis, usually in a common trunk with the inferior gluteal artery. The internal pudendal artery gives rise to small muscular branches to the muscles of the pelvic floor and then leaves the pelvis through the greater sciatic foramen between the piriformis and coccygeus muscles. It provides small and inconstant muscular branches in the gluteal region and then enters the perineum, crossing the ischial spine with the pudendal nerve to its medial side and the nerve to the obturator internus to its lateral side. Accompanied by these structures and by its vein, the internal pudendal artery traverses the pudendal canal (fig. 403), distributing to the same areas and having essentially the same branches as the pudendal nerve. The inferior rectal branch

pierces the wall of the pudendal canal and breaks up into several branches which accompany the branches of the inferior rectal nerve across the ischiorectal fossa to the muscles and skin of the anal region and to the region of the lower edge of the gluteus maximus muscle. This artery anastomoses with the superior rectal, middle rectal, and perineal arteries, and the vessel of the opposite side. The perineal artery, like the perineal nerve, arises toward the end of the pudendal canal and turns forward over the superficial transverse perineus muscle into the interval between the bulbospongiosus and ischiocavernosus muscles. It supplies these three muscles and gives off a posterior scrotal branch to the skin and tunica dartos of the scrotum and also a transverse perineal branch. The transverse perineal branch follows the superficial transverse perineus muscle medialward and anastomoses with the vessel of the opposite side and with the perineal and inferior rectal arteries.

The internal pudendal artery passes above the perineal membrane and immediately gives off the artery of the bulb of the penis (or artery of the bulb of the vestibule in the female). This large vessel passes medialward through the posterior edge of the sphincter urethrae muscle, pierces the perineal membrane, and enters the bulb of the penis (or bulb of the vestibule). It supplies the bulb, the bulbourethral gland, and the posterior part of the corpus spongiosum penis. The urethral artery arises a short distance anterior to the artery of the bulb; it also passes medialward and pierces the perineal membrane to enter the corpus spongiosum penis. It follows forward in the corpus spongiosum to the glans penis. The terminal branches of the internal pudendal artery are the deep artery of the penis (or clitoris) and the dorsal artery of the penis (or clitoris). The deep artery of the penis (or clitoris) is usually the larger. It pierces the perineal membrane and, entering the corpus cavernosum penis (or corpus cavernosum clitoridis), runs forward in its center. The dorsal artery of the penis (or clitoris) pierces the anterior end of the perineal membrane and runs forward between the two layers of the suspensory ligament of the penis. It follows along the dorsum of the penis (or clitoris) to the glans, where it divides into two branches to the glans and prepuce. On the penis it has the dorsal nerve to its lateral side and the deep dorsal vein to its medial side. It supplies the superficial tissues of the penis (or clitoris) and sends branches into the corpus cavernosum penis (or corpus cavernosum clitoridis).

Veins of the Perineum

The veins of the perineum correspond with the branches of the internal pudendal artery. The **internal pudendal vein** begins in the deep vein of the penis (or clitoris) and has other tributaries which are venae comitantes of the artery. The internal pudendal vein is a tributary of the internal iliac vein. The veins of the penis (or clitoris) end to only a minor degree in the internal pudendal system. The **superficial dorsal vein of the penis** runs backward immediately under the skin to the pubic symphysis, where it divides into right and left branches which end in the superficial external pudendal veins (p. 370). The **deep dorsal vein of the penis** lies in the midline under the deep fascia of the penis. It begins behind the glans, receiving numerous tributaries from the glans and the corpora cavernosa penis. The vein runs proximalward in the dorsal sulcus between the corpora cavernosa penis from which it receives many tributaries. At the root of the penis it passes between the two layers of the suspensory ligament and then ascends between the arcuate pubic ligament and the transverse ligament of the perineum. The deep dorsal vein ends by dividing into two branches which join the prostatic plexus within the pelvis (p. 513). At the root of the penis, it communicates with the internal pudendal vein. The corresponding **deep dorsal vein of the clitoris** is small but is otherwise similar to that of the penis. It ends in the lower part of the vesical plexus of veins within the pelvis.

THE DEEP SPACE OF THE PERINEUM

The deep space of the perineum lies deep (superior) to the perineal membrane. In the male (fig. 384) the space contains the sphincter urethrae and deep transverse perineus muscles, the membranous portion of the urethra, and the bulbourethral glands. In the female the muscular layer is traversed by both the urethra and the vagina. The greater vestibular glands of the female, comparable to the bulbourethral glands of the male, occupy the superficial space of the perineum. The pudendal nerve and the internal pudendal vessels fill the lateral limits of the space.

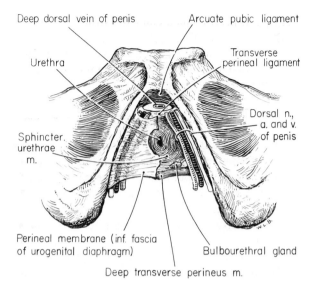

Fig. 384 The muscles, vessels, and nerves of the deep space of the perineum. On the right side, the perineal membrane has been removed; on the left side, it is intact (based on T. M. Oelrich '80).

THE PERINEAL MEMBRANE (figs. 384, 387, 395)

The current concept of a urogenital diaphragm consisting of superior and inferior fascial layers enclosing the sphincter urethrae and deep transverse perineus muscles is erroneous. The strong **perineal membrane** (inferior fascia) is the final support of the pelvic viscera. The sphincter urethrae muscle does not constitute a flat plane stretched between the ischiopubic rami (see its description), and the only superior fascia is the intrinsic fascia of the sphincter urethrae muscle (T. M. Oelrich). The **perineal membrane,** triangular in form, fills the interval between the two ischiopubic rami. It is attached laterally to the medial borders of the ischiopubic rami as far back as the ischial tuberosities. It contributes to and is attached to the central tendinous point of the perineum. At its anterior limit the perineal membrane forms a fibrous band which stretches across the subpubic angle just behind the deep dorsal vein of the penis; this is the **transverse ligament of the perineum.** This ligament is separated from the arcuate pubic ligament by a small interval occupied by the deep dorsal vein. The perineal membrane is perforated by the urethra in the male and by both the urethra and the vagina in the female. It blends with their walls along the line of perforation. It is also perforated by the ducts of the bulbourethral glands and by the arteries, veins, and nerves which enter the structures of the superficial space of the perineum from above. The bulb of the penis is attached to the under surface of the perineal membrane, and the bulbospongiosus muscle fibers insert into it.

Muscles of the Deep Space (figs. 384, 385, 387)

In both sexes the posterior division of this layer is the deep transverse perineus muscle. In front of it, in the male, is the sphincter urethrae muscle; in the female this portion of the muscle layer invests both the urethra and the vagina.

In both sexes the **deep transverse perineus muscle** arises from the ischial ramus and passes medialward to interlace in a tendinous raphe with its fellow of the opposite side. In this interlacement, fibers are received from the deep portion of the external sphincter ani muscle and from the puborectal portion of the pelvic diaphragm. In the female the deep transverse perineus muscle partly passes behind the vagina and partly blends with its wall. The **sphincter urethrae muscle** of the male forms a circular investment (a true sphincter) for the urethra inferior to the prostate gland. It is a striated muscle but its fibers are 25% smaller than usual in striated muscles. It also extends vertically

to the base of the bladder, investing the prostate gland anteriorly and anterolaterally (especially superiorly). The muscular primordium is laid down around the whole length of the urethra prior to the development of the prostate gland, and as development of the prostate from urethral glands proceeds, there is displacement and some atrophy later of fascicles of the urethral sphincter. No muscle remains posteriorly or posterolaterally on the prostate in the adult. (Oelrich '64, '80, figs. 384, 385, 390, 395). The sphincter urethrae muscle, in the female as in the male, provides an annular sphincter for the urethra, and its more superior portion rises to the base of the bladder. At perineal levels the muscle has a thin extension to the ischial ramus of each side (compressor urethrae m.) and a discrete band which encircles both the vagina and the urethra (fig. 387—urethrovaginal sphincter—Oelrich, in press). The muscles of the deep space of the perineum act as sphincters, compressing the urethra in the male and both the urethra and the vagina in the female. The sphincter urethrae muscle is the essential muscle in voluntary control of micturition. The muscles of the deep space of the perineum are innervated by the perineal branch of the pudendal nerve (p. 474).

The Bulbourethral Glands

The pea-size bulbourethral glands of the male are embedded in the fibers of the sphincter urethrae muscle posterolateral to the membranous urethra and overlie the bulb of the penis. The ducts of these glands pass downward through the perineal membrane into the bulb. After a course of about 2.5 cm., they open into the lumen of the urethra by minute apertures.

THE SPECIAL ANATOMY OF THE UROGENITAL TRIANGLE IN THE FEMALE

Embryologically, the external genital organs in the two sexes develop along relatively parallel lines, and there are many homologies between the parts. The labia majora are the homologues of the scrotum, and the labia minora represent skin and subcutaneous tissue at the side of the original perineal urethra. The penis has its homologue in the clitoris of the female.

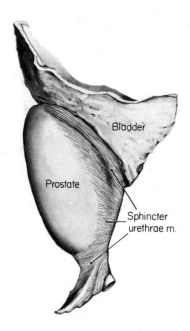

Fig. 385 The prostate gland and the sphincter urethrae muscle in a 21-year-old male. (From the original of figure 32, T.M. Oelrich, *Am. J. Anat.,* 158:224, 1980.)

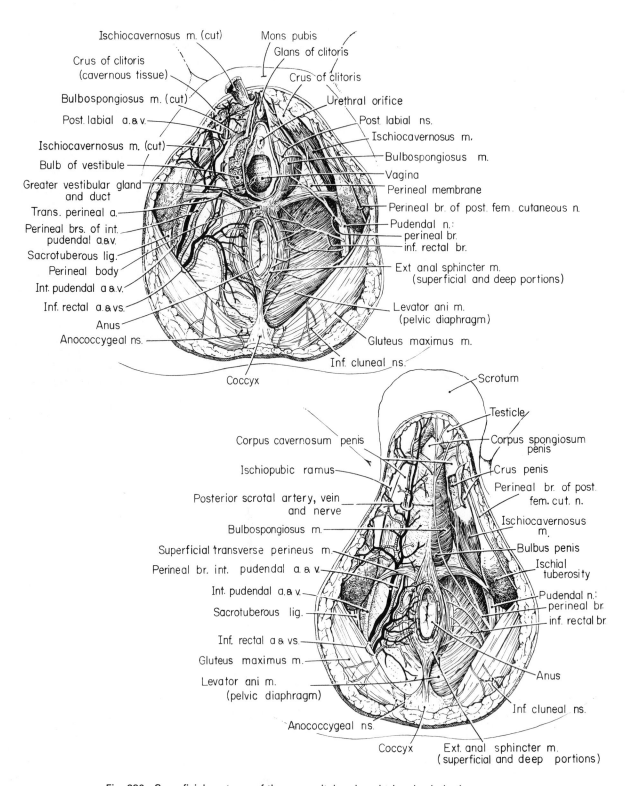

Ischiocavernosus m. (cut)
Crus of clitoris (cavernous tissue)
Bulbospongiosus m. (cut)
Post. labial a. & v.
Ischiocavernosus m. (cut)
Bulb of vestibule
Greater vestibular gland and duct
Trans. perineal a.
Perineal brs. of int. pudendal a. & v.
Sacrotuberous lig.
Perineal body
Int. pudendal a. & v.
Inf. rectal a. & vs.
Anus
Anococcygeal ns.

Mons pubis
Glans of clitoris
Crus of clitoris
Urethral orifice
Post. labial ns.
Ischiocavernosus m.
Bulbospongiosus m.
Vagina
Perineal membrane
Perineal br. of post. fem. cutaneous n.
Pudendal n.:
perineal br.
inf. rectal br.
Ext anal sphincter m. (superficial and deep portions)
Levator ani m. (pelvic diaphragm)
Gluteus maximus m.
Inf. cluneal ns.
Coccyx

Corpus cavernosum penis
Ischiopubic ramus
Posterior scrotal artery, vein and nerve
Bulbospongiosus m.
Superficial transverse perineus m.
Perineal br. int. pudendal a. & v.
Int. pudendal a. & v.
Sacrotuberous lig.
Inf. rectal a. & vs.
Gluteus maximus m.
Levator ani m. (pelvic diaphragm)
Anococcygeal ns.
Coccyx

Scrotum
Testicle
Corpus spongiosum penis
Crus penis
Perineal br. of post. fem. cut. n.
Ischiocavernosus m.
Bulbus penis
Ischial tuberosity
Pudendal n.:
perineal br.
inf. rectal br.
Anus
Inf. cluneal ns.
Ext. anal sphincter m. (superficial and deep portions)

Fig. 386 Superficial anatomy of the urogenital and anal triangles in both sexes.

478

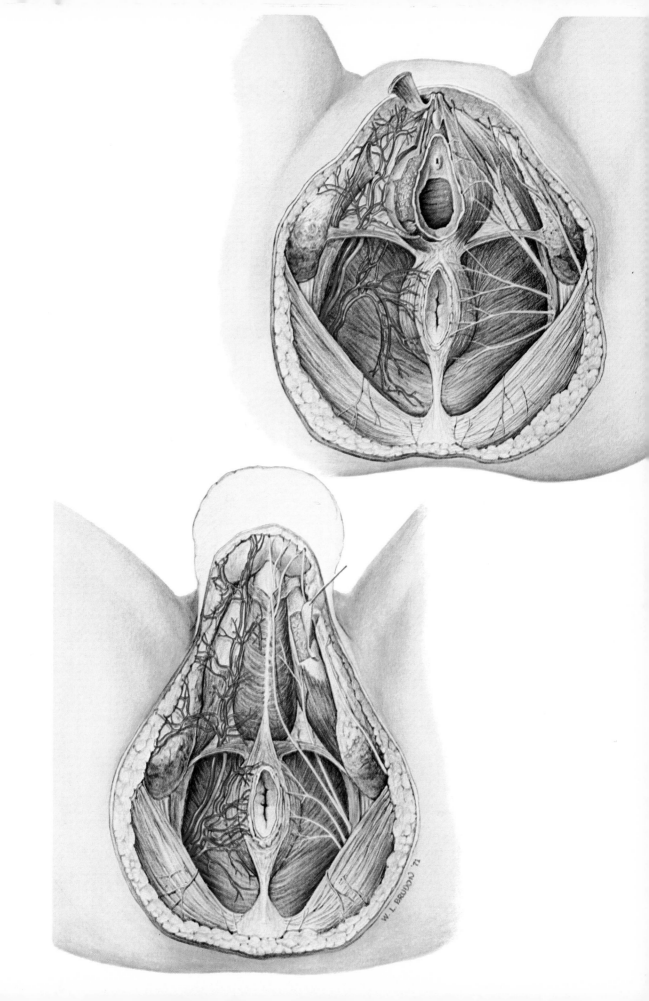

The Clitoris (fig. 386)

The clitoris is attached to the ischiopubic rami by crura which are no different from and only slightly smaller than those of the penis. The crura of the clitoris are invested by paired **ischiocavernosus muscles** equivalent to those of the penis. Significant differences in the median line are related to the persisting vestibular cleft in the female and to closure of the perineum in the male. The bilateral halves of the bulbospongiosus muscle of the male are united in a midline raphe, but in the female they are separate. The single fused bulb of the penis is represented in the separated bulbs of the vestibule.

Bulbs of the Vestibule (fig. 386)

These are the homologues of the bulb of the penis and the adjoining part of the corpus spongiosum. They are elongated masses of erectile tissue, thicker posteriorly and hollowed medially where they fit against the wall of the vagina. Thus, each has somewhat the shape of a half-pear and is about 2.5 cm. long. The tapered anterior ends of the bulbs are united to one another in front of the urethra by a narrow band, or commissure, which is, in turn, united to the glans clitoridis by a slender band of erectile tissue. The bulbs of the vestibule are invested by the two bulbospongiosus muscles and lie deep to the line of attachment of the labia minora.

The Bulbospongiosus Muscles

In the female the bulbospongiosus muscles are thin and overlie the bulbs of the vestibule. Arising posteriorly from the perineal body, the fibers of each muscle conform to the curvature of the outer surface of the bulb and insert into the perineal membrane and into the corpus cavernosum clitoridis. A few fascicles of the muscle pass to the dorsum of the clitoris. The muscle compresses the bulb of the vestibule and acts as a sphincter of the vaginal orifice. The **superficial transverse perineus muscle** of the female is not different from that of the male.

The Greater Vestibular Glands

The greater vestibular glands are the homologues of the bulbourethral glands of the male, but they lie in the superficial space of the perineum.

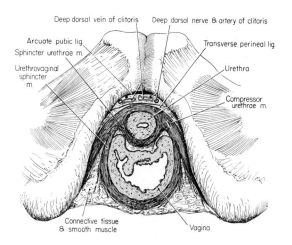

Fig. 387 The sphincter urethrae muscle in the female (based on unpublished research of T. M. Oelrich).

They are situated, one on either side of the vagina, in contact with and partly under the posterior end of each vestibular bulb. The greater vestibular glands are somewhat larger than pea-size, and their ducts, about 2 cm. long, open in the grooves between the hymen and the labia minora.

The Perineal Membrane

The female perineal membrane is somewhat broader than that of the male due to the greater spread of the ischiopubic rami. It is perforated by both the urethra and the vagina, and the muscle layer of the deep space is, therefore, broken into smaller parts with less evident continuity at the midline. In the male the perineal membrane is rather tightly stretched across the urogenital triangle, whereas in the female it is broader, somewhat more lax, and projects toward the surface arounds its apertures. Thus, the bulbospongiosus muscles and the bulbs of the vestibule lie on a more inclined plane than do their homologues in the male. The muscles of the deep perineal space in the female are described on p. 477. The blood vessels and nerves of the female perineum, branches of the pudendal nerve and internal pudendal arteries, correspond with those of the male perineum. They are described on p. 475.

INNERVATION OF THE EXTERNAL GENITALIA IN BOTH SEXES

The pudendal nerve is the somatic nerve of the perineum. It has a sensory distribution to all the skin and epithelial areas of the external genitalia, and its afferent impulses are thus important in sexual excitation. Sympathetic and parasympathetic fibers pass to the structures of the perineum from the prostatic plexus in the male. The nerve fibers concerned are derived from the upper three lumbar segments and from the pelvic splanchnic nerves (sacral segments two to four) respectively. Terminal fibers communicate with the pudendal nerves, and offshoots from both sets of nerves innervate the penile vessels, the erectile tissue of the corpora cavernosa and corpus spongiosum penis, the membranous and penile portions of the urethra, and the bulbourethral glands. Both the sympathetic and parasympathetic nerve mechanisms are necessary to sexual activity. Erection depends on stimuli through the parasympathetic nerves, leading to engorgement of the corpora cavernosa by dilatation of their arteries or, possibly, by relaxation of their vascular tone. Contraction of the bulbospongiosus and ischiocavernosus muscles helps maintain erection by interference with the venous return of blood from the cavernous tissues. Stimulation of sympathetic nerves leads to constriction of the arteries and subsidence of erection. Ejaculation consists of two phases, emission and ejaculation proper. Emission, or delivery of semen to the membranous urethra, follows reflex peristalsis in the ductus deferentes and seminal vesicles and contraction of the smooth muscle of the prostate gland. This is a sympathetic response. Ejaculation, or expulsion of the seminal fluid, follows parasympathetic stimulation and is accompanied by clonic spasm of the bulbospongiosus and ischiocavernosus muscles. In the female the comparable parasympathetic stimulation leads to increased vaginal secretion, erection of the clitoris, and engorgement of the erectile tissue of the vestibule. Afferent fibers from the viscera accompany both sympathetic and parasympathetic fibers.

In the female the innervation of the genital organs is mediated through the vesical plexuses and the vaginal nerves. Filaments follow the vaginal arteries; communications are made with pudendal nerve branches; and terminals are distributed to the vaginal wall, the urethra, the erectile tissue of the vestibular bulbs, the greater vestibular glands, and the erectile tissue of the clitoris.

THE ANAL TRIANGLE

The **anal triangle** of the perineum (fig. 381) is limited behind by the tip of the coccyx and in front by the posterior border of the urogenital diaphragm (line between the anterior parts of the ischial tuberosities). The sacrotuberous ligaments form the posterolateral limits of the triangle. Overriding the

ligaments, the gluteus maximus muscles encroach on the triangle. The anal triangle contains the aperture of the anus, the external sphincter ani muscle, and the ischiorectal fossae. The skin around the anal orifice is pigmented and thrown into radiating folds by the underlying involuntary musculature (corrugator cutis ani). The skin of the anus is provided with hair follicles, and with large sebaceous and sweat glands.

THE CUTANEOUS NERVES

The **cutaneous nerves** of the anal triangle are the inferior rectal nerve, the perineal branch of the fourth sacral nerve, and the anococcygeal nerves. The **inferior rectal nerve** is described on p. 474. It supplies the skin around the anus as far lateralward as the ischial tuberosity. The **perineal branch of the fourth sacral nerve** arises from the lower portion of the loop between the third and fourth sacral nerves which provides the nerves to the coccygeus and levator ani muscles. The perineal branch perforates the coccygeus muscle and appears in the posterior angle of the ischiorectal fossa. It innervates the posterior part of the external sphincter ani muscle and supplies the skin between the coccyx and the anus. The **anococcygeal nerves** arise from the coccygeal nerve with communications from the fourth and fifth sacral nerves. The anococcygeal nerves constitute a few delicate filaments which pierce the sacrotuberous ligament and supply the skin in the region of the coccyx and behind the anus.

THE EXTERNAL SPHINCTER ANI MUSCLE (figs. 386, 388, 392, 403)

The **external sphincter ani muscle** constitutes the large voluntary sphincter of the anal canal and aperture. Forming a broad band on either side of the canal, the muscle consists of a superficial and a deep portion. The **superficial portion** is distinctly fusiform, ending posteriorly in the subcutaneous tissue over the tip of the coccyx and inserting into its dorsum; anteriorly, it ends in the subcutaneous tissue over the bulbospongiosus muscle. This portion of the external sphincter is characteristically infiltrated by the spreading terminal fibers of the longitudinal muscular layer of the anal canal, which end in the subcutaneous tissue and skin around the anus. The superficial portion of the external sphincter caps the distal end of the internal sphincter ani muscle and, there, lies directly under the submucosal layer of the anal verge. A fascicle or two of circular fibers cross posterior to the anal canal. These fascicles constitute what have sometimes been designated as the subcutaneous portion of the external anal sphincter. There are seldom any circular fascicles at the anterior border of the canal.

The **deep portion** of the external sphincter ani muscle is generally circular in disposition. It encircles the anal canal both anteriorly and posteriorly, although a few fibers join the fusiform superficial portion. The deep portion is also intermingled with the inferior fibers of the puborectalis portion of the pelvic diaphragm, and the puborectalis fibers swing posterior to the anal canal with the external anal sphincter. When the superficial transverse perineus muscle is present, fibers are contributed to it from the deep portion of the external sphincter and others contribute to the deep transverse perineus muscle. Some of its fibers also decussate anteriorly into the bulbospongiosus muscle.

The external sphincter ani muscle is in a continuous state of tonic contraction; having no antagonist,

it keeps the anal canal and orifice closed. It can, of course, be relaxed voluntarily to allow defecation to take place, and it can be more tightly contracted under the action of the will. It helps to form the central tendinous point of the perineum, giving stability for the action of the bulbospongiosus muscle. It is also a support for the pelvic floor. The external sphincter ani muscle overlaps the internal sphincter ani; it is immediately inferior to the pelvic diaphragm and blends with inferior fibers of its puborectal portion. The external anal sphincter is supplied by the inferior rectal nerves and blood vessels (p. 474). In addition, a twig from the perineal branch of the fourth sacral nerve reaches its posterior portion.

THE ISCHIORECTAL FOSSAE (figs. 386, 388, 403)

The **ischiorectal fossae** are spaces at either side of the anal canal which, containing only fat and areolar tissue, accommodate to the enlargement of the canal during the passage of feces. In a frontal section of the perineum these spaces are wedge-shaped, with their bases at the skin surface and their apices deep under the pelvic diaphragm. The lateral wall of the wedge-shaped fossa is the fascia of the extrapelvic part of the obturator internus muscle and the tuberosity of the ischium; its sloping medial wall is the inferior fascia of the pelvic diaphragm. More superficially, the external sphincter ani muscle is also a medial boundary. The sloping medial wall follows the inferior surface of the pelvic diaphragm forward superior to the perineal membrane, and thus a small **anterior recess** of the space is prolonged forward. Lateroposteriorly, the gluteus maximus muscle overhangs the space and creates a **posterior recess** (fig. 403). The ischiorectal fossae of the two sides communicate over the anococcygeal raphe. The adipose tissue of each fossa is traversed by fibrous bands and septa and the space is crossed by the inferior rectal nerves and vessels on their way to the anal region. Posteriorly, the perineal branch of the fourth sacral nerve emerges through the ischiorectal fossa; anteriorly, the perineal and posterior scrotal (or posterior labial) branches of the pudendal nerve come to the surface through it. The pudendal canal for the pudendal vessels and nerve is formed by a split in the obturator internus fascia in the lateral wall of the space. This canal lies about 4 cm. superior to the lower border of the ischial tuberosity.

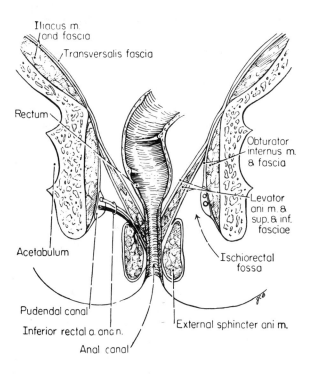

Fig. 388 A schematic frontal section of the rectum, pelvic diaphragm, and anal canal.

Clinical Note   The ischiorectal fossa is occasionally the site of an abscess, in certain cases arising in the anal canal. An abscess of one side can extend to the other fossa by way of the communication over the anococcygeal raphe.

THE LYMPHATICS OF THE PERINEUM

The lymphatics of the perineum and the external genital structures consist of subcutaneous lymphatic networks and lymphatic channels, all of which drain to the nodes of the **superficial inguinal group** which lie on the medial side of the greater saphenous vein (p. 527).

In the Male

The scrotum has a very abundant cutaneous lymphatic plexus; it is especially well developed in the region of the raphe where numerous connections are made across the midline. Collecting vessels pass lateralward and forward and join channels from the skin of the buttocks and the anal triangle which run forward in the genitofemoral sulcus. These vessels proceed to the more medial superficial inguinal nodes. The cutaneous lymphatics of the penis converge onto the dorsum of the penis, uniting there with channels from the glans and the prepuce. They accompany the deep dorsal vein to the root of the penis and then turn left and right to reach the superficial inguinal lymphatic nodes of either side. There are numerous anastomoses across the midline. The lymphatic vessels of the erectile bodies of the penis also follow the deep dorsal vein to the root of the penis and there turn lateralward to end in the superficial inguinal nodes. The lymphatic vessels of the testis, the epididymis, and the ductus deferens take routes entirely different from the drainage paths of the cutaneous and cavernous portions of the penis and of the scrotum (p. 495).

In the Female

The lymphatics of the labia, the vestibule, and the clitoris correspond in arrangement and drainage to those of the homologous parts in the male. The mucocutaneous covering of the vulva (vestibule, labia minora, and medial surfaces of labia majora) contains an exceedingly copious and intricate network of lymphatic channels. These channels lead lateralward and forward through the coarser meshwork of the lateral surfaces of the labia majora and the mons pubis to the superficial inguinal nodes. They are accompanied in the genitofemoral sulcus by coarse lymphatic channels from the anal and buttock regions. The lymphatics of the clitoris correspond to those of the penis. The drainage of the greater vestibular glands is also to the superficial inguinal lymph nodes.

VIII

The Pelvis

The **pelvis** (basin) is the reduced inferior portion of the abdomen. Its cavity is directly continuous with the abdomen although angulated backward from it. The pelvis is also the region of transition from the trunk to the lower limbs, and the bony pelvis forms a pelvic girdle for the attachment of the limbs. Thus, the walls of the pelvis are bony (os coxae, sacrum, and coccyx), ligamentous, and muscular. A greater and a lesser pelvis can be recognized, the greater pelvis being bounded laterally by the alae of the ilia and represented by the iliac fossae. The lesser pelvis is the true pelvis. It is bounded above by the terminal line of the pelvis. This consists, bilaterally, of the pecten of the pubis, the arcuate line of the ilium, and the sacral promontory. The pelvis is closed inferiorly by the pelvic diaphragm, reinforced below by the perineal membrane. It contains the urinary bladder, the rectum, the internal genital organs, and blood vessels and nerves. The pelvis also lodges the most dependent coils of the small intestine.

THE PERITONEUM IN THE PELVIS

The peritoneal cavity (figs. 388, 389, 396, 400) falls short of the inferior limit of the pelvic cavity; thus the viscera of the pelvis are only partially covered by peritoneum and lie largely below it. Under the reflections of the peritoneum, the viscera are embedded in rather abundant extraperitoneal connective tissue, the **endopelvic fascia.** From the anterior abdominal wall the peritoneum reflects onto the superior surface of the bladder and descends a short distance over the posterior curvature of its fundus. When distended, the bladder bulges the peritoneum into the lower abdomen; but when the bladder is collapsed, the peritoneum sweeps smoothly from the abdominal wall onto the bladder.

When the bladder is distended, shallow peritoneal pouches appear at its sides, the **paravesical fossae.** In the male the peritoneum posterior to the bladder curves onto the rectum with an occasional intermediate elevation over the superior extremities of the seminal vesicles. The **rectovesical pouch** is the hollow in the peritoneal cavity between the bladder and the rectum in the male. In the female this hollow is separated into two fossae by the elevation from the pelvic floor of the **broad ligaments** of the peritoneum over the uterus, the uterine tubes, and adnexa. The peritoneum lies in contact with only the upper two-thirds of the rectum, and the rectum is distinguished from the sigmoid colon by its lack of a complete peritoneal investment. In its upper one-third the rectum is covered anteriorly and on its sides by peritoneum; in its middle one-third it is covered only anteriorly as the peritoneum sweeps away from it. The lower one-third of the rectum is retroperitoneal. Small peritoneal fossae lie on either side of the rectum in its upper one-third; these are the **pararectal fossae.** The lateral pelvic walls are peritonealized only to the extent of the rectovesical pouch and the fossae mentioned.

In the female the peritoneum reflected from the posterior border of the bladder turns onto the vesical surface of the uterus at the level of the uterine isthmus. It covers the vesical surface, the fundus, and the intestinal surface of the uterus; in addition, it is continued across the posterior fornix of the vagina for from 1 to 2 cm. Leaving the vaginal fornix, it is then reflected posteriorly onto the rectum. The interposition of the uterus and its peritoneal covering in the female forms two peritoneal pouches instead of the one in the male, a **vesicouterine pouch** anteriorly and a **rectouterine pouch** posteriorly. The two layers of peritoneum that cover the two surfaces of the uterus come together

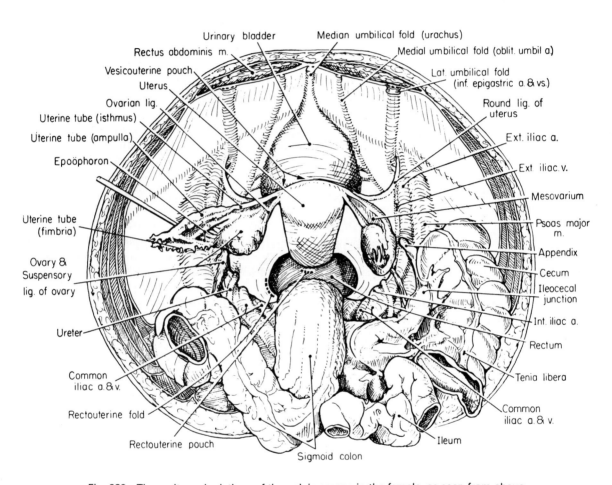

Urinary bladder

Rectus abdominis m.

Vesicouterine pouch

Uterus

Ovarian lig.

Uterine tube (isthmus)

Uterine tube (ampulla)

Epoöphoron

Uterine tube (fimbria)

Ovary & Suspensory lig. of ovary

Ureter

Common iliac a. & v.

Rectouterine fold

Rectouterine pouch

Sigmoid colon

Median umbilical fold (urachus)

Medial umbilical fold (oblit. umbil a.)

Lat. umbilical fold (inf. epigastric a. & vs.)

Round lig. of uterus

Ext. iliac a.

Ext. iliac. v.

Mesovarium

Psoas major m.

Appendix

Cecum

Ileocecal junction

Int. iliac a.

Rectum

Tenia libera

Common iliac a. & v.

Ileum

Fig. 389   The peritoneal relations of the pelvic organs in the female, as seen from above.

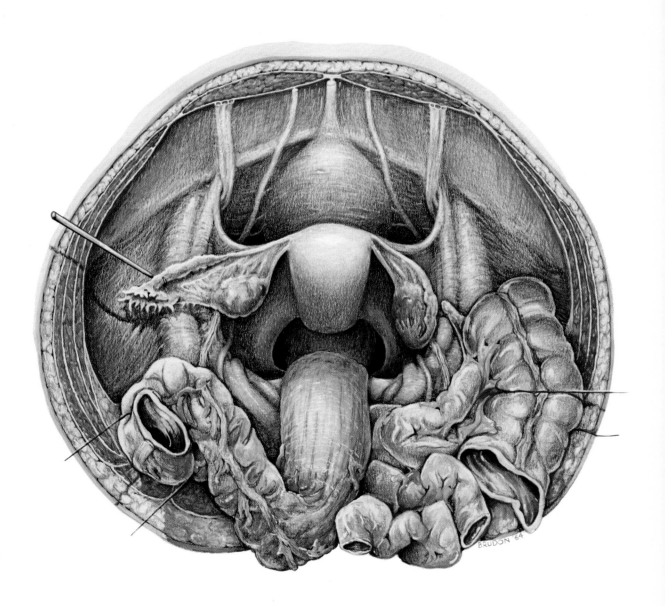

at the margins of the uterus and extend lateralward therefrom as far as the side walls of the pelvis. These double-layered sheets are the **broad ligaments of the uterus.** Enclosed in the free upper edge of each broad ligament is a uterine tube. The ends of the round ligament of the uterus and the ovarian ligament are attached into the side of the uterus under the peritoneum just below the entrance of the uterine tube. The round ligament lies under the anterior leaf of the broad ligament in its course to the deep inguinal ring. The ovarian ligament attaches laterally to the ovary, and the peritoneum covering the ligament is prolonged posteriorly as the **mesovarium;** that part which is superior to the mesovarium and surrounds the uterine tube is the **mesosalpinx.** The mesovarium and the mesosalpinx are continuous at the pelvic brim with the **suspensory ligament of the ovary.** The lateral part of the mesosalpinx is free, and the end of the uterine tube curves downward along the posterior border of the ovary. This, together with a ridging of the parietal peritoneum by the ureter behind and below the ovary, forms the shallow **ovarian fossa.** The remainder of the broad ligament, concerned in attaching the uterus to the lateral walls and floor of the pelvis, is the **mesometrium.** Inferiorly, at the base of the broad ligament its layers diverge, and endopelvic fascia is inserted between them. The posterior layer of the mesometrium is elevated by the **rectouterine fold,** a variable but sometimes sharp shelf-like fold, which passes from the isthmus of the uterus to the posterior wall of the pelvis lateral to the rectum. It corresponds to the sacrogenital fold in the male.

THE RECTUM AND THE ANAL CANAL

THE RECTUM (figs. 389, 390, 391, 403)

Of the two terminal portions of the gastrointestinal tract, only the rectum lies in the pelvis. It is, however, convenient to consider the two parts together. The **rectum** is continuous above with the sigmoid colon and begins where the complete peritoneal investment of the latter ends. This is about the level of the third sacral vertebra. Curving downward and forward in the concavity of the sacrum and coccyx and onto the pelvic diaphragm, the rectum ends about 4 cm. below and in front of the tip of the

coccyx. The rectum also winds slightly to the right and then to the left in its descent. The length of the tube is about 12 cm. It abuts inferiorly against the prostate in the male and the vagina in the female and then turns abruptly backward and downward to pierce the pelvic diaphragm and continue as the anal canal. Generally smaller in caliber than the sigmoid colon, the rectum is dilated just above the pelvic diaphragm in the **rectal ampulla.**

The side to side sinuosity of the rectum is a reflection of the **transverse rectal folds** in its interior. These are composed of an infolding of the mucous and submucous coats and of the greater part of the circular muscle layer. Three folds are usually present. The most prominent one, about 8 cm. from the anal aperture, lies on the right anterior side of the tube. The other two folds, projecting inward from the left posterior aspect of the lumen, lie from 2 to 3 cm. above and below the right anterior fold. These are permanent folds maintained by the relative shortness of the longitudinal muscular layer which, in the rectum, is concentrated into anterior longitudinal and posterior longitudinal bands. The transverse rectal folds appear to function in the support of the fecal mass in the rectum. They may interfere with instrumentation or digital examination of the rectum.

The Rectal Fusion Fascia

The rectum does not have a complete peritoneal investment in the adult. In its upper one-third it is covered on the sides and in front; in its middle one-third, only in front. The lower one-third of the rectum is completely below the peritoneal reflection which passes forward onto the bladder or the vagina. In the adult the peritoneum does not reach the pelvic floor, although it did during developmental stages. In the adult its deeper-lying layers have become apposed; the epithelial lining has disappeared; and the connective tissue components have remained as a fusion fascia similar to fusion fasciae in the abdomen proper. The **rectal fusion fascia,** then, lies on either side of the rectum below the peritoneal reflections and has the position and attachments of the original peritoneum of the rectum. Thus, the rectal fusion fascia begins on either side of the rectum on the under surface of the sloping peritoneal gutter of the pararectal fossae. It encloses the rectum, together with its blood vessels, and the fasciae of the two sides end in a midline attachment over the sacrum. The peritoneum obliterated in front of the lower one-third of the rectum is the peritoneum of the inferior limit of the rectovesical pouch of the male. Here the fusion fascia, continuous behind with the rectal fusion fascia at the sides of the rectum, is

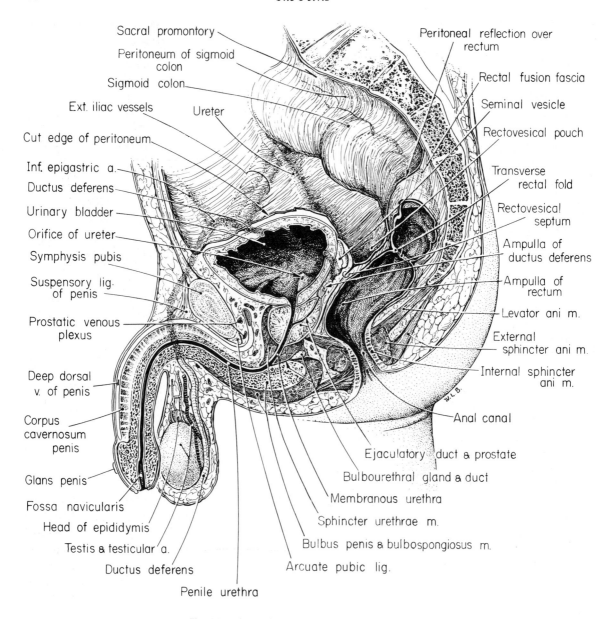

Sacral promontory

Peritoneum of sigmoid colon

Sigmoid colon

Ext. iliac vessels

Ureter

Cut edge of peritoneum

Inf. epigastric a.

Ductus deferens

Urinary bladder

Orifice of ureter

Symphysis pubis

Suspensory lig. of penis

Prostatic venous plexus

Deep dorsal v. of penis

Corpus cavernosum penis

Glans penis

Fossa navicularis

Head of epididymis

Testis & testicular a.

Ductus deferens

Penile urethra

Peritoneal reflection over rectum

Rectal fusion fascia

Seminal vesicle

Rectovesical pouch

Transverse rectal fold

Rectovesical septum

Ampulla of ductus deferens

Ampulla of rectum

Levator ani m.

External sphincter ani m.

Internal sphincter ani m.

Anal canal

Ejaculatory duct & prostate

Bulbourethral gland & duct

Membranous urethra

Sphincter urethrae m.

Bulbus penis & bulbospongiosus m.

Arcuate pubic lig.

Fig. 390   A median section of the male pelvis.

the **rectovesical septum.** It separates the ampullary portion of the rectum from the fundus of the bladder and from the ductus deferentes, the seminal vesicles, and the posterior surface of the prostate gland. The comparable **rectovaginal septum** of the female lies between the ampulla of the rectum and the upper half of the posterior vaginal wall. The pelvic fusion fasciae are less distinct at the sides of the bladder, but the rectovesical and rectovaginal fusion fasciae have continuities anteriorly with a less deep and relatively inconspicuous vesical fusion fascia.

Muscular Layers

The inner circular layer of involuntary muscle of the rectum is complete, as in the other parts of the intestines. The outer longitudinal layer is incomplete, being concentrated into an anterior and a posterior band on the corresponding surfaces of the rectum. This is a modification of the three teniae coli of the sigmoid colon. Certain fascicles separate from the anterior rectal band just above the pelvic diaphragm. These descend in the genital hiatus of the diaphragm and end in the midline raphe of the urethral sphincter muscle. This is the **rectourethral muscle.**

Blood Vessels and Nerves

The superior rectal artery supplies the rectum with the assistance of the middle and inferior rectal branches of the internal iliac artery. The superior rectal artery and vein are described on p. 438; the middle rectal and inferior rectal vessels, on pp. 507 and 475, respectively. The nerve supply of the rectum is discussed in connection with the autonomic nervous system of the pelvis (p. 513).

Lymphatics of the Rectum and the Anal Canal (fig. 391)

The lymph vessels of the rectum and the anal canal arise in a plexus in the mucous layer which is continuous with the cutaneous plexus of the perineum. For the most part, the collecting vessels follow the blood vessels supplying the rectum and the anal canal. Five channels are usually listed: (1) The most important drainage is by channels which accompany the superior rectal vessels to nodes located in the angle of division of the superior rectal artery opposite the third sacral vertebra and from there ascend to the inferior mesenteric nodes at the root of the inferior mesenteric artery. These ascending lymphatic channels receive lymph from the entire length of the rectum and the anal canal. **Pararectal lymph nodules** may interrupt any of the ascending channels. (2) From the lower part of the rectum, lymphatic channels course lateralward along the middle rectal vessels to the internal iliac nodes (p. 508). (3) Small channels pass under the sacrogenital folds to the sacral nodes (p. 509). (4) Drainage channels arising below the pelvic diaphragm follow the inferior rectal vessels across the ischiorectal fossae, perforate the lateral edges of the diaphragm, and empty into the internal iliac nodes. (5) From the terminal part of the anal canal and from the cutaneous plexus around the anus, lymphatic channels follow the genitofemoral sulcus forward to the superficial inguinal nodes.

THE ANAL CANAL (figs. 386, 388, 390–92)

The ampullary portion of the rectum rests on the pelvic diaphragm. At this point the tube turns backward and downward at about a 90° angle and

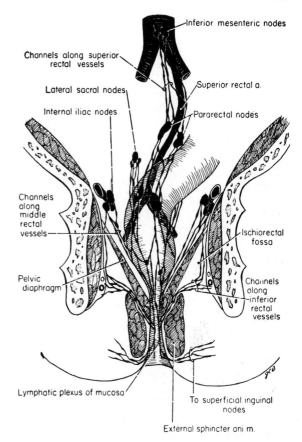

Fig. 391 The lymphatic drainage of the rectum and anal canal. Semischematic.

passes through the narrow interval between the medial borders of the levator ani muscles (p. 515). The anal canal is only from 2.5 to 3.5 cm. long and, under the tonic contraction of its sphincter muscles, maintains a contracted collapsed lumen except during the passage of feces. The anal canal lies below the pelvic diaphragm within the anal triangle of the perineum and ends at the anus. It has the fibrofatty connective tissue of the ischiorectal fossae on either side. In front of it is the central tendinous point of the perineum (the more bulky perineal body in the female).

The Muscular Relations (fig. 392)

The circular layer of the rectum continues into the anal canal. At the passage of the tube through the pelvic diaphragm, this layer of involuntary muscle undergoes a sudden and considerable thickening to form the **internal sphincter ani muscle.** The sphincter surrounds the anal canal for from 2.5 to 3 cm., reaching toward the anal orifice, and terminates at

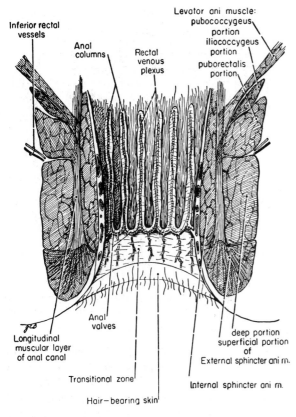

Fig. 392 The anal canal is seen in a semischematic vertical section.

the level of the transitional cutaneous zone. It is overlapped distally by the superficial portion of the external anal sphincter, and an intermuscular groove separating the muscles lies at the distal edge of the transitional cutaneous zone. The external sphincter ani is a voluntary muscle of the perineum and is described on p. 480. The longitudinal muscular layer of the rectum, with some admixture of fibers of the levator ani muscle as the rectum pierces the pelvic diaphragm, continues over the anal canal as far as the anal aperture, separating the internal sphincter ani and external sphincter ani muscles. Its terminal fibers filter through the superficial portion of the external anal sphincter to end in the subcutaneous connective tissue and skin at the anal orifice.

The Mucous Membrane (fig. 392)

The mucous membrane of the anal canal exhibits differentiation at various levels. The skin of the margin of the anus is pigmented, contains hair follicles and glands, and covers the **cutaneous zone** of the canal. Internal to it is the **intermediate zone** (transitional zone), about 1 cm. wide, which is characterized by smooth hairless skin. The line of change from the cutaneous zone to the intermediate zone is referred to clinically as the 'white line.' True mucous membrane begins at the level of the **anal valves,** which are remnants of the proctodeal sphincter, the end of the anal portion of the cloaca in embryonic development. The anal valves and the anal columns characterize the upper 1.5 to 2 cm. of the anal canal. The **anal columns** are from five to ten vertical folds of mucous membrane, separated by grooves, which overlie veins (and sometimes small arteries). The columns are connected by mucosal folds, the **anal valves,** under which pass communicating veins. At the underside of the valves the mucous membrane becomes continuous with the smooth skin of the transitional zone along an irregular line; this is the **pectinate line.**

Clinical Note The characteristics of carcinomas differ above and below the pectinate line, and the line also separates the lymphatic, vascular, and nerve supply of the anal canal. Afferent nerves are visceral above the line; somatic, below. The lymphatic drainage ascends above the line (p. 487) but, from the terminal canal, follows the genitofemoral sulcus to the superficial inguinal nodes. Dilated and redundant veins of the anal columns may form 'internal hemorrhoids' above the line in distinction to 'external hemorrhoids' below the pectinate line (see below).

The Rectal Venous Plexus (figs. 392, 407)

The submucosa of the anal canal represents an important area of junction between the veins draining to the portal and systemic systems. Longitudinal veins under the anal columns are interrelated by cross-connecting veins under the anal valves, and both types of veins communicate with veins under the skin of the transitional and cutaneous zones. This venous network, the rectal plexus, drains in several directions. Its inferior radicles pierce the external sphincter ani muscle and join other radicles in the formation of the inferior rectal veins. Above the pelvic diaphragm emergent veins of the plexus pierce the wall of the canal and form the middle rectal veins. The inferior rectal and middle rectal veins are ultimately tributary to the internal iliac vein and the inferior vena cava. The veins ascending under the anal columns communicate with one another and unite to form several larger vessels which ascend within the rectal wall. These vessels pierce the rectal wall at about its midlength and join to form the superior rectal vein. As a tributary of the inferior mesenteric vein, the superior rectal vein flows into the portal system.

Arteries and Nerves

The **inferior rectal arteries** and the **inferior rectal nerves** supply the anal canal (p. 474). The **middle rectal artery,** the branches of which assist in the supply of the upper part of the canal, is described with the internal iliac artery (p.507). The autonomic nerves to the anal canal are similar to those of the rectum and are conducted to it along its blood vessels (p. 513).

THE URINARY BLADDER

The **urinary bladder** (figs. 378, 389, 390, 393, 395, 397, 400) is a hollow muscular organ which serves as a reservoir for the urinary system. It lies in the anterior half of the pelvis, occupying a triangular space bounded in front and at the sides by the symphysis pubis and the diverging walls of the pelvis and behind by the rectovesical septum (p. 486 & fig. 390). The structures of the pelvic walls which form the space of the bladder are the obturator internus muscles above and the pelvic diaphragm (levator ani muscle) below. Posteriorly, the male bladder is closely related to the ductus deferentes and the seminal vesicles and, through the rectovesical septum, to the rectum. The superior surface of the bladder is flat when empty and relaxed in death or when pressed on by other viscera; it is generally rounded in life, and the whole bladder is globular when distended. When filled, it contains about 500 cc. of urine. The superior surface and the uppermost one or two centimeters of the posterior aspect of the bladder are covered by peritoneum, which sweeps off the bladder into the rectovesical (or vesicouterine) pouch. Inferiorly, the bladder rests on and is firmly attached to the base of the prostate gland in the male; in the female it lies on the pelvic diaphragm and, in the genital hiatus, on the sphincter urethrae muscle. The bladder is enveloped in endopelvic fascia. Sweeping from the neck of the bladder and the base of the prostate, a sheet of this fascia attaches to the parietal fascia on the back of the pubis and to the superior fascia of the pelvic diaphragm. This is the **puboprostatic ligament** (fig. 395), or lateral true ligament of the

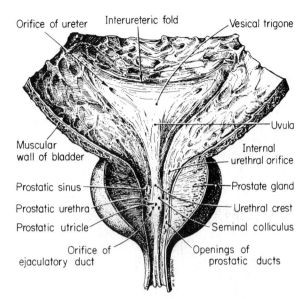

Fig. 393 A frontal section of the bladder and prostate gland to show the vesical trigone and the prostatic urethra.

bladder. In the female the puboprostatic ligament becomes the **pubovesical ligament.** The inferior region of the bladder is its most fixed part. In addition to the fascial fixation, there is continuity of structure between the neck of the bladder and the prostatic urethra in the male; in both sexes the urethra passes inferiorly into the structures of the perineum, providing a stable attachment.

In spite of its rounded form, the bladder is peaked anteriorly at its apex by the remains of its original connection to the allantois, the **urachus.** In the adult this is reduced to a fibrous strand which connects the apex of the bladder to the connective tissue of the umbilicus; it is known as the **median umbilical ligament.** The relation of the **fundus** (posterior-inferior portion) of the bladder with the seminal vesicles and the ductus deferentes is particularly close (fig. 397). In the male the two seminal vesicles form a V-shaped figure. These are irregular, convoluted structures, approximately the size of a small finger, which converge to ducts opposite the base of the prostate. The deferent ducts descend on the back of the bladder within the limbs of the V of the seminal vesicles. Their ampullated portions lie against the seminal vesicles and join the ducts of the seminal vesicles to form the **ejaculatory ducts** opposite the base of the prostate. The seminal vesicles are enveloped in abundant endopelvic fascia and are held firmly thereby to the fundus of the bladder. The ureters enter the bladder 5 or 6 cm. apart at the superolateral portion of the fundus, each being crossed at its point of entrance by the ductus deferens of the same side. In the female (fig. 400) the fundus is loosely attached to the anterior wall of the vagina.

Inferolaterally, the bladder is separated from the pelvic diaphragm, the obturator internus muscle, and the back of the pubis by loose areolar connective tissue and fat. The space occupied by this loose material is the **prevesical space.** Around the neck of the bladder, the dense vesical plexus of veins lies in this areolar tissue.

Interior of the Bladder (fig. 393)

The mucous membrane lining the bladder is loosely attached to its musculature over most of its surface area and appears wrinkled or folded except when the bladder is distended. There is in the fundus, however, a small triangular area, the **vesical trigone,** where the mucous membrane is firmly bound to the muscular coat and is always smooth. The vesical trigone is bounded by the openings of the ureters at its posterolateral angles and by the urethral aperture at its inferior (anterior) angle. A ridge stands out between the ureteral openings; this **interureteric fold** is formed by an underlying transverse bundle of involuntary muscle prolonged medialward from the terminations of the ureters. Other fascicles of ureteral muscle fan out to underlie the mucous membrane of the whole vesical trigone; they end in the mucous membrane and in the underlying detrusor muscle. The two ureteral openings are slit-like and, in an undistended organ, lie about 3 cm. apart. The internal urethral opening is an equal distance from the ureters, so that the vesical trigone is equilateral. The distances between the three openings increase somewhat in distention of the organ. A small tongue-like elevation of the trigone above and behind the urethral aperture is the **uvula of the bladder.** It is due to the middle lobe of the prostate. The urethral opening is at the most dependent part of the bladder and is usually somewhat semilunar in form, indented from behind by the uvula.

Musculature of the Bladder

The involuntary musculature of the bladder is plexiform in arrangement, forming a meshwork. However, a fairly common orientation of the muscle fibers results in the frequent designation

of three layers. The **external stratum** is principally longitudinal in direction. A slender band of its fibers constitutes the **pubovesical muscle,** which underlies the medial portion of the puboprostatic ligament; its fibers end in the retropublic connective tissue. The fibers of the **middle layer** are generally transverse to the long axis of the bladder. An **internal stratum** consists of largely longitudinally oriented muscle fibers which inferiorly converge toward the urethral aperture. The term 'detrusor urinae muscle' is properly applied to all layers, for they contract together. Further, there is continuity of single fiber bundles through several layers, the circular stratum being composed of bundles which continue into it from the longitudinal layer and vica versa.

Under the mucous membrane of the vesical trigone lies an extension of the muscular layers of the ureters. Muscular bands curve medialward from the ureteral orifices and constitute the interureteric fold. Other laminae descend with a medialward inclination and crisscross in a thin submucosal sheet throughout the trigone area. Similar fasciculi of the **trigonal muscle** descend at the margins of the trigone to approach the urethral aperture. This muscular complex is entirely an extension of ureteral muscle and can be separated from the bladder wall by prolonged maceration.

The muscular arrangements at the neck of the bladder are of interest. A vesical sphincter had traditionally been recognized although no annular sphincter exists. A sphincteric action has been attributed to opposing arcades of muscle in the neck region, in spite of the logical difficulty of the detrusor muscle being called upon to both empty the bladder and close its orifice. Fasciculi of both the longitudinal and circular layers of bladder muscle intermingle at the bladder neck. In addition, muscular fibers of both layers turn downward along the urethra and form the longitudinal muscle layer of the urethra.

The neck of the bladder and the urethral wall are characterized by much circularly disposed elastic tissue, which doubtless aids in maintaining a collapsed lumen in the urethra.

Blood Vessels

The arteries of the bladder are the **superior vesical** and **inferior vesical arteries** from the anterior trunk of the internal iliac artery (p. 506). A dense plexus of veins surrounds the neck of the bladder in the endopelvic areolar tissue. This **vesical plexus** extends over the seminal vesicles and the deferent ducts to the regions of entrance of the ureters. It communicates with the prostatic plexus in the male, and from it emerge several **vesical veins** on each side which are tributary to the internal iliac vein.

Lymphatics of the Bladder

The lymphatics of the bladder arise from a submucous plexus which is most abundant in the region of the trigone. Piercing the muscular coat, the vessels organize in two groups: from the superior and inferolateral surfaces they pass to the external iliac nodes; from the posterior surface they pass to both the external iliac and internal iliac lymph nodes. Lymph vessels from the neck of the bladder, in association with some from the prostate, pass to the sacral nodes and to the median common iliac nodes.

Nerves of the Bladder

The **vesical plexus** is a continuation of the anterior portion of the inferior hypogastric plexus and contains postganglionic sympathetic and preganglionic parasympathetic fibers. The latter are branches of the pelvic splanchnic nerves which join the inferior hypogastric plexus from sacral nerves, two, three, and four. The vesical plexus of each side invests the terminal portion of the ureter and lies against the posterolateral aspect of the bladder (fig. 409). It supplies the bladder, the lower portion of the ureter, the seminal vesicle, and the ductus deferens as far as the epididymis. The parasympathetic innervation of the bladder serves the emptying reflex, causing contraction of the musculature of the wall of the bladder (detrusor urinae muscle). The sympathetic supply of the bladder reaches the trigonal muscle and the blood vessels. Afferent impulses are conducted along both the sympathetic and parasympathetic sets of nerves, impulses of pain due to an overstretched bladder traveling with the sympathetic fibers. The more important proprioceptive impulses from the muscular wall, initiated by normal stretching of the muscle layers as the viscus fills, travel over the parasympathetic fibers. Their stimulation results in reflex emptying of the bladder.

Clinical Note The Mechanisms of Micturition As urine accumulates in the bladder, the tonicity of the bladder musculature adjusts to the increased volume of fluid until the stretch receptors in the muscle are sufficiently stimulated to set off the emptying reflex and to initiate the desire to void. This occurs at urine volumes of from 400 to 500 cc. The efferent portion of the emptying reflex begins in the second, third, and fourth sacral segments of the spinal cord and both the afferent and efferent nerve fibers are carried over the pelvic splanchnic nerves; this is a parasympathetic reflex. Normal bladder control requires the superposition over this reflex of inhibiting impulses from higher brain centers and positive voluntary control of the urethral sphincter muscle. The sympathetic innervation of the bladder reaches the trigonal muscle and the blood vessels and includes afferents of pain. This innervation has little to do with normal voiding.

Just as there is only one innervation which serves the emptying reflex, so there is only one muscular mechanism involved in the bladder. All the musculature of the wall and the neck of the bladder contracts at once, and there is no evidence for a vesical sphincter or its inhibition. The descending muscular bundles of the bladder which run longitudinally in the urethra serve to open the internal urethral orifice in a funnel-shaped opening. These bundles curve downward into the urethra from the surrounding musculature and, with ballooning out of the bladder, are increasingly oblique to the urethral lumen. Their traction, as the bladder muscle contracts, draws radially on the mouth of the internal urethral orifice and assists the pressure being exerted on the orifice by the detrusor muscle in carrying out the voiding reaction. Voluntary relaxation of the skeletal urethral sphincter accompanies the initiation of voiding. Relaxation of the pelvic diaphragm, contributing to dropping the bladder and shortening the urethra, supplements the response.

It is equally important to consider the closing mechanism. Attention has been called to circularly and spirally disposed elastic tissue interspersed in the largely longitudinal muscle layer of the vesical neck and urethra. These components aid in maintaining a collapsed urethral lumen during the intervals between voidings. Lengthening of the urethra by the ascending traction of the pelvic diaphragm and the downward and backward pull of the urethral sphincter are actions important in closing the urethra (fig. 394). The urethral sphincter muscle (striated) is also capable of voluntary abrupt termination of urination.

THE PELVIC URETER (figs. 378, 393, 397)

The abdominal portion of the ureter is described in connection with the kidney (p. 444). The pelvic portion is about 12.5 cm. long and begins at the point where the ureter crosses the bifurcation of the common iliac artery or the beginning of the external

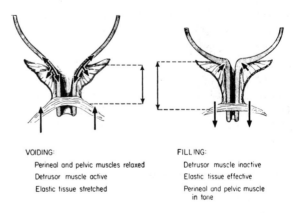

VOIDING:
Perineal and pelvic muscles relaxed
Detrusor muscle active
Elastic tissue stretched

FILLING:
Detrusor muscle inactive
Elastic tissue effective
Perineal and pelvic muscle in tone

Fig. 394 Schema of structural relations in voiding and filling of the bladder (from the original of fig. 6, R. T. Woodburne, *Anat. Rec.,* 141:11, 1961).

iliac artery. The pelvic ureter descends retroperitoneally on the side of the pelvic wall, with a
curvature which is concave forward and medialward. It lies below and in front of the internal iliac
artery and crosses medially the obturator nerve and
vessels and the umbilical artery. It then inclines
medialward and reaches the posterolateral aspect of
the bladder in front of the upper end of the seminal
vesicle. Here the ductus deferens turns over the
posterosuperior aspect of the bladder and passes
downward over the fundus, crossing the termination
of the ureter. The end of the ureter is surrounded
by the vesical plexus of veins. The ureters run
obliquely through the wall of the bladder for about
1.5 cm., so that their internal openings are much
closer together than are their external penetrations
of the bladder wall. Their oblique passage allows
the contraction of the bladder musculature to act as
a sphincter of the ureter, preventing reflux of urine
into the ureters as the bladder is emptied. The
musculature of the ureter does not blend with that
of the bladder, the ureter passing through a gap in
the muscular wall of the bladder. Under the
mucous membrane of the vesical trigone the
musculature of the ureter fans out, and its marginal
fibers descend toward the urethra. The lumen of
the ureter is not uniform throughout; it is slightly
constricted where it crosses the common iliac
artery; it is narrowest in its passage through the
bladder wall.

The musculature of the ureter is arranged
generally in two layers; these are inner longitudinal
and outer circular, with oblique fascicles intermingled. In the middle and lower parts of its length
the circular fibers form a thicker stratum than exists
in the superior part, but, terminally, the musculature
is wholly longitudinal. Near the termination of the
ureter in the bladder, coarse bundles of longitudinal
fibers are applied to it from the external stratum of
musculature of the bladder.

In the female the ureter has several relationships
of special significance. As it descends and then turns
forward along the lateral wall of the pelvis, it
underlies the parietal peritoneum behind and below
the ovary and forms the posterior and inferior
limits of the **ovarian fossa** (figs. 388, 400). It then
runs forward and medialward toward the bladder,
passing the cervix of the uterus and the lateral
fornix of the vagina at a distance of from 1 to 2 cm.
In this portion of its course the ureter is crossed
obliquely by the uterine artery, which provides

a small arterial twig to it. This close relationship of ureter and uterine artery must always be in
mind when ligating the artery so that the ureter
not be damaged.

Vessels and Nerves

The pelvic portion of the ureter receives branches of
the superior vesical **artery** of the internal iliac system
(p. 506). Ureteric branches may also arise from the
common iliac or the internal iliac artery directly. In the
female the uterine artery provides a small ureteric
branch as it crosses the ureter. The blood supply of the
abdominal portion of the ureter is described on p. 444,
and the various vessels anastomose along the length of
the ureter. The **veins** of the ureter correspond with its
arteries. They form, along the ureter, an anastomosing
system which drains into the inferior vena cava inferiorly by way of the internal iliac vein and, superiorly,
by way of the testicular (or ovarian) or renal veins. The
ureteric veins may be concerned in a circumadrenorenal
anastomosis (p. 449). The lymphatics are considered
with the kidney (p. 446); the nerves, with the description of the abdominal ureter (p. 444) and the subsidiary
sympathetic plexuses of the abdomen (p. 459).

THE MALE URETHRA

The **male urethra** (figs. 390, 393, 395, 398) traverses
the prostate gland, the sphincter urethrae muscle
and the perineal membrane, and the penis. Its
length is about 20 cm., and its parts are prostatic,
membranous, and spongy. In its passage from the
internal to the external urethral orifice, the urethra
has a sinuous course, first downward, then forward, and again downward. It conducts urine
from the bladder but also receives the ejaculatory ducts and the prostatic ducts in its passage through the prostate gland. Farther along,
the ducts of the bulbourethral glands empty into
the urethra, and numerous minute, mucous urethral
glands add their secretion to its contents.

The **prostatic portion** of the urethra is from 3 to 4
cm. long and traverses the prostate in a gentle
curve, the concavity of which is forward (fig. 389).
The lumen is somewhat spindle-shaped, wider in its
middle than above or below (figs. 393, 398). A
narrow, longitudinal ridge in the posterior wall, the
urethral crest, indents the lumen, so that it is
crescentic in cross section. The grooves on either
side of the crest are the **prostatic sinuses,** and it is
here that most of the ducts of the prostate gland
open. A minority open along the sides of the
urethral crest. Beyond the midlength of the crest

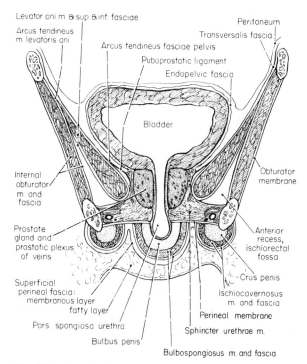

Fig. 395 A semischematic frontal section of the pelvis and perineum through the bladder and prostate gland.

there is a rounded eminence, the **seminal colliculus.** In the median plane of the colliculus a small slit leads into a 'blind' pouch, less than a centimeter in length, which is directed backward and upward into the prostate; this is the **prostatic utricle.** It is formed from the fused ends of the paramesonephric ducts and is the male homologue of the uterus and the vagina. On each side of the mouth of the prostatic utricle there is the more minute opening of the **ejaculatory duct.** The urethral crest diminishes in height both above and below the seminal colliculus but may have continuity with the uvula of the bladder above. Inferiorly, it frequently divides into two inconspicuous elevations in the membranous portion of the urethra. Internal longitudinal and external circular layers of involuntary muscle characterize the prostatic urethra. The circular lamina is largely lacking in the female urethra. In the male it may be regarded as a sphincter preventing entrance of seminal fluid into the bladder.

The **membranous portion** of the urethra (figs. 390, 395) traverses the annular part of the sphincter urethrae muscle and the perineal membrane and is the shortest part of the urethra; its

length is about 1 cm. It pierces the perineal membrane about 2.5 cm. behind the pubic symphysis. The bulbourethral glands lie behind and to either side of the membranous portion of the urethra; their ducts perforate the perineal membrane and enter the spongy portion of the urethra. The wall of the membranous portion is thin.

The **spongy portion** of the urethra (figs. 390, 395), about 15 cm. long, enters the bulb of the penis at the underside of the perineal membrane, traverses the full length of the corpus spongiosum, and terminates at the external urethral orifice. The lumen, about 5 mm. in diameter through most of the length of the penis, is larger in the bulb and is again widened in the glans penis as the **fossa navicularis.** Throughout most of the spongy portion of the urethra the lumen is a transverse slit but becomes a vertical aperture at the external urethral opening. The mucosa presents a number of minute openings for the ducts of the mucous **urethral glands.** A number of small, pitlike recesses, the **urethral lacunae,** also open into the spongy portion; their openings are directed distalward along the urethra.

The **blood vessels** of the urethra are those of the structures it traverses. Branches of prostatic vessels supply it within the prostate; the artery of the bulb and the urethral artery, in its more distal portions. The nerves are branches of the pudendal nerve and autonomic contributions from the cavernous plexuses.

Clinical Note The size and the distensibility of the lumen of the male urethra are of both anatomical and clinical interest. The external urethral opening, the vertical slit at the end of the glans penis, is both the narrowest part of the urethra and its least distensible portion. An instrument passing through this opening should pass through any of the other parts of the lumen. The membranous portion of the urethra is its next narrowest portion. This narrowing is, however, due to the contraction of the sphincter urethrae muscle, so that the membranous urethra is distensible. Immediately below the perineal membrane, the urethra is well covered below and behind by the erectile tissue of the bulb, but its superior surface is unprotected for a short distance. Here the wall is thin and distensible and is vulnerable to rough instrumentation. Enlargement of the prostate is common with advancing years and is a cause of obstruction of the urethra. The hyperplastic prostatic tissue is commonly from a posterior midline component (middle lobe, fig. 398) but may also occur in the lateral lobes. The urethral canal is developmentally first a groove on the underside of the penis. Normally this groove is closed over, but in certain cases the closing does not extend to the end of the glans and an opening is left on the underside of the penis. This is hypospadias. Minor degrees are not important but too proximal an opening is associated with faulty development of the corpus spongiosum as well, and surgical correction should then be considered. The opposite abnormality of development, an opening on the dorsum of the penis is epispadias, a very rare condition.

THE FEMALE URETHRA

The **female urethra** (fig. 400), about 4 cm. long, corresponds to the prostatic and membranous portions of the male urethra. Its lumen, about 5 mm. in diameter, is slightly curved, its concavity being directed forward. Its external orifice is immediately in front of that of the vagina and is about 2.5 cm. behind the glans clitoridis. The urethra lies in front of the lower half of the vagina; its upper part is separated from the vagina by loose areolar tissue, but its lower portion is so intimately related to the vaginal wall that it appears to be embedded in it. The muscular wall is composed of oblique and longitudinal muscle fibers prolonged downward from the bladder. In its lower portion some muscle fibers blend with the anterior wall of the vagina. The tube is characterized by urethral glands and **urethral lacunae** as in the male. One group of glands opens by the **paraurethral ducts** in the vestibule at the sides of the external urethral orifice (p. 472). A thin layer of spongy erectile tissue containing a plexus of large veins, intermingled with involuntary muscle fibers, lies in the submucosa. The female urethra is surrounded by the voluntary **sphincter urethrae muscle**; some of the fibers of the muscle enclosing the urethra and the vagina together as a urethrovaginal sphincter. There is also a compressor urethrae component of the muscular complex (p. 477 & fig. 387). The urethra then perforates the perineal membrane. The blood supply of the female urethra is through the inferior vesical and vaginal branches of the internal iliac system of vessels.

THE MALE GENITAL ORGANS

The **male genital organs** include the testes, the ductus deferentes, the seminal vesicles, the ejaculatory ducts, and the penis. The prostate and bulbourethral glands are accessory glandular structures. The male semen is a mixture of spermatozoa and a thick, alkaline, mucous fluid. The addition to the spermatozoa of the mucous vehicle and the emission of the semen are contributions of the tubular system and the glands associated with the testes.

THE TESTES (figs. 390, 396)

The **testes** are paired, oval bodies which elaborate the spermatozoa; they are located in the scrotum. Each testis is from 4 to 5 cm. long, 3 cm. wide, and about 2 cm. thick and has its long axis directed upward and slightly forward and lateralward. The lateral and medial surfaces of the testis are flattened; its rounded anterior border is free; whereas its posterior border provides an attachment for the epididymis.

The Epididymis (fig. 396)

The epididymis is a comma-shaped structure curved over the back and upper end of the testis and bulging onto its lateral surface posteriorly. It is essentially an irregularly twisted tube, the **duct of the epididymis,** having a total uncoiled length of from 15 to 20 feet and a diameter of about 1 mm. The head of the comma is represented by the **head of the epididymis.** The mass of coils is somewhat reduced at the back of the testis, where it is called the **body of the epididymis.** This portion is separated from the posterior part of the lateral surface of the testis by a recess of the visceral layer of the tunica vaginalis testis which constitutes the **sinus of the epididymis.**

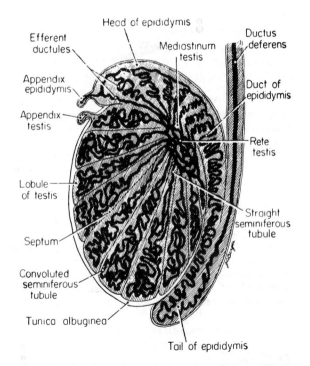

Fig. 396 A schematic vertical section of the testis and epididymis.

The smaller inferior end is the **tail of the epididymis.** It is attached to the inferior pole of the testis by loose areolar connective tissue and by the tunica vaginalis testis. In the tail portion the duct of the epididymis increases in thickness and diameter and becomes continuous with the ductus deferens.

On the upper extremity of the testis, usually emerging from under the head of the epididymis, is a minute oval remnant of the upper end of the paramesonephric duct; this is the **appendix of the testis.** On the head of the epididymis is a second small appendage, the stalked **appendix of the epididymis,** usually regarded as a detached efferent duct.

The Tunica Vaginalis Testis

The tunica vaginalis testis is described with the scrotum and the coverings of the cord and the testis (p. 385). An invaginated serous sac covering most of the testis, the epididymis, and the lower end of the spermatic cord, it represents the lowermost pinched-off portion of the processus vaginalis of the peritoneum. Its visceral layer dips between the posterolateral surface of the testis and the body of the epididymis as the **sinus of the epididymis.** The tunica vaginalis reflects on itself to leave the posterior border of the testis uncovered; here the blood vessels and nerves of the testis enter from the spermatic cord.

General Structure of the Testis (fig. 396)

The thick external covering of the testis is the **tunica albuginea,** composed of dense, white fibrous connective tissue. At the posterior border of the gland the tunica albuginea turns into the substance of the testis and is broken into the more open mesh-like **mediastinum testis.** The mediastinum is traversed by the major ducts of the testis and by its arteries, veins, and lymphatics. From the internal aspect of the mediastinum, numerous septa radiate to the surface of the organ, where they blend with the tunica albuginea. The **tunica vasculosa,** a delicate network of vessels, is formed on the septa and on the deep surface of the tunica albuginea. In the compartments between the septa lie a large number of fine thread-like **convoluted seminiferous tubules.** There are estimated to be eight hundred or more tubules in the testis. The convoluted tubes unite to form a smaller number of **straight seminiferous tubules** which pass into the mediastinum. These anastomose in a network of epithelial-lined channels in the fibrous stroma which constitute the **rete testis.** At the upper end of the mediastinum the channels of the rete testis coalesce into from twelve to fifteen efferent ducts. These perforate the tunica albuginea and form convoluted masses, the **coni epididymis,** which together constitute the head of the epididymis. The efferent ducts, each from 15 to 20 cm. in length if uncoiled, open opposite the bases of the cones into a single duct which constitutes the **duct of the epididymis.**

Vessels and Nerves of the Testis

The **testicular artery,** described on p. 460, breaks into two main branches at the posterior border of the testis. These enter the testis on either side and divide into terminals that form the tunica vasculosa and ramify to all the constituent parts of the testis and the epididymis. The **veins** emerge from the testis and form the dense **pampiniform plexus** from which is derived the testicular vein (p. 464). The **lymphatics** of the testis emerge at the mediastinum and join channels draining the epididymis. The collecting vessels ascend in the spermatic cord, accompany the testicular vessels through the inguinal canal, and terminate in the lumbar chain of nodes at levels below the renal blood vessels. The autonomic **nerves** of the testis and the epididymis arise from the same upper abdominal levels as do their vessels and lymphatics and as reflected in the abdominal origin of the parts. The testicular plexus is described on p. 459; it contains vagal parasympathetic fibers and sympathetic neurons from the tenth thoracic segment of the spinal cord.

THE DUCTUS DEFERENS (figs. 390, 396, 397)

The **ductus deferens** begins in the tail of the epididymis as the continuation of the duct of the epididymis. About 45 cm. long, it ascends in the spermatic cord, traverses the inguinal canal and then, running backward and medialward, descends on the fundus of the bladder. Tortuous like the epididymis at its beginning, it straightens out behind the testis and ascends on the medial aspect of the epididymis. The ductus deferens lies in the center of the other constituents of the spermatic cord and can be identified by its hard and cord-like character when rolled between the fingers, due to its small lumen and very thick muscular coat. It is this thick muscular layer which, by its peristaltic action, provides for delivery of the semen to the prostatic urethra. Emerging from the deep inguinal ring, the ductus deferens passes on the lateral side of the inferior epigastric artery and ascends obliquely across the external iliac arteries. Reaching the pelvic brim, it descends between the peritoneum and the lateral wall of the pelvis medial to the obliterated umbilical artery and the obturator nerve and vessels. It then curves onto the back of the bladder, crossing to the medial side of the ureter. The ductus deferens descends on the fundus of the bladder medial to the ureter and the seminal vesicle, where it becomes dilated and tortuous in its ampullary portion. This enlarged portion narrows markedly just above the base of the prostate and, joined there by the duct of the seminal vesicle, forms the **ejaculatory duct.**

THE SEMINAL VESICLES (fig. 397)

The **seminal vesicles,** each roughly the size and shape of a small finger, are lobulated 'blind' pouches which secrete an alkaline constituent of the seminal fluid. They lie against the fundus of the bladder, diverging like the limbs of a V. They are enclosed by endopelvic fascia, and the ductus deferentes descend along their medial surfaces. The upper ends of the vesicles are rounded and they taper toward their lower ends. Above the base of the prostate, each vesicle is constricted to form a short duct which joins the lateral side of the narrowed ductus deferens at an acute angle. The common duct formed by this union is the **ejaculatory duct.** The seminal vesicles are related to the rectum behind through the rectovesical septum. They are more intimately connected to the bladder by the rectovesical portion of the endopelvic fascia. The seminal vesicle and the ampulla of the ductus deferens are similar in structure. They are thin-walled and their

mucous membranes present a honeycomb appearance.

The **artery** of the ductus deferens is a minute vessel which is closely applied to its surface. It arises from the umbilical artery (p. 506) and ends by anastomosing with the testicular artery at the back of the testis. The seminal vesicle is supplied by twigs of the inferior vesical artery (p. 506). The **lymphatics** of the ductus deferens which arise from parts distal to the brim of the pelvis join the lymphatic vessels of the testis. The lymphatics of the pelvic part of the ductus deferens and of the seminal vesicles pass, with those of the prostate and the bladder, to the internal iliac lymph nodes and the posteromedial nodes of the external iliac group. The **nerves** of the ductus deferens and the seminal vesicle are offshoots of the inferior hypogastric plexus (p. 512).

THE EJACULATORY DUCTS (figs. 390, 397, 398)

The **ejaculatory ducts** are formed just above the base of the prostate by the union of the ducts of the seminal vesicles with the narrowed ends of the ductus deferentes. About 2 cm. long, each ejaculatory duct lies almost completely within the prostate. Close to its fellow of the opposite side, each duct passes downward and forward through the prostate. The ducts open by slit-like apertures into the prostatic urethra on the colliculus seminalis at either side of the opening of the prostatic utricle. The walls of the ejaculatory ducts are extremely thin.

THE PROSTATE (figs. 385, 390, 393, 395, 397, 398)

The **prostate** is an accessory gland in the seminal tract. It underlies the male bladder, resting by its **apex** on the perineal membrane. The ampulla of the rectum abuts against it behind, the rectovesical septum intervening. The greatest breadth of the prostate is superior in relation to the bladder; it is narrow inferiorly and thus has the shape of a blunted cone. The prostate measures about 4 cm. transversely across its **base**; it is about 3 cm. in its vertical diameter and about 2 cm. in its anteroposterior diameter. Most of the base of the prostate is structurally continuous with the bladder wall. The **lateral surfaces** of the prostate are convex and are related to the superior fascia of the pelvic diaphragm.

The urethra enters the prostate near the middle of its base and leaves it on its **anterior surface** immediately above and in front of the apex of the gland. The ejaculatory ducts enter at the posterior border

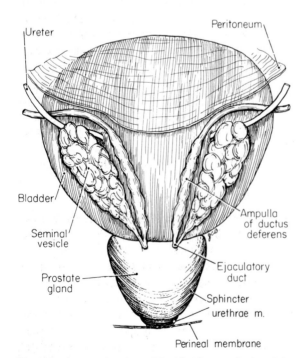

Fig. 397 A posterior view of the bladder, seminal vesicle, and prostate gland.

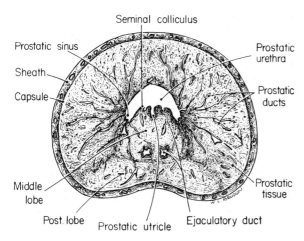

Seminal colliculus

Prostatic sinus

Sheath

Capsule

Prostatic urethra

Prostatic ducts

Prostatic tissue

Middle lobe

Post. lobe   Prostatic utricle   Ejaculatory duct

Fig. 398 A semischematic cross section of the prostate gland.

of the base and run obliquely downward and forward to open at either side of the prostatic utricle. The gland is formed by enlargement of urethral glands and is developmentally divisible into two lateral lobes which surround the urethra and fuse anteriorly in an isthmus which is largely muscular. Although the prostate has been regarded as homogeneous histologically, McNeal has recently pointed out that the area surrounding the course of the ejaculatory ducts has the histologic character of the seminal vesicles. This 'central zone' may constitute about 25% of the total mass.

Fascial Covering

A **sheath** for the prostate is supplied by a condensation of endopelvic fascia. The sheath is continuous with the puboprostatic ligaments, two fibrous sheets which unite the anterior and lateral surfaces of the gland with the back of the pubis and the superior fascia of the pelvic diaphragm along the arcus tendineus fasciae pelvis (fig. 395).

Structure

The prostate consists of a fibromuscular stroma enclosing the glandular elements. A condensation of the stroma forms the capsule of the gland from which septa pass into it, dividing it into about fifty lobules. The glandular tissue is composed of minute, slightly branched tubules which lead into from twenty to thirty **prostatic ducts.** Most of these empty into the prostatic sinuses at either side of the urethral crest of the posterior urethral wall. The urethral crest, the seminal colliculus, and other features of the prostatic portion of the urethra are described on p. 493. The prostate is formed

from enlarged urethral glands and so its fibromuscular stroma is derived from the urethral wall. Its involuntary musculature is also continuous from the neck of the bladder. Peripherally the fibromuscular prostate is not sharply separable from surrounding striated muscle fibers of the sphincter urethrae muscle (figs. 385, 390, 397). This muscle is applied to the anterolateral surface of the gland superiorly, is more anterior at middle levels of the gland, and then, immediately inferior to the prostate, becomes an annular sphincter for the urethra.

Vessels and Nerves

The **arteries** of the prostate are branches of the inferior vesical and middle rectal arteries of the internal iliac system (p. 507). The wide, thin-walled **veins** form the prostatic plexus embedded in the prostatic sheath in front of and at the sides of the gland. This plexus receives the deep dorsal vein of the penis, communicates with the vesical plexus, and empties into the internal iliac veins. The **lymph** vessels are associated with those of the seminal vesicles and the neck of the bladder. They end in the sacral and internal iliac nodes. The **nerves** of the prostate are branches of the prostatic plexus, an offshoot of the inferior hypogastric plexus. For a functional analysis of genital innervation, see p. 480.

THE FEMALE GENITAL ORGANS

The **female genital organs** consist of an internal and an external set; the external group is described in the chapter on the Perineum (p. 471). The internal organs are the ovaries, the uterine tubes, the uterus, and the vagina. The ovaries are the female gonads, homologous with the testes of the male. The uterine tube conducts the ovum to the uterus, where, if fertilization has taken place, the zygote becomes implanted. Development into the embryo and fetus takes place in the uterus. The vagina is the connection of the uterus to the exterior and serves for introduction of the sperm and as the birth canal.

THE OVARIES (figs. 389, 399, 400)

The **ovaries** lie against the lateral pelvic walls, each attached to the mesovarium of the broad ligament. The ovary is an oval body, about 3 cm. long, 2 cm. wide, and about 1 cm. thick. It lies in the **ovarian fossa,** a shallow depression of parietal peritoneum bounded postero-inferiorly by the ureter; superiorly, by the external iliac vessels; and anteriorly, by the pelvic attachment of the broad ligament. In the erect posture the long axis of the ovary is vertical, and its superior extremity is overlain by the fimbriated end of the uterine tube. From

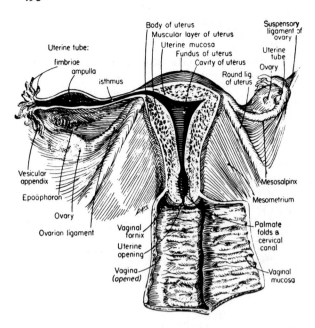

Fig. 399　The internal genital organs in the female, partly in frontal section.

this **tubal extremity,** a fold of peritoneum, the **suspensory ligament of the ovary,** passes upward across the iliac vessels and the psoas muscles. It conducts the ovarian vessels, nerves, and lymphatics to the tubal end of the gonad. The **uterine,** or inferior, **extremity** of the ovary gives rise to the **ovarian ligament,** a round fibromuscular cord which attaches the ovary to the uterus just below the entrance of the uterine tube into the uterus. The ovarian ligament is enclosed within a posteriorly directed fold of the broad ligament, the **mesovarium.** The anterior border of the ovary is attached to the mesovarium; its posterior border is free. Vessels and nerves enter the attached border at its hilum. The surface epithelium of the ovary is columnar; it is modified from the developmental peritoneal covering of the ovary. The uterine tube arches over the ovary; ascending along its mesovarial border, it curves over the superior pole of the ovary and passes downward in contact with its free border and its medial surface. Vestigial tubular structures, epoöphoron (above the ovary) and paraöphoron (beside the ovary), lie in the mesosalpinx between the uterine tube and the ovarian ligament.

Vessels and Nerves (fig. 402)

The **ovarian artery,** a branch of the abdominal aorta (p. 460), reaches the hilum of the ovary by way of its suspensory ligament. Besides branches to the ovary, the vessel gives off a **tubal branch,** which passes medialward in the mesosalpinx for the supply of the tube, and a branch which follows the ovarian ligament to the side of the uterus. Both branches anastomose with offshoots of the uterine artery. Venous blood is returned by venae comitantes of the ovarian arteries which, becoming confluent as **ovarian veins,** terminate in the inferior vena cava on the right side and the renal vein on the left side. The **lymphatic drainage** of the ovary is by channels which ascend in the suspensory ligament to the lumbar chain of nodes. Its **nerves** descend along the vessels from the abdominal sympathetic plexus; the ovarian plexus is described on p. 459. Afferent fibers from the ovary enter the spinal cord through the tenth thoracic nerve.

The Uterine Tubes (figs. 389, 399, 400, 401)

The **uterine tubes** extend from the uterus lateralward to the vicinity of the ovary. About 10 cm. long, each tube is divisible into an isthmus, an ampulla, and an infundibulum. The **isthmus**—narrow, short, and thick-walled—attaches the tube to the superolateral angle of the uterus. Its lumen is about 1 mm. in diameter. The widest and longest part of the tube is the **ampulla.** Slightly tortuous, it extends lateralward to cap the superior pole of the ovary. The ampulla ascends along the attached margin of the ovary, arches over its pole, and turns down its free border. Here the ampulla ends in the **infundibulum,** a funnel-like termination formed of numerous, irregular fringed and branched processes, the **fimbriae.** The fimbriae spread over most of the medial surface of the ovary, one large process, the **ovarian fimbria,** being attached to the superior pole of the ovary. Deep in the funnel of the infundibulum lies the pelvic opening of the uterine tube; it has a diameter of about 2 mm., and it is into this opening that an ovum discharged from the ovary is swept by the cilia of the mucosal lining of the fimbriae.

Vessels and Nerves (fig. 402)

Tubal branches of both the ovarian and uterine **arteries** supply the uterine tubes. The **veins** also empty into both the uterine and ovarian veins. The **lymphatics** join the ovarian channels, ending in the

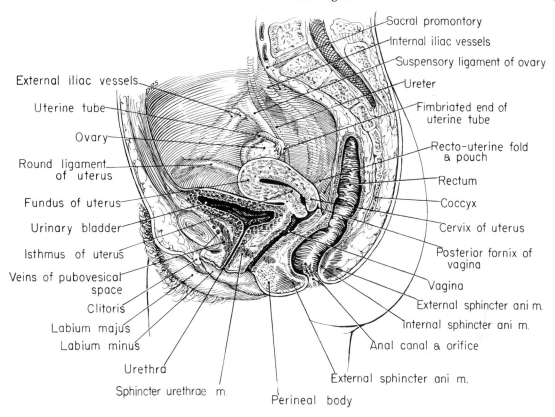

External iliac vessels

Uterine tube

Ovary

Round ligament of uterus

Fundus of uterus

Urinary bladder

Isthmus of uterus

Veins of pubovesical space

Clitoris

Labium majus

Labium minus

Urethra

Sphincter urethrae m.

Perineal body

Sacral promontory

Internal iliac vessels

Suspensory ligament of ovary

Ureter

Fimbriated end of uterine tube

Recto-uterine fold & pouch

Rectum

Coccyx

Cervix of uterus

Posterior fornix of vagina

Vagina

External sphincter ani m.

Internal sphincter ani m.

Anal canal & orifice

External sphincter ani m.

Fig. 400   A semischematic median section of the female pelvis.

lumbar nodes. The **nerve supply** of the uterine tubes is provided partly by the ovarian plexus and partly by the uterine plexus. Afferent fibers are contained in the eleventh and twelfth thoracic nerves and the first lumbar nerve.

THE UTERUS (figs. 389, 399–401)

The **uterus** is the pear-shaped, highly muscular organ in which the fertilized ovum becomes implanted and development of the embryo and fetus takes place. The uterus is about 7.5 cm. long, 5 cm. broad, and 2.5 cm. thick. It is mainly horizontal in orientation, its **vesical,** or under, **surface** resting on the urinary bladder. Its more rounded **intestinal surface** faces the cavity of the pelvis and is frequently in contact with coils of the small intestine. The uterus is not completely straight, for there is an angulation between its body and its cervix (anteflexion). It also forms an angle of 100 to 110° with the vagina (anteversion). The uterus is, however, readily displaced upward by elevation of the superior surface of the bladder as urine accumulates, so that uterine angles and relations change constantly. The

parts of the uterus are the fundus, the body, the isthmus, and the cervix.

The **fundus** of the uterus is the bluntly rounded free extremity of the organ. It is the part above a line joining the points of entrance of the two uterine tubes. The **body** narrows from the fundus to the isthmus. Its vesical surface is somewhat flattened, but its intestinal surface is distinctly rounded. Its lateral borders are thick, and here the mesometrial portions of the broad ligaments sweep lateralward from the uterus. At the upper end of each lateral border the body of the uterus is pierced by the uterine tube. Below and anterior to the end of the uterine tube, the round ligament is attached to the uterus, while behind it is the point of fixation of the ovarian ligament. The **isthmus** of the uterus is the 1 cm. long constricted region between the body and the cervix. It includes a part previously known as the upper portion of the cervix, together with its opening to the body of the uterus. At the vesical surface of the isthmus the peritoneum reflects from the superior surface of the bladder onto the uterus. The **cervix,** or neck of the uterus, is its tapered vaginal end. It is approximately the lower 2 cm. of

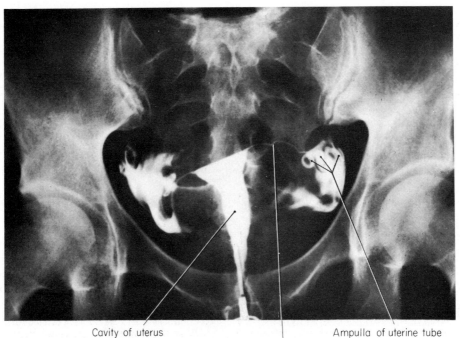

Cavity of uterus   Ampulla of uterine tube

Isthmus portion, uterine tube

Fig. 401   A normal uterosalpingogram.

the organ and is nearly cylindrical in form. Around its circumference the vagina is attached along an oblique line, and so the cervix has a supravaginal and a vaginal portion. The **supravaginal portion** of the cervix is separated from the bladder and adjacent lateral structures by fibrous connective tissue. The uterine arteries reach the margins of the cervix in this fibrous tissue after crossing the ureters 1 to 2 cm. from the cervix (fig. 402). The **vaginal portion** of the cervix projects freely into the anterior wall of the vagina. On its rounded extremity, the **uterine ostium** appears as a small, depressed aperture.

Cavity of the Uterus

The cavity of the body of the uterus is a mere slit in its anteroposterior dimension (fig. 400). In the mediolateral direction, the cavity is triangular (figs. 399, 401), the three points of the triangle being at the ostia of the uterine tubes and at the constriction of the isthmus. The **canal of the cervix** is a spindle-shaped passage, narrowed at the cervical ostium and at its junction with the canal of the isthmus. The cervical canal is characterized by an anterior and a posterior longitudinal fold, or ridge, from which

arises a series of secondary, oblique **palmate folds.** These folds are directed laterally and toward the isthmus of the uterus.

Clinical Note   Supports of the Uterus   The principal supports of the uterus are the pelvic floor and the viscera surrounding the uterus. In the erect posture the uterus lies directly on the superior surface of the bladder, and the cervix and the upper vagina are supported from below and behind by the ampulla of the rectum. A healthy, intact pelvic floor is important in the support of the uterus. Acting as a central attachment for the perineal muscles, a sound **perineal body** is essential to a properly supportive pelvic floor. The ligaments of the uterus maintain the orientation of the organ but are not capable of supporting it. The **round ligaments** of the uterus hold the fundus forward. These flattened bands attach to the lateral border of the uterus just below and anterior to the end of the uterine tubes. They pass anterolaterally in the broad ligaments, forming a ridge in their anterior leaves, pass over the external iliac vessels, and enter the inguinal canal of each side at the deep inguinal ring. Having traversed the inguinal canal, the fibers of each round ligament spread out and end in the skin and connective tissue of the labium majus of the same side. The **rectouterine ligaments** hold the cervix back and upward, and in the erect posture these ligaments are almost vertically oriented. They are attached to the deep fascia and periosteum of the sacrum and sometimes elevate crescentic shelves of peritoneum (rectouterine folds) which delimit the rectouterine pouch. Each ligament consists of a slender band of white fibrous connective tissue and a bundle of smooth muscle, the rectouterine muscle. The cardinal ligaments, long regarded as important supporting elements of the uterus, have been shown to be no more than the connective tissue around the uterine blood vessels as they come from the side wall near the floor of the pelvis. There is certainly some support from the connective tissue here, but discrete ligaments in the traditional sense are entirely lacking. The broad ligaments, by their peritoneal continuations, assist in maintaining the proper orientation of the uterus.

Blood Vessels (fig. 402)

The arteries of the uterus are the uterine from the internal iliac artery and the ovarian from the abdominal aorta (p. 460). The **uterine artery** approaches the organ from the lateral pelvic wall in company with veins, nerves, and lymphatics and is embedded in areolar connective tissue. One to two centimeters from the cervix, the ureter, passing obliquely forward to the bladder, courses under the uterine artery, and a twig of the artery is supplied to it. Reaching the side of the uterus at the isthmus, the uterine artery divides into a larger branch, which supplies the body and fundus, and a smaller descending **vaginal branch** for the cervix and the vagina. The larger branch has a tortuous course

along the side of the body and, at the junctions of the ovarian ligament and the uterine tube with the uterus, anastomoses with branches of the ovarian artery which lie along these structures. From the main vessel ascending at the lateral margin of the uterus, branches pass on to its vesical and intestinal surfaces, penetrate the muscular wall, and make numerous anastomoses. Thin-walled **veins** form a pattern like the arteries, descend along the lateral margins of the uterus, and drain into the internal iliac vein.

Lymphatics (fig. 402)

The lymphatic collecting vessels of the uterus arise in the subserous plexus; to this plexus the deeper lymphatic channels drain. The lymphatic drainage of the uterus is directed to various node groups, principally on a regional basis: (1) From the body of the uterus collecting vessels pass lateralward between the layers of the broad ligaments to end in the external iliac nodes. (2) Collecting vessels from the body of the uterus also follow the uterine vessels, terminating in the internal iliac nodes. (3) Lymph vessels from the fundus join ovarian collecting channels and follow the ovarian vessels to the lumbar nodes. (4) A few lymph channels from the fundus and the upper body accompany the round ligaments of the uterus and end in the superficial inguinal nodes. (5) Lymphatic channels from the cervical portion of the uterus end partly in the internal and external iliac nodes but also pass backward under the rectouterine ligament to the sacral and the median common iliac nodes.

Nerves

The **uterovaginal plexus,** located against the supravaginal portion of the cervix on either side of the uterus, is an extension of the inferior hypogastric plexus. Through the uterovaginal plexus pass sympathetic, parasympathetic, and afferent fibers. The autonomic fibers are largely vasomotor. Most of the afferent neurons ascend through the hypogastric plexus and reach the spinal cord by way of the tenth to twelfth thoracic nerves and the first lumbar nerve.

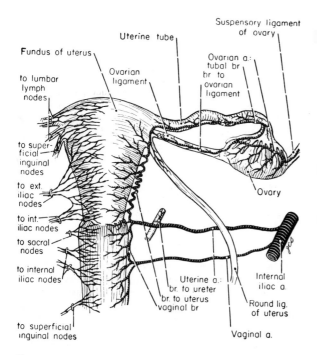

Fig. 402 The arterial supply and lymphatic drainage of the internal genital organs of the female; blood vessels on right, lymphatic channels on left.

THE VAGINA (figs. 399, 400, 403)

The **vagina** extends from the vestibule to the cervix of the uterus. Its direction is parallel to that of the

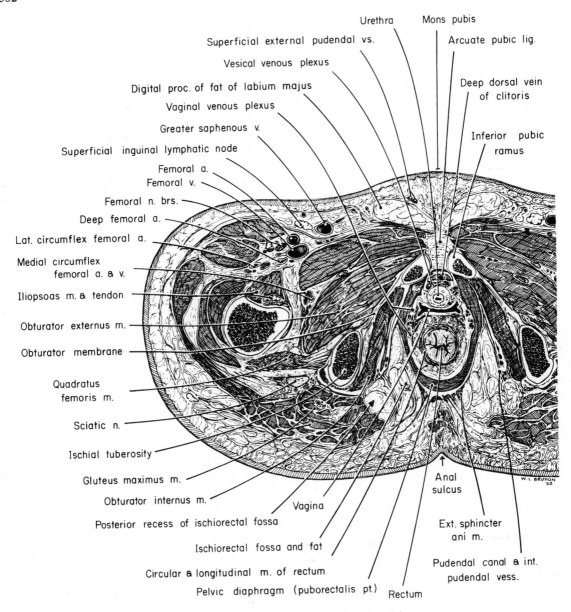

Urethra

Mons pubis

Superficial external pudendal vs.

Arcuate pubic lig.

Vesical venous plexus

Digital proc. of fat of labium majus

Deep dorsal vein
of clitoris

Vaginal venous plexus

Greater saphenous v.

Inferior pubic
ramus

Superficial inguinal lymphatic node

Femoral a.

Femoral v.

Femoral n. brs.

Deep femoral a.

Lat. circumflex femoral a.

Medial circumflex
femoral a. & v.

Iliopsoas m. & tendon

Obturator externus m.

Obturator membrane

Quadratus
femoris m.

Sciatic n.

Ischial tuberosity

Gluteus maximus m.

Obturator internus m.

Posterior recess of ischiorectal fossa

Ischiorectal fossa and fat

Circular & longitudinal m. of rectum

Pelvic diaphragm (puborectalis pt.)

Vagina

Anal
sulcus

Ext. sphincter
ani m.

Pudendal canal & int.
pudendal vess.

Rectum

W. L. BRUDON
'62

Fig. 403 A cross section of the female pelvis.

pelvic inlet, and it describes an angle of 100 to 110°
with the axis of the uterus. Its upper end encloses
the vaginal portion of the cervix, and, consequently,
its walls are unequal in length—about 7.5 cm. for
the anterior wall and 9 cm. for the posterior wall.
The culs-de-sac around the cervix are known as
fornices; and an anterior fornix, a posterior fornix,
and lateral fornices are designated. The walls of the
vagina are in contact except where its lumen is held
open by the cervix uteri. The vestibular entrance is
a sagittal cleft; but throughout its midregion the

vagina has an H-shaped lumen, the principal dimension being transverse. The upper end of the
vagina, clasping the cervix, is circular.

Anteriorly, the vagina is related to the fundus
of the bladder from which it is separated by loose
connective tissue. Lower, it is intimately connected to the wall of the urethra. Posteriorly, the
vagina is related to the rectouterine pouch for 1 or
2 cm., then to the ampulla of the rectum and,
inferiorly, to the perineal body. The rectovaginal
septum intervenes between the vagina and the

rectum. The terminal portions of the ureters pass close to the lateral fornices of the vagina, and the vaginal plexuses of veins lie on either side between the vagina and the pelvic diaphragm.

The vagina has a **mucous coat** continuous with that of the uterus. It is thick and is corrugated by transverse elevations, the **vaginal rugae.** Longitudinal ridges form the **anterior rugal column** and the **posterior rugal column** in the corresponding walls. The female urethra is closely applied to the anterior column in its lower part. The **musculature** of the vaginal wall is principally longitudinal in arrangement, and a thin layer of erectile tissue exists between the mucosal and muscular coats.

Vessels and Nerves (fig. 402)

The vagina is supplied by the **vaginal artery** and the vaginal branch of the **uterine artery.** Vaginal twigs of the middle rectal artery reach it and, in the perineum, branches of the internal pudendal artery. The vaginal, uterine, and middle rectal arteries are described on p. 507. The **veins** form a vaginal plexus along the lateral aspect of the vagina; the efferent vessels of the plexus are tributary to the internal iliac veins. The **lymphatics** of the vagina differ regionally. From the upper portion of the vagina the collecting channels join those from the cervix uteri and reach both the external iliac and internal iliac nodes. The middle and lower regions drain particularly along the vaginal vessels to the internal iliac groups. The lowest part of the vagina (below the hymen) drains, as does the perineum, to the superficial inguinal node groups. The **nerves** of the vagina are derived from the uterovaginal plexus (p. 513).

THE VESSELS AND NERVES OF THE PELVIS

THE INTERNAL ILIAC ARTERY (figs. 404, 431)

The **internal iliac artery** of each side arises at the bifurcation of the common iliac artery. It is the artery of the pelvis but also has a number of branches which supply the buttocks, the medial thigh regions, and the perineum. The origin of the internal iliac artery is opposite the lumbosacral disk and in front of the sacroiliac articulation. The vessel is thick and short (from 3 to 4 cm.) and descends to the upper margin of the greater sciatic foramen, where it

commonly divides into an anterior and a posterior division. The most common pattern of branching of the internal iliac artery is only an approximate guide to the identification of the vessels in any specimen, for it is one of the most variable arterial systems of the body. It is useful, therefore, to classify the branches rather than to describe them by the order of their origin. The **posterior division** of the artery passes backward and provides parietal branches only—two to the pelvic wall and one to the gluteal region. These are the iliolumbar, the lateral sacral, and the superior gluteal arteries. The **anterior division** is the direct continuation of the main stem of the internal iliac. It gives rise to three parietal branches to the thigh and the perineum—obturator, internal pudendal, and inferior gluteal—and provides all the branches to the pelvic viscera. These are the umbilical, the inferior vesical, the middle rectal and, in the female, the uterine and vaginal arteries.

The Iliolumbar Artery

The iliolumbar artery, a branch of the posterior division of the internal iliac artery, passes upward to the iliac fossa. Its course is anterior to the sacroiliac joint and posterior to the lower part of the common iliac vessels and the psoas major muscle, and it ascends between the lumbosacral trunk and the obturator nerve. The vessel divides in the iliac fossa into an iliac and a lumbar branch.

The **iliac branch** supplies the iliacus muscle, provides a large nutrient artery to the ilium, and sends branches into the abdominal wall muscles at the iliac crest. It anastomoses with branches of the obturator and deep circumflex iliac arteries. The **lumbar branch** supplies the psoas major and quadratus lumborum muscles and anastomoses with the fourth lumbar artery. Like the lumbar arteries, it sends a small **spinal branch** into the intervertebral foramen between the fifth lumbar vertebra and the sacrum for the supply of the nerves of the cauda equina.

The Lateral Sacral Arteries

The lateral sacral arteries are usually two, a superior and an inferior. The large **superior artery** passes medialward, anastomoses with the median sacral artery, and enters the first or the second pelvic sacral foramen to supply the contents of the sacral canal. It then emerges through the dorsal sacral foramen and ends in the muscles on the dorsum of the sacrum, anastomosing with the superior gluteal branches. The **inferior artery** passes

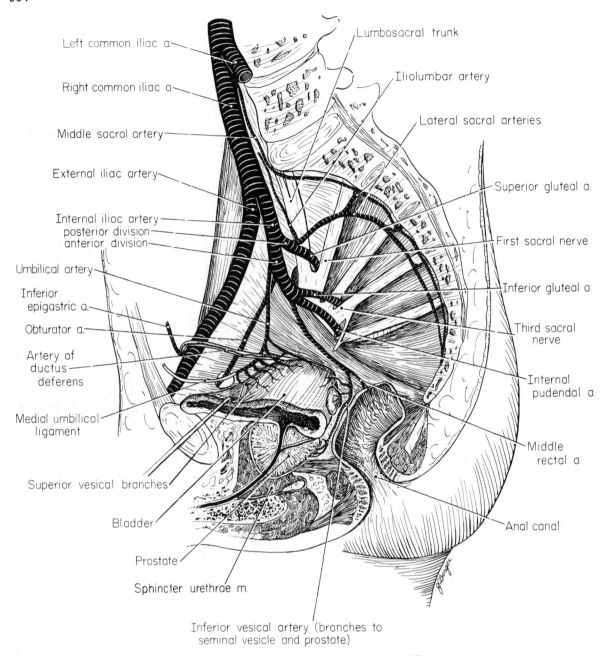

Left common iliac a.

Right common iliac a.

Middle sacral artery

External iliac artery

Internal iliac artery
posterior division
anterior division

Umbilical artery

Inferior
epigastric a.

Obturator a.

Artery of
ductus
deferens

Medial umbilical
ligament

Superior vesical branches

Bladder

Prostate

Sphincter urethrae m.

Lumbosacral trunk

Iliolumbar artery

Lateral sacral arteries

Superior gluteal a.

First sacral nerve

Inferior gluteal a.

Third sacral
nerve

Internal
pudendal a.

Middle
rectal a.

Anal canal

Inferior vesical artery (branches to
seminal vesicle and prostate)

Fig. 404   The common pattern of division of the internal iliac artery.

over the piriformis muscle and the sacral nerves and descends on the pelvic surface of the sacrum between the sacral sympathetic trunk and the sacral foramina. Over the coccyx the artery ends by anastomosing with the median sacral artery and with its fellow of the opposite side. In its descent the inferior artery sends branches into the pelvic sacral foramina which partly end in the sacral canal and partly emerge on the dorsum.

The Superior Gluteal Artery (figs. 406, 431, 476)

The large superior gluteal artery appears to be the continuation of the posterior division of the internal iliac artery. Usually passing between the lumbosacral trunk and the first sacral nerve, the artery leaves the pelvis through the upper part of the greater sciatic foramen above the piriformis muscle. Before leaving the pelvis, a few muscular branches and a nutrient artery to the ilium are given off.

Along the upper border of the piriformis muscle, the artery divides into a superficial and a deep branch.

The **superficial branch** enters the gluteus maximus muscle and divides into a number of branches to the muscle, to the skin overlying its origin, and to the sacrum. The branches anastomose with branches of the inferior gluteal and lateral sacral arteries. The **deep branch** arches forward in the intermuscular plane between the gluteus medius and gluteus minimus muscles. An upper radicle of the deep branch lies over the upper border of the gluteus minimus and reaches as far as the anterior superior iliac spine, anastomosing with the deep circumflex iliac artery and the ascending branch of the lateral circumflex femoral artery. It supplies branches to the gluteus medius, gluteus minimus, and tensor fasciae latae muscles. A lower radicle, accompanied by the superior gluteal nerve, is directed toward the greater trochanter of the femur. It supplies the gluteal muscles and the hip joint and anastomoses with the lateral circumflex femoral artery.

The Obturator Artery (fig. 405)

The obturator artery, frequently the first branch of the anterior division of the internal iliac artery, runs downward and forward to the obturator foramen, passing along the lateral pelvic wall near its brim. Traversing the upper part of the foramen,

the artery enters the thigh and immediately divides into an anterior and a posterior branch. In its pelvic course the obturator nerve lies above the obturator artery and vein (fig. 406). Medial to the artery is the ureter, the ductus deferens, and the peritoneum. Muscular branches supply adjacent muscles, a nutrient branch supplies the ilium, and a vesical branch passes to the bladder. A **pubic branch** (fig. 406) arises just before the obturator artery leaves the pelvis. It ramifies on the pelvic surface of the pubis, anastomosing with its fellow of the opposite side and with the pubic branch of the inferior epigastric artery.

The **anterior branch** of the obturator artery passes forward and downward on the external surface of the obturator membrane. It supplies the external and internal obturator muscles and the origins of the adductor magnus and brevis muscles and anastomoses with the medial circumflex femoral artery. The **posterior branch** follows the posterior margin of the obturator foramen, sending one offshoot along the ischial ramus to anastomose with the anterior branch. The remainder of the vessel supplies the origin of the adductor magnus muscle and the hip joint. Its **acetabular branch** enters the joint through the acetabular notch and reaches the head of the femur by way of the ligamentum capitis femoris (see also acetabular branch of medial circumflex femoral artery, p. 551).

Clinical Note An 'abnormal' obturator artery occurs in a high percentage of cases (30 per cent of sides—Pick, Anson, & Ashley, '42). Such a vessel takes its origin from the inferior epigastric or, rarely, the external iliac artery and descends across the brim of the pelvis to the obturator foramen. Its presence is due to the enlargement of the normal anastomosis of the pubic branches of the inferior epigastric and obturator arteries (fig. 406), so that a vessel of considerable size is formed. The particular significance of the 'abnormal' obturator artery is that it may pass toward the brim of the pelvis along the free margin of the lacunar ligament and thus lie at the medial edge of the femoral ring; it is in danger if that ring has to be enlarged to free the sac of a femoral hernia.

The Internal Pudendal Artery (figs. 404, 431)

The internal pudendal and the inferior gluteal arteries are terminals of the anterior division of the internal iliac artery and frequently arise from a common trunk either within or outside the pelvis. The internal pudendal artery (or the common trunk) leaves the pelvis at the lower border of the greater sciatic foramen, passing between the piriformis and the coccygeus muscles. It then crosses the ischial spine and enters the perineum through the lesser sciatic foramen. Its further course and branches are described in the chapter on the Perineum (p. 475).

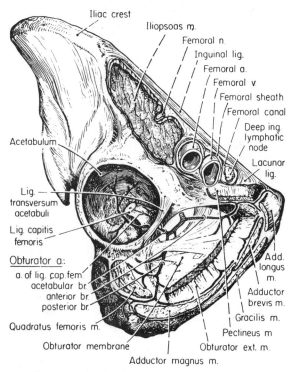

Fig. 405 The distribution of the obturator artery.

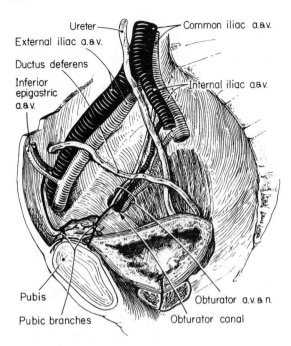

Ureter
External iliac a.&v.
Ductus deferens
Inferior epigastric a.&v.
Common iliac a.&v.
Internal iliac a.&v.
Pubis
Pubic branches
Obturator a.v. & n.
Obturator canal

Fig. 406 The connection of the inferior epigastric and obturator arteries through their pubic branches.

The Inferior Gluteal Artery (figs. 404, 431)

The inferior gluteal is the larger of the two terminal branches of the anterior division of the internal iliac artery. It frequently passes between the first and the second or the second and the third sacral nerves and leaves the pelvis just below the piriformis muscle. Small irregular branches to the viscera and muscles may arise in the pelvis. In the gluteal region the artery descends on the postero-medial side of the sciatic nerve deep to the gluteus maximus muscle, and its lowermost terminals accompany the sciatic and posterior femoral cutaneous nerves.

Large **muscular branches** are given off to the gluteus maximus and the muscles attached to the tuberosity of the ischium; they anastomose with the internal pudendal, the medial circumflex femoral, and the obturator arteries. One or two **coccygeal branches** pass medialward, pierce the sacrotuberous ligament, and supply gluteus maximus and the skin and subcutaneous tissues over the coccyx. An **anastomotic branch** descends across the lateral rotator muscles of the hip and contributes to the 'cruciate anastomosis' of the back of the thigh by anastomosing with the ascending branch of the first perforating artery and with the medial and lateral circumflex femoral arteries (p. 551). The anastomotic branch supplies an articular branch to the back of the hip joint. **Cutaneous branches** accompany the posterior

femoral cutaneous nerve, and the **accompanying artery of the sciatic nerve** descends along the sciatic nerve. This is a long, slender vessel which passes downward along or in the substance of the sciatic nerve as far as the lower part of the thigh. It is developmentally the major artery of the lower limb.

The Umbilical Artery

The umbilical arteries of the fetus are the principal channels for fetal blood flowing from the heart and the aorta to the placenta. With tying of the umbilical cord and abandonment of the placental circulation, the portion of the umbilical artery between its last pelvic branch and the umbilicus becomes atrophic and degenerates into a fibrous strand. This is the **medial umbilical ligament** (obliterated umbilical artery). The proximal part of the umbilical artery, concerned in the supply of the pelvic viscera, persists in the adult, although it is reduced in size from that of the fetal period. The medial umbilical ligament, beginning just beyond the origin of the superior vesical artery, passes retroperitoneally over the bladder toward its apex and then ascends on the posterior surface of the anterior abdominal wall in the medial umbilical fold of the peritoneum. In the pelvis it is crossed by the ductus deferens in the male and by the round ligament of the uterus in the female. The umbilical artery is the first of the visceral branches of the anterior division of the internal iliac artery. It passes forward along the lateral pelvic wall and along the side of the bladder, giving rise to the artery of the ductus deferens and the superior vesical artery.

The **artery of the ductus deferens** is a long, slender vessel that accompanies the ductus deferens as far as the testis. It anastomoses in the spermatic cord with the testicular and cremasteric arteries. In the pelvis it provides twigs to the ureter and the seminal vesicles. The artery of the ductus deferens may be a branch of either the superior or the inferior vesical artery.

Several **superior vesical arteries** arise from the umbilical artery at the side of the bladder. They pass medially onto the superior surface of the urinary bladder, supplying to it numerous branches which anastomose with the other vesical arteries. They also provide small branches to the urachus and to the terminal part of the ureter.

The Inferior Vesical Artery

The inferior vesical artery may arise in common with the middle rectal artery or separately. It runs medialward on the upper surface of the pelvic diaphragm to the underside of the bladder and distributes branches to the fundus of the bladder, the prostate, and the seminal vesicles. The inferior vesical anastomoses with the other vesical arteries and the middle rectal artery. Its prostatic branch anastomoses, within the prostatic sheath, with the artery of the opposite side.

The Uterine Artery (fig. 402)

The uterine artery is the homologue of the artery of the ductus deferens in the male. However, it usually arises separately from the anterior division of the internal iliac, or it may originate in common with the vaginal or the middle rectal artery. The uterine artery runs medialward and slightly forward upon the upper surface of the pelvic diaphragm and, entering the base of the broad ligament, passes to the uterus between its two layers. Accompanied by large veins, it is surrounded by endopelvic fascia. The further course, relations, and distribution of this vessel are described with the uterus (p. 501).

The Vaginal Artery (fig. 402)

The vaginal artery may be represented by one or several branches and may arise directly from the internal iliac or in common with the uterine artery. It runs medialward on the floor of the pelvis and divides into numerous branches for both the anterior and posterior walls of the vagina. Their longitudinal anastomoses form the **azygos arteries** of the vagina which run vertically on both the anterior and posterior surfaces. The vaginal artery anastomoses with the vaginal branch of the uterine artery and with the perineal branch of the internal pudendal artery. It has branches for the bulb of the vestibule, the base of the bladder, and the rectum.

The Middle Rectal Artery

The middle rectal artery may be a separate branch of the anterior division of the internal iliac, or it may arise in common with the inferior vesical or the internal pudendal arteries. From its origin it passes medialward to the middle portion of the rectum where its branches anastomose with those of the superior rectal, inferior rectal, and inferior vesical arteries. It also provides branches to the prostate and the seminal vesicles in the male and to the vagina in the female.

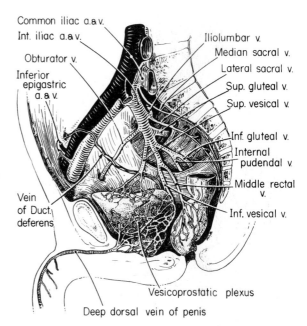

Fig. 407   The internal iliac vein and its tributaries.

THE INTERNAL ILIAC VEIN (figs. 407, 431)

The **internal iliac vein** is formed by the confluence of the veins corresponding to branches of the internal iliac artery. The iliolumbar veins empty into the common iliac veins, and the median sacral vein empties into the left common iliac vein. The internal iliac vein receives extrapelvic tributaries—superior gluteal, inferior gluteal, obturator, and internal pudendal veins—as well as pelvic tributaries. Of the pelvic tributaries, the lateral sacral vein is parietal; others arise in the venous plexuses associated with the pelvic viscera. These are the middle rectal, vesical, prostatic, uterine, and vaginal veins. The internal iliac vein is usually a short, thick trunk which extends from the upper part of the greater sciatic foramen to the sacroiliac articulation. Here it joins the external iliac to form the common iliac vein. The vein lies a little behind and medial to its corresponding artery and contains no valves.

The Superior Gluteal Veins

The superior gluteal veins, venae comitantes of the superior gluteal artery, are formed by tributaries from the gluteal muscles. Entering the pelvis through the greater sciatic foramen with the artery, they end in the internal iliac vein either separately or as a single vein.

The Inferior Gluteal Veins

These are venae comitantes of the inferior gluteal

artery. They begin in the posterior hip area, in venous radicles which follow the pattern of the artery, anastomosing with the medial circumflex femoral, lateral circumflex femoral, and first perforating veins. The inferior gluteal veins enter the pelvis through the lower part of the greater sciatic foramen and combine to form a single vein which empties into the lower part of the internal iliac vein.

The Obturator Vein

The obturator vein, arising in the muscles of the upper medial aspect of the thigh, enters the pelvis through the obturator canal. On the pelvic wall it runs backward and upward below the obturator artery, lateral to the ureter and the descending branches of the obturator artery, and ends in the internal iliac vein.

The Internal Pudendal Veins

The internal pudendal veins begin in the deep veins of the penis, follow the internal pudendal artery, and receive tributaries corresponding to its branches. They usually communicate with the deep dorsal vein of the penis which, however, ends primarily in the prostatic plexus.

The Lateral Sacral Veins

These veins arise from the **sacral venous plexus** located in front of the sacrum. The plexus receives tributaries from the sacral canal and also gives origin to the median sacral vein. The lateral sacral veins ascend in front of the sacrum and end in the internal iliac vein.

The Pelvic Venous Plexuses

The thick, endopelvic fascial investment of the pelvic viscera supports and encloses dense networks of thin-walled veins. These lie adjacent to the borders and surfaces of the rectum, the urinary bladder, the prostate, the uterus, and the vagina. They communicate freely with one another, and from them arise the visceral tributaries of the internal iliac vein.

The rectal plexuses and the middle rectal veins. The **rectal plexuses** (fig. 396) lie in the submucous coat of the rectum and the anal canal and on the outer surface of their muscular layers. The two plexuses intercommunicate freely and, as more fully described with the anal canal (p. 488), are drained by the superior rectal, middle rectal, and inferior rectal veins. They form a peripheral link between the portal and inferior vena caval systems. The **middle rectal veins** arise by tributaries from both the internal and external plexuses of the lower rectum. A right middle rectal and a left middle rectal vein run lateralward on the pelvic diaphragm to end in the internal iliac veins. In the male each vein receives tributaries from the seminal vesicle and the ductus deferens of the same side.

The vesical plexus and the vesical veins. In the male the **vesical plexus** lies against the muscular coat of the bladder. The plexus is particularly dense around the neck of the bladder, where it is continuous with the prostatic plexus, and on the fundus, where it has ex-

tensions around the seminal vesicles, the ends of the ureters, and the ductus deferentes. In the female the **vesical plexus** surrounds the upper part of the urethra and the neck of the bladder. It receives the deep dorsal vein of the clitoris and communicates with the vaginal plexus. From the vesical plexus, in either sex, several **vesical veins** from each side pass to the internal iliac vein.

The prostatic plexus and veins. The **prostatic plexus** (figs. 395, 398, 407) is a dense network of communicating veins which lie partly within the sheath of the prostate gland and partly between its sheath and capsule. The plexus receives the deep dorsal vein of the penis in front and, behind, communicates with the vesical plexus, particularly that part on the fundus of the bladder. The prostatic plexus gives rise to **prostatic veins** which end in the internal iliac vein, but it drains mainly through the vesical plexus.

The uterine plexuses and uterine veins. The **uterine plexuses** are formed of veins along the sides and superior angles of the uterus between the layers of the broad ligaments. They communicate above with the ovarian pampiniform plexus and below with the vaginal plexus. The **uterine veins,** usually two on each side, arise from the lower part of the uterine plexus above the lateral vaginal fornix. They accompany the uterine artery lateralward to terminate in the internal iliac vein.

The vaginal plexuses and the vaginal vein. The **vaginal plexuses** lie at the sides of the vagina and receive tributaries from its walls. They communicate with the uterine, vesical, and rectal plexuses and with the veins of the bulb of the vestibule. A **vaginal vein** arises from the plexus on either side and, accompanying the corresponding artery, ends in the internal iliac vein.

LYMPHATICS OF THE PELVIS

The lymphatic drainage of the pelvic viscera is given with the description of each organ. Considerable anterior and lateral drainage reaches the external iliac nodes, but the internal iliac and sacral nodes receive a large proportion of the lymph from both the viscera and the pelvic wall.

The Internal Iliac Nodes (fig. 371)

The internal iliac nodes, from four to eight in number, lie against or in the angles formed by the branches of the internal iliac artery as these diverge from the parent trunk. Thus, the nodes are found between the origins of the umbilical and obturator arteries as well as in relation to the lower branches of the internal iliac system. The internal iliac nodes receive part of the lymphatic drainage of all the pelvic organs—bladder, ductus deferentes, seminal vesicles, prostate, membranous and prostatic portions of urethra, uterus, vagina, and rectum. They also receive lymph vessels from the gluteal region and the thigh. Small and variable lymph nodes are found intercalated along afferent channels—a middle rectal node, vesical nodes, and, in the female, parauterine nodes. The efferent channels of the internal iliac nodes pass to both the external iliac and common iliac nodes, the latter being the final common path for both of the more peripheral groups.

The Sacral Nodes

The sacral nodes comprise two or three nodes located along the lateral sacral arteries opposite the second sacral and third sacral foramina. They receive afferent channels from the prostate gland, the uterus and the vagina, the rectum, and the posterior pelvic wall. The sacral nodes send their efferent channels to the more medial of the common iliac nodes.

NERVES OF THE PELVIS

Various nerves are represented in the pelvis. The somatic nerves are those of the sacral and coccygeal plexuses, formed by the ventral rami of the sacral and coccygeal nerves. Autonomic nerves and their afferent fibers are numerous. The sacral sympathetic trunk and ganglionic chain descends on the sacrum and the coccyx (p. 511). The abdominal aortic plexus continues over the sacral promontory as the superior hypogastric plexus and divides into the hypogastric nerves for the sympathetic supply of the pelvic viscera. The pelvic splanchnic nerves represent the sacral portion of the parasympathetic division of the autonomic nervous system, and the distribution of these nerves is largely pelvic.

The Sacral Plexus (figs. 408, 431)

Like the other major nerve plexuses of the body, the sacral plexus is formed by the union, division, and regrouping of the ventral rami of spinal nerves. Into the sacral plexus combine the ventral ramus of a part of the fourth lumbar nerve, the ventral rami of the fifth lumbar and the first, second, and third sacral nerves, and the ventral ramus of a part of the fourth sacral nerve.

Formation of the plexus. The sacral plexus is located on the dorsal wall of the pelvis; the nerves forming it converge toward the lower part of the greater sciatic foramen where the plexus forms a broad triangular band, the apex of which passes

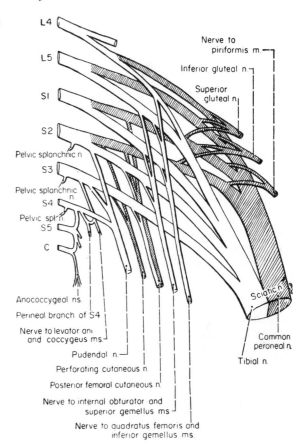

Fig. 408   The formation of the nerves of the sacral and coccygeal plexuses.

into the gluteal region as the sciatic nerve. The descending portion of the fourth lumbar nerve joins the ventral ramus of the fifth lumbar nerve over the ala of the sacrum to form the **lumbosacral trunk.** This trunk descends across the sacroiliac articulation, emerges from the psoas muscle on the medial side of the obturator nerve, and joins the first sacral nerve. The lumbosacral trunk contains anterior and posterior branches of the ventral rami of the fourth and fifth lumbar nerves. The ventral rami of the first, second, and third sacral nerves pass lateralward from the pelvic sacral foramina and divide in front of the piriformis muscle into anterior and posterior parts. The fourth sacral nerve divides between the sacral and coccygeal plexuses. The largest nerves of the sacral plexus are the fifth lumbar and the first sacral, and the superior gluteal artery usually leaves the pelvis by passing between them. All nerves except the fourth sacral divide into anterior and posterior branches; like the other plexuses, the anterior branches form **pre-axial** nerves related to pre-axial muscle masses and skin areas, and the posterior branches form similarly related **postaxial nerves.**

Nerves of the plexus. The principal nerve of the sacral plexus is the **sciatic.** It is composed of a preaxial nerve, the tibial, and a postaxial nerve, the common peroneal, enclosed within a single sheath. In 10 per cent of cases (Beaton & Anson, '38) these are separated in the greater sciatic foramen by all or part of the piriformis muscle. They are occasionally separated through the full length of the thigh. The nerves of the plexus and their sources are tabulated as follows:

	Anterior Branches	Posterior Branches
Sciatic nerve	(tibial)—L4, 5, S1, 2, 3	(common peroneal)—L4, 5, S1, 2
Muscular branches to piriformis		S1, 2
to levator ani and coccygeus	S3, 4	
Superior gluteal nerve		L4, 5, S1
Inferior gluteal nerve		L5, S1, 2
Nerve to quadratus femoris and inferior gemellus muscles	L4, 5, S1	
Nerve to obturator internus and superior gemellus muscles	L5, S1, 2	
Posterior femoral cutaneous nerve	S2, 3	S1, 2
Perforating cutaneous nerve		S2, 3
Pudendal nerve	S2, 3, 4	
Pelvic splanchnic nerves	S2, 3, 4	
Perineal branch of fourth sacral nerve	S4	

The **sciatic nerve** emerges from the pelvis at the lower border of the piriformis muscle and enters the thigh in the hollow between the tuberosity of the ischium and the greater trochanter of the femur. Its distribution is entirely within the lower limb, and its detailed description falls within that chapter (p. 556).

The **nerve to the piriformis muscle** may be represented by separate contributions from S1 and 2. It arises from the dorsal aspect of these nerves and immediately enters the pelvic surface of the muscle.

The **nerves to the coccygeus and levator ani muscles** arise from a loop between the third and fourth sacral nerves and descend to enter the pelvic surface of these muscles.

The **superior gluteal nerve,** formed from the posterior branches of L4 and 5 and S1, passes from the pelvis above the piriformis muscle. Deep to the gluteus maximus and gluteus medius muscles, the nerve accompanies the superior gluteal artery and vein over the surface of the gluteus minimus muscle. It supplies the gluteus medius and gluteus minimus muscles. It also innervates the tensor fasciae latae muscle by a branch which accompanies the lower branch of the deep division of the superior gluteal artery (p. 505).

The **inferior gluteal nerve,** formed from the posterior branches of L5 and S1 and 2, passes from the pelvis below the piriformis muscle. In the gluteal region it is superficial to the sciatic nerve and, after a short course, breaks up into a number of branches which enter the deep surface of the gluteus maximus muscle. It is the sole supply of this muscle.

The **nerve to the quadratus femoris and inferior gemellus muscles** is formed from the anterior branches of nerves L4 and 5 and S1. The nerve leaves the pelvis through the lower part of the greater sciatic foramen deep to the sciatic nerve. It descends over the back of the ischium anterior to the gemelli and obturator internus muscles. In its course it provides an articular branch to the hip joint and a branch to the inferior gemellus muscle and ends in the anterior surface of the quadratus femoris.

The **nerve to the obturator internus and superior gemellus muscles** arises from the junction of the anterior branches of nerves L5 and S1 and 2. In the gluteal region it lies inferomedial to the sciatic nerve and on the lateral side of the internal pudendal vessels. As it crosses the superior gemellus, it gives off a small nerve to this muscle. The nerve to the internal obturator crosses the ischial spine and enters the ischiorectal fossa through the lesser sciatic foramen. It enters the perineal surface of the obturator internus muscle.

The **posterior femoral cutaneous nerve** is a mixed nerve, formed by posterior branches from nerves S1 and 2 and anterior branches from nerves S2 and 3. It leaves the pelvis below the piriformis muscle and, in the gluteal region, lies alongside the sciatic nerve. It descends to the back of the knee and beyond as a cutaneous nerve of the back of the thigh, but at the lower border of gluteus maximus it gives rise to gluteal and perineal branches. The gluteal branches are several **inferior cluneal nerves** which turn around the lower border of the gluteus maximus and supply the skin over the lower and lateral part of the muscle. The **perineal branches** also arise at the lower border of the gluteus maximus muscle. They run medialward across the origins of muscles arising from the ischial tuberosity, pierce the fascia lata, and cross the ischiopubic ramus to the perineum. They are more completely described in the chapter on the Perineum (p. 471).

The **perforating cutaneous nerve** arises from posterior branches of nerves S2 and 3 and is associated at its origin with the lower roots of the posterior femoral cutaneous nerve. It pierces the sacrotuberous ligament and the lower fibers of the gluteus maximus muscle and distributes to the skin over the medial part of the fold of the buttock.

The **pudendal nerve,** formed from the anterior branches of nerves S2, 3, and 4, is the principal nerve of the perineum. In the gluteal region it is inferomedial to the sciatic nerve. It crosses the spine of the ischium and the sacrospinous ligament medial to the internal pudendal vessels and enters the pudendal canal. Its distribution in the perineum is fully covered in the chapter on the Perineum (p. 474).

The **pelvic splanchnic nerves** arise from the anterior aspect of nerves S2, 3, and 4. They pass forward to join the inferior hypogastric plexus as more fully described on p. 513.

The **perineal branch of the fourth sacral nerve** arises from the lower part of the loop between the third and fourth sacral nerves which supplies the levator ani and coccygeus muscles. The perineal branch descends through the coccygeus muscle, enters the ischiorectal fossa, and runs forward to end in the posterior part of the external sphincter ani muscle. It also supplies cutaneous nerves to the overlying skin.

The Coccygeal Plexus

The coccygeal plexus is formed from the junction of part of the ventral ramus of the fourth sacral nerve with all the fifth sacral and coccygeal nerves. This delicate plexus provides the **anococcygeal nerves** which pierce the sacrotuberous ligament and supply the skin in the region of the coccyx.

The Sympathetic Trunk (fig. 409)

The sacral sympathetic trunk is directly continuous with the lumbar portion of the trunk behind the common iliac vessels. Smaller than the lumbar trunk, it descends on the pelvic surface of the sacrum just medial to the pelvic sacral foramina. The trunks of the two sides converge inferiorly and end in front of the coccyx in the small **ganglion impar.** The trunks descend behind the rectum in the retroperitoneal connective tissue. There are usually four ganglia in the sacral region, one opposite each of the upper three sacral segments and one between the fourth and fifth segments. The ganglia are small, but each gives rise to one or more gray rami communicantes which join the adjacent sacral and coccygeal nerves for distribution to blood vessels, sweat glands, the arrector pili muscles, and bones and joints. Direct visceral branches (sacral splanchnic nerves), usually two or three in number, arise from the second and third ganglia. They pass to the inferior hypogastric plexus and distribute through it.

The Hypogastric Plexuses and Nerves (fig. 409)

The **superior hypogastric plexus** of the abdomen continues into the pelvis as the hypogastric nerves and the inferior hypogastric plexuses. The superior hypogastric plexus is continuous with the intermesenteric plexus; it lies against the lower part of the abdominal aorta, its bifurcation, and the median sacral vessels. Its extent is from the lower border of the third lumbar vertebra to the middle of the first sacral segment. The superior hypogastric plexus consists of a broad, flattened band of intercommunicating nerve bundles which descend over the aortic bifurcation. In addition to its continuity with the intermesenteric plexus, it receives the lower two lumbar splanchnic nerves. Broadening below, the plexus divides opposite the first sacral segment into the right and left hypogastric nerves. *There are few, if any, parasympathetic neurons in the superior hypogastric plexus, but the afferent fibers it contains are sometimes clinically more important than its efferent bundles.*

The right and left **hypogastric nerves** are formed from the bifurcation of the superior hypogastric plexus. These relatively solid nerve trunks diverge on either side of the rectum and curve outward, downward, and backward into the pelvis. They are from 7.5 to 10 cm. long and contain no ganglia. They interconnect the superior hypogastric and inferior hypogastric plexuses and are the principal sympathetic roots of the latter. The hypogastric nerves lie in the extraperitoneal connective tissue lateral to the rectum and medial to the internal iliac vessels; they are in the base of the rectouterine fold in the female (or the rectovesical fold in the male). Near their upper extremities, the hypogastric nerves provide branches to the sigmoid colon and the descending colon (p. 513). A prominent vascular branch of each hypogastric nerve joins the internal

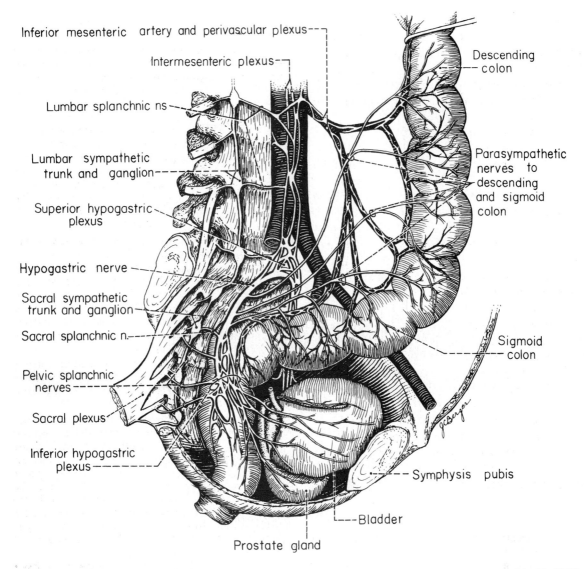

Inferior mesenteric artery and perivascular plexus

Intermesenteric plexus

Descending colon

Lumbar splanchnic ns

Lumbar sympathetic trunk and ganglion

Superior hypogastric plexus

Parasympathetic nerves to descending and sigmoid colon

Hypogastric nerve

Sacral sympathetic trunk and ganglion

Sacral splanchnic n.

Sigmoid colon

Pelvic splanchnic nerves

Sacral plexus

Inferior hypogastric plexus

Symphysis pubis

Bladder

Prostate gland

Fig. 409   The autonomic nerves of the pelvis (from original of fig. 2 R. T. Woodburne, *Anat. Rec.,* **124:** 67, 1956).

iliac or the external iliac artery, and there may also be branches to the testis (or ovary) and the ureter.

The **inferior hypogastric plexuses** are fan-like expansions of the hypogastric nerves which are located, in the male, on either side of the rectum, the prostate, and the seminal vesicles, and against the inferolateral surface of the bladder. In the female the relations are similar, the cervix of the uterus and the lateral vaginal fornices taking the place of the seminal vesicles and the prostate. The intercommunicating nerves and ganglia that constitute

the plexus form a thin meshwork, the dimensions of which are about 6 cm. in the anteroposterior direction and about 4 cm. from above downward. The plexus lies medial to the internal iliac vessels and their radicles and is separated from the viscera by endopelvic fascia. The ureter crosses the superior border of the plexus from without inward, and the terminal ureter is related to the inner surface of the anterior part of the plexus. The hypogastric nerves which join the inferior plexuses at their superolateral angles are their principal sympathetic roots, but the sacral splanchnic branches also convey sympathetic filaments to each plexus. The pelvic splanchnic nerves contribute the parasympathetic supply for the pelvic viscera. The inferior hypogastric plexuses may be divided into rectal and vesical parts in the male, and rectal, uterovaginal, and vesical parts in the female.

The **pelvic splanchnic nerves** represent the sacral portion of the craniosacral (parasympathetic) portion of the autonomic nervous system (p. 30). With cells of origin in the second, third, and fourth sacral spinal cord segments, these preganglionic nerves supply the parasympathetic innervation of all the pelvic and perineal viscera and of the abdominal viscera supplied by the inferior mesenteric artery. The pelvic splanchnic nerves spring from the ventral rami of the second, third, and fourth sacral nerves shortly after their emergence from the pelvic sacral foramina. The contribution from the third sacral nerve is usually the largest. From three to ten strands of nerves pass forward to and become incorporated in the inferior hypogastric plexus. The fibers become indistinguishable from other components in the plexus and distribute through its branches. They synapse in the ganglia of the inferior hypogastric plexus and in minute ganglia in the muscular walls of the pelvic viscera.

The **subsidiary plexuses** of the inferior hypogastric plexus are the middle rectal, the vesical, the deferential, and the prostatic (or the uterovaginal).

(1) The **middle rectal plexus** is an offshoot of the posterior portion of the inferior hypogastric plexus. From four to eight nerves run from the inferior hypogastric plexus to the rectum, mostly directly, but one or two in company with the middle rectal artery. They penetrate the walls of the rectum and provide enteric plexuses like those found in the other parts of the intestines. The parasympathetic fibers of the middle rectal plexus are derived from the pelvic splanchnic nerves.

The parasympathetic supply of the large intestine from its left flexure to the lower rectum and the anal canal is separate from its sympathetic supply. The sympathetic supply is carried through the perivascular inferior mesenteric plexus (p. 439). The **parasympathetic innervation of the distal part of the colon** (fig. 409) is by long, slender nerves which, bilaterally, run courses independent of the sympathetic nerves. Branches emerge from the superior aspect of the inferior hypogastric plexus (occasionally one or more are direct branches of the pelvic splanchnic nerves) and ascend parallel to the rectum and the sigmoid colon. They

pass across the course of the blood vessels in the mesocolon, joining the perivascular plexuses only along the vasa recta. From the upper end of each hypogastric nerve, just below the sacral promontory, a branch passes into the sigmoid mesocolon and runs diagonally toward the descending colon. Present bilaterally, this nerve passes across the sigmoid and left colic arteries and communicates with but does not end in their perivascular plexuses. The nerve ascends along the descending colon, turning into the perivascular nerve plexuses along the vasa recta, and is presumed to be the parasympathetic supply of the descending colon and the left colic flexure. The separate course of the parasympathetic innervation of the distal part of the colon reduces the constitution of the superior hypogastric plexus to sympathetic and afferent fibers.

(2) The **vesical plexus** is the anterior portion of the inferior hypogastric plexus. It forms loops about the terminal portion of the ureter and is disposed along the inferolateral surface of the bladder. It includes both sympathetic and parasympathetic fibers which sink into the muscular walls. The fibers of the plexus supplying the fundus of the bladder and the periureteral nerve loops have offshoots to the seminal vesicle, the ductus deferens, and the ejaculatory duct.

(3) The **deferential plexus** consists of several nerves which accompany and supply the ductus deferens as far as the epididymis. They are off-shoots of the inferior portion of the vesical plexus.

(4) The **prostatic plexus** is derived from the larger nerves of the anterior inferior part of the inferior hypogastric plexus. It lies along the side of the prostate gland, and its nerves communicate with twigs to the neck of the bladder and to the seminal vesicle. It supplies the prostate gland, the prostatic urethra, and the ejaculatory ducts. Terminals communicate with the pudendal nerve, and branches of both sets innervate the penile vessels, the corpora cavernosa, the corpus spongiosum, the membranous and penile portions of the urethra, and the bulbourethral glands. This distribution is occasionally by way of definite **greater and lesser cavernous nerves** which descend along the urethra. The former accompanies the dorsal nerve of the penis and supplies the corpora cavernosa; the latter supplies the corpus spongiosum and the urethra.

(5) The **uterovaginal plexus** is the anterior and intermediate part of the inferior hypogastric plexus in the female, especially those parts lying between the layers of the broad ligament medial to the uterine vessels. Its upper branches innervate the uterus, several nerves ascending with the artery along the lateral border of the uterus. Some filaments reach the uterine tube, innervating its inner end and communicating with the ovarian nerves. The numerous lower uterine branches supply the cervix uteri and the upper vagina. The **vaginal nerves** follow vaginal arteries and end in the vaginal wall, the urethra, the erectile tissue of the vestibular bulbs, the greater vestibular glands, and the erectile tissue of the clitoris. The uterovaginal plexus of the female is comparable to the prostatic plexus of the male.

THE MUSCLES AND FASCIAE OF THE PELVIC WALL

The muscles of the lateral pelvic wall are the piriformis and the obturator internus. These are muscles that move the lower limb but have their origins on the inner aspect of the pelvic girdle. The pelvic diaphragm closes the pelvis posteroinferiorly.

PIRIFORMIS (figs. 411, 431, 442)

The **piriformis muscle,** flat and triangular in form, arises from the pelvic surfaces of the second, third, and fourth sacral vertebrae between and lateral to the pelvic sacral foramina. The muscle is directed toward the greater sciatic foramen which it largely fills. Its rounded tendon inserts into the upper border and medial side of the greater trochanter of

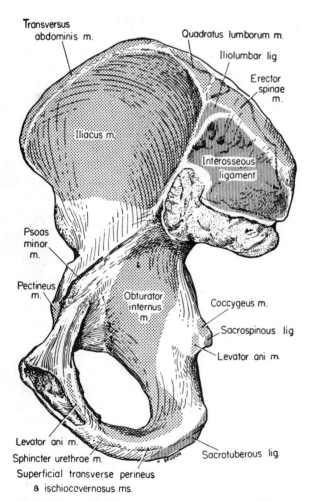

Fig. 410 The internal surface of the coxal bone. (Black dots—origin; white dots on black—insertions; lines—ligaments.)

the femur (fig. 442). The sacral plexus is formed on the pelvic surface of the piriformis. In the subgluteal region of the hip this muscle is an important landmark, for the superior gluteal vessels and nerve emerge above the piriformis, and the sciatic and other nerves of the sacral plexus and the inferior gluteal and internal pudendal vessels emerge at its lower border. The muscle is frequently pierced by the common peroneal nerve in variations of formation of the sciatic nerve. The piriformis is a lateral rotator of the thigh. It is innervated by postaxial branches of the first and second sacral nerves.

OBTURATOR INTERNUS (figs. 388, 395, 403, 410, 411)

The **internal obturator muscle** arises from the entire bony margin of the obturator foramen (except the obturator groove), the inner surface of the obturator membrane, and the pelvic surface of the os coxae behind and above the obturator foramen. The overlying obturator fascia also provides an origin for muscle fibers. The **obturator membrane** completely closes the obturator foramen except at the obturator groove, where the obturator vessels and nerve pass into the thigh. Its interlacing fiber bundles are mainly transverse in direction; they attach to the sharp bony margins of the foramen. The two obturator muscles arise from the opposite surfaces of this membrane. The fibers of the fan-shaped obturator internus muscle converge to pass through the lesser sciatic foramen. Four or five tendinous slips arise from the deep surface of the muscle and, in the foramen, are separated from the bone of the lesser sciatic notch by a large bursa. The tendon turns a 90° angle in passing the lesser sciatic notch and inserts into the medial surface of the greater trochanter of the femur above the trochanteric fossa. In the gluteal region the tendons of the superior gemellus and inferior gemellus muscles blend with the tendon of the internal obturator muscle (fig. 431). The obturator internus is a lateral rotator of the thigh and also assists in abduction. It is innervated, in association with the superior gemellus, by a pre-axial nerve composed of fibers from the fifth lumbar and the first and second sacral nerves.

The **obturator fascia** is thick and strong and provides an origin for part of the fibers of the muscle. It blends with the periosteum of the bony rim of the obturator foramen. A thickening of the fascia which is stretched between the spine of the ischium and the superior pubic ramus anterior to the exit of the obturator nerve and vessels gives origin to a part of the levator ani muscle. This is the arcus tendineus m. levatoris ani (fig. 411). Superior to this thickening, the obturator fascia is strong and represents the true obturator fascia blended with the fascial remnant of the levator ani between the arcuate line of the ilium and the arcus tendineus.

THE PELVIC DIAPHRAGM

The **pelvic diaphragm** is a sheet of muscle covered on both surfaces by fasciae, which closes the pelvic cavity postero-inferiorly and assists in the support of the abdominopelvic viscera. It is stretched somewhat like a hammock between the pubis in front and the coccyx behind and is attached along the lateral pelvic wall to a thickened band in the obturator fascia—**arcus tendineus m. levatoris ani** (fig. 395). The passage of the urethra and the anal canal in the male (or urethra, vagina, and anal canal in the female) necessitates a separation of the two halves of the diaphragm in front of the rectum. In this **genital hiatus** the medial margins of the fasciae of the pelvic diaphragm blend with the adventitia of the perforating structures. The pelvic diaphragm

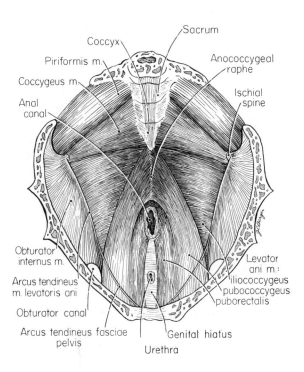

Fig. 411 The muscles of the pelvis as seen from above.

consists of the levator ani and coccygeus muscles and the superior and inferior fasciae.

Levator Ani (figs. 395, 403, 410, 411)

The levator ani muscle consists of three principal parts—puborectalis, pubococcygeus, and iliococcygeus. The levator ani arises, in part, from the dorsal surface of the pubis along an oblique line from the lower part of the symphysis to the obturator canal. The more medial portion of this origin (from the dorsum of the pubis) constitutes the puborectalis; the more lateral portion pertains to the pubococcygeus. The **puborectalis portion** passes backward along the edge of the genital hiatus in contact with the side of the prostate (or the vagina) and, turning behind the rectum, joins the bundle from the other side at the midline. The two sides thus form a U-shaped rectal sling which holds forward the recto-anal junction and assists in producing the 90° angle of its turn. This is a relatively thick portion of the levator ani. Some of the most medial fibers of the puborectalis, those in contact with the prostate (or vagina), turn into the sheath of the prostate, the vagina, and the perineal body in front of the anal canal. Such fibers may be designated as the **levator prostatae** and the **pubovaginalis.** Others end in the anterior and lateral walls of the anal canal. The more inferior fascicles of puborectalis intermingle with the deep portion of the external sphincter ani muscle and reinforce this muscle at the sides and posterior to the anal canal. Others turn into the deep transverse perineus muscle. The **pubococcygeus portion** of the levator ani arises from the back of the superior pubic ramus as far lateralward as the obturator canal. Its fibers pass backward, converging behind the rectum, and insert into the anococcygeal raphe and into the front and sides of the coccyx. The raphe represents the fibrous line of junction of the fibers of the two sides. The **iliococcygeus portion** of the levator ani arises from the arcus tendineus m. levatoris ani and the spine of the ischium. It inserts into the anococcygeal raphe and the coccyx, and at the level of the anal canal its fibers blend with the longitudinal coat of the anal canal. The iliococcygeus is a very thin layer, and its muscular fasciculi are often separated by membranous intervals. Its origin from the tendinous arch over the obturator internus muscle is due to degeneration of the upper portion of the muscle. In many mammals the arcuate line is the origin of this muscle. In Man its superior portion is reduced to a fascial remnant which is blended with the upper portion of the obturator internus fascia, and the muscle takes origin from a thickened line of union of the fascia of the levator ani with that of the obturator internus. This is the **arcus tendineus m. levatoris ani** which extends in a curved line from the spine of the ischium forward and upward to the superior pubic ramus anterior to the obturator canal.

The levator ani muscle supports and raises the pelvic floor and resists increased internal pressure as in forced expiration and defecation. The puborectalis draws the recto-anal junction toward the pubis and reinforces the action of the sphincters of the anal canal and the vagina.

Coccygeus (figs. 410, 411)

The coccygeus muscle abuts against the posterior border of the iliococcygeus. The muscle is triangular in form and arises by its apex from the spine of the ischium and the sacrospinous ligament.

It inserts by its base into the borders of the coccyx and the lowest segment of the sacrum. The coccygeus muscle draws the coccyx forward, elevating the pelvic floor. The entire pelvic diaphragm is innervated by anterior branches of the ventral rami of the third and fourth sacral nerves.

The Fasciae (figs. 396, 412)

The **inferior fascia** of the pelvic diaphragm is an extension of the fascia of the obturator internus muscle. Anteriorly, it is attached to the superior pubic ramus and, at the genital hiatus, it blends with the superior fascia of the pelvic diaphragm. It is continuous with the fascia of the external sphincter ani muscle and, deep to this muscle, attaches to the perineal body anteriorly and covers the anococcygeal raphe posteriorly. The inferior fascia of the pelvic diaphragm forms the superomedial boundary of the ischiorectal fossa.

The **superior fascia** of the pelvic diaphragm is a continuity of the **transversalis fascia**, the parietal

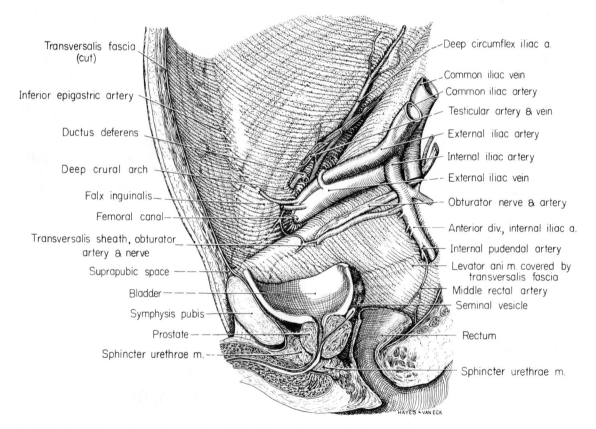

Fig. 412 Semischematic median section of the pelvis and lower abdomen to show the continuity of parietal abdominal and parietal pelvic fasciae (from original of fig. 11, Mark A. Hayes, *Am. J. Anat.*, 87:145, 1950).

Labels on figure:
Transversalis fascia (cut)
Inferior epigastric artery
Ductus deferens
Deep crural arch
Falx inguinalis
Femoral canal
Transversalis sheath, obturator artery & nerve
Suprapubic space
Bladder
Symphysis pubis
Prostate
Sphincter urethrae m.

Deep circumflex iliac a.
Common iliac vein
Common iliac artery
Testicular artery & vein
External iliac artery
Internal iliac artery
External iliac vein
Obturator nerve & artery
Anterior div., internal iliac a.
Internal pudendal artery
Levator ani m. covered by transversalis fascia
Middle rectal artery
Seminal vesicle
Rectum
Sphincter urethrae m.

HAYES + VAN ECK

abdominopelvic fascia (p. 378). From its attachment to the periosteum of the bony rim of the pelvis, the transversalis fascia sweeps down over the obturator internus fascia and onto the superior surface of the pelvic diaphragm. The layer also covers the piriformis muscle. In the genital hiatus it is continuous with the inferior diaphragmatic fascia.

Clinical Note   A prolongation of the transversalis fascia along the obturator nerve and vessels is common; a small hernia of fat and connective tissue frequently occupies this diverticulum.

THE PELVIC FASCIAE

The **parietal fasciae** of the pelvis are the fascia of the obturator internus muscle and the superior fascia of the pelvic diaphragm. These are described on pp. 514 and 516. Fasciae associated with the pelvic viscera consist of the enveloping extraperitoneal connective tissue, the endopelvic fascia, and the fusion fasciae associated with the peritoneum of the pelvis—the rectal fusion fascia and the rectovesical (or rectovaginal) septum and the vesical fusion fascia (p. 485).

The **endopelvic fascia** invests the pelvic viscera, forming a subserous covering for them and enclosing their vascular pedicles. It is continuous with and is an elaboration of extraperitoneal connective tissue. The sheath of the prostate and the connective tissue embedding the seminal vesicles are thick investments of this tissue. The bladder, the rectum, and the female genital organs receive an endopelvic fascial covering, and the thin-walled veins of the pelvis are contained within its meshes. In addition to its pervasive character, the endopelvic fascia has an attachment to the superior fascia of the pelvic diaphragm at the sides and in front of the bladder and the prostate. This attachment is along a line of thickening of the superior fascia designated as the **arcus tendineus fasciae pelvis.** It separates from the arcus tendineus m. levatoris ani as the latter turns toward the obturator canal and, continuing a forward course on the pubococcygeus muscle, ends near the median line on the back of the lower border of the pubis. The sheet of endopelvic fascia which sweeps medialward from the arcus tendineus fasciae pelvis to invest the prostate and the bladder is the **puboprostatic ligament** or **pubovesical ligament** in the female (lateral true ligament of the bladder).

THE BONY PELVIS

Os Coxae (The Hip Bone) (figs. 403, 405, 410, 413, 414, 428)

The hip bones of the two sides form most of the bony pelvis. They are united to one another in front in the pubic symphysis and, behind, are joined to the sacrum, thus forming the ring of bone which encloses the pelvis and unites the trunk and the lower limbs. The hip bone is large and irregular. It has been likened to a propeller, for it has two expanded and oppositely bent blades, and its acetabulum approximates a central axis for the shaft. The os coxae is formed of three separate bones, the ilium, the ischium, and the pubis. These come together in the acetabulum and, fusing during the middle teens, are indistinguishably joined in the adult. In orienting the hip bone for examination,

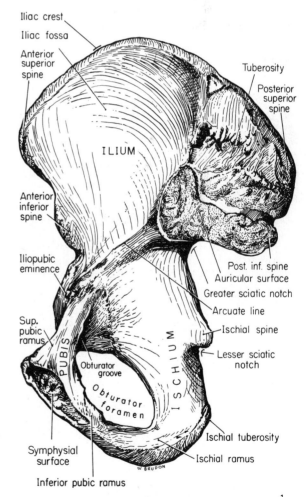

Fig. 413   The internal aspect of the coxal bone. ($\times \frac{1}{2}$)

the acetabulum is directed outward and downward, and the acetabular notch should point directly downward. The anterior superior spine of the ilium and the pubic tubercle lie on the same vertical frontal plane.

The **ilium** is fan-shaped, the ala representing the spread of the fan; the body, the rather broad handle. The body forms about two-fifths of the acetabulum. The **crest** of the ilium is the rim of the fan. It exhibits a reverse curve following the contour of the ala; it is concave in its forward part as the iliac fossa but convex behind as the **auricular surface** of the ilium. The **crest** has a slightly overhanging **external lip,** to the whole length of which is attached the fascia lata. About 6 cm. behind the anterior limit of the crest is a thickened **tubercle.** An **internal lip** marks the upper limit of the iliac fossa; it provides attachment for the iliac fascia. The iliac crest is palpable throughout its entire length and ends anteriorly in the prominent **anterior superior spine** to which is fixed the inguinal ligament. At the posterior end of the crest is the **posterior superior spine** for the attachment of the sacrotuberous ligament and the posterior sacroiliac ligament. A **posterior inferior iliac spine** marks the inferior extermity of the auricular part of the sacropelvic surface of the bone. This spine is rounded like the lobule of an ear. Below it, the posterior surface of the bone curves sharply inward as the **greater sciatic notch.** Below the anterior superor spine is a shallow notch, succeeded by the **anterior inferior spine.**

The **iliac fossa** is the smooth internal concavity of the ala. It narrows below, ending at the roughened **iliopubic eminence,** the line of junction of the ilium and the pubis. The pelvic brim, marked by the **arcuate line** of the ilium, is the postero-inferior limit of the iliac fossa. Behind and below the iliac fossa and the arcuate line is the sacropelvic surface of the ilium. Posterior to the iliac fossa, this region exhibits the **auricular surface** for articulation with the first two segments of the sacrum and, behind and above it, the **tuberosity.** The roughened tuberosity provides attachment for the short posterior sacroiliac ligaments and for fibers of the erector spinae and multifidus muscles.

The **gluteal surface** of the ala is convex anteriorly and concave posteriorly; it presents three gluteal lines. The **posterior gluteal line** begins at the iliac crest in front of the posterior superior spine and arches downward, ending in the greater sciatic notch. The **anterior gluteal line** begins at the crest several centimeters behind the anterior superior spine. It arches backward and downward across the gluteal surface, ending in the greater sciatic notch. The **inferior gluteal line** begins in the notch above the anterior inferior spine and, running almost horizontally, terminates in the anterior part of the greater sciatic notch. The gluteal surface is succeeded inferiorly by the acetabular portion of the ilium, although the reflected tendon of the rectus femoris muscle intervenes just above the acetabular rim.

The **ischium** is the postero-inferior V-shaped member of the hip bone. The blunt apex of the V is heavy and rounded as the ischial tuberosity. In the sitting position the weight of the body rests on the two ischial tuberosities. The ischium has a **body** and a **ramus.** The body forms a little more than two-fifths of the acetabulum posterior to the acetabular notch. The body is

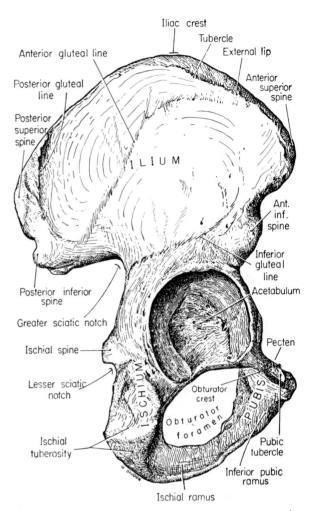

Fig. 414   The external aspect of the coxal bone. ($\times \frac{1}{2}$)

heavy in response to its weight-bearing function. It bears a blunt projection which is pointed posteromedially; this is the **spine of the ischium.** Above the spine the **greater sciatic notch** is spanned by the sacrospinous ligament which converts this notch into the greater sciatic foramen. Below the ischial spine the **lesser sciatic notch** is smooth except for shallow hollows across which play the tendinous slips of the obturator internus muscle. The notch is converted into a foramen by the crossing of the sacrotuberous and sacrospinous ligaments. The **ischial tuberosity** has an upper smooth quadrilateral portion and a lower rough triangular portion. The **pelvic surface** of the body of the ischium is smooth and forms part of the lateral wall of the pelvis. It borders the obturator foramen and gives rise to some of the fibers of the internal obturator muscles.

The **ramus** of the ischium is flattened from side to side and everted externally along its inferior border. It joins the inferior ramus of the pubis to form the ischiopubic ramus, the junction being indicated by a narrowing of the ramus.

The **pubis** is an angulated bone. Its **body** forms one-fifth of the acetabulum, completing, with the body portions of both the ilium and the ischium, the construction of the fossa for the head of the femur. The ilio-pubic eminence marks the junction of the body of the pubis with that of the ilium. The **superior ramus** of the pubis extends to the median plane, where it is joined to its fellow of the opposite side in the symphysis pubis. The lateral portion of the ramus carries on its upper surface an oblique ridge, the **pecten** of the pubis, which begins at the pubic tubercle and is confluent with the arcuate line of the ilium laterally. The pecten provides attachment for the falx inguinalis, the lacunar ligament, the reflected inguinal ligament, and the pectineal fascia. In front of the pecten the superior ramus is smooth; the pectineus muscle and femoral vessels overlie this area. This surface is limited below by the **obturator crest,** which runs from the pubic tubercle to the acetabulum and gives origin to the pubo-femoral ligament. The underside of the superior ramus bounds the obturator foramen and presents the deep **obturator groove** for the passage of the obturator nerve and vessels.

The compressed medial portion of the superior ramus (formerly called the body) is rough along its superior margin (pubic crest) and, laterally, presents the projecting **pubic tubercle**. The inguinal ligament attaches to the tubercle. The posterior aspect of the medial part of the superior ramus is smooth and underlies the urinary bladder. The oval **symphyseal surface** is crossed by eight or nine transverse ridges to which the fibrocartilage of the symphysis pubis is attached.

The **inferior ramus** of the pubis extends inferiorly and posterolaterally into continuity with the ramus of the ischium. The inferior pubic ramus is thin and flattened. The medial border is thick, rough, and everted. Its ridges, continuous with those of the ramus of the ischium, give attachment to the perineal membrane and the membranous layer of the subcutaneous connective tissue of the perineum.

The **acetabulum** (figs. 405, 414) is a deep hemispherical cavity which receives the head of the femur. Its heavy wall consists of a semilunar articular portion, open below, and a deep, central nonarticular portion. One-fifth of the smooth **lunate surface** of the articular part of the acetabulum is formed from the body of the pubis; two-fifths, from the body of the ilium; and two-fifths, from the body of the ischium. Its prominent rim gives attachment to the glenoid labrum of the hip joint; its uneven internal edge provides an attachment for the synovial membrane of the joint. The **acetabular fossa** is the deep, central, pitted nonarticular part of the acetabulum. It is formed mainly from the ischium, and its wall is frequently thin. It lodges a mass of fat. The acetabular fossa is open inferiorly as the **acetabular notch.** This notch is bridged by the transverse ligament, and the margins of the notch give attachment to the ligamentum capitis femoris.

The **obturator foramen** is a large aperture surrounded by the bodies and the rami of the ischium and the pubis. It has an oval form in the male pelvis and a more triangular form in the female. Its thin margin provides attachment for the obturator membrane except superiorly where the underside of the superior ramus of the pubis presents the obturator groove for the obturator vessels and nerve. The groove runs obliquely forward and downward from the pelvis into the thigh and is converted into a canal by a specialization of the obturator fascia.

Ossification

As the bony support of the pelvic organs and the connection between the trunk and the lower limb, the sequences in ossification of the coxal bone have much importance. The hip bone is ossified from eight centers: three primary centers for the ilium, the ischium, and the pubis; and five secondary centers for the iliac crest and the anterior inferior spine, the ischial tuberosity, the pubic symphysis, and the triradiate piece at the center of the acetabulum. The primary centers appear during the third, fourth, and fifth months of development, but at birth the bones are quite separate and the secondary centers have not yet appeared. About the thirteenth or the fourteenth year, the major portions of the ilium, the ischium, and the pubis are completely bony, but they are still separated by a Y-shaped (triradiate) cartilage at the centre of the acetabulum. Ossification of this area proceeds to fusion of the three bones between the ages of fifteen and sixteen. The other secondary ossification centers begin to exhibit bone at puberty and unite with the major bones between the ages of twenty and twenty-two.

ARTICULATIONS OF THE BONY PELVIS

The articulations of the bony pelvis include the sacroiliac joint and the symphysis pubis. These are joints where free motion is sacrificed in favor of stability in the union between the trunk and the pelvic girdle. Other articulations of the pelvis are the lumbosacral and the sacrococcygeal joints.

The Lumbosacral Joint

This joint unites the fifth lumbar vertebra with the first segment of the sacrum. These parts are joined, like all typical vertebrae, by an intervertebral disk, anterior and posterior longitudinal ligaments, synovial joints between their articular processes, ligamenta flava, and interspinal and supraspinal ligaments. In addition, an iliolumbar ligament helps to stabilize the lumbosacral joint. The strong iliolumbar ligament (fig. 416) passes lateralward from the tranverse process of the fifth lumbar vertebra to the posterior part of the inner lip of the iliac crest. A partially separate lumbosacral portion of the ligament descends obliquely

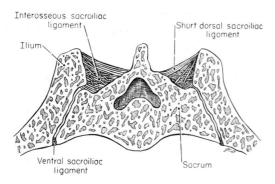

Fig. 415 A cross section of the sacrum and ilium to show the sacroiliac articulations and their ligaments.

to the base and the ala of the sacrum, intermingling with the ventral sacroiliac ligament.

The Sacrococcygeal Joint

This joint exhibits an intervertebral disk, reinforced all around by longitudinal strands designated as the **sacrococcygeal ligaments** (fig. 416).

The Sacroiliac Articulation (figs. 415, 416)

The sacroiliac joint is formed between the **auricular surfaces** of the sacrum and the ilium. These are covered by cartilage, and their irregularities and sinuosities result in a partial interlocking of the facets. The joint is a synovial one, but the movement permitted is limited by the interlocking and by the thick firm dorsal sacroiliac ligaments. The dorsal sacroiliac ligaments are supplemented by the ventral sacroiliac and interosseous sacroiliac ligaments.

The **ventral sacroiliac ligaments** consist of numerous thin bands which close the joint on its pelvic aspect. They connect the pelvic surface of the lateral part of the sacrum to the margin of the auricular surface of the ilium. The **dorsal sacroiliac ligaments** fill the deep depression between the sacrum and the tuberosity of the ilium. Numerous fasciculi pass in various directions between the bones. The **short dorsal sacroiliac ligaments** of the upper part of the space pass horizontally between the first and second transverse tubercles of the sacrum and the iliac tuberosity. The **long dorsal sacroiliac ligament** is oblique; its fibers descend from the posterior superior iliac spine to the third and fourth transverse tubercles of the sacrum. The **interosseous sacroiliac ligaments** are deep to the dorsal ligaments. They consist of a thick group of short, strong fibers that fill the narrow cleft between the rough areas on the bones immediately above and behind the auricular surfaces.

Accessory ligaments of the sacroiliac articulation are the sacrotuberous and sacrospinous ligaments

(fig. 417). The **sacrotuberous ligament** is long, flat, and triangular. Its superior attachments are the posterior superior and posterior inferior iliac spines, the back and the side of the lower part of the sacrum, and the side of the coccyx. The fibers converge below to end on the ischial tuberosity, some of the fibers continuing along its medial margin and on to the ischial ramus. The ligament is in line with the long head of the biceps femoris muscle and may be considered as being derived therefrom. The posterior surface of the ligament gives origin to the gluteus maximus muscle. The **sacrospinous ligament,** triangular in form, is attached by its apex to the ischial spine. Its broader base arises from the side of the lower sacral and coccygeal segments. This ligament converts the greater sciatic notch into the greater sciatic foramen. With the sacrotuberous ligament, which crosses it dorsally, it likewise converts the lesser sciatic notch into the lesser sciatic foramen. On the pelvic

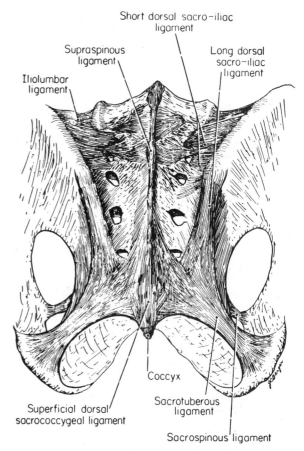

Fig. 416 The posterior ligaments of the sacrum.

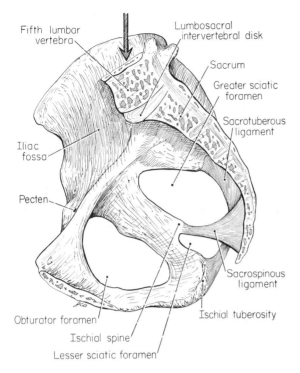

Fifth lumbar vertebra

Lumbosacral intervertebral disk

Sacrum

Greater sciatic foramen

Sacrotuberous ligament

Iliac fossa

Pecten

Sacrospinous ligament

Obturator foramen

Ischial tuberosity

Ischial spine

Lesser sciatic foramen

Fig. 417 A median section of the bony pelvis and the sacrotuberous and sacrospinous ligaments. The arrow indicates the line of gravity in the erect posture.

surface of the sacrospinous ligament lies the coccygeus muscle; the ligament may be regarded as a degenerated part of this muscle.

The sacroiliac joints permit but little movement. Slight gliding and rotary movements between the auricular surfaces take place, but any appreciable action here would lead to instability in the erect posture. The arterial supply of the joint is by branches of the superior gluteal, iliolumbar, and lateral sacral arteries. Nerves reach the joint from the superior gluteal nerve and from branches of the dorsal rami of the first and second sacral nerves.

The Symphysis Pubis

The pubic symphysis is a median joint between the superior rami of the pubic bones. Each articular surface is covered with a thin layer of hyaline cartilage. Each surface is irregularly ridged and grooved, and the irregularities of the two sides fit closely together. The apposed cartilaginous layers are united by an **interpubic disk** of fibrocartilage, thicker in the female than in the male. There may be a cavity in its center. Overlying the disk anteriorly are decussating tendinous fibers of the rectus

abdominis and external abdominal oblique muscles; these strengthen the joint. A **superior pubic ligament** connects the rami along their superior surfaces and extends as far lateralward as the pubic tubercles. The **arcuate pubic ligament** (figs. 384, 403) arches across the joint inferiorly and rounds off the subpubic angle. Between this thick triangular arch of ligamentous fibers and the transverse ligament of the perineum there is a small gap through which the deep dorsal vein of the penis or clitoris passes into the pelvis.

THE MECHANISM OF THE PELVIS (fig. 417)

The principal function of the bony pelvis is that of a bony member intermediate between the trunk and the lower limbs, and the services it renders include transmitting the weight of the body to the limbs and absorbing the stresses of muscular activity in the erect posture. The weight of the body above the hips may be thought of as being concentrated in the first segment of the sacrum and being distributed from there into the pelvic girdle. In transferring it to the heads of the femurs, the lines of weight-bearing diverge in a sacrofemoral arch. For the sitting posture weight is shifted to the ischial tuberosities through a hypothetical sacroischial arch. The greatest thickness and strength of the hip bones lie through the body portions of the ilium and the ischium along the lines of these arches. The tie-rods of the arches are the superior pubic and ischiopubic rami, which resist inward collapse of the bones. The resilient weight support afforded by the curvatures of the vertebral column imposes additional stresses in the pelvis, for the weight of the body reaches the sacrum anterior to the sacroiliac articulation. This results in a turning or rotary force, so that the lower sacrum and the coccyx tend to tilt backward and the sacral promontory tends to drop forward. These tendencies are resisted by irregular articular planes and by ligaments. The central portion of the auricular surface of the sacrum is concave and is lipped on each edge; conversely, the auricular surface of the ilium is convex in the corresponding central area. Furthermore, the lower parts of the auricular surfaces are sloped, so that the widest portion of this surface of the sacrum is on its pelvic side. The effect of this slight wedge shape is to resist backward displacement of the lower sacrum. A similar eversion of the anterior edge of the auricular surface of the ilium in its upper portion prevents the sacral promontory from dropping forward. These small but important reciprocal bony features are effective only as long as the bones are held in intimate and firm apposition. The very strong interosseous and dorsal sacroiliac ligaments normally maintain such apposition. The tendency to rotation of the sacrum and the coccyx is also countered by the strong sacrotuberous and sacrospinous ligaments (fig. 417). They hold the lower sacral segments forward and resist any tendency for them to rotate backward. The iliolumbar ligaments help to prevent the last lumbar vertebra from slipping forward

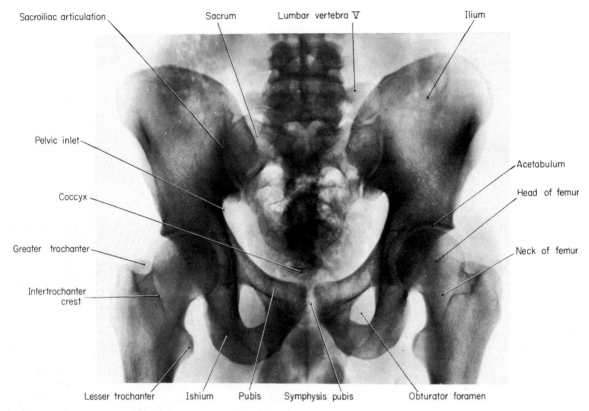

Sacroiliac articulation   Sacrum   Lumbar vertebra Ⅴ   Ilium

Pelvic inlet

Acetabulum

Head of femur

Coccyx

Neck of femur

Greater trochanter

Intertrochanter
crest

Lesser trochanter   Ishium   Pubis   Symphysis pubis   Obturator foramen

Fig. 418   A normal pelvic radiogram, the sacroiliac and hip joints.

on the oblique upper surface of the first sacral segment.

The overlapping of the articular processes of the fifth lumbar vertebra in relation to the sacrum is also a factor in stabilizing the lumbosacral junction. The importance of this feature is seen in the abnormal condition of spondylolisthesis, a poorly understood separation of the neural arch of the fifth lumbar vertebra from its body. Such a separation permits forward displacement of the vertebral column on the sacrum. The pelvis is divided into the greater and lesser pelves by an oblique plane which passes through the sacral promontory, the arcuate line of the ilium, the pectineal line of the pubis, and the upper margin of the pubic symphysis. The whole line is called the **terminal line,** or pelvic brim. The **greater pelvis** is related to the lower abdominal cavity; it is largely the iliac fossae. The **lesser pelvis** is the true pelvis and contains the pelvic viscera. It has a pelvic inlet, or **superior pelvic aperture,** circumscribed by the terminal line.

Clinical Note   Special Features of the Female Pelvis   The female pelvis shows differences from the male which are mainly related to the childbearing function. The principal difference is a modification which enlarges the pelvic outlet; thus the subpubic angle is wider, the ischial tuberosities are everted, and the symphysis pubis is shallower. With eversion below, there is a straightening above, so that the alae of the ilia are more upright in the female; they have a greater flare in

the male. As a consequence, the male iliac fossa is deeper than the female fossa. The cavity of the true pelvis is shallower but wider in the female; the greater pelvis is relatively narrower. In addition, in the female the sacrum is broader and less curved. Other differences are the greater delicacy of the bones in the female, the greater angle of the sciatic notch, the triangularity and smaller size of the obturator foramina, and the smaller size and more forward attitude of the acetabulae.

In making a judgment of the capacity of the female pelvis for childbearing, the diameters of the true pelvis may be determined radiographically. The **conjugate diameter** extends in the median line from the sacral promontory to the symphysis pubis; it averages 11 cm. The **transverse diameter** is the greatest width of the superior aperture from the middle of the brim on one side to the other; it is usually 13.5 cm. The **oblique diameter** is measured between the iliopubic eminence on one side and the sacroiliac articulation on the other; it averages 12.5 cm. The pelvic outlet, or **inferior pelvic aperture,** is completely described in the chapter on the Perineum (p. 469). It is diamond-shaped and is bounded by the back of the pubis, the ischiopubic rami, the ischial tuberosities, the sacrotuberous ligaments, and the tip of the coccyx. In the female its anteroposterior diameter ranges from 9 to 11.5 cm.; it varies with the mobility of the coccyx. The transverse diameter of the pelvic outlet, measured across the posterior parts of the ischial tuberosities, is from 11 to 12 cm. In comparison with the pelves of lower animals, the human pelvis is characterized by its breadth, shallowness, and great capacity.

IX

The Lower Limb

THE SUPERFICIAL ASPECTS OF THE LIMB

The lower limb is separated from the pelvic portion of the trunk by the external borders of the pelvic bones together with their associated ligaments. Thus the uppermost limits of the limb are the iliac crest, the inguinal ligament, the symphysis pubis, the ischiopubic ramus, the ischial tuberosity, the sacrotuberous ligament, and the dorsum of the sacrum and the coccyx. The limb is divisible into the hip and the thigh, the knee, the leg, the ankle, and the foot.

SURFACE ANATOMY

The Hip and the Thigh (figs. 419, 427, 431)

In the thigh the skin is relatively thin and fine and is loosely attached to the underlying fascia. It is especially fine on the medial side of the thigh, where it contains only delicate hairs and evenly distributed sebaceous glands. The skin of the gluteal region, on the other hand, is thick and coarse and is a frequent site for boils. It is poorly vascularized, but its nerve supply is rich.

The **iliac crest** is palpable throughout its entire length. The highest point of its curvature is at the level of the fourth lumbar vertebra. The anterior limit of the crest, the **anterior superior spine of the ilium,** is readily felt and is visible in thin individuals. It is an important landmark and gives attachment to the inguinal ligament. About 6 cm. behind the anterior spine the iliac crest has a lateral projection known as the **tubercle of the crest.** The crest ends posteriorly in the **posterior superior iliac spine.** This spine is often in the depths of a dimple of skin and lies at the level of the second sacral spine and opposite the middle of the sacroiliac articulation. The **inguinal ligament** is a tense band which connects the anterior superior spine of the ilium and the tubercle of the pubis. It becomes visible as a subcutaneous cord when, lying on one's back on the edge of a bed, one leg is dropped to the floor. The pubic tubercle is palpable at the lower end of the inguinal ligament. In the erect posture the pubic tubercle and the anterior

superior iliac spine lie in the same vertical plane frontal to the body. Posteriorly, the **gluteal fold** marks the lower border of the thick gluteus maximus muscle. The **ischial tuberosity** is palpable under the middle of this fold. The tuberosity is not covered by the gluteus maximus muscle when the thigh is flexed. Anterior to the tuberosity, the entire length of the **ischiopubic ramus** is subcutaneous; it is the lateral boundary of the urogenital triangle of the perineum.

A hand-breadth below the tubercle of the crest of the ilium lies the **greater trochanter of the femur.** It is palpable. The tip of the greater trochanter is normally located on a line carried around the thigh from the anterior superior iliac spine to the ischial tuberosity (Nelaton's line). The **sciatic nerve** leaves the gluteal region for the thigh midway between the greater trochanter and the ischial tuberosity. The nerve descends from this point toward the middle of the back of the knee.

The femur is enclosed by the muscles of the thigh; its epicondyles become palpable at the knee. The **sartorius muscle** is visible on the front of the thigh (especially when the thigh is flexed, abducted, and laterally rotated), for the muscle arises on the anterior superior iliac spine and inserts on the medial aspect of the tibia. Lateral to and below the sartorius is the bulky **quadriceps femoris muscle** which tapers into continuity with the subcutaneous **patella** at the knee. Medial to the sartorius lies the **adductor group of muscles** of the thigh. The **hamstring muscles** cover the femur posteriorly. They arise from the ischial tuberosity, and their tendons form the prominent cords (hamstrings) at the back of the knee. Medial to the sartorius muscle and inferior to the inguinal ligament is the **femoral triangle;** its medial limit is the adductor longus muscle. Within this triangle the center of the **saphenous opening** lies about 4 cm. below and lateral to the pubic tubercle. The femoral triangle (p. 533) locates the femoral vessels and the termination of the **greater saphenous** vein; the latter passes through the saphenous opening to end in the femoral vein. The **femoral artery** can be projected to the surface on a line which begins superiorly midway between the anterior superior iliac spine and the symphysis pubis and ends below at the upper end of the medial epicondyle of the femur (adductor tubercle). The course of the greater saphenous vein can be indicated by a line drawn from the adductor tubercle to the lower edge of the saphenous opening. Within the subcutaneous tissues of the femoral

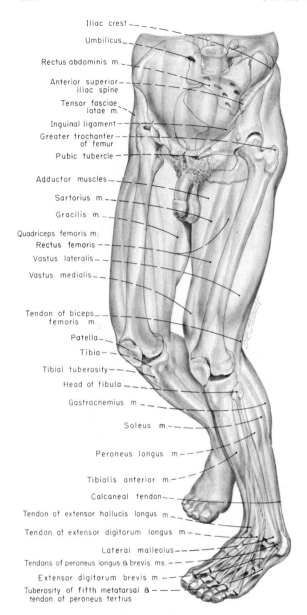

Iliac crest
Umbilicus
Rectus abdominis m.
Anterior superior
iliac spine
Tensor fasciae
latae m.
Inguinal ligament
Greater trochanter
of femur
Pubic tubercle
Adductor muscles
Sartorius m.
Gracilis m.
Quadriceps femoris m:
Rectus femoris
Vastus lateralis
Vastus medialis
Tendon of biceps
femoris m.
Patella
Tibia
Tibial tuberosity
Head of fibula
Gastrocnemius m.
Soleus m.
Peroneus longus m.
Tibialis anterior m.
Calcaneal tendon
Tendon of extensor hallucis longus m.
Tendon of extensor digitorum longus m
Lateral malleolus
Tendons of peroneus longus & brevis ms.
Extensor digitorum brevis m.
Tuberosity of fifth metatarsal &
tendon of peroneus tertius

Fig. 419 The surface anatomy and underlying bones of the lower limbs.

triangle lie the **superficial inguinal lymphatic nodes.** These nodes may be enlarged and palpable in disease. They include a horizontal set below the inguinal ligament and a vertical set along the termination of the greater saphenous vein. The vertical groove on the lateral aspect of the thigh is due to the tight **iliotibial tract** of the fascia lata.

The Knee (fig. 427)

About the knee the skin is thick. In front of the knee it is lax and freely movable, contributing to easy movement of the joint. The bony prominence anteriorly is the **patella,** a sesamoid bone in the quadriceps femoris tendon. In the

erect posture, with the quadriceps femoris relaxed, the patella can be grasped between the fingers and moved from side to side. When the quadriceps muscle is contracting, the patella is firmly fixed, and the broad tendon of the muscle above it and the **ligamentum patellae,** by which the patella is attached to the front of the tibia, become tense and prominent. In front of the lower part of the patella and the upper part of the ligamentum patellae is the **prepatellar bursa,** the inflamed condition of which is commonly called 'housemaids knee.' On either side of the patella and forming the width of the knee are the **condyles** of the femur. At the superior limit of the medial condyle is the sharp **adductor tubercle** which receives the cord-like tendon of the adductor magnus. Below the medial condyle of the femur the **medial condyle of the tibia** is readily felt, and on the lateral side of the knee the iliotibial tract can be followed to the **lateral tibial condyle.** The **head of the fibula** and the tendon of the biceps femoris muscle, which ends on it, are readily apparent in semiflexion of the knee. The head of the fibula lies a little below the most prominent part of the lateral femoral condyle and is on a level with the **tibial tuberosity**, which is the point of insertion of the ligamentum patellae and the upper end of the anterior margin of the tibia.

The hollow behind the knee is the **popliteal fossa.** The heavy cord-like tendons which form its boundaries are the **tendon of the biceps femoris muscle** laterally and the **tendons of the semitendinosus** and **semimembranosus muscles** medially. The popliteal fascia covers the interval between these tendons. Vertically through the subfascial fossa courses the tibial portion of the sciatic nerve superficially, followed more deeply by the popliteal vein and then the popliteal artery. The common peroneal nerve follows the tendon of the biceps femoris muscle to the head of the fibula and enters the leg by curving superficial to the neck of the bone.

The Leg

The skin of the leg is more closely adherent to deeper tissues than is the skin of the thigh. Over the medial (subcutaneous) surface of the tibia the skin is scarcely separated from the periosteum of the bone, and this is a common region for ulcers. The contours of the leg are reflections of the underlying bones and muscles. The medial surface of the **tibia** is subcutaneous throughout, and its **anterior margin** forms a sharp crest which terminates superiorly in the **tibial tuberosity.** The shaft of the **fibula** is enclosed by the muscles of the leg. The head of the fibula can be palpated at the knee, and the lateral malleolus is subcutaneous at the ankle. Most of the bulky musculature of the leg is placed posteriorly, but lateral to the anterior margin of the tibia, a fullness represents the **extensor** and **peroneal groups of muscles** of the leg. A furrow between this fullness and the bulging calf muscles locates the **posterior peroneal septum** extending from the back of the head of the fibula to the hollow behind the lateral malleolus. Behind the septum the **soleus** and **gastrocnemius muscles** form the mass of the calf, the margins of the soleus bulging on both sides of the gastrocnemius. These two muscles insert by means of the **calcaneal tendon** which descends as a prominent ridge on the back of the ankle and ends on the tuberosity of the calcaneus. The **greater saphenous vein** lies along the medial margin of the tibia. It enters the leg in front of the medial malleolus and leaves it by passing behind the condyles of the knee. The **lesser saphenous vein** leaves the foot behind the lateral malleolus, enters the leg on the lateral side of the calcaneal tendon, and ascends over the middle of the calf. It perforates the crural fascia, ending in the popliteal vein.

The Ankle and the Foot (fig. 453)

Over the ankle and the dorsum of the foot the skin is thin and loosely attached to the tissues beneath. There is very little fat here, and the skin is separated only slightly from the malleoli and the bones of the foot. On the contrary, the skin is dense and thick on the sole of the foot, especially on those parts which make contact with the ground. The subcutaneous tissue is thick and holds fat in discrete loculi. On the heel this fatty layer may be over 1 cm. thick and serves a useful function as a cushion for the jars taken by the heel in walking. As in the hand, but somewhat less in degree, the nerve supply of the sole of the foot is rich and sensation is acute. The malleoli of the tibia and the fibula are the two prominent bones at the sides of the ankle. The tip of the **lateral malleolus** is 1.5 cm. lower than the tip of the **medial malleolus.** Several centimeters below the tip of the medial malleolus one can palpate the **sustentaculum tali of the calcaneus** and, an equal distance in front of the sustentaculum, the **tuberosity of the navicular bone.** The line of the ankle joint can be determined by palpation on either side of the tendons of the digital extensor muscles, and when the foot is turned downward in plantar flexion, the anterior part of the upper surface of the body of the **talus** forms a prominence below the distal edge of the tibia. The tendons which pass over the dorsum of the foot to insert in the tarsal and phalangeal bones are, from medial to lateral, those of the prominent **tibialis anterior muscle** and the **extensor hallucis longus, extensor digitorum longus,** and **peroneus tertius muscles.** The fleshy pad on the lateral and proximal surface of the foot is the belly of the **extensor digitorum brevis muscle.** On the lateral border of the foot the base of the **fifth metatarsal bone** makes a prominence to which can be traced the tendon of the **peroneus brevis muscle.** The two peroneal tendons enter the foot behind and below the lateral malleolus, but the tendon of the **peroneus longus** runs a deep course across the sole of the foot.

SUPERFICIAL VEINS (figs. 420, 421, 423–25)

As in the upper limb, the superficial veins of the lower limb arise from the ends of a venous arch on the dorsum of the foot which crosses the distal parts of the shafts of the metatarsal bones. **Dorsal digital veins** run along the two dorsal margins of each digit; these unite at the webs of the toes into

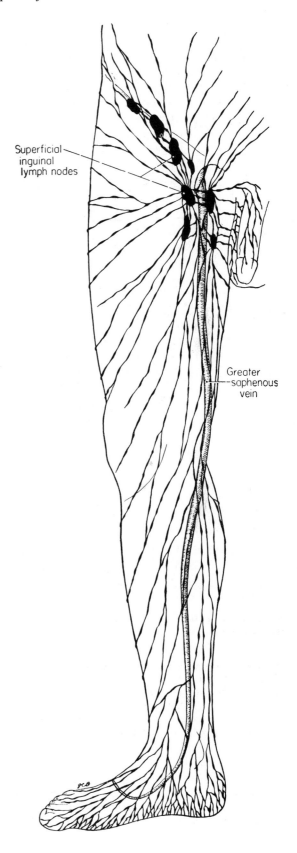

Superficial inguinal lymph nodes

Greater saphenous vein

Fig. 420 The superficial lymphatic vessels of the lower limb.

short **dorsal metatarsal veins** which empty into the venous arch. From the plantar side of the toes **plantar digital veins** communicate between the heads of the metatarsals with the dorsal metatarsal veins. The **dorsal venous arch** of the foot lies in the subcutaneous fat superficial to the cutaneous nerves. Medially, it receives the **medial dorsal digital vein** of the great toe to form the greater saphenous vein and, laterally, the **lateral dorsal digital vein** of the little toe to form the lesser saphenous vein.

The **greater saphenous vein,** the longest vein of the body, begins at the junction of the medial end of the dorsal venous arch and the medial dorsal vein of the great toe. It turns up past the ankle anterior to the medial malleolus and, ascending immediately posterior to the medial margin of the tibia, passes the knee against the posterior border of the medial condyle of the femur. It is thus on the flexion side of the axis of movement at both the ankle and the knee and is not put on a stretch in the action of these joints. In the thigh the greater saphenous vein inclines anteriorly and lateralward, and in the femoral triangle it turns deeply through the saphenous opening to empty into the femoral vein. In the leg the greater saphenous vein accompanies a branch of the saphenous nerve; in the thigh an anterior femoral cutaneous nerve lies next to it.

The origin and communications of the greater saphenous vein in the foot have been listed; in the leg this vein is joined by tributaries from the dorsum of the foot, the heel, the front of the leg, and the calf. It also communicates with radicles of the lesser saphenous vein. The vein receives tributaries from the anterior and lateral portions of the thigh. Into it also empties a large **accessory saphenous vein** which collects the superficial radicles from the medial and posterior parts of the thigh. Just before it turns through the saphenous opening, the greater saphenous vein receives the **superficial epigastric, superficial circumflex iliac,** and **superficial external pudendal** veins. These veins, more particularly related to the abdominal wall, are described on p. 370. The valves in the greater saphenous vein vary from ten to twenty in number; they are more numerous in the leg than in the thigh. These valves assist in the support of the long column of blood that fills this vein. **Perforating communications** pierce the deep fascia and interconnect the superficial and deep veins at all levels of the limb. The blood flows from the superficial to the deep veins through these communications, and the valves in the communicating veins prevent deep to superficial drainage.

The **lesser saphenous vein** is formed by the union of the lateral part of the dorsal venous arch with the lateral dorsal digital vein of the little toe. Receiving lateral marginal veins in the foot, the lesser saphenous vein passes backward along the lateral border of the foot in company with the sural nerve. It turns upward behind the lateral malleolus and against the calcaneal tendon and ascends through the middle of the calf. The vein pierces the crural fascia, mostly in the middle third of the leg but frequently in the upper third, and ascends deep to or in a split of the deep fascia. It usually (72%, Moosman & Hartwell,'64) terminates in the popliteal vein, directly or through muscular or thigh tributaries. The lesser saphenous vein receives tributaries from the lateral side of the foot and the heel and from the back of the leg and communicates, around the medial side of the leg, with the greater saphenous vein. It contains from six to twelve valves and interconnects with the deep veins of the leg. Just before it passes through the deep fascia, the lesser saphenous vein may give rise to a branch which passes toward the medial side of the thigh and joins the accessory saphenous vein.

Clinical Note Varicose veins are frequently seen in the lower limb. Such veins are dilated and tortuous, and their valves tend to be incompetent due to failure of the valve cusps to meet in a grossly enlarged vein. The superficial veins of the limb have to support a long column of blood, are not supported by surrounding muscles, and may be overpowered by failure of the perforating veins which normally shunt blood to the deeper venous system. Varicose veins may be painful and, in the extreme case, may lead to cutaneous ulceration. If the deep veins are patent, the varicosed superficial veins may be removed by 'stripping.' The greater saphenous vein is a favorable source for venous segments commonly used as shunting vessels in coronary bypass operations.

SUPERFICIAL LYMPHATICS OF THE LIMB (fig. 420)

The superficial lymphatic vessels of the foot are similar in arrangement to those of the hand. From plexuses on the plantar side of the toes and the foot, collecting channels pass through the interdigital clefts to the dorsum of the foot where they are united with collecting vessels from the dorsal aspect of the toes. The collecting vessels of the dorsum, for the most part, accompany the greater saphenous vein; those of the lateral part of the foot follow the course of the lesser saphenous vein. The principal stream of ascending lymphatic vessels runs with the **greater saphenous vein** toward which collecting vessels flow from the lateral and medial borders and from the front and the back of the leg and the thigh. The lymphatic vessels accompanying the greater saphenous vein end above in the **superficial inguinal lymph nodes** to which also pass collecting vessels from the lower parts of the abdomen, the perineum, the scrotum and penis (or vulva in the female), and the gluteal region. The area of venous drainage of the lesser saphenous vein gives rise to lymph vessels which accompany that vein, pierce the popliteal fascia with it, and end in the popliteal lymph nodes (fig. 427). Lymph collected by the popliteal nodes ascends by deep lymphatic channels to the deep inguinal nodes.

The **superficial inguinal nodes** are from twelve to twenty lymph nodes arranged in the form of a T in the subcutaneous tissue of the groin. The horizontal part of the T accounts for most of the nodes; they form a chain parallel to and about 1 cm. below the inguinal ligament. These nodes receive lymph from the lower abdominal wall, the buttocks, the penis and scrotum (or the labia majora), and the perineum. The fewer, larger nodes of the vertical limb of the T lie along the termination of the greater saphenous vein. Their afferent vessels drain chiefly the superficial lymphatic plexus of the lower limb, although they also receive some vessels from the skin of the penis and scrotum, the perineum, and the buttocks. The superficial inguinal nodes send their efferent channels mainly within the femoral sheath to the external iliac nodes. Only a few vessels end in the deep inguinal nodes.

Deep Lymphatics of the Limb

The deep lymph vessels of the lower limb accompany the deep blood vessels. In the leg they form three sets with the anterior tibial, posterior tibial, and peroneal vessels; these end in the popliteal lymph nodes. Certain vessels of the gluteal region follow the gluteal blood vessels to the internal iliac nodes.

The **popliteal lymph nodes** (fig. 427) are usually small, six or seven in number, and lie in the fat of the popliteal fossa. One lies at the termination of the lesser saphenous vein and receives the lymph vessels which accompany that vein. Deeper nodes receive the channels accompanying the anterior

tibial, posterior tibial, and peroneal vessels and the lymphatics from the capsule of the knee joint. A node is frequently interposed between the popliteal artery and the posterior surface of the knee joint. The efferents of the popliteal nodes follow the femoral vessels to the deep inguinal nodes.

The **deep inguinal lymph nodes** (fig. 405) are from one to three nodes on the medial side of the femoral vein. If three are present, one is usually situated in the femoral canal, one at its upper end in the femoral ring, and one below the junction of the greater saphenous and femoral veins. These nodes receive the deep lymphatic drainage of the limb, some of the lymphatics from the penis (or clitoris), and a few of the efferents of the superficial inguinal nodes. They discharge to the external iliac nodes.

Cutaneous Nerves (figs. 421–23, 427, 453)

The cutaneous nerves of the lower limb, with the exception of a few of the uppermost nerves, are branches of the lumbar and sacral plexuses.

Cutaneous Nerves of the Hip and the Thigh

The **cluneal nerves** are superior, middle, and inferior. The **superior cluneal nerves** are the lateral cutaneous branches of the dorsal rami of the upper three lumbar nerves (p. 49). They pierce the thoracolumbar fascia at the lateral border of the erector spinae muscle, cross the iliac crest a short distance in advance of the posterior superior spine of the ilium, and distribute to the skin of the gluteal region as far as the greater trochanter. The **middle cluneal nerves** are lateral branches of the dorsal rami of the upper three sacral nerves. They become cutaneous on a line connecting the posterior superior iliac spine with the tip of the coccyx and supply the skin and the subcutaneous tissue over the back of the sacrum and the adjacent area of the gluteal region. The **inferior cluneal nerves** (fig. 423) are the gluteal branches of the posterior femoral cutaneous nerves (p. 510). These branches become cutaneous by turning around the lower border of the gluteus maximus muscle and supply the skin of the lower half of the buttock. Their lateral branches reach as far as the greater trochanter; their medial branches distribute almost to the coccyx.

The **perforating cutaneous nerve** of the sacral plexus (S2, 3) pierces the sacrotuberous ligament and the lower

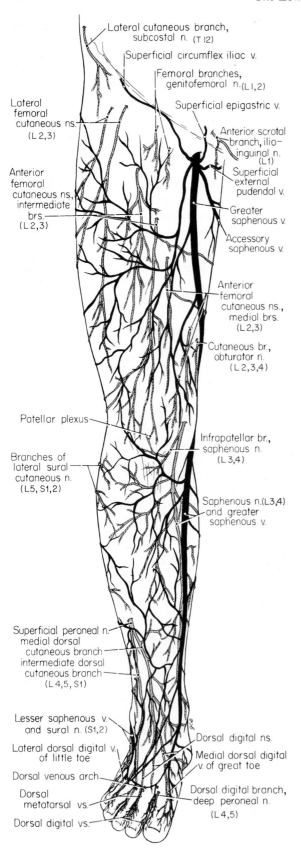

Lateral cutaneous branch, subcostal n. (T 12)

Superficial circumflex iliac v.

Femoral branches, genitofemoral n.(L1,2)

Superficial epigastric v.

Lateral femoral cutaneous ns. (L 2,3)

Anterior scrotal branch, ilio-inguinal n.(L1)

Superficial external pudendal v.

Anterior femoral cutaneous ns., intermediate brs. (L 2,3)

Greater saphenous v.

Accessory saphenous v.

Anterior femoral cutaneous ns., medial brs. (L 2,3)

Cutaneous br., obturator n. (L 2,3,4)

Patellar plexus

Infrapatellar br., saphenous n. (L 3,4)

Branches of lateral sural cutaneous n. (L5, S1,2)

Saphenous n.(L3,4) and greater saphenous v.

Superficial peroneal n. medial dorsal cutaneous branch intermediate dorsal cutaneous branch (L 4,5, S1)

Lesser saphenous v. and sural n. (S1,2)

Lateral dorsal digital v. of little toe

Dorsal venous arch

Dorsal metatarsal vs.

Dorsal digital vs.

Dorsal digital ns.

Medial dorsal digital v. of great toe

Dorsal digital branch, deep peroneal n. (L 4,5)

fibers of the gluteus maximus muscle. It supplies the skin over the lower half of the buttock and the medial part of the fold of the buttock. This distribution corresponds with part of the area of supply of the inferior cluneal nerve; on occasion, the perforating cutaneous nerve is combined with the inferior cluneal.

The **lateral cutaneous branches of the subcostal and iliohypogastric nerves** descend across the crest of the ilium into the upper thigh. The **lateral cutaneous branch of the subcostal nerve** crosses the crest several centimeters behind the anterior superior iliac spine. It supplies the skin and the subcutaneous tissue of the thigh as low as but in front of the greater trochanter of the femur. The **lateral cutaneous branch of the iliohypogastric nerve** crosses the iliac crest at the junction of its anterior and middle thirds and supplies the skin of the gluteal region posterior to the region supplied by the lateral cutaneous branch of the subcostal nerve. These nerves tend to be reciprocal in size and distribution, and one or the other may be lacking.

The **ilioinguinal nerve** has a small femoral distribution through its anterior scrotal (or anterior labial) branch. Twigs of this branch terminate in the skin of the proximal and medial part of the femoral triangle adjacent to the scrotum.

The **femoral branch of the genitofemoral nerve** (p. 467) arises from the lumbar plexus and enters the thigh behind the inguinal ligament and on the anterior surface of the femoral artery. It pierces the femoral sheath and passes through the saphenous opening (or directly through the fascia lata) to supply the skin over the femoral triangle lateral to the area covered by the ilioinguinal nerve. In the triangle it supplies a small twig to the femoral artery and communicates with the anterior cutaneous branch of the femoral nerve.

The **anterior cutaneous branches of the femoral nerve** are multiple. Several anterior cutaneous branches descend vertically over the quadriceps femoris muscle, becoming cutaneous by piercing the fascia lata toward the apex of the femoral triangle. These nerves (intermediate femoral cutaneous nerves) supply the skin of the distal three-fourths of the front of the thigh and extend to the front of the patella, where they assist in forming the patellar plexus. A more medially distributing ramus crosses the femoral vessels and divides into branches which descend deep to the fascia lata in the groove between the sartorius and the adductor muscles. These branches (medial femoral cutaneous) distribute to the skin and the subcutaneous tissue of the distal two-thirds of the medial portion of the thigh.

Fig. 421 The superficial veins and cutaneous nerves of the anterior aspect of the lower limb.

The anterior cutaneous branches usually arise from the femoral nerve in the femoral triangle on the lateral aspect of the femoral vessels. They may, however, arise in the iliac fossa, or they may be separate branches of the lumbar plexus.

The **lateral femoral cutaneous nerve** is a direct branch of the lumbar plexus (p. 468). Passing under the lateral end of the inguinal ligament, either superficial or deep to the sartorius muscle, it descends at first under the fascia lata. It becomes subcutaneous about 10 cm. below the anterior superior spine of the ilium and divides into an anterior and a posterior branch. The larger anterior branch distributes on the lateral aspect of the front of the thigh as far as the knee, where it may communicate with the patellar plexus. The smaller posterior branch supplies the skin and the subcutaneous tissue over the lateral part of the buttock

distal to the greater trochanter and on the proximal two-thirds of the lateral aspect of the thigh.

The **posterior femoral cutaneous nerve** is a branch of the sacral plexus (p. 509), and its perineal and inferior cluneal branches have been described (pp. 471, 510). The nerve descends in the posterior midline of the thigh deep to the fascia lata, giving off small branches from both sides which pierce the fascia lata individually. They distribute to the skin of the back of the thigh and over the popliteal fossa. The posterior femoral cutaneous nerve finally pierces the fascia lata and ends as one or two cutaneous branches over the calf.

The **cutaneous branch of the obturator nerve** is a variable offshoot of the anterior branch of that nerve (p. 468). The anterior branch of the obturator nerve, mainly muscular and articular, communicates with an anterior

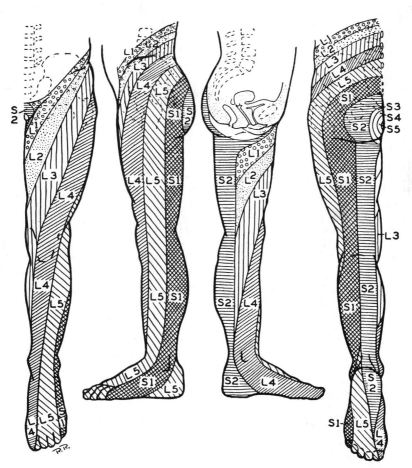

Fig. 422 Dermatome chart of the lower limb of man outlined by the pattern of hyposensitivity from loss of function of a single nerve root (from original of fig. 7, J. J. Keegan and F. D. Garrett, *Anat. Rec.*, 102:415, 1948).

cutaneous branch and the saphenous branch of the femoral nerve. From this anastomosis, a cutaneous branch of the obturator contribution becomes superficial between the gracilis and adductor longus muscles. It is distributed, when present, to the skin of the distal one-third of the thigh on its medial side.

Cutaneous Nerves of the Leg, the Ankle, and the Foot

The **saphenous nerve** is the terminal branch of the femoral nerve (p. 555). Arising from the femoral nerve in the femoral triangle, it enters the adductor canal, where it crosses the femoral vessels anteriorly from their lateral to their medial side. At the distal end of the canal, the nerve pierces its fascial covering in company with the saphenous branch of the descending genicular artery. Opposite the knee joint, the nerve becomes cutaneous, emerging between the tendons of the sartorius and gracilis muscles. It descends in the leg in company with the greater saphenous vein. A branch arising in the adductor canal communicates with the obturator nerve and the anterior cutaneous branch of the femoral nerve.

An **infrapatellar branch**, given off at the medial side of the knee, pierces the sartorius muscle and fascia lata and curves downward below the patella and over the medial condyle of the tibia to the front of the knee and the upper part of the leg. It forms the **patellar plexus** with communicating branches of the anterior cutaneous branch of the femoral nerve and branches of the lateral femoral cutaneous nerve. The saphenous nerve continues along the medial surface of the leg, distributing branches to the skin and the subcutaneous tissue of the front and the medial portion of the leg and the posterior half of the dorsum and the medial side of the foot.

The **lateral sural cutaneous nerve** is a branch of the common peroneal nerve which arises in the popliteal fossa. It pierces the deep fascia over the lateral head of the gastrocnemius muscle and distributes to the skin and the subcutaneous connective tissue on the lateral part of the back of the leg in its proximal two-thirds.

The **peroneal communicating branch** is a small nerve which usually takes origin from the lateral sural cutaneous nerve. It may arise in the popliteal fossa or over the calf. It joins the medial sural cutaneous nerve in the middle one-third of the leg to form the sural nerve, usually in the subcutaneous tissue but sometimes under the deep fascia.

The **medial sural cutaneous nerve** arises from the tibial nerve in the popliteal fossa. It descends imme-

diately under the deep fascia in the groove between the two heads of the gastrocnemius muscle as far as the middle of the leg. Here it pierces the deep fascia and is joined by the peroneal communicating branch to form the sural nerve. The level of junction of these two contributing nerves is quite variable, and in about 20 per cent of cases they do not unite. In such cases, the usual area of distribution of the sural nerve is divided between them—the peroneal communicating branches ending in the leg and the medial sural cutaneous nerve supplying the heel and foot areas (Huelke, '57).

The **sural nerve,** formed as just described, usually becomes superficial at the middle of the length of the leg. It descends in company with the lesser saphenous vein, and with this vein it turns under the lateral malleolus onto the side of the foot. The nerve is cutaneous to the lateral side and the back of the distal one-third of the leg and to the ankle and the heel (**lateral calcaneal branches**). On the side of the foot it is known as the **lateral dorsal cutaneous nerve.** It supplies the lateral aspect of the foot and the little toe, provides articular branches to the ankle and the tarsal joints, and communicates with the intermediate dorsal cutaneous branch of the superficial peroneal nerve.

The **superficial peroneal nerve,** a branch of the common peroneal, descends through the lateral muscular compartment of the leg and then, piercing the deep fascia in the distal one-third of the leg, terminates in cutaneous branches. It emerges from under the deep fascia anterolaterally between the peroneus longus and extensor digitorum longus muscles and almost immediately divides into two terminal branches.

The **medial dorsal cutaneous nerve** descends across the extensor retinaculum of the ankle, supplying cutaneous twigs in the distal one-third of the leg. Over the dorsum of the foot it divides into two or three branches for the supply of the dorsum and the sides of the medial two and a half toes. The web and adjacent sides of the great toe and the second toe receive, from beneath the extensor digitorum brevis muscle, the cutaneous terminals of the deep peroneal nerve. The twigs of the dorsal cutaneous nerve of the foot in this interval simply communicate with these branches of the deep peroneal nerve. The **intermediate dorsal cutaneous nerve** passes more laterally over the dorsum of the foot. It supplies cutaneous branches to the lateral side of the ankle and the foot and terminates in **dorsal digital branches** for the adjacent sides of the third and fourth and the fourth and fifth toes. The more lateral branches communicate with terminals of the lateral dorsal cutaneous nerve. As in the fingers, the dorsal digital

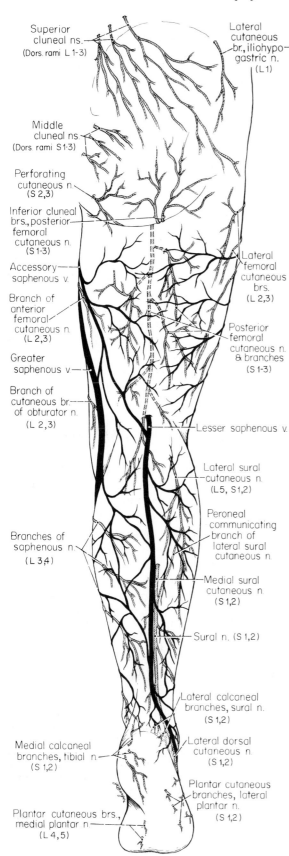

Superior cluneal ns. (Dors. rami L 1-3)

Lateral cutaneous br., iliohypogastric n. (L1)

Middle cluneal ns (Dors. rami S 1-3)

Perforating cutaneous n. (S 2,3)

Inferior cluneal brs., posterior femoral cutaneous n. (S 1-3)

Accessory saphenous v.

Branch of anterior femoral cutaneous n. (L 2,3)

Greater saphenous v.

Branch of cutaneous br. of obturator n. (L 2,3)

Branches of saphenous n. (L 3,4)

Medial calcaneal branches, tibial n. (S 1,2)

Plantar cutaneous brs., medial plantar n. (L 4,5)

Lateral femoral cutaneous brs. (L 2,3)

Posterior femoral cutaneous n. & branches (S 1-3)

Lesser saphenous v.

Lateral sural cutaneous n. (L5, S1,2)

Peroneal communicating branch of lateral sural cutaneous n.

Medial sural cutaneous n. (S 1,2)

Sural n. (S 1,2)

Lateral calcaneal branches, sural n. (S 1,2)

Lateral dorsal cutaneous n. (S 1,2)

Plantar cutaneous branches, lateral plantar n. (S 1,2)

branches to the toes are smaller than the corresponding plantar nerves and distribute only over the proximal and middle phalanges. The distal parts of the toes receive terminals of the plantar nerves.

The **dorsal digital branches of the deep peroneal nerve** supply the web and adjacent sides of the great and second toes. These nerves become superficial between the tendons of the extensor digitorum brevis muscle to the corresponding digits and communicate with terminals of the medial dorsal cutaneous nerve.

The **medial calcaneal** and **plantar digital branches of the tibial nerve.** The tibial nerve supplies the musculature of the back of the leg and continues into the foot behind the medial malleolus. It is subfascial at the ankle where its medial calcaneal branches pierce the flexor retinaculum to distribute to the skin and the subcutaneous tissue of the heel and the posterior part of the sole of the foot. Deep to the origin of the abductor hallucis muscle, the tibial nerve divides into medial and lateral plantar nerves which, corresponding to the median and ulnar nerves in the hand, provide the muscular and cutaneous nerves of the sole of the foot and the toes. The **medial plantar nerve** provides a proper digital branch and three common digital branches. The **proper digital nerve** of the great toe pierces the plantar aponeurosis behind the ball of the great toe. This nerve, having given a branch to the flexor hallucis muscle, distributes to the medial aspect of the great toe. The three **common digital nerves** (fig. 453) pass between the division of the plantar aponeurosis, and each splits into **two proper digital nerves.** Those of the first common digital supply the adjacent sides of the great and second toes; those of the second common digital, the adjacent sides of the second and third toes; and those of the third common digital, the surfaces of the third and fourth toes. The first common digital nerve supplies the first lumbrical muscle.

The **superficial branch of the lateral plantar nerve** provides a **common digital nerve** which, in turn, divides into proper digital nerves to the adjacent sides of the fourth and fifth toes and the **proper digital nerve** to the lateral side of the little toe. The superficial branch also innervates the flexor digiti

Fig. 423 The superficial veins and cutaneous nerves of the posterior aspect of the lower limb.

minimi brevis muscle and the two interosseous muscles of the fourth interosseous space; its common digital branch communicates with the third common digital branch of the medial plantar nerve. All the plantar proper digital nerves supply the whole plantar surface of the toes and, in addition, furnish small dorsal twigs for the supply of the nail bed and the tip of each toe.

THE HIP AND THE THIGH

The region of the hip and the thigh includes the entire area from the iliac crest to the knee. The uppermost part from the iliac crest to the greater trochanter, representing the part of the limb at the side of the pelvic girdle, is designated as the hip.

THE SUBCUTANEOUS CONNECTIVE TISSUE

The **subcutaneous connective tissue** contains a considerable amount of fat over the hip and the thigh, although it varies in different regions. In the gluteal region the fat is deposited in a thick layer which contributes to the contour of the buttock and the formation of the transverse fold of the buttock. The subcutaneous tissue of the thigh is continuous with the similar layer of the abdomen, the back, the perineum, and the leg. At the fold of the groin the subcutaneous tissue is separable into a superficial fat-laden layer and a deeper membranous layer. The superficial inguinal lymph nodes and the subcutaneous blood vessels and nerves lie between these layers. The membranous layer is attached, along with the similar membranous layer of the abdominal subcutaneous tissue, to the fascia lata a short distance below the inguinal ligament and, medially, to the pubic tubercle. It is attached to the margins of the saphenous opening and fills the opening itself with a spongy layer which has a number of perforations for the passage of the greater saphenous vein and other blood and lymphatic vessels. A large **subcutaneous prepatellar bursa** lies in the subcutaneous connective tissue over the patella.

THE DEEP FASCIA (fig. 424)

The **fascia lata** is the uppermost subdivision of a complete stocking-like investment of the soft parts of the lower limb. Its continuities are the crural

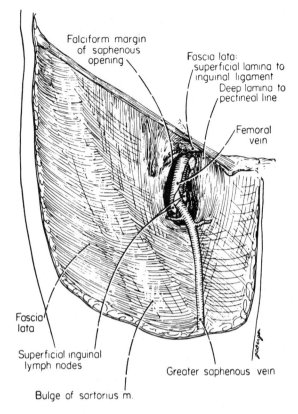

Fig. 424   The saphenous opening and the fascia lata.

fascia of the leg and the plantar and dorsal fasciae of the foot. The fascia lata is a strong membranous fascia, thicker where it is reinforced by tendinous contributions and thinner in the gluteal region. It has a continuous bony and ligamentous attachment superiorly. Medially, it is attached to the pubic crest, the pubic symphysis, the ischiopubic ramus, and the tuberosity of the ischium. From here an attachment to the sacrotuberous ligament carries it to the dorsum of the sacrum and to the coccyx. It is continued to the posterior superior iliac spine and then along the external lip of the iliac crest to the anterior superior iliac spine, the inguinal ligament, and the pubic tubercle. A deep lamina follows the pecten of the pubis behind the femoral vein. Between the iliac crest and the superior border of the gluteus maximus muscle, the fascia is thickened by vertical tendinous fibers; this **gluteal aponeurosis** (fig. 431) provides part of the origin of the gluteus medius muscle. The adductor portion and the remaining gluteal portion of the fascia lata are thin, but a lateral band, the **iliotibial tract,** is an especially strong tendinous part. Arising from the tubercle of the crest of the ilium, the iliotibial tract receives tendinous reinforcements from the tensor fasciae latae and gluteus maximus muscles and serves as the tendon of insertion for the former muscle and as part of the insertion for the latter muscle. The iliotibial tract ends on the lateral condyle of the tibia, where it blends with the fibrous expansions of the vastus lateralis and biceps femoris muscles. The fascia lata forms a compartment superiorly enclosing the tensor fasciae latae muscle. Above the knee the fascia lata is thickened by the addition of tendinous expansions of the vastus muscles and thus forms the lateral and medial retinacula of the patella, which attach to the borders of the patella and to the condyles of the tibia. The fascia is also attached in the knee region to the tuberosity of the tibia and the head of the fibula. On the back of the thigh and over the popliteal fossa it is reinforced by transverse fibers, so that the popliteal fascia is specially thick.

Intermuscular septa unite the fascia lata to the periosteum of the femur. A **lateral intermuscular septum** extends from the iliotibial tract to the lateral epicondylar line and the lateral lip of the linea aspera; it separates the vastus lateralis from the biceps femoris and serves as an attachment for the fibers of both muscles. The **medial intermuscular septum** lies between the vastus medialis and the adductor group of muscles and attaches to the femur along the medial lip of the linea aspera. It contributes to the formation of the adductor canal and, superficially, splits to enclose the sartorius muscle.

The **saphenous opening** provides for the passage of the great saphenous vein under the fascia lata to its termination in the femoral vein. Its center is located 4 cm. below the inguinal ligament lateral to the pubic tubercle. The opening, about 4 cm. long, has the form of an oval fossa as seen from in front, but it is formed by superficial and deep laminae of the fascia lata and thus is a slit between these laminae when viewed from the medial side. The deep lamina of the fascia lata attaches to the pecten of the pubis and is continuous with that part of the fascia lata ending on the pubic crest and the pubic tubercle. The deep lamina thus passes under the femoral vein to reach its attachment to bone. The superficial lamina is that portion of the fascia lata which sweeps medialward down the inguinal ligament to end at the pubic tubercle. This lamina crosses the femoral vein and presents a free sharp **falciform margin** on the lateral side of the opening. Inferiorly, the falciform margin blends with the flat surface of the deep lamina, and across the junction of the laminae, the greater saphenous vein passes deeply to reach the femoral vein. The **fascia cribrosa,** derived from subcutaneous connective tissue, fills the saphenous opening.

THE SUBFASCIAL SPACES OF THE THIGH

The Femoral Triangle (Fig. 425)

The femoral triangle is the subfascial space of the upper one-third of the thigh. It contains the first portion of the femoral vessels and many of their important branches and is the principal site of division of the femoral nerve. The triangle is bounded above by the inguinal ligament; laterally, by the medial border of the sartorius muscle; and medially, by the medial border of the adductor longus muscle. The crossing of the adductor longus by the sartorius closes the triangle below. The floor of the triangle is also muscular; it is represented medially by the pectineus and adductor longus muscles and, laterally, by the iliopsoas. The juxtaposition of the pectineus and iliopsoas muscles forms a deep groove in the floor of the triangle, and here the medial circumflex femoral artery passes to the back of the thigh. The **femoral artery** bisects the triangle in a vertical direction. It enters the thigh at the midpoint between the anterior superior spine of the ilium and the symphysis pubis and leaves the femoral triangle at its apex by entering the adductor canal under the margin of the sartorius muscle. The artery descends over the tendon of the psoas major muscle. The **femoral vein** enters the thigh medial to the artery and descends over the pectineus muscle. The two vessels twist on one another as they descend, so that at the apex of the triangle they enter the adductor canal with the artery anterior and the vein posterior. The **femoral nerve** passes under the inguinal ligament in the groove between the rounded psoas muscle and the flat iliacus muscle (figs. 405, 425). Lying on the iliacus muscle, it divides into many of its muscular and sensory branches while in the triangle. Only the saphenous nerve and one of the nerves to the vastus medialis muscle continue into the adductor canal with the femoral vessels.

The **femoral sheath** (figs. 405, 425) is a diverticulum of the transversalis fascia which is prolonged along the femoral vessels and covers them on all sides for about 2 to 3 cm. beyond the inguinal ligament. At its termination, its fibers blend with the adventitia of the blood vessels. The extraperitoneal connective tissue of the abdomen extends along the vessels and contributes to the subdivision of the sheath into three compartments—a lateral one for the artery; a middle one for the vein; and a medial one which lodges one or more deep inguinal

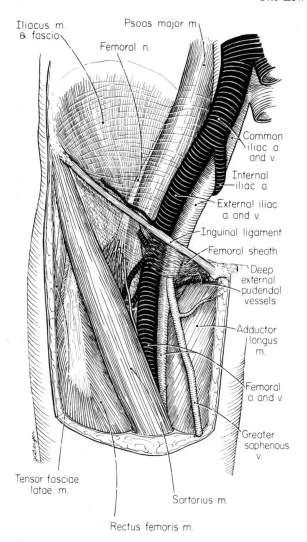

Iliacus m. & fascia

Psoas major m.

Femoral n.

Common iliac a. and v.

Internal iliac a.

External iliac a. and v.

Inguinal ligament

Femoral sheath

Deep external pudendal vessels

Adductor longus m.

Femoral a. and v.

Greater saphenous v.

Tensor fasciae latae m.

Sartorius m.

Rectus femoris m.

Fig. 425 The boundaries and contents of the femoral triangle.

hernia. From within outward, these are the parietal peritoneum, the extraperitoneal connective tissue including the compressed contents of the femoral canal, the anterior layer of the femoral sheath, the fascia cribrosa, and the skin. A femoral hernia presents lateral and inferior to the pubic tubercle, whereas an inguinal hernia emerges superior and medial to the pubic tubercle.

The Adductor Canal (fig. 425)

The adductor canal is a fascial compartment which conducts the femoral vessels through the middle one-third of the thigh. It begins about 15 cm. below the inguinal ligament at the crossing of the sartorius muscle over the adductor longus muscle. The canal ends at the upper limit of the adductor hiatus—a separation in the tendinous insertion of the adductor magnus muscle which allows the femoral vessels to pass to the back of the knee. On the anteromedial aspect of the thigh, the quadriceps femoris and adductor groups of muscles are separated by a deep groove. This groove, covered by the sartorius muscle and thus converted into a triangular channel, is the adductor canal. The muscular fasciae are the immediate boundaries of the canal —the fascia of vastus medialis is anterolateral; the fasciae of the adductor longus and adductor magnus muscles limit the canal posteromedially. Its anteromedial, or superficial, boundary is the deep layer of the fascial sheath of the sartorius muscle. Within the adductor canal the femoral artery and vein descend through the

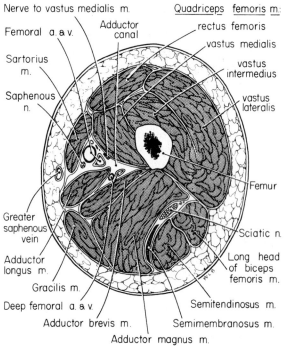

Nerve to vastus medialis m.

Femoral a. & v.

Sartorius m.

Saphenous n.

Adductor canal

Quadriceps femoris m.:

rectus femoris

vastus medialis

vastus intermedius

vastus lateralis

Femur

Greater saphenous vein

Adductor longus m.

Gracilis m.

Deep femoral a. & v.

Adductor brevis m.

Adductor magnus m.

Sciatic n.

Long head of biceps femoris m.

Semitendinosus m.

Semimembranosus m.

Fig. 426 A cross section of the thigh through levels of the adductor canal.

lymph nodes and fat. The medial compartment is designated as the **femoral canal,** and its small abdominal opening is known as the **femoral ring.** The femoral sheath is bounded medially by the free concave margin of the lacunar ligament.

Clinical Note   The small femoral canal is the usual site of a **femoral hernia.** The canal represents a point of possible exit from the abdominal cavity, and its contents are readily compressed to allow passage for herniating material. Further, the lower end of the femoral canal reaches the level of the upper part of the saphenous opening, a weakness in the fascial covering of the thigh. Thus, it is common for a femoral hernia to bulge forward under the skin over the saphenous opening. Consideration of the anatomy of the femoral sheath and the saphenous opening indicates the coverings of a femoral middle of the thigh, the artery anterior to the vein. Behind the main vessels course the deep femoral vessels, perforating branches of which pierce the tendons of insertion of the adductor muscles to reach the back of the thigh. The saphenous nerve enters the adductor canal lateral to the vessels, crosses them anteriorly, and lies medial to them at the lower end of the canal. The lower nerve to the vastus medialis also traverses the canal, entering the muscle toward its lower end.

The Popliteal Fossa (fig. 427)

The popliteal fossa lies behind the knee and is covered by the dense popliteal fascia. The space is diamond-shaped; the lateral and medial heads of the gastrocnemius muscle of the leg arise from the lateral and medial epicondyles of the femur and, bulging into the midline as they descend, form the inferior sides of the diamond. The superior borders of the figure are produced by the hamstring tendons diverging from the posterior midline of the thigh. Laterally, the tendon of the biceps femoris muscle crosses the lateral head of the gastrocnemius at the lateral angle of the diamond and ends below by inserting into the head of the fibula. Medially, the superimposed tendons of the semitendinosus and semimembranosus muscles cross the medial head of the gastrocnemius on the way to their insertions. The lesser saphenous vein ascends to end in the popliteal vein. Directly under the fascia, the posterior femoral cutaneous nerve and the tibial and common peroneal divisions of the sciatic nerve cross the fossa. The **tibial nerve** continues the midline course of the sciatic nerve from the thigh and thus runs through the fossa in a vertical direction from its superior to its inferior angle. Within the space the tibial nerve gives off genicular branches to the knee and sural branches to the gastrocnemius and plantaris muscles. The posterior femoral cutaneous nerve also descends across the fossa in the midline. It sends cutaneous branches through the fascia lata to the skin of the back of the knee and the upper part of the calf. The **common peroneal nerve** follows the tendon of the biceps femoris muscle downward and lateralward along the margin of the fossa. This nerve provides genicular branches and gives off the lateral sural cutaneous nerve (p. 530) within the space. The peroneal communicating branch also frequently arises here. The common peroneal nerve passes subfascially into the leg by curving around the neck of the fibula.

The **femoral artery** and the **femoral vein** migrate, in their course down the thigh, from an anterior position in the femoral triangle to a location opposite the posterior border of the femur at the lower end of the adductor canal. Here they lie against the tendon of the adductor magnus and, continuing their backward inclination, slip through a division in the tendon into the popliteal fossa (fig. 427). The adductor magnus has a long insertion on the medial lip of the linea aspera of the femur and a lower insertion by a heavy rounded tendon into a tubercle on the upper border of the medial epicondyle of the femur (adductor tubercle). The elongated interval of separation of these tendons is the **adductor hiatus;** through it the femoral vessels pass to the back of the knee, where they become the **popliteal vessels.** In the adductor canal the femoral artery lies anterior to the femoral vein, and the popliteal vessels retain this relationship. This places the popliteal artery next to the femur, well protected from external trauma but vulnerable to supracondylar fractures of the femur. The popliteal vessels provide genicular branches to the knee and large sural branches to the muscles of the calf. The popliteal vein receives the lesser saphenous vein. The vessels are embedded in a considerable amount of fat and areolar connective tissue which also lodges the popliteal lymph nodes.

THE MUSCLES OF THE HIP AND THE THIGH

The muscles of the hip and the thigh are divided into four groups, the muscles within each group being generally classified by location but having certain common actions. The **lateral femoral muscles** lie in the gluteal region and include strong abductors and lateral rotators of the thigh. The **medial femoral muscles** are also known as the adductor group and have adduction as their principal function. They are medial rotators but do not have strong rotary actions. The **anterior femoral muscles** occupy the entire anterior half of the thigh and almost completely envelop the femur. They powerfully extend the leg at the knee and have a weaker action in flexion of the thigh at the hip. The latter action is the principal function of the psoas major and iliacus muscles which are considered on p. 465. The **posterior femoral muscles** are the hamstring muscles. These muscles flex the leg at the knee and extend the thigh at the hip.

The anterior femoral group is entirely postaxial in classification and innervation, and the medial femoral group is pre-axial. However, the other groups contain muscles of both types. The

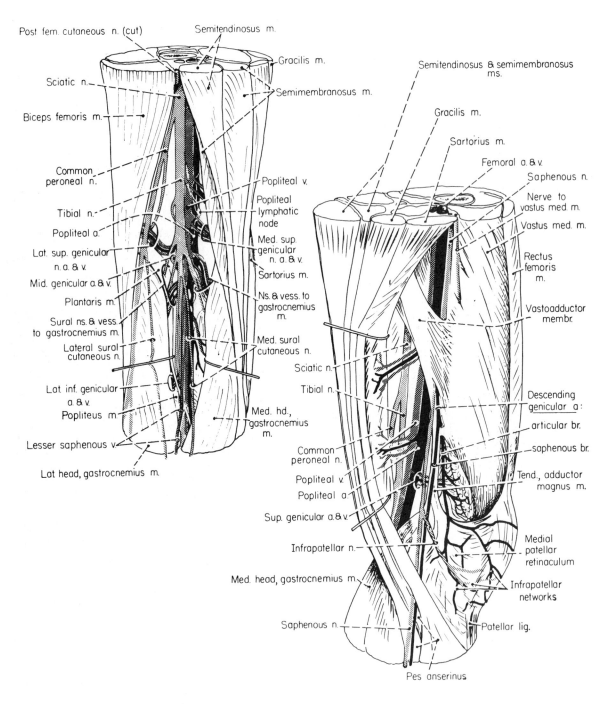

Post. fem. cutaneous n. (cut)
Semitendinosus m.
Gracilis m.
Sciatic n.
Semimembranosus m.
Biceps femoris m.
Semitendinosus & semimembranosus ms.
Gracilis m.
Sartorius m.
Femoral a. & v.
Saphenous n.
Common peroneal n.
Popliteal v.
Nerve to vastus med. m.
Popliteal lymphatic node
Vastus med. m.
Tibial n.
Med. sup. genicular n. a. & v.
Rectus femoris m.
Popliteal a.
Sartorius m.
Lat. sup. genicular n. a. & v.
Vastoadductor membr.
Ns. & vess. to gastrocnemius m.
Mid. genicular a. & v.
Plantaris m.
Med. sural cutaneous n.
Sural ns. & vess. to gastrocnemius m.
Sciatic n.
Descending genicular a.:
Lateral sural cutaneous n.
articular br.
Tibial n.
Lat. inf. genicular a. & v.
saphenous br.
Popliteus m.
Med. hd., gastrocnemius m.
Common peroneal n.
Tend., adductor magnus m.
Lesser saphenous v.
Popliteal v.
Lat. head, gastrocnemius m.
Popliteal a.
Medial patellar retinaculum
Sup. genicular a. & v.
Infrapatellar n.
Infrapatellar networks
Med. head, gastrocnemius m.
Saphenous n.
Patellar lig.
Pes anserinus

Fig. 427   The left popliteal space and the major nerves and vessels at the knee.

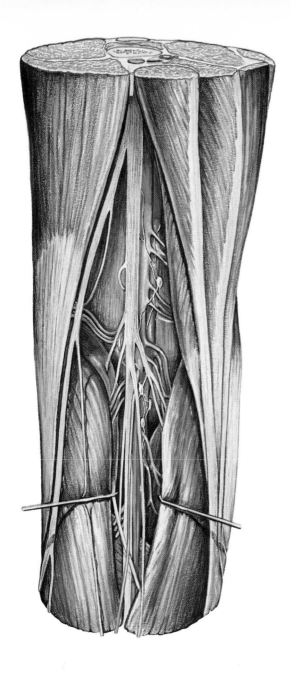
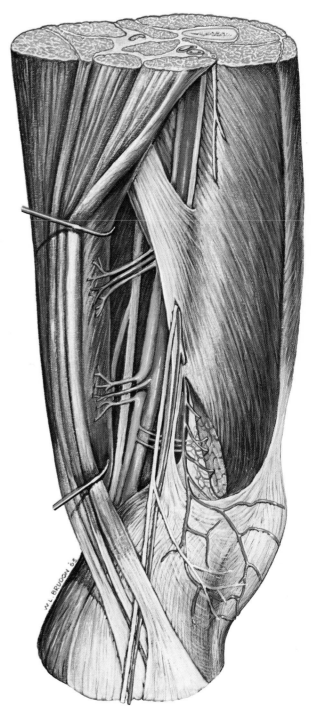

subdivision of the total muscle mass follows the orientation of the pelvic girdle in four-footed animals—the ilium is dorsal; the pubis and the ischium are ventral. *Thus, muscles arising from the pubis or the ischium are pre-axial and are innervated by either the obturator or the tibial nerve; muscles arising from the ilium or the femur are postaxial and are supplied by either the femoral or the common peroneal nerve.*

The Lateral Femoral (Gluteal) Muscles (figs. 428–33, 442)

The muscles of the lateral femoral group are the following:

Postaxial muscles	Pre-axial muscles
Gluteus maximus	Obturator internus
Gluteus medius	Gemellus superior
Gluteus minimus	Gemellus inferior
Tensor fasciae latae	Quadratus femoris
Piriformis	(Obturator externus)

The **gluteus maximus** is a heavy, coarsely fasciculated muscle, superficially situated in the buttock. It is quadrilateral in form and its fasciculi are directed downward and outward. The muscle arises from the posterior gluteal line of the ilium and the area of bone above and behind it, from the aponeurosis of the erector spinae muscle, from the posterior surface of the sacrum and the coccyx, from the sacrotuberous ligament, and from the deep fascia (gluteal aponeurosis) overlying the gluteus medius. The larger upper portion and the superficial fibers of the lower portion of the muscle insert into the iliotibial tract of the fascia lata; the deeper fibers of the lower portion reach the gluteal tuberosity and the lateral intermuscular septum. The upper fibers of the muscle play across the greater trochanter of the femur and are separated from it by the large **trochanteric bursa.** A bursa also separates the tendon of the gluteus maximus from the origin of the vastus lateralis portion of the quadriceps femoris muscle. Another bursa is frequently located between the lower portion of the muscle and the ischial tuberosity. The gluteus maximus is distinctly a muscle of the erect posture but becomes active only under conditions of effort. It is a powerful extensor of the thigh and, acting from its insertion, is a strong extensor of the trunk. Its insertion into the iliotibial tract may promote stability of the femur on the tibia but the

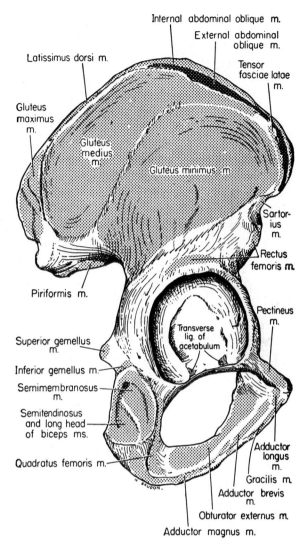

Fig. 428 The muscular attachments to the external surface of the coxal bone. (Black dots—origin; white dots on black—insertion.)

extensive blending of this tract with the lateral intermuscular septum of the thigh would prevent significant action of the muscle on the tibia. The muscle is a lateral rotator, and its superior fibers come into play in forcible abduction of the thigh. The **inferior gluteal nerve,** a postaxial branch of the sacral plexus with fibers from L5 and S1 and 2, is the sole supply of the gluteus maximus muscle. It enters the muscle on its deep surface accompanied by branches of the inferior gluteal artery.

The **gluteus medius** lies largely anterior to the gluteus maximus muscle under the strong vertical fibers of the gluteal aponeurosis, but it also partly underlies the gluteus maximus. The gluteus medius arises from the external aspect of the ilium between the anterior and posterior gluteal lines and from the gluteal aponeurosis. The strong flattened tendon of the muscle inserts into the posterosuperior angle of the greater trochanter and into a diagonal ridge on its lateral surface. A bursa separates the tendon from the trochanter in front of the insertion.

The **gluteus minimus** immediately underlies the gluteus medius, arising from the outer surface of the ilium between the anterior gluteal and inferior gluteal lines. Its fibers converge to an insertion on the anterosuperior angle of the greater trochanter. A bursa lies between the tendon and the medial part of the anterior surface of the greater trochanter. The gluteus medius and gluteus minimus muscles abduct the femur and rotate the thigh medialward. Both muscles are innervated by the superior gluteal nerve (L4, 5, S1) which, emerging from the pelvis at the upper border of the greater sciatic foramen, proceeds lateralward between the two muscles (fig. 442). It is accompanied by branches of the superior gluteal vessels.

Clinical Note The actions of the gluteus medius and gluteus minimus are important in walking, for when one foot is raised off the ground, the glutei, holding the pelvic bone of the opposite side down to the greater trochanter, prevent the collapse of the pelvis on the side which is not supported. The same muscles, by their rotary action, swing the pelvis forward as the step is taken.

The **tensor fasciae latae** is a fusiform muscle enclosed between two layers of the fascia lata. It arises from the anterior part of the external lip of the iliac crest, from the outer surface of the anterior superior iliac spine, and from the notch below the spine. The muscle inserts into the iliotibial tract with which blend both layers of its investing fascia. The tensor fasciae latae lies between the gluteus medius and sartorius muscles. It assists in flexion, abduction, and medial rotation (weakly) of the thigh. The superior gluteal nerve ends in the tensor fasciae latae muscle together with an inferior branch of the superior gluteal artery.

The **piriformis muscle** arises within the pelvis from the front of the sacrum between the first, second, third, and fourth pelvic sacral foramina. It is described in the chapter on the Pelvis (p. 514 & figs. 411, 431). The muscle passes into the thigh through the greater sciatic foramen and inserts into the upper border of the greater trochanter. Its

tendon may be partly blended with the common tendon of the obturator internus and gemellus muscles. The piriformis is a lateral rotator of the thigh and assists in abduction. It is innervated by one or two branches from the first and second sacral nerves. The muscle is occasionally (10 per cent of cases) divided by the aberrant passage of the common peroneal division of the sciatic nerve through it, or the nerve may emerge above the muscle.

The **obturator internus muscle** arises from the circumference of the obturator foramen and from the obturator membrane. It is more completely described in the chapter on the Pelvis (p. 514). The tendon of the obturator internus enters the thigh by passing through the lesser sciatic foramen, and, running horizontally, inserts into the medial surface of the greater trochanter above the trochanteric fossa. The muscle is a lateral rotator of the thigh and is also capable of some abduction of the femur. It is innervated by a pre-axial nerve of the sacral plexus from the fifth lumbar and the first and second sacral nerves. This nerve (figs. 431, 442) also supplies the superior gemellus muscle and, crossing the spine of the ischium, enters the pudendal canal in company with the internal pudendal vessels and the pudendal nerve. The tendon of the obturator internus muscle receives the tendon of the superior gemellus muscle along its superior margin and superficial surface and the tendon of the inferior gemellus along its inferior margin. The bellies of these muscles may overlie and obscure the tendon of the internal obturator.

The **superior gemellus** is a small, tapered muscular fasciculus which arises from the ischial spine. It passes lateralward above the tendon of the internal obturator and then joins that tendon to insert with it. The superior gemellus muscle is innervated by the nerve to the internal obturator.

The **inferior gemellus** is also accessory to the internal obturator muscle. Arising from the ischial tuberosity, this muscle lies along the underside of the obturator tendon and unites with it toward the greater trochanter. The nerve to the inferior gemellus muscle is a branch of the nerve to the quadratus femoris muscle.

The **quadratus femoris** is a thick, quadrilateral muscle located inferior to the inferior gemellus. Its origin is the upper part of the lateral border of the ischial tuberosity (fig. 402); its insertion is the quadrate line, which extends vertically downward from the intertrochanteric crest. This muscle is a strong lateral rotator of the thigh. Its nerve is a pre-axial branch of the sacral plexus which contains fibers from nerves L4 and 5 and S1. The nerve descends anterior to the inferior gemellus and quadratus femoris muscles and enters them on their anterior aspects.

The **obturator externus** is another lateral rotator muscle of the thigh, the tendon of which ends in the trochanteric fossa of the femur. Its origin is deep to those muscles of the medial femoral group which arise from the rami of the pubis and the ischium, and its main description is given under the medial femoral muscles (p. 544).

The Anterior Femoral Muscles (figs. 426, 427–30, 432, 433, 440, 441)

The anterior femoral muscles are all *postaxial* and are *innervated by the femoral nerve.* They are the following:

Sartorius

Quadriceps femoris $\begin{cases} \text{Rectus femoris} \\ \text{Vastus lateralis} \\ \text{Vastus medialis} \\ \text{Vastus intermedius} \end{cases}$

Articularis genu

The **sartorius** is the longest muscle in the body. Ribbon-like in form, it arises from the anterior superior spine of the ilium and the upper half of the notch just below the spine. It inserts into the medial surface of the tibia below the tuberosity and nearly as far forward as the anterior crest of the bone. This insertion is by a common aponeurosis, formed in association with the tendons of the gracilis and semitendinosus muscles (pes anserinus), which is separated from the surface of the tibia by a bursa. A slip of the aponeurosis blends above with the capsule of the knee joint, and another slip blends below with the crural fascia. The sartorius muscle lies in a loose facial sheath formed by the fascia lata and the medial intermuscular septum. Within this superficial sheath the muscle lies diagonal to the thigh, arising anterolaterally and inserting on the medial side of the leg. Its contraction thus produces flexion, abduction, and lateral rotation of the thigh. The muscle also flexes the leg, for its tendon passes the knee behind its transverse axis. These actions led to its name, for 'sartorius' means 'tailor', and the muscle places both the thigh and the leg in the correct position for the traditional cross-legged sitting posture of the tailor. Electromyography indicates that the muscle is more active in flexion

of the thigh than in abduction and lateral rotation.

The femoral nerve innervates the sartorius by two branches which conduct fibers from spinal cord segments L2 and 3—an upper branch arises in common with one of the anterior femoral cutaneous branches and distributes to the proximal part of the muscle; a lower branch reaches its distal portion. The sartorius muscle forms the lateral border of the femoral triangle in the upper one-third of the thigh; in the middle one-third it forms the roof of the adductor canal.

The **quadriceps femoris muscle** is composed of four parts—rectus femoris, vastus lateralis, vastus medialis, and vastus intermedius. The rectus femoris arises from the ilium and thus has an action at the hip joint as well as at the knee. The vasti arise from the femur. All four heads end in the tibia—its tuberosity and its condyles.

(1) The **rectus femoris muscle,** as its name implies, runs straight down the thigh. Somewhat fusiform in shape, its superficial fibers have a bipennate arrangement. It arises by two tendons. The **straight head** takes origin from the anterior inferior spine of the ilium; the **reflected head** arises from a groove above the rim of the acetabulum. The two tendons unite at an acute angle and continue into a central aponeurosis which is expanded downward onto the muscle. From this the muscle fibers arise, turn around its margins, and end in the tendon of insertion. The tendon of insertion broadens to attach to the proximal border of the patella and spreads over its surface to the tuberosity of the tibia. Two branches of the **femoral nerve** which carry nerve fibers from lumbar segments three and four innervate the muscle—one enters the deep surface of the muscle in its proximal fourth; the other enters the medial margin of the muscle below its midlength.

(2) The **vastus lateralis** is the largest part of the quadriceps femoris muscle. It arises from the femur by a broad aponeurosis which is attached to the upper part of the intertrochanteric line, to the anterior and inferior borders of the greater trochanter, to the gluteal tuberosity, to the immediately adjacent portion of the lateral lip of the linea aspera, and to the lateral intermuscular septum throughout the length of the septum. This aponeurosis covers the superior portion of the muscle, and from its deep surface many muscle fibers take origin. The flat tendon of the large, fleshy, muscular mass arises on the deep surface of the lower part of the muscle. The tendon of the vastus lateralis inserts into the superolateral border of the patella and into the lateral condyle of the tibia (lateral retinaculum of the patella, p. 541). The large branch of the femoral nerve (L3, 4) to the vastus lateralis accompanies the descending branch of the lateral circumflex femoral artery, and its radicles enter the muscle at various levels.

(3) The **vastus medialis muscle** arises from the whole extent of the medial lip of the linea aspera, from the distal half of the intertrochanteric line, and from the medial intermuscular septum. The fibers of the vastus medialis muscle are directed downward and forward toward the knee, the lowest fibers being almost horizontal. The strong aponeu-

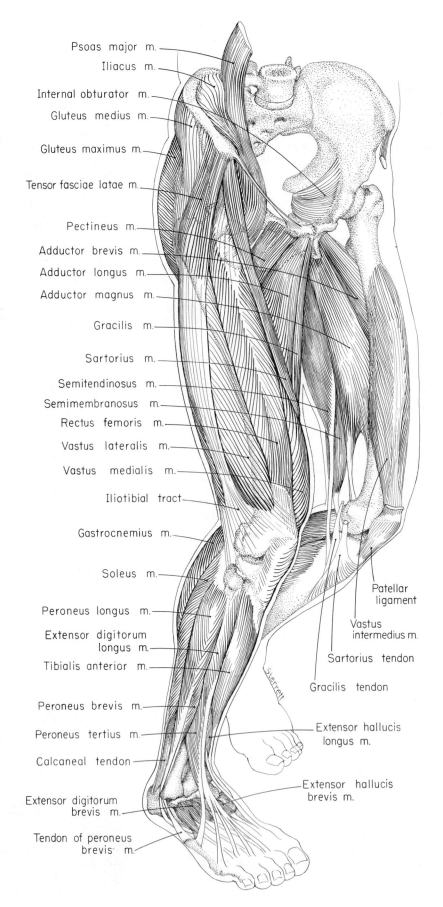

Psoas major m.

Iliacus m.

Internal obturator m.

Gluteus medius m.

Gluteus maximus m.

Tensor fasciae latae m.

Pectineus m.

Adductor brevis m.

Adductor longus m.

Adductor magnus m.

Gracilis m.

Sartorius m.

Semitendinosus m.

Semimembranosus m.

Rectus femoris m.

Vastus lateralis m.

Vastus medialis m.

Iliotibial tract

Gastrocnemius m.

Soleus m.

Peroneus longus m.

Extensor digitorum longus m.

Tibialis anterior m.

Peroneus brevis m.

Peroneus tertius m.

Calcaneal tendon

Extensor digitorum brevis m.

Tendon of peroneus brevis m.

Patellar ligament

Vastus intermedius m.

Sartorius tendon

Gracilis tendon

Extensor hallucis longus m.

Extensor hallucis brevis m.

Fig. 429 The muscles of the anterolateral and anteromedial aspects of the thigh and leg.

rotic tendon, organized on the deep aspect of the muscle, inserts into the tendon of the rectus femoris, the superomedial border of the patella, and the medial condyle of the tibia (medial retinaculum of the patella). Two branches of the femoral nerve (L3, 4) supply the vastus medialis—a proximal branch enters the upper part of the muscle, and a distal branch passes part way along the adductor canal to enter the medial aspect of the lower portion of the muscle.

(4) The **vastus intermedius** portion of the quadriceps femoris arises from the anterior and lateral surfaces of the upper two-thirds of the shaft of the femur, from the lower half of the lateral lip of the linea aspera, and from the lateral intermuscular septum. Its fibers end in a superficial aponeurosis which blends with the deep surface of the tendons of the rectus femoris and the other vastus muscles. The vastus intermedius is innervated superficially by a branch of the **femoral nerve** (L3, 4) which passes through the muscle and supplies also the articularis genu muscle. The vastus intermedius is also supplied by the upper nerve to the vastus medialis.

The four parts of the quadriceps femoris muscle converge on the patella, but this bone is actually a sesamoid bone which is developed in the quadriceps femoris tendon. It bears against the condyles of the femur and serves as both a lever and a roller-bearing for the tendon, its lever function resulting in an improved angle of pull on the tibial tuberosity (Haxton, '45). The **patellar ligament,** ending in the tuberosity of the tibia, is actually the terminal part of the tendinous insertion of the quadriceps femoris muscle. Expansions of the aponeuroses of the vastus muscles insert into the condyles of the tibia. These are the **medial** and **lateral retinacula of the patella;** of these, the medial retinaculum is broader and better developed.

The quadriceps femoris muscle is the great extensor of the leg at the knee. All parts of the muscle contribute to this action. The line of traction, which is along the axis of the femur, is not directly in the line of the tibia. This results in a tendency to displace the patella lateralward as the muscle contracts. The numerous low, almost horizontal fibers of the vastus medialis counter this displacing force. The rectus femoris, having an origin from the ilium, also acts in flexion of the thigh at the hip. Its two, almost right-angled tendons of origin combine to serve the full range of flexion action. As the thigh is raised, the straight tendon has a progressively less effective attachment for the action. As this tendon loses its effectiveness, the flexed thigh becomes aligned with the reflected tendon, and its attachment then forms

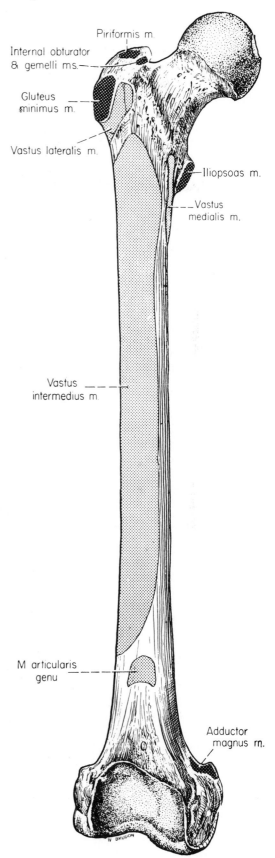

Fig. 430 The muscular attachments on the anterior aspect of the right femur. (Code as fig. 428).

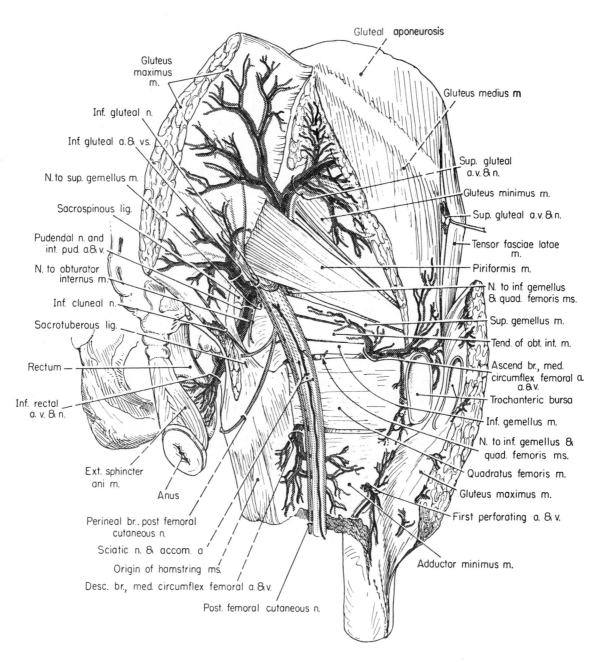

Gluteal **aponeurosis**

Gluteus maximus m.

Inf. gluteal n.

Inf. gluteal a. & vs.

N. to sup. gemellus m.

Sacrospinous lig.

Pudendal n. and int. pud. a. & v.

N. to obturator internus m.

Inf. cluneal n.

Sacrotuberous lig.

Rectum —

Inf. rectal a. v. & n.

Ext. sphincter ani m.

Anus

Perineal br., post femoral cutaneous n.

Sciatic n. & accom. a.

Origin of hamstring ms.

Desc. br., med. circumflex femoral a. & v.

Post. femoral cutaneous n.

Gluteus medius **m**

Sup. gluteal a.v. & n.

Gluteus minimus m.

Sup. gluteal a.v. & n.

Tensor fasciae latae m.

Piriformis m.

N. to inf. gemellus & quad. femoris ms.

Sup. gemellus m.

Tend. of obt. int. m.

Ascend br., med. circumflex femoral a. a. & v.

Trochanteric bursa

Inf. gemellus m.

N. to inf. gemellus & quad. femoris ms.

Quadratus femoris m.

Gluteus maximus m.

First perforating a. & v.

Adductor minimus m.

Fig. 431   The anatomy of the subgluteal area.

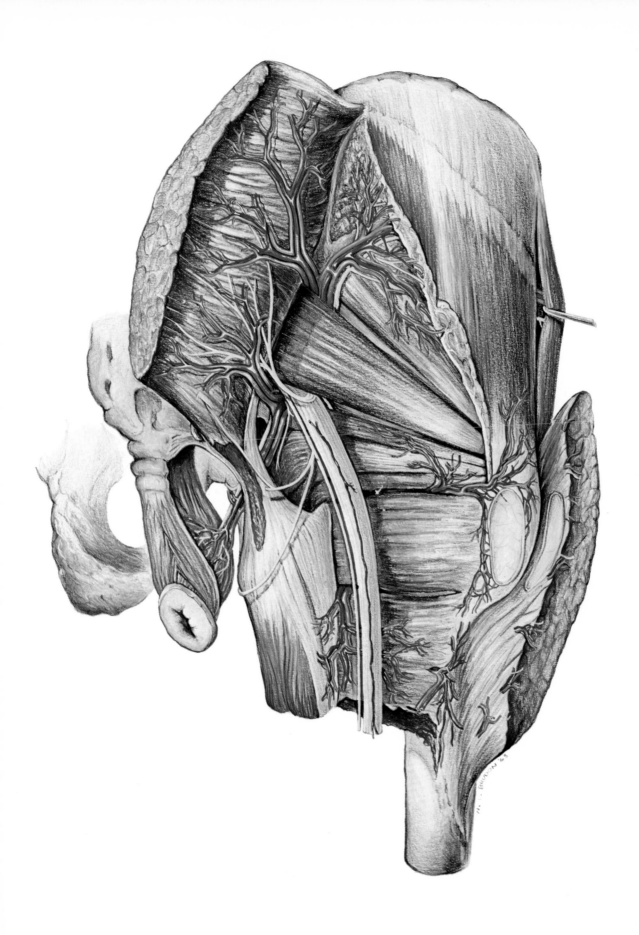

an optimal site for traction of the muscle. The vasti are generally quiescent during relaxed standing.

The **articularis genu** consists of a number of small muscular bundles that arise from the lower one-fourth of the front of the femur. Deep to the vastus intermedius, they insert into the upper part of the synovial membrane of the knee joint. These bundles are innervated by a branch of the nerve to the vastus intermedius.

The Medial Femoral Muscles (figs. 405, 426–29, 431, 432)

The medial femoral muscles arise from the external surfaces of the pubic rami and the ramus of the ischium. *They are all pre-axial adductors and, except for variability in the innervation of the pectineus, are supplied by the obturator nerve.* The muscles are the following:

Gracilis   Adductor longus   Adductor magnus
Pectineus   Adductor brevis   Obturator externus

The **gracilis** is a long, thin muscle, superficially placed on the medial aspect of the thigh. It arises by a thin tendon which attaches to the whole length of the inferior pubic ramus and along the edge of the symphysis pubis. The slender muscle lies along the medial side of the thigh and tapers below into a tendon which, at the knee, is placed between the tendons of the sartorius and semitendinosus muscles. The tendon of the gracilis inserts into the upper part of the shaft of the tibia as part of the pes anserinus (p.539). A **bursa** deep to the tendons of the gracilis and the semitendinosus separate them from the tibial collateral ligament of the knee joint.

The gracilis muscle adducts the thigh and assists in flexing the leg at the knee. It also participates in flexion and medial rotation of the thigh at the hip.

Its nerve, a branch of the anterior division of the obturator nerve containing fibers from the second and third lumbar segments, enters the deep surface of the muscle near the junction of its upper and middle thirds.

The **pectineus** is a flat, quadrangular muscle which forms the medial part of the floor of the femoral triangle. It arises from the pecten of the pubis and from the surface of the bone below the pecten between the iliopubic eminence laterally and the pubic tubercle medially. The fibers of the muscle pass downward, backward, and lateralward and insert by a tendon, which is about 5 cm. wide, into the pectineal line of the femur.

The pectineus muscle adducts the thigh, rotates it medially, and assists in its flexion. It is innervated by a branch of the femoral nerve which passes behind the femoral artery and vein and enters the lateral portion of the muscle. It is also supplied by the accessory obturator nerve when this nerve is present (9 per cent of cases). There is also a variable division of the muscle into ventromedial and dorsolateral portions. These variabilities make for uncertainty in the classification and nerve assignment of the muscle; nor is it established whether the muscle develops phylogenetically from the primitive dorsal or ventral musculature. The muscle appears to the writer to be best handled for Man in the preaxial classification; but in those cases where no obturator nerve supply reaches the muscle, it must be assumed that some pre-axial nerve fibers are carried to it in the femoral nerve.

The **adductor longus muscle** lies in the same plane as the pectineus and forms the medial boundary of the femoral triangle. It arises by a flat, narrow tendon from the medial portion of the superior ramus of the pubis from which it expands into a triangular muscular belly. The muscle inserts by a very thin, wide tendon into the middle one-third of the medial lip of the linea aspera between the tendons of the vastus medialis and adductor magnus muscles.

The adductor longus muscle adducts the thigh and assists in its flexion and medial rotation. It is innervated by a branch of the anterior division of the obturator nerve (L2, 3) which reaches it on the deep surface of its middle one-third.

The **adductor brevis muscle** lies deep to the pectineus and adductor longus muscles. It has a narrow origin from the inferior pubic ramus between the attachments of the gracilis and obturator externus muscles. Like the adductor longus, the muscular fibers of the adductor brevis fan out and end in an aponeurosis which inserts into the lower two-thirds of the pectineal line and the upper half of the medial lip of the linea aspera. Its fibers end dorsal to those of the pectineus and the adductor longus.

The adductor brevis muscle is an adductor of the thigh; to a lesser degree, it assists in its flexion and its medial rotation. The anterior division of the obturator nerve descends on the superficial surface of the adductor brevis. Branches of this nerve (L2, 3) enter the middle one-third of the muscle near its proximal border. The tendon of the adductor brevis muscle is perforated by from one to three perforating branches of the deep femoral artery and their accompanying veins.

The **adductor magnus** is the largest muscle of the medial femoral group. It is triangular in shape and is formed by the combination of two muscles with differing innervations. The adductor magnus muscle arises from the lower part of the inferior pubic ramus, from the ramus of the ischium, and from the ischial tuberosity. Its muscular fibers fan out to the whole length of the linea aspera of the femur—the uppermost fibers are horizontal; the lowermost, vertical. The upper horizontal fibers, sometimes designated as the **adductor minimus muscle,** insert into the medial side of the gluteal ridge and the uppermost part of the linea aspera. Below this, the aponeurosis of the muscle ends in the whole length of the medial lip of the linea aspera and the supracondylar line of the femur. The most medial portion of the muscle (ischiocondylar portion) arises from the ischial tuberosity and forms a round tendon which ends in the adductor tubercle of the medial epicondyle of the femur.

In the lower one-third of the thigh, tendinous fibers spread lateralward from the rounded tendon of the muscle toward the vastus medialis and end in the medial intermuscular septum. This strong aponeurotic sheet (vastoadductor membrane, fig. 427) overlies the femoral vessels in the more distal segment of the adductor canal and may be pierced by the saphenous nerve and the descending genicular artery. The aponeurosis of insertion of the adductor magnus muscle provides four openings adjacent to the femur for the passage of the perforating vessels, and a large gap exists between the lower end of the aponeurosis and the rounded tendon to the adductor tubercle. Through this **adductor hiatus** the femoral vessels pass to the popliteal fossa (p. 551). The adductor magnus muscle is a very powerful adductor of the thigh. The accessory actions and the nerve supply differ between its two parts. The upper portion of the muscle assists in flexion and **medial** rotation of the thigh. It is innervated by the posterior division of the obturator nerve (L3, 4) which descends upon the muscle under its fascia. Branches pierce the muscle on its anterior surface about midway between its pelvic and femoral attachments. The ischiocondylar portion of the muscle is one of the hamstring muscles of the back of the thigh. It extends the thigh and rotates it medialward. It is supplied on its dorsal surface by a branch from the tibial division of the sciatic nerve (L4, 5, S1) which separates from the nerve to the semimembranosus muscle.

The **obturator externus muscle** arises from the external aspect of the superior pubic and inferior pubic rami and the ramus of the ischium and from the external surface of the obturator membrane (fig. 405). The muscle is deep to the origins of the adductor group of muscles. The tendon of the obtura-

tor externus muscle passes across the back of the neck of the femur and the lower part of the posterior capsule of the hip joint to insert into the trochanteric fossa of the femur. The synovial membrane of the hip joint descends below the edge of the posterior capsule and acts as a bursa separating the tendon from the neck of the femur.

The external obturator muscle is a lateral rotator of the thigh. The muscle is innervated by the obturator nerve, which, while in the obturator canal, gives off a branch that enters the superior border and the anterior surface of the muscle (fibers from L3, 4). The anterior branch of the obturator nerve reaches the thigh by passing anterior to the muscle; the posterior branch, by piercing it.

The Posterior Femoral Muscles (figs. 426–28, 431, 433)

The posterior femoral muscles compose the 'hamstring' group. *They arise from the ischial tuberosity (pre-axial) and are supplied by the tibial division of the sciatic nerve.* The biceps femoris muscle, however, has a supplemental origin from the femur, and this portion of the muscle (postaxial) is correspondingly supplied by the common peroneal portion of the sciatic nerve. The muscles are as follows:

Semitendinosus	Biceps femoris
Semimembranosus	(Ischiocondylar portion of adductor magnus)

The **semitendinosus muscle** is aptly named, for it is tendinous in about half its length. It arises from the lower and medial impression on the tuberosity of the ischium in common with the long head of the biceps femoris muscle. The adjacent surfaces of the two muscles are joined by an aponeurosis for about 7.5 cm. from this origin, and additional muscle fibers arise from this aponeurosis. The tendon of the semitendinosus muscle forms the medial margin of the popliteal fossa; it then curves around the medial condyle of the tibia and inserts as part of the pes anserinus into the upper part of the medial surface of the tibia. Along with the gracilis, it is separated from the tibial collateral ligament of the knee joint by a bursa.

Two branches of the tibial division of the sciatic nerve (L4, 5, S1, 2) usually reach the muscle. The upper nerve, frequently also supplying the long head of the biceps femoris, enters the middle one-third of the semitendinosus muscle on its deep surface. The lower nerve enters the deep surface of the distal half of the muscle after arising in common with the nerve to the semimembranosus.

The **semimembranosus muscle** arises by a long flat tendon that lies beneath the proximal half of the semitendinosus muscle and attaches to the upper and outer impression on the tuberosity of the ischium. This tendon, at first adherent to the tendon of the adductor magnus in front and to that of the biceps femoris and semitendinosus muscles behind, descends to the middle of the muscle. The tendon of insertion of the semimembranosus ends mainly in the horizontal groove on the posteromedial aspect of the medial condyle of the tibia. Fibrous expansions of the tendon reach the lateral condyle of the femur (oblique popliteal ligament of the knee joint, p. 587), the fascia of the popliteus muscle, and the tibial collateral ligament.

The semimembranosus muscle is innervated by a branch of the tibial division of the sciatic nerve which arises with the lower branch to the semitendinosus. This branch divides into several radicles which enter the deep surface of the muscle at about its midlength.

The **biceps femoris** is a secondary combination of one pre-axial muscle (the long head) and one post-axial muscle (the short head). The **long head** arises from the lower and medial impression on the ischial tuberosity by a tendon in common with the semitendinosus and from the lower part of the sacrotuberous ligament. The **short head** arises from the lateral lip of the linea aspera from the middle of the shaft down to its bifurcation, from the proximal two-thirds of the supracondylar line, and from the lateral intermuscular septum. From this line of origin, the muscle fibers join the tendon of the long head to form a heavy, round tendon which constitutes the lateral margin of the popliteal fossa. At the knee the tendon divides around the fibular collateral ligament and ends on the lateral aspect of the head of the fibula, the lateral condyle of the tibia, and in the deep fascia of the lateral aspect of the leg.

The biceps femoris, as a combined muscle, has a double innervation. The long head usually receives two branches of the **tibial** portion of the sciatic nerve (S1, 2, 3), one to the upper one-third and one to the middle one-third of the

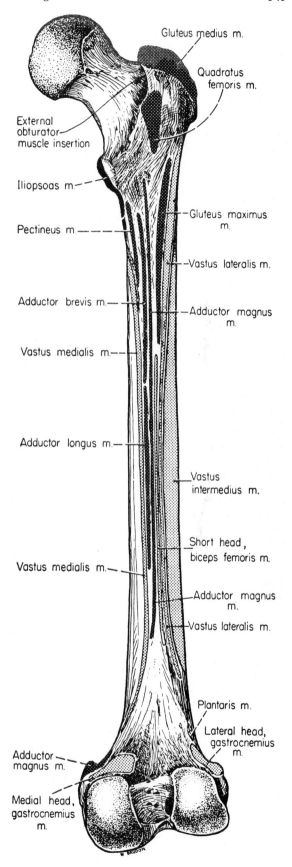

Fig. 432 The muscular attachments on the posterior surface of the right femur. (Code as fig. 428).

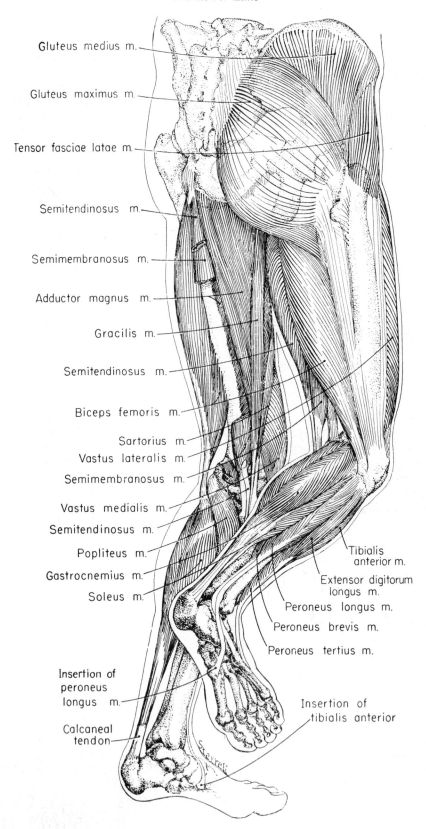

Gluteus medius m.

Gluteus maximus m.

Tensor fasciae latae m.

Semitendinosus m.

Semimembranosus m.

Adductor magnus m.

Gracilis m.

Semitendinosus m.

Biceps femoris m.

Sartorius m.

Vastus lateralis m.

Semimembranosus m.

Vastus medialis m.

Semitendinosus m.

Popliteus m.

Gastrocnemius m.

Soleus m.

Insertion of peroneus longus m.

Calcaneal tendon

Tibialis anterior m.

Extensor digitorum longus m.

Peroneus longus m.

Peroneus brevis m.

Peroneus tertius m.

Insertion of tibialis anterior

Fig. 433 The muscles of the posterolateral and posteromedial aspects of the thigh and leg.

deep surface of the muscle. The nerve to the short head is a branch of the **common peroneal** division of the sciatic (L5, S1, 2). It enters the superficial surface of the muscle near its lateral margin.

The hamstring muscles flex the leg and extend the thigh, and these actions are carried out together in the usual movement patterns of the lower limb. Since the muscles usually produce a combined action, maximal excursion at one joint (such as extension of the thigh) limits the movement at the other joint to less than maximal. Their ligamentous or protective action at the hip is important. Additionally, the "semi" muscles rotate the flexed leg medially, biceps femoris rotates it laterally. These are minimal actions.

THE FEMUR

The **femur** (figs. 430, 432, 434, 435) is the longest and strongest bone in the body. Its **shaft** is nearly cylindrical and fairly uniform in caliber, but its two ends are quite irregular. The inferior extremity of the bone is broadened to form the knee joint and has smooth articular surfaces for the condyles of the tibia and the facets of the patella. The superior extremity of the bone has a nearly spherical head mounted on an angulated neck and prominent trochanters provide for muscle attachments.

The **head** of the femur is smooth and forms about two-thirds of a sphere. It is received into the acetabulum of the os coxae in the formation of the hip joint. The articular surface of the head is largest above and anteriorly; it is interrupted medially by a depression, the **fovea capitis femoris,** into which attaches the **ligamentum capitis femoris.**

The **neck** is a strong buttress of bone whch connects the head with the shaft in the region of the trochanters. It is compressed anteroposteriorly and contains a large number of prominent pits, especially on its posterosuperior aspect, for the entrance of blood vessels.

The neck is about 5 cm. in length and allows the femur free movement without the interference of contact with the os coxae. The neck forms an angle with the shaft, which varies in the normal individual from 115 to 140°. The head also points somewhat forward as the result of a forward angle between the neck and shaft, which averages 8°. The superior border of the neck leads to the greater trochanter; the inferior border slopes down and approaches the direction of the shaft as it ends in the lesser trochanter.

The **greater trochanter** is a large, square prominence on the upper end of the shaft of the femur. It is separated from the smooth shaft below by a horizontal ridge. The greater trochanter is the bony prominence of the hip and is palpable 12 or 14 cm. below the iliac crest. The

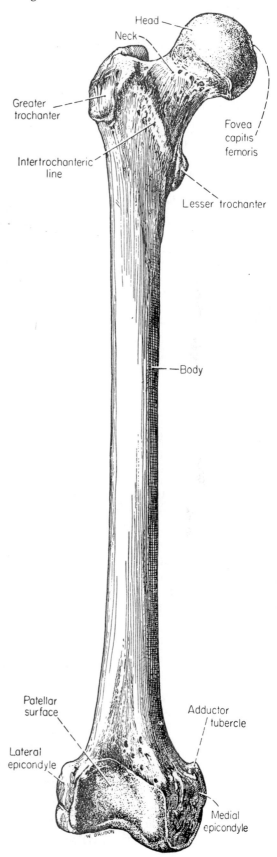

Fig. 434 The anterior surface of the right femur. ($\times \frac{1}{2}$)

lateral quadrilateral surface of the greater trochanter is divided by an oblique ridge which runs between its posterosuperior and antero-inferior angles. Its posterior rounded border bounds the trochanteric fossa and contributes to the intertrochanteric crest. The **trochanteric fossa** is a deep pit on the medial aspect of the greater trochanter.

The **lesser trochanter** is a blunt conical projection which is located at the junction of the inferior border of the neck and the shaft of the femur. The trochanters are joined behind by a thick **intertrochanteric crest.** On the anterior surface of the femur, the junction of the neck and the shaft is ridged; this is the **intertrochanteric line.** This begins in the superior border of the greater trochanter, provides attachment for the capsule of the hip joint across the front of the bone, and is continued below the lesser trochanter as a spiral line which winds backward to blend into the medial lip of the linea aspera.

The **shaft** of the bone, fairly uniform in caliber, nevertheless broadens slightly at its extremities. Bowing somewhat forward, its surface is smooth, except along the linea aspera, and it is covered by muscles. The almost circular cross section of the shaft is modified by a thickened ridge along its posterior aspect; this is the **linea aspera.** The linea aspera is especially prominent in the middle one-third of the bone, where it exhibits a **lateral lip** and a **medial lip.** The lateral and medial lips diverge below and form the supracondylar ridges of the femur. Superiorly, the linea aspera has three rough lines: the lateral lip blends with the prominent **gluteal tuberosity** (for gluteus maximus) and is prolonged to the base of the greater trochanter; the intermediate lip extends upward as the **pectineal line** to the posterior border of the lesser trochanter. The medial lip of the linea aspera continues in a spiral line to the front of the femur, where the spiral line ends in the intertrochanteric line. The **nutrient foramen** of the femur lies on the linea aspera about two-fifths of the distance down the bone. It is directed upward.

The **lower extremity** of the femur is broadened about threefold for articulation with the tibia at the knee. Except at the sides, the surfaces are articular. Two oblong **condyles** for articulation with the tibia, separated by a deep **intercondylar fossa,** occupy the inferior and posterior parts of the articular surface.

These wheel-like surfaces are curved from side to side and are only indistinctly separated from the anteriorly placed **patellar surface.** The contact surface for the patella includes parts of both condyles but is derived mostly from the lateral condyle. The **lateral condyle** is broader than the medial condyle. The intercondylar fossa is especially deep posteriorly and is separated by a ridge from the popliteal surface of the femur above. The **medial condyle** is longer than the lateral condyle, and, if the femur is oriented vertically, it extends farther inferiorly. In life, however, the two condyles rest on the horizontal plane of the tibial condyles, and the shaft of the femur inclines

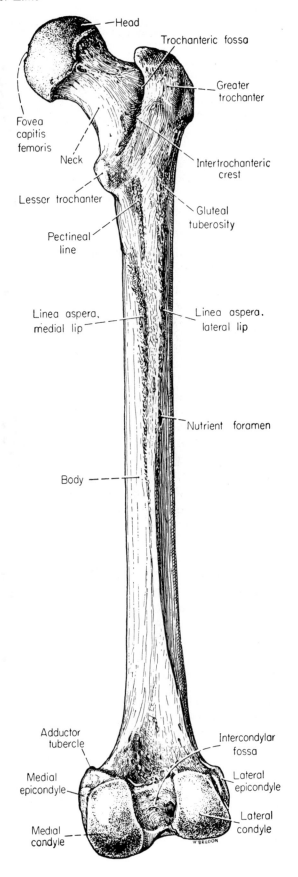

Fig. 435 The posterior surface of the right femur. ($\times \frac{1}{2}$)

downward and inward. This inclination is an expression of the greater breadth of the body at the hips than at the knees.

The **epicondyles** of the femur bulge above and within the curvature of the condyles. The **medial epicondyle** is the more prominent. It gives attachment to the tibial collateral ligament and bears a pointed projection, the **adductor tubercle,** on which the tendon of the adductor magnus muscle inserts. The **lateral epicondyle** gives rise to the fibular collateral ligament. Immediately below the lateral epicondyle is an oblique groove which borders the articular surface of the condyle. This groove lodges the tendon of the popliteus muscle.

The structure of the femur shows an excellent adaptation to the mechanical conditions existing at all points along the bone. Its internal architecture appears to be determined by the requirements of mechanical laws which result in a maximum of strength with a minimum of material. A more detailed description of the structure of bone is given under the general description of bone (p. 40).

The femur is ossified from five centers: one for the shaft, one each for the head and the inferior extremity, and one for each trochanter. The shaft begins to ossify during the seventh week of development and is ossified at birth; ossification extends into the neck after birth. The center of ossification for the lower end of the bone appears during the ninth month of intrauterine life; that for the head, during the first year of postnatal life. The ossification center in the greater trochanter appears during the third to the fifth year; that for the lesser trochanter, about the ninth or the tenth year. The epiphyses at the upper end of the bone—head and trochanters—fuse with the shaft during the fourteenth to the seventeenth year. The epiphysis at the knee fuses with the shaft at about seventeen and a half years of age in the male and by the fifteenth year in the female. Individual variations may be as great as two years earlier or later.

THE PATELLA

The **patella** (fig. 436) is a large sesamoid bone developed in the tendon of the quadriceps femoris muscle. It bears against the anterior articular surface of the inferior extremity of the femur and, by holding the tendon off the lower end of the femur, improves the angle of approach of the tendon to the tibial tuberosity.

The **anterior surface** of the patella is convex and is vertically striated by the tendon fibers. The **base** (superior border) is thick and gives attachment to the tendinous fibers of the rectus femoris and vastus intermedius muscles. The **lateral** and **medial borders** are thinner and receive the tendinous fibers of the vastus lateralis and vastus medialis muscles. These two borders converge below into the pointed **apex** which gives attachment to the patellar ligament. The **articular surface** of the patella presents a smooth, oval articular area, divided into two facets by a vertical ridge. The ridge occupies the groove on the patellar surface of the

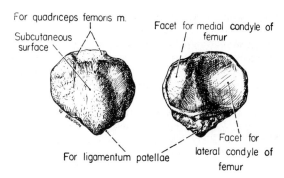

Fig. 436 The patella; anterior surface on left, posterior surface on right. ($\times \frac{1}{2}$)

femur; the medial and lateral facets correspond to the similar parts of the patellar surface of the femur. The lateral facet is broader and deeper than the medial facet. Inferior to the articular surface is a rough nonarticular area, from the lower half of which the patellar ligament arises. The patella arises from a single ossification center which appears early in the third year of life in the male and at about two and a half years of age in the female. Ossification is complete about the thirteenth year in the male and about the tenth year in the female.

BLOOD VESSELS OF THE HIP AND THE THIGH

The blood vessels of the hip region—superior and inferior gluteal—are fully described in the chapter on the Pelvis (pp. 504, 506). Supplemented by these and the obturator vessels, the **femoral vessels,** the continuation of the external iliac vessels, supply the major part of the lower limb. The femoral vessels become the popliteal vessels within the popliteal fossa; and the popliteal vessels, in the upper one-third of the leg, divide into the anterior tibial and posterior tibial vessels. From these are derived the dorsalis pedis and the lateral plantar and medial plantar vessels of the foot.

The Obturator Artery (figs. 405, 440)

The obturator artery is a branch of the internal iliac artery which distributes largely outside the pelvis. Its origin, its pelvic branches, and its frequently abnormal course are considered in the description of the internal iliac artery (p. 505). Having passed through the obturator canal, the obturator artery divides at the upper margin of the obturator foramen into anterior and posterior branches.

The **anterior branch** runs forward and downward on the outer surface of the obturator membrane and gives off muscular arteries to the obturator externus and internus and to the origins of adductor magnus and brevis. It anastomoses with a radicle of the posterior branch and with the medial circumflex femoral artery. The **posterior branch** follows the posterior margin of the obturator foramen and sends an offshoot anteriorly along the ramus of the ischium to anastomose with the anterior branch. The posterior branch also supplies the origin of the adductor magnus muscle and gives off an **acetabular branch** which enters the hip joint through the acetabular notch and provides the small artery of the ligamentum capitis femoris.

The Femoral Artery (figs. 405, 417, 427, 437)

The femoral artery enters the lower limb by passing under the inguinal ligament midway between the anterior superior spine of the ilium and the pubic symphysis. It descends through the femoral triangle and, at its apex, enters the adductor canal. At the **adductor hiatus** in the tendon of the adductor magnus muscle, the femoral artery passes into the popliteal space and becomes the popliteal artery. The relations of the vessel within these areas are discussed with the spaces concerned (pp. 533, 534, and 535).

The **branches** of the femoral artery are as follows:

> Superficial epigastric
> Superficial circumflex iliac
> Superficial external pudendal
> Deep external pudendal
> Deep femoral
> Descending genicular

The **superficial epigastric, superficial circumflex iliac,** and **superficial external pudendal arteries** arise from the anterior aspect of the femoral artery within 2 cm. of the inguinal ligament. They are subcutaneous vessels which are related to the lower abdominal wall and the pudendal area. The **deep external pudendal artery** arises from the medial aspect of the femoral artery or as a branch of the medial circumflex femoral artery. It passes medialward over the pectineus and adductor longus muscles, and pierces the fascia lata to become subcutaneous in the scrotum (or labium majus). These arteries are described more fully on p. 369.

The **deep femoral artery,** the largest branch of the femoral artery, arises from its lateral side about 5 cm. below the inguinal ligament. The deep femoral artery sinks deeply into the thigh as it descends, so that it lies behind the femoral artery and vein on the medial side of the femur. It crosses the tendon of the adductor brevis muscle and, at the lower border of this tendon, passes deep to the tendon of the adductor longus muscle. In the lower one-third of the thigh the deep femoral artery ends as the fourth perforating artery which pierces the

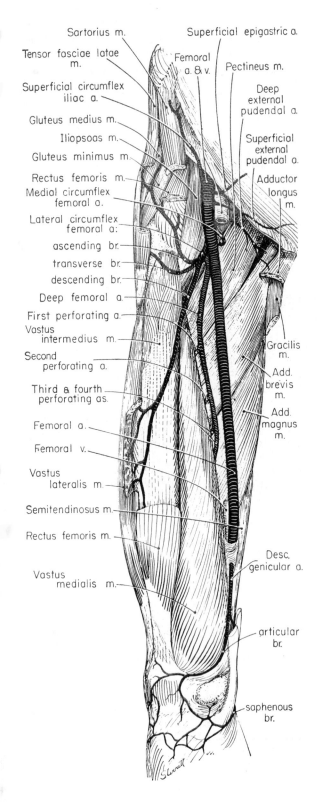

Fig. 437   The femoral artery and its branches.

adductor magnus and distributes to the hamstring muscles on the back of the thigh. In the femoral triangle the deep femoral artery gives off the medial circumflex femoral, lateral circumflex femoral, and muscular branches; in the adductor canal it provides three perforating branches. The circumflex femoral arteries are variable, both arising from the deep femoral in only 56 per cent of cases (Williams, Martin, & McIntire, '34).

(1) The **medial circumflex femoral artery** arises from the medial and posterior aspect of the deep femoral in most cases; it arises from the femoral artery above the origin of the deep branch in 20 per cent of specimens. It sinks deeply into the femoral triangle, passing between the pectineus and iliopsoas muscles, and continues backward under the neck of the femur. The medial circumflex femoral artery supplies branches to the upper portion of the adductor muscles but distributes principally in the back of the hip and thigh. Deep to adductor brevis an **acetabular branch** enters the hip joint beneath the transverse ligament of the acetabulum. Several **muscular branches** supply adductor magnus, one distributing adjacent to the obturator nerve. Anterior to the quadratus femoris muscle the medial circumflex femoral artery divides into **ascending** and **descending branches**. The former passes upward and lateralward to the trochanteric fossa. The descending branch crosses the upper border of adductor minimus and descends into the hamstring muscles beyond the ischial tuberosity.

(2) The **lateral circumflex femoral artery** arises from the lateral side of the deep femoral, passes lateralward deep to the sartorius and rectus femoris muscles, and divides into ascending, transverse, and descending branches. It has an origin from the femoral artery in 14 per cent of cases. The **ascending branch** passes upward, beneath the tensor fasciae latae muscle, to the adjacent anterior borders of the gluteus medius and gluteus minimus muscles and passes between them to anastomose with terminals of the superior gluteal artery. The small **transverse branch** enters the substance of the vastus lateralis muscle, winds around the femur below the greater trochanter, and anastomoses on the back of the thigh with the medial circumflex femoral, inferior gluteal, and first perforating arteries. An articular artery for the hip joint may arise from either the ascending or the transverse branch. The **descending branch** passes downward on the vastus lateralis accompanied by the branch of the femoral nerve to this muscle. It supplies the muscle and anastomoses with the descending genicular branch of the femoral artery and with the lateral superior genicular branch of the popliteal.

(3) The **perforating arteries** are usually three in number. Many muscular branches of the deep femoral artery arise in the adductor canal, but the perforating arteries come from the posterior aspect of the artery, pass directly against the linea aspera of the femur, and pierce the tendons of the adductor muscles to reach the posterior muscular compartment of the thigh. They supply the muscles of the back of the thigh. The first perforating artery passes immediately below the pec-

tineus and through the middle of the adductor brevis muscle; the second, through its lower 3 or 4 cm.; and the third, just below the lowest fibers of the muscle. The **first perforating artery,** after piercing the adductor brevis and magnus muscles, anastomoses with the terminals of the inferior gluteal, the medial circumflex femoral, and the lateral circumflex femoral arteries. The first perforating artery supplies the gluteus maximus and vastus lateralis muscles and sometimes others. The **second perforating artery,** on the back of the thigh, passes between the gluteus maximus and the short head of the biceps femoris and enters the vastus lateralis. It supplies the vastus lateralis, usually provides the nutrient artery of the femur, and may also provide branches to the long head of the biceps and the semitendinosus muscles. The **third perforating artery** passes through the adductor magnus and supplies the vastus lateralis and the upper portion of the short head of the biceps femoris muscles. The deep femoral artery passes under the tendon of the adductor longus and ends by piercing the adductor magnus muscle as the **fourth perforating artery.** This vessel descends in and ends in the short head of the biceps femoris muscle.

The **descending genicular artery** arises from the femoral artery just before the femoral artery passes through the adductor hiatus. The descending genicular artery immediately divides into saphenous and articular branches. The **saphenous branch** pierces the aponeurotic covering of the adductor canal and passes between the tendons of the gracilis and sartorius muscles in company with the saphenous nerve. With the nerve, the artery perforates the fascia lata and supplies the skin and the superficial tissues in the upper medial part of the leg. The **articular branch** descends in the substance of the vastus medialis muscle, anterior to the adductor magnus tendon, to the medial side of the knee. It supplies branches to the vastus medialis muscle and anastomoses with the medial superior genicular and the anterior tibial recurrent arteries. A branch passes lateralward over the patellar surface of the femur, supplies the knee joint, and anastomoses with the descending branch of the lateral circumflex femoral artery and with the lateral superior genicular artery.

The Popliteal Artery (figs. 427, 438)

The popliteal artery is the direct continuation of the femoral artery at the adductor hiatus. It descends through the popliteal space with a slight mediolateral inclination and ends opposite the lower border of the popliteus muscle by dividing into the anterior tibial and posterior tibial arteries. The artery is the deepest of the vascular structures in the popliteal space, lying in the intercondylar fossa of the femur and against the posterior portion of the capsule of the knee joint. Below, it has the popliteus muscle anterior to it in the upper fourth of the leg. The popliteal vein and the tibial nerve are superficial to the artery. The branches of the popliteal artery are as follows:

Lateral superior genicular
Medial superior genicular
Middle genicular
Sural
Lateral inferior genicular
Medial inferior genicular

The **lateral superior genicular artery** passes above the lateral condyle of the femur against the plantaris muscle and the lateral head of the gastrocnemius muscle. Lying deep to the tendon of the biceps femoris muscle, the artery winds around the femur immediately above its condyle. By a superficial branch, it supplies the vastus lateralis muscle and anastomoses with the descending branch of the lateral circumflex femoral and the lateral inferior genicular arteries. Its deep branch supplies the knee joint and forms an anastomotic arch across the front of the femur with the descending genicular and the medial superior genicular arteries.

The **medial superior genicular artery** passes medial-ward on the medial head of the gastrocnemius anterior to the tendons of the semimembranosus and semi-tendinosus muscles. It has two branches, one of which supplies the vastus medialis muscle and anastomoses with the descending genicular and the medial inferior genicular arteries. The other branch supplies the knee joint and, by the anastomotic arch across the femur, anastomoses with the lateral superior genicular artery.

The **middle genicular artery** is a small, unpaired branch which arises from the anterior aspect of the popliteal artery opposite the back of the knee joint. It pierces the oblique popliteal ligament and supplies the cruciate ligaments and the synovial membrane within the joint cavity.

The **sural arteries** are usually two large muscular branches which enter the proximal part of the heads of the gastrocnemius muscle and send other branches into the plantaris and the upper part of the soleus muscle. The sural arteries are the distal muscular branches of the popliteal. There are also proximal unnamed muscular branches to the lower end of the hamstring muscles and cutaneous branches to the skin over the popliteal fossa. One cutaneous branch descends along the middle of the back of the calf with the lesser saphenous vein.

The **lateral inferior genicular artery** passes across the popliteus muscle anterior to the lateral head of the gastrocnemius and plantaris muscles. Turning at the side of the knee, it lies deep to the fibular collateral ligament and ends by dividing into branches which anastomose with the lateral superior genicular, the anterior tibial recurrent, and the medial inferior genicular arteries.

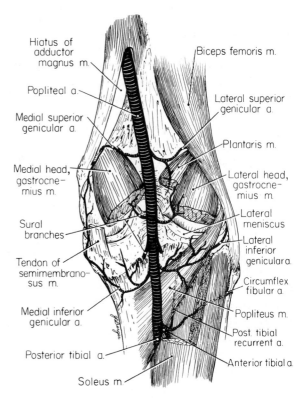

Fig. 438 The popliteal artery and its branches.

The **medial inferior genicular artery** passes medial-ward along the upper border of the popliteus muscle anterior to the medial head of the gastrocnemius muscle. On the medial side of the knee the artery runs deep to the tibial collateral ligament. At the anterior border of the ligament branches ascend to anastomose with the descending genicular and medial superior genicular arteries. Other branches pass across the tibia under the patellar ligament and anastomose with the lateral inferior genicular and the anterior tibial recurrent arteries. For collateral circulation in the femoral and popliteal systems, see p. 602.

Veins of the Hip and the Thigh

In general, the veins of the hip and the thigh are venae comitantes of the arteries. The **superior gluteal veins** receive tributaries from the buttock, especially from the gluteal muscles. They accompany the artery through the greater sciatic foramen and end, usually as a single trunk, in the internal iliac vein. The **inferior gluteal veins** arise, similarly, in the subcutaneous tissue and muscles of the buttock. They anastomose with the medial circumflex femoral and first perforating veins. Entering the pelvis through the greater sciatic foramen below the piriformis muscle, the inferior gluteal veins unite into a single vessel which ends in the lower part of the internal iliac vein. The **obturator vein** is formed from tributaries which arise in the hip joint and the muscles of the adductor region of the thigh. It enters the pelvis through the obturator canal, passes backward and upward on the lateral pelvic wall below the corresponding artery, and ends in the internal iliac vein.

The **popliteal vein** (fig. 427) is a large, single vein or is represented by several veins which ascend through the popliteal fossa superficial to the artery and between it and the tibial nerve. It is bound to the artery by a dense fascial sheath within which it lies somewhat medial to the artery inferiorly but against its lateral side above the knee joint. Three or four bicuspid valves prevent a descending flow in the vein. One valve is constantly found just distal to the adductor hiatus. The popliteal vein is formed at various levels by the union of the venae comitantes of the anterior tibial, posterior tibial, and peroneal arteries. A single popliteal vein is only to be expected with certainty above a point about 5 cm. above the level of the knee joint (Williams, '53). Tributaries within the popliteal fossa are veins corresponding to the branches of the popliteal artery and the lesser saphenous vein (p. 526). The popliteal vein becomes continuous with the femoral vein at the adductor hiatus.

The **femoral vein** (figs. 427, 439) accompanies the femoral artery through the adductor canal and the femoral triangle. Beginning at the adductor hiatus in continuity with the popliteal vein, the femoral vein is posterolateral to its artery. As it ascends through the adductor canal, it becomes posterior and is finally medial to the artery in the femoral triangle. It occupies the middle compartment of the femoral sheath and, as it passes under the inguinal ligament, becomes the external iliac vein. The femoral vein contains two or three bicuspid valves; usually one lies just inferior to the entrance of the deep femoral vein, and one is located at the level of the inguinal ligament (Powell & Lynn, '51).

The femoral vein receives the deep femoral and greater saphenous veins and the venae comitantes of the descending genicular, lateral circumflex femoral, medial circumflex femoral, and deep external pudendal arteries. The superficial epigastric, superficial circumflex iliac, and superficial external pudendal veins enter the greater saphenous vein which passes through the saphenous opening to empty into the femoral vein at the lower end of the femoral sheath. The deep external pudendal vein

is usually the highest tributary of the femoral vein. The venae comitantes of the descending genicular artery may enter the lower end of the femoral vein or may, with muscular tributaries, form a long collateral vessel which may not end in the femoral vein until the femoral triangle is reached. The deep femoral vein receives muscular and perforating tributaries. The **perforating veins** also help form, on the back of the thigh, a long anastomotic column formed of their interconnections and of connections with radicles of the popliteal vein below and the inferior gluteal vein above. The medial circumflex femoral and lateral circumflex femoral veins are only infrequently tributary to the deep femoral vein. Charles et al. ('30) found that the circumflex femoral veins were tributary to the femoral vein above the entrance of the deep femoral in more than three-fourths of the specimens examined. The circumflex femoral veins have the same course and regions of supply as do the corresponding arteries and make numerous anastomoses with the obturator, gluteal, and genicular veins, and with perforating and muscular veins of the femoral system.

Clinical Note The **deep femoral vein** enters the femoral vein at a much less regular level than characterizes the origin of the deep femoral artery. The deep femoral vein joins the femoral vein, on an average, 8 cm. below the inguinal ligament, and the junction may be expected within the range of from 5 to 11.5 cm. from the ligament (Edwards & Robuck, '47).

NERVES OF THE HIP AND THE THIGH

The cutaneous nerves of the hip and the thigh are described with the superficial structures (p. 527). The nerves to the muscles and the joints of the area are the obturator and femoral nerves from the lumbar plexus and the muscular branches and sciatic nerve of the sacral plexus.

The Obturator Nerve (fig. 440)

The obturator nerve is the principal pre-axial nerve of the lumbar plexus. It arises from the anterior branches of lumbar nerves two, three, and four and descends along the medial border of the psoas muscle. Its further course within the pelvis is described under the lumbar plexus (p. 468). As it enters the thigh through the obturator canal, the nerve divides into an anterior and a posterior branch.

The **anterior branch** communicates with the accessory obturator nerve, when this nerve is present, and descends on the surface of the adductor brevis muscle, behind the pectineus and adductor longus muscles. An **articular branch** to the hip joint is given off near the obturator foramen and **muscular branches** are supplied by the anterior branch to the adductor longus, gracilis,

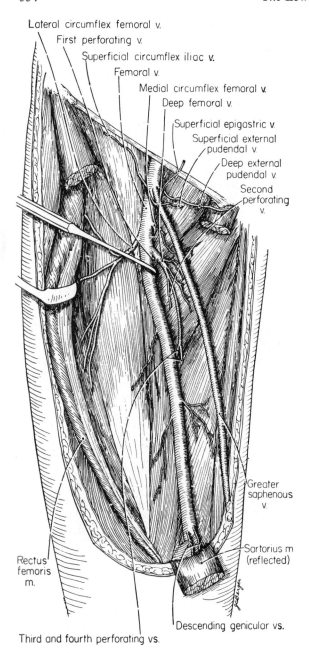

Lateral circumflex femoral v.

First perforating v.

Superficial circumflex iliac v.

Femoral v.

Medial circumflex femoral v.

Deep femoral v.

Superficial epigastric v.

Superficial external pudendal v.

Deep external pudendal v.

Second perforating v.

Greater saphenous v.

Sartorius m (reflected)

Rectus femoris m.

Descending genicular vs.

Third and fourth perforating vs.

Fig. 439   The femoral vein.

under the muscular fascia of the adductor magnus. It supplies the obturator externus and adductor magnus muscles and sometimes the adductor brevis muscle. The nerve continues through the substance of the adductor magnus and ends in an articular branch to the knee joint. This **articular branch** accompanies the popliteal artery in the popliteal fossa, where it penetrates the oblique popliteal ligament to supply the posterior part of the knee joint.

The Accessory Obturator Nerve

The variable accessory obturator nerve is described with the lumbar plexus (p. 468).

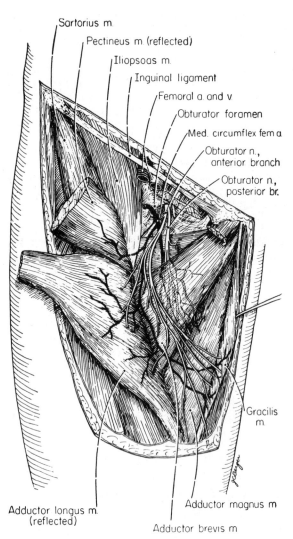

Sartorius m.

Pectineus m. (reflected)

Iliopsoas m.

Inguinal ligament

Femoral a. and v.

Obturator foramen

Med. circumflex fem a.

Obturator n., anterior branch

Obturator n., posterior br.

Gracilis m.

Adductor longus m. (reflected)

Adductor magnus m

Adductor brevis m

Fig. 440   The branches and distribution of the obturator nerve.

and adductor brevis muscles. Rarely it gives a branch to the pectineus. Terminal radicles of the anterior branch are a twig to the femoral artery, which ramifies on the artery in the adductor canal, and a cutaneous branch. A **cutaneous branch** becomes superficial between gracilis and adductor longus and is described with the cutaneous nerves of the thigh (p. 529). **The posterior branch** descends behind the adductor brevis muscle and

The Femoral Nerve (fig. 441)

The femoral nerve is the largest branch of the lumbar plexus. It is a postaxial nerve formed by the posterior branches of the second, third, and fourth lumbar nerves (p.468). It passes under the inguinal ligament in the groove formed by the adjacent margins of the psoas and iliacus muscles and thus enters the femoral triangle. Here the terminal muscular, articular, and cutaneous branches arise as a large bundle of nerves. The cutaneous branches, the anterior femoral cutaneous nerve and the saphenous nerve, are fully described with the cutaneous nerves of the thigh and the hip (p.530). Muscular branches supply the pectineus, sartorius, and quadriceps femoris muscles.

The **nerve to the pectineus muscle** arises close to the inguinal ligament and passes behind the femoral vessels to enter the muscle near its lateral border. There are two sets of **nerves to the sartorius muscle.** An upper set of short branches arises in common with an anterior femoral cutaneous nerve and enters the proximal part of the sartorius muscle; a set of longer branches enters the middle of the muscle. The **nerves to the quadriceps femoris muscle** are multiple. Branches to the rectus femoris enter the deep surface of the muscle both in its proximal one-fourth and below its midlength. The **nerve to the vastus lateralis** accompanies the descending branch of the lateral circumflex femoral artery and sends branches into the muscle along its length. Two branches supply the **vastus medialis muscle.** A proximal branch enters its upper part, and a distal branch passes part way along the adductor canal to enter the muscle in its medial aspect. The **vastus intermedius** is innervated superficially by a branch which passes through the muscle and also supplies the articularis genu. The vastus intermedius is also innervated by the upper nerve to the vastus medialis muscle. The **articular branches** of the femoral nerve reach both the hip and the knee. The nerve to the hip joint arises from the nerve to the rectus femoris muscle and is accompanied by branches from the lateral circumflex femoral artery. Three or four articular branches reach the knee joint. A continuation of the nerve to the vastus lateralis muscle penetrates the joint capsule on its anterior aspect. A filament derived from the lower nerve to the vastus medialis muscle accompanies the descending genicular artery to the knee joint on its medial side. The nerve to the articularis genu muscle also supplies the joint. A fourth articular branch sometimes arises from the saphenous nerve.

The Muscular Branches of the Sacral Plexus

The muscular branches of the sacral plexus are described under that plexus (p.510). They include the nerve to the piriformis muscle, the superior gluteal nerve, the inferior gluteal nerve, the nerve

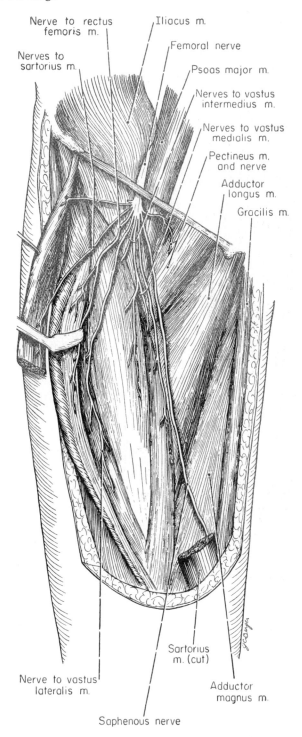

Fig. 441 The branches and distribution of the femoral nerve in the thigh.

to the quadratus femoris and inferior gemellus muscles, and the nerve to the obturator internus and superior gemellus muscles.

The Sciatic Nerve (figs. 431, 442)

The sciatic nerve is actually two nerves contained within a common connective tissue sheath—the pre-axial tibial nerve and the postaxial common peroneal nerve. In 10 per cent of cases they are separated in the greater sciatic foramen by all or part of the piriformis muscle (Beaton & Anson, '38). Occasionally the separate identity of the two nerves is clear throughout the thigh. The sciatic nerve arises from spinal cord segments L4 through S3, with the tibial division being formed from the anterior branches of the entire series of nerves. The common peroneal division of the sciatic is formed from the posterior branches of nerves L4 through S2. The tibial division is the medial part of the sciatic nerve; the common peroneal portion is lateral in the nerve. The sciatic nerve leaves the pelvis through the lower part of the greater sciatic foramen and extends from the inferior border of the piriformis muscle to the lower one-third of the thigh, where its two parts separate. In the upper part of its course it descends over the internal obturator tendon and the gemellus muscles and across the quadratus femoris muscle between the greater trochanter of the femur and the ischial tuberosity. Inferiorly, the nerve lies against the posterior surface of the adductor magnus muscle and is crossed obliquely by the long head of the biceps femoris muscle. The branches of the sciatic nerve in the thigh are articular branches to the hip joint and muscular branches to the hamstring muscles.

The **articular branches** arise from the proximal part of the nerve and perforate the posterior capsule of the hip joint. The **nerves to the hamstrings** (except the short head of the biceps femoris) leave the medial aspect of the sciatic nerve below the quadratus femoris muscle. They are branches of the tibial division of the sciatic nerve. The branch to the semitendinosus muscle is double; the branch to the part above the tendinous intersection arises with the nerve to the long head of the biceps femoris muscle. The nerve to the ischial portion of the adductor magnus muscle arises in common with the nerve to the semimembranosus muscle, which, in turn, branches from a trunk common to it and the lower nerve of the semitendinosus muscle. The **nerve to the short head of the biceps femoris muscle** (L5, S1, 2) arises from the lateral side of the common peroneal division of the sciatic in the middle one-third of the thigh and enters the superficial surface of the muscle near its lateral margin. This nerve continues to the knee

Fig. 442 The branches and distribution of the sciatic nerve.

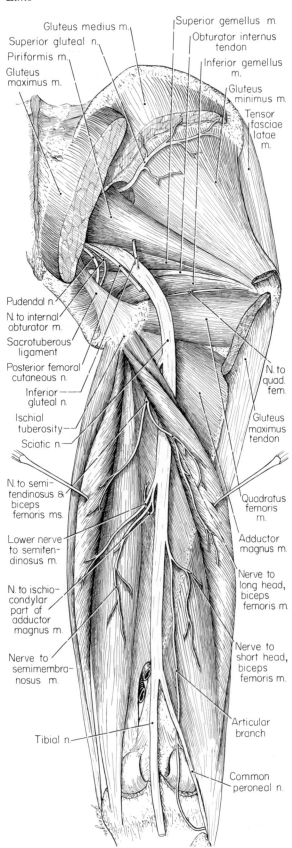

as an **articular branch.** In the popliteal fossa it divides into proximal and distal branches which accompany the lateral superior genicular and lateral inferior genicular arteries to the knee joint.

The **tibial nerve** is the larger of the two parts of the sciatic nerve. Separating from the common peroneal nerve at the upper limit of the popliteal fossa, the tibial nerve continues the straight course of the sciatic and runs vertically through the popliteal fossa directly under the popliteal fascia. Inferiorly, it passes between the heads of the gastrocnemius and under the soleus muscle. Three **articular branches** arise in the popliteal fossa. They accompany the medial superior genicular, the middle genicular, and medial inferior genicular arteries to the knee joint. There is also a vascular branch to the popliteal artery which arises here. Branches of the tibial nerve concerned with the leg are the medial sural cutaneous nerve (p. 530) and muscular branches to the calf muscles.

The **common peroneal nerve** follows the tendon of the biceps femoris muscle along the upper lateral margin of the popliteal fossa. Thus conducted to the back of the head of the fibula, the nerve winds around the neck of the bone, passes deep to the peroneus longus muscle, and divides into the superficial peroneal and deep peroneal nerves. Its muscular branch to the short head of the biceps femoris muscle and its articular nerves to the knee joint arise while in combination in the sciatic nerve (p. 556). In the popliteal fossa the common peroneal nerve gives off the lateral sural cutaneous nerve and, occasionally, the peroneal communicating branch. These are described with the cutaneous nerves of the lower limb (p. 530).

THE LEG

Fasciae of the Leg, the Ankle, and the Foot

The subcutaneous tissue of the leg contains a large amount of fat over the muscles but little over the bones and joints. Subcutaneous bursae are located over the tuberosity of the tibia and over both the medial and lateral malleoli.

The fascia lata of the thigh is continuous onto the leg where it is designated as the **crural fascia** (fig. 443). In the region of the knee, the crural fascia is attached anteriorly to the patella, the patellar ligament, and the tibial tuberosity. Medially and laterally, it attaches to the condyles of the tibia and to the head of the fibula and, reinforced by tendinous fibers of the vastus muscles, forms the **retinacula of the patella.** The crural fascia is strengthened in the region of the knee by expansions of the tendons of the sartorius, gracilis, semitendinosus, and biceps femoris muscles. The fascia envelops the soft parts and the bones of the leg, is connected to the bones by intermuscular septa, and gives origin to superficial muscle fibers. It blends with the periosteum of the subcutaneous medial surface of the tibia and, above the ankle, with the periosteum of the lower part of the shaft of the fibula. The crural fascia is attached distally to the medial and lateral malleoli and to the posterior surface of the calcaneus.

Two septa pass from the deep surface of the crural fascia and define the lateral muscular compartment of the leg (fig. 443). The **anterior intermuscular septum** is attached to the anterior border of the fibula and separates the extensor muscles of the anterior compartment from the peroneal muscles of the lateral compartment. The **posterior intermuscular septum** is attached to the posterior border of the fibula and separates the peroneal muscles from the flexors of the back of the leg. From the posterior intermuscular septum, a broad **transverse intermuscular septum** extends medialward, separating the superficial and deep groups of posterior, or calf, muscles. It ends medially on the tibia and in the crural fascia covering the flexor digitorum longus muscle. This sheet is thick over the popliteus muscle, where it is reinforced by fibers from the semimembranosus tendon; in the lower part of the leg it is strong anterior to the calcaneal tendon.

In the region of the ankle the crural fascia is reinforced by transverse and oblique fibers and forms thickenings, or retinacula, which hold the tendons close to the bones, preventing them from bowstringing (fig. 453). Anteriorly, immediately above the ankle, the **superior extensor retinaculum** is attached laterally to the lower end of the fibula and, medially, to the tibia. Blending indistinctly at its borders into the crural fascia, it covers the tendons of the tibialis anterior, extensor hallucis, and extensor digitorum longus muscles. From its deep surface a strong septum passes to the tibia and divides the underlying space into two canals, a medial one for tibialis anterior and a lateral one for the long extensor muscles. The **inferior extensor retinaculum** is a well-defined, Y-shaped band which overlies the dorsum of the foot and the front of the ankle. The stem of the Y arises from the upper surface of the calcaneus and passes upward onto the dorsum of the foot as two laminae, one superficial and one deep to the tendons of the peroneus tertius and extensor digitorum longus muscles. At the medial border of the tendon of the extensor digitorum longus muscle the two laminae blend, and

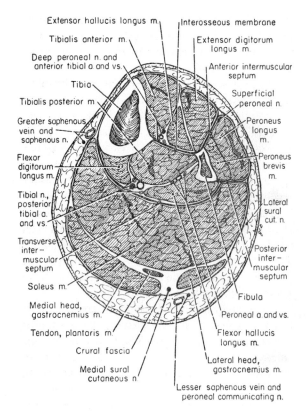

Fig. 443   A cross section of the leg at its midlength.

the two limbs of the Y begin to diverge. One limb is directed upward and medialward and attaches to the medial malleolus. It passes over the tendon of the extensor hallucis longus muscle, the dorsalis pedis vessels, and the deep peroneal nerve but splits to enclose the tendon of the tibialis anterior muscle. The lower limb of the Y passes medialward and downward across the medial border of the foot and is lost in the deep fascia of the sole. It passes over the tendons of the tibialis anterior and extensor hallucis longus muscles, the dorsalis pedis vessels, and the deep peroneal nerve. The **flexor retinaculum** stretches between the medial malleolus and the medial tubercle of the calcaneus. From its deep surface, septa pass to the back of the lower end of the tibia and to the capsule of the ankle joint. The four canals defined by these septa, enumerated from the medial side, transmit the tendon of the tibialis posterior, the tendon of the flexor digitorum longus muscle, the posterior tibial vessels and tibial nerve, and the tendon of the flexor hallucis longus muscle. The upper border of the flexor retinaculum is con-

tinuous with the transverse intermuscular septum which covers these muscles in the calf. Its lower border is continuous with the deep fascia of the sole and gives origin to the abductor hallucis muscle. The peroneal retinacula are thickenings of the fascia that hold the tendons of the peroneus longus and peroneus brevis muscles in place. The **superior peroneal retinaculum** extends from the lateral malleolus into the fascia on the back of the leg and to the lateral surface of the calcaneus. The **inferior peroneal retinaculum** overlies the peroneal tendons on the lateral surface of the calcaneus and is attached to that bone on either side of them. It is connected superiorly with the stem of the Y of the inferior extensor retinaculum.

The **fascia of the dorsum** of the foot is thin. It is continuous with the extensor retinacula, curves over the margins of the foot, and becomes the **fascia of the sole.** The latter is greatly thickened and specialized in its central portion as the **plantar aponeurosis;** it is more completely described with the sole of the foot (p. 576).

COMPARTMENTS OF THE LEG (fig. 443)

The tibia and the fibula are joined along their interosseous borders by an interosseous membrane. Externally, the bones make connection with the crural fascia—the tibia through its periosteum and the fibula through the anterior intermuscular and posterior intermuscular septa. The bones and the fasciae complete the separation of the leg into anterior and posterior portions, each of which is further subdivided. The anterior portion of the leg contains the anterior and lateral compartments separated by the anterior intermuscular septum. The posterior portion of the leg is subdivided by the transverse intermuscular septum into superficial posterior and deep posterior compartments. The two posterior compartments are pre-axial in classification. They contain the flexor muscles of the foot and the toes and the pre-axial tibial nerve. The muscles of the **superficial posterior compartment**—gastrocnemius, plantaris, and soleus—insert by the calcaneal tendon on the tuberosity of the calcaneus. Their contraction produces plantar flexion of the foot. The muscles of the **deep posterior compartment** are the tibialis posterior, the flexor digitorum longus, and the flexor hallucis longus. These muscles act in flexion of the toes and in plantar flexion and inversion (medial rotation) of the foot.

The deep posterior compartment also includes the popliteus muscle, a flexor of the knee. The tibial nerve and posterior tibial vessels descend under the transverse intermuscular septum and serve both of these compartments.

The anterior and lateral compartments are postaxial. The **anterior compartment** lies in front of the interosseous membrane, the fibula, and the anterior intermuscular septum. It contains the tibialis anterior, extensor hallucis longus, and extensor digitorum longus muscles. These muscles extend the toes and dorsiflex the foot. In addition, the tibialis anterior muscle inverts the sole of the foot. The muscles of this compartment are supplied by the anterior tibial vessels and the deep peroneal nerve. The **lateral compartment,** bounded by the anterior intermuscular and posterior intermuscular septa, the crural fascia, and the fibula, contains the peroneus longus and peroneus brevis muscles. It also contains and its muscles are innervated by the superficial peroneal nerve. The anterior and lateral compartments of the leg are postaxial in spite of their forward position—a consequence of the phylogenetic twisting of the limb which places the sole of the foot flat on the ground in the erect posture. The placing of the hands palm down on a desk or table induces a similar temporary twist in the upper limb with the same change in position of the pre-axial and postaxial compartments of the forearm and the hand.

Muscles of the Anterior Compartment (figs. 442–47, 453)

The muscles of the anterior compartment of the leg are as follows:

> Tibialis anterior
> Extensor hallucis longus
> Extensor digitorum longus
> Peroneus tertius

The **tibialis anterior muscle** lies against the lateral surface of the tibia. It arises from the lateral condyle and the upper half of the lateral surface of the tibia. Fibers also take origin from the interosseous membrane, the intermuscular septum between it and the extensor digitorum longus muscle, and the overlying crural fascia. The tendon appears in the lower one-third of the leg and passes through the compartments formed by the superior extensor and inferior extensor retinacula surrounded by a syno-

vial sheath. The tendon inserts on the medial side of the foot into the medial surface of the medial cuneiform bone and the base of the first metatarsal bone (figs. 447, 475).

The **extensor hallucis longus** is a thin muscle which comes to the surface between the tibialis anterior and the extensor digitorum longus in the lower half of the leg. The extensor hallucis longus muscle arises from the middle two-fourths of the anterior surface of the fibula and from the adjacent interosseous membrane for the same distance. The anterior tibial vessels and the deep peroneal nerve lie on the interosseous membrane between it and the tibialis anterior muscle. The tendon of the extensor hallucis longus muscle arises on its superficial surface, passes under the superior extensor and inferior extensor retinacula, and inserts into the base of the distal phalanx of the great toe (fig. 475).

The **extensor digitorum longus** is a pennate muscle which occupies the fibular portion of the anterior compartment of the leg. The muscle arises from the lateral condyle of the tibia, the anterior surface of the fibula, the lateral margin of the interosseous membrane, the intermuscular septum between it and the tibialis anterior, the septum between it and the peroneus longus, and the fascia of the leg near the tibial origin. The tendon begins at about the middle of the leg and receives fibers nearly to the ankle. Passing under the superior extensor retinaculum, the tendon divides into two parts. The two parts descend under the inferior extensor retinaculum, where they are enclosed within a synovial sheath. Below the inferior retinaculum they divide, so that four tendons pass across the dorsum of the foot to the lateral four toes. The tendons insert, as do the tendons of the extensor digitorum muscle to the fingers, by each dividing into two lateral slips and one central slip. The central slip ends on the dorsum of the base of the middle phalanx of each digit, and the lateral slips converge beyond it to the dorsum of the base of the distal phalanx (fig. 475). Extensor expansions are formed by the tendons of the extensor digitorum longus similar to but shorter than those of the tendons of the extensor digitorum over the fingers (p. 97).

The tendons of the dorsal interosseous muscles join the expansions alongside the digits; the tendons of the lumbrical muscles unite with them on the great toe sides. The tendons of the extensor digitorum brevis muscle unite with the extensor expansions of the second, third, and fourth toes on their fibular sides.

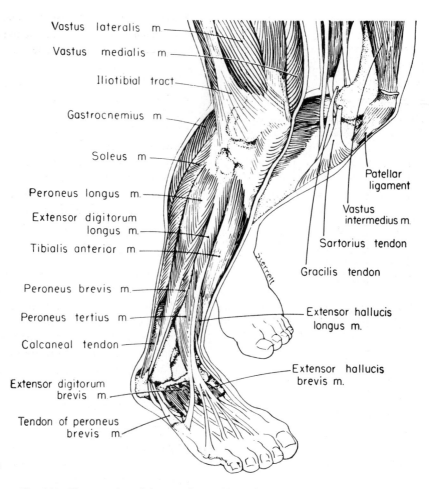

Vastus lateralis m

Vastus medialis m

Iliotibial tract

Gastrocnemius m

Soleus m

Peroneus longus m.

Extensor digitorum longus m.

Tibialis anterior m

Peroneus brevis m.

Peroneus tertius m

Calcaneal tendon

Extensor digitorum brevis m.

Tendon of peroneus brevis m.

Patellar ligament

Vastus intermedius m.

Sartorius tendon

Gracilis tendon

Extensor hallucis longus m.

Extensor hallucis brevis m.

Fig. 444   The muscles of the anterior and lateral compartments of the leg.

The **peroneus tertius** is essentially a lateral slip of the extensor digitorum longus muscle from which it is seldom completely separated. The peroneus tertius muscle arises from the distal one-third of the anterior surface of the fibula, the adjacent part of the interosseous membrane, and the anterior intermuscular septum. The tendon passes under the extensor retinacula in the compartments for extensor digitorum longus and then turns lateralward to end on the dorsum of the shaft of the fifth metatarsal bone (fig. 475).

Muscles of the Lateral Compartment (figs. 444–47, 451–53)

The muscles of the lateral compartment are the following:

Peroneus longus   Peroneus brevis

The **peroneus longus muscle,** bipennate in form, arises higher and is more superficial than the peroneus brevis. It takes its origin from the head and the upper two-thirds of the lateral surface of the body of the fibula, from the anterior intermuscular and posterior intermuscular septa, and from the crural fascia overlying the muscle. Its tendon begins high on the superficial surface of the muscle and receives fibers (posteriorly) to within a few centimeters of the lateral malleolus. Behind the lateral malleolus, the tendon lies posterior to the tendon of the peroneus brevis; both are enclosed within a common synovial sheath and are crossed by the superior peroneal retinaculum. The tendon of the peroneus longus passes diagonally forward below the tendon of the peroneus brevis and turns into the foot against the smooth cartilage-covered anterior slope of the tuberosity of the cuboid bone. The tendon is thickened by the development of a sesamoid fibrocartilage which protects it over the tuberosity of the cuboid. It almost never lies in the so-called groove for the tendon in front of the tuberosity. Crossing the sole of the foot deep to its intrinsic muscles, the tendon of the peroneus longus muscle ends on the inferolateral surface of the medial cuneiform bone and on the base and inferolateral border of the first metatarsal bone (figs. 456, 475).

The **peroneus brevis muscle** lies deep to the peroneus longus and is a shorter and smaller muscle. It arises from the lower two-thirds of the lateral surface of the shaft of the fibula and from the anterior intermuscular and posterior intermuscular septa. Its tendon grooves the back of the lateral malleolus, where it lies in a common synovial sheath with the tendon of the peroneus longus muscle. The tendon of the peroneus brevis passes under the superior peroneal retinaculum anterior to the long tendon and then runs diagonally downward and forward deep to the inferior peroneal retinaculum. It inserts into the tuberosity on the base of the fifth metatarsal bone (fig. 475).

Muscles of the Superficial Posterior Compartment
(figs. 427, 443, 446–48, 452)

The muscles of the superficial posterior compartment end by means of the calcaneal tendon on the tuberosity of the calcaneus, and their principal actions are alike. Supplemented by the almost vestigial plantaris muscle, the two heads of the gastrocnemius muscle and the soleus muscle form the triceps surae.

The **gastrocnemius muscle** is the most superficial of the group and forms the greater bulk of the calf. It arises by two heads. The larger **medial head** takes origin from the rough area of the popliteal surface of the femur immediately above the medial femoral condyle. The **lateral head** arises from an impression on the upper and posterior part of the lateral surface of the lateral femoral condyle and from the distal end of the lateral supracondylar line of the femur. Both heads also receive fibers from the back of the capsule of the knee joint, and a bursa lies deep to each tendon of origin. The bursa of the medial head frequently opens into the knee joint and may also communicate with the bursa deep to the tendon of the semimembranosus muscle. Extending downward for about two-thirds of the length of the muscle, aponeurotic bands cover the

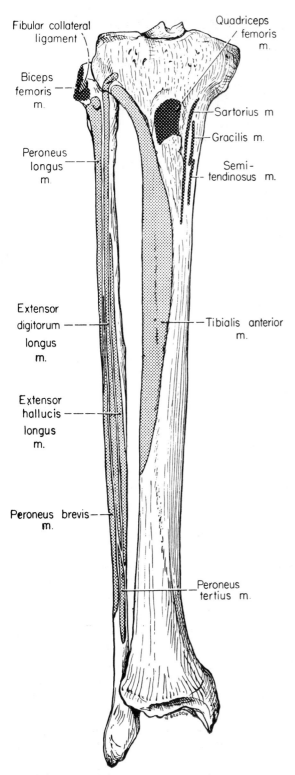

Fibular collateral ligament

Biceps femoris m.

Peroneus longus m.

Extensor digitorum longus m.

Extensor hallucis longus m.

Peroneus brevis m.

Quadriceps femoris m.

Sartorius m.

Gracilis m.

Semi-tendinosus m.

Tibialis anterior m.

Peroneus tertius m.

Fig. 445 The muscular attachments on the anterior aspects of the tibia and fibula. (Black dots—origin; white dots on black—insertion.)

outer margins and posterior surfaces of the two heads and provide further attachment for the muscular fibers. The fibers of both heads of origin and those from their aponeuroses converge toward the midline of the leg and unite at about the midlength of the muscle in a tendinous raphe which broadens into an aponeurosis on the anterior aspect of the muscle. This aponeurosis, narrowing below, fuses with the tendon of the soleus muscle and, with it, forms the calcaneal tendon. Generally, the medial head is the broader and thicker, and its muscular fibers descend farther toward the heel.

The **soleus** is a broad, fleshy muscle lying immediately anterior to the gastrocnemius but arising entirely below the knee. It has a triple origin from (1) the posterior surfaces of the head of the fibula and the upper one-third of its shaft; (2) a **tendinous arch** that represents the upper part of the transverse intermuscular septum and extends between the tibia and the fibula posterior to the popliteal vessels and the tibial nerve; and (3) the **soleal line** of the tibia and the middle one-third of the medial border of this bone. The muscular fibers descend into a broad membranous tendon which fuses below with the deep surface of the tendon of the gastrocnemius muscle to form the calcaneal tendon. The **calcaneal tendon** is the thickest and strongest in the body. It is about 15 cm. long, begins at about the middle of the leg, and receives muscular fibers almost to its termination. Narrowing below, it inserts into the middle part of the posterior surface of the calcaneus; a bursa lies deep to the tendon and separates it from the upper part of the posterior surface of the calcaneus.

The **plantaris muscle** has a short fleshy belly (about 10 cm. long) which arises above the lateral head of the gastrocnemius from the lower 2 to 3 cm. of the lateral supracondylar line of the femur and from the adjacent part of its popliteal surface. The muscle also arises from the oblique popliteal ligament of the knee joint. Its belly ends in a long slender tendon which descends obliquely between the gastrocnemius and soleus muscles and then along the medial border of the calcaneal tendon. It is inserted into the calcaneus in front of or at the side of the calcaneal tendon.

Muscles of the Deep Posterior Compartment (figs. 443, 447, 448, 452, 455)

The muscles of the deep posterior compartment include a weak flexor of the knee—popliteus—and

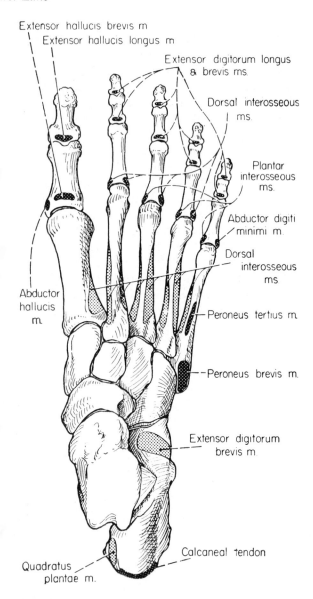

Fig. 446 The muscular attachments on the dorsum of the foot (Code as fig. 445.)

three muscles which produce plantar flexion of the foot and the toes. They are as follows:

Popliteus	Flexor digitorum longus
Flexor hallucis longus	Tibialis posterior

The **popliteus** is a flat, thin muscle of triangular outline which lies in the floor of the lower part of the popliteal fossa. It arises by a stout tendon from the rough pit at the anterior end of the groove on the lateral aspect of the lateral condyle of the femur close to the articular margin of the bone. The tendon passes between the lateral meniscus and the capsule of the knee joint and has a bursa under it which regularly communicates with the cavity of the joint (subpopliteal recess, p. 590). Its fibers also come in part from the arcuate popliteal ligament and, in lesser amount, from the lateral meniscus. The muscle inserts by fleshy fibers into the triangular area of the back of the tibia above the soleal line and into the fascia covering the muscle. The popliteus flexes the leg at the knee and rotates it medialward. With the foot placed firmly on the ground, the contraction of the muscle leads to lateral rotation of the femur and starts the flexion of the leg. The popliteus muscle is innervated by a branch of the tibial nerve (L4, 5, S1) which winds around the lower border of the muscle and enters its deep surface.

The **flexor hallucis longus muscle** lies on the fibular rather than on the tibial side of the leg. It arises from the lower two-thirds of the posterior surface of the shaft of the fibula and from the intermuscular septa between it and the posterior tibial and peroneal muscles. The fiber bundles of the flexor hallucis longus converge on a tendon which appears on the posteromedial margin of the muscle in its distal half. This tendon, passing deep to the flexor retinaculum, is enclosed in a separate synovial sheath. It grooves the posterior surface of the lower end of the tibia, the posterior surface of the talus, and the under surface of the sustentaculum tali of the calcaneus. In the sole of the foot the flexor hallucis longus tendon crosses diagonally above and gives a tendinous slip to the tendon of the flexor digitorum longus muscle. The flexor hallucis longus tendon then passes forward between the two heads of the flexor hallucis brevis muscle and inserts into the base of the distal phalanx of the great toe. The grooves on the talus and the cal-

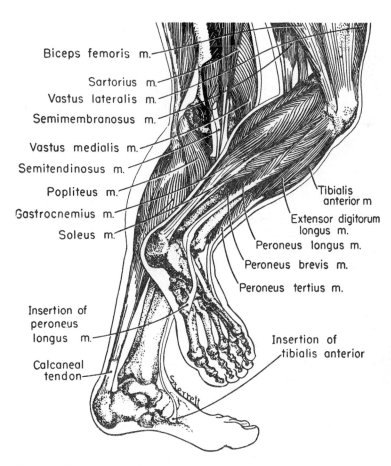

Biceps femoris m.
Sartorius m.
Vastus lateralis m.
Semimembranosus m.
Vastus medialis m.
Semitendinosus m.
Popliteus m.
Gastrocnemius m.
Soleus m.
Tibialis anterior m
Extensor digitorum longus m.
Peroneus longus m.
Peroneus brevis m.
Peroneus tertius m.
Insertion of peroneus longus m.
Insertion of tibialis anterior
Calcaneal tendon

Fig. 447   The muscles of the posterior and lateral compartments of the leg.

The Lower Limb

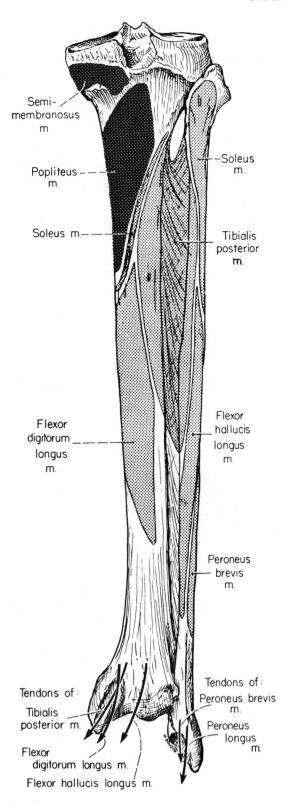

Semi-
membranosus
m.

Popliteus
m.

Soleus m.

Flexor
digitorum
longus
m.

Soleus
m.

Tibialis
posterior
m.

Flexor
hallucis
longus
m.

Peroneus
brevis
m.

Tendons of :

Tibialis
posterior m.

Flexor
digitorum longus m.

Flexor hallucis longus m.

Tendons of :
Peroneus brevis
m.

Peroneus
longus
m.

Fig. 448 The muscular attachments and the course of tendons on the posterior surfaces of the tibia and fibula. (Code as fig. 439.)

caneus are converted into canals by ligamentous fibers, and within these canals the tendon is enclosed by a continuation of its synovial sheath.

The **flexor digitorum longus muscle** is located on the tibial side of the leg. It arises from the medial side of the posterior surface of the middle three-fifths of the tibia (below the soleal line) and from the intermuscular septum between it and the tibialis posterior muscle. The pennate fibers converge on a tendon which runs along nearly the whole length of the medial margin of the muscle, receiving muscle fibers almost to the medial malleolus. The tendon of the tibialis posterior muscle crosses anterior to the tendon of the flexor digitorum longus muscle, and the two tendons enter the foot by passing in a groove on the back of the medial malleolus in separate synovial sheaths and separated from each other by a fibrous septum derived from the flexor retinaculum. The tendon of the flexor digitorum longus muscle passes diagonally into the sole of the foot, crossing the deltoid ligament of the ankle joint (p. 592) and running superficial to the tendon of the flexor hallucis longus muscle. It receives a slip from the tendon of the flexor hallucis longus muscle. In the middle of the sole the tendon of the flexor digitorum longus receives the insertion of the quadratus plantae muscle and then divides into four tendons which insert into the bases of the distal phalanges of the second, third, fourth, and fifth toes.

Like the tendons of the flexor digitorum profundus of the forearm the tendons of the flexor digitorum longus give rise, in the foot, to the lumbrical muscles. They also pass through divisions of the tendons of the flexor digitorum brevis muscle in order to reach the distal phalanges. The tendons of the flexor digitorum longus and flexor digitorum brevis muscles are conducted, in each digit, through fibrous digital sheaths within which they are invested by synovial membranes. These relations are given in greater detail in the description of the foot (p. 579).

The **tibialis posterior,** the most deeply situated muscle of the compartment, lies between the flexor digitorum longus and flexor hallucis longus muscles. Beginning above in two pointed processes, the tibialis posterior arises from all except the lowest part of the posterior surface of the interosseous membrane, from the lateral part of the posterior surface of the tibia in its upper two-thirds but below the soleal line, from the upper two-thirds of the medial surface of the fibula, from the intermuscular septa between it and the muscles on either side, and from the transverse intermuscular septum of the leg. The fiber bundles converge on a tendon which, about the middle of the leg, emerges on the medial side of the muscle and continues to receive muscle fibers almost to the medial malleolus. The tendon of the tibialis posterior passes in front of the flexor digitorum longus muscle in the lower one-fourth of the leg and then behind the medial malleolus and under the flexor retinaculum. Enclosed within a synovial sheath, it is the most antero-medial of the structures which groove the malleolus. Continuing, the tibialis posterior tendon crosses the deltoid ligament of the ankle, passes inferior to the plantar calcaneonavicular ligament (p. 597), and inserts into the tuberosity of the navicular and into the underside of the medial cuneiform bone (fig. 455). Expansions of the tendon continue forward and lateralward to the intermediate and lateral cuneiform bones and to the plantar surfaces of the bases of the second, third, and fourth meta-tarsal bones. One expansion passes backward to the medial border of the sustentaculum tali of the calcaneus.

INNERVATIONS

The tibial and common peroneal nerves supply the muscles of the leg. The tibial nerve distributes to the muscles of both the superficial and deep portions of the posterior (pre-axial) compartment and the common peroneal nerve to those of the anterior and lateral (postaxial) compartments. The superficial peroneal supplies the peroneus longus and brevis muscles of the lateral compartment and the deep peroneal the muscles of the anterior compartment. The full description of these nerves includes their course and branches (p. 571, 572).

ACTIONS OF THE MUSCLES OF THE LEG

The muscles of the leg are predominantly movers of the foot at the ankle, though some have actions on the joints of the foot and toes and a few muscles are concerned with action of the knee. The muscles of the superficial posterior compartment all insert through the calcaneal tendon onto the tuberosity of the calcaneus. They act together to produce plantar flexion of the foot, accompanied by some inversion. Thus they raise the heel against the weight of the body in walking. The shorter fibers of the soleus act more powerfully over a shorter range; the longer fibers of the gastrocnemius have a longer

excursion. In standing, these muscles draw back on the leg, stabilizing the ankle joint and preventing dorsiflexion of the foot.

The gastrocnemius arises above the knee and produces flexion at the knee as well. Like other muscles crossing two joints, however, it does not have sufficient length to permit, at both joints, movements opposite in direction to those produced by its own traction. Hence, extension of the leg restricts dorsiflexion of the foot, and dorsiflexion of the foot is freer in combination with flexion of the leg at the knee. Gastrocnemius may show intermittent activity in relaxed standing since the ankle joint is not in the 'close-packed' position.

In the deep posterior compartment, popliteus is a weak flexor of the leg at the knee. Its full action is given under the description of the muscle. The other muscles assist the muscles of the superficial compartment in plantar flexion and inversion of the foot at the ankle but their particular functions are otherwise. Tibialis posterior acts powerfully in adduction and inversion of the foot. It distributes the body weight among the heads of the metatarsals, serving to reduce flat-foot and shifting the body's weight toward the lateral side of the foot. The flexor hallucis longus muscle flexes the distal phalanx of the great toe and shows its greatest activity at heel-off in walking. Flexor digitorum longus similarly flexes the distal phalanges of the lateral four toes.

The muscles of the anterior compartment of the leg dorsiflex and invert the foot at the ankle. Tibialis anterior is powerful in this action; with the foot on the ground, it draws the tibia forward (as in walking). Additionally the extensor hallucis longus dorsiflexes the great toe, and extensor digitorum longus dorsiflexes the proximal and middle phalanges of the lateral four toes. Peroneus tertius assists in dorsiflexion of the foot at the ankle but operates in eversion, not inversion. The peroneus longus and brevis muscles of the lateral compartment evert and abduct the foot and assist in its plantar flexion.

The support of the longitudinal and transverse arches of the foot is assisted during muscular action by the tendons of certain muscles of the leg. Peroneus longus and tibialis anterior end on the same bones at the medial border of the foot, the tendon of peroneus longus passing under the tarsal bones en route. The broad grasp of the tendinous fibers of the tibialis posterior muscle on the bones of the medial side and underside of the foot combine with the traction of tibialis anterior and peroneus longus to lift the arches and to give aid, during muscular activity, to the ligaments which maintain the arched orientation of the bones. They are inactive during relaxed standing in a normally strong foot characterized by normal ligaments (fig. 475).

BONES OF THE LEG

The Tibia (figs. 445, 448–50)

The tibia is the weight-bearing bone of the leg, the fibula serving for muscular attachments and completing the ankle joint on its lateral side. The tibia is one of the largest bones of the body. Occupying the anteromedial aspect of the leg, the bone can be felt through the skin. It is a long bone with a body and two extremities.

The **upper extremity** of the tibia is expanded to receive the condyles of the femur. The shaft of the bone flares out into lateral and medial buttresses, which form the **medial** and **lateral condyles.**

The **superior articular surface** presents two facets. The medial facet is oval in shape and has a slight concavity. The lateral facet is nearly round; concave from side to side, it is convex from in front backward. The rims of the facets are in contact with the menisci of the knee joint, but the central portions receive the condyles of the femur. An **intercondylar eminence** is raised between the two articular facets, surmounted on either side by the **medial intercondylar tubercle** and the **lateral intercondylar tubercle.** The articular surfaces continue on to the adjacent sides of these tubercles. Anterior to the intercondylar eminence is the **anterior intercondylar area;** it provides attachment for the anterior extremities of the medial and lateral menisci and for the anterior cruciate ligament. The **posterior intercondylar area** is a broad groove separating the posterior aspects of the condyles. It lodges the posterior cruciate ligament and, behind the eminence, gives attachment to the posterior extremities of the medial and lateral menisci. Anteriorly, the two condyles blend in a triangular area leading below to the tibial tuberosity. This triangle has large vascular foramina and is sharply marked at its borders by oblique lines where the fascia lata attaches. The **tibial tuberosity** contains a smooth upper portion for the termination of the ligamentum patellae and a rough lower portion. They are separated by an irregular line. Posteriorly, the medial condyle is marked by a transverse groove which accommodates the insertion of the tendon of the semimembranosus muscle. The rough medial surface of the condyle gives attachment to the tibial collateral ligament. The **lateral condyle** has, on its postero-inferior surface, a nearly circular facet for articulation with the head of the fibula. Anterior to the facet, at the junction of the anterior and lateral surfaces of the condyle, is an oblique line on which the iliotibial tract ends.

The **body of the tibia** is somewhat expanded at its extremities but is otherwise fairly uniform in size. In cross section it is triangular with medial, lateral, and posterior surfaces and anterior, medial, and interosseous borders.

The **anterior border** is subcutaneous and prominent. It is slightly sinuous, beginning at the lateral margin of the tuberosity above and ending on the medial malleolus below; it is sharpest in its middle one-third. The **medial border** extends from the posterior aspect of the medial condyle to the posterior border of the medial malleolus. It is sharper in its lower half. In its upper portion it provides for attachment of the tibial collateral ligament. The **interosseous border** is on the fibular side of the bone and is sharp throughout its length. It begins superiorly below and anterior to the facet for the head of the fibula and provides, throughout the length of the bone, for the attachment of the interosseous membrane. Just above the ankle, the interosseous border bifurcates and encloses a triangular area for the attachment of the ligamentous tissue which unites the tibia and fibula distally.

The **medial surface** of the body of the tibia is smooth and convex. The **lateral surface** has a shallow groove in its upper two-thirds. Its lower one-third is smooth and spirals anteriorly. A prominent marking of the **posterior surface** of the bone is the **soleal line.** Beginning behind the facet for the head of the fibula, it extends obliquely downward across the back of the tibia, ending on the medial border of the bone at the junction of the upper and middle thirds of the shaft. On the posterior surface of the tibia below the soleal line is the nutrient foramen. It is large and is directed downward.

The **inferior extremity** of the tibia projects medialward and downward as the **medial malleolus** which forms the subcutaneous prominence on the medial aspect of the ankle.

The **malleolar groove** of its posterior surface accommodates the tendons of the tibialis posterior and flexor digitorum longus muscles; the tendon of the flexor hallucis longus muscle may produce a shallow groove on the posterior surface of the inferior extremity of the bone lateral to the malleolar groove. The lateral surface of the distal extremity forms the triangular **fibular notch** which receives the distal end of the shaft of the fibula and is roughened by the interosseous ligaments uniting the two bones. The borders of the fibular notch are sharp for the attachments of the anterior tibiofibular and posterior tibiofibular ligaments. The distal extremity forms a quadrilateral **inferior articular surface** for articulation with the body of the talus. Its surface is wider anteriorly than posteriorly and is concave anteroposteriorly. It is continuous medially with the **malleolar articular surface.** The malleolar articular surface, on the internal surface of the medial malleolus, lies almost at right angles to the articular surface of the shaft and extends beyond it

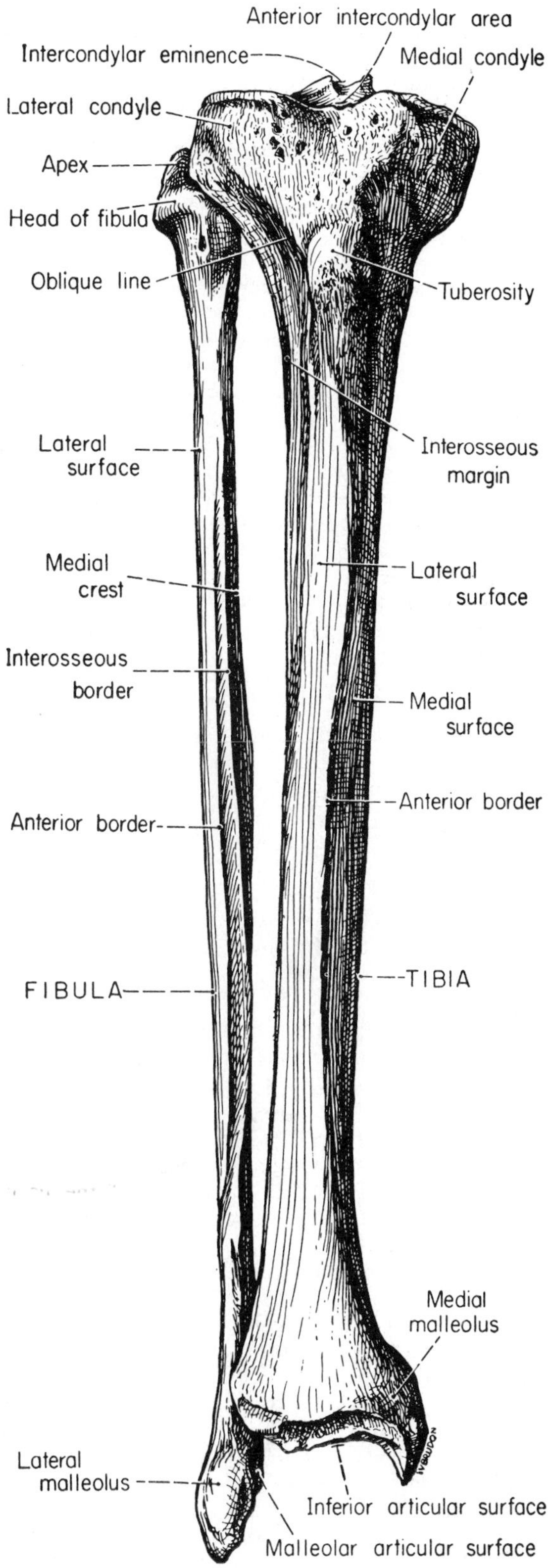

Fig. 449 The anterior surfaces of the right tibia and fibula. ($\times\frac{1}{2}$)

about 1.5 cm. The lower edge of the malleolus is notched, and from its tip, anterior border, and notch the deltoid ligament passes to the bones of the foot.

The tibia is ossified from three centers—one for the body and one for each extremity. They appear in the seventh week of intrauterine life for the body, in the upper epiphysis shortly before birth, and in the lower end during the second year. The lower epiphysis joins the body at about the age of sixteen and a half in the male and fourteen and a half in the female; the upper one, at approximately seventeen and a half years in the male and at fifteen in the female.

The Fibula (figs. 445, 448–50)

The fibula is a long, slender bone which lies parallel to the tibia. It does not participate in weight-bearing but serves for muscle attachments and aids in forming the ankle joint. It has a long body and two somewhat expanded extremities.

The superior extremity, or **head of the fibula,** is knob-like. On its superior aspect and slanted toward the tibia is the almost circular **articular surface of the head** which participates in the tibiofibular articulation. At the posterolateral limit of the articular facet, the **apex of the head** projects upward. This gives attachment to the fibular collateral ligament of the knee joint and, on its lateral aspect, to the tendon of the biceps femoris muscle. The circumference of the head is rough for muscle attachments and has anterior and posterior tubercles.

The **body** of the fibula has anterior, interosseous, and posterior borders and medial, lateral, and posterior surfaces.

The sharp **anterior border** begins just below the head and terminates distally by dividing to enclose a triangular subcutaneous surface immediately above the lateral malleolus. This border gives attachment to the anterior intermuscular septum of the leg which separates the anterior and lateral muscular compartments. The **interosseous border** lies close to the anterior border. It begins in front of the head above, is directed anteromedially on the shaft, and terminates by dividing into two parts which enclose medially the rough area above the lateral malleolus for the tibiofibular syndesmosis. The **medial crest** of the medial surface of the fibula begins on the medial border of the head and curves forward to join the interosseous border in the lower one-third of the shaft. The nutrient foramen of the fibula is located posterior to the distal end of the medial crest. A relatively indistinct **posterior border** runs from the posterior aspect of the head down the posterolateral surface of the shaft, ending just above the **fossa of the lateral malleolus.** The posterior border gives attachment to the posterior intermuscular septum separating the lateral and posterior compartments of the leg. The surfaces of the fibula are medial, posterior, and lateral. They provide origins for various leg muscles.

The **lateral malleolus** is the pointed distal extremity of the bone. Its tip descends about 1.5 cm. below that of the medial malleolus of the tibia. Its lateral surface, convex and subcutaneous, is continuous with the subcutaneous area of the shaft.

The medial surface of the malleolus consists of the triangular **malleolar articular surface** and the **fossa of the lateral malleolus.** The malleolar articular surface, angulated outward below, makes contact with the corresponding surface of the lateral side of the talus. The fossa provides attachment for the transverse tibio-fibular and the posterior talofibular ligaments.

The fibula has three centers of ossification. A center appears in the middle of the shaft in the eighth week of intrauterine life, and only the ends are cartilaginous at birth. Ossification of its lower extremity begins at the end of the first year or the beginning of the second; ossification of its upper extremity begins in the fourth year in the male and early in the third year in the female. Fusion of the epiphysis of the lower extremity takes place at approximately fourteen and a half (female) and sixteen and a half years (male). In the upper extremity of the bone the epiphysis fuses during the eighteenth year in the male and at fifteen and a half years in the female. Individual variation may alter the fusion times by several years.

BLOOD VESSELS OF THE LEG (figs. 427, 438, 443, 451–53)

The popliteal artery (p. 551) descends across the popliteus muscle and, at its lower border, divides into the anterior tibial and posterior tibial arteries. The posterior tibial continues the course of the popliteal artery down the back of the leg. The anterior tibial artery passes into the anterior compartment of the leg and descends within that compartment.

The Anterior Tibial Artery (figs. 443, 451, 453)

The anterior tibial artery, arising by the division of the popliteal artery at the lower border of the popliteus muscle, passes forward between the two pointed origins of the tibialis posterior muscle, above the upper margin of the interosseous membrane, and into the anterior compartment of the leg. Here it lies against the medial side of the neck of the fibula. The artery descends on the interosseous membrane, at first between the tibialis anterior and extensor digitorum longus muscles and

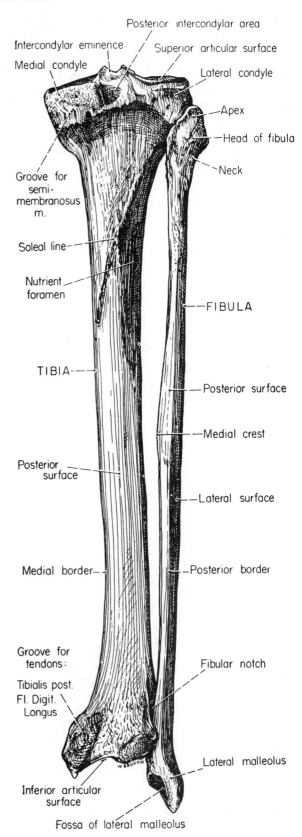

Fig. 450 The posterior surfaces of the right tibia and fibula. ($\times\frac{1}{2}$)

then, beginning in the middle one-third of the leg, between the tibialis anterior and extensor hallucis longus muscles. The artery is accompanied by venae comitantes which lie on either side of it. It is joined, on its lateral side, by the deep peroneal nerve after the nerve turns around the neck of the fibula and enters the compartment. At the ankle the vessels and nerve are crossed diagonally from the lateral to the medial side by the tendon of the extensor hallucis longus muscle, and then they lie between this tendon and the tendons of the extensor digitorum longus muscle. The artery crosses the ankle midway between the lateral and medial malleoli and enters the dorsum of the foot as the dorsalis pedis artery. The named branches of the anterior tibial artery, like those of the posterior tibial, are concentrated at the levels of the knee and the ankle. Between these regions many unnamed muscular branches arise. The named branches are as follows:

> Posterior tibial recurrent
> Circumflex fibular artery
> Anterior tibial recurrent
> Anterior medial malleolar
> Anterior lateral malleolar

The **posterior tibial recurrent artery**, sometimes a branch of the posterior tibial artery, usually arises from the anterior tibial in the posterior compartment of the leg. It ascends between the popliteus muscle and the back of the knee, supplies the popliteus muscle and the tibiofibular joint, and anastomoses with the lateral inferior genicular artery.

The small **circumflex fibular artery** also arises before the anterior tibial artery pierces the interosseous membrane but may, at times, be a branch of the posterior tibial. It passes lateralward around the neck of the fibula through the soleus muscle, the fibers of which it supplies. The artery anastomoses with the lateral inferior genicular artery.

The **anterior tibial recurrent artery** is given off by the anterior tibial as soon as it enters the anterior compartment. It ascends among the deep fibers of the tibialis anterior muscle and then branches over the front and sides of the knee joint. Participating in the patellar plexus, it anastomoses with the genicular branches of the popliteal artery and with the descending genicular and the lateral circumflex femoral branches of the femoral artery.

The **anterior medial malleolar artery** arises at the level of the ankle joint. It passes medialward deep to the tendons of the tibialis anterior and extensor hallucis longus muscles. The vessel ramifies on the medial aspect of the ankle and supplies the skin and the ankle joint, anastomosing with the malleolar branches of the posterior tibial artery.

The **anterior lateral malleolar artery** takes origin opposite the anterior medial malleolar and passes lateralward deep to the tendons of the extensor digitorum longus muscle. It supplies the lateral side of the ankle and the ankle joint and anastomoses with the perforating branch of the peroneal artery and with ascending branches of the lateral tarsal branch of the dorsalis pedis artery.

The Posterior Tibial Artery (figs. 443, 452)

The posterior tibial artery is the direct continuation of the popliteal artery and begins at the division of the latter at the lower border of the popliteus muscle. The posterior tibial artery, accompanied by two venae comitantes and the tibial nerve, descends in the deep posterior compartment of the leg immediately under the transverse intermuscular septum. It runs, at first, diagonally toward the fibula, in conformity with the diagonal course of the popliteal artery. Then, after giving off the peroneal artery, the posterior tibial inclines medialward and passes behind the medial malleolus at the ankle. Entering the foot, the artery divides beneath the origin of the abductor hallucis muscle into the medial plantar and lateral plantar arteries. Separated from the gastrocnemius and soleus muscles by the transverse intermuscular septum, the posterior tibial artery lies successively on the tibialis posterior and the flexor digitorum longus muscles, the tibia, and the back of the ankle joint. In the lower one-third of the leg, the artery is covered only by skin and fascia and runs parallel to and along the medial side of the calcaneal tendon. Like the anterior tibial artery, the posterior tibial gives rise to many muscular branches along its length. The several named branches are near the knee and at the level of the ankle. They are as follows:

> Peroneal   Communicating
> Nutrient   Posterior medial malleolar
>   Medial calcaneal

The large **nutrient branch** of the posterior tibial artery arises from its upper part. It pierces the tibialis posterior muscle, to which it supplies muscular twigs, and enters the nutrient foramen in the proximal one-third of the posterior surface of the tibia.

The **communicating branch** unites the posterior tibial and peroneal arteries a centimeter or more above the tibiofibular syndesmosis.

The **posterior medial malleolar branch** of the posterior tibial passes onto the medial malleolus, where it anastomoses with the anterior medial malleolar branch of the anterior tibial artery.

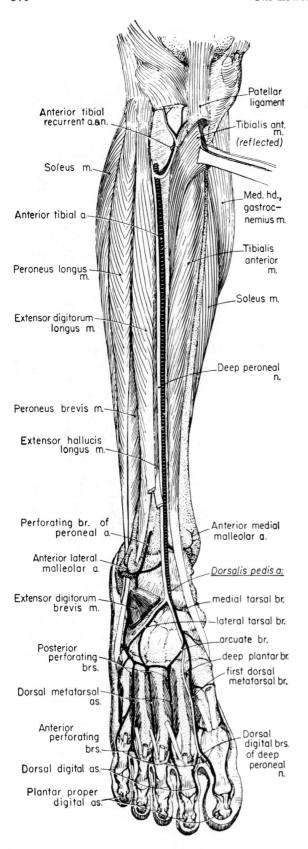

Patellar ligament

Anterior tibial recurrent a.an.

Soleus m.

Anterior tibial a.

Peroneus longus m.

Extensor digitorum longus m.

Peroneus brevis m.

Extensor hallucis longus m.

Perforating br. of peroneal a.

Anterior lateral malleolar a.

Extensor digitorum brevis m.

Posterior perforating brs.

Dorsal metatarsal as.

Anterior perforating brs.

Dorsal digital as.

Plantar proper digital as.

Tibialis ant. m. (reflected)

Med. hd., gastrocnemius m.

Tibialis anterior m.

Soleus m.

Deep peroneal n.

Anterior medial malleolar a.

Dorsalis pedis a.

medial tarsal br.

lateral tarsal br.

arcuate br.

deep plantar br.

first dorsal metatarsal br.

Dorsal digital brs. of deep peroneal n.

The **medial calcaneal branches** arise from the posterior tibial artery just proximal to its division. They penetrate the flexor retinaculum and spread into the fat and the areolar tissue on the medial side and the back of the heel. Here they supply the skin and the deeper tissues and anastomose with the lateral calcaneal branches of the peroneal artery and with the posterior medial malleolar branch of the posterior tibial artery.

The **peroneal artery** is the largest branch of the posterior tibial and developmentally is the major artery of the leg. It is the muscular artery of the fibular side of the leg. It also serves as a large collateral vessel, for near the ankle it is connected by a horizontal communicating branch with the posterior tibial artery and by a perforating ramus with the anterior tibial artery. The peroneal artery arises from the posterior tibial, from 2 to 3 cm. below the lower border of the popliteus muscle, and descends near the fibula within the substance of the flexor hallucis longus muscle or in the intermuscular septum between it and the tibialis posterior. At the ankle it lies behind the tibiofibular syndesmosis and ends in branches to the ankle and the heel. The peroneal artery gives rise to muscular branches in its descent through the leg. Its named branches are the following:

Nutrient | Communicating
Perforating | Posterior lateral malleolar

(1) The **nutrient branch** of the peroneal artery enters the nutrient foramen of the fibula.

(2) The **perforating branch** passes foward at the distal border of the interosseous membrane, immediately above the tibiofibular syndesmosis, to enter the anterior compartment of the leg. It supplies the tibiofibular syndesmosis and the ankle joint and anastomoses with the anterior lateral malleolar branch of the anterior tibial and with the lateral tarsal and arcuate branches of the dorsalis pedis artery.

(3) The **communicating branch,** arising at about the same level as the perforating branch, runs transversely across the back of the leg deep to the tendon of the flexor hallucis longus muscle and joins the communicating branch of the posterior tibial artery.

(4) The **posterior lateral malleolar branch** is the terminal part of the peroneal. It descends posterior to tibiofibular syndesmosis, supplies branches to the malleolus, and anastomoses with the anterior lateral malleolar branch of the anterior tibial. The **lateral calcaneal branches** of the posterior lateral malleolar artery reach the lateral side and the back of the heel,

Fig. 451 The anterior tibial and dorsalis pedis arteries, the deep peroneal nerve, and related structures.

where they anastomose with the medial calcaneal branches of the posterior tibial artery.

For collateral circulation of the arteries of the leg, see p. 602.)

Veins of the Leg

The veins of the leg are paired accompanying veins of the anterior tibial, peroneal, and posterior tibial arteries. They have tributaries and continuities which are like the arteries, and they are supplied with numerous valves. They receive many perforating communications from the superficial veins of the leg. The peroneal veins unite with the posterior tibial veins. At a variable level in the popliteal fossa, the anterior tibial, peroneal, and posterior tibial veins combine to form the large, single popliteal vein (p. 553).

NERVES OF THE LEG (figs. 421–23, 427, 442, 443, 451, 452)

The cutaneous nerves of the leg are described with the superficial structures of the lower limb (p. 530). The muscular and articular nerves are branches of the common peroneal and tibial nerves.

The Common Peroneal Nerve (figs. 427, 442, 443, 451)

The common peroneal nerve takes its origin from the posterior branches of the fourth and fifth lumbar nerves and the first and second sacral nerves. Separating from the tibial portion of the sciatic nerve at the apex of the popliteal fossa, it follows the tendon of the biceps femoris muscle along the lateral margin of the space and crosses the lateral head of the gastrocnemius muscle to reach the back of the head of the fibula. In the lower thigh the nerve gives rise to **muscular branches** to the short head of the biceps femoris muscle and articular branches to the back of the knee (p. 557). In the popliteal fossa arise the lateral sural cutaneous and the peroneal communicating branches (p. 530). The common peroneal nerve enters the leg by turning forward around the neck of the fibula beneath the uppermost fibers of the peroneus longus muscle, where it divides into the superficial

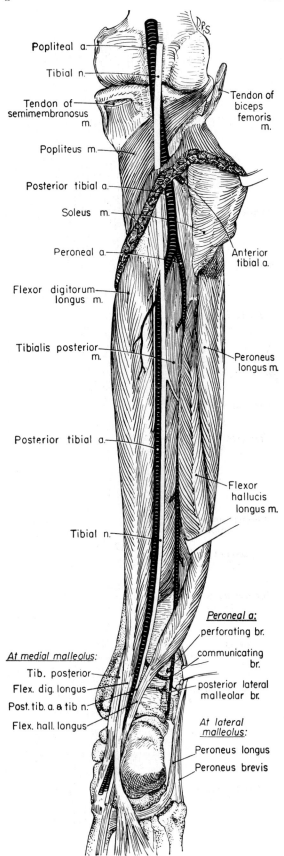

Fig. 452 The posterior tibial and peroneal arteries, the tibial nerve, and related structures.

peroneal and deep peroneal nerves. Just proximal to its terminal division, the **recurrent articular branch** arises and continues forward through the fibers of the peroneus longus and extensor digitorum longus muscles. Ascending with the anterior tibial recurrent artery, the recurrent articular nerve supplies the uppermost fibers of the tibialis anterior, the tibiofibular articulation, and the knee joint.

The **superficial peroneal nerve** arises between the peroneus longus muscle and the neck of the fibula. It descends in the anterior intermuscular septum and, there, supplies branches to the peroneus longus and peroneus brevis muscles and cutaneous twigs to the skin of the lower part of the front of the leg. At about the junction of the middle and lower thirds of the leg, the superficial peroneal nerve pierces the deep fascia and, in the subcutaneous tissue, divides into a medial dorsal cutaneous and an intermediate dorsal cutaneous nerve of the foot (p. 530).

The **deep peroneal nerve** also arises from the division of the common peroneal nerve between the peroneus longus muscle and the neck of the fibula. It continues the forward and downward course of the main nerve, passing through the upper part of the origin of the extensor digitorum longus muscle to the lateral border of the tibialis anterior muscle. The nerve takes up a position along the lateral aspect of the anterior tibial artery and descends with it, first between the tibialis anterior and extensor digitorum longus muscles and then between the tibialis anterior and extensor hallucis longus muscles. In the region of the ankle the tendon of the extensor hallucis longus muscles crosses to the medial side of the nerve and artery. The deep peroneal nerve supplies, in its course, the tibialis anterior, extensor digitorum longus, extensor hallucis longus, and peroneus tertius muscles and provides articular branches to the tibiofibular syndesmosis and the ankle joint. After crossing the ankle, the deep peroneal nerve divides into medial and lateral branches which supply structures of the dorsum of the foot (p. 575).

The Tibial Nerve (figs. 427, 442, 443, 452)

The tibial nerve arises from the anterior branches of the fourth and fifth lumbar nerves and the first, second, and third sacral nerves. The tibial nerve, as a component of the sciatic nerve, supplies the hamstring muscles of the thigh (p. 557) and, in the popliteal fossa, descends in the midline of the limb across the back of the knee.

In the popliteal fossa the tibial nerve gives origin to the articular branches which accompany the medial superior genicular, middle genicular, and medial inferior genicular arteries (p. 557). Also arising here are the medial sural cutaneous nerve (p. 530) and **muscular branches** to both heads of the gastrocnemius muscle and the plantaris, soleus, and popliteus muscles. The large **nerve to the soleus** passes between the lateral head of the gastrocnemius muscle and the plantaris muscle in its course to the soleus. The **nerve to the popliteus muscle** turns forward at the distal border of the muscle and ascends on its anterior surface, supplying it from that aspect. This nerve also supplies articular branches to the knee and the tibiofibular articulation, a medullary branch to the shaft of the tibia which accompanies the nutrient artery, and gives rise to the interosseous crural nerve. The **interosseous crural nerve** descends on the interosseous membrane, supplying it and the periosteum of the tibia, and ends in the tibiofibular syndesmosis.

The tibial nerve passes under the tendinous arch of the soleus muscle and descends immediately beneath the transverse intermuscular septum, where it overlies the tibialis posterior muscle. The nerve is superficial to the posterior tibial artery and is also at first medial and then lateral to the artery. This is due to an initial shift of the artery toward the fibula, to give off the peroneal artery, and then an inclination toward the tibia below. From its position deep to the transverse intermuscular septum, the tibial nerve provides **muscular branches** to the tibialis posterior, flexor digitorum longus, soleus, and flexor hallucis longus muscles. Twigs are provided to the posterior tibial and peroneal arteries. The **medial calcaneal branches** arise in the lower part of the leg. They pierce the flexor retinaculum and distribute to the skin of the heel and of the posterior part of the sole of the foot. Small **articular branches** penetrate the deltoid ligament to supply the ankle joint. The tibial nerve passes through the ankle deep to the flexor retinaculum, lateral to the posterior tibial artery and veins and between them and the tendon of the flexor hallucis longus muscle. Under the origin of the abductor hallucis muscle, the nerve divides into the medial plantar and lateral plantar nerves of the foot (pp. 578, 582).

THE ANKLE

The **ankle** is the region of transition from the leg to the foot and contains the ankle joint. Here the shift from the vertical orientation of the bones and muscles of the leg to a horizontal orientation in the foot entails a turning forward of all the tendons, nerves, and vessels entering the foot. Provision is made by the various retinacula for holding these structures close to the bones of the ankle and for the prevention of bowstringing by the tendons. The retinacula are areas of reinforcement and thickening of the deep fascia and are described with the deep fascia of the leg (p. 557). They define the tendon compartments at the ankle.

Tendon Compartments and Synovial Sheaths at the Ankle (fig. 453)

The **superior extensor** and **inferior extensor retinacula** (p. 557) overlie the tendons of the anterior compartment of the leg and define the compartments through which they run. The tendon of the tibialis anterior muscle runs within a separate compartment beneath the medial portions of the two retinacula. The tendon of the extensor hallucis longus lies in a separate compartment of the upper limb of the Y of the inferior extensor retinaculum but is not separated from adjacent tendons by the other bands. The tendons of the extensor digitorum longus and peroneus tertius muscles lie under the stem of the Y of the inferior extensor retinaculum, where they are enclosed between its superficial and deep laminae.

The tendons are invested by synovial sheaths as they run under the extensor retinacula. The sheath of the tibialis anterior rises to the upper limit of the superior extensor retinaculum and covers the tendon beyond the lower limit of the inferior extensor retinaculum. The sheath of the extensor hallucis longus rises a little above the upper limb of the Y of the inferior extensor retinaculum and is prolonged downward to the proximal phalanx of the great toe. A common synovial sheath invests the tendons of the extensor digitorum longus and peroneus tertius muscles. It begins at the lower border of the superior extensor retinaculum and, widening below, ends at about the middle of the dorsum of the foot.

The **flexor retinaculum** covers a number of structures which pass behind the medial malleolus, and four canals are defined by deep extensions of its fibrous tissue to the back of the lower end of the tibia and to the capsule of the ankle joint. Enumerated from the medial side, these canals transmit the tendon of the tibialis posterior muscle, the tendon of the flexor digitorum longus muscle, the posterior tibial artery and veins and the tibial nerve, and the tendon of the flexor hallucis longus muscle.

The three tendons are provided with **synovial sheaths.** The sheath of the tibialis posterior begins about 5 cm. above the tip of the medial malleolus and ends at the insertion into the navicular bone. The sheath of the flexor digitorum longus has an almost equal superior extent and, inferiorly, reaches the middle of the foot. The sheath of the flexor hallucis longus rises just above the superior limit of the flexor retinaculum and ends distally at about the middle of the first metatarsal bone.

Deep to the **peroneal retinacula** pass the tendons of the peroneus longus and peroneus brevis muscles, the peroneus brevis tendon being anterior behind the lateral malleolus and superior to the peroneus longus tendon beneath the inferior peroneal retinaculum.

These tendons are enclosed, at first, within a common synovial sheath which begins behind the lateral malleolus and then bifurcates below the superior peroneal reti-

naculum. The sheaths are separate beneath the inferior peroneal retinaculum. One covers the tendon of the peroneus brevis muscle to its insertion; the other is prolonged across the sole of the foot to the insertion of the peroneus longus. This sheath may be separated into two parts as the tendon turns at the margin of the cuboid bone.

The Arteries and Nerves at the Ankle (figs. 451–53, 458)

The **anterior tibial artery** leaves the anterior compartment of the leg and crosses the ankle midway between the medial and lateral malleoli. The **deep peroneal nerve** lies on the lateral side of this artery and both are crossed by the tendon of the extensor hallucis longus muscle. They pass under the inferior extensor retinaculum between the tendons of the extensor hallucis longus and extensor digitorum longus muscles. The anterior tibial artery becomes the dorsalis pedis artery in front of the ankle joint.

The **posterior tibial artery** passes through the third of the four canals formed by the flexor retinaculum behind the medial malleolus. Its position, as projected to the surface, is midway between the medial malleolus and the medial tubercle of the calcaneus. The tibial nerve lies on the lateral aspect of the artery and an accompanying vein. The vessels and the nerve enter the foot by passing deep to the origin of the abductor hallucis muscle and, under that muscle, divide into medial plantar and lateral plantar vessels and nerves.

The malleolar branches of the anterior tibial, posterior tibial, and peroneal arteries form arterial networks over the medial and lateral malleoli. These are described more completely in connection with the collateral circulation of the lower limb (p. 560).

THE FOOT

THE DORSUM OF THE FOOT

The **dorsum of the foot** exhibits structures that are extensions of those of the anterior compartment of the leg. The skin of the dorsum is thin, and there is relatively little subcutaneous fat beneath it; the tendons are usually visible, especially during dorsiflexion of the foot and the toes. The deep fascia is thin. It is continuous with the extensor retinacula and curves over the margins of the foot to become the fascia of the sole. Anteriorly on the dorsum, the fascia encloses the extensor tendons as it does on the back of the hand. Associated with the ten-

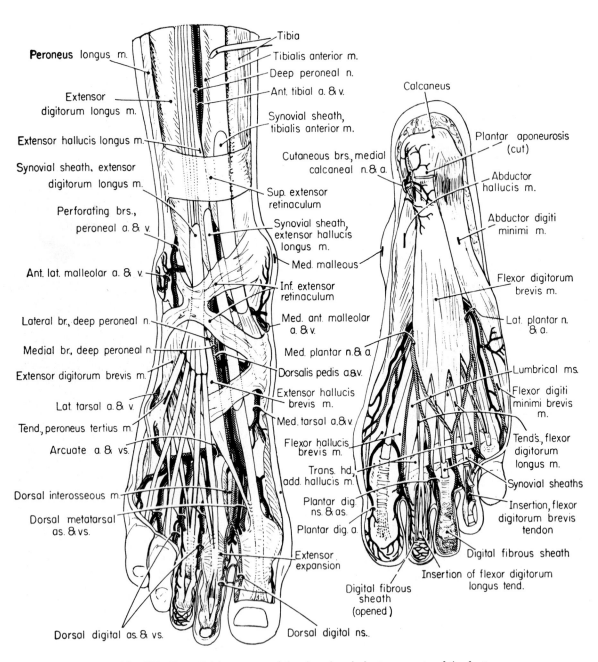

Peroneus longus m.

Extensor digitorum longus m.

Extensor hallucis longus m.

Synovial sheath, extensor digitorum longus m.

Perforating brs., peroneal a. & v.

Ant. lat. malleolar a. & v.

Lateral br., deep peroneal n.

Medial br., deep peroneal n.

Extensor digitorum brevis m.

Lat. tarsal a. & v.

Tend., peroneus tertius m.

Arcuate a. & vs.

Dorsal interosseous m.

Dorsal metatarsal as. & vs.

Dorsal digital as. & vs.

Tibia

Tibialis anterior m.

Deep peroneal n.

Ant. tibial a. & v.

Synovial sheath, tibialis anterior m.

Cutaneous brs., medial calcaneal n.& a.

Sup. extensor retinaculum

Synovial sheath, extensor hallucis longus m.

Med. malleous

Inf. extensor retinaculum

Med. ant. malleolar a. & v.

Med. plantar n.& a.

Dorsalis pedis a.&v.

Extensor hallucis brevis m.

Med. tarsal a.&v.

Flexor hallucis brevis m.

Trans. hd, add. hallucis m.

Plantar dig. ns. & as.

Plantar dig. a.

Extensor expansion

Digital fibrous sheath (opened)

Dorsal digital ns..

Calcaneus

Plantar aponeurosis (cut)

Abductor hallucis m.

Abductor digiti minimi m.

Flexor digitorum brevis m.

Lat. plantar n. & a.

Lumbrical ms.

Flexor digiti minimi brevis m.

Tend's, flexor digitorum longus m.

Synovial sheaths

Insertion, flexor digitorum brevis tendon

Digital fibrous sheath

Insertion of flexor digitorum longus tend.

Fig. 453 Superficial anatomy of the dorsal and plantar aspects of the foot.

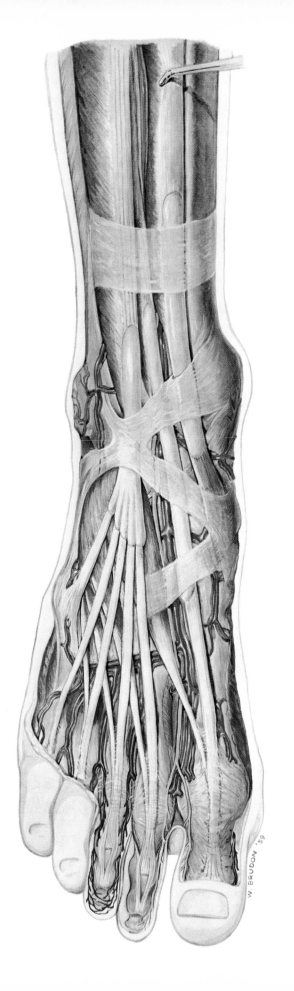
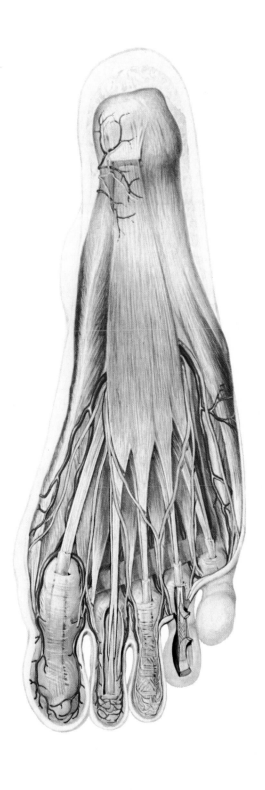

W. BRUDON '59

dons of the extensor hallucis longus and extensor digitorum longus muscles is one intrinsic muscle, the extensor digitorum brevis. It underlies the tendons of the extensor digitorum longus muscle and largely covers the branches of the dorsalis pedis artery and the deep peroneal nerve.

The Extensor Digitorum Brevis Muscle (figs. 451, 453, 459)

The extensor digitorum brevis muscle is broad and thin. It arises from the distal part of the superior and lateral surfaces of the calcaneus and from the stem of the inferior extensor retinaculum. The muscle passes diagonally across the dorsum, dividing into four tendons to the medial four toes. The medialmost and largest tendon inserts into the dorsal surface of the base of the first phalanx of the great toe. This tendon, together with its belly, is frequently designated separately as the **extensor hallucis brevis muscle.** The other three tendons join the lateral sides of the tendons of the extensor digitorum longus muscle to the second, third, and fourth toes and assist in forming the extensor expansions on these digits. The muscle assists the long extensor muscles in extending the proximal phalanges of the medial four toes. During extension, it draws the toes somewhat lateralward. The **deep peroneal nerve** (S1, 2) passes under the medial belly of the muscle and gives off a lateral branch which, passing lateralward deep to the muscle, supplies it.

The Dorsalis Pedis Artery (figs. 451, 453, 459)

The dorsalis pedis artery, the continuation of the anterior tibial artery opposite the ankle joint, is directed forward across the dorsum of the foot to the proximal end of the first intermetatarsal space. Here the artery divides into the first dorsal metatarsal and deep plantar arteries. The dorsalis pedis artery lies against the bones and ligaments of the dorsum and is crossed near its termination by the tendon of the extensor hallucis brevis muscle. On the lateral side of the artery lies the medial branch of the deep peroneal nerve; the artery is accompanied by two venae comitantes. Its branches are as follows:

Lateral tarsal	Arcuate
Medial tarsal	First dorsal metatarsal
	Deep plantar

The **lateral tarsal artery** arises over the navicular bone and passes lateralward and distally deep to the extensor digitorum brevis muscle. It supplies this muscle and the tarsal articulations and anastomoses with branches of the arcuate, anterior lateral malleolar, and lateral plantar arteries, and with the perforating branch of the peroneal artery.

The **medial tarsal arteries** are two or three small branches which ramify on the medial border of the foot. They anastomose with the medial malleolar arteries.

The **arcuate artery** arises at the level of the bases of the metatarsal bones and runs lateralward across the proximal ends of these bones beneath the tendons of the extensor digitorum longus and the extensor digitorum brevis. The artery has a slight curve, with its convexity forward, and ends on the lateral aspect of the foot by anastomosing with the lateral tarsal and lateral plantar arteries. Three **dorsal metatarsal arteries** arise from the arcuate and pass forward over the dorsal interosseous muscles to the clefts of the toes. Here each artery divides into two small **dorsal digital arteries** for the adjacent sides of the toes on either side of the cleft. Like the dorsal digital arteries of the fingers, they do not reach the distal phalanx. The fourth metatarsal artery provides an additional small branch to the lateral side of the fifth toe. Each dorsal metatarsal artery gives off a small **posterior perforating branch,** which descends between the heads of origin of the dorsal interosseous muscles to anastomose with the plantar arch, and an **anterior perforating branch,** which passes deeply through the anterior part of the interosseous space to anastomose with a corresponding plantar metatarsal artery.

The **first dorsal metatarsal artery** is like the other dorsal metatarsal arteries in its course and in its bifurcation for the supply of the adjacent sides of the first and second toes. It also gives off a branch which crosses the first metatarsal bone beneath the tendon of the extensor hallucis longus muscle to supply the medial side of the great toe. Its posterior perforating branch is enlarged as the deep plantar artery.

The **deep plantar artery,** in size the continuation of the dorsalis pedis, descends into the sole of the foot by passing between the two heads of origin of the first dorsal interosseous muscle. The deep plantar artery unites with the lateral plantar artery to form the plantar arterial arch. Its course and destination are like those of the radial artery in the hand which completes the formation of the deep palmar arterial arch.

The anastomosis between the vessels on the lateral side of the dorsum of the foot—perforating branch of the peroneal, lateral anterior malleolar, lateral tarsal, and arcuate arteries—is sometimes enlarged, and the dorsalis pedis artery is reduced in size. This configuration results in the absence or the great reduction of the dorsalis pedis artery in about 5 per cent of feet (Huber, '41). In most of these variations the arteries of the dorsum are branches of the perforating branch of the peroneal artery prolonged onto the foot.

The Deep Peroneal Nerve (figs. 451, 453, 459)

The deep peroneal nerve divides at about the lower border of the inferior extensor retinaculum into medial and lateral terminal branches. The

medial branch passes downward lateral to the dorsalis pedis artery and divides into two **dorsal digital branches** which supply the adjacent sides of the first and second digits. The medial branch also gives twigs to the metatarsophalangeal and interphalangeal articulations of the great toe and one to the first dorsal interosseous muscle. The **lateral branch** passes lateralward deep to the extensor digitorum brevis muscle. It ends in an enlargement from which branches distribute to this muscle, the tarsal joints, and the three lateral intermetatarsal spaces for the supply of the periosteum and the joints.

THE SOLE OF THE FOOT

The skin of the sole of the foot is thin on the toes and over the instep. It is thickened over the heel and the heads of the metatarsal bones in response to weight-bearing and friction. There is much fat in the subcutaneous tissue of the sole of the foot and the plantar aspect of the toes. It is intermingled with fibrous connective tissue, and the firmly supported fat over the sole and the heel forms a cushioning pad on the weight-bearing parts of the foot.

Cutaneous Nerves of the Sole (figs. 453, 454)

Sensory nerves turn into the sole from nearby nerves and pierce the deep fascia to reach the skin and the subcutaneous tissue. The **medial calcaneal branches** of the tibial nerve (p. 531) arise in the lower part of the leg. They pierce the flexor retinaculum and distribute to the skin of the heel and the posterior part of the sole of the foot. The **lateral calcaneal branches** of the sural nerve (p. 530) are distributed to the lateral side of the heel. The **plantar cutaneous branches** of the medial plantar nerve pierce the plantar fascia between the abductor hallucis and flexor digitorum brevis muscles and distribute to a large part of the sole of the foot. They are supplemented laterally by small plantar cutaneous twigs of the lateral plantar nerve which perforate the deep fascia. The **common digital branches of the medial plantar nerve** divide into **proper digital nerves** for the supply of the medial three and a half toes. The **lateral plantar nerve** distributes to the lateral one and a half toes. These nerves are described more completely with the cutaneous nerves of the limb (p. 531). Like the palmar digital nerves of the fingers, the plantar

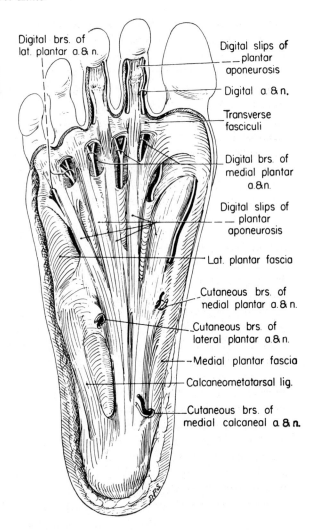

Fig. 454 The cutaneous nerves and vessels of the sole of the foot; the plantar fascia.

digital nerves of the toes are larger than the dorsal digital nerves and, terminally, send twigs onto the dorsum of the last phalanx.

The Plantar Fascia (figs. 454, 458, 459)

The plantar fascia is continuous with the fascia of the dorsum of the foot after an attachment to the periosteum of the sides of the first and fifth metatarsal bones. It exhibits, in the sole, thinner membranous sheets medially and laterally over the compartments of the great and little toes and a thickened plantar aponeurosis over the central compartment. The **plantar aponeurosis** consists largely of glistening, longitudinally arranged bands of white fibrous connective tissue which diverge toward the toes from the medial process of the tuberosity of the calcaneus. Over the sole of the foot the aponeurosis is undivided, but toward the ball of the foot **digital slips** separate and are directed onto the plantar surface of each toe, the broader bands passing to the middle three toes. Deeper transverse fibers form a reinforcing band over the heads of the metatarsal bones, interconnecting the digital slips. This is the **superficial transverse metatarsal ligament.** **Transverse fasciculi** reinforce the webs of the toes. Marginal fibers of the digital slips at the webs of the toes pass deeply to blend with the proximal ends of the fibrous sheaths of the flexor tendons and to attach to the **deep transverse metatarsal and plantar ligaments** (Hicks, '54). The superficial central fibers of the digital slips end largely in the flexion creases of the skin between the toes and the sole, although some fibers continue into the toes. At the lateral and medial margins of the plantar aponeurosis, fibers radiate onto the thinner membranous fasciae of the side compartments, and the deep intermuscular septa penetrate the soft parts of the sole of the foot. These septa separate the central compartment from the lateral and medial compartments and end in the plantar interosseous fascia and in the bones and ligaments deep in the sole of the foot. They contribute to the intermuscular fasciae which separate the muscular laminae of the central compartment and provide attachments for muscle fibers in the various compartments.

The thinner **medial plantar fascia** covers the intrinsic muscles of the great toe. The **lateral plantar fascia** is thick and well developed near the heel and thinner toward the little toe. A dense fibrous band, the **calcaneometatarsal ligament,** extends within it from the lateral process of the tuberosity of the calcaneus to the tuberosity of the fifth metatarsal bone.

Compartments of the Sole (figs. 453, 459)

The muscles of the sole of the foot and the compartments within which they lie have many similarities with those in the hand. A **great toe compartment,** located medially beneath the medial plantar fascia, is bounded at its sides by the medial intermuscular septum from the plantar aponeurosis and by the attachment of the fascia of the dorsum of the foot to the side of the first metatarsal bone. A **compartment** for the soft parts associated with the **small toe** is similarly formed on the lateral side of the sole. *Each compartment contains an abductor*

and a flexor for the toe in question. Opponens muscles are only infrequently and minimally represented in the foot and may be neglected. The **central compartment,** deep to the plantar aponeurosis, is bounded by the lateral intermuscular and medial intermuscular septa at its borders. It is occupied by the flexor digitorum brevis muscle, the muscles associated with the tendons of the flexor digitorum longus muscle—quadratus plantae and the four lumbrical muscles. The **interosseous-adductor compartment** is defined by the plantar interosseous and dorsal interosseous fasciae and contains the deeper muscles of the central compartment—adductor hallucis, dorsal interosseous, and plantar interosseous muscles.

The muscular nerves of the sole of the foot are the medial plantar and lateral plantar nerves. These branches of the tibial nerve correspond in their distribution to the median and ulnar nerves of the hand respectively, and the rule of innervation for the muscles of the sole of the foot is similar to that for the muscles of the hand. The medial plantar nerve, like the median, has the greater cutaneous and the lesser muscular distribution. The lateral plantar nerve, like the ulnar nerve of the hand, has the lesser cutaneous and the greater muscular distribution.

The Compartment of the Great Toe (figs. 453–59)

This compartment contains the abductor hallucis and flexor hallucis brevis muscles, the medial plantar nerve and vessels, and the first metatarsal bone.

The **abductor hallucis** lies superficially along the medial border of the sole. It arises from the medial process of the tuberosity of the calcaneus, the flexor retinaculum, the plantar aponeurosis, and the intermuscular septum between the abductor hallucis and flexor digitorum brevis muscles. Its tendon inserts into the medial side of the base of the proximal phalanx of the great toe, partly blending wth the medial head of the flexor hallucis brevis muscle.

The **flexor hallucis brevis** is a two-bellied muscle; between its bellies the tendon of the flexor hallucis longus muscle passes to the great toe. The flexor hallucis brevis muscle arises by a pointed extremity from the medial part of the plantar surface of the cuboid bone, from the adjacent part of the lateral cuneiform bone, from the tibialis posterior tendon, and from the medial aspect of the first metatarsal bone. The fibers of the two bellies also arise in front of this attachment from a tendinous raphe. The

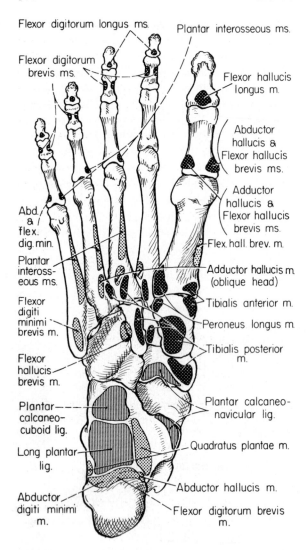

Flexor digitorum longus ms.

Plantar interosseous ms.

Flexor digitorum brevis ms.

Flexor hallucis longus m.

Abductor hallucis & Flexor hallucis brevis ms.

Adductor hallucis & Flexor hallucis brevis ms.

Flex. hall. brev. m.

Abd. & flex. dig. min.

Plantar interosseous ms.

Flexor digiti minimi brevis m.

Flexor hallucis brevis m.

Plantar calcaneocuboid lig.

Long plantar lig.

Abductor digiti minimi m.

Adductor hallucis m. (oblique head)

Tibialis anterior m.

Peroneus longus m.

Tibialis posterior m.

Plantar calcaneonavicular lig.

Quadratus plantae m.

Abductor hallucis m.

Flexor digitorum brevis m.

Fig. 455 The muscular and ligamentous attachments on the bones of the sole of the foot. (Black dots—origin; white dots on black—insertion; lines—ligaments.)

medial belly, blended with the tendon of the abductor hallucis muscle, inserts into the medial side of the base of the proximal phalanx of the great toe. The **lateral belly,** combining with the tendon of the adductor hallucis muscle, inserts into the lateral side of the base of the same phalanx. A sesamoid bone in each tendon of the flexor hallucis brevis muscle acts as a roller-bearing for these tendons and plays against the underside of the head of the first metatarsal bone. The lateral belly of this muscle is derived from the first plantar interosseous muscle and, reflecting that derivation, occasionally receives an innervation from the lateral plantar nerve. Its

derivation, its insertion with the adductor hallucis, and its lateral plantar innervation correspond to similar relations of a deep portion of the flexor pollicis brevis muscle of the hand (p. 124).

The **medial plantar nerve** is the larger of the two terminal branches of the tibial nerve. It arises by the division of the tibial nerve beneath the posterior portion of abductor hallucis and passes forward, accompanied by the small medial plantar artery, in the medial intermuscular septum between the abductor hallucis and flexor digitorum brevis muscles. **The muscular branches** to the abductor hallucis and the flexor digitorum brevis arise in this location and enter the deep surfaces of the muscles, and the **articular branches** supply the joints of the tarsal and metatarsal bones. The **plantar cutaneous branches** perforate the deep fascia to supply the skin of the medial part of the sole. The medial plantar nerve becomes superficial at the middle of the length of the sole and divides into the **proper digital branch** for the medial side of the great toe and three **common digital branches.** These are more completely described with the cutaneous nerves of the lower limb (p. 531). The proper digital nerve of the great toe supplies the flexor hallucis brevis muscle, and the first lumbrical muscle is innervated by a branch from the first common digital nerve.

The **medial plantar artery,** the smaller branch of the posterior tibial artery, does not, as a rule, form an arch in the foot like the superficial arterial arch of the palm. The artery accompanies the medial plantar nerve and lies first under the abductor hallucis muscle and then between it and the flexor digitorum brevis muscle. It sends branches into these muscles. Opposite the head of the first metatarsal bone, the medial plantar artery anastomoses with the digital branch of the first dorsal metatarsal artery on the medial side of the great toe. In addition to small branches to the skin, the muscles, and the joints, the artery provides three **digital branches** which anastomose with the three plantar metatarsal arteries of the plantar arch (p. 581) at the bases of the first three interdigital clefts. These vessels accompany the common digital branches of the medial plantar nerve.

The Compartment of the Small Toe (figs. 453–59)

The compartment of the small toe underlies the lateral plantar fascia and is bounded by the lateral intermuscular septum from the plantar aponeurosis medially and by the attachment of the fascia of the dorsum to the side of the fifth metatarsal bone laterally. It includes an abductor and a short flexor muscle for this digit and the fifth metatarsal bone.

The **abductor digiti minimi muscle** lies along the lateral border of the foot. It arises from both the medial and lateral processes of the tuberosity of the calcaneus, from the lateral plantar fascia, and from the intermuscular septum between it and the flexor digitorum brevis muscle. Its tendon crosses a smooth depression on the plantar surface of the base of the fifth metatarsal bone and inserts on the lateral side of the base of the proximal phalanx of the fifth toe.

The **flexor digiti minimi brevis muscle** underlies and is medial to the tendon of the abductor digiti minimi muscle. Its origin is the base of the fifth metatarsal bone and the fibrous sheath of the peroneus longus tendon. The tendon of the flexor digiti minimi brevis muscle inserts into the lateral side of the base of the proximal phalanx of the fifth digit.

The Central Compartment (figs. 453–59)

The central compartment of the sole of the foot lies deep to the plantar aponeurosis. It is bounded on either side by the medial and lateral intermuscular septa which pass from the margins of the aponeurosis to the plantar interosseous fascia, to the underside of the bones of the foot, and to the plantar ligaments. It contains the flexor digitorum brevis muscle, the tendon of the flexor digitorum longus and its associated muscles—quadratus plantae and the four lumbrical muscles—a portion of the tendon of the flexor hallucis longus muscle, and the lateral plantar vessels and nerve.

The Flexor Digitorum Brevis Muscle This muscle lies immediately beneath the plantar aponeurosis. It arises from the medial process of the tuberosity of the calcaneus, the posterior one-third of the plantar aponeurosis, and the medial and lateral intermuscular septa. The muscle divides into four tendons, one for each of the four lateral toes. Opposite the base of each proximal phalanx, each tendon divides into two slips between which passes a tendon of the flexor digitorum longus muscle (on the way to its insertion into the base of the distal phalanx). Turning under the tendon of the flexor digitorum longus, the two slips of each tendon of the flexor digitorum brevis muscle unite and insert into the base of the middle phalanx. The detail of this arrangement is like that of the flexor digitorum superficialis in the hand (p. 100).

Fibrous and Synovial Sheaths of the Digits of the Foot (fig. 453) Fibro-osseous canals enclose the flexor tendons of the toes, as in the fingers (p. 123). The **digital fibrous sheaths** begin over the heads of the metatarsal bones and extend as far distalward as the bases of the distal phalanges. They arch over the tendons, attaching to the capsules of the joints and to the margins of the proximal and middle phalanges. Over the shafts of the proximal and middle phalanges the sheaths are strong and their fibers are transverse; but over the joints the sheaths are much thinner and most of their constituent fiber bundles are oblique. Slender transverse bands cross the joint spaces. The marginal fibers of the digital slips of the plantar aponeurosis terminate in the proximal end of the fibrous sheaths. **Synovial**

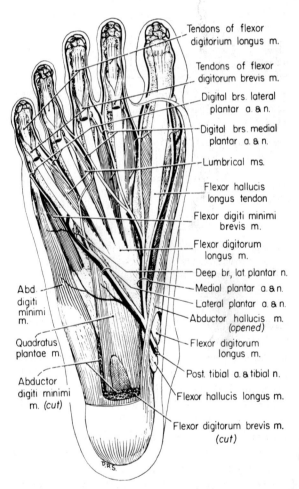

Fig. 456   The intermediate stratum of the sole of the foot; the muscles, vessels, and nerves related to the tendon of the flexor digitorum longus muscle.

sheaths occupy the fibrous sheaths of all five toes. They enclose the tendon of the flexor hallucis longus muscle and the tendons of the flexor digitorum longus and flexor digitorum brevis muscles of the four lateral toes. The synovial sheaths extend from the heads of the metatarsals, just proximal to the opening of the fibrous sheaths, to the bases of the distal phalanges.

The Long Tendons in the Sole of the Foot (figs. 456, 458, 459) The **tendon of the flexor digitorum longus muscle** passes the ankle in the second compartment under the flexor retinaculum and enters the foot deep to the abductor hallucis muscle. It passes diagonally toward the center of the sole where it expands and receives, from behind, the broad insertion of the quadratus plantae muscle. Here the tendon divides into four slips which are directed to the four lateral toes. They enter the digital fibrous sheaths surrounded by synovial sheaths, pass through the separation of the tendons of the flexor digitorum brevis muscle, and terminate by inserting into the bases of the distal phalanges. From the sides of the four tendons, the lumbrical muscles arise and pass toward the toes.

The **tendon of the flexor hallucis longus muscle** passes through the lateralmost compartment under the flexor retinaculum and turns into the foot under the sustentaculum tali of the calcaneus. Directed toward the great toe, it passes superior to the tendon of the flexor digitorum longus muscle, to which it contributes a tendinous slip, and then lies in the groove between the two bellies of the flexor hallucis brevis muscle. It enters the fibrous sheath of the first digit and inserts on the base of the distal phalanx.

The Quadratus Plantae (figs. 456, 458) This is an accessory muscle to the long digital flexor. It arises by two heads which are separated from each other by the long plantar ligament. Its tendinous lateral head arises from the lateral border of the plantar surface of the calcaneus and from the long plantar ligament. The fleshy medial head takes origin from the medial surface of the calcaneus and from the medial border of the long plantar ligament. The two heads join at an acute angle to form a flattened band which inserts into the lateral margin and into both surfaces of the tendon of the flexor digitorum longus muscle. The quadratus plantae lies on the superior aspect of the flexor digitorum brevis muscle, and the lateral plantar vessels and nerve are situated between them. The vessels and nerve are

behind and parallel to the diagonally directed tendon of the flexor digitorum longus muscle.

The Lumbrical Muscles (fig. 456) These are four small cylindrical muscles which arise from the four tendons of the flexor digitorum longus muscle. They are comparable to the lumbrical muscles in the hand—associated there with the tendons of the flexor digitorum profundus (p. 123). Except for the first one, each of the lumbrical muscles of the foot arises from the two adjacent tendons of the flexor digitorum longus muscle; the first muscle springs from the medial side of the first tendon alone. The tendons of the lumbrical muscles pass forward on the plantar side of the deep transverse metatarsal ligament and end in the medial aspect of the extensor expansion over the four lateral toes. The lumbrical muscles flex the proximal phalanges at the metatarsophalangeal joints and extend the middle and distal phalanges.

The Lateral Plantar Nerve (figs. 453, 456, 459) The lateral plantar nerve, the smaller of the two plantar nerves arising from the tibial nerve under the abductor hallucis muscle, has a distribution like the ulnar nerve in the hand. Medial to the lateral plantar vessels, the nerve passes diagonally into the sole of the foot between the flexor digitorum brevis, and the quadratus plantae muscles.

In this interval it provides **muscular branches** to the abductor digiti minimi and quadratus plantae muscles and **articular branches** to the calcaneocuboid joint. At the lateral margin of the quadratus plantae muscle the nerve divides into a superficial and a deep branch. The **deep branch** descends around the margin of the muscle into the deeper interosseous-adductor compartment of sole. The **superficial branch** splits into a common digital nerve and a nerve which supplies a proper digital branch for the lateral side of the little toe and muscular branches to the flexor digiti minimi brevis muscle and (sometimes) the two interosseous muscles of the fourth intermetatarsal space. The **common digital nerve** communicates with the third common digital branch of the medial plantar nerve and divides into two **proper digital nerves** for the adjacent sides of the fourth and fifth toes.

The Lateral Plantar Vessels (figs. 453, 456, 459) The **lateral plantar artery** is the larger of the two terminal branches of the posterior tibial artery. Its course and relations are like those of the ulnar artery of the hand. Accompanied on either side by venae comitantes, the lateral plantar artery passes diagonally across the sole of the foot toward the base of the fifth metatarsal bone. It lies lateral to the lateral plantar nerve in the interval between the flexor digitorum brevis and the quadratus plantae muscles. At the medial side of the base of the fifth metatarsal bone, the artery turns medialward around the margin of the quadratus plantae muscle and sinks into a deeper plane in the foot. It perforates the plantar interosseous fascia and passes medialward across the proximal ends of the second, third, and fourth metatarsal bones and the corresponding interosseous muscles. Here it forms the **plantar arterial arch,** anastomosing in the first interosseous space with the deep plantar branch of the dorsalis pedis artery.

Between its origin and the base of the fifth metatarsal bone, the lateral plantar artery gives off the **calcaneal branch** to the skin and subcutaneous tissue of the heel, **muscular branches** to the adjacent muscles, and **cutaneous branches** to the skin of the lateral side of the sole of the foot. The branches of the plantar arch are described under the interosseous-adductor compartment. The **lateral plantar veins** lie on either side of the artery. Their course and distribution are the same as for the artery.

The Interosseous-adductor Compartment (figs. 457, 459)

This is the deepest plane of the sole of the foot. It is enclosed dorsally by the dorsal interosseous fascia and its attachments to the periosteum of the metatarsal bones. It is limited on its plantar aspect by the plantar interosseous fascia which covers the superficial surface of the adductor hallucis muscle and blends with the medial intermuscular septum. Within the compartment are enclosed the dorsal interosseous, plantar interosseous, and adductor hallucis muscles, the plantar arterial arch, the deep branch of the lateral plantar nerve, and the dorsal metatarsal branches of the dorsalis pedis artery.

The Adductor Hallucis Muscle (figs. 457, 459) The adductor hallucis arises by oblique and transverse heads. Occupying the hollow on the plantar aspect of the metatarsal bones, the **oblique head** arises from the bases of the second, third, and fourth metatarsal bones and from the sheath of the tendon of the peroneus longus muscle. The **transverse head** takes origin from the plantar metatarsophalangeal ligaments of the third, fourth, and fifth digits and from the deep transverse metatarsal ligament. The tendons of both heads, together with the lateral portion of the flexor hallucis brevis muscle, insert into the lateral side of the base of the proximal phalanx of the great toe.

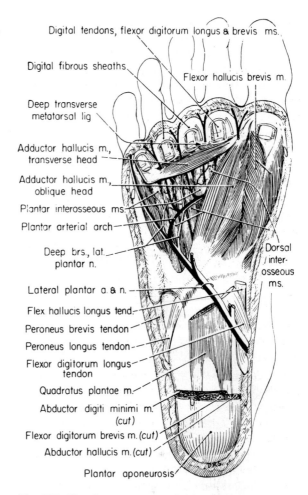

Fig. 457 The deepest stratum of the sole of the foot; the plantar arterial arch and associated muscles and nerves.

The Plantar Arterial Arch (figs. 456, 457) This arch is comparable to the deep palmar arterial arch (p. 124). It is continuous with the lateral plantar artery where that vessel sinks deeply into the sole of the foot opposite the base of the fifth metatarsal bone. The arch is completed medially by its union with the deep plantar branch of the dorsalis pedis artery which reaches the sole through the proximal end of the first intermetatarsal space. In 81 per cent of his cases, Vann ('43) considered this arch to be formed primarily by the deep plantar branch of the dorsalis pedis artery. The plantar arch is convex forward and lies across the bases of the central metatarsal bones and the origin of the interosseous muscles. Viewed from the plantar side, it is deep to the adductor hallucis muscle. The plantar arch gives off four plantar metatarsal arteries, three perforating branches, and twigs to the tarsal joints and the muscles of the compartment.

The **plantar metatarsal arteries** run forward between the metatarsal bones on the plantar surface of the interosseous muscles. Each artery divides into a pair of **plantar digital arteries** which supply the adjacent sides of the toes. The first plantar metatarsal artery arises from the junction of the plantar arch and the deep plantar artery. The proper digital artery to the lateral aspect of the fifth toe is a branch of the lateral plantar artery as it becomes the plantar arch opposite the base of the fifth metatarsal bone. Each plantar metatarsal artery gives off, near its point of division, an **anterior perforating branch** which passes upward through the interosseous space to anastomose with a corresponding branch of a dorsal metatarsal artery (p. 575). Like the corresponding arteries of fingers, the plantar digital arteries are larger than the dorsal digital arteries and provide terminal branches which pass dorsally to supply the nail beds and the skin over the distal phalanges.

The **perforating branches** of the arch ascend through the proximal ends of the lateral three intermetatarsal spaces and between the heads of the dorsal interosseous muscles and join the posterior perforating branches of the dorsal metatarsal arteries.

The Deep Branch of the Lateral Plantar Nerve
The deep branch of the lateral plantar nerve sinks into the interosseous-adductor compartment with the lateral plantar artery and swings medialward across the bases of the metatarsal bones posterior to the plantar arterial arch. It supplies **muscular branches** to the lateral three lumbrical muscles, the interosseous muscles of each space (except the fourth space in some cases), and both heads of the adductor hallucis muscle. Articular branches also arise from the deep branch of the lateral plantar nerve for the supply of the intertarsal and tarsometatarsal joints.

The Interossoeus Muscles (figs. 446, 451, 453, 455, 457) The dorsal interosseous and plantar interosseous muscles lie side by side in the intermetatarsal spaces. There are four dorsal and four plantar interossei (one of the latter is combined with flexor hallucis brevis); the dorsal interossei are abductors of the digits and the plantar interossei are adductors. The plane of reference for abduction and adduction of the toes is the midplane of the second digit. The requirements of abduction and adduction for the five toes are met in ten muscles. Of these, three specific abductors and adductors exist in the abductor hallucis, the adductor hallucis, and the abductor digiti minimi muscles. The seven other muscles are the four dorsal interosseous and three plantar interosseous muscles. (The first plantar

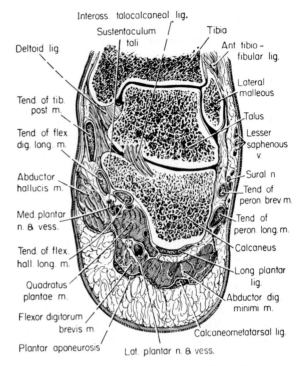

Fig. 458   A vertical section of the ankle.

interosseous muscle, like the corresponding muscle in the hand (p. 126), is incorporated with the lateral head of the flexor hallucis brevis, inserts with the adductor hallucis, and accounts for the occasional twigs of the lateral plantar nerve which enter the lateral head of the flexor hallucis brevis muscle. Like the first palmar interosseous muscle in the hand, it is not separately dissectible.)

The **dorsal interosseous muscles** are bipennate in form and arise from the adjacent sides of both metatarsal bones of the space in which they lie. They have a longer origin from the metatarsal of the digit into which they insert. The **plantar interosseous muscles** arise from the bases and medial sides of the metatarsal bones of the digits into which they insert.

The first and second dorsal interosseous muscles lie on the medial and lateral sides of the second metatarsal bone, respectively, and insert into the corresponding sides of the base of the proximal phalanx of this digit. The third and fourth dorsal interossei lie on the lateral aspects of the third and fourth metatarsal bones, respectively, and insert into the lateral sides of the bases of the proximal phalanges of the corresponding digits. The tendons also send slips into the capsules of the metatarsophalangeal joints. The second, third, and fourth plantar interossei lie on the medial sides of the third, fourth, and fifth metatarsal bones, respectively, and insert into the medial sides of the proximal phalanges of these digits. They provide traction on these digits toward the midplane of the second digit.

Only the dorsal interossei have an insertion into the extensor expansion of the extensor digitorum longus tendons, and it is a variable and minor attachment.

The Dorsal Metatarsal Arteries (figs. 457, 453) These are branches of the dorsalis pedis artery and are described with that vessel (p. 575). They underlie the dorsal interosseous fascia and thus occupy the interosseous-adductor compartment, but are more logically considered with the arterial system to which they belong.

INNERVATIONS

The nerves of the sole of the foot correspond in distribution and function with those of the hand; thus, the medial plantar nerve mirrors the median nerve and the lateral plantar nerve the ulnar nerve in the hand. The medial plantar nerve supplies the abductor hallucis brevis, flexor hallucis brevis, first lumbrical and flexor digitorum brevis muscles. (The latter muscle is the homologue of the flexor digitorum superficialis of the forearm with a median nerve innervation). The lateral plantar nerve supplies all the other pre-axial musculature of the foot—the abductor digiti minimi, flexor digiti minimi brevis, quadratus plantae, adductor hallucis, the dorsal and plantar interosseous and the lateral three lumbrical muscles.

ACTIONS OF THE MUSCLES

The great and small toe compartments each lodge an abductor and a short flexor muscle. The abductors abduct and flex the proximal phalanges of the digit in question; the flexors flex the same phalanx. Flexor digitorum brevis flexes the middle phalanges of the lateral four toes and assists in metatarsophalangeal flexion of the same digits. Its contraction gives assistance in supporting the longitudinal arch of the foot. Quadratus plantae (old name—flexor accessorius) assists the tendon of the flexor digitorum longus in flexing the toes and makes the line of traction of that tendon more nearly parallel with the long axis of the foot. The adductor hallucis muscle adducts the great toe and aids in maintaining the arches of the foot. The lumbrical muscles flex the proximal phalanges at the metatarsophalangeal joints and extend the middle and distal phalanges. The interosseous muscles, as in the hand, are abductors and adductors of the digits and also serve in flexion of the metatarsophalangeal joints. Their action in extension of the middle and distal phalanges (as in the hand, p. 126) is uncertain in the foot.

THE ARTICULATIONS OF THE LOWER LIMB

The articulations of the lower limb include those of the pelvic girdle as well as those of the limb proper. The articulations of the pelvic girdle are, however, described in the chapter on the Pelvis—the sacroiliac joint on p. 520 and the symphysis pubis on p. 521 The remaining joints are as follows:

Hip	Tarsometatarsal
Knee	Intermetatarsal
Tibiofibular	Metatarsophalangeal
Ankle	Interphalangeal
Intertarsal	

THE HIP JOINT (figs. 28, 460, 461)

The **hip joint** is a synovial joint of spheroidal type.

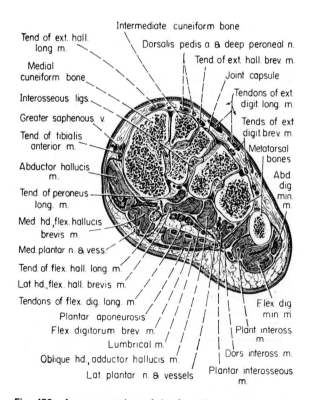

Tend. of ext. hall. long. m.
Medial cuneiform bone
Interosseous ligs.
Greater saphenous v.
Tend. of tibialis anterior m.
Abductor hallucis m.
Tend. of peroneus long. m.
Med. hd, flex. hallucis brevis m.
Med. plantar n. & vess.
Tend. of flex. hall. long. m.
Lat. hd., flex. hall. brevis m.
Tendons of flex. dig. long. m.
Plantar aponeurosis
Flex. digitorum brev. m.
Lumbrical m.
Oblique hd., adductor hallucis m.
Lat. plantar n. & vessels

Intermediate cuneiform bone
Dorsalis pedis a. & deep peroneal n.
Tend. of ext. hall. brev. m.
Joint capsule
Tendons of ext. digit. long. m.
Tends. of ext. digit. brev. m.
Metatarsal bones
Abd. dig. min. m.
Flex. dig. min. m.
Plant. inteross. m.
Dors. inteross. m.
Plantar interosseous m.

Fig. 459 A cross section of the foot through the cuneiform bones.

It consists of the articulation of the globular head of the femur in the cup-like acetabulum of the coxal bone. This is the truest example of a 'ball and socket' joint in the body. In comparison with the shoulder joint, it is modified in the direction of greater stability and some decrease in freedom of movement. The head of the femur forms two-thirds of a sphere. It is covered by an articular cartilage which is thickest above, in the line of weight-bearing, and thins to an irregular line of termination at the junction of the head of the bone with its neck. The articular cartilage stops at the pit for the **ligamentum capitis femoris.** The articular surface of the acetabulum is horseshoe-shaped and arches around the acetabular fossa. Its cartilage is thick above and is thinner at the borders of the fossa. The acetabular fossa lodges a mass of fat covered by the synovial membrane, and the **transverse ligament of the acetabulum** closes the fossa below. Articular vessels and nerves enter the fossa under the deep border of this ligament. The acetabulum is deepened by the fibrocartilaginous **acetabular labrum** attached to the bony rim and to the ligament. The free thin edge of the lip cups around the head of the femur and holds it firmly.

The **articular capsule** of the hip joint is strong. It is attached above to the bony rim of the acetabulum, along the outer border of the acetabular labrum on the anterior side of the joint, but 5 mm. beyond the labrum behind. Inferiorly, the capsule is attached to the transverse ligament of the acetabulum. On the femur the capsule is attached anteriorly to the intertrochanteric line and to the junction of the neck and the trochanters. Behind the femur the capsule has an arched free border and covers only about two-thirds of the neck from the head toward the intertrochanteric crest. Most of the fibers of the capsule are longitudinal, running from the pelvis to the femur, but some deeper fibers run circularly. These **zona orbicularis** fibers are most marked in the posterior part of the capsule and help to hold the head in the acetabulum. The fibers of the capsule wind downward and forward anteriorly in a spiral course that is continued on to the posterior part of the capsule and provides for a 'screw-home' effect in full extension of the joint. As the thigh becomes fully extended, the capsule is twisted and shortened, and the head of the femur is guided like a screw into its socket. In hyperextension, the head is in its most completely congruent 'close-packed' position in the acetabulum, and stability is at its maximum.

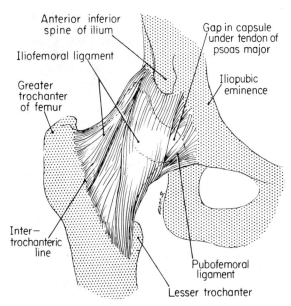

Fig. 460 The anterior aspect of the capsule of the hip joint.

Adding strength to the articular capsule are three accessory bands: the iliofemoral, ischiofemoral, and pubofemoral ligaments. The **iliofemoral ligament** lies on the anterior surface of the capsule. It is very strong and has the form of an inverted Y. The stem of the inverted Y (Λ) is attached by its apex to the lower part of the anterior inferior spine of the ilium. The diverging fibers of the ligament attach to the whole length of the intertrochanteric line, the thicker marginal bundles reaching the ends of the line. The iliofemoral ligament becomes taut in full extension and thus helps to maintain the erect posture, for in this position the weight of the body tends to roll the pelvis backward on the femoral heads. The **pubofemoral ligament** is applied to the medial and inferior part of the capsule. It arises from the pubic part of the acetabular rim and from the obturator crest of the superior ramus of the pubis. Below, the fibers of the ligament reach the neck of the femur, blending with the lower fibers of the iliofemoral ligament. The pubofemoral ligament becomes tight in extension and also limits abduction. The capsule of the joint is thinnest between the iliofemoral and pubofemoral ligaments but is crossed here by the strong tendon of the iliopsoas muscle. Between the tendon and the capsule lies the **iliopectineal bursa,** and in those cases in which the capsule is perforated here, the bursa is open to the joint cavity. The **ischiofemoral ligament** lies on the posterior aspect of the capsule. It arises from the ischial portion of the acetabulum and spirals lateralward and upward across the posterior part of the neck of the femur. The fibers insert into the superior portion of the neck of the femur medial to the root of the greater trochanter. This ligament forms the posterior free margin of the capsule.

The **ligamentum capitis femoris** is intracapsular. It arises from the two margins of the acetabular notch, intermediate fibers springing from the lower border of the transverse ligament of the acetabulum. About 3.5 cm. long, the ligament ends in the fovea of the head of the femur. It is covered by a sleeve of synovial membrane and becomes taut in adduction of the femur.

The **synovial membrane** of the hip joint lines the articular capsule, covers the acetabular labrum, and is extended, sleeve-like, over the ligament of the head of the femur. The membrane surrounds the fat of the acetabular notch and, at the femoral attachment of the capsule, is reflected back along the femoral neck. Blood vessels to the head and the neck of the femur course under these synovial reflections. A prolongation of the synovial membrane beyond the free margin of the capsule at the back of the neck of the femur serves as a bursa for the tendon of the obturator externus muscle.

The **arteries** of the hip joint are branches of the medial and lateral circumflex femoral arteries, the deep division of the superior gluteal artery, and the inferior gluteal artery. The posterior branch of the obturator artery gives rise to the artery of the ligamentum capitis femoris. This artery is of variable size but represents a significant part of the blood supply of the head of the femur. The **nerves** of the hip joint are derived from the nerve to the rectus femoris muscle, the nerve to the quadratus femoris muscle, the anterior division of the obturator nerve, the accessory obturator nerve (when present), and the superior gluteal nerve (Gardner, '48).

The movements of the femur may be expressed as extension and flexion through a transverse axis, abduction and adduction through an anteroposterior axis, and medial and lateral rotation through a vertical axis. Circumduction is also allowed. The three axes intersect in the center of the head of the femur. Since the head is at an angle to the shaft, all movements involve conjoint rotation of the head. Flexion of the hip joint is very free and is only limited by contact of the thigh with the abdomen if the knee is flexed. Flexion is limited by tension of the hamstring muscles when the knee is extended as can be appreciated in the attempt to touch the fingers to the floor in the stiff-knee bend. Extension of the hip is limited by tension in the iliofemoral ligament and can be carried to only about 30° beyond the straight limb position of the erect posture. Abduction and adduction are free in all positions of the lower limb. Abduction is limited by the pubofemoral ligament; adduction, by the opposite limb and by tension in the ischiofemoral ligament and the ligament of the head. Lateral rotation is freer than medial rotation and is a more powerful movement. Medial rotation is the weakest movement at the joint.

The description of the actions of the muscles operating on the joint is supplemented by the following group listing:

Flexion	**Extension**
Iliopsoas	Gluteus maximus
Pectineus	Semitendinosus
Rectus femoris	Semimembranosus
Sartorius	Biceps femoris (long head)
Tensor fasciae latae	Ischiocondylar part of adductor magnus
Adductor muscles	
Anterior part of gluteus medius and minimus	

Abduction	**Adduction**
Gluteus medius	Adductor longus
Gluteus minimus	Adductor brevis
Piriformis	Pectineus
Tensor fasciae latae	Adductor magnus
Sartorius	Gracilis
	The hamstrings

Lateral Rotation	**Medial Rotation**
Gluteus maximus	Gluteus minimus (anterior part)
Piriformis	Gluteus medius (anterior part)
Quadratus femoris	Tensor fasciae latae
Obturator internus and the gemelli	Semitendinosus
Obturator externus	Semimembranosus
	Pectineus
	Adductor longus
	Adductor brevis
	Adductor magnus

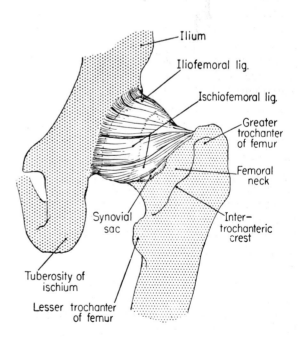

Fig. 461   The capsule of the hip joint from behind.

The Knee Joint (figs. 29, 462–66)

The **knee joint** satisfies, although not always perfectly, the requirements of a weight-bearing joint in which there is free motion in one primary plane combined with considerable stability. The support of the weight of the body on the vertically apposed ends of two long bones is intrinsically an unstable arrangement, but security at the knee joint is attained by a number of compensating mechanisms. Among these are the twofold to threefold expansion of the bearing surfaces of the femur and the tibia, the application to the joint of strong collateral ligaments, its reinforcing aponeuroses and tendons, a strong capsule, and intraarticular ligaments. The knee joint is functionally a hinge-joint, or ginglymus, but also permits a small amount of rotation of the leg, especially in the flexed position of the knee. It is divisible into three articulations: the femoropatellar articulation and two tibiofemoral joints. The separation between the two tibiofemoral joints is made by the intra-articular cruciate ligaments and the infrapatellar synovial fold. The three joint cavities are not separate in Man but are connected by restricted openings.

The Articular Surfaces

The cartilage-covered articular surfaces of the femur are its medial and lateral condyles. They have the form of rollers or thick wheels which are not entirely parallel but diverge somewhat inferiorly and posteriorly. The medial condylar surface is longer than the lateral, and the curvatures gradually change from a flatter curvature anteriorly to a tighter curvature posteriorly. The condylar surfaces of the femur are separated from the patellar surface by a shallow groove and the patellar surface projects more laterally than medially.

On the tibia there are, in the superior articular surface, two entirely separate cartilage-covered areas. The surface of the medial condyle of the tibia is the larger; it is oval and slightly concave. The surface of the lateral condyle of the tibia is approximately circular and is concave from side to side but concavoconvex from before backward. The fossae of the articular surfaces of the tibia are deepened by disk-like menisci.

The Articular Capsule

The articular capsule of the knee joint is inseparable from the ligaments and aponeuroses apposed to it. Posteriorly, its vertical fibers arise from the

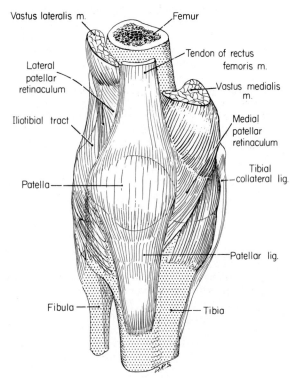

Fig. 462   The anterior aspect of the knee and its ligaments.

condyles and the intercondylar fossa of the femur. Inferiorly, they are overlain by the oblique popliteal ligament. The capsule attaches below to the tibial condyles. The external ligaments which reinforce the capsule are the fascia lata and the iliotibial tract, the medial patellar and lateral patellar retinacula, and the patellar, oblique popliteal, and arcuate popliteal ligaments.

The aponeurotic tendons of the vastus medialis and vastus lateralis muscles are attached to the medial and lateral margins of the patella as far as the attachment of the patellar ligament. Expanding over the sides of the capsule as the **medial patellar** and **lateral patellar retinacula,** they also insert into the front of the tibial condyles and into the oblique lines of the condyles as far to the sides as the collateral ligaments. Medially, the retinaculum blends with the periosteum of the shaft of the tibia; laterally, it blends with the overlying iliotibal tract. Superficial to the retinacula, the **fascia lata** covers the front and the sides of the knee joint. As it descends to attach to the tibial tuberosity and to the oblique lines of the condyles, it overlies and blends with the patellar retinacula. On the lateral side its strong **iliotibial tract,** which, in the thigh, is attached deeply to the lateral intermuscular septum, attaches to the oblique line of the lateral condyle after blending anteriorly with the capsule of the joint. Thinner on the medial side of the patella, the fascia lata sends some longitudinal fibers inferiorly to blend with the fibrous expansion of the sartorius muscle.

The **ligamentum patellae** is the continuation of the tendon of the quadriceps femoris muscle to the tibial tuberosity. Extremely strong, it is a flat band attached above to the inferior border of the patella and is continuous over the front of the patella with fibers of the quadriceps femoris tendon. The ligament ends somewhat obliquely on the tibia, being prolonged downward several centimeters farther on the lateral than on the medial side. A **deep infrapatellar bursa** intervenes between the ligament and the bone immediately superior to the insertion. A large **subcutaneous infrapatellar bursa** is developed in the subcutaneous tissue over the ligament.

The **oblique popliteal ligament** is a posterior reinforcement of the capsule provided by the tendon of the semimembranosus muscle. As this tendon inserts into the groove on the posterior aspect of the medial condyle of the tibia, it sends an oblique expansion lateralward and superiorly across the posterior surface of the capsule toward the lateral condyle of the femur. Large foramina for vessels and nerves perforate it, and the popliteal artery lies against it. The lower lateral part of the knee joint is strengthened posteriorly by the **arcuate popliteal ligament.** This arises from the back of the head of the fibula, arches upward and medialward over the tendon of the popliteus muscle, and spreads out over the posterior surface of the joint.

The Collateral Ligaments

The collateral ligaments are two strong bands at the sides of the knee. Their points of attachment lie slightly behind the vertical axis of the bones, so that they become taut in extension and prevent hyperextension. They also prevent an abduction or adduction angulation of the bones. The inferior genicular blood vessels pass between them and the capsule of the joint. The **tibial collateral ligament** is a strong, flat band that extends from the tubercle on the medial condyle of the femur to the medial condyle of the tibia and to the medial surface of its shaft. The ligament is from 8 to 9 cm. long and is well defined anteriorly where it blends with the medial patellar retinaculum though bursae sometimes separate them in part. Its deeper fibers end on the condyle and also spread triangularly (fig. 465) to attach to the medial meniscus. Only the **bursa anserina** separates the tibial collateral ligament from the overlying insertions of the sartorius, gracilis, and semitendinosus muscles. The **fibular collateral ligament** is a rounded, pencil-like cord about 5 cm. long. It is attached to the tubercle on the lateral condyle of the femur above and behind the groove for the popliteus muscle. It ends below on the lateral surface of the head of the fibula about 1 cm. anterior to its apex. The tendon of the popliteus muscle passes deep to the fibular collateral ligament, and the tendon of the biceps femoris muscle divides on either side of its lower attachment.

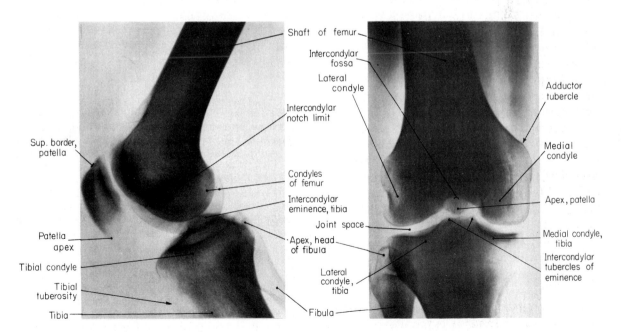

Fig. 463   The knee joint; posteroanterior and lateral views.

The small **inferior subtendinous bursa of the biceps femoris muscle** is interposed between the tendon of the biceps femoris muscle and the fibular collateral ligament. Another bursa lies nearer the upper end of the ligament, separating it from the tendon of the popliteus muscle. To complete the sequence, it should be noted that the synovial membrane of the joint protrudes as the **subpopliteal recess** and separates the tendon of the popliteus muscle from the lateral meniscus.

The Intraarticular Cruciate Ligaments

The cruciate ligaments lie within the capsule of the knee joint. They are situated in the vertical plane between the condyles somewhat closer to the posterior than to the anterior surface of the joint. They are strong, rounded cords which cross each other like the limbs of an X. Their designation of anterior and posterior is taken from their relation to the intercondylar eminence of the tibia. The **anterior cruciate ligament** arises from the rough, nonarticular area in front of the intercondylar eminence of the tibia and extends upward and backward to the posterior part of the medial aspect of the lateral femoral condyle. The **posterior cruciate ligament** is directed upward and forward on the medial side of the anterior ligament. It extends from the area behind the tibial intercondylar eminence to the lateral side of the medial condyle of the femur. These ligaments prevent movement of the tibia forward or backward under the femoral condyles, the anterior cruciate ligament preventing anterior displacement of the tibia and the posterior ligament preventing its posterior displacement. They are somewhat taut in all positions of flexion but become tightest in full extension and full flexion.

The Menisci

These are two crescent-shaped wafers of fibrocartilage which rest on the peripheral parts of the articular surfaces of the tibia and deepen the articular fossae for the reception of the condyles of the femur. They are thicker at their external margins and taper to thin unattached edges in the interior of the articulation. They are attached along the outer borders of the condyles of the tibia and, at their ends, in front of and behind its intercondylar eminence. Their upper surfaces are slightly concave for the condyles of the femur; their under surfaces are flatter. The **medial meniscus** is larger and is more nearly oval in outline. It is broader posteriorly; anteriorly, it is thin and pointed as it attaches in the anterior intercondylar area of the tibia in front of

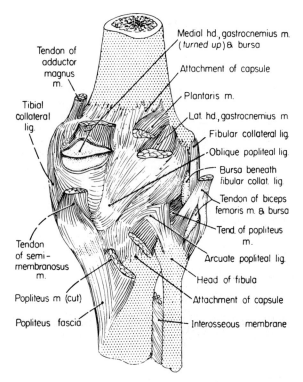

Fig. 464   The capsule of the right knee joint from behind.

the anterior cruciate ligament. Its posterior attachment is in the corresponding posterior fossa in front of the origin of the posterior cruciate ligament. The **lateral meniscus** is more nearly circular and covers a somewhat greater proportion of the tibial surface than does the medial meniscus. Its anterior end is attached in the anterior intercondylar area lateral to and behind the end of the anterior cruciate ligament. Its posterior termination is in the posterior intercondylar area in front of the end of the medial meniscus. The lateral meniscus is weakly attached around the margin of the lateral condyle of the tibia, and it lacks an attachment where it is crossed and notched by the popliteus tendon. At the back of the joint, the lateral meniscus gives rise to some of the fibers of the popliteus muscle. Close to its posterior attachment, it frequently sends off a collection of fibers, the **posterior meniscofemoral ligament,** which either joins or lies behind the posterior cruciate ligament.

The posterior meniscofemoral ligament ends in the medial condyle of the femur immediately behind the attachment of the posterior cruciate ligament. An occasional **anterior meniscofemoral ligament** has a

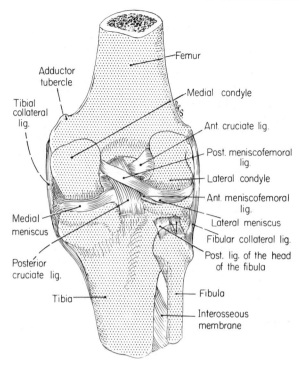

Fig. 465 The right knee joint opened from behind; the menisci and the cruciate ligaments.

similar but anterior relationship to the posterior cruciate ligament. The lateral meniscus is thus loosely attached to the tibia and has frequent attachments to the femur. It tends to move forward and backward with the lateral femoral condyle. The **transverse genicular ligament** connects the anterior convex margin of the lateral meniscus to the anterior end of the medial meniscus.

Clinical Note Athletic injuries, particularly football injuries, are the most frequent causes of damage to the menisci, collateral ligaments, and cruciate ligaments of the knee. Hit from the side and behind while the foot is firmly planted on the ground and the knee is semiflexed, both rotary and abduction strains are placed on the medial side of the knee and both the tibial collateral ligament and the medial meniscus may be torn. Violence to the medial meniscus tends to separate its thinner inner portion from its outer portion—'bucket handle rupture.'

The Synovial Membrane and Cavity

The articular cavity of the knee is the largest joint space of the body. Aside from the space intervening between and around the condyles, it extends upward behind the patella to include the femoropatellar articulation and then goes into free communication with the suprapatellar bursa between the tendon of the quadriceps femoris muscle and

the femur. The synovial membrane lines the articular capsule and reflects onto the bones as far as the edges of their cartilages. It lines the suprapatellar bursa and extends at the sides of the patella under the aponeuroses of the vastus muscles.

There are also recesses of the cavity lined by synovial membrane behind the posterior part of each femoral condyle. At the upper end of the medial recess, the bursa under the medial head of the gastrocnemius muscle occasionally opens into the joint cavity. In the **subpopliteal recess** the cavity and its lining membrane extend beyond the capsule into relation with the tendon of the popliteus muscle. The synovial membrane covers the cruciate ligaments except behind where the posterior cruciate is connected to the back of the capsule. Thus, the cruciate ligaments are excluded from the synovial cavity. The **infrapatellar fat pad** represents an anterior part of the median septum of tissue which, with the cruciate ligaments, separates the two tibiofemoral articulations. It lies below the patella. From the synovial surface of the infrapatellar fat pad, a vertical crescentic fold, the **infrapatellar fold** of synovial membrane, passes in the median plane toward the cruciate ligaments. It is attached in the intercondylar fossa of the femur anterior to the anterior cruciate ligament and lateral to the posterior ligament. From the medial and lateral borders of the articular surface of the patella, reduplications of synovial membrane project into the interior of the joint. These form two fringe-like **alar folds** which also cover collections of fat.

The Bursae

Since all tendons at the knee lie parallel to the bones and pull lengthwise across the joint, bursae are numerous. A number of them have been mentioned in the foregoing account. Besides the **suprapatellar bursa**, three others are associated with the patella and its ligament. The large **subcutaneous prepatellar bursa** is interposed between the patella and the overlying skin, and the **subcutaneous infrapatellar bursa** overlies the patellar ligament. A **deep infrapatellar bursa** occupies the interval between the patellar ligament and the tibia

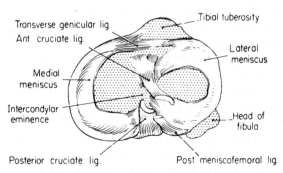

Fig. 466 The superior articular surface of the tibia, the menisci, and the attachments of the cruciate ligaments.

just above the insertion of the ligament. On the lateral side of the joint the **inferior subtendinous bursa of the biceps femoris muscle** (p. 545) lies between the tendon of this muscle and the fibular collateral ligament, and the **subpopliteal recess** of synovial membrane underlies the tendon of the popliteus muscle. Another bursa may separate the popliteus tendon from the fibular collateral ligament, or the membrane of the subpopliteal recess may wrap around the tendon and separate them. Also belonging to the lateral group is the **subtendinous bursa of the lateral head of the gastrocnemius muscle**. It lies beneath the tendon of origin of this muscle and occasionally communicates with the knee joint. Medially, the **bursa anserina** is deep to the pes anserinus tendons (sartorius, gracilis, and semitendinosus) and separates them from the tibial collateral ligament. The **bursa of the semimembranosus muscle** lies between that muscle and the tibia. The **subtendinous bursa of the medial head of the gastrocnemius** underlies the tendon of origin of the medial head and separates the tendon from the femur. It sometimes communicates with the knee joint.

Vessels and Nerves

The **arteries** of the knee joint are the ten vessels which form the genicular anastomosis (p. 602). Its **lymphatics** drain to the popliteal and inguinal node groups. The **nerves** of the knee joint are numerous. From the **femoral nerve,** articular branches reach the joint by way of the **nerves to the vastus muscles** and through the **saphenous nerve.** The posterior division of the **obturator nerve,** after supplying the adductor magnus muscle, ends in the knee joint. Articular branches arise from both the **tibial** and **common peroneal nerves.** Those from the tibial nerve follow the two medial genicular arteries and the middle genicular artery, those from the common peroneal nerve, the two lateral genicular arteries and the anterior tibial recurrent artery (Gardner, '48).

Actions

The knee joint is primarily a hinge joint. In full flexion, the joint is loose enough to allow some rotation of the leg; and in terminal extension, there is normally a small amount of rotation which increases the tension in the ligaments. Biological hinge joints do not operate around a fixed axis like the pin of a door hinge, and the knee joint clearly demonstrates a gliding component in its transition from a flexed to an extended position. The femoral condyles both roll and glide on the relatively flat tibial articular surfaces and progressively approach the relatively greater stability which results from the contact of the flatter anterior femoral surfaces with the superior articular surfaces of the tibia. As the extended position is approached, the smaller lateral meniscus is displaced forward on the tibia and becomes firmly seated in a groove on the lateral femoral condyle. This stops extension, but

the medial condyle of the femur is still capable of gliding backward on the medial tibial plateau so as to bring its flatter, more anterior surface into full contact with the tibia. The two movements constitute a medial rotation of the femur on the tibia (foot on the ground) which brings the cruciate ligaments into a taut, or 'locked,' position. Barnett ('53) describes several cycles of partial locking and shifting of the menisci. The collateral ligaments are attached posterior to the vertical axis of the extended bones, and they, too, become maximally tensed in full extension. The tension of the ligaments and the approximation of the flatter parts of the condyles (close-packed position) make the erect position relatively easy to maintain with only minor attention from the muscles, the quadriceps femoris being relaxed. The sequence of actions in flexion are the reverse of those in extension, the pull of the popliteus muscle beginning the movement by reversing the rotary movements of terminal extension. Flexion may be carried through about 130° and is finally limited by contact between calf and thigh. The collateral ligaments are loose in flexion and rotary displacements can occur in this position.

The patella maintains a shifting contact with the femur in all positions of the knee. It serves as a protection to the front of the joint and improves the angle of traction of the tendon of the quadriceps femoris muscle. As the knee shifts from a fully flexed to a fully extended position, first the superior, then the middle, and lastly the inferior parts of the articular surface of the patella are brought into contact with the patellar surface of the femur. On some patellae these surfaces are relatively plane and are indistinctly marked off from one another.

The menisci may have a function in improving the congruity of the joint surfaces. It is also likely that they function in producing efficient utilization of the synovial fluid as a joint lubricant (p. 36).

The **muscles** acting at the knee joint are principally flexors and extensors of the leg. The extensors are the quadriceps femoris, gluteus maximus, and tensor fasciae latae muscles. The flexors are the semimembranosus, semitendinosus, biceps femoris, popliteus, gracilis, sartorius, and gastrocnemius muscles. Medial rotation of the leg is the function of popliteus assisted by the pes anserinus muscles and by the semimembranosus. Lateral rotation is produced by the biceps femoris and tensor fasciae latae muscles. The rotary movements of the leg are slight.

Articulations between the Tibia and the Fibula (figs. 448, 464, 465, 467–69)

The union of the tibia and the fibula is in three parts: the tibiofibular articulation at their proximal ends, a connection between their shafts by means of an interosseous membrane, and the distal tibiofibular syndesmosis.

The Tibiofibular Articulation (figs. 464, 465)

This is a plane joint between the circular facet on the head of the fibula and a similarly shaped surface on the posterolateral aspect of the underside of the lateral condyle of the tibia. The **articular capsule** is attached at the margins of the facets on the tibia and the fibula and is strengthened by accessory ligaments anteriorly and posteriorly.

The **anterior ligament of the head of the fibula** consists of fibrous bands which pass obliquely from the front of the head of the fibula to the front of the lateral condyle of the tibia. The **posterior ligament of the head of the fibula** is a single broad band running obliquely between the head and back of the lateral tibial condyle. It is crossed by the tendon of the popliteus muscle; the subpopliteal recess of the synovial cavity of the knee joint occasionally communicates here with the cavity of the tibiofibular articulation. The joint receives its **arterial supply** from the lateral inferior genicular and anterior tibial recurrent arteries. Its **lymphatics** drain into the popliteal nodes. **Nerves** to the articulation are derived from the common peroneal nerve, the nerve to the popliteus muscle, and the anterior tibial recurrent nerve. Movement is slight at this joint, but the articulation imparts a certain flexibility in the relationship between the tibia and the fibula during ankle movement and in response to the action of the muscles attached to the fibula.

The Interosseus Membrane

The interosseous membrane of the leg (fig. 448) extends between the interosseous borders of the tibia and the fibula and consists largely of fibers which pass from the tibia lateralward and downward toward the fibula. The upper margin of the membrane does not reach the tibiofibular articulation, and the anterior tibial vessels pass over the upper edge of the membrane to the anterior compartment of the leg. The membrane is continuous below with the interosseous ligament of the tibiofibular syndesmosis. It is perforated by a number of vessels, including the perforating branch of the peroneal artery. The interosseous membrane separates the muscles of the anterior and posterior compartments of the leg and gives origin to muscles of both groups.

The Tibiofibular Syndesmosis (figs. 467–69)

This is a fibrous joint in which the rough convex surface of the medial aspect of the lower end of the fibula is attached by the interosseous ligament to the corresponding rough concave surface on the lateral side of the tibia. The **interosseous ligament,** continuous with the interosseous membrane above, consists of short, strong fibrous bands which form the principal connection of the two bones. It is supplemented by anterior, posterior, and transverse tibiofibular ligaments.

The **anterior tibiofibular ligament** and the **posterior tibiofibular ligament** pass from the borders of the fibular notch of the tibia to the anterior and posterior surfaces of the lateral malleolus of the fibula respectively. Each is inclined downward and lateralward. The **transverse tibiofibular ligament** is largely deep to the posterior tibiofibular ligament. It arises from nearly the entire inferior border of the posterior surface of the tibia and attaches on the upper end of the malleolar fossa of the fibula. This strong thick ligament projects below the margin of the bones and forms part of the articulating surface for the talus. A recess of the articular cavity of the ankle joint, lined by a synovial membrane, usually extends upward between the tibia and the fibula as far as the distal end of the interosseous ligament (1 cm. or less). The **blood supply** of the tibiofibular syndesmosis is from the peroneal artery and its perforating branch, assisted sometimes by the anterior tibial artery or its lateral anterior malleolar branch. The joint is supplied by twigs from the **deep peroneal** and **tibial nerves.** This articulation forms the firm union of tibia and the fibula required in the box-like mortise of the ankle. Its fibrous tissue allows a slight yielding of the bones for the accommodation of the talus in the movements of the ankle.

The Ankle Joint (Talocrural articulation) (figs. 467–69)

The **ankle** is a synovial joint of the hinge (ginglymus) type. Its form is that of a mortise and tenon, the box-like mortise being constructed by the ends of the bones of the leg. The sides of the mortise are formed by the cartilage-covered area of the distal surface of the tibia and the lateral aspect of its medial malleolus and by the triangular facet on the medial side of the lateral malleolus of the fibula. Posteriorly, the transverse tibiofibular ligament completes the socket between the tibia and the fibula. The tenon of the joint is the **trochlea** of the body of the talus. It is smoothly convex from before backward and slightly concave from side to side. Its medial margin is straight anteroposteriorly, but

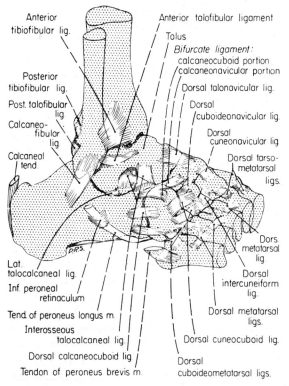

Fig. 467 The lateral ligaments of the ankle and of the tibiofibular syndesmosis; the dorsal ligaments of the tarsal joints.

and the anterior and posterior parts of the lateral ligament are blended with the articular capsule. The middle lateral band is free.

The **deltoid ligament** is a strong, triangular ligament which is attached by its apex to the anterior and posterior borders and the tip of the medial malleolus. The ligament broadens below to form a continuous attachment to the bones of the foot, its four parts being designated according to their separate distal attachments. The most anterior fibers compose the **anterior tibiotalar ligament** to the medial part of the neck of the talus. They are partly overlain by the superficial **tibionavicular ligament** to the upper and medial part of the navicular bone. Continued backward, this ligament blends below with the medial margin of the plantar calcaneonavicular ligament and is succeeded by the tibiocalcaneal ligament. The fibers of the **tibiocalcaneal ligament** descend almost vertically to attach to the whole length of the sustentaculum tali of the calcaneus. The posterior and thickest part of the deltoid is the **posterior tibiotalar ligament**, the fibers of which run lateralward and backward to the medial side of the talus and to the medial tubercle of its posterior process.

The **lateral ligaments** are the anterior talofibular, the calcaneofibular, and the posterior talofibular. These are separated ligaments and do not constitute so strong a ligamentous connection as does the deltoid ligament on the medial side. This is evident in the fact that an ankle sprain is almost always due to a turning of the ankle outward and a consequent straining of the lateral ligaments (inversion stress). The **anterior talofibular ligament**

its lateral margin is oblique, so that the trochlea is broader in front than behind. A small articular surface on the anteromedial aspect of the trochlea articulates with the medial malleolus. The lateral articular surface comprises the whole side of the trochlea. Triangular in shape, it articulates with the lateral malleolus of the fibula.

The Articular Capsule

The capsule, in correspondence with the requirements of free movement in flexion and extension at the ankle, is weak anteriorly and posteriorly. The stability of the joint, however, is assured by very strong collateral ligaments. The thin anterior and posterior parts of the capsule are attached above to the margins of the tibia and the fibula and, below, to the talus both in front of and behind the superior surface of its trochlea. The collateral ligaments of the joint are the exceedingly strong deltoid ligament on the medial side of the ankle and the separated ligaments on the lateral side. Both sets of ligaments radiate downward from the malleoli and both have a middle band to the calcaneus and anterior and posterior bands to the talus. The deltoid ligament

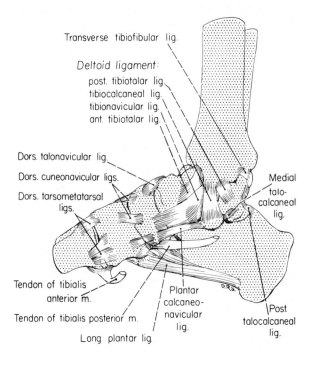

Fig. 468 The medial ligaments of the ankle and of the tarsal joints.

stretches from the anterior border and tip of the lateral malleolus to the neck of the talus. The **calcaneofibular ligament** is a narrow, rounded cord that descends, with a slight backward inclination, from the tip of the malleolus to a tubercle at the middle of the lateral surface of the calcaneus. The almost horizontal **posterior talofibular ligament** is strong and thick. It arises in the malleolar fossa of the lateral malleolus and passes medially and backward to the upper surface of the posterior process of the talus.

The Synovial Membrane

The synovial membrane of the ankle joint is loose and capacious, and the synovial cavity extends upward between the apposed surfaces of the ends of the tibia and the fibula as far as the interosseous ligament of the tibiofibular syndesmosis. The ankle joint receives its **blood supply** from the four malleolar branches of the anterior tibial, peroneal, and posterior tibial arteries. Its **nerve supply** is provided by twigs from the tibial nerve and the lateral branch of the deep peroneal nerve.

Actions

The ankle joint permits dorsiflexion and plantar flexion of the foot through a range of 90°. In dorsiflexion, the broader anterior portion of the trochlea of the talus occupies and completely fills the mortise of the joint, and stability of the foot at the ankle is greatest in this close-packed position. Conversely, in full plantar flexion the narrower part of the trochlea occupies the joint space, and a little rotation and abduction-adduction are then permitted. This is the position of least stability at the ankle. Adjustments at the tibiofibular syndesmosis contribute to stability in extreme dorsiflexion, for then the malleoli are spread apart and the lateral malleolus is everted. The joint space is reduced in positions in which the talus is less forcibly wedged between the bones.

All the muscles which enter the foot behind the malleoli, those of the lateral and both posterior compartments of the leg, produce plantar flexion at the ankle. Dorsiflexion of the foot follows the contraction of the muscles of the anterior compartment of the leg. In the erect position the line of gravity of the body falls in front of the transverse axis of the ankle joint, and as a result, the body tends to fall forward at this joint. In a normally strong foot, ligaments take most of the strain of standing but some activity in soleus (and perhaps gastrocnemius) may be needed for easy standing and will certainly come into play with activity. The tendons of the muscles of the deep posterior and lateral compartments of the leg turn sharply at the ankle to enter the foot, and part of their traction is such as to draw the foot backward under the malleoli. This posteriorly directed displacing force on the foot is resisted by the collateral ligaments at the ankle, the strongest of them running posteriorly between the malleoli and the tarsal bones.

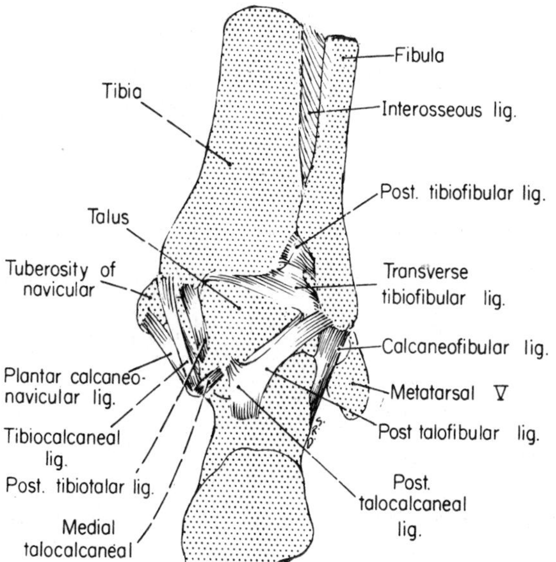

THE BONES OF THE FOOT (figs. 470–72)

The bones of the foot have fundamental similarities to the bones of the wrist and the hand. They are the seven tarsal bones, the five metatarsal bones, and the fourteen phalanges. The tarsal bones are the talus, the calcaneus, the navicular, the medial cuneiform, the intermediate cuneiform, the lateral cuneiform, and the cuboid. The weight of the body is transmitted into the foot through the talus, and this bone constitutes the summit of the longitudinal arch of the foot. The arrangement of the bones of the foot provides for a limited independence between the bones forming the medial three digits and those forming the lateral two digits. The bones of the medial digits are the talus, the navicular, the three cuneiform bones, and the metatarsal bones and phalanges of the first three toes. The bones of the lateral digits are the calcaneus, the cuboid, and the metatarsal bones and phalanges of the lateral two toes.

Fig. 469   The ligaments of the ankle and of the tibiofibular syndesmosis from behind.

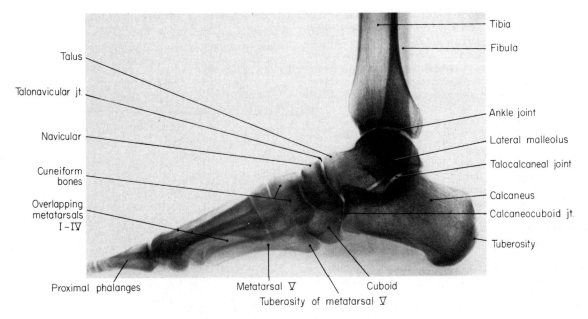

Fig. 470 A lateral radiogram of the foot.

The Talus

The talus articulates with many bones: the tibia and the fibula above and at the sides, the calcaneus below, and the navicular in front. It does not provide attachment for muscles but does receive ligaments. It is divisible for description into a head, a neck, and a body.

The **head** of the talus is the rounded anterior end of the bone. It is directed forward and medialward and has three articular surfaces. The large **navicular articular surface** is convex and oval. As this surface is followed to the underside of the head, it is succeeded by a flat, triangular **anterior calcaneal articular surface** through which the head bears on the anterior facet of the calcaneus and on the plantar calcaneonavicular ligament. The third facet, most posteriorly located, is the oval **middle calcaneal articular surface** which makes contact with the upper surface of the sustentaculum tali of the calcaneus. The **neck** of the talus is the somewhat constricted part between the head and the body. On the upper side of the talus, the neck is rough for ligaments and shows a number of vascular foramina. Inferiorly, the bone is deeply grooved as the **sulcus tali.** With the talus and the calcaneus articulated, the sulcus tali is the roof of the tarsal sinus which is filled in life by the interosseous talocalcaneal ligament.

The **body** of the talus is roughly quadrilateral. Its upper portion, the **trochlea,** enters into the formation of the ankle joint. The superior surface of the trochlea is smoothly convex from before backward and slightly concave from side to side. It articulates with the distal end of the tibia. The trochlea is wide anteriorly and narrow behind, and its lateral surface is oblique to the anteropos-

terior axis of the talus. The **lateral malleolar surface** of the trochlea is almost entirely articular for contact with the lateral malleolus. This surface is triangular in shape; it is broad above and narrows below to become the **lateral process** of the talus. The **medial malleolar surface** of the trochlea has the form of a broad crescent and extends over only one-third of the side of the talus. The region inferior to it is rough for the deltoid ligament and is perforated by vascular foramina. Prominent on the underside of the body is an oblong articular facet, deeply concave from side to side. This is the **posterior calcaneal articular surface.** The **posterior process** of the talus is grooved by the tendon of the flexor hallucis longus muscle. Medial to the groove is the **medial tubercle** for the medial talocalcaneal and posterior tibiotalar ligaments, and on the lateral side of the groove is a **lateral tubercle** to which attaches the **posterior talofibular ligament** of the ankle joint.

The Calcaneus

This is the largest and strongest bone of the foot. It is long, flattened from side to side, and bulbous posteriorly and inferiorly where it forms the heel. The calcaneus is shaped somewhat like a pistol grip, the thumb gliding naturally into the hollow under the sustentaculum tali.

Held thus, the **superior surface** exhibits anteriorly three articular facets for the talus. The largest is the **posterior talar articular surface,** triangular in form and convex from behind forward. Anterior to it is a deep depression which leads onto the sustentaculum as the **calcaneal sulcus.** This underlies the sulcus of the talus, and the tarsal sinus formed by the superimposed sulci lodges the interosseous talocalcaneal ligament. The upper surface of the **sustentaculum tali** carries the **middle talar articular surface.** At the anterior extremity of the calcaneus, the superior surface has a small, oval **anterior talar articular surface.** Toward the posterior end of the superior surface, the calcaneus is rough and is in relation with the fat in front of the calcaneal tendon. The **inferior surface** of the calcaneus is narrow and uneven. It provides attachment for the long plantar ligament and ends posteriorly in the tuberosity of the calcaneus. The **tuberosity** of the calcaneus is the postero-inferior part of the bone. It is rough and striated for the attachment of the calcaneal tendon. On its inferior aspect it has two processes—a larger **medial process** and a narrow **lateral process.**

The **medial surface** of the calcaneus is smoothly concave and is overhung by the prominent **sustentaculum tali.** The sustentaculum tali has a **sulcus** on its underside for the tendon of the flexor hallucis longus muscle. The **lateral surface** of the calcaneus is rough. At about its middle, it has a small swelling for attachment of the calcaneofibular ligament and, inferior to this, a **peroneal trochlea** which separates the tendons of the peroneus brevis and the longus muscles. The anterior extremity of the calcaneus is the **cuboidal articular surface.** Roughly triangular in form, it carries a saddle-shaped articulation for the cuboid bone.

The Navicular

The navicular is a flattened, oval bone located between the head of the talus behind and the three cuneiform bones in front.

It has a large, oval, concave articular facet on its posterior surface for the head of the talus and a rounded eminence at its medial extremity which projects plantarward and gives attachment to the tendon of the tibialis posterior muscle. This is the **tuberosity** of the navicular. The anterior surface of the bone is subdivided by two vertical ridges into three slightly convex, triangular facets for articulation with the three cuneiform bones.

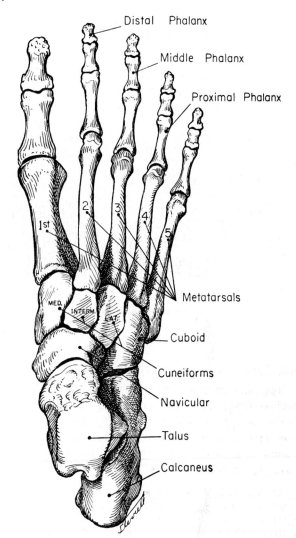

Fig. 471   The dorsal aspect of the bones of the foot. ($\times \frac{1}{2}$)

The Cuneiform Bones

The three cuneiform bones are all wedge-shaped, but the broad side of the wedge faces plantarward on the medial cuneiform and dorsalward on the other two. The posteriorly directed end of each bone is concave and articulates with one of the facets of the navicular bone. Each bone has a distal articular extremity which enters into the formation of the tarsometatarsal joint of the first, second, or third digit. There are articular surfaces between the adjacent cuneiform bones and one between the lateral cuneiform and the cuboid. The dorsal and plantar surfaces are rough for the attachment of ligaments and tendons. The **medial cuneiform**

bone is the largest of the three, and the **intermediate cuneiform** is shorter than the others.

The Cuboid Bone

The cuboid bone lies on the lateral side of the foot between the calcaneus behind and the fourth and fifth metatarsal bones in front.

Its plantar surface has a prominent ridge which receives the long plantar ligament and which ends laterally in the **tuberosity.** The tuberosity exhibits a convex cartilage-covered facet over which the tendon of the peroneus longus plays as it enters the foot. This tendon almost never lies in the **groove** anterior to the tuberosity. Laterally the cuboid is short and concave for the passage of the peroneus longus tendon, but the medial side of the bone is extensive. The posterior surface of the cuboid bone is entirely articular; it is saddle-shaped for participation in the calcaneocuboid joint. The distal articular surface has two slightly concave facets for the bases of the fourth and fifth metatarsal bones respectively.

The Metatarsal Bones

The metatarsal bones, like the metacarpal bones of the hand, are long bones and each consists of a base, a body, and a head. They are from six to eight centimeters in length. The metatarsal bones are relatively flat dorsally but are concave longitudinally on their plantar sides. The **bases** are smooth on their ends for articulation with the three cuneiform bones and the cuboid bone and carry articular facets and pits for ligaments on their sides. The **bodies** are narrow and tend to be triangular in cross section. The **heads** of the metatarsal bones present convex articular surfaces, flattened from side to side, for contact with the proximal phalanges. The flattened sides of the heads are surmounted by tubercles dorsally for the attachment of the collateral ligaments of the metatarsophalangeal joints.

The **first metatarsal bone** is the shortest, broadest, and most massive of the series. The base projects downward and lateralward to form the tuberosity on which the tendon of the peroneus longus inserts. The head of the first metatarsal is broad, and its plantar surface is marked by two deep grooves separated by a ridge. In these grooves the sesamoid bones of the tendons of the flexor hallucis brevis muscle play. The **second metatarsal bone** is the longest, and its base fits into a recess formed by the three cuneiform bones. The **third, fourth, and fifth metatarsal bones** conform to the general type. The base of the fifth bone is expanded lateralward into a rough tuberosity which gives insertion to the peroneus brevis tendon.

The Phalanges

The phalanges of the toes, like the fingers, are fourteen in number—three for each digit except the great toe which has two. They are designated as proximal, middle, and distal.

Except for that of the great toe which is broad and thick, the **proximal phalanges** are expanded at their extremities and narrow throughout their bodies. The **bases** have single round or oval, cup-like facets for reception of the heads of the corresponding metatarsal bones. The **heads** of the proximal phalanges present pulley-like surfaces, grooved in the middle and raised at the edges, for articulation with the bases of the middle phalanges. The **middle phalanges** are short, but their bodies are proportionately broader than those of the proximal phalanges. Both ends of the middle phalanges have trochlear surfaces. The **distal phalanges** are also short. They have broadened bases with trochlear surfaces and rough, broadened **distal phalangeal tuberosities** for the support of the nails and the pulp of the toes.

Ossification of the Bones of the Foot

The **tarsal bones** are ossified from a single center for each bone, excepting the calcaneus which has a separate epiphysis for its tuberosity. The center of ossification for the calcaneus appears at the sixth fetal month; that for the talus, during the seventh fetal month; that for the cuboid, at the time of birth; that for the lateral cuneiform, during the first year; and that for the medial cuneiform, during the third year. The ossification centers for the intermediate cuneiform and navicular bones appear in the fourth year. The epiphysis for the tuberosity of the calcaneus appears about the eighth to the tenth year and unites with the rest of the bone at puberty. Ossification of all the tarsal bones is complete shortly after puberty.

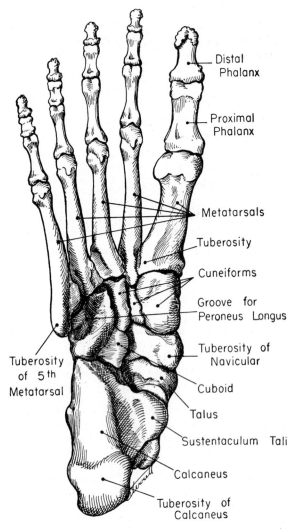

Fig. 472   The plantar aspect of the bones of the foot. ($\times \frac{1}{2}$)

In each of the **metatarsal bones,** a primary center of ossification for the body and the base (except the first where it is in the head rather than the base) appears about the ninth week of fetal life, and these bones are well ossified by birth. A secondary center of ossification for each of the heads of the four lateral bones (and one for the base of the first metatarsal) appears in the third year and fuses to the shaft from the fourteenth to the seventeenth year.

The **phalanges** are each ossified from two centers, one for the body and head and one for the base. Those for the bodies and heads appear from between the tenth fetal week to the time of birth—in the distal phalanges first and in the middle phalanges last. The secondary ossification centers for the bases appear during the third year and unite with the shafts from the fourteenth to the seventeenth year.

THE INTERTARSAL JOINTS (figs. 467–73)

The **intertarsal joints** are the subtalar, the talocalcaneonavicular, the calcaneocuboid, the transverse tarsal, the cuneonavicular, the intercuneiform, and the cuneocuboid joints. These are all characterized by interosseous, dorsal, and plantar ligaments. In order to resist the stress of the body weight, the plantar ligaments are much stronger than the dorsal. The bones and ligaments receive blood vessels from adjacent branches of the dorsalis pedis, medial plantar, and lateral plantar arteries and nerves from the deep peroneal, medial plantar, and lateral plantar nerves.

The Subtalar Joint

This joint is formed between the large concave facet on the under surface of the body of the talus and the convex posterior articular surface on the superior aspect of the calcaneus. A loose, thin-walled articular capsule unites the two bones by attaching to the margins of the articular surfaces. Slightly stronger portions are designated as the posterior, medial, and lateral talocalcaneal ligaments. The **medial talocalcaneal ligament** connects the medial tubercle of the posterior process of the talus with the posterior portion of the sustentaculum tali; the **lateral talocalcaneal ligament** is parallel to and deeper than the calcaneofibular ligament. The **posterior talocalcaneal ligament** is a short band, the fibers of which radiate from a narrow attachment on the lateral tubercle of the talus to the upper and medial part of the calcaneus. The **interosseous talocalcaneal ligament,** located in the tarsal sinus, is blended with the anterior part of the articular capsule of the subtalar joint and with the posterior part of the capsule of the talocalcaneonavicular joint. It is a strong band, composed of several laminae of fibers with fatty tissue between, which connects the adjacent surfaces of the talus and the calcaneus along the oblique tarsal grooves. Support for the subtalar

articulation is also provided by certain ligaments of the ankle joint which, passing from the tibia and the fibula to the calcaneus, span the talus.

The Talocalcaneonavicular Joint

This joint is formed by the articular surfaces of the head of the talus which are in contact with the navicular bone, the plantar calcaneonavicular ligament, the sustentaculum tali, and the adjacent part of the anterior articular surface of the calcaneus. The articular capsule is thin and encloses the common articular cavity. It blends with the interosseous talocalcaneal ligament posteriorly. The capsule is reinforced on its dorsum between the neck of the talus and the dorsal surface of the navicular bone by the broad **dorsal talonavicular ligament.** The **plantar calcaneonavicular ligament** is a thick dense, fibroelastic ligament. This ligament extends from the sustentaculum tali and the distal surface of the calcaneus to the entire width of the inferior surface of the navicular and to its medial surface behind the tuberosity. Medially, it is blended with the deltoid ligament of the ankle and, laterally, with the lower border of the calcaneonavicular part of the bifurcate ligament. Its upper surface is smooth and faceted and contains a fibrocartilaginous plate on which the head of the talus bears. Because of its elasticity under the pressure of the head of the talus it is also called the 'spring ligament.' The **calcaneonavicular part of the bifurcate ligament** completes the socket on the lateral side. Its short fibers pass from the upper surface of the anterior end of the calcaneus to the adjacent lateral surface of the navicular bone. The synovial cavity of the joint does not communicate with adjacent articulations.

The Calcaneocuboid Joint

The calcaneocuboid joint is the articulation between the saddle-shaped facets of the anterior surface of the calcaneus and of the posterior surface of the cuboid bone. The joint cavity is separate from adjacent ones, and an articular capsule completely surrounds it. A thin, broad band in the dorsal part of the capsule is the **dorsal calcaneocuboid ligament.** The **bifurcate ligament** is a strong band which is attached to the deep hollow on the upper surface of the calcaneus lateral to the anterior articular surface. It divides into two parts. The **calcaneocuboid** part ends on the dorsomedial angle of the cuboid and is one of the main connections between the first and the second row of tarsal bones. The **calcaneonavicular** part is attached to the lateral side of the navicular; it is concerned in the talocalcaneonavicular articulation.

The calcaneocuboid joint receives the weight of the body as it is transmitted to the lateral portion of the longitudinal arch of the foot and therefore requires strong plantar ligaments. These are provided in the long plantar and plantar calcaneocuboid ligaments. The **long plantar ligament** stretches from the plantar surface of the calcaneus in front of its tuberosity to the tuberosity of the cuboid bone, its more superficial fibers spreading forward to the bases of the third, fourth, and fifth metatarsal bones. Partially concealed by the long plantar ligament, the **plantar calcaneocuboid ligament** is attached to the rounded eminence at the anterior end of the inferior surface of the calcaneus and, forward of this, to the plantar surface of the cuboid bone behind the tuberosity and the oblique ridge. The fibers of this wide band are short and strong. These two ligaments and the plantar calcaneonavicular ligament underlie the summit of the longitudinal arch of the foot and are of prime importance in the support of that arch.

The Transverse Tarsal Joint

This is a designation for the irregular articular plane which extends from side to side across the foot and is composed of the talonavicular articulation medially and the calcaneocuboid joint laterally. These joints are separate but combine in a distinctive movement pattern which is an important contribution to the action of the foot. At this plane primarily and at the other tarsal joints to a lesser degree are produced the movements of inversion and eversion of the foot. With inversion is combined adduction and plantar flexion; with eversion, abduction and dorsiflexion. The contribution of the subtalar and talocalcaneonavicular joints is that of movement around an axis that passes through the tarsal sinus. These movements allow the foot to be placed firmly on slanting and irregular surfaces and still to serve as a firm basis of support for the body. Many ligaments are concerned in maintaining the transverse tarsal joint; these ligaments are described under its parts. Additional ligaments are concerned in uniting the navicular and cuboid bones. The **dorsal cuboideonavicular**

and **plantar cuboideonavicular ligaments** unite the adjacent surfaces of these two bones, and a strong **interosseous cuboideonavicular ligament** connects the rough nonarticular portions of their adjacent surfaces. A small joint cavity frequently exists between the posterior medial angle of the cuboid bone and the lateral margin of the navicular bone. This **cuboideonavicular articulation** is continuous with the cuneonavicular joint in front of it.

The Distal Intertarsal Joints

The distal intertarsal joints are the **cuneonavicular,** the **intercuneiform,** and the **cuneocuboid articulations.** These have a common articular cavity enclosed by a common articular capsule. The synovial cavity of these joints extends forward between the cuneiform bones to include the tarsometatarsal joints between the intermediate cuneiform and the second and third metatarsal bones, and the intermetatarsal joints between the second and third metatarsal bones and between the third and fourth metatarsal bones. The ligaments of the distal intertarsal joints are weak dorsal bands and stronger plantar and interosseous ligaments. Adjacent bones are united dorsally by the **dorsal cuneonavicular ligaments** for each of the cuneiform bones, the **dorsal intercuneiform ligaments,** and the **dorsal cuneocuboid ligament.** Plantar ligaments are the **plantar cuneonavicular ligaments,** the **plantar intercuneiform ligaments,** and the **plantar cuneocuboid ligament.** The interosseous ligaments are the **intercuneiform** and the **cuneocuboid.** The movement permitted between these joints is a slight gliding action which contributes to the flexibility and adaptability of the foot.

THE TARSOMETATARSAL JOINTS (figs. 467, 473)

The **tarsometatarsal joints** are plane joints between the adjacent surfaces of the distal row of tarsal bones and the bases of the metatarsal bones. The first metatarsal bone articulates with the medial cuneiform only. The base of the second metatarsal is contained in a mortise formed by the three cuneiform bones and articulates with all of them. The third metatarsal makes contact with the lateral cuneiform bone. The fourth and fifth metatarsal bones articulate with the slanted distal surface of the cuboid bone, and the fourth metatarsal also has a small area of contact with the lateral cuneiform bone.

There are three tarsometatarsal joint cavities. The medial one is a separate cavity for the joint between the first metatarsal and the medial cuneiform bone. The intermediate joint includes the articulations of the second and third metatarsal bones and is an extension of the distal intertarsal joint space. The lateral articular cavity includes the articulations of the fourth and fifth

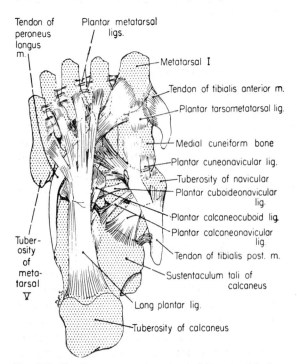

Tendon of peroneus longus m.

Plantar metatarsal ligs.

Metatarsal I

Tendon of tibialis anterior m.

Plantar tarsometatarsal lig.

Medial cuneiform bone

Plantar cuneonavicular lig.

Tuberosity of navicular

Plantar cuboideonavicular lig.

Plantar calcaneocuboid lig.

Plantar calcaneonavicular lig.

Tendon of tibialis post. m.

Sustentaculum tali of calcaneus

Long plantar lig.

Tuberosity of calcaneus

Tuberosity of metatarsal V

Fig. 473   The ligaments and tendons of the sole of the foot.

metatarsal bones with the cuboid bone. Weak **dorsal tarsometatarsal ligaments** pass between adjacent surfaces of the bones concerned, each metatarsal bone receiving a slip from each tarsal bone with which it articulates. **Plantar tarsometatarsal ligaments** are less regularly arranged and consist of both longitudinal and oblique bands. Two or three **interosseous cuneometatarsal ligaments** are present. The first ligament passes from the lateral surface of the medial cuneiform to the adjacent angle of the second metatarsal bone, and the second ligament, small and inconstant, connects the lateral aspect of the second metatarsal bone with the lateral cuneiform bone. The third interosseous ligament connects the lateral angle of the lateral cuneiform to the adjacent side of the base of the third metatarsal bone.

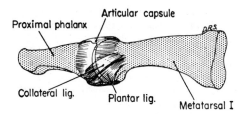

Fig. 474   The capsule and ligaments of the metatarsophalangeal joint.

THE INTERMETATARSAL JOINTS (figs. 467, 473)

The **intermetatarsal joints** unite the bases of the metatarsal bones. The first is joined to the second metatarsal bone by interosseous fibers only but articular cavities exist between the four lateral metatarsal bones. The joint cavity between the bases of the second and third metatarsal bones and the third and fourth metatarsal bones is a forward extension of the common distal intertarsal and intermediate tarsometatarsal articular cavity. That between the fourth and fifth metatarsal bones is continuous from the joint between their bases and the cuboid bone.

These joint spaces are closed by **dorsal metatarsal** and **plantar metatarsal ligaments** and are limited distalward by **interosseous metatarsal ligaments**. The interosseous ligaments are strong and help to maintain the transverse arch of the foot. Except at the first joint, the intermetatarsal joints provide only a slight, gliding movement which contributes to the flexibility of the foot. The first joint permits slight rotary movements of the great toe, and a small amount of flexion accompanies the gliding movement of the fifth metatarsal bone and, to a lesser degree, the fourth metatarsal bone.

THE METATARSOPHALANGEAL JOINTS (fig. 474)

The **metatarsophalangeal joints** are condyloid joints between the rounded heads of the metatarsal bones and the cupped posterior extremities of the proximal phalanges. They are very similar to the metacarpophalangeal joints of the fingers. Each joint is enclosed by an articular capsule, reinforced by the plantar ligament and by collateral ligaments.

The **articular capsule** is loose and is attached closer to the articular borders dorsally than plantarward. It is reinforced by

fibers from the extensor tendons. The **plantar ligament,** like its palmar counterpart, is a dense, fibrocartilaginous plate that is attached firmly to the proximal plantar border of each phalanx and serves as part of the bearing surface for the head of each metatarsal bone. At the sides, it is attached to the collateral and deep transverse metatarsal ligaments. The strong **collateral ligaments** of each joint pass from the tubercles on each side of the head of the metatarsal bone to the sides of the proximal end of the phalanx and to the sides of the plantar ligament. For the great toe the sesamoid bones and their interconnecting ligamentous band replace the plantar ligament. All the plantar ligaments are interconnected by the **deep transverse metatarsal ligament** which connects the heads and the joint capsules of all the metatarsal bones. Its plantar surface is grooved by the flexor tendons. The lumbrical tendons also cross its plantar aspect, whereas the interosseous tendons cross it dorsally. The movements permitted in the metatarsophalangeal joints are dorsiflexion, plantar flexion, abduction, adduction, and circumduction.

THE INTERPHALANGEAL JOINTS

The **interphalangeal joints** are similar to the metatarsophalangeal articulations except that their trochlear surfaces permit only dorsiflexion and plantar flexion. Each joint is provided with an articular capsule and with plantar and collateral ligaments. The blood vessels and nerves of the metatarsophalangeal and interphalangeal joints are branches of the digital vessels and nerves.

THE ARCHES OF THE FOOT

The foot is strong to support the weight of the body, but it is also flexible and resilient to absorb the shocks transmitted to it and to provide spring and lift for the activities of the body. It has an arched structure composed of a number of bones and their interconnecting joints. The joints and ligaments, together with muscle action, provide for spring,

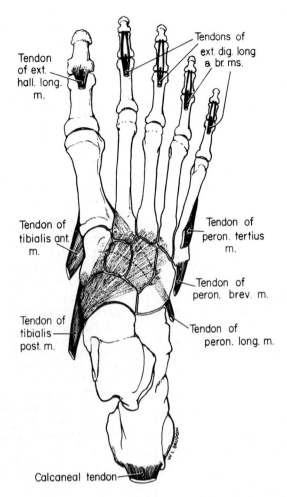

Tendon of ext. hall. long. m.

Tendons of ext. dig. long & br. ms.

Tendon of tibialis ant. m.

Tendon of peron. tertius m.

Tendon of peron. brev. m.

Tendon of tibialis post. m.

Tendon of peron. long. m.

Calcaneal tendon

Fig. 475 The insertion in the foot of the tendons of certain of the leg muscles.

for they yield when weight is applied and recoil when the weight is removed.

The bones of the foot are arranged in longitudinal and transverse arches. The **longitudinal arch** is supported posteriorly on the tuberosity of the calcaneus and, anteriorly, on the heads of the metatarsal bones. The **transverse arch** results from the shape of the tarsal bones of the distal row and of the bases of the metatarsal bones. Generally broader dorsally, these bones articulate in a domed curve, forming the transverse arch of the foot. The talus is at the summit of the longitudinal arch. In the fundamental construction of the foot, the talus is primarily related to the navicular, the three cuneiform bones, and the medial three metatarsal bones. Likewise, the calcaneus is more directly related

to the cuboid and the lateral two metatarsal bones. These differences appear in the function of the foot as well, for the medial segment of the longitudinal arch is characterized by its greater curvature and by its remarkable elasticity. The flatter, more rigid, lateral segment of the arch makes contact with the ground and provides a firm basis for support in the upright position. In walking, the heel touches the ground first, and then contact is made by the lateral margin of the foot. Finally, the heel rising, the weight of the body is borne on the heads of the metatarsal bones and the toes. A recent alternative suggestion (MacConaill) would relate the skeletal and joint structure of the foot to a twisted plate spanning from metatarsal heads to the calcaneus with the twisted feature exemplified by the high medial arch (instep) and the flattened border laterally.

The arches of the foot are completed and maintained by strong ligaments and tendons. They are also sustained by both the intrinsic muscles of the foot and the long muscles of the leg. In standing, the weight of the body is distributed equally between the heel and the ball of the foot and is shared equally by the two feet. However, in walking, the raising of the body weight on the ball of each foot alternately increases the stress applied at that point by four (the entire body weight). The gastrocnemius and soleus muscles provide the muscular power required to raise the weight of the body on the toes. The ligaments are primary in static standing (Manter, '46; Basmajian & Bentzon, '54), but the muscular support of the foot becomes significant both in the minor movements of shifting the weight in the erect posture and in the more extensive bodily actions. The plantar ligaments of the joints of the foot are the strongest, and they are supplemented by robust interosseous ligaments which keep the bones from spreading apart. Notable on the sole of the foot are the long plantar ligament and the plantar calcaneocuboid and plantar calcaneonavicular ligaments. The elasticity of the latter and its support of the head of the talus have led to its being called the 'spring ligament'. The plantar aponeurosis may be likened to a 'tie rod' for the longitudinal arch, firmly connecting its ends and preventing their spread.

The long muscles of the leg are of only limited significance in supporting the arches of the foot and are inactive in relaxed standing. The insertion of the tibialis posterior muscle on the tuberosity of the navicular and the spread of its fibers to all the other tarsal bones except the talus and to the middle three metatarsal bones combines with the transverse course and medial insertion of the peroneus longus to form a tendinous 'sling,' the parts of which interlace under the longitudinal and transverse arches (fig. 475). The tibialis anterior, inserting on the medial margin, and the peroneus brevis, inserting on the lateral margin of the foot, draw upward on the bones to which they are attached. The flexor hallucis longus and flexor digitorum longus muscles also contribute to support of the arches during movement. Probably more significant in foot function, however, are the eversion and inversion actions of these muscles. The stresses of standing are born at the ball of the foot between the head of the first metatarsal bone and those of the second to the fifth metatarsal bones in the ratio of 1:2. Function is interfered with if there is significant alteration of this ratio by muscle imbalance (Jones, '41).

The toes add to the 'grasp' of the foot on the ground and the great toe is of special importance. The foot is raised against the contact of the great toe with the ground, and its bones and muscles constitute a possible 'push-off' mechanism. The unique contribution of the arches, ligaments, and muscles of the foot to an elastic, springy tread is clearest when such a tread is absent—when the lameness of an artificial limb is substituted for it, or in the shuffling, hobbling gait which may result from flat feet or deformed toes.

Clinical Note Deformities of the foot commonly result from muscle imbalance. Flatfoot (pes planus) shows depression or collapse of the medial longitudinal arch with eversion and abduction of the forefoot. The talus is forced downward and ligaments are severely stretched. Clubfoot is a term used for a congenitally deformed foot, most commonly talipes equinovarus. This defect is characterized by plantar flexion, inversion, and adduction of the foot.

APPLIED ANATOMY OF THE LOWER LIMB

The student is reminded that the general principles involved in dislocations and fractures have been covered under the comparable section for the Upper Limb (p. 141). Surface Anatomy (p. 523) should also be reviewed.

Dislocations Congenital dislocation of the hip is thought to be due to faulty development of the acetabulum, especially of its upper rim. The femoral head may also be poorly developed. The condition is much more common in girls.

Traumatic dislocation of the hip is not common except as the sequel of an automobile accident. The capsule of the joint is very strong anteriorly but posteriorly it only covers a portion of the neck of the bone, so posterior dislocation is more common, favored also by the usual direction of the dislocating force. The capsule will be ruptured, the head of the femur will be in the posterior iliac fossa and the limb will be shortened, adducted, and medially rotated, the knee on the affected side tending to overlie the normal knee. The position is almost diagnostic and is to be contrasted with the position in fracture of the femoral neck, when the limb will be shortened also but will be laterally rotated (toes pointing outward).

Dislocation of the knee is rare due to the great strength of the ligaments and muscles here but can occur anteriorly, posteriorly, or laterally. Dislocation of the patella is not uncommon, favored by the somewhat lateralward displacing force of the pull of the quadriceps femoris muscle and the changing relation of the patella to the femur.

Fractures Fractures of the femur are relatively common and displacements of the fragments with shortening of the limb usually follow. Of particular concern is fracture of the femoral neck, occurring frequently in older women. Such fractures tend to be intracapsular and thus realigning of the head and neck fragments can be a problem. Furthermore, the blood vessels serving the head and neck enter at the capsular attachments and run under the synovial membrane to foramina of entrance into the bone. These constitute the blood supply of the separated fragments except for the small artery of the ligament of the head of the femur, and their rupture can result in ischemic degeneration of the head fragment. The blood vessels of the head and neck of the femur are branches of the medial circumflex femoral and lateral circumflex femoral arteries and of the posterior branch of the obturator artery (see descriptions). Fracture of the upper third of the shaft of the femur exemplifies the effects of muscle pull. The proximal fragment receives the tendon of the iliopsoas muscle in its lesser trochanter and, on and adjacent to the greater trochanter, the glutei and other muscles of the posterior femoral group. This fragment is then strongly flexed, abducted, and laterally rotated while the distal fragment is displaced upward and medially by the adductor and hamstring muscles. Realignment requires some sort of traction apparatus and surgical intervention may be necessary in the adult. A supracondylar fracture of the lower end of the femur is of interest because the distal fragment is drawn backward due to the attachments of the gastrocnemius muscle, and the jagged edges of bone can do great harm to the popliteal artery.

Fractures of the tibia and fibula are common. If the fracture is due to indirect injury, the tibia usually fractures at the junction of its middle and lower thirds, and the fibula at the junction of its middle and upper thirds. In a direct injury, the fracture is usually at the same level in both bones. Because of the subcutaneous position of much of both these bones, compound fracture is common. When only one bone is fractured the integrity of the other provides a splinting function and shortening and overriding is minimal.

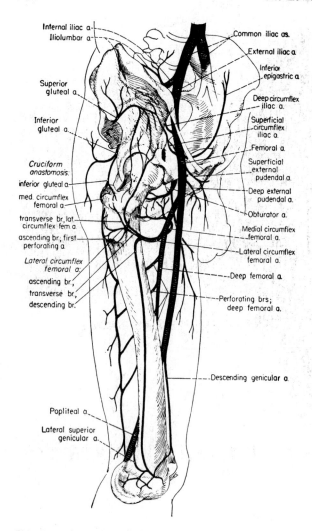

Internal iliac a.
Iliolumbar a.

Common iliac as.

External iliac a.

Inferior epigastric a.

Superior gluteal a.

Deep circumflex iliac a.

Superficial circumflex iliac a.

Inferior gluteal a.

Femoral a.

Superficial external pudendal a.

Cruciform anastomosis:

inferior gluteal a.

Deep external pudendal a.

med. circumflex femoral a.

Obturator a.

transverse br; lat. circumflex fem a.

Medial circumflex femoral a.

ascending br; first perforating a.

Lateral circumflex femoral a.

Lateral circumflex femoral a.

Deep femoral a.

ascending br;
transverse br;
descending br.

Perforating brs; deep femoral a.

Descending genicular a.

Popliteal a.

Lateral superior genicular a.

Fig. 476   The collateral circulation in the hip and thigh.

Collateral Circulation

The general features of a collateral circulation are considered in the chapter on the Upper Limb (p. 142). The collateral circulation in the lower limb (figs. 476–78) is poorer than in the upper limb. The tissue masses which are supplied are large, and the major vessels are large and single.

In obstruction of the external iliac artery, collateral circulation is established through abdominal and other trunk vessels and by connections of the terminal

branches of the internal iliac artery. There are anastomoses between the ililolumbar and lumbar arteries and the superficial circumflex iliac and deep circumflex iliac vessels and between the superior epigastric branch of the internal thoracic artery and the inferior epigastric artery. The inferior gluteal branch of the internal iliac artery anastomoses with the medial circumflex femoral and lateral circumflex femoral arteries and with the ascending branch of the first perforating artery. This is known as the **cruciate anastomosis** of the thigh. The obturator artery anastomoses with the medial circumflex femoral artery, and its pubic branch connects with the pubic branch of the inferior epigastric artery. The external pudendal branches of the femoral artery anastomose with the terminals of the internal pudendal artery.

The principal anastomoses which continue the flow of blood into the lower limb after obstruction of the femoral artery in the proximal part of the thigh are among those previously listed. The cruciate anastomosis, the anastomosis of the obturator and medial circumflex femoral arteries, and the anastomosis of the external pudendal and internal pudendal arteries aid in carrying blood distally into the thigh. In obstruction of the femoral artery below its deep branch, blood is supplied to the leg through the genicular anastomosis at the knee by the lateral circumflex femoral artery and through the anastomoses of the perforating branches of the deep femoral artery in the back of the thigh.

In the region of the knee there is an important **genicular anastomosis.** This consists of superficial plexuses above and below the patella and a deep plexus on the capsule of the knee joint and on the adjacent surfaces of the condyles of the femur and the tibia. The genicular anastomosis consists of the terminal interconnections of ten vessels. Two of these descend from above—the descending branch of the lateral circumflex femoral artery and the descending genicular branch of the femoral artery. Five are branches of the popliteal artery at the level of the knee joint: the medial superior genicular, lateral superior genicular, middle genicular, medial inferior genicular, and lateral inferior genicular arteries. Three branches of vessels of the leg ascend to join in the anastomosis; these are the posterior tibial recurrent artery, the circumflex fibular artery, and the anterior tibial recurrent artery. These are listed as branches of the anterior tibial artery, but the first two occasionally arise from the posterior tibial artery.

The arteries of the leg may be considered as forming long loops with one another by reason of their distal interconnections through branches of the peroneal artery. A short distance above the tibiofibular syndesmosis, the communicating branch of the peroneal artery passes across the back of the leg and connects with the posterior tibial artery, and the perforating branch of the peroneal artery passes into the anterior compartment to communicate with the anterior lateral malleolar branch of the anterior tibial artery.

The arteries around the ankle joint anastomose freely by small branches which form the **malleolar net-**

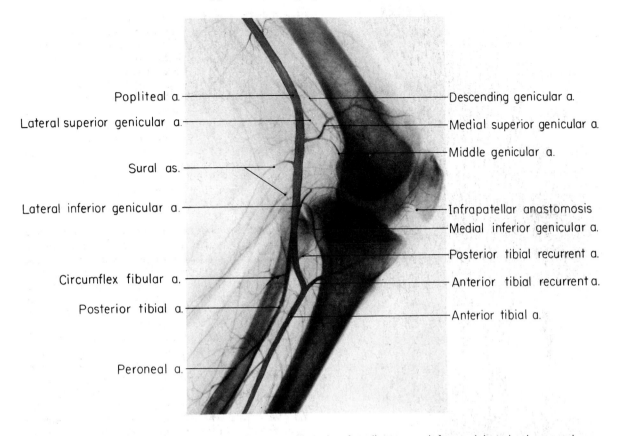

Popliteal a.

Lateral superior genicular a.

Sural as.

Lateral inferior genicular a.

Circumflex fibular a.

Posterior tibial a.

Peroneal a.

Descending genicular a.

Medial superior genicular a.

Middle genicular a.

Infrapatellar anastomosis

Medial inferior genicular a.

Posterior tibial recurrent a.

Anterior tibial recurrent a.

Anterior tibial a.

Fig. 477   The collateral circulation at the knee. From a positive print of a radiogram made from an injected cadaver specimen.

works over and below the medial and lateral malleoli. Participating in these anastomoses are the anterior medial malleolar and anterior lateral malleolar branches of the anterior tibial artery, the posterior medial malleolar branch of the posterior tibial artery, and the posterior lateral malleolar branch of the peroneal artery. The medial malleolar network is augmented by the medial tarsal branches of the dorsalis pedis artery and by medial calcaneal branches of the posterior tibial artery. The lateral malleolar network receives, in addition to the 'malleolar' vessels, branches of the lateral tarsal branch of the dorsalis pedis artery and twigs from the lateral calcaneal branches of the peroneal artery.

Important arterial connections in the foot include the union of the deep plantar branch of the dorsalis pedis with the plantar arch of the lateral plantar artery and the union of the dorsal metatarsal arteries with the plantar arch through the posterior perforating arteries. The dorsal metatarsal arteries also connect with the plantar metatarsal arteries through the anterior perforating arteries. Minor anastomoses are made at the margins of the foot between the lateral plantar and medial plantar arteries, the lateral tarsal and medial tarsal branches of the dorsalis pedis artery, and the metatarsal vessels of the borders of the foot. The digital arteries anastomose over the middle and distal phalanges and on the sides of the toes.

On the back and sole of the heel the medial calcaneal and lateral calcaneal arteries of the posterior tibial and peroneal arteries form a network. Contributing to this network are branches from the malleolar anastomoses.

Nerve Injuries In posterior dislocation of the hip, the sciatic and other nerves of the posterior hip region may be stretched over the head of the femur or otherwise damaged. The nerves supplying the hip joint are the femoral, obturator, and sciatic nerves. Primary disease of the hip joint

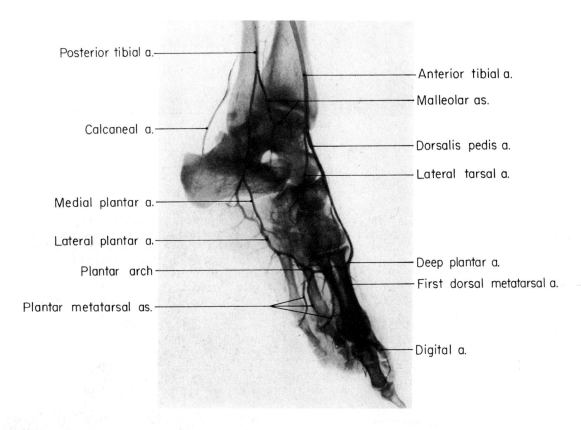

Posterior tibial a.
Calcaneal a.
Medial plantar a.
Lateral plantar a.
Plantar arch
Plantar metatarsal as.

Anterior tibial a.
Malleolar as.
Dorsalis pedis a.
Lateral tarsal a.
Deep plantar a.
First dorsal metatarsal a.
Digital a.

Fig. 478　The vessels of the ankle and foot. From a positive print of a radiogram made from an injected cadaver specimen.

frequently manifests itself in pain referred to the knee because the same nerves provide branches to that joint and a cutaneous branch of the obturator nerve distributes on the medial aspect of the knee. Injury to the superior gluteal nerve supplying gluteus medius and minimus results in collapse of the pelvis on the opposite side (as the leg is raised) from paralysis of these muscles (positive Trendelenberg test).

At the knee the tibial and common peroneal nerves may be damaged, as in supracondylar fracture of the femur, and the common peroneal can be damaged as it grooves the neck of the fibula to enter the leg. In the paralysis following a

lesion of the common peroneal nerve, the foot hangs down from its own weight (foot drop with loss of the extensors of the foot and toes) and turns inward. These effects are due to paralysis of the muscles of the anterior and lateral compartments of the leg. With a lesion of the tibial nerve, the actions of the muscles of the posterior compartment of the leg and of the sole of the foot are lost. The patient cannot plantar flex the ankle; the toes cannot be flexed or moved sidewise. There will be sensory loss in areas appropriate to the distribution of the tibial and common peroneal nerves when these nerves are damaged.

Glossary

This text follows the Nomina Anatomica adopted in Paris in 1955, together with its revisions and additions of 1960, 1965, and 1977. For some years there will be certain confusions with the previously established anatomical terminology and the eponymous terms heretofore widely used. Eponyms are now completely dropped, and the P.N.A. changes (as revised) are generally in the direction of logic and simplicity. The more commonly used earlier terms and eponyms are here related to revised terminology.

Adenoids: hypertrophy of the pharyngeal tonsil
Alcock's canal: pudendal canal
Angle, medial, of scapula: superior angle of scapula
Ansa hypoglossi: ansa cervicalis
Antibrachium: antebrachium
Aponeurosis, epicranial: galea aponeurotica
Appendage, auricular: auricle (of heart)
Arantius, bodies or nodules of: nodules of semilunar valvules of heart
Arnold's nerve: auricular branch of vagus
Artery, auditory, internal: labyrinthine
  cervical, superficial: transverse cervical
  frontal: supratrochlear
  genicular, highest: descending genicular
  hemorrhoidal: rectal
  hypogastric: internal iliac
  innominate: brachiocephalic
  mammary, internal: internal thoracic
  maxillary, external: facial
  maxillary, internal: maxillary
  scapular, transverse: suprascapular
  spermatic, external: cremasteric
  spermatic, internal: testicular
Arthrodial joint: plane joint
Auditory: acoustic
Auerbach's plexus and ganglia: myenteric portion of the enteric nerve plexus

Bartholin's ducts: sublingual ducts that join submandibular duct
  gland: greater vestibular gland

Bigelow, 'Y' ligament of: iliofemoral lig.
Bone, astragalus: talus
  cuneiform, first, second, and third: medial, intermediate, and lateral cuneiforms
  innominate: coxal bone
  malar: zygomatic
  multangular, greater and lesser: trapezium and trapezoid
  navicular, of hand: scaphoid
  scaphoid, of foot: navicular
  turbinate: nasal concha
Brunner's glands: duodenal glands
Buck's fascia: deep fascia of penis
Burn's space: suprasternal fascial space

Calot, triangle of: cystohepatic triangle, *see* Triangle
Camper's fascia: fatty layer of subcutaneous connective tissue of lower abdomen
Canal, pterygopalatine: major palatine canal
  semicircular, membranous: semicircular duct
  subsartorial: adductor
Canthi: angles at which the eyelids meet
Cartilage, lower nasal: greater alar
  sesamoid, nasal: accessory
  upper nasal: lateral
Chain, sympathetic: sympathetic trunk
Cisterna magna: cerebellomedullary cistern
Cloquet's septum; node: femoral septum; lymph node at the femoral ring
Colles' fascia: membranous layer of subcutaneous connective tissue of perineum
  ligament: reflected inguinal ligament
Column, rectal: anal column
Cooper's ligament: pectineal lig., or 'suspensory ligaments' (retinacula cutis) of breast
Corpus cavernosum urethrae: corpus spongiosum penis
Corti, ganglion and organ of: spiral ganglion and spiral organ of ear
Cowper's glands: bulbourethral gland
Cuvier, duct of: common cardinal vein

Denonvillier's fascia: rectovesical septum
Diarthrosis: synovial joint
Digiti quinti: digiti minimi
Dorsal: posterior, usually, except in hand and for nerve roots

Douglas, cavity or pouch of: rectouterine excavation or pouch
  fold of: rectouterine fold
  line or fold of: arcuate line of sheath of rectus abdominis
Duct, submaxillary: submandibular duct

Eustachian tube: auditory tube
  valve: valve of inferior vena cava

Fallopian aqueduct or canal: facial canal
  foramen: hiatus of facial canal
  tube: uterine tube
Fascia bulbi: sheath of bulb of eye
  lumbodorsal: thoracolumbar
Fenestra ovalis: fenestra vestibuli
  rotunda: fenestra cochleae
Filum terminale, internal: filum terminale (of cord)
  external: filum of the spinal dura mater
Flack's node: sinuatrial node
Flexure, hepatic: right colic
  lienal: left colic
Fold, epigastric: lateral umbilical
  ileocecal, inferior: ileocecal
  lateral umbilical: medial umbilical
  medial umbilical: median umbilical
  superior: vascular cecal fold
Fontanna, spaces of: spaces of iridocorneal angle
Foramen, of Magendie: median aperture of fourth ventricle
  optic: optic canal
Foramina of Luschka: lateral apertures of fourth ventricle
Fossa, antecubital: cubital fossa
  ovalis femoris: saphenous opening
Foveae, inguinal and supravesical: fossae, lateral and medial inguinal and supravesical

Galen, vein of: great cerebral vein
Ganglion, sphenopalatine: pterygopalatine
Gartner's duct: longitudinal duct of epoöphoron, remains of mesonephric duct in female
Gasserian ganglion: semilunar ganglion of trigeminal nerve
Gerota's capsule or fascia: renal fascia
Gimbernat's ligament: lacunar ligament
Gland, adrenal: suprarenal
  lymph: lymphatic node
  submaxillary: submandibular
Glaserian fissure: petrotympanic fissure
Glisson's capsule: fibrous capsule of liver
Groove, bicipital, of humerus: intertubercular

Hasner, fold or valve of: lacrimal fold
Haversian canals: canals for vessels in compact bone
Heister, spiral valve of: spiral fold of cystic duct
Henle's ligament: part of conjoined tendon forming medial margin of femoral ring
Hensen's duct: ductus reuniens
Hepatic duct: common hepatic duct
Hering, nerve of: carotid sinus branch of N. IX

Herophilus, torcular of: confluence of the cranial venous sinuses
Hesselbach's ligament: interfoveolar lig.
  triangle: inguinal triangle, *see* Triangle
Hiatus, adductor: tendinous hiatus of adductor magnus
Highmore, antrum of: maxillary sinus
Hilton's line: the white line (mucocutaneous junction) of the anal canal
His, bundle of: atrioventricular bundle
Horner's muscle: pars lacrimalis of orbicularis oculi
Houston's valves or folds: transverse rectal folds
Humphrey, ligament of: anterior meniscofemoral lig.
Hunter's canal: adductor canal

Inscription, tendinous: tendinous intersection
Ischiadic: sciatic

Jacobson's cartilage, organ: vomeronasal cartilage and organ
  nerve: tympanic br. of glossopharyngeal nerve

Labrum, glenoid (of hip): acetabular labrum
Lacertus fibrosus: bicipital aponeurosis
Langer's lines: cleavage lines of skin
Langerhans, islets of: pancreatic islets
Langley's ganglion: ganglion cells in hilus of submandibular gland
Ligament, arcuate, of diaphragm: lumbocostal arch
  cruciate crural: inferior extensor retinaculum of ankle
  dorsal carpal: extensor retinaculum at wrist
  iliopectineal: iliopectineal arch
  laciniate: flexor retinaculum at ankle
  transverse carpal: flexor retinaculum at wrist
  transverse crural: superior extensor retinaculum of leg and ankle
Ligamentum teres (of femur): lig. of the head of the femur
Line, pectinate: line of the anal valves at the bases of the anal columns
Linea semicircularis: arcuate line of sheath of rectus abdominis
Lister's tubercle: dorsal radial tubercle
Littre, glands of: urethral glands
Lockwood's ligament: sling for eyeball formed by its muscular fascias (suspensory lig. of eyeball)
Louis, angle of: sternal angle
Lumbodorsal fascia: thoracolumbar fascia
Luschka's foramina: lateral apertures of fourth ventricle

Mackenrodt's ligament: cardinal (lateral cervical) lig. of uterus
Magendie, foramen of: median aperture of fourth ventricle
Marshall, fold of: fold of the left superior vena cava in the pericardial sac
  vein of: oblique vein of left atrium
Maxillary antrum (of Highmore): maxillary sinus
Mayo, vein of: prepyloric vein
Meatus, urethral: external urethral ostium in male

Meckel's cartilage: cartilage of the first branchial arch
  cave: trigeminal cavum, the subarachnoid space around N.V. as it lies in the middle cranial fossa
  diverticulum: diverticulum of the ileum, a persistent proximal part of the connection to the yolk sac
  ganglion: pterygopalatine ganglion
Meibomian glands: tarsal glands
Meissner's plexus and ganglia: submucosal portion of enteric nerve plexus
Membrane, basilar (of ear): lamina basilaris
  vestibular or Reissner's: vestibular wall of cochlear duct
Moll, glands of: sudoriferous ciliary glands
Monro, foramina of: interventricular foramina
Montgomery's tubercles or glands: sebaceous glands of the areola of the nipple
Morgagni, appendix or hydatid of: appendix testis or vesicular appendix of epoöphoron
  columns of: anal columns
  foramen of: foramen cecum linguae
  fossa of: fossa navicularis urethrae
  sinus of: laryngeal ventricle, or anal sinus, or space between superior constrictor and base of skull
Morison, pouch of: hepatorenal recess
Muller's duct: paramesonephric duct
  muscle: smooth muscle related to orbit, most commonly the superior tarsal m.
Muscle, caninus: levator anguli oris
  compressor naris: pars transversa, m. nasalis
  dilatator naris: pars alaris, m. nasalis
  extensor digitorum communis: extensor digitorum
  flexor digitorum sublimis: fl. digit. superficialis
  quadratus labii inferioris: depressor labii inferioris
  quadratus labii superioris: zygomaticus minor, levator labii superioris and lev. lab. sup. alaeque nasi
  sacrospinalis: erector spinae
  triangularis: depressor anguli oris

Nerve, anterior thoracic: pectoral
  buccinator (N.V.): buccal
  cutaneous cervical: transverse cervical
  dorsal, in forearm: posterior
  external spermatic: genital branch of genitofemoral
  hemorrhoidal: rectal
  lateral brachial cutaneous: superior lateral brachial cutaneous
  lumboinguinal: femoral branch of genitofemoral
  petrosal, superficial, greater and lesser: greater and lesser petrosals
  posterior antebrachial cutaneous, upper branch: inferior lateral brachial cutaneous
  sacral parasympathetic: pelvic splanchnic
  sphenopalatine: pterygopalatine
  spinal accessory: accessory (N. XI)
  statoacustic: vestibulocochlear
  volar: in forearm, anterior; in hand, palmar
Notch, semilunar: trochlear notch
Nuck, canal of: persistent processus vaginalis in female

Oddi, sphincter of: sphincter of hepatopancreatic ampulla

Pacchionian bodies or granulations: arachnoidal granulations
Passavant, fold or ridge of: a fold produced in the posterior pharyngeal wall which helps close the nasopharynx
Petit, triangle of: lumbar triangle
Peyer's patches: aggregated lymphatic follicles in the ileum
Plexus, hypogastric: pelvic plexus, superior and inferior hypogastric plexuses
Poupart's ligament: inguinal lig.
Purkinje fibers: modified muscle fibers of the conduction system of the heart

Ranvier, nodes of: constrictions of the myelin of nerve fibers
Rathke's pouch: craniobuccal or neurobuccal pouch, from which the anterior lobe of the hypophysis develops
Reissner's membrane: vestibular wall of cochlear duct
Retzius, cave of: retropubic space
  veins of: retroperitoneal vv. connecting portal and inferior caval systems
Riolan, arc of: usually an anastomosis between the left and middle colic aa. at the base of the mesocolon
Rivinus, ducts of: lesser sublingual ducts
Rolandic fissure: central sulcus of cerebral hemisphere
Rosenmuller, fossa of: pharyngeal recess

Sac, lesser: omental bursa
Santorini, cartilage of: corniculate cart.
  duct of: accessory pancreatic duct
Sappey's veins: small vv. in the falciform lig.
Scala media: cochlear duct
Scarpa's fascia: membranous layer of subcutaneous connective tissue of lower abdomen
  ganglion: vestibular ganglion
  triangle: femoral triangle
Schlemm, canal of: sinus venosus sclerae
Schwann, cells of: neurilemma cells
Shrapnell's membrane: pars flaccida of the tympanic membrane
Sibson's fascia: suprapleural membrane
Sinus, piriform: piriform recess
  pleural: costodiaphragmatic and costomediastinal recesses of the pleura
  rectal: anal sinus
Skene's ducts or glands: paraurethral ducts or glands
Spieghel's lobe: caudate lobe of liver
Stensen, canal of: incisive canal
  duct of: parotid duct
Sulcus, longitudinal, anterior and posterior: anterior and posterior interventricular
Sylvian aqueduct: cerebral aqueduct
  fissure: lateral sulcus of cerebral hemisphere

Tawara, node of: atrioventricular node
Tenon's capsule or fascia: the fascial sheath of the eyeball
Thebesian valve: valve of coronary sinus
  veins: least cardiac veins

Treitz, muscle or ligament of: suspensory muscle of duodenum

Treves, bloodless fold of: ileocecal fold

Triangle, cystohepatic: triangle between cystic duct, common hepatic duct, and liver

  inguinal: triangle containing the medial inguinal fossa, i.e., between the inf. epigastric vessels and the rectus abdominis

Tubercle, dorsal radial: a small tubercle on distal end of radius, serving as a pulley for the ext. poll. longus muscle

Tuberosity, bicipital: radial tuberosity

  of humerus: tubercle

Tunica vaginalis communis: internal spermatic fascia

Tunnel, carpal: carpal canal

Turbinate: concha

Urethra, anterior, spongy, or spongy and membranous, part of male urethra

  posterior: prostatic, or prostatic and membranous, part of male urethra

Uvea, or uveal tract: the choroid, ciliary body, and iris of the eyeball

Valsalva, sinus of: aortic sinus

Vas deferens: ductus deferens

Vasa efferentia testis: ductuli efferentes

Vater, ampulla of: hepatopancreatic ampulla

Vein, coronary: left gastric mainly

  facial, anterior: facial

  facial, common: lower end of facial v. after being joined by retromandibular vein

  facial, posterior: retromandibular

  frontal: supratrochlear

  hemorrhoidal: rectal

  hypogastric: internal iliac

  innominate: brachiocephalic

  pyloric: right gastric vein

  sacral, medial: median sacral

  transverse scapular: suprascapular

Verumontanum: colliculus seminalis

Vidian canal, nerve: pterygoid canal, nerve of pterygoid canal

Volar: in forearm, anterior; in hand, palmar

Waldeyer's ring: lymphatic ring in the pharynx

Wharton's duct: submandibular duct

Willis, circle of: circulus arteriosus cerebri

Window: *see* Fenestra

Winslow, foramen of: epiploic foramen

Wirsung, duct of: pancreatic duct

Wolffian body and duct: mesonephros and mesonephric duct

Wormian bones: sutural bones of skull

Wrisberg, cartilage of: cuneiform cartilage

  ganglion of: ganglion in the superficial part of the cardiac plexus

  ligament of: posterior meniscofemoral lig.

Zeis, glands of: sebaceous ciliary glands

Zinn, annulus of: common ring tendon

Zuckerkandl, organs of: paired chromaffin cell bodies on the aorta near origin of inferior mesenteric a.

Selected References

GENERAL

Clark, W. E. Le Gros 1971 *The Tissues of the Body. An Introduction to the Study of Anatomy.* 6th ed., Clarendon Press, Oxford.

Corliss, C. E. 1976 *Patten's Human Embryology.* McGraw-Hill Book Co., New York.

Cox, H. T. 1941 'The cleavage lines of skin.' *Brit. J. Surg.,* 29:234.

Cummins, H. and C. Midlo 1943 *Finger Prints, Palms and Soles: An Introduction to Dermatoglyphics.* Blakiston, Philadelphia.

Jones, Frederic Wood 1941 *The Principles of Anatomy as seen in the Hand.* 2nd ed., Bailliere, Tindall & Cox, London.

Kaplan, E. B. 1953 *Functional and Surgical Anatomy of the Hand.* Lippincott, Philadelphia.

Krogman, W. 1941 *A Bibliography of Human Morphology, 1914-1939.* Univ. of Chicago Press, Chicago.

Lockhart, R. D. 1949 *Living Anatomy. A Photographic Atlas of Muscles in Action and Surface Contours.* 2nd ed., The University Press, Glasgow.

Steindler, A. 1935 *Mechanics of Normal and Pathological Locomotion in Man.* Thomas, Springfield.

BONES AND JOINTS

Atkinson, W. B. and H. Elftman 1945 'The carrying angle of the human arm as a secondary sex character.' *Anat. Rec.,* 91:49.

Coventry, M. B., R. K. Gormley, and J. W. Kernohan 1945 'The intervertebral disc; its microscopic anatomy and pathology.' *J. Bone & Joint Surg.,* 27:105, 233 and 460.

Dempster, W. T. and J. C. Finerty 1947 'Relative activity of wrist moving muscles in static support of the wrist joint: An electromyographic study.' *Am. J. Physiol.,* 150:596.

Dempster, W. T. and R. T. Liddicoat 1952 'Compact bone as a non-isotropic material.' *Am. J. Anat.,* 91:331.

Enlow, D. H. 1963 *Principles of Bone Remodeling.* Chas. C. Thomas, Springfield.

Evans, F. G. and H. R. Lissner 1948 ' "Stresscoat" deformation studies of the femur under static vertical loading.' *Anat. Rec.,* 100:159.

Felts, W. J. L. 1954 'The prenatal development of the human femur.' *Am. J. Anat.,* 94:1.

Frazer, J. E. 1958 *The Anatomy of the Human Skeleton.* 5th ed., Churchill, London.

Gardner, E 1952 'The Anatomy of the Joints. Development of Joints.' *Amer. Acad. Orthoped. Surg., Instructional Course Lectures,* vol. 9.

Greulich, W. W. and S. I. Pyle 1950 *Radiographic Atlas of Skeletal Development of the Hand and Wrist.* Stanford Univ. Press, Stanford.

Inman, V. T., J. B. DeC. M. Saunders, and L. C. Abbott 1944 'Observations on the function of the shoulder joint.' *J. Bone & Joint Surg.,* 26:1.

Koch, J. C. 1917 'The laws of bone architecture.' *Am. J. Anat.,* 21:177.

MacConaill, M. A. 1932 'The function of intra-articular fibrocartilages, with special reference to the knee and inferior radio-ulnar joints.' *J. Anat.,* 66:211.

Manter, J. T. 1946 'Distribution of compression forces in the joints of the human foot.' *Anat. Rec.,* 96:313.

Murray, P. D. F. 1936 *Bones; A Study of the Development and Structure of the Vertebrate Skeleton.* Cambridge Univ. Press, London.

Payton, C. G. 1932 'The growth in length of the long bones in the madder-fed pig.' *J. Anat.,* 66:414.

Pyle, S. I. and L. W. Sontag 1943 'Variability in onset of ossification in epiphyses and short bones of the extremities.' *Am. J. Roentgenol. & Rad. Therapy,* 49:795.

Sarnat, B. G. 1951 *The Temporomandibular Joint.* Thomas, Springfield.

Stevenson, P. H. 1924 'Age order of epiphyseal union in Man.' *Am. J. Physical Anthrop.,* 7:53.

Todd, T. W. 1937 *Atlas of Skeletal Maturation.* Part I. *Hand.* Mosby, St. Louis.

FASCIAS, TENDONS, AND SPACES

Coller, F. A. and L. Yglesias 1937 'The relation of the spread of infection to fascial planes in the neck and thorax.' *Surgery,* 1:323.

Gallaudet, B. B. 1931 *A Description of the Planes of Fascia of the Human Body.* Columbia Univ. Press, New York.

Grodinsky, M. 1929 'A study of the fascial spaces of the foot and their bearing on infections.' *Surg., Gyn., and Obs.,* 49:737.

Grodinsky, M. and E. A. Holyoke 1938 'The fasciae and fascial spaces of the head, neck and adjacent regions.' *Am. J. Anat.,* 63:367.

Grodinsky, M. and E. A. Holyoke 1941 'The fasciae and fascial spaces of the palm.' *Anat. Rec.,* 79:435.

Hayes, M. A. 1950 'Abdominopelvic fasciae.' *Am. J. Anat.,* 87:119.

Hicks, J. H. 1954 'The mechanics of the foot. II. The plantar aponeurosis and the arch.' *J. Anat.,* 88:25.

Kanavel, A. B. 1939 *Infections of the Hand.* 7th ed., Lea & Febiger, Philadelphia.

Scheldrup, E. W. 1951 'Tendon sheath patterns in the hand.' *Surg., Gyn., and Obs.,* 93:16.

Tobin, C. E. and J. A. Benjamin 1945 'Anatomical and surgical restudy of Denonvilliers' fascia.' *Surg., Gyn., and Obs.,* 80:373.

MUSCLES

Abd-el-Malek, S. 1955 'The part played by the tongue in mastication and deglutition.' *Am. J. Anat.,* 89:250.

Basmajian, J. V. 1974 *Muscles Alive.* Williams & Wilkins, Baltimore.

Basmajian, J. V. and J. W. Bentzon 1954 'An electromyographic study of certain muscles of the leg and foot in the standing position.' *Surg., Gyn., and Obs.,* 98:662.

Campbell, E. J. M. 1955 'The role of scalene and stenomastoid muscles in breathing in normal subjects.' *J. Anat.,* 89:378.

Doty, R. W. and J. F. Bosma 1956 'An electromyographic analysis of reflex deglutition.' *J. Neurophysiol.,* 19:44.

Huber, E. 1931 *Evolution of Facial Musculature and Facial Expression.* Johns Hopkins Press, Baltimore.

Jit, I. 1952 'The development and the structure of the suspensory muscle of the duodenum.' *Anat. Rec.,* 113:395.

Jones, D. S., R. J. Beargie, and J. E. Pauly 1953 'An electromyographic study of some muscles of costal respiration in man.' *Anat. Rec.,* 117:17.

Jones, R. L. 1941 'The human foot. An experimental study of its mechanics, and the role of its muscles and ligaments in the support of the arch.' *Am. J. Anat.,* 68:1.

MacConaill, M. A. and J. V. Basmajian 1977 *Muscles and Movements: A basis for human kinesiology.* Robt. Krieger Publishing Co., Huntington, N.Y.

Markee, J. E., J. T. Logue, M. Williams, W. B. Stanton, R. N. Wrenn, and L. B. Walker 1955 'Two-joint muscles of the thigh.' *J. Bone and Joint Surg.,* 37:125.

McMurrich, J. P. 1903 'The phylogeny of the forearm flexors.' *Am. J. Anat.,* 2:177.

McMurrich, J. P. 1903 'The phylogeny of the palmar musculature.' *Am. J. Anat.,* 2:463.

Morris, J. M., G. Benner, and D. B. Lucas 1962 'An electromyographic study of the intrinsic muscles of the back in man.' *J. Anat.,* 96:509.

Moyers, R. E. 1950 'An electromyographic analysis of certain muscles involved in temporomandibular movement.' *Am. J. Orthodont.,* 36:481.

Oelrich, T. M. 1980 'The urethral sphincter muscle in the male. *Am. J. Anat.,* 158:229–246.

Sullivan, W. E., O. A. Mortensen, M. Miles, and L. S. Greene 1950 'Electromyographic studies of M. biceps brachii during normal voluntary movement at the elbow.' *Anat. Rec.,* 107:243.

Sunderland, S. 1945 'The actions of the extensor digitorum communis, interosseous and lumbrical muscles.' *Am. J. Anat.,* 77:189.

Uhlenhuth, E. 1953 *Problems in the Anatomy of the Pelvis.* Lippincott Co., Philadelphia.

BLOOD AND LYMPHATIC VESSELS

Anson, B. J., E. W. Cauldwell, J. W. Pick, and L. E. Beaton 1947 'Blood supply of kidney, suprarenal gland, and associated structures.' *Surg., Gyn., and Obs.,* 84:313.

Basmajian, J. V. 1952 'The distribution of valves in the femoral, external iliac, and common iliac veins and their relationship to varicose veins.' *Surg., Gyn., and Obs.,* 95:537.

Basmajian, J. V. 1954 'The marginal anastomoses of the arteries to the large intestine.' *Surg., Gyn., and Obs.,* 99:614.

Batson, O. V. 1940 'The function of the vertebral veins and their role in the spread of metastases.' *Ann. Surg.,* 112:138.

Beaton, L. E. and B. J. Anson 1942 'The arterial supply of the small intestine.' *Quart. Bull., Northwestern Univ. Med. Sch.,* 16:114.

Browning, H. C. 1953 'The confluence of dural venous sinuses.' *Am. J. Anat.,* 93:542.

Clark, E. R. 1938 'Arterio-venous anastomoses.' *Physiol. Rev.,* 18:229.

Cauldwell, E. W. and B. J. Anson 1943 'The visceral branches of the abdominal aorta: topographical relationships.' *Am. J. Anat.,* 73:27.

DeGaris, C. F. 1941 'The aortic arch in primates.' *Am. J. Phys. Anthrop.,* 28:41.

Drinker, C. K. and J. M. Yoffey 1941 *Lymphatics, Lymph and Lymphoid Tissue.* Harvard Univ. Press, Cambridge.

Edwards, E. A. and J. D. Robuck, Jr. 1947 'Applied anatomy of the femoral vein and its tributaries.' *Surg., Gyn., and Obs.,* 85:547.

Faller, A. 1946 'Zur Kenntnis der Gefässverhältnisse der Carotisteilungstelle.' *Schweizerische Medizinische Wochenschrift,* 76:1156.

Franklin, K. J. 1937 *A Monograph on Veins.* Thomas, Springfield.

Franklin, K. J. 1948 *Cardiovascular Studies*. Vol. 16, Blackwell Scientific Publications, Oxford.

Franklin, K. J. 1949 'The history of research upon the renal circulation.' *Proc. Roy. Soc. Med.*, 42:721.

Gagnon, R. 1956 'The venous drainage of the human adrenal gland.' *Rev. Canad. de Biol.*, 14:350.

Graves, F. T. 1954 'The anatomy of the intrarenal arteries and its application to segmental resection of the kidney.' *Brit. J. Surg.*, 42:132.

Gross, L. 1921 *The Blood Supply to the Heart*. Hoeber, New York.

Healey, J., P. Schroy, and R. Sorensen 1953 'The intrahepatic distribution of the hepatic artery in man.' *J. Internat. Coll. Surg.*, 20:133.

Huber, J. F. 1941 'The arterial network supplying the dorsum of the foot.' *Anat. Rec.*, 80:373.

Huelke, D. F. 1958 'A study of the transverse cervical and dorsal scapular arteries.' *Anat. Rec.*, 132:233.

Huelke, D. F. 1959 'Variation in the origins of the branches of the axillary artery.' *Anat. Rec.*, 135:33.

Kampmeier, O. F. and C. Birch 1927 'The origin and development of the venous valves, with particular reference to the saphenous district.' *Am. J. Anat.*, 38:451.

Kay, E. B. 1942 'Regional lymphatic metastases of carcinoma of the gastrointestinal tract.' *Surgery*, 12:553.

Keen, J. A. 1961 'A study of the arterial variations in the limbs, with special reference to symmetry of vascular patterns.' *Am. J. Anat.*, 108:245.

Latarjet, M. 1954 'La vascularisation sanguine des bronches.' *Les Bronches*, 4:145.

Lurje, A. 1946 'On the topographical anatomy of the internal maxillary artery.' *Acta Anatom.*, 2:219.

McConnell, E. M. 1953 'The arterial blood supply of the human hypophysis cerebri.' *Anat. Rec.*, 115:175.

McCormack, L. J., E. W. Cauldwell, and Barry J. Anson 1953 'Brachial and antebrachial arterial patterns.' *Surg., Gyn., and Obs.*, 96:43.

Michels, N. A. 1942 'The variational anatomy of the spleen and splenic artery.' *Am. J. Anat.*, 70:21.

Michels, N. A. 1955 *Blood Supply and Anatomy of the Upper Abdominal Organs with a Descriptive Atlas*. Lippincott, Philadelphia.

Moosman, D. A. and S. Hartwell 1964 'The surgical significance of the subfascial course of the lesser saphenous vein.' *Surg., Gyn., and Obs.*, 118:761.

Noer, R. J. 1943 'The blood vessels of the jejunum and ileum: A comparative study of man and certain laboratory animals.' *Am. J. Anat.*, 73:293.

Patek, P. R. 1939 'The morphology of the lymphatics of the mammalian heart.' *Am. J. Anat.*, 64:203.

Petren, T. 1929 'Die Arterien und Venen des Duodenums und des Pankreaskopfes beim Menschen.' *Z. Anat. Entwickl. Gesch.*, 90:234.

Pick, J. W., B. J. Anson, and F. J. Ashley 1942 'The origin of the obturator artery.' *Am. J. Anat.*, 70:317.

Powell, T. and R. B. Lynn 1951 'The valves of the external iliac, femoral, and upper third of the popliteal veins.' *Surg., Gyn., and Obs.*, 92:453.

Röhlich, Karl 1934 'Uber die Arteria transversa colli des Menschen.' *Anat. Anz.*, 79:37.

Rouvière, H. 1932 *Anatomie des lymphatiques de l'homme*. Masson et Cie, Paris.

Sabin, F. R. 1916 'The origin and development of the lymphatic system.' *Johns Hopkins Hosp. Rep.*, 17:347.

Seib, G. A. 1934 'The azygos system of veins in American whites and American negroes, including observations on the inferior caval system.' *Am. J. Phys. Anthropol.*, 19:39.

Shapiro, A. L. and G. L. Robillard 1946 'Morphology and variations of the duodenal vasculature.' *Arch. Surg.*, 52:571.

Shapiro, A. L. and G. L. Robillard 1950 'The esophageal arteries. Their configurational anatomy and variations in relations to surgery.' *Ann. of Surg.*, 131:671.

Simer, P. H. 1952 'Drainage of pleural lymphatics.' *Anat. Rec.*, 113:269.

Streeter, G. L. 1915 'The development of the venous sinuses of the dura mater in the human embryo.' *Am. J. Anat.*, 18:145.

Swigart, La V. I., R. G. Siekert, W. C. Hambley, and B. J. Anson 1950 'The esophageal arteries. An anatomic study of 150 specimens.' *Surg., Gyn., and Obs.*, 90:234.

Tobin, C. E. 1952 'The bronchial arteries and their connections with other vessels in the human lung.' *Surg., Gyn., and Obs.*, 95:741.

Tobin, C. E. and M. O. Zariquiey 1953 'Some observations on the blood supply of the human lung.' *Med. Radiog. and Photog.*, 29:9.

Trueta, J., A. E. Barclay, P. M. Daniel, K. J. Franklin, and M. M. L. Prichard 1947 *Studies on the Renal Circulation*. Blackwell, Oxford.

Vann, H. M. 1943 'A note on the formation of the plantar arterial arch of the human foot.' *Anat. Rec.*, 85:269.

Van Pernis, P. A. 1949 'Variations of the thoracic duct.' *Surg.*, 26:806.

Weathersby, Hal T. 1955 'The artery of the index finger.' *Anat. Rec.*, 122:57.

Weathersby, H. T. 1956 'The valves of the axillary, subclavian, and internal jugular veins.' *Anat. Rec.*, 124:379.

Wilkie, D. P. D. 1911 'The blood supply of the duodenum.' *Surg., Gyn., and Obs.*, 13:399.

Williams, G. D., C. H. Martin, and L. R. McIntire 1934 'Origin of the deep and circumflex femoral group of arteries.' *Anat. Rec.*, 60:189.

Wolff, H. G. 1938 'The cerebral blood vessels—anatomical principles.' *A. Res. Nerv. & Ment. Dis., Proc.*, 18:29.

Woodburne, R. T. and L. L. Olsen 1951 'The arteries of the pancreas.' *Anat. Rec.*, 111:255.

NERVES

Ashley, F. L. and B. J. Anson 1946 'The pelvic autonomic nerves in the male.' *Surg., Gyn., and Obs.*, 82:598.

Corbin, K. N. and F. Harrison 1939 'The sensory innervation of the spinal accessory and tongue musculature in the rhesus monkey.' *Brain*, 62:191.

Edward, L. F. and R. C. Baker 1940 'Variations in the formation of the splanchnic nerves in man.' *Anat. Rec.*, 77:335.

Eisler, P. 1891 'Der Plexus lumbosacralis des Menschen.' *Anat. Anz.*, 6:274.

Foerster, O. 1933 'The dermatomes in Man.' *Brain*, 56:1.

Foley, J. O. 1945 'The sensory and motor axons of the chorda tympani.' *Proc. Soc. Exper. Biol. & Med.*, 60:262.

Gardner, E. 1948 'The innervation of the knee joint.' *Anat. Rec.*, 101:109.

Gillilan, L. A. 1954 *Clinical Aspects of the Autonomic Nervous System*. Little, Brown and Company, Boston.

Harris, W. 1939 *The Morphology of the Brachial Plexus*. Oxford Univ. Press, London.

Hollinshead, W. H. and J. E. Markee 1946 'The multiple innervation of limb muscles in Man.' *J. Bone & Joint Surg.*, 28:721.

Huelke, D. 1957 'A study of the formation of the sural nerve in Man.' *Am. J. Phys. Anthrop.*, 15:137.

Jackson, R. G. 1949 'Anatomy of the vagus nerves in the region of the lower esophagus and the stomach.' *Anat. Rec.*, 103:1.

Jamieson, R. W., D. B. Smith, and B. J. Anson 1952 'The cervical sympathetic ganglia. An anatomical study of 100 cervico-thoracic dissections.' *Quart. Bull. Northwestern Univ. Med. Sch.*, 26:219.

Keegan, J. J. and F. D. Garrett 1948 'The segmental distribution of cutaneous nerves in the limbs of man.' *Anat. Rec.*, 102:409.

McCormack, L. J., E. W. Cauldwell, and B. J. Anson 1945 'The surgical anatomy of the facial nerve.' *Surg., Gyn., and Obs.*, 80:620.

Mitchell, G. A. G. 1935 'The innervation of the distal colon.' *Edinb. Med. J.*, 42:11.

Mitchell, G. A. G. 1938 'The innervation of the ovary, uterine tube, testis and epididymis.' *J. Anat.*, 72:508.

Mitchell, G. A. G. 1953 *Anatomy of the Autonomic Nervous System*. Williams & Wilkins, Baltimore.

Mitchell, G. A. G. 1953 'The innervation of the heart.' *Brit. Heart J.*, 15:159.

Nonidez, J. N. 1939 'Studies on the innervation of the heart. I. Distribution of the cardiac nerves, with special reference to the identification of the sympathetic and parasympathetic postganglionics.' *Am. J. Anat.*, 65:361.

Pick, J. and D. Sheehan 1946 'Sympathetic rami in Man.' *J. Anat.*, 80:12.

Reed, A. F. 1943 'The relations of the inferior laryngeal nerve to the inferior thyroid artery.' *Anat. Rec.*, 85:17.

Reed, A. F. 1951 'The origins of the splanchnic nerves.' *Anat. Rec.*, 109:341.

Richins, C. A. 1945 'The innervation of the pancreas.' *J. Comp. Neurol.*, 83:223.

Sheehan, D., J. H. Mulholland, and B. Shafiroff 1941 'Surgical anatomy of the carotid sinus nerve.' *Anat. Rec.*, 80:431.

Sheehan, D. 1941 'Spinal autonomic outflows in man and monkey.' *J. Comp. Neurol.*, 75:341.

Sherrington, C. S. 1920 *The Integrative Action of the Nervous System*. Yale University Press, New Haven.

Sprague, J. M. 1944 'The innervation of the pharynx in the rhesus monkey, and the formation of the pharyngeal plexus in primates.' *Anat. Rec.*, 90:197.

Tsouras, S. 1949 'Anatomical observations on the lumbar chain ganglia and the sympathetic system.' *Greek Medical Progress*, 3:3.

White, J. C. and R. H. Smithwick 1942 *The Autonomic Nervous System*. 2nd ed., Henry Kimpton, London.

Woodburne, R. T. 1960 'The accessory obturator nerve and the innervation of the pectineus muscle.' *Anat. Rec.*, 136:367.

Woolard, H. H. 1926 'The innervation of the heart.' *J. Anat.*, 60:345.

VISCERA

Anson, B. J., J. W. Pick, and E. W. Cauldwell 1942 'The anatomy of commoner renal anomalies. Ectopic and horseshoe kidneys.' *J. Urol.*, 47:112.

Bast, T. H. and B. J. Anson 1949 *The Temporal Bone and the Ear*. Thomas, Springfield.

Boyden, E. A. 1932 'Congenital absence of the kidney.' *Anat. Rec.*, 52:325.

Boyden, E. A. 1953 'Humoral vs. neural regulation of the extra-hepatic biliary tract.' *Minn. Med.*, 36:720.

Boyden, E. A. 1955 *Segmental Anatomy of the Lungs*. McGraw-Hill Book Co., New York.

Boyden, E. A. 1957 'The anatomy of the choledochoduodenal junction in man.' *Surg. Gyn. and Obs.*, 104:641.

Colton, E. J., B. J. Anson, E. W. Gibbs, L. J. McCormack, A. F. Reimann, and E. H. Daseler 1947 'Positions of abdominal viscera.' *Quart. Bull., Northwestern Univ. Med. Sch.*, 21:154.

Healey, J. E. and P. C. Schroy 1953 'Anatomy of the biliary ducts within the human liver.' *A.M.A. Arch. Surg.*, 66:599.

Heinbach, W. F. Jr. 1933 'A study of the number and location of the parathyroid glands in man.' *Anat. Rec.*, 57:251.

Hollinshead, W. H. 1952 'Anatomy of the endocrine glands.' *Surg. Clin. North Amer.*, 32:1115.

Huber, G. C. 1905 'On the development and shape of uriniferous tubules of certain of the higher mammals.' *Am. J. Anat. Suppl.*, 4:1.

Huber, J. F. 1949 'Practical correlative anatomy of the bronchial tree and lungs.' *J. Nat'l Med. Assn.*, 41:49.

Jackson, C. L. and J. F. Huber 1943 'Correlated applied anatomy of the bronchial tree and lungs with a system of nomenclature.' *Dis. of Chest*, 9:319.

Johnstone, A. S. 1942 'A radiological study of deglutition.' *J. Anat.*, 77:97.

Knisely, M. 1936 'Spleen studies: I Microscopic observations of the circulatory system of living unstimulated mammalian spleens.' *Anat. Rec.*, 65:23.

Kramer, T. C. 1942 'The partitioning of the truncus and conus and the formation of the membranous portion of the interventricular septum in the human heart.' *Am. J. Anat.*, 71:343.

Lachman, E. 1942 'A comparison of the posterior boundaries of lungs and pleurae as demonstrated on the cadaver and on the roentgenogram of the living.' *Anat. Rec.*, 83:521.

Langworthy, O. R., L. C. Kolb, and L. G. Lewis 1940 *Physiology of Micturition*. Williams and Wilkins Co., Baltimore.

Lapides, J. 1958 'Structure and function of the internal vesical sphincter.' *J. Urol.*, 80:341.

Lawrence, M. 1950 'Recent investigations of sound conduction. I. The normal ear.' *Ann. Otol., Rhin. & Laryng.*, 59:1020.

Lerche, W. 1950 *The Esophagus and Pharynx in Action: A Study of Structure in Relation to Function*. Thomas, Springfield.

Macklin, C. C. 1932 'The dynamic bronchial tree.' *Am. Rev. Tuberc.*, 25:393.

Miller, W. S. 1921 *The Lung*. 2nd ed., Thomas, Springfield.

Moosman, D. A. and F. A. Coller 1951 'Prevention of traumatic injury to the bile ducts.' *Am. J. Surg.*, 82:132.

Patten, B. M. 1938 'Developmental defects at the foramen ovale.' *Am. J. Pathol.*, 14:135.

Polyak, S. L. 1941 *The Retina*. Univ. of Chic. Press, Chicago.

Robb, J. S. and R. C. Robb 1942 'The normal heart; anatomy and physiology of the structural units.' *Am. Heart J.*, 23:455.

Schaeffer, J. P. 1920 *The Nose, Paranasal Sinuses, Nasolacrimal Passageways, and Olfactory Organ in Man*. Blakiston, Philadelphia.

Shapiro, H. H. and R. C. Truex 1943 'The temporomandibular joint and the auditory function.' *J. Am. Dental Assn.*, 30:1147.

Smith, H. W. 1951 *The Kidney: Structure and Function in Health and Disease*. Oxford University Press, New York.

Sterling, J. A. 1954 'The common channel for bile and pancreatic ducts.' *Surg., Gyn., and Obs.*, 98:420.

Taussig, H. B. 1947 *Congenital Malformations of the Heart*. Vol. 36, The Commonwealth Fund, New York.

Truex, R. C. and W. M. Copenhaver 1947 'Histology of the moderator band in man and other mammals with special reference to the conduction system.' *Am. J. Anat.*, 80:173.

Wells, L. J. 1943 'Descent of the testis: anatomical and hormonal considerations.' *Surg.*, 14:436.

Wolff, E. 1954 *The Anatomy of the Eye and Orbit*. 4th ed., H. K. Lewis & Co., Ltd., London.

Woodburne, R. T. 1947. 'The costomediastinal border of the left pleura in the precordial area.' *Anat. Rec.*, 97:197.

Woodburne, R. T. 1961 'The sphincter mechanism of the urinary bladder and urethra.' *Anat. Rec.*, 141:11.

Woodburne, R. T. 1968 'Anatomy of the bladder and bladder outlet.' *J. Urol.*, 100:474.

Index

Items are indexed both by description and by illustration. Bold face numbers refer to pages of description, light face type indicates pages on which items are illustrated.